treatment of a bacterial infection, 626
virus spread, 212, 311, 444
Members of a population, heights, 12
Natural history museum, 516
Optometry, near point and far point of the eye, 603
Pest management, 66
Physiology
   blood flow, 402
   body surface area, 248
   brain weight, 67
   child gender, 672
   femur length, 80
   Hardy-Weinberg Law, 593
   heart rate, 7
   red blood cell volume, 310
   weights and body surface areas of infants, 598
Wheelchair ramp, 58, 66
Zoology, surface area of a pond, 461

Social and Behavioral Sciences

Alcohol use
   among eighth-graders, 38
   among tenth-graders, 38
Assistance for needy families, 537
Assisted living, 109
Attending college, 672
Child support, 537
   cases, 133
Choice of newscasts, 542
Choosing a month to take a vacation, 694
Civil rights cases, 222
College enrollment, 67
Comparing populations, 542
Consumer awareness
   ambulance transport cost, 111
   average prices of prescription drugs, 35
   change in beef prices, 424
   dormitory charges, 497
   electricity cost, 690
   fuel economy, 270, 661
   fuel mileage, 690
   gasoline price, 7, 424
   ground beef price, 157
   health club fees, 103
   health insurance fraud, 671
   ice cream price, 180
   medical index, 545
   U.S. Postal Service first class mail rates, 103
   vitamins cost, 111
Consumer trends
   agricultural imports, 42, 312
   airline complaints, 256
   annual energy production
      from crude oil, 51
      from wind, 51
   annual geothermal energy consumption, 54

basic cable television subscr
beef production, 410
cars per household, 695
cellular telephone subscribers, 41
citrus fruits production, 173
coupons used in a grocery store, 690
dental services expenditures, 108
egg production, 7
grain and feed imports, 42
health care expenditures, 67
households that have indoor houseplants, 257
lumber use, 470
magazine subscribers, 180, 470
malt beverage imports, 42
pork product imports, 42
prescription drugs (amount spent), 35
prescriptions, numbers of, 497
public expenditures, 583
shrimp supplied to the United States, 626
surfing the Internet, 288
textbook spending, 55
TV usage (hours), 254
visitors to a national park, 122, 144
wine imports, 42
Consumption of
   energy, 453
   fruits and vegetables, 166
   geothermal energy, 247
   milk, 564, 583
   pineapples, 410
   softwoods, 123, 136, 626
Department of Defense manpower, 241
Education, 673
   ACT
      scores, 690
      testing, 497
   average grade on a calculus final, 669
   exam
      essay questions, 694
      scores, 689
      true-false question, 671
   freshman classes, 695
   math lab usage, 279
   quiz scores, 694
   SAT testing, 497
Employment, 278
   amusement park workers, 361, 455
   construction workers, 345, 361, 453
   dentist office and nursing care facility, 603
   farm work force, 55
   health care and social assistance, 256
   medical and surgical hospitals, 314
   mineral production workers, 627
   unemployed workers, 108
   unemployment insurance, 313
   women in the labor force, 383, 627
Fatal crashes, 66

space, and other technology, 305
Federally funded programs for disabled children, 435
Financial aid, 497
Food stamps, 41
Food supply
   corn products, 185
   rice, 185
Head Start enrollment, 108
Health insurance for children, 178, 179
Health insurance coverage status, 673
Health insurance survey, 12
High school dropouts, 212
Higher education enrollment, 311
Households, number of, 407
Internet users, rate of increase, 374
Living arrangements, 374
Medical degrees, number of, 192
Medicare enrollees, 173
National health expenditures, 410
National monuments, 41
Newspaper circulation, 254
Nonprofit associations, social welfare, 313
Overseas international travelers to the U.S., 426
Poll, 671, 672
Population
   of Alaska, 542
   of Arizona, 50
   of Australia, 304
   of California, 108
   of Canada, 304
   center of, 145
   of the District of Columbia, 222
   of France, 296
   of Kentucky, 109
   of Las Vegas, Nevada, 270
   of Maryland, 50, 112
   of Missouri, 68, 112
   of Oklahoma, 54
   of Oregon, 54
   of the Philippines, 304
   of South Africa, 304
   of South Carolina, 67, 109
   of Tennessee, 50
   of Texas, 108
   of Turkey, 304
   of the United States, 202, 305, 424
   of Vermont, 542
Population density, 616, 619
   contour map of New York, 625
Population growth, 18, 311, 313, 314, 374, 654, 656
   Bulgaria, 267
   Horry County, South Carolina, 374
   Houston, Texas, 287
   India, 267
   Japan, 145

Orlando, Florida, 287
rural area, 145
suburban, 383
United States, 263, 445
world, 312, 602
Population per square mile of the United States, 18
Presidential election, 546
Private schools enrollment, 374
Psychology
autistic children, 485, 600
Ebbinghaus Model, 278
learning curve, 232, 305, 648
learning theory, 270, 288, 297, 642, 654, 680
memory model, 32, 255, 285, 435
migraine prevalence, 136
rate of change, 292
sleep patterns, 425
Stanford-Binet Test (IQ test), 583
Public school enrollment in Puerto Rico, 181
Queuing model, 571
Recycling
aluminum cans, 111
fluorescent bulbs, 67, 100
paper products, 180
Rehabilitation counseling, 93
Research and development, 143
School enrollment, 304
Hispanic students in public schools (percent), 435
Seizing drugs, 157, 232
Social Security beneficiaries, 133
Social Security benefits, 241
Stay-at-home mothers, 371
Students served by disability programs, 600
Treatment facilities, substance abuse, 305
Veteran benefits, 212
Vital statistics
children in families (numbers), 695
divorced adults, rate of increase, 371
married couples, rate of increase, 374
median age, 470
people 65 years old and over, 311
Volunteers, rate of change, 402
Voter poll, 11, 12
Voting preferences, 528
Work groups, 691

Physical Sciences

Acceleration, 169
Acceleration due to gravity
on Earth, 170
on the moon, 170
Adiabatic expansion, A15
Airplane speed, 497
Arc length, 323, 470
Architecture, 555
Area, 221, 248, A14
of a pasture, 219
Average elevation, 627
Average velocity, 139
Beam strength, 221
Biomechanics, Froude number, 625
Catenary, 274
Changing area, A11
Changing surface area, A13
Changing volume, A13
Chemistry
acid mixture, 506, 542
Apparent Temperature Index, 583
carbon dating, 287, 304
chemical mixture, 517, 653, 655, 662
chemical reaction, 649, 655, 662
copper (melting point), 12
dating organic material, 260
evaporation rate, 656, 662
finding concentrations, 24
hydrogen orbitals, 696
Ideal Gas Law, 571
molecular velocity, 192
Newton's Law of Cooling, 650, 655, 661, 663
pH levels, 313
radioactive decay, 264, 297, 299, 304, 313, 314, 642, 662
rainwater acidity, 625
rate of change, 294
raw materials, 545
solution mixture, 546
temperature, 297
in a steel plate, 583
water (boiling temperature), 296
Circuit analysis, 537
Digital camera tripod leg length, 362
Distance, 323, 334, 360
across a lake, 335
Earthquake intensity, Richter scale, 297
Electricity, A15
Engineering, 602
Erosion, 18
Exercise ball, equation of, 624
Geology
crystals, 555, 624
seismic amplitudes contour map, 572
Greenhouse dimensions, 41
Heat loss, 593
Height, 360
of a broadcasting tower, 334
of a building, 329
of the Empire State Building, 334
of a mountain, 334
of a streetlight, 322
of a tree, 360
Ice sculptures, 24
Instantaneous rate of change, 140
Length, 254, 334
of a guy wire, 323, 359
of a spring, 82
of a walk, 37
Machine shop calculations, 334
Material minimum amount, 220
Maximum area, 219, 220, 221, 222
Maximum volume, 213, 219, 221, 590, 593
Measurement errors, 246
Metallurgy, iron in ore samples, 689
Meteorology
atmospheric pressure, 288, 572, 603
average temperature, 451
barometric pressure, 656, 662
contour map of average precipitation for Iowa, 625
daylight in New Orleans, Louisiana, (hours), 353
dew point, 572
isotherms, 584
monthly normal high and low temperatures for Erie, Pennsylvania, 345
monthly normal temperature
for New York City, 253
for Pittsburgh, Pennsylvania, 241
monthly precipitation (average), 156
for Bismarck, North Dakota, 454
for Sacramento, California, 453
for San Francisco, California, 486
monthly rainfall, 696
monthly temperature and precipitation for Honolulu, Hawaii, (mean), 345
monthly temperature (average)
in Duluth, Minnesota, 115
in Honolulu, Hawaii, 136
normal average daily temperature, 353
normal daily maximum temperature, 181
rainfall (amount), 680
snow accumulation, 422
temperature change, 351
tornado wind speed, 314
Minimum area, 221
Minimum distance, 216, 221
Minimum length, 220, 221
Minimum perimeter, 219, 254
Minimum surface area, 219, 220, 221
Peripheral vision, 329
Physical science
apple pie temperature, 232
Earth and its shape, 564
Earth as a sphere (equation), 584
Food (temperature) in
a freezer, 181
a refrigerator, 156
Physics
wave properties, 345
wind chill, 144, 569
Position function, 370
Position (net change), 397, 422
Projectile motion, 173
Ramp angle, 335
Revolution speed, 323
Ripples in a pond, 81
River width, 334

(continued on back endsheets)

Applied Calculus
for the Life and Social Sciences

Applied Calculus
for the Life and Social Sciences

RON LARSON

The Pennsylvania State University
The Behrend College

with the assistance of
DAVID C. FALVO

The Pennsylvania State University
The Behrend College

HOUGHTON MIFFLIN
HARCOURT PUBLISHING
COMPANY
Boston   New York

Publisher: Richard Stratton
Senior Sponsoring Editor: Cathy Cantin
Senior Marketing Manager: Jennifer Jones
Development Editor: Peter Galuardi
Associate Editor: Jeannine Lawless
Project Editor: Margaret M. Kearney/Sarah S. Evans
Senior Media Producer: Douglas Winicki
Senior Content Manager: Maren Kunert
Art and Design Manager: Jill Haber
Cover Design Manager: Anne S. Katzeff
Senior Photo Editor: Jennifer Meyer Dare
Senior Composition Buyer: Chuck Dutton
Manager of New Title Project Management: Pat O'Neill
Editorial Assistant: Amy Haines

Cover photo: Multiple jellyfish. Copyright © iStock International Inc.

Copyright © 2009 by Houghton Mifflin Harcourt Publishing Company. All rights reserved.

No part of this work may be reproduced or transmitted in any form or by any means, electronic or mechanical, including photocopying and recording, or by any information storage or retrieval system without the prior written permission of Houghton Mifflin Harcourt Publishing Company unless such copying is expressly permitted by federal copyright law. Address inquiries to College Permissions, Houghton Mifflin Harcourt Publishing Company, 222 Berkeley Street, Boston, MA 02116-3764.

Printed in the U.S.A.

Library of Congress Control Number: 2008924205

Instructor's examination copy
ISBN-10: 0-547-20367-5
ISBN-13: 978-0-547-20367-6

For orders, use student text ISBNs
ISBN-10: 0-618-96259-X
ISBN-13: 978-0-618-96259-4

123456789–DOW–12 11 10 09 08

Contents

From the Desk of Ron Larson   ix
Goals for this Text   x
A Plan for You as a Student   xi
Supplements   xii
Acknowledgments   xiii
Features   xiv

0 A Precalculus Review   1

- **0.1** The Real Number Line and Order   2
- **0.2** Absolute Value and Distance on the Real Number Line   8
- **0.3** Exponents and Radicals   13
- **0.4** Factoring Polynomials   19
- **0.5** Fractions and Rationalization   25

1 Functions, Graphs, and Limits   33

- **1.1** The Cartesian Plane and the Distance Formula   34
- **1.2** Graphs of Equations   43
- **1.3** Lines in the Plane and Slope   56
- Mid-Chapter Quiz   68
- **1.4** Functions   69
  - *Life Science Capsule: Nursing Careers*   81
- **1.5** Limits   82
- **1.6** Continuity   94
- Chapter 1 Algebra Review   104
- Chapter Summary and Study Strategies   106
- Review Exercises   108
- Chapter Test   112

2 Differentiation   113

- **2.1** The Derivative and the Slope of a Graph   114
- **2.2** Some Rules for Differentiation   125
- **2.3** Rates of Change   137
- Mid-Chapter Quiz   146
- **2.4** The Product and Quotient Rules   147
  - *Social Science Capsule: Social Work*   157
- **2.5** The Chain Rule   158
- **2.6** Higher-Order Derivatives   167
- Chapter 2 Algebra Review   174
- Chapter Summary and Study Strategies   176
- Review Exercises   178
- Chapter Test   182

3 Applications of the Derivative — 183

- **3.1** Increasing and Decreasing Functions 184
- **3.2** Extrema and the First-Derivative Test 193
- **3.3** Concavity and the Second-Derivative Test 203
  - *Life Science Capsule: Biomedical Science* 212
- **3.4** Optimization Problems 213
- Mid-Chapter Quiz 222
- **3.5** Asymptotes 223
- **3.6** Curve Sketching: A Summary 233
- **3.7** Differentials: Linear Approximation 242
- Chapter 3 Algebra Review 249
- Chapter Summary and Study Strategies 251
- Review Exercises 253
- Chapter Test 257

4 Exponential and Logarithmic Functions — 258

- **4.1** Exponential Functions 259
- **4.2** Natural Exponential Functions 265
- **4.3** Derivatives of Exponential Functions 271
- Mid-Chapter Quiz 279
- **4.4** Logarithmic Functions 280
- **4.5** Derivatives of Logarithmic Functions 289
  - *Life Science Capsule: Agriculture and Food Science* 297
- **4.6** Exponential Growth and Decay 298
- Chapter 4 Algebra Review 306
- Chapter Summary and Study Strategies 308
- Review Exercises 310
- Chapter Test 314

5 Trigonometric Functions — 315

- **5.1** Radian Measure of Angles 316
- **5.2** The Trigonometric Functions 324
- Mid-Chapter Quiz 335
- **5.3** Graphs of Trigonometric Functions 336
- **5.4** Derivatives of Trigonometric Functions 346
  - *Social Science Capsule: Landscape Architecture* 354
- Chapter 5 Algebra Review 355
- Chapter Summary and Study Strategies 357
- Review Exercises 359
- Chapter Test 362

6 Integration and Its Applications — 363

- **6.1** Antiderivatives and Indefinite Integrals 364
- **6.2** Integration by Substitution and The General Power Rule 375
- **6.3** Exponential and Logarithmic Integrals 384
- Mid-Chapter Quiz 391
- **6.4** Area and the Fundamental Theorem of Calculus 392
- **6.5** The Area of a Region Bounded by Two Graphs 403
  - *Social Science Capsule: USDA* 410
- **6.6** Volumes of Solids of Revolution 411
- Chapter 6 Algebra Review 418
- Chapter Summary and Study Strategies 420
- Review Exercises 422
- Chapter Test 426

Contents   vii

7 Techniques of Integration   427

- **7.1** Integration by Parts   428
- **7.2** Partial Fractions and Logistic Growth   436
- **7.3** Integrals of Trigonometric Functions   446
- Mid-Chapter Quiz   455
- **7.4** The Definite Integral as the Limit of a Sum   456
- **7.5** Numerical Integration   462
- **7.6** Improper Integrals   471
  - *Life Science Capsule: Orthodontics*   480
- Chapter 7 Algebra Review   481
- Chapter Summary and Study Strategies   483
- Review Exercises   485
- Chapter Test   488

8 Matrices   489

- **8.1** Systems of Linear Equations in Two Variables   490
- **8.2** Systems of Linear Equations in More Than Two Variables   498
  - *Life Science Capsule: Chemical and Material Science*   506
- **8.3** Matrices and Systems of Linear Equations   507
- Mid-Chapter Quiz   517
- **8.4** Operations with Matrices   518
- **8.5** The Inverse of a Matrix   529
- Chapter 8 Algebra Review   538
- Chapter Summary and Study Strategies   540
- Review Exercises   542
- Chapter Test   546

9 Functions of Several Variables   547

- **9.1** The Three-Dimensional Coordinate System   548
- **9.2** Surfaces in Space   556
- **9.3** Functions of Several Variables   565
- **9.4** Partial Derivatives   573
  - *Social Science Capsule: Empirisoft*   583
- Mid-Chapter Quiz   584
- **9.5** Extrema of Functions of Two Variables   585
- **9.6** Least Squares Regression Analysis   594
- **9.7** Double Integrals and Area in the Plane   604
- **9.8** Applications of Double Integrals   612
- Chapter 9 Algebra Review   620
- Chapter Summary and Study Strategies   622
- Review Exercises   624
- Chapter Test   628

10 Differential Equations — 629

- **10.1** Solutions of Differential Equations   630
- **10.2** Separation of Variables   636
    - *Life Science Capsule: Radiation Oncology*   642
- Mid-Chapter Quiz   643
- **10.3** First-Order Linear Differential Equations   644
- **10.4** Applications of Differential Equations   649
- Chapter 10 Algebra Review   657
- Chapter Summary and Study Strategies   659
- Review Exercises   660
- Chapter Test   663

11 Probability and Calculus — 664

- **11.1** Discrete Probability   665
- **11.2** Continuous Random Variables   674
    - *Life Science Capsule: Occupational Health and Safety*   680
- **11.3** Expected Value and Variance   681
- Chapter 11 Algebra Review   691
- Chapter Summary and Study Strategies   693
- Review Exercises   694
- Chapter Test   697

Appendices

Appendix A: Differentiation and Integration Formulas   A2

Appendix B: Additional Topics in Differentiation   A3
- **B.1** Implicit Differentiation   A3
- **B.2** Related Rates   A10

Appendix C: Probability and Probability Distributions (web only)*
- **C.1** Probability
- **C.2** Probability Computations
- **C.3** Conditional Probability
- **C.4** Tree Diagrams and Bayes' Theorem
- **C.5** Probability Distributions
- **C.6** Normal Distribution
- **C.7** Binomial Distribution

Appendix D: Properties and Measurement (web only)*
- **D.1** Review of Algebra, Geometry, and Trigonometry
- **D.2** Units of Measurements

Appendix E: Graphing Utility Programs (web only)*

Answers to Selected Exercises   A17
Answers to Checkpoints   A109
Index   A125

*Available at the text-specific website:
college.hmco.com/pic/larsonACLS

From the Desk of Ron Larson

This text was written in response to requests from several users of my *Calculus: An Applied Approach* book to write a book intended for students majoring in the Life and Social Sciences. Enrollments in environmental science, biology, nursing, and social science programs are increasing each year. Most students in these programs are required to take one or more applied calculus courses. However, most traditional applied calculus texts have a strong business focus, which many students may find dull or irrelevant. It was my goal to write a textbook that would teach calculus concepts within a context best suited for these students.

I'm excited about this new textbook because it acknowledges where students are when they enter the course…and where they should be when they complete it. We review the precalculus that students have studied previously (in Chapter 0, skills reviews, notes, study tips, and algebra reviews throughout the text), *and* present a solid applied calculus course that balances understanding of concepts with the practicality of skills.

I am also excited about this textbook program because it is being published as part of a whole series of textbooks tailored to the needs of college algebra and applied calculus students majoring in business, biology, and related courses:

College Algebra with Applications for Business and the Life Sciences
College Algebra and Calculus: An Applied Approach (©2010)
Calculus: An Applied Approach, 8th edition
Brief Calculus: An Applied Approach, 8th edition
Applied Calculus for the Life and Social Sciences

This new textbook program helps students learn the math in the ways I have found most effective for my students, by **practicing their problem-solving skills and reinforcing their understanding in the context of actual problems they may encounter in their lives and careers.**

I hope you and your students enjoy *Applied Calculus for the Life and Social Sciences*. Feel free to tell me what you think about it. Over the years, I have received many useful comments from both instructors and students, and I value these comments very much.

Ron Larson

Ron Larson

Goals for This Text

Establish a Solid Foundation in Calculus

Many effective tools are incorporated throughout the text to help students master calculus concepts. These features help students evaluate and reinforce their understanding of the math.

- After each worked Example, a **Checkpoint** offers the opportunity for immediate practice.
- At the end of each section before the Section Exercises, a **Concept Check** poses noncomputational questions designed to test students' basic understanding of that section's concepts.
- Each exercise set begins with a **Skills Review** of cumulative exercises that test prerequisite skills from earlier sections.
- The **Mid-Chapter Quiz** offers frequent opportunities for self-assessment so students can discover any topics they might need to study further before they progress too far into the chapter.
- The **Chapter Summary** summarizes skills presented in the chapter and correlates each skill to **Review Exercises** for extra practice.
- The **Study Strategies** provide invaluable tips for overcoming common study obstacles.
- The **Chapter Test** enables students to identify and strengthen any weaknesses before their exam.

Present Real-World Problems to Motivate Interest and Understanding

Great care was taken in choosing the applications for this text, drawing them from news sources, current events, industry data, world events, and government data. If students can see calculus as it applies to the world around them, they will see calculus' relevance.

Ample Review

Students coming into the course approach from many different paths. Students who have not taken a math course recently may have difficulty at first. This book contains a number of features designed to get students up to speed quickly and keep them there, including:

- **Algebra Reviews** appear in each chapter to review algebra concepts relevant to the calculus being studied.
- **Prerequisite Skills Reviews** appear before each exercise set and review important material students will need to use in order to successfully complete the exercises.

Enhance Understanding Using Technology

Students can visualize the math by using powerful technology, such as graphing calculators and spreadsheet software, and so develop a deeper comprehension of mathematical concepts.

- Optional **Technology boxes** feature exercises that offer students opportunities to practice using these tools.
- The (T) icon in the exercises suggests when a graphing calculator or other technology tool can be used.
- The (S) icon appears when spreadsheet software, such as Microsoft's Excel®, can be incorporated.

A Plan for You as a Student

Study Strategies

Your success in mathematics depends on your active participation both in class and outside of class. Because the material you learn each day builds on the material you have learned previously, it is important that you keep up with your course work every day and develop a clear plan of study. This set of guidelines highlights key study strategies to help you learn how to study mathematics.

Preparing for Class   The syllabus your instructor provides is an invaluable resource that outlines the major topics to be covered in the course. Use it to help you prepare. As a general rule, you should set aside two to four hours of study time for each hour spent in class. Being prepared is the first step toward success. Before class:

- Review your notes from the previous class.
- Read the portion of the text that will be covered in class.

Keeping Up   Another important step toward success in mathematics involves your ability to keep up with the work. It is very easy to fall behind, especially if you miss a class. To keep up with the course work, be sure to:

- Attend every class. Bring your text, a notebook, a pen or pencil, and a calculator (scientific or graphing). If you miss a class, get the notes from a classmate as soon as possible and review them carefully.
- Participate in class. As mentioned above, if there is a topic you do not understand, ask about it before the instructor moves on to a new topic.
- Take notes in class. After class, read through your notes and add explanations so that your notes make sense to *you*. Fill in any gaps and note any questions you might have.

Getting Extra Help   It can be very frustrating when you do not understand concepts and are unable to complete homework assignments. However, there are many resources available to help you with your studies.

- Your instructor may have office hours. If you are feeling overwhelmed and need help, make an appointment to discuss your difficulties with your instructor.
- Find a study partner or a study group. Sometimes it helps to work through problems with another person.
- Special assistance with algebra appears in the *Algebra Reviews*, which are located throughout each chapter. These short reviews are tied together in the larger *Algebra Review* section at the end of each chapter.

Preparing for an Exam   The last step toward success in mathematics lies in how you prepare for and complete exams. If you have followed the suggestions given above, then you are almost ready for exams. Do not assume that you can cram for the exam the night before—this seldom works. As a final preparation for the exam:

- When you study for an exam, first look at all definitions, properties, and formulas until you know them. Review your notes and the portion of the text that will be covered on the exam. Then work as many exercises as you can, especially any kinds of exercises that have given you trouble in the past, reworking homework problems as necessary.
- Start studying for your exam well in advance (at least a week). The first day or two, study only about two hours. Gradually increase your study time each day. Be completely prepared for the exam two days in advance. Spend the final day just building confidence so you can be relaxed during the exam.

For a more comprehensive list of study strategies, please visit *college.hmco.com/pic/larsonACLS*.

Get more value from your textbook!

Supplements for the Instructor

Online Complete Solutions Manual
Found on the instructor website, this manual contains the complete, worked-out solutions for all the exercises in the text.

Supplements for the Student

Student Solutions Manual
This manual contains worked-out solutions to all odd-numbered exercises in the text.

Excel Made Easy CD
This CD offers easy-to-follow videos to help students use Microsoft Excel to master mathematical concepts introduced in class. Electronic spreadsheets and detailed tutorials are included.

Instructor and Student Websites
The Instructor and Student websites at *college.hmco.com/pic/larsonACLS* contain an abundance of resources for teaching and learning, such as Note Taking Guides, Digital Art, ACE Practice Quizzes and a graphing calculator simulator.

Instructional DVDs
Hosted by Dana Mosely and captioned for the hearing-impaired, these DVDs cover all sections in the text. Ideal for promoting individual study and review, these comprehensive DVDs also support students in online courses or those who have missed a lecture.

HM Testing (powered by Diploma®)
HM Testing (powered by Diploma) provides instructors with a wide array of algorithmic items along with improved functionality and ease of use. HM Testing offers all the tools needed to create, deliver, and customize multiple types of tests—including authoring and editing algorithmic questions. In addition to producing an unlimited number of tests for each chapter, including cumulative tests and final exams, HM Testing also offers instructors the ability to deliver tests online, or by paper and pencil. Diploma is currently in use at thousands of college and university campuses throughout the United States and Canada.

SMARTHINKING
Houghton Mifflin's unique partnership with SMARTHINKING brings students live, online tutorial support when they need it most. SMARTHINKING provides an easy-to-use and effective online, text-specific tutoring service. A dynamic *Whiteboard* and *Graphing Calculator* function enables students and e-structors to collaborate easily. Visit *www.smarthinking.com* for more information.

Online Course Content for Blackboard/WebCT®
Houghton Mifflin can provide you with valuable content to include in your existing Blackboard and WebCT systems. This text-specific content enables instructors to teach all or part of their course online. Contact your Houghton Mifflin sales rep for cartridge availability.

WebAssign®
Developed by teachers for teachers, WebAssign allows instructors to focus on teaching rather than grading. Instructors can create assignments from a ready-to-use database of algorithmic questions based on end-of-section exercises or write and customize their own. With WebAssign, students can access homework, quizzes, and tests anytime of day or night.

Acknowledgments

I would like to thank the many people who have helped at the various stages of this project. Their encouragement, criticisms, and suggestions have been invaluable.

Thank you to all of the instructors who took the time to provide feedback on the material for this book. I would also like to thank the reviewers of the other books in this series.

Reviewers

Dr. Mohammad Z. Abu-Sbeih, *King Fahd University of Petroleum and Minerals;* Lateef Adelani, *Harris–Stowe State University;* Frederick Adkins, *Indiana University of Pennsylvania;* Polly Amstutz, *University of Nebraska at Kearney;* Judy Barclay, *Cuesta College, Saint Louis;* Jean Michelle Benedict, *Augusta State University;* Ben Brink, *Wharton County Junior College;* Jimmy Chang, *St. Petersburg College;* Derron Coles, *Oregon State University;* Steve Comer, *The Citadel–Military College of SC;* David French, *Tidewater Community College;* Randy Gallaher, *Lewis & Clark Community College;* Perry Gillespie, *Fayetteville State University;* Walter J. Gleason, *Bridgewater State College;* Larry Hoehn, *Austin Peay State University;* Raja Khoury, *Collin County Community College;* Ivan Loy, *Front Range Community College;* Lewis D. Ludwig, *Denison University;* Augustine Maison, *Eastern Kentucky University;* Dr. Eric Marland, *Appalachain State University;* John Nardo, *Oglethorpe University;* Darla Ottman, *Elizabethtown Community & Technical College;* William Parzynski, *Montclair State University;* Laurie Poe, *Santa Clara University;* Adelaida Quesada, *Miami Dade College–Kendall;* Brooke P. Quinlan, *Hillsborough Community College;* David Ray, *University of Tennessee at Martin;* Carol Rychly, *Augusta State University;* Mike Shirazi, *Germanna Community College;* Rick Simon, *University of La Verne;* Marvin Stick, *University of Massachusetts–Lowell;* Devki Talwar, *Indiana University of Pennsylvania;* Linda Taylor, *Northern Virginia Community College;* Stephen Tillman, *Wilkes University;* Blake Thornton, *Washington University in St. Louis;* Donna Tupper, *CCBC-Essex;* Jay Wiestling, *Palomar College;* John Williams, *St. Petersburg College;* Ted Williamson, *Montclair State University.*

My thanks to David Falvo, The Behrend College, The Pennsylvania State University, for his contributions to this project. My thanks also to Robert Hostetler, The Behrend College, The Pennsylvania State University, and Bruce Edwards, University of Florida, for their significant contributions.

I would also like to thank the staff at Larson Texts, Inc. who assisted with proofreading the manuscript, preparing and proofreading the art package, and checking and typesetting the supplements.

On a personal level, I am grateful to my spouse, Deanna Gilbert Larson, for her love, patience, and support. Also, a special thanks goes to R. Scott O'Neil.

If you have suggestions for improving this text, please feel free to write to me. Over the years I have received many useful comments from both instructors and students, and I value these comments highly.

Ron Larson

Ron Larson

How to get the most out of your textbook...

Establish a Solid Foundation in Applied Calculus

CHAPTER OPENERS

Each opener has an applied example of a core topic from the chapter. The section outline provides a comprehensive overview of the material being presented.

Demography, a field within sociology, studies changes in population. Derivatives are used to determine the rates of change of populations over time. Recently, growth of Japan's population has been slowing down. (See Section 2.3, Exercise 23.)

Applications

Differentiation has many real-life applications. The applications listed below represent a sample of the applications in this chapter.

- Psychology: Migraine Prevalence, Exercise 64, page 136
- Center of Population, Exercise 28, page 145
- Biology: Wildlife Management, Exercise 60, page 156
- Make a Decision: Environment, Exercise 63, page 157
- Fruit and Vegetable Consumption, Exercise 73, page 166

2.1 The Derivative and the Slope of a Graph
2.2 Some Rules for Differentiation
2.3 Rates of Change
2.4 The Product and Quotient Rules
2.5 The Chain Rule
2.6 Higher-Order Derivatives

Section 2.1

The Derivative and the Slope of a Graph

- Identify tangent lines to a graph at a point.
- Approximate the slopes of tangent lines to graphs at points.
- Use the limit definition to find the slopes of graphs at points.
- Use the limit definition to find the derivatives of functions.
- Describe the relationship between differentiability and continuity.

SECTION OBJECTIVES

A bulleted list of learning objectives allows you the opportunity to preview what will be presented in the upcoming section.

Tangent Line to a Graph

Calculus is a branch of mathematics that studies rates of change of functions. In this course, you will learn that rates of change have many applications in real life. In Section 1.3, you learned how the slope of a line indicates the rate at which the line rises or falls. For a line, this rate (or slope) is the same at every point on the line. For graphs other than lines, the rate at which the graph rises or falls changes from point to point. For instance, in Figure 2.1, the parabola is rising more quickly at the point (x_1, y_1) than it is at the point (x_2, y_2). At the vertex (x_3, y_3), the graph levels off, and at the point (x_4, y_4), the graph is falling.

To determine the rate at which a graph rises or falls at a *single point*, you can find the slope of the **tangent line** at the point. In simple terms, the tangent line to the graph of a function f at a point $P(x_1, y_1)$ is the line that best approximates the graph at that point, as shown in Figure 2.1. Figure 2.2 shows other examples of tangent lines.

FIGURE 2.1 The slope of a nonlinear graph changes from one point to another.

DEFINITIONS AND THEOREMS

All definitions and theorems are highlighted for emphasis and easy reference.

Definition of Instantaneous Rate of Change

The **instantaneous rate of change** (or simply **rate of change**) of $y = f(x)$ at x is the limit of the average rate of change on the interval $[x, x + \Delta x]$, as Δx approaches 0.

$$\lim_{\Delta x \to 0} \frac{\Delta y}{\Delta x} = \lim_{\Delta x \to 0} \frac{f(x + \Delta x) - f(x)}{\Delta x}$$

If y is a distance and x is time, then the rate of change is a **velocity**.

The Sum and Difference Rules

The derivative of the sum or difference of two differentiable functions is the sum or difference of their derivatives.

$$\frac{d}{dx}[f(x) + g(x)] = f'(x) + g'(x) \qquad \text{Sum Rule}$$

$$\frac{d}{dx}[f(x) - g(x)] = f'(x) - g'(x) \qquad \text{Difference Rule}$$

---

154 CHAPTER 2 Differentiation

Application

Example 9 Rate of Change of Systolic Blood Pressure Ⓡ

As blood moves from the heart through the major arteries out to the capillaries and back through the veins, the systolic blood pressure continuously drops. Consider a person whose systolic blood pressure P (in millimeters of mercury) is given by

$$P = \frac{25t^2 + 125}{t^2 + 1}, \quad 0 \le t \le 10$$

where t is measured in seconds. At what rate is the blood pressure changing 5 seconds after blood leaves the heart?

SOLUTION Begin by applying the Quotient Rule.

$$\frac{dP}{dt} = \frac{(t^2 + 1)(50t) - (25t^2 + 125)(2t)}{(t^2 + 1)^2} \qquad \text{Quotient Rule}$$

$$= \frac{50t^3 + 50t - 50t^3 - 250t}{(t^2 + 1)^2}$$

$$= -\frac{200t}{(t^2 + 1)^2} \qquad \text{Simplify.}$$

When $t = 5$, the rate of change is

$$-\frac{200(5)}{26^2} \approx -1.48 \text{ millimeters per second.}$$

So, the pressure is *dropping* at a rate of 1.48 millimeters per second when $t = 5$ seconds.

✓**CHECKPOINT 9**

In Example 9, find the rate at which systolic blood pressure is changing at each time shown in the table below. Describe the changes in blood pressure as the blood moves away from the heart.

t	0	1	2	3	4	5	6	7
$\frac{dP}{dt}$								

EXAMPLES

There are a wide variety of relevant examples in the text, each titled for easy reference. Many of the solutions are presented graphically, analytically, and/or numerically to provide further insight into mathematical concepts. Examples using a real-life situation are identified with an icon. Ⓡ

CHECKPOINT

After each example, a similar problem is presented to allow for immediate practice, and to further reinforce your understanding of the concepts just learned.

Features

ALGEBRA REVIEWS

These appear throughout each chapter and offer algebraic support at point of use. Many of the reviews are then revisited in the Algebra Review at the end of the chapter, where additional details of examples with solutions and explanations are provided.

---

174 CHAPTER 2 Differentiation

Algebra Review

Simplifying Algebraic Expressions

To be successful in using derivatives, you must be good at simplifying algebraic expressions. Here are some helpful simplification techniques.

1. Combine *like terms*. This may involve expanding an expression by multiplying factors.
2. Divide out *like factors* in the numerator and denominator of an expression.
3. Factor an expression.
4. Rationalize a denominator.
5. Add, subtract, multiply, or divide fractions.

TECHNOLOGY

Symbolic algebra systems can simplify algebraic expressions. If you have access to such a system, try using it to simplify the expressions in this Algebra Review.

Example 1 Simplifying Fractional Expressions

a. $\dfrac{(x + \Delta x)^2 - x^2}{\Delta x} = \dfrac{x^2 + 2x(\Delta x) + (\Delta x)^2 - x^2}{\Delta x}$ Expand expression.

$= \dfrac{2x(\Delta x) + (\Delta x)^2}{\Delta x}$ Combine like terms.

$= \dfrac{\Delta x(2x + \Delta x)}{\Delta x}$ Factor.

$= 2x + \Delta x, \quad \Delta x \neq 0$ Divide out like factors.

b. $\dfrac{(x^2 - 1)(-2 - 2x) - (3 - 2x - x^2)(2)}{(x^2 - 1)^2}$

$= \dfrac{(-2x^2 - 2x^3 + 2 + 2x) - (6 - 4x - 2x^2)}{(x^2 - 1)^2}$ Expand expression.

$= \dfrac{-2x^2 - 2x^3 + 2 + 2x - 6 + 4x + 2x^2}{(x^2 - 1)^2}$ Remove parentheses.

$= \dfrac{-2x^3 + 6x - 4}{(x^2 - 1)^2}$ Combine like terms.

c. $2\left(\dfrac{2x + 1}{3x}\right)\left[\dfrac{3x(2) - (2x + 1)(3)}{(3x)^2}\right]$

$= 2\left(\dfrac{2x + 1}{3x}\right)\left[\dfrac{6x - (6x + 3)}{(3x)^2}\right]$ Multiply factors.

$= \dfrac{2(2x + 1)(6x - 6x - 3)}{(3x)^3}$ Multiply fractions and remove parentheses.

$= \dfrac{2(2x + 1)(-3)}{3(9)x^3}$ Combine like terms and factor.

$= \dfrac{-2(2x + 1)}{9x^3}$ Divide out like factors.

Algebra Review

When applying the Quotient Rule, it is suggested that you enclose all factors and derivatives in symbols of grouping, such as parentheses. Also, pay special attention to the subtraction required in the numerator. For help in evaluating expressions such as the one in Example 4, see the *Chapter 2 Algebra Review* on page 175, Example 2(d).

STUDY TIP

An interpretation of the Constant Rule says that the tangent line to a constant function is the function itself. Find an equation of the tangent line to $f(x) = -4$ at $x = 3$.

STUDY TIPS

Scattered throughout the text, study tips address special cases, expand on concepts, and help you to avoid common errors.

STUDY TIP

In real-life problems, it is important to list the units of measure for a rate of change. The units for $\Delta y/\Delta x$ are "*y*-units" per "*x*-units." For example, if y is measured in miles and x is measured in hours, then $\Delta y/\Delta x$ is measured in *miles per hour*.

LIFE AND SOCIAL SCIENCE CAPSULES

Life and Social Science Capsules appear at the ends of several sections. These capsules and their accompanying exercises discuss life and social science careers and companies that are related to the mathematical concepts covered in the chapter.

Social Science Capsule

© Bill Aron/PhotoEdit

Social work is a profession for those with a strong desire to help improve people's lives. About 90% of social workers are in health care and social assistance industries, as well as state and local government agencies. Most states require 3000 hours of supervised clinical experience for licensure. Employment of social workers is expected to increase faster than the average for all occupations through 2014.

70. **Research Project** Use your school's library, the Internet, or some other reference source to find information about the many career paths possible for someone with a degree in social work. Choose one career path and write a short paper about it.

CONCEPT CHECK

1. Complete the following: When determining the rate of change in the concentration of a drug over a 2-minute time interval, you are finding an _____ rate of change.

2. Complete the following: When determining the rate of change in the concentration of a drug exactly 1 hour after the drug was administered, you are finding an _____ rate of change.

3. You are asked to find the rate of change of a function over a certain interval. Should you find the average rate of change or the instantaneous rate of change?

4. You are asked to find the rate of change of a function at a certain instant. Should you find the average rate of change or the instantaneous rate of change?

CONCEPT CHECK

These non-computational questions appear at the end of each section and are designed to check your understanding of the concepts covered in that section.

Enhance Your Understanding Using Technology

DISCOVERY

These projects appear before selected topics and allow you to explore concepts on your own. These boxed features are optional, so they can be omitted with no loss of continuity in the coverage of material.

DISCOVERY

Use a graphing utility to graph $f(x) = 2x^3 - 4x^2 + 3x - 5$. On the same screen, sketch the graphs of $y = x - 5$, $y = 2x - 5$, and $y = 3x - 5$. Which of these lines, if any, appears to be tangent to the graph of f at the point $(0, -5)$? Explain your reasoning.

TECHNOLOGY

There are several ways to use technology to find relative extrema of a function. One way is to use a graphing utility to graph the function, and then use the *zoom* and *trace* features to find the relative minimum and relative maximum points. For instance, consider the graph of

$$f(x) = 3.1x^3 - 7.3x^2 + 1.2x + 2.5$$

as shown below.

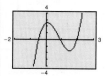

From the graph, you can see that the function has one relative maximum one relative minimum. You can approximate these values by zooming using the *trace* feature, as shown below.

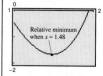 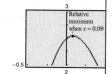

A second way to use technology to find relative extrema is to perform First-Derivative Test with a symbolic differentiation utility. You can utility to differentiate the function, set the derivative equal to zero, and solve the resulting equation. After obtaining the critical numbers, 1.4 and 0.0870148, you can graph the function and observe that the first a relative minimum and the second yields a relative maximum. Comp two ways shown above with doing the calculations by hand, as show below.

$f(x) = 3.1x^3 - 7.3x^2 + 1.2x + 2.5$ Write original

$f'(x) = \dfrac{d}{dx}[3.1x^3 - 7.3x^2 + 1.2x + 2.5]$ Differentiate w respect to x.

$f'(x) = 9.3x^2 - 14.6x + 1.2$ First derivative

$9.3x^2 - 14.6x + 1.2 = 0$ Set derivative

$x = \dfrac{73 \pm \sqrt{4213}}{93}$ Solve for x.

$x \approx 1.48288, x \approx 0.0870148$ Approximate.

TECHNOLOGY BOXES

These boxes appear throughout the text and provide guidance on using technology to ease lengthy calculations, present a graphical solution, or discuss where using technology can lead to misleading or wrong solutions.

TECHNOLOGY

You can use a graphing utility to confirm the result given in Example 7. One way to do this is to choose a point on the graph of $y = 2/t$, such as $(1, 2)$, and find the equation of the tangent line at that point. Using the derivative found in the example, you know that the slope of the tangent line when $t = 1$ is $m = -2$. This means that the tangent line at the point $(1, 2)$ is

$y - y_1 = m(t - t_1)$

$y - 2 = -2(t - 1)$ or

$y = -2t + 4$.

By graphing $y = 2/t$ and $y = -2t + 4$ in the same viewing window, as shown below, you can confirm that the line is tangent to the graph at the point $(1, 2)$.*

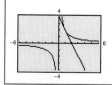

Practice

SKILLS REVIEW

These exercises at the beginning of each exercise set help you review skills covered in previous sections. The answers are provided at the back of the text to reinforce your understanding of the skill sets learned.

EXERCISE SETS

These exercises offer opportunities for practice and review. They progress in difficulty from skill-development problems to more challenging problems, to build confidence and understanding.

MAKE A DECISION

Multi-step exercises reinforce your problem-solving skills and mastery of concepts, as well as taking a real-life application further by testing what you know about a given problem to make a decision within the context of the problem.

---

63. MAKE A DECISION: ENVIRONMENT The predicted cost C (in hundreds of thousands of dollars) for a company to remove $p\%$ of a chemical from its waste water can be modeled by

$$C = \frac{124p}{(10 + p)(100 - p)}, \quad 0 \leq p < 100.$$

(a) Find the rates of change in the cost when $p = 25\%$ and $p = 75\%$.

---

55. MAKE A DECISION: SOCIAL SECURITY The table lists the average monthly Social Security benefits B (in dollars) for retired workers aged 62 and over from 1998 through 2005. A model for the data is

$$B = \frac{582.6 + 38.38t}{1 + 0.025t - 0.0009t^2}, \quad 8 \leq t \leq 15$$

where t is the year, with $t = 8$ corresponding to 1998. *(Source: U.S. Social Security Administration)*

t	8	9	10	11	12	13	14	15
B	780	804	844	874	895	922	955	1002

---

78. U.S. AIDS Epidemic The number of cases f (in thousands) of AIDS reported in the years 1995 through 2005 can be modeled by

$$f = \sqrt{-1.8840t^4 + 66.325t^3 - 739.92t^2 + 2307.0t + 5125}$$

where t represents the year, with $t = 5$ corresponding to 1995. *(Source: U.S. Centers for Disease Control and Prevention)*

(a) Find the derivative of the model. Which differentiation rule(s) did you use?

(b) Use a graphing utility to graph the derivative on the interval $5 \leq t \leq 15$.

TECHNOLOGY EXERCISES

Many exercises in the text can be solved with or without technology. The (T) symbol identifies exercises for which you are specifically instructed to use a graphing calculator or a computer algebra system to solve the problem. Additionally, the (S) symbol denotes exercises best solved by using a spreadsheet.

---

68. Human Memory Model Consider the learning curve modeled by

$$P = \frac{0.55 + 0.87(n - 1)}{1 + 0.87(n - 1)}, \quad n > 0$$

where P is the percent of the correct responses after n trials.

(a) Use a spreadsheet software pro table for the model.

n	1	2	3	4	5	6
P						

(b) Find the limit of P as n approach

---

83. Modeling Data The table shows the numbers n (in thousands) of students enrolled in public schools in Puerto Rico for the years 2000 through 2005, where t is the year, with $t = 0$ corresponding to 2000. *(Source: Puerto Rico Planning Board)*

t	0	1	2	3	4	5
n	612.3	610.8	603.5	596.3	584.9	575.9

Study and Review

MID-CHAPTER QUIZ

Appearing in the middle of each chapter, this one-page test allows you to practice skills and concepts learned in the chapter. This opportunity for self-assessment will uncover any potential weak areas that might require further review of the material.

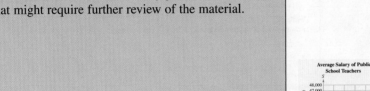

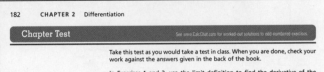

CHAPTER TEST

Appearing at the end of the chapter, this test is designed to simulate an in-class exam. Taking these tests will help you to determine what concepts require further study and review.

CHAPTER SUMMARY AND STUDY STRATEGIES

The *Chapter Summary* reviews the skills covered in the chapter and correlates each skill to the Review Exercises that test the skill. Following each *Chapter Summary* is a short list of Study Strategies for addressing topics or situations in the chapter.

CHAPTER 2 Differentiation

Chapter Summary and Study Strategies

After studying this chapter, you should have acquired the following skills. The exercise numbers are keyed to the Review Exercises that begin on page 178. Answers to odd-numbered Review Exercises are given in the back of the text.*

Section 2.1	Review Exercises
■ Approximate the slope of the tangent line to a graph at a point.	1–4
■ Interpret the slope of a graph in a real-life setting.	5–8
■ Use the limit definition to find the derivative of a function and the slope of a graph at a point.	9–16
$\quad f'(x) = \lim_{\Delta x \to 0} \dfrac{f(x + \Delta x) - f(x)}{\Delta x}$	
■ Use the derivative to find the slope of a graph at a point.	17–24
■ Use the graph of a function to recognize points at which the function is not differentiable.	25–28

Section 2.2	
■ Use the Constant Multiple Rule for differentiation.	29, 30
$\quad \dfrac{d}{dx}[cf(x)] = cf'(x)$	
■ Use the Sum and Difference Rules for differentiation.	31–38
$\quad \dfrac{d}{dx}[f(x) \pm g(x)] = f'(x) \pm g'(x)$	

Section 2.3	
■ Find the average rate of change of a function over an interval and the instantaneous rate of change at a point.	39, 40
$\quad \text{Average rate of change} = \dfrac{f(b) - f(a)}{b - a}$	
$\quad \text{Instantaneous rate of change} = \lim_{\Delta x \to 0} \dfrac{f(x + \Delta x) - f(x)}{\Delta x}$	
■ Find the average and instantaneous rates of change of a quantity in a real-life problem.	41, 42
■ Use rates of change to solve real-life problems.	43–48

---

* Use a wide range of valuable study aids to help you master the material in this chapter. The *Student Solutions Guide* includes step-by-step solutions to all odd-numbered exercises to help you review and prepare. The student website at *college.hmco.com/info/larsonapplied* offers algebra help and a *Graphing Technology Guide*. The *Graphing Technology Guide* contains step-by-step commands and instructions for a wide variety of graphing calculators, including the most recent models.

Applied Calculus
for the Life and Social Sciences

A Precalculus Review

One way erosion occurs is by running water. Running water transports particles such as soil or fragments of rock. These particles strike against the bedrock of the stream channel, grind it away, and eventually settle in the channel or find their way to the sea. (See Section 0.3, Exercise 58.)

Applications

Topics in precalculus have many real-life applications. The applications listed below represent a sample of the applications in this chapter.

- Biology: pH Values, Exercise 33, page 7
- Life Expectancy, Exercise 46, page 12
- Chemistry: Finding Concentrations, Exercise 77, page 24
- Plants, Exercise 46, page 32

0.1 The Real Number Line and Order

0.2 Absolute Value and Distance on the Real Number Line

0.3 Exponents and Radicals

0.4 Factoring Polynomials

0.5 Fractions and Rationalization

Section 0.1

The Real Number Line and Order

- Represent, classify, and order real numbers.
- Use inequalities to represent sets of real numbers.
- Solve inequalities.
- Use inequalities to model and solve real-life problems.

The Real Number Line

Real numbers can be represented with a coordinate system called the **real number line** (or x-axis), as shown in Figure 0.1. The **positive direction** (to the right) is denoted by an arrowhead and indicates the direction of increasing values of x. The real number corresponding to a particular point on the real number line is called the **coordinate** of the point. As shown in Figure 0.1, it is customary to label those points whose coordinates are integers.

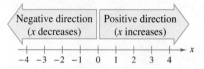

FIGURE 0.1 The Real Number Line

The point on the real number line corresponding to zero is called the **origin**. Numbers to the right of the origin are **positive,** and numbers to the left of the origin are **negative**. The term **nonnegative** describes a number that is either positive or zero.

The importance of the real number line is that it provides you with a conceptually perfect picture of the real numbers. That is, each point on the real number line corresponds to one and only one real number, and each real number corresponds to one and only one point on the real number line. This type of relationship is called a **one-to-one correspondence** and is illustrated in Figure 0.2.

Each of the four points in Figure 0.2 corresponds to a real number that can be expressed as the ratio of two integers.

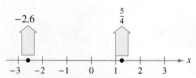

Every point on the real number line corresponds to one and only one real number.

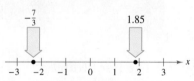

Every real number corresponds to one and only one point on the real number line.

FIGURE 0.2

$$-2.6 = -\frac{13}{5} \qquad \frac{5}{4} \qquad -\frac{7}{3} \qquad 1.85 = \frac{37}{20}$$

Such numbers are called **rational**. Rational numbers have either terminating or infinitely repeating decimal representations.

Terminating Decimals

$\dfrac{2}{5} = 0.4$

$\dfrac{7}{8} = 0.875$

Infinitely Repeating Decimals

$\dfrac{1}{3} = 0.333\ldots = 0.\overline{3}$*

$\dfrac{12}{7} = 1.714285714285\ldots = 1.\overline{714285}$

Real numbers that are not rational are called **irrational,** and they cannot be represented as the ratio of two integers (or as terminating or infinitely repeating decimals). So, a decimal approximation is used to represent an irrational number. Some irrational numbers occur so frequently in applications that mathematicians have invented special symbols to represent them. For example, the symbols $\sqrt{2}$, π, and e represent irrational numbers whose decimal approximations are as shown. (See Figure 0.3.)

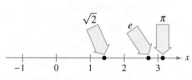

FIGURE 0.3

$$\sqrt{2} \approx 1.4142135623 \qquad \pi \approx 3.1415926535 \qquad e \approx 2.7182818284$$

---

*The bar indicates which digit or digits repeat infinitely.

Order and Intervals on the Real Number Line

One important property of the real numbers is that they are **ordered**: 0 is less than 1, -3 is less than -2.5, π is less than $\frac{22}{7}$, and so on. You can visualize this property on the real number line by observing that a is less than b if and only if a lies to the left of b on the real number line. Symbolically, "a is less than b" is denoted by the inequality $a < b$. For example, the inequality $\frac{3}{4} < 1$ follows from the fact that $\frac{3}{4}$ lies to the left of 1 on the real number line, as shown in Figure 0.4.

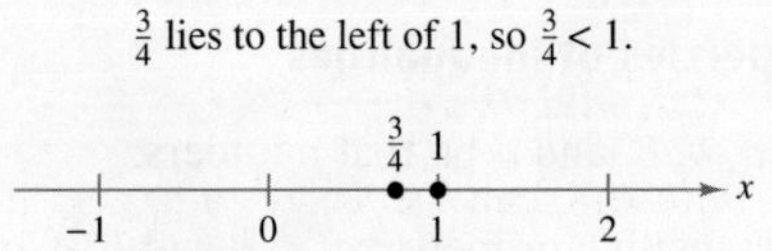

$\frac{3}{4}$ lies to the left of 1, so $\frac{3}{4} < 1$.

FIGURE 0.4

When three real numbers a, x, and b are ordered such that $a < x$ and $x < b$, we say that x is **between** a and b and write

$a < x < b.$ x is between a and b.

The set of *all* real numbers between a and b is called the **open interval** between a and b and is denoted by (a, b). An interval of the form (a, b) does not contain the "endpoints" a and b. Intervals that include their endpoints are called **closed** and are denoted by $[a, b]$. Intervals of the form $[a, b)$ and $(a, b]$ are neither open nor closed. Figure 0.5 shows the nine types of intervals on the real number line.

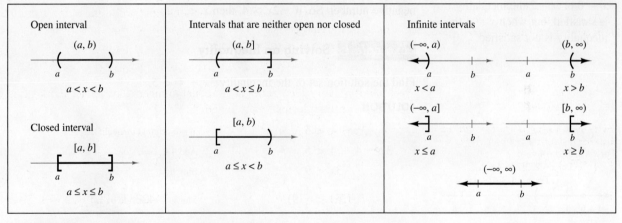

FIGURE 0.5 Intervals on the Real Number Line

> **STUDY TIP**
> Note that a square bracket is used to denote "less than or equal to" ($\leq$) or "greater than or equal to" ($\geq$). Furthermore, the symbols ∞ and $-\infty$ denote **positive** and **negative infinity**. These symbols do not denote real numbers; they merely let you describe unbounded conditions more concisely. For instance, the interval $[b, \infty)$ is unbounded to the right because it includes *all* real numbers that are greater than or equal to b.

CHAPTER 0 A Precalculus Review

Solving Inequalities

In calculus, you are frequently required to "solve inequalities" involving variable expressions such as $3x - 4 < 5$. The number a is a **solution** of an inequality if the inequality is true when a is substituted for x. The set of all values of x that satisfy an inequality is called the **solution set** of the inequality. The following properties are useful for solving inequalities. (Similar properties are obtained if $<$ is replaced by $\le$ and $>$ is replaced by $\ge$.)

STUDY TIP

Notice the differences between Properties 3 and 4. For example,
$$-3 < 4 \Longrightarrow (-3)(2) < (4)(2)$$
and
$$-3 < 4 \Longrightarrow (-3)(-2) > (4)(-2).$$

Properties of Inequalities

Let a, b, c, and d be real numbers.

1. Transitive property: $a < b$ and $b < c$ $\Longrightarrow$ $a < c$
2. Adding inequalities: $a < b$ and $c < d$ $\Longrightarrow$ $a + c < b + d$
3. Multiplying by a (positive) constant: $a < b$ $\Longrightarrow$ $ac < bc$, $c > 0$
4. Multiplying by a (negative) constant: $a < b$ $\Longrightarrow$ $ac > bc$, $c < 0$
5. Adding a constant: $a < b$ $\Longrightarrow$ $a + c < b + c$
6. Subtracting a constant: $a < b$ $\Longrightarrow$ $a - c < b - c$

STUDY TIP

Once you have solved an inequality, it is a good idea to check some x-values in your solution set to see whether they satisfy the original inequality. You might also check some values outside your solution set to verify that they do *not* satisfy the inequality. For example, Figure 0.6 shows that when $x = 0$ or $x = 2$ the inequality is satisfied, but when $x = 4$ the inequality is not satisfied.

Note that you *reverse the inequality* when you multiply by a negative number. For example, if $x < 3$, then $-4x > -12$. This principle also applies to division by a negative number. So, if $-2x > 4$, then $x < -2$.

Example 1 Solving an Inequality

Find the solution set of the inequality $3x - 4 < 5$.

SOLUTION

$3x - 4 < 5$	Write original inequality.
$3x - 4 + 4 < 5 + 4$	Add 4 to each side.
$3x < 9$	Simplify.
$\frac{1}{3}(3x) < \frac{1}{3}(9)$	Multiply each side by $\frac{1}{3}$.
$x < 3$	Simplify.

So, the solution set is the interval $(-\infty, 3)$, as shown in Figure 0.6.

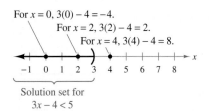

Solution set for $3x - 4 < 5$

FIGURE 0.6

✓ CHECKPOINT 1

Find the solution set of the inequality $2x - 3 < 7$. ■

In Example 1, all five inequalities listed as steps in the solution have the same solution set, and they are called **equivalent inequalities**.

SECTION 0.1 The Real Number Line and Order

The inequality in Example 1 involves a first-degree polynomial. To solve inequalities involving polynomials of higher degree, you can use the fact that a polynomial can change signs *only* at its real zeros (the real numbers that make the polynomial zero). Between two consecutive real zeros, a polynomial must be entirely positive or entirely negative. This means that when the real zeros of a polynomial are put in order, they divide the real number line into **test intervals** in which the polynomial has no sign changes. That is, if a polynomial has the factored form

$$(x - r_1)(x - r_2), \ldots, (x - r_n), \quad r_1 < r_2 < r_3 < \cdots < r_n$$

then the test intervals are

$$(-\infty, r_1), \quad (r_1, r_2), \quad \ldots, \quad (r_{n-1}, r_n), \quad \text{and} \quad (r_n, \infty).$$

For example, the polynomial

$$x^2 - x - 6 = (x - 3)(x + 2)$$

can change signs only at $x = -2$ and $x = 3$. To determine the sign of the polynomial in the intervals $(-\infty, -2)$, $(-2, 3)$, and $(3, \infty)$, you need to test only *one value* from each interval.

Example 2 Solving a Polynomial Inequality

Find the solution set of the inequality $x^2 < x + 6$.

SOLUTION

$$x^2 < x + 6 \qquad \text{Write original inequality.}$$
$$x^2 - x - 6 < 0 \qquad \text{Polynomial form}$$
$$(x - 3)(x + 2) < 0 \qquad \text{Factor.}$$

So, the polynomial $x^2 - x - 6$ has $x = -2$ and $x = 3$ as its zeros. You can solve the inequality by testing the sign of the polynomial in each of the following intervals.

$$x < -2, \quad -2 < x < 3, \quad x > 3$$

To test an interval, choose a representative number in the interval and compute the sign of each factor. For example, for any $x < -2$, both of the factors $(x - 3)$ and $(x + 2)$ are negative. Consequently, the product (of two negative numbers) is positive, and the inequality is *not* satisfied in the interval

$$x < -2.$$

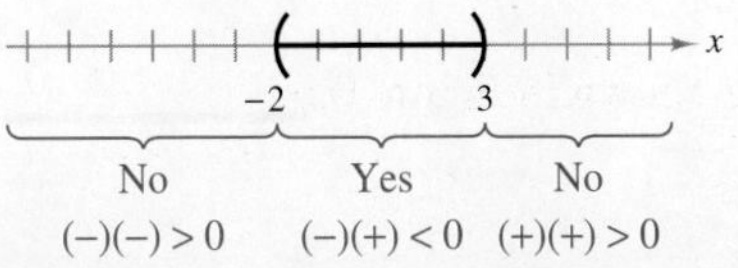

A convenient testing format is shown in Figure 0.7. Because the inequality is satisfied only by the center test interval, you can conclude that the solution set is given by the interval

$$-2 < x < 3. \qquad \text{Solution set}$$

✓ CHECKPOINT 2

Find the solution set of the inequality $x^2 > 3x + 10$. ■

Sign of $(x - 3)(x + 2)$

x	Sign	<0?
-3	$(-)(-)$	No
-2	$(-)(0)$	No
-1	$(-)(+)$	Yes
0	$(-)(+)$	Yes
1	$(-)(+)$	Yes
2	$(-)(+)$	Yes
3	$(0)(+)$	No
4	$(+)(+)$	No

```
++++++(++++)++++→ x
     -2      3
   No    Yes    No
 (-)(-)>0  (-)(+)<0  (+)(+)>0
```

FIGURE 0.7 Is $(x - 3)(x + 2) < 0$?

Application

Inequalities are frequently used to describe conditions that occur in business and science. For instance, the inequality

$$144 \le W \le 180$$

describes the recommended weight W for a man whose height is 5 feet 10 inches. Example 3 shows how an inequality can be used to describe temperatures at which a lizard will be comfortable during the day.

Example 3 MAKE A DECISION Temperature

© Danita Delimont/Alamy

Crested geckos are native to Southern Grand Terre, New Caledonia and at least one nearby island (Isle of Pines). At temperatures of 85°F or warmer, crested geckos will become stressed, possibly leading to illness or death.

According to a pet care guide, your pet lizard should be kept in a room where the daytime temperature ranges from 71.6°F to 77°F. Keeping the lizard in higher or lower temperatures could make the lizard ill. Find the high and low temperatures in degrees Celsius. Will a daytime temperature of 24.5°C be harmful to the lizard?

SOLUTION The range of temperatures F in degrees Fahrenheit is

$$71.6 \le F \le 77.0.$$

Use the temperature conversion formula $F = 1.8C + 32$ to write the following.

$71.6 \le$	F	≤ 77	Write original inequality.
$71.6 \le$	$1.8C + 32$	≤ 77	Substitute $1.8C + 32$ for F.
$71.6 - 32 \le$	$1.8C + 32 - 32$	$\le 77 - 32$	Subtract 32 from each part.
$39.6 \le$	$1.8C$	≤ 45	Simplify.
$\dfrac{39.6}{1.8} \le$	$\dfrac{1.8C}{1.8}$	$\le \dfrac{45}{1.8}$	Divide each part by 1.8.
$22 \le$	C	≤ 25	Simplify.

So, the daytime temperature ranges from 22°C to 25°C. No, a daytime temperature of 24.5°C will not be harmful to the lizard, as shown in Figure 0.8.

FIGURE 0.8

✓ CHECKPOINT 3

During nighttime, the lizard in Example 3 should be kept in a room where the temperature ranges from 68°F to 75.2°F. Find the high and low temperatures in degrees Celsius. ∎

The symbol ® indicates an example that uses or is derived from real-life data.

Exercises 0.1

See www.CalcChat.com for worked-out solutions to odd-numbered exercises.

In Exercises 1–10, determine whether the real number is rational or irrational.

*1. 0.25
2. -3678
3. $\dfrac{3\pi}{2}$
4. $3\sqrt{2} - 1$
5. $4.3\overline{451}$
6. $\dfrac{22}{7}$
7. $\sqrt[3]{64}$
8. $0.\overline{8177}$
9. $\sqrt[3]{60}$
10. $2e$

In Exercises 11–14, determine whether each given value of x satisfies the inequality.

11. $5x - 12 > 0$
   (a) $x = 3$ (b) $x = -3$ (c) $x = \tfrac{5}{2}$

12. $x + 1 < \dfrac{x}{3}$
   (a) $x = 0$ (b) $x = 4$ (c) $x = -4$

13. $0 < \dfrac{x-2}{4} < 2$
   (a) $x = 4$ (b) $x = 10$ (c) $x = 0$

14. $-1 < \dfrac{3-x}{2} \leq 1$
   (a) $x = 0$ (b) $x = 1$ (c) $x = 5$

In Exercises 15–28, solve the inequality and sketch the graph of the solution on the real number line.

15. $x - 5 \geq 7$
16. $2x > 3$
17. $4x + 1 < 2x$
18. $2x + 7 < 3$
19. $4 - 2x < 3x - 1$
20. $x - 4 \leq 2x + 1$
21. $-4 < 2x - 3 < 4$
22. $0 \leq x + 3 < 5$
23. $\dfrac{3}{4} > x + 1 > \dfrac{1}{4}$
24. $-1 < -\dfrac{x}{3} < 1$
25. $\dfrac{x}{2} + \dfrac{x}{3} > 5$
26. $\dfrac{x}{2} - \dfrac{x}{3} > 5$
27. $2x^2 - x < 6$
28. $2x^2 + 1 < 9x - 3$

In Exercises 29–32, use inequality notation to describe the subset of real numbers.

29. A gas station expects the cost per gallon g of 87-octane gasoline for the next week to be no less than \$2.39 per gallon and no more than \$3.25 per gallon.

30. The average length l of a fish is no less than 7 inches and no more than 12 inches.

31. A person expects their weight loss w for the next three months to be no less than 18 pounds.

32. To grow cabbage properly, gardeners keep the pH level p of the soil at no more than 7.5.

Ⓑ 33. **Biology: pH Values** The pH scale measures the concentration of hydrogen ions in a solution. Strong acids produce low pH values, while strong bases produce high pH values. Represent the following approximate pH values on a real number line: hydrochloric acid, 0.0; lemon juice, 2.0; oven cleaner, 13.0; baking soda, 9.0; pure water, 7.0; black coffee, 5.0. *(Source: Adapted from Levine/Miller, Biology: Discovering Life, Second Edition)*

34. **Physiology** The maximum heart rate of a person in normal health is related to the person's age by the equation

$$r = 220 - A$$

where r is the maximum heart rate in beats per minute and A is the person's age in years. Some physiologists recommend that during physical activity a person should strive to increase his or her heart rate to at least 60% of the maximum heart rate for sedentary people and at most 90% of the maximum heart rate for highly fit people. Express as an interval the range of the target heart rate for a 20-year-old.

35. **Terrarium Temperature** Your pet frog should be kept in a terrarium where the temperature ranges from 68°F to 75°F. The current temperature of the terrarium is 18.3°C. Does this temperature fall within the acceptable range?

36. **Egg Production** The number E of eggs (in billions) produced in the United States from 2000 through 2005 can be modeled by

$$E = 1.02t + 84.9, \quad 0 \leq t \leq 5$$

where t represents the year, with $t = 0$ corresponding to 2000. According to this model, when was the annual egg production at least 86 billion, but no more than 88 billion? *(Source: U.S. Department of Agriculture)*

In Exercises 37 and 38, determine whether each statement is true or false, given $a < b$.

37. (a) $-2a < -2b$
    (b) $a + 2 < b + 2$
    (c) $6a < 6b$
    (d) $\dfrac{1}{a} < \dfrac{1}{b}$

38. (a) $a - 4 < b - 4$
    (b) $4 - a < 4 - b$
    (c) $-3b < -3a$
    (d) $\dfrac{a}{4} < \dfrac{b}{4}$

---

*The answers to the odd-numbered and selected even-numbered exercises are given in the back of the text. Worked-out solutions to the odd-numbered exercises are given in the *Student Solutions Guide*.

Section 0.2

Absolute Value and Distance on the Real Number Line

- Find the absolute values of real numbers and understand the properties of absolute value.
- Find the distance between two numbers on the real number line.
- Define intervals on the real number line.
- Find the midpoint of an interval and use intervals to model and solve real-life problems.

Absolute Value of a Real Number

TECHNOLOGY

Absolute value expressions can be evaluated on a graphing utility. When an expression such as $|3 - 8|$ is evaluated, parentheses should surround the expression, as in abs(3 − 8).

Definition of Absolute Value

The **absolute value** of a real number a is

$$|a| = \begin{cases} a, & \text{if } a \geq 0 \\ -a, & \text{if } a < 0. \end{cases}$$

At first glance, it may appear from this definition that the absolute value of a real number can be negative, but this is not possible. For example, let $a = -3$. Then, because $-3 < 0$, you have

$$|a| = |-3| = -(-3) = 3.$$

The following properties are useful for working with absolute values.

Properties of Absolute Value

1. Multiplication: $|ab| = |a||b|$
2. Division: $\left|\dfrac{a}{b}\right| = \dfrac{|a|}{|b|}, \quad b \neq 0$
3. Power: $|a^n| = |a|^n$
4. Square root: $\sqrt{a^2} = |a|$

Be sure you understand the fourth property in this list. A common error in algebra is to imagine that by squaring a number and then taking the square root, you come back to the original number. But this is true only if the original number is nonnegative. For instance, if $a = 2$, then

$$\sqrt{2^2} = \sqrt{4} = 2$$

but if $a = -2$, then

$$\sqrt{(-2)^2} = \sqrt{4} = 2.$$

The reason for this is that (by definition) the square root symbol $\sqrt{}$ denotes only the nonnegative root.

Distance on the Real Number Line

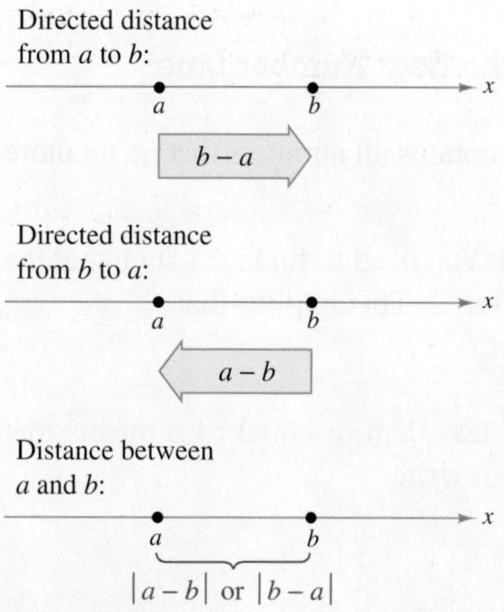

Consider two distinct points on the real number line, as shown in Figure 0.9.

1. The **directed distance from a to b** is $b - a$.
2. The **directed distance from b to a** is $a - b$.
3. The **distance between a and b** is $|a - b|$ or $|b - a|$.

In Figure 0.9, note that because b is to the right of a, the directed distance from a to b (moving to the right) is positive. Moreover, because a is to the left of b, the directed distance from b to a (moving to the left) is negative. The distance *between* two points on the real number line can never be negative.

Directed distance from a to b:

Directed distance from b to a:

Distance between a and b:

FIGURE 0.9

Distance Between Two Points on the Real Number Line

The distance d between points x_1 and x_2 on the real number line is given by

$$d = |x_2 - x_1| = \sqrt{(x_2 - x_1)^2}.$$

Note that the order of subtraction with x_1 and x_2 does not matter because

$$|x_2 - x_1| = |x_1 - x_2| \quad \text{and} \quad (x_2 - x_1)^2 = (x_1 - x_2)^2.$$

Example 1 Finding Distance on the Real Number Line

Determine the distance between -3 and 4 on the real number line. What is the directed distance from -3 to 4? What is the directed distance from 4 to -3?

SOLUTION The distance between -3 and 4 is given by

$$|-3 - 4| = |-7| = 7 \quad \text{or} \quad |4 - (-3)| = |7| = 7 \qquad |a - b| \text{ or } |b - a|$$

as shown in Figure 0.10.

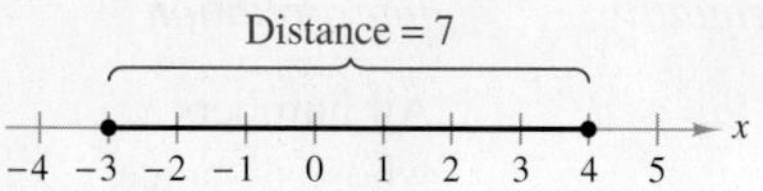

FIGURE 0.10

The directed distance from -3 to 4 is

$$4 - (-3) = 7. \qquad b - a$$

The directed distance from 4 to -3 is

$$-3 - 4 = -7. \qquad a - b$$

✓ CHECKPOINT 1

Determine the distance between -2 and 6 on the real number line. What is the directed distance from -2 to 6? What is the directed distance from 6 to -2? ■

Intervals Defined by Absolute Value

Example 2 Defining an Interval on the Real Number Line

Find the interval on the real number line that contains all numbers that lie no more than two units from 3.

SOLUTION Let x be any point in this interval. You need to find all x such that the distance between x and 3 is less than or equal to 2. This implies that

$$|x - 3| \leq 2.$$

Requiring the absolute value of $x - 3$ to be less than or equal to 2 means that $x - 3$ must lie between -2 and 2. So, you can write

$$-2 \leq x - 3 \leq 2.$$

Solving this pair of inequalities, you have

$$-2 + 3 \leq x - 3 + 3 \leq 2 + 3$$
$$1 \leq x \leq 5. \quad \text{Solution set}$$

So, the interval is $[1, 5]$, as shown in Figure 0.11.

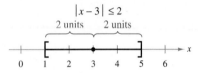

FIGURE 0.11

✓ **CHECKPOINT 2**

Find the interval on the real number line that contains all numbers that lie no more than four units from 6. ∎

Two Basic Types of Inequalities Involving Absolute Value

Let a and d be real numbers, where $d > 0$.

$|x - a| \leq d$ if and only if $a - d \leq x \leq a + d$.

$|x - a| \geq d$ if and only if $x \leq a - d$ or $a + d \leq x$.

Inequality	Interpretation	Graph
$\|x - a\| \leq d$	All numbers x whose distance from a is less than or equal to d.	
$\|x - a\| \geq d$	All numbers x whose distance from a is greater than or equal to d.	

STUDY TIP

Be sure you see that inequalities of the form $|x - a| \geq d$ have solution sets consisting of two intervals. To describe the two intervals without using absolute values, you must use *two* separate inequalities, connected by an "or" to indicate union.

Application

Example 3 Voter Poll

In a poll for an upcoming mayoral election, 40% of likely voters said they would vote for Candidate A. The poll has a margin of error of ±3%.

a. Write and solve an inequality that represents the percent of likely voters who said they would vote for Candidate A.

b. Assume there are 50,000 likely voters. Of these voters, how many does the poll predict would vote for Candidate A?

SOLUTION

a. Because the poll has a margin of error of ±3%, Candidate A expects to get 40% ± 3% of likely voters. Let A be the percent (written in decimal form) of likely voters for Candidate A. You know that A will differ from 0.40 by at most 0.03.

$$0.40 - 0.03 \leq A \leq 0.40 + 0.03$$
$$0.37 \leq A \leq 0.43 \qquad \text{Figure 0.12(a)}$$

b. Letting V be the number of people out of 50,000 who vote for Candidate A, it follows that $V = 50{,}000A$. So, the number of people who would vote for Candidate A would be given by

$$0.37(50{,}000) \leq 50{,}000A \leq 0.43(50{,}000)$$
$$18{,}500 \leq V \leq 21{,}500. \qquad \text{Figure 0.12(b)}$$

According to the poll, Candidate A can expect at least 18,500 votes and at most 21,500 votes.

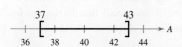

(a) Percent of likely voters

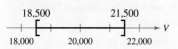

(b) Number of likely voters

FIGURE 0.12

✓ CHECKPOINT 3

Repeat Example 3 for Candidate B, where 37% of likely voters said they would vote for Candidate B with a margin of error of ±3%. ■

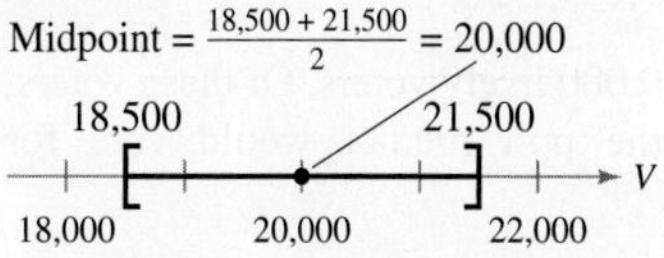

In Example 3(b), the candidate should expect to receive between 18,500 and 21,500 votes. Of course, a conservative expectation would be the lower of these estimates. However, from a statistical point of view, the most representative estimate would be the average of these two extremes. Graphically, the average of two numbers is the **midpoint** of the interval with the two numbers as endpoints, as shown in Figure 0.13.

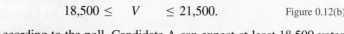

FIGURE 0.13

Midpoint of an Interval

The **midpoint** of the interval with endpoints a and b is found by taking the average of the endpoints.

$$\text{Midpoint} = \frac{a+b}{2}$$

Exercises 0.2

In Exercises 1–6, find (a) the directed distance from a to b, (b) the directed distance from b to a, and (c) the distance between a and b.

1. $a = 126, b = 75$
2. $a = -126, b = -75$
3. $a = 9.34, b = -5.65$
4. $a = -2.05, b = 4.25$
5. $a = \frac{16}{5}, b = \frac{112}{75}$
6. $a = -\frac{18}{5}, b = \frac{61}{15}$

In Exercises 7–18, use absolute values to describe the given interval (or pair of intervals) on the real number line.

7. $[-2, 2]$
8. $(-3, 3)$
9. $(-\infty, -2) \cup (2, \infty)$
10. $(-\infty, -3] \cup [3, \infty)$
11. $[2, 8]$
12. $(-7, -1)$
13. $(-\infty, 0) \cup (4, \infty)$
14. $(-\infty, 20) \cup (24, \infty)$
15. All numbers *less than* three units from 5
16. All numbers *more than* five units from 2
17. y is *at most* two units from a.
18. y is *less than* h units from c.

In Exercises 19–34, solve the inequality and sketch the graph of the solution on the real number line.

19. $|x| < 4$
20. $|2x| < 6$
21. $\left|\frac{x}{2}\right| > 3$
22. $|3x| > 12$
23. $|x - 5| < 2$
24. $|3x + 1| \geq 4$
25. $\left|\frac{x-3}{2}\right| \geq 5$
26. $|2x + 1| < 5$
27. $|10 - x| > 4$
28. $|25 - x| \geq 20$
29. $|9 - 2x| < 1$
30. $\left|1 - \frac{2x}{3}\right| < 1$
31. $|x - a| \leq b, b > 0$
32. $|2x - a| \geq b, b > 0$
33. $\left|\frac{3x-a}{4}\right| < 2b, b > 0$
34. $\left|a - \frac{5x}{2}\right| > b, b > 0$

In Exercises 35–40, find the midpoint of the given interval.

35. $[8, 24]$
36. $[7.3, 12.7]$
37. $[-6.85, 9.35]$
38. $[-4.6, -1.3]$
39. $\left[-\frac{1}{2}, \frac{3}{4}\right]$
40. $\left[\frac{5}{6}, \frac{5}{2}\right]$

41. **Chemistry** Copper has a melting point M within 0.2°C of 1083.4°C. Use absolute values to write the range as an inequality.

42. **Body Temperature** Normal body temperature T in humans is within 1°F of 98.6°F. Use absolute values to write the range as an inequality.

43. **Heights of a Population** The heights h of two-thirds of the members of a population satisfy the inequality

$$\left|\frac{h - 68.5}{2.7}\right| \leq 1$$

where h is measured in inches. Determine the interval on the real number line in which these heights lie.

44. **Biology** The American Kennel Club has developed guidelines for judging the features of various breeds of dogs. For collies, the guidelines specify that the weights for males satisfy the inequality

$$\left|\frac{w - 67.5}{7.5}\right| \leq 1$$

where w is measured in pounds. Determine the interval on the real number line in which these weights lie.

45. **Landscaping** The amount of grass contained in one square foot of a typical lawn is given by

$$|n - 850| \leq 100$$

where n is measured in blades of grass. Determine the high and low amounts of grass per square foot.

46. **Life Expectancy** Estimates of the life expectancy of a loggerhead sea turtle are given by $|x - 67.5| \leq 7.5$, where x is measured in years. Determine the high and low life expectancy estimates.

47. **Voter Poll** In a poll for an upcoming election, 45% of likely voters said they would vote for Candidate X. The poll has a margin of error of ±3%.

    (a) Write and solve an inequality that represents the percent of likely voters who said they would vote for Candidate X.

    (b) Assume there are 80,000 likely voters. Of these voters, how many does the poll predict would vote for Candidate X?

48. **Survey** In a survey on health insurance, 92% of persons covered by a particular plan said they were satisfied with the coverage they were receiving. The survey has a margin of error of ±5%.

    (a) Write and solve an inequality that represents the percent of persons who said they were satisfied with their coverage.

    (b) Assume there are 120,000 persons covered by the plan. According to the survey, how many said they were satisfied with their coverage?

Section 0.3
Exponents and Radicals

- Evaluate expressions involving exponents or radicals.
- Simplify expressions with exponents.
- Find the domains of algebraic expressions.

Expressions Involving Exponents or Radicals

Properties of Exponents

1. Whole-number exponents: $x^n = \underbrace{x \cdot x \cdot x \cdots x}_{n \text{ factors}}$

2. Zero exponent: $x^0 = 1, \quad x \neq 0$

3. Negative exponents: $x^{-n} = \dfrac{1}{x^n}, \quad x \neq 0$

4. Radicals (principal nth root): $\sqrt[n]{x} = a \implies x = a^n$

5. Rational exponents $(1/n)$: $x^{1/n} = \sqrt[n]{x}$

6. Rational exponents (m/n): $x^{m/n} = (x^{1/n})^m = \left(\sqrt[n]{x}\right)^m$

   $x^{m/n} = (x^m)^{1/n} = \sqrt[n]{x^m}$

7. Special convention (square root): $\sqrt[2]{x} = \sqrt{x}$

STUDY TIP
If n is even, then the principal nth root is positive. For example, $\sqrt{4} = +2$ and $\sqrt[4]{81} = +3$.

Example 1 Evaluating Expressions

	Expression	x-Value	Substitution
a.	$y = -2x^2$	$x = 4$	$y = -2(4^2) = -2(16) = -32$
b.	$y = 3x^{-3}$	$x = -1$	$y = 3(-1)^{-3} = \dfrac{3}{(-1)^3} = \dfrac{3}{-1} = -3$
c.	$y = (-x)^2$	$x = \dfrac{1}{2}$	$y = \left(-\dfrac{1}{2}\right)^2 = \dfrac{1}{4}$
d.	$y = \dfrac{2}{x^{-2}}$	$x = 3$	$y = \dfrac{2}{3^{-2}} = 2(3^2) = 18$

✓ CHECKPOINT 1
Evaluate $y = 4x^{-2}$ for $x = 3$. ∎

Example 2 Evaluating Expressions

	Expression	x-Value	Substitution
a.	$y = 2x^{1/2}$	$x = 4$	$y = 2\sqrt{4} = 2(2) = 4$
b.	$y = \sqrt[3]{x^2}$	$x = 8$	$y = 8^{2/3} = (8^{1/3})^2 = 2^2 = 4$

✓ CHECKPOINT 2
Evaluate $y = 4x^{1/3}$ for $x = 8$. ∎

Operations with Exponents

> **TECHNOLOGY**
>
> Graphing utilities perform the established order of operations when evaluating an expression. To see this, try entering the expressions
>
> $$1200\left(1 + \frac{0.09}{12}\right)^{12 \cdot 6}$$
>
> and
>
> $$1200 \times 1 + \left(\frac{0.09}{12}\right)^{12 \cdot 6}$$
>
> into your graphing utility to see that the expressions result in different values.*

Operations with Exponents

1. Multiplying like bases: $\quad x^n x^m = x^{n+m}$ $\quad$ Add exponents.

2. Dividing like bases: $\quad \dfrac{x^n}{x^m} = x^{n-m}$ $\quad$ Subtract exponents.

3. Removing parentheses: $\quad (xy)^n = x^n y^n$

   $\quad \left(\dfrac{x}{y}\right)^n = \dfrac{x^n}{y^n}$

   $\quad (x^n)^m = x^{nm}$

4. Special conventions: $\quad -x^n = -(x^n), \quad -x^n \neq (-x)^n$

   $\quad cx^n = c(x^n), \quad cx^n \neq (cx)^n$

   $\quad x^{n^m} = x^{(n^m)}, \quad x^{n^m} \neq (x^n)^m$

Example 3 Simplifying Expressions with Exponents

Simplify each expression.

a. $2x^2(x^3)$ $\quad$ **b.** $(3x)^2 \sqrt[3]{x}$ $\quad$ **c.** $\dfrac{3x^2}{(x^{1/2})^3}$

d. $\dfrac{5x^4}{(x^2)^3}$ $\quad$ **e.** $x^{-1}(2x^2)$ $\quad$ **f.** $\dfrac{-\sqrt{x}}{5x^{-1}}$

SOLUTION

a. $2x^2(x^3) = 2x^{2+3} = 2x^5$ $\qquad\qquad\qquad\qquad\qquad x^n x^m = x^{n+m}$

b. $(3x)^2 \sqrt[3]{x} = 9x^2 x^{1/3} = 9x^{2+(1/3)} = 9x^{7/3}$ $\qquad x^n x^m = x^{n+m}$

c. $\dfrac{3x^2}{(x^{1/2})^3} = 3\left(\dfrac{x^2}{x^{3/2}}\right) = 3x^{2-(3/2)} = 3x^{1/2}$ $\qquad (x^n)^m = x^{nm}, \dfrac{x^n}{x^m} = x^{n-m}$

d. $\dfrac{5x^4}{(x^2)^3} = \dfrac{5x^4}{x^6} = 5x^{4-6} = 5x^{-2} = \dfrac{5}{x^2}$ $\qquad (x^n)^m = x^{nm}, \dfrac{x^n}{x^m} = x^{n-m}$

e. $x^{-1}(2x^2) = 2x^{-1}x^2 = 2x^{2-1} = 2x$ $\qquad\qquad\qquad x^n x^m = x^{n+m}$

f. $\dfrac{-\sqrt{x}}{5x^{-1}} = -\dfrac{1}{5}\left(\dfrac{x^{1/2}}{x^{-1}}\right) = -\dfrac{1}{5}x^{(1/2)+1} = -\dfrac{1}{5}x^{3/2}$ $\qquad \dfrac{x^n}{x^m} = x^{n-m}$

✓ CHECKPOINT 3

Simplify each expression.

a. $3x^2(x^4)$ $\quad$ **b.** $(2x)^3 \sqrt{x}$ $\quad$ **c.** $\dfrac{4x^2}{(x^{1/3})^2}$ ∎

---

*Specific calculator keystroke instructions for operations in this and other technology boxes can be found at *college.hmco.com/info/larsonapplied.*

Note in Example 3 that one characteristic of simplified expressions is the absence of negative exponents. Another characteristic of simplified expressions is that sums and differences are written in *factored form*. To do this, you can use the **Distributive Property**.

$$abx^n + acx^{n+m} = ax^n(b + cx^m)$$

Study the next example carefully to be sure that you understand the concepts involved in the factoring process.

Example 4 Simplifying by Factoring

Simplify each expression by factoring.

a. $2x^2 - x^3$ **b.** $2x^3 + x^2$ **c.** $2x^{1/2} + 4x^{5/2}$ **d.** $2x^{-1/2} + 3x^{5/2}$

SOLUTION

a. $2x^2 - x^3 = x^2(2 - x)$

b. $2x^3 + x^2 = x^2(2x + 1)$

c. $2x^{1/2} + 4x^{5/2} = 2x^{1/2}(1 + 2x^2)$

d. $2x^{-1/2} + 3x^{5/2} = x^{-1/2}(2 + 3x^3) = \dfrac{2 + 3x^3}{\sqrt{x}}$

✓ **CHECKPOINT 4**

Simplify each expression by factoring.

a. $x^3 - 2x$

b. $2x^{1/2} + 8x^{3/2}$ ∎

STUDY TIP

To check that the simplified expression is equivalent to the original expression, try substituting values for x into each expression.

Many algebraic expressions obtained in calculus occur in unsimplified form. For instance, the two expressions shown in the following example are the result of an operation in calculus called **differentiation**. [The first is the derivative of $2(x + 1)^{3/2}(2x - 3)^{5/2}$, and the second is the derivative of $2(x + 1)^{1/2}(2x - 3)^{5/2}$.]

Example 5 Simplifying by Factoring

Simplify each expression by factoring.

a. $3(x + 1)^{1/2}(2x - 3)^{5/2} + 10(x + 1)^{3/2}(2x - 3)^{3/2}$

$= (x + 1)^{1/2}(2x - 3)^{3/2}[3(2x - 3) + 10(x + 1)]$

$= (x + 1)^{1/2}(2x - 3)^{3/2}(6x - 9 + 10x + 10)$

$= (x + 1)^{1/2}(2x - 3)^{3/2}(16x + 1)$

b. $(x + 1)^{-1/2}(2x - 3)^{5/2} + 10(x + 1)^{1/2}(2x - 3)^{3/2}$

$= (x + 1)^{-1/2}(2x - 3)^{3/2}[(2x - 3) + 10(x + 1)]$

$= (x + 1)^{-1/2}(2x - 3)^{3/2}(2x - 3 + 10x + 10)$

$= (x + 1)^{-1/2}(2x - 3)^{3/2}(12x + 7)$

$= \dfrac{(2x - 3)^{3/2}(12x + 7)}{(x + 1)^{1/2}}$

✓ **CHECKPOINT 5**

Simplify the expression by factoring.

$(x + 2)^{1/2}(3x - 1)^{3/2}$
$+ 4(x + 2)^{-1/2}(3x - 1)^{5/2}$ ∎

CHAPTER 0 A Precalculus Review

Example 6 shows some additional types of expressions that can occur in calculus. [The expression in Example 6(d) is an antiderivative of $(x + 1)^{2/3}(2x + 3)$, and the expression in Example 6(e) is the derivative of $(x + 2)^3/(x − 1)^3$.]

Example 6 Factors Involving Quotients

Simplify each expression by factoring.

a. $\dfrac{3x^2 + x^4}{2x}$

b. $\dfrac{\sqrt{x} + x^{3/2}}{x}$

c. $(9x + 2)^{-1/3} + 18(9x + 2)$

d. $\dfrac{3}{5}(x + 1)^{5/3} + \dfrac{3}{4}(x + 1)^{8/3}$

e. $\dfrac{3(x + 2)^2(x − 1)^3 − 3(x + 2)^3(x − 1)^2}{[(x − 1)^3]^2}$

SOLUTION

a. $\dfrac{3x^2 + x^4}{2x} = \dfrac{x^2(3 + x^2)}{2x} = \dfrac{x^{2-1}(3 + x^2)}{2} = \dfrac{x(3 + x^2)}{2}$

b. $\dfrac{\sqrt{x} + x^{3/2}}{x} = \dfrac{x^{1/2}(1 + x)}{x} = \dfrac{1 + x}{x^{1-(1/2)}} = \dfrac{1 + x}{\sqrt{x}}$

c. $(9x + 2)^{-1/3} + 18(9x + 2) = (9x + 2)^{-1/3}[1 + 18(9x + 2)^{4/3}]$
$= \dfrac{1 + 18(9x + 2)^{4/3}}{\sqrt[3]{9x + 2}}$

d. $\dfrac{3}{5}(x + 1)^{5/3} + \dfrac{3}{4}(x + 1)^{8/3} = \dfrac{12}{20}(x + 1)^{5/3} + \dfrac{15}{20}(x + 1)^{8/3}$
$= \dfrac{3}{20}(x + 1)^{5/3}[4 + 5(x + 1)]$
$= \dfrac{3}{20}(x + 1)^{5/3}(4 + 5x + 5)$
$= \dfrac{3}{20}(x + 1)^{5/3}(5x + 9)$

e. $\dfrac{3(x + 2)^2(x − 1)^3 − 3(x + 2)^3(x − 1)^2}{[(x − 1)^3]^2}$
$= \dfrac{3(x + 2)^2(x − 1)^2[(x − 1) − (x + 2)]}{(x − 1)^6}$
$= \dfrac{3(x + 2)^2(x − 1 − x − 2)}{(x − 1)^{6-2}}$
$= \dfrac{-9(x + 2)^2}{(x − 1)^4}$

TECHNOLOGY

A graphing utility offers several ways to calculate rational exponents and radicals. You should be familiar with the x-squared key ⌐x²⌐. This key squares the value of an expression.

For rational exponents or exponents other than 2, use the ⌐^⌐ key.

For radical expressions, you can use the square root key, the cube root key, or the xth root key. Consult your graphing utility user's guide for specific keystrokes you can use to evaluate rational exponents and radical expressions.

Use a graphing utility to evaluate each expression.

a. $(-8)^{2/3}$ b. $(16 − 5)^4$

c. $\sqrt{576}$ d. $\sqrt[3]{729}$

e. $\sqrt[4]{(16)^3}$

✓ CHECKPOINT 6

Simplify the expression by factoring.

$\dfrac{5x^3 + x^6}{3x}$ ■

Domain of an Algebraic Expression

When working with algebraic expressions involving x, you face the potential difficulty of substituting a value of x for which the expression is not defined (does not produce a real number). For example, the expression $\sqrt{2x+3}$ is *not defined* when $x = -2$ because $\sqrt{2(-2)+3}$ is not a real number.

The set of all values for which an expression is defined is called its **domain**. So, the domain of $\sqrt{2x+3}$ is the set of all values of x such that $\sqrt{2x+3}$ is a real number. In order for $\sqrt{2x+3}$ to represent a real number, it is necessary that $2x + 3 \geq 0$. In other words, $\sqrt{2x+3}$ is defined only for those values of x that lie in the interval $\left[-\frac{3}{2}, \infty\right)$, as shown in Figure 0.14.

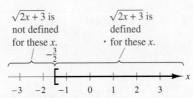

FIGURE 0.14

Example 7 Finding the Domain of an Expression

Find the domain of each expression.

a. $\sqrt{3x-2}$

b. $\dfrac{1}{\sqrt{3x-2}}$

c. $\sqrt[3]{9x+1}$

SOLUTION

a. The domain of $\sqrt{3x-2}$ consists of all x such that

$$3x - 2 \geq 0 \qquad \text{Expression must be nonnegative.}$$

which implies that $x \geq \frac{2}{3}$. So, the domain is $\left[\frac{2}{3}, \infty\right)$.

b. The domain of $1/\sqrt{3x-2}$ is the same as the domain of $\sqrt{3x-2}$, except that $1/\sqrt{3x-2}$ is not defined when $3x - 2 = 0$. Because this occurs when $x = \frac{2}{3}$, the domain is $\left(\frac{2}{3}, \infty\right)$.

c. Because $\sqrt[3]{9x+1}$ is defined for all real numbers, its domain is $(-\infty, \infty)$.

✓ CHECKPOINT 7

Find the domain of each expression.

a. $\sqrt{x-2}$

b. $\dfrac{1}{\sqrt{x-2}}$

c. $\sqrt[3]{x-2}$

Exercises 0.3

See www.CalcChat.com for worked-out solutions to odd-numbered exercises.

In Exercises 1–20, evaluate the expression for the given value of x.

	Expression	x-Value		Expression	x-Value
1.	$-2x^3$	$x = 3$	2.	$\dfrac{x^2}{3}$	$x = 6$
3.	$4x^{-3}$	$x = 2$	4.	$7x^{-2}$	$x = 5$
5.	$\dfrac{1 + x^{-1}}{x^{-1}}$	$x = 3$	6.	$x - 4x^{-2}$	$x = 3$
7.	$3x^2 - 4x^3$	$x = -2$	8.	$5(-x)^3$	$x = 3$
9.	$6x^0 - (6x)^0$	$x = 10$	10.	$\dfrac{1}{(-x)^{-3}}$	$x = 4$
11.	$\sqrt[3]{x^2}$	$x = 27$	12.	$\sqrt{x^3}$	$x = \tfrac{1}{9}$
13.	$x^{-1/2}$	$x = 4$	14.	$x^{-3/4}$	$x = 16$
15.	$x^{-2/5}$	$x = -32$	16.	$(x^{2/3})^3$	$x = 10$
17.	$500x^{60}$	$x = 1.01$	18.	$\dfrac{10{,}000}{x^{120}}$	$x = 1.075$
19.	$\sqrt[3]{x}$	$x = -154$	20.	$\sqrt[6]{x}$	$x = 325$

In Exercises 21–30, simplify the expression.

21. $6y^{-2}(2y^4)^{-3}$
22. $z^{-3}(3z^4)$
23. $10(x^2)^2$
24. $(4x^3)^2$
25. $\dfrac{7x^2}{x^{-3}}$
26. $\dfrac{x^{-3}}{\sqrt{x}}$
27. $\dfrac{10(x+y)^3}{4(x+y)^{-2}}$
28. $\left(\dfrac{12s^2}{9s}\right)^3$
29. $\dfrac{3x\sqrt{x}}{x^{1/2}}$
30. $\left(\sqrt[3]{x^2}\right)^3$

In Exercises 31–36, simplify by removing all possible factors from the radical.

31. $\sqrt{8}$
32. $\sqrt[3]{\dfrac{16}{27}}$
33. $\sqrt[3]{54x^5}$
34. $\sqrt[4]{(3x^2y^3)^4}$
35. $\sqrt[3]{144x^9y^{-4}z^5}$
36. $\sqrt[4]{32xy^5z^{-8}}$

In Exercises 37–44, simplify each expression by factoring.

37. $3x^3 - 12x$
38. $8x^4 - 6x^2$
39. $2x^{5/2} + x^{-1/2}$
40. $5x^{3/2} - x^{-3/2}$
41. $3x(x+1)^{3/2} - 6(x+1)^{1/2}$
42. $2x(x-1)^{5/2} - 4(x-1)^{3/2}$
43. $\dfrac{(x+1)(x-1)^2 - (x-1)^3}{(x+1)^2}$
44. $(x^4 + 2)^3(x+3)^{-1/2} + 4x^3(x^4+2)^2(x+3)^{1/2}$

In Exercises 45–52, find the domain of the given expression.

45. $\sqrt{x-4}$
46. $\sqrt{5-2x}$
47. $\sqrt{x^2+3}$
48. $\sqrt{4x^2+1}$
49. $\dfrac{1}{\sqrt[3]{x-4}}$
50. $\dfrac{1}{\sqrt[3]{x+4}}$
51. $\dfrac{\sqrt{x+2}}{1-x}$
52. $\dfrac{1}{\sqrt{2x+3}} + \sqrt{6-4x}$

(T) **Body Surface Area** In Exercises 53–56, the body surface area (BSA) of a person is given by the formula

$$B = \sqrt{\dfrac{hw}{3600}}$$

where B is the BSA (in square meters), h is the height (in centimeters), and w is the weight (in kilograms). Enter the formula into a graphing utility and use it to find the BSA for the given height and weight.

53. $h = 147$ cm, $w = 48$ kg
54. $h = 162$ cm, $w = 50$ kg
55. $h = 180$ cm, $w = 80$ kg
56. $h = 189$ cm, $w = 84$ kg

57. **Population Growth** The population P of a species of mollusk is given by the formula

$$P = N(1+r)^t$$

where N is the initial population, r is the annual percentage growth rate of the population (expressed as a decimal), and t is the time period (in years). Find the population in 4 years if the initial population is 2 million and the growth rate is 3% per year.

58. **Erosion** A stream of water moving at the rate of v feet per second can carry particles of size $0.03\sqrt{v}$ inches. Find the size of the largest particle that can be carried by a stream flowing at the rate of $\tfrac{1}{2}$ foot per second.

59. **Extended Application** To work an extended application analyzing the population per square mile of the United States, visit this text's website at *college.hmco.com/info/larsonapplied*. (Data Source: U.S. Census Bureau)

The symbol (T) indicates when to use graphing technology or a symbolic computer algebra system to solve a problem or an exercise. The solutions to other problems or exercises may also be facilitated by use of appropriate technology.

Section 0.4
Factoring Polynomials

- Use special products and factorization techniques to factor polynomials.
- Find the domains of radical expressions.
- Use synthetic division to factor polynomials of degree three or more.
- Use the Rational Zero Theorem to find the real zeros of polynomials.

Factorization Techniques

The **Fundamental Theorem of Algebra** states that every nth-degree polynomial

$$a_n x^n + a_{n-1} x^{n-1} + \cdots + a_1 x + a_0, \quad a_n \neq 0$$

has precisely n **zeros.** (The zeros may be repeated or imaginary.) The problem of finding the zeros of a polynomial is equivalent to the problem of factoring the polynomial into linear factors.

Special Products and Factorization Techniques

Quadratic Formula

$ax^2 + bx + c = 0 \implies x = \dfrac{-b \pm \sqrt{b^2 - 4ac}}{2a}$

Example

$x^2 + 3x - 1 = 0 \implies x = \dfrac{-3 \pm \sqrt{13}}{2}$

Special Products

$x^2 - a^2 = (x - a)(x + a)$

$x^3 - a^3 = (x - a)(x^2 + ax + a^2)$

$x^3 + a^3 = (x + a)(x^2 - ax + a^2)$

$x^4 - a^4 = (x - a)(x + a)(x^2 + a^2)$

Examples

$x^2 - 9 = (x - 3)(x + 3)$

$x^3 - 8 = (x - 2)(x^2 + 2x + 4)$

$x^3 + 64 = (x + 4)(x^2 - 4x + 16)$

$x^4 - 16 = (x - 2)(x + 2)(x^2 + 4)$

Binomial Theorem

$(x + a)^2 = x^2 + 2ax + a^2$

$(x - a)^2 = x^2 - 2ax + a^2$

$(x + a)^3 = x^3 + 3ax^2 + 3a^2 x + a^3$

$(x - a)^3 = x^3 - 3ax^2 + 3a^2 x - a^3$

$(x + a)^4 = x^4 + 4ax^3 + 6a^2 x^2 + 4a^3 x + a^4$

$(x - a)^4 = x^4 - 4ax^3 + 6a^2 x^2 - 4a^3 x + a^4$

$(x + a)^n = x^n + nax^{n-1} + \dfrac{n(n-1)}{2!} a^2 x^{n-2} + \dfrac{n(n-1)(n-2)}{3!} a^3 x^{n-3} + \cdots + na^{n-1} x + a^n$ *

$(x - a)^n = x^n - nax^{n-1} + \dfrac{n(n-1)}{2!} a^2 x^{n-2} - \dfrac{n(n-1)(n-2)}{3!} a^3 x^{n-3} + \cdots \pm na^{n-1} x \mp a^n$

Examples

$(x + 3)^2 = x^2 + 6x + 9$

$(x^2 - 5)^2 = x^4 - 10x^2 + 25$

$(x + 2)^3 = x^3 + 6x^2 + 12x + 8$

$(x - 1)^3 = x^3 - 3x^2 + 3x - 1$

$(x + 2)^4 = x^4 + 8x^3 + 24x^2 + 32x + 16$

$(x - 4)^4 = x^4 - 16x^3 + 96x^2 - 256x + 256$

Factoring by Grouping

$acx^3 + adx^2 + bcx + bd = ax^2(cx + d) + b(cx + d)$
$ = (ax^2 + b)(cx + d)$

Example

$3x^3 - 2x^2 - 6x + 4 = x^2(3x - 2) - 2(3x - 2)$
$ = (x^2 - 2)(3x - 2)$

---

* The factorial symbol ! is defined as follows: $0! = 1$, $1! = 1$, $2! = 2 \cdot 1 = 2$, $3! = 3 \cdot 2 \cdot 1 = 6$, $4! = 4 \cdot 3 \cdot 2 \cdot 1 = 24$, and so on.

Example 1 Applying the Quadratic Formula

Use the Quadratic Formula to find all real zeros of each polynomial.

a. $4x^2 + 6x + 1$ **b.** $x^2 + 6x + 9$ **c.** $2x^2 - 6x + 5$

SOLUTION

a. Using $a = 4$, $b = 6$, and $c = 1$, you can write

$$x = \frac{-b \pm \sqrt{b^2 - 4ac}}{2a} = \frac{-6 \pm \sqrt{36 - 16}}{8}$$

$$= \frac{-6 \pm \sqrt{20}}{8}$$

$$= \frac{-6 \pm 2\sqrt{5}}{8}$$

$$= \frac{2(-3 \pm \sqrt{5})}{2(4)}$$

$$= \frac{-3 \pm \sqrt{5}}{4}.$$

So, there are two real zeros:

$$x = \frac{-3 - \sqrt{5}}{4} \approx -1.309 \quad \text{and} \quad x = \frac{-3 + \sqrt{5}}{4} \approx -0.191.$$

b. In this case, $a = 1$, $b = 6$, and $c = 9$, and the Quadratic Formula yields

$$x = \frac{-b \pm \sqrt{b^2 - 4ac}}{2a} = \frac{-6 \pm \sqrt{36 - 36}}{2} = -\frac{6}{2} = -3.$$

So, there is one (repeated) real zero: $x = -3$.

c. For this quadratic equation, $a = 2$, $b = -6$, and $c = 5$. So,

$$x = \frac{-b \pm \sqrt{b^2 - 4ac}}{2a} = \frac{6 \pm \sqrt{36 - 40}}{4} = \frac{6 \pm \sqrt{-4}}{4}.$$

Because $\sqrt{-4}$ is imaginary, there are no real zeros.

STUDY TIP

Try solving Example 1(b) by factoring. Do you obtain the same answer?

✓ CHECKPOINT 1

Use the Quadratic Formula to find all real zeros of each polynomial.

a. $2x^2 + 4x + 1$ **b.** $x^2 - 8x + 16$ **c.** $2x^2 - x + 5$ ■

The zeros in Example 1(a) are irrational, and the zeros in Example 1(c) are imaginary. In both of these cases the quadratic is said to be **irreducible** because it cannot be factored into linear factors with rational coefficients. The next example shows how to find the zeros associated with *reducible* quadratics. In this example, factoring is used to find the zeros of each quadratic. Try using the Quadratic Formula to obtain the same zeros.

Example 2 Factoring Quadratics

Find the zeros of each quadratic polynomial.

a. $x^2 - 5x + 6$ **b.** $x^2 - 6x + 9$ **c.** $2x^2 + 5x - 3$

SOLUTION

a. Because
$$x^2 - 5x + 6 = (x - 2)(x - 3)$$
the zeros are $x = 2$ and $x = 3$.

b. Because
$$x^2 - 6x + 9 = (x - 3)^2$$
the only zero is $x = 3$.

c. Because
$$2x^2 + 5x - 3 = (2x - 1)(x + 3)$$
the zeros are $x = \frac{1}{2}$ and $x = -3$.

STUDY TIP

The zeros of a polynomial in x are the values of x that make the polynomial zero. To find the zeros, factor the polynomial into linear factors and set each factor equal to zero. For instance, the zeros of $(x - 2)(x - 3)$ occur when $x - 2 = 0$ and $x - 3 = 0$.

✓ **CHECKPOINT 2**

Find the zeros of each quadratic polynomial.

a. $x^2 - 2x - 15$ **b.** $x^2 + 2x + 1$ **c.** $2x^2 - 7x + 6$ ■

Example 3 Finding the Domain of a Radical Expression

Find the domain of $\sqrt{x^2 - 3x + 2}$.

SOLUTION Because
$$x^2 - 3x + 2 = (x - 1)(x - 2)$$
you know that the zeros of the quadratic are $x = 1$ and $x = 2$. So, you need to test the sign of the quadratic in the three intervals $(-\infty, 1)$, $(1, 2)$, and $(2, \infty)$, as shown in Figure 0.15. After testing each of these intervals, you can see that the quadratic is negative in the center interval and positive in the outer two intervals. Moreover, because the quadratic is zero when $x = 1$ and $x = 2$, you can conclude that the domain of $\sqrt{x^2 - 3x + 2}$ is

$$(-\infty, 1] \cup [2, \infty). \quad \text{Domain}$$

Values of $\sqrt{x^2 - 3x + 2}$

x	$\sqrt{x^2 - 3x + 2}$
0	$\sqrt{2}$
1	0
1.5	Undefined
2	0
3	$\sqrt{2}$

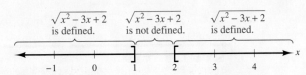

FIGURE 0.15

✓ **CHECKPOINT 3**

Find the domain of
$\sqrt{x^2 + x - 2}$. ■

Factoring Polynomials of Degree Three or More

It can be difficult to find the zeros of polynomials of degree three or more. However, if one of the zeros of a polynomial is known, then you can use that zero to reduce the degree of the polynomial. For example, if you know that $x = 2$ is a zero of $x^3 - 4x^2 + 5x - 2$, then you know that $(x - 2)$ is a factor, and you can use long division to factor the polynomial as shown.

$$x^3 - 4x^2 + 5x - 2 = (x - 2)(x^2 - 2x + 1)$$
$$= (x - 2)(x - 1)(x - 1)$$

As an alternative to long division, many people prefer to use **synthetic division** to reduce the degree of a polynomial.

Synthetic Division for a Cubic Polynomial

Given: $x = x_1$ is a zero of $ax^3 + bx^2 + cx + d$.

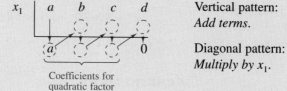

Vertical pattern: Add terms.

Diagonal pattern: Multiply by x_1.

Performing synthetic division on the polynomial

$$x^3 - 4x^2 + 5x - 2$$

using the given zero, $x = 2$, produces the following.

$$\begin{array}{c|cccc} 2 & 1 & -4 & 5 & -2 \\ & & 2 & -4 & 2 \\ \hline & 1 & -2 & 1 & 0 \end{array}$$

$(x - 2)(x^2 - 2x + 1) = x^3 - 4x^2 + 5x - 2$

When you use synthetic division, remember to take *all* coefficients into account—*even if some of them are zero*. For instance, if you know that $x = -2$ is a zero of $x^3 + 3x + 14$, you can apply synthetic division as shown.

$$\begin{array}{c|cccc} -2 & 1 & 0 & 3 & 14 \\ & & -2 & 4 & -14 \\ \hline & 1 & -2 & 7 & 0 \end{array}$$

$(x + 2)(x^2 - 2x + 7) = x^3 + 3x + 14$

STUDY TIP

The algorithm for synthetic division given above works *only* for divisors of the form $x - x_1$. Remember that $x + x_1 = x - (-x_1)$.

The Rational Zero Theorem

There is a systematic way to find the *rational* zeros of a polynomial. You can use the **Rational Zero Theorem** (also called the Rational Root Theorem).

Rational Zero Theorem

If a polynomial

$$a_n x^n + a_{n-1} x^{n-1} + \cdots + a_1 x + a_0$$

has integer coefficients, then every *rational* zero is of the form $x = p/q$, where p is a factor of a_0, and q is a factor of a_n.

Example 4 — Using the Rational Zero Theorem

Find all real zeros of the polynomial.

$$2x^3 + 3x^2 - 8x + 3$$

SOLUTION

$$\underset{}{\boxed{2}} x^3 + 3x^2 - 8x + \underset{}{\boxed{3}}$$

Factors of constant term: $\pm 1, \pm 3$
Factors of leading coefficient: $\pm 1, \pm 2$

The possible rational zeros are the factors of the constant term divided by the factors of the leading coefficient.

$$1, -1, 3, -3, \frac{1}{2}, -\frac{1}{2}, \frac{3}{2}, -\frac{3}{2}$$

By testing these possible zeros, you can see that $x = 1$ works.

$$2(1)^3 + 3(1)^2 - 8(1) + 3 = 2 + 3 - 8 + 3 = 0$$

Now, by synthetic division you have the following.

```
1 | 2   3   -8    3
  |     2    5   -3
  |_____
    2   5   -3    0
```

$(x - 1)(2x^2 + 5x - 3) = 2x^3 + 3x^2 - 8x + 3$

Finally, by factoring the quadratic, $2x^2 + 5x - 3 = (2x - 1)(x + 3)$, you have

$$2x^3 + 3x^2 - 8x + 3 = (x - 1)(2x - 1)(x + 3)$$

and you can conclude that the zeros are $x = 1$, $x = \frac{1}{2}$, and $x = -3$.

STUDY TIP

In Example 4, you can check that the zeros are correct by substituting into the original polynomial.

Check that $x = 1$ is a zero.

$2(1)^3 + 3(1)^2 - 8(1) + 3$
$= 2 + 3 - 8 + 3$
$= 0$

Check that $x = \frac{1}{2}$ is a zero.

$2\left(\frac{1}{2}\right)^3 + 3\left(\frac{1}{2}\right)^2 - 8\left(\frac{1}{2}\right) + 3$
$= \frac{1}{4} + \frac{3}{4} - 4 + 3$
$= 0$

Check that $x = -3$ is a zero.

$2(-3)^3 + 3(-3)^2 - 8(-3) + 3$
$= -54 + 27 + 24 + 3$
$= 0$

✓ CHECKPOINT 4

Find all real zeros of the polynomial.

$$2x^3 - 3x^2 - 3x + 2 \quad \blacksquare$$

Exercises 0.4

In Exercises 1–8, use the Quadratic Formula to find all real zeros of the second-degree polynomial.

1. $6x^2 - 7x + 1$
2. $8x^2 - 2x - 1$
3. $4x^2 - 12x + 9$
4. $9x^2 + 12x + 4$
5. $y^2 + 4y + 1$
6. $y^2 + 5y - 2$
7. $2x^2 + 3x - 4$
8. $3x^2 - 8x - 4$

In Exercises 9–18, write the second-degree polynomial as the product of two linear factors.

9. $x^2 - 4x + 4$
10. $x^2 + 10x + 25$
11. $4x^2 + 4x + 1$
12. $9x^2 - 12x + 4$
13. $3x^2 - 4x + 1$
14. $2x^2 - x - 1$
15. $3x^2 - 5x + 2$
16. $x^2 - xy - 2y^2$
17. $x^2 - 4xy + 4y^2$
18. $a^2b^2 - 2abc + c^2$

In Exercises 19–34, completely factor the polynomial.

19. $81 - y^4$
20. $x^4 - 16$
21. $x^3 - 8$
22. $y^3 - 64$
23. $y^3 + 64$
24. $z^3 + 125$
25. $x^3 - y^3$
26. $(x - a)^3 + b^3$
27. $x^3 - 4x^2 - x + 4$
28. $x^3 - x^2 - x + 1$
29. $2x^3 - 3x^2 + 4x - 6$
30. $x^3 - 5x^2 - 5x + 25$
31. $2x^3 - 4x^2 - x + 2$
32. $x^3 - 7x^2 - 4x + 28$
33. $x^4 - 15x^2 - 16$
34. $2x^4 - 49x^2 - 25$

In Exercises 35–54, find all real zeros of the polynomial.

35. $x^2 - 5x$
36. $2x^2 - 3x$
37. $x^2 - 9$
38. $x^2 - 25$
39. $x^2 - 3$
40. $x^2 - 8$
41. $(x - 3)^2 - 9$
42. $(x + 1)^2 - 36$
43. $x^2 + x - 2$
44. $x^2 + 5x + 6$
45. $x^2 - 5x - 6$
46. $x^2 + x - 20$
47. $3x^2 + 5x + 2$
48. $2x^2 - x - 1$
49. $x^3 + 64$
50. $x^3 - 216$
51. $x^4 - 16$
52. $x^4 - 625$
53. $x^3 - x^2 - 4x + 4$
54. $2x^3 + x^2 + 6x + 3$

In Exercises 55–60, find the interval (or intervals) on which the given expression is defined.

55. $\sqrt{x^2 - 4}$
56. $\sqrt{4 - x^2}$
57. $\sqrt{x^2 - 7x + 12}$
58. $\sqrt{x^2 - 8x + 15}$
59. $\sqrt{5x^2 + 6x + 1}$
60. $\sqrt{3x^2 - 10x + 3}$

In Exercises 61–64, use synthetic division to complete the indicated factorization.

61. $x^3 - 3x^2 - 6x - 2 = (x + 1)(\quad)$
62. $x^3 - 2x^2 - x + 2 = (x - 2)(\quad)$
63. $2x^3 - x^2 - 2x + 1 = (x + 1)(\quad)$
64. $x^4 - 16x^3 + 96x^2 - 256x + 256 = (x - 4)(\quad)$

In Exercises 65–74, use the Rational Zero Theorem as an aid in finding all real zeros of the polynomial.

65. $x^3 - x^2 - 10x - 8$
66. $x^3 - 7x - 6$
67. $x^3 - 6x^2 + 11x - 6$
68. $x^3 + 2x^2 - 5x - 6$
69. $6x^3 - 11x^2 - 19x - 6$
70. $18x^3 - 9x^2 - 8x + 4$
71. $x^3 - 3x^2 - 3x - 4$
72. $2x^3 - x^2 - 13x - 6$
73. $4x^3 + 11x^2 + 5x - 2$
74. $3x^3 + 4x^2 - 13x + 6$

75. Ice Sculptures Some ice sculptures are made by filling a mold with water and then freezing it. A pyramid shaped mold has a height that is 1 foot greater than the length of each side of its square base. The volume of the ice sculpture is 4 cubic feet. What are the dimensions of the mold?

76. Liver Transplants The number L of liver transplant procedures from 1995 through 2005 is given by

$$L = 12.41t^2 + 21.4t + 3356$$

where t represents the year, with $t = 5$ corresponding to 1995. In which year was the number of transplant procedures greater than 5300? *(Source: U.S. Department of Health and Human Services)*

Ⓑ 77. Chemistry: Finding Concentrations Use the Quadratic Formula to solve the expression

$$1.8 \times 10^{-5} = \frac{x^2}{1.0 \times 10^{-4} - x}$$

which is needed to determine the quantity of hydrogen ions ($[H^+]$) in a solution of 1.0×10^{-4}M acetic acid. Because x represents a concentration of $[H^+]$, only positive values of x are possible solutions. *(Source: Adapted from Zumdahl, Chemistry, Seventh Edition)*

The symbol Ⓑ indicates an exercise that contains material from textbooks in other disciplines.

Section 0.5
Fractions and Rationalization

- Add and subtract rational expressions.
- Simplify rational expressions involving radicals.
- Rationalize numerators and denominators of rational expressions.

Operations with Fractions

In this section, you will review operations involving fractional expressions such as

$$\frac{2}{x}, \quad \frac{x^2 + 2x - 4}{x + 6}, \quad \text{and} \quad \frac{1}{\sqrt{x^2 + 1}}.$$

The first two expressions have polynomials as both numerator and denominator and are called **rational expressions.** A rational expression is **proper** if the degree of the numerator is less than the degree of the denominator. For example,

$$\frac{x}{x^2 + 1}$$

is proper. If the degree of the numerator is greater than or equal to the degree of the denominator, then the rational expression is **improper.** For example,

$$\frac{x^2}{x^2 + 1}, \quad \text{and} \quad \frac{x^3 + 2x + 1}{x + 1}$$

are both improper.

Operations with Fractions

1. Add fractions (find a common denominator):

$$\frac{a}{b} + \frac{c}{d} = \frac{a}{b}\left(\frac{d}{d}\right) + \frac{c}{d}\left(\frac{b}{b}\right) = \frac{ad}{bd} + \frac{bc}{bd} = \frac{ad + bc}{bd}, \quad b \neq 0, d \neq 0$$

2. Subtract fractions (find a common denominator):

$$\frac{a}{b} - \frac{c}{d} = \frac{a}{b}\left(\frac{d}{d}\right) - \frac{c}{d}\left(\frac{b}{b}\right) = \frac{ad}{bd} - \frac{bc}{bd} = \frac{ad - bc}{bd}, \quad b \neq 0, d \neq 0$$

3. Multiply fractions:

$$\left(\frac{a}{b}\right)\left(\frac{c}{d}\right) = \frac{ac}{bd}, \quad b \neq 0, d \neq 0$$

4. Divide fractions (invert and multiply):

$$\frac{a/b}{c/d} = \left(\frac{a}{b}\right)\left(\frac{d}{c}\right) = \frac{ad}{bc}, \quad \frac{a/b}{c} = \frac{a/b}{c/1} = \left(\frac{a}{b}\right)\left(\frac{1}{c}\right) = \frac{a}{bc}, \quad b \neq 0,$$
$$c \neq 0, d \neq 0$$

5. Divide out like factors:

$$\frac{\cancel{a}b}{\cancel{a}c} = \frac{b}{c}, \quad \frac{ab + ac}{ad} = \frac{\cancel{a}(b + c)}{\cancel{a}d} = \frac{b + c}{d}, \quad a \neq 0, c \neq 0, d \neq 0$$

CHAPTER 0 A Precalculus Review

Example 1 Adding and Subtracting Rational Expressions

Perform each indicated operation and simplify.

a. $x + \dfrac{1}{x}$ **b.** $\dfrac{1}{x+1} - \dfrac{2}{2x-1}$

SOLUTION

a. $x + \dfrac{1}{x} = \dfrac{x^2}{x} + \dfrac{1}{x}$ Write with common denominator.

$\phantom{x + \dfrac{1}{x}} = \dfrac{x^2 + 1}{x}$ Add fractions.

b. $\dfrac{1}{x+1} - \dfrac{2}{2x-1} = \dfrac{(2x-1)}{(x+1)(2x-1)} - \dfrac{2(x+1)}{(x+1)(2x-1)}$

$\phantom{\dfrac{1}{x+1} - \dfrac{2}{2x-1}} = \dfrac{2x - 1 - 2x - 2}{2x^2 + x - 1} = \dfrac{-3}{2x^2 + x - 1}$

✓ CHECKPOINT 1

Perform each indicated operation and simplify.

a. $x + \dfrac{2}{x}$ **b.** $\dfrac{2}{x+1} - \dfrac{1}{2x+1}$ ■

In adding (or subtracting) fractions whose denominators have no common factors, it is convenient to use the following pattern.

$$\dfrac{a}{b} + \dfrac{c}{d} = \dfrac{a \cdot d}{b \cdot d} + \dfrac{c \cdot b}{d \cdot b} = \dfrac{ad + bc}{bd}$$

For instance, in Example 1(b), you could have used this pattern as shown.

$\dfrac{1}{x+1} - \dfrac{2}{2x-1} = \dfrac{(2x-1) - 2(x+1)}{(x+1)(2x-1)}$

$\phantom{\dfrac{1}{x+1} - \dfrac{2}{2x-1}} = \dfrac{2x - 1 - 2x - 2}{(x+1)(2x-1)} = \dfrac{-3}{2x^2 + x - 1}$

In Example 1, the denominators of the rational expressions have no common factors. When the denominators do have common factors, it is best to find the least common denominator before adding or subtracting. For instance, when adding $1/x$ and $2/x^2$, you can recognize that the least common denominator is x^2 and write

$\dfrac{1}{x} + \dfrac{2}{x^2} = \dfrac{x}{x^2} + \dfrac{2}{x^2}$ Write with common denominator.

$\phantom{\dfrac{1}{x} + \dfrac{2}{x^2}} = \dfrac{x+2}{x^2}.$ Add fractions.

This is further demonstrated in Example 2.

SECTION 0.5 Fractions and Rationalization

Example 2 Adding and Subtracting Rational Expressions

Perform each indicated operation and simplify.

a. $\dfrac{x}{x^2 - 1} + \dfrac{3}{x + 1}$ **b.** $\dfrac{1}{2(x^2 + 2x)} - \dfrac{1}{4x}$

SOLUTION

a. Because $x^2 - 1 = (x + 1)(x - 1)$, the least common denominator is $x^2 - 1$.

$$\dfrac{x}{x^2 - 1} + \dfrac{3}{x + 1} = \dfrac{x}{(x - 1)(x + 1)} + \dfrac{3}{x + 1} \qquad \text{Factor.}$$

$$= \dfrac{x}{(x - 1)(x + 1)} + \dfrac{3(x - 1)}{(x - 1)(x + 1)} \qquad \text{Write with common denominator.}$$

$$= \dfrac{x + 3x - 3}{(x - 1)(x + 1)} \qquad \text{Add fractions.}$$

$$= \dfrac{4x - 3}{x^2 - 1} \qquad \text{Simplify.}$$

b. In this case, the least common denominator is $4x(x + 2)$.

$$\dfrac{1}{2(x^2 + 2x)} - \dfrac{1}{4x} = \dfrac{1}{2x(x + 2)} - \dfrac{1}{2(2x)} \qquad \text{Factor.}$$

$$= \dfrac{2}{2(2x)(x + 2)} - \dfrac{x + 2}{2(2x)(x + 2)} \qquad \text{Write with common denominator.}$$

$$= \dfrac{2 - x - 2}{4x(x + 2)} \qquad \text{Subtract fractions.}$$

$$= \dfrac{-\cancel{x}}{4\cancel{x}(x + 2)} \qquad \text{Divide out like factor.}$$

$$= \dfrac{-1}{4(x + 2)}, \quad x \neq 0 \qquad \text{Simplify.}$$

✓ CHECKPOINT 2

Perform each indicated operation and simplify.

a. $\dfrac{x}{x^2 - 4} + \dfrac{2}{x - 2}$ **b.** $\dfrac{1}{3(x^2 + 2x)} - \dfrac{1}{3x}$ ■

STUDY TIP

To add more than two fractions, you must find a denominator that is common to all the fractions. For instance, to add $\tfrac{1}{2}, \tfrac{1}{3},$ and $\tfrac{1}{5}$, use a (least) common denominator of 30 and write

$$\dfrac{1}{2} + \dfrac{1}{3} + \dfrac{1}{5} = \dfrac{15}{30} + \dfrac{10}{30} + \dfrac{6}{30} \qquad \text{Write with common denominator.}$$

$$= \dfrac{31}{30}. \qquad \text{Add fractions.}$$

To add more than two rational expressions, use a similar procedure, as shown in Example 3. (Expressions such as those shown in this example are used in calculus to perform an integration technique called integration by partial fractions.)

Example 3 Adding More than Two Rational Expressions

Perform each indicated addition of rational expressions.

a. $\dfrac{A}{x+2} + \dfrac{B}{x-3} + \dfrac{C}{x+4}$

b. $\dfrac{A}{x+2} + \dfrac{B}{(x+2)^2} + \dfrac{C}{x-1}$

SOLUTION

a. The least common denominator is $(x+2)(x-3)(x+4)$.

$$\dfrac{A}{x+2} + \dfrac{B}{x-3} + \dfrac{C}{x+4}$$

$$= \dfrac{A(x-3)(x+4) + B(x+2)(x+4) + C(x+2)(x-3)}{(x+2)(x-3)(x+4)}$$

$$= \dfrac{A(x^2+x-12) + B(x^2+6x+8) + C(x^2-x-6)}{(x+2)(x-3)(x+4)}$$

$$= \dfrac{Ax^2 + Bx^2 + Cx^2 + Ax + 6Bx - Cx - 12A + 8B - 6C}{(x+2)(x-3)(x+4)}$$

$$= \dfrac{(A+B+C)x^2 + (A+6B-C)x + (-12A+8B-6C)}{(x+2)(x-3)(x+4)}$$

b. Here the least common denominator is $(x+2)^2(x-1)$.

$$\dfrac{A}{x+2} + \dfrac{B}{(x+2)^2} + \dfrac{C}{x-1}$$

$$= \dfrac{A(x+2)(x-1) + B(x-1) + C(x+2)^2}{(x+2)^2(x-1)}$$

$$= \dfrac{A(x^2+x-2) + B(x-1) + C(x^2+4x+4)}{(x+2)^2(x-1)}$$

$$= \dfrac{Ax^2 + Cx^2 + Ax + Bx + 4Cx - 2A - B + 4C}{(x+2)^2(x-1)}$$

$$= \dfrac{(A+C)x^2 + (A+B+4C)x + (-2A-B+4C)}{(x+2)^2(x-1)}$$

✓ CHECKPOINT 3

Perform each indicated addition of rational expressions.

a. $\dfrac{A}{x+1} + \dfrac{B}{x-1} + \dfrac{C}{x+2}$

b. $\dfrac{A}{x+1} + \dfrac{B}{(x+1)^2} + \dfrac{C}{x-2}$

SECTION 0.5 Fractions and Rationalization

Expressions Involving Radicals

In calculus, the operation of differentiation tends to produce "messy" expressions when applied to fractional expressions. This is especially true when the fractional expressions involve radicals. When differentiation is used, it is important to be able to simplify these expressions so that you can obtain more manageable forms. All of the expressions in Examples 4 and 5 are the results of differentiation. In each case, note how much *simpler* the simplified form is than the original form.

Example 4 Simplifying an Expression with Radicals

Simplify each expression.

a. $\dfrac{\sqrt{x+1} - \dfrac{x}{2\sqrt{x+1}}}{x+1}$ b. $\left(\dfrac{1}{x+\sqrt{x^2+1}}\right)\left(1 + \dfrac{2x}{2\sqrt{x^2+1}}\right)$

SOLUTION

a. $\dfrac{\sqrt{x+1} - \dfrac{x}{2\sqrt{x+1}}}{x+1} = \dfrac{\dfrac{2(x+1)}{2\sqrt{x+1}} - \dfrac{x}{2\sqrt{x+1}}}{x+1}$ Write with common denominator.

$= \dfrac{\dfrac{2x+2-x}{2\sqrt{x+1}}}{\dfrac{x+1}{1}}$ Subtract fractions.

$= \dfrac{x+2}{2\sqrt{x+1}}\left(\dfrac{1}{x+1}\right)$ To divide, invert and multiply

$= \dfrac{x+2}{2(x+1)^{3/2}}$ Multiply.

b. $\left(\dfrac{1}{x+\sqrt{x^2+1}}\right)\left(1 + \dfrac{2x}{2\sqrt{x^2+1}}\right)$

$= \left(\dfrac{1}{x+\sqrt{x^2+1}}\right)\left(1 + \dfrac{x}{\sqrt{x^2+1}}\right)$

$= \left(\dfrac{1}{x+\sqrt{x^2+1}}\right)\left(\dfrac{\sqrt{x^2+1}}{\sqrt{x^2+1}} + \dfrac{x}{\sqrt{x^2+1}}\right)$

$= \left(\dfrac{1}{x+\sqrt{x^2+1}}\right)\left(\dfrac{x+\sqrt{x^2+1}}{\sqrt{x^2+1}}\right)$

$= \dfrac{1}{\sqrt{x^2+1}}$

✓ **CHECKPOINT 4**

Simplify each expression.

a. $\dfrac{\sqrt{x+2} - \dfrac{x}{4\sqrt{x+2}}}{x+2}$ b. $\left(\dfrac{1}{x+\sqrt{x^2+4}}\right)\left(1 + \dfrac{x}{\sqrt{x^2+4}}\right)$

Example 5 Simplifying an Expression with Radicals

Simplify the expression.

$$\frac{-x\left(\dfrac{2x}{2\sqrt{x^2+1}}\right) + \sqrt{x^2+1}}{x^2} + \left(\dfrac{1}{x+\sqrt{x^2+1}}\right)\left(1 + \dfrac{2x}{2\sqrt{x^2+1}}\right)$$

SOLUTION From Example 4(b), you already know that the second part of this sum simplifies to $1/\sqrt{x^2+1}$. The first part simplifies as shown.

$$\frac{-x\left(\dfrac{2x}{2\sqrt{x^2+1}}\right) + \sqrt{x^2+1}}{x^2} = \frac{-x^2}{x^2\sqrt{x^2+1}} + \frac{\sqrt{x^2+1}}{x^2}$$

$$= \frac{-x^2}{x^2\sqrt{x^2+1}} + \frac{x^2+1}{x^2\sqrt{x^2+1}}$$

$$= \frac{-x^2 + x^2 + 1}{x^2\sqrt{x^2+1}}$$

$$= \frac{1}{x^2\sqrt{x^2+1}}$$

So, the sum is

$$\frac{-x\left(\dfrac{2x}{2\sqrt{x^2+1}}\right) + \sqrt{x^2+1}}{x^2} + \left(\dfrac{1}{x+\sqrt{x^2+1}}\right)\left(1 + \dfrac{2x}{2\sqrt{x^2+1}}\right)$$

$$= \frac{1}{x^2\sqrt{x^2+1}} + \frac{1}{\sqrt{x^2+1}}$$

$$= \frac{1}{x^2\sqrt{x^2+1}} + \frac{x^2}{x^2\sqrt{x^2+1}}$$

$$= \frac{x^2+1}{x^2\sqrt{x^2+1}}$$

$$= \frac{\sqrt{x^2+1}}{x^2}.$$

✓ CHECKPOINT 5

Simplify the expression.

$$\frac{-x\left(\dfrac{3x}{3\sqrt{x^2+4}}\right) + \sqrt{x^2+4}}{x^2} + \left(\dfrac{1}{x+\sqrt{x^2+4}}\right)\left(1 + \dfrac{3x}{3\sqrt{x^2+4}}\right) \quad\blacksquare$$

STUDY TIP

To check that the simplified expression in Example 5 is equivalent to the original expression, try substituting values of x into each expression. For instance, when you substitute $x = 1$ into each expression, you obtain $\sqrt{2}$.

Rationalization Techniques

In working with quotients involving radicals, it is often convenient to move the radical expression from the denominator to the numerator, or vice versa. For example, you can move $\sqrt{2}$ from the denominator to the numerator in the following quotient by multiplying by $\sqrt{2}/\sqrt{2}$.

$$\text{Radical in Denominator} \qquad \text{Rationalize} \qquad \text{Radical in Numerator}$$

$$\frac{1}{\sqrt{2}} \quad \Longrightarrow \quad \frac{1}{\sqrt{2}}\left(\frac{\sqrt{2}}{\sqrt{2}}\right) \quad \Longrightarrow \quad \frac{\sqrt{2}}{2}$$

This process is called **rationalizing the denominator.** A similar process is used to **rationalize the numerator.**

STUDY TIP

The success of the second and third rationalizing techniques stems from the following.

$$(\sqrt{a} - \sqrt{b})(\sqrt{a} + \sqrt{b})$$
$$= a - b$$

Rationalizing Techniques

1. If the denominator is $\sqrt{a}$, multiply by $\dfrac{\sqrt{a}}{\sqrt{a}}$.

2. If the denominator is $\sqrt{a} - \sqrt{b}$, multiply by $\dfrac{\sqrt{a} + \sqrt{b}}{\sqrt{a} + \sqrt{b}}$.

3. If the denominator is $\sqrt{a} + \sqrt{b}$, multiply by $\dfrac{\sqrt{a} - \sqrt{b}}{\sqrt{a} - \sqrt{b}}$.

The same guidelines apply to rationalizing numerators.

Example 6 Rationalizing Denominators and Numerators

Rationalize the denominator or numerator.

a. $\dfrac{3}{\sqrt{12}}$ b. $\dfrac{\sqrt{x+1}}{2}$ c. $\dfrac{1}{\sqrt{5} + \sqrt{2}}$ d. $\dfrac{1}{\sqrt{x} - \sqrt{x+1}}$

SOLUTION

a. $\dfrac{3}{\sqrt{12}} = \dfrac{3}{2\sqrt{3}} = \dfrac{3}{2\sqrt{3}}\left(\dfrac{\sqrt{3}}{\sqrt{3}}\right) = \dfrac{3\sqrt{3}}{2(3)} = \dfrac{\sqrt{3}}{2}$

b. $\dfrac{\sqrt{x+1}}{2} = \dfrac{\sqrt{x+1}}{2}\left(\dfrac{\sqrt{x+1}}{\sqrt{x+1}}\right) = \dfrac{x+1}{2\sqrt{x+1}}$

c. $\dfrac{1}{\sqrt{5} + \sqrt{2}} = \dfrac{1}{\sqrt{5} + \sqrt{2}}\left(\dfrac{\sqrt{5} - \sqrt{2}}{\sqrt{5} - \sqrt{2}}\right) = \dfrac{\sqrt{5} - \sqrt{2}}{5 - 2} = \dfrac{\sqrt{5} - \sqrt{2}}{3}$

d. $\dfrac{1}{\sqrt{x} - \sqrt{x+1}} = \dfrac{1}{\sqrt{x} - \sqrt{x+1}}\left(\dfrac{\sqrt{x} + \sqrt{x+1}}{\sqrt{x} + \sqrt{x+1}}\right)$

$\qquad = \dfrac{\sqrt{x} + \sqrt{x+1}}{x - (x+1)}$

$\qquad = -\sqrt{x} - \sqrt{x+1}$

✓CHECKPOINT 6

Rationalize the denominator or numerator.

a. $\dfrac{5}{\sqrt{8}}$

b. $\dfrac{\sqrt{x+2}}{4}$

c. $\dfrac{1}{\sqrt{6} - \sqrt{3}}$

d. $\dfrac{1}{\sqrt{x} + \sqrt{x+2}}$ ∎

Exercises 0.5

In Exercises 1–16, perform the indicated operations and simplify your answer.

1. $\dfrac{x}{x-2} + \dfrac{3}{x-2}$

2. $\dfrac{2x-1}{x+3} + \dfrac{1-x}{x+3}$

3. $\dfrac{2x}{x^2+2} - \dfrac{1-3x}{x^2+2}$

4. $\dfrac{5x+10}{2x-1} - \dfrac{2x+10}{2x-1}$

5. $\dfrac{2}{x^2-4} - \dfrac{1}{x-2}$

6. $\dfrac{x}{x^2+x-2} - \dfrac{1}{x+2}$

7. $\dfrac{5}{x-3} + \dfrac{3}{3-x}$

8. $\dfrac{x}{2-x} + \dfrac{2}{x-2}$

9. $\dfrac{A}{x-1} + \dfrac{B}{(x-1)^2} + \dfrac{C}{x+2}$

10. $\dfrac{A}{x-5} + \dfrac{B}{x+5} + \dfrac{C}{(x+5)^2}$

11. $\dfrac{A}{x-6} + \dfrac{Bx+C}{x^2+3}$

12. $\dfrac{Ax+B}{x^2+2} + \dfrac{C}{x-4}$

13. $-\dfrac{2}{x} + \dfrac{1}{x^2+2}$

14. $\dfrac{2}{x+1} + \dfrac{1-x}{x^2-2x+3}$

15. $\dfrac{1}{x^2-x-2} - \dfrac{x}{x^2-5x+6}$

16. $\dfrac{x-1}{x^2+5x+4} + \dfrac{2}{x^2-x-2} + \dfrac{10}{x^2+2x-8}$

In Exercises 17–26, simplify each expression.

17. $\dfrac{-x}{(x+1)^{3/2}} + \dfrac{2}{(x+1)^{1/2}}$

18. $2\sqrt{x}(x-2) + \dfrac{(x-2)^2}{2\sqrt{x}}$

19. $\dfrac{2-t}{2\sqrt{1+t}} - \sqrt{1+t}$

20. $-\dfrac{\sqrt{x^2+1}}{x^2} + \dfrac{1}{\sqrt{x^2+1}}$

21. $\left(2x\sqrt{x^2+1} - \dfrac{x^3}{\sqrt{x^2+1}}\right) \div (x^2+1)$

22. $\left(\sqrt{x^3+1} - \dfrac{3x^3}{2\sqrt{x^3+1}}\right) \div (x^3+1)$

23. $\dfrac{(x^2+2)^{1/2} - x^2(x^2+2)^{-1/2}}{x^2}$

24. $\dfrac{x(x+1)^{-1/2} - (x+1)^{1/2}}{x^2}$

25. $\dfrac{-x^2}{(2x+3)^{3/2}} + \dfrac{2x}{(2x+3)^{1/2}}$

26. $\dfrac{-x}{2(3+x^2)^{3/2}} + \dfrac{3}{(3+x^2)^{1/2}}$

In Exercises 27–42, rationalize the numerator or denominator and simplify.

27. $\dfrac{2}{\sqrt{10}}$

28. $\dfrac{3}{\sqrt{21}}$

29. $\dfrac{4x}{\sqrt{x-1}}$

30. $\dfrac{5y}{\sqrt{y+7}}$

31. $\dfrac{49(x-3)}{\sqrt{x^2-9}}$

32. $\dfrac{10(x+2)}{\sqrt{x^2-x-6}}$

33. $\dfrac{5}{\sqrt{14}-2}$

34. $\dfrac{13}{6+\sqrt{10}}$

35. $\dfrac{2x}{5-\sqrt{3}}$

36. $\dfrac{x}{\sqrt{2}+\sqrt{3}}$

37. $\dfrac{1}{\sqrt{6}+\sqrt{5}}$

38. $\dfrac{\sqrt{15}+3}{12}$

39. $\dfrac{2}{\sqrt{x}+\sqrt{x-2}}$

40. $\dfrac{10}{\sqrt{x}+\sqrt{x+5}}$

41. $\dfrac{\sqrt{x+2}-\sqrt{2}}{x}$

42. $\dfrac{\sqrt{x+1}-1}{x}$

In Exercises 43 and 44, perform the indicated operations and rationalize as needed.

43. $\dfrac{\dfrac{\sqrt{4-x^2}}{x^4} - \dfrac{2}{x^2\sqrt{4-x^2}}}{4-x^2}$

44. $\dfrac{\dfrac{\sqrt{x^2+1}}{x^2} - \dfrac{1}{x\sqrt{x^2+1}}}{x^2+1}$

(T) 45. Memory Model Psychologists have developed mathematical models to predict memory performance based on the number of trials n of a certain task. One such model is

$$P = \dfrac{0.6 + 0.85(n-1)}{1 + 0.85(n-1)}, \quad n > 0$$

where P is the percent of correct responses (in decimal form) after n trials. Enter the formula into a graphing utility and use it to complete the table.

n	1	2	3	4	5	6	7	8	9	10
P										

46. Plants The numbers of endangered and threatened plant species in the United States from 2000 through 2005 are given by

Endangered plants: $E = \dfrac{2342.52t^2 + 565}{3.91t^2 + 1}$ and

Threatened plants: $T = \dfrac{243.48t^2 + 139}{1.65t^2 + 1}$

where t represents the year, with $t = 0$ corresponding to 2000. Find a rational model that represents the total number of endangered and threatened plant species. *(Source: U.S. Fish and Wildlife Service)*

Functions, Graphs, and Limits

During the first 12 months, the growth of a killer whale calf can be modeled by a linear equation. (See Section 1.3, Example 3.)

Applications

Functions and limit concepts have many real-life applications. The applications listed below represent a sample of the applications in this chapter.

- State Populations, Exercise 61, page 54
- Fatal Crashes, Exercise 89, page 66
- Femur Length, Exercise 67, page 80
- Rehabilitation Counseling, Exercise 72, page 93
- Population Growth, Exercise 65, page 103

1.1 The Cartesian Plane and the Distance Formula
1.2 Graphs of Equations
1.3 Lines in the Plane and Slope
1.4 Functions
1.5 Limits
1.6 Continuity

Section 1.1

The Cartesian Plane and the Distance Formula

- Plot points in a coordinate plane and read data presented graphically.
- Find the distance between two points in a coordinate plane.
- Find the midpoints of line segments connecting two points.
- Translate points in a coordinate plane.

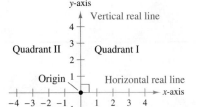

FIGURE 1.1 The Cartesian Plane

FIGURE 1.2

FIGURE 1.3

The Cartesian Plane

Just as you can represent real numbers by points on a real number line, you can represent ordered pairs of real numbers by points in a plane called the **rectangular coordinate system,** or the **Cartesian plane,** after the French mathematician René Descartes (1596–1650).

The Cartesian plane is formed by using two real number lines intersecting at right angles, as shown in Figure 1.1. The horizontal real number line is usually called the ***x*-axis,** and the vertical real number line is usually called the ***y*-axis.** The point of intersection of these two axes is the **origin,** and the two axes divide the plane into four parts called **quadrants.**

Each point in the plane corresponds to an **ordered pair** (x, y) of real numbers x and y, called **coordinates** of the point. The ***x*-coordinate** represents the directed distance from the y-axis to the point, and the ***y*-coordinate** represents the directed distance from the x-axis to the point, as shown in Figure 1.2.

$$(x, y)$$

Directed distance from y-axis ↑ ↑ Directed distance from x-axis

STUDY TIP

The notation (x, y) denotes both a point in the plane and an open interval on the real number line. The context will tell you which meaning is intended.

Example 1 Plotting Points in the Cartesian Plane

Plot the points $(-1, 2)$, $(3, 4)$, $(0, 0)$, $(3, 0)$, and $(-2, -3)$.

SOLUTION To plot the point

$$(-1, 2)$$

x-coordinate ↑ ↑ y-coordinate

imagine a vertical line through -1 on the x-axis and a horizontal line through 2 on the y-axis. The intersection of these two lines is the point $(-1, 2)$. The other four points can be plotted in a similar way and are shown in Figure 1.3.

✓ CHECKPOINT 1

Plot the points $(-3, 2)$, $(4, -2)$, $(3, 1)$, $(0, -2)$, and $(-1, -2)$. ■

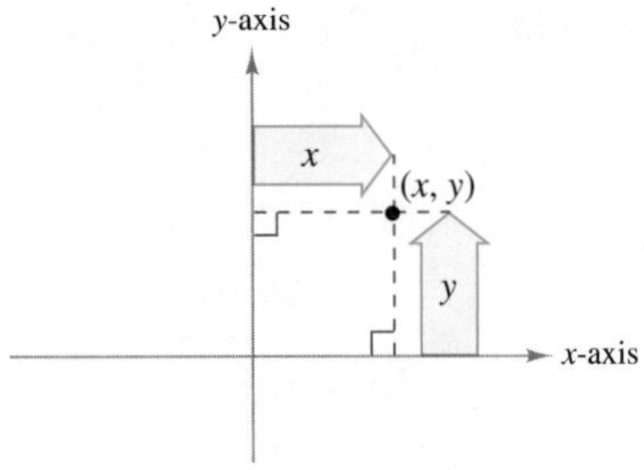

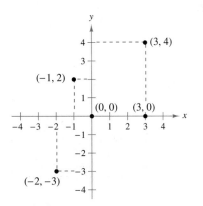

SECTION 1.1 The Cartesian Plane and the Distance Formula

Using a rectangular coordinate system allows you to visualize relationships between two variables. In Example 2, notice how much your intuition is enhanced by the use of a graphical presentation.

Example 2 Sketching a Scatter Plot

The amounts A (in billions of dollars) spent on prescription drugs in the United States in the years 1998 through 2005 are shown in the table, where t represents the year. Sketch a scatter plot of the data. *(Source: National Association of Chain Drug Stores)*

t	1998	1999	2000	2001	2002	2003	2004	2005
A	108.7	125.8	145.6	164.1	182.7	203.1	221.0	230.3

SOLUTION To sketch a *scatter plot* of the data given in the table, you simply represent each pair of values by an ordered pair (t, A), and plot the resulting points, as shown in Figure 1.4. For instance, the first pair of values is represented by the ordered pair (1998, 108.7). Note that the break in the t-axis indicates that the numbers between 0 and 1997 have been omitted.

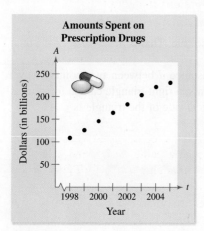

FIGURE 1.4

STUDY TIP

In Example 2, you could let $t = 1$ represent the year 1998. In that case, the horizontal axis would not have been broken, and the tick marks would have been labeled 1 through 8 (instead of 1998 through 2005).

✓ CHECKPOINT 2

From 1998 through 2005, the average prices P (in dollars) of prescription drugs (excluding mail order) in the United States are shown in the table, where t represents the year. Sketch a scatter plot of the data. *(Source: National Association of Chain Drug Stores)*

t	1998	1999	2000	2001	2002	2003	2004	2005
P	38.43	42.42	45.79	50.06	55.37	59.52	63.59	64.86

TECHNOLOGY

The scatter plot in Example 2 is only one way to represent the given data graphically. Two other techniques are shown at the right. The first is a *bar graph* and the second is a *line graph*. All three graphical representations were created with a computer. If you have access to computer graphing software, try using it to represent graphically the data given in Example 2.

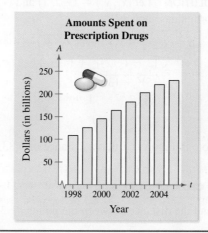
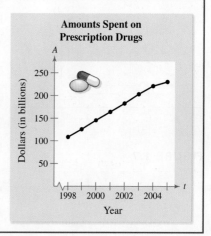

The symbol ® indicates an example that uses or is derived from real-life data.

The Distance Formula

Recall from the Pythagorean Theorem that, for a right triangle with hypotenuse of length c and sides of lengths a and b, you have

$$a^2 + b^2 = c^2 \qquad \text{Pythagorean Theorem}$$

as shown in Figure 1.5. (The converse is also true. That is, if $a^2 + b^2 = c^2$, then the triangle is a right triangle.)

Suppose you want to determine the distance d between two points (x_1, y_1) and (x_2, y_2) in the plane. With these two points, a right triangle can be formed, as shown in Figure 1.6. The length of the vertical side of the triangle is

$$|y_2 - y_1|$$

and the length of the horizontal side is

$$|x_2 - x_1|.$$

By the Pythagorean Theorem, you can write

$$d^2 = |x_2 - x_1|^2 + |y_2 - y_1|^2$$
$$d = \sqrt{|x_2 - x_1|^2 + |y_2 - y_1|^2}$$
$$d = \sqrt{(x_2 - x_1)^2 + (y_2 - y_1)^2}.$$

This result is the **Distance Formula**.

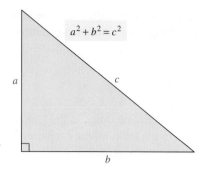

FIGURE 1.5 Pythagorean Theorem

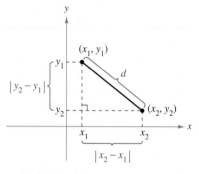

FIGURE 1.6 Distance Between Two Points

The Distance Formula

The distance d between the points (x_1, y_1) and (x_2, y_2) in the plane is

$$d = \sqrt{(x_2 - x_1)^2 + (y_2 - y_1)^2}.$$

Example 3 Finding a Distance

Find the distance between the points $(-2, 1)$ and $(3, 4)$.

SOLUTION Let $(x_1, y_1) = (-2, 1)$ and $(x_2, y_2) = (3, 4)$. Then apply the Distance Formula as shown.

$$d = \sqrt{(x_2 - x_1)^2 + (y_2 - y_1)^2} \qquad \text{Distance Formula}$$
$$= \sqrt{[3 - (-2)]^2 + (4 - 1)^2} \qquad \text{Substitute for } x_1, y_1, x_2, \text{ and } y_2.$$
$$= \sqrt{(5)^2 + (3)^2} \qquad \text{Simplify.}$$
$$= \sqrt{34}$$
$$\approx 5.83 \qquad \text{Use a calculator.}$$

Note in Figure 1.7 that a distance of 5.83 looks about right.

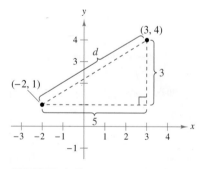

FIGURE 1.7

✓ CHECKPOINT 3

Find the distance between the points $(-2, 1)$ and $(2, 4)$. ■

SECTION 1.1   The Cartesian Plane and the Distance Formula   37

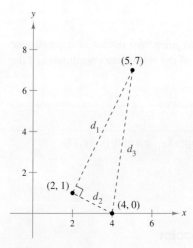

FIGURE 1.8

Example 4   Verifying a Right Triangle

Use the Distance Formula to show that the points $(2, 1)$, $(4, 0)$, and $(5, 7)$ are vertices of a right triangle.

SOLUTION The three points are plotted in Figure 1.8. Using the Distance Formula, you can find the lengths of the three sides as shown below.

$$d_1 = \sqrt{(5-2)^2 + (7-1)^2} = \sqrt{9 + 36} = \sqrt{45}$$
$$d_2 = \sqrt{(4-2)^2 + (0-1)^2} = \sqrt{4 + 1} = \sqrt{5}$$
$$d_3 = \sqrt{(5-4)^2 + (7-0)^2} = \sqrt{1 + 49} = \sqrt{50}$$

Because

$$d_1^2 + d_2^2 = 45 + 5 = 50 = d_3^2$$

you can apply the converse of the Pythagorean Theorem to conclude that the triangle must be a right triangle.

✓ CHECKPOINT 4

Use the Distance Formula to show that the points $(2, -1)$, $(5, 5)$, and $(6, -3)$ are vertices of a right triangle. ■

The figures provided with Examples 3 and 4 were not really essential to the solution. *Nevertheless,* we strongly recommend that you develop the habit of including sketches with your solutions—even if they are not required.

Example 5   Finding the Length of a Walk

Figure 1.9 shows part of a trail system at a nature preserve. Each unit represents 0.1 mile. You walk from the visitor center at *V* to the observation stand at *S*. How far do you walk?

SOLUTION You can find the length of the walk by finding the distance between the points $(3, 0)$ and $(6, 4)$.

$$d = \sqrt{(6-3)^2 + (4-0)^2} \quad \text{Distance Formula}$$
$$= \sqrt{9 + 16}$$
$$= 5 \quad \text{Simplify.}$$

Because each unit represents 0.1 mile, the walk is $5(0.1) = 0.5$ mile long.

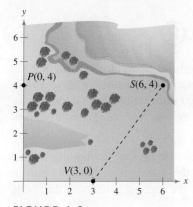

FIGURE 1.9

✓ CHECKPOINT 5

After walking to the observation stand in Example 5, you walk from *S* to the picnic area at *P*. How far do you walk? ■

STUDY TIP

When you use coordinate geometry to solve real-life problems, you are free to place the coordinate system in any way that is convenient for the solution of the problem.

The Midpoint Formula

To find the **midpoint** of the line segment that joins two points in a coordinate plane, you can simply find the average values of the respective coordinates of the two endpoints.

> **The Midpoint Formula**
>
> The midpoint of the segment joining the points (x_1, y_1) and (x_2, y_2) is
> $$\text{Midpoint} = \left(\frac{x_1 + x_2}{2}, \frac{y_1 + y_2}{2}\right).$$

Example 6 Finding a Segment's Midpoint

Find the midpoint of the line segment joining the points $(-5, -3)$ and $(9, 3)$, as shown in Figure 1.10.

SOLUTION Let $(x_1, y_1) = (-5, -3)$ and $(x_2, y_2) = (9, 3)$.

$$\text{Midpoint} = \left(\frac{x_1 + x_2}{2}, \frac{y_1 + y_2}{2}\right) = \left(\frac{-5 + 9}{2}, \frac{-3 + 3}{2}\right) = (2, 0)$$

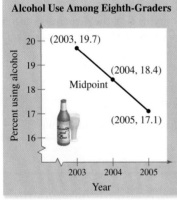

FIGURE 1.10

✓ CHECKPOINT 6

Find the midpoint of the line segment joining $(-6, 2)$ and $(2, 8)$. ■

Example 7 Estimating Alcohol Use ®

A survey in 2003 found that 19.7% of eighth-graders used alcohol, and in 2005 another survey found that 17.1% of eighth-graders used alcohol. Without knowing any additional information, what would you estimate the alcohol use among 2004 eighth-graders to have been? *(Source: National Institute of Health, National Institute on Drug Abuse)*

SOLUTION One solution to the problem is to assume that alcohol use among eighth-graders followed a linear pattern. With this assumption, you can estimate the 2004 alcohol use by finding the midpoint of the segment connecting the points (2003, 19.7) and (2005, 17.1).

$$\text{Midpoint} = \left(\frac{2003 + 2005}{2}, \frac{19.7 + 17.1}{2}\right) = (2004, 18.4)$$

So, you would estimate that 18.4% of eighth-graders used alcohol in 2004, as shown in Figure 1.11. (The actual survey results found that 18.6% of eighth-graders used alcohol.)

FIGURE 1.11

✓ CHECKPOINT 7

A survey in 2003 found that 35.4% of tenth-graders used alcohol, and a survey in 2005 found that 33.2% of tenth-graders used alcohol. What would you estimate the alcohol use among 2004 tenth-graders to have been? *(Source: National Institute of Health, National Institute on Drug Abuse)* ■

SECTION 1.1 The Cartesian Plane and the Distance Formula

Translating Points in the Plane

Example 8 **Translating Points in the Plane**

Figure 1.12(a) shows the vertices of a parallelogram. Find the vertices of the parallelogram after it has been translated two units down and four units to the right.

SOLUTION To translate each vertex two units down, subtract 2 from each y-coordinate. To translate each vertex four units to the right, add 4 to each x-coordinate.

Original Point	*Translated Point*
$(1, 0)$	$(1 + 4, 0 - 2) = (5, -2)$
$(3, 2)$	$(3 + 4, 2 - 2) = (7, 0)$
$(3, 6)$	$(3 + 4, 6 - 2) = (7, 4)$
$(1, 4)$	$(1 + 4, 4 - 2) = (5, 2)$

The translated parallelogram is shown in Figure 1.12(b).

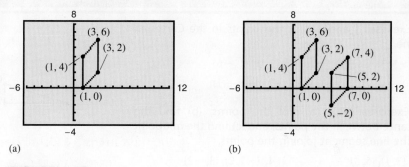

FIGURE 1.12

Courtesy of Masumi Yajima and the Sun Center of Excellence for Visual Genomics

Scientists at the University of Calgary have created a computer-generated three-dimensional display of the human body as an aid to both doctors and patients. Much of computer graphics, including this image, consists of transformations of points in two- and three-dimensional space. One type of transformation, a translation, is illustrated in Example 8. Other types include reflections, rotations, and stretches.

✓ **CHECKPOINT 8**

Find the vertices of the parallelogram in Example 8 after it has been translated two units to the left and four units down. ■

CONCEPT CHECK

1. What is the y-coordinate of any point on the x-axis? What is the x-coordinate of any point on the y-axis?

2. Describe the signs of the x- and y-coordinates of points that lie in the first and second quadrants.

3. To divide a line segment into four equal parts, how many times is the Midpoint Formula used?

4. When finding the distance between two points, does it matter which point is chosen as (x_1, y_1)? Explain.

Skills Review 1.1

The following warm-up exercises involve skills that were covered in earlier sections. You will use these skills in the exercise set for this section. For additional help, review Section 0.3.

In Exercises 1–6, simplify each expression.

1. $\sqrt{(3-6)^2 + [1-(-5)]^2}$
2. $\sqrt{(-2-0)^2 + [-7-(-3)]^2}$
3. $\dfrac{5 + (-4)}{2}$
4. $\dfrac{-3 + (-1)}{2}$
5. $\sqrt{27} + \sqrt{12}$
6. $\sqrt{8} - \sqrt{18}$

In Exercises 7–10, solve for x or y.

7. $\sqrt{(3-x)^2 + (7-4)^2} = \sqrt{45}$
8. $\sqrt{(6-2)^2 + (-2-y)^2} = \sqrt{52}$
9. $\dfrac{x + (-5)}{2} = 7$
10. $\dfrac{-7 + y}{2} = -3$

Exercises 1.1

See www.CalcChat.com for worked-out solutions to odd-numbered exercises.

In Exercises 1 and 2, plot the points in the Cartesian plane.

1. $(-5, 3), (1, -1), (-2, -4), (2, 0), (1, -6)$
2. $(0, -4), (5, 1), (-3, 5), (2, -2), (-6, -1)$

In Exercises 3–12, (a) plot the points, (b) find the distance between the points, and (c) find the midpoint of the line segment joining the points.

3. $(3, 1), (5, 5)$
4. $(-3, 2), (3, -2)$
5. $\left(\tfrac{1}{2}, 1\right), \left(-\tfrac{3}{2}, -5\right)$
6. $\left(\tfrac{2}{3}, -\tfrac{1}{3}\right), \left(\tfrac{5}{6}, 1\right)$
7. $(2, 2), (4, 14)$
8. $(-3, 7), (1, -1)$
9. $\left(1, \sqrt{3}\right), (-1, 1)$
10. $(-2, 0), \left(0, \sqrt{2}\right)$
11. $(0, -4.8), (0.5, 6)$
12. $(5.2, 6.4), (-2.7, 1.8)$

In Exercises 13–16, (a) find the length of each side of the right triangle and (b) show that these lengths satisfy the Pythagorean Theorem.

13.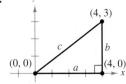
14.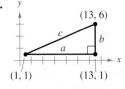

15.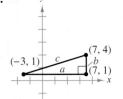
16.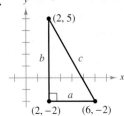

In Exercises 17–20, show that the points form the vertices of the given figure. (A rhombus is a quadrilateral whose sides have the same length.)

Vertices	Figure
17. $(0, 1), (3, 7), (4, -1)$	Right triangle
18. $(1, -3), (3, 2), (-2, 4)$	Isosceles triangle
19. $(0, 0), (1, 2), (2, 1), (3, 3)$	Rhombus
20. $(0, 1), (3, 7), (4, 4), (1, -2)$	Parallelogram

In Exercises 21 and 22, find x such that the distance between the points is 5.

21. $(1, 0), (x, -4)$
22. $(2, -1), (x, 2)$

In Exercises 23 and 24, find y such that the distance between the points is 8.

23. $(0, 0), (3, y)$
24. $(5, 1), (5, y)$

The answers to the odd-numbered and selected even-numbered exercises are given in the back of the text. Worked-out solutions to the odd-numbered exercises are given in the *Student Solutions Guide*.

25. Building Dimensions The base and height of the trusses for the roof of a greenhouse are 32 feet and 5 feet, respectively (see figure).

(a) Find the distance from the eaves to the peak of the roof.

(b) The length of the greenhouse is 40 feet. Use the result of part (a) to find the number of square feet of roofing.

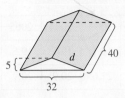

Figure for 25

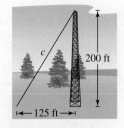

Figure for 26

26. Wire Length A guy wire is stretched from a communications tower in a nature preserve at a point 200 feet above the ground to an anchor 125 feet from the base (see figure). How long is the wire?

(T) In Exercises 27 and 28, use a graphing utility to sketch a scatter plot, a bar graph, or a line graph to represent the data. Describe any trends that appear.

27. Consumer Trends The numbers (in millions) of basic cable television subscribers in the United States for 1996 through 2005 are shown in the table. *(Source: National Cable & Telecommunications Association)*

Year	1996	1997	1998	1999	2000
Subscribers	62.3	63.6	64.7	65.5	66.3

Year	2001	2002	2003	2004	2005
Subscribers	66.7	66.5	66.0	65.7	65.3

28. Consumer Trends The numbers (in millions) of cellular telephone subscribers in the United States for 1996 through 2005 are shown in the table. *(Source: Cellular Telecommunications & Internet Association)*

Year	1996	1997	1998	1999	2000
Subscribers	44.0	55.3	69.2	86.0	109.5

Year	2001	2002	2003	2004	2005
Subscribers	128.4	140.8	158.7	182.1	207.9

National Monuments In Exercises 29 and 30, use the figure below showing the numbers of recreational visits (in millions) to national monuments. *(Source: U.S. National Park Service)*

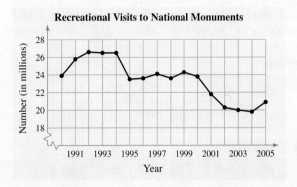

29. Estimate the number of recreational visits (in millions) to national monuments for each year.

(a) 1992   (b) 1997   (c) 2001   (d) 2004

30. Estimate the percent increase or decrease in the number of recreational visits (a) from 1991 to 1998 and (b) from 2000 to 2005.

Food Stamps In Exercises 31 and 32, use the figure, which shows the numbers of participants (in millions) in federal food stamp programs in the United States from 1990 through 2005. *(Source: U.S. Department of Agriculture, Food and Nutrition Service)*

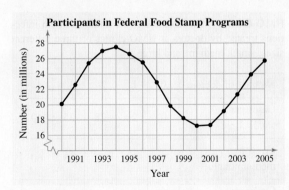

31. Estimate the number of participants in federal food stamp programs for each year.

(a) 1991   (b) 1994   (c) 1997   (d) 2005

32. Estimate the percent increase or decrease in the number of participants in federal food stamp programs (a) from 1996 to 1997 and (b) from 2003 to 2004.

The symbol (T) indicates an exercise in which you are instructed to use graphing technology or a symbolic computer algebra system. The solutions of other exercises may also be facilitated by use of appropriate technology.

Agricultural Imports In Exercises 33–36, (a) use the Midpoint Formula to estimate the value of the commodity in 2003. (b) Then use your school's library, the Internet, or some other reference source to find the actual value of the commodity for 2003. (c) Did the value increase in a linear pattern from 2001 to 2005? Explain your reasoning. *(Source: U.S. Department of Agriculture, Economic Research Service)*

33. Pork

Year	2001	2003	2005
Value (millions of $)	1048		1281

34. Grains and feeds

Year	2001	2003	2005
Value (millions of $)	3320		4526

35. Wine

Year	2001	2003	2005
Value (millions of $)	2250		3763

36. Malt beverages

Year	2001	2003	2005
Value (millions of $)	2348		3096

Ⓑ 37. HMO Clinics The table shows the numbers of ear infections treated by doctors at HMO clinics of three different sizes: small, medium, and large.

Number of doctors	0	1	2	3	4
Cases per small clinic	0	20	28	35	40
Cases per medium clinic	0	30	42	53	60
Cases per large clinic	0	35	49	62	70

(a) Show the relationship between doctors and treated ear infections using *three* curves, where the number of doctors is on the horizontal axis and the number of ear infections treated is on the vertical axis.

(b) Compare the three relationships. *(Source: Adapted from Taylor, Economics, Fifth Edition)*

38. Health The percents of adults who are considered drinkers or smokers are shown in the table. Drinkers were defined as those who had five or more drinks in 1 day at least once during a recent year. Smokers were defined as those who smoked more than 100 cigarettes in their lifetime and smoked daily or semi-daily. *(Source: National Health Interview Survey)*

Year	2001	2002	2003	2004	2005
Drinkers	20.0	19.9	19.1	19.1	19.5
Smokers	22.7	22.4	21.6	20.9	20.9

(a) Sketch a line graph of each data set.

(b) Describe any trends that appear.

Ⓣ Computer Graphics In Exercises 39 and 40, the red figure is translated to a new position in the plane to form the blue figure. (a) Find the vertices of the transformed figure. (b) Then use a graphing utility to draw both figures.

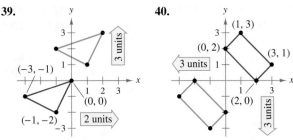

41. Use the Midpoint Formula repeatedly to find the three points that divide the segment joining (x_1, y_1) and (x_2, y_2) into four equal parts.

42. Show that $\left(\frac{1}{3}[2x_1 + x_2], \frac{1}{3}[2y_1 + y_2]\right)$ is one of the points of trisection of the line segment joining (x_1, y_1) and (x_2, y_2). Then, find the second point of trisection by finding the midpoint of the segment joining

$$\left(\frac{1}{3}[2x_1 + x_2], \frac{1}{3}[2y_1 + y_2]\right) \text{ and } (x_2, y_2).$$

43. Use Exercise 41 to find the points that divide the line segment joining the given points into four equal parts.

(a) $(1, -2), (4, -1)$ (b) $(-2, -3), (0, 0)$

44. Use Exercise 42 to find the points of trisection of the line segment joining the given points.

(a) $(1, -2), (4, 1)$ (b) $(-2, -3), (0, 0)$

The symbol Ⓑ indicates an exercise that contains material from textbooks in other disciplines.

Section 1.2
Graphs of Equations

- Sketch graphs of equations by hand.
- Find the *x*- and *y*-intercepts of graphs of equations.
- Write the standard forms of equations of circles.
- Find the points of intersection of two graphs.
- Use mathematical models to model and solve real-life problems.

The Graph of an Equation

In Section 1.1, you used a coordinate system to represent graphically the relationship between two quantities. There, the graphical picture consisted of a collection of points in a coordinate plane (see Example 2 in Section 1.1).

Frequently, a relationship between two quantities is expressed as an equation. For instance, degrees on the Fahrenheit scale are related to degrees on the Celsius scale by the equation $F = \frac{9}{5}C + 32$. In this section, you will study some basic procedures for sketching the graphs of such equations. The **graph of an equation** is the set of all points that are solutions of the equation.

Example 1 Sketching the Graph of an Equation

Sketch the graph of $y = 7 - 3x$.

SOLUTION The simplest way to sketch the graph of an equation is the *point-plotting method*. With this method, you construct a table of values that consists of several solution points of the equation, as shown in the table below. For instance, when $x = 0$

$$y = 7 - 3(0) = 7$$

which implies that $(0, 7)$ is a solution point of the graph.

x	0	1	2	3	4
$y = 7 - 3x$	7	4	1	-2	-5

From the table, it follows that $(0, 7)$, $(1, 4)$, $(2, 1)$, $(3, -2)$, and $(4, -5)$ are solution points of the equation. After plotting these points, you can see that they appear to lie on a line, as shown in Figure 1.13. The graph of the equation is the line that passes through the five plotted points.

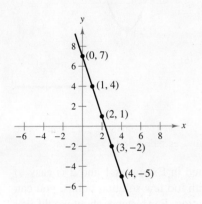

FIGURE 1.13 Solution Points for $y = 7 - 3x$

✓ **CHECKPOINT 1**

Sketch the graph of $y = 2x + 1$. ∎

> **STUDY TIP**
>
> Even though we refer to the sketch shown in Figure 1.13 as the graph of $y = 7 - 3x$, it actually represents only a *portion* of the graph. The entire graph is a line that would extend off the page.

44 CHAPTER 1 Functions, Graphs, and Limits

Example 2 Sketching the Graph of an Equation

Sketch the graph of $y = x^2 - 2$.

SOLUTION Begin by constructing a table of values, as shown below.

x	-2	-1	0	1	2	3
$y = x^2 - 2$	2	-1	-2	-1	2	7

Next, plot the points given in the table, as shown in Figure 1.14(a). Finally, connect the points with a smooth curve, as shown in Figure 1.14(b).

> **STUDY TIP**
>
> The graph shown in Example 2 is a **parabola**. The graph of any second-degree equation of the form
>
> $y = ax^2 + bx + c, \quad a \neq 0$
>
> has a similar shape. If $a > 0$, the parabola opens upward, as in Figure 1.14(b), and if $a < 0$, the parabola opens downward.

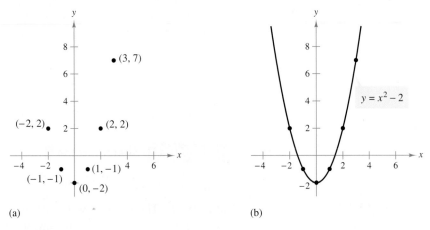

FIGURE 1.14

✓ CHECKPOINT 2

Sketch the graph of $y = x^2 - 4$. ∎

The point-plotting technique demonstrated in Examples 1 and 2 is easy to use, but it does have some shortcomings. With too few solution points, you can badly misrepresent the graph of a given equation. For instance, how would you connect the four points in Figure 1.15? Without further information, any one of the three graphs in Figure 1.16 would be reasonable.

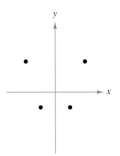

FIGURE 1.15

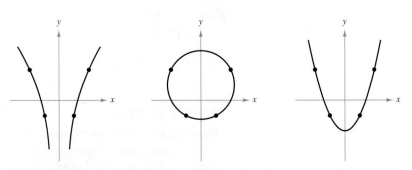

FIGURE 1.16

Intercepts of a Graph

> **Algebra Review**
>
> Finding intercepts involves solving equations. For a review of some techniques for solving equations, see page 105.

It is often easy to determine the solution points that have zero as either the *x*-coordinate or the *y*-coordinate. These points are called **intercepts** because they are the points at which the graph intersects the *x*- or *y*-axis.

Some texts denote the *x*-intercept as the *x*-coordinate of the point $(a, 0)$ rather than the point itself. Unless it is necessary to make a distinction, we will use the term *intercept* to mean either the point or the coordinate.

A graph may have no intercepts or several intercepts, as shown in Figure 1.17.

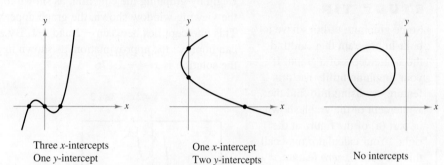

No *x*-intercept
One *y*-intercept

Three *x*-intercepts
One *y*-intercept

One *x*-intercept
Two *y*-intercepts

No intercepts

FIGURE 1.17

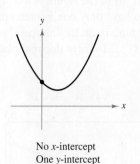

FIGURE 1.18

> **Finding Intercepts**
>
> 1. To find **x-intercepts**, let *y* be zero and solve the equation for *x*.
> 2. To find **y-intercepts**, let *x* be zero and solve the equation for *y*.

Example 3 Finding x- and y-Intercepts

Find the *x*- and *y*-intercepts of the graph of each equation.

a. $y = x^3 - 4x$ **b.** $x = y^2 - 3$

SOLUTION

a. Let $y = 0$. Then $0 = x(x^2 - 4) = x(x + 2)(x - 2)$ has solutions $x = 0$ and $x = \pm 2$. Let $x = 0$. Then $y = (0)^3 - 4(0) = 0$.

   x-intercepts: $(0, 0), (2, 0), (-2, 0)$ *y*-intercept: $(0, 0)$ See Figure 1.18.

b. Let $y = 0$. Then $x = (0)^2 - 3 = -3$. Let $x = 0$. Then $y^2 - 3 = 0$ has solutions $y = \pm\sqrt{3}$.

   x-intercept: $(-3, 0)$ *y*-intercepts: $(0, \sqrt{3}), (0, -\sqrt{3})$ See Figure 1.19.

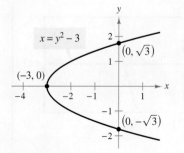

FIGURE 1.19

✓CHECKPOINT 3

Find the *x*- and *y*-intercepts of the graph of each equation.

a. $y = x^2 - 2x - 3$ **b.** $y^2 - 4 = x$ ∎

TECHNOLOGY

Zooming in to Find Intercepts

You can use the *zoom* feature of a graphing utility to approximate the x-intercepts of a graph. Suppose you want to approximate the x-intercept(s) of the graph of

$$y = 2x^3 - 3x + 2.$$

STUDY TIP

Some graphing utilities have a built-in program that can find the x-intercepts of a graph. If your graphing utility has this feature, try using it to find the x-intercept of the graph shown in part (a) of the figure at the right. (Your calculator may call this the *root* or *zero* feature.)*

Begin by graphing the equation, as shown in part (a) of the figure below. From the viewing window shown, the graph appears to have only one x-intercept. This intercept lies between -2 and -1. By zooming in on the intercept, you can improve the approximation, as shown in part (b). To three decimal places, the solution is $x \approx -1.476$.

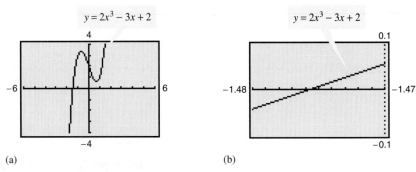

(a)   (b)

Here are some suggestions for using the *zoom* feature.

1. With each successive zoom-in, adjust the x-scale so that the viewing window shows at least one tick mark on each side of the x-intercept.

2. The error in your approximation will be less than the distance between two scale marks.

3. The *trace* feature can usually be used to add one more decimal place of accuracy without changing the viewing window.

Part (a) of the figure below shows the graph of $y = x^2 - 5x + 3$. Parts (b) and (c) show "zoom-in views" of the two intercepts. From these views, you can approximate the x-intercepts to be $x \approx 0.697$ and $x \approx 4.303$.

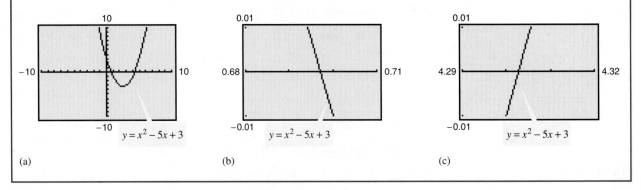

(a)   (b)   (c)

---

*Specific calculator keystroke instructions for operations in this and other technology boxes can be found at *college.hmco.com/info/larsonapplied*.

Circles

Throughout this course, you will learn to recognize several types of graphs from their equations. For instance, you should recognize that the graph of a second-degree equation of the form

$$y = ax^2 + bx + c, \quad a \neq 0$$

is a parabola (see Example 2). Another easily recognized graph is that of a **circle**.

Consider the circle shown in Figure 1.20. A point (x, y) is on the circle if and only if its distance from the center (h, k) is r. By the Distance Formula,

$$\sqrt{(x - h)^2 + (y - k)^2} = r.$$

By squaring both sides of this equation, you obtain the **standard form of the equation of a circle.**

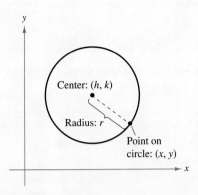

FIGURE 1.20

Standard Form of the Equation of a Circle

The point (x, y) lies on the circle of **radius** r and **center** (h, k) if and only if

$$(x - h)^2 + (y - k)^2 = r^2.$$

From this result, you can see that the standard form of the equation of a circle *with its center at the origin*, $(h, k) = (0, 0)$, is simply

$$x^2 + y^2 = r^2. \qquad \text{Circle with center at origin}$$

Example 4 Finding the Equation of a Circle

The point $(3, 4)$ lies on a circle whose center is at $(-1, 2)$, as shown in Figure 1.21. Find the standard form of the equation of this circle.

SOLUTION The radius of the circle is the distance between $(-1, 2)$ and $(3, 4)$.

$$r = \sqrt{[3 - (-1)]^2 + (4 - 2)^2} \qquad \text{Distance Formula}$$
$$= \sqrt{16 + 4} \qquad \text{Simplify.}$$
$$= \sqrt{20} \qquad \text{Radius}$$

Using $(h, k) = (-1, 2)$, the standard form of the equation of the circle is

$$(x - h)^2 + (y - k)^2 = r^2$$
$$[x - (-1)]^2 + (y - 2)^2 = \left(\sqrt{20}\right)^2 \qquad \text{Substitute for } h, k, \text{ and } r.$$
$$(x + 1)^2 + (y - 2)^2 = 20. \qquad \text{Write in standard form.}$$

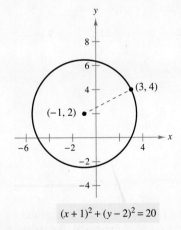

FIGURE 1.21

✓ CHECKPOINT 4

The point $(1, 5)$ lies on a circle whose center is at $(-2, 1)$. Find the standard form of the equation of this circle. ■

TECHNOLOGY

To graph a circle on a graphing utility, you can solve its equation for y and graph the top and bottom halves of the circle separately. For instance, you can graph the circle $(x + 1)^2 + (y - 2)^2 = 20$ by graphing the following equations.

$$y = 2 + \sqrt{20 - (x + 1)^2}$$
$$y = 2 - \sqrt{20 - (x + 1)^2}$$

If you want the result to appear circular, you need to use a square setting, as shown below.

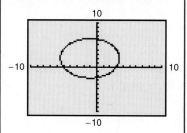

Standard setting

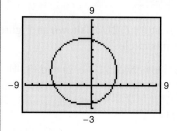

Square setting

General Form of the Equation of a Circle

$$Ax^2 + Ay^2 + Dx + Ey + F = 0, \quad A \neq 0$$

To change from general form to standard form, you can use a process called **completing the square,** as demonstrated in Example 5.

Example 5 Completing the Square

Sketch the graph of the circle whose general equation is

$$4x^2 + 4y^2 + 20x - 16y + 37 = 0.$$

SOLUTION First divide by 4 so that the coefficients of x^2 and y^2 are both 1.

$4x^2 + 4y^2 + 20x - 16y + 37 = 0$ Write original equation.

$x^2 + y^2 + 5x - 4y + \frac{37}{4} = 0$ Divide each side by 4.

$(x^2 + 5x +) + (y^2 - 4y +) = -\frac{37}{4}$ Group terms.

$\left(x^2 + 5x + \frac{25}{4}\right) + (y^2 - 4y + 4) = -\frac{37}{4} + \frac{25}{4} + 4$ Complete the square.

$\qquad\qquad$(Half)^2 $\qquad\qquad$(Half)^2

$\left(x + \frac{5}{2}\right)^2 + (y - 2)^2 = 1$ Write in standard form.

From the standard form, you can see that the circle is centered at $\left(-\frac{5}{2}, 2\right)$ and has a radius of 1, as shown in Figure 1.22.

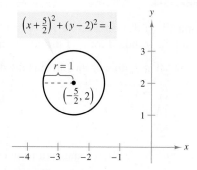

FIGURE 1.22

The general equation $Ax^2 + Ay^2 + Dx + Ey + F = 0$ may not always represent a circle. In fact, such an equation will have no solution points if the procedure of completing the square yields the impossible result

$$(x - h)^2 + (y - k)^2 = \text{negative number.} \qquad \text{No solution points}$$

✓ CHECKPOINT 5

Write the equation of the circle $x^2 + y^2 - 4x + 2y + 1 = 0$ in standard form and sketch its graph. ∎

Points of Intersection

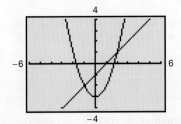

FIGURE 1.23

A **point of intersection** of two graphs is an ordered pair that is a solution point of both graphs. For instance, Figure 1.23 shows that the graphs of

$$y = x^2 - 3 \quad \text{and} \quad y = x - 1$$

have two points of intersection: $(2, 1)$ and $(-1, -2)$. To find the points analytically, set the two y-values equal to each other and solve the equation

$$x^2 - 3 = x - 1$$

for x as shown in Example 6.

Example 6 Finding Points of Intersection

Find the points of intersection of the graphs of

$$y = x^2 - 3 \quad \text{and} \quad y = x - 1.$$

SOLUTION Set the two y-values equal to each other and solve the following equation

$x^2 - 3 = x - 1$	Set the y-values equal to each other.
$x^2 - x - 2 = 0$	Write in general form.
$(x - 2)(x + 1) = 0$	Factor.
$x - 2 = 0 \implies x = 2$	Set 1st factor equal to 0.
$x + 1 = 0 \implies x = -1$	Set 2nd factor equal to 0.

Use either of the original equations to find y. When $x = 2$, $y = 1$ and when $x = -1$, $y = -2$. So, there are two points of intersection: $(2, 1)$ and $(-1, -2)$. (See Figure 1.24.)

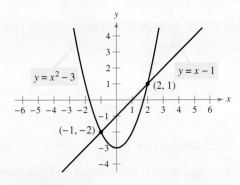

FIGURE 1.24

✓ CHECKPOINT 6

Find the points of intersection (if any) of the graphs of

$$y = x^2 - 5 \quad \text{and} \quad y = x + 1. \ \blacksquare$$

STUDY TIP

The *Technology* note on page 46 describes how to use a graphing utility to find the x-intercepts of a graph. A similar procedure can be used to find the points of intersection of two graphs. (Your calculator may call this the *intersect* feature.) Use this procedure to verify the results of Example 6.

Example 7 State Populations

From 2000 through 2006, the population of Arizona increased more rapidly than the population of Tennessee. Two models that approximate the populations P (in thousands) are

$$P = 5129 + 164.4t \qquad \text{Arizona}$$
$$P = 5687 + 54.4t \qquad \text{Tennessee}$$

where t represents the year, with $t = 0$ corresponding to 2000. *(Source: U.S. Census Bureau)*

a. According to these two models, when would you expect the population of Arizona to have exceeded the population of Tennessee?

b. Use the two models to predict the populations of both states in 2010.

SOLUTION

a. Set the two P-values equal to each other and solve the following equation.

$5129 + 164.4t = 5687 + 54.4t$	Set the P-values equal to each other.
$5129 + 110t = 5687$	Subtract $54.4t$ from each side.
$110t = 558$	Subtract 5129 from each side.
$t \approx 5.07$	Divide each side by 110.

So, from the given models, you would expect that the population of Arizona would have exceeded the population of Tennessee after $t \approx 5.07$ years, which was sometime during 2005. (See Figure 1.25.)

b. To predict the populations of both states in 2010, substitute $t = 10$ into each model and evaluate.

Arizona	Tennessee
$P = 5129 + 164.4t$	$P = 5687 + 54.4t$
$= 5129 + 164.4(10)$	$= 5687 + 54.4(10)$
$= 6773$	$= 6231$

So, according to the models, you can predict that Arizona's population in 2010 will be 6,773,000 and Tennessee's population will be 6,231,000.

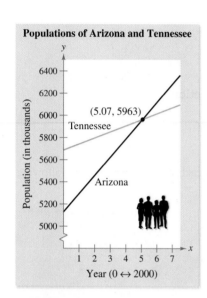

FIGURE 1.25

✓ CHECKPOINT 7

The population P (in thousands) of Maryland from 2000 through 2006 can be modeled by

$$P = 5331 + 51.6t$$

where t represents the year, with $t = 0$ corresponding to 2000. *(Source: U.S. Census Bureau)*

a. According to this model and the model for the population of Arizona in Example 7, when would you expect the population of Arizona to have exceeded the population of Maryland?

b. Use the model to predict Maryland's population in 2010. ■

STUDY TIP

In Example 7(a), an analytical method was used to answer the question. You can also use a numerical method to find the answer. For instance, you could use a spreadsheet to create a table of values for each model to determine when the population of Arizona exceeded the population of Tennessee. Use a numerical method to solve Example 7(a) and compare results.

Mathematical Models

In this text, you will see many examples of the use of equations as **mathematical models** of real-life phenomena. In developing a mathematical model to represent actual data, you should strive for two (often conflicting) goals—accuracy and simplicity.

Example 8 Using a Mathematical Model

The table shows the annual amounts of energy (in quadrillion Btu) produced from crude oil in the United States in the years 2000 through 2005. Projected 2006 energy production from crude oil is 10.9 quadrillion Btu. How do you think this projection was obtained? *(Source: U.S. Energy Information Association)*

Algebra Review

For help in evaluating the expression in Example 8, see the review of order of operations on page 104.

Year	2000	2001	2002	2003	2004	2005
t	0	1	2	3	4	5
Crude oil	12.358	12.282	12.163	12.026	11.503	10.963

SOLUTION The projection was obtained by using past production to predict future production. The past production was modeled by an equation that was found by a statistical procedure called *least squares regression analysis*.

$$E = -0.07029t^2 + 0.0815t + 12.323 \quad \text{Crude oil energy production}$$

Using $t = 6$ to represent 2006, you can predict the 2006 energy production from crude oil to be

$$E = -0.07029(6)^2 + 0.0815(6) + 12.323 \approx 10.282$$

This projection is close to the given projection. The graph of the model and the actual data are shown in Figure 1.26.

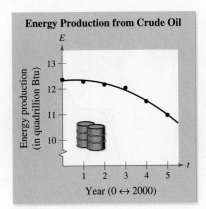

FIGURE 1.26

✓ CHECKPOINT 8

The table shows the annual amounts of energy (in quadrillion Btu) produced from wind in the United States in the years 2000 through 2005. Projected 2006 energy production from wind is 0.258 quadrillion Btu. How does this projection compare with the projection obtained using the model below? *(Source: U.S. Energy Information Association)*

$$E = 0.00148t^2 + 0.0163t + 0.057$$

Year	2000	2001	2002	2003	2004	2005
t	0	1	2	3	4	5
Wind	0.057	0.070	0.105	0.115	0.142	0.178

STUDY TIP

To test the accuracy of a model, you can compare the actual data with the values given by the model. For instance, the table below compares the actual amounts of energy produced from crude oil with those given by the model found in Example 8.

Year	2000	2001	2002	2003	2004	2005
Actual	12.358	12.282	12.163	12.026	11.503	10.963
Model	12.323	12.334	12.205	11.935	11.524	10.973

Much of your study of calculus will center around the behavior of the graphs of mathematical models. Figure 1.27 shows the graphs of six basic algebraic equations. Familiarity with these graphs will help you in the creation and use of mathematical models.

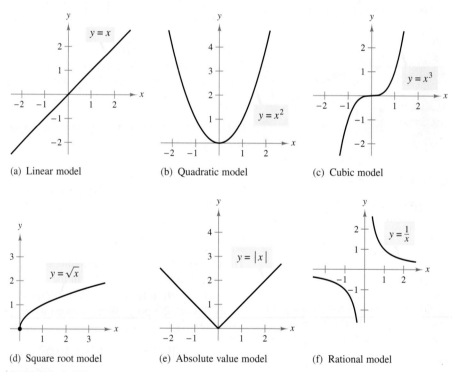

(a) Linear model (b) Quadratic model (c) Cubic model

(d) Square root model (e) Absolute value model (f) Rational model

FIGURE 1.27

CONCEPT CHECK

1. What does the graph of an equation represent?
2. Describe how to find the *x*- and *y*-intercepts of the graph of an equation.
3. How can you check that an ordered pair is a point of intersection of two graphs?
4. Can the graph of an equation have more than one *y*-intercept?

SECTION 1.2 Graphs of Equations

Skills Review 1.2

The following warm-up exercises involve skills that were covered in earlier sections. You will use these skills in the exercise set for this section. For additional help, review Sections 0.3 and 0.4.

In Exercises 1–6, solve for y.

1. $5y - 12 = x$
2. $-y = 15 - x$
3. $x^3y + 2y = 1$
4. $x^2 + x - y^2 - 6 = 0$
5. $(x - 2)^2 + (y + 1)^2 = 9$
6. $(x + 6)^2 + (y - 5)^2 = 81$

In Exercises 7–10, complete the square to write the expression as a perfect square trinomial.

7. $x^2 - 4x +$ ____
8. $x^2 + 6x +$ ____
9. $x^2 - 5x +$ ____
10. $x^2 + 3x +$ ____

In Exercises 11–14, factor the expression.

11. $x^2 - 3x + 2$
12. $x^2 + 5x + 6$
13. $y^2 - 3y + \frac{9}{4}$
14. $y^2 - 7y + \frac{49}{4}$

Exercises 1.2

See www.CalcChat.com for worked-out solutions to odd-numbered exercises.

In Exercises 1–4, determine whether the points are solution points of the given equation.

1. $2x - y - 3 = 0$
   (a) $(1, 2)$ (b) $(1, -1)$ (c) $(4, 5)$
2. $7x + 4y - 6 = 0$
   (a) $(6, -9)$ (b) $(-5, 10)$ (c) $\left(\frac{1}{2}, \frac{5}{8}\right)$
3. $x^2 + y^2 = 4$
   (a) $(1, -\sqrt{3})$ (b) $\left(\frac{1}{2}, -1\right)$ (c) $\left(\frac{3}{2}, \frac{7}{2}\right)$
4. $x^2y + x^2 - 5y = 0$
   (a) $\left(0, \frac{1}{5}\right)$ (b) $(2, 4)$ (c) $(-2, -4)$

In Exercises 5–10, match the equation with its graph. Use a graphing utility, set for a square setting, to confirm your result. [The graphs are labeled (a)–(f).]

5. $y = x - 2$
6. $y = -\frac{1}{2}x + 2$
7. $y = x^2 + 2x$
8. $y = \sqrt{9 - x^2}$
9. $y = |x| - 2$
10. $y = x^3 - x$

(a)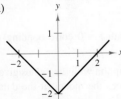

(b)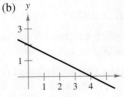

(c)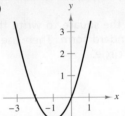

(d)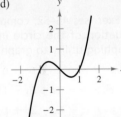

(e)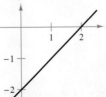

(f)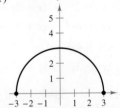

In Exercises 11–20, find the x- and y-intercepts of the graph of the equation.

11. $2x - y - 3 = 0$
12. $4x - 2y - 5 = 0$
13. $y = x^2 + x - 2$
14. $y = x^2 - 4x + 3$
15. $y = \sqrt{4 - x^2}$
16. $y^2 = x^3 - 4x$
17. $y = \dfrac{x^2 - 4}{x - 2}$
18. $y = \dfrac{x^2 + 3x}{2x}$
19. $x^2y - x^2 + 4y = 0$
20. $2x^2y + 8y - x^2 = 1$

In Exercises 21–36, sketch the graph of the equation and label the intercepts. Use a graphing utility to verify your results.

21. $y = 2x + 3$
22. $y = -3x + 2$
23. $y = x^2 - 3$
24. $y = x^2 + 6$
25. $y = (x - 1)^2$
26. $y = (5 - x)^2$
27. $y = x^3 + 2$
28. $y = 1 - x^3$
29. $y = -\sqrt{x - 1}$
30. $y = \sqrt{x + 1}$
31. $y = |x + 1|$
32. $y = -|x - 2|$
33. $y = 1/(x - 3)$
34. $y = 1/(x + 1)$
35. $x = y^2 - 4$
36. $x = 4 - y^2$

In Exercises 37–44, write the general form of the equation of the circle.

37. Center: $(0, 0)$; radius: 4
38. Center: $(0, 0)$; radius: 5
39. Center: $(2, -1)$; radius: 3
40. Center: $(-4, 3)$; radius: 2
41. Center: $(-1, 2)$; solution point: $(0, 0)$
42. Center: $(3, -2)$; solution point: $(-1, 1)$
43. Endpoints of a diameter: $(0, 0), (6, 8)$
44. Endpoints of a diameter: $(-4, -1), (4, 1)$

(T) In Exercises 45–52, complete the square to write the equation of the circle in standard form. Then use a graphing utility to graph the circle.

45. $x^2 + y^2 - 2x + 6y + 6 = 0$
46. $x^2 + y^2 - 2x + 6y - 15 = 0$
47. $x^2 + y^2 - 4x - 2y + 1 = 0$
48. $x^2 + y^2 - 4x + 2y + 3 = 0$
49. $2x^2 + 2y^2 - 2x - 2y - 3 = 0$
50. $4x^2 + 4y^2 - 4x + 2y - 1 = 0$
51. $4x^2 + 4y^2 + 12x - 24y + 41 = 0$
52. $3x^2 + 3y^2 - 6y - 1 = 0$

In Exercises 53–60, find the points of intersection (if any) of the graphs of the equations. Use a graphing utility to check your results.

53. $x + y = 2, 2x - y = 1$
54. $x + y = 7, 3x - 2y = 11$
55. $x^2 + y^2 = 25, 2x + y = 10$
56. $x^2 - y = -2, x - y = -4$
57. $y = x^3, y = 2x$
58. $y = \sqrt{x}, y = x$
59. $y = x^4 - 2x^2 + 1, y = 1 - x^2$
60. $y = x^3 - 2x^2 + x - 1, y = -x^2 + 3x - 1$

61. **State Populations** The populations P (in thousands) of Oregon and Oklahoma for the years 2000 through 2006 can be modeled by

$P = 3431 + 43.0t$ Oregon

$P = 3448 + 20.1t$ Oklahoma

where t represents the year, with $t = 0$ corresponding to 2000. *(Source: U.S. Census Bureau)*

(a) According to these two models, when would you expect the population of Oregon to have exceeded the population of Oklahoma?

(b) Use the two models to estimate the populations of both states in 2010.

62. **Biology** The body length L (in centimeters) and weight W (in kilograms) of a walrus calf can be approximated by

$L = 0.42t + 109$

$W = 0.8t + 60$

where t is the age of the calf in days.

(a) Use the given equations to complete the table.

Age	0			45	
Length		111.1		117.82	
Weight			68		108

(b) One calf is 163 centimeters long and another calf weighs 163 kilograms. Is it possible that the calves are the same age? Explain your reasoning.

63. **Geothermal Energy Consumption** The table shows the annual amounts of U.S. consumption G (in trillion Btu) of geothermal energy in the commercial sector for the years 2000 through 2005. *(Source: U.S. Energy Information Association)*

Year	2000	2001	2002
Geothermal	7.600	8.270	8.753

Year	2003	2004	2005
Geothermal	11.000	12.000	13.600

A mathematical model for the data is given by

$G = 0.120t^2 + 0.64t + 7.5$

where t represents the year, with $t = 0$ corresponding to 2000.

(a) Compare the actual consumption with the values given by the model. How well does the model fit the data? Explain your reasoning.

(b) Use the model to predict the consumption in 2013.

64. Textbook Spending The amounts of money y (in millions of dollars) spent on college textbooks in the United States in the years 2000 through 2005 are shown in the table. *(Source: Book Industry Study Group, Inc.)*

Year	2000	2001	2002	2003	2004	2005
Expense	4265	4571	4899	5086	5479	5703

A mathematical model for the data is given by $y = 0.796t^3 - 8.65t^2 + 312.9t + 4268$, where t represents the year, with $t = 0$ corresponding to 2000.

(a) Compare the actual expenses with those given by the model. How well does the model fit the data? Explain your reasoning.

(b) Use the model to predict the expense in 2013.

65. Federal Education Spending The federal outlays y (in billions of dollars) for elementary, secondary, and vocational education in the United States in the years 2000 through 2005 are shown in the table. *(Source: U.S. Office of Management and Budget)*

Year	2000	2001	2002	2003	2004	2005
Outlay	20.6	22.9	25.9	31.5	34.4	38.3

A mathematical model for the data is given by

$y = 0.136t^2 + 3.00t + 20.2$

where t represents the year, with $t = 0$ corresponding to 2000.

(a) Compare the actual outlays with those given by the model. How well does the model fit the data? Explain.

(b) Use the model to predict the outlay in 2012.

66. MAKE A DECISION: WEEKLY SALARY A mathematical model for the average weekly salary y of a person in home health care services is given by

$y = -0.80t^2 + 18.16t + 350$

where t represents the year, with $t = 0$ corresponding to 2000. *(Source: U.S. Bureau of Labor Statistics)*

(a) Use the model to complete the table.

Year	2001	2002	2003	2004	2005	2008
Salary						

(b) This model was created using actual data from 2000 through 2006. How accurate do you think the model is in predicting the 2008 average weekly salary? Explain your reasoning.

(c) What does this model predict the average weekly salary will be in 2010? Do you think this prediction is valid?

67. MAKE A DECISION: KIDNEY TRANSPLANTS A mathematical model for the numbers of kidney transplants performed in the United States in the years 2001 through 2005 is given by

$y = 30.57t^2 + 381.4t + 13{,}852$

where y is the number of transplants and t represents the year, with $t = 1$ corresponding to 2001. *(Source: United Network for Organ Sharing)*

(a) Use a graphing utility or a spreadsheet to complete the table.

Year	2001	2002	2003	2004	2005
Transplants					

(b) Use your school's library, the Internet, or some other reference source to find the actual numbers of kidney transplants in the years 2001 through 2005. Compare the actual numbers with those given by the model. How well does the model fit the data? Explain your reasoning.

(c) Using this model, what is the prediction for the number of transplants in the year 2011? Do you think this prediction is valid? What factors could affect this model's accuracy?

68. Use a graphing utility to graph the equation $y = cx + 1$ for $c = 1, 2, 3, 4,$ and 5. Then make a conjecture about the constant term c and the graph of the equation.

69. Use a graphing utility to graph the equation $y = x + c$ for $c = 0, 1, 2, 3,$ and 4. Then make a conjecture about the constant term c and the graph of the equation.

In Exercises 70–75, use a graphing utility to graph the equation and approximate the x- and y-intercepts of the graph.

70. $y = 0.24x^2 + 1.32x + 5.36$

71. $y = -0.56x^2 - 5.34x + 6.25$

72. $y = \sqrt{0.3x^2 - 4.3x + 5.7}$

73. $y = \sqrt{-1.21x^2 + 2.34x + 5.6}$

74. $y = \dfrac{0.2x^2 + 1}{0.1x + 2.4}$

75. $y = \dfrac{0.4x - 5.3}{0.4x^2 + 5.3}$

76. Extended Application To work an extended application analyzing the numbers of workers in the farm work force in the United States from 1955 through 2005, visit this text's website at *college.hmco.com/info/larsonapplied*. *(Data Source: U.S. Bureau of Labor Statistics)*

Section 1.3

Lines in the Plane and Slope

- Use the slope-intercept form of a linear equation to sketch graphs.
- Find slopes of lines passing through two points.
- Use the point-slope form to write equations of lines.
- Find equations of parallel and perpendicular lines.
- Use linear equations to model and solve real-life problems.

Using Slope

The simplest mathematical model for relating two variables is the **linear equation** $y = mx + b$. The equation is called *linear* because its graph is a line. (In this text, the term *line* is used to mean *straight line*.) By letting $x = 0$, you can see that the line crosses the y-axis at $y = b$, as shown in Figure 1.28. In other words, the y-intercept is $(0, b)$. The steepness or slope of the line is m.

$$y = \underset{\text{Slope}}{m}x + \underset{y\text{-intercept}}{b}$$

The **slope** of a line is the number of units the line rises (or falls) vertically for each unit of horizontal change from left to right, as shown in Figure 1.28.

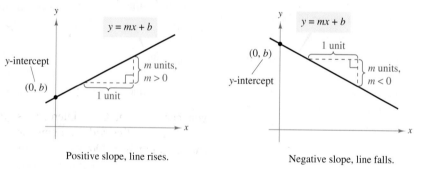

Positive slope, line rises. Negative slope, line falls.

FIGURE 1.28

A linear equation that is written in the form $y = mx + b$ is said to be written in **slope-intercept form.**

The Slope-Intercept Form of the Equation of a Line

The graph of the equation

$$y = mx + b$$

is a line whose slope is m and whose y-intercept is $(0, b)$.

TECHNOLOGY

On most graphing utilities, the display screen is two-thirds as high as it is wide. On such screens, you can obtain a graph with a true geometric perspective by using a **square setting**—one in which

$$\frac{Y_{max} - Y_{min}}{X_{max} - X_{min}} = \frac{2}{3}.$$

One such setting is shown below. Notice that the x and y tick marks are equally spaced on a square setting, but not on a standard setting.

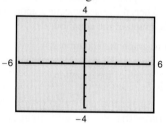

DISCOVERY

Use a graphing utility to compare the slopes of the lines $y = mx$, where $m = 0.5, 1, 2,$ and 4. Which line rises most quickly? Now, let $m = -0.5, -1, -2,$ and -4. Which line falls most quickly? Let $m = 0.01, 0.001,$ and 0.0001. What is the slope of a horizontal line? Use a square setting to obtain a true geometric perspective.

SECTION 1.3 Lines in the Plane and Slope

Once you have determined the slope and the y-intercept of a line, it is a relatively simple matter to sketch its graph.

In the following example, note that none of the lines is vertical. A vertical line has an equation of the form

$x = a.$ Vertical line

Because such an equation cannot be written in the form $y = mx + b$, it follows that the slope of a vertical line is undefined, as indicated in Figure 1.29.

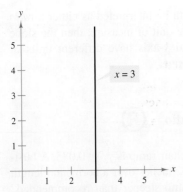

FIGURE 1.29 When the line is vertical, the slope is undefined.

Example 1 Graphing Linear Equations

Sketch the graph of each linear equation.

a. $y = 2x + 1$

b. $y = 2$

c. $x + y = 2$

SOLUTION

a. Because $b = 1$, the y-intercept is $(0, 1)$. Moreover, because the slope is $m = 2$, the line *rises* two units for each unit the line moves to the right, as shown in Figure 1.30(a).

b. By writing this equation in the form $y = (0)x + 2$, you can see that the y-intercept is $(0, 2)$ and the slope is zero. A zero slope implies that the line is horizontal—that is, it doesn't rise *or* fall, as shown in Figure 1.30(b).

c. By writing this equation in slope-intercept form

$x + y = 2$ Write original equation.

$y = -x + 2$ Subtract x from each side.

$y = (-1)x + 2$ Write in slope-intercept form.

you can see that the y-intercept is $(0, 2)$. Moreover, because the slope is $m = -1$, this line *falls* one unit for each unit the line moves to the right, as shown in Figure 1.30(c).

✓ CHECKPOINT 1

Sketch the graph of each linear equation.

a. $y = 4x - 2$

b. $x = 1$ ■

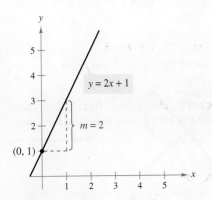

(a) When m is positive, the line rises.

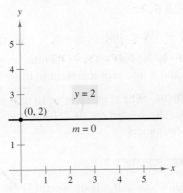

(b) When m is zero, the line is horizontal.

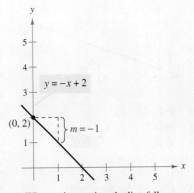

(c) When m is negative, the line falls.

FIGURE 1.30

In real-life problems, the slope of a line can be interpreted as either a *ratio* or a *rate*. If the *x*-axis and *y*-axis have the same unit of measure, then the slope has no units and is a **ratio.** If the *x*-axis and *y*-axis have different units of measure, then the slope is a **rate** or **rate of change.**

Example 2 — MAKE A DECISION: Using Slope as a Ratio

The maximum recommended slope of a wheelchair ramp is $\frac{1}{12} \approx 0.083$. A business is installing a wheelchair ramp that rises 22 inches over a horizontal length of 24 feet, as shown in Figure 1.31. Is the ramp steeper than recommended? *(Source: American Disabilities Act Handbook)*

SOLUTION The horizontal length of the ramp is 24 feet or $12(24) = 288$ inches. So, the slope of the ramp is

$$\text{Slope} = \frac{\text{vertical change}}{\text{horizontal change}}$$

$$= \frac{22 \text{ in.}}{288 \text{ in.}}$$

$$\approx 0.076.$$

So, the slope is not steeper than recommended.

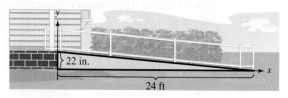

FIGURE 1.31

✓**CHECKPOINT 2**

If the ramp in Example 2 rises 27 inches over a horizontal length of 26 feet, is it steeper than recommended? ■

Example 3 Using Slope as a Rate of Change

A killer whale calf's body length L (in centimeters) during its first year can be modeled by

$$L = 10t + 260$$

where t is the calf's age in months, as shown in Figure 1.32. Describe what the slope and *y*-intercept represent in this situation.

SOLUTION The slope of $m = 10$ tells you that the calf grows 10 centimeters per month. The *y*-intercept $(0, 260)$ tells you that the calf's body length at birth was 260 centimeters.

✓**CHECKPOINT 3**

A killer whale calf's body weight W (in kilograms) during its first year can be modeled by $W = 37t + 158$, where t is the calf's age in months. Describe what the slope and *y*-intercept represent in this situation. ■

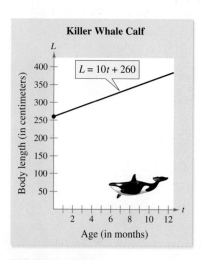

FIGURE 1.32

Finding the Slope of a Line

Given an equation of a nonvertical line, you can find its slope by writing the equation in slope-intercept form. If you are not given an equation, you can still find the slope of a line. For instance, suppose you want to find the slope of the line passing through the points (x_1, y_1) and (x_2, y_2), as shown in Figure 1.33. As you move from left to right along this line, a change of $(y_2 - y_1)$ units in the vertical direction corresponds to a change of $(x_2 - x_1)$ units in the horizontal direction. These two changes are denoted by the symbols

$$\Delta y = y_2 - y_1 = \text{the change in } y$$

and

$$\Delta x = x_2 - x_1 = \text{the change in } x.$$

(The symbol Δ is the Greek capital letter delta, and the symbols Δy and Δx are read as "delta y" and "delta x.") The ratio of Δy to Δx represents the slope of the line that passes through the points (x_1, y_1) and (x_2, y_2).

$$\text{Slope} = \frac{\Delta y}{\Delta x} = \frac{y_2 - y_1}{x_2 - x_1}$$

Be sure you see that Δx represents a single number, not the product of two numbers (Δ and x). The same is true for Δy.

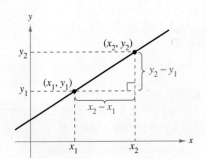

FIGURE 1.33

The Slope of a Line Passing Through Two Points

The **slope** m of the line passing through (x_1, y_1) and (x_2, y_2) is

$$m = \frac{\Delta y}{\Delta x} = \frac{y_2 - y_1}{x_2 - x_1}$$

where $x_1 \neq x_2$.

When this formula is used for slope, the *order of subtraction* is important. Given two points on a line, you are free to label either one of them as (x_1, y_1) and the other as (x_2, y_2). However, once you have done this, you must form the numerator and denominator using the same order of subtraction.

$$m = \frac{y_2 - y_1}{x_2 - x_1} \qquad m = \frac{y_1 - y_2}{x_1 - x_2} \qquad m = \frac{y_2 - y_1}{x_1 - x_2}$$

Correct   Correct   Incorrect

For instance, the slope of the line passing through the points $(3, 4)$ and $(5, 7)$ can be calculated as

$$m = \frac{7 - 4}{5 - 3} = \frac{3}{2}$$

or

$$m = \frac{4 - 7}{3 - 5} = \frac{-3}{-2} = \frac{3}{2}.$$

60 CHAPTER 1 Functions, Graphs, and Limits

DISCOVERY

The line in Example 4(b) is a horizontal line. Find an equation for this line. The line in Example 4(d) is a vertical line. Find an equation for this line.

Example 4 Finding Slopes of Lines

Find the slope of the line passing through each pair of points.

a. $(-2, 0)$ and $(3, 1)$ **b.** $(-1, 2)$ and $(2, 2)$

c. $(0, 4)$ and $(1, -1)$ **d.** $(3, 4)$ and $(3, 1)$

SOLUTION

a. Letting $(x_1, y_1) = (-2, 0)$ and $(x_2, y_2) = (3, 1)$, you obtain a slope of

$$m = \frac{y_2 - y_1}{x_2 - x_1} = \frac{1 - 0}{3 - (-2)} = \frac{1}{5} \quad \longleftarrow \text{ Difference in } y\text{-values} \\ \longleftarrow \text{ Difference in } x\text{-values}$$

as shown in Figure 1.34(a).

b. The slope of the line passing through $(-1, 2)$ and $(2, 2)$ is

$$m = \frac{2 - 2}{2 - (-1)} = \frac{0}{3} = 0. \qquad \text{See Figure 1.34(b).}$$

c. The slope of the line passing through $(0, 4)$ and $(1, -1)$ is

$$m = \frac{-1 - 4}{1 - 0} = \frac{-5}{1} = -5. \qquad \text{See Figure 1.34(c).}$$

d. The slope of the vertical line passing through $(3, 4)$ and $(3, 1)$ is not defined because division by zero is undefined. [See Figure 1.34(d).]

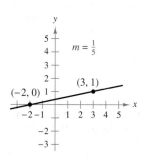

(a) Positive slope, line rises.

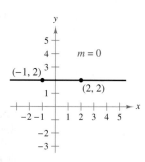

(b) Zero slope, line is horizontal.

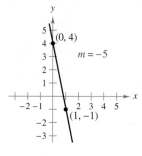

(c) Negative slope, line falls.

(d) Vertical line, undefined slope.

FIGURE 1.34

✓ **CHECKPOINT 4**

Find the slope of the line passing through each pair of points.

a. $(-3, 2)$ and $(5, 18)$

b. $(-2, 1)$ and $(-4, 2)$ ∎

Writing Linear Equations

If (x_1, y_1) is a point lying on a nonvertical line of slope m and (x, y) is *any other* point on the line, then

$$\frac{y - y_1}{x - x_1} = m.$$

This equation, involving the variables x and y, can be rewritten in the form $y - y_1 = m(x - x_1)$, which is the **point-slope form** of the equation of a line.

Point-Slope Form of the Equation of a Line

The equation of the line with slope m passing through the point (x_1, y_1) is

$$y - y_1 = m(x - x_1).$$

The point-slope form is most useful for *finding* the equation of a nonvertical line. You should remember this formula—it is used throughout the text.

Example 5 Using the Point-Slope Form

Find the equation of the line that has a slope of 3 and passes through the point $(1, -2)$.

SOLUTION Use the point-slope form with $m = 3$ and $(x_1, y_1) = (1, -2)$.

$y - y_1 = m(x - x_1)$	Point-slope form
$y - (-2) = 3(x - 1)$	Substitute for m, x_1, and y_1.
$y + 2 = 3x - 3$	Simplify.
$y = 3x - 5$	Write in slope-intercept form.

The slope-intercept form of the equation of the line is $y = 3x - 5$. The graph of this line is shown in Figure 1.35.

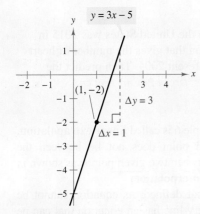

FIGURE 1.35

✓CHECKPOINT 5

Find the equation of the line that has a slope of 2 and passes through the point $(-1, 2)$. ■

The point-slope form can be used to find an equation of the line passing through points (x_1, y_1) and (x_2, y_2). To do this, first find the slope of the line

$$m = \frac{y_2 - y_1}{x_2 - x_1}, \quad x_1 \neq x_2$$

and then use the point-slope form to obtain the equation

$$y - y_1 = \frac{y_2 - y_1}{x_2 - x_1}(x - x_1). \quad \text{Two-point form}$$

This is sometimes called the **two-point form** of the equation of a line.

STUDY TIP

The two-point form of a line is similar to the slope-intercept form. What is the slope of a line given in two-point form

$$y - y_1 = \frac{y_2 - y_1}{x_2 - x_1}(x - x_1)?$$

Example 6 Predicting Organ Transplants

The number of liver transplants performed in the United States was 6169 in 2004 and 6443 in 2005. Write a linear equation that gives the number of liver transplants in terms of the year. Then predict the number of liver transplants in 2006. *(Source: Organ Procurement and Transplantation Network)*

SOLUTION Let $t = 4$ represent 2004. Then the two given values are represented by the data points $(4, 6169)$ and $(5, 6443)$. The slope of the line through these points is

$$m = \frac{6443 - 6169}{5 - 4} = 274.$$

Using the point-slope form, you can find the equation that relates the number of liver transplants y and the year t to be $y = 274t + 5073$. According to this equation, the number of liver transplants performed in 2006 was 6717, as shown in Figure 1.36. (In this case, the prediction is fairly good—the actual number of liver transplants performed in 2006 was 6650.)

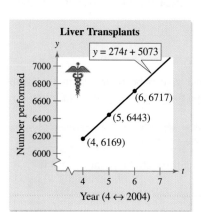

FIGURE 1.36

✓CHECKPOINT 6

The number of heart transplants performed in the United States was 2015 in 2004 and 2125 in 2005. Write a linear equation that gives the number of heart transplants in terms of the year. Let $t = 4$ represent 2004. Then predict the number of heart transplants in 2006. *(Source: Organ Procurement and Transplantation Network)* ■

The prediction method illustrated in Example 6 is called **linear extrapolation**. Note in Figure 1.37(a) that an extrapolated point does not lie between the given points. When the estimated point lies between two given points, as shown in Figure 1.37(b), the procedure is called **linear interpolation**.

Because the slope of a vertical line is not defined, its equation cannot be written in slope-intercept form. However, every line has an equation that can be written in the **general form**

$$Ax + By + C = 0 \qquad \text{General form}$$

where A and B are not both zero. For instance, the vertical line given by $x = a$ can be represented by the general form $x - a = 0$. The five most common forms of equations of lines are summarized below.

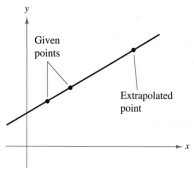

(a) Linear Extrapolation

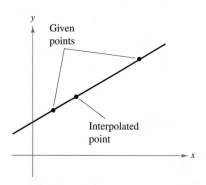

(b) Linear Interpolation

FIGURE 1.37

Equations of Lines	
1. General form:	$Ax + By + C = 0$
2. Vertical line:	$x = a$
3. Horizontal line:	$y = b$
4. Slope-intercept form:	$y = mx + b$
5. Point-slope form:	$y - y_1 = m(x - x_1)$

SECTION 1.3 Lines in the Plane and Slope

Parallel and Perpendicular Lines

Slope can be used to decide whether two nonvertical lines in a plane are parallel, perpendicular, or neither.

Parallel and Perpendicular Lines

1. Two distinct nonvertical lines are **parallel** if and only if their slopes are equal. That is, $m_1 = m_2$.

2. Two nonvertical lines are **perpendicular** if and only if their slopes are negative reciprocals of each other. That is, $m_1 = -1/m_2$.

Example 7 Finding Parallel and Perpendicular Lines

Find equations of the lines that pass through the point $(2, -1)$ and are

a. parallel to the line $2x - 3y = 5$.

b. perpendicular to the line $2x - 3y = 5$.

SOLUTION By writing the given equation in slope-intercept form

$2x - 3y = 5$	Write original equation.
$-3y = -2x + 5$	Subtract $2x$ from each side.
$y = \frac{2}{3}x - \frac{5}{3}$	Write in slope-intercept form.

you can see that it has a slope of $m = \frac{2}{3}$, as shown in Figure 1.38.

a. Any line parallel to the given line must also have a slope of $\frac{2}{3}$. So, the line through $(2, -1)$ that is parallel to the given line has the following equation.

$y - (-1) = \frac{2}{3}(x - 2)$	Write in point-slope form.
$3(y + 1) = 2(x - 2)$	Multiply each side by 3.
$3y + 3 = 2x - 4$	Distributive Property
$2x - 3y - 7 = 0$	Write in general form.
$y = \frac{2}{3}x - \frac{7}{3}$	Write in slope-intercept form.

b. Any line perpendicular to the given line must have a slope of

$$\frac{-1}{\frac{2}{3}} \text{ or } -\frac{3}{2}.$$

So, the line through $(2, -1)$ that is perpendicular to the given line has the following equation.

$y - (-1) = -\frac{3}{2}(x - 2)$	Write in point-slope form.
$2(y + 1) = -3(x - 2)$	Multiply each side by 2.
$2y + 2 = -3x + 6$	Distributive Property
$3x + 2y - 4 = 0$	Write in general form.
$y = -\frac{3}{2}x + 2$	Write in slope-intercept form.

TECHNOLOGY

On a graphing utility, lines will not appear to have the correct slope unless you use a viewing window that has a "square setting." For instance, try graphing the lines in Example 7 using the standard setting $-10 \leq x \leq 10$ and $-10 \leq y \leq 10$. Then reset the viewing window with the square setting $-9 \leq x \leq 9$ and $-6 \leq y \leq 6$. On which setting do the lines $y = \frac{2}{3}x - \frac{5}{3}$ and $y = -\frac{3}{2}x + 2$ appear to be perpendicular?

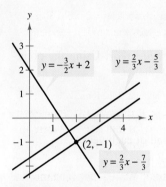

FIGURE 1.38

✓ CHECKPOINT 7

Find equations of the lines that pass through the point $(2, 1)$ and are

a. parallel to the line $2x - 4y = 5$.

b. perpendicular to the line $2x - 4y = 5$.

Extended Application: Linear Depreciation

Most business expenses can be deducted the same year they occur. One exception to this is the cost of property that has a useful life of more than 1 year, such as buildings, cars, or equipment. Such costs must be **depreciated** over the useful life of the property. If the *same amount* is depreciated each year, the procedure is called **linear depreciation** or **straight-line depreciation.** The *book value* is the difference between the original value and the total amount of depreciation accumulated to date.

Example 8 Depreciating Medical Equipment

Your medical lab has purchased a $20,000 analytical lab balance scale that has a useful life of 8 years. The salvage value at the end of 8 years is $4000. Write a linear equation that describes the book value of the machine each year.

SOLUTION Let V represent the value of the machine at the end of year t. You can represent the initial value of the machine by the ordered pair (0, 20,000) and the salvage value of the machine by the ordered pair (8, 4000). The slope of the line is

$$m = \frac{4000 - 20{,}000}{8 - 0} = -\$2000 \qquad m = \frac{V_2 - V_1}{t_2 - t_1}$$

which represents the annual depreciation in *dollars per year*. Using the point-slope form, you can write the equation of the line as shown.

$$V - 20{,}000 = -2000(t - 0) \qquad \text{Write in point-slope form.}$$
$$V = -2000t + 20{,}000 \qquad \text{Write in slope-intercept form.}$$

The table shows the book value of the scale at the end of each year.

t	0	1	2	3	4
V	20,000	18,000	16,000	14,000	12,000

t	5	6	7	8
V	10,000	8000	6000	4000

The graph of the equation is shown in Figure 1.39.

✓ CHECKPOINT 8

Write a linear equation for the machine in Example 8 if the salvage value at the end of 8 years is $2000. ∎

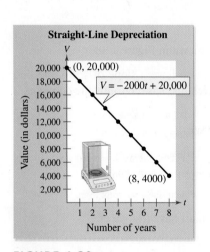

FIGURE 1.39

CONCEPT CHECK

1. In the form $y = mx + b$, what does the m represent? What does the b represent?
2. Can any pair of points on a line be used to calculate the slope of the line? Explain.
3. The slopes of two lines are -4 and $\frac{5}{2}$. Which is steeper? Explain your reasoning.
4. Is it possible for two lines with positive slopes to be perpendicular? Why or why not?

Skills Review 1.3

The following warm-up exercises involve skills that were covered in earlier sections. You will use these skills in the exercise set for this section. For additional help, review Sections 0.3 and 0.5.

In Exercises 1 and 2, simplify the expression.

1. $\dfrac{5 - (-2)}{-3 - 4}$

2. $\dfrac{-7 - (-10)}{4 - 1}$

3. Evaluate $-\dfrac{1}{m}$ when $m = -3$.

4. Evaluate $-\dfrac{1}{m}$ when $m = \dfrac{6}{7}$.

In Exercises 5–10, solve for y in terms of x.

5. $-4x + y = 7$

6. $3x - y = 7$

7. $y - 2 = 3(x - 4)$

8. $y - (-5) = -1[x - (-2)]$

9. $y - (-3) = \dfrac{4 - (-3)}{2 - 1}(x - 2)$

10. $y - 1 = \dfrac{-3 - 1}{-7 - (-1)}[x - (-1)]$

Exercises 1.3

See www.CalcChat.com for worked-out solutions to odd-numbered exercises.

In Exercises 1–4, estimate the slope of the line.

1.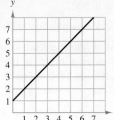

2.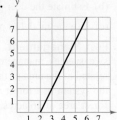

3.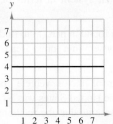

4.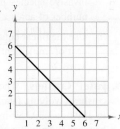

In Exercises 5–18, plot the points and find the slope of the line passing through the pair of points.

5. $(0, -3), (9, 0)$
6. $(-2, 0), (1, 4)$
7. $(3, -4), (5, 2)$
8. $(1, 2), (-2, 2)$
9. $\left(\tfrac{1}{2}, 2\right), (6, 2)$
10. $\left(\tfrac{11}{3}, -2\right), \left(\tfrac{11}{3}, -10\right)$
11. $(-8, -3), (-8, -5)$
12. $(2, -1), (-2, -5)$
13. $(-2, 1), (4, -3)$
14. $(3, -5), (-2, -5)$
15. $\left(\tfrac{1}{4}, -2\right), \left(-\tfrac{3}{8}, 1\right)$
16. $\left(-\tfrac{3}{2}, -5\right), \left(\tfrac{5}{6}, 4\right)$
17. $\left(\tfrac{2}{3}, \tfrac{5}{2}\right), \left(\tfrac{1}{4}, -\tfrac{5}{6}\right)$
18. $\left(\tfrac{7}{8}, \tfrac{3}{4}\right), \left(\tfrac{5}{4}, -\tfrac{1}{4}\right)$

In Exercises 19–26, use the point on the line and the slope of the line to find three additional points through which the line passes. (There are many correct answers.)

Point	Slope		Point	Slope
19. $(2, 1)$	$m = 0$		20. $(-3, -1)$	$m = 0$
21. $(6, -4)$	$m = \tfrac{2}{3}$		22. $(-1, -6)$	$m = -\tfrac{1}{2}$
23. $(1, 7)$	$m = -3$		24. $(7, -2)$	$m = 2$
25. $(-8, 1)$	m is undefined.			
26. $(-3, 4)$	m is undefined.			

In Exercises 27–36, find the slope and y-intercept (if possible) of the equation of the line.

27. $x + 5y = 20$
28. $2x + y = 40$
29. $7x + 6y = 30$
30. $2x + 3y = 9$
31. $3x - y = 15$
32. $2x - 3y = 24$
33. $x = 4$
34. $x + 5 = 0$
35. $y - 4 = 0$
36. $y + 1 = 0$

In Exercises 37–48, write an equation of the line that passes through the points. Then use the equation to sketch the line.

37. $(4, 3), (0, -5)$
38. $(-3, -4), (1, 4)$
39. $(0, 0), (-1, 3)$
40. $(-3, 6), (1, 2)$
41. $(2, 3), (2, -2)$
42. $(6, 1), (10, 1)$
43. $(3, -1), (-2, -1)$
44. $(2, 5), (2, -10)$
45. $\left(-\tfrac{1}{3}, 1\right), \left(-\tfrac{2}{3}, \tfrac{5}{6}\right)$
46. $\left(\tfrac{7}{8}, \tfrac{3}{4}\right), \left(\tfrac{5}{4}, -\tfrac{1}{4}\right)$
47. $\left(-\tfrac{1}{2}, 4\right), \left(\tfrac{1}{2}, 8\right)$
48. $(4, -1), \left(\tfrac{1}{4}, -5\right)$

In Exercises 49–58, write an equation of the line that passes through the given point and has the given slope. Then use a graphing utility to graph the line.

Point	Slope	Point	Slope
49. $(0, 3)$	$m = \frac{3}{4}$	50. $(0, 0)$	$m = \frac{2}{3}$
51. $(-1, 2)$	m is undefined.		
52. $(0, 4)$	m is undefined.		
53. $(-2, 7)$	$m = 0$	54. $(-2, 4)$	$m = 0$
55. $(0, -2)$	$m = -4$	56. $(-1, -4)$	$m = -2$
57. $\left(0, \frac{2}{3}\right)$	$m = \frac{3}{4}$	58. $\left(0, -\frac{2}{3}\right)$	$m = \frac{1}{6}$

In Exercises 59–62, explain how to use the concept of slope to determine whether the three points are collinear.

59. $(-2, 1), (-1, 0), (2, -2)$ 60. $(0, 4), (7, -6), (-5, 11)$
61. $(-2, -1), (0, 3), (2, 7)$ 62. $(4, 1), (-2, -2), (8, 3)$

63. Write an equation of the vertical line with x-intercept at 3.
64. Write an equation of the horizontal line through $(0, -5)$.
65. Write an equation of the line with y-intercept at -10 and parallel to all horizontal lines.
66. Write an equation of the line with x-intercept at -5 and parallel to all vertical lines.

In Exercises 67–74, write equations of the lines through the given point (a) parallel to the given line and (b) perpendicular to the given line. Then use a graphing utility to graph all three equations in the same viewing window.

Point	Line	Point	Line
67. $(-3, 2)$	$x + y = 7$	68. $(2, 1)$	$4x - 2y = 3$
69. $\left(-\frac{2}{3}, \frac{7}{8}\right)$	$3x + 4y = 7$	70. $\left(\frac{7}{8}, \frac{3}{4}\right)$	$5x + 3y = 0$
71. $(-1, 0)$	$y + 3 = 0$	72. $(2, 5)$	$y + 4 = 0$
73. $(1, 1)$	$x - 2 = 0$	74. $(12, -3)$	$x + 4 = 0$

In Exercises 75–84, sketch the graph of the equation. Use a graphing utility to verify your result.

75. $y = -2$
76. $y = -4$
77. $2x - y - 3 = 0$
78. $x + 2y + 6 = 0$
79. $y = -2x + 1$
80. $4x + 5y = 20$
81. $3x + 5y + 15 = 0$
82. $-5x + 2y - 20 = 0$
83. $y + 2 = -4(x + 1)$
84. $y - 1 = 3(x + 4)$

Wheelchair Ramp In Exercises 85 and 86, use the following information. According to the *Americans with Disabilities Act Handbook*, the recommended slope of a wheelchair ramp is $\frac{1}{12}$. *(Source: Americans with Disabilities Act Handbook)*

85. A business is installing a wheelchair ramp that rises 36 inches over a horizontal length of 32 feet. Is the ramp steeper than recommended?

86. A city is modifying its sidewalks with curb ramps for wheelchair access. The height of each curb is 6 inches. Determine the horizontal length x of the curb ramp (see figure) so that the slope is the maximum recommended value.

87. **Deer Population** The deer population P in a forest region can be modeled by

$$P = 60t + 1300$$

where t is the year, with $t = 0$ corresponding to 2000.

(a) Describe what the slope and y-intercept mean in this situation.

(b) Estimate the deer population in 2005.

(c) Use the model to predict the deer population in 2012.

88. **Pest Management** The cost of implementing an invasive species management system in a forest is linearly related to the area of the forest. It costs $740 to implement the system in a forest area of 10 acres. It costs $1130 in a forest area of 16 acres.

(a) Write a linear equation expressing the cost C of the invasive species management system in terms of the number of acres x of forest.

(b) Use the linear equation you found in part (a) to find the cost of implementing the system in a forest area of 40 acres.

89. **Fatal Crashes** The number of motor vehicle traffic crashes involving fatalities in the United States was 37,107 in 1998 and 39,189 in 2005. Assume that the relationship between the number of crashes y and the time t (in years) is linear. Let $t = 8$ correspond to 1998. *(Source: National Highway Traffic Safety Administration)*

(a) Write a linear equation expressing y in terms of t.

(b) *Linear Interpolation* Use the linear equation you found in part (a) to estimate the number of fatal crashes in 2002.

(c) *Linear Extrapolation* Use the linear equation you found in part (a) to estimate the number of fatal crashes in 2008.

90. College Enrollment The enrollment at a small college was 3076 students in 2006 and 3488 students in 2008. Assume that the relationship between the enrollment y and the time t (in years) is linear. Let $t = 6$ correspond to 2006.

(a) Write a linear equation expressing y in terms of t. Describe what the slope and y-intercept mean in this situation.

(b) *Linear Interpolation* Use the linear equation you found in part (a) to estimate the enrollment in 2007.

(c) *Linear Extrapolation* Use the linear equation you found in part (a) to predict the enrollment in 2012.

91. Population The resident population of South Carolina (in thousands) was 4024 in 2000 and 4255 in 2005. Assume that the relationship between the population y and the year t is linear. Let $t = 0$ represent 2000. *(Source: U.S. Census Bureau)*

(a) Write a linear model for the data. What is the slope and what does it tell you about the population?

(b) Estimate the population in 2002.

(c) Use your model to estimate the population in 2004.

(d) Use your school's library, the Internet, or some other reference source to find the actual populations in 2002 and 2004. How close were your estimates?

(e) Do you think your model could be used to predict the population in 2009? Explain.

92. Brain Weight The average weight of a male child's brain is 970 grams at age 1 and 1270 grams at age 3. Assume that the relationship between brain weight y and age t is linear. *(Source: American Neurological Association)*

(a) Write a linear model for the data. What is the slope and what does it tell you about brain weight?

(b) Use your model to estimate the average brain weight at age 2.

(c) Use your school's library, the Internet, or some other reference source to find the actual average brain weight at age 2. How close was your estimate?

(d) Do you think your model could be used to determine the average brain weight of an adult? Explain.

93. Temperature Conversion Write a linear equation that expresses the relationship between the temperature in degrees Celsius C and degrees Fahrenheit F. Use the fact that water freezes at 0°C (32°F) and boils at 100°C (212°F).

94. Chemistry Use the result of Exercise 93 to answer the following:

(a) A person has a temperature of 100.4°F. What is this temperature on the Celsius scale?

(b) If the temperature in a room is 72°F, what is this temperature on the Celsius scale?

(Source: Adapted from Zumdahl, Chemistry, Seventh Edition)

95. Linear Depreciation A hospital purchases a piece of medical equipment for $45,000. After 5 years the equipment will be outdated, having no value.

(a) Write a linear equation giving the value y of the equipment in terms of the time t in years, $0 \leq t \leq 5$.

(b) Use a graphing utility to graph the equation.

(c) Move the cursor along the graph and estimate (to two-decimal-place accuracy) the value of the equipment when $t = 3$.

(d) Move the cursor along the graph and estimate (to two-decimal-place accuracy) the time when the value of the equipment will be $28,000.

96. Recycling The maintenance department of an office building purchases a machine that crushes used fluorescent lamps in preparation for recycling. The machine is purchased for $2750 and has a useful life of 5 years, after which time the machine will have no value.

(a) Write a linear equation giving the value y of the machine in terms of the time t in years, $0 \leq t \leq 5$.

(b) Use a graphing utility to graph the equation.

(c) Move the cursor along the graph and estimate the value of the machine when $t = 2$.

(d) Move the cursor along the graph and estimate (to two-decimal-place accuracy) the time when the value of the machine will be $1000.

97. Health Care The data below show the total yearly health care expenditures E (in billions of dollars) in the United States for the years 2001 through 2005, where $t = 1$ represents 2001. *(Source: U.S. Centers for Medicare and Medicaid Services)*

t	1	2	3	4	5
E	1470	1603	1733	1859	1988

(a) Use a graphing utility to create a scatter plot of the data.

(b) Use the *regression* feature of a graphing utility to find a linear model for the data.

(c) What is the rate of change in health care expenditures from 2001 through 2005?

(d) Use your model to predict the health care expenditure in the year 2012. Does this prediction seem reasonable? Explain.

Mid-Chapter Quiz

Take this quiz as you would take a quiz in class. When you are done, check your work against the answers given in the back of the book.

In Exercises 1–3, (a) plot the points, (b) find the distance between the points, and (c) find the midpoint of the line segment joining the points.

1. $(3, -2), (-3, 1)$
2. $\left(\frac{1}{4}, -\frac{3}{2}\right), \left(\frac{1}{2}, 2\right)$
3. $(0, -4), (\sqrt{3}, 0)$

4. Use the Distance Formula to show that the points $(4, 0)$, $(2, 1)$, and $(-1, -5)$ are vertices of a right triangle.

5. The resident population of Missouri (in thousands) was 5719 in 2003 and 5800 in 2005. Use the Midpoint Formula to estimate the population in 2004. *(Source: U.S. Census Bureau)*

In Exercises 6–8, sketch the graph of the equation and label the intercepts.

6. $y = 5x + 2$
7. $y = x^2 + x - 6$
8. $y = |x - 3|$

In Exercises 9 and 10, write the general form of the equation of the circle.

9. Center: $(-1, 0)$; radius: $\sqrt{37}$
10. Center: $(2, -2)$; solution point: $(-1, 2)$

Ⓣ In Exercises 11 and 12, write the equation of the circle in standard form. Then use a graphing utility to graph the circle.

11. $x^2 + y^2 + 8x - 6y + 16 = 0$
12. $4x^2 + 4y^2 - 8x + 4y - 11 = 0$

13. Retail sales S of mail order prescription drugs (in billions of dollars) in the United States for the years 2000 through 2005 can be modeled by

$$S = 4.42t + 22.3$$

where t is the year, with $t = 0$ corresponding to 2000. Use the model to complete the table. *(Source: National Association of Chain Drug Stores)*

Year	2000	2001	2002
Sales			

Year	2003	2004	2005
Sales			

Table for 13

In Exercises 14–16, write an equation of the line that passes through the points. Then use the equation to sketch the line.

14. $(1, -1), (-4, 5)$
15. $(-2, 3), (-2, 2)$
16. $\left(\frac{5}{2}, 2\right), (0, 2)$

17. Find equations of the lines that pass through the point $(3, -5)$ and are
    (a) parallel to the line $x + 4y = -2$.
    (b) perpendicular to the line $x + 4y = -2$.

18. The pressure (in atmospheres) exerted on a scuba diver's body has a linear relationship with the diver's depth. At sea level (or a depth of 0 feet), the pressure exerted on a diver is 1 atmosphere. At a depth of 132 feet, the pressure exerted on a diver is 5 atmospheres.

    (a) Write a linear equation expressing the pressure p (in atmospheres) in terms of depth d (in feet).

    (b) What is the rate of change in pressure with respect to the diver's depth?

Section 1.4
Functions

- Decide whether relations between two variables are functions.
- Find the domains and ranges of functions.
- Use function notation and evaluate functions.
- Combine functions to create other functions.
- Find inverse functions algebraically.

Functions

In many common relationships between two variables, the value of one of the variables depends on the value of the other variable. For example, the sales tax on an item depends on its selling price, the distance a nerve impulse travels in a given amount of time depends on its speed, the weight of a human brain depends on body weight, and the area of a circle depends on its radius.

Consider the relationship between the area of a circle and its radius. This relationship can be expressed by the equation

$$A = \pi r^2.$$

In this equation, the value of A depends on the choice of r. Because of this, A is the **dependent variable** and r is the **independent variable**.

Most of the relationships that you will study in this course have the property that for a given value of the independent variable, there corresponds exactly one value of the dependent variable. Such a relationship is a **function**.

> **Definition of Function**
>
> A **function** is a relationship between two variables such that to each value of the independent variable there corresponds exactly one value of the dependent variable.
>
> The **domain** of the function is the set of all values of the independent variable for which the function is defined. The **range** of the function is the set of all values taken on by the dependent variable.

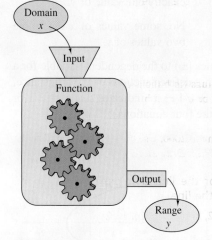

FIGURE 1.40

In Figure 1.40, notice that you can think of a function as a machine that inputs values of the independent variable and outputs values of the dependent variable.

Although functions can be described by various means such as tables, graphs, and diagrams, they are most often specified by formulas or equations. For instance, the equation

$$y = 4x^2 + 3$$

describes y as a function of x. For this function, x is the independent variable and y is the dependent variable.

70   CHAPTER 1   Functions, Graphs, and Limits

TECHNOLOGY

The procedure used in Example 1, isolating the dependent variable on the left side, is also useful for graphing equations with a graphing utility. In fact, the standard graphing program on most graphing utilities is called a "function grapher." To graph an equation in which y is not a function of x, such as a circle, you usually have to enter two or more equations into the graphing utility.

Example 1 Deciding Whether Relations Are Functions

Which of the equations below define y as a function of x?

a. $x + y = 1$ **b.** $x^2 + y^2 = 1$

c. $x^2 + y = 1$ **d.** $x + y^2 = 1$

SOLUTION To decide whether an equation defines a function, it is helpful to isolate the dependent variable on the left side. For instance, to decide whether the equation $x + y = 1$ defines y as a function of x, write the equation in the form

$$y = 1 - x.$$

From this form, you can see that for any value of x, there is exactly one value of y. So, y is a function of x.

Original Equation	Rewritten Equation	Test: Is y a function of x?
a. $x + y = 1$	$y = 1 - x$	Yes, each value of x determines exactly one value of y.
b. $x^2 + y^2 = 1$	$y = \pm\sqrt{1 - x^2}$	No, some values of x determine two values of y.
c. $x^2 + y = 1$	$y = 1 - x^2$	Yes, each value of x determines exactly one value of y.
d. $x + y^2 = 1$	$y = \pm\sqrt{1 - x}$	No, some values of x determine two values of y.

Note that the equations that assign two values ($\pm$) to the dependent variable for a given value of the independent variable do not define functions of x. For instance, in part (b), when $x = 0$, the equation $y = \pm\sqrt{1 - x^2}$ indicates that $y = +1$ or $y = -1$. Figure 1.41 shows the graphs of the four equations.

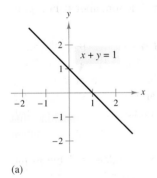
(a)

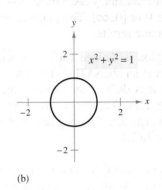
(b)

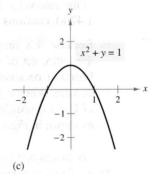
(c)

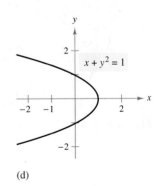
(d)

FIGURE 1.41

✓ CHECKPOINT 1

Which of the equations below define y as a function of x? Explain your answer.

a. $x - y = 1$ **b.** $x^2 + y^2 = 4$ **c.** $y^2 + x = 2$ **d.** $x^2 - y = 0$ ■

The Graph of a Function

When the graph of a function is sketched, the standard convention is to let the horizontal axis represent the independent variable. When this convention is used, the test described in Example 1 has a nice graphical interpretation called the **Vertical Line Test.** This test states that if every vertical line intersects the graph of an equation at most once, then the equation defines y as a function of x. For instance, in Figure 1.41, the graphs in parts (a) and (c) pass the Vertical Line Test, but those in parts (b) and (d) do not.

The domain of a function may be described explicitly, or it may be *implied* by an equation used to define the function. For example, the function given by

$$y = \frac{1}{x^2 - 4}$$

has an implied domain that consists of all real x except $x = \pm 2$. These two values are excluded from the domain because division by zero is undefined.

Another type of implied domain is that used to avoid even roots of negative numbers, as indicated in Example 2.

Example 2 Finding the Domains and Ranges of Functions

Find the domain and range of each function.

a. $y = \sqrt{x - 1}$ **b.** $y = \begin{cases} 1 - x, & x < 1 \\ \sqrt{x - 1}, & x \geq 1 \end{cases}$

SOLUTION

a. Because $\sqrt{x - 1}$ is not defined for $x - 1 < 0$ (that is, for $x < 1$), it follows that the domain of the function is the interval $x \geq 1$ or $[1, \infty)$. To find the range, observe that $\sqrt{x - 1}$ is never negative. Moreover, as x takes on the various values in the domain, y takes on all nonnegative values. So, the range is the interval $y \geq 0$ or $[0, \infty)$. The graph of the function, shown in Figure 1.42(a), confirms these results.

b. Because this function is defined for $x < 1$ *and* for $x \geq 1$, the domain is the entire set of real numbers. This function is called a **piecewise-defined function** because it is defined by two or more equations over a specified domain. When $x \geq 1$, the function behaves as in part (a). For $x < 1$, the value of $1 - x$ is positive, and therefore the range of the function is $y \geq 0$ or $[0, \infty)$, as shown in Figure 1.42(b).

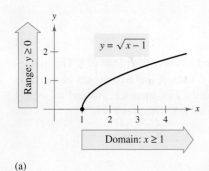
(a)

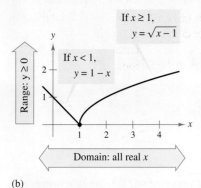
(b)

FIGURE 1.42

A function is **one-to-one** if to each value of the dependent variable in the range there corresponds exactly one value of the independent variable. For instance, the function in Example 2(a) is one-to-one, whereas the function in Example 2(b) is not one-to-one.

Geometrically, a function is one-to-one if every horizontal line intersects the graph of the function at most once. This geometrical interpretation is the **Horizontal Line Test** for one-to-one functions. So, a graph that represents a one-to-one function must satisfy *both* the Vertical Line Test and the Horizontal Line Test.

✓ CHECKPOINT 2

Find the domain and range of each function.

a. $y = \sqrt{x + 1}$

b. $y = \begin{cases} x^2, & x \leq 0 \\ \sqrt{x}, & x > 0 \end{cases}$

Function Notation

When using an equation to define a function, you generally isolate the dependent variable on the left. For instance, writing the equation $x + 2y = 1$ as

$$y = \frac{1-x}{2}$$

indicates that y is the dependent variable. In **function notation,** this equation has the form

$$f(x) = \frac{1-x}{2}. \qquad \text{Function notation}$$

The independent variable is x, and the name of the function is "f." The symbol $f(x)$ is read as "f of x," and it denotes the value of the dependent variable. For instance, the value of f when $x = 3$ is

$$f(3) = \frac{1-(3)}{2} = \frac{-2}{2} = -1.$$

The value $f(3)$ is called a **function value,** and it lies in the range of f. This means that the point $(3, f(3))$ lies on the graph of f. One of the advantages of function notation is that it allows you to be less wordy. For instance, instead of asking "What is the value of y when $x = 3$?" you can ask "What is $f(3)$?"

Example 3 Evaluating a Function

Find the values of the function $f(x) = 2x^2 - 4x + 1$ when x is -1, 0, and 2. Is f one-to-one?

SOLUTION When $x = -1$, the value of f is

$$f(-1) = 2(-1)^2 - 4(-1) + 1 = 2 + 4 + 1 = 7.$$

When $x = 0$, the value of f is

$$f(0) = 2(0)^2 - 4(0) + 1 = 0 - 0 + 1 = 1.$$

When $x = 2$, the value of f is

$$f(2) = 2(2)^2 - 4(2) + 1 = 8 - 8 + 1 = 1.$$

Because two different values of x yield the same value of $f(x)$, the function is *not* one-to-one, as shown in Figure 1.43.

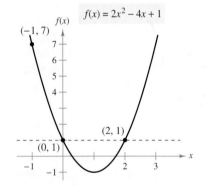

FIGURE 1.43

✓ CHECKPOINT 3

Find the values of $f(x) = x^2 - 5x + 1$ when x is 0, 1, and 4. Is f one-to-one? ∎

STUDY TIP

You can use the Horizontal Line Test to determine whether the function in Example 3 is one-to-one. Because the line $y = 1$ intersects the graph of the function twice, the function is *not* one-to-one.

Example 3 suggests that the role of the variable x in the equation

$$f(x) = 2x^2 - 4x + 1$$

is simply that of a placeholder. Informally, f could be defined by the equation

$$f(\quad) = 2(\quad)^2 - 4(\quad) + 1.$$

To evaluate $f(-2)$, simply place -2 in each set of parentheses.

$$f(-2) = 2(-2)^2 - 4(-2) + 1 = 8 + 8 + 1 = 17$$

The ratio in Example 4(b) is called a *difference quotient*. In Section 2.1, you will see that it has special significance in calculus.

Example 4 Evaluating a Function

Let $f(x) = x^2 - 4x + 7$, and find

a. $f(x + \Delta x)$ **b.** $\dfrac{f(x + \Delta x) - f(x)}{\Delta x}$.

SOLUTION

a. To evaluate f at $x + \Delta x$, substitute $x + \Delta x$ for x in the original function, as shown.

$$\begin{aligned} f(x + \Delta x) &= (x + \Delta x)^2 - 4(x + \Delta x) + 7 \\ &= x^2 + 2x\,\Delta x + (\Delta x)^2 - 4x - 4\,\Delta x + 7 \end{aligned}$$

b. Using the result of part (a), you can write

$$\begin{aligned} \frac{f(x + \Delta x) - f(x)}{\Delta x} &= \frac{[(x + \Delta x)^2 - 4(x + \Delta x) + 7] - [x^2 - 4x + 7]}{\Delta x} \\ &= \frac{x^2 + 2x\,\Delta x + (\Delta x)^2 - 4x - 4\,\Delta x + 7 - x^2 + 4x - 7}{\Delta x} \\ &= \frac{2x\,\Delta x + (\Delta x)^2 - 4\,\Delta x}{\Delta x} \\ &= 2x + \Delta x - 4, \quad \Delta x \neq 0. \end{aligned}$$

✓ CHECKPOINT 4

Let $f(x) = x^2 - 2x + 3$, and find (a) $f(x + \Delta x)$ and (b) $\dfrac{f(x + \Delta x) - f(x)}{\Delta x}$. ∎

Although f is often used as a convenient function name and x as the independent variable, you can use other symbols. For instance, the following equations all define the same function.

$$\begin{aligned} f(x) &= x^2 - 4x + 7 \\ f(t) &= t^2 - 4t + 7 \\ g(s) &= s^2 - 4s + 7 \end{aligned}$$

TECHNOLOGY

Most graphing utilities can be programmed to evaluate functions. The program depends on the calculator used. The pseudocode below can be translated into a program for a graphing utility. (Appendix E lists the program for several models of graphing utilities.)

Program

- Label a.
- Input x.
- Display function value.
- Goto a.

To use this program, enter a function. Then run the program—it will allow you to evaluate the function at several values of x.

74   CHAPTER 1   Functions, Graphs, and Limits

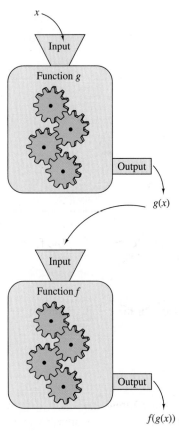

FIGURE 1.44

Combinations of Functions

Two functions can be combined in various ways to create new functions. For instance, if $f(x) = 2x - 3$ and $g(x) = x^2 + 1$, you can form the following functions.

$f(x) + g(x) = (2x - 3) + (x^2 + 1) = x^2 + 2x - 2$ Sum

$f(x) - g(x) = (2x - 3) - (x^2 + 1) = -x^2 + 2x - 4$ Difference

$f(x)g(x) = (2x - 3)(x^2 + 1) = 2x^3 - 3x^2 + 2x - 3$ Product

$\dfrac{f(x)}{g(x)} = \dfrac{2x - 3}{x^2 + 1}$ Quotient

You can combine two functions in yet another way called a **composition**. The resulting function is a **composite function**.

Definition of Composite Function

The function given by $(f \circ g)(x) = f(g(x))$ is the **composite** of f with g. The **domain** of $(f \circ g)$ is the set of all x in the domain of g such that $g(x)$ is in the domain of f, as indicated in Figure 1.44.

The composite of f with g may not be equal to the composite of g with f, as shown in the next example.

Example 5 Forming Composite Functions

Let $f(x) = 2x - 3$ and $g(x) = x^2 + 1$, and find

a. $f(g(x))$ **b.** $g(f(x))$.

SOLUTION

a. The composite of f with g is given by

$f(g(x)) = 2(g(x)) - 3$ Evaluate f at $g(x)$.

$ = 2(x^2 + 1) - 3$ Substitute $x^2 + 1$ for $g(x)$.

$ = 2x^2 - 1.$ Simplify.

b. The composite of g with f is given by

$g(f(x)) = (f(x))^2 + 1$ Evaluate g at $f(x)$.

$ = (2x - 3)^2 + 1$ Substitute $2x - 3$ for $f(x)$.

$ = 4x^2 - 12x + 10.$ Simplify.

✓ CHECKPOINT 5

Let $f(x) = 2x + 1$ and $g(x) = x^2 + 2$, and find

a. $f(g(x))$ **b.** $g(f(x))$. ∎

STUDY TIP

The results of $f(g(x))$ and $g(f(x))$ are different in Example 5. You can verify this by substituting specific values of x into each function and comparing the results.

Inverse Functions

Informally, the inverse function of f is another function g that "undoes" what f has done.

Definition of Inverse Function

Let f and g be two functions such that

$$f(g(x)) = x \text{ for each } x \text{ in the domain of } g$$

and

$$g(f(x)) = x \text{ for each } x \text{ in the domain of } f.$$

Under these conditions, the function g is the **inverse function** of f. The function g is denoted by f^{-1}, which is read as "f-inverse." So,

$$f(f^{-1}(x)) = x \quad \text{and} \quad f^{-1}(f(x)) = x.$$

The domain of f must be equal to the range of f^{-1}, and the range of f must be equal to the domain of f^{-1}.

STUDY TIP

Don't be confused by the use of the superscript -1 to denote the inverse function f^{-1}. In this text, whenever f^{-1} is written, it *always* refers to the inverse function of f and *not* to the reciprocal of $f(x)$.

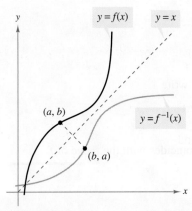

FIGURE 1.45 The graph of f^{-1} is a reflection of the graph of f in the line $y = x$.

Example 6 Finding Inverse Functions

Several functions and their inverse functions are shown below. In each case, note that the inverse function "undoes" the original function. For instance, to undo multiplication by 2, you should divide by 2.

a. $f(x) = 2x$ $f^{-1}(x) = \frac{1}{2}x$

b. $f(x) = \frac{1}{3}x$ $f^{-1}(x) = 3x$

c. $f(x) = x + 4$ $f^{-1}(x) = x - 4$

d. $f(x) = 2x - 5$ $f^{-1}(x) = \frac{1}{2}(x + 5)$

e. $f(x) = x^3$ $f^{-1}(x) = \sqrt[3]{x}$

f. $f(x) = \dfrac{1}{x}$ $f^{-1}(x) = \dfrac{1}{x}$

STUDY TIP

You can verify that the functions in Example 6 are inverse functions by showing that $f(f^{-1}(x)) = x$ and $f^{-1}(f(x)) = x$. For Example 6(a), you obtain the following.

$f(f^{-1}(x)) = f\left(\frac{1}{2}x\right) = 2\left(\frac{1}{2}x\right) = x$

$f^{-1}(f(x)) = f^{-1}(2x) = \frac{1}{2}(2x) = x$

✓ CHECKPOINT 6

Informally find the inverse function of each function.

a. $f(x) = \frac{1}{5}x$ b. $f(x) = 3x + 2$ ∎

The graphs of f and f^{-1} are mirror images of each other (with respect to the line $y = x$), as shown in Figure 1.45. Try using a graphing utility to confirm this for each of the functions given in Example 6.

The functions in Example 6 are simple enough so that their inverse functions can be found by inspection. The next example demonstrates a strategy for finding the inverse functions of more complicated functions.

Example 7 Finding an Inverse Function

Find the inverse function of $f(x) = \sqrt{2x - 3}$.

SOLUTION Begin by replacing $f(x)$ with y. Then, interchange x and y and solve for y.

$$f(x) = \sqrt{2x - 3} \qquad \text{Write original function.}$$
$$y = \sqrt{2x - 3} \qquad \text{Replace } f(x) \text{ with } y.$$
$$x = \sqrt{2y - 3} \qquad \text{Interchange } x \text{ and } y.$$
$$x^2 = 2y - 3 \qquad \text{Square each side.}$$
$$x^2 + 3 = 2y \qquad \text{Add 3 to each side.}$$
$$\frac{x^2 + 3}{2} = y \qquad \text{Divide each side by 2.}$$

So, the inverse function has the form

$$f^{-1}() = \frac{()^2 + 3}{2}.$$

Using x as the independent variable, you can write

$$f^{-1}(x) = \frac{x^2 + 3}{2}, \quad x \geq 0.$$

In Figure 1.46, note that the domain of f^{-1} coincides with the range of f.

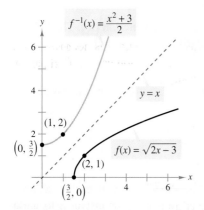

FIGURE 1.46

✓ CHECKPOINT 7

Find the inverse function of $f(x) = x^2 + 2$ for $x \geq 0$. ∎

After you have found an inverse function, you should check your results. You can check your results *graphically* by observing that the graphs of f and f^{-1} are reflections of each other in the line $y = x$. You can check your results *algebraically* by evaluating $f(f^{-1}(x))$ and $f^{-1}(f(x))$—both should be equal to x. The algebraic check for Example 7 is shown below.

Check that $f(f^{-1}(x)) = x$

$$f(f^{-1}(x)) = f\left(\frac{x^2 + 3}{2}\right)$$
$$= \sqrt{2\left(\frac{x^2 + 3}{2}\right) - 3}$$
$$= \sqrt{x^2}$$
$$= x, \quad x \geq 0$$

Check that $f^{-1}(f(x)) = x$

$$f^{-1}(f(x)) = f^{-1}\left(\sqrt{2x - 3}\right)$$
$$= \frac{\left(\sqrt{2x - 3}\right)^2 + 3}{2}$$
$$= \frac{2x}{2}$$
$$= x, \quad x \geq \frac{3}{2}$$

TECHNOLOGY

A graphing utility can help you check that the graphs of f and f^{-1} are reflections of each other in the line $y = x$. To do this, graph $y = f(x)$, $y = f^{-1}(x)$, and $y = x$ in the same viewing window, using a square setting.

Not every function has an inverse function. In fact, for a function to have an inverse function, it must be one-to-one.

Example 8 A Function That Has No Inverse Function

Show that the function

$$f(x) = x^2 - 1$$

has no inverse function. (Assume that the domain of f is the set of all real numbers.)

SOLUTION Begin by sketching the graph of f, as shown in Figure 1.47. Note that

$$f(2) = (2)^2 - 1 = 3$$

and

$$f(-2) = (-2)^2 - 1 = 3.$$

So, f does not pass the Horizontal Line Test, which implies that it is not one-to-one, and therefore has no inverse function. The same conclusion can be obtained by trying to find the inverse function of f algebraically.

$f(x) = x^2 - 1$	Write original function.
$y = x^2 - 1$	Replace $f(x)$ with y.
$x = y^2 - 1$	Interchange x and y.
$x + 1 = y^2$	Add 1 to each side.
$\pm\sqrt{x + 1} = y$	Take square root of each side.

The last equation does not define y as a function of x, and so f has no inverse function.

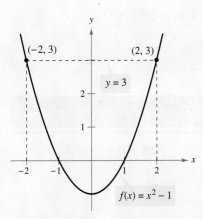

FIGURE 1.47 f is not one-to-one and has no inverse function.

✓ CHECKPOINT 8

Show that the function

$$f(x) = x^2 + 4$$

has no inverse function. ■

CONCEPT CHECK

1. Explain the difference between a relation and a function.
2. In your own words, explain the meanings of *domain* and *range*.
3. Is every relation a function? Explain.
4. Describe how to find the inverse of a function given by an equation in x and y.

CHAPTER 1 Functions, Graphs, and Limits

Skills Review 1.4

The following warm-up exercises involve skills that were covered in earlier sections. You will use these skills in the exercise set for this section. For additional help, review Sections 0.3 and 0.5.

In Exercises 1–6, simplify the expression.

1. $5(-1)^2 - 6(-1) + 9$
2. $(-2)^3 + 7(-2)^2 - 10$
3. $(x - 2)^2 + 5x - 10$
4. $(3 - x) + (x + 3)^3$
5. $\dfrac{1}{1 - (1 - x)}$
6. $1 + \dfrac{x - 1}{x}$

In Exercises 7–12, solve for y in terms of x.

7. $2x + y - 6 = 11$
8. $5y - 6x^2 - 1 = 0$
9. $(y - 3)^2 = 5 + (x + 1)^2$
10. $y^2 - 4x^2 = 2$
11. $x = \dfrac{2y - 1}{4}$
12. $x = \sqrt[3]{2y - 1}$

Exercises 1.4

See www.CalcChat.com for worked-out solutions to odd-numbered exercises.

In Exercises 1–8, decide whether the equation defines y as a function of x.

1. $x^2 + y^2 = 4$
2. $x + y^2 = 4$
3. $\tfrac{1}{2}x - 6y = -3$
4. $3x - 2y + 5 = 0$
5. $x^2 + y = 4$
6. $x^2 + y^2 + 2x = 0$
7. $y = |x + 2|$
8. $x^2y - x^2 + 4y = 0$

(T) In Exercises 9–16, use a graphing utility to graph the function. Then determine the domain and range of the function.

9. $f(x) = 2x^2 - 5x + 1$
10. $f(x) = 5x^3 + 6x^2 - 1$
11. $f(x) = \dfrac{|x|}{x}$
12. $f(x) = \sqrt{9 - x^2}$
13. $f(x) = \dfrac{x}{\sqrt{x - 4}}$
14. $f(x) = \begin{cases} 3x + 2, & x < 0 \\ 2 - x, & x \geq 0 \end{cases}$
15. $f(x) = \dfrac{x - 2}{x + 4}$
16. $f(x) = \dfrac{x^2}{1 - x}$

In Exercises 17–20, find the domain and range of the function. Use interval notation to write your result.

17. $f(x) = x^3$
18. $f(x) = \sqrt{2x - 3}$

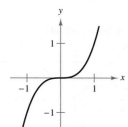
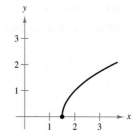

19. $f(x) = 4 - x^2$
20. $f(x) = |x - 2|$

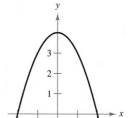
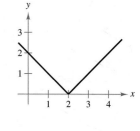

In Exercises 21–24, evaluate the function at the specified values of the independent variable. Simplify the result.

21. $f(x) = 3x - 2$
   (a) $f(0)$ (b) $f(x - 1)$ (c) $f(x + \Delta x)$
22. $f(x) = x^2 - 4x + 1$
   (a) $f(-1)$ (b) $f(c + 2)$ (c) $f(x + \Delta x)$
23. $g(x) = 1/x$
   (a) $g\left(\tfrac{1}{4}\right)$ (b) $g(x + 4)$ (c) $g(x + \Delta x) - g(x)$
24. $f(x) = |x| + 4$
   (a) $f(-2)$ (b) $f(x + 2)$ (c) $f(x + \Delta x) - f(x)$

In Exercises 25–30, evaluate the difference quotient and simplify the result.

25. $f(x) = x^2 - 5x + 2$
   $\dfrac{f(x + \Delta x) - f(x)}{\Delta x}$
26. $h(x) = x^2 + x + 3$
   $\dfrac{h(2 + \Delta x) - h(2)}{\Delta x}$

27. $g(x) = \sqrt{x+1}$

$$\frac{g(x + \Delta x) - g(x)}{\Delta x}$$

28. $f(x) = \dfrac{1}{\sqrt{x}}$

$$\frac{f(x) - f(2)}{x - 2}$$

29. $f(x) = \dfrac{1}{x-2}$

$$\frac{f(x + \Delta x) - f(x)}{\Delta x}$$

30. $f(x) = \dfrac{1}{x+4}$

$$\frac{f(x + \Delta x) - f(x)}{\Delta x}$$

In Exercises 31–34, use the Vertical Line Test to determine whether y is a function of x.

31. $x^2 + y^2 = 9$

32. $x - xy + y + 1 = 0$

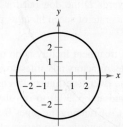
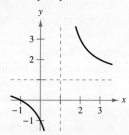

33. $x^2 = xy - 1$

34. $x = |y|$

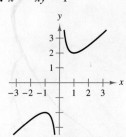
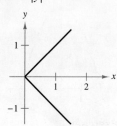

In Exercises 35–38, find (a) $f(x) + g(x)$, (b) $f(x) \cdot g(x)$, (c) $f(x)/g(x)$, (d) $f(g(x))$, and (e) $g(f(x))$, if defined.

35. $f(x) = 2x - 5$
$g(x) = 5$

36. $f(x) = x^2 + 5$
$g(x) = \sqrt{1-x}$

37. $f(x) = x^2 + 1$
$g(x) = x - 1$

38. $f(x) = \dfrac{x}{x+1}$
$g(x) = x^3$

39. Given $f(x) = \sqrt{x}$ and $g(x) = x^2 - 1$, find the composite functions.

(a) $f(g(1))$
(b) $g(f(1))$
(c) $g(f(0))$
(d) $f(g(-4))$
(e) $f(g(x))$
(f) $g(f(x))$

40. Given $f(x) = 1/x$ and $g(x) = x^2 - 1$, find the composite functions.

(a) $f(g(2))$
(b) $g(f(2))$
(c) $f(g(1/\sqrt{2}))$
(d) $g(f(1/\sqrt{2}))$
(e) $f(g(x))$
(f) $g(f(x))$

In Exercises 41–44, show that f and g are inverse functions by showing that $f(g(x)) = x$ and $g(f(x)) = x$. Then sketch the graphs of f and g on the same coordinate axes.

41. $f(x) = 5x + 1$, $\qquad g(x) = \dfrac{x-1}{5}$

42. $f(x) = \dfrac{1}{x}$, $\qquad g(x) = \dfrac{1}{x}$

43. $f(x) = 9 - x^2$, $x \geq 0$, $\quad g(x) = \sqrt{9-x}$, $x \leq 9$

44. $f(x) = 1 - x^3$, $\qquad g(x) = \sqrt[3]{1-x}$

(T) In Exercises 45–52, find the inverse function of f. Then use a graphing utility to graph f and f^{-1} on the same coordinate axes.

45. $f(x) = 2x - 3$
46. $f(x) = 7 - x$
47. $f(x) = x^5$
48. $f(x) = x^3$
49. $f(x) = \sqrt{9 - x^2}$, $0 \leq x \leq 3$
50. $f(x) = \sqrt{x^2 - 4}$, $x \geq 2$
51. $f(x) = x^{2/3}$, $x \geq 0$ $\qquad$ **52.** $f(x) = x^{3/5}$

(T) In Exercises 53–58, use a graphing utility to graph the function. Then use the Horizontal Line Test to determine whether the function is one-to-one. If it is, find its inverse function.

53. $f(x) = 3 - 7x$
54. $f(x) = \sqrt{x - 2}$
55. $f(x) = x^2$
56. $f(x) = x^4$
57. $f(x) = |x + 3|$
58. $f(x) = -5$

59. Use the graph of $f(x) = \sqrt{x}$ below to sketch the graph of each function.

(a) $y = \sqrt{x} + 2$
(b) $y = -\sqrt{x}$
(c) $y = \sqrt{x - 2}$
(d) $y = \sqrt{x + 3}$
(e) $y = \sqrt{x - 4}$
(f) $y = 2\sqrt{x}$

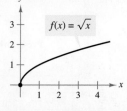

60. Use the graph of $f(x) = |x|$ below to sketch the graph of each function.

(a) $y = |x| + 3$
(b) $y = -\frac{1}{2}|x|$
(c) $y = |x - 2|$
(d) $y = |x + 1| - 1$
(e) $y = 2|x|$

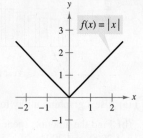

61. Use the graph of $f(x) = x^2$ to write an equation for each function whose graph is shown.

(a) (b)

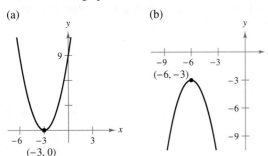

62. Use the graph of $f(x) = x^3$ to write an equation for each function whose graph is shown.

(a) (b)

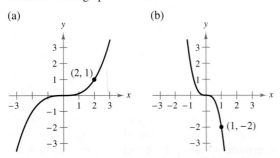

63. Prescription Drugs The amount d (in billions of dollars) spent on prescription drugs in the United States in the years 1991 through 2005 (see figure) can be approximated by the model

$$d(t) = \begin{cases} y = 0.68t^2 - 0.3t + 45, & 1 \le t \le 8 \\ y = 16.7t - 45, & 9 \le t \le 15 \end{cases}$$

where t represents the year, with $t = 1$ corresponding to 1991. *(Source: U.S. Centers for Medicare & Medicaid Services)*

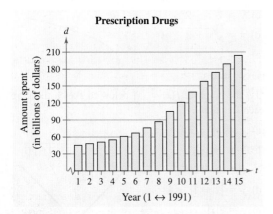

(T) (a) Use a graphing utility to graph the function.

(b) Find the amounts spent on prescription drugs in 1997, 2000, and 2004.

64. Cadavers The number of cat cadavers purchased for dissection in a biology class in the years 2000 through 2007 can be modeled by the function

$$C = \begin{cases} 2t + 48, & 0 \le t \le 3 \\ 4t + 42, & 4 \le t \le 7 \end{cases}$$

with $t = 0$ corresponding to 2000.

(T) (a) Use a graphing utility to graph the function.

(b) Find the numbers of cat cadavers purchased for dissection in a biology class in 2001, 2003, and 2006.

65. Dog Race A greyhound ran a total of 1100 yards in two races. In the third race, the greyhound ran for 35 seconds. Express the total number of yards D the dog ran for all three races as a function of r, its rate of speed in yards per second.

66. Water Treatment A water treatment facility is on one side of a river that is $\frac{1}{2}$ mile wide. A factory is 3 miles downstream on the other side of the river (see figure). It costs \$170/ft to run the pipes on land and \$230/ft to run them under water. Express the cost C of running the pipes from the water treatment facility to the factory as a function of x.

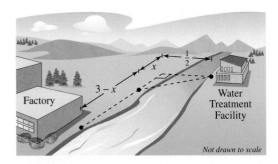

67. Femur Length Anthropologists can estimate a person's height from the length of certain bones. The height h (in inches) of an adult human female can be modeled by the function

$$h(l) = 1.95l + 28.7, \quad 15 \le l \le 24$$

where l is the length (in inches) of the femur, or thigh bone.

(a) Find the domain and range of the function.

(b) Suppose a female's femur was 16 inches long. About how tall was she?

(c) If an anthropologist estimates a female's height to be 5 feet 10 inches, about how long was her femur?

68. Path of a Salmon Part of the life cycle of a salmon is to migrate for reproduction. Salmon are anadromous fish. This means they swim from the ocean to fresh water streams to lay their eggs. During migration, salmon must jump waterfalls to reach their destination. The path of a jumping salmon is given by

$$h = -0.42x^2 + 2.52x$$

where h is the height (in feet) and x is the horizontal distance (in feet) from where the salmon left the water. Will the salmon clear a waterfall that is 3 feet high if it leaves the water 4 feet from the waterfall?

69. Bacteria The number N of bacteria is given by $N(T) = 8T^2 - 14T + 200$, where T is the temperature. The temperature is given by $T(t) = 2t + 2$, where t is the time in hours. Find and interpret $(N \circ T)(t)$.

70. Ecology A square concrete foundation is prepared as a base for a large cylindrical aquatic tank that is to be used in ecology experiments (see figure).

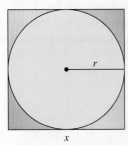

(a) Write the radius r of the tank as a function of the length x of the sides of the square.

(b) Write the area A of the circular base of the tank as a function of the radius r.

(c) Find and interpret $(A \circ r)(x)$.

71. Ripples in a Pond A stone is thrown into the middle of a calm pond causing ripples to form in concentric circles. The radius r of the outermost ripple increases at the rate of 0.75 foot per second.

(a) Write a function for the radius r (in feet) of the circle formed by the outermost ripple in terms of time t (in seconds).

(b) Write a function in terms of t for the area A enclosed by the outermost ripple. Complete the table.

Time, t	1	2	3	4	5
Radius, r					
Area, A					

(c) Compare the ratios $A(2)/A(1)$ and $A(4)/A(2)$. What do you observe? Based on your observations, predict the area when $t = 8$. Verify your prediction by checking $t = 8$ in the area function.

72. Medicine The temperature of a patient after being given a fever-reducing drug is given by

$$F(t) = 98 + \frac{3}{t + 1}$$

where F is the temperature (in degrees Fahrenheit) and t is the time (in hours) since the drug was administered. Use a graphing utility to graph the function. Be sure to choose an appropriate viewing window. For what values of t do you think this function would be valid? Explain.

In Exercises 73–78, use a graphing utility to graph the function. Then use the *zoom* and *trace* features to find the zeros of the function. Is the function one-to-one?

73. $f(x) = 9x - 4x^2$

74. $f(x) = 2\left(3x^2 - \dfrac{6}{x}\right)$

75. $g(t) = \dfrac{t + 3}{1 - t}$

76. $h(x) = 6x^3 - 12x^2 + 4$

77. $g(x) = x^2\sqrt{x^2 - 4}$

78. $g(x) = \left|\dfrac{1}{2}x^2 - 4\right|$

Life Science Capsule

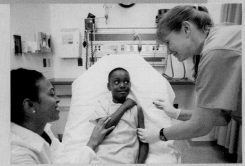

ERproductions Ltd/Getty Images

By 2020, there will be an estimated shortage of 800,000 nurses. Because of the increasing need for nurses, some hospitals are offering signing bonuses of up to $14,000. As of August 2006, students pursuing a Baccalaureate in Nursing at the School of Nursing - University of Pittsburgh are required to complete 124 course credits.

79. Research Project Use your school's library, the Internet, or some other reference source to find information about the needs for and requirements of a life science career, such as nursing. Write a short paper about this career.

Section 1.5

Limits

- Find limits of functions graphically and numerically.
- Use the properties of limits to evaluate limits of functions.
- Use different analytic techniques to evaluate limits of functions.
- Evaluate one-sided limits.
- Recognize unbounded behavior of functions.

The Limit of a Function

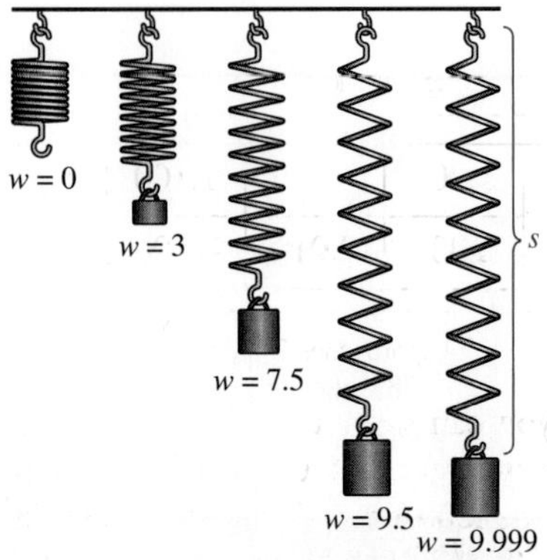

In everyday language, people refer to a speed limit, a wrestler's weight limit, the limit of one's endurance, or stretching a spring to its limit. These phrases all suggest that a limit is a bound, which on some occasions may not be reached but on other occasions may be reached or exceeded.

Consider a spring that will break only if a weight of 10 pounds or more is attached. To determine how far the spring will stretch without breaking, you could attach increasingly heavier weights and measure the spring length s for each weight w, as shown in Figure 1.48. If the spring length approaches a value of L, then it is said that "the limit of s as w approaches 10 is L." A mathematical limit is much like the limit of a spring. The notation for a limit is

$$\lim_{x \to c} f(x) = L$$

which is read as "the limit of $f(x)$ as x approaches c is L."

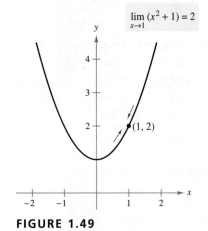

FIGURE 1.48 What is the limit of s as w approaches 10 pounds?

Example 1 Finding a Limit

Find the limit: $\lim_{x \to 1} (x^2 + 1)$.

SOLUTION Let $f(x) = x^2 + 1$. From the graph of f in Figure 1.49, it appears that $f(x)$ approaches 2 as x approaches 1 from either side, and you can write

$$\lim_{x \to 1} (x^2 + 1) = 2.$$

The table yields the same conclusion. Notice that as x gets closer and closer to 1, $f(x)$ gets closer and closer to 2.

	x approaches 1.				x approaches 1.	
x	0.900	0.990	0.999	1.000	1.001	1.010	1.100
$f(x)$	1.810	1.980	1.998	2.000	2.002	2.020	2.210

$f(x)$ approaches 2. $f(x)$ approaches 2.

FIGURE 1.49

✓ CHECKPOINT 1

Find the limit: $\lim_{x \to 1} (2x + 4)$. ■

SECTION 1.5 Limits

Example 2 Finding Limits Graphically and Numerically

Find the limit: $\lim_{x \to 1} f(x)$.

a. $f(x) = \dfrac{x^2 - 1}{x - 1}$ **b.** $f(x) = \dfrac{|x - 1|}{x - 1}$ **c.** $f(x) = \begin{cases} x, & x \neq 1 \\ 0, & x = 1 \end{cases}$

SOLUTION

a. From the graph of f, in Figure 1.50(a), it appears that $f(x)$ approaches 2 as x approaches 1 from either side. A missing point is denoted by the open dot on the graph. This conclusion is reinforced by the table. Be sure you see that *it does not matter that $f(x)$ is undefined when $x = 1$. The limit depends only on values of $f(x)$ near 1, not at 1.*

	x approaches 1.				*x* approaches 1.		
x	0.900	0.990	0.999	1.000	1.001	1.010	1.100
$f(x)$	1.900	1.990	1.999	?	2.001	2.010	2.100
	$f(x)$ approaches 2.				$f(x)$ approaches 2.		

b. From the graph of f, in Figure 1.50(b), you can see that $f(x) = -1$ for all values to the left of $x = 1$ and $f(x) = 1$ for all values to the right of $x = 1$. So, $f(x)$ is approaching a different value from the left of $x = 1$ than it is from the right of $x = 1$. In such situations, we say that *the limit does not exist.* This conclusion is reinforced by the table.

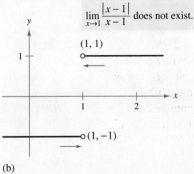

	x approaches 1.				*x* approaches 1.		
x	0.900	0.990	0.999	1.000	1.001	1.010	1.100
$f(x)$	−1.000	−1.000	−1.000	?	1.000	1.000	1.000
	$f(x)$ approaches -1.				$f(x)$ approaches 1.		

c. From the graph of f, in Figure 1.50(c), it appears that $f(x)$ approaches 1 as x approaches 1 from either side. This conclusion is reinforced by the table. It does not matter that $f(1) = 0$. The limit depends only on values of $f(x)$ near 1, not at 1.

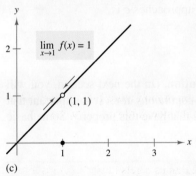

FIGURE 1.50

	x approaches 1.				*x* approaches 1.		
x	0.900	0.990	0.999	1.000	1.001	1.010	1.100
$f(x)$	0.900	0.990	0.999	?	1.001	1.010	1.100
	$f(x)$ approaches 1.				$f(x)$ approaches 1.		

✓ CHECKPOINT 2

Find the limit: $\lim_{x \to 2} f(x)$.

a. $f(x) = \dfrac{x^2 - 4}{x - 2}$

b. $f(x) = \dfrac{|x - 2|}{x - 2}$

c. $f(x) = \begin{cases} x^2, & x \neq 2 \\ 0, & x = 2 \end{cases}$

> **TECHNOLOGY**
>
> Try using a graphing utility to determine the following limit.
>
> $$\lim_{x \to 1} \frac{x^3 + 4x - 5}{x - 1}$$
>
> You can do this by graphing
>
> $$f(x) = \frac{x^3 + 4x - 5}{x - 1}$$
>
> and zooming in near $x = 1$. From the graph, what does the limit appear to be?

There are three important ideas to learn from Examples 1 and 2.

1. Saying that the limit of $f(x)$ approaches L as x approaches c means that the value of $f(x)$ may be made *arbitrarily close* to the number L by choosing x closer and closer to c.

2. For a limit to exist, you must allow x to approach c from *either side* of c. If $f(x)$ approaches a different number as x approaches c from the left than it does as x approaches c from the right, then the limit *does not exist*. [See Example 2(b).]

3. The value of $f(x)$ when $x = c$ has no bearing on the existence or nonexistence of the limit of $f(x)$ as x approaches c. For instance, in Example 2(a), the limit of $f(x)$ exists as x approaches 1 even though the function f is not defined at $x = 1$.

> **Definition of the Limit of a Function**
>
> If $f(x)$ becomes arbitrarily close to a single number L as x approaches c from either side, then
>
> $$\lim_{x \to c} f(x) = L$$
>
> which is read as "the **limit** of $f(x)$ as x approaches c is L."

Properties of Limits

Many times the limit of $f(x)$ as x approaches c is simply $f(c)$, as shown in Example 1. Whenever the limit of $f(x)$ as x approaches c is

$$\lim_{x \to c} f(x) = f(c) \qquad \text{Substitute } c \text{ for } x.$$

the limit can be evaluated by **direct substitution.** (In the next section, you will learn that a function that has this property is *continuous at c*.) It is important that you learn to recognize the types of functions that have this property. Some basic ones are given in the following list.

> **Properties of Limits**
>
> Let b and c be real numbers, and let n be a positive integer.
>
> 1. $\lim_{x \to c} b = b$
> 2. $\lim_{x \to c} x = c$
> 3. $\lim_{x \to c} x^n = c^n$
> 4. $\lim_{x \to c} \sqrt[n]{x} = \sqrt[n]{c}$
>
> In Property 4, if n is even, then c must be positive.

By combining the properties of limits with the rules for operating with limits shown below, you can find limits for a wide variety of algebraic functions.

TECHNOLOGY

Symbolic computer algebra systems are capable of evaluating limits. Try using a computer algebra system to evaluate the limit given in Example 3.

Operations with Limits

Let b and c be real numbers, let n be a positive integer, and let f and g be functions with the following limits.

$$\lim_{x \to c} f(x) = L \quad \text{and} \quad \lim_{x \to c} g(x) = K$$

1. Scalar multiple: $\lim_{x \to c} [bf(x)] = bL$
2. Sum or difference: $\lim_{x \to c} [f(x) \pm g(x)] = L \pm K$
3. Product: $\lim_{x \to c} [f(x) \cdot g(x)] = LK$
4. Quotient: $\lim_{x \to c} \dfrac{f(x)}{g(x)} = \dfrac{L}{K}$, provided $K \neq 0$
5. Power: $\lim_{x \to c} [f(x)]^n = L^n$
6. Radical: $\lim_{x \to c} \sqrt[n]{f(x)} = \sqrt[n]{L}$

In Property 6, if n is even, then L must be positive.

DISCOVERY

Use a graphing utility to graph $y_1 = 1/x^2$. Does y_1 approach a limit as x approaches 0? Evaluate $y_1 = 1/x^2$ at several positive and negative values of x near 0 to confirm your answer. Does $\lim_{x \to 1} 1/x^2$ exist?

Example 3 Finding the Limit of a Polynomial Function

Find the limit: $\lim_{x \to 2} (x^2 + 2x - 3)$.

$$\lim_{x \to 2} (x^2 + 2x - 3) = \lim_{x \to 2} x^2 + \lim_{x \to 2} 2x - \lim_{x \to 2} 3 \quad \text{Apply Property 2.}$$
$$= 2^2 + 2(2) - 3 \quad \text{Use direct substitution.}$$
$$= 4 + 4 - 3 \quad \text{Simplify.}$$
$$= 5$$

✓ **CHECKPOINT 3**

Find the limit: $\lim_{x \to 1} (2x^2 - x + 4)$. ■

Example 3 is an illustration of the following important result, which states that the limit of a polynomial function can be evaluated by direct substitution.

The Limit of a Polynomial Function

If p is a polynomial function and c is any real number, then

$$\lim_{x \to c} p(x) = p(c).$$

Techniques for Evaluating Limits

Many techniques for evaluating limits are based on the following important theorem. Basically, the theorem states that if two functions agree at all but a single point c, then they have identical limit behavior at $x = c$.

> **The Replacement Theorem**
>
> Let c be a real number and let $f(x) = g(x)$ for all $x \neq c$. If the limit of $g(x)$ exists as $x \to c$, then the limit of $f(x)$ also exists and
>
> $$\lim_{x \to c} f(x) = \lim_{x \to c} g(x).$$

To apply the Replacement Theorem, you can use a result from algebra which states that for a polynomial function p, $p(c) = 0$ if and only if $(x - c)$ is a factor of $p(x)$. This concept is demonstrated in Example 4.

Example 4 Finding the Limit of a Function

Find the limit: $\lim\limits_{x \to 1} \dfrac{x^3 - 1}{x - 1}$.

SOLUTION Note that the numerator and denominator are zero when $x = 1$. This implies that $x - 1$ is a factor of both, and you can divide out this like factor.

$$\dfrac{x^3 - 1}{x - 1} = \dfrac{(x - 1)(x^2 + x + 1)}{x - 1} \quad \text{Factor numerator.}$$

$$= \dfrac{\cancel{(x - 1)}(x^2 + x + 1)}{\cancel{x - 1}} \quad \text{Divide out like factor.}$$

$$= x^2 + x + 1, \quad x \neq 1 \quad \text{Simplify.}$$

So, the rational function $(x^3 - 1)/(x - 1)$ and the polynomial function $x^2 + x + 1$ agree for all values of x other than $x = 1$, and you can apply the Replacement Theorem.

$$\lim_{x \to 1} \dfrac{x^3 - 1}{x - 1} = \lim_{x \to 1} (x^2 + x + 1) = 1^2 + 1 + 1 = 3$$

Figure 1.51 illustrates this result graphically. Note that the two graphs are identical except that the graph of g contains the point $(1, 3)$, whereas this point is missing on the graph of f. (In the graph of f in Figure 1.51, the missing point is denoted by an open dot.)

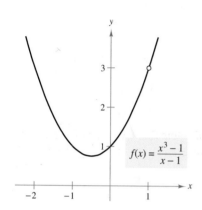

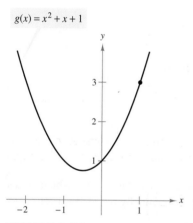

FIGURE 1.51

✓ CHECKPOINT 4

Find the limit: $\lim\limits_{x \to 2} \dfrac{x^3 - 8}{x - 2}$. ■

The technique used to evaluate the limit in Example 4 is called the **dividing out** technique. This technique is further demonstrated in the next example.

DISCOVERY

Use a graphing utility to graph

$$y = \frac{x^2 + x - 6}{x + 3}.$$

Is the graph a line? Why or why not?

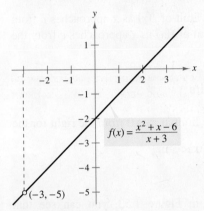

FIGURE 1.52 f is undefined when $x = -3$.

Example 5 Using the Dividing Out Technique

Find the limit: $\lim\limits_{x \to -3} \dfrac{x^2 + x - 6}{x + 3}$.

SOLUTION Direct substitution fails because both the numerator and the denominator are zero when $x = -3$.

$$\lim_{x \to -3} \frac{x^2 + x - 6}{x + 3} \qquad \begin{array}{l} \leftarrow \lim\limits_{x \to -3}(x^2 + x - 6) = 0 \\ \leftarrow \lim\limits_{x \to -3}(x + 3) = 0 \end{array}$$

However, because the limits of both the numerator and denominator are zero, you know that they have a *common factor* of $x + 3$. So, for all $x \ne -3$, you can divide out this factor to obtain the following.

$$\begin{aligned}
\lim_{x \to -3} \frac{x^2 + x - 6}{x + 3} &= \lim_{x \to -3} \frac{(x - 2)(x + 3)}{x + 3} && \text{Factor numerator.} \\
&= \lim_{x \to -3} \frac{(x - 2)\cancel{(x + 3)}}{\cancel{x + 3}} && \text{Divide out like factor.} \\
&= \lim_{x \to -3} (x - 2) && \text{Simplify.} \\
&= -5 && \text{Direct substitution}
\end{aligned}$$

This result is shown graphically in Figure 1.52. Note that the graph of f coincides with the graph of $g(x) = x - 2$, except that the graph of f has a hole at $(-3, -5)$.

✓ **CHECKPOINT 5**

Find the limit: $\lim\limits_{x \to 3} \dfrac{x^2 + x - 12}{x - 3}$. ∎

Example 6 Finding a Limit of a Function

Find the limit: $\lim\limits_{x \to 0} \dfrac{\sqrt{x + 1} - 1}{x}$.

SOLUTION Direct substitution fails because both the numerator and the denominator are zero when $x = 0$. In this case, you can rewrite the fraction by rationalizing the numerator.

$$\begin{aligned}
\frac{\sqrt{x + 1} - 1}{x} &= \left(\frac{\sqrt{x + 1} - 1}{x}\right)\left(\frac{\sqrt{x + 1} + 1}{\sqrt{x + 1} + 1}\right) \\
&= \frac{(x + 1) - 1}{x(\sqrt{x + 1} + 1)} \\
&= \frac{\cancel{x}}{\cancel{x}(\sqrt{x + 1} + 1)} = \frac{1}{\sqrt{x + 1} + 1}, \quad x \ne 0
\end{aligned}$$

Now, using the Replacement Theorem, you can evaluate the limit as shown.

$$\lim_{x \to 0} \frac{\sqrt{x + 1} - 1}{x} = \lim_{x \to 0} \frac{1}{\sqrt{x + 1} + 1} = \frac{1}{1 + 1} = \frac{1}{2}$$

STUDY TIP

When you try to evaluate a limit and both the numerator and denominator are zero, remember that you must rewrite the fraction so that the new denominator does not have 0 as its limit. One way to do this is to divide out like factors, as shown in Example 5. Another technique is to rationalize the numerator, as shown in Example 6.

✓ **CHECKPOINT 6**

Find the limit: $\lim\limits_{x \to 0} \dfrac{\sqrt{x + 4} - 2}{x}$. ∎

One-Sided Limits

In Example 2(b), you saw that one way in which a limit can fail to exist is when a function approaches a different value from the left of c than it approaches from the right of c. This type of behavior can be described more concisely with the concept of a **one-sided limit**.

$$\lim_{x \to c^-} f(x) = L \quad \text{Limit from the left}$$

$$\lim_{x \to c^+} f(x) = L \quad \text{Limit from the right}$$

The first of these two limits is read as "the limit of $f(x)$ as x approaches c from the left is L." The second is read as "the limit of $f(x)$ as x approaches c from the right is L."

Example 7 Finding One-Sided Limits

Find the limit as $x \to 0$ from the left and the limit as $x \to 0$ from the right for the function

$$f(x) = \frac{|2x|}{x}.$$

SOLUTION From the graph of f, shown in Figure 1.53, you can see that $f(x) = -2$ for all $x < 0$. So, the limit from the left is

$$\lim_{x \to 0^-} \frac{|2x|}{x} = -2. \quad \text{Limit from the left}$$

Because $f(x) = 2$ for all $x > 0$, the limit from the right is

$$\lim_{x \to 0^+} \frac{|2x|}{x} = 2. \quad \text{Limit from the right}$$

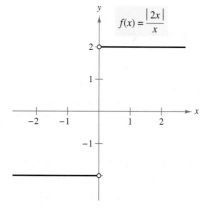

FIGURE 1.53

TECHNOLOGY

On most graphing utilities, the absolute value function is denoted by *abs*. You can verify the result in Example 7 by graphing

$$y = \frac{\text{abs}(2x)}{x}$$

in the viewing window $-3 \le x \le 3$ and $-3 \le y \le 3$.

✓ CHECKPOINT 7

Find each limit. (a) $\lim_{x \to 2^-} \frac{|x-2|}{x-2}$ (b) $\lim_{x \to 2^+} \frac{|x-2|}{x-2}$

In Example 7, note that the function approaches different limits from the left and from the right. In such cases, the limit of $f(x)$ as $x \to c$ does not exist. For the limit of a function to exist as $x \to c$, *both* one-sided limits must exist and must be equal.

Existence of a Limit

If f is a function and c and L are real numbers, then

$$\lim_{x \to c} f(x) = L$$

if and only if both the left and right limits are equal to L

SECTION 1.5 Limits

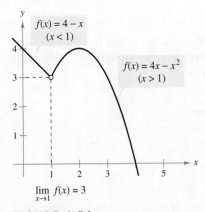

FIGURE 1.54

Example 8 Finding One-Sided Limits

Find the limit of $f(x)$ as x approaches 1.

$$f(x) = \begin{cases} 4 - x, & x < 1 \\ 4x - x^2, & x > 1 \end{cases}$$

SOLUTION Remember that you are concerned about the value of f near $x = 1$ rather than at $x = 1$. So, for $x < 1$, $f(x)$ is given by $4 - x$, and you can use direct substitution to obtain

$$\lim_{x \to 1^-} f(x) = \lim_{x \to 1^-} (4 - x)$$
$$= 4 - 1 = 3.$$

For $x > 1$, $f(x)$ is given by $4x - x^2$, and you can use direct substitution to obtain

$$\lim_{x \to 1^+} f(x) = \lim_{x \to 1^+} (4x - x^2)$$
$$= 4(1) - 1^2 = 4 - 1 = 3.$$

Because both one-sided limits exist and are equal to 3, it follows that

$$\lim_{x \to 1} f(x) = 3.$$

The graph in Figure 1.54 confirms this conclusion.

✓ CHECKPOINT 8

Find the limit of $f(x)$ as x approaches 0.

$$f(x) = \begin{cases} x^2 + 1, & x < 0 \\ 2x + 1, & x > 0 \end{cases}$$

Example 9 Recommended Dosage

The recommended dosage of a children's pain reliever is based on the child's age and is administered every 4 hours as needed. For infants up to 3 months, the dosage is 40 milligrams. For more than 3 months and up to 12 months, the dosage is 80 milligrams. For more than 12 months and up to 24 months, the dosage is 120 milligrams. Let x represent the age of the child (in months) and let $f(x)$ represent the dosage (in milligrams).

$$f(x) = \begin{cases} 40, & 0 < x \leq 3 \\ 80, & 3 < x \leq 12 \\ 120, & 12 < x \leq 24 \end{cases}$$

Show that the limit of $f(x)$ as $x \to 12$ does not exist.

SOLUTION The graph of f is shown in Figure 1.55. The limit of $f(x)$ as x approaches 12 from the left is

$$\lim_{x \to 12^-} f(x) = 80$$

whereas the limit of $f(x)$ as x approaches 12 from the right is

$$\lim_{x \to 12^+} f(x) = 120.$$

Because these one-sided limits are not equal, the limit of $f(x)$ as $x \to 12$ does not exist.

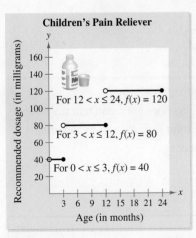

FIGURE 1.55

✓ CHECKPOINT 9

Show that the limit of $f(x)$ as $x \to 3$ does not exist in Example 9.

Unbounded Behavior

Example 9 shows a limit that fails to exist because the limits from the left and right differ. Another important way in which a limit can fail to exist is when $f(x)$ increases or decreases without bound as x approaches c.

Example 10 An Unbounded Function

Find the limit (if possible).

$$\lim_{x \to 2} \frac{3}{x - 2}$$

SOLUTION From Figure 1.56, you can see that $f(x)$ decreases without bound as x approaches 2 from the left and $f(x)$ increases without bound as x approaches 2 from the right. Symbolically, you can write this as

$$\lim_{x \to 2^-} \frac{3}{x - 2} = -\infty$$

and

$$\lim_{x \to 2^+} \frac{3}{x - 2} = \infty.$$

Because f is unbounded as x approaches 2, the limit does not exist.

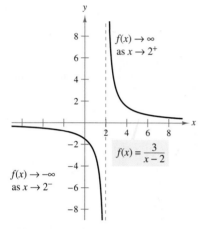

FIGURE 1.56

✓ CHECKPOINT 10

Find the limit (if possible): $\lim_{x \to -2} \dfrac{5}{x + 2}$.

> **STUDY TIP**
>
> The equal sign in the statement $\lim_{x \to c^+} f(x) = \infty$ does not mean that the limit exists. On the contrary, it tells you how the limit *fails to exist* by denoting the unbounded behavior of $f(x)$ as x approaches c.

CONCEPT CHECK

1. If $\lim_{x \to c^-} f(x) \neq \lim_{x \to c^+} f(x)$, what can you conclude about $\lim_{x \to c} f(x)$?

2. Describe how to find the limit of a polynomial function $p(x)$ as x approaches c.

3. Is the limit of $f(x)$ as x approaches c always equal to $f(c)$? Why or why not?

4. If f is undefined at $x = c$, can you conclude that the limit of $f(x)$ as x approaches c does not exist? Explain.

SECTION 1.5 Limits

Skills Review 1.5

The following warm-up exercises involve skills that were covered in earlier sections. You will use these skills in the exercise set for this section. For additional help, review Section 1.4.

In Exercises 1–4, evaluate the expression and simplify.

1. $f(x) = x^2 - 3x + 3$
   (a) $f(-1)$ (b) $f(c)$ (c) $f(x + h)$

2. $f(x) = \begin{cases} 2x - 2, & x < 1 \\ 3x + 1, & x \geq 1 \end{cases}$
   (a) $f(-1)$ (b) $f(3)$ (c) $f(t^2 + 1)$

3. $f(x) = x^2 - 2x + 2$ $\dfrac{f(1 + h) - f(1)}{h}$

4. $f(x) = 4x$ $\dfrac{f(2 + h) - f(2)}{h}$

In Exercises 5–8, find the domain and range of the function and sketch its graph.

5. $h(x) = -\dfrac{5}{x}$

6. $g(x) = \sqrt{25 - x^2}$

7. $f(x) = |x - 3|$

8. $f(x) = \dfrac{|x|}{x}$

In Exercises 9 and 10, determine whether y is a function of x.

9. $9x^2 + 4y^2 = 49$

10. $2x^2 y + 8x = 7y$

Exercises 1.5

See www.CalcChat.com for worked-out solutions to odd-numbered exercises.

In Exercises 1–8, complete the table and use the result to estimate the limit. Use a graphing utility to graph the function to confirm your result.

1. $\lim\limits_{x \to 2} (2x + 5)$

x	1.9	1.99	1.999	2	2.001	2.01	2.1
$f(x)$				?			

2. $\lim\limits_{x \to 2} (x^2 - 3x + 1)$

x	1.9	1.99	1.999	2	2.001	2.01	2.1
$f(x)$				?			

3. $\lim\limits_{x \to 2} \dfrac{x - 2}{x^2 - 4}$

x	1.9	1.99	1.999	2	2.001	2.01	2.1
$f(x)$				?			

4. $\lim\limits_{x \to 2} \dfrac{x - 2}{x^2 - 3x + 2}$

x	1.9	1.99	1.999	2	2.001	2.01	2.1
$f(x)$				?			

5. $\lim\limits_{x \to 0} \dfrac{\sqrt{x + 1} - 1}{x}$

x	−0.1	−0.01	−0.001	0	0.001	0.01	0.1
$f(x)$				?			

6. $\lim\limits_{x \to 0} \dfrac{\sqrt{x + 2} - \sqrt{2}}{x}$

x	−0.1	−0.01	−0.001	0	0.001	0.01	0.1
$f(x)$				?			

7. $\displaystyle\lim_{x\to 0^-} \frac{\frac{1}{x+4}-\frac{1}{4}}{x}$

x	-0.5	-0.1	-0.01	-0.001	0
$f(x)$					?

8. $\displaystyle\lim_{x\to 0^+} \frac{\frac{1}{2+x}-\frac{1}{2}}{2x}$

x	0.5	0.1	0.01	0.001	0
$f(x)$					?

In Exercises 9–12, use the graph to find the limit (if it exists).

9.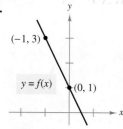

(a) $\displaystyle\lim_{x\to 0} f(x)$

(b) $\displaystyle\lim_{x\to -1} f(x)$

10.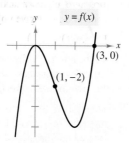

(a) $\displaystyle\lim_{x\to 1} f(x)$

(b) $\displaystyle\lim_{x\to 3} f(x)$

11.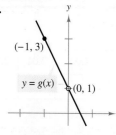

(a) $\displaystyle\lim_{x\to 0} g(x)$

(b) $\displaystyle\lim_{x\to -1} g(x)$

12.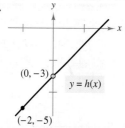

(a) $\displaystyle\lim_{x\to -2} h(x)$

(b) $\displaystyle\lim_{x\to 0} h(x)$

In Exercises 13 and 14, find the limit of (a) $f(x) + g(x)$, (b) $f(x)g(x)$, and (c) $f(x)/g(x)$, as x approaches c.

13. $\displaystyle\lim_{x\to c} f(x) = 3$

$\displaystyle\lim_{x\to c} g(x) = 9$

14. $\displaystyle\lim_{x\to c} f(x) = \frac{3}{2}$

$\displaystyle\lim_{x\to c} g(x) = \frac{1}{2}$

In Exercises 15 and 16, find the limit of (a) $\sqrt{f(x)}$, (b) $[3f(x)]$, and (c) $[f(x)]^2$, as x approaches c.

15. $\displaystyle\lim_{x\to c} f(x) = 16$

16. $\displaystyle\lim_{x\to c} f(x) = 9$

In Exercises 17–22, use the graph to find the limit (if it exists).

(a) $\displaystyle\lim_{x\to c^+} f(x)$ (b) $\displaystyle\lim_{x\to c^-} f(x)$ (c) $\displaystyle\lim_{x\to c} f(x)$

17.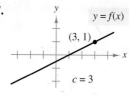

18.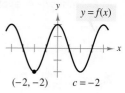

19.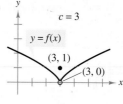

20.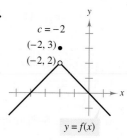

21.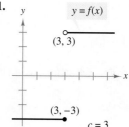

22.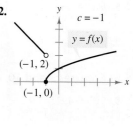

In Exercises 23–40, find the limit.

23. $\displaystyle\lim_{x\to 2} x^2$

24. $\displaystyle\lim_{x\to -2} x^3$

25. $\displaystyle\lim_{x\to -3} (2x+5)$

26. $\displaystyle\lim_{x\to 0} (3x-2)$

27. $\displaystyle\lim_{x\to 1} (1-x^2)$

28. $\displaystyle\lim_{x\to 2} (-x^2+x-2)$

29. $\displaystyle\lim_{x\to 3} \sqrt{x+6}$

30. $\displaystyle\lim_{x\to 4} \sqrt[3]{x+4}$

31. $\displaystyle\lim_{x\to -3} \frac{2}{x+2}$

32. $\displaystyle\lim_{x\to -2} \frac{3x+1}{2-x}$

33. $\displaystyle\lim_{x\to -2} \frac{x^2-1}{2x}$

34. $\displaystyle\lim_{x\to -1} \frac{4x-5}{3-x}$

35. $\displaystyle\lim_{x\to 7} \frac{5x}{x+2}$

36. $\displaystyle\lim_{x\to 3} \frac{\sqrt{x+1}}{x-4}$

37. $\displaystyle\lim_{x\to 3} \frac{\sqrt{x+1}-1}{x}$

38. $\displaystyle\lim_{x\to 5} \frac{\sqrt{x+4}-2}{x}$

39. $\displaystyle\lim_{x\to 1} \frac{\frac{1}{x+4}-\frac{1}{4}}{x}$

40. $\displaystyle\lim_{x\to 2} \frac{\frac{1}{x+2}-\frac{1}{2}}{x}$

In Exercises 41–60, find the limit (if it exists).

41. $\lim_{x \to 1} \dfrac{x^2 - 1}{x - 1}$

42. $\lim_{x \to -1} \dfrac{2x^2 - x - 3}{x + 1}$

43. $\lim_{x \to 2} \dfrac{x - 2}{x^2 - 4x + 4}$

44. $\lim_{x \to 2} \dfrac{2 - x}{x^2 - 4}$

45. $\lim_{t \to 4} \dfrac{t + 4}{t^2 - 16}$

46. $\lim_{t \to 1} \dfrac{t^2 + t - 2}{t^2 - 1}$

47. $\lim_{x \to -2} \dfrac{x^3 + 8}{x + 2}$

48. $\lim_{x \to -1} \dfrac{x^3 - 1}{x + 1}$

49. $\lim_{x \to -2} \dfrac{|x + 2|}{x + 2}$

50. $\lim_{x \to 2} \dfrac{|x - 2|}{x - 2}$

51. $\lim_{x \to 2} f(x)$, where $f(x) = \begin{cases} 4 - x, & x \neq 2 \\ 0, & x = 2 \end{cases}$

52. $\lim_{x \to 1} f(x)$, where $f(x) = \begin{cases} x^2 + 2, & x \neq 1 \\ 1, & x = 1 \end{cases}$

53. $\lim_{x \to 3} f(x)$, where $f(x) = \begin{cases} \frac{1}{3}x - 2, & x \leq 3 \\ -2x + 5, & x > 3 \end{cases}$

54. $\lim_{s \to 1} f(s)$, where $f(s) = \begin{cases} s, & s \leq 1 \\ 1 - s, & s > 1 \end{cases}$

55. $\lim_{\Delta x \to 0} \dfrac{2(x + \Delta x) - 2x}{\Delta x}$

56. $\lim_{\Delta x \to 0} \dfrac{4(x + \Delta x) - 5 - (4x - 5)}{\Delta x}$

57. $\lim_{\Delta x \to 0} \dfrac{\sqrt{x + 2 + \Delta x} - \sqrt{x + 2}}{\Delta x}$

58. $\lim_{\Delta x \to 0} \dfrac{\sqrt{x + \Delta x} - \sqrt{x}}{\Delta x}$

59. $\lim_{\Delta t \to 0} \dfrac{(t + \Delta t)^2 - 5(t + \Delta t) - (t^2 - 5t)}{\Delta t}$

60. $\lim_{\Delta t \to 0} \dfrac{(t + \Delta t)^2 - 4(t + \Delta t) + 2 - (t^2 - 4t + 2)}{\Delta t}$

(T) Graphical, Numerical, and Analytic Analysis In Exercises 61–64, use a graphing utility to graph the function and estimate the limit. Use a table to reinforce your conclusion. Then find the limit by analytic methods.

61. $\lim_{x \to 1^-} \dfrac{2}{x^2 - 1}$

62. $\lim_{x \to 1^+} \dfrac{5}{1 - x}$

63. $\lim_{x \to -2^-} \dfrac{1}{x + 2}$

64. $\lim_{x \to 0^-} \dfrac{x + 1}{x}$

(T) In Exercises 65–68, use a graphing utility to estimate the limit (if it exists).

65. $\lim_{x \to 2} \dfrac{x^2 - 5x + 6}{x^2 - 4x + 4}$

66. $\lim_{x \to 1} \dfrac{x^2 + 6x - 7}{x^3 - x^2 + 2x - 2}$

67. $\lim_{x \to -4} \dfrac{x^3 + 4x^2 + x + 4}{2x^2 + 7x - 4}$

68. $\lim_{x \to -2} \dfrac{4x^3 + 7x^2 + x + 6}{3x^2 - x - 14}$

69. **Medicine** Consider the function $F(t)$ given in Section 1.4, Exercise 72, for the temperature of a patient after being given a fever-reducing drug. Show that the limit of $F(t)$ as t approaches 4 is 98.6. Explain the significance of this limit.

70. **Environment** The cost (in dollars) of removing $p\%$ of the pollutants from the water in a small lake is given by

$$C = \dfrac{25,000p}{100 - p}, \quad 0 \leq p < 100$$

where C is the cost and p is the percent of pollutants.

(a) Find the cost of removing 50% of the pollutants.

(b) What percent of the pollutants can be removed for $100,000?

(c) Evaluate $\lim_{p \to 100^-} C$. Explain your results.

71. **Population Growth** The population of a species of mosquito in a wetland area is 5000. After 10 days, the population P is given by

$$P = 5000(1 + r)^{10}$$

where r is the daily percentage growth rate of the population (expressed as a decimal). Does the limit of P exist as the growth rate approaches 6%? If so, what is the limit?

72. **Rehabilitation Counseling** The number of patients $f(x)$ receiving counseling at a rehabilitation facility varies over a one-year period according to the function

$$f(x) = \begin{cases} 68, & 1 \leq x < 4 \\ 60, & 4 \leq x < 8 \\ 64, & 8 \leq x \leq 12 \end{cases}$$

where x is the month, with $x = 1$ corresponding to January. Find the following limits (if they exist).

(a) $\lim_{x \to 3} f(x)$ (b) $\lim_{x \to 4^+} f(x)$

(c) $\lim_{x \to 4^-} f(x)$ (d) $\lim_{x \to 4} f(x)$

73. The limit of

$$f(x) = (1 + x)^{1/x}$$

is a natural base for many real-life applications, as you will see in Section 4.2.

$$\lim_{x \to 0} (1 + x)^{1/x} = e \approx 2.718$$

(a) Show the reasonableness of this limit by completing the table.

x	-0.01	-0.001	-0.0001	0	0.0001	0.001	0.01
$f(x)$							

(T) (b) Use a graphing utility to graph f and to confirm the answer in part (a).

(c) Find the domain and range of the function.

Section 1.6
Continuity

- Determine the continuity of functions.
- Determine the continuity of functions on a closed interval.
- Use the greatest integer function to model and solve real-life problems.

Continuity

In mathematics, the term "continuous" has much the same meaning as it does in everyday use. To say that a function is continuous at $x = c$ means that there is no interruption in the graph of f at c. The graph of f is unbroken at c, and there are no holes, jumps, or gaps. As simple as this concept may seem, its precise definition eluded mathematicians for many years. In fact, it was not until the early 1800s that a precise definition was finally developed.

Before looking at this definition, consider the function whose graph is shown in Figure 1.57. This figure identifies three values of x at which the function f is not continuous.

1. At $x = c_1$, $f(c_1)$ is not defined.
2. At $x = c_2$, $\lim_{x \to c_2} f(x)$ does not exist.
3. At $x = c_3$, $f(c_3) \neq \lim_{x \to c_3} f(x)$.

At all other points in the interval (a, b), the graph of f is uninterrupted, which implies that the function f is continuous at all other points in the interval (a, b).

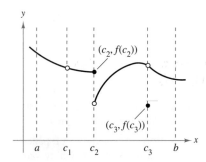

FIGURE 1.57 f is not continuous when $x = c_1, c_2, c_3$.

Definition of Continuity

Let c be a number in the interval (a, b), and let f be a function whose domain contains the interval (a, b). The function f is **continuous at the point c** if the following conditions are true.

1. $f(c)$ is defined.
2. $\lim_{x \to c} f(x)$ exists.
3. $\lim_{x \to c} f(x) = f(c)$.

If f is continuous at every point in the interval (a, b), then it is **continuous on an open interval (a, b)**.

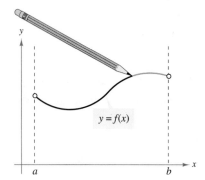

FIGURE 1.58 On the interval (a, b), the graph of f can be traced with a pencil.

Roughly, you can say that a function is continuous on an interval if its graph on the interval can be traced using a pencil and paper without lifting the pencil from the paper, as shown in Figure 1.58.

SECTION 1.6 Continuity

TECHNOLOGY

Most graphing utilities can draw graphs in two different modes: *connected mode* and *dot mode*. The *connected mode* works well as long as the function is continuous on the entire interval represented by the viewing window. If, however, the function is not continuous at one or more x-values in the viewing window, then the *connected mode* may try to "connect" parts of the graph that should not be connected. For instance, try graphing the function $y_1 = (x + 3)/(x - 2)$ on the viewing window $-8 \leq x \leq 8$ and $-6 \leq y \leq 6$. Do you notice any problems?

In Section 1.5, you studied several types of functions that meet the three conditions for continuity. Specifically, if *direct substitution* can be used to evaluate the limit of a function at c, then the function is continuous at c. Two types of functions that have this property are polynomial functions and rational functions.

Continuity of Polynomial and Rational Functions

1. A polynomial function is continuous at every real number.
2. A rational function is continuous at every number in its domain.

Example 1 Determining Continuity of Polynomial Functions

Discuss the continuity of each function.

a. $f(x) = x^2 - 2x + 3$

b. $f(x) = x^3 - x$

SOLUTION   Each of these functions is a *polynomial function*. So, each is continuous on the entire real line, as indicated in Figure 1.59.

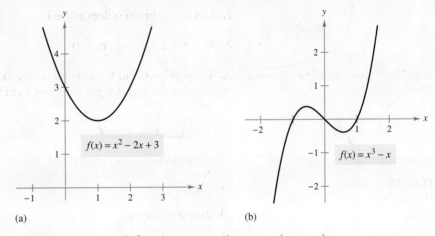

FIGURE 1.59   Both functions are continuous on $(-\infty, \infty)$.

✓ CHECKPOINT 1

Discuss the continuity of each function.

a. $f(x) = x^2 + x + 1$ **b.** $f(x) = x^3 + x$ ∎

Polynomial functions are one of the most important types of functions used in calculus. Be sure you see from Example 1 that the graph of a polynomial function is continuous on the entire real line, and therefore has no holes, jumps, or gaps. Rational functions, on the other hand, need not be continuous on the entire real line, as shown in Example 2.

STUDY TIP

A graphing utility can give misleading information about the continuity of a function. Graph the function

$$f(x) = \frac{x^3 + 8}{x + 2}$$

in the standard viewing window. Does the graph appear to be continuous? For what values of x is the function continuous?

Example 2 Determining Continuity of Rational Functions

Discuss the continuity of each function.

a. $f(x) = 1/x$ **b.** $f(x) = (x^2 - 1)/(x - 1)$ **c.** $f(x) = 1/(x^2 + 1)$

SOLUTION Each of these functions is a rational function and is therefore continuous at every number in its domain.

a. The domain of $f(x) = 1/x$ consists of all real numbers except $x = 0$. So, this function is continuous on the intervals $(-\infty, 0)$ and $(0, \infty)$. [See Figure 1.60(a).]

b. The domain of $f(x) = (x^2 - 1)/(x - 1)$ consists of all real numbers except $x = 1$. So, this function is continuous on the intervals $(-\infty, 1)$ and $(1, \infty)$. [See Figure 1.60(b).]

c. The domain of $f(x) = 1/(x^2 + 1)$ consists of all real numbers. So, this function is continuous on the entire real line. [See Figure 1.60(c).]

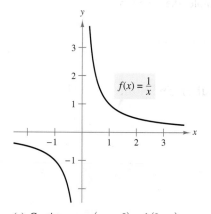
(a) Continuous on $(-\infty, 0)$ and $(0, \infty)$.

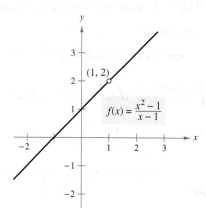
(b) Continuous on $(-\infty, 1)$ and $(1, \infty)$.

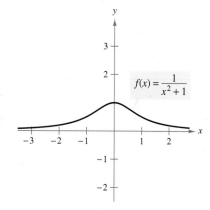
(c) Continuous on $(-\infty, \infty)$.

FIGURE 1.60

✓ CHECKPOINT 2

Discuss the continuity of each function.

a. $f(x) = \dfrac{1}{x - 1}$ **b.** $f(x) = \dfrac{x^2 - 4}{x - 2}$ **c.** $f(x) = \dfrac{1}{x^2 + 2}$ ∎

Consider an open interval I that contains a real number c. If a function f is defined on I (except possibly at c), and f is not continuous at c, then f is said to have a **discontinuity** at c. Discontinuities fall into two categories: **removable** and **nonremovable**. A discontinuity at c is called removable if f can be made continuous by appropriately defining (or redefining) $f(c)$. For instance, the function in Example 2(b) has a removable discontinuity at $(1, 2)$. To remove the discontinuity, all you need to do is redefine the function so that $f(1) = 2$.

A discontinuity at $x = c$ is nonremovable if the function cannot be made continuous at $x = c$ by defining or redefining the function at $x = c$. For instance, the function in Example 2(a) has a nonremovable discontinuity at $x = 0$.

Continuity on a Closed Interval

The intervals discussed in Examples 1 and 2 are open. To discuss continuity on a closed interval, you can use the concept of one-sided limits, as defined in Section 1.5.

Definition of Continuity on a Closed Interval

Let f be defined on a closed interval $[a, b]$. If f is continuous on the open interval (a, b) and

$$\lim_{x \to a^+} f(x) = f(a) \quad \text{and} \quad \lim_{x \to b^-} f(x) = f(b)$$

then f is **continuous on the closed interval** $[a, b]$. Moreover, f is **continuous from the right** at a and **continuous from the left** at b.

Similar definitions can be made to cover continuity on intervals of the form $(a, b]$ and $[a, b)$, or on infinite intervals. For example, the function

$$f(x) = \sqrt{x}$$

is continuous on the infinite interval $[0, \infty)$.

Example 3 Examining Continuity at an Endpoint

Discuss the continuity of

$$f(x) = \sqrt{3 - x}.$$

SOLUTION Notice that the domain of f is the set $(-\infty, 3]$. Moreover, f is continuous from the left at $x = 3$ because

$$\lim_{x \to 3^-} f(x) = \lim_{x \to 3^-} \sqrt{3 - x}$$
$$= 0$$
$$= f(3).$$

For all $x < 3$, the function f satisfies the three conditions for continuity. So, you can conclude that f is continuous on the interval $(-\infty, 3]$, as shown in Figure 1.61.

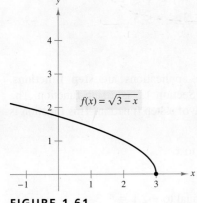

FIGURE 1.61

✓ CHECKPOINT 3

Discuss the continuity of $f(x) = \sqrt{x - 2}$. ∎

STUDY TIP

When working with radical functions of the form

$$f(x) = \sqrt{g(x)}$$

remember that the domain of f coincides with the solution of $g(x) \geq 0$.

Example 4 Examining Continuity on a Closed Interval

Discuss the continuity of $g(x) = \begin{cases} 5 - x, & -1 \leq x \leq 2 \\ x^2 - 1, & 2 < x \leq 3 \end{cases}$.

SOLUTION The polynomial functions $5 - x$ and $x^2 - 1$ are continuous on the intervals $[-1, 2]$ and $(2, 3]$, respectively. So, to conclude that g is continuous on the entire interval $[-1, 3]$, you only need to check the behavior of g when $x = 2$. You can do this by taking the one-sided limits when $x = 2$.

$$\lim_{x \to 2^-} g(x) = \lim_{x \to 2^-} (5 - x) = 3 \qquad \text{Limit from the left}$$

and

$$\lim_{x \to 2^+} g(x) = \lim_{x \to 2^+} (x^2 - 1) = 3 \qquad \text{Limit from the right}$$

Because these two limits are equal,

$$\lim_{x \to 2} g(x) = g(2) = 3.$$

So, g is continuous at $x = 2$ and, consequently, it is continuous on the entire interval $[-1, 3]$. The graph of g is shown in Figure 1.62.

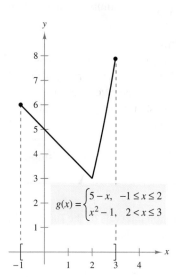

FIGURE 1.62

✓ CHECKPOINT 4

Discuss the continuity of $f(x) = \begin{cases} x + 2, & -1 \leq x < 3 \\ 14 - x^2, & 3 \leq x \leq 5 \end{cases}$. ■

The Greatest Integer Function

Some functions that are used in real-life applications are **step functions**. For instance, the function in Example 9 in Section 1.5 is a step function. The **greatest integer function** is another example of a step function. This function is denoted by

$[\![x]\!]$ = greatest integer less than or equal to x.

For example,

$[\![-2.1]\!]$ = greatest integer less than or equal to $-2.1 = -3$
$[\![-2]\!]$ = greatest integer less than or equal to $-2 = -2$
$[\![1.5]\!]$ = greatest integer less than or equal to $1.5 = 1$.

Note that the graph of the greatest integer function (Figure 1.63) jumps up one unit at each integer. This implies that the function is not continuous at each integer.

In real-life applications, the domain of the greatest integer function is often restricted to nonnegative values of x. In such cases this function serves the purpose of **truncating** the decimal portion of x. For example, 1.345 is truncated to 1 and 3.57 is truncated to 3. That is,

$[\![1.345]\!] = 1$ and $[\![3.57]\!] = 3.$

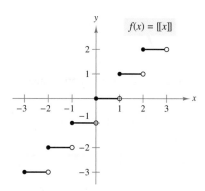

FIGURE 1.63 Greatest Integer Function

TECHNOLOGY

Use a graphing utility to calculate the following.

a. $[\![3.5]\!]$ **b.** $[\![-3.5]\!]$ **c.** $[\![0]\!]$

TECHNOLOGY

Step Functions and Compound Functions

To graph a step function or compound function with a graphing utility, you must be familiar with the utility's programming language. For instance, different graphing utilities have different "integer truncation" functions. One is IPart(x), and it yields the truncated integer part of x. For example, IPart(-1.2) = -1 and IPart(3.4) = 3. Another function is Int(x), which is the greatest integer function. The graphs of these two functions are shown below. When graphing a step function, you should set your graphing utility to *dot mode*.

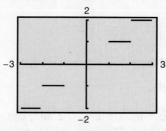

Graph of $f(x) = $ IPart(x)

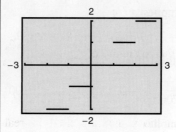

Graph of $f(x) = $ Int(x)

On some graphing utilities, you can graph a piecewise-defined function such as

$$f(x) = \begin{cases} x^2 - 4, & x \leq 2 \\ -x + 2, & 2 < x \end{cases}.$$

The graph of this function is shown below.

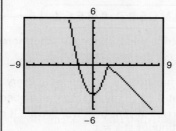

Consult the user's guide for your graphing utility for specific keystrokes you can use to graph these functions.

Example 5 Recycling Costs

A county recycling program purchases a fluorescent bulb eater machine for $2750. The disposal of 700 pounds of recycled material costs $290. The total cost C (in dollars) of removing up to 2100 pounds can be modeled by

$$C = 2750 + 290\left(1 + \left[\!\left[\frac{x-1}{700}\right]\!\right]\right), \quad 0 < x \leq 2100$$

where x is the weight (to the nearest pound) of the bulbs recycled. Sketch a graph of this function.

SOLUTION For the first 700 pounds recycled

$$\left[\!\left[\frac{x-1}{700}\right]\!\right] = 0, \quad 0 < x \leq 700$$

which implies

$$C = 2750 + 290(1 + 0) = 3040.$$

For the second 700 pounds recycled

$$\left[\!\left[\frac{x-1}{700}\right]\!\right] = 1, \quad 700 < x \leq 1400$$

which implies

$$C = 2750 + 290(1 + 1) = 3330.$$

For the third 700 pounds recycled

$$\left[\!\left[\frac{x-1}{700}\right]\!\right] = 2, \quad 1400 < x \leq 2100$$

which implies

$$C = 2750 + 290(1 + 2) = 3620.$$

The graph of C is shown in Figure 1.64. Note the graph's discontinuities.

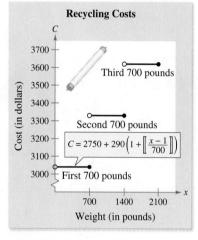

FIGURE 1.64

✓ CHECKPOINT 5

Use a graphing utility to graph the function in Example 5. ■

CONCEPT CHECK

1. Describe the continuity of a polynomial function.
2. Describe the continuity of a rational function.
3. If a function f is continuous at every point in the interval (a, b), then what can you say about f on an open interval (a, b)?
4. Describe in your own words what it means to say that a function f is continuous at $x = c$.

SECTION 1.6 Continuity

Skills Review 1.6

The following warm-up exercises involve skills that were covered in earlier sections. You will use these skills in the exercise set for this section. For additional help, review Sections 0.4, 0.5, and 1.5.

In Exercises 1–4, simplify the expression.

1. $\dfrac{x^2 + 6x + 8}{x^2 - 6x - 16}$

2. $\dfrac{x^2 - 5x - 6}{x^2 - 9x + 18}$

3. $\dfrac{2x^2 - 2x - 12}{4x^2 - 24x + 36}$

4. $\dfrac{x^3 - 16x}{x^3 + 2x^2 - 8x}$

In Exercises 5–8, solve for x.

5. $x^2 + 7x = 0$

6. $x^2 + 4x - 5 = 0$

7. $3x^2 + 8x + 4 = 0$

8. $x^3 + 5x^2 - 24x = 0$

In Exercises 9 and 10, find the limit.

9. $\lim\limits_{x \to 3} (2x^2 - 3x + 4)$

10. $\lim\limits_{x \to -2} (3x^3 - 8x + 7)$

Exercises 1.6

See www.CalcChat.com for worked-out solutions to odd-numbered exercises.

In Exercises 1–10, determine whether the function is continuous on the entire real line. Explain your reasoning.

1. $f(x) = 5x^3 - x^2 + 2$

2. $f(x) = (x^2 - 1)^3$

3. $f(x) = \dfrac{1}{x^2 - 4}$

4. $f(x) = \dfrac{1}{9 - x^2}$

5. $f(x) = \dfrac{1}{4 + x^2}$

6. $f(x) = \dfrac{3x}{x^2 + 1}$

7. $f(x) = \dfrac{2x - 1}{x^2 - 8x + 15}$

8. $f(x) = \dfrac{x + 4}{x^2 - 6x + 5}$

9. $g(x) = \dfrac{x^2 - 4x + 4}{x^2 - 4}$

10. $g(x) = \dfrac{x^2 - 9x + 20}{x^2 - 16}$

In Exercises 11–34, describe the interval(s) on which the function is continuous. Explain why the function is continuous on the interval(s). If the function has a discontinuity, identify the conditions of continuity that are not satisfied.

11. $f(x) = \dfrac{x^2 - 1}{x}$

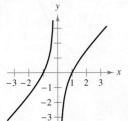

12. $f(x) = \dfrac{1}{x^2 - 4}$

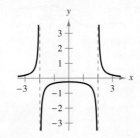

13. $f(x) = \dfrac{x^2 - 1}{x + 1}$

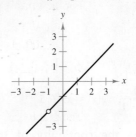

14. $f(x) = \dfrac{x^3 - 8}{x - 2}$

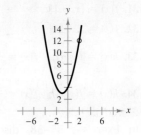

15. $f(x) = x^2 - 2x + 1$

16. $f(x) = 3 - 2x - x^2$

17. $f(x) = \dfrac{x}{x^2 - 1}$

18. $f(x) = \dfrac{x - 3}{x^2 - 9}$

19. $f(x) = \dfrac{x}{x^2 + 1}$

20. $f(x) = \dfrac{1}{x^2 + 1}$

21. $f(x) = \dfrac{x - 5}{x^2 - 9x + 20}$

22. $f(x) = \dfrac{x - 1}{x^2 + x - 2}$

23. $f(x) = [\![2x]\!] + 1$ **24.** $f(x) = \dfrac{[\![x]\!]}{2} + x$

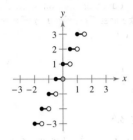
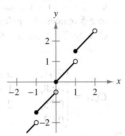

25. $f(x) = \begin{cases} -2x + 3, & x < 1 \\ x^2, & x \geq 1 \end{cases}$

26. $f(x) = \begin{cases} 3 + x, & x \leq 2 \\ x^2 + 1, & x > 2 \end{cases}$

27. $f(x) = \begin{cases} \frac{1}{2}x + 1, & x \leq 2 \\ 3 - x, & x > 2 \end{cases}$

28. $f(x) = \begin{cases} x^2 - 4, & x \leq 0 \\ 3x + 1, & x > 0 \end{cases}$

29. $f(x) = \dfrac{|x + 1|}{x + 1}$

30. $f(x) = \dfrac{|4 - x|}{4 - x}$

31. $f(x) = [\![x - 1]\!]$

32. $f(x) = x - [\![x]\!]$

33. $h(x) = f(g(x))$, $f(x) = \dfrac{1}{\sqrt{x}}$, $g(x) = x - 1, x > 1$

34. $h(x) = f(g(x))$, $f(x) = \dfrac{1}{x - 1}$, $g(x) = x^2 + 5$

In Exercises 35–38, discuss the continuity of the function on the closed interval. If there are any discontinuities, determine whether they are removable.

Function	Interval
35. $f(x) = x^2 - 4x - 5$	$[-1, 5]$
36. $f(x) = \dfrac{5}{x^2 + 1}$	$[-2, 2]$
37. $f(x) = \dfrac{1}{x - 2}$	$[1, 4]$
38. $f(x) = \dfrac{x}{x^2 - 4x + 3}$	$[0, 4]$

In Exercises 39–44, sketch the graph of the function and describe the interval(s) on which the function is continuous.

39. $f(x) = \dfrac{x^2 - 16}{x - 4}$ **40.** $f(x) = \dfrac{2x^2 + x}{x}$

41. $f(x) = \dfrac{x^3 + x}{x}$

42. $f(x) = \dfrac{x - 3}{4x^2 - 12x}$

43. $f(x) = \begin{cases} x^2 + 1, & x < 0 \\ x - 1, & x \geq 0 \end{cases}$

44. $f(x) = \begin{cases} x^2 - 4, & x \leq 0 \\ 2x + 4, & x > 0 \end{cases}$

In Exercises 45 and 46, find the constant a (Exercise 45) and the constants a and b (Exercise 46) such that the function is continuous on the entire real line.

45. $f(x) = \begin{cases} x^3, & x \leq 2 \\ ax^2, & x > 2 \end{cases}$

46. $f(x) = \begin{cases} 2, & x \leq -1 \\ ax + b, & -1 < x < 3 \\ -2, & x \geq 3 \end{cases}$

(T) In Exercises 47–52, use a graphing utility to graph the function. Use the graph to determine any *x*-value(s) at which the function is not continuous. Explain why the function is not continuous at the *x*-value(s).

47. $h(x) = \dfrac{1}{x^2 - x - 2}$

48. $k(x) = \dfrac{x - 4}{x^2 - 5x + 4}$

49. $f(x) = \begin{cases} 2x - 4, & x \leq 3 \\ x^2 - 2x, & x > 3 \end{cases}$

50. $f(x) = \begin{cases} 3x - 1, & x \leq 1 \\ x + 1, & x > 1 \end{cases}$

51. $f(x) = x - 2[\![x]\!]$

52. $f(x) = [\![2x - 1]\!]$

In Exercises 53–56, describe the interval(s) on which the function is continuous.

53. $f(x) = \dfrac{x}{x^2 + 1}$ **54.** $f(x) = x\sqrt{x + 3}$

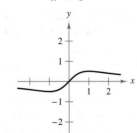
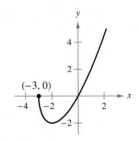

55. $f(x) = \frac{1}{2}[\![2x]\!]$ 56. $f(x) = \frac{x+1}{\sqrt{x}}$

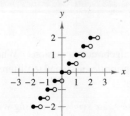

Writing In Exercises 57 and 58, use a graphing utility to graph the function on the interval $[-4, 4]$. Does the graph of the function appear to be continuous on this interval? Is the function in fact continuous on $[-4, 4]$? Write a short paragraph about the importance of examining a function analytically as well as graphically.

57. $f(x) = \frac{x^2 + x}{x}$ 58. $f(x) = \frac{x^3 - 8}{x - 2}$

59. **Health Club Fees** A health club charges a yearly fee of $225 for a student membership. To offset the cost of implementing its new Yoga, Pilates, and flexibility programs, all membership fees will increase 2% compounded yearly. The amount A paid for a student membership after t years is

$A = 225(1.02)^{[\![t]\!]}, \quad t \geq 0.$

(a) Sketch the graph of A. Is the graph continuous? Explain your reasoning.

(b) What is the cost of a student membership after 18 months?

60. **Environmental Cost** The cost C (in millions of dollars) of removing x percent of the pollutants emitted from the smokestack of a factory can be modeled by

$C = \frac{2x}{100 - x}.$

(a) What is the implied domain of C? Explain your reasoning.

(b) Use a graphing utility to graph the cost function. Is the function continuous on its domain? Explain your reasoning.

(c) Find the cost of removing 75% of the pollutants from the smokestack.

61. **Health Food** A co-op health food store charges $3.50 for the first pound of organically grown peanuts and $1.90 for each additional pound or fraction thereof. Use the greatest integer function to create a model for the charge C for x pounds of organically grown peanuts. Use a graphing utility to graph the function, and discuss its continuity.

62. **Consumer Awareness** The United States Postal Service first class mail rates are $0.41 for the first ounce and $0.17 for each additional ounce or fraction thereof up to 3.5 ounces. A model for the cost C (in dollars) of a first class mailing that weighs 3.5 ounces or less is given below. *(Source: United States Postal Service)*

$C(x) = \begin{cases} 0.41, & 0 \leq x \leq 1 \\ 0.58, & 1 < x \leq 2 \\ 0.75, & 2 < x \leq 3 \\ 0.92, & 3 < x \leq 3.5 \end{cases}$

(a) Use a graphing utility to graph the function and discuss its continuity. At what values is the function not continuous? Explain your reasoning.

(b) Find the cost of mailing a 2.5-ounce letter.

63. **Salary Contract** A union contract guarantees a 9% yearly increase for 5 years. For a current salary of $28,500, the salaries for the next 5 years are given by

$S = 28,500(1.09)^{[\![t]\!]}$

where $t = 0$ represents the present year.

(a) Use the greatest integer function of a graphing utility to graph the salary function, and discuss its continuity.

(b) Find the salary during the fifth year (when $t = 5$).

64. **Sporting Goods** The number of fishing poles in inventory at a small company is

$N = 25\left(2\left[\!\left[\frac{t+2}{2}\right]\!\right] - t\right), \quad 0 \leq t \leq 12$

where the real number t is the time in months.

(a) Use the greatest integer function of a graphing utility to graph this function, and discuss its continuity.

(b) How often must the company replenish its inventory?

65. **Population Growth** A biologist determined a linear model for the growth of a deer population as a function of time. Is the model a continuous function? Would the actual growth of the population be a continuous function of time? Explain your reasoning.

66. **Biology** The gestation period of rabbits is about 29 to 35 days. Therefore, the population of a form (rabbits' home) can increase dramatically in a short period of time. The table gives the populations of a form, where t is the time in months and N is the rabbit population.

t	0	1	2	3	4	5	6
N	2	8	10	14	10	15	12

Graph the population as a function of time. Find any points of discontinuity in the function. Explain your reasoning.

Algebra Review

Order of Operations

Much of the algebra in this chapter involves evaluation of algebraic expressions. When you evaluate an algebraic expression, you need to know the priorities assigned to different operations. These priorities are called the *order of operations*.

1. Perform operations inside *symbols of grouping or absolute value symbols*, starting with the innermost symbol.
2. Evaluate all *exponential* expressions.
3. Perform all *multiplications* and *divisions* from left to right.
4. Perform all *additions* and *subtractions* from left to right.

Example 1 Using Order of Operations

Evaluate each expression.

a. $7 - [(5 \cdot 3) + 2^3]$

b. $[36 \div (3^2 \cdot 2)] + 6$

c. $36 - [3^2 \cdot (2 \div 6)]$

d. $10 - 2(8 + |5 - 7|)$

SOLUTION

a.
$$7 - [(5 \cdot 3) + 2^3] = 7 - [15 + 2^3] \quad \text{Multiply inside parentheses.}$$
$$= 7 - [15 + 8] \quad \text{Evaluate exponential expression.}$$
$$= 7 - 23 \quad \text{Add inside brackets.}$$
$$= -16 \quad \text{Subtract.}$$

b.
$$[36 \div (3^2 \cdot 2)] + 6 = [36 \div (9 \cdot 2)] + 6 \quad \text{Evaluate exponential expression inside parentheses.}$$
$$= [36 \div 18] + 6 \quad \text{Multiply inside parentheses.}$$
$$= 2 + 6 \quad \text{Divide inside brackets.}$$
$$= 8 \quad \text{Add.}$$

c.
$$36 - [3^2 \cdot (2 \div 6)] = 36 - [3^2 \cdot \tfrac{1}{3}] \quad \text{Divide inside parentheses.}$$
$$= 36 - [9 \cdot \tfrac{1}{3}] \quad \text{Evaluate exponential expression.}$$
$$= 36 - 3 \quad \text{Multiply inside brackets.}$$
$$= 33 \quad \text{Subtract.}$$

d.
$$10 - 2(8 + |5 - 7|) = 10 - 2(8 + |-2|) \quad \text{Subtract inside absolute value symbols.}$$
$$= 10 - 2(8 + 2) \quad \text{Evaluate absolute value.}$$
$$= 10 - 2(10) \quad \text{Add inside parentheses.}$$
$$= 10 - 20 \quad \text{Multiply.}$$
$$= -10 \quad \text{Subtract.}$$

TECHNOLOGY

Most scientific and graphing calculators use the same order of operations listed above. Try entering the expressions in Example 1 into your calculator. Do you get the same results?

Algebra Review

Solving Equations

A second algebraic skill that is central to this chapter is solving an equation in one variable.

1. To solve a *linear equation*, you can add or subtract the same quantity from each side of the equation. You can also multiply or divide each side of the equation by the same *nonzero* quantity.
2. To solve a *quadratic equation*, you can take the square root of each side, use factoring, or use the Quadratic Formula.
3. To solve a *radical equation*, isolate the radical on one side of the equation and square each side of the equation.
4. To solve an *absolute value equation*, use the definition of absolute value to rewrite the equation as two equations.

TECHNOLOGY

The equations in Example 2 are solved algebraically. Most graphing utilities have a "solve" key that allows you to solve equations graphically. If you have a graphing utility, try using it to solve graphically the equations in Example 2.

STUDY TIP

You should be aware that solving radical equations can sometimes lead to *extraneous solutions* (those that do not satisfy the original equation). For example, squaring both sides of the following equation yields two possible solutions, one of which is extraneous.

$\sqrt{x} = x - 2$
$x = x^2 - 4x + 4$
$0 = x^2 - 5x + 4$
$ = (x - 4)(x - 1)$
$x - 4 = 0 \implies x = 4$ (solution)
$x - 1 = 0 \implies x = 1$ (extraneous)

Example 2 — Solving Equations

Solve each equation.

a. $3x - 3 = 5x - 7$
b. $2x^2 = 10$
c. $2x^2 + 5x - 6 = 6$
d. $\sqrt{2x - 7} = 5$

SOLUTION

a. $3x - 3 = 5x - 7$ Write original (linear) equation.
$ -3 = 2x - 7$ Subtract $3x$ from each side.
$ 4 = 2x$ Add 7 to each side.
$ 2 = x$ Divide each side by 2.

b. $2x^2 = 10$ Write original (quadratic) equation.
$x^2 = 5$ Divide each side by 2.
$x = \pm\sqrt{5}$ Take the square root of each side.

c. $2x^2 + 5x - 6 = 6$ Write original (quadratic) equation.
$2x^2 + 5x - 12 = 0$ Write in general form.
$(2x - 3)(x + 4) = 0$ Factor.

$2x - 3 = 0 \implies x = \frac{3}{2}$ Set first factor equal to zero.
$x + 4 = 0 \implies x = -4$ Set second factor equal to zero.

d. $\sqrt{2x - 7} = 5$ Write original (radical) equation.
$2x - 7 = 25$ Square each side.
$2x = 32$ Add 7 to each side.
$x = 16$ Divide each side by 2.

Chapter Summary and Study Strategies

After studying this chapter, you should have acquired the following skills. The exercise numbers are keyed to the Review Exercises that begin on page 108. Answers to odd-numbered Review Exercises are given in the back of the text.*

Section 1.1   Review Exercises

- Plot points in a coordinate plane.   1–4
- Read data presented graphically.   5–8
- Find the distance between two points in a coordinate plane.   9–12

  $d = \sqrt{(x_2 - x_1)^2 + (y_2 - y_1)^2}$

- Find the midpoints of line segments connecting two points.   13–16

  $\text{Midpoint} = \left(\dfrac{x_1 + x_2}{2}, \dfrac{y_1 + y_2}{2} \right)$

- Interpret real-life data that are presented graphically.   17, 18
- Translate points in a coordinate plane.   19, 20

Section 1.2

- Sketch graphs of equations by hand.   21–30
- Find the x- and y-intercepts of graphs of equations algebraically and graphically using a graphing utility.   31, 32
- Write the standard forms of equations of circles, given the center and a point on the circle.   33, 34

  $(x - h)^2 + (y - k)^2 = r^2$

- Convert equations of circles from general form to standard form by completing the square, and sketch the circles.   35, 36
- Find the points of intersection of two graphs algebraically and graphically using a graphing utility.   37–40
- Solve real-life problems by finding the points of intersection of two graphs.   41
- Use mathematical modeling to solve real-life problems.   42

Section 1.3

- Use the slope-intercept form of a linear equation to sketch graphs of lines.   43–48

  $y = mx + b$

---

* Use a wide range of valuable study aids to help you master the material in this chapter. The *Student Solutions Guide* includes step-by-step solutions to all odd-numbered exercises to help you review and prepare. The student website at *college.hmco.com/info/larsonapplied* offers algebra help and a *Graphing Technology Guide*. The *Graphing Technology Guide* contains step-by-step commands and instructions for a wide variety of graphing calculators, including the most recent models.

Chapter Summary and Study Strategies

Section 1.3 (continued)   Review Exercises

- Find slopes of lines passing through two points.   49–52

$$m = \frac{y_2 - y_1}{x_2 - x_1}$$

- Use the point-slope form to write equations of lines and graph equations using a graphing utility.   53–56

$$y - y_1 = m(x - x_1)$$

- Find equations of parallel and perpendicular lines.   57, 58

Parallel lines: $m_1 = m_2$ Perpendicular lines: $m_1 = -\dfrac{1}{m_2}$

- Use linear equations to solve real-life problems.   59, 60

Section 1.4

- Use the Vertical Line Test to decide whether equations define functions.   61–64
- Use function notation to evaluate functions.   65, 66
- Use a graphing utility to graph functions and find the domains and ranges of functions.   67–72
- Combine functions to create other functions.   73, 74
- Use the Horizontal Line Test to determine whether functions have inverse functions. If they do, find the inverse functions.   75–78

Section 1.5

- Determine whether limits exist. If they do, find the limits.   79–96
- Use a table to estimate one-sided limits.   97, 98
- Determine whether statements about limits are true or false.   99–104

Section 1.6

- Determine whether functions are continuous at a point, on an open interval, and on a closed interval.   105–112
- Determine the value of an unknown constant in a function such that the function is continuous.   113, 114
- Use analytic and graphical models of real-life data to solve real-life problems.   115–119

Study Strategies

- **Use a Graphing Utility** A graphing calculator or graphing software for a computer can help you in this course in two important ways. As an *exploratory device*, a graphing utility allows you to learn concepts by allowing you to compare graphs of equations. For instance, sketching the graphs of $y = x^2$, $y = x^2 + 1$, and $y = x^2 - 1$ helps confirm that adding (or subtracting) a constant to (or from) a function shifts the graph of the function vertically. As a *problem-solving tool*, a graphing utility frees you of some of the drudgery of sketching complicated graphs by hand. The time that you save can be spent using mathematics to solve real-life problems.

- **Use the Skills Review Exercises** Each exercise set in this text begins with a set of skills review exercises. We urge you to begin each homework session by quickly working all of these exercises (all are answered in the back of the text). The "old" skills covered in these exercises are needed to master the "new" skills in the section exercise set. The skills review exercises remind you that mathematics is cumulative—to be successful in this course, you must retain "old" skills.

- **Use the Additional Study Aids** The additional study aids were prepared specifically to help you master the concepts discussed in the text. They are the *Student Solutions Guide*, the student website, and the *Graphing Technology Guide*.

Review Exercises

See www.CalcChat.com for worked-out solutions to odd-numbered exercises.

In Exercises 1–4, plot the points

1. $(2, 3), (0, 6)$
2. $(-5, 1), (4, -3)$
3. $(0.5, -4), (-1, -2)$
4. $(-1.5, 0), (6, -5)$

In Exercises 5–8, match the data with the real-life situation that they represent. [The graphs are labeled (a)–(d).]

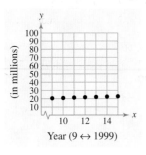

(a)

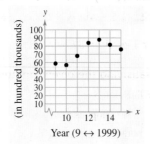

(b)

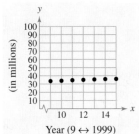

(c)

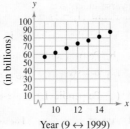

(d)

5. Population of Texas
6. Population of California
7. Number of unemployed workers in the United States
8. National health expenditures for dental services

In Exercises 9–12, find the distance between the two points.

9. $(0, 0), (5, 2)$
10. $(1, 2), (4, 3)$
11. $(-1, 3), (-4, 6)$
12. $(6, 8), (-3, 7)$

In Exercises 13–16, find the midpoint of the line segment connecting the two points.

13. $(5, 6), (9, 2)$
14. $(0, 0), (-4, 8)$
15. $(-10, 4), (-6, 8)$
16. $(7, -9), (-3, 5)$

In Exercises 17 and 18, use the graph below showing the numbers of children enrolled in Head Start for the years 1995 through 2005. (Head Start is a federal program that focuses on assisting children from low-income families.) *(Source: U.S. Administration for Children and Families)*

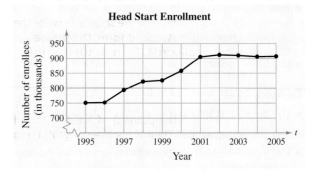

17. Estimate the number of children (in thousands) enrolled in Head Start for each year.

    (a) 1996 (b) 1999
    (c) 2002 (d) 2005

18. Estimate the percent increase or decrease in the number of children (in thousands) enrolled in Head Start.

    (a) from 1995 to 1999
    (b) from 2000 to 2005

19. Translate the triangle whose vertices are $(1, 3), (2, 4)$, and $(5, 6)$ three units to the right and four units up. Find the coordinates of the translated vertices.

20. Translate the rectangle whose vertices are $(-2, 1), (-1, 2), (1, 0)$, and $(0, -1)$ four units to the right and one unit down.

In Exercises 21–30, sketch the graph of the equation.

21. $y = 4x - 12$
22. $y = 4 - 3x$
23. $y = x^2 + 5$
24. $y = 1 - x^2$
25. $y = |4 - x|$
26. $y = |2x - 3|$
27. $y = x^3 + 4$
28. $y = 2x^3 - 1$
29. $y = \sqrt{4x + 1}$
30. $y = \sqrt{2x}$

In Exercises 31 and 32, find the *x*- and *y*-intercepts of the graph of the equation algebraically. Use a graphing utility to verify your results.

31. $4x + y + 3 = 0$
32. $y = (x - 1)^3 + 2(x - 1)^2$

In Exercises 33 and 34, write the standard form of the equation of the circle.

33. Center: $(0, 0)$
    Solution point: $(2, \sqrt{5})$
34. Center: $(2, -1)$
    Solution point: $(-1, 7)$

In Exercises 35 and 36, complete the square to write the equation of the circle in standard form. Determine the radius and center of the circle. Then sketch the circle.

35. $x^2 + y^2 - 6x + 8y = 0$
36. $x^2 + y^2 + 10x + 4y - 7 = 0$

In Exercises 37–40, find the point(s) of intersection of the graphs algebraically. Then use a graphing utility to verify your results.

37. $2x - 3y = 13$, $5x + 3y = 1$
38. $x^2 + y^2 = 5$, $x - y = 1$
39. $y = x^3$, $y = x$
40. $x^2 + y = 4$, $2x - y = 1$

41. **State Populations** The populations *P* (in thousands) of Kentucky and South Carolina from 2000 through 2006 can be modeled by

    $P = 4042 + 26.1t$ Kentucky
    $P = 4010 + 48.5t$ South Carolina

    where *t* represents the year, with $t = 0$ corresponding to 2000. *(Source: U.S. Census Bureau)*

    (a) According to these two models, when would you expect the population of South Carolina to have exceeded the population of Kentucky?

    (b) Use the two models to estimate the populations of both states in 2010.

42. **Fish Population** A lake is stocked with 700 fish. The fish population *P* can be modeled by $P = 175t + 700$, where *t* is the number of months since the lake was stocked.

    (a) Describe what the slope and *y*-intercept mean in this situation. Then determine the fish population after 3 months.

    (b) Is a linear model accurate for long-term analysis of the fish population? Explain.

In Exercises 43–48, find the slope and *y*-intercept (if possible) of the linear equation. Then sketch the graph of the equation.

43. $3x + y = -2$
44. $-\frac{1}{3}x + \frac{5}{6}y = 1$
45. $y = -\frac{5}{3}$
46. $x = -3$
47. $-2x - 5y - 5 = 0$
48. $3.2x - 0.8y + 5.6 = 0$

In Exercises 49–52, find the slope of the line passing through the two points.

49. $(0, 0), (7, 6)$
50. $(-1, 5), (-5, 7)$
51. $(10, 17), (-11, -3)$
52. $(-11, -3), (-1, -3)$

In Exercises 53–56, find an equation of the line that passes through the point and has the given slope. Then use a graphing utility to graph the line.

Point	Slope	Point	Slope
53. $(3, -1)$	$m = -2$	54. $(-3, -3)$	$m = \frac{1}{2}$
55. $(1.5, -4)$	$m = 0$	56. $(8, 2)$	*m* is undefined

In Exercises 57 and 58, find the general form of the equation of the line passing through the point and satisfying the given condition.

57. Point: $(-3, 6)$
    (a) Slope is $\frac{7}{8}$
    (b) Parallel to the line $4x + 2y = 7$
    (c) Passes through the origin
    (d) Perpendicular to the line $3x - 2y = 2$

58. Point: $(1, -3)$
    (a) Parallel to the *x*-axis
    (b) Perpendicular to the *x*-axis
    (c) Parallel to the line $-4x + 5y = -3$
    (d) Perpendicular to the line $5x - 2y = 3$

59. **Assisted Living** The number of residents at an assisted living facility was 429 in 2005 and 564 in 2008. Assume that the relationship between the number of residents *L* and the time *t* (in years) is linear. Let $t = 5$ correspond to 2005.

    (a) Write a linear function expressing *L* in terms of *t*.

    (b) *Linear Interpolation* Use the linear equation you found in part (a) to estimate the number of residents in 2006.

    (c) *Linear Extrapolation* Use the linear equation you found in part (a) to predict the number of residents in 2013.

60. **Linear Depreciation** A municipality purchases an ambulance for $125,000. After 10 years the ambulance will be obsolete and have no value.

    (a) Write a linear equation giving the value *v* of the ambulance in terms of the time *t*.

    (b) Use a graphing utility to graph the function.

    (c) Use a graphing utility to estimate the value of the ambulance after 4 years.

    (d) Use a graphing utility to estimate the time when the ambulance's value will be $84,000.

In Exercises 61–64, use the Vertical Line Test to determine whether y is a function of x.

61. $y = -x^2 + 2$

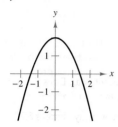

62. $x^2 + y^2 = 4$

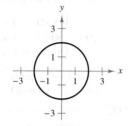

63. $y^2 - \frac{1}{4}x^2 = 4$

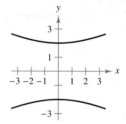

64. $y = |x + 4|$

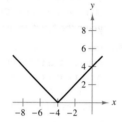

In Exercises 65 and 66, evaluate the function at the specified values of the independent variable. Simplify the result.

65. $f(x) = 3x + 4$
 (a) $f(1)$ (b) $f(x + 1)$ (c) $f(2 + \Delta x)$

66. $f(x) = x^2 + 4x + 3$
 (a) $f(0)$ (b) $f(x - 1)$ (c) $f(x + \Delta x) - f(x)$

(T) In Exercises 67–72, use a graphing utility to graph the function. Then find the domain and range of the function.

67. $f(x) = x^3 + 2x^2 - x + 2$

68. $f(x) = 2$

69. $f(x) = \sqrt{x + 1}$

70. $f(x) = \dfrac{x - 3}{x^2 + x - 12}$

71. $f(x) = -|x| + 3$

72. $f(x) = -\frac{12}{13}x - \frac{7}{8}$

In Exercises 73 and 74, use f and g to find the combinations of the functions.

(a) $f(x) + g(x)$ (b) $f(x) - g(x)$ (c) $f(x)g(x)$

(d) $\dfrac{f(x)}{g(x)}$ (e) $f(g(x))$ (f) $g(f(x))$

73. $f(x) = 1 + x^2$, $g(x) = 2x - 1$

74. $f(x) = 2x - 3$, $g(x) = \sqrt{x + 1}$

In Exercises 75–78, find the inverse function of f (if it exists).

75. $f(x) = \frac{3}{2}x$

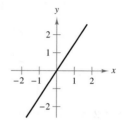

76. $f(x) = |x + 1|$

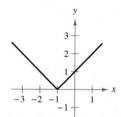

77. $f(x) = -x^2 + \frac{1}{2}$

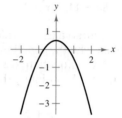

78. $f(x) = x^3 - 1$

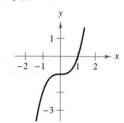

In Exercises 79–96, find the limit (if it exists).

79. $\lim\limits_{x \to 2} (5x - 3)$

80. $\lim\limits_{x \to 2} (2x + 9)$

81. $\lim\limits_{x \to 2} (5x - 3)(2x + 3)$

82. $\lim\limits_{x \to 2} \dfrac{5x - 3}{2x + 9}$

83. $\lim\limits_{t \to 3} \dfrac{t^2 + 1}{t}$

84. $\lim\limits_{t \to 0} \dfrac{t^2 + 1}{t}$

85. $\lim\limits_{t \to 1} \dfrac{t + 1}{t - 2}$

86. $\lim\limits_{t \to 2} \dfrac{t + 1}{t - 2}$

87. $\lim\limits_{x \to -2} \dfrac{x + 2}{x^2 - 4}$

88. $\lim\limits_{x \to 3^-} \dfrac{x^2 - 9}{x - 3}$

89. $\lim\limits_{x \to 0^+} \left(x - \dfrac{1}{x}\right)$

90. $\lim\limits_{x \to 1/2} \dfrac{2x - 1}{6x - 3}$

91. $\lim\limits_{x \to 0} \dfrac{[1/(x - 2)] - 1}{x}$

92. $\lim\limits_{x \to 0} \dfrac{[1/(x - 4)] - (1/4)}{x}$

93. $\lim\limits_{t \to 0} \dfrac{(1/\sqrt{t + 4}) - (1/2)}{t}$

94. $\lim\limits_{s \to 0} \dfrac{(1/\sqrt{1 + s}) - 1}{s}$

95. $\lim\limits_{\Delta x \to 0} \dfrac{(x + \Delta x)^3 - (x + \Delta x) - (x^3 - x)}{\Delta x}$

96. $\lim\limits_{\Delta x \to 0} \dfrac{1 - (x + \Delta x)^2 - (1 - x^2)}{\Delta x}$

In Exercises 97 and 98, use a table to estimate the limit.

97. $\lim\limits_{x \to 1^+} \dfrac{\sqrt{2x + 1} - \sqrt{3}}{x - 1}$

98. $\lim\limits_{x \to 1^+} \dfrac{1 - \sqrt[3]{x}}{x - 1}$

True or False? In Exercises 99–104, determine whether the statement is true or false. If it is false, explain why or give an example that shows it is false.

99. $\lim\limits_{x \to 0} \dfrac{|x|}{x} = 1$

100. $\lim\limits_{x \to 0} x^3 = 0$

101. $\lim\limits_{x \to 0} \sqrt{x} = 0$

102. $\lim\limits_{x \to 0} \sqrt[3]{x} = 0$

103. $\lim\limits_{x \to 2} f(x) = 3, \quad f(x) = \begin{cases} 3, & x \le 2 \\ 0, & x > 2 \end{cases}$

104. $\lim\limits_{x \to 3} f(x) = 1, \quad f(x) = \begin{cases} x - 2, & x \le 3 \\ -x^2 + 8x - 14, & x > 3 \end{cases}$

In Exercises 105–112, describe the interval(s) on which the function is continuous. Explain why the function is continuous on the interval(s). If the function has a discontinuity, identify the conditions of continuity that are not satisfied.

105. $f(x) = \dfrac{1}{(x+4)^2}$

106. $f(x) = \dfrac{x+2}{x}$

107. $f(x) = \dfrac{3}{x+1}$

108. $f(x) = \dfrac{x+1}{2x+2}$

109. $f(x) = [\![x+3]\!]$

110. $f(x) = [\![x]\!] - 2$

111. $f(x) = \begin{cases} x, & x \le 0 \\ x+1, & x > 0 \end{cases}$

112. $f(x) = \begin{cases} x, & x \le 0 \\ x^2, & x > 0 \end{cases}$

In Exercises 113 and 114, find the constant a such that f is continuous on the entire real line.

113. $f(x) = \begin{cases} -x+1, & x \le 3 \\ ax - 8, & x > 3 \end{cases}$

114. $f(x) = \begin{cases} x+1, & x < 1 \\ 2x+a, & x \ge 1 \end{cases}$

115. **Consumer Awareness** The cost C (in dollars) of purchasing x bottles of vitamins at a whole foods store is shown below.

$$C(x) = \begin{cases} 5.99x, & 0 < x \le 5 \\ 4.99x, & 5 < x \le 10 \\ 3.99x, & 10 < x \le 15 \\ 2.99x, & x > 15 \end{cases}$$

(T) (a) Use a graphing utility to graph the function and discuss its continuity. At what values is the function not continuous? Explain your reasoning.

(b) Find the cost of purchasing 10 bottles.

116. **Salary Contract** A union contract guarantees a 10% yearly increase for 3 years. For a current salary of $28,000, the salaries S (in thousands of dollars) for the next 3 years are given by

$$S(t) = \begin{cases} 28.00, & 0 < t \le 1 \\ 30.80, & 1 < t \le 2 \\ 33.88, & 2 < t \le 3 \end{cases}$$

where $t = 0$ represents the present year. Does the limit of S exist as t approaches 2? Explain your reasoning.

(T) 117. **Consumer Awareness** A company charges $5 for the first mile a patient is transported by an ambulance. Each additional mile or fraction thereof costs $4. Use the greatest integer function to create a model for the cost C of an ambulance trip lasting x miles. Use a graphing utility to graph the function, and discuss its continuity.

118. **Recycling** A recycling center pays $0.50 for each pound of aluminum cans. Twenty-four aluminum cans weigh one pound. A mathematical model for the amount A paid by the recycling center is

$$A = \dfrac{1}{2}\left[\!\!\left[\dfrac{x}{24}\right]\!\!\right]$$

where x is the number of cans.

(T) (a) Use a graphing utility to graph the function and then discuss its continuity.

(b) How much does the recycling center pay out for 1500 cans?

119. **National Debt** The table shows the national debt D (in billions of dollars) for selected years. A mathematical model for the national debt is

$$D = 4.2845t^3 - 97.655t^2 + 861.14t + 2571.1,$$
$$2 \le t \le 15$$

where $t = 2$ represents 1992. *(Source: U.S. Department of the Treasury)*

t	2	3	4	5	6
D	4001.8	4351.0	4643.3	4920.6	5181.5

t	7	8	9	10	11
D	5369.2	5478.2	5605.5	5628.7	5769.9

t	12	13	14	15
D	6198.4	6760.0	7354.7	7905.3

(T) (a) Use a graphing utility to graph the model.

(b) Create a table that compares the values given by the model with the actual data.

(c) Use the model to estimate the national debt in 2010.

CHAPTER 1 Functions, Graphs, and Limits

Chapter Test

See www.CalcChat.com for worked-out solutions to odd-numbered exercises.

Take this test as you would take a test in class. When you are done, check your work against the answers given in the back of the book.

In Exercises 1–3, (a) find the distance between the points, (b) find the midpoint of the line segment joining the points, and (c) find the slope of the line passing through the points.

1. $(1, -1), (-4, 4)$
2. $\left(\frac{5}{2}, 2\right), (0, 2)$
3. $(3\sqrt{2}, 2), (\sqrt{2}, 1)$

4. Sketch the graph of the circle whose general equation is $x^2 + y^2 - 4x - 2y - 4 = 0$.

5. The populations P (in thousands) of Missouri and Maryland from 2000 through 2006 can be modeled by

   $P = 5603 + 38.9t$ Missouri

   $P = 5331 + 51.6t$ Maryland

   where t represents the year, with $t = 0$ corresponding to 2000. Assuming these two models are accurate for long-term analysis, when would you expect the population of Maryland to exceed the population of Missouri? *(Source: U.S. Census Bureau)*

In Exercises 6–8, find the slope and y-intercept (if possible) of the linear equation. Then sketch the graph of the equation.

6. $y = \frac{1}{5}x - 2$
7. $x - \frac{7}{4} = 0$
8. $-x - 0.4y + 2.5 = 0$

In Exercises 9–11, (a) graph the function and label the intercepts, (b) determine the domain and range of the function, (c) find the value of the function when x is -3, -2, and 3, and (d) determine whether the function is one-to-one.

9. $f(x) = 2x + 5$
10. $f(x) = x^2 - x - 2$
11. $f(x) = |x| - 4$

In Exercises 12 and 13, find the inverse function of f. Then check your results algebraically by showing that $f(f^{-1}(x)) = x$ and $f^{-1}(f(x)) = x$.

12. $f(x) = 4x + 6$
13. $f(x) = \sqrt[3]{8 - 3x}$

In Exercises 14–17, find the limit (if it exists).

14. $\lim\limits_{x \to 0} \dfrac{x + 5}{x - 5}$
15. $\lim\limits_{x \to 5} \dfrac{x + 5}{x - 5}$
16. $\lim\limits_{x \to -3} \dfrac{x^2 + 2x - 3}{x^2 + 4x + 3}$
17. $\lim\limits_{x \to 0} \dfrac{\sqrt{x + 9} - 3}{x}$

In Exercises 18–20, describe the interval(s) on which the function is continuous. Explain why the function is continuous on the interval(s). If the function has a discontinuity at a point, identify all conditions of continuity that are not satisfied.

18. $f(x) = \dfrac{x^2 - 16}{x - 4}$
19. $f(x) = \sqrt{5 - x}$
20. $f(x) = \begin{cases} 1 - x, & x < 1 \\ x - x^2, & x \geq 1 \end{cases}$

21. The table lists the numbers of farms y (in thousands) in the United States for selected years. A mathematical model for the data is given by $y = 0.54t^2 - 15.4t + 2166$, where t represents the year, with $t = 0$ corresponding to 2000. *(Source: U.S. Department of Agriculture)*

    (a) Compare the values given by the model with the actual data. How well does the model fit the data? Explain your reasoning.

    (b) Use the model to predict the number of farms in 2009.

t	0	1	2
y	2167	2149	2135

t	3	4	5
y	2127	2113	2101

Table for 21

2 Differentiation

Demography, a field within sociology, studies changes in population. Derivatives are used to determine the rates of change of populations over time. Recently, growth of Japan's population has been slowing down. (See Section 2.3, Exercise 23.)

Applications

Differentiation has many real-life applications. The applications listed below represent a sample of the applications in this chapter.

- Psychology: Migraine Prevalence, Exercise 64, page 136
- Center of Population, Exercise 28, page 145
- Biology: Wildlife Management, Exercise 60, page 156
- Make a Decision: Environment, Exercise 63, page 157
- Fruit and Vegetable Consumption, Exercise 73, page 166

2.1 The Derivative and the Slope of a Graph

2.2 Some Rules for Differentiation

2.3 Rates of Change

2.4 The Product and Quotient Rules

2.5 The Chain Rule

2.6 Higher-Order Derivatives

Section 2.1
The Derivative and the Slope of a Graph

- Identify tangent lines to a graph at a point.
- Approximate the slopes of tangent lines to graphs at points.
- Use the limit definition to find the slopes of graphs at points.
- Use the limit definition to find the derivatives of functions.
- Describe the relationship between differentiability and continuity.

Tangent Line to a Graph

Calculus is a branch of mathematics that studies rates of change of functions. In this course, you will learn that rates of change have many applications in real life. In Section 1.3, you learned how the slope of a line indicates the rate at which the line rises or falls. For a line, this rate (or slope) is the same at every point on the line. For graphs other than lines, the rate at which the graph rises or falls changes from point to point. For instance, in Figure 2.1, the parabola is rising more quickly at the point (x_1, y_1) than it is at the point (x_2, y_2). At the vertex (x_3, y_3), the graph levels off, and at the point (x_4, y_4), the graph is falling.

To determine the rate at which a graph rises or falls at a *single point*, you can find the slope of the **tangent line** at the point. In simple terms, the tangent line to the graph of a function f at a point $P(x_1, y_1)$ is the line that best approximates the graph at that point, as shown in Figure 2.1. Figure 2.2 shows other examples of tangent lines.

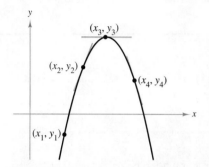

FIGURE 2.1 The slope of a nonlinear graph changes from one point to another.

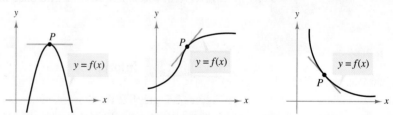

FIGURE 2.2 Tangent Line to a Graph at a Point

When Isaac Newton (1642–1727) was working on the "tangent line problem," he realized that it is difficult to define precisely what is meant by a tangent to a general curve. From geometry, you know that a line is tangent to a circle if the line intersects the circle at only one point, as shown in Figure 2.3. Tangent lines to a noncircular graph, however, can intersect the graph at more than one point. For instance, in the second graph in Figure 2.2, if the tangent line were extended, it would intersect the graph at a point other than the point of tangency. In this section, you will see how the notion of a limit can be used to define a general tangent line.

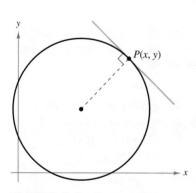

FIGURE 2.3 Tangent Line to a Circle

DISCOVERY

Use a graphing utility to graph $f(x) = 2x^3 - 4x^2 + 3x - 5$. On the same screen, sketch the graphs of $y = x - 5$, $y = 2x - 5$, and $y = 3x - 5$. Which of these lines, if any, appears to be tangent to the graph of f at the point $(0, -5)$? Explain your reasoning.

Slope of a Graph

Because a tangent line approximates the graph at a point, the problem of finding the slope of a graph at a point becomes one of finding the slope of the tangent line at the point.

Example 1 Approximating the Slope of a Graph

Use the graph in Figure 2.4 to approximate the slope of the graph of $f(x) = x^2$ at the point $(1, 1)$.

SOLUTION From the graph of $f(x) = x^2$, you can see that the tangent line at $(1, 1)$ rises approximately two units for each unit change in x. So, the slope of the tangent line at $(1, 1)$ is given by

$$\text{Slope} = \frac{\text{change in } y}{\text{change in } x} \approx \frac{2}{1} = 2.$$

Because the tangent line at the point $(1, 1)$ has a slope of about 2, you can conclude that the graph has a slope of about 2 at the point $(1, 1)$.

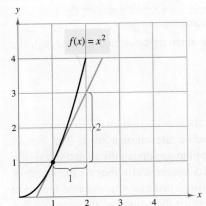

FIGURE 2.4

STUDY TIP
When visually approximating the slope of a graph, note that the scales on the horizontal and vertical axes may differ. When this happens (as it frequently does in applications), the slope of the tangent line is distorted, and you must be careful to account for the difference in scales.

✓ **CHECKPOINT 1**

Use the graph to approximate the slope of the graph of $f(x) = x^3$ at the point $(1, 1)$.

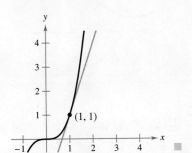

Example 2 Interpreting Slope

Figure 2.5 graphically depicts the average monthly temperature (in degrees Fahrenheit) in Duluth, Minnesota. Estimate the slope of this graph at the indicated point and give a physical interpretation of the result. *(Source: National Oceanic and Atmospheric Administration)*

SOLUTION From the graph, you can see that the tangent line at the given point falls approximately 28 units for each two-unit change in x. So, you can estimate the slope at the given point to be

$$\text{Slope} = \frac{\text{change in } y}{\text{change in } x} \approx \frac{-28}{2}$$

$$= -14 \text{ degrees per month.}$$

This means that you can expect the average daily temperatures in November to be about 14 degrees *lower* than the corresponding temperatures in October.

✓ **CHECKPOINT 2**

For which months do the slopes of the tangent lines appear to be positive? Negative? Interpret these slopes in the context of the problem. ■

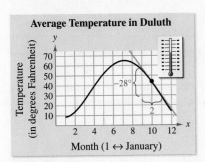

FIGURE 2.5

Slope and the Limit Process

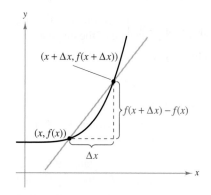

FIGURE 2.6 The Secant Line Through the Two Points $(x, f(x))$ and $(x + \Delta x, f(x + \Delta x))$

In Examples 1 and 2, you approximated the slope of a graph at a point by making a careful graph and then "eyeballing" the tangent line at the point of tangency. A more precise method of approximating the slope of a tangent line makes use of a **secant line** through the point of tangency and a second point on the graph, as shown in Figure 2.6. If $(x, f(x))$ is the point of tangency and $(x + \Delta x, f(x + \Delta x))$ is a second point on the graph of f, then the slope of the secant line through the two points is

$$m_{\text{sec}} = \frac{f(x + \Delta x) - f(x)}{\Delta x}. \qquad \text{Slope of secant line}$$

The right side of this equation is called the **difference quotient.** The denominator Δx is the **change in x,** and the numerator is the **change in y.** The beauty of this procedure is that you obtain more and more accurate approximations of the slope of the tangent line by choosing points closer and closer to the point of tangency, as shown in Figure 2.7. Using the limit process, you can find the *exact* slope of the tangent line at $(x, f(x))$, which is also the slope of the graph of f at $(x, f(x))$.

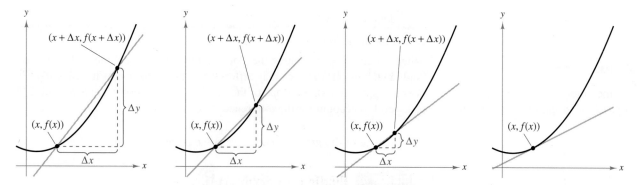

FIGURE 2.7 As Δx approaches 0, the secant lines approach the tangent line.

> **Definition of the Slope of a Graph**
>
> The **slope** m of the graph of f at the point $(x, f(x))$ is equal to the slope of its tangent line at $(x, f(x))$, and is given by
>
> $$m = \lim_{\Delta x \to 0} m_{\text{sec}} = \lim_{\Delta x \to 0} \frac{f(x + \Delta x) - f(x)}{\Delta x}$$
>
> provided this limit exists.

> **STUDY TIP**
>
> Δx is used as a variable to represent the change in x in the definition of the slope of a graph. Other variables may also be used. For instance, this definition is sometimes written as
>
> $$m = \lim_{h \to 0} \frac{f(x + h) - f(x)}{h}.$$

SECTION 2.1 The Derivative and the Slope of a Graph

Algebra Review

For help in evaluating the expressions in Examples 3–6, see the review of simplifying fractional expressions on page 174.

Example 3 Finding Slope by the Limit Process

Find the slope of the graph of $f(x) = x^2$ at the point $(-2, 4)$.

SOLUTION Begin by finding an expression that represents the slope of a secant line at the point $(-2, 4)$.

$$m_{sec} = \frac{f(-2 + \Delta x) - f(-2)}{\Delta x} \quad \text{Set up difference quotient.}$$

$$= \frac{(-2 + \Delta x)^2 - (-2)^2}{\Delta x} \quad \text{Use } f(x) = x^2.$$

$$= \frac{4 - 4\Delta x + (\Delta x)^2 - 4}{\Delta x} \quad \text{Expand terms.}$$

$$= \frac{-4\Delta x + (\Delta x)^2}{\Delta x} \quad \text{Simplify.}$$

$$= \frac{\Delta x(-4 + \Delta x)}{\Delta x} \quad \text{Factor and divide out.}$$

$$= -4 + \Delta x, \quad \Delta x \neq 0 \quad \text{Simplify.}$$

Next, take the limit of m_{sec} as $\Delta x \to 0$.

$$m = \lim_{\Delta x \to 0} m_{sec} = \lim_{\Delta x \to 0} (-4 + \Delta x) = -4$$

So, the graph of f has a slope of -4 at the point $(-2, 4)$, as shown in Figure 2.8.

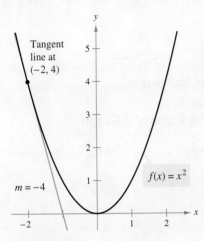

FIGURE 2.8

✓ CHECKPOINT 3

Find the slope of the graph of $f(x) = x^2$ at the point $(2, 4)$. ■

Example 4 Finding the Slope of a Graph

Find the slope of the graph of $f(x) = -2x + 4$.

SOLUTION You know from your study of linear functions that the line given by $f(x) = -2x + 4$ has a slope of -2, as shown in Figure 2.9. This conclusion is consistent with the limit definition of slope.

$$m = \lim_{\Delta x \to 0} \frac{f(x + \Delta x) - f(x)}{\Delta x}$$

$$= \lim_{\Delta x \to 0} \frac{[-2(x + \Delta x) + 4] - [-2x + 4]}{\Delta x}$$

$$= \lim_{\Delta x \to 0} \frac{-2x - 2\Delta x + 4 + 2x - 4}{\Delta x}$$

$$= \lim_{\Delta x \to 0} \frac{-2\Delta x}{\Delta x} = -2$$

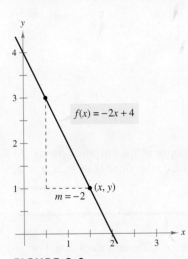

FIGURE 2.9

✓ CHECKPOINT 4

Find the slope of the graph of $f(x) = 2x + 5$. ■

DISCOVERY

Use a graphing utility to graph the function $y_1 = x^2 + 1$ and the three lines $y_2 = 3x - 1$, $y_3 = 4x - 3$, and $y_4 = 5x - 5$. Which of these lines appears to be tangent to y_1 at the point $(2, 5)$? Confirm your answer by showing that the graphs of y_1 and its tangent line have only one point of intersection, whereas the graphs of y_1 and the other lines each have two points of intersection.

It is important that you see the distinction between the ways the difference quotients were set up in Examples 3 and 4. In Example 3, you were finding the slope of a graph at a specific point $(c, f(c))$. To find the slope, you can use the following form of a difference quotient.

$$m = \lim_{\Delta x \to 0} \frac{f(c + \Delta x) - f(c)}{\Delta x} \quad \text{Slope at specific point}$$

In Example 4, however, you were finding a formula for the slope at *any* point on the graph. In such cases, you should use x, rather than c, in the difference quotient.

$$m = \lim_{\Delta x \to 0} \frac{f(x + \Delta x) - f(x)}{\Delta x} \quad \text{Formula for slope}$$

Except for linear functions, this form will always produce a function of x, which can then be evaluated to find the slope at any desired point.

Example 5 Finding a Formula for the Slope of a Graph

Find a formula for the slope of the graph of $f(x) = x^2 + 1$. What are the slopes at the points $(-1, 2)$ and $(2, 5)$?

SOLUTION

$$\begin{aligned}
m_{\text{sec}} &= \frac{f(x + \Delta x) - f(x)}{\Delta x} &&\text{Set up difference quotient.} \\
&= \frac{[(x + \Delta x)^2 + 1] - (x^2 + 1)}{\Delta x} &&\text{Use } f(x) = x^2 + 1. \\
&= \frac{x^2 + 2x\,\Delta x + (\Delta x)^2 + 1 - x^2 - 1}{\Delta x} &&\text{Expand terms.} \\
&= \frac{2x\,\Delta x + (\Delta x)^2}{\Delta x} &&\text{Simplify.} \\
&= \frac{\Delta x(2x + \Delta x)}{\Delta x} &&\text{Factor and divide out.} \\
&= 2x + \Delta x, \quad \Delta x \neq 0 &&\text{Simplify.}
\end{aligned}$$

Next, take the limit of m_{sec} as $\Delta x \to 0$.

$$\begin{aligned}
m &= \lim_{\Delta x \to 0} m_{\text{sec}} \\
&= \lim_{\Delta x \to 0} (2x + \Delta x) \\
&= 2x
\end{aligned}$$

Using the formula $m = 2x$, you can find the slopes at the specified points. At $(-1, 2)$ the slope is $m = 2(-1) = -2$, and at $(2, 5)$ the slope is $m = 2(2) = 4$. The graph of f is shown in Figure 2.10.

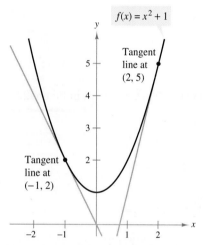

FIGURE 2.10

✓ CHECKPOINT 5

Find a formula for the slope of the graph of $f(x) = 4x^2 + 1$. What are the slopes at the points $(0, 1)$ and $(1, 5)$?

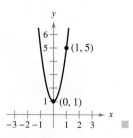

STUDY TIP

The slope of the graph of $f(x) = x^2 + 1$ varies for different values of x. For what value of x is the slope equal to 0?

The Derivative of a Function

In Example 5, you started with the function $f(x) = x^2 + 1$ and used the limit process to derive another function, $m = 2x$, that represents the slope of the graph of f at the point $(x, f(x))$. This derived function is called the **derivative** of f at x. It is denoted by $f'(x)$, which is read as "f prime of x."

> **STUDY TIP**
>
> The notation dy/dx is read as "the derivative of y with respect to x," and using limit notation, you can write
>
> $$\frac{dy}{dx} = \lim_{\Delta x \to 0} \frac{\Delta y}{\Delta x}$$
>
> $$= \lim_{\Delta x \to 0} \frac{f(x + \Delta x) - f(x)}{\Delta x}$$
>
> $$= f'(x).$$

Definition of the Derivative

The **derivative of f at x** is given by

$$f'(x) = \lim_{\Delta x \to 0} \frac{f(x + \Delta x) - f(x)}{\Delta x}$$

provided this limit exists. A function is **differentiable** at x if its derivative exists at x. The process of finding derivatives is called **differentiation.**

In addition to $f'(x)$, other notations can be used to denote the derivative of $y = f(x)$. The most common are

$$\frac{dy}{dx}, \quad y', \quad \frac{d}{dx}[f(x)], \quad \text{and} \quad D_x[y].$$

Example 6 Finding a Derivative

Find the derivative of $f(x) = 3x^2 - 2x$.

SOLUTION

$$f'(x) = \lim_{\Delta x \to 0} \frac{f(x + \Delta x) - f(x)}{\Delta x}$$

$$= \lim_{\Delta x \to 0} \frac{[3(x + \Delta x)^2 - 2(x + \Delta x)] - (3x^2 - 2x)}{\Delta x}$$

$$= \lim_{\Delta x \to 0} \frac{3x^2 + 6x\,\Delta x + 3(\Delta x)^2 - 2x - 2\,\Delta x - 3x^2 + 2x}{\Delta x}$$

$$= \lim_{\Delta x \to 0} \frac{6x\,\Delta x + 3(\Delta x)^2 - 2\,\Delta x}{\Delta x}$$

$$= \lim_{\Delta x \to 0} \frac{\Delta x(6x + 3\,\Delta x - 2)}{\Delta x}$$

$$= \lim_{\Delta x \to 0} (6x + 3\,\Delta x - 2)$$

$$= 6x - 2$$

So, the derivative of $f(x) = 3x^2 - 2x$ is $f'(x) = 6x - 2$.

✓ CHECKPOINT 6

Find the derivative of $f(x) = x^2 - 5x$. ■

TECHNOLOGY

You can use a graphing utility to confirm the result given in Example 7. One way to do this is to choose a point on the graph of $y = 2/t$, such as $(1, 2)$, and find the equation of the tangent line at that point. Using the derivative found in the example, you know that the slope of the tangent line when $t = 1$ is $m = -2$. This means that the tangent line at the point $(1, 2)$ is

$$y - y_1 = m(t - t_1)$$
$$y - 2 = -2(t - 1) \text{ or}$$
$$y = -2t + 4.$$

By graphing $y = 2/t$ and $y = -2t + 4$ in the same viewing window, as shown below, you can confirm that the line is tangent to the graph at the point $(1, 2)$.*

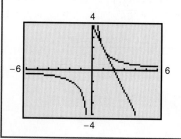

✓ CHECKPOINT 7

Find the derivative of y with respect to t for the function $y = 4/t$. ■

In many applications, it is convenient to use a variable other than x as the independent variable. Example 7 shows a function that uses t as the independent variable.

Example 7 Finding a Derivative

Find the derivative of y with respect to t for the function

$$y = \frac{2}{t}.$$

SOLUTION Consider $y = f(t)$, and use the limit process as shown.

$$\frac{dy}{dt} = \lim_{\Delta t \to 0} \frac{f(t + \Delta t) - f(t)}{\Delta t} \quad \text{Set up difference quotient.}$$

$$= \lim_{\Delta t \to 0} \frac{\frac{2}{t + \Delta t} - \frac{2}{t}}{\Delta t} \quad \text{Use } f(t) = 2/t.$$

$$= \lim_{\Delta t \to 0} \frac{\frac{2t - 2t - 2\Delta t}{t(t + \Delta t)}}{\Delta t} \quad \text{Expand terms.}$$

$$= \lim_{\Delta t \to 0} \frac{-2\Delta t}{t(\Delta t)(t + \Delta t)} \quad \text{Factor and divide out.}$$

$$= \lim_{\Delta t \to 0} \frac{-2}{t(t + \Delta t)} \quad \text{Simplify.}$$

$$= -\frac{2}{t^2} \quad \text{Evaluate the limit.}$$

So, the derivative of y with respect to t is

$$\frac{dy}{dt} = -\frac{2}{t^2}.$$

Remember that the derivative of a function gives you a formula for finding the slope of the tangent line at any point on the graph of the function. For example, the slope of the tangent line to the graph of f at the point $(1, 2)$ is given by

$$f'(1) = -\frac{2}{1^2} = -2.$$

To find the slopes of the graph at other points, substitute the t-coordinate of the point into the derivative, as shown below.

Point	t-Coordinate	Slope
$(2, 1)$	$t = 2$	$m = f'(2) = -\dfrac{2}{2^2} = -\dfrac{1}{2}$
$(-2, -1)$	$t = -2$	$m = f'(-2) = -\dfrac{2}{(-2)^2} = -\dfrac{1}{2}$

---

*Specific calculator keystroke instructions for operations in this and other technology boxes can be found at *college.hmco.com/info/larsonapplied*.

Differentiability and Continuity

Not every function is differentiable. Figure 2.11 shows some common situations in which a function will not be differentiable at a point—vertical tangent lines, discontinuities, and sharp turns in the graph. Each of the functions shown in Figure 2.11 is differentiable at every value of x *except* $x = 0$.

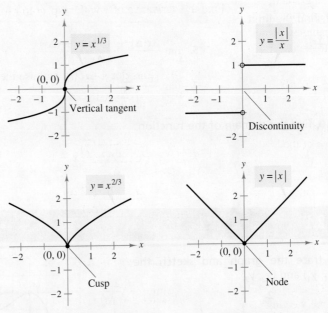

FIGURE 2.11 Functions That Are Not Differentiable at $x = 0$

In Figure 2.11, you can see that all but one of the functions are continuous at $x = 0$ but none are differentiable there. This shows that continuity is not a strong enough condition to guarantee differentiability. On the other hand, if a function is differentiable at a point, then it must be continuous at that point. This important result is stated in the following theorem.

Differentiability Implies Continuity

If a function f is differentiable at $x = c$, then f is continuous at $x = c$.

CONCEPT CHECK

1. What is the name of the line that best approximates the slope of a graph at a point?
2. What is the name of a line through the point of tangency and a second point on the graph?
3. Sketch a graph of a function whose derivative is always negative.
4. Sketch a graph of a function whose derivative is always positive.

Skills Review 2.1

The following warm-up exercises involve skills that were covered in earlier sections. You will use these skills in the exercise set for this section. For additional help, review Sections 1.3, 1.4, and 1.5.

In Exercises 1–3, find an equation of the line containing P and Q.

1. $P(2, 1)$, $Q(2, 4)$
2. $P(2, 2)$, $Q(-5, 2)$
3. $P(2, 0)$, $Q(3, -1)$

In Exercises 4–7, find the limit.

4. $\lim\limits_{\Delta x \to 0} \dfrac{2x\Delta x + (\Delta x)^2}{\Delta x}$

5. $\lim\limits_{\Delta x \to 0} \dfrac{3x^2\Delta x + 3x(\Delta x)^2 + (\Delta x)^3}{\Delta x}$

6. $\lim\limits_{\Delta x \to 0} \dfrac{1}{x(x + \Delta x)}$

7. $\lim\limits_{\Delta x \to 0} \dfrac{(x + \Delta x)^2 - x^2}{\Delta x}$

In Exercises 8–10, find the domain of the function.

8. $f(x) = \dfrac{1}{x - 1}$
9. $f(x) = \dfrac{1}{5}x^3 - 2x^2 + \dfrac{1}{3}x - 1$
10. $f(x) = \dfrac{6x}{x^3 + x}$

Exercises 2.1

See www.CalcChat.com for worked-out solutions to odd-numbered exercises.

In Exercises 1–4, trace the graph and sketch the tangent lines at (x_1, y_1) and (x_2, y_2).

1.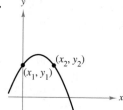
2.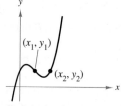
3.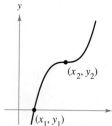
4.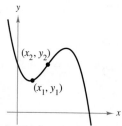

In Exercises 5–10, estimate the slope of the graph at the point (x, y). (Each square on the grid is 1 unit by 1 unit.)

5.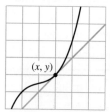
6.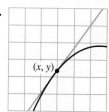
7.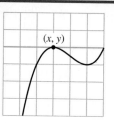
8.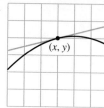
9.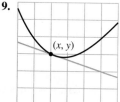
10.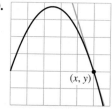

11. **Consumer Trends** The graph shows the number of visitors V (in hundreds of thousands) to a national park during a one-year period, where $t = 1$ corresponds to January. Estimate the slopes of the graph at $t = 2, 8,$ and 11.

12. Uranium Concentrate The graph represents the amount U of uranium (in thousands of pounds per year) produced in mines for the years 2000 through 2005, where t represents the year, with $t = 0$ corresponding to 2000. Estimate the slopes of the graph for the years 2001 and 2004. *(Source: U.S. Department of Energy)*

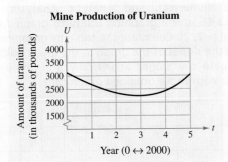

13. Timber Products The graph represents the amount L of softwood products (in millions of cubic feet per year) consumed from 2000 through 2005, where t represents the year, with $t = 0$ corresponding to 2000. Estimate the slopes of the graph for the years 2002 and 2004. *(Source: U.S. Forest Service)*

14. Attendance The attendance for four high school basketball games is given by $s = f(t)$ and the attendance for four high school football games is given by $s = g(t)$, where $t = 1$ corresponds to the first game.

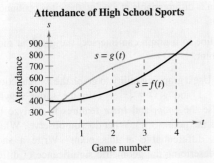

(a) Which attendance rate is greater at game 1?

(b) What conclusion can you make regarding the attendance rates at game 3?

(c) What conclusion can you make regarding the attendance rates at game 4?

In Exercises 15–24, use the limit definition to find the slope of the tangent line to the graph of f at the given point.

15. $f(x) = 6 - 2x$, $(2, 2)$
16. $f(x) = 2x + 4$, $(1, 6)$
17. $f(x) = -1$, $(0, -1)$
18. $f(x) = 6$, $(-2, 6)$
19. $f(x) = x^2 - 1$, $(2, 3)$
20. $f(x) = 4 - x^2$, $(2, 0)$
21. $f(x) = x^3 - x$, $(2, 6)$
22. $f(x) = x^3 + 2x$, $(1, 3)$
23. $f(x) = 2\sqrt{x}$, $(4, 4)$
24. $f(x) = \sqrt{x + 1}$, $(8, 3)$

In Exercises 25–38, use the limit definition to find the derivative of the function.

25. $f(x) = 3$
26. $f(x) = -2$
27. $f(x) = -5x$
28. $f(x) = 4x + 1$
29. $g(s) = \frac{1}{3}s + 2$
30. $h(t) = 6 - \frac{1}{2}t$
31. $f(x) = x^2 - 4$
32. $f(x) = 1 - x^2$
33. $h(t) = \sqrt{t - 1}$
34. $f(x) = \sqrt{x + 2}$
35. $f(t) = t^3 - 12t$
36. $f(t) = t^3 + t^2$
37. $f(x) = \dfrac{1}{x + 2}$
38. $g(s) = \dfrac{1}{s - 1}$

In Exercises 39–46, use the limit definition to find an equation of the tangent line to the graph of f at the given point. Then verify your results by using a graphing utility to graph the function and its tangent line at the point.

39. $f(x) = \frac{1}{2}x^2$, $(2, 2)$
40. $f(x) = -x^2$, $(-1, -1)$
41. $f(x) = (x - 1)^2$, $(-2, 9)$
42. $f(x) = 2x^2 - 1$, $(0, -1)$
43. $f(x) = \sqrt{x} + 1$, $(4, 3)$
44. $f(x) = \sqrt{x + 2}$, $(7, 3)$
45. $f(x) = \dfrac{1}{x}$, $(1, 1)$
46. $f(x) = \dfrac{1}{x - 1}$, $(2, 1)$

In Exercises 47–50, find an equation of the line that is tangent to the graph of f and parallel to the given line.

Function	Line
47. $f(x) = -\frac{1}{4}x^2$	$x + y = 0$
48. $f(x) = x^2 + 1$	$2x + y = 0$
49. $f(x) = -\frac{1}{2}x^3$	$6x + y + 4 = 0$
50. $f(x) = x^2 - x$	$x + 2y - 6 = 0$

In Exercises 51–58, describe the x-values at which the function is differentiable. Explain your reasoning.

51. $y = |x + 3|$

52. $y = |x^2 - 9|$

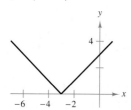
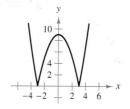

53. $y = (x - 3)^{2/3}$

54. $y = x^{2/5}$

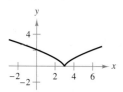
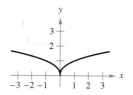

55. $y = \sqrt{x - 1}$

56. $y = \dfrac{x^2}{x^2 - 4}$

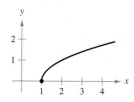
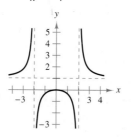

57. $y = \begin{cases} x^3 + 3, & x < 0 \\ x^3 - 3, & x \geq 0 \end{cases}$

58. $y = \begin{cases} x^2, & x \leq 1 \\ -x^2, & x > 1 \end{cases}$

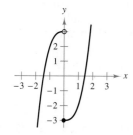
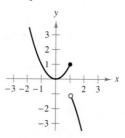

In Exercises 59 and 60, describe the x-values at which f is differentiable.

59. $f(x) = \dfrac{1}{x - 1}$

60. $f(x) = \begin{cases} x^2 - 3, & x \leq 0 \\ 3 - x^2, & x > 0 \end{cases}$

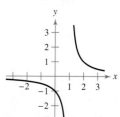
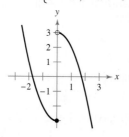

In Exercises 61 and 62, identify a function f that has the given characteristics. Then sketch the function.

61. $f(0) = 2; f'(x) = -3, -\infty < x < \infty$

62. $f(-2) = f(4) = 0; f'(1) = 0, f'(x) < 0$ for $x < 1$; $f'(x) > 0$ for $x > 1$

(T) **Graphical, Numerical, and Analytic Analysis** In Exercises 63–66, use a graphing utility to graph f on the interval $[-2, 2]$. Complete the table by graphically estimating the slopes of the graph at the given points. Then evaluate the slopes analytically and compare your results with those obtained graphically.

x	-2	$-\frac{3}{2}$	-1	$-\frac{1}{2}$	0	$\frac{1}{2}$	1	$\frac{3}{2}$	2
$f(x)$									
$f'(x)$									

63. $f(x) = \frac{1}{4}x^3$

64. $f(x) = \frac{1}{2}x^2$

65. $f(x) = -\frac{1}{2}x^3$

66. $f(x) = -\frac{3}{2}x^2$

(T) In Exercises 67–70, find the derivative of the given function f. Then use a graphing utility to graph f and its derivative in the same viewing window. What does the x-intercept of the derivative indicate about the graph of f?

67. $f(x) = x^2 - 4x$

68. $f(x) = 2 + 6x - x^2$

69. $f(x) = x^3 - 3x$

70. $f(x) = x^3 - 6x^2$

True or False? In Exercises 71–74, determine whether the statement is true or false. If it is false, explain why or give an example that shows it is false.

71. The slope of the graph of $f(x) = x^2$ is different at every point on the graph of f.

72. If a function is continuous at a point, then it is differentiable at that point.

73. If a function is differentiable at a point, then it is continuous at that point.

74. A tangent line to a graph can intersect the graph at more than one point.

(T) **75. Writing** Use a graphing utility to graph the two functions $f(x) = x^2 + 1$ and $g(x) = |x| + 1$ in the same viewing window. Use the *zoom* and *trace* features to analyze the graphs near the point (0, 1). What do you observe? Which function is differentiable at this point? Write a short paragraph describing the geometric significance of differentiability at a point.

Section 2.2
Some Rules for Differentiation

- Find the derivatives of functions using the Constant Rule.
- Find the derivatives of functions using the Power Rule.
- Find the derivatives of functions using the Constant Multiple Rule.
- Find the derivatives of functions using the Sum and Difference Rules.
- Use derivatives to answer questions about real-life situations.

The Constant Rule

In Section 2.1, you found derivatives by the limit process. This process is tedious, even for simple functions, but fortunately there are rules that greatly simplify differentiation. These rules allow you to calculate derivatives without the *direct* use of limits.

The Constant Rule

The derivative of a constant function is zero. That is,

$$\frac{d}{dx}[c] = 0, \quad c \text{ is a constant.}$$

PROOF Let $f(x) = c$. Then, by the limit definition of the derivative, you can write

$$f'(x) = \lim_{\Delta x \to 0} \frac{f(x + \Delta x) - f(x)}{\Delta x} = \lim_{\Delta x \to 0} \frac{c - c}{\Delta x} = \lim_{\Delta x \to 0} 0 = 0.$$

So, $\frac{d}{dx}[c] = 0.$

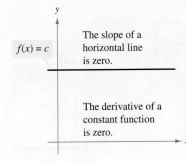

FIGURE 2.12

STUDY TIP

Note in Figure 2.12 that the Constant Rule is equivalent to saying that the slope of a horizontal line is zero.

STUDY TIP

An interpretation of the Constant Rule says that the tangent line to a constant function is the function itself. Find an equation of the tangent line to $f(x) = -4$ at $x = 3$.

Example 1 Finding Derivatives of Constant Functions

a. $\dfrac{d}{dx}[7] = 0$

b. If $f(x) = 0$, then $f'(x) = 0$.

c. If $y = 2$, then $\dfrac{dy}{dx} = 0$.

d. If $g(t) = -\dfrac{3}{2}$, then $g'(t) = 0$.

✓ CHECKPOINT 1

Find the derivative of each function.

a. $f(x) = -2$ **b.** $y = \pi$ **c.** $g(w) = \sqrt{5}$ **d.** $s(t) = 320.5$

The Power Rule

The binomial expansion process is used to prove the Power Rule.

$$(x + \Delta x)^2 = x^2 + 2x\,\Delta x + (\Delta x)^2$$
$$(x + \Delta x)^3 = x^3 + 3x^2\,\Delta x + 3x(\Delta x)^2 + (\Delta x)^3$$
$$(x + \Delta x)^n = x^n + nx^{n-1}\,\Delta x + \underbrace{\frac{n(n-1)x^{n-2}}{2}(\Delta x)^2 + \cdots + (\Delta x)^n}_{(\Delta x)^2 \text{ is a factor of these terms.}}$$

The (Simple) Power Rule

$$\frac{d}{dx}[x^n] = nx^{n-1}, \qquad n \text{ is any real number.}$$

PROOF We prove only the case in which n is a positive integer. Let $f(x) = x^n$. Using the binomial expansion, you can write

$$f'(x) = \lim_{\Delta x \to 0} \frac{f(x + \Delta x) - f(x)}{\Delta x} \qquad \text{Definition of derivative}$$
$$= \lim_{\Delta x \to 0} \frac{(x + \Delta x)^n - x^n}{\Delta x}$$
$$= \lim_{\Delta x \to 0} \frac{x^n + nx^{n-1}\,\Delta x + \frac{n(n-1)x^{n-2}}{2}(\Delta x)^2 + \cdots + (\Delta x)^n - x^n}{\Delta x}$$
$$= \lim_{\Delta x \to 0} \left[nx^{n-1} + \frac{n(n-1)x^{n-2}}{2}(\Delta x) + \cdots + (\Delta x)^{n-1} \right]$$
$$= nx^{n-1} + 0 + \cdots + 0 = nx^{n-1}.$$

For the Power Rule, the case in which $n = 1$ is worth remembering as a separate differentiation rule. That is,

$$\frac{d}{dx}[x] = 1. \qquad \text{The derivative of } x \text{ is 1.}$$

This rule is consistent with the fact that the slope of the line given by $y = x$ is 1. (See Figure 2.13.)

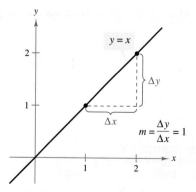

FIGURE 2.13 The slope of the line $y = x$ is 1.

SECTION 2.2 Some Rules for Differentiation

Example 2 Applying the Power Rule

Find the derivative of each function.

Function	Derivative
a. $f(x) = x^3$	$f'(x) = 3x^2$
b. $y = \dfrac{1}{x^2} = x^{-2}$	$\dfrac{dy}{dx} = (-2)x^{-3} = -\dfrac{2}{x^3}$
c. $g(t) = t$	$g'(t) = 1$
d. $R = x^4$	$\dfrac{dR}{dx} = 4x^3$

✓ **CHECKPOINT 2**

Find the derivative of each function.

a. $f(x) = x^4$ **b.** $y = \dfrac{1}{x^3}$

c. $g(w) = w^2$ **d.** $s(t) = \dfrac{1}{t}$

In Example 2(b), note that *before* differentiating, you should rewrite $1/x^2$ as x^{-2}. Rewriting is the first step in *many* differentiation problems.

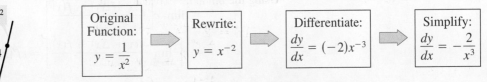

Original Function: $y = \dfrac{1}{x^2}$ ⇒ Rewrite: $y = x^{-2}$ ⇒ Differentiate: $\dfrac{dy}{dx} = (-2)x^{-3}$ ⇒ Simplify: $\dfrac{dy}{dx} = -\dfrac{2}{x^3}$

Remember that the derivative of a function f is another function that gives the slope of the graph of f at any point at which f is differentiable. So, you can use the derivative to find slopes, as shown in Example 3.

Example 3 Finding the Slope of a Graph

Find the slopes of the graph of

$f(x) = x^2$ Original function

when $x = -2, -1, 0, 1,$ and 2.

SOLUTION Begin by using the Power Rule to find the derivative of f.

$f'(x) = 2x$ Derivative

You can use the derivative to find the slopes of the graph of f, as shown.

x-Value	Slope of Graph of f
$x = -2$	$m = f'(-2) = 2(-2) = -4$
$x = -1$	$m = f'(-1) = 2(-1) = -2$
$x = 0$	$m = f'(0) = 2(0) = 0$
$x = 1$	$m = f'(1) = 2(1) = 2$
$x = 2$	$m = f'(2) = 2(2) = 4$

The graph of f is shown in Figure 2.14.

FIGURE 2.14

✓ **CHECKPOINT 3**

Find the slopes of the graph of $f(x) = x^3$ when $x = -1, 0,$ and 1.

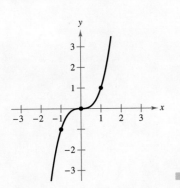

The Constant Multiple Rule

To prove the Constant Multiple Rule, the following property of limits is used.

$$\lim_{x \to a} cg(x) = c\left[\lim_{x \to a} g(x)\right]$$

> **The Constant Multiple Rule**
>
> If f is a differentiable function of x, and c is a real number, then
>
> $$\frac{d}{dx}[cf(x)] = cf'(x), \qquad c \text{ is a constant.}$$

PROOF Apply the definition of the derivative to produce

$$\frac{d}{dx}[cf(x)] = \lim_{\Delta x \to 0} \frac{cf(x + \Delta x) - cf(x)}{\Delta x} \qquad \text{Definition of derivative}$$

$$= \lim_{\Delta x \to 0} c\left[\frac{f(x + \Delta x) - f(x)}{\Delta x}\right]$$

$$= c\left[\lim_{\Delta x \to 0} \frac{f(x + \Delta x) - f(x)}{\Delta x}\right] = cf'(x).$$

Informally, the Constant Multiple Rule states that constants can be factored out of the differentiation process.

$$\frac{d}{dx}[cf(x)] = c\frac{d}{dx}[f(x)] = cf'(x)$$

The usefulness of this rule is often overlooked, especially when the constant appears in the denominator, as shown below.

$$\frac{d}{dx}\left[\frac{f(x)}{c}\right] = \frac{d}{dx}\left[\frac{1}{c}f(x)\right] = \frac{1}{c}\left(\frac{d}{dx}[f(x)]\right) = \frac{1}{c}f'(x)$$

To use the Constant Multiple Rule efficiently, look for constants that can be factored out *before* differentiating. For example,

$$\frac{d}{dx}[5x^2] = 5\frac{d}{dx}[x^2] \qquad \text{Factor out 5.}$$
$$= 5(2x) \qquad \text{Differentiate.}$$
$$= 10x \qquad \text{Simplify.}$$

and

$$\frac{d}{dx}\left[\frac{x^2}{5}\right] = \frac{1}{5}\left(\frac{d}{dx}[x^2]\right) \qquad \text{Factor out } \tfrac{1}{5}.$$
$$= \frac{1}{5}(2x) \qquad \text{Differentiate.}$$
$$= \frac{2}{5}x. \qquad \text{Simplify.}$$

SECTION 2.2 Some Rules for Differentiation

TECHNOLOGY

If you have access to a symbolic differentiation utility, try using it to confirm the derivatives shown in this section.

Example 4 Using the Power and Constant Multiple Rules

Differentiate each function.

a. $y = 2x^{1/2}$ **b.** $f(t) = \dfrac{4t^2}{5}$

SOLUTION

a. Using the Constant Multiple Rule and the Power Rule, you can write

$$\frac{dy}{dx} = \frac{d}{dx}[2x^{1/2}] = 2\underbrace{\frac{d}{dx}[x^{1/2}]}_{\text{Constant Multiple Rule}} = 2\underbrace{\left(\frac{1}{2}x^{-1/2}\right)}_{\text{Power Rule}} = x^{-1/2} = \frac{1}{\sqrt{x}}.$$

b. Begin by rewriting $f(t)$ as

$$f(t) = \frac{4t^2}{5} = \frac{4}{5}t^2.$$

Then, use the Constant Multiple Rule and the Power Rule to obtain

$$f'(t) = \frac{d}{dt}\left[\frac{4}{5}t^2\right] = \frac{4}{5}\left[\frac{d}{dt}(t^2)\right] = \frac{4}{5}(2t) = \frac{8}{5}t.$$

✓ **CHECKPOINT 4**

Differentiate each function.

a. $y = 4x^2$
b. $f(x) = 16x^{1/2}$

You may find it helpful to combine the Constant Multiple Rule and the Power Rule into one combined rule.

$$\frac{d}{dx}[cx^n] = cnx^{n-1}, \quad n \text{ is a real number, } c \text{ is a constant.}$$

For instance, in Example 4(b), you can apply this combined rule to obtain

$$\frac{d}{dt}\left[\frac{4}{5}t^2\right] = \left(\frac{4}{5}\right)(2)(t) = \frac{8}{5}t.$$

The three functions in the next example are simple, yet errors are frequently made in differentiating functions involving constant multiples of the first power of x. Keep in mind that

$$\frac{d}{dx}[cx] = c, \quad c \text{ is a constant.}$$

Example 5 Applying the Constant Multiple Rule

Find the derivative of each function.

✓ **CHECKPOINT 5**

Find the derivative of each function.

a. $y = \dfrac{t}{4}$

b. $y = -\dfrac{2x}{5}$

Original Function	Derivative
a. $y = -\dfrac{3x}{2}$	$y' = -\dfrac{3}{2}$
b. $y = 3\pi x$	$y' = 3\pi$
c. $y = -\dfrac{x}{2}$	$y' = -\dfrac{1}{2}$

130 CHAPTER 2 Differentiation

Parentheses can play an important role in the use of the Constant Multiple Rule and the Power Rule. In Example 6, be sure you understand the mathematical conventions involving the use of parentheses.

Example 6 Using Parentheses When Differentiating

Find the derivative of each function.

a. $y = \dfrac{5}{2x^3}$ **b.** $y = \dfrac{5}{(2x)^3}$ **c.** $y = \dfrac{7}{3x^{-2}}$ **d.** $y = \dfrac{7}{(3x)^{-2}}$

SOLUTION

Function	Rewrite	Differentiate	Simplify
a. $y = \dfrac{5}{2x^3}$	$y = \dfrac{5}{2}(x^{-3})$	$y' = \dfrac{5}{2}(-3x^{-4})$	$y' = -\dfrac{15}{2x^4}$
b. $y = \dfrac{5}{(2x)^3}$	$y = \dfrac{5}{8}(x^{-3})$	$y' = \dfrac{5}{8}(-3x^{-4})$	$y' = -\dfrac{15}{8x^4}$
c. $y = \dfrac{7}{3x^{-2}}$	$y = \dfrac{7}{3}(x^2)$	$y' = \dfrac{7}{3}(2x)$	$y' = \dfrac{14x}{3}$
d. $y = \dfrac{7}{(3x)^{-2}}$	$y = 63(x^2)$	$y' = 63(2x)$	$y' = 126x$

✓ CHECKPOINT 6

Find the derivative of each function.

a. $y = \dfrac{9}{4x^2}$ **b.** $y = \dfrac{9}{(4x)^2}$ ■

Example 7 Differentiating Radical Functions

Find the derivative of each function.

a. $y = \sqrt{x}$ **b.** $y = \dfrac{1}{2\sqrt[3]{x^2}}$ **c.** $y = \sqrt{2x}$

SOLUTION

Function	Rewrite	Differentiate	Simplify
a. $y = \sqrt{x}$	$y = x^{1/2}$	$y' = \left(\dfrac{1}{2}\right)x^{-1/2}$	$y' = \dfrac{1}{2\sqrt{x}}$
b. $y = \dfrac{1}{2\sqrt[3]{x^2}}$	$y = \dfrac{1}{2}x^{-2/3}$	$y' = \dfrac{1}{2}\left(-\dfrac{2}{3}\right)x^{-5/3}$	$y' = -\dfrac{1}{3x^{5/3}}$
c. $y = \sqrt{2x}$	$y = \sqrt{2}(x^{1/2})$	$y' = \sqrt{2}\left(\dfrac{1}{2}\right)x^{-1/2}$	$y' = \dfrac{1}{\sqrt{2x}}$

STUDY TIP

When differentiating functions involving radicals, you should rewrite the function with rational exponents. For instance, you should rewrite $y = \sqrt[3]{x}$ as $y = x^{1/3}$, and you should rewrite

$y = \dfrac{1}{\sqrt[3]{x^4}}$ as $y = x^{-4/3}$.

✓ CHECKPOINT 7

Find the derivative of each function.

a. $y = \sqrt{5x}$
b. $y = \sqrt[3]{x}$ ■

The Sum and Difference Rules

The next two rules are ones that you might expect to be true, and you may have used them without thinking about it. For instance, if you were asked to differentiate $y = 3x + 2x^3$, you would probably write

$$y' = 3 + 6x^2$$

without questioning your answer. The validity of differentiating a sum or difference of functions term by term is given by the Sum and Difference Rules.

The Sum and Difference Rules

The derivative of the sum or difference of two differentiable functions is the sum or difference of their derivatives.

$$\frac{d}{dx}[f(x) + g(x)] = f'(x) + g'(x) \qquad \text{Sum Rule}$$

$$\frac{d}{dx}[f(x) - g(x)] = f'(x) - g'(x) \qquad \text{Difference Rule}$$

PROOF Let $h(x) = f(x) + g(x)$. Then, you can prove the Sum Rule as shown.

$$h'(x) = \lim_{\Delta x \to 0} \frac{h(x + \Delta x) - h(x)}{\Delta x} \qquad \text{Definition of derivative}$$

$$= \lim_{\Delta x \to 0} \frac{f(x + \Delta x) + g(x + \Delta x) - f(x) - g(x)}{\Delta x}$$

$$= \lim_{\Delta x \to 0} \frac{f(x + \Delta x) - f(x) + g(x + \Delta x) - g(x)}{\Delta x}$$

$$= \lim_{\Delta x \to 0} \left[\frac{f(x + \Delta x) - f(x)}{\Delta x} + \frac{g(x + \Delta x) - g(x)}{\Delta x} \right]$$

$$= \lim_{\Delta x \to 0} \frac{f(x + \Delta x) - f(x)}{\Delta x} + \lim_{\Delta x \to 0} \frac{g(x + \Delta x) - g(x)}{\Delta x}$$

$$= f'(x) + g'(x)$$

So,

$$\frac{d}{dx}[f(x) + g(x)] = f'(x) + g'(x).$$

The Difference Rule can be proved in a similar manner.

The Sum and Difference Rules can be extended to the sum or difference of any finite number of functions. For instance, if $y = f(x) + g(x) + h(x)$, then $y' = f'(x) + g'(x) + h'(x)$.

STUDY TIP

Look back at Example 6 on page 119. Notice that the example asks for the derivative of the difference of two functions. Verify this result by using the Difference Rule.

CHAPTER 2 Differentiation

With the four differentiation rules listed in this section, you can differentiate *any* polynomial function.

Example 8 Using the Sum and Difference Rules

Find the slope of the graph of $f(x) = x^3 - 4x + 2$ at the point $(1, -1)$.

SOLUTION The derivative of $f(x)$ is

$$f'(x) = 3x^2 - 4.$$

So, the slope of the graph of f at $(1, -1)$ is

$$\text{Slope} = f'(1) = 3(1)^2 - 4 = -1$$

as shown in Figure 2.15.

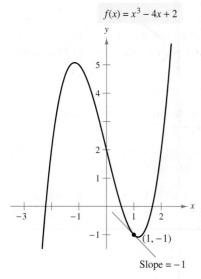

FIGURE 2.15

✓ **CHECKPOINT 8**

Find the slope of the graph of $f(x) = x^2 - 5x + 1$ at the point $(2, -5)$. ■

Example 8 illustrates the use of the derivative for determining the shape of a graph. A rough sketch of the graph of $f(x) = x^3 - 4x + 2$ might lead you to think that the point $(1, -1)$ is the minimum point of the graph. After finding the slope at this point to be -1, however, you can conclude that the minimum point (where the slope is 0) is farther to the right. (You will study techniques for finding minimum and maximum points in Section 3.2.)

Example 9 Using the Sum and Difference Rules

Find an equation of the tangent line to the graph of

$$g(x) = -\frac{1}{2}x^4 + 3x^3 - 2x$$

at the point $\left(-1, -\frac{3}{2}\right)$.

SOLUTION The derivative of $g(x)$ is $g'(x) = -2x^3 + 9x^2 - 2$, which implies that the slope of the graph at the point $\left(-1, -\frac{3}{2}\right)$ is

$$\text{Slope} = g'(-1) = -2(-1)^3 + 9(-1)^2 - 2$$
$$= 2 + 9 - 2$$
$$= 9$$

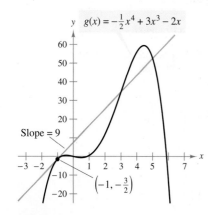

FIGURE 2.16

as shown in Figure 2.16. Using the point-slope form, you can write the equation of the tangent line at $\left(-1, -\frac{3}{2}\right)$ as shown.

$$y - \left(-\frac{3}{2}\right) = 9[x - (-1)] \qquad \text{Point-slope form}$$

$$y = 9x + \frac{15}{2} \qquad \text{Equation of tangent line}$$

✓ **CHECKPOINT 9**

Find an equation of the tangent line to the graph of $f(x) = -x^2 + 3x - 2$ at the point $(2, 0)$. ■

Application

Example 10 Modeling Social Security Beneficiaries

From 2000 through 2005, the number of Social Security beneficiaries (in thousands) in the United States can be modeled by

$$N = 31.27t^2 + 447.06t + 45{,}412, \quad 0 \le t \le 5$$

where t represents the year, with $t = 0$ corresponding to 2000. At what rate was the number of beneficiaries changing in 2002? *(Source: U.S. Social Security Administration)*

SOLUTION One way to answer this question is to find the derivative of the model with respect to time.

$$\frac{dN}{dt} = 62.54t + 447.06, \quad 0 \le t \le 5$$

In 2002 (when $t = 2$), the rate of change of the number of beneficiaries with respect to time is given by

$$62.54(2) + 447.06 = 572.14.$$

Because N is measured in thousands of beneficiaries and t is measured in years, it follows that the derivative dN/dt is measured in thousands of beneficiaries per year. So, at the end of 2002, the number of Social Security beneficiaries was increasing at a rate of 572.14 thousand per year, as shown in Figure 2.17.

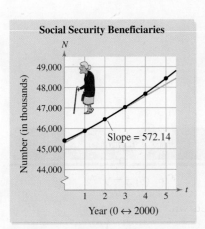

FIGURE 2.17

✓ CHECKPOINT 10

From 2002 through 2006, the number of child support cases (in thousands) in the United States can be modeled by

$$N = 23.43t^2 - 238.0t + 16{,}440, \quad 2 \le t \le 6$$

where t represents the year, with $t = 2$ corresponding to 2002. At what rate was the number of cases changing in 2004? *(Source: U.S. Department of Health and Human Services, Office of Child Support Enforcement)* ∎

CONCEPT CHECK

1. What is the derivative of any constant function?
2. Write a verbal description of the Power Rule.
3. According to the Sum Rule, the derivative of the sum of two differentiable functions is equal to what?
4. According to the Difference Rule, the derivative of the difference of two differentiable functions is equal to what?

134 CHAPTER 2 Differentiation

Skills Review 2.2

The following warm-up exercises involve skills that were covered in earlier sections. You will use these skills in the exercise set for this section. For additional help, review Sections 0.3 and 0.4.

In Exercises 1 and 2, evaluate each expression when $x = 2$.

1. (a) $2x^2$ (b) $(2x)^2$ (c) $2x^{-2}$
2. (a) $\dfrac{1}{(3x)^2}$ (b) $\dfrac{1}{4x^3}$ (c) $\dfrac{(2x)^{-3}}{4x^{-2}}$

In Exercises 3–6, simplify the expression.

3. $4(3)x^3 + 2(2)x$
4. $\frac{1}{2}(3)x^2 - \frac{3}{2}x^{1/2}$
5. $\left(\frac{1}{4}\right)x^{-3/4}$
6. $\frac{1}{3}(3)x^2 - 2\left(\frac{1}{2}\right)x^{-1/2} + \frac{1}{3}x^{-2/3}$

In Exercises 7–10, solve the equation.

7. $3x^2 + 2x = 0$
8. $x^3 - x = 0$
9. $x^2 + 8x - 20 = 0$
10. $x^2 - 10x - 24 = 0$

Exercises 2.2

See www.CalcChat.com for worked-out solutions to odd-numbered exercises.

In Exercises 1–4, find the slope of the tangent line to $y = x^n$ at the point $(1, 1)$.

1. (a) $y = x^2$ (b) $y = x^{1/2}$

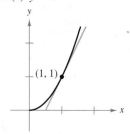

2. (a) $y = x^{3/2}$ (b) $y = x^3$

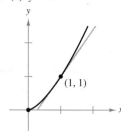

3. (a) $y = x^{-1}$ (b) $y = x^{-1/3}$

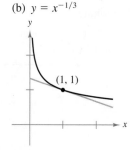

4. (a) $y = x^{-1/2}$ (b) $y = x^{-2}$

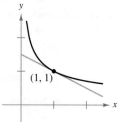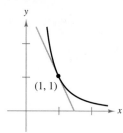

In Exercises 5–22, find the derivative of the function.

5. $y = 3$
6. $f(x) = -2$
7. $y = x^4$
8. $h(x) = 2x^5$
9. $f(x) = 4x + 1$
10. $g(x) = 3x - 1$
11. $g(x) = x^2 + 5x$
12. $y = t^2 - 6$
13. $f(t) = -3t^2 + 2t - 4$
14. $y = x^3 - 9x^2 + 2$
15. $s(t) = t^3 - 2t + 4$
16. $y = 2x^3 - x^2 + 3x - 1$
17. $y = 4t^{4/3}$
18. $h(x) = x^{5/2}$
19. $f(x) = 4\sqrt{x}$
20. $g(x) = 4\sqrt[3]{x} + 2$
21. $y = 4x^{-2} + 2x^2$
22. $s(t) = 4t^{-1} + 1$

In Exercises 23–28, use Example 6 as a model to find the derivative.

Function	Rewrite	Differentiate	Simplify
23. $y = \dfrac{1}{x^3}$			
24. $y = \dfrac{2}{3x^2}$			
25. $y = \dfrac{1}{(4x)^3}$			
26. $y = \dfrac{\pi}{(3x)^2}$			
27. $y = \dfrac{\sqrt{x}}{x}$			
28. $y = \dfrac{4x}{x^{-3}}$			

In Exercises 29–34, find the value of the derivative of the function at the given point.

Function	Point
29. $f(x) = \dfrac{1}{x}$	$(1, 1)$
30. $f(t) = 4 - \dfrac{4}{3t}$	$\left(\dfrac{1}{2}, \dfrac{4}{3}\right)$
31. $f(x) = -\dfrac{1}{2}x(1 + x^2)$	$(1, -1)$
32. $y = 3x\left(x^2 - \dfrac{2}{x}\right)$	$(2, 18)$
33. $y = (2x + 1)^2$	$(0, 1)$
34. $f(x) = 3(5 - x)^2$	$(5, 0)$

In Exercises 35–48, find $f'(x)$.

35. $f(x) = x^2 - \dfrac{4}{x} - 3x^{-2}$

36. $f(x) = x^2 - 3x - 3x^{-2} + 5x^{-3}$

37. $f(x) = x^2 - 2x - \dfrac{2}{x^4}$

38. $f(x) = x^2 + 4x + \dfrac{1}{x}$

39. $f(x) = x(x^2 + 1)$

40. $f(x) = (x^2 + 2x)(x + 1)$

41. $f(x) = (x + 4)(2x^2 - 1)$

42. $f(x) = (3x^2 - 5x)(x^2 + 2)$

43. $f(x) = \dfrac{2x^3 - 4x^2 + 3}{x^2}$

44. $f(x) = \dfrac{2x^2 - 3x + 1}{x}$

45. $f(x) = \dfrac{4x^3 - 3x^2 + 2x + 5}{x^2}$

46. $f(x) = \dfrac{-6x^3 + 3x^2 - 2x + 1}{x}$

47. $f(x) = x^{4/5} + x$

48. $f(x) = x^{1/3} - 1$

In Exercises 49–52, (a) find an equation of the tangent line to the graph of the function at the given point, (b) use a graphing utility to graph the function and its tangent line at the point, and (c) use the *derivative* feature of a graphing utility to confirm your results.

Function	Point
49. $y = -2x^4 + 5x^2 - 3$	$(1, 0)$
50. $y = x^3 + x$	$(-1, -2)$
51. $f(x) = \sqrt[3]{x} + \sqrt[5]{x}$	$(1, 2)$
52. $f(x) = \dfrac{1}{\sqrt[3]{x^2}} - x$	$(-1, 2)$

In Exercises 53–56, determine the point(s), if any, at which the graph of the function has a horizontal tangent line.

53. $y = -x^4 + 3x^2 - 1$

54. $y = x^3 + 3x^2$

55. $y = \frac{1}{2}x^2 + 5x$

56. $y = x^2 + 2x$

In Exercises 57 and 58, (a) sketch the graphs of f and g, (b) find $f'(1)$ and $g'(1)$, (c) sketch the tangent line to each graph when $x = 1$, and (d) explain the relationship between f' and g'.

57. $f(x) = x^3$
    $g(x) = x^3 + 3$

58. $f(x) = x^2$
    $g(x) = 3x^2$

59. Use the Constant Rule, the Constant Multiple Rule, and the Sum Rule to find $h'(1)$ given that $f'(1) = 3$.
   (a) $h(x) = f(x) - 2$
   (b) $h(x) = 2f(x)$

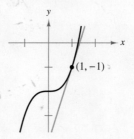

(c) $h(x) = -f(x)$

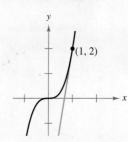

(d) $h(x) = -1 + 2f(x)$

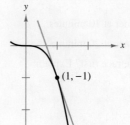
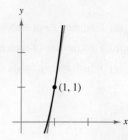

60. Timber Products The amount L (in millions of cubic feet per year) of softwood products consumed from 2000 through 2005 can be modeled by

$$L = -38.8333t^4 + 302.944t^3 - 538.50t^2 + 209.7t + 12{,}652$$

where t is the year, with $t = 0$ corresponding to 2000. (Source: U.S. Forest Service)

(a) Use the model to find the slopes of the graph of L for the years 2002 and 2004.

(b) Compare your results with those obtained in Exercise 13 in Section 2.1.

(c) What are the units for the slope of the graph?

61. Uranium Concentrate The amount U (in thousands of pounds per year) of uranium produced in mines from 2000 through 2005 can be modeled by

$$U = 1.6875t^4 - 0.255t^3 + 57.91t^2 - 507.9t + 3118$$

where t is the year, with $t = 0$ corresponding to 2000. (Source: U.S. Department of Energy)

(a) Use the model to find the slopes of the graph of U for the years 2001 and 2004.

(b) Compare your results with those obtained in Exercise 12 in Section 2.1.

(c) What are the units for the slope of the graph?

62. Athletics Runner f and runner g start out side by side on a 10,000-meter run. Their distances (in thousands of meters) are given by $f(t) = 1.58\sqrt{t}$ and $g(t) = 0.0063t^2$, respectively, where t is measured in minutes.

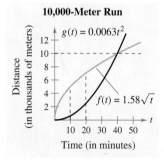

10,000-Meter Run

(a) Find the rate of each runner at 10 minutes.

(b) Find the rate of each runner at 20 minutes.

(c) Which runner finishes the race first? Explain.

63. Meteorology The mean monthly temperature T (in degrees Fahrenheit) for Honolulu, Hawaii can be modeled by

$$T = -0.0418t^3 + 0.605t^2 - 1.16t + 73.4$$

where t is the month, with $t = 1$ corresponding to January.

(a) Use the model to find the slopes of the graph of T at $t = 1, 8$, and 12.

(b) What are the units for the slope of the graph?

64. Psychology: Migraine Prevalence The graph illustrates the prevalence of migraine headaches in males and females in selected income groups. (Source: Adapted from Sue/Sue/Sue, Understanding Abnormal Behavior, Seventh Edition)

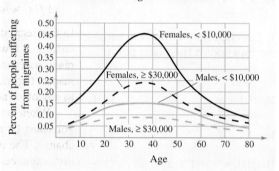

Prevalence of Migraine Headaches

(a) Write a short paragraph describing your general observations about the prevalence of migraines in females and males with respect to age group and income bracket.

(b) Describe the graph of the derivative of each curve, and explain the significance of each derivative. Include an explanation of the units of the derivatives, and indicate the time intervals in which the derivatives would be positive and negative.

In Exercises 65 and 66, use a graphing utility to graph f and f' over the given interval. Determine any points at which the graph of f has horizontal tangents.

Function	Interval
65. $f(x) = 4.1x^3 - 12x^2 + 2.5x$	$[0, 3]$
66. $f(x) = x^3 - 1.4x^2 - 0.96x + 1.44$	$[-2, 2]$

True or False? In Exercises 67 and 68, determine whether the statement is true or false. If it is false, explain why or give an example that shows it is false.

67. If $f'(x) = g'(x)$, then $f(x) = g(x)$.

68. If $f(x) = g(x) + c$, then $f'(x) = g'(x)$.

Section 2.3
Rates of Change

- Find the average rates of change of functions over intervals.
- Find the instantaneous rates of change of functions at points.
- Use rates of change to solve real-life problems.

Average Rate of Change

In Sections 2.1 and 2.2, you studied the two primary applications of derivatives.

1. **Slope** The derivative of f is a function that gives the slope of the graph of f at a point $(x, f(x))$.

2. **Rate of Change** The derivative of f is a function that gives the rate of change of $f(x)$ with respect to x at the point $(x, f(x))$.

In this section, you will see that there are many real-life applications of rates of change. A few are velocity, acceleration, population growth rates, unemployment rates, and water flow rates. Although rates of change often involve change with respect to time, you can investigate the rate of change of one variable with respect to any other related variable.

When determining the rate of change of one variable with respect to another, you must be careful to distinguish between *average* and *instantaneous* rates of change. The distinction between these two rates of change is comparable to the distinction between the slope of the secant line through two points on a graph and the slope of the tangent line at one point on the graph.

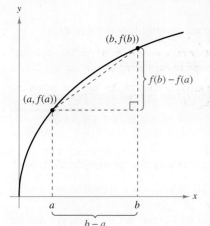

FIGURE 2.18

Definition of Average Rate of Change

If $y = f(x)$, then the **average rate of change** of y with respect to x on the interval $[a, b]$ is

$$\text{Average rate of change} = \frac{f(b) - f(a)}{b - a}$$

$$= \frac{\Delta y}{\Delta x}.$$

Note that $f(a)$ is the value of the function at the *left* endpoint of the interval, $f(b)$ is the value of the function at the *right* endpoint of the interval, and $b - a$ is the width of the interval, as shown in Figure 2.18.

STUDY TIP

In real-life problems, it is important to list the units of measure for a rate of change. The units for $\Delta y / \Delta x$ are "y-units" per "x-units." For example, if y is measured in miles and x is measured in hours, then $\Delta y / \Delta x$ is measured in *miles per hour*.

Example 1 Medicine

The concentration C (in milligrams per milliliter) of a drug in a patient's bloodstream is monitored over 10-minute intervals for 2 hours, where t is measured in minutes, as shown in the table. Find the average rate of change over each interval.

a. $[0, 10]$ **b.** $[0, 20]$ **c.** $[100, 110]$

t	0	10	20	30	40	50	60	70	80	90	100	110	120
C	0	2	17	37	55	73	89	103	111	113	113	103	68

STUDY TIP

In Example 1, the average rate of change is positive when the concentration increases and negative when the concentration decreases, as shown in Figure 2.19.

SOLUTION

a. For the interval $[0, 10]$, the average rate of change is

$$\frac{\Delta C}{\Delta t} = \frac{2 - 0}{10 - 0} = \frac{2}{10} = 0.2 \text{ milligram per milliliter per minute.}$$

where the numerator is the value of C at right endpoint minus value of C at left endpoint, and the denominator is the width of interval.

b. For the interval $[0, 20]$, the average rate of change is

$$\frac{\Delta C}{\Delta t} = \frac{17 - 0}{20 - 0} = \frac{17}{20} = 0.85 \text{ milligram per milliliter per minute.}$$

c. For the interval $[100, 110]$, the average rate of change is

$$\frac{\Delta C}{\Delta t} = \frac{103 - 113}{110 - 100} = \frac{-10}{10} = -1 \text{ milligram per milliliter per minute.}$$

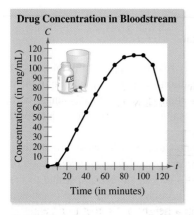

FIGURE 2.19

✓ CHECKPOINT 1

Use the table in Example 1 to find the average rate of change over each interval.

a. $[0, 120]$ **b.** $[90, 100]$ **c.** $[90, 120]$ ■

The rates of change in Example 1 are in milligrams per milliliter per minute because the concentration is measured in milligrams per milliliter and the time is measured in minutes.

$$\frac{\Delta C}{\Delta t} = \frac{2 - 0}{10 - 0} = \frac{2}{10} = 0.2 \text{ milligram per milliliter per minute}$$

where concentration is measured in milligrams per milliliter, rate of change is measured in milligrams per milliliter per minute, and time is measured in minutes.

SECTION 2.3 Rates of Change

A common application of an average rate of change is to find the **average velocity** of an object that is moving in a straight line. That is,

$$\text{Average velocity} = \frac{\text{change in distance}}{\text{change in time}}.$$

This formula is demonstrated in Example 2.

Example 2 Finding an Average Velocity

If a free-falling object is dropped from a height of 100 feet, and *air resistance is neglected*, the height h (in feet) of the object at time t (in seconds) is given by

$$h = -16t^2 + 100. \quad \text{(See Figure 2.20.)}$$

Find the average velocity of the object over each interval.

a. $[1, 2]$ **b.** $[1, 1.5]$ **c.** $[1, 1.1]$

SOLUTION You can use the position equation $h = -16t^2 + 100$ to determine the heights at $t = 1$, $t = 1.1$, $t = 1.5$, and $t = 2$, as shown in the table.

t (in seconds)	0	1	1.1	1.5	2
h (in feet)	100	84	80.64	64	36

a. For the interval $[1, 2]$, the object falls from a height of 84 feet to a height of 36 feet. So, the average velocity is

$$\frac{\Delta h}{\Delta t} = \frac{36 - 84}{2 - 1} = \frac{-48}{1} = -48 \text{ feet per second}.$$

b. For the interval $[1, 1.5]$, the average velocity is

$$\frac{\Delta h}{\Delta t} = \frac{64 - 84}{1.5 - 1} = \frac{-20}{0.5} = -40 \text{ feet per second}.$$

c. For the interval $[1, 1.1]$, the average velocity is

$$\frac{\Delta h}{\Delta t} = \frac{80.64 - 84}{1.1 - 1} = \frac{-3.36}{0.1} = -33.6 \text{ feet per second}.$$

✓ CHECKPOINT 2

The height h (in feet) of a free-falling object at time t (in seconds) is given by $h = -16t^2 + 180$. Find the average velocity of the object over each interval.

a. $[0, 1]$ **b.** $[1, 2]$ **c.** $[2, 3]$ ∎

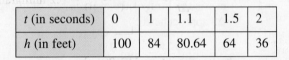

FIGURE 2.20 Some falling objects have considerable air resistance. Other falling objects have negligible air resistance. When modeling a falling-body problem, you must decide whether to account for air resistance or neglect it.

STUDY TIP

In Example 2, the average velocities are negative because the object is moving downward.

Instantaneous Rate of Change and Velocity

Suppose in Example 2 you wanted to find the rate of change of h at the instant $t = 1$ second. Such a rate is called an **instantaneous rate of change**. You can approximate the instantaneous rate of change at $t = 1$ by calculating the average rate of change over smaller and smaller intervals of the form $[1, 1 + \Delta t]$, as shown in the table. From the table, it seems reasonable to conclude that the instantaneous rate of change of the height when $t = 1$ is -32 feet per second.

Δt approaches 0.

Δt	1	0.5	0.1	0.01	0.001	0.0001	0
$\dfrac{\Delta h}{\Delta t}$	-48	-40	-33.6	-32.16	-32.016	-32.0016	-32

$\dfrac{\Delta h}{\Delta t}$ approaches -32.

STUDY TIP

The limit in this definition is the same as the limit in the definition of the derivative of f at x. This is the second major interpretation of the derivative—as an *instantaneous rate of change in one variable with respect to another*. Recall that the first interpretation of the derivative is as the slope of the graph of f at x.

Definition of Instantaneous Rate of Change

The **instantaneous rate of change** (or simply **rate of change**) of $y = f(x)$ at x is the limit of the average rate of change on the interval $[x, x + \Delta x]$, as Δx approaches 0.

$$\lim_{\Delta x \to 0} \frac{\Delta y}{\Delta x} = \lim_{\Delta x \to 0} \frac{f(x + \Delta x) - f(x)}{\Delta x}$$

If y is a distance and x is time, then the rate of change is a **velocity**.

Example 3 Finding an Instantaneous Rate of Change

Find the velocity of the object in Example 2 when $t = 1$.

SOLUTION From Example 2, you know that the height of the falling object is given by

$h = -16t^2 + 100.$ Position function

By taking the derivative of this position function, you obtain the velocity function.

$h'(t) = -32t$ Velocity function

The velocity function gives the velocity at *any* time. So, when $t = 1$, the velocity is

$h'(1) = -32(1)$
$= -32$ feet per second.

✓ CHECKPOINT 3

Find the velocities of the object in Checkpoint 2 when $t = 1.75$ and $t = 2$. ∎

SECTION 2.3 Rates of Change

DISCOVERY

Graph the polynomial function $h = -16t^2 + 16t + 32$ from Example 4 on the domain $0 \leq t \leq 2$. What is the maximum value of h? What is the derivative of h at this maximum point? In general, discuss how the derivative can be used to find the maximum or minimum values of a function.

The general **position function** for a free-falling object, neglecting air resistance, is

$$h = -16t^2 + v_0 t + h_0 \quad \text{Position function}$$

where h is the height (in feet), t is the time (in seconds), v_0 is the initial velocity (in feet per second), and h_0 is the initial height (in feet). Remember that the model assumes that positive velocities indicate upward motion and negative velocities indicate downward motion. The derivative $h' = -32t + v_0$ is the **velocity function**. The absolute value of the velocity is the **speed** of the object.

Example 4 Finding the Velocity of a Diver

At time $t = 0$, a diver jumps from a diving board that is 32 feet high, as shown in Figure 2.21. Because the diver's initial velocity is 16 feet per second, his position function is

$$h = -16t^2 + 16t + 32. \quad \text{Position function}$$

a. When does the diver hit the water?

b. What is the diver's velocity at impact?

SOLUTION

a. To find the time at which the diver hits the water, let $h = 0$ and solve for t.

$$-16t^2 + 16t + 32 = 0 \quad \text{Set } h \text{ equal to 0.}$$
$$-16(t^2 - t - 2) = 0 \quad \text{Factor out common factor.}$$
$$-16(t + 1)(t - 2) = 0 \quad \text{Factor.}$$
$$t = -1 \text{ or } t = 2 \quad \text{Solve for } t.$$

The solution $t = -1$ does not make sense in the problem because it would mean the diver hits the water 1 second before he jumps. So, you can conclude that the diver hits the water when $t = 2$ seconds.

b. The velocity at time t is given by the derivative

$$h' = -32t + 16. \quad \text{Velocity function}$$

The velocity at time $t = 2$ is $-32(2) + 16 = -48$ feet per second.

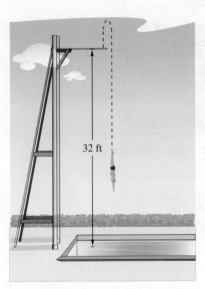

FIGURE 2.21

✓ CHECKPOINT 4

Give the position function of a diver who jumps from a board 12 feet high with initial velocity 16 feet per second. Then find the diver's velocity function. ∎

In Example 4, note that the diver's initial velocity is $v_0 = 16$ feet per second (upward) and his initial height is $h_0 = 32$ feet.

$$h = -16t^2 + \underset{\underset{\text{Initial velocity is 16 feet per second.}}{\downarrow}}{16}t + \underset{\underset{\text{Initial height is 32 feet.}}{\downarrow}}{32}$$

Application

Example 5 Holstein Calf Growth

The weights w (in pounds) for a Holstein calf during the first 12 months of its life are shown in the table, where t is the time (in months) since birth.

t	0	1	2	4	6	12
w	96	118	161	272	396	714

The calf's weight can be modeled by

$$w = 0.48t^2 + 47.3t + 80, \quad 0 \le t \le 12.$$

a. Find the average rate of change of the calf's weight during the first 12 months of its life.

b. Find the rate of change of the calf's weight when $t = 4$.

SOLUTION

a. The first 12 months of the calf's life can be represented by the interval $[0, 12]$. For the interval $[0, 12]$, the calf's weight increases from 96 pounds to 714 pounds. So, the average rate of change is

$$\frac{\Delta w}{\Delta t} = \frac{714 - 96}{12 - 0} = \frac{618}{12} = 51.5 \text{ pounds per month.}$$

b. The rate of change of the calf's weight is given by the derivative

$$w' = 0.96t + 47.3.$$

When $t = 4$, the rate of change of the calf's weight is

$$w' = 0.96(4) + 47.3 = 3.84 + 47.3 = 51.14 \text{ pounds per month.}$$

Thiriet Claudius/Peter Arnold Inc.
An average Holstein calf weighs about 90 pounds at birth.

✓ CHECKPOINT 5

Use the information in Example 5 to answer the following.

a. Find the average rate of change of the calf's weight during the first 6 months of its life.

b. Find the rate of change of the calf's weight when $t = 6$.

CONCEPT CHECK

1. Complete the following: When determining the rate of change in the concentration of a drug over a 2-minute time interval, you are finding an _____ rate of change.

2. Complete the following: When determining the rate of change in the concentration of a drug exactly 1 hour after the drug was administered, you are finding an _____ rate of change.

3. You are asked to find the rate of change of a function over a certain interval. Should you find the average rate of change or the instantaneous rate of change?

4. You are asked to find the rate of change of a function at a certain instant. Should you find the average rate of change or the instantaneous rate of change?

SECTION 2.3 Rates of Change

Skills Review 2.3

The following warm-up exercises involve skills that were covered in earlier sections. You will use these skills in the exercise set for this section. For additional help, review Sections 2.1 and 2.2.

In Exercises 1 and 2, evaluate the expression.

1. $\dfrac{-63 - (-105)}{21 - 7}$

2. $\dfrac{-37 - 54}{16 - 3}$

In Exercises 3–10, find the derivative of the function.

3. $y = 4x^2 - 2x + 7$

4. $y = -3t^3 + 2t^2 - 8$

5. $s = -16t^2 + 24t + 30$

6. $y = -16x^2 + 54x + 70$

7. $A = \frac{1}{10}(-2r^3 + 3r^2 + 5r)$

8. $y = \frac{1}{9}(6x^3 - 18x^2 + 63x - 15)$

9. $y = 12x - \dfrac{x^2}{5000}$

10. $y = 138 + 74x - \dfrac{x^3}{10{,}000}$

Exercises 2.3

See www.CalcChat.com for worked-out solutions to odd-numbered exercises.

1. **Research and Development** The table shows the amounts A (in billions of dollars per year) spent on R&D in the United States from 1980 through 2004, where t is the year, with $t = 0$ corresponding to 1980. Approximate the average rate of change of A during each period. *(Source: U.S. National Science Foundation)*

   (a) 1980–1985 (b) 1985–1990 (c) 1990–1995
   (d) 1995–2000 (e) 1980–2004 (f) 1990–2004

t	0	1	2	3	4	5	6
A	63	72	81	90	102	115	120

t	7	8	9	10	11	12
A	126	134	142	152	161	165

t	13	14	15	16	17	18
A	166	169	184	197	212	228

t	19	20	21	22	23	24
A	245	267	277	276	292	312

2. **Trade Deficit** The graph shows the values I (in billions of dollars per year) of goods imported to the United States and the values E (in billions of dollars per year) of goods exported from the United States from 1980 through 2005. Approximate each indicated average rate of change. *(Source: U.S. International Trade Administration)*

   (a) Imports: 1980–1990 (b) Exports: 1980–1990
   (c) Imports: 1990–2000 (d) Exports: 1990–2000
   (e) Imports: 1980–2005 (f) Exports: 1980–2005

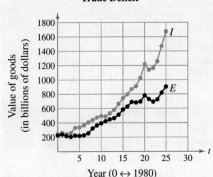

Figure for 2

 In Exercises 3–12, use a graphing utility to graph the function and find its average rate of change on the interval. Compare this rate with the instantaneous rates of change at the endpoints of the interval.

3. $f(t) = 3t + 5$, $[1, 2]$

4. $h(x) = 2 - x$, $[0, 2]$

5. $h(x) = x^2 - 4x + 2$, $[-2, 2]$

6. $f(x) = x^2 - 6x - 1$, $[-1, 3]$

7. $f(x) = 3x^{4/3}$, $[1, 8]$

8. $f(x) = x^{3/2}$, $[1, 4]$

9. $f(x) = \dfrac{1}{x}$, $[1, 4]$

10. $f(x) = \dfrac{1}{\sqrt{x}}$, $[1, 4]$

11. $g(x) = x^4 - x^2 + 2$, $[1, 3]$

12. $g(x) = x^3 - 1$, $[-1, 1]$

13. Medicine The graph shows the estimated number of milligrams of a pain medication M in the bloodstream t hours after a 1000-milligram dose of the drug has been given.

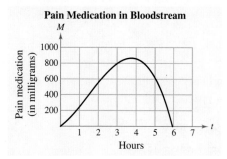

Pain Medication in Bloodstream

(a) Estimate the one-hour interval over which the average rate of change is the greatest.

(b) Over what interval is the average rate of change approximately equal to the rate of change at $t = 4$? Explain your reasoning.

14. Consumer Trends The graph shows the number of visitors V to a national park (in hundreds of thousands) during a one-year period, where $t = 1$ represents January.

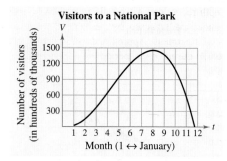

Visitors to a National Park

(a) Estimate the rate of change of V over the interval $[9, 12]$ and explain your results.

(b) Over what interval is the average rate of change approximately equal to the rate of change at $t = 8$? Explain your reasoning.

15. Health The temperature (in degrees Fahrenheit) of a person during an illness can be modeled by

$$T = -0.0375t^2 + 0.3t + 100.4$$

where t is time in hours since the person started to show signs of a fever.

(T) (a) Use a graphing utility to graph the function. Be sure to choose an appropriate window.

(b) Do the slopes of the tangent lines appear to be positive or negative? What does this tell you?

(c) Evaluate the function for $t = 0, 4, 8,$ and 12.

(d) Find dT/dt and explain its meaning in this situation.

(e) Evaluate dT/dt for $t = 0, 4, 8,$ and 12.

16. Biomedical Inventory The annual inventory cost C for a biomedical manufacturer can be modeled by

$$C = 1{,}008{,}000/Q + 6.3Q$$

where Q is the order size when the inventory is replenished. Find the change in annual cost when Q is increased from 350 to 351, and compare this with the instantaneous rate of change when $Q = 350$.

17. Physics: Wind Chill At 0° Celsius, the heat loss H (in kilocalories per square meter per hour) from a person's body can be modeled by $H = 33(10\sqrt{v} - v + 10.45)$, where v is the wind speed (in meters per second).

(a) Find $\dfrac{dH}{dv}$ and interpret its meaning in this situation.

(b) Find the rates of change of H when $v = 2$ and when $v = 5$.

18. Medicine The effectiveness E (on a scale from 0 to 1) of a pain-killing drug t hours after entering the bloodstream is given by

$$E = \frac{1}{27}(9t + 3t^2 - t^3), \quad 0 \leq t \leq 4.5.$$

Find the average rate of change of E on each indicated interval and compare this rate with the instantaneous rates of change at the endpoints of the interval.

(a) $[0, 1]$ (b) $[1, 2]$ (c) $[2, 3]$ (d) $[3, 4]$

(B) **19. Physics: Velocity** A racecar travels northward on a straight, level track at a constant speed, traveling 0.750 kilometer in 20.0 seconds. The return trip over the same track is made in 25.0 seconds.

(a) What is the average velocity of the car in meters per second for the first leg of the run?

(b) What is the average velocity for the total trip?

(Source: Shipman/Wilson/Todd, An Introduction to Physical Science, Eleventh Edition)

20. Velocity The height s (in feet) at time t (in seconds) of a silver dollar dropped from the top of the Washington Monument is given by

$$s = -16t^2 + 555.$$

(a) Find the average velocity on the interval $[2, 3]$.

(b) Find the instantaneous velocities when $t = 2$ and when $t = 3$.

(c) How long will it take the dollar to hit the ground?

(d) Find the velocity of the dollar when it hits the ground.

(T) In Exercises 21 and 22, use a graphing utility to graph f and f'. Then determine the points (if any) at which f has a horizontal tangent.

21. $f(x) = \frac{1}{4}x^3, \quad -2 \leq x \leq 2$

22. $f(x) = x^4 - 12x^3 + 52x^2 - 96x + 64, \quad 1 \leq x \leq 5$

23. Population Growth The population P (in thousands) of Japan can be modeled by

$$P = -14.71t^2 + 785.5t + 117{,}216$$

where t is time in years, with $t = 0$ corresponding to 1980. *(Source: U.S. Census Bureau)*

(a) Evaluate P for $t = 0, 10, 15, 20,$ and 25. Explain these values.

(b) Determine the population growth rate, dP/dt.

(c) Evaluate dP/dt for the same values as in part (a). Explain your results.

24. Population Growth The population of a developing rural area has been growing according to the model $P = 22t^2 + 52t + 10{,}000$, where t is time in years, with $t = 0$ corresponding to 1990.

(a) Evaluate P for $t = 0, 10, 15, 20,$ and 25. Explain these values.

(b) Determine the population growth rate, dP/dt.

(c) Evaluate dP/dt for the same values as in part (a). Explain your results.

25. MAKE A DECISION: FUEL COST A car is driven 15,000 miles a year and gets x miles per gallon. Assume that the average fuel cost is \$2.95 per gallon. Find the annual cost of fuel C as a function of x and use this function to complete the table.

x	10	15	20	25	30	35	40
C							
dC/dx							

Who would benefit more from a 1 mile per gallon increase in fuel efficiency—the driver who gets 15 miles per gallon or the driver who gets 35 miles per gallon? Explain.

26. Gasoline Sales The number N of gallons of regular unleaded gasoline sold by a gasoline station at a price of p dollars per gallon is given by $N = f(p)$.

(a) Describe the meaning of $f'(2.959)$.

(b) Is $f'(2.959)$ usually positive or negative? Explain.

27. Consider the function given by $f(x) = \dfrac{4}{x}$, $0 < x \leq 5$.

(a) Use a graphing utility to graph f and f' in the same viewing window.

(b) Complete the table.

x	$\frac{1}{8}$	$\frac{1}{4}$	$\frac{1}{2}$	1	2	3	4	5
$f(x)$								
$f'(x)$								

(c) Find the average rate of change of f over the intervals determined by consecutive x-values in the table.

28. Center of Population Since 1790, the center of population of the United States has been gradually moving westward. Use the figure to estimate the rate (in miles per year) at which the center of population was moving westward during the given period. *(Source: U.S. Census Bureau)*

(a) From 1790 to 1900   (b) From 1900 to 2000

(c) From 1790 to 2000

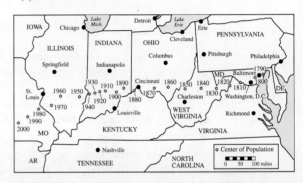

29. Periodontal Disease The table shows the average pocket depth reductions A (in millimeters) of periodontal pockets under teeth caused by periodontal disease after t months of treatment. Approximate the average rate of change of A during each period. *(Submitted by Terry Maher, Clinical Dental Researcher)*

t	0	1	3	6	9	12
A	0	1.5	1.5	1.7	1.9	1.7

(a) 0–1 month   (b) 3–6 months

(c) 6–9 months   (d) 9–12 months

30. Biology Many populations in nature exhibit logistic growth, which consists of four phases, as shown in the figure. Describe the rate of growth of the population in each phase, and give possible reasons as to why the rates might be changing from phase to phase. *(Source: Adapted from Levine/Miller, Biology: Discovering Life, Second Edition)*

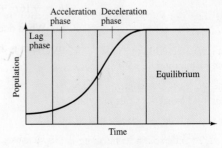

Mid-Chapter Quiz

Take this test as you would take a test in class. When you are done, check your work against the answers given in the back of the book.

In Exercises 1–3, use the limit definition to find the derivative of the function. Then find the slope of the tangent line to the graph of f at the given point.

1. $f(x) = -x + 2$, $(2, 0)$
2. $f(x) = \sqrt{x + 3}$, $(1, 2)$
3. $f(x) = \dfrac{4}{x}$, $(1, 4)$

In Exercises 4–9, find the derivative of the function.

4. $f(x) = 12$
5. $f(x) = 19x + 9$
6. $f(x) = 5 - 3x^2$
7. $f(x) = 12x^{1/4}$
8. $f(x) = 4x^{-2}$
9. $f(x) = 2\sqrt{x}$

(T) In Exercises 10–13, use a graphing utility to graph the function and find its average rate of change on the interval. Compare this rate with the instantaneous rates of change at the endpoints of the interval.

10. $f(x) = x^2 - 3x + 1$, $[0, 3]$
11. $f(x) = 2x^3 + x^2 - x + 4$, $[-1, 1]$
12. $f(x) = \dfrac{1}{2x}$, $[2, 5]$
13. $f(x) = \sqrt[3]{x}$, $[8, 27]$

14. The graph represents the average salary S (in dollars) of classroom teachers in public schools for the years 2000 through 2005, where t represents the year, with $t = 0$ corresponding to 2000. Estimate the slopes of the graph for the years 2002 and 2004. *(Source: Educational Research Service)*

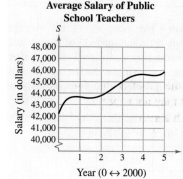

Figure for 14

(T) In Exercises 15 and 16, find an equation of the tangent line to the graph of f at the given point. Then use a graphing utility to graph the function and the tangent line in the same viewing window.

15. $f(x) = 5x^2 + 6x - 1$, $(-1, -2)$
16. $f(x) = (x - 1)(x + 1)$, $(0, -1)$

17. At time $t = 0$, a diver jumps from a diving board that is 25 feet high. Because the diver's initial velocity is 15 feet per second, her position function is
$$h = -16t^2 + 15t + 25.$$
(a) When does the diver hit the water?
(b) What is the diver's velocity at impact?

Section 2.4
The Product and Quotient Rules

- Find the derivatives of functions using the Product Rule.
- Find the derivatives of functions using the Quotient Rule.
- Simplify derivatives.
- Use derivatives to answer questions about real-life situations.

The Product Rule

In Section 2.2, you saw that the derivative of a sum or difference of two functions is simply the sum or difference of their derivatives. The rules for the derivative of a product or quotient of two functions are not as simple.

STUDY TIP

Rather than trying to remember the formula for the Product Rule, it can be more helpful to remember its verbal statement: *the first function times the derivative of the second plus the second function times the derivative of the first.*

The Product Rule

The derivative of the product of two differentiable functions is equal to the first function times the derivative of the second plus the second function times the derivative of the first.

$$\frac{d}{dx}[f(x)g(x)] = f(x)g'(x) + g(x)f'(x)$$

PROOF Some mathematical proofs, such as the proof of the Sum Rule, are straightforward. Others involve clever steps that may not appear to follow clearly from a prior step. The proof below involves such a step—adding and subtracting the same quantity. (This step is shown in color.) Let $F(x) = f(x)g(x)$.

$$F'(x) = \lim_{\Delta x \to 0} \frac{F(x + \Delta x) - F(x)}{\Delta x}$$

$$= \lim_{\Delta x \to 0} \frac{f(x + \Delta x)g(x + \Delta x) - f(x)g(x)}{\Delta x}$$

$$= \lim_{\Delta x \to 0} \frac{f(x + \Delta x)g(x + \Delta x) - f(x + \Delta x)g(x) + f(x + \Delta x)g(x) - f(x)g(x)}{\Delta x}$$

$$= \lim_{\Delta x \to 0} \left[f(x + \Delta x) \frac{g(x + \Delta x) - g(x)}{\Delta x} + g(x) \frac{f(x + \Delta x) - f(x)}{\Delta x} \right]$$

$$= \lim_{\Delta x \to 0} f(x + \Delta x) \frac{g(x + \Delta x) - g(x)}{\Delta x} + \lim_{\Delta x \to 0} g(x) \frac{f(x + \Delta x) - f(x)}{\Delta x}$$

$$= \left[\lim_{\Delta x \to 0} f(x + \Delta x) \right] \left[\lim_{\Delta x \to 0} \frac{g(x + \Delta x) - g(x)}{\Delta x} \right]$$

$$+ \left[\lim_{\Delta x \to 0} g(x) \right] \left[\lim_{\Delta x \to 0} \frac{f(x + \Delta x) - f(x)}{\Delta x} \right]$$

$$= f(x)g'(x) + g(x)f'(x)$$

Example 1 Finding the Derivative of a Product

Find the derivative of $y = (3x - 2x^2)(5 + 4x)$.

SOLUTION Using the Product Rule, you can write

$$\frac{dy}{dx} = \overbrace{(3x - 2x^2)}^{\text{First}} \overbrace{\frac{d}{dx}[5 + 4x]}^{\substack{\text{Derivative} \\ \text{of second}}} + \overbrace{(5 + 4x)}^{\text{Second}} \overbrace{\frac{d}{dx}[3x - 2x^2]}^{\substack{\text{Derivative} \\ \text{of first}}}$$

$$= (3x - 2x^2)(4) + (5 + 4x)(3 - 4x)$$
$$= (12x - 8x^2) + (15 - 8x - 16x^2)$$
$$= 15 + 4x - 24x^2.$$

✓ CHECKPOINT 1

Find the derivative of $y = (4x + 3x^2)(6 - 3x)$. ∎

STUDY TIP

In general, the derivative of the product of two functions is not equal to the product of the derivatives of the two functions. To see this, compare the product of the derivatives of $f(x) = 3x - 2x^2$ and $g(x) = 5 + 4x$ with the derivative found in Example 1.

In the next example, notice that the first step in differentiating is *rewriting the original function*.

TECHNOLOGY

(T) If you have access to a symbolic differentiation utility, try using it to confirm several of the derivatives in this section. The form of the derivative can depend on how you use software.

Example 2 Finding the Derivative of a Product

Find the derivative of

$$f(x) = \left(\frac{1}{x} + 1\right)(x - 1). \qquad \text{Original function}$$

SOLUTION Rewrite the function. Then use the Product Rule to find the derivative.

$$f(x) = (x^{-1} + 1)(x - 1) \qquad \text{Rewrite function.}$$

$$f'(x) = (x^{-1} + 1)\frac{d}{dx}[x - 1] + (x - 1)\frac{d}{dx}[x^{-1} + 1] \qquad \text{Product Rule}$$

$$= (x^{-1} + 1)(1) + (x - 1)(-x^{-2})$$

$$= \frac{1}{x} + 1 - \frac{x - 1}{x^2}$$

$$= \frac{x + x^2 - x + 1}{x^2} \qquad \text{Write with common denominator.}$$

$$= \frac{x^2 + 1}{x^2} \qquad \text{Simplify.}$$

✓ CHECKPOINT 2

Find the derivative of

$$f(x) = \left(\frac{1}{x} + 1\right)(2x + 1).$$ ∎

SECTION 2.4 The Product and Quotient Rules

You now have two differentiation rules that deal with products—the Constant Multiple Rule and the Product Rule. The difference between these two rules is that the Constant Multiple Rule deals with the product of a constant and a variable quantity:

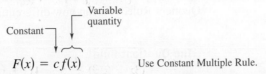

$$F(x) = c f(x)$$ Use Constant Multiple Rule.

whereas the Product Rule deals with the product of two variable quantities:

$$F(x) = f(x) g(x).$$ Use Product Rule.

The next example compares these two rules.

Example 3 Comparing Differentiation Rules

Find the derivative of each function.

a. $y = 2x(x^2 + 3x)$

b. $y = 2(x^2 + 3x)$

SOLUTION

a. By the Product Rule,

$$\frac{dy}{dx} = (2x)\frac{d}{dx}[x^2 + 3x] + (x^2 + 3x)\frac{d}{dx}[2x] \quad \text{Product Rule}$$
$$= (2x)(2x + 3) + (x^2 + 3x)(2)$$
$$= 4x^2 + 6x + 2x^2 + 6x$$
$$= 6x^2 + 12x.$$

b. By the Constant Multiple Rule,

$$\frac{dy}{dx} = 2\frac{d}{dx}[x^2 + 3x] \quad \text{Constant Multiple Rule}$$
$$= 2(2x + 3)$$
$$= 4x + 6.$$

The Product Rule can be extended to products that have more than two factors. For example, if f, g, and h are differentiable functions of x, then

$$\frac{d}{dx}[f(x)g(x)h(x)] = f'(x)g(x)h(x) + f(x)g'(x)h(x) + f(x)g(x)h'(x).$$

STUDY TIP

You could calculate the derivatives in Example 3 without the Product Rule. For Example 3(a),

$$y = 2x(x^2 + 3x) = 2x^3 + 6x^2$$

and

$$\frac{dy}{dx} = 6x^2 + 12x.$$

✓ CHECKPOINT 3

Find the derivative of each function.

a. $y = 3x(2x^2 + 5x)$

b. $y = 3(2x^2 + 5x)$ ∎

The Quotient Rule

In Section 2.2, you saw that by using the Constant Rule, the Power Rule, the Constant Multiple Rule, and the Sum and Difference Rules, you were able to differentiate any polynomial function. By combining these rules with the Quotient Rule, you can now differentiate any *rational* function.

The Quotient Rule

The derivative of the quotient of two differentiable functions is equal to the denominator times the derivative of the numerator minus the numerator times the derivative of the denominator, all divided by the square of the denominator.

$$\frac{d}{dx}\left[\frac{f(x)}{g(x)}\right] = \frac{g(x)f'(x) - f(x)g'(x)}{[g(x)]^2}, \quad g(x) \neq 0$$

STUDY TIP

From this differentiation rule, you can see that the derivative of a quotient is not, in general, the quotient of the derivatives. That is,

$$\frac{d}{dx}\left[\frac{f(x)}{g(x)}\right] \neq \frac{f'(x)}{g'(x)}.$$

PROOF Let $F(x) = f(x)/g(x)$. As in the proof of the Product Rule, a key step in this proof is adding and subtracting the same quantity.

$$F'(x) = \lim_{\Delta x \to 0} \frac{F(x + \Delta x) - F(x)}{\Delta x}$$

$$= \lim_{\Delta x \to 0} \frac{\frac{f(x + \Delta x)}{g(x + \Delta x)} - \frac{f(x)}{g(x)}}{\Delta x}$$

$$= \lim_{\Delta x \to 0} \frac{g(x)f(x + \Delta x) - f(x)g(x + \Delta x)}{\Delta x g(x)g(x + \Delta x)}$$

$$= \lim_{\Delta x \to 0} \frac{g(x)f(x + \Delta x) - f(x)g(x) + f(x)g(x) - f(x)g(x + \Delta x)}{\Delta x g(x)g(x + \Delta x)}$$

$$= \frac{\lim_{\Delta x \to 0} \frac{g(x)[f(x + \Delta x) - f(x)]}{\Delta x} - \lim_{\Delta x \to 0} \frac{f(x)[g(x + \Delta x) - g(x)]}{\Delta x}}{\lim_{\Delta x \to 0} [g(x)g(x + \Delta x)]}$$

$$= \frac{g(x)\left[\lim_{\Delta x \to 0} \frac{f(x + \Delta x) - f(x)}{\Delta x}\right] - f(x)\left[\lim_{\Delta x \to 0} \frac{g(x + \Delta x) - g(x)}{\Delta x}\right]}{\lim_{\Delta x \to 0} [g(x)g(x + \Delta x)]}$$

$$= \frac{g(x)f'(x) - f(x)g'(x)}{[g(x)]^2}$$

STUDY TIP

As suggested for the Product Rule, it can be more helpful to remember the verbal statement of the Quotient Rule rather than trying to remember the formula for the rule.

SECTION 2.4 The Product and Quotient Rules

Algebra Review

When applying the Quotient Rule, it is suggested that you enclose all factors and derivatives in symbols of grouping, such as parentheses. Also, pay special attention to the subtraction required in the numerator. For help in evaluating expressions such as the one in Example 4, see the *Chapter 2 Algebra Review* on page 175, Example 2(d).

Example 4 Finding the Derivative of a Quotient

Find the derivative of $y = \dfrac{x-1}{2x+3}$.

SOLUTION Apply the Quotient Rule, as shown.

$$\frac{dy}{dx} = \frac{(2x+3)\dfrac{d}{dx}[x-1] - (x-1)\dfrac{d}{dx}[2x+3]}{(2x+3)^2}$$

$$= \frac{(2x+3)(1) - (x-1)(2)}{(2x+3)^2}$$

$$= \frac{2x + 3 - 2x + 2}{(2x+3)^2}$$

$$= \frac{5}{(2x+3)^2}$$

✓ CHECKPOINT 4

Find the derivative of $y = \dfrac{x+4}{5x-2}$. ∎

Example 5 Finding an Equation of a Tangent Line

Find an equation of the tangent line to the graph of

$$y = \frac{2x^2 - 4x + 3}{2 - 3x}$$

when $x = 1$.

SOLUTION Apply the Quotient Rule, as shown.

$$\frac{dy}{dx} = \frac{(2-3x)\dfrac{d}{dx}[2x^2 - 4x + 3] - (2x^2 - 4x + 3)\dfrac{d}{dx}[2-3x]}{(2-3x)^2}$$

$$= \frac{(2-3x)(4x-4) - (2x^2 - 4x + 3)(-3)}{(2-3x)^2}$$

$$= \frac{-12x^2 + 20x - 8 - (-6x^2 + 12x - 9)}{(2-3x)^2}$$

$$= \frac{-12x^2 + 20x - 8 + 6x^2 - 12x + 9}{(2-3x)^2}$$

$$= \frac{-6x^2 + 8x + 1}{(2-3x)^2}$$

When $x = 1$, the value of the function is $y = -1$ and the slope is $m = 3$. Using the point-slope form of a line, you can find the equation of the tangent line to be $y = 3x - 4$. The graph of the function and the tangent line is shown in Figure 2.22.

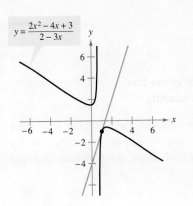

$y = \dfrac{2x^2 - 4x + 3}{2 - 3x}$

FIGURE 2.22

✓ CHECKPOINT 5

Find an equation of the tangent line to the graph of

$$y = \frac{x^2 - 4}{2x + 5} \quad \text{when } x = 0.$$

Sketch the line tangent to the graph at $x = 0$. ∎

Example 6 — Finding the Derivative of a Quotient

Find the derivative of

$$y = \frac{3 - (1/x)}{x + 5}.$$

SOLUTION Begin by rewriting the original function. Then apply the Quotient Rule and simplify the result.

$$y = \frac{3 - (1/x)}{x + 5} \qquad \text{Write original function.}$$

$$= \frac{3x - 1}{x(x + 5)} \qquad \text{Multiply numerator and denominator by } x.$$

$$= \frac{3x - 1}{x^2 + 5x} \qquad \text{Rewrite.}$$

$$\frac{dy}{dx} = \frac{(x^2 + 5x)(3) - (3x - 1)(2x + 5)}{(x^2 + 5x)^2} \qquad \text{Apply Quotient Rule.}$$

$$= \frac{(3x^2 + 15x) - (6x^2 + 13x - 5)}{(x^2 + 5x)^2}$$

$$= \frac{-3x^2 + 2x + 5}{(x^2 + 5x)^2} \qquad \text{Simplify.}$$

✓ CHECKPOINT 6

Find the derivative of $y = \dfrac{3 - (2/x)}{x + 4}$. ■

Not every quotient needs to be differentiated by the Quotient Rule. For instance, each of the quotients in the next example can be considered as the product of a constant and a function of x. In such cases, the Constant Multiple Rule is more efficient than the Quotient Rule.

STUDY TIP
To see the efficiency of using the Constant Multiple Rule in Example 7, try using the Quotient Rule to find the derivatives of the four functions.

Example 7 — Rewriting Before Differentiating

Find the derivative of each function.

	Original Function	Rewrite	Differentiate	Simplify
a.	$y = \dfrac{x^2 + 3x}{6}$	$y = \dfrac{1}{6}(x^2 + 3x)$	$y' = \dfrac{1}{6}(2x + 3)$	$y' = \dfrac{1}{3}x + \dfrac{1}{2}$
b.	$y = \dfrac{5x^4}{8}$	$y = \dfrac{5}{8}x^4$	$y' = \dfrac{5}{8}(4x^3)$	$y' = \dfrac{5}{2}x^3$
c.	$y = \dfrac{-3(3x - 2x^2)}{7x}$	$y = -\dfrac{3}{7}(3 - 2x)$	$y' = -\dfrac{3}{7}(-2)$	$y' = \dfrac{6}{7}$
d.	$y = \dfrac{9}{5x^2}$	$y = \dfrac{9}{5}(x^{-2})$	$y' = \dfrac{9}{5}(-2x^{-3})$	$y' = -\dfrac{18}{5x^3}$

✓ CHECKPOINT 7

Find the derivative of each function.

a. $y = \dfrac{x^2 + 4x}{5}$ **b.** $y = \dfrac{3x^4}{4}$ ■

Simplifying Derivatives

Example 8 Combining the Product and Quotient Rules

Find the derivative of
$$y = \frac{(1 - 2x)(3x + 2)}{5x - 4}.$$

SOLUTION This function contains a product within a quotient. You could first multiply the factors in the numerator and then apply the Quotient Rule. However, to gain practice in using the Product Rule within the Quotient Rule, try differentiating as shown.

$$y' = \frac{(5x - 4)\frac{d}{dx}[(1 - 2x)(3x + 2)] - (1 - 2x)(3x + 2)\frac{d}{dx}[5x - 4]}{(5x - 4)^2}$$

$$= \frac{(5x - 4)[(1 - 2x)(3) + (3x + 2)(-2)] - (1 - 2x)(3x + 2)(5)}{(5x - 4)^2}$$

$$= \frac{(5x - 4)(-12x - 1) - (1 - 2x)(15x + 10)}{(5x - 4)^2}$$

$$= \frac{(-60x^2 + 43x + 4) - (-30x^2 - 5x + 10)}{(5x - 4)^2}$$

$$= \frac{-30x^2 + 48x - 6}{(5x - 4)^2}$$

✓ CHECKPOINT 8

Find the derivative of $y = \dfrac{(1 + x)(2x - 1)}{x - 1}.$ ■

In the examples in this section, much of the work in obtaining the final form of the derivative occurs *after* the differentiation. As summarized in the list below, direct application of differentiation rules often yields results that are not in simplified form. Note that two characteristics of simplified form are the absence of negative exponents and the combining of like terms.

	$f'(x)$ After Differentiating	$f'(x)$ After Simplifying
Example 1	$(3x - 2x^2)(4) + (5 + 4x)(3 - 4x)$	$15 + 4x - 24x^2$
Example 2	$(x^{-1} + 1)(1) + (x - 1)(-x^{-2})$	$\dfrac{x^2 + 1}{x^2}$
Example 5	$\dfrac{(2 - 3x)(4x - 4) - (2x^2 - 4x + 3)(-3)}{(2 - 3x)^2}$	$\dfrac{-6x^2 + 8x + 1}{(2 - 3x)^2}$
Example 8	$\dfrac{(5x - 4)[(1 - 2x)(3) + (3x + 2)(-2)] - (1 - 2x)(3x + 2)(5)}{(5x - 4)^2}$	$\dfrac{-30x^2 + 48x - 6}{(5x - 4)^2}$

Application

Example 9 Rate of Change of Systolic Blood Pressure

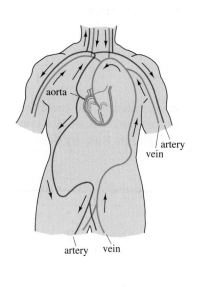

As blood moves from the heart through the major arteries out to the capillaries and back through the veins, the systolic blood pressure continuously drops. Consider a person whose systolic blood pressure P (in millimeters of mercury) is given by

$$P = \frac{25t^2 + 125}{t^2 + 1}, \quad 0 \leq t \leq 10$$

where t is measured in seconds. At what rate is the blood pressure changing 5 seconds after blood leaves the heart?

SOLUTION Begin by applying the Quotient Rule.

$$\frac{dP}{dt} = \frac{(t^2 + 1)(50t) - (25t^2 + 125)(2t)}{(t^2 + 1)^2} \quad \text{Quotient Rule}$$

$$= \frac{50t^3 + 50t - 50t^3 - 250t}{(t^2 + 1)^2}$$

$$= -\frac{200t}{(t^2 + 1)^2} \quad \text{Simplify.}$$

When $t = 5$, the rate of change is

$$-\frac{200(5)}{26^2} \approx -1.48 \text{ millimeters per second.}$$

So, the pressure is *dropping* at a rate of 1.48 millimeters per second when $t = 5$ seconds.

✓ CHECKPOINT 9

In Example 9, find the rate at which systolic blood pressure is changing at each time shown in the table below. Describe the changes in blood pressure as the blood moves away from the heart.

t	0	1	2	3	4	5	6	7
$\dfrac{dP}{dt}$								

CONCEPT CHECK

1. Write a verbal statement that represents the Product Rule.
2. Write a verbal statement that represents the Quotient Rule.
3. Is it possible to find the derivative of $f(x) = \dfrac{x^3 + 5x}{2}$ without using the Quotient Rule? If so, what differentiation rule can you use to find f'? (You do not need to find the derivative.)
4. Complete the following: In general, you can use the Product Rule to differentiate the _____ of two variable quantities and the Quotient Rule to differentiate any _____ function.

SECTION 2.4 The Product and Quotient Rules

Skills Review 2.4

The following warm-up exercises involve skills that were covered in earlier sections. You will use these skills in the exercise set for this section. For additional help, review Sections 0.4, 0.5, and 2.2.

In Exercises 1–10, simplify the expression.

1. $(x^2 + 1)(2) + (2x + 7)(2x)$
2. $(2x - x^3)(8x) + (4x^2)(2 - 3x^2)$
3. $x(4)(x^2 + 2)^3(2x) + (x^2 + 4)(1)$
4. $x^2(2)(2x + 1)(2) + (2x + 1)^4(2x)$
5. $\dfrac{(2x + 7)(5) - (5x + 6)(2)}{(2x + 7)^2}$
6. $\dfrac{(x^2 - 4)(2x + 1) - (x^2 + x)(2x)}{(x^2 - 4)^2}$
7. $\dfrac{(x^2 + 1)(2) - (2x + 1)(2x)}{(x^2 + 1)^2}$
8. $\dfrac{(1 - x^4)(4) - (4x - 1)(-4x^3)}{(1 - x^4)^2}$
9. $(x^{-1} + x)(2) + (2x - 3)(-x^{-2} + 1)$
10. $\dfrac{(1 - x^{-1})(1) - (x - 4)(x^{-2})}{(1 - x^{-1})^2}$

In Exercises 11–14, find $f'(2)$.

11. $f(x) = 3x^2 - x + 4$
12. $f(x) = -x^3 + x^2 + 8x$
13. $f(x) = \dfrac{1}{x}$
14. $f(x) = x^2 - \dfrac{1}{x^2}$

Exercises 2.4

See www.CalcChat.com for worked-out solutions to odd-numbered exercises.

In Exercises 1–16, find the value of the derivative of the function at the given point. State which differentiation rule you used to find the derivative.

Function	Point
1. $f(x) = x(x^2 + 3)$	$(2, 14)$
2. $g(x) = (x - 4)(x + 2)$	$(4, 0)$
3. $f(x) = x^2(3x^3 - 1)$	$(1, 2)$
4. $f(x) = (x^2 + 1)(2x + 5)$	$(-1, 6)$
5. $f(x) = \frac{1}{3}(2x^3 - 4)$	$(0, -\frac{4}{3})$
6. $f(x) = \frac{1}{7}(5 - 6x^2)$	$(1, -\frac{1}{7})$
7. $g(x) = (x^2 - 4x + 3)(x - 2)$	$(4, 6)$
8. $g(x) = (x^2 - 2x + 1)(x^3 - 1)$	$(1, 0)$
9. $h(x) = \dfrac{x}{x - 5}$	$(6, 6)$
10. $h(x) = \dfrac{x^2}{x + 3}$	$\left(-1, \frac{1}{2}\right)$
11. $f(t) = \dfrac{2t^2 - 3}{3t + 1}$	$\left(3, \frac{3}{2}\right)$
12. $f(x) = \dfrac{3x}{x^2 + 4}$	$\left(-1, -\frac{3}{5}\right)$
13. $g(x) = \dfrac{2x + 1}{x - 5}$	$(6, 13)$
14. $f(x) = \dfrac{x + 1}{x - 1}$	$(2, 3)$
15. $f(t) = \dfrac{t^2 - 1}{t + 4}$	$(1, 0)$
16. $g(x) = \dfrac{4x - 5}{x^2 - 1}$	$(0, 5)$

In Exercises 17–24, find the derivative of the function. Use Example 7 as a model.

Function	Rewrite	Differentiate	Simplify
17. $y = \dfrac{x^2 + 2x}{x}$			
18. $y = \dfrac{4x^{3/2}}{x}$			
19. $y = \dfrac{7}{3x^3}$			
20. $y = \dfrac{4}{5x^2}$			
21. $y = \dfrac{4x^2 - 3x}{8\sqrt{x}}$			
22. $y = \dfrac{3x^2 - 4x}{6x}$			
23. $y = \dfrac{x^2 - 4x + 3}{x - 1}$			
24. $y = \dfrac{x^2 - 4}{x + 2}$			

In Exercises 25–40, find the derivative of the function. State which differentiation rule(s) you used to find the derivative.

25. $f(x) = (x^3 - 3x)(2x^2 + 3x + 5)$
26. $h(t) = (t^5 - 1)(4t^2 - 7t - 3)$
27. $g(t) = (2t^3 - 1)^2$
28. $h(p) = (p^3 - 2)^2$
29. $f(x) = \sqrt[3]{x}(\sqrt{x} + 3)$
30. $f(x) = \sqrt[3]{x}(x + 1)$
31. $f(x) = \dfrac{3x - 2}{2x - 3}$
32. $f(x) = \dfrac{x^3 + 3x + 2}{x^2 - 1}$
33. $f(x) = \dfrac{3 - 2x - x^2}{x^2 - 1}$
34. $f(x) = (x^5 - 3x)\left(\dfrac{1}{x^2}\right)$
35. $f(x) = x\left(1 - \dfrac{2}{x + 1}\right)$
36. $h(t) = \dfrac{t + 2}{t^2 + 5t + 6}$
37. $g(s) = \dfrac{s^2 - 2s + 5}{\sqrt{s}}$
38. $f(x) = \dfrac{x + 1}{\sqrt{x}}$
39. $g(x) = \left(\dfrac{x - 3}{x + 4}\right)(x^2 + 2x + 1)$
40. $f(x) = (3x^3 + 4x)(x - 5)(x + 1)$

(T) In Exercises 41–46, find an equation of the tangent line to the graph of the function at the given point. Then use a graphing utility to graph the function and the tangent line in the same viewing window.

Function	Point
41. $f(x) = (x - 1)^2(x - 2)$	$(0, -2)$
42. $h(x) = (x^2 - 1)^2$	$(-2, 9)$
43. $f(x) = \dfrac{x - 2}{x + 1}$	$(1, -\tfrac{1}{2})$
44. $f(x) = \dfrac{2x + 1}{x - 1}$	$(2, 5)$
45. $f(x) = \left(\dfrac{x + 5}{x - 1}\right)(2x + 1)$	$(0, -5)$
46. $g(x) = (x + 2)\left(\dfrac{x - 5}{x + 1}\right)$	$(0, -10)$

In Exercises 47–50, find the point(s), if any, at which the graph of f has a horizontal tangent.

47. $f(x) = \dfrac{x^2}{x - 1}$
48. $f(x) = \dfrac{x^2}{x^2 + 1}$
49. $f(x) = \dfrac{x^4}{x^3 + 1}$
50. $f(x) = \dfrac{x^4 + 3}{x^2 + 1}$

(T) In Exercises 51–54, use a graphing utility to graph f and f' on the interval $[-2, 2]$.

51. $f(x) = x(x + 1)$
52. $f(x) = x^2(x + 1)$
53. $f(x) = x(x + 1)(x - 1)$
54. $f(x) = x^2(x + 1)(x - 1)$

55. **Environment** The model
$$f(t) = \dfrac{t^2 - t + 1}{t^2 + 1}$$
measures the level of oxygen in a pond, where t is the time (in weeks) after organic waste is dumped into the pond. Find the rates of change of f with respect to t when (a) $t = 0.5$, (b) $t = 2$, and (c) $t = 8$.

56. **Physical Science** The temperature T (in degrees Fahrenheit) of food placed in a refrigerator is modeled by
$$T = 10\left(\dfrac{4t^2 + 16t + 75}{t^2 + 4t + 10}\right)$$
where t is the time (in hours). What is the initial temperature of the food? Find the rates of change of T with respect to t when (a) $t = 1$, (b) $t = 3$, (c) $t = 5$, and (d) $t = 10$.

57. **Population Growth** A population of bacteria is introduced into a culture. The number P of bacteria can be modeled by
$$P = 500\left(1 + \dfrac{4t}{50 + t^2}\right)$$
where t is the time (in hours). Find the rate of change of the population when $t = 2$.

58. **Population Growth** The population P (in millions) of microbes in a contaminated water sample can be modeled by
$$P = (t - 12)(3t^2 - 20t) + 250$$
where t is the time (in hours). Find the rate of change of the population when $t = 2$.

59. **Precipitation** The average monthly precipitation P (in inches) for a city in the southwest United States can be modeled by
$$P = 0.0059t^3 + 0.0083(t - 133.13)(t - 5.10),$$
$$1 \le t \le 12$$
where t is the month, with $t = 1$ corresponding to January. Find the rate of change in the monthly precipitation for the month of June and interpret your result.

60. **Biology: Wildlife Management** A state game commission introduces 30 elk into a new state park. The population N of the herd is modeled by
$$N = \dfrac{10(3 + 4t)}{1 + 0.1t}$$
where t is the time in years.

(a) Find the rates of change of the population of the herd after 5, 10, and 25 years.

(T) (b) Use a graphing utility to graph N.

(c) What can you say about the rate of change of the population as t increases?

61. Body Temperature The temperature T (in degrees Fahrenheit) of a person during an illness can be modeled by

$$T = \frac{12t}{t^2 + 8} + 98.6$$

where t is time in hours since the person started to show signs of a fever.

(a) Find the rates of change of T with respect to t when $t = 0.5$ and $t = 4$.

(b) Use a graphing utility to graph T.

(c) Use your graph to determine when the person's temperature starts to decrease.

(d) Normal body temperature is within 1°F of 98.6°F. After how many hours does the person have a body temperature that is considered normal? Explain.

62. MAKE A DECISION: SEIZING DRUGS The annual cost C (in millions of dollars) for a government agency to seize $p\%$ of an illegal drug can be modeled by

$$C = \frac{528p}{100 - p}, \quad 0 \le p < 100.$$

(a) Find the rates of change in the cost when $p = 30\%$ and $p = 60\%$.

(b) Use a graphing utility to graph C.

(c) What happens to C and C' as p approaches 100?

(d) Is it possible for the agency to seize 100% of the drug? Explain.

63. MAKE A DECISION: ENVIRONMENT The predicted cost C (in hundreds of thousands of dollars) for a company to remove $p\%$ of a chemical from its waste water can be modeled by

$$C = \frac{124p}{(10 + p)(100 - p)}, \quad 0 \le p < 100.$$

(a) Find the rates of change in the cost when $p = 25\%$ and $p = 75\%$.

(b) Use a graphing utility to graph C.

(c) What happens to C and C' as p approaches 100?

(d) What would you say is a reasonable percent of the chemical for the company to remove? Explain.

64. Consumer Awareness The price P per pound of lean and extra lean ground beef in the United States from 1998 through 2005 can be modeled by

$$P = \frac{1.755 - 0.2079t + 0.00673t^2}{1 - 0.1282t + 0.00434t^2}, \quad 8 \le t \le 15$$

where t is the year, with $t = 8$ corresponding to 1998. Find dP/dt and evaluate it for $t = 8, 10, 12,$ and 14. Interpret the meaning of these values. *(Source: U.S. Bureau of Labor Statistics)*

65. Sales Analysis The monthly sales $M(t)$ of memberships at a newly built fitness center are modeled by

$$M(t) = \frac{300t}{t^2 + 1} + 8$$

where t is the number of months since the center opened.

(a) Find $M'(t)$.

(b) Find $M(3)$ and $M'(3)$ and interpret the results.

(c) Find $M(24)$ and $M'(24)$ and interpret the results.

In Exercises 66–69, use the given information to find $f'(2)$.

$g(2) = 3$ and $g'(2) = -2$

$h(2) = -1$ and $h'(2) = 4$

66. $f(x) = 2g(x) + h(x)$

67. $f(x) = 3 - g(x)$

68. $f(x) = g(x) + h(x)$

69. $f(x) = \dfrac{g(x)}{h(x)}$

Social Science Capsule

© Bill Aron/PhotoEdit

Social work is a profession for those with a strong desire to help improve people's lives. About 90% of social workers are in health care and social assistance industries, as well as state and local government agencies. Most states require 3000 hours of supervised clinical experience for licensure. Employment of social workers is expected to increase faster than the average for all occupations through 2014.

70. Research Project Use your school's library, the Internet, or some other reference source to find information about the many career paths possible for someone with a degree in social work. Choose one career path and write a short paper about it.

Section 2.5
The Chain Rule

- Find derivatives using the Chain Rule.
- Find derivatives using the General Power Rule.
- Write derivatives in simplified form.
- Use derivatives to answer questions about real-life situations.
- Use the differentiation rules to differentiate algebraic functions.

The Chain Rule

In this section, you will study one of the most powerful rules of differential calculus—the **Chain Rule.** This differentiation rule deals with composite functions and adds versatility to the rules presented in Sections 2.2 and 2.4. For example, compare the functions below. Those on the left can be differentiated without the Chain Rule, whereas those on the right are best done with the Chain Rule.

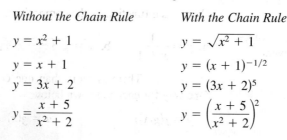

Without the Chain Rule	With the Chain Rule
$y = x^2 + 1$	$y = \sqrt{x^2 + 1}$
$y = x + 1$	$y = (x + 1)^{-1/2}$
$y = 3x + 2$	$y = (3x + 2)^5$
$y = \dfrac{x + 5}{x^2 + 2}$	$y = \left(\dfrac{x + 5}{x^2 + 2}\right)^2$

The Chain Rule

If $y = f(u)$ is a differentiable function of u, and $u = g(x)$ is a differentiable function of x, then $y = f(g(x))$ is a differentiable function of x, and

$$\frac{dy}{dx} = \frac{dy}{du} \cdot \frac{du}{dx}$$

or, equivalently,

$$\frac{d}{dx}[f(g(x))] = f'(g(x))g'(x).$$

Basically, the Chain Rule states that if y changes dy/du times as fast as u, and u changes du/dx times as fast as x, then y changes

$$\frac{dy}{du} \cdot \frac{du}{dx}$$

times as fast as x, as illustrated in Figure 2.23. One advantage of the dy/dx notation for derivatives is that it helps you remember differentiation rules, such as the Chain Rule. For instance, in the formula

$$dy/dx = (dy/du)(du/dx)$$

you can imagine that the du's divide out.

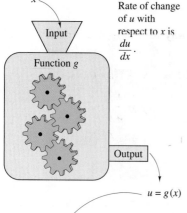

FIGURE 2.23

SECTION 2.5 The Chain Rule

When applying the Chain Rule, it helps to think of the composite function $y = f(g(x))$ or $y = f(u)$ as having two parts—an *inside* and an *outside*—as illustrated below.

$$y = f(\underbrace{g(x)}_{\text{Inside}}) = f(u)$$
$$\underbrace{}_{\text{Outside}}$$

The Chain Rule tells you that the derivative of $y = f(u)$ is the derivative of the outer function (at the inner function u) *times* the derivative of the inner function. That is,

$$y' = f'(u) \cdot u'.$$

Example 1 Decomposing Composite Functions

✓ **CHECKPOINT 1**

Write each function as the composition of two functions, where $y = f(g(x))$.

a. $y = \dfrac{1}{\sqrt{x+1}}$

b. $y = (x^2 + 2x + 5)^3$ ∎

Write each function as the composition of two functions.

a. $y = \dfrac{1}{x+1}$ **b.** $y = \sqrt{3x^2 - x + 1}$

SOLUTION There is more than one correct way to decompose each function. One way for each is shown below.

$y = f(g(x))$	$u = g(x)$ (inside)	$y = f(u)$ (outside)
a. $y = \dfrac{1}{x+1}$	$u = x + 1$	$y = \dfrac{1}{u}$
b. $y = \sqrt{3x^2 - x + 1}$	$u = 3x^2 - x + 1$	$y = \sqrt{u}$

Example 2 Using the Chain Rule

Find the derivative of $y = (x^2 + 1)^3$.

SOLUTION To apply the Chain Rule, you need to identify the inside function u.

$$y = (\underbrace{x^2 + 1}_{u})^3 = u^3$$

By the Chain Rule, you can write the derivative as shown.

$$\frac{dy}{dx} = \underbrace{3(x^2 + 1)^2}_{\frac{dy}{du}} \underbrace{(2x)}_{\frac{du}{dx}} = 6x(x^2 + 1)^2$$

STUDY TIP

Try checking the result of Example 2 by expanding the function to obtain

$$y = x^6 + 3x^4 + 3x^2 + 1$$

and finding the derivative. Do you obtain the same answer?

✓ **CHECKPOINT 2**

Find the derivative of $y = (x^3 + 1)^2$. ∎

The General Power Rule

The function in Example 2 illustrates one of the most common types of composite functions—a power function of the form

$$y = [u(x)]^n.$$

The rule for differentiating such functions is called the **General Power Rule,** and it is a special case of the Chain Rule.

The General Power Rule

If $y = [u(x)]^n$, where u is a differentiable function of x and n is a real number, then

$$\frac{dy}{dx} = n[u(x)]^{n-1}\frac{du}{dx}$$

or, equivalently,

$$\frac{d}{dx}[u^n] = nu^{n-1}u'.$$

PROOF Apply the Chain Rule and the Simple Power Rule as shown.

$$\frac{dy}{dx} = \frac{dy}{du} \cdot \frac{du}{dx}$$

$$= \frac{d}{du}[u^n]\frac{du}{dx}$$

$$= nu^{n-1}\frac{du}{dx}$$

Example 3 Using the General Power Rule

Find the derivative of

$$f(x) = (3x - 2x^2)^3.$$

SOLUTION The inside function is $u = 3x - 2x^2$. So, by the General Power Rule,

$$f'(x) = \overset{n}{3}(\underbrace{3x - 2x^2}_{u^{n-1}})^2\underbrace{\frac{d}{dx}[3x - 2x^2]}_{u'}$$

$$= 3(3x - 2x^2)^2(3 - 4x)$$

$$= (9 - 12x)(3x - 2x^2)^2.$$

✓ CHECKPOINT 3

Find the derivative of $y = (x^2 + 3x)^4$. ∎

TECHNOLOGY

If you have access to a symbolic differentiation utility, try using it to confirm the result of Example 3.

Example 4 Rewriting Before Differentiating

Find the tangent line to the graph of

$$y = \sqrt[3]{(x^2 + 4)^2} \qquad \text{Original function}$$

when $x = 2$.

SOLUTION Begin by rewriting the function in rational exponent form.

$$y = (x^2 + 4)^{2/3} \qquad \text{Rewrite original function.}$$

Then, using the inside function, $u = x^2 + 4$, apply the General Power Rule.

$$\frac{dy}{dx} = \overset{n}{\tfrac{2}{3}} \overbrace{(x^2 + 4)^{-1/3}}^{u^{n-1}} \overbrace{(2x)}^{u'} \qquad \text{Apply General Power Rule.}$$

$$= \frac{4x(x^2 + 4)^{-1/3}}{3}$$

$$= \frac{4x}{3\sqrt[3]{x^2 + 4}} \qquad \text{Simplify.}$$

When $x = 2$, $y = 4$ and the slope of the line tangent to the graph at $(2, 4)$ is $\tfrac{4}{3}$. Using the point-slope form, you can find the equation of the tangent line to be $y = \tfrac{4}{3}x + \tfrac{4}{3}$. The graph of the function and the tangent line is shown in Figure 2.24.

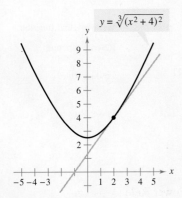

FIGURE 2.24

✓ CHECKPOINT 4

Find the tangent line to the graph of $y = \sqrt[3]{(x + 4)^2}$ when $x = 4$. Sketch the line tangent to the graph at $x = 4$. ■

Example 5 Finding the Derivative of a Quotient

Find the derivative of each function.

a. $y = \dfrac{3}{x^2 + 1}$ **b.** $y = \dfrac{3}{(x + 1)^2}$

SOLUTION

a. Begin by rewriting the function as

$$y = 3(x^2 + 1)^{-1}. \qquad \text{Rewrite original function.}$$

Then apply the General Power Rule to obtain

$$\frac{dy}{dx} = -3(x^2 + 1)^{-2}(2x) = -\frac{6x}{(x^2 + 1)^2}. \qquad \text{Apply General Power Rule.}$$

b. Begin by rewriting the function as

$$y = 3(x + 1)^{-2}. \qquad \text{Rewrite original function.}$$

Then apply the General Power Rule to obtain

$$\frac{dy}{dx} = -6(x + 1)^{-3}(1) = -\frac{6}{(x + 1)^3}. \qquad \text{Apply General Power Rule.}$$

STUDY TIP

The derivative of a quotient can sometimes be found more easily with the General Power Rule than with the Quotient Rule. This is especially true when the numerator is a constant, as shown in Example 5.

✓ CHECKPOINT 5

Find the derivative of each function.

a. $y = \dfrac{4}{2x + 1}$

b. $y = \dfrac{2}{(x - 1)^3}$ ■

CHAPTER 2 Differentiation

Simplification Techniques

Throughout this chapter, writing derivatives in simplified form has been emphasized. The reason for this is that most applications of derivatives require a simplified form. The next two examples illustrate some useful simplification techniques.

Algebra Review

In Example 6, note that you subtract exponents when factoring. That is, when $(1 - x^2)^{-1/2}$ is factored out of $(1 - x^2)^{1/2}$, the *remaining* factor has an exponent of $\frac{1}{2} - \left(-\frac{1}{2}\right) = 1$. So,

$$(1 - x^2)^{1/2} = (1 - x^2)^{-1/2} (1 - x^2)^1.$$

For help in evaluating expressions such as the one in Example 6, see the *Chapter 2 Algebra Review* on pages 174 and 175.

Example 6 Simplifying by Factoring Out Least Powers

Find the derivative of $y = x^2\sqrt{1 - x^2}$.

$$\begin{aligned}
y &= x^2\sqrt{1 - x^2} & &\text{Write original function.}\\
&= x^2(1 - x^2)^{1/2} & &\text{Rewrite function.}\\
y' &= x^2 \frac{d}{dx}\left[(1 - x^2)^{1/2}\right] + (1 - x^2)^{1/2} \frac{d}{dx}[x^2] & &\text{Product Rule}\\
&= x^2\left[\frac{1}{2}(1 - x^2)^{-1/2}(-2x)\right] + (1 - x^2)^{1/2}(2x) & &\text{Power Rule}\\
&= -x^3(1 - x^2)^{-1/2} + 2x(1 - x^2)^{1/2} & &\\
&= x(1 - x^2)^{-1/2}\left[-x^2(1) + 2(1 - x^2)\right] & &\text{Factor.}\\
&= x(1 - x^2)^{-1/2}(2 - 3x^2) & &\\
&= \frac{x(2 - 3x^2)}{\sqrt{1 - x^2}} & &\text{Simplify.}
\end{aligned}$$

✓**CHECKPOINT 6**

Find and simplify the derivative of $y = x^2\sqrt{x^2 + 1}$. ■

STUDY TIP

In Example 7, try to find $f'(x)$ by applying the Quotient Rule to

$$f(x) = \frac{(3x - 1)^2}{(x^2 + 3)^2}.$$

Which method do you prefer?

Example 7 Differentiating a Quotient Raised to a Power

Find the derivative of

$$f(x) = \left(\frac{3x - 1}{x^2 + 3}\right)^2.$$

SOLUTION

$$\begin{aligned}
f'(x) &= \overbrace{2}^{n}\overbrace{\left(\frac{3x - 1}{x^2 + 3}\right)}^{u^{n-1}} \overbrace{\frac{d}{dx}\left[\frac{3x - 1}{x^2 + 3}\right]}^{u'}\\
&= \left[\frac{2(3x - 1)}{x^2 + 3}\right]\left[\frac{(x^2 + 3)(3) - (3x - 1)(2x)}{(x^2 + 3)^2}\right]\\
&= \frac{2(3x - 1)(3x^2 + 9 - 6x^2 + 2x)}{(x^2 + 3)^3}\\
&= \frac{2(3x - 1)(-3x^2 + 2x + 9)}{(x^2 + 3)^3}
\end{aligned}$$

✓**CHECKPOINT 7**

Find the derivative of

$$f(x) = \left(\frac{x + 1}{x - 5}\right)^2.$$ ■

Example 8 Finding Rates of Change

From 2000 through 2005, the amounts A (in millions of dollars) of private investment in medical research are shown in the table, where t is the year, with $t = 0$ corresponding to 2000. *(Source: U.S. Centers for Medicare & Medicaid Services)*

t	0	1	2	3	4	5
A	2.5	2.8	3.1	3.4	3.5	3.7

The amounts invested can be modeled by

$$A = (-0.0079t^2 + 0.108t + 1.58)^2, \quad 0 \le t \le 5.$$

a. Find the average rate of change of the amounts invested from 2000 through 2005.

b. Find the rate of change of the amounts invested in 2005.

SOLUTION

a. The years 2000 through 2005 can be represented by the interval $[0, 5]$. For the interval $[0, 5]$, the amount invested increases from \$2.5 million to \$3.7 million. So, the average rate of change is

$$\frac{\Delta A}{\Delta t} = \frac{3.7 - 2.5}{5 - 0} = \frac{1.2}{5} = \$0.24 \text{ million per year.}$$

b. The rate of change of A is given by the derivative dA/dt. You can use the General Power Rule to find the derivative.

$$\frac{dA}{dt} = 2(-0.0079t^2 + 0.108t + 1.58)^1(-0.0158t + 0.108)$$

$$= (-0.0316t + 0.216)(-0.0079t^2 + 0.108t + 1.58)$$

In 2005, the amount of private investment in medical research was changing at a rate of

$$[-0.0316(5) + 0.216][-0.0079(5)^2 + 0.108(5) + 1.58] \approx \$0.11$$

million per year.

The graph of the model is shown in Figure 2.25.

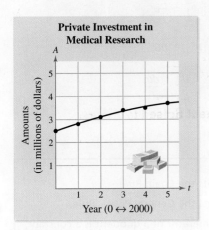

FIGURE 2.25

✓ CHECKPOINT 8

Use the information in Example 8 to answer the following.

a. Find the average rate of change of the amounts invested from 2002 through 2005.

b. Find the rate of change of the amounts invested when $t = 2$.

Summary of Differentiation Rules

You now have all the rules you need to differentiate *any* algebraic function. For your convenience, they are summarized below.

Summary of Differentiation Rules

Let u and v be differentiable functions of x.

1. **Constant Rule** $\quad \dfrac{d}{dx}[c] = 0, \quad c$ is a constant.

2. **Constant Multiple Rule** $\quad \dfrac{d}{dx}[cu] = c\dfrac{du}{dx}, \quad c$ is a constant.

3. **Sum and Difference Rules** $\quad \dfrac{d}{dx}[u \pm v] = \dfrac{du}{dx} \pm \dfrac{dv}{dx}$

4. **Product Rule** $\quad \dfrac{d}{dx}[uv] = u\dfrac{dv}{dx} + v\dfrac{du}{dx}$

5. **Quotient Rule** $\quad \dfrac{d}{dx}\left[\dfrac{u}{v}\right] = \dfrac{v\dfrac{du}{dx} - u\dfrac{dv}{dx}}{v^2}$

6. **Power Rules** $\quad \dfrac{d}{dx}[x^n] = nx^{n-1}$

    $\quad \dfrac{d}{dx}[u^n] = nu^{n-1}\dfrac{du}{dx}$

7. **Chain Rule** $\quad \dfrac{dy}{dx} = \dfrac{dy}{du} \cdot \dfrac{du}{dx}$

CONCEPT CHECK

1. Write a verbal statement that represents the Chain Rule.
2. Write a verbal statement that represents the General Power Rule.
3. Complete the following: When the numerator of a quotient is a constant, you may be able to find the derivative of the quotient more easily with the _____ _____ Rule than with the Quotient Rule.
4. In the expression $f(g(x))$, f is the outer function and g is the inner function. Write a verbal statement of the Chain Rule using the words "inner" and "outer."

SECTION 2.5 The Chain Rule

Skills Review 2.5

The following warm-up exercises involve skills that were covered in earlier sections. You will use these skills in the exercise set for this section. For additional help, review Sections 0.3 and 0.4.

In Exercises 1–6, rewrite the expression with rational exponents.

1. $\sqrt[5]{(1-5x)^2}$
2. $\sqrt[4]{(2x-1)^3}$
3. $\dfrac{1}{\sqrt{4x^2+1}}$
4. $\dfrac{1}{\sqrt[3]{x-6}}$
5. $\dfrac{\sqrt{x}}{\sqrt[3]{1-2x}}$
6. $\dfrac{\sqrt{(3-7x)^3}}{2x}$

In Exercises 7–10, factor the expression.

7. $3x^3 - 6x^2 + 5x - 10$
8. $5x\sqrt{x} - x - 5\sqrt{x} + 1$
9. $4(x^2+1)^2 - x(x^2+1)^3$
10. $-x^5 + 3x^3 + x^2 - 3$

Exercises 2.5

See www.CalcChat.com for worked-out solutions to odd-numbered exercises.

In Exercises 1–8, identify the inside function, $u = g(x)$, and the outside function, $y = f(u)$.

	$y = f(g(x))$	$u = g(x)$	$y = f(u)$
1.	$y = (6x - 5)^4$		
2.	$y = (x^2 - 2x + 3)^3$		
3.	$y = (4 - x^2)^{-1}$		
4.	$y = (x^2 + 1)^{4/3}$		
5.	$y = \sqrt{5x - 2}$		
6.	$y = \sqrt{1 - x^2}$		
7.	$y = (3x + 1)^{-1}$		
8.	$y = (x + 2)^{-1/2}$		

In Exercises 9–14, find dy/du, du/dx, and dy/dx.

9. $y = u^2,\ u = 4x + 7$
10. $y = u^3,\ u = 3x^2 - 2$
11. $y = \sqrt{u},\ u = 3 - x^2$
12. $y = 2\sqrt{u},\ u = 5x + 9$
13. $y = u^{2/3},\ u = 5x^4 - 2x$
14. $y = u^{-1},\ u = x^3 + 2x^2$

In Exercises 15–22, match the function with the rule that you would use to find the derivative *most efficiently*.

(a) Simple Power Rule   (b) Constant Rule
(c) General Power Rule   (d) Quotient Rule

15. $f(x) = \dfrac{2}{1 - x^3}$
16. $f(x) = \dfrac{2x}{1 - x^3}$
17. $f(x) = \sqrt[3]{8^2}$
18. $f(x) = \sqrt[3]{x^2}$
19. $f(x) = \dfrac{x^2 + 2}{x}$
20. $f(x) = \dfrac{x^4 - 2x + 1}{\sqrt{x}}$
21. $f(x) = \dfrac{2}{x - 2}$
22. $f(x) = \dfrac{5}{x^2 + 1}$

In Exercises 23–40, use the General Power Rule to find the derivative of the function.

23. $y = (2x - 7)^3$
24. $y = (2x^3 + 1)^2$
25. $g(x) = (4 - 2x)^3$
26. $h(t) = (1 - t^2)^4$
27. $h(x) = (6x - x^3)^2$
28. $f(x) = (4x - x^2)^3$
29. $f(x) = (x^2 - 9)^{2/3}$
30. $f(t) = (9t + 2)^{2/3}$
31. $f(t) = \sqrt{t + 1}$
32. $g(x) = \sqrt{5 - 3x}$
33. $s(t) = \sqrt{2t^2 + 5t + 2}$
34. $y = \sqrt[3]{3x^3 + 4x}$
35. $y = \sqrt[3]{9x^2 + 4}$
36. $y = 2\sqrt{4 - x^2}$
37. $f(x) = -3\sqrt[4]{2 - 9x}$
38. $f(x) = (25 + x^2)^{-1/2}$
39. $h(x) = (4 - x^3)^{-4/3}$
40. $f(x) = (4 - 3x)^{-5/2}$

In Exercises 41–46, find an equation of the tangent line to the graph of f at the point $(2, f(2))$. Use a graphing utility to check your result by graphing the original function and the tangent line in the same viewing window.

41. $f(x) = 2(x^2 - 1)^3$
42. $f(x) = 3(9x - 4)^4$
43. $f(x) = \sqrt{4x^2 - 7}$
44. $f(x) = x\sqrt{x^2 + 5}$
45. $f(x) = \sqrt{x^2 - 2x + 1}$
46. $f(x) = (4 - 3x^2)^{-2/3}$

(T) In Exercises 47–50, use a symbolic differentiation utility to find the derivative of the function. Graph the function and its derivative in the same viewing window. Describe the behavior of the function when the derivative is zero.

47. $f(x) = \dfrac{\sqrt{x} + 1}{x^2 + 1}$
48. $f(x) = \sqrt{\dfrac{2x}{x + 1}}$
49. $f(x) = \sqrt{\dfrac{x + 1}{x}}$
50. $f(x) = \sqrt{x}(2 - x^2)$

In Exercises 51–66, find the derivative of the function. State which differentiation rule(s) you used to find the derivative.

51. $y = \dfrac{1}{x-2}$

52. $s(t) = \dfrac{1}{t^2 + 3t - 1}$

53. $y = -\dfrac{4}{(t+2)^2}$

54. $f(x) = \dfrac{3}{(x^3 - 4)^2}$

55. $f(x) = \dfrac{1}{(x^2 - 3x)^2}$

56. $y = \dfrac{1}{\sqrt{x+2}}$

57. $g(t) = \dfrac{1}{t^2 - 2}$

58. $g(x) = \dfrac{3}{\sqrt[3]{x^3 - 1}}$

59. $f(x) = x(3x - 9)^3$

60. $f(x) = x^3(x - 4)^2$

61. $y = x\sqrt{2x+3}$

62. $y = t\sqrt{t+1}$

63. $y = t^2\sqrt{t-2}$

64. $y = \sqrt{x}(x-2)^2$

65. $y = \left(\dfrac{6-5x}{x^2-1}\right)^2$

66. $y = \left(\dfrac{4x^2}{3-x}\right)^3$

(T) In Exercises 67–72, find an equation of the tangent line to the graph of the function at the given point. Then use a graphing utility to graph the function and the tangent line in the same viewing window.

Function	Point
67. $f(t) = \dfrac{36}{(3-t)^2}$	$(0, 4)$
68. $s(x) = \dfrac{1}{\sqrt{x^2 - 3x + 4}}$	$(3, \tfrac{1}{2})$
69. $f(t) = (t^2 - 9)\sqrt{t+2}$	$(-1, -8)$
70. $y = \dfrac{2x}{\sqrt{x+1}}$	$(3, 3)$
71. $f(x) = \dfrac{x+1}{\sqrt{2x-3}}$	$(2, 3)$
72. $y = \dfrac{x}{\sqrt{25 + x^2}}$	$(0, 0)$

73. **Fruit and Vegetable Consumption** The amount A (in billions of pounds) of vegetables and melons consumed from 2000 through 2005 can be modeled by

$$A = \sqrt{7.54t^2 + 29.8t + 2348}$$

where t represents the year, with $t = 0$ corresponding to 2000. Find the rates of change of A with respect to t when $t = 2$ and $t = 4$. (Source: U.S. Department of Agriculture)

74. **Environment** An environmental study indicates that the average daily level P of a certain pollutant in the air (in parts per million) can be modeled by the equation

$$P = 0.25\sqrt{0.5n^2 + 5n + 25}$$

where n is the number of residents of the community (in thousands). Find the rate at which the level of pollutant is increasing when the population of the community is 12,000.

75. **Biology** The number N of bacteria in a culture after t days is modeled by

$$N = 400\left[1 - \dfrac{3}{(t^2 + 2)^2}\right].$$

Complete the table. What can you conclude?

t	0	1	2	3	4
dN/dt					

76. **Depreciation** The value V of a security x-ray screening system t years after it was purchased is inversely proportional to the square root of $t + 1$. The initial value of the security x-ray screening system was $10,000.

(a) Write V as a function of t.

(b) Find the rate of depreciation when $t = 1$.

(c) Find the rate of depreciation when $t = 3$.

77. **Depreciation** Repeat Exercise 76 given that the value of the machine t years after it was purchased is inversely proportional to the cube root of $t + 1$.

(T) 78. **U.S. AIDS Epidemic** The number of cases f (in thousands) of AIDS reported in the years 1995 through 2005 can be modeled by

$$f = \sqrt{-1.8840t^4 + 66.325t^3 - 739.92t^2 + 2307.0t + 5125}$$

where t represents the year, with $t = 5$ corresponding to 1995. (Source: U.S. Centers for Disease Control and Prevention)

(a) Find the derivative of the model. Which differentiation rule(s) did you use?

(b) Use a graphing utility to graph the derivative on the interval $5 \le t \le 15$.

(c) Use the *trace* feature to find the years during which the number of AIDS cases was changing the most.

(d) Use the *trace* feature to find the years during which the number of AIDS cases was changing the least.

True or False? In Exercises 79 and 80, determine whether the statement is true or false. If it is false, explain why or give an example that shows it is false.

79. If $y = (1 - x)^{1/2}$, then $y' = \tfrac{1}{2}(1 - x)^{-1/2}$.

80. If y is a differentiable function of u, u is a differentiable function of v, and v is a differentiable function of x, then

$$\dfrac{dy}{dx} = \dfrac{dy}{du} \cdot \dfrac{du}{dv} \cdot \dfrac{dv}{dx}.$$

81. Given that $f(x) = h(g(x))$, find $f'(2)$ for each of the following.

(a) $g(2) = -6$ and $g'(2) = 5$, $h(5) = 4$ and $h'(-6) = 3$

(b) $g(2) = -1$ and $g'(2) = -2$, $h(2) = 4$ and $h'(-1) = 5$

Section 2.6
Higher-Order Derivatives

- Find higher-order derivatives.
- Find and use the position function to determine the velocity and acceleration of moving objects.

Second, Third, and Higher-Order Derivatives

The derivative of f' is the **second derivative** of f and is denoted by f''.

$$\frac{d}{dx}[f'(x)] = f''(x) \qquad \text{Second derivative}$$

The derivative of f'' is the **third derivative** of f and is denoted by f'''.

$$\frac{d}{dx}[f''(x)] = f'''(x) \qquad \text{Third derivative}$$

By continuing this process, you obtain **higher-order derivatives** of f. Higher-order derivatives are denoted as follows.

STUDY TIP

In the context of higher-order derivatives, the "standard" derivative f' is often called the **first derivative** of f.

DISCOVERY

For each function, find the indicated higher-order derivative.

a. $y = x^2$ b. $y = x^3$
   y'' y'''

c. $y = x^4$ d. $y = x^n$
   $y^{(4)}$ $y^{(n)}$

Notation for Higher-Order Derivatives

1. 1st derivative: y', $f'(x)$, $\dfrac{dy}{dx}$, $\dfrac{d}{dx}[f(x)]$, $D_x[y]$

2. 2nd derivative: y'', $f''(x)$, $\dfrac{d^2y}{dx^2}$, $\dfrac{d^2}{dx^2}[f(x)]$, $D_x^2[y]$

3. 3rd derivative: y''', $f'''(x)$, $\dfrac{d^3y}{dx^3}$, $\dfrac{d^3}{dx^3}[f(x)]$, $D_x^3[y]$

4. 4th derivative: $y^{(4)}$, $f^{(4)}(x)$, $\dfrac{d^4y}{dx^4}$, $\dfrac{d^4}{dx^4}[f(x)]$, $D_x^4[y]$

5. nth derivative: $y^{(n)}$, $f^{(n)}(x)$, $\dfrac{d^ny}{dx^n}$, $\dfrac{d^n}{dx^n}[f(x)]$, $D_x^n[y]$

Example 1   Finding Higher-Order Derivatives

Find the first five derivatives of $f(x) = 2x^4 - 3x^2$.

$f(x) = 2x^4 - 3x^2$ Write original function.
$f'(x) = 8x^3 - 6x$ First derivative
$f''(x) = 24x^2 - 6$ Second derivative
$f'''(x) = 48x$ Third derivative
$f^{(4)}(x) = 48$ Fourth derivative
$f^{(5)}(x) = 0$ Fifth derivative

✓ CHECKPOINT 1

Find the first four derivatives of $f(x) = 6x^3 - 2x^2 + 1$. ∎

Example 2 Finding Higher-Order Derivatives

Find the value of $g'''(2)$ for the function

$$g(t) = -t^4 + 2t^3 + t + 4.$$ Original function

SOLUTION Begin by differentiating three times.

$$g'(t) = -4t^3 + 6t^2 + 1$$ First derivative

$$g''(t) = -12t^2 + 12t$$ Second derivative

$$g'''(t) = -24t + 12$$ Third derivative

Then, evaluate the third derivative of g at $t = 2$.

$$g'''(2) = -24(2) + 12$$
$$= -36$$ Value of third derivative

✓ **CHECKPOINT 2**

Find the value of $g'''(1)$ for $g(x) = x^4 - x^3 + 2x$. ∎

Examples 1 and 2 show how to find higher-order derivatives of *polynomial* functions. Note that with each successive differentiation, the degree of the polynomial drops by one. Eventually, higher-order derivatives of polynomial functions degenerate to a constant function. Specifically, the nth-order derivative of an nth-degree polynomial function

$$f(x) = a_n x^n + a_{n-1} x^{n-1} + \cdots + a_1 x + a_0$$

is the constant function

$$f^{(n)}(x) = n! a_n$$

where $n! = 1 \cdot 2 \cdot 3 \cdots n$. Each derivative of order higher than n is the zero function. Polynomial functions are the *only* functions with this characteristic. For other functions, successive differentiation never produces a constant function.

Example 3 Finding Higher-Order Derivatives

Find the first four derivatives of $y = x^{-1}$.

$$y = x^{-1} = \frac{1}{x}$$ Write original function.

$$y' = (-1)x^{-2} = -\frac{1}{x^2}$$ First derivative

$$y'' = (-1)(-2)x^{-3} = \frac{2}{x^3}$$ Second derivative

$$y''' = (-1)(-2)(-3)x^{-4} = -\frac{6}{x^4}$$ Third derivative

$$y^{(4)} = (-1)(-2)(-3)(-4)x^{-5} = \frac{24}{x^5}$$ Fourth derivative

TECHNOLOGY

Higher-order derivatives of nonpolynomial functions can be difficult to find by hand. If you have access to a symbolic differentiation utility, try using it to find higher-order derivatives.

✓ **CHECKPOINT 3**

Find the fourth derivative of

$$y = \frac{1}{x^2}.$$ ∎

SECTION 2.6 Higher-Order Derivatives

STUDY TIP
Acceleration is measured in units of length per unit of time squared. For instance, if the velocity is measured in feet per second, then the acceleration is measured in "feet per second squared," or, more formally, in "feet per second per second."

Acceleration

In Section 2.3, you saw that the velocity of an object moving in a straight path (neglecting air resistance) is given by the derivative of its position function. In other words, the rate of change of the position with respect to time is defined to be the velocity. In a similar way, the rate of change of the velocity with respect to time is defined to be the **acceleration** of the object.

$$s = f(t) \quad \text{Position function}$$

$$\frac{ds}{dt} = f'(t) \quad \text{Velocity function}$$

$$\frac{d^2s}{dt^2} = f''(t) \quad \text{Acceleration function}$$

To find the position, velocity, or acceleration at a particular time t, substitute the given value of t into the appropriate function, as illustrated in Example 4.

Example 4 Finding Acceleration

A ball is thrown upward from the top of a 160-foot cliff, as shown in Figure 2.26. The initial velocity of the ball is 48 feet per second, which implies that the position function is

$$s = -16t^2 + 48t + 160$$

where the time t is measured in seconds. Find the height, the velocity, and the acceleration of the ball when $t = 3$.

SOLUTION Begin by differentiating to find the velocity and acceleration functions.

$$s = -16t^2 + 48t + 160 \quad \text{Position function}$$

$$\frac{ds}{dt} = -32t + 48 \quad \text{Velocity function}$$

$$\frac{d^2s}{dt^2} = -32 \quad \text{Acceleration function}$$

To find the height, velocity, and acceleration when $t = 3$, substitute $t = 3$ into each of the functions above.

Height $= -16(3)^2 + 48(3) + 160 = 160$ feet

Velocity $= -32(3) + 48 = -48$ feet per second

Acceleration $= -32$ feet per second squared

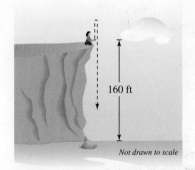

FIGURE 2.26

✓ CHECKPOINT 4

A ball is thrown upward from the top of an 80-foot cliff with an initial velocity of 64 feet per second. Give the position function. Then find the velocity and acceleration functions. ■

In Example 4, notice that the acceleration of the ball is -32 feet per second squared at any time t. This constant acceleration is due to the gravitational force of Earth and is called the **acceleration due to gravity.** Note that the negative value indicates that the ball is being pulled *down*—toward Earth.

Although the acceleration exerted on a falling object is relatively constant near Earth's surface, it varies greatly throughout our solar system. Large planets exert a much greater gravitational pull than do small planets or moons. The next example describes the motion of a free-falling object on the moon.

Example 5 Finding Acceleration on the Moon

An astronaut standing on the surface of the moon throws a rock into the air. The height s (in feet) of the rock is given by

$$s = -\frac{27}{10}t^2 + 27t + 6$$

where t is measured in seconds. How does the acceleration due to gravity on the moon compare with that on Earth?

SOLUTION

$$s = -\frac{27}{10}t^2 + 27t + 6 \qquad \text{Position function}$$

$$\frac{ds}{dt} = -\frac{27}{5}t + 27 \qquad \text{Velocity function}$$

$$\frac{d^2s}{dt^2} = -\frac{27}{5} \qquad \text{Acceleration function}$$

So, the acceleration at any time is

$$-\frac{27}{5} = -5.4 \text{ feet per second squared}$$

—about one-sixth of the acceleration due to gravity on Earth.

The position function described in Example 5 neglects air resistance, which is appropriate because the moon has no atmosphere—and *no air resistance.* This means that the position function for any free-falling object on the moon is given by

$$s = -\frac{27}{10}t^2 + v_0 t + h_0$$

where s is the height (in feet), t is the time (in seconds), v_0 is the initial velocity, and h_0 is the initial height. For instance, the rock in Example 5 was thrown upward with an initial velocity of 27 feet per second and had an initial height of 6 feet. This position function is valid for all objects, whether they are heavy ones such as hammers or light ones such as feathers.

In 1971, astronaut David R. Scott demonstrated the lack of atmosphere on the moon by dropping a hammer and a feather from the same height. Both took exactly the same time to fall to the ground. If they were dropped from a height of 6 feet, how long did each take to hit the ground?

NASA

The acceleration due to gravity on the surface of the moon is only about one-sixth that exerted by Earth. So, if you were on the moon and threw an object into the air, it would rise to a greater height than it would on Earth's surface.

✓ CHECKPOINT 5

The position function on Earth, where s is measured in meters, t is measured in seconds, v_0 is the initial velocity in meters per second, and h_0 is the initial height in meters, is

$$s = -4.9t^2 + v_0 t + h_0.$$

If the initial velocity is 2.2 and the initial height is 3.6, what is the acceleration due to gravity on Earth in meters per second per second? ■

Example 6 Finding Velocity and Acceleration

The velocity v (in feet per second) of a certain automobile starting from rest is

$$v = \frac{80t}{t+5} \qquad \text{Velocity function}$$

where t is the time (in seconds). The positions of the automobile at 10-second intervals are shown in Figure 2.27. Find the velocity and acceleration of the automobile at 10-second intervals from $t = 0$ to $t = 60$.

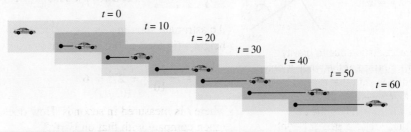

FIGURE 2.27

SOLUTION To find the acceleration function, differentiate the velocity function.

$$\frac{dv}{dt} = \frac{(t+5)(80) - (80t)(1)}{(t+5)^2}$$

$$= \frac{400}{(t+5)^2} \qquad \text{Acceleration function}$$

t (seconds)	0	10	20	30	40	50	60
v (ft/sec)	0	53.5	64.0	68.6	71.1	72.7	73.8
$\frac{dv}{dt}$ (ft/sec²)	16	1.78	0.64	0.33	0.20	0.13	0.09

✓ **CHECKPOINT 6**

Use a graphing utility to graph the velocity function and acceleration function in Example 6 in the same viewing window. Compare the graphs with the table at the right. As the velocity levels off, what does the acceleration approach? ∎

In the table, note that the acceleration approaches zero as the velocity levels off. This observation should agree with your experience—when riding in an accelerating automobile, you do not feel the velocity, but you do feel the acceleration. In other words, you feel changes in velocity.

CONCEPT CHECK

1. Use mathematical notation to write the third derivative of $f(x)$.
2. Give a verbal description of what is meant by $\frac{d^2y}{dx^2}$.
3. Complete the following: If $f(x)$ is an nth-degree polynomial, then $f^{(n+1)}(x)$ is equal to _____.
4. If the velocity of an object is constant, what is its acceleration?

CHAPTER 2 Differentiation

Skills Review 2.6

The following warm-up exercises involve skills that were covered in earlier sections. You will use these skills in the exercise set for this section. For additional help, review Sections 1.4 and 2.4.

In Exercises 1–4, solve the equation.

1. $-16t^2 + 24t = 0$
2. $-16t^2 + 80t + 224 = 0$
3. $-16t^2 + 128t + 320 = 0$
4. $-16t^2 + 9t + 1440 = 0$

In Exercises 5–8, find *dy/dx*.

5. $y = x^2(2x + 7)$
6. $y = (x^2 + 3x)(2x^2 - 5)$
7. $y = \dfrac{x^2}{2x + 7}$
8. $y = \dfrac{x^2 + 3x}{2x^2 - 5}$

In Exercises 9 and 10, find the domain and range of *f*.

9. $f(x) = x^2 - 4$
10. $f(x) = \sqrt{x - 7}$

Exercises 2.6

See www.CalcChat.com for worked-out solutions to odd-numbered exercises.

In Exercises 1–16, find the second derivative of the function.

1. $f(x) = 9 - 2x$
2. $f(x) = 4x + 15$
3. $f(x) = x^2 + 7x - 4$
4. $f(x) = 3x^2 + 4x$
5. $g(t) = \frac{1}{3}t^3 - 4t^2 + 2t$
6. $f(x) = 4(x^2 - 1)^2$
7. $f(t) = \dfrac{3}{4t^2}$
8. $g(t) = 32t^{-2}$
9. $f(x) = 3(2 - x^2)^3$
10. $f(x) = x\sqrt[3]{x}$
11. $y = (x^3 - 2x)^4$
12. $y = 4(x^2 + 5x)^3$
13. $f(x) = \dfrac{x + 1}{x - 1}$
14. $g(t) = -\dfrac{4}{(t + 2)^2}$
15. $y = x^2(x^2 + 4x + 8)$
16. $h(s) = s^3(s^2 - 2s + 1)$

In Exercises 17–22, find the third derivative of the function.

17. $f(x) = x^5 - 3x^4$
18. $f(x) = x^4 - 2x^3$
19. $f(x) = 5x(x + 4)^3$
20. $f(x) = (x^3 - 6)^4$
21. $f(x) = \dfrac{3}{16x^2}$
22. $f(x) = \dfrac{1}{x}$

In Exercises 23–28, find the given value.

Function	Value
23. $g(t) = 5t^4 + 10t^2 + 3$	$g''(2)$
24. $f(x) = 9 - x^2$	$f''(-\sqrt{5})$
25. $f(x) = \sqrt{4 - x}$	$f'''(-5)$
26. $f(t) = \sqrt{2t + 3}$	$f'''(\frac{1}{2})$
27. $f(x) = x^2(3x^2 + 3x - 4)$	$f'''(-2)$
28. $g(x) = 2x^3(x^2 - 5x + 4)$	$g'''(0)$

In Exercises 29–34, find the higher-order derivative.

Given	Derivative
29. $f'(x) = 2x^2$	$f''(x)$
30. $f''(x) = 20x^3 - 36x^2$	$f'''(x)$
31. $f'''(x) = (3x - 1)/x$	$f^{(4)}(x)$
32. $f'''(x) = 2\sqrt{x - 1}$	$f^{(4)}(x)$
33. $f^{(4)}(x) = (x^2 + 1)^2$	$f^{(6)}(x)$
34. $f''(x) = 2x^2 + 7x - 12$	$f^{(5)}(x)$

In Exercises 35–42, find the second derivative and solve the equation $f''(x) = 0$.

35. $f(x) = x^3 - 9x^2 + 27x - 27$
36. $f(x) = 3x^3 - 9x + 1$
37. $f(x) = (x + 3)(x - 4)(x + 5)$
38. $f(x) = (x + 2)(x - 2)(x + 3)(x - 3)$
39. $f(x) = x\sqrt{x^2 - 1}$
40. $f(x) = x\sqrt{4 - x^2}$
41. $f(x) = \dfrac{x}{x^2 + 3}$
42. $f(x) = \dfrac{x}{x - 1}$

43. Velocity and Acceleration A ball is propelled straight upward from ground level with an initial velocity of 144 feet per second.

(a) Write the position, velocity, and acceleration functions for the ball.

(b) When is the ball at its highest point? How high is this point?

(c) How fast is the ball traveling when it hits the ground? How is this speed related to the initial velocity?

44. Velocity and Acceleration A brick becomes dislodged from the top of the Empire State Building (at a height of 1250 feet) and falls to the sidewalk below.

(a) Write the position, velocity, and acceleration functions for the brick.

(b) How long does it take the brick to hit the sidewalk?

(c) How fast is the brick traveling when it hits the sidewalk?

45. Velocity and Acceleration The velocity (in feet per second) of an automobile starting from rest is modeled by

$$\frac{ds}{dt} = \frac{90t}{t+10}.$$

Create a table showing the velocity and acceleration at 10-second intervals during the first minute of travel. What can you conclude?

46. Stopping Distance A car is traveling at a rate of 66 feet per second (45 miles per hour) when the brakes are applied. The position function for the car is given by $s = -8.25t^2 + 66t$, where s is measured in feet and t is measured in seconds. Create a table showing the position, velocity, and acceleration for each given value of t. What can you conclude?

In Exercises 47 and 48, use a graphing utility to graph f, f', and f'' in the same viewing window. What is the relationship among the degree of f and the degrees of its successive derivatives? In general, what is the relationship among the degree of a polynomial function and the degrees of its successive derivatives?

47. $f(x) = x^2 - 6x + 6$ **48.** $f(x) = 3x^3 - 9x$

In Exercises 49 and 50, the graphs of f, f', and f'' are shown on the same set of coordinate axes. Which is which? Explain your reasoning.

49.

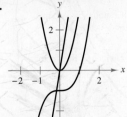

50.

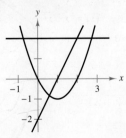

51. Modeling Data The table shows the utilized productions y of citrus fruits (in millions of pounds) in the United States for the years 2000 through 2005, where t is the year, with $t = 0$ corresponding to 2000. *(Source: U.S. Department of Agriculture)*

t	0	1	2	3	4	5
y	8355	8331	8256	8442	8156	7366

(a) Use a graphing utility to find a cubic model for the data.

(b) Use a graphing utility to graph the model and plot the data in the same viewing window. How well does the model fit the data?

(c) Find the first and second derivatives of the function.

(d) Show that the utilized production was decreasing from 2003 to 2005.

(e) Find the year in which the utilized production was increasing at the greatest rate by solving $y''(t) = 0$.

(f) Explain the relationship among your answers for parts (c), (d), and (e).

52. Projectile Motion An object is thrown upward from the top of a 64-foot building with an initial velocity of 48 feet per second.

(a) Write the position, velocity, and acceleration functions for the object.

(b) When will the object hit the ground?

(c) When is the velocity of the object zero?

(d) How high does the object go?

(e) Use a graphing utility to graph the position, velocity, and acceleration functions in the same viewing window. Write a short paragraph that describes the relationship among these functions.

True or False? In Exercises 53–56, determine whether the statement is true or false. If it is false, explain why or give an example that shows it is false.

53. If $y = f(x)g(x)$, then $y' = f'(x)g'(x)$.

54. If $y = (x+1)(x+2)(x+3)(x+4)$, then $\dfrac{d^5y}{dx^5} = 0$.

55. If $f'(c)$ and $g'(c)$ are zero and $h(x) = f(x)g(x)$, then $h'(c) = 0$.

56. The second derivative represents the rate of change of the first derivative.

57. Finding a Pattern Develop a general rule for $[xf(x)]^{(n)}$ where f is a differentiable function of x.

58. Extended Application To work an extended application analyzing the numbers of Medicare enrollees in the years 1980 through 2005, visit this text's website at *college.hmco.com/info/larsonapplied*. *(Data Source: U.S. Centers for Medicare and Medicaid Services)*

Algebra Review

Simplifying Algebraic Expressions

To be successful in using derivatives, you must be good at simplifying algebraic expressions. Here are some helpful simplification techniques.

1. Combine *like terms*. This may involve expanding an expression by multiplying factors.
2. Divide out *like factors* in the numerator and denominator of an expression.
3. Factor an expression.
4. Rationalize a denominator.
5. Add, subtract, multiply, or divide fractions.

TECHNOLOGY

Symbolic algebra systems can simplify algebraic expressions. If you have access to such a system, try using it to simplify the expressions in this Algebra Review.

Example 1 Simplifying Fractional Expressions

a. $\dfrac{(x + \Delta x)^2 - x^2}{\Delta x} = \dfrac{x^2 + 2x(\Delta x) + (\Delta x)^2 - x^2}{\Delta x}$ Expand expression.

$= \dfrac{2x(\Delta x) + (\Delta x)^2}{\Delta x}$ Combine like terms.

$= \dfrac{\Delta x(2x + \Delta x)}{\Delta x}$ Factor.

$= 2x + \Delta x, \quad \Delta x \neq 0$ Divide out like factors.

b. $\dfrac{(x^2 - 1)(-2 - 2x) - (3 - 2x - x^2)(2)}{(x^2 - 1)^2}$

$= \dfrac{(-2x^2 - 2x^3 + 2 + 2x) - (6 - 4x - 2x^2)}{(x^2 - 1)^2}$ Expand expression.

$= \dfrac{-2x^2 - 2x^3 + 2 + 2x - 6 + 4x + 2x^2}{(x^2 - 1)^2}$ Remove parentheses.

$= \dfrac{-2x^3 + 6x - 4}{(x^2 - 1)^2}$ Combine like terms.

c. $2\left(\dfrac{2x + 1}{3x}\right)\left[\dfrac{3x(2) - (2x + 1)(3)}{(3x)^2}\right]$

$= 2\left(\dfrac{2x + 1}{3x}\right)\left[\dfrac{6x - (6x + 3)}{(3x)^2}\right]$ Multiply factors.

$= \dfrac{2(2x + 1)(6x - 6x - 3)}{(3x)^3}$ Multiply fractions and remove parentheses.

$= \dfrac{2(2x + 1)(-3)}{3(9)x^3}$ Combine like terms and factor.

$= \dfrac{-2(2x + 1)}{9x^3}$ Divide out like factors.

Example 2 Simplifying Expressions with Powers or Radicals

a. $(2x + 1)^2(6x + 1) + (3x^2 + x)(2)(2x + 1)(2)$

$= (2x + 1)[(2x + 1)(6x + 1) + (3x^2 + x)(2)(2)]$ Factor.

$= (2x + 1)[12x^2 + 8x + 1 + (12x^2 + 4x)]$ Multiply factors.

$= (2x + 1)(12x^2 + 8x + 1 + 12x^2 + 4x)$ Remove parentheses.

$= (2x + 1)(24x^2 + 12x + 1)$ Combine like terms.

b. $(-1)(6x^2 - 4x)^{-2}(12x - 4)$

$= \dfrac{(-1)(12x - 4)}{(6x^2 - 4x)^2}$ Rewrite as a fraction.

$= \dfrac{(-1)(4)(3x - 1)}{(6x^2 - 4x)^2}$ Factor.

$= \dfrac{-4(3x - 1)}{(6x^2 - 4x)^2}$ Multiply factors.

c. $(x)\left(\dfrac{1}{2}\right)(2x + 3)^{-1/2} + (2x + 3)^{1/2}(1)$

$= (2x + 3)^{-1/2}\left(\dfrac{1}{2}\right)[x + (2x + 3)(2)]$ Factor.

$= \dfrac{x + 4x + 6}{(2x + 3)^{1/2}(2)}$ Rewrite as a fraction.

$= \dfrac{5x + 6}{2(2x + 3)^{1/2}}$ Combine like terms.

d. $\dfrac{x^2\left(\frac{1}{2}\right)(2x)(x^2 + 1)^{-1/2} - (x^2 + 1)^{1/2}(2x)}{x^4}$

$= \dfrac{(x^3)(x^2 + 1)^{-1/2} - (x^2 + 1)^{1/2}(2x)}{x^4}$ Multiply factors.

$= \dfrac{(x^2 + 1)^{-1/2}(x)[x^2 - (x^2 + 1)(2)]}{x^4}$ Factor.

$= \dfrac{x[x^2 - (2x^2 + 2)]}{(x^2 + 1)^{1/2}x^4}$ Write with positive exponents.

$= \dfrac{x^2 - 2x^2 - 2}{(x^2 + 1)^{1/2}x^3}$ Divide out like factors and remove parentheses.

$= \dfrac{-x^2 - 2}{(x^2 + 1)^{1/2}x^3}$ Combine like terms.

All but one of the expressions in this Algebra Review are derivatives. Can you see what the original function is? Explain your reasoning.

Chapter Summary and Study Strategies

After studying this chapter, you should have acquired the following skills. The exercise numbers are keyed to the Review Exercises that begin on page 178. Answers to odd-numbered Review Exercises are given in the back of the text.*

Section 2.1   Review Exercises

- Approximate the slope of the tangent line to a graph at a point.   1–4
- Interpret the slope of a graph in a real-life setting.   5–8
- Use the limit definition to find the derivative of a function and the slope of a graph at a point.   9–16

$$f'(x) = \lim_{\Delta x \to 0} \frac{f(x + \Delta x) - f(x)}{\Delta x}$$

- Use the derivative to find the slope of a graph at a point.   17–24
- Use the graph of a function to recognize points at which the function is not differentiable.   25–28

Section 2.2

- Use the Constant Multiple Rule for differentiation.   29, 30

$$\frac{d}{dx}[cf(x)] = cf'(x)$$

- Use the Sum and Difference Rules for differentiation.   31–38

$$\frac{d}{dx}[f(x) \pm g(x)] = f'(x) \pm g'(x)$$

Section 2.3

- Find the average rate of change of a function over an interval and the instantaneous rate of change at a point.   39, 40

$$\text{Average rate of change} = \frac{f(b) - f(a)}{b - a}$$

$$\text{Instantaneous rate of change} = \lim_{\Delta x \to 0} \frac{f(x + \Delta x) - f(x)}{\Delta x}$$

- Find the average and instantaneous rates of change of a quantity in a real-life problem.   41, 42
- Use rates of change to solve real-life problems.   43–48

---

* Use a wide range of valuable study aids to help you master the material in this chapter. The *Student Solutions Guide* includes step-by-step solutions to all odd-numbered exercises to help you review and prepare. The student website at *college.hmco.com/info/larsonapplied* offers algebra help and a *Graphing Technology Guide*. The *Graphing Technology Guide* contains step-by-step commands and instructions for a wide variety of graphing calculators, including the most recent models.

Section 2.4

Review Exercises

- Use the Product Rule for differentiation.   49–52

$$\frac{d}{dx}[f(x)g(x)] = f(x)g'(x) + g(x)f'(x)$$

- Use the Quotient Rule for differentiation.   53, 54

$$\frac{d}{dx}\left[\frac{f(x)}{g(x)}\right] = \frac{g(x)f'(x) - f(x)g'(x)}{[g(x)]^2}, \quad g(x) \neq 0$$

Section 2.5

- Use the General Power Rule for differentiation.   55–58

$$\frac{d}{dx}[u^n] = nu^{n-1}u'$$

- Use differentiation rules efficiently to find the derivative of any algebraic function, then simplify the result.   59–68
- Use derivatives to answer questions about real-life situations. (Sections 2.1–2.5)   69–72

Section 2.6

- Find higher-order derivatives.   73–80
- Find and use the position function to determine the velocity and acceleration of a moving object.   81, 82
- Use higher-order derivatives to answer questions about real–life situations.   83

Study Strategies

- **Simplify Your Derivatives**   Often our students ask if they have to simplify their derivatives. Our answer is "Yes, if you expect to use them." In the next chapter, you will see that almost all applications of derivatives require that the derivatives be written in simplified form. It is not difficult to see the advantage of a derivative in simplified form. Consider, for instance, the derivative of

$$f(x) = x/\sqrt{x^2 + 1}.$$

The "raw form" produced by the Quotient and Chain Rules

$$f'(x) = \frac{(x^2 + 1)^{1/2}(1) - (x)(\frac{1}{2})(x^2 + 1)^{-1/2}(2x)}{\left(\sqrt{x^2 + 1}\right)^2}$$

is obviously much more difficult to use than the simplified form

$$f'(x) = \frac{1}{(x^2 + 1)^{3/2}}.$$

- **List Units of Measure in Applied Problems**   When using derivatives in real-life applications, be sure to list the units of measure for each variable. For instance, if P is measured in millimeters and t is measured in seconds, then the derivative dP/dt is measured in millimeters per second.

Review Exercises

See www.CalcChat.com for worked-out solutions to odd-numbered exercises.

In Exercises 1–4, approximate the slope of the tangent line to the graph at (x, y).

1.
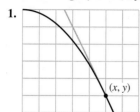

2.
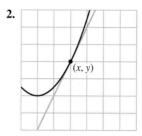

3.
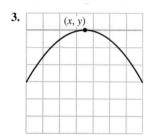

4.
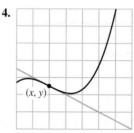

5. Space Launches The graph approximates the number S of successful space launches world-wide for the years 2000 through 2005, where t is the year, with $t = 0$ corresponding to 2000. Estimate the slopes of the graph when $t = 1$, $t = 2$, and $t = 4$. Interpret each slope in the context of the problem. *(Source: Library of Congress)*

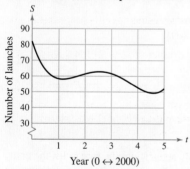

World-Wide Successful Space Launches

6. Health Insurance The graph approximates the number of children enrolled E (in thousands per year) in the State Children's Health Insurance Program (SCHIP) for the years 1998 through 2005, where t is the year, with $t = 8$ corresponding to 1998. Estimate the slopes of the graph when $t = 9$, $t = 12$, and $t = 14$. Interpret each slope in the context of the problem. *(Source: Congressional Budget Office)*

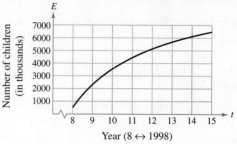

State Children's Health Insurance Program

Figure for 6

7. Medicine The graph shows the estimated number of milligrams of a pain medication M in the bloodstream t hours after a 1000-milligram dose of the drug has been given. Estimate the slopes of the graph at $t = 1$, 4, and 5.

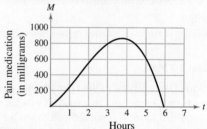

Pain Medication in Bloodstream

8. White-Water Rafting Two white-water rafters leave a campsite simultaneously and start downstream on a 9-mile trip. Their distances from the campsite are given by $s = f(t)$ and $s = g(t)$, where s is measured in miles and t is measured in hours.

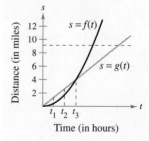

White-Water Rafting

(a) Which rafter is traveling at a greater rate at t_1?

(b) What can you conclude about their rates at t_2?

(c) What can you conclude about their rates at t_3?

(d) Which rafter finishes the trip first? Explain your reasoning.

In Exercises 9–16, use the limit definition to find the derivative of the function. Then use the limit definition to find the slope of the tangent line to the graph of f at the given point.

9. $f(x) = -3x - 5$, $(-2, 1)$
10. $f(x) = 7x + 3$, $(-1, 4)$
11. $f(x) = x^2 - 4x$, $(1, -3)$
12. $f(x) = x^2 + 10$, $(2, 14)$
13. $f(x) = \sqrt{x + 9}$, $(-5, 2)$
14. $f(x) = \sqrt{x - 1}$, $(10, 3)$
15. $f(x) = \dfrac{1}{x - 5}$, $(6, 1)$
16. $f(x) = \dfrac{1}{x + 4}$, $(-3, 1)$

In Exercises 17–24, find the slope of the graph of f at the given point.

17. $f(x) = 5 - 3x$, $(1, -2)$
18. $f(x) = 1 - 4x$, $(2, -7)$
19. $f(x) = -\tfrac{1}{2}x^2 + 2x$, $(2, 2)$
20. $f(x) = 4 - x^2$, $(-1, 3)$
21. $f(x) = \sqrt{x} + 2$, $(9, 5)$
22. $f(x) = 2\sqrt{x} + 1$, $(4, 5)$
23. $f(x) = \dfrac{5}{x}$, $(1, 5)$
24. $f(x) = \dfrac{2}{x} - 1$, $\left(\dfrac{1}{2}, 3\right)$

In Exercises 25–28, determine the x-value at which the function is not differentiable.

25. $y = \dfrac{x + 1}{x - 1}$
26. $y = -|x| + 3$

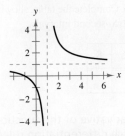
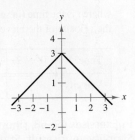

27. $y = \begin{cases} -x - 2, & x \le 0 \\ x^3 + 2, & x > 0 \end{cases}$
28. $y = (x + 1)^{2/3}$

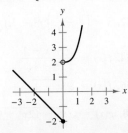
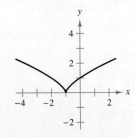

Ⓣ In Exercises 29–38, find the equation of the tangent line at the given point. Then use a graphing utility to graph the function and the tangent line in the same viewing window.

29. $g(t) = \dfrac{2}{3t^2}$, $\left(1, \dfrac{2}{3}\right)$
30. $h(x) = \dfrac{2}{(3x)^2}$, $\left(2, \dfrac{1}{18}\right)$

31. $f(x) = x^2 + 3$, $(1, 4)$
32. $f(x) = 2x^2 - 3x + 1$, $(2, 3)$
33. $y = 11x^4 - 5x^2 + 1$, $(-1, 7)$
34. $y = x^3 - 5 + \dfrac{3}{x^3}$, $(-1, -9)$
35. $f(x) = \sqrt{x} - \dfrac{1}{\sqrt{x}}$, $(1, 0)$
36. $f(x) = 2x^{-3} + 4 - \sqrt{x}$, $(1, 5)$
37. $f(x) = \dfrac{x^2 + 3}{x}$, $(1, 4)$
38. $f(x) = -x^2 - 4x - 4$, $(-4, -4)$

In Exercises 39 and 40, find the average rate of change of the function over the indicated interval. Then compare the average rate of change with the instantaneous rates of change at the endpoints of the interval.

39. $f(x) = x^2 + 3x - 4$, $[0, 1]$
40. $f(x) = x^3 + x$, $[-2, 2]$

41. **Space Launches** The number of world-wide successful space launches S for the years 2000 through 2005 can be modeled by

$$S = 0.9792t^4 - 10.819t^3 + 38.98t^2 - 52.8t + 82$$

where t is the time in years, with $t = 0$ corresponding to 2000. A graph of this model appears in Exercise 5. *(Source: Library of Congress)*

(a) Find the average rate of change from 2000 through 2005.

(b) Find the instantaneous rates of change of the model for 2000 and 2005.

(c) Interpret the results of parts (a) and (b) in the context of the problem.

42. **Health Insurance** The number of children enrolled E (in thousands per year) in the State Children's Health Insurance Program (SCHIP) for 1998 through 2005 can be modeled by

$$E = \dfrac{21{,}892 - 2812.9t}{1 - 0.2759t}$$

where t is the time in years, with $t = 8$ corresponding to 1998. A graph of this model appears in Exercise 6. *(Source: Congressional Budget Office)*

(a) Find the average rate of change from 2000 through 2005.

(b) Find the instantaneous rates of change of the model for 2000 and 2005.

(c) Interpret the results of parts (a) and (b) in the context of the problem.

43. Magazine Subscribers The number N (in millions) of subscribers to a magazine from 2001 through 2008 can be modeled by the equation

$$N = 2\sqrt{t} + t + 2$$

where t is the year, with $t = 1$ corresponding to 2001.

(a) Find the rate of change of the number of subscribers with respect to the year.

(T) (b) Use a graphing utility to graph the function for $1 \leq t \leq 8$. During which year does the rate of change appear to be greatest?

(c) Create a table that shows the rates of change for $1 \leq t \leq 8$. For what year is the rate of change the greatest?

(d) Compare your answers for parts (b) and (c).

44. Retail Price The average retail price P (in dollars) of a half-gallon of prepackaged ice cream from 1992 through 2006 can be modeled by the equation

$$P = -0.00149t^3 + 0.0340t^2 - 0.086t + 2.53$$

where t is the year, with $t = 2$ corresponding to 1992. (Source: U.S. Bureau of Labor Statistics)

(a) Find the rate of change of the price with respect to the year.

(b) At what rate was the price of a half-gallon of prepackaged ice cream changing in 1997? in 2003? in 2005?

(T) (c) Use a graphing utility to graph the function for $2 \leq t \leq 16$. During which years was the price increasing? decreasing?

(d) For which years do the slopes of the tangent lines appear to be positive? negative?

(e) Compare your answers for parts (c) and (d).

45. Population Growth The mosquito population P in a wetland area during a 10-day period can be modeled by the equation

$$P = (0.035t^2 + 2.04t + 70.7)^2, \quad 0 \leq t \leq 10$$

where t is measured in days.

(T) (a) Use a graphing utility to graph the equation. Be sure to choose an appropriate window.

(b) Determine dP/dt. Evaluate dP/dt for $t = 0, 5,$ and 10.

(c) Is dP/dt positive for $t \geq 0$? Does this agree with the graph of the function? What does this tell you about this situation? Explain your reasoning.

46. Recycling The amount T of recycled paper products (in millions of tons) from 1997 through 2005 can be modeled by the equation

$$T = \sqrt{1.3150t^3 - 42.747t^2 + 522.28t - 885.2}$$

where t is the year, with $t = 7$ corresponding to 1997. (Source: Franklin Associates, Ltd.)

(T) (a) Use a graphing utility to graph the equation. Be sure to choose an appropriate window.

(b) Determine dT/dt. Evaluate dT/dt for 1997, 2002, and 2005.

(c) Is dT/dt positive for $t \geq 7$? Does this agree with the graph of the function? What does this tell you about this situation? Explain your reasoning.

47. Velocity A rock is dropped from a tower on the Brooklyn Bridge, 276 feet above the East River. Let t represent the time in seconds.

(a) Write a model for the position function (assume that air resistance is negligible).

(b) Find the average velocity during the first 2 seconds.

(c) Find the instantaneous velocities when $t = 2$ and $t = 3$.

(d) How long will it take for the rock to hit the water?

(e) When it hits the water, what is the rock's speed?

48. Velocity The straight-line distance s (in feet) traveled by an accelerating bicyclist can be modeled by

$$s = 2t^{3/2}, \quad 0 \leq t \leq 8$$

where t is the time (in seconds). Complete the table, showing the velocity of the bicyclist at two-second intervals.

Time, t	0	2	4	6	8
Velocity					

In Exercises 49–68, find the derivative of the function. Simplify your result. State which differentiation rule(s) you used to find the derivative.

49. $f(x) = x^3(5 - 3x^2)$

50. $y = (3x^2 + 7)(x^2 - 2x)$

51. $y = (4x - 3)(x^3 - 2x^2)$

52. $s = \left(4 - \dfrac{1}{t^2}\right)(t^2 - 3t)$

53. $f(x) = \dfrac{6x - 5}{x^2 + 1}$

54. $f(x) = \dfrac{x^2 + x - 1}{x^2 - 1}$

55. $f(x) = (5x^2 + 2)^3$

56. $f(x) = \sqrt[3]{x^2 - 1}$

57. $h(x) = \dfrac{2}{\sqrt{x + 1}}$

58. $g(x) = \sqrt{x^6 - 12x^3 + 9}$

59. $g(x) = x\sqrt{x^2 + 1}$

60. $g(t) = \dfrac{t}{(1 - t)^3}$

61. $f(x) = x(1 - 4x^2)^2$

62. $f(x) = \left(x^2 + \dfrac{1}{x}\right)^5$

63. $h(x) = [x^2(2x + 3)]^3$

64. $f(x) = [(x - 2)(x + 4)]^2$ 65. $f(x) = x^2(x - 1)^5$

66. $f(s) = s^3(s^2 - 1)^{5/2}$

67. $h(t) = \dfrac{\sqrt{3t + 1}}{(1 - 3t)^2}$

68. $g(x) = \dfrac{(3x + 1)^2}{(x^2 + 1)^2}$

69. **Population Growth** A population of 400 bacteria is introduced into a culture and grows in number according to the equation

$$P(t) = 400\left(1 + \dfrac{3t}{40 + t^2}\right)$$

where t is the time (in hours).

(a) Find the rates of change of the population when $t = 2$, $t = 4$, $t = 6$, and $t = 20$.

(T) (b) Graph the model on a graphing utility and describe the rate at which the population is changing.

70. **Physical Science** The temperature T (in degrees Fahrenheit) of food placed in a freezer can be modeled by

$$T = \dfrac{1300}{t^2 + 2t + 25}$$

where t is the time (in hours).

(a) Find the rates of change of T when $t = 1$, $t = 3$, $t = 5$, and $t = 10$.

(T) (b) Graph the model on a graphing utility and describe the rate at which the temperature is changing.

71. **Forestry** According to the *Doyle Log Rule*, the volume V (in board-feet) of a log of length L (feet) and diameter D (inches) at the small end is

$$V = \left(\dfrac{D - 4}{4}\right)^2 L.$$

Find the rates at which the volume is changing with respect to D for a 12-foot-long log whose smallest diameter is (a) 8 inches, (b) 16 inches, (c) 24 inches, and (d) 36 inches.

(T) 72. **Temperature** The normal daily maximum temperature T (in degrees Fahrenheit) for a midwest city can be modeled by

$$T = \sqrt{5.9918t^4 - 171.015t^3 + 1469.25t^2 - 3560.6t + 4292}$$

where t is the time in months, with $t = 1$ corresponding to January.

(a) Find the derivative of this model. Which differentiation rule(s) did you use?

(b) Use a graphing utility to graph the derivative on the interval $1 \leq t \leq 12$.

(c) Use the *trace* feature to find the month(s) during which the temperature was changing the most.

(d) Use the *trace* feature to find the month(s) during which the temperature was changing the least.

In Exercises 73–80, find the higher-order derivative.

73. Given $f(x) = 3x^2 + 7x + 1$, find $f''(x)$.

74. Given $f'(x) = 5x^4 - 6x^2 + 2x$, find $f'''(x)$.

75. Given $f'''(x) = -\dfrac{6}{x^4}$, find $f^{(5)}(x)$.

76. Given $f(x) = \sqrt{x}$, find $f^{(4)}(x)$.

77. Given $f'(x) = 7x^{5/2}$, find $f''(x)$.

78. Given $f(x) = x^2 + \dfrac{3}{x}$, find $f''(x)$.

79. Given $f''(x) = 6\sqrt[3]{x}$, find $f'''(x)$.

80. Given $f'''(x) = 20x^4 - \dfrac{2}{x^3}$, find $f^{(5)}(x)$.

81. **Athletics** A person dives from a 30-foot platform with an initial velocity of 5 feet per second (upward).

(a) Find the position function for the diver.

(b) How long will it take for the diver to hit the water?

(c) What is the diver's velocity at impact?

(d) What is the diver's acceleration at impact?

82. **Velocity and Acceleration** The position function for a particle is given by

$$s = \dfrac{1}{t^2 + 2t + 1}$$

where s is the height (in feet) and t is the time (in seconds). Find the velocity and acceleration functions.

(T) 83. **Modeling Data** The table shows the numbers n (in thousands) of students enrolled in public schools in Puerto Rico for the years 2000 through 2005, where t is the year, with $t = 0$ corresponding to 2000. *(Source: Puerto Rico Planning Board)*

t	0	1	2	3	4	5
n	612.3	610.8	603.5	596.3	584.9	575.9

(a) Use a graphing utility to find a cubic model for the data.

(b) Use a graphing utility to graph the model and plot the data in the same viewing window. How well does the model fit the data?

(c) Find the first and second derivatives of the function.

(d) Show that the number of students enrolled in public schools was decreasing from 2003 through 2005.

(e) Find the year in which enrollment was decreasing at the greatest rate by solving $n''(t) = 0$.

(f) Explain the relationship among your answers for parts (c), (d), and (e).

Chapter Test

See www.CalcChat.com for worked-out solutions to odd-numbered exercises.

Take this test as you would take a test in class. When you are done, check your work against the answers given in the back of the book.

In Exercises 1 and 2, use the limit definition to find the derivative of the function. Then find the slope of the tangent line to the graph of f at the given point.

1. $f(x) = x^2 + 1$, $(2, 5)$
2. $f(x) = \sqrt{x} - 2$, $(4, 0)$

In Exercises 3–11, find the derivative of the function. Simplify your result.

3. $f(t) = t^3 + 2t$
4. $f(x) = 4x^2 - 8x + 1$
5. $f(x) = x^{3/2}$
6. $f(x) = (x + 3)(x - 3)$
7. $f(x) = -3x^{-3}$
8. $f(x) = \sqrt{x}(5 + x)$
9. $f(x) = (3x^2 + 4)^2$
10. $f(x) = \sqrt{1 - 2x}$
11. $f(x) = \dfrac{(5x - 1)^3}{x}$

(T) 12. Find an equation of the tangent line to the graph of $f(x) = x - \dfrac{1}{x}$ at the point $(1, 0)$. Then use a graphing utility to graph the function and the tangent line in the same viewing window.

13. The number N of bone graft procedures (in thousands) in the United States for the years 2000 through 2005 can be modeled by $N = 12.95t^2 + 110.3t + 780$, where t represents the year, with $t = 0$ corresponding to 2000. *(Source: U.S. Department of Health and Human Services)*

    (a) Find the average rate of change from 2001 through 2004.

    (b) Find the instantaneous rates of change of the model for 2001 and 2004.

    (c) Interpret the results of parts (a) and (b) in the context of the problem.

In Exercises 14–16, find the third derivative of the function. Simplify your result.

14. $f(x) = 2x^2 + 3x + 1$
15. $f(x) = \sqrt{3 - x}$
16. $f(x) = \dfrac{2x + 1}{2x - 1}$

(T) 17. The table shows the numbers N of bald eagle breeding pairs in the lower 48 states for selected years from 1996 through 2006. *(Source: U.S. Fish & Wildlife Service)*

t	6	7	8	9	10	15	16
N	5094	5295	5748	6404	6471	7066	9789

A model for the data is $N = 0.00263t^6 - 1.035t^4 + 124.5t^2 + 1590$, where t is the year, with $t = 6$ corresponding to 1996.

(a) Use a graphing utility to graph the model and the data in the same viewing window. How well does the model fit the data?

(b) Find the first and second derivatives of the function.

(c) Use the model to show that the number of breeding pairs was decreasing in the years 2000 through 2002.

(d) Find the year in which the number of breeding pairs was decreasing at the greatest rate by solving $N''(t) = 0$.

Applications of the Derivative

3

Ultraviolet (UV) radiation is emitted by the sun. Small amounts of UV are essential for the production of vitamin D in humans. However, overexposure may result in acute and chronic health effects on the skin, eyes, and immune system. (See Section 3.5, Exercise 64.)

Applications

Derivatives have many real-life applications. The applications listed below represent a sample of the applications in this chapter.

- Threatened and Endangered Species, Exercise 41, page 192
- Biology: Fertility Rates, Exercise 51, page 202
- Learning Curve, Exercise 63, page 232
- Meteorology, Exercise 57, page 241
- Physiology: Body Surface Area, Exercise 30, page 248

3.1 Increasing and Decreasing Functions

3.2 Extrema and the First-Derivative Test

3.3 Concavity and the Second-Derivative Test

3.4 Optimization Problems

3.5 Asymptotes

3.6 Curve Sketching: A Summary

3.7 Differentials: Linear Approximation

Section 3.1

Increasing and Decreasing Functions

- Test for increasing and decreasing functions.
- Find the critical numbers of functions and find the open intervals on which functions are increasing or decreasing.
- Use increasing and decreasing functions to model and solve real-life problems.

Increasing and Decreasing Functions

A function is **increasing** if its graph moves up as x moves to the right and **decreasing** if its graph moves down as x moves to the right. The following definition states this more formally.

Definition of Increasing and Decreasing Functions

A function f is **increasing** on an interval if for any two numbers x_1 and x_2 in the interval

$$x_2 > x_1 \quad \text{implies} \quad f(x_2) > f(x_1).$$

A function f is **decreasing** on an interval if for any two numbers x_1 and x_2 in the interval

$$x_2 > x_1 \quad \text{implies} \quad f(x_2) < f(x_1).$$

The function in Figure 3.1 is decreasing on the interval $(-\infty, a)$, constant on the interval (a, b), and increasing on the interval (b, ∞). Actually, from the definition of increasing and decreasing functions, the function shown in Figure 3.1 is decreasing on the interval $(-\infty, a]$ and increasing on the interval $[b, \infty)$. This text restricts the discussion to finding *open* intervals on which a function is increasing or decreasing.

The derivative of a function can be used to determine whether the function is increasing or decreasing on an interval.

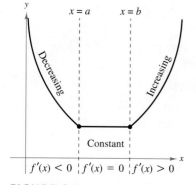

FIGURE 3.1

Test for Increasing and Decreasing Functions

Let f be differentiable on the interval (a, b).

1. If $f'(x) > 0$ for all x in (a, b), then f is increasing on (a, b).
2. If $f'(x) < 0$ for all x in (a, b), then f is decreasing on (a, b).
3. If $f'(x) = 0$ for all x in (a, b), then f is constant on (a, b).

STUDY TIP

The conclusions in the first two cases of testing for increasing and decreasing functions are valid even if $f'(x) = 0$ at a finite number of x-values in (a, b).

SECTION 3.1 Increasing and Decreasing Functions

Example 1 Testing for Increasing and Decreasing Functions

Show that the function

$$f(x) = x^2$$

is decreasing on the open interval $(-\infty, 0)$ and increasing on the open interval $(0, \infty)$.

SOLUTION The derivative of f is

$$f'(x) = 2x.$$

On the open interval $(-\infty, 0)$, the fact that x is negative implies that $f'(x) = 2x$ is also negative. So, by the test for a decreasing function, you can conclude that f is *decreasing* on this interval. Similarly, on the open interval $(0, \infty)$, the fact that x is positive implies that $f'(x) = 2x$ is also positive. So, it follows that f is *increasing* on this interval, as shown in Figure 3.2.

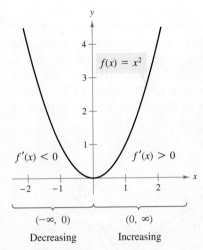

FIGURE 3.2

✓ CHECKPOINT 1

Show that the function $f(x) = x^4$ is decreasing on the open interval $(-\infty, 0)$ and increasing on the open interval $(0, \infty)$. ∎

Example 2 Modeling a Food Supply

From 1998 through 2005, the food supply S of corn products in the United States (in pounds per person per year) can be modeled by

$$S = -0.004t^2 + 0.71t + 21.8, \quad 8 \leq t \leq 15$$

where t is the year, with $t = 8$ corresponding to 1998 (see Figure 3.3). Show that the supply of corn products was increasing from 1998 through 2005. *(Source: U.S. Department of Agriculture)*

SOLUTION The derivative of this model is $dS/dt = -0.008t + 0.71$. On the open interval $(8, 15)$, the derivative is positive. So, the function is increasing, which implies that the supply of corn products was increasing during the given time period.

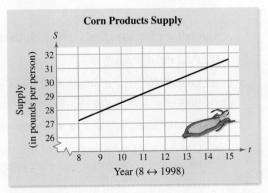

FIGURE 3.3

DISCOVERY

Use a graphing utility to graph $f(x) = 2 - x^2$ and $f'(x) = -2x$ in the same viewing window. On what interval is f increasing? On what interval is f' positive? Describe how the first derivative can be used to determine where a function is increasing and decreasing. Repeat this analysis for $g(x) = x^3 - x$ and $g'(x) = 3x^2 - 1$.

✓ CHECKPOINT 2

From 1997 through 2005, the food supply S of rice in the United States (in pounds per person per year) can be modeled by

$$S = -0.011t^2 + 0.68t + 13.4,$$

$$7 \leq t \leq 15$$

where t is the year, with $t = 7$ corresponding to 1997. Show that the supply of rice was increasing from 1997 through 2005. *(Source: U.S. Department of Agriculture)* ∎

Critical Numbers and Their Use

In Example 1, you were given two intervals: one on which the function was decreasing and one on which it was increasing. Suppose you had been asked to determine these intervals. To do this, you could have used the fact that for a continuous function, $f'(x)$ can change signs only at x-values where $f'(x) = 0$ or at x-values where $f'(x)$ is undefined, as shown in Figure 3.4. These two types of numbers are called the **critical numbers** of f.

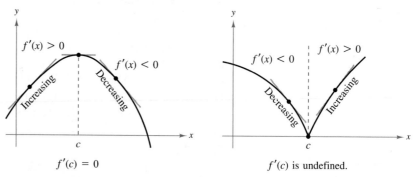

FIGURE 3.4

Definition of Critical Number

If f is defined at c, then c is a critical number of f if $f'(c) = 0$ or if $f'(c)$ is undefined.

STUDY TIP

This definition requires that a critical number be in the domain of the function. For example, $x = 0$ is not a critical number of the function $f(x) = 1/x$.

To determine the intervals on which a continuous function is increasing or decreasing, you can use the guidelines below.

Guidelines for Applying Increasing/Decreasing Test

1. Find the derivative of f.
2. Locate the critical numbers of f and use these numbers to determine test intervals. That is, find all x for which $f'(x) = 0$ or $f'(x)$ is undefined.
3. Test the sign of $f'(x)$ at an arbitrary number in each of the test intervals.
4. Use the test for increasing and decreasing functions to decide whether f is increasing or decreasing on each interval.

SECTION 3.1 Increasing and Decreasing Functions

Example 3 Finding Increasing and Decreasing Intervals

Find the open intervals on which the function is increasing or decreasing.

$$f(x) = x^3 - \frac{3}{2}x^2$$

SOLUTION Begin by finding the derivative of f. Then set the derivative equal to zero and solve for the critical numbers.

$f'(x) = 3x^2 - 3x$	Differentiate original function.
$3x^2 - 3x = 0$	Set derivative equal to 0.
$3(x)(x - 1) = 0$	Factor.
$x = 0, x = 1$	Critical numbers

Because there are no x-values for which f' is undefined, it follows that $x = 0$ and $x = 1$ are the *only* critical numbers. So, the intervals that need to be tested are $(-\infty, 0)$, $(0, 1)$, and $(1, \infty)$. The table summarizes the testing of these three intervals.

Interval	$-\infty < x < 0$	$0 < x < 1$	$1 < x < \infty$
Test value	$x = -1$	$x = \frac{1}{2}$	$x = 2$
Sign of $f'(x)$	$f'(-1) = 6 > 0$	$f'(\frac{1}{2}) = -\frac{3}{4} < 0$	$f'(2) = 6 > 0$
Conclusion	Increasing	Decreasing	Increasing

The graph of f is shown in Figure 3.5. Note that the test values in the intervals were chosen for convenience—other x-values could have been used.

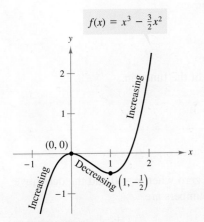

FIGURE 3.5

✓ **CHECKPOINT 3**

Find the open intervals on which the function $f(x) = x^3 - 12x$ is increasing or decreasing. ∎

TECHNOLOGY

You can use the *trace* feature of a graphing utility to confirm the result of Example 3. Begin by graphing the function, as shown at the right. Then activate the *trace* feature and move the cursor from left to right. In intervals on which the function is increasing, note that the y-values increase as the x-values increase, whereas in intervals on which the function is decreasing, the y-values decrease as the x-values increase.*

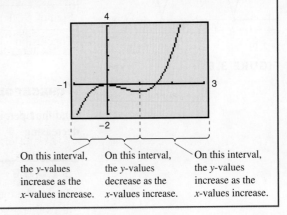

On this interval, the y-values increase as the x-values increase.

On this interval, the y-values decrease as the x-values increase.

On this interval, the y-values increase as the x-values increase.

---

*Specific calculator keystroke instructions for operations in this and other technology boxes can be found at *college.hmco.com/info/larsonapplied*.

188   CHAPTER 3   Applications of the Derivative

Not only is the function in Example 3 continuous on the entire real line, it is also differentiable there. For such functions, the only critical numbers are those for which $f'(x) = 0$. The next example considers a continuous function that has *both* types of critical numbers—those for which $f'(x) = 0$ and those for which $f'(x)$ is undefined.

> **Algebra Review**
>
> For help with the algebra in Example 4, see Example 2(d) in the *Chapter 3 Algebra Review*, on page 250.

Example 4   Finding Increasing and Decreasing Intervals

Find the open intervals on which the function

$$f(x) = (x^2 - 4)^{2/3}$$

is increasing or decreasing.

SOLUTION   Begin by finding the derivative of the function.

$$f'(x) = \frac{2}{3}(x^2 - 4)^{-1/3}(2x) \qquad \text{Differentiate.}$$

$$= \frac{4x}{3(x^2 - 4)^{1/3}} \qquad \text{Simplify.}$$

From this, you can see that the derivative is zero when $x = 0$ and the derivative is undefined when $x = \pm 2$. So, the critical numbers are

$$x = -2, \quad x = 0, \quad \text{and} \quad x = 2. \qquad \text{Critical numbers}$$

This implies that the test intervals are

$$(-\infty, -2), \quad (-2, 0), \quad (0, 2), \quad \text{and} \quad (2, \infty). \qquad \text{Test intervals}$$

The table summarizes the testing of these four intervals, and the graph of the function is shown in Figure 3.6.

Interval	$-\infty < x < -2$	$-2 < x < 0$	$0 < x < 2$	$2 < x < \infty$
Test value	$x = -3$	$x = -1$	$x = 1$	$x = 3$
Sign of $f'(x)$	$f'(-3) < 0$	$f'(-1) > 0$	$f'(1) < 0$	$f'(3) > 0$
Conclusion	Decreasing	Increasing	Decreasing	Increasing

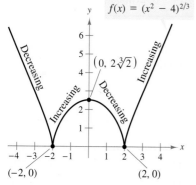

FIGURE 3.6

✓ CHECKPOINT 4

Find the open intervals on which the function $f(x) = x^{2/3}$ is increasing or decreasing. ■

> **STUDY TIP**
>
> To test the intervals in the table, it is not necessary to *evaluate* $f'(x)$ at each test value—you only need to determine its sign. For example, you can determine the sign of $f'(-3)$ as shown.
>
> $$f'(-3) = \frac{4(-3)}{3(9-4)^{1/3}} = \frac{\text{negative}}{\text{positive}} = \text{negative}$$

SECTION 3.1 Increasing and Decreasing Functions

The functions in Examples 1 through 4 are continuous on the entire real line. If there are isolated x-values at which a function is not continuous, then these x-values should be used along with the critical numbers to determine the test intervals. For example, the function

$$f(x) = \frac{x^4 + 1}{x^2}$$

is not continuous at $x = 0$. Because the derivative of f

$$f'(x) = \frac{2(x^4 - 1)}{x^3}$$

is zero when $x = \pm 1$, you should use the following numbers to determine the test intervals.

$x = -1, x = 1$ Critical numbers

$x = 0$ Discontinuity

After testing $f'(x)$, you can determine that the function is decreasing on the intervals $(-\infty, -1)$ and $(0, 1)$, and increasing on the intervals $(-1, 0)$ and $(1, \infty)$, as shown in Figure 3.7.

The converse of the test for increasing and decreasing functions is *not* true. For instance, it is possible for a function to be increasing on an interval even though its derivative is not positive at every point in the interval.

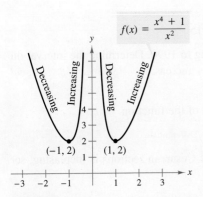

FIGURE 3.7

Example 5 Testing an Increasing Function

Show that

$$f(x) = x^3 - 3x^2 + 3x$$

is increasing on the entire real line.

SOLUTION From the derivative of f

$$f'(x) = 3x^2 - 6x + 3 = 3(x - 1)^2$$

you can see that the only critical number is $x = 1$. So, the test intervals are $(-\infty, 1)$ and $(1, \infty)$. The table summarizes the testing of these two intervals. From Figure 3.8, you can see that f is increasing on the entire real line, even though $f'(1) = 0$. To convince yourself of this, look back at the definition of an increasing function.

Interval	$-\infty < x < 1$	$1 < x < \infty$
Test value	$x = 0$	$x = 2$
Sign of $f'(x)$	$f'(0) = 3(-1)^2 > 0$	$f'(2) = 3(1)^2 > 0$
Conclusion	Increasing	Increasing

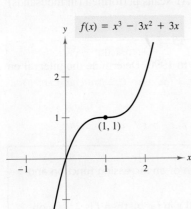

FIGURE 3.8

✓CHECKPOINT 5

Show that $f(x) = -x^3 + 2$ is decreasing on the entire real line. ■

Application

Example 6 Cesarean Section Procedures

From 1990 through 2005, the number N of Cesarean section procedures (in thousands) can be modeled by

$$N = 6.51t^2 - 71.9t + 954, \quad 0 \le t \le 15$$

where t is the year, with $t = 0$ corresponding to 1990. Determine the interval on which the number of Cesarean sections is increasing. *(Source: U.S. National Center for Health Statistics)*

SOLUTION Begin by finding the derivative of the function.

$N' = 13.02t - 71.9$ *Differentiate.*

To find the interval on which the number of Cesarean sections is increasing, set N' equal to zero and solve for t.

$$13.02t - 71.9 = 0 \quad \text{Set } N' \text{ equal to 0.}$$
$$13.02t = 71.9 \quad \text{Add 71.9 to each side.}$$
$$t = \frac{71.9}{13.02} \quad \text{Divide each side by 13.02.}$$
$$t \approx 5.5 \quad \text{Simplify.}$$

On the interval $(0, 5.5)$, N' is negative and the number of Cesarean sections is *decreasing*. On the interval $(5.5, 15)$, N' is positive and the number of Cesarean sections is *increasing*. So, the number of Cesarean sections has been increasing since about the middle of 1995. The graph of the function is shown in Figure 3.9.

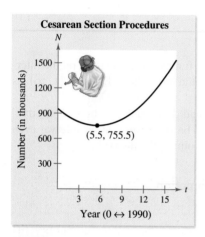

FIGURE 3.9

✓ CHECKPOINT 6

From 1990 through 2005, the number N of CAT scans performed (in thousands) can be modeled by

$$N = 6.64t^2 - 141.4t + 1507, \quad 0 \le t \le 15$$

where t is the year, with $t = 0$ corresponding to 1990. Determine the interval on which the number of CAT scans is increasing. *(Source: U.S. National Center for Health Statistics)* ∎

CONCEPT CHECK

1. Write a verbal description of (a) the graph of an increasing function and (b) the graph of a decreasing function.
2. Complete the following: If $f'(x) > 0$ for all x in (a, b), then f is _____ on (a, b). [Assume f is differentiable on (a, b).]
3. If f is defined at c, under what condition(s) is c a critical number of f?
4. In your own words, state the guidelines for determining the intervals on which a continuous function is increasing or decreasing.

SECTION 3.1   Increasing and Decreasing Functions

Skills Review 3.1

The following warm-up exercises involve skills that were covered in earlier sections. You will use these skills in the exercise set for this section. For additional help, review Sections 0.3 and 1.4.

In Exercises 1–4, solve the equation.

1. $x^2 = 8x$

2. $15x = \dfrac{5}{8}x^2$

3. $\dfrac{x^2 - 25}{x^3} = 0$

4. $\dfrac{2x}{\sqrt{1 - x^2}} = 0$

In Exercises 5–8, find the domain of the expression.

5. $\dfrac{x + 3}{x - 3}$

6. $\dfrac{2}{\sqrt{1 - x}}$

7. $\dfrac{2x + 1}{x^2 - 3x - 10}$

8. $\dfrac{3x}{\sqrt{9 - 3x^2}}$

In Exercises 9–12, evaluate the expression for $x = -2$, 0, and 2.

9. $-2(x + 1)(x - 1)$

10. $4(2x + 1)(2x - 1)$

11. $\dfrac{2x + 1}{(x - 1)^2}$

12. $\dfrac{-2(x + 1)}{(x - 4)^2}$

Exercises 3.1

See www.CalcChat.com for worked-out solutions to odd-numbered exercises.

In Exercises 1–4, evaluate the derivative of the function at the indicated points on the graph.

1. $f(x) = \dfrac{x^2}{x^2 + 4}$

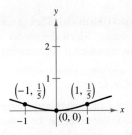

2. $f(x) = x + \dfrac{32}{x^2}$

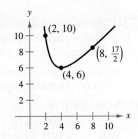

3. $f(x) = (x + 2)^{2/3}$

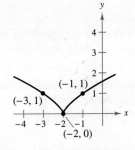

4. $f(x) = -3x\sqrt{x + 1}$

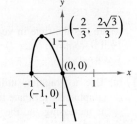

In Exercises 5–8, use the derivative to identify the open intervals on which the function is increasing or decreasing. Verify your result with the graph of the function.

5. $f(x) = -(x + 1)^2$

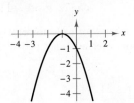

6. $f(x) = \dfrac{x^3}{4} - 3x$

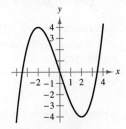

7. $f(x) = x^4 - 2x^2$

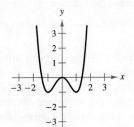

8. $f(x) = \dfrac{x^2}{x + 1}$

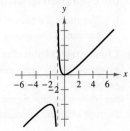

In Exercises 9–32, find the critical numbers and the open intervals on which the function is increasing or decreasing. Then use a graphing utility to graph the function.

9. $f(x) = 2x - 3$
10. $f(x) = 5 - 3x$
11. $g(x) = -(x - 1)^2$
12. $g(x) = (x + 2)^2$
13. $y = x^2 - 6x$
14. $y = -x^2 + 2x$
15. $y = x^3 - 6x^2$
16. $y = (x - 2)^3$
17. $f(x) = \sqrt{x^2 - 1}$
18. $f(x) = \sqrt{9 - x^2}$
19. $y = x^{1/3} + 1$
20. $y = x^{2/3} - 4$
21. $g(x) = (x - 1)^{1/3}$
22. $g(x) = (x - 1)^{2/3}$
23. $f(x) = -2x^2 + 4x + 3$
24. $f(x) = x^2 + 8x + 10$
25. $y = 3x^3 + 12x^2 + 15x$
26. $y = x^3 - 3x + 2$
27. $f(x) = x\sqrt{x + 1}$
28. $h(x) = x\sqrt[3]{x - 1}$
29. $f(x) = x^4 - 2x^3$
30. $f(x) = \frac{1}{4}x^4 - 2x^2$
31. $f(x) = \dfrac{x}{x^2 + 4}$
32. $f(x) = \dfrac{x^2}{x^2 + 4}$

In Exercises 33–38, find the critical numbers and the open intervals on which the function is increasing or decreasing. (*Hint:* Check for discontinuities.) Sketch the graph of the function.

33. $f(x) = \dfrac{2x}{16 - x^2}$
34. $f(x) = \dfrac{x}{x + 1}$
35. $y = \begin{cases} 4 - x^2, & x \le 0 \\ -2x, & x > 0 \end{cases}$
36. $y = \begin{cases} 2x + 1, & x \le -1 \\ x^2 - 2, & x > -1 \end{cases}$
37. $y = \begin{cases} 3x + 1, & x \le 1 \\ 5 - x^2, & x > 1 \end{cases}$
38. $y = \begin{cases} -x^3 + 1, & x \le 0 \\ -x^2 + 2x, & x > 0 \end{cases}$

39. **Emergency Room** The number N of patients per day in a hospital's emergency room over a 31-day period is modeled by
$$N = 100\left(\frac{1}{t} + \frac{t}{t + 3}\right), \quad 1 \le t \le 31$$
where t is measured in days.
(a) Find the open intervals on which N is increasing or decreasing.
(b) Use a graphing utility to graph the function.
(c) Use the *trace* feature to determine when the number of patients is the lowest.

40. **Chemistry: Molecular Velocity** Plots of the relative numbers of N_2 (nitrogen) molecules that have a given velocity at each of three temperatures (in degrees Kelvin) are shown in the figure. Identify the differences in the average velocities (indicated by the peaks of the curves) for the three temperatures, and describe the intervals on which velocity is increasing and decreasing for each of the three temperatures. (*Source: Adapted from Zumdahl, Chemistry, Seventh Edition*)

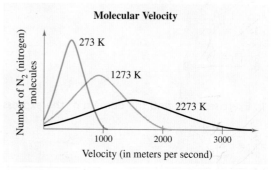

Figure for 40

41. **Threatened and Endangered Species** The number N of threatened and endangered mammals in the world from 2001 through 2006 can be modeled by $N = 0.75t^2 - 2.1t + 342$, $1 \le t \le 6$, where t is the time in years, with $t = 1$ corresponding to 2001. Determine the interval on which the number of threatened and endangered mammals is increasing. (*Source: U.S. Fish and Wildlife Service*)

42. **Medical Degrees** The number y of medical degrees conferred in the United States from 1970 through 2004 can be modeled by
$$y = 0.813t^3 - 55.70t^2 + 1185.2t + 7752, \quad 0 \le t \le 34$$
where t is the time in years, with $t = 0$ corresponding to 1970. (*Source: U.S. National Center for Education Statistics*)
(a) Use a graphing utility to graph the model. Then graphically estimate the years during which the model is increasing and the years during which it is decreasing.
(b) Use the test for increasing and decreasing functions to verify the result of part (a).

43. **Multiple Births** The number y of triplets and higher-order multiple births (per 100,000 live births) in the United States from 1971 through 2004 can be modeled by
$$y = -0.00165t^4 + 0.1064t^3 - 1.941t^2 + 13.02t + 9.0,$$
$$1 \le t \le 34$$
where t is the time in years, with $t = 1$ corresponding to 1971. (*Source: U.S. Department of Health and Human Services*)
(a) Use a graphing utility to graph the model. Then graphically estimate the years during which the model is increasing and the years during which the model is decreasing.
(b) Use the test for increasing and decreasing functions to verify the result of part (a).

Section 3.2
Extrema and the First-Derivative Test

- Recognize the occurrence of relative extrema of functions.
- Use the First-Derivative Test to find the relative extrema of functions.
- Find absolute extrema of continuous functions on closed intervals.
- Find minimum and maximum values of real-life models and interpret the results in context.

Relative Extrema

You have used the derivative to determine the intervals on which a function is increasing or decreasing. In this section, you will examine the points at which a function changes from increasing to decreasing, or vice versa. At such a point, the function has a **relative extremum**. (The plural of extremum is *extrema*.) The **relative extrema** of a function include the **relative minima** and **relative maxima** of the function. For instance, the function shown in Figure 3.10 has a relative maximum at the left point and a relative minimum at the right point.

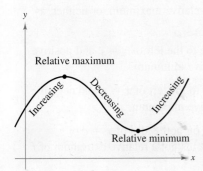

FIGURE 3.10

Definition of Relative Extrema

Let f be a function defined at c.

1. $f(c)$ is a **relative maximum** of f if there exists an interval (a, b) containing c such that $f(x) \leq f(c)$ for all x in (a, b).
2. $f(c)$ is a **relative minimum** of f if there exists an interval (a, b) containing c such that $f(x) \geq f(c)$ for all x in (a, b).

If $f(c)$ is a relative extremum of f, then the relative extremum is said to occur at $x = c$.

For a continuous function, the relative extrema must occur at critical numbers of the function, as shown in Figure 3.11.

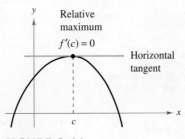

FIGURE 3.11

Occurrences of Relative Extrema

If f has a relative minimum or relative maximum when $x = c$, then c is a critical number of f. That is, either $f'(c) = 0$ or $f'(c)$ is undefined.

The First-Derivative Test

The discussion on the preceding page implies that in your search for relative extrema of a continuous function, you only need to test the critical numbers of the function. Once you have determined that c is a critical number of a function f, the **First-Derivative Test** for relative extrema enables you to classify $f(c)$ as a relative minimum, a relative maximum, or neither.

DISCOVERY

Use a graphing utility to graph the function $f(x) = x^2$ and its first derivative $f'(x) = 2x$ in the same viewing window. Where does f have a relative minimum? What is the sign of f' to the left of this relative minimum? What is the sign of f' to the right? Describe how the sign of f' can be used to determine the relative extrema of a function.

First-Derivative Test for Relative Extrema

Let f be continuous on the interval (a, b) in which c is the only critical number. If f is differentiable on the interval (except possibly at c), then $f(c)$ can be classified as a relative minimum, a relative maximum, or neither, as shown.

1. On the interval (a, b), if $f'(x)$ is negative to the left of $x = c$ and positive to the right of $x = c$, then $f(c)$ is a relative minimum.

2. On the interval (a, b), if $f'(x)$ is positive to the left of $x = c$ and negative to the right of $x = c$, then $f(c)$ is a relative maximum.

3. On the interval (a, b), if $f'(x)$ is positive on both sides of $x = c$ or negative on both sides of $x = c$, then $f(c)$ is not a relative extremum of f.

A graphical interpretation of the First-Derivative Test is shown in Figure 3.12.

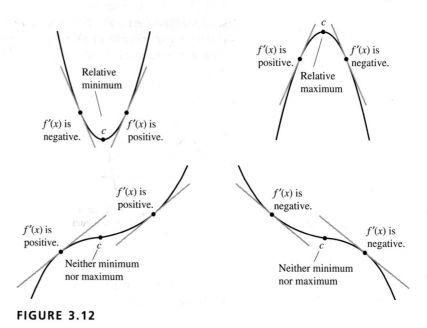

FIGURE 3.12

SECTION 3.2 Extrema and the First-Derivative Test

Example 1 Finding Relative Extrema

Find all relative extrema of the function

$$f(x) = 2x^3 - 3x^2 - 36x + 14.$$

SOLUTION Begin by finding the critical numbers of f.

$f'(x) = 6x^2 - 6x - 36$	Find derivative of f.
$6x^2 - 6x - 36 = 0$	Set derivative equal to 0.
$6(x^2 - x - 6) = 0$	Factor out common factor.
$6(x - 3)(x + 2) = 0$	Factor.
$x = -2, x = 3$	Critical numbers

Because $f'(x)$ is defined for all x, the only critical numbers of f are $x = -2$ and $x = 3$. Using these numbers, you can form the three test intervals $(-\infty, -2)$, $(-2, 3)$, and $(3, \infty)$. The testing of the three intervals is shown in the table.

Interval	$-\infty < x < -2$	$-2 < x < 3$	$3 < x < \infty$
Test value	$x = -3$	$x = 0$	$x = 4$
Sign of $f'(x)$	$f'(-3) = 36 > 0$	$f'(0) = -36 < 0$	$f'(4) = 36 > 0$
Conclusion	Increasing	Decreasing	Increasing

Using the First-Derivative Test, you can conclude that the critical number -2 yields a relative maximum [$f'(x)$ changes sign from positive to negative], and the critical number 3 yields a relative minimum [$f'(x)$ changes sign from negative to positive].

STUDY TIP

In Section 2.2, Example 8, you examined the graph of the function $f(x) = x^3 - 4x + 2$ and discovered that it does *not* have a relative minimum at the point $(1, -1)$. Try using the First-Derivative Test to find the point at which the graph *does* have a relative minimum.

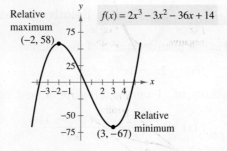

FIGURE 3.13

The graph of f is shown in Figure 3.13. The relative maximum is $f(-2) = 58$ and the relative minimum is $f(3) = -67$.

✓CHECKPOINT 1

Find all relative extrema of $f(x) = 2x^3 - 6x + 1$. ■

Algebra Review

For help with the algebra in Example 2, see Example 2(c) in the *Chapter 3 Algebra Review*, on page 250.

In Example 1, both critical numbers yielded relative extrema. In the next example, only one of the two critical numbers yields a relative extremum.

Example 2 Finding Relative Extrema

Find all relative extrema of the function $f(x) = x^4 - x^3$.

SOLUTION From the derivative of the function

$$f'(x) = 4x^3 - 3x^2 = x^2(4x - 3)$$

you can see that the function has only two critical numbers: $x = 0$ and $x = \frac{3}{4}$. These numbers produce the test intervals $(-\infty, 0)$, $(0, \frac{3}{4})$, and $(\frac{3}{4}, \infty)$, which are tested in the table.

Interval	$-\infty < x < 0$	$0 < x < \frac{3}{4}$	$\frac{3}{4} < x < \infty$
Test value	$x = -1$	$x = \frac{1}{2}$	$x = 1$
Sign of $f'(x)$	$f'(-1) = -7 < 0$	$f'(\frac{1}{2}) = -\frac{1}{4} < 0$	$f'(1) = 1 > 0$
Conclusion	Decreasing	Decreasing	Increasing

By the First-Derivative Test, it follows that f has a relative minimum when $x = \frac{3}{4}$, as shown in Figure 3.14. The relative minimum is $f(\frac{3}{4}) = -\frac{27}{256}$. Note that the critical number $x = 0$ does not yield a relative extremum.

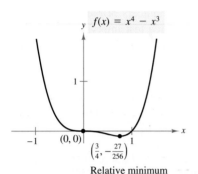

FIGURE 3.14

✓ CHECKPOINT 2

Find all relative extrema of $f(x) = x^4 - 4x^3$. ∎

Example 3 Finding Relative Extrema

Find all relative extrema of the function

$$f(x) = 2x - 3x^{2/3}.$$

SOLUTION From the derivative of the function

$$f'(x) = 2 - \frac{2}{x^{1/3}} = \frac{2(x^{1/3} - 1)}{x^{1/3}}$$

you can see that $f'(1) = 0$ and f' is undefined at $x = 0$. So, the function has two critical numbers: $x = 1$ and $x = 0$. These numbers produce the test intervals $(-\infty, 0)$, $(0, 1)$, and $(1, \infty)$. By testing these intervals, you can conclude that f has a relative maximum at $(0, 0)$ and a relative minimum at $(1, -1)$, as shown in Figure 3.15.

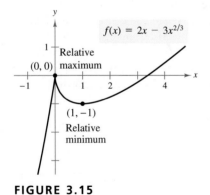

FIGURE 3.15

✓ CHECKPOINT 3

Find all relative extrema of $f(x) = 3x^{2/3} - 2x$. ∎

TECHNOLOGY

There are several ways to use technology to find relative extrema of a function. One way is to use a graphing utility to graph the function, and then use the *zoom* and *trace* features to find the relative minimum and relative maximum points. For instance, consider the graph of

$$f(x) = 3.1x^3 - 7.3x^2 + 1.2x + 2.5$$

as shown below.

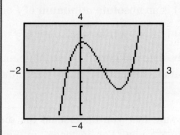

From the graph, you can see that the function has one relative maximum and one relative minimum. You can approximate these values by zooming in and using the *trace* feature, as shown below.

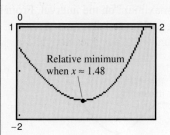
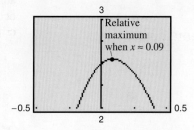

A second way to use technology to find relative extrema is to perform the First-Derivative Test with a symbolic differentiation utility. You can use the utility to differentiate the function, set the derivative equal to zero, and then solve the resulting equation. After obtaining the critical numbers, 1.48288 and 0.0870148, you can graph the function and observe that the first yields a relative minimum and the second yields a relative maximum. Compare the two ways shown above with doing the calculations by hand, as shown below.

$f(x) = 3.1x^3 - 7.3x^2 + 1.2x + 2.5$ Write original function.

$f'(x) = \dfrac{d}{dx}[3.1x^3 - 7.3x^2 + 1.2x + 2.5]$ Differentiate with respect to x.

$f'(x) = 9.3x^2 - 14.6x + 1.2$ First derivative

$9.3x^2 - 14.6x + 1.2 = 0$ Set derivative equal to 0.

$x = \dfrac{73 \pm \sqrt{4213}}{93}$ Solve for x.

$x \approx 1.48288,\ x \approx 0.0870148$ Approximate.

STUDY TIP

Some graphing calculators have a special feature that allows you to find the minimum or maximum of a function on an interval. Consult the user's manual for information on the *minimum value* and *maximum value* features of your graphing utility.

Absolute Extrema

The terms *relative minimum* and *relative maximum* describe the *local* behavior of a function. To describe the *global* behavior of the function on an entire interval, you can use the terms **absolute maximum** and **absolute minimum**.

> **Definition of Absolute Extrema**
>
> Let f be defined on an interval I containing c.
>
> 1. $f(c)$ is an **absolute minimum of f** on I if $f(c) \leq f(x)$ for every x in I.
> 2. $f(c)$ is an **absolute maximum of f** on I if $f(c) \geq f(x)$ for every x in I.
>
> The absolute minimum and absolute maximum values of a function on an interval are sometimes simply called the **minimum** and **maximum** of f on I.

Be sure that you understand the distinction between relative extrema and absolute extrema. For instance, in Figure 3.16, the function has a relative minimum that also happens to be an absolute minimum on the interval $[a, b]$. The relative maximum of f, however, is not the absolute maximum on the interval $[a, b]$. The next theorem points out that if a continuous function has a closed interval as its domain, then it *must* have both an absolute minimum and an absolute maximum on the interval. From Figure 3.16, note that these extrema can occur at endpoints of the interval.

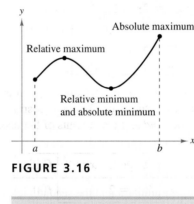

FIGURE 3.16

> **Extreme Value Theorem**
>
> If f is continuous on $[a, b]$, then f has both a minimum value and a maximum value on $[a, b]$.

Although a continuous function has just one minimum and one maximum value on a closed interval, either of these values can occur for more than one x-value. For instance, on the interval $[-3, 3]$, the function $f(x) = 9 - x^2$ has a minimum value of zero when $x = -3$ *and* when $x = 3$, as shown in Figure 3.17.

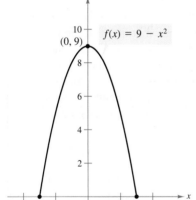

FIGURE 3.17

SECTION 3.2 Extrema and the First-Derivative Test

TECHNOLOGY

A graphing utility can help you locate the extrema of a function on a closed interval. For instance, try using a graphing utility to confirm the results of Example 4. (Set the viewing window at $-1 \leq x \leq 6$ and $-8 \leq y \leq 4$.) Use the *trace* feature to check that the minimum y-value occurs when $x = 3$ and the maximum y-value occurs when $x = 0$.

When looking for extrema of a function on a *closed* interval, remember that you must consider the values of the function at the endpoints as well as at the critical numbers of the function. You can use the guidelines below to find extrema on a closed interval.

Guidelines for Finding Extrema on a Closed Interval

To find the extrema of a continuous function f on a closed interval $[a, b]$, use the steps below.

1. Evaluate f at each of its critical numbers in (a, b).
2. Evaluate f at each endpoint, a and b.
3. The least of these values is the minimum, and the greatest is the maximum.

Example 4 Finding Extrema on a Closed Interval

Find the minimum and maximum values of

$$f(x) = x^2 - 6x + 2$$

on the interval $[0, 5]$.

SOLUTION Begin by finding the critical numbers of the function.

$$f'(x) = 2x - 6 \quad \text{Find derivative of } f.$$
$$2x - 6 = 0 \quad \text{Set derivative equal to 0.}$$
$$2x = 6 \quad \text{Add 6 to each side.}$$
$$x = 3 \quad \text{Solve for } x.$$

From this, you can see that the only critical number of f is $x = 3$. Because this number lies in the interval under question, you should test the values of $f(x)$ at this number *and* at the endpoints of the interval, as shown in the table.

x-value	Endpoint: $x = 0$	Critical number: $x = 3$	Endpoint: $x = 5$
$f(x)$	$f(0) = 2$	$f(3) = -7$	$f(5) = -3$
Conclusion	Maximum is 2	Minimum is -7	Neither maximum nor minimum

From the table, you can see that the minimum of f on the interval $[0, 5]$ is $f(3) = -7$. Moreover, the maximum of f on the interval $[0, 5]$ is $f(0) = 2$. This is confirmed by the graph of f, as shown in Figure 3.18.

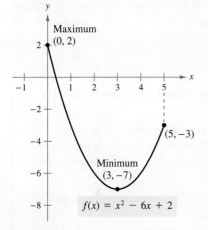

FIGURE 3.18

✓ CHECKPOINT 4

Find the minimum and maximum values of $f(x) = x^2 - 8x + 10$ on the interval $[0, 7]$. Sketch the graph of $f(x)$ and label the minimum and maximum values. ∎

CHAPTER 3 Applications of the Derivative

Applications of Extrema

Finding the minimum and maximum values of a function is one of the most common applications of calculus.

Example 5 Oxygen Level of a Pond

When organic waste is dumped into a pond, the decomposition of the waste consumes oxygen. A model for the oxygen level L (where $L = 1$ is the normal unpolluted level) of a pond as waste material oxidizes is

$$L = \frac{t^2 - t + 1}{t^2 + 1}, \quad t \geq 0$$

where t is the time in weeks and organic waste is dumped into the pond when $t = 0$. When is the oxygen level lowest?

SOLUTION Begin by finding the critical numbers of the function.

$$L = \frac{t^2 - t + 1}{t^2 + 1} \qquad \text{Write original function.}$$

$$\frac{dL}{dt} = \frac{(t^2 + 1)(2t - 1) - (t^2 - t + 1)(2t)}{(t^2 + 1)^2} \qquad \text{Quotient Rule}$$

$$= \frac{2t^3 - t^2 + 2t - 1 - 2t^3 + 2t^2 - 2t}{(t^2 + 1)^2}$$

$$= \frac{t^2 - 1}{(t^2 + 1)^2} \qquad \text{Simplify.}$$

From this and the domain restriction $t \geq 0$, you can see that the derivative is zero when $t = 1$. From Figure 3.19, you can see that the critical number $t = 1$ corresponds to the time when the oxygen level of the pond is lowest. The oxygen level is lowest 1 week after the organic waste is dumped.

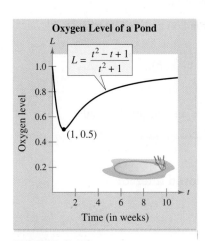

FIGURE 3.19

✓ CHECKPOINT 5

Verify the results of Example 5 by completing the table.

t	0	1	2	3	4
L					

t	5	6	7	8
L				

CONCEPT CHECK

1. Complete the following: The relative extrema of a function include the relative _____ and the relative _____.

2. Let f be continuous on the open interval (a, b) in which c is the only critical number, and assume that f is differentiable on the interval (except possibly at c). According to the First-Derivative Test, what are the three possible classifications for $f(c)$?

3. Let f be defined on an interval I containing c. The value $f(c)$ is an absolute minimum of f on I under what condition?

4. In your own words, state the guidelines for finding the extrema of a continuous function f on a closed interval $[a, b]$.

Skills Review 3.2

The following warm-up exercises involve skills that were covered in earlier sections. You will use these skills in the exercise set for this section. For additional help, review Sections 2.2, 2.4, and 3.1.

In Exercises 1–6, solve the equation $f'(x) = 0$.

1. $f(x) = 4x^4 - 2x^2 + 1$
2. $f(x) = \frac{1}{3}x^3 - \frac{3}{2}x^2 - 10x$
3. $f(x) = 5x^{4/5} - 4x$
4. $f(x) = \frac{1}{2}x^2 - 3x^{5/3}$
5. $f(x) = \frac{x+4}{x^2+1}$
6. $f(x) = \frac{x-1}{x^2+4}$

In Exercises 7–10, use $g(x) = -x^5 - 2x^4 + 4x^3 + 2x - 1$ to determine the sign of the derivative.

7. $g'(-4)$
8. $g'(0)$
9. $g'(1)$
10. $g'(3)$

In Exercises 11 and 12, decide whether the function is increasing or decreasing on the given interval.

11. $f(x) = 2x^2 - 11x - 6$, $(3, 6)$
12. $f(x) = x^3 + 2x^2 - 4x - 8$, $(-2, 0)$

Exercises 3.2

See www.CalcChat.com for worked-out solutions to odd-numbered exercises.

In Exercises 1–4, use a table similar to that in Example 1 to find all relative extrema of the function.

1. $f(x) = -2x^2 + 4x + 3$
2. $f(x) = x^2 + 8x + 10$
3. $f(x) = x^2 - 6x$
4. $f(x) = -4x^2 + 4x + 1$

In Exercises 5–12, find all relative extrema of the function.

5. $g(x) = 6x^3 - 15x^2 + 12x$
6. $g(x) = \frac{1}{5}x^5 - x$
7. $h(x) = -(x+4)^3$
8. $h(x) = 2(x-3)^3$
9. $f(x) = x^3 - 6x^2 + 15$
10. $f(x) = x^4 - 32x + 4$
11. $f(x) = x^4 - 2x^3 + x + 1$
12. $f(x) = x^4 - 12x^3$

(T) In Exercises 13–18, use a graphing utility to graph the function. Then find all relative extrema of the function.

13. $f(x) = (x-1)^{2/3}$
14. $f(t) = (t-1)^{1/3}$
15. $g(t) = t - \frac{1}{2t^2}$
16. $f(x) = x + \frac{1}{x}$
17. $f(x) = \frac{x}{x+1}$
18. $h(x) = \frac{4}{x^2+1}$

In Exercises 19–30, find the absolute extrema of the function on the closed interval. Use a graphing utility to verify your results.

19. $f(x) = 2(3 - x)$, $[-1, 2]$
20. $f(x) = \frac{1}{3}(2x + 5)$, $[0, 5]$
21. $f(x) = 5 - 2x^2$, $[0, 3]$
22. $f(x) = x^2 + 2x - 4$, $[-1, 1]$
23. $f(x) = x^3 - 3x^2$, $[-1, 3]$
24. $f(x) = x^3 - 12x$, $[0, 4]$
25. $h(s) = \frac{1}{3-s}$, $[0, 2]$
26. $h(t) = \frac{t}{t-2}$, $[3, 5]$
27. $f(x) = 3x^{2/3} - 2x$, $[-1, 2]$
28. $g(t) = \frac{t^2}{t^2+3}$, $[-1, 1]$
29. $h(t) = (t-1)^{2/3}$, $[-7, 2]$
30. $g(x) = 4\left(1 + \frac{1}{x} + \frac{1}{x^2}\right)$, $[-4, 5]$

In Exercises 31 and 32, approximate the critical numbers of the function shown in the graph. Determine whether the function has a relative maximum, a relative minimum, an absolute maximum, an absolute minimum, or none of these at each critical number on the interval shown.

31.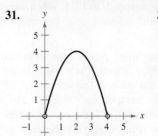

32.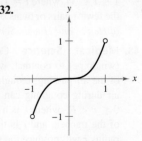

In Exercises 33–36, use a graphing utility to find graphically the absolute extrema of the function on the closed interval.

33. $f(x) = 0.4x^3 - 1.8x^2 + x - 3$, $[0, 5]$
34. $f(x) = 3.2x^5 + 5x^3 - 3.5x$, $[0, 1]$
35. $f(x) = \frac{4}{3}x\sqrt{3-x}$, $[0, 3]$
36. $f(x) = 4\sqrt{x} - 2x + 1$, $[0, 6]$

In Exercises 37–40, find the absolute extrema of the function on the interval $[0, \infty)$.

37. $f(x) = \dfrac{4x}{x^2+1}$
38. $f(x) = \dfrac{8}{x+1}$
39. $f(x) = \dfrac{2x}{x^2+4}$
40. $f(x) = 8 - \dfrac{4x}{x^2+1}$

In Exercises 41 and 42, find the maximum value of $|f''(x)|$ on the closed interval. (You will use this skill in Section 7.5 to estimate the error in the Trapezoidal Rule.)

41. $f(x) = \sqrt{1+x^3}$, $[0, 2]$ 42. $f(x) = \dfrac{1}{x^2+1}$, $[0, 3]$

In Exercises 43 and 44, find the maximum value of $|f^{(4)}(x)|$ on the closed interval. (You will use this skill in Section 7.5 to estimate the error in Simpson's Rule.)

43. $f(x) = (x+1)^{2/3}$, $[0, 2]$
44. $f(x) = \dfrac{1}{x^2+1}$, $[-1, 1]$

In Exercises 45 and 46, graph a function on the interval $[-2, 5]$ having the given characteristics.

45. Absolute maximum at $x = -2$
Absolute minimum at $x = 1$
Relative maximum at $x = 3$

46. Relative minimum at $x = -1$
Critical number, but no extremum, at $x = 0$
Absolute maximum at $x = 2$
Absolute minimum at $x = 5$

47. **Twins** The number T of pairs of twins born (per 1000 live births) in the United States from 1971 through 2004 can be modeled by $T = 0.0143t^2 - 0.074t + 18$, $1 \leq t \leq 34$, where $t = 1$ corresponds to 1971. When were the fewest pairs of twins born? *(Source: U.S. Department of Health and Human Services)*

48. **Medical Science** Coughing forces the trachea (windpipe) to contract, which in turn affects the velocity of the air passing through the trachea. The velocity of the air during coughing can be modeled by $v = k(R - r)r^2$, $0 \leq r < R$, where k is a constant, R is the normal radius of the trachea, and r is the radius during coughing. What radius r will produce the maximum air velocity?

49. **Bloodstream** The concentration C (in milligrams per milliliter) of a chemical in the bloodstream t hours after injection into muscle tissue can be modeled by

$$C = \dfrac{3t}{27+t^3}, \quad t \geq 0.$$

(a) Complete the table and use it to approximate the time when the concentration reached a maximum.

t	0	0.5	1	1.5	2	2.5	3
$C(t)$							

(b) Use a graphing utility to graph the concentration function. Use the *trace* feature to approximate the time when the concentration reached a maximum.

(c) Determine analytically the time when the concentration reached a maximum.

50. **Population** The resident population P (in millions) of the United States from 1790 through 2000 can be modeled by $P = 0.00000583t^3 + 0.005003t^2 + 0.13776t + 4.658$, $-10 \leq t \leq 200$, where $t = 0$ corresponds to 1800. *(Source: U.S. Census Bureau)*

(a) Make a conjecture about the maximum and minimum populations from 1790 through 2000.

(b) Analytically find the maximum and minimum populations over the interval.

51. **Biology: Fertility Rates** The graph of the United States fertility rate shows the number of births per 1000 women in their lifetime according to the birth rate in that particular year. *(Source: U.S. National Center for Health Statistics)*

(a) In which year was the fertility rate the highest, and to how many births per 1000 women did this rate correspond?

(b) During which time period was the fertility rate decreasing most rapidly? Most slowly?

(c) Give some possible real-life reasons for fluctuations in the fertility rate.

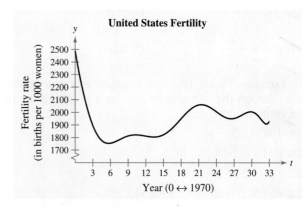

Section 3.3
Concavity and the Second-Derivative Test

- Determine the intervals on which the graphs of functions are concave upward or concave downward.
- Find the points of inflection of the graphs of functions.
- Use the Second-Derivative Test to find the relative extrema of functions.
- Use the Second-Derivative Test to answer questions about real-life situations.

Concavity

You already know that locating the intervals over which a function f increases or decreases is helpful in determining its graph. In this section, you will see that locating the intervals on which f' increases or decreases can determine where the graph of f is curving upward or curving downward. This property of curving upward or downward is defined formally as the **concavity** of the graph of the function.

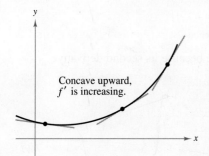

Definition of Concavity

Let f be differentiable on an open interval I. The graph of f is

1. **concave upward** on I if f' is increasing on the interval.
2. **concave downward** on I if f' is decreasing on the interval.

From Figure 3.20, you can observe the following graphical interpretation of concavity.

1. A curve that is concave upward lies *above* its tangent lines.
2. A curve that is concave downward lies *below* its tangent lines.

This visual test for concavity is useful when the graph of a function is given. To determine concavity without seeing a graph, you need an analytic test. It turns out that you can use the second derivative to determine these intervals in much the same way that you use the first derivative to determine the intervals on which f is increasing or decreasing.

FIGURE 3.20

Test for Concavity

Let f be a function whose second derivative exists on an open interval I.

1. If $f''(x) > 0$ for all x in I, then f is concave upward on I.
2. If $f''(x) < 0$ for all x in I, then f is concave downward on I.

204 CHAPTER 3 Applications of the Derivative

For a *continuous* function f, you can find the open intervals on which the graph of f is concave upward and concave downward as follows. [For a function that is not continuous, the test intervals should be formed using points of discontinuity, along with the points at which $f''(x)$ is zero or undefined.]

DISCOVERY

Use a graphing utility to graph the function $f(x) = x^3 - x$ and its second derivative $f''(x) = 6x$ in the same viewing window. On what interval is f concave upward? On what interval is f'' positive? Describe how the second derivative can be used to determine where a function is concave upward and concave downward. Repeat this analysis for the functions $g(x) = x^4 - 6x^2$ and $g''(x) = 12x^2 - 12$.

Guidelines for Applying Concavity Test

1. Locate the x-values at which $f''(x) = 0$ or $f''(x)$ is undefined.
2. Use these x-values to determine the test intervals.
3. Test the sign of $f''(x)$ in each test interval.

Example 1 Applying the Test for Concavity

a. The graph of the function

$$f(x) = x^2 \qquad \text{Original function}$$

is concave upward on the entire real line because its second derivative

$$f''(x) = 2 \qquad \text{Second derivative}$$

is positive for all x. (See Figure 3.21.)

b. The graph of the function

$$f(x) = \sqrt{x} \qquad \text{Original function}$$

is concave downward for $x > 0$ because its second derivative

$$f''(x) = -\frac{1}{4}x^{-3/2} \qquad \text{Second derivative}$$

is negative for all $x > 0$. (See Figure 3.22.)

✓ CHECKPOINT 1

a. Find the second derivative of $f(x) = -2x^2$ and discuss the concavity of its graph.

b. Find the second derivative of $f(x) = -2\sqrt{x}$ and discuss the concavity of its graph. ∎

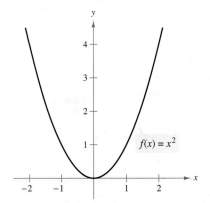

FIGURE 3.21 Concave Upward

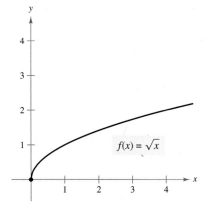

FIGURE 3.22 Concave Downward

SECTION 3.3 Concavity and the Second-Derivative Test

Example 2 Determining Concavity

Determine the open intervals on which the graph of the function is concave upward or concave downward.

$$f(x) = \frac{6}{x^2 + 3}$$

SOLUTION Begin by finding the second derivative of f.

$$f(x) = 6(x^2 + 3)^{-1} \qquad \text{Rewrite original function.}$$
$$f'(x) = (-6)(2x)(x^2 + 3)^{-2} \qquad \text{Chain Rule}$$
$$= \frac{-12x}{(x^2 + 3)^2} \qquad \text{Simplify.}$$
$$f''(x) = \frac{(x^2 + 3)^2(-12) - (-12x)(2)(2x)(x^2 + 3)}{(x^2 + 3)^4} \qquad \text{Quotient Rule}$$
$$= \frac{-12(x^2 + 3) + (48x^2)}{(x^2 + 3)^3} \qquad \text{Simplify.}$$
$$= \frac{36(x^2 - 1)}{(x^2 + 3)^3} \qquad \text{Simplify.}$$

From this, you can see that $f''(x)$ is defined for all real numbers and $f''(x) = 0$ when $x = \pm 1$. So, you can test the concavity of f by testing the intervals $(-\infty, -1)$, $(-1, 1)$, and $(1, \infty)$, as shown in the table. The graph of f is shown in Figure 3.23.

Interval	$-\infty < x < -1$	$-1 < x < 1$	$1 < x < \infty$
Test value	$x = -2$	$x = 0$	$x = 2$
Sign of $f'(x)$	$f''(-2) > 0$	$f''(0) < 0$	$f''(2) > 0$
Conclusion	Concave upward	Concave downward	Concave upward

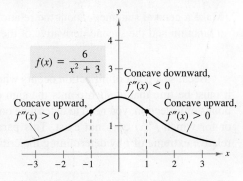

FIGURE 3.23

Algebra Review

For help with the algebra in Example 2, see Example 1(a) in the *Chapter 3 Algebra Review*, on page 249.

STUDY TIP

In Example 2, f' is increasing on the interval $(1, \infty)$ even though f is decreasing there. Be sure you see that the increasing or decreasing of f' does not necessarily correspond to the increasing or decreasing of f.

✓ CHECKPOINT 2

Determine the intervals on which the graph of the function is concave upward or concave downward.

$$f(x) = \frac{12}{x^2 + 4}$$

Points of Inflection

If the tangent line to a graph exists at a point at which the concavity changes, then the point is a **point of inflection.** Three examples of inflection points are shown in Figure 3.24. (Note that the third graph has a vertical tangent line at its point of inflection.)

> **STUDY TIP**
> As shown in Figure 3.24, a graph crosses its tangent line at a point of inflection.

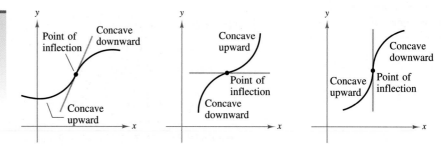

FIGURE 3.24   The graph *crosses* its tangent line at a point of inflection.

> **Definition of Point of Inflection**
>
> If the graph of a continuous function has a tangent line at a point where its concavity changes from upward to downward (or downward to upward), then the point is a **point of inflection.**

> ### DISCOVERY
>
> Use a graphing utility to graph
>
> $$f(x) = x^3 - 6x^2 + 12x - 6 \quad \text{and} \quad f''(x) = 6x - 12$$
>
> in the same viewing window. At what x-value does $f''(x) = 0$? At what x-value does the point of inflection occur? Repeat this analysis for
>
> $$g(x) = x^4 - 5x^2 + 7 \quad \text{and} \quad g''(x) = 12x^2 - 10.$$
>
> Make a general statement about the relationship of the point of inflection of a function and the second derivative of the function.

Because a point of inflection occurs where the concavity of a graph changes, it must be true that at such points the sign of f'' changes. So, to locate possible points of inflection, you need to determine only the values of x for which $f''(x) = 0$ or for which $f''(x)$ does not exist. This parallels the procedure for locating the relative extrema of f by determining the critical numbers of f.

> **Property of Points of Inflection**
>
> If $(c, f(c))$ is a point of inflection of the graph of f, then either $f''(c) = 0$ or $f''(c)$ is undefined.

SECTION 3.3 Concavity and the Second-Derivative Test

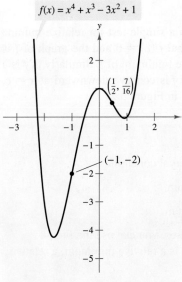

FIGURE 3.25 Two Points of Inflection

Example 3 Finding Points of Inflection

Discuss the concavity of the graph of f and find its points of inflection.

$$f(x) = x^4 + x^3 - 3x^2 + 1$$

SOLUTION Begin by finding the second derivative of f.

$f(x) = x^4 + x^3 - 3x^2 + 1$	Write original function.
$f'(x) = 4x^3 + 3x^2 - 6x$	Find first derivative.
$f''(x) = 12x^2 + 6x - 6$	Find second derivative.
$= 6(2x - 1)(x + 1)$	Factor.

From this, you can see that the possible points of inflection occur at $x = \frac{1}{2}$ and $x = -1$. After testing the intervals $(-\infty, -1)$, $\left(-1, \frac{1}{2}\right)$, and $\left(\frac{1}{2}, \infty\right)$, you can determine that the graph is concave upward on $(-\infty, -1)$, concave downward on $\left(-1, \frac{1}{2}\right)$, and concave upward on $\left(\frac{1}{2}, \infty\right)$. Because the concavity changes at $x = -1$ and $x = \frac{1}{2}$, you can conclude that the graph has points of inflection at these x-values, as shown in Figure 3.25.

✓ CHECKPOINT 3

Discuss the concavity of the graph of f and find its points of inflection.

$$f(x) = x^4 - 2x^3 + 1$$

It is possible for the second derivative to be zero at a point that is *not* a point of inflection. For example, compare the graphs of $f(x) = x^3$ and $g(x) = x^4$, as shown in Figure 3.26. Both second derivatives are zero when $x = 0$, but only the graph of f has a point of inflection at $x = 0$. This shows that before concluding that a point of inflection exists at a value of x for which $f''(x) = 0$, you must test to be certain that the concavity actually changes at that point.

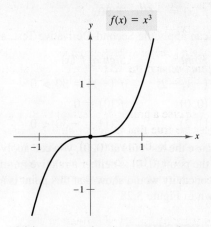

$f''(0) = 0$, and $(0, 0)$ is a point of inflection.

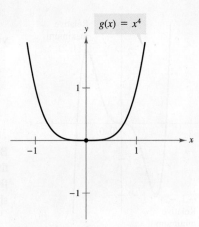

$g''(0) = 0$, but $(0, 0)$ is not a point of inflection.

FIGURE 3.26

The Second-Derivative Test

The second derivative can be used to perform a simple test for relative minima and relative maxima. If f is a function such that $f'(c) = 0$ and the graph of f is concave upward at $x = c$, then $f(c)$ is a relative minimum of f. Similarly, if f is a function such that $f'(c) = 0$ and the graph of f is concave downward at $x = c$, then $f(c)$ is a relative maximum of f, as shown in Figure 3.27.

Second-Derivative Test

Let $f'(c) = 0$, and let f'' exist on an open interval containing c.

1. If $f''(c) > 0$, then $f(c)$ is a relative minimum.
2. If $f''(c) < 0$, then $f(c)$ is a relative maximum.
3. If $f''(c) = 0$, then the test fails. In such cases, you can use the First-Derivative Test to determine whether $f(c)$ is a relative minimum, a relative maximum, or neither.

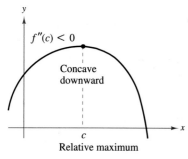
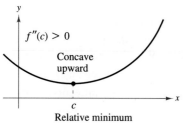

FIGURE 3.27

Algebra Review

For help with the algebra in Example 4, see Example 2(b) in the *Chapter 3 Algebra Review*, on page 250.

Example 4 Using the Second-Derivative Test

Find the relative extrema of
$$f(x) = -3x^5 + 5x^3.$$

SOLUTION Begin by finding the first derivative of f.
$$f'(x) = -15x^4 + 15x^2$$
$$= 15x^2(1 - x^2)$$

From this derivative, you can see that $x = 0$, $x = -1$, and $x = 1$ are the only critical numbers of f. Using the second derivative
$$f''(x) = -60x^3 + 30x$$
you can apply the Second-Derivative Test, as shown.

Point	Sign of $f''(x)$	Conclusion
$(-1, -2)$	$f''(-1) = 30 > 0$	Relative minimum
$(0, 0)$	$f''(0) = 0$	Test fails.
$(1, 2)$	$f''(1) = -30 < 0$	Relative maximum

Because the test fails at $(0, 0)$, you can apply the First-Derivative Test to conclude that the point $(0, 0)$ is neither a relative minimum nor a relative maximum—a test for concavity would show that this point is a point of inflection. The graph of f is shown in Figure 3.28.

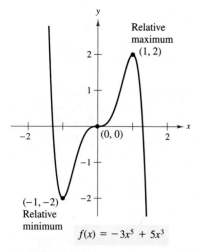

FIGURE 3.28

✓**CHECKPOINT 4**

Find all relative extrema of $f(x) = x^4 - 4x^3 + 1$. ∎

Application

Example 5 Death Rate from HIV

Since 1990, the annual death rate r (per 100,000 persons) from Human Immunodeficiency Virus (HIV) in the United States for 15- to 24-year-olds can be modeled by

$$r = 0.0032t^3 - 0.073t^2 + 0.32t + 1.5, \quad 0 \le t \le 15$$

where t is the year, with $t = 0$ corresponding to 1990. Find the point of inflection of the graph of r and interpret its meaning. *(Source: U.S. National Center for Health Statistics)*

SOLUTION Begin by finding the first and second derivatives.

$r' = 0.0096t^2 - 0.146t + 0.32$ First derivative

$r'' = 0.0192t - 0.146$ Second derivative

The second derivative is zero only when

$$t = \frac{0.146}{0.0192} \approx 7.6.$$

By testing the intervals $(0, 7.6)$ and $(7.6, 15)$, you can conclude that the graph has a point of inflection when $t = 7.6$, as shown in Figure 3.29. The death rate from HIV for 15- to 24-year-olds has been declining in recent years, but since 1997 the rate of this decline has been slowing down.

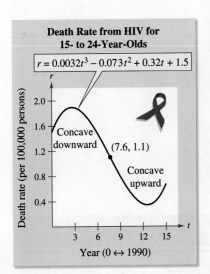

FIGURE 3.29

✓ CHECKPOINT 5

Find and interpret any relative extrema of the model given in Example 5. ■

CONCEPT CHECK

1. Let f be differentiable on an open interval I. If the graph of f is concave upward on I, what can you conclude about the behavior of f' on the interval?
2. Let f be a function whose second derivative exists on an open interval I, and let $f''(x) > 0$ for all x in I. Is f concave upward or concave downward on I?
3. Let $f'(c) = 0$, and let f'' exist on an open interval containing c. According to the Second-Derivative Test, what are the possible classifications for $f(c)$?
4. A newspaper headline states that "The rate of growth of the national deficit is decreasing." What does this mean? What does it imply about the graph of the deficit as a function of time?

210 CHAPTER 3 Applications of the Derivative

Skills Review 3.3

The following warm-up exercises involve skills that were covered in earlier sections. You will use these skills in the exercise set for this section. For additional help, review Sections 2.4, 2.5, 2.6, and 3.1.

In Exercises 1–6, find the second derivative of the function.

1. $f(x) = 4x^4 - 9x^3 + 5x - 1$
2. $g(s) = (s^2 - 1)(s^2 - 3s + 2)$
3. $g(x) = (x^2 + 1)^4$
4. $f(x) = (x - 3)^{4/3}$
5. $h(x) = \dfrac{4x + 3}{5x - 1}$
6. $f(x) = \dfrac{2x - 1}{3x + 2}$

In Exercises 7–10, find the critical numbers of the function.

7. $f(x) = 5x^3 - 5x + 11$
8. $f(x) = x^4 - 4x^3 - 10$
9. $g(t) = \dfrac{16 + t^2}{t}$
10. $h(x) = \dfrac{x^4 - 50x^2}{8}$

Exercises 3.3

See www.CalcChat.com for worked-out solutions to odd-numbered exercises.

In Exercises 1–8, find the open intervals on which the graph is concave upward and those on which it is concave downward.

1. $y = x^2 - x - 2$
2. $y = -x^3 + 3x^2 - 2$

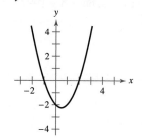
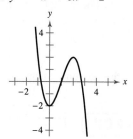

3. $f(x) = \dfrac{x^2 - 1}{2x + 1}$
4. $f(x) = \dfrac{x^2 + 4}{4 - x^2}$

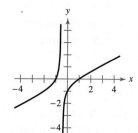
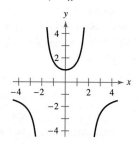

5. $f(x) = \dfrac{24}{x^2 + 12}$
6. $f(x) = \dfrac{x^2}{x^2 + 1}$

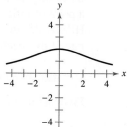
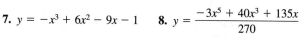

7. $y = -x^3 + 6x^2 - 9x - 1$
8. $y = \dfrac{-3x^5 + 40x^3 + 135x}{270}$

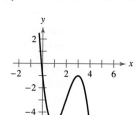
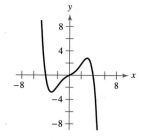

In Exercises 9–22, find all relative extrema of the function. Use the Second-Derivative Test when applicable.

9. $f(x) = 6x - x^2$
10. $f(x) = (x - 5)^2$
11. $f(x) = x^3 - 5x^2 + 7x$
12. $f(x) = x^4 - 4x^3 + 2$
13. $f(x) = x^{2/3} - 3$
14. $f(x) = x + \dfrac{4}{x}$
15. $f(x) = \sqrt{x^2 + 1}$
16. $f(x) = \sqrt{2x^2 + 6}$

17. $f(x) = \sqrt{9 - x^2}$ 18. $f(x) = \sqrt{4 - x^2}$

19. $f(x) = \dfrac{8}{x^2 + 2}$ 20. $f(x) = \dfrac{18}{x^2 + 3}$

21. $f(x) = \dfrac{x}{x - 1}$ 22. $f(x) = \dfrac{x}{x^2 - 1}$

(T) In Exercises 23–26, use a graphing utility to estimate graphically all relative extrema of the function.

23. $f(x) = \tfrac{1}{2}x^4 - \tfrac{1}{3}x^3 - \tfrac{1}{2}x^2$ 24. $f(x) = -\tfrac{1}{3}x^5 - \tfrac{1}{2}x^4 + x$

25. $f(x) = 5 + 3x^2 - x^3$ 26. $f(x) = 3x^3 + 5x^2 - 2$

In Exercises 27–30, state the signs of $f'(x)$ and $f''(x)$ on the interval $(0, 2)$.

27.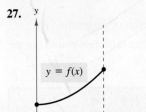

28.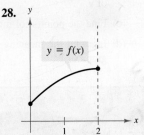

29.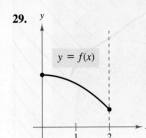

30.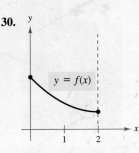

In Exercises 31–38, find the point(s) of inflection of the graph of the function.

31. $f(x) = x^3 - 9x^2 + 24x - 18$
32. $f(x) = x(6 - x)^2$
33. $f(x) = (x - 1)^3(x - 5)$ 34. $f(x) = x^4 - 18x^2 + 5$
35. $g(x) = 2x^4 - 8x^3 + 12x^2 + 12x$
36. $f(x) = -4x^3 - 8x^2 + 32$
37. $h(x) = (x - 2)^3(x - 1)$
38. $f(t) = (1 - t)(t - 4)(t^2 - 4)$

(T) In Exercises 39–50, use a graphing utility to graph the function and identify all relative extrema and points of inflection.

39. $f(x) = x^3 - 12x$ 40. $f(x) = x^3 - 3x$
41. $f(x) = x^3 - 6x^2 + 12x$ 42. $f(x) = x^3 - \tfrac{3}{2}x^2 - 6x$
43. $f(x) = \tfrac{1}{4}x^4 - 2x^2$ 44. $f(x) = 2x^4 - 8x + 3$

45. $g(x) = (x - 2)(x + 1)^2$ 46. $g(x) = (x - 6)(x + 2)^3$
47. $g(x) = x\sqrt{x + 3}$ 48. $g(x) = x\sqrt{9 - x}$
49. $f(x) = \dfrac{4}{1 + x^2}$ 50. $f(x) = \dfrac{2}{x^2 - 1}$

In Exercises 51–54, sketch a graph of a function f having the given characteristics.

51. $f(2) = f(4) = 0$
    $f'(x) < 0$ if $x < 3$
    $f'(3) = 0$
    $f'(x) > 0$ if $x > 3$
    $f''(x) > 0$

52. $f(2) = f(4) = 0$
    $f'(x) > 0$ if $x < 3$
    $f'(3)$ is undefined.
    $f'(x) < 0$ if $x > 3$
    $f''(x) > 0, x \neq 3$

53. $f(0) = f(2) = 0$
    $f'(x) > 0$ if $x < 1$
    $f'(1) = 0$
    $f'(x) < 0$ if $x > 1$
    $f''(x) < 0$

54. $f(0) = f(2) = 0$
    $f'(x) < 0$ if $x < 1$
    $f'(1) = 0$
    $f'(x) > 0$ if $x > 1$
    $f''(x) > 0$

In Exercises 55 and 56, use the graph to sketch the graph of f'. Find the intervals on which (a) $f'(x)$ is positive, (b) $f'(x)$ is negative, (c) f' is increasing, and (d) f' is decreasing. For each of these intervals, describe the corresponding behavior of f.

55.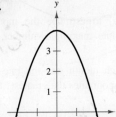

56.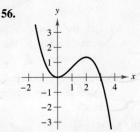

In Exercises 57–60, you are given f'. Find the intervals on which (a) $f'(x)$ is increasing or decreasing and (b) the graph of f is concave upward or concave downward. (c) Find the relative extrema and inflection points of f. (d) Then sketch a graph of f.

57. $f'(x) = 2x + 5$ 58. $f'(x) = 3x^2 - 2$
59. $f'(x) = -x^2 + 2x - 1$ 60. $f'(x) = x^2 + x - 6$

Productivity In Exercises 61 and 62, consider a college student who works from 7 P.M. to 11 P.M. assembling mechanical components. The number N of components assembled after t hours is given by the function. At what time is the student assembling components at the greatest rate?

61. $N = -0.12t^3 + 0.54t^2 + 8.22t$, $0 \le t \le 4$

62. $N = \dfrac{20t^2}{4 + t^2}$, $0 \le t \le 4$

In Exercises 63–66, use a graphing utility to graph f, f', and f'' in the same viewing window. Graphically locate the relative extrema and points of inflection of the graph of f. State the relationship between the behavior of f and the signs of f' and f''.

63. $f(x) = \frac{1}{2}x^3 - x^2 + 3x - 5$, $[0, 3]$

64. $f(x) = -\frac{1}{20}x^5 - \frac{1}{12}x^2 - \frac{1}{3}x + 1$, $[-2, 2]$

65. $f(x) = \frac{2}{x^2 + 1}$, $[-3, 3]$ 66. $f(x) = \frac{x^2}{x^2 + 1}$, $[-3, 3]$

67. **Phishing** Phishing is a criminal activity used by an individual or group to fraudulently acquire information by masquerading as a trustworthy person or business in an electronic communication. Criminals create spoof sites on the Internet to trick victims into giving them information. A model for the number of reported spoof sites from November 2005 through October 2006 is

$f(t) = 88.253t^3 - 1116.16t^2 + 4541.4t + 4161$, $0 \le t \le 11$

where t represents the number of months since November 2005. *(Source: Anti-Phishing Working Group)*

(a) Use a graphing utility to graph the model on the interval $[0, 11]$.

(b) Use the graph in part (a) to estimate the month corresponding to the absolute minimum number of spoof sites.

(c) Use the graph in part (a) to estimate the month corresponding to the absolute maximum number of spoof sites.

(d) During approximately which month was the rate of increase of the number of spoof sites the greatest? the least?

68. **Medicine** The spread of a virus can be modeled by

$N = -t^3 + 12t^2$, $0 \le t \le 12$

where N is the number of people infected (in hundreds), and t is the time (in weeks).

(a) What is the maximum number of people projected to be infected?

(b) When will the virus be spreading most rapidly?

(c) Use a graphing utility to graph the model and to verify your results.

69. **High School Dropouts** From 2000 through 2005, the number d of high school dropouts not in the labor force (in thousands) can be modeled by

$d = -20.444t^3 + 152.33t^2 - 266.6t + 1162$

where t is the year, with $t = 0$ corresponding to 2000. *(Source: U.S. Bureau of Labor Statistics)*

(a) Use a graphing utility to graph the model.

(b) Use the second derivative to determine the concavity of d.

(c) Find the point(s) of inflection of the graph of d.

(d) Interpret the meaning of the inflection point(s) of the graph of d.

70. **Veteran Benefits** From 1995 through 2005, the number v of veterans (in thousands) receiving compensation and pension benefits for service in the armed forces can be modeled by

$v = -0.1731t^4 + 7.405t^3 - 106.50t^2 + 607.2t + 2133$

where t is the year, with $t = 5$ corresponding to 1995. *(Source: U.S. Department of Veterans Affairs)*

(a) Use a graphing utility to graph the model.

(b) Use the second derivative to determine the concavity of v.

(c) Find the point(s) of inflection of the graph of v.

(d) Interpret the meaning of the inflection point(s) of the graph of v.

Life Science Capsule

© Owen Franken/CORBIS

Biomedical scientists have made worthwhile contributions to the benefit of humankind. For example, James D. Watson is best known for his discovery of the structure of DNA. Also, Fredrick Sanger determined the complete amino acid sequence of insulin. According to a recent survey from ASEE (American Society of Engineering Education), the number of biomedical bachelor's degrees has doubled since 2001.

71. **Research Project** Use your school's library, the Internet, or some other reference source to research the number of students earning bachelor's, master's, or doctoral degrees like the one above. Gather the data for the number of degrees earned in a particular field over a period of time, and use a graphing utility to graph a scatter plot of the data. Fit models to the data. Do the models appear to be concave upward or downward? Do they appear to be increasing or decreasing? Discuss the implications of your answers.

Section 3.4
Optimization Problems

- Solve real-life optimization problems.

Solving Optimization Problems

One of the most common applications of calculus is the determination of optimum (minimum or maximum) values. Before learning a general method for solving optimization problems, consider the next example.

Example 1   Finding the Maximum Volume

A manufacturer wants to design an open box that has a square base and a surface area of 108 square inches, as shown in Figure 3.30. What dimensions will produce a box with a maximum volume?

SOLUTION   Because the base of the box is square, the volume is

$$V = x^2 h. \quad \text{Primary equation}$$

This equation is called the **primary equation** because it gives a formula for the quantity to be optimized. The surface area of the box is

$$S = \text{(area of base)} + \text{(area of four sides)}$$
$$108 = x^2 + 4xh. \quad \text{Secondary equation}$$

Because V is to be optimized, it helps to express V as a function of just one variable. To do this, solve the secondary equation for h in terms of x to obtain

$$h = \frac{108 - x^2}{4x}$$

and substitute this expression into the primary equation.

$$V = x^2 h = x^2 \left(\frac{108 - x^2}{4x} \right) = 27x - \frac{1}{4}x^3 \quad \text{Function of one variable}$$

Before finding which x-value yields a maximum value of V, you need to determine the *feasible domain* of the function. That is, what values of x make sense in the problem? Because x must be nonnegative and the area of the base ($A = x^2$) is at most 108, you can conclude that the feasible domain is

$$0 \leq x \leq \sqrt{108}. \quad \text{Feasible domain}$$

Using the techniques described in the first three sections of this chapter, you can determine that (on the interval $0 \leq x \leq \sqrt{108}$) this function has an absolute maximum when $x = 6$ inches and $h = 3$ inches.

✓ CHECKPOINT 1

Use a graphing utility to graph the volume function $V = 27x - \frac{1}{4}x^3$ on $0 \leq x \leq \sqrt{108}$ from Example 1. Verify that the function has an absolute maximum when $x = 6$. What is the maximum volume? ∎

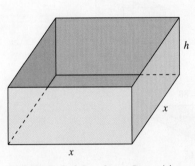

FIGURE 3.30   Open Box with Square Base: $S = x^2 + 4xh = 108$

In studying Example 1, be sure that you understand the basic question that it asks. Some students have trouble with optimization problems because they are too eager to start solving the problem by using a standard formula. For instance, in Example 1, you should realize that there are infinitely many open boxes having 108 square inches of surface area. You might begin to solve this problem by asking yourself which basic shape would seem to yield a maximum volume. Should the box be tall, squat, or nearly cubical? You might even try calculating a few volumes, as shown in Figure 3.31, to see if you can get a good feeling for what the optimum dimensions should be.

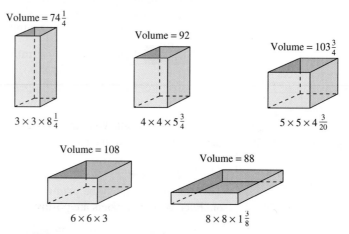

FIGURE 3.31 Which box has the greatest volume?

STUDY TIP

Remember that you are not ready to begin solving an optimization problem until you have clearly identified what the problem is. Once you are sure you understand what is being asked, you are ready to begin considering a method for solving the problem.

There are several steps in the solution of Example 1. The first step is to sketch a diagram and identify all *known* quantities and all quantities *to be determined*. The second step is to write a primary equation for the quantity to be optimized. Then, a secondary equation is used to rewrite the primary equation as a function of one variable. Finally, calculus is used to determine the optimum value. These steps are summarized below.

STUDY TIP

When performing Step 5, remember that to determine the maximum or minimum value of a continuous function f on a closed interval, you need to compare the values of f at its critical numbers with the values of f at the endpoints of the interval. The greatest of these values is the desired maximum and the least is the desired minimum.

Guidelines for Solving Optimization Problems

1. Identify all given quantities and all quantities to be determined. If possible, make a sketch.

2. Write a **primary equation** for the quantity that is to be maximized or minimized. (A summary of several common formulas is given in Appendix D.)

3. Reduce the primary equation to one having a single independent variable. This may involve the use of a **secondary equation** that relates the independent variables of the primary equation.

4. Determine the feasible domain of the primary equation. That is, determine the values for which the stated problem makes sense.

5. Determine the desired maximum or minimum value by the calculus techniques discussed in Sections 3.1 through 3.3.

SECTION 3.4 Optimization Problems

Example 2 Finding a Minimum Sum

The product of two positive numbers is 288. Minimize the sum of the second number and twice the first number.

SOLUTION

1. Let x be the first number, y the second, and S the sum to be minimized.

2. Because you want to minimize S, the primary equation is

$$S = 2x + y. \quad \text{Primary equation}$$

3. Because the product of the two numbers is 288, you can write the secondary equation as

$$xy = 288 \quad \text{Secondary equation}$$

$$y = \frac{288}{x}.$$

Using this result, you can rewrite the primary equation as a function of one variable.

$$S = 2x + \frac{288}{x} \quad \text{Function of one variable}$$

4. Because the numbers are positive, the feasible domain is

$$x > 0. \quad \text{Feasible domain}$$

5. To find the minimum value of S, begin by finding its critical numbers.

$$\frac{dS}{dx} = 2 - \frac{288}{x^2} \quad \text{Find derivative of } S.$$

$$0 = 2 - \frac{288}{x^2} \quad \text{Set derivative equal to 0.}$$

$$x^2 = 144 \quad \text{Simplify.}$$

$$x = \pm 12 \quad \text{Critical numbers}$$

Choosing the positive x-value, you can use the First-Derivative Test to conclude that S is decreasing on the interval $(0, 12)$ and increasing on the interval $(12, \infty)$, as shown in the table. So, $x = 12$ yields a minimum, and the two numbers are

$$x = 12 \quad \text{and} \quad y = \frac{288}{12} = 24.$$

Interval	$0 < x < 12$	$12 < x < \infty$
Test value	$x = 1$	$x = 13$
Sign of $\frac{dS}{dx}$	$\frac{dS}{dx} < 0$	$\frac{dS}{dx} > 0$
Conclusion	S is decreasing.	S is increasing.

Algebra Review

For help with the algebra in Example 2, see Example 1(b) in the Chapter 3 Algebra Review, on page 249.

TECHNOLOGY

After you have written the primary equation as a function of a single variable, you can estimate the optimum value by graphing the function. For instance, the graph of

$$S = 2x + \frac{288}{x}$$

shown below indicates that the minimum value of S occurs when x is about 12.

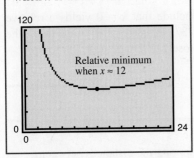

Relative minimum when $x \approx 12$

✓CHECKPOINT 2

The product of two numbers is 72. Minimize the sum of the second number and twice the first number. ■

Example 3 Finding a Minimum Distance

Find the points on the graph of

$$y = 4 - x^2$$

that are closest to $(0, 2)$.

SOLUTION

1. Figure 3.32 indicates that there are two points at a minimum distance from the point $(0, 2)$.

2. You are asked to minimize the distance d. So, you can use the Distance Formula to obtain a primary equation.

$$d = \sqrt{(x - 0)^2 + (y - 2)^2} \qquad \text{Primary equation}$$

3. Using the secondary equation $y = 4 - x^2$, you can rewrite the primary equation as a function of a single variable.

$$d = \sqrt{x^2 + (4 - x^2 - 2)^2} \qquad \text{Substitute } 4 - x^2 \text{ for } y.$$
$$= \sqrt{x^4 - 3x^2 + 4} \qquad \text{Simplify.}$$

Because d is smallest when the expression under the radical is smallest, you simplify the problem by finding the minimum value of $f(x) = x^4 - 3x^2 + 4$.

4. The domain of f is the entire real line.

5. To find the minimum value of $f(x)$, first find the critical numbers of f.

$$f'(x) = 4x^3 - 6x \qquad \text{Find derivative of } f.$$
$$0 = 4x^3 - 6x \qquad \text{Set derivative equal to 0.}$$
$$0 = 2x(2x^2 - 3) \qquad \text{Factor.}$$
$$x = 0, \, x = \sqrt{\tfrac{3}{2}}, \, x = -\sqrt{\tfrac{3}{2}} \qquad \text{Critical numbers}$$

By the First-Derivative Test, you can conclude that $x = 0$ yields a relative maximum, whereas both $\sqrt{3/2}$ and $-\sqrt{3/2}$ yield a minimum. So, on the graph of $y = 4 - x^2$, the points that are closest to the point $(0, 2)$ are

$$\left(\sqrt{\tfrac{3}{2}}, \tfrac{5}{2}\right) \quad \text{and} \quad \left(-\sqrt{\tfrac{3}{2}}, \tfrac{5}{2}\right).$$

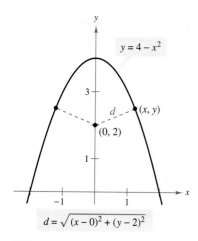

$d = \sqrt{(x - 0)^2 + (y - 2)^2}$

FIGURE 3.32

✓ **CHECKPOINT 3**

Find the points on the graph of $y = 4 - x^2$ that are closest to $(0, 3)$. ∎

Algebra Review

For help with the algebra in Example 3, see Example 1(c) in the *Chapter 3 Algebra Review*, on page 249.

STUDY TIP

To confirm the result in Example 3, try computing the distances between several points on the graph of $y = 4 - x^2$ and the point $(0, 2)$. For instance, the distance between $(1, 3)$ and $(0, 2)$ is

$$d = \sqrt{(0 - 1)^2 + (2 - 3)^2} = \sqrt{2} \approx 1.414.$$

Note that this is greater than the distance between $\left(\sqrt{3/2}, 5/2\right)$ and $(0, 2)$, which is

$$d = \sqrt{\left(0 - \sqrt{\tfrac{3}{2}}\right)^2 + \left(2 - \tfrac{5}{2}\right)^2} = \sqrt{\tfrac{7}{4}} \approx 1.323.$$

Example 4 Fencing a Study Plot

An ecologist has 500 meters of fencing to enclose a rectangular study plot. What should the dimensions of the plot be to maximize the enclosed area?

SOLUTION

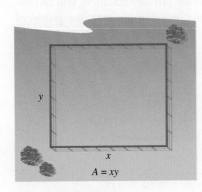

FIGURE 3.33

1. A diagram of the plot is shown in Figure 3.33.

2. Letting A be the area to be maximized, the primary equation is

   $A = xy.$ Primary equation

3. The perimeter of the rectangle is given by

   $500 = 2x + 2y.$ Secondary equation

   Solving this equation for y produces

   $y = 250 - x.$

   By substituting this into the primary equation, you obtain

   $A = x(250 - x)$ Write as a function of one variable.

   $= 250x - x^2$ Multiply.

4. Because y must be positive, $250 - x > 0$ or $250 > x$. Also, x must be positive. So, the feasible domain is

   $0 < x < 250.$

5. To find the maximum area, begin by finding the critical numbers of A.

   $\dfrac{dA}{dx} = 250 - 2x$ Find derivative of A.

   $0 = 250 - 2x$ Set derivative equal to 0.

   $2x = 250$ Add $2x$ to each side.

   $x = 125$ Divide each side by 2.

   So, the only critical number is $x = 125$. Using the First-Derivative Test, it follows that A is a maximum when

   $x = 125$ and $y = 250 - 125 = 125.$

   The dimensions of the study plot should be 125 meters by 125 meters.

✓ CHECKPOINT 4

Suppose the ecologist in Example 4 has 600 meters of fencing to enclose the rectangular study plot. What should the dimensions of the plot be to maximize the enclosed area? ■

STUDY TIP

Note that the shape of the study plot obtained in Example 4 is a square. For a rectangle with a given perimeter, the maximum area occurs when the rectangle is a square.

218 **CHAPTER 3** Applications of the Derivative

As applications go, the four examples described in this section are fairly simple, and yet the resulting primary equations are quite complicated. Real-life applications often involve equations that are at least as complex as these four. Remember that one of the main goals of this course is to enable you to use the power of calculus to analyze equations that at first glance seem formidable.

Also remember that once you have found the primary equation, you can use the graph of the equation to help solve the problem. For instance, the graphs of the primary equations in Examples 1 through 4 are shown in Figure 3.34.

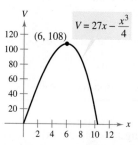
Example 1

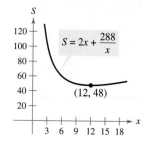
Example 2

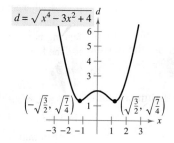
Example 3

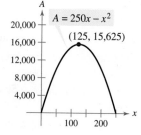
Example 4

FIGURE 3.34

CONCEPT CHECK

1. Complete the following: In an optimization problem, the formula that represents the quantity to be optimized is called the _____ _____.
2. Explain what is meant by the term *feasible domain*.
3. Explain the difference between a primary equation and a secondary equation.
4. In your own words, state the guidelines for solving an optimization problem.

SECTION 3.4 Optimization Problems

Skills Review 3.4

The following warm-up exercises involve skills that were covered in earlier sections. You will use these skills in the exercise set for this section. For additional help, review Section 3.1.

In Exercises 1–4, write a formula for the written statement.

1. The sum of one number and half a second number is 12.
2. The product of one number and twice another is 24.
3. The area of a rectangle is 24 square units.
4. The distance between two points is 10 units.

In Exercises 5–10, find the critical numbers of the function.

5. $y = x^2 + 6x - 9$
6. $y = 2x^3 - x^2 - 4x$
7. $y = 5x + \dfrac{125}{x}$
8. $y = 3x + \dfrac{96}{x^2}$
9. $y = \dfrac{x^2 + 1}{x}$
10. $y = \dfrac{x}{x^2 + 9}$

Exercises 3.4

See www.CalcChat.com for worked-out solutions to odd-numbered exercises.

In Exercises 1–6, find two positive numbers satisfying the given requirements.

1. The sum is 120 and the product is a maximum.
2. The sum is S and the product is a maximum.
3. The sum of the first and twice the second is 36 and the product is a maximum.
4. The sum of the first and twice the second is 100 and the product is a maximum.
5. The product is 192 and the sum is a minimum.
6. The product is 192 and the sum of the first plus three times the second is a minimum.

In Exercises 7 and 8, find the length and width of a rectangle that has the given perimeter and a maximum area.

7. Perimeter: 100 meters
8. Perimeter: P units

In Exercises 9 and 10, find the length and width of a rectangle that has the given area and a minimum perimeter.

9. Area: 64 square feet
10. Area: A square centimeters

11. **Maximum Area** A rancher has 200 feet of fencing to enclose two adjacent rectangular corrals (see figure). What dimensions should be used so that the enclosed area will be a maximum?

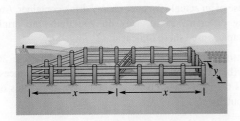

12. **Area** A dairy farmer plans to enclose a rectangular pasture adjacent to a river. To provide enough grass for the herd, the pasture must contain 180,000 square meters. No fencing is required along the river. What dimensions will use the least amount of fencing?

13. **Maximum Volume**
    (a) Verify that each of the rectangular solids shown in the figure has a surface area of 150 square inches.
    (b) Find the volume of each solid.
    (c) Determine the dimensions of a rectangular solid (with a square base) of maximum volume if its surface area is 150 square inches.

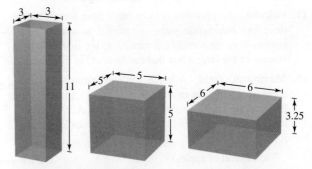

14. **Maximum Volume** A rectangular solid with a square base has a surface area of 337.5 square centimeters.
    (a) Determine the dimensions that yield the maximum volume.
    (b) Find the maximum volume.

15. **Minimum Surface Area** A rectangular solid with a square base has a volume of 8000 cubic inches.
    (a) Determine the dimensions that yield the minimum surface area.
    (b) Find the minimum surface area.

16. **Maximum Area** A Norman window is constructed by adjoining a semicircle to the top of an ordinary rectangular window (see figure). Find the dimensions of a Norman window of maximum area if the total perimeter is 16 feet.

17. **Minimum Surface Area** A net enclosure for golf practice is open at one end (see figure). The volume of the enclosure is $83\frac{1}{3}$ cubic meters. Find the dimensions that require the least amount of netting.

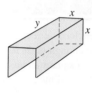 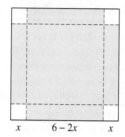

Figure for 17   Figure for 18

18. **Volume** An open box is to be made from a six-inch by six-inch square piece of material by cutting equal squares from the corners and turning up the sides (see figure). Find the volume of the largest box that can be made.

19. **Volume** An open box is to be made from a two-foot by three-foot rectangular piece of material by cutting equal squares from the corners and turning up the sides. Find the volume of the largest box that can be made in this manner.

20. **Maximum Yield** A home gardener estimates that 16 apple trees will have an average yield of 80 apples per tree. But because of the size of the garden, for each additional tree planted, the yield will decrease by four apples per tree.

    (a) How many trees should be planted to maximize the total yield of apples?

    (b) What is the maximum yield?

21. **Maximum Yield** A citrus grower estimates that 90 orange trees per acre will have an average yield of 700 oranges per tree. For each additional tree per acre, the yield will decrease by 25 oranges per tree.

    (a) How many trees should be planted per acre to maximize the yield of oranges?

    (b) What is the maximum yield per acre?

22. **Maximum Area** A rectangle is bounded by the x- and y-axes and the graph of $y = (6 - x)/2$ (see figure). What length and width should the rectangle have so that its area is a maximum?

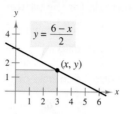 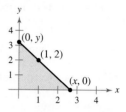

Figure for 22   Figure for 23

23. **Minimum Length** A right triangle is formed in the first quadrant by the x- and y-axes and a line through the point $(1, 2)$ (see figure).

    (a) Write the length L of the hypotenuse as a function of x.

    (T) (b) Use a graphing utility to approximate x graphically such that the length of the hypotenuse is a minimum.

    (c) Find the vertices of the triangle such that its area is a minimum.

24. **Maximum Area** A rectangle is bounded by the x-axis and the semicircle

    $$y = \sqrt{25 - x^2}$$

    (see figure). What length and width should the rectangle have so that its area is a maximum?

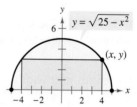

25. **Area** Find the dimensions of the largest rectangle that can be inscribed in a semicircle of radius r. (See Exercise 24.)

26. **MAKE A DECISION: MINIMUM AMOUNT OF MATERIAL** An energy drink container has the shape of a right circular cylinder. The container holds 12 fluid ounces (1 fluid ounce is approximately 1.80469 cubic inches).

    (a) What container dimensions use a minimum amount of construction material?

    (b) What is the minimum amount of construction material?

    (c) Compare the dimensions you obtained in part (a) with those of an actual 12-ounce energy drink container. Are they different? If so, why do you think the drink company used these dimensions?

27. **Minimum Amount of Material** Repeat Exercise 26 (a) and (b) if the container has a volume of 16 fluid ounces. What is the relationship between the height and radius of each container?

In Exercises 28–31, find the points on the graph of the function that are closest to the given point.

28. $f(x) = x^2$, $\left(2, \dfrac{1}{2}\right)$

29. $f(x) = (x + 1)^2$, $(5, 3)$

30. $f(x) = \sqrt{x}$, $(4, 0)$

31. $f(x) = \sqrt{x - 8}$, $(2, 0)$

32. Maximum Volume A rectangular package to be sent by a postal service can have a maximum combined length and girth (perimeter of a cross section) of 108 inches. Find the dimensions of the package with maximum volume. Assume that the package's dimensions are x by x by y (see figure).

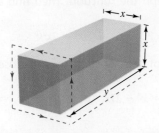

33. Minimum Surface Area A solid is formed by adjoining a hemisphere to one end of a right circular cylinder. The total volume of the solid is 12 cubic inches. Find the radius of the cylinder that produces the minimum surface area.

34. Maximum Volume A solid of the shape described in Exercise 33 has a total surface area of 1000 square centimeters. Find the radius of the cylinder that produces the maximum volume.

35. Minimum Area The sum of the perimeters of a circle and a square is 16. Find the dimensions of the circle and square that produce a minimum total area.

36. Minimum Area The sum of the perimeters of an equilateral triangle and a square is 10. Find the dimensions of the triangle and square that produce a minimum total area.

37. Minimum Time You are in a boat 2 miles from the nearest point on the coast. You are to go to point Q, located 3 miles down the coast and 1 mile inland (see figure). You can row at a rate of 2 miles per hour and you can walk at a rate of 4 miles per hour. Toward what point on the coast should you row in order to reach point Q in the least time?

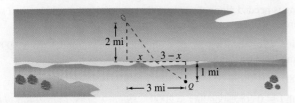

38. Maximum Area An indoor physical fitness room consists of a rectangular region with a semicircle on each end. The perimeter of the room is to be a 200-meter running track. Find the dimensions that will make the area of the rectangular region as large as possible.

39. Farming A strawberry farmer will receive $30 per bushel of strawberries during the first week of harvesting. Each week after that, the value will drop $0.80 per bushel. The farmer estimates that there are approximately 120 bushels of strawberries in the fields, and that the crop is increasing at a rate of four bushels per week. When should the farmer harvest the strawberries to maximize their value? How many bushels of strawberries will yield the maximum value? What is the maximum value of the strawberries?

40. Beam Strength A wooden beam has a rectangular cross section of height h and width w (see figure). The strength S of the beam is directly proportional to its width and the square of its height. What are the dimensions of the strongest beam that can be cut from a round log of diameter 24 inches? (*Hint:* $S = kh^2w$, where $k > 0$ is the proportionality constant.)

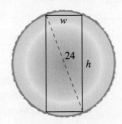

41. Area Four feet of wire is to be used to form a square and a circle.

(a) Express the sum of the areas of the square and the circle as a function A of the side of the square x.

(b) What is the domain of A?

(c) Use a graphing utility to graph A on its domain.

(d) How much wire should be used for the square and how much for the circle in order to enclose the least total area? the greatest total area?

42. Minimum Length Two posts, one 12 feet high and the other 28 feet high, stand 30 feet apart. They are to be secured by two wires, attached to a single stake, running from ground level to the top of each post (see figure). Where should the stake be placed to use the least amount of wire?

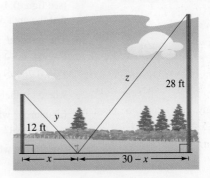

Mid-Chapter Quiz

Take this quiz as you would take a quiz in class. When you are done, check your work against the answers given in the back of the book.

In Exercises 1–3, find the critical numbers of the function and the open intervals on which the function is increasing or decreasing.

1. $f(x) = x^2 - 6x + 1$ **2.** $f(x) = 2x^3 + 12x^2$ **3.** $f(x) = \dfrac{1}{x^2 + 2}$

(T) In Exercises 4–6, use a graphing utility to graph the function. Then find all relative extrema of the function.

4. $f(x) = x^3 + 3x^2 - 5$ **5.** $f(x) = x^4 - 8x^2 + 3$ **6.** $f(x) = 2x^{2/3}$

In Exercises 7–9, find the absolute extrema of the function on the closed interval.

7. $f(x) = x^2 + 2x - 8$, $[-2, 1]$ **8.** $f(x) = x^3 - 27x$, $[-4, 4]$

9. $f(x) = \dfrac{x}{x^2 + 1}$, $[0, 2]$

In Exercises 10 and 11, find the point(s) of inflection of the graph of the function. Then determine the open intervals on which the graph of the function is concave upward or concave downward.

10. $f(x) = x^3 - 6x^2 + 7x$ **11.** $f(x) = x^4 - 24x^2$

In Exercises 12 and 13, use the Second-Derivative Test to find all relative extrema of the function.

12. $f(x) = 2x^3 + 3x^2 - 12x + 16$ **13.** $f(x) = \dfrac{x^2 + 1}{x}$

14. The number C of civil rights cases pending in the U.S. district courts from 2000 through 2005 can be modeled by

$$C = -166.472t^3 + 1013.92t^2 - 1642.9t + 44{,}316, \quad 0 \le t \le 5$$

where t is the year, with $t = 0$ corresponding to 2000. *(Source: Administrative Office of the U. S. Courts)*

(T) (a) Use a graphing utility to graph the model.

(b) Use the second derivative to determine the concavity of C.

(c) Find the point of inflection of the graph of C and interpret its meaning.

15. A gardener has 200 feet of fencing to enclose a rectangular garden adjacent to a river (see figure). No fencing is needed along the river. What dimensions should be used so that the area of the garden will be a maximum?

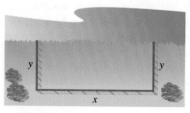

Figure for 15

16. The resident population P of the District of Columbia (in thousands) from 1999 through 2005 can be modeled by

$$P = 0.2694t^3 - 2.048t^2 - 0.73t + 571.9, \quad -1 \le t \le 5$$

where t is the year, with $t = 0$ corresponding to 2000. *(Source: U.S. Census Bureau)*

(a) During which year, from 1999 through 2005, was the population the greatest? the least?

(b) During which year(s) was the population increasing? decreasing?

Section 3.5
Asymptotes

- Find the vertical asymptotes of functions and find infinite limits.
- Find the horizontal asymptotes of functions and find limits at infinity.
- Use asymptotes to answer questions about real-life situations.

Vertical Asymptotes and Infinite Limits

In the first three sections of this chapter, you studied ways in which you can use calculus to help analyze the graph of a function. In this section, you will study another valuable aid to curve sketching: the determination of vertical and horizontal asymptotes.

Recall from Section 1.5, Example 10, that the function

$$f(x) = \frac{3}{x-2}$$

is unbounded as x approaches 2 (see Figure 3.35). This type of behavior is described by saying that the line $x = 2$ is a **vertical asymptote** of the graph of f. The type of limit in which $f(x)$ approaches infinity (or negative infinity) as x approaches c from the left or from the right is an **infinite limit**. The infinite limits for the function $f(x) = 3/(x-2)$ can be written as

$$\lim_{x \to 2^-} \frac{3}{x-2} = -\infty$$

and

$$\lim_{x \to 2^+} \frac{3}{x-2} = \infty.$$

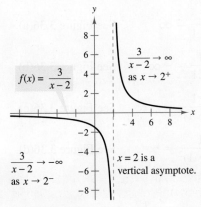

FIGURE 3.35

Definition of Vertical Asymptote

If $f(x)$ approaches infinity (or negative infinity) as x approaches c from the right or from the left, then the line $x = c$ is a **vertical asymptote** of the graph of f.

TECHNOLOGY

When you use a graphing utility to graph a function that has a vertical asymptote, the utility may try to connect separate branches of the graph. For instance, the figure at the right shows the graph of

$$f(x) = \frac{3}{x-2}$$

on a graphing calculator.

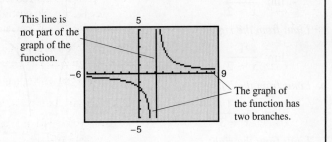

224 **CHAPTER 3** Applications of the Derivative

TECHNOLOGY

Use a spreadsheet or table to verify the results shown in Example 1. (Consult the user's manual for a spreadsheet software program for specific instructions on how to create a table.) For instance, in Example 1(a), notice that the values of $f(x) = 1/(x - 1)$ decrease and increase without bound as x gets closer and closer to 1 from the left and the right.

x Approaches 1 from the Left

x	$f(x) = 1/(x - 1)$
0	-1
0.9	-10
0.99	-100
0.999	-1000
0.9999	$-10{,}000$

x Approaches 1 from the Right

x	$f(x) = 1/(x - 1)$
2	1
1.1	10
1.01	100
1.001	1000
1.0001	10,000

✓ **CHECKPOINT 1**

Find each limit.

a. *Limit from the left*

$$\lim_{x \to 2^-} \frac{1}{x - 2}$$

Limit from the right

$$\lim_{x \to 2^+} \frac{1}{x - 2}$$

b. *Limit from the left*

$$\lim_{x \to -3^-} \frac{-1}{x + 3}$$

Limit from the right

$$\lim_{x \to -3^+} \frac{-1}{x + 3}$$

One of the most common instances of a vertical asymptote is the graph of a *rational function*—that is, a function of the form $f(x) = p(x)/q(x)$, where $p(x)$ and $q(x)$ are polynomials. If c is a real number such that $q(c) = 0$ and $p(c) \neq 0$, the graph of f has a vertical asymptote at $x = c$. Example 1 shows four cases.

Example 1 Finding Infinite Limits

Find each limit.

Limit from the left   *Limit from the right*

a. $\lim\limits_{x \to 1^-} \dfrac{1}{x - 1} = -\infty$ $\lim\limits_{x \to 1^+} \dfrac{1}{x - 1} = \infty$ See Figure 3.36(a).

b. $\lim\limits_{x \to 1^-} \dfrac{-1}{x - 1} = \infty$ $\lim\limits_{x \to 1^+} \dfrac{-1}{x - 1} = -\infty$ See Figure 3.36(b).

c. $\lim\limits_{x \to 1^-} \dfrac{-1}{(x - 1)^2} = -\infty$ $\lim\limits_{x \to 1^+} \dfrac{-1}{(x - 1)^2} = -\infty$ See Figure 3.36(c).

d. $\lim\limits_{x \to 1^-} \dfrac{1}{(x - 1)^2} = \infty$ $\lim\limits_{x \to 1^+} \dfrac{1}{(x - 1)^2} = \infty$ See Figure 3.36(d).

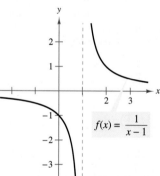

(a)   (b)

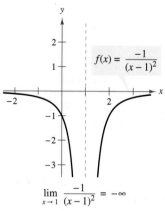

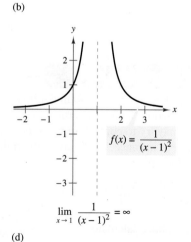

(c)   (d)

FIGURE 3.36

SECTION 3.5 Asymptotes 225

Each of the graphs in Example 1 has only one vertical asymptote. As shown in the next example, the graph of a rational function can have more than one vertical asymptote.

Example 2 Finding Vertical Asymptotes

Find the vertical asymptotes of the graph of

$$f(x) = \frac{x + 2}{x^2 - 2x}.$$

SOLUTION The possible vertical asymptotes correspond to the x-values for which the denominator is zero.

$x^2 - 2x = 0$ Set denominator equal to 0.
$x(x - 2) = 0$ Factor.
$x = 0, x = 2$ Zeros of denominator

Because the numerator of f is not zero at either of these x-values, you can conclude that the graph of f has two vertical asymptotes—one at $x = 0$ and one at $x = 2$, as shown in Figure 3.37.

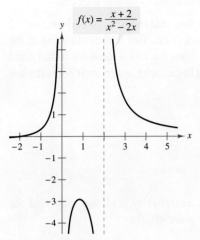

FIGURE 3.37 Vertical Asymptotes at $x = 0$ and $x = 2$

✓ **CHECKPOINT 2**

Find the vertical asymptote(s) of the graph of

$$f(x) = \frac{x + 4}{x^2 - 4x}.$$ ■

Example 3 Finding Vertical Asymptotes

Find the vertical asymptotes of the graph of

$$f(x) = \frac{x^2 + 2x - 8}{x^2 - 4}.$$

SOLUTION First factor the numerator and denominator. Then divide out like factors.

$f(x) = \dfrac{x^2 + 2x - 8}{x^2 - 4}$ Write original function.

$= \dfrac{(x + 4)(x - 2)}{(x + 2)(x - 2)}$ Factor numerator and denominator.

$= \dfrac{(x + 4)\cancel{(x - 2)}}{(x + 2)\cancel{(x - 2)}}$ Divide out like factors.

$= \dfrac{x + 4}{x + 2}, \quad x \neq 2$ Simplify.

For all values of x other than $x = 2$, the graph of this simplified function is the same as the graph of f. So, you can conclude that the graph of f has only one vertical asymptote. This occurs at $x = -2$, as shown in Figure 3.38.

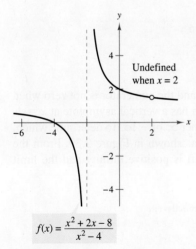

FIGURE 3.38 Vertical Asymptote at $x = -2$

✓ **CHECKPOINT 3**

Find the vertical asymptote(s) of the graph of

$$f(x) = \frac{x^2 + 4x + 3}{x^2 - 9}.$$ ■

From Example 3, you know that the graph of

$$f(x) = \frac{x^2 + 2x - 8}{x^2 - 4}$$

has a vertical asymptote at $x = -2$. This implies that the limit of $f(x)$ as $x \to -2$ from the right (or from the left) is either ∞ or $-\infty$. But without looking at the graph, how can you determine that the limit from the left is *negative* infinity and the limit from the right is *positive* infinity? That is, why is the limit from the left

$$\lim_{x \to -2^-} \frac{x^2 + 2x - 8}{x^2 - 4} = -\infty \qquad \text{Limit from the left}$$

and why is the limit from the right

$$\lim_{x \to -2^+} \frac{x^2 + 2x - 8}{x^2 - 4} = \infty? \qquad \text{Limit from the right}$$

It is cumbersome to determine these limits analytically, and you may find the graphical method shown in Example 4 to be more efficient.

Example 4 Determining Infinite Limits

Find the limits.

$$\lim_{x \to 1^-} \frac{x^2 - 3x}{x - 1} \quad \text{and} \quad \lim_{x \to 1^+} \frac{x^2 - 3x}{x - 1}$$

SOLUTION Begin by considering the function

$$f(x) = \frac{x^2 - 3x}{x - 1}.$$

Because the denominator is zero when $x = 1$ and the numerator is not zero when $x = 1$, it follows that the graph of the function has a vertical asymptote at $x = 1$. This implies that each of the given limits is either ∞ or $-\infty$. To determine which, use a graphing utility to graph the function, as shown in Figure 3.39. From the graph, you can see that the limit from the left is positive infinity and the limit from the right is negative infinity. That is,

$$\lim_{x \to 1^-} \frac{x^2 - 3x}{x - 1} = \infty \qquad \text{Limit from the left}$$

and

$$\lim_{x \to 1^+} \frac{x^2 - 3x}{x - 1} = -\infty. \qquad \text{Limit from the right}$$

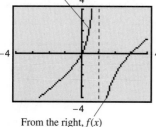

From the left, $f(x)$ approaches positive infinity.

From the right, $f(x)$ approaches negative infinity.

FIGURE 3.39

STUDY TIP

In Example 4, try evaluating $f(x)$ at x-values that are just barely to the left of 1. You will find that you can make the values of $f(x)$ arbitrarily large by choosing x sufficiently close to 1. For instance, $f(0.99999) = 199,999$.

✓ CHECKPOINT 4

Find the limits.

$$\lim_{x \to 2^-} \frac{x^2 - 4x}{x - 2} \quad \text{and} \quad \lim_{x \to 2^+} \frac{x^2 - 4x}{x - 2}$$

Then verify your solution by graphing the function. ∎

SECTION 3.5 Asymptotes

Horizontal Asymptotes and Limits at Infinity

Another type of limit, called a **limit at infinity,** specifies a finite value approached by a function as x increases (or decreases) without bound.

> **Definition of Horizontal Asymptote**
>
> If f is a function and L_1 and L_2 are real numbers, the statements
>
> $$\lim_{x \to \infty} f(x) = L_1 \quad \text{and} \quad \lim_{x \to -\infty} f(x) = L_2$$
>
> denote **limits at infinity.** The lines $y = L_1$ and $y = L_2$ are **horizontal asymptotes** of the graph of f.

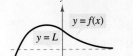

FIGURE 3.40

Figure 3.40 shows two ways in which the graph of a function can approach one or more horizontal asymptotes. Note that it is possible for the graph of a function to cross its horizontal asymptote.

Limits at infinity share many of the properties of limits discussed in Section 1.5. When finding horizontal asymptotes, you can use the property that

$$\lim_{x \to \infty} \frac{1}{x^r} = 0, \quad r > 0 \quad \text{and} \quad \lim_{x \to -\infty} \frac{1}{x^r} = 0, \quad r > 0.$$

(The second limit assumes that x^r is defined when $x < 0$.)

Example 5 Finding Limits at Infinity

Find the limit: $\lim_{x \to \infty} \left(5 - \dfrac{2}{x^2}\right)$.

SOLUTION

$$\begin{aligned}
\lim_{x \to \infty} \left(5 - \frac{2}{x^2}\right) &= \lim_{x \to \infty} 5 - \lim_{x \to \infty} \frac{2}{x^2} && \lim_{x \to \infty}[f(x) - g(x)] = \lim_{x \to \infty} f(x) - \lim_{x \to \infty} g(x) \\
&= \lim_{x \to \infty} 5 - 2\left(\lim_{x \to \infty} \frac{1}{x^2}\right) && \lim_{x \to \infty} cf(x) = c \lim_{x \to \infty} f(x) \\
&= 5 - 2(0) \\
&= 5
\end{aligned}$$

You can verify this limit by sketching the graph of

$$f(x) = 5 - \frac{2}{x^2}$$

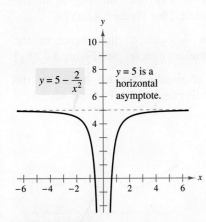

FIGURE 3.41

as shown in Figure 3.41. Note that the graph has $y = 5$ as a horizontal asymptote to the right. By evaluating the limit of $f(x)$ as $x \to -\infty$, you can show that this line is also a horizontal asymptote to the left.

✓ CHECKPOINT 5

Find the limit: $\lim_{x \to \infty} \left(2 + \dfrac{5}{x^2}\right)$.

TECHNOLOGY

Some functions have two horizontal asymptotes: one to the right and one to the left. For instance, try sketching the graph of

$$f(x) = \frac{x}{\sqrt{x^2 + 1}}.$$

What horizontal asymptotes does the function appear to have?

Horizontal Asymptotes of Rational Functions

Let $f(x) = p(x)/q(x)$ be a rational function.

1. If the degree of the numerator is less than the degree of the denominator, then $y = 0$ is a horizontal asymptote of the graph of f (to the left and to the right).

2. If the degree of the numerator is equal to the degree of the denominator, then $y = a/b$ is a horizontal asymptote of the graph of f (to the left and to the right), where a and b are the leading coefficients of $p(x)$ and $q(x)$, respectively.

3. If the degree of the numerator is greater than the degree of the denominator, then the graph of f has no horizontal asymptote.

Example 6 Finding Horizontal Asymptotes

Find the horizontal asymptote of the graph of each function.

a. $y = \dfrac{-2x + 3}{3x^2 + 1}$ b. $y = \dfrac{-2x^2 + 3}{3x^2 + 1}$ c. $y = \dfrac{-2x^3 + 3}{3x^2 + 1}$

SOLUTION

a. Because the degree of the numerator is less than the degree of the denominator, $y = 0$ is a horizontal asymptote. [See Figure 3.42(a).]

b. Because the degree of the numerator is equal to the degree of the denominator, the line $y = -\frac{2}{3}$ is a horizontal asymptote. [See Figure 3.42(b).]

c. Because the degree of the numerator is greater than the degree of the denominator, the graph has no horizontal asymptote. [See Figure 3.42(c).]

✓ CHECKPOINT 6

Find the horizontal asymptote of the graph of each function.

a. $y = \dfrac{2x + 1}{4x^2 + 5}$

b. $y = \dfrac{2x^2 + 1}{4x^2 + 5}$

c. $y = \dfrac{2x^3 + 1}{4x^2 + 5}$

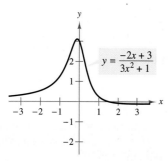
(a) $y = 0$ is a horizontal asymptote.

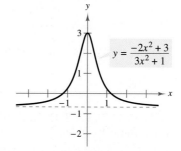
(b) $y = -\frac{2}{3}$ is a horizontal asymptote.

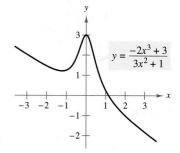
(c) No horizontal asymptote

FIGURE 3.42

SECTION 3.5 Asymptotes

Since the 1980s, industries in the United States have spent billions of dollars to reduce air pollution.

Application

Example 7 Modeling Smokestack Emission

A manufacturing plant has determined that the cost C (in dollars) of removing $p\%$ of the smokestack pollutants of its main smokestack is modeled by

$$C = \frac{80{,}000p}{100 - p}, \quad 0 \le p < 100.$$

Find the vertical asymptote of this function and interpret its meaning.

SOLUTION The graph of the cost function is shown in Figure 3.43. From the graph, you can see that $p = 100$ is the vertical asymptote. This means that as the plant attempts to remove higher and higher percents of the pollutants, the cost increases dramatically. For instance, the cost of removing 85% of the pollutants is

$$C = \frac{80{,}000(85)}{100 - 85} \approx \$453{,}333 \qquad \text{Cost of 85\% removal}$$

but the cost of removing 90% is

$$C = \frac{80{,}000(90)}{100 - 90} = \$720{,}000. \qquad \text{Cost of 90\% removal}$$

✓ CHECKPOINT 7

According to the cost function in Example 7, is it possible to remove 100% of the smokestack pollutants? Why or why not? ∎

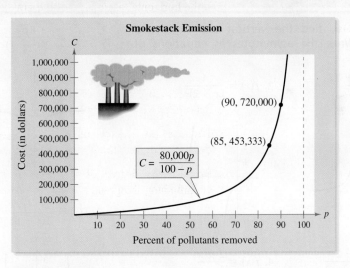

FIGURE 3.43

CONCEPT CHECK

1. Complete the following: If $f(x) \to \pm\infty$ as $x \to c$ from the right or the left, then the line $x = c$ is a _____ _____ of the graph of f.
2. Describe in your own words what is meant by $\lim\limits_{x \to \infty} f(x) = 4$.
3. Describe in your own words what is meant by $\lim\limits_{x \to -\infty} f(x) = 2$.
4. Complete the following: Given a rational function f, if the degree of the numerator is less than the degree of the denominator, then _____ is a horizontal asymptote of the graph of f (to the left and to the right).

Skills Review 3.5

The following warm-up exercises involve skills that were covered in earlier sections. You will use these skills in the exercise set for this section. For additional help, review Sections 1.4 and 1.5.

In Exercises 1–4, determine the domain and range of the function.

1. $f(x) = \dfrac{1}{x-5}$

2. $f(x) = \dfrac{1}{x^2 - 4}$

3. $f(x) = \sqrt{x+3}$

4. $f(x) = \dfrac{1}{\sqrt{x-3}}$

In Exercises 5–12, find the limit.

5. $\lim\limits_{x \to 2} (x+1)$

6. $\lim\limits_{x \to -1} (3x+4)$

7. $\lim\limits_{x \to -3} \dfrac{2x^2 + x - 15}{x+3}$

8. $\lim\limits_{x \to 2} \dfrac{3x^2 - 8x + 4}{x-2}$

9. $\lim\limits_{x \to 2^+} \dfrac{x^2 - 5x + 6}{x^2 - 4}$

10. $\lim\limits_{x \to 1^-} \dfrac{x^2 - 6x + 5}{x^2 - 1}$

11. $\lim\limits_{x \to 0^+} \sqrt{x}$

12. $\lim\limits_{x \to 1^+} \left(x + \sqrt{x-1}\right)$

Exercises 3.5

See www.CalcChat.com for worked-out solutions to odd-numbered exercises.

In Exercises 1–8, find the vertical and horizontal asymptotes. Write the asymptotes as equations of lines.

1. $f(x) = \dfrac{x^2 + 1}{x^2}$

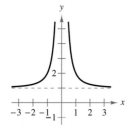

2. $f(x) = \dfrac{4}{(x-2)^3}$

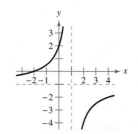

3. $f(x) = \dfrac{x^2 - 2}{x^2 - x - 2}$

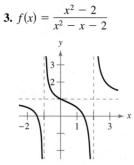

4. $f(x) = \dfrac{2+x}{1-x}$

5. $f(x) = \dfrac{3x^2}{2(x^2+1)}$

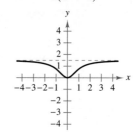

6. $f(x) = \dfrac{-4x}{x^2 + 4}$

7. $f(x) = \dfrac{x^2 - 1}{2x^2 - 8}$

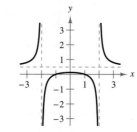

8. $f(x) = \dfrac{x^2 + 1}{x^3 - 8}$

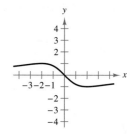

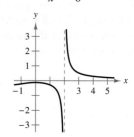

In Exercises 9–12, match the function with its graph. Use horizontal asymptotes as an aid. [The graphs are labeled (a)–(d).]

(a)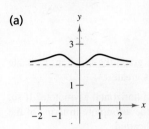

(b)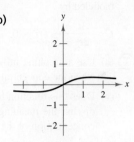

(c)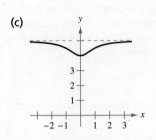

(d)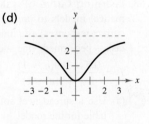

9. $f(x) = \dfrac{3x^2}{x^2 + 2}$

10. $f(x) = \dfrac{x}{x^2 + 2}$

11. $f(x) = 2 + \dfrac{x^2}{x^4 + 1}$

12. $f(x) = 5 - \dfrac{1}{x^2 + 1}$

In Exercises 13–20, find the limit.

13. $\lim\limits_{x \to -2^-} \dfrac{1}{(x + 2)^2}$

14. $\lim\limits_{x \to -2^-} \dfrac{1}{x + 2}$

15. $\lim\limits_{x \to 3^+} \dfrac{x - 4}{x - 3}$

16. $\lim\limits_{x \to 1^+} \dfrac{2 + x}{1 - x}$

17. $\lim\limits_{x \to 4^-} \dfrac{x^2}{x^2 - 16}$

18. $\lim\limits_{x \to 4} \dfrac{x^2}{x^2 + 16}$

19. $\lim\limits_{x \to 0^-} \left(1 + \dfrac{1}{x}\right)$

20. $\lim\limits_{x \to 0^-} \left(x^2 - \dfrac{1}{x}\right)$

(T)(S) In Exercises 21–24, use a graphing utility or spreadsheet software program to complete the table. Then use the result to estimate the limit of $f(x)$ as x approaches infinity.

x	10^0	10^1	10^2	10^3	10^4	10^5	10^6
$f(x)$							

21. $f(x) = \dfrac{x + 1}{x\sqrt{x}}$

22. $f(x) = \dfrac{2x^2}{x + 1}$

23. $f(x) = \dfrac{x^2 - 1}{0.02x^2}$

24. $f(x) = \dfrac{3x^2}{0.1x^2 + 1}$

(T)(S) In Exercises 25 and 26, use a graphing utility or a spreadsheet software program to complete the table and use the result to estimate the limit of $f(x)$ as x approaches infinity and as x approaches negative infinity.

x	-10^6	-10^4	-10^2	10^0	10^2	10^4	10^6
$f(x)$							

25. $f(x) = \dfrac{2x}{\sqrt{x^2 + 4}}$

26. $f(x) = x - \sqrt{x(x - 1)}$

In Exercises 27 and 28, find $\lim\limits_{x \to \infty} h(x)$, if possible.

27. $f(x) = 5x^3 - 3$

(a) $h(x) = \dfrac{f(x)}{x^2}$ (b) $h(x) = \dfrac{f(x)}{x^3}$ (c) $h(x) = \dfrac{f(x)}{x^4}$

28. $f(x) = 3x^2 + 7$

(a) $h(x) = \dfrac{f(x)}{x}$ (b) $h(x) = \dfrac{f(x)}{x^2}$ (c) $h(x) = \dfrac{f(x)}{x^3}$

In Exercises 29 and 30, find each limit, if possible.

29. (a) $\lim\limits_{x \to \infty} \dfrac{x^2 + 2}{x^3 - 1}$

(b) $\lim\limits_{x \to \infty} \dfrac{x^2 + 2}{x^2 - 1}$

(c) $\lim\limits_{x \to \infty} \dfrac{x^2 + 2}{x - 1}$

30. (a) $\lim\limits_{x \to \infty} \dfrac{3 - 2x}{3x^3 - 1}$

(b) $\lim\limits_{x \to \infty} \dfrac{3 - 2x}{3x - 1}$

(c) $\lim\limits_{x \to \infty} \dfrac{3 - 2x^2}{3x - 1}$

In Exercises 31–40, find the limit.

31. $\lim\limits_{x \to \infty} \dfrac{4x - 3}{2x + 1}$

32. $\lim\limits_{x \to \infty} \dfrac{5x^3 + 1}{10x^3 - 3x^2 + 7}$

33. $\lim\limits_{x \to \infty} \dfrac{3x}{4x^2 - 1}$

34. $\lim\limits_{x \to -\infty} \dfrac{2x^2 - 5x - 12}{1 - 6x - 8x^2}$

35. $\lim\limits_{x \to -\infty} \dfrac{5x^2}{x + 3}$

36. $\lim\limits_{x \to \infty} \dfrac{x^3 - 2x^2 + 3x + 1}{x^2 - 3x + 2}$

37. $\lim\limits_{x \to \infty} (2x - x^{-2})$

38. $\lim\limits_{x \to \infty} (2 - x^{-3})$

39. $\lim\limits_{x \to -\infty} \left(\dfrac{2x}{x - 1} + \dfrac{3x}{x + 1}\right)$

40. $\lim\limits_{x \to \infty} \left(\dfrac{2x^2}{x - 1} + \dfrac{3x}{x + 1}\right)$

In Exercises 41–58, sketch the graph of the equation. Use intercepts, extrema, and asymptotes as sketching aids.

41. $y = \dfrac{3x}{1 - x}$

42. $y = \dfrac{x - 3}{x - 2}$

43. $f(x) = \dfrac{x^2}{x^2 + 9}$

44. $f(x) = \dfrac{x}{x^2 + 4}$

45. $g(x) = \dfrac{x^2}{x^2 - 16}$

46. $g(x) = \dfrac{x}{x^2 - 4}$

47. $xy^2 = 4$

48. $x^2y = 4$

49. $y = \dfrac{2x}{1-x}$ **50.** $y = \dfrac{2x}{1-x^2}$

51. $y = 1 - 3x^{-2}$ **52.** $y = 1 + x^{-1}$

53. $f(x) = \dfrac{1}{x^2 - x - 2}$

54. $f(x) = \dfrac{x - 2}{x^2 - 4x + 3}$

55. $g(x) = \dfrac{x^2 - x - 2}{x - 2}$

56. $g(x) = \dfrac{x^2 - 9}{x + 3}$

57. $y = \dfrac{2x^2 - 6}{(x - 1)^2}$

58. $y = \dfrac{x}{(x + 1)^2}$

59. Temperature The graph shows the temperature T (in degrees Fahrenheit) of an apple pie t seconds after it is removed from an oven and placed on a cooling rack.

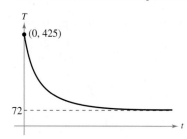

(a) Find $\lim\limits_{t \to 0^+} T$. What does this limit represent?

(b) Find $\lim\limits_{t \to \infty} T$. What does this limit represent?

60. Seizing Drugs The cost C (in millions of dollars) for the federal government to seize $p\%$ of an illegal drug as it enters the country is modeled by

$C = 528p/(100 - p)$, $0 \le p < 100$.

(a) Find the costs of seizing 25%, 50%, and 75%.

(b) Find the limit of C as $p \to 100^-$. Interpret the limit in the context of the problem. Use a graphing utility to verify your result.

61. Removing Pollutants The cost C (in dollars) of removing $p\%$ of the air pollutants in the stack emission of a utility company that burns coal is modeled by

$C = 80{,}000p/(100 - p)$, $0 \le p < 100$.

(a) Find the costs of removing 15%, 50%, and 90%.

(b) Find the limit of C as $p \to 100^-$. Interpret the limit in the context of the problem. Use a graphing utility to verify your result.

62. Drug Concentration The concentration C (in milligrams per milliliter) of a medication in the bloodstream t minutes after sublingual (under the tongue) application is modeled by

$C = \dfrac{3t - 1}{2t^2 + 5}$, $t > 0$.

(a) Use a graphing utility to graph the function. Estimate when the concentration is greatest.

(b) Does this function have a horizontal asymptote? If so, discuss the meaning of the asymptote in terms of the concentration of the medication.

63. Learning Curve Psychologists have developed mathematical models to predict performance P (the percent of correct responses) as a function of n, the number of times a task is performed. One such model is

$P = \dfrac{0.5 + 0.9(n - 1)}{1 + 0.9(n - 1)}$, $n > 0$.

(a) Use a spreadsheet software program to complete the table for the model.

n	1	2	3	4	5	6	7	8	9	10
P										

(b) Find the limit as n approaches infinity.

(c) Use a graphing utility to graph this learning curve and interpret the graph in the context of the problem.

64. Ultraviolet Radiation The amount of time T (in hours) a person with sensitive skin can be exposed to the sun with minimal burning can be modeled by

$T = \dfrac{0.37s + 23.8}{s}$, $0 < s \le 120$

where s is the Sunsor Scale reading. The Sunsor Scale is based on the level of intensity of UVB rays. *(Source: Sunsor, Inc.)*

(a) Find the amounts of time a person with sensitive skin can be exposed to the sun with minimal burning when $s = 10$, $s = 25$, and $s = 100$.

(b) If the model were valid for all $s > 0$, what would be the horizontal asymptote of this function, and what would it represent?

65. Biology: Wildlife Management The state game commission introduces 30 elk into a new state park. The population N of the herd is modeled by

$N = [10(3 + 4t)]/(1 + 0.1t)$

where t is the time in years.

(a) Find the size of the herd after 5, 10, and 25 years.

(b) According to this model, what is the limiting size of the herd as time progresses?

Section 3.6
Curve Sketching: A Summary

- Analyze the graphs of functions.
- Recognize the graphs of simple polynomial functions.

Summary of Curve-Sketching Techniques

It would be difficult to overstate the importance of using graphs in mathematics. Descartes's introduction of analytic geometry contributed significantly to the rapid advances in calculus that began during the mid-seventeenth century.

So far, you have studied several concepts that are useful in analyzing the graph of a function.

- x-intercepts and y-intercepts (Section 1.2)
- Domain and range (Section 1.4)
- Continuity (Section 1.6)
- Differentiability (Section 2.1)
- Relative extrema (Section 3.2)
- Concavity (Section 3.3)
- Points of inflection (Section 3.3)
- Vertical asymptotes (Section 3.5)
- Horizontal asymptotes (Section 3.5)

When you are sketching the graph of a function, either by hand or with a graphing utility, remember that you cannot normally show the *entire* graph. The decision as to which part of the graph to show is crucial. For instance, which of the viewing windows in Figure 3.44 better represents the graph of

$$f(x) = x^3 - 25x^2 + 74x - 20?$$

The lower viewing window gives a more complete view of the graph, but the context of the problem might indicate that the upper view is better. Here are some guidelines for analyzing the graph of a function.

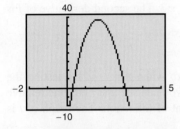
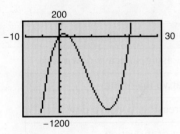

FIGURE 3.44

Guidelines for Analyzing the Graph of a Function

1. Determine the domain and range of the function. If the function models a real-life situation, consider the context.
2. Determine the intercepts and asymptotes of the graph.
3. Locate the x-values where $f'(x)$ and $f''(x)$ are zero or undefined. Use the results to determine where the relative extrema and the points of inflection occur.

In these guidelines, note the importance of *algebra* (as well as calculus) for solving the equations $f(x) = 0$, $f'(x) = 0$, and $f''(x) = 0$.

TECHNOLOGY

Which of the viewing windows best represents the graph of the function

$$f(x) = \frac{x^3 + 8x^2 - 33x}{5}?$$

a. Xmin = -15, Xmax = 1,
Ymin = -10, Ymax = 60
b. Xmin = -10, Xmax = 10,
Ymin = -10, Ymax = 10
c. Xmin = -13, Xmax = 5,
Ymin = -10, Ymax = 60

Example 1 Analyzing a Graph

Analyze the graph of

$$f(x) = x^3 + 3x^2 - 9x + 5.$$ Original function

SOLUTION Begin by finding the intercepts of the graph. This function factors as

$$f(x) = (x - 1)^2(x + 5).$$ Factored form

So, the x-intercepts occur when $x = 1$ and $x = -5$. The derivative is

$$f'(x) = 3x^2 + 6x - 9$$ First derivative
$$= 3(x - 1)(x + 3).$$ Factored form

So, the critical numbers of f are $x = 1$ and $x = -3$. The second derivative of f is

$$f''(x) = 6x + 6$$ Second derivative
$$= 6(x + 1)$$ Factored form

which implies that the second derivative is zero when $x = -1$. By testing the values of $f'(x)$ and $f''(x)$, as shown in the table, you can see that f has one relative minimum, one relative maximum, and one point of inflection. The graph of f is shown in Figure 3.45.

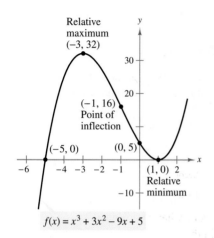

FIGURE 3.45

	$f(x)$	$f'(x)$	$f''(x)$	Characteristics of graph
x in $(-\infty, -3)$		+	−	Increasing, concave downward
$x = -3$	32	0	−	Relative maximum
x in $(-3, -1)$		−	−	Decreasing, concave downward
$x = -1$	16	−	0	Point of inflection
x in $(-1, 1)$		−	+	Decreasing, concave upward
$x = 1$	0	0	+	Relative minimum
x in $(1, \infty)$		+	+	Increasing, concave upward

✓**CHECKPOINT 1**

Analyze the graph of $f(x) = -x^3 + 3x^2 + 9x - 27$.

> **TECHNOLOGY**
>
> In Example 1, you are able to find the zeros of f, f', and f'' algebraically (by factoring). When this is not feasible, you can use a graphing utility to find the zeros. For instance, the function
>
> $$g(x) = x^3 + 3x^2 - 9x + 6$$
>
> is similar to the function in the example, but it does not factor with integer coefficients. Using a graphing utility, you can determine that the function has only one x-intercept, $x \approx -5.0275$.

SECTION 3.6 Curve Sketching: A Summary

Example 2 Analyzing a Graph

Analyze the graph of

$$f(x) = x^4 - 12x^3 + 48x^2 - 64x.$$ Original function

SOLUTION Begin by finding the intercepts of the graph. This function factors as

$$f(x) = x(x^3 - 12x^2 + 48x - 64)$$
$$= x(x - 4)^3.$$ Factored form

So, the x-intercepts occur when $x = 0$ and $x = 4$. The derivative is

$$f'(x) = 4x^3 - 36x^2 + 96x - 64$$ First derivative
$$= 4(x - 1)(x - 4)^2.$$ Factored form

So, the critical numbers of f are $x = 1$ and $x = 4$. The second derivative of f is

$$f''(x) = 12x^2 - 72x + 96$$ Second derivative
$$= 12(x - 4)(x - 2).$$ Factored form

which implies that the second derivative is zero when $x = 2$ and $x = 4$. By testing the values of $f'(x)$ and $f''(x)$, as shown in the table, you can see that f has one relative minimum and two points of inflection. The graph is shown in Figure 3.46.

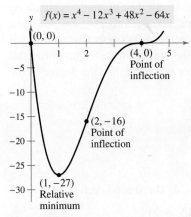

FIGURE 3.46

	$f(x)$	$f'(x)$	$f''(x)$	Characteristics of graph
x in $(-\infty, 1)$		$-$	$+$	Decreasing, concave upward
$x = 1$	-27	0	$+$	Relative minimum
x in $(1, 2)$		$+$	$+$	Increasing, concave upward
$x = 2$	-16	$+$	0	Point of inflection
x in $(2, 4)$		$+$	$-$	Increasing, concave downward
$x = 4$	0	0	0	Point of inflection
x in $(4, \infty)$		$+$	$+$	Increasing, concave upward

✓ **CHECKPOINT 2**

Analyze the graph of $f(x) = x^4 - 4x^3 + 5$. ■

DISCOVERY

A polynomial function of degree n can have at most $n - 1$ relative extrema and at most $n - 2$ points of inflection. For instance, the third-degree polynomial in Example 1 has two relative extrema and one point of inflection. Similarly, the fourth-degree polynomial function in Example 2 has one relative extremum and two points of inflection. Is it possible for a third-degree function to have no relative extrema? Is it possible for a fourth-degree function to have no relative extrema?

DISCOVERY

Show that the function in Example 3 can be rewritten as

$$f(x) = \frac{x^2 - 2x + 4}{x - 2}$$

$$= x + \frac{4}{x - 2}.$$

Use a graphing utility to graph f together with the line $y = x$. How do the two graphs compare as you zoom out? Describe what is meant by a "slant asymptote." Find the slant asymptote of the function $g(x) = \dfrac{x^2 - x - 1}{x - 1}$.

Example 3 Analyzing a Graph

Analyze the graph of

$$f(x) = \frac{x^2 - 2x + 4}{x - 2}.$$ Original function

SOLUTION The y-intercept occurs at $(0, -2)$. Using the Quadratic Formula on the numerator, you can see that there are no x-intercepts. Because the denominator is zero when $x = 2$ (and the numerator is not zero when $x = 2$), it follows that $x = 2$ is a vertical asymptote of the graph. There are no horizontal asymptotes because the degree of the numerator is greater than the degree of the denominator. The derivative is

$$f'(x) = \frac{(x-2)(2x-2) - (x^2 - 2x + 4)}{(x-2)^2}$$ First derivative

$$= \frac{x(x-4)}{(x-2)^2}.$$ Factored form

So, the critical numbers of f are $x = 0$ and $x = 4$. The second derivative is

$$f''(x) = \frac{(x-2)^2(2x-4) - (x^2 - 4x)(2)(x-2)}{(x-2)^4}$$ Second derivative

$$= \frac{(x-2)(2x^2 - 8x + 8 - 2x^2 + 8x)}{(x-2)^4}$$

$$= \frac{8}{(x-2)^3}.$$ Factored form

Because the second derivative has no zeros and because $x = 2$ is not in the domain of the function, you can conclude that the graph has no points of inflection. By testing the values of $f'(x)$ and $f''(x)$, as shown in the table, you can see that f has one relative minimum and one relative maximum. The graph of f is shown in Figure 3.47.

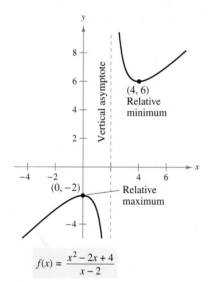

FIGURE 3.47

	$f(x)$	$f'(x)$	$f''(x)$	Characteristics of graph
x in $(-\infty, 0)$		+	−	Increasing, concave downward
$x = 0$	−2	0	−	Relative maximum
x in $(0, 2)$		−	−	Decreasing, concave downward
$x = 2$	Undef.	Undef.	Undef.	Vertical asymptote
x in $(2, 4)$		−	+	Decreasing, concave upward
$x = 4$	6	0	+	Relative minimum
x in $(4, \infty)$		+	+	Increasing, concave upward

✓ CHECKPOINT 3

Analyze the graph of $f(x) = \dfrac{x^2}{x - 1}$. ■

SECTION 3.6 Curve Sketching: A Summary

Example 4 Analyzing a Graph

Analyze the graph of

$$f(x) = \frac{2(x^2 - 9)}{x^2 - 4}.$$ Original function

SOLUTION Begin by writing the function in factored form.

$$f(x) = \frac{2(x - 3)(x + 3)}{(x - 2)(x + 2)}$$ Factored form

The y-intercept is $\left(0, \frac{9}{2}\right)$, and the x-intercepts are $(-3, 0)$ and $(3, 0)$. The graph of f has vertical asymptotes at $x = \pm 2$ and a horizontal asymptote at $y = 2$. The first derivative is

$$f'(x) = \frac{2[(x^2 - 4)(2x) - (x^2 - 9)(2x)]}{(x^2 - 4)^2}$$ First derivative

$$= \frac{20x}{(x^2 - 4)^2}.$$ Factored form

So, the critical number of f is $x = 0$. The second derivative of f is

$$f''(x) = \frac{(x^2 - 4)^2(20) - (20x)(2)(2x)(x^2 - 4)}{(x^2 - 4)^4}$$ Second derivative

$$= \frac{20(x^2 - 4)(x^2 - 4 - 4x^2)}{(x^2 - 4)^4}$$

$$= -\frac{20(3x^2 + 4)}{(x^2 - 4)^3}.$$ Factored form

Because the second derivative has no zeros and $x = \pm 2$ are not in the domain of the function, you can conclude that the graph has no points of inflection. By testing the values of $f'(x)$ and $f''(x)$, as shown in the table, you can see that f has one relative minimum. The graph of f is shown in Figure 3.48.

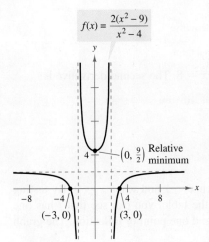

FIGURE 3.48

	$f(x)$	$f'(x)$	$f''(x)$	Characteristics of graph
x in $(-\infty, -2)$		$-$	$-$	Decreasing, concave downward
$x = -2$	Undef.	Undef.	Undef.	Vertical asymptote
x in $(-2, 0)$		$-$	$+$	Decreasing, concave upward
$x = 0$	$\frac{9}{2}$	0	$+$	Relative minimum
x in $(0, 2)$		$+$	$+$	Increasing, concave upward
$x = 2$	Undef.	Undef.	Undef.	Vertical asymptote
x in $(2, \infty)$		$+$	$-$	Increasing, concave downward

✓ CHECKPOINT 4

Analyze the graph of $f(x) = \dfrac{x^2 + 1}{x^2 - 1}$. ∎

TECHNOLOGY

Some graphing utilities will not graph the function in Example 5 properly if the function is entered as

$f(x) = 2x\wedge(5/3) - 5x\wedge(4/3)$.

To correct for this, you can enter the function as

$f(x) = 2(\sqrt[3]{x})\wedge 5 - 5(\sqrt[3]{x})\wedge 4$.

Try entering both functions into a graphing utility to see whether both functions produce correct graphs.

Algebra Review

For help with the algebra in Example 5, see Example 2(a) in the *Chapter 3 Algebra Review*, on page 250.

Example 5 Analyzing a Graph

Analyze the graph of

$$f(x) = 2x^{5/3} - 5x^{4/3}.$$ Original function

SOLUTION Begin by writing the function in factored form.

$$f(x) = x^{4/3}(2x^{1/3} - 5)$$ Factored form

One of the intercepts is $(0, 0)$. A second x-intercept occurs when $2x^{1/3} = 5$.

$$2x^{1/3} = 5$$
$$x^{1/3} = \tfrac{5}{2}$$
$$x = \left(\tfrac{5}{2}\right)^3$$
$$x = \tfrac{125}{8}$$

The first derivative is

$$f'(x) = \tfrac{10}{3}x^{2/3} - \tfrac{20}{3}x^{1/3}$$ First derivative
$$= \tfrac{10}{3}x^{1/3}(x^{1/3} - 2).$$ Factored form

So, the critical numbers of f are $x = 0$ and $x = 8$. The second derivative is

$$f''(x) = \tfrac{20}{9}x^{-1/3} - \tfrac{20}{9}x^{-2/3}$$ Second derivative
$$= \tfrac{20}{9}x^{-2/3}(x^{1/3} - 1)$$
$$= \frac{20(x^{1/3} - 1)}{9x^{2/3}}.$$ Factored form

So, possible points of inflection occur when $x = 1$ and when $x = 0$. By testing the values of $f'(x)$ and $f''(x)$, as shown in the table, you can see that f has one relative maximum, one relative minimum, and one point of inflection. The graph of f is shown in Figure 3.49.

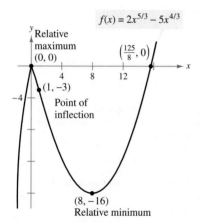

FIGURE 3.49

	$f(x)$	$f'(x)$	$f''(x)$	Characteristics of graph
x in $(-\infty, 0)$		+	−	Increasing, concave downward
$x = 0$	0	0	Undef.	Relative maximum
x in $(0, 1)$		−	−	Decreasing, concave downward
$x = 1$	−3	−	0	Point of inflection
x in $(1, 8)$		−	+	Decreasing, concave upward
$x = 8$	−16	0	+	Relative minimum
x in $(8, \infty)$		+	+	Increasing, concave upward

✓ CHECKPOINT 5

Analyze the graph of

$$f(x) = 2x^{3/2} - 6x^{1/2}.$$

Summary of Simple Polynomial Graphs

A summary of the graphs of polynomial functions of degrees 0, 1, 2, and 3 is shown in Figure 3.50. Because of their simplicity, lower-degree polynomial functions are commonly used as mathematical models.

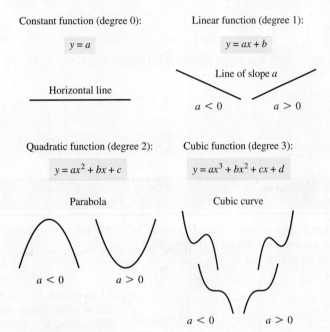

FIGURE 3.50

STUDY TIP

The graph of any cubic polynomial has one point of inflection. The slope of the graph at the point of inflection may be zero or nonzero.

CONCEPT CHECK

1. A fourth-degree polynomial can have at most how many relative extrema?
2. A fourth-degree polynomial can have at most how many points of inflection?
3. Complete the following: A polynomial function of degree *n* can have at most _____ relative extrema.
4. Complete the following: A polynomial function of degree *n* can have at most _____ points of inflection.

Skills Review 3.6

The following warm-up exercises involve skills that were covered in earlier sections. You will use these skills in the exercise set for this section. For additional help, review Sections 3.1 and 3.5.

In Exercises 1–4, find the vertical and horizontal asymptotes of the graph.

1. $f(x) = \dfrac{1}{x^2}$ **2.** $f(x) = \dfrac{8}{(x-2)^2}$ **3.** $f(x) = \dfrac{40x}{x+3}$ **4.** $f(x) = \dfrac{x^2 - 3}{x^2 - 4x + 3}$

In Exercises 5–10, determine the open intervals on which the function is increasing or decreasing.

5. $f(x) = x^2 + 4x + 2$ **6.** $f(x) = -x^2 - 8x + 1$ **7.** $f(x) = x^3 - 3x + 1$

8. $f(x) = \dfrac{-x^3 + x^2 - 1}{x^2}$ **9.** $f(x) = \dfrac{x-2}{x-1}$ **10.** $f(x) = -x^3 - 4x^2 + 3x + 2$

Exercises 3.6

See www.CalcChat.com for worked-out solutions to odd-numbered exercises.

In Exercises 1–22, sketch the graph of the function. Choose a scale that allows all relative extrema and points of inflection to be identified on the graph.

1. $y = -x^2 - 2x + 3$
2. $y = 2x^2 - 4x + 1$
3. $y = x^3 - 4x^2 + 6$
4. $y = -x^3 + x - 2$
5. $y = 2 - x - x^3$
6. $y = x^3 + 3x^2 + 3x + 2$
7. $y = 3x^3 - 9x + 1$
8. $y = -4x^3 + 6x^2$
9. $y = 3x^4 + 4x^3$
10. $y = x^4 - 2x^2$
11. $y = x^3 - 6x^2 + 3x + 10$
12. $y = -x^3 + 3x^2 + 9x - 2$
13. $y = x^4 - 8x^3 + 18x^2 - 16x + 5$
14. $y = x^4 - 4x^3 + 16x - 16$
15. $y = x^4 - 4x^3 + 16x$
16. $y = x^5 + 1$
17. $y = x^5 - 5x$
18. $y = (x-1)^5$
19. $y = \dfrac{x^2 + 1}{x}$
20. $y = \dfrac{x+2}{x}$
21. $y = \begin{cases} x^2 + 1, & x \le 0 \\ 1 - 2x, & x > 0 \end{cases}$
22. $y = \begin{cases} x^2 + 4, & x < 0 \\ 4 - x, & x \ge 0 \end{cases}$

(T) In Exercises 23–34, use a graphing utility to graph the function. Choose a window that allows all relative extrema and points of inflection to be identified on the graph.

23. $y = \dfrac{x^2}{x^2 + 3}$
24. $y = \dfrac{x}{x^2 + 1}$
25. $y = 3x^{2/3} - 2x$
26. $y = 3x^{2/3} - x^2$
27. $y = 1 - x^{2/3}$
28. $y = (1 - x)^{2/3}$
29. $y = x^{1/3} + 1$
30. $y = x^{-1/3}$
31. $y = x^{5/3} - 5x^{2/3}$
32. $y = x^{4/3}$
33. $y = x\sqrt{x^2 - 9}$
34. $y = \dfrac{x}{\sqrt{x^2 - 4}}$

In Exercises 35–44, sketch the graph of the function. Label the intercepts, relative extrema, points of inflection, and asymptotes. Then state the domain of the function.

35. $y = \dfrac{5 - 3x}{x - 2}$
36. $y = \dfrac{x^2 + 1}{x^2 - 9}$
37. $y = \dfrac{2x}{x^2 - 1}$
38. $y = \dfrac{x^2 - 6x + 12}{x - 4}$
39. $y = x\sqrt{4 - x}$
40. $y = x\sqrt{4 - x^2}$
41. $y = \dfrac{x - 3}{x}$
42. $y = x + \dfrac{32}{x^2}$
43. $y = \dfrac{x^3}{x^3 - 1}$
44. $y = \dfrac{x^4}{x^4 - 1}$

In Exercises 45 and 46, find values of a, b, c, and d such that the graph of $f(x) = ax^3 + bx^2 + cx + d$ will resemble the given graph. Then use a graphing utility to verify your result. (There are many correct answers.)

45. 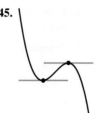   **46.**

In Exercises 47–50, use the graph of f' or f'' to sketch the graph of f. (There are many correct answers.)

47.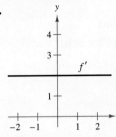

48.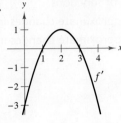

49.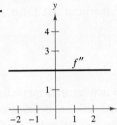

50.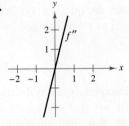

In Exercises 51 and 52, sketch a graph of a function f having the given characteristics. (There are many correct answers.)

51. $f(-2) = f(0) = 0$
    $f'(x) > 0$ if $x < -1$
    $f'(x) < 0$ if $-1 < x < 0$
    $f'(x) > 0$ if $x > 0$
    $f'(-1) = f'(0) = 0$

52. $f(-1) = f(3) = 0$
    $f'(1)$ is undefined.
    $f'(x) < 0$ if $x < 1$
    $f'(x) > 0$ if $x > 1$
    $f''(x) < 0$, $x \neq 1$
    $\lim_{x \to \infty} f(x) = 4$

In Exercises 53 and 54, create a function whose graph has the given characteristics. (There are many correct answers.)

53. Vertical asymptote: $x = 5$
    Horizontal asymptote: $y = 0$

54. Vertical asymptote: $x = -3$
    Horizontal asymptote: None

55. **MAKE A DECISION: SOCIAL SECURITY** The table lists the average monthly Social Security benefits B (in dollars) for retired workers aged 62 and over from 1998 through 2005. A model for the data is

$$B = \frac{582.6 + 38.38t}{1 + 0.025t - 0.0009t^2}, \quad 8 \leq t \leq 15$$

where t is the year, with $t = 8$ corresponding to 1998.
(Source: U.S. Social Security Administration)

t	8	9	10	11	12	13	14	15
B	780	804	844	874	895	922	955	1002

(a) Use a graphing utility to create a scatter plot of the data and graph the model in the same viewing window. How well does the model fit the data?

(b) Use the model to predict the average monthly benefit in 2008.

(c) Should this model be used to predict the average monthly Social Security benefits in future years? Why or why not?

56. **Bacteria** The data in the table show the number N of bacteria in a culture at time t, where t is measured in days. A model for these data is

$$N = \frac{24{,}670 - 35{,}153t + 13{,}250t^2}{100 - 39t + 7t^2}, \quad 1 \leq t \leq 8.$$

t	1	2	3	4	5	6	7	8
N	25	200	804	1756	2296	2434	2467	2473

(a) Use a graphing utility to create a scatter plot of the data and graph the model in the same viewing window. How well does the model fit the data?

(b) Use the model to predict the number of bacteria in the culture after 10 days.

(c) Should this model be used to predict the number of bacteria in the culture in a few months? Why or why not?

57. **Meteorology** The monthly normal temperature T (in degrees Fahrenheit) for Pittsburgh, Pennsylvania can be modeled by

$$T = \frac{22.329 - 0.7t + 0.029t^2}{1 - 0.203t + 0.014t^2}, \quad 1 \leq t \leq 12$$

where t is the month, with $t = 1$ corresponding to January. Use a graphing utility to graph the model and find all absolute extrema. Interpret the meaning of these values in the context of the problem. (Source: National Climatic Data Center)

Writing In Exercises 58 and 59, use a graphing utility to graph the function. Explain why there is no vertical asymptote when a superficial examination of the function may indicate that there should be one.

58. $h(x) = \dfrac{6 - 2x}{3 - x}$

59. $g(x) = \dfrac{x^2 + x - 2}{x - 1}$

60. **Extended Application** To work an extended application analyzing the Department of Defense manpower from 1980 through 2005, visit this text's website at *college.hmco.com*. (Data Source: U.S. Department of Defense)

Section 3.7
Differentials: Linear Approximation

- Find the differentials of functions.
- Use differentials to approximate changes in functions.
- Use differentials to approximate changes in real-life models.

Differentials

When the derivative was defined in Section 2.1 as the limit of the ratio $\Delta y / \Delta x$, it seemed natural to retain the quotient symbolism for the limit itself. So, the derivative of y with respect to x was denoted by

$$\frac{dy}{dx} = \lim_{\Delta x \to 0} \frac{\Delta y}{\Delta x}$$

even though we did not interpret dy/dx as the quotient of two separate quantities. In this section, you will see that the quantities dy and dx can be assigned meanings in such a way that their quotient, when $dx \neq 0$, is equal to the derivative of y with respect to x.

> **STUDY TIP**
>
> In this definition, dx can have any nonzero value. In most applications, however, dx is chosen to be small and this choice is denoted by $dx = \Delta x$.

Definition of Differentials

Let $y = f(x)$ represent a differentiable function. The **differential of x** (denoted by dx) is any nonzero real number. The **differential of y** (denoted by dy) is $dy = f'(x)\, dx$.

One use of differentials is in approximating the change in $f(x)$ that corresponds to a change in x, as shown in Figure 3.51. This change is denoted by

$$\Delta y = f(x + \Delta x) - f(x). \qquad \text{Change in } y$$

In Figure 3.51, notice that as Δx gets smaller and smaller, the values of dy and Δy get closer and closer. That is, when Δx is small, $dy \approx \Delta y$.

> **STUDY TIP**
>
> Note in Figure 3.51 that near the point of tangency, the graph of f is very close to the tangent line. This is the essence of the approximations used in this section. In other words, near the point of tangency, $dy \approx \Delta y$.

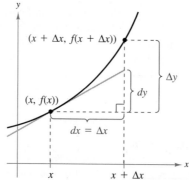

FIGURE 3.51

This **tangent line approximation** is the basis for most applications of differentials.

SECTION 3.7 Differentials: Linear Approximation

Example 1 Interpreting Differentials Graphically

Consider the function given by

$$f(x) = x^2.$$ Original function

Find the value of dy when $x = 1$ and $dx = 0.01$. Compare this with the value of Δy when $x = 1$ and $\Delta x = 0.01$. Interpret the results graphically.

SOLUTION Begin by finding the derivative of f.

$$f'(x) = 2x$$ Derivative of f

When $x = 1$ and $dx = 0.01$, the value of the differential dy is

$$\begin{aligned} dy &= f'(x)\,dx & &\text{Differential of } y \\ &= f'(1)(0.01) & &\text{Substitute 1 for } x \text{ and 0.01 for } dx. \\ &= 2(1)(0.01) & &\text{Use } f'(x) = 2x. \\ &= 0.02. & &\text{Simplify.} \end{aligned}$$

When $x = 1$ and $\Delta x = 0.01$, the value of Δy is

$$\begin{aligned} \Delta y &= f(x + \Delta x) - f(x) & &\text{Change in } y \\ &= f(1.01) - f(1) & &\text{Substitute 1 for } x \text{ and 0.01 for } \Delta x. \\ &= (1.01)^2 - (1)^2 \\ &= 0.0201. & &\text{Simplify.} \end{aligned}$$

Note that $dy \approx \Delta y$, as shown in Figure 3.52.

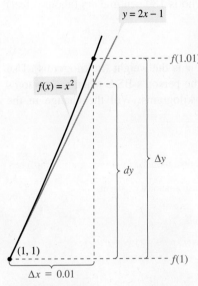

FIGURE 3.52

✓ CHECKPOINT 1

Find the value of dy when $x = 2$ and $dx = 0.01$ for $f(x) = x^4$. Compare this with the value of Δy when $x = 2$ and $\Delta x = 0.01$. ■

The validity of the approximation

$$dy \approx \Delta y, \quad dx \neq 0$$

stems from the definition of the derivative. That is, the existence of the limit

$$f'(x) = \lim_{\Delta x \to 0} \frac{f(x + \Delta x) - f(x)}{\Delta x}$$

implies that when Δx is close to zero, then $f'(x)$ is close to the difference quotient. So, you can write

$$\frac{f(x + \Delta x) - f(x)}{\Delta x} \approx f'(x)$$

$$f(x + \Delta x) - f(x) \approx f'(x)\,\Delta x$$

$$\Delta y \approx f'(x)\,\Delta x.$$

Substituting dx for Δx and dy for $f'(x)\,dx$ produces

$$\Delta y \approx dy.$$

STUDY TIP

Find an equation of the tangent line $y = g(x)$ to the graph of $f(x) = x^2$ at the point $x = 1$. Evaluate $g(1.01)$ and $f(1.01)$.

Differentials can be used in the life sciences to approximate changes in populations, drug concentrations, or body surface areas, as shown in Example 2.

Example 2
MAKE A DECISION   Body Surface Area

The body surface area (BSA) of a person who is 180 centimeters (about 6 feet) tall is modeled by

$$B = 0.1\sqrt{5w}$$

where B is the BSA (in square meters) and w is the weight (in kilograms). Use differentials to approximate the change in the person's BSA when the person's weight changes from 90 kilograms to 95 kilograms. Will the change in the person's BSA be less than 0.1 square meter?

SOLUTION   Begin by finding dB/dw.

$$B = 0.1\sqrt{5w} \qquad \text{Write original formula.}$$
$$= 0.1(5w)^{1/2} \qquad \text{Rewrite using rational exponent.}$$
$$\frac{dB}{dw} = 0.1\left(\frac{1}{2}\right)(5w)^{-1/2}(5) \qquad \text{Power Rule}$$
$$= 0.05\frac{5}{\sqrt{5w}} \qquad \text{Simplify.}$$
$$= 0.05\sqrt{\frac{5}{w}} \qquad \text{Rationalize.}$$

When $w = 90$ and $dw = 5$, you can approximate the change in the person's BSA to be

$$\Delta B \approx dB = 0.05\sqrt{\frac{5}{w}}\,dw$$
$$= 0.05\left(\sqrt{\frac{5}{90}}\right)(5)$$
$$\approx 0.06$$

So, the change in the person's BSA is approximately 0.06 square meter. Yes, the change in the person's BSA will be less than 0.1 square meter.

✓CHECKPOINT 2

Use the differential dB to approximate the change in a person's BSA for the model in Example 2 when the person's weight changes from 100 kilograms to 103 kilograms. Will the change in the person's BSA be less than 0.1 square meter? ■

Formulas for Differentials

You can use the definition of differentials to rewrite each differentiation rule in **differential form**. For example, if u and v are differentiable functions of x, then $du = (du/dx)\, dx$ and $dv = (dv/dx)\, dx$, which implies that you can write the Product Rule in the following differential form.

$$d[uv] = \frac{d}{dx}[uv]\, dx \qquad \text{Differential of } uv$$

$$= \left[u\frac{dv}{dx} + v\frac{du}{dx}\right] dx \qquad \text{Product Rule}$$

$$= u\frac{dv}{dx}\, dx + v\frac{du}{dx}\, dx$$

$$= u\, dv + v\, du \qquad \text{Differential form of Product Rule}$$

The following summary gives the differential forms of the differentiation rules presented so far in the text.

Differential Forms of Differentiation Rules

Constant Multiple Rule:	$d[cu] = c\, du$
Sum or Difference Rule:	$d[u \pm v] = du \pm dv$
Product Rule:	$d[uv] = u\, dv + v\, du$
Quotient Rule:	$d\left[\dfrac{u}{v}\right] = \dfrac{v\, du - u\, dv}{v^2}$
Constant Rule:	$d[c] = 0$
Power Rule:	$d[x^n] = nx^{n-1}\, dx$

The next example compares the derivatives and differentials of several simple functions.

Example 3 Finding Differentials

Find the differential dy of each function.

Function	Derivative	Differential
a. $y = x^2$	$\dfrac{dy}{dx} = 2x$	$dy = 2x\, dx$
b. $y = \dfrac{3x + 2}{5}$	$\dfrac{dy}{dx} = \dfrac{3}{5}$	$dy = \dfrac{3}{5}\, dx$
c. $y = 2x^2 - 3x$	$\dfrac{dy}{dx} = 4x - 3$	$dy = (4x - 3)\, dx$
d. $y = \dfrac{1}{x}$	$\dfrac{dy}{dx} = -\dfrac{1}{x^2}$	$dy = -\dfrac{1}{x^2}\, dx$

✓ CHECKPOINT 3

Find the differential dy of each function.

a. $y = 4x^3$

b. $y = \dfrac{2x + 1}{3}$

c. $y = 3x^2 - 2x$

d. $y = \dfrac{1}{x^2}$

Error Propagation

A common use of differentials is the estimation of errors that result from inaccuracies of physical measuring devices. This is shown in Example 4.

Example 4 Estimating Measurement Errors

The radius of a ball bearing is measured to be 0.7 inch, as shown in Figure 3.53. This implies that the volume of the ball bearing is $\frac{4}{3}\pi(0.7)^3 \approx 1.4368$ cubic inches. You are told that the measurement of the radius is correct to within 0.01 inch. How far off could the calculation of the volume be?

SOLUTION Because the value of r can be off by 0.01 inch, it follows that

$$-0.01 \leq \Delta r \leq 0.01. \qquad \text{Possible error in measuring}$$

Using $\Delta r = dr$, you can estimate the possible error in the volume.

$$V = \tfrac{4}{3}\pi r^3 \qquad \text{Formula for volume}$$

$$dV = \frac{dV}{dr}\,dr = 4\pi r^2\,dr \qquad \text{Formula for differential of } V$$

The possible error in the volume is

$$4\pi r^2\,dr = 4\pi(0.7)^2(\pm 0.01) \qquad \text{Substitute for } r \text{ and } dr.$$

$$\approx \pm 0.0616 \text{ cubic inch.} \qquad \text{Possible error}$$

So, the volume of the ball bearing could range between

$$(1.4368 - 0.0616) = 1.3752 \text{ cubic inches}$$

and

$$(1.4368 + 0.0616) = 1.4984 \text{ cubic inches.}$$

In Example 4, the **relative error** in the volume is defined to be the ratio of dV to V. This ratio is

$$\frac{dV}{V} \approx \frac{\pm 0.0616}{1.4368} \approx \pm 0.0429.$$

This corresponds to a **percentage error** of 4.29%.

FIGURE 3.53

✓ CHECKPOINT 4

Find the surface area of the ball bearing in Example 4. How far off could your calculation of the surface area be? The surface area of a sphere is given by $S = 4\pi r^2$. ∎

CONCEPT CHECK

1. Given a differentiable function $y = f(x)$, what is the differential of x?
2. Given a differentiable function $y = f(x)$, write an expression for the differential of y.
3. Write the differential form of the Quotient Rule.
4. When using differentials, what is meant by the terms *relative error* and *percentage error*?

Skills Review 3.7

The following warm-up exercises involve skills that were covered in earlier sections. You will use these skills in the exercise set for this section. For additional help, review Sections 2.2 and 2.4.

In Exercises 1–12, find the derivative.

1. $C = 44 + 0.09x^2$
2. $C = 250 + 0.15x$
3. $R = x(1.25 + 0.02\sqrt{x})$
4. $R = x(15.5 - 1.55x)$
5. $P = -0.03x^{1/3} + 1.4x - 2250$
6. $P = -0.02x^2 + 25x - 1000$
7. $A = \frac{1}{4}\sqrt{3}x^2$
8. $A = 6x^2$
9. $C = 2\pi r$
10. $P = 4w$
11. $S = 4\pi r^2$
12. $P = 2x + \sqrt{2}x$

In Exercises 13–16, write a formula for the quantity.

13. Area A of a circle of radius r
14. Area A of a square of side x
15. Volume V of a cube of edge x
16. Volume V of a sphere of radius r

Exercises 3.7

See www.CalcChat.com for worked-out solutions to odd-numbered exercises.

In Exercises 1–6, find the differential dy.

1. $y = 3x^2 - 4$
2. $y = 3x^{2/3}$
3. $y = (4x - 1)^3$
4. $y = \dfrac{x + 1}{2x - 1}$
5. $y = \sqrt{9 - x^2}$
6. $y = \sqrt[3]{6x^2}$

In Exercises 7–10, let $x = 1$ and $\Delta x = 0.01$. Find Δy.

7. $f(x) = 5x^2 - 1$
8. $f(x) = \sqrt{3x}$
9. $f(x) = \dfrac{4}{\sqrt[3]{x}}$
10. $f(x) = \dfrac{x}{x^2 + 1}$

In Exercises 11–14, compare the values of dy and Δy.

11. $y = 0.5x^3$, $x = 2$, $\Delta x = dx = 0.1$
12. $y = 1 - 2x^2$, $x = 0$, $\Delta x = dx = -0.1$
13. $y = x^4 + 1$, $x = -1$, $\Delta x = dx = 0.01$
14. $y = 2x + 1$, $x = 2$, $\Delta x = dx = 0.01$

In Exercises 15–20, let $x = 2$ and complete the table for the function.

$dx = \Delta x$	dy	Δy	$\Delta y - dy$	$dy/\Delta y$
1.000				
0.500				
0.100				
0.010				
0.001				

15. $y = x^2$
16. $y = x^5$
17. $y = \dfrac{1}{x^2}$
18. $y = \dfrac{1}{x}$
19. $y = \sqrt[4]{x}$
20. $y = \sqrt{x}$

In Exercises 21–24, find an equation of the tangent line to the function at the given point. Then find the function values and the tangent line values at $f(x + \Delta x)$ and $y(x + \Delta x)$ for $\Delta x = -0.01$ and 0.01.

Function	Point
21. $f(x) = 2x^3 - x^2 + 1$	$(-2, -19)$
22. $f(x) = 3x^2 - 1$	$(2, 11)$
23. $f(x) = \dfrac{x}{x^2 + 1}$	$(0, 0)$
24. $f(x) = \sqrt{25 - x^2}$	$(3, 4)$

25. **Bacteria** The number N of bacteria present on an unwashed fork can be modeled by $N = 32t^2 + 36t + 204$, where t is the time in hours. Use differentials to approximate the change in the number of bacteria from $t = 12$ to $t = 16$.

26. **Geothermal Energy Consumption** The annual geothermal energy consumption y (in trillion Btu) in the commercial sector of the United States for the years 2000 through 2005 can be modeled by

$$y = 0.120t^2 + 0.64t + 7.5, \quad 0 \le t \le 5$$

where t represents the year, with $t = 0$ corresponding to 2000. Use differentials to approximate the change in consumption from 2004 to 2005. *(Source: U.S. Energy Information Association)*

27. Insect Control The percent p of households that use lawn and garden insect control can be modeled by
$$p = 0.537t^3 - 4.46t^2 + 9.4t + 27, \quad 0 \leq t \leq 5$$
where t is the year, with $t = 0$ corresponding to 2000. *(Source: National Gardening Association)*

(a) Use differentials to approximate the percent change from 2001 to 2002.

(b) Use differentials to approximate the percent change from 2004 to 2005.

28. Biology: Wildlife Management A state game commission introduces 50 deer into newly acquired state game lands. The population N of the herd can be modeled by
$$N = \frac{10(5 + 3t)}{1 + 0.04t}$$
where t is the time in years. Use differentials to approximate the change in the herd size from $t = 5$ to $t = 6$.

29. Medical Science The concentration C (in milligrams per milliliter) of a drug in a patient's bloodstream t hours after injection into muscle tissue is modeled by
$$C = \frac{3t}{27 + t^3}.$$
Use differentials to approximate the change in the concentration when t changes from $t = 1$ to $t = 1.5$.

30. Physiology: Body Surface Area The body surface area (BSA) of a person who is 160 centimeters (about 5 feet 3 inches) tall is modeled by
$$B = \frac{1}{15}\sqrt{10w}$$
where B is the BSA (in square meters) and w is the weight (in kilograms). Use differentials to approximate the change in the person's BSA when the person's weight changes from 52 kilograms to 54 kilograms.

31. Systolic Blood Pressure Consider a person whose systolic blood pressure P (in millimeters of mercury) is given by
$$P = \frac{25t^2 + 125}{t^2 + 1}, \quad 0 \leq t \leq 10$$
where t is the number of seconds that have passed since blood has left the heart.

(a) Use differentials to approximate the change in pressure from $t = 8$ to $t = 9$.

(T) (b) Use a graphing utility to graph the model, and use the *trace* feature to find the change in pressure from $t = 8$ to $t = 9$.

(c) Compare the results of parts (a) and (b).

32. Bicycling The distance D (in kilometers) a bicyclist travels during a race can be modeled by
$$D = 0.1t^{4/3}, \quad 0 \leq t \leq 30$$
where t is the time in minutes.

(a) Use differentials to approximate the change in the distance traveled from $t = 14$ to $t = 15$.

(T) (b) Use a graphing utility to graph the model, and use the *trace* feature to find the change in the distance traveled from $t = 14$ to $t = 15$.

(c) Compare the results of parts (a) and (b).

33. Area The area A of a square of side x is $A = x^2$.

(a) Compare dA and ΔA in terms of x and Δx.

(b) In the figure, identify the region whose area is dA.

(c) Identify the region whose area is $\Delta A - dA$.

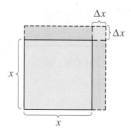

34. Area The side of a square is measured to be 12 inches, with a possible error of $\frac{1}{64}$ inch. Use differentials to approximate the possible error and the relative error in computing the area of the square.

35. Area The radius of a circle is measured to be 10 inches with a possible error of $\frac{1}{8}$ inch. Use differentials to approximate the possible error and the relative error in computing the area of the circle.

36. Volume and Surface Area The edge of a cube is measured to be 12 inches, with a possible error of 0.03 inch. Use differentials to approximate the possible error and the relative error in computing (a) the volume of the cube and (b) the surface area of the cube.

37. Volume The radius of a sphere is measured to be 6 inches, with a possible error of 0.02 inch. Use differentials to approximate the possible error and the relative error in computing the volume of the sphere.

True or False? In Exercises 38 and 39, determine whether the statement is true or false. If it is false, explain why or give an example that shows it is false.

38. If $y = x + c$, then $dy = dx$.

39. If $y = ax + b$, then $\Delta y / \Delta x = dy/dx$.

Algebra Review

Solving Equations

Much of the algebra in Chapter 3 involves simplifying algebraic expressions (see pages 174 and 175) and solving algebraic equations (see page 105). The Algebra Review on page 105 illustrates some of the basic techniques for solving equations. On these two pages, you can review some of the more complicated techniques for solving equations.

When solving an equation, remember that your basic goal is to isolate the variable on one side of the equation. To do this, you use inverse operations. For instance, to get rid of the *subtract* 2 in

$$x - 2 = 0$$

you *add* 2 to each side of the equation. Similarly, to get rid of the *square root* in

$$\sqrt{x + 3} = 2$$

you *square* both sides of the equation.

Example 1 Solving an Equation

Solve each equation.

a. $\dfrac{36(x^2 - 1)}{(x^2 + 3)^3} = 0$ **b.** $0 = 2 - \dfrac{288}{x^2}$ **c.** $0 = 2x(2x^2 - 3)$

SOLUTION

a.
$\dfrac{36(x^2 - 1)}{(x^2 + 3)^3} = 0$	Example 2, page 205
$36(x^2 - 1) = 0$	A fraction is zero only if its numerator is zero.
$x^2 - 1 = 0$	Divide each side by 36.
$x^2 = 1$	Add 1 to each side.
$x = \pm 1$	Take the square root of each side.

b.
$0 = 2 - \dfrac{288}{x^2}$	Example 2, page 215
$-2 = -\dfrac{288}{x^2}$	Subtract 2 from each side.
$1 = \dfrac{144}{x^2}$	Divide each side by -2.
$x^2 = 144$	Multiply each side by x^2.
$x = \pm 12$	Take the square root of each side.

c.
$0 = 2x(2x^2 - 3)$	Example 3, page 216
$2x = 0 \implies x = 0$	Set first factor equal to zero.
$2x^2 - 3 = 0 \implies x = \pm\sqrt{\dfrac{3}{2}}$	Set second factor equal to zero.

Example 2 Solve an Equation

Solve each equation.

a. $\dfrac{20(x^{1/3} - 1)}{9x^{2/3}} = 0$ b. $15x^2(1 - x^2) = 0$

c. $x^2(4x - 3) = 0$ d. $\dfrac{4x}{3(x^2 - 4)^{1/3}} = 0$

e. $g'(x) = 0$, where $g(x) = (x - 2)(x + 1)^2$

SOLUTION

a. $\dfrac{20(x^{1/3} - 1)}{9x^{2/3}} = 0$ Example 5, page 238

$\qquad 20(x^{1/3} - 1) = 0$ A fraction is zero only if its numerator is zero.

$\qquad\qquad x^{1/3} - 1 = 0$ Divide each side by 20.

$\qquad\qquad\qquad x^{1/3} = 1$ Add 1 to each side.

$\qquad\qquad\qquad\quad x = 1$ Cube each side.

b. $\quad 15x^2(1 - x^2) = 0$ Example 4, page 208

$15x^2(1 + x)(1 - x) = 0$ Factor completely.

$\qquad\qquad 15x^2 = 0 \;\Rightarrow\; x = 0$ Set first factor equal to zero.

$\qquad\qquad 1 + x = 0 \;\Rightarrow\; x = -1$ Set second factor equal to zero.

$\qquad\qquad 1 - x = 0 \;\Rightarrow\; x = 1$ Set third factor equal to zero.

c. $x^2(4x - 3) = 0$ Example 2, page 196

$\qquad x^2 = 0 \;\Rightarrow\; x = 0$ Set first factor equal to zero.

$\qquad 4x - 3 = 0 \;\Rightarrow\; x = \tfrac{3}{4}$ Set second factor equal to zero.

d. $\dfrac{4x}{3(x^2 - 4)^{1/3}} = 0$ Example 4, page 188

$\qquad 4x = 0$ A fraction is zero only if its numerator is zero.

$\qquad\; x = 0$ Divide each side by 4.

e. $g(x) = (x - 2)(x + 1)^2$ Exercise 45, page 211

$(x - 2)(2)(x + 1) + (x + 1)^2(1) = 0$ Find derivative and set equal to zero.

$(x + 1)[2(x - 2) + (x + 1)] = 0$ Factor.

$(x + 1)(2x - 4 + x + 1) = 0$ Multiply factors.

$(x + 1)(3x - 3) = 0$ Combine like terms.

$\qquad x + 1 = 0 \;\Rightarrow\; x = -1$ Set first factor equal to zero.

$\qquad 3x - 3 = 0 \;\Rightarrow\; x = 1$ Set second factor equal to zero.

Chapter Summary and Study Strategies

After studying this chapter, you should have acquired the following skills. The exercise numbers are keyed to the Review Exercises that begin on page 253. Answers to odd-numbered Review Exercises are given in the back of the text.*

Section 3.1 — Review Exercises

- Find the critical numbers of a function. — 1–4
    - c is a critical number of f if $f'(c) = 0$ or $f'(c)$ is undefined.
- Find the open intervals on which a function is increasing or decreasing. — 5–8
    - Increasing if $f'(x) > 0$
    - Decreasing if $f'(x) < 0$
- Find intervals on which a real-life model is increasing or decreasing, and interpret the results in context. — 9, 10, 51, 77

Section 3.2

- Use the First-Derivative Test to find the relative extrema of a function. — 11–20
- Find the absolute extrema of a continuous function on a closed interval. — 21–30
- Find minimum and maximum values of a real-life model and interpret the results in context. — 31, 32, 48, 83

Section 3.3

- Find the open intervals on which the graph of a function is concave upward or concave downward. — 33–36
    - Concave upward if $f''(x) > 0$
    - Concave downward if $f''(x) < 0$
- Find the points of inflection of the graph of a function. — 37–40
- Use the Second-Derivative Test to find the relative extrema of a function. — 41–44
- Use the Second-Derivative Test to solve real-life problems. — 45, 50, 52, 78

Section 3.4

- Solve real-life optimization problems. — 46, 47, 49

Section 3.5

- Find the vertical and horizontal asymptotes of a function and sketch its graph. — 53–58
- Find infinite limits and limits at infinity. — 59–66
- Use asymptotes to answer questions about real life. — 67, 68

---

* Use a wide range of valuable study aids to help you master the material in this chapter. The *Student Solutions Guide* includes step-by-step solutions to all odd-numbered exercises to help you review and prepare. The student website at *college.hmco.com/info/larsonapplied* offers algebra help and a *Graphing Technology Guide*. The *Graphing Technology Guide* contains step-by-step commands and instructions for a wide variety of graphing calculators, including the most recent models.

Section 3.6
- Analyze the graph of a function.

Review Exercises 69–76

Section 3.7
- Find the differential of a function. 79–82
- Use differentials to approximate changes in real-life models. 84–87

Study Strategies

- **Solve Problems Graphically, Analytically, and Numerically** When analyzing the graph of a function, use a variety of problem-solving strategies. For instance, if you were asked to analyze the graph of

$$f(x) = x^3 - 4x^2 + 5x - 4$$

you could begin *graphically*. That is, you could use a graphing utility to find a viewing window that appears to show the important characteristics of the graph. From the graph shown below, the function appears to have one relative minimum, one relative maximum, and one point of inflection.

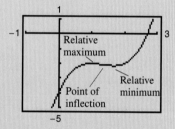

Next, you could use calculus to *analyze* the graph. Because the derivative of f is

$$f'(x) = 3x^2 - 8x + 5 = (3x - 5)(x - 1)$$

the critical numbers of f are $x = \frac{5}{3}$ and $x = 1$. By the First-Derivative Test, you can conclude that $x = \frac{5}{3}$ yields a relative minimum and $x = 1$ yields a relative maximum. Because

$$f''(x) = 6x - 8$$

you can conclude that $x = \frac{4}{3}$ yields a point of inflection. Finally, you could analyze the graph *numerically*. For instance, you could construct a table of values and observe that f is increasing on the interval $(-\infty, 1)$, decreasing on the interval $\left(1, \frac{5}{3}\right)$, and increasing on the interval $\left(\frac{5}{3}, \infty\right)$.

- **Problem-Solving Strategies** If you get stuck when trying to solve an optimization problem, consider the strategies below.

1. *Draw a Diagram.* If feasible, draw a diagram that represents the problem. Label all known values and unknown values on the diagram.

2. *Solve a Simpler Problem.* Simplify the problem, or write several simple examples of the problem. For instance, if you are asked to find the dimensions that will produce a maximum area, try calculating the areas of several examples.

3. *Rewrite the Problem in Your Own Words.* Rewriting a problem can help you understand it better.

4. *Guess and Check.* Try guessing the answer, then check your guess in the statement of the original problem. By refining your guesses, you may be able to think of a general strategy for solving the problem.

Review Exercises

See www.CalcChat.com for worked-out solutions to odd-numbered exercises.

In Exercises 1–4, find the critical numbers of the function.

1. $f(x) = -x^2 + 2x + 4$
2. $g(x) = (x-1)^2(x-3)$
3. $h(x) = \sqrt{x}(x-3)$
4. $f(x) = (x+3)^2$

In Exercises 5–8, determine the open intervals on which the function is increasing or decreasing. Solve the problem analytically and graphically.

5. $f(x) = x^2 + x - 2$
6. $g(x) = (x+2)^3$
7. $h(x) = \dfrac{x^2 - 3x - 4}{x - 3}$
8. $f(x) = -x^3 + 6x^2 - 2$

9. **Meteorology** The normal monthly temperature T (in degrees Fahrenheit) in New York City can be modeled by
$$T = 0.0380t^4 - 1.092t^3 + 9.23t^2 - 19.6t + 44$$
where $1 \le t \le 12$ and $t = 1$ corresponds to January. *(Source: National Climatic Data Center)*

   (a) Find the interval(s) on which the model is increasing.
   (b) Find the interval(s) on which the model is decreasing.
   (c) Interpret the results of parts (a) and (b).
   (T) (d) Use a graphing utility to graph the model.

10. **CD Shipments** The number S of manufacturer unit shipments (in millions) of CDs in the United States from 2000 through 2005 can be modeled by
$$S = -4.17083t^4 + 40.3009t^3 - 110.524t^2 + 19.40t + 941.6, \quad 0 \le t \le 5$$
where t is the year, with $t = 0$ corresponding to 2000. *(Source: Recording Industry Association of America)*

   (a) Find the interval(s) on which the model is increasing.
   (b) Find the interval(s) on which the model is decreasing.
   (c) Interpret the results of parts (a) and (b).
   (T) (d) Use a graphing utility to graph the model.

In Exercises 11–20, use the First-Derivative Test to find the relative extrema of the function. Then use a graphing utility to verify your result.

11. $f(x) = 4x^3 - 6x^2 - 2$
12. $f(x) = \tfrac{1}{4}x^4 - 8x$
13. $g(x) = x^2 - 16x + 12$
14. $h(x) = 4 + 10x - x^2$
15. $h(x) = 2x^2 - x^4$
16. $s(x) = x^4 - 8x^2 + 3$
17. $f(x) = \dfrac{6}{x^2 + 1}$
18. $f(x) = \dfrac{2}{x^2 - 1}$
19. $h(x) = \dfrac{x^2}{x - 2}$
20. $g(x) = x - 6\sqrt{x}, \quad x > 0$

In Exercises 21–30, find the absolute extrema of the function on the closed interval. Then use a graphing utility to confirm your result.

21. $f(x) = x^2 + 5x + 6; \quad [-3, 0]$
22. $f(x) = x^4 - 2x^3; \quad [0, 2]$
23. $f(x) = x^3 - 12x + 1; \quad [-4, 4]$
24. $f(x) = x^3 + 2x^2 - 3x + 4; \quad [-3, 2]$
25. $f(x) = 4\sqrt{x} - x^2; \quad [0, 3]$
26. $f(x) = 2\sqrt{x} - x; \quad [0, 9]$
27. $f(x) = \dfrac{x}{\sqrt{x^2 + 1}}; \quad [0, 2]$
28. $f(x) = -x^4 + x^2 + 2; \quad [0, 2]$
29. $f(x) = \dfrac{2x}{x^2 + 1}; \quad [-1, 2]$
30. $f(x) = \dfrac{8}{x} + x; \quad [1, 4]$

(T) 31. **Surface Area** The radius r (in inches) and slant height l (in inches) of a right circular cone are related by $l = \dfrac{1}{r^2}$. The total surface area of the cone in terms of r is given by
$$S = \pi r^2 + \dfrac{\pi}{r}.$$
Use a graphing utility to graph S and S' and to find the value of r that yields the minimum surface area.

(T) 32. **Surface Area** A right circular cylinder of radius r and height h has a volume of 25 cubic inches. The total surface area of the cylinder in terms of r is given by
$$S = 2\pi r\left(r + \dfrac{25}{\pi r^2}\right).$$
Use a graphing utility to graph S and S' and to find the value of r that yields the minimum surface area.

In Exercises 33–36, analytically find the intervals on which the graph is concave upward and those on which it is concave downward. Verify your results using the graph of the function.

33. $f(x) = (x - 2)^3$

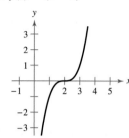

34. $h(x) = x^5 - 10x^2$

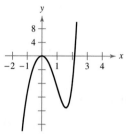

35. $g(x) = \frac{1}{4}(-x^4 + 8x^2)$

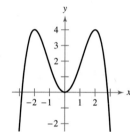

36. $h(x) = x^3 - 6x$

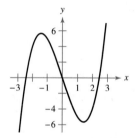

In Exercises 37–40, find the points of inflection of the graph of the function.

37. $f(x) = \frac{1}{2}x^4 - 4x^3$

38. $f(x) = \frac{1}{4}x^4 - 2x^2 - x$

39. $f(x) = x^3(x - 3)^2$

40. $f(x) = (x - 1)^2(x - 3)$

In Exercises 41–44, use the Second-Derivative Test to find the relative extrema of the function.

41. $f(x) = x^5 - 5x^3$

42. $f(x) = x(x^2 - 3x - 9)$

43. $f(x) = 2x^2(1 - x^2)$

44. $f(x) = x - 4\sqrt{x + 1}$

45. Lightning Fatalities From 1995 through 2006, the number d of fatalities in the United States due to lightning can be modeled by

$$d = 0.0793t^4 - 3.463t^3 + 54.74t^2 - 370.8t + 953,$$
$$5 \leq t \leq 16$$

where t is the year, with $t = 5$ corresponding to 1995. (*Source: U.S. National Oceanic and Atmospheric Administration*)

(T) (a) Use a graphing utility to graph the model.

(b) Use the second derivative to determine the concavity of d.

(c) Find the point(s) of inflection of the graph of d.

(d) Interpret the meaning of the inflection point(s) of the graph of d.

(T) **46. Minimum Sum** Find two positive numbers whose product is 169 and whose sum is a minimum. Solve the problem analytically, and use a graphing utility to solve the problem graphically.

47. Length The wall of a building is to be braced by a beam that must pass over a five-foot fence that is parallel to the building and 4 feet from the building. Find the length of the shortest beam that can be used.

48. Newspaper Circulation The total number N of daily newspapers in circulation (in millions) in the United States from 1970 through 2005 can be modeled by

$$N = 0.022t^3 - 1.27t^2 + 9.7t + 1746, \quad 0 \leq t \leq 35$$

where t is the year, with $t = 0$ corresponding to 1970. (*Source: Editor and Publisher Company*)

(a) Find the absolute maximum and minimum circulation over the time period.

(b) Find the year in which the circulation was changing at the greatest rate.

(c) Briefly explain your results for parts (a) and (b).

49. Minimum Perimeter A fence is to be built to enclose a rectangular region of 4900 square feet.

(a) Find the dimensions of the region that will minimize the amount of fencing required.

(b) How would the result of part (a) change if the area of the region were increased to 10,000 square feet?

(T) **50. Biology** The growth of a red oak tree is approximated by the model

$$y = -0.003x^3 + 0.137x^2 + 0.458x - 0.839,$$
$$2 \leq x \leq 34$$

where y is the height of the tree (in feet) and x is its age (in years). Find the age of the tree when it is growing most rapidly. Then use a graphing utility to graph the function and to verify your result. (*Hint:* Use the viewing window $2 \leq x \leq 34$ and $-10 \leq y \leq 60$.)

51. Consumer Trends The average number of hours N (per person per year) of TV usage in the United States from 2000 through 2005 can be modeled by

$$N = -0.382t^3 - 0.97t^2 + 30.5t + 1466, \quad 0 \leq t \leq 5$$

where t is the year, with $t = 0$ corresponding to 2000. (*Source: Veronis Suhler Stevenson*)

(a) Find the intervals on which dN/dt is increasing and decreasing.

(b) Find the limit of N as $t \to \infty$.

(c) Briefly explain your results for parts (a) and (b).

52. Medicine: Poiseuille's Law The speed of blood that is r centimeters from the center of an artery is modeled by

$$s(r) = c(R^2 - r^2), \quad c > 0$$

where c is a constant, R is the radius of the artery, and s is measured in centimeters per second. Show that the speed is a maximum at the center of an artery.

In Exercises 53–58, find the vertical and horizontal asymptotes of the graph. Then use a graphing utility to graph the function.

53. $h(x) = \dfrac{2x + 3}{x - 4}$

54. $g(x) = \dfrac{3}{x} - 2$

55. $f(x) = \dfrac{\sqrt{9x^2 + 1}}{x}$

56. $h(x) = \dfrac{3x}{\sqrt{x^2 + 2}}$

57. $f(x) = \dfrac{4}{x^2 + 1}$

58. $h(x) = \dfrac{2x^2 + 3x - 5}{x - 1}$

In Exercises 59–66, find the limit, if it exists.

59. $\lim\limits_{x \to 0^+} \left(x - \dfrac{1}{x^3}\right)$

60. $\lim\limits_{x \to 0^-} \left(3 + \dfrac{1}{x}\right)$

61. $\lim\limits_{x \to -1^+} \dfrac{x^2 - 2x + 1}{x + 1}$

62. $\lim\limits_{x \to 3^-} \dfrac{3x^2 + 1}{x^2 - 9}$

63. $\lim\limits_{x \to \infty} \dfrac{2x^2}{3x^2 + 5}$

64. $\lim\limits_{x \to \infty} \dfrac{3x^2 - 2x + 3}{x + 1}$

65. $\lim\limits_{x \to -\infty} \dfrac{3x}{x^2 + 1}$

66. $\lim\limits_{x \to -\infty} \left(\dfrac{x}{x - 2} + \dfrac{2x}{x + 2}\right)$

67. Health For a person with sensitive skin, the maximum amount T (in hours) of exposure to the sun that can be tolerated before skin damage occurs can be modeled by

$$T = \dfrac{-0.03s + 33.6}{s}, \quad 0 < s \le 120$$

where s is the Sunsor Scale reading. *(Source: Sunsor, Inc.)*

(a) Use a graphing utility to graph the model. Compare your result with the graph below.

(b) Describe the value of T as s increases.

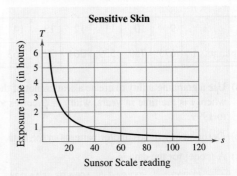

Sensitive Skin

68. Human Memory Model Consider the learning curve modeled by

$$P = \dfrac{0.55 + 0.87(n - 1)}{1 + 0.87(n - 1)}, \quad n > 0$$

where P is the percent of the correct responses after n trials.

(a) Use a spreadsheet software program to complete the table for the model.

n	1	2	3	4	5	6	7	8	9	10
P										

(b) Find the limit of P as n approaches infinity.

In Exercises 69–76, use a graphing utility to graph the function. Use the graph to approximate any intercepts, relative extrema, points of inflection, and asymptotes. State the domain of the function.

69. $f(x) = 4x - x^2$

70. $f(x) = 4x^3 - x^4$

71. $f(x) = x\sqrt{16 - x^2}$

72. $f(x) = x^2\sqrt{9 - x^2}$

73. $f(x) = \dfrac{x + 1}{x - 1}$

74. $f(x) = \dfrac{x - 1}{3x^2 + 1}$

75. $f(x) = x^2 + \dfrac{2}{x}$

76. $f(x) = x^{4/5}$

CHAPTER 3 Applications of the Derivative

77. Airline Complaints The numbers A of flight overbooking complaints against U.S. airlines for the years 1999 through 2005 are shown in the table. *(Source: U.S. Department of Transportation)*

Year, t	9	10	11	12	13	14	15
Complaints, A	673	759	539	364	223	263	283

(a) Use a graphing utility to create a scatter plot of the data, where t is the time in years, with $t = 9$ corresponding to 1999.

(b) Describe any trends and/or patterns in the data.

(c) A model for the data is
$$A = 11.722t^3 - 409.64t^2 + 4596.9t - 16{,}036,$$
$9 \le t \le 15$.

Graph the model and the data in the same viewing window.

(d) Find the years in which the number of complaints was increasing and decreasing.

(e) Find the years in which the rate of change of the number of complaints was increasing and decreasing.

78. Land in Farms The total amounts L of land in U.S. farms (in millions of acres) for the years 2000 through 2005 are shown in the table. *(Source: U.S. Department of Agriculture)*

Year, t	0	1	2	3	4	5
Land, L	945	942	940	939	936	933

(a) Use a graphing utility to create a scatter plot of the data, where t is the time in years, with $t = 0$ corresponding to 2000.

(b) Describe any trends and/or patterns in the data.

(c) Use the *regression* feature of a graphing utility to find a cubic model for the data. Graph the model and the data in the same viewing window.

(d) Find the first and second derivatives of L.

(e) Find the relative extrema and inflection point(s) of L.

(f) Interpret the meanings of the values from part (e) in the context of the problem.

In Exercises 79–82, find the differential dy.

79. $y = x(1 - x)$

80. $y = (3x^2 - 2)^3$

81. $y = \sqrt{36 - x^2}$

82. $y = \dfrac{2 - x}{x + 5}$

83. Medicine The effectiveness E of a pain-killing drug t hours after entering the bloodstream is modeled by
$$E = 22.5t + 7.5t^2 - 2.5t^3, \quad 0 \le t \le 4.5.$$

(a) Use a graphing utility to graph the equation. Choose an appropriate window.

(b) Find the maximum effectiveness the pain-killing drug attains over the interval $[0, 4.5]$.

84. Employment The employment E (in thousands) in health care and social assistance in the United States for the years 2000 to 2005 can be modeled by
$$E = 2.6389t^3 - 36.131t^2 + 475.9t + 12{,}712,$$
$0 \le t \le 5$

where t is the year, with $t = 0$ corresponding to 2000. Use differentials to approximate the change in employment from 2004 to 2005. *(Source: U.S. Bureau of Labor Statistics)*

85. Aquaculture The recommended daily percent p of biomass (plant matter) to be included in a fish's diet can be modeled by
$$p = 0.000235w^2 - 0.054w + 7.1$$
where w is the weight of the fish in grams. *(Source: Food and Agriculture Organization of the United Nations)*

(a) Use differentials to approximate the change in the recommended percent of biomass when the fish's weight changes from 10 grams to 20 grams.

(b) Use differentials to approximate the change in the recommended percent of biomass when the fish's weight changes from 40 grams to 60 grams.

86. Death Rate The number D of deaths in the United States (per 1000 persons) for the years 2000 through 2005 can be modeled by
$$D = 0.0148t^3 - 0.111t^2 + 0.11t + 8.7, \quad 0 \le t \le 5$$
where t is the year, with $t = 0$ corresponding to 2000. *(Source: U.S. Census Bureau)*

(a) Use differentials to approximate the change in the death rate from 2002 to 2004.

(b) Use a graphing utility to graph the model, and use the *trace* feature to find the change in the death rate from 2002 to 2004.

(c) Compare the results of parts (a) and (b).

87. Surface Area and Volume The diameter of a sphere is measured to be 18 inches with a possible error of 0.05 inch. Use differentials to approximate the possible error in the surface area and the volume of the sphere.

Chapter Test

See www.CalcChat.com for worked-out solutions to odd-numbered exercises.

Take this test as you would take a test in class. When you are done, check your work against the answers given in the back of the book.

In Exercises 1–3, find the critical numbers of the function and the open intervals on which the function is increasing or decreasing.

1. $f(x) = 3x^2 - 4$ **2.** $f(x) = x^3 - 12x$ **3.** $f(x) = (x - 5)^4$

In Exercises 4–6, use the First-Derivative Test to find all relative extrema of the function. Then use a graphing utility to verify your result.

4. $f(x) = \frac{1}{3}x^3 - 9x + 4$ **5.** $f(x) = 2x^4 - 4x^2 - 5$ **6.** $f(x) = \frac{5}{x^2 + 2}$

In Exercises 7–9, find the absolute extrema of the function on the closed interval.

7. $f(x) = x^2 + 6x + 8$, $[-4, 0]$ **8.** $f(x) = 12\sqrt{x} - 4x$, $[0, 5]$

9. $f(x) = \frac{6}{x} + \frac{x}{2}$, $[1, 6]$

In Exercises 10 and 11, determine the open intervals on which the graph of the function is concave upward or concave downward.

10. $f(x) = x^5 - 4x^2$ **11.** $f(x) = \frac{20}{3x^2 + 8}$

In Exercises 12 and 13, find the point(s) of inflection of the graph of the function.

12. $f(x) = x^4 + 6$ **13.** $f(x) = \frac{1}{5}x^5 - 4x^2$

In Exercises 14 and 15, use the Second-Derivative Test to find all relative extrema of the function.

14. $f(x) = x^3 - 6x^2 - 24x + 12$ **15.** $f(x) = \frac{3}{5}x^5 - 9x^3$

(T) In Exercises 16–18, find the vertical and horizontal asymptotes of the graph. Then use a graphing utility to graph the function.

16. $f(x) = \frac{3x + 2}{x - 5}$ **17.** $f(x) = \frac{2x^2}{x^2 + 3}$ **18.** $f(x) = \frac{2x^2 - 5}{x - 1}$

In Exercises 19–21, find the limit, if it exists.

19. $\lim_{x \to \infty} \left(\frac{3}{x} + 1\right)$ **20.** $\lim_{x \to \infty} \frac{3x^2 - 4x + 1}{x - 7}$ **21.** $\lim_{x \to -\infty} \frac{6x^2 + x - 5}{2x^2 - 5x}$

In Exercises 22–24, find the differential dy.

22. $y = 5x^2 - 3$ **23.** $y = \frac{1 - x}{x + 3}$ **24.** $y = (x + 4)^3$

25. The percent p of households that have indoor houseplants can be modeled by

$$p = 0.500t^3 - 3.71t^2 + 5.8t + 43, \quad 1 \le t \le 5$$

where t is the year, with $t = 1$ corresponding to 2001. When was the percent the highest? the lowest? *(Source: National Gardening Association)*

4 Exponential and Logarithmic Functions

4.1 Exponential Functions
4.2 Natural Exponential Functions
4.3 Derivatives of Exponential Functions
4.4 Logarithmic Functions
4.5 Derivatives of Logarithmic Functions
4.6 Exponential Growth and Decay

After nearly disappearing from most of the United States, the bald eagle is now flourishing. According to the U.S. Fish and Wildlife Service, the bald eagle has recovered from an all-time low of 417 nesting pairs in 1963 to an estimated 9789 breeding pairs in 2006. (See Section 4.1, Exercise 35.)

Applications

Exponential and logarithmic functions have many real-life applications. The applications listed below represent a sample of the applications in this chapter.

- Radioactive Decay, Exercises 31 and 32, page 264
- Forest Defoliation, Exercise 31, page 270
- Ebbinghaus Model, Exercise 39, page 278
- Carbon Dating: Exercises 77–80, page 287
- Make a Decision: Treatment Facilities, Exercise 35, page 305

Section 4.1
Exponential Functions

- Use the properties of exponents to evaluate and simplify exponential expressions.
- Sketch the graphs of exponential functions.

Exponential Functions

You are already familiar with the behavior of algebraic functions such as

$$f(x) = x^2, \quad g(x) = \sqrt{x} = x^{1/2}, \quad \text{and} \quad h(x) = \frac{1}{x} = x^{-1}$$

each of which involves a variable raised to a constant power. By interchanging roles and raising a constant to a variable power, you obtain another important class of functions called **exponential functions**. Some simple examples are

$$f(x) = 2^x, \quad g(x) = \left(\frac{1}{10}\right)^x = \frac{1}{10^x}, \quad \text{and} \quad h(x) = 3^{2x} = 9^x.$$

In general, you can use any positive base $a \neq 1$ as the base of an exponential function.

Definition of Exponential Function

If $a > 0$ and $a \neq 1$, then the **exponential function** with base a is given by

$$f(x) = a^x.$$

STUDY TIP

In the definition above, the base $a = 1$ is excluded because it yields $f(x) = 1^x = 1$. This is a constant function, not an exponential function.

When working with exponential functions, the properties of exponents, shown below, are useful.

Properties of Exponents

Let a and b be positive numbers.

1. $a^0 = 1$
2. $a^x a^y = a^{x+y}$
3. $\dfrac{a^x}{a^y} = a^{x-y}$
4. $(a^x)^y = a^{xy}$
5. $(ab)^x = a^x b^x$
6. $\left(\dfrac{a}{b}\right)^x = \dfrac{a^x}{b^x}$
7. $a^{-x} = \dfrac{1}{a^x}$

Example 1 Applying Properties of Exponents

Simplify each expression using the properties of exponents.

a. $(2^2)(2^3)$ **b.** $(2^2)(2^{-3})$ **c.** $(3^2)^3$

d. $\left(\dfrac{1}{3}\right)^{-2}$ **e.** $\dfrac{3^2}{3^3}$ **f.** $(2^{1/2})(3^{1/2})$

SOLUTION

a. $(2^2)(2^3) = 2^{2+3} = 2^5 = 32$ Apply Property 2.

b. $(2^2)(2^{-3}) = 2^{2-3} = 2^{-1} = \dfrac{1}{2}$ Apply Properties 2 and 7.

c. $(3^2)^3 = 3^{2(3)} = 3^6 = 729$ Apply Property 4.

d. $\left(\dfrac{1}{3}\right)^{-2} = \dfrac{1}{(1/3)^2} = \left(\dfrac{1}{1/3}\right)^2 = 3^2 = 9$ Apply Properties 6 and 7.

e. $\dfrac{3^2}{3^3} = 3^{2-3} = 3^{-1} = \dfrac{1}{3}$ Apply Properties 3 and 7.

f. $(2^{1/2})(3^{1/2}) = [(2)(3)]^{1/2} = 6^{1/2} = \sqrt{6}$ Apply Property 5.

✓ CHECKPOINT 1

Simplify each expression using the properties of exponents.

a. $(3^2)(3^3)$ **b.** $(3^2)(3^{-1})$

c. $(2^3)^2$ **d.** $\left(\dfrac{1}{2}\right)^{-3}$

e. $\dfrac{2^2}{2^3}$ **f.** $(2^{1/2})(5^{1/2})$ ∎

Although Example 1 demonstrates the properties of exponents with integer and rational exponents, it is important to realize that the properties hold for *all* real exponents. With a calculator, you can obtain approximations of a^x for any base a and any real exponent x. Here are some examples.

$$2^{-0.6} \approx 0.660, \qquad \pi^{0.75} \approx 2.360, \qquad (1.56)^{\sqrt{2}} \approx 1.876$$

Example 2 Dating Organic Material

In living organic material, the ratio of radioactive carbon isotopes to the total number of carbon atoms is about 1 to 10^{12}. When organic material dies, its radioactive carbon isotopes begin to decay, with a half-life of about 5715 years. This means that after 5715 years, the ratio of isotopes to atoms will have decreased to one-half the original ratio, after a second 5715 years the ratio will have decreased to one-fourth of the original, and so on. Figure 4.1 shows this decreasing ratio. The formula for the ratio R of carbon isotopes to carbon atoms is

$$R = \left(\dfrac{1}{10^{12}}\right)\left(\dfrac{1}{2}\right)^{t/5715}$$

where t is the time in years. Find the value of R for each period of time.

a. 10,000 years **b.** 20,000 years **c.** 25,000 years

SOLUTION

a. $R = \left(\dfrac{1}{10^{12}}\right)\left(\dfrac{1}{2}\right)^{10,000/5715} \approx 2.973 \times 10^{-13}$ Ratio for 10,000 years

b. $R = \left(\dfrac{1}{10^{12}}\right)\left(\dfrac{1}{2}\right)^{20,000/5715} \approx 8.842 \times 10^{-14}$ Ratio for 20,000 years

c. $R = \left(\dfrac{1}{10^{12}}\right)\left(\dfrac{1}{2}\right)^{25,000/5715} \approx 4.821 \times 10^{-14}$ Ratio for 25,000 years

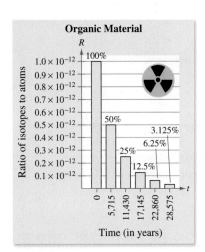

FIGURE 4.1

✓ CHECKPOINT 2

Use the formula for the ratio of carbon isotopes to carbon atoms in Example 2 to find the value of R for each period of time.

a. 5000 years

b. 15,000 years

c. 30,000 years ∎

Graphs of Exponential Functions

The basic nature of the graph of an exponential function can be determined by the point-plotting method or by using a graphing utility.

Example 3 Graphing Exponential Functions

Sketch the graph of each exponential function.

a. $f(x) = 2^x$ **b.** $g(x) = \left(\frac{1}{2}\right)^x = 2^{-x}$ **c.** $h(x) = 3^x$

SOLUTION To sketch these functions by hand, you can begin by constructing a table of values, as shown below.

x	-3	-2	-1	0	1	2	3	4
$f(x) = 2^x$	$\frac{1}{8}$	$\frac{1}{4}$	$\frac{1}{2}$	1	2	4	8	16
$g(x) = 2^{-x}$	8	4	2	1	$\frac{1}{2}$	$\frac{1}{4}$	$\frac{1}{8}$	$\frac{1}{16}$
$h(x) = 3^x$	$\frac{1}{27}$	$\frac{1}{9}$	$\frac{1}{3}$	1	3	9	27	81

The graphs of the three functions are shown in Figure 4.2. Note that the graphs of $f(x) = 2^x$ and $h(x) = 3^x$ are increasing, whereas the graph of $g(x) = 2^{-x}$ is decreasing.

STUDY TIP

Note that a graph of the form $f(x) = a^x$, as shown in Example 3(a), is a reflection in the y-axis of the graph of the form $f(x) = a^{-x}$, as shown in Example 3(b).

✓ **CHECKPOINT 3**

Complete the table of values for $f(x) = 5^x$. Sketch the graph of the exponential function.

x	-3	-2	-1	0
$f(x)$				

x	1	2	3
$f(x)$			

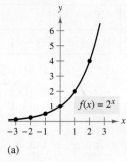
(a)

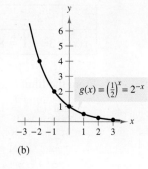
(b)

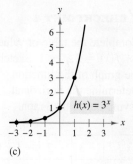
(c)

FIGURE 4.2

TECHNOLOGY

Try graphing the functions $f(x) = 2^x$ and $h(x) = 3^x$ in the same viewing window, as shown at the right. From the display, you can see that the graph of h is increasing more rapidly than the graph of f.*

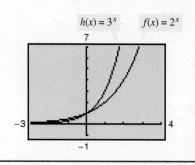

---

*Specific calculator keystroke instructions for operations in this and other technology boxes can be found at *college.hmco.com/info/larsonapplied*.

262   CHAPTER 4   Exponential and Logarithmic Functions

The forms of the graphs in Figure 4.2 are typical of the graphs of the exponential functions $y = a^{-x}$ and $y = a^x$, where $a > 1$. The basic characteristics of such graphs are summarized in Figure 4.3.

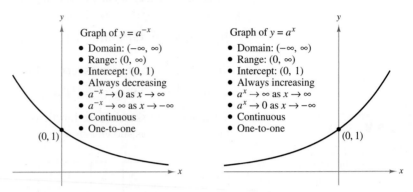

FIGURE 4.3   Characteristics of the Exponential Functions $y = a^{-x}$ and $y = a^x$ ($a > 1$)

Example 4   Graphing an Exponential Function

Sketch the graph of $f(x) = 3^{-x} - 1$.

SOLUTION   Begin by creating a table of values, as shown below.

x	-2	-1	0	1	2
$f(x)$	$3^2 - 1 = 8$	$3^1 - 1 = 2$	$3^0 - 1 = 0$	$3^{-1} - 1 = -\frac{2}{3}$	$3^{-2} - 1 = -\frac{8}{9}$

From the limit

$$\lim_{x \to \infty} (3^{-x} - 1) = \lim_{x \to \infty} 3^{-x} - \lim_{x \to \infty} 1$$

$$= \lim_{x \to \infty} \frac{1}{3^x} - \lim_{x \to \infty} 1$$

$$= 0 - 1$$

$$= -1$$

you can see that $y = -1$ is a horizontal asymptote of the graph. The graph is shown in Figure 4.4.

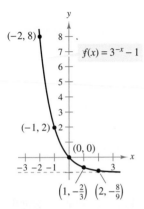

FIGURE 4.4

✓ CHECKPOINT 4

Complete the table of values for $f(x) = 2^{-x} + 1$. Sketch the graph of the function. Determine the horizontal asymptote of the graph.

x	-3	-2	-1	0
$f(x)$				

x	1	2	3
$f(x)$			

CONCEPT CHECK

1. Complete the following: If $a > 0$ and $a \neq 1$, then $f(x) = a^x$ is a(n) _____ function.
2. Identify the domains and ranges of the exponential functions (a) $y = a^{-x}$ and (b) $y = a^x$. (Assume $a > 1$.)
3. As x approaches ∞, what does a^{-x} approach? (Assume $a > 1$.)
4. Explain why 1^x is *not* an exponential function.

SECTION 4.1 Exponential Functions

Skills Review 4.1

The following warm-up exercises involve skills that were covered in earlier sections. You will use these skills in the exercise set for this section. For additional help, review Sections 1.4 and 1.6.

In Exercises 1–6, describe how the graph of g is related to the graph of f.

1. $g(x) = f(x + 2)$
2. $g(x) = -f(x)$
3. $g(x) = -1 + f(x)$
4. $g(x) = f(-x)$
5. $g(x) = f(x - 1)$
6. $g(x) = f(x) + 2$

In Exercises 7–10, discuss the continuity of the function.

7. $f(x) = \dfrac{x^2 + 2x - 1}{x + 4}$
8. $f(x) = \dfrac{x^2 - 3x + 1}{x^2 + 2}$
9. $f(x) = \dfrac{x^2 - 3x - 4}{x^2 - 1}$
10. $f(x) = \dfrac{x^2 - 5x + 4}{x^2 + 1}$

In Exercises 11–16, solve for x.

11. $2x - 6 = 4$
12. $3x + 1 = 5$
13. $(x + 4)^2 = 25$
14. $(x - 2)^2 = 8$
15. $x^2 + 4x - 5 = 0$
16. $2x^2 + 3x + 1 = 0$

Exercises 4.1

See www.CalcChat.com for worked-out solutions to odd-numbered exercises.

In Exercises 1 and 2, evaluate each expression.

1. (a) $5(5^3)$ (b) $27^{2/3}$
   (c) $64^{3/4}$ (d) $81^{1/2}$
   (e) $25^{3/2}$ (f) $32^{2/5}$

2. (a) $\left(\frac{1}{5}\right)^3$ (b) $\left(\frac{1}{8}\right)^{1/3}$
   (c) $64^{2/3}$ (d) $\left(\frac{5}{8}\right)^2$
   (e) $100^{3/2}$ (f) $4^{5/2}$

In Exercises 3–6, use the properties of exponents to simplify the expression.

3. (a) $(5^2)(5^3)$ (b) $(5^2)(5^{-3})$
   (c) $(5^2)^2$ (d) 5^{-3}

4. (a) $\dfrac{5^3}{5^6}$ (b) $\left(\dfrac{1}{5}\right)^{-2}$
   (c) $(8^{1/2})(2^{1/2})$ (d) $(32^{3/2})\left(\dfrac{1}{2}\right)^{3/2}$

5. (a) $\dfrac{5^3}{25^2}$ (b) $(9^{2/3})(3)(3^{2/3})$
   (c) $[(25^{1/2})(5^2)]^{1/3}$ (d) $(8^2)(4^3)$

6. (a) $(4^3)(4^2)$ (b) $\left(\dfrac{1}{4}\right)^2(4^2)$
   (c) $(4^6)^{1/2}$ (d) $[(8^{-1})(8^{2/3})]^3$

(T) In Exercises 7–10, evaluate the function. If necessary, use a graphing utility, rounding your answers to three decimal places.

7. $f(x) = 2^{x-1}$
   (a) $f(3)$ (b) $f\left(\frac{1}{2}\right)$ (c) $f(-2)$ (d) $f\left(-\frac{3}{2}\right)$

8. $f(x) = 3^{x+2}$
   (a) $f(-4)$ (b) $f\left(-\frac{1}{2}\right)$ (c) $f(2)$ (d) $f\left(-\frac{5}{2}\right)$

9. $g(x) = 1.05^x$
   (a) $g(-2)$ (b) $g(120)$ (c) $g(12)$ (d) $g(5.5)$

10. $g(x) = 1.075^x$
    (a) $g(1.2)$ (b) $g(180)$ (c) $g(60)$ (d) $g(12.5)$

11. **Biology: Endangered Species** The population P of an endangered species can be modeled by the exponential function $P(t) = 2184(0.956)^t$, where t is the time in years, with $t = 0$ corresponding to 2000. Use the model to estimate the populations in the years (a) 2009 and (b) 2013.

12. **Population Growth** The population P (in millions) of the United States from 1992 through 2005 can be modeled by the exponential function

    $P(t) = 252.12(1.011)^t$

    where t is the time in years, with $t = 2$ corresponding to 1992. Use the model to estimate the populations in the years (a) 2008 and (b) 2012. *(Source: U.S. Census Bureau)*

In Exercises 13–18, match the function with its graph. [The graphs are labeled (a)–(f).]

(a)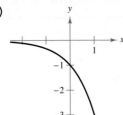

(b)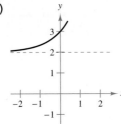

(c)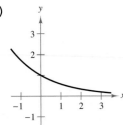

(d)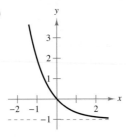

(e)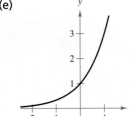

(f)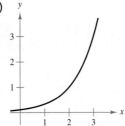

13. $f(x) = 3^x$
14. $f(x) = 3^{-x/2}$
15. $f(x) = -3^x$
16. $f(x) = 3^{x-2}$
17. $f(x) = 3^{-x} - 1$
18. $f(x) = 3^x + 2$

(T) In Exercises 19–30, use a graphing utility to graph the function.

19. $f(x) = 6^x$
20. $f(x) = 4^x$
21. $f(x) = \left(\frac{1}{5}\right)^x = 5^{-x}$
22. $f(x) = \left(\frac{1}{4}\right)^x = 4^{-x}$
23. $y = 2^{x-1}$
24. $y = 4^x + 3$
25. $y = -2^x$
26. $y = -5^x$
27. $y = 3^{-x^2}$
28. $y = 6^{-x^2}$
29. $s(t) = \frac{1}{4}(3^{-t})$
30. $s(t) = 2^{-t} + 3$

(T) **31. Radioactive Decay** After t years, the remaining mass y (in grams) of 16 grams of a radioactive element whose half-life is 30 years is given by

$$y = 16\left(\frac{1}{2}\right)^{t/30}, \quad t \geq 0.$$

(a) Use a graphing utility to graph the function.
(b) How much of the initial mass remains after 150 years?
(c) Use the *zoom* and *trace* features of a graphing utility to find the time required for the mass to decay to an amount of 1 gram.

(T) **32. Radioactive Decay** After t years, the remaining mass y (in grams) of 23 grams of a radioactive element whose half-life is 45 years is given by

$$y = 23\left(\frac{1}{2}\right)^{t/45}, \quad t \geq 0.$$

(a) Use a graphing utility to graph the function.
(b) How much of the initial mass remains after 90 years?
(c) Use the *zoom* and *trace* features of a graphing utility to find the time required for the mass to decay to an amount of 1 gram.

33. Depreciation After t years, the value of a wheelchair conversion van that originally cost $30,500 depreciates so that each year it is worth $\frac{7}{8}$ of its value for the previous year.

(a) Find a model for $V(t)$, the value of the van after t years.
(b) Determine the value of the van 4 years after it was purchased.

34. Drug Concentration Immediately following an injection, the concentration of a drug in the bloodstream is 300 milligrams per milliliter. After t hours, the concentration is 75% of the level of the previous hour.

(a) Find a model for $C(t)$, the concentration of the drug after t hours.
(b) Determine the concentration of the drug after 8 hours.

(T) **35. Bald Eagle Population** The number $N(t)$ of bald eagle breeding pairs in the lower 48 states for the years 1963 through 2006 can be modeled by

$$N(t) = 315.5(1.078)^t, \quad 3 \leq t \leq 46$$

where t is the year, with $t = 3$ corresponding to 1963. (Source: U.S. Fish and Wildlife Service)

(a) Use a graphing utility to graph the function.
(b) Use the model to estimate the number of breeding pairs in 1990.
(c) Use the model to estimate the number of breeding pairs in 2000.
(d) Use the *zoom* and *trace* features of a graphing utility to predict the year the number of breeding pairs will reach 15,000.
(e) Does your answer to part (d) seem reasonable? Explain.

Section 4.2
Natural Exponential Functions

- Evaluate and graph functions involving the natural exponential function.
- Solve real-life problems involving the natural exponential function.

Natural Exponential Functions

In Section 4.1, exponential functions were introduced using an unspecified base a. In calculus, the most convenient (or natural) choice for a base is the irrational number e, whose decimal approximation is

$$e \approx 2.71828182846.$$

Although this choice of base may seem unusual, its convenience will become apparent as the rules for differentiating exponential functions are developed in Section 4.3. In that development, you will encounter the limit used in the definition of e.

TECHNOLOGY

Try graphing $y = (1 + x)^{1/x}$ with a graphing utility. Then use the *zoom* and *trace* features to find values of y near $x = 0$. You will find that the y-values get closer and closer to the number $e \approx 2.71828$.

Limit Definition of e

The irrational number e is defined to be the limit of $(1 + x)^{1/x}$ as $x \to 0$. That is,

$$\lim_{x \to 0} (1 + x)^{1/x} = e.$$

Example 1 Investigating the Number e

Describe the behavior of the function

$$f(x) = (1 + x)^{1/x}$$

at values of x that are close to 0.

SOLUTION One way to examine the values of $f(x)$ near 0 is to construct a table.

x	-0.01	-0.001	-0.0001	0.0001	0.001	0.01
$(1 + x)^{1/x}$	2.7320	2.7196	2.7184	2.7181	2.7169	2.7048

From the table, it appears that the closer x gets to 0, the closer $(1 + x)^{1/x}$ gets to e. This can be confirmed by the graph of the function, as shown in Figure 4.5.

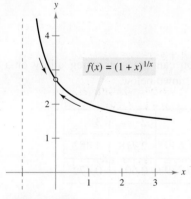

FIGURE 4.5

✓ CHECKPOINT 1

Complete the following table. Do the values of $(1 + x)^{1/x}$ get closer and closer to the number $e \approx 2.71828182846$?

x	0.00001	0.000001	0.0000001	0.00000001
$(1 + x)^{1/x}$				

266 **CHAPTER 4** Exponential and Logarithmic Functions

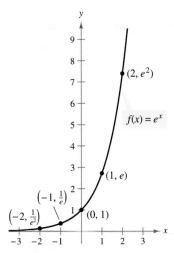

FIGURE 4.6

Example 2 Graphing the Natural Exponential Function

Sketch the graph of $f(x) = e^x$.

SOLUTION Begin by evaluating the function for several values of x, as shown in the table.

x	-2	-1	0	1	2
$f(x)$	$e^{-2} \approx 0.135$	$e^{-1} \approx 0.368$	$e^0 \approx 1$	$e^1 \approx 2.718$	$e^2 \approx 7.389$

The graph of $f(x) = e^x$ is shown in Figure 4.6. Note that e^x is positive for all values of x. Moreover, the graph has the x-axis as a horizontal asymptote to the left. That is, $\lim_{x \to -\infty} e^x = 0$.

✓ CHECKPOINT 2

Complete the table of values for $f(x) = e^{-x}$. Sketch the graph of the function.

x	-2	-1	0	1	2
$f(x)$					

Example 3 Graphing Natural Exponential Functions

Sketch the graph of each natural exponential function.

a. $f(x) = e^{0.5x}$

b. $g(x) = e^{-0.25x}$

SOLUTION To sketch these two graphs, you can use a graphing utility or a spreadsheet to construct a table of values, as shown below.

x	-3	-2	-1	0	1	2	3
$f(x)$	0.223	0.368	0.607	1.000	1.649	2.718	4.482
$g(x)$	2.117	1.649	1.284	1.000	0.779	0.607	0.472

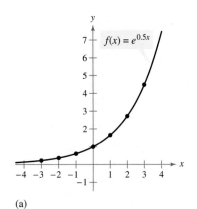

(a)

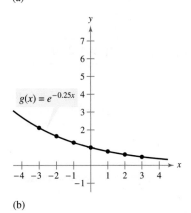

(b)

FIGURE 4.7

The graphs of f and g are shown in Figure 4.7(a) and 4.7(b), respectively. Note that the graph of f is increasing, whereas the graph of g is decreasing.

✓ CHECKPOINT 3

Sketch the graph of each natural exponential function.

a. $f(x) = e^{2x}$ **b.** $g(x) = e^{-2x}$

Applications

Example 4 MAKE A DECISION Modeling a Population

A model for the population P (in millions) of India is given by

$$P = 839.61e^{0.0177t}$$

where t is the year, with $t = 0$ corresponding to 1990. *(Source: U.S. Census Bureau)*

a. Sketch the graph of the model.

b. According to the model, was the population of India increasing from 2000 to 2006?

SOLUTION

a. The graph of the model is shown in Figure 4.8.

b. From the graph in Figure 4.8, the population was increasing from 2000 to 2006. You can verify this using a table of values, starting with 2000 ($t = 10$).

t	10	11	12	13	14	15	16
P	1002	1020	1038	1057	1076	1095	1114

Yes, according to the model, the population of India was increasing from 2000 to 2006.

Exponential functions are often used to model the growth of a quantity or a population. When the quantity's growth *is not* restricted, an exponential model is often used. When the quantity's growth *is* restricted, the best model is often a **logistic growth function** of the form

$$f(t) = \frac{a}{1 + be^{-kt}}.$$

Graphs of both types of population growth models are shown in Figure 4.9.

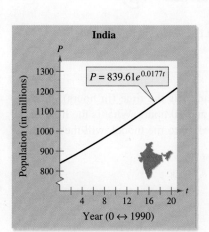

FIGURE 4.8

✓CHECKPOINT 4

A model for the population P (in millions) of Bulgaria is given by

$$P = 8.79e^{-0.0114t}$$

Where t is the year, with $t = 0$ corresponding to 1990. *(Source: U.S. Census Bureau)*

a. Sketch a graph of the model.

b. According to the model, was the population of Bulgaria increasing from 2000 to 2006? ■

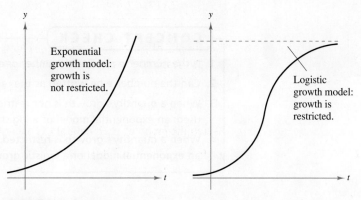

FIGURE 4.9

Example 5 MAKE A DECISION Modeling a Population

A bacterial culture is growing according to the *logistic growth model*

$$y = \frac{1.25}{1 + 0.25e^{-0.4t}}, \quad t \geq 0$$

where y is the culture weight (in grams) and t is the time (in hours). Find the weight of the culture after 0 hours, 1 hour, and 10 hours. What is the limit of the model as t increases without bound? According to the model, will the weight of the culture reach 1.5 grams?

SOLUTION

$y = \dfrac{1.25}{1 + 0.25e^{-0.4(0)}} = 1$ gram   Weight when $t = 0$

$y = \dfrac{1.25}{1 + 0.25e^{-0.4(1)}} \approx 1.071$ grams   Weight when $t = 1$

$y = \dfrac{1.25}{1 + 0.25e^{-0.4(10)}} \approx 1.244$ grams   Weight when $t = 10$

As t approaches infinity, the limit of y is

$$\lim_{t \to \infty} \frac{1.25}{1 + 0.25e^{-0.4t}} = \lim_{t \to \infty} \frac{1.25}{1 + (0.25/e^{0.4t})} = \frac{1.25}{1 + 0} = 1.25.$$

So, as t increases without bound, the weight of the culture approaches 1.25 grams. According to the model, the weight of the culture will not reach 1.5 grams. The graph of the model is shown in Figure 4.10.

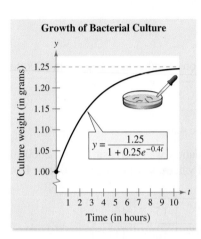

FIGURE 4.10 When a culture is grown in a dish, the size of the dish and the available food limit the culture's growth.

✓ CHECKPOINT 5

A bacterial culture is growing according to the model

$$y = \frac{1.50}{1 + 0.2e^{-0.5t}}, \quad t \geq 0$$

where y is the culture weight (in grams) and t is the time (in hours). Find the weight of the culture after 0 hours, 1 hour, and 10 hours. What is the limit of the model as t increases without bound? ∎

CONCEPT CHECK

1. Is the number e a rational number or an irrational number?
2. Can the number e be written as the ratio of two integers? Explain.
3. When a quantity's growth is not restricted, which model is more often used: an exponential model or a logistic growth model?
4. When a quantity's growth is restricted, which model is more often used: an exponential model or a logistic growth model?

SECTION 4.2 Natural Exponential Functions

Skills Review 4.2
The following warm-up exercises involve skills that were covered in earlier sections. You will use these skills in the exercise set for this section. For additional help, review Sections 1.6 and 3.5.

In Exercises 1–4, discuss the continuity of the function.

1. $f(x) = \dfrac{3x^2 + 2x + 1}{x^2 + 1}$

2. $f(x) = \dfrac{x + 1}{x^2 - 4}$

3. $f(x) = \dfrac{x^2 - 6x + 5}{x^2 - 3}$

4. $g(x) = \dfrac{x^2 - 9x + 20}{x - 4}$

In Exercises 5–12, find the limit.

5. $\lim\limits_{x \to \infty} \dfrac{25}{1 + 4x}$

6. $\lim\limits_{x \to \infty} \dfrac{16x}{3 + x^2}$

7. $\lim\limits_{x \to \infty} \dfrac{8x^3 + 2}{2x^3 + x}$

8. $\lim\limits_{x \to \infty} \dfrac{x}{2x}$

9. $\lim\limits_{x \to \infty} \dfrac{3}{2 + (1/x)}$

10. $\lim\limits_{x \to \infty} \dfrac{6}{1 + x^{-2}}$

11. $\lim\limits_{x \to \infty} 2^{-x}$

12. $\lim\limits_{x \to \infty} \dfrac{7}{1 + 5x}$

Exercises 4.2

See www.CalcChat.com for worked-out solutions to odd-numbered exercises.

In Exercises 1–4, use the properties of exponents to simplify the expression.

1. (a) $(e^3)(e^4)$ (b) $(e^3)^4$
   (c) $(e^3)^{-2}$ (d) e^0

2. (a) $\left(\dfrac{1}{e}\right)^{-2}$ (b) $\left(\dfrac{e^5}{e^2}\right)^{-1}$
   (c) $\dfrac{e^5}{e^3}$ (d) $\dfrac{1}{e^{-3}}$

3. (a) $(e^2)^{5/2}$ (b) $(e^2)(e^{1/2})$
   (c) $(e^{-2})^{-3}$ (d) $\dfrac{e^5}{e^{-2}}$

4. (a) $(e^{-3})^{2/3}$ (b) $\dfrac{e^4}{e^{-1/2}}$
   (c) $(e^{-2})^{-4}$ (d) $(e^{-4})(e^{-3/2})$

In Exercises 5–10, match the function with its graph. [The graphs are labeled (a)–(f).]

(a)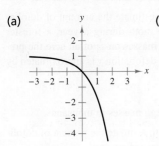

(b)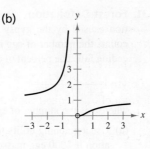

(c)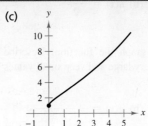

(d)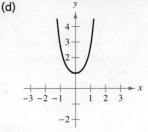

(e)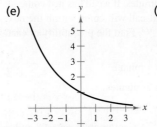

(f)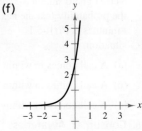

5. $f(x) = e^{2x+1}$

6. $f(x) = e^{-x/2}$

7. $f(x) = e^{x^2}$

8. $f(x) = e^{-1/x}$

9. $f(x) = e^{\sqrt{x}}$

10. $f(x) = -e^x + 1$

In Exercises 11–14, sketch the graph of the function.

11. $h(x) = e^{x-3}$

12. $f(x) = e^{3x}$

13. $g(x) = e^{1-x}$

14. $j(x) = e^{-x+2}$

In Exercises 15–18, use a graphing utility to graph the function. Be sure to choose an appropriate viewing window.

15. $N(t) = 500e^{-0.2t}$

16. $A(t) = 500e^{0.15t}$

17. $g(x) = \dfrac{2}{1 + e^{x^2}}$

18. $g(x) = \dfrac{10}{1 + e^{-x}}$

In Exercises 19–22, use a graphing utility to graph the function. Determine whether the function has any horizontal asymptotes and discuss the continuity of the function.

19. $f(x) = \dfrac{e^x + e^{-x}}{2}$

20. $f(x) = \dfrac{e^x - e^{-x}}{2}$

21. $f(x) = \dfrac{2}{1 + e^{1/x}}$

22. $f(x) = \dfrac{2}{1 + 2e^{-0.2x}}$

23. Use a graphing utility to graph $f(x) = e^x$ and the given function in the same viewing window. How are the two graphs related?

   (a) $g(x) = e^{x-2}$

   (b) $h(x) = -\dfrac{1}{2}e^x$

   (c) $q(x) = e^x + 3$

24. Use a graphing utility to graph the function. Describe the shape of the graph for very large and very small values of x.

   (a) $f(x) = \dfrac{8}{1 + e^{-0.5x}}$

   (b) $g(x) = \dfrac{8}{1 + e^{-0.5/x}}$

25. **Probability** The average time between incoming calls at a coast guard station is 3 minutes. If a call has just come in, the probability that the next call will come within the next t minutes is $P(t) = 1 - e^{-t/3}$. Find the probability of each situation.

   (a) A call comes in within $\frac{1}{2}$ minute.

   (b) A call comes in within 2 minutes.

   (c) A call comes in within 5 minutes.

26. **Consumer Awareness** An automobile gets 28 miles per gallon at speeds of up to and including 50 miles per hour. At speeds greater than 50 miles per hour, the number of miles per gallon drops at the rate of 12% for each 10 miles per hour. If s is the speed (in miles per hour) and y is the number of miles per gallon, then $y = 28e^{0.6 - 0.012s}$, $s > 50$. Use this information and a spreadsheet to complete the table. What can you conclude?

Speed (s)	50	55	60	65	70
Miles per gallon (y)					

27. **Population** The population P (in thousands) of Las Vegas, Nevada from 1960 through 2005 can be modeled by $P = 68.4e^{0.0467t}$, where t is the time in years, with $t = 0$ corresponding to 1960. *(Source: U.S. Census Bureau)*

   (a) Find the populations in 1960, 1970, 1980, 1990, 2000, and 2005.

   (b) Explain why the data do not fit a linear model.

   (c) Use the model to estimate when the population will exceed 900,000.

28. **Biology** The population y of a bacterial culture can be modeled by the logistic growth function $y = 925/(1 + e^{-0.3t})$, where t is the time in days.

   (a) Use a graphing utility to graph the model.

   (b) Does the population have a limit as t increases without bound? Explain your answer.

   (c) How would the limit change if the model were $y = 1000/(1 + e^{-0.3t})$? Explain your answer. Draw some conclusions about this type of model.

29. **Biology: Cell Division** Suppose that you have a single imaginary bacterium able to divide to form two new cells every 30 seconds. Make a table of values for the number of individuals in the population over 30-second intervals up to 5 minutes. Graph the points and use a graphing utility to fit an exponential model to the data. *(Source: Adapted from Levine/Miller, Biology: Discovering Life, Second Edition)*

30. **Learning Theory** In a learning theory project, the proportion P of correct responses after n trials can be modeled by

$$P = \dfrac{0.83}{1 + e^{-0.2n}}.$$

   (a) Use a graphing utility to estimate the proportion of correct responses after 10 trials. Verify your result analytically.

   (b) Use a graphing utility to estimate the number of trials required to have a proportion of correct responses of 0.75.

   (c) Does the proportion of correct responses have a limit as n increases without bound? Explain your answer.

31. **Forest Defoliation** To estimate the amount of defoliation caused by the gypsy moth during a year, a forester counts the number of egg masses on $\frac{1}{40}$ of an acre the preceding fall. The percent of defoliation y is approximated by

$$y = \dfrac{300}{3 + 17e^{-0.0625x}}$$

where x is the number of egg masses in thousands.

   (a) Use a graphing utility to estimate the percent of defoliation if 2000 egg masses are counted.

   (b) Use a graphing utility to estimate the number of egg masses that existed if you observed that approximately 66.67% of a forest was defoliated.

Section 4.3
Derivatives of Exponential Functions

- Find the derivatives of natural exponential functions.
- Use calculus to analyze the graphs of functions that involve the natural exponential function.
- Explore the normal probability density function.

DISCOVERY

Use a spreadsheet software program to compare the expressions $e^{\Delta x}$ and $1 + \Delta x$ for values of Δx near 0.

Δx	$e^{\Delta x}$	$1 + \Delta x$
0.1		
0.01		
0.001		

What can you conclude? Explain how this result is used in the development of the derivative of $f(x) = e^x$.

Derivatives of Exponential Functions

In Section 4.2, it was stated that the most convenient base for exponential functions is the irrational number e. The convenience of this base stems primarily from the fact that the function $f(x) = e^x$ *is its own derivative*. You will see that this is not true of other exponential functions of the form $y = a^x$ where $a \neq e$. To verify that $f(x) = e^x$ is its own derivative, notice that the limit

$$\lim_{\Delta x \to 0} (1 + \Delta x)^{1/\Delta x} = e$$

implies that for small values of Δx, $e \approx (1 + \Delta x)^{1/\Delta x}$, or $e^{\Delta x} \approx 1 + \Delta x$. This approximation is used in the following derivation.

$$f'(x) = \lim_{\Delta x \to 0} \frac{f(x + \Delta x) - f(x)}{\Delta x} \quad \text{Definition of derivative}$$

$$= \lim_{\Delta x \to 0} \frac{e^{x + \Delta x} - e^x}{\Delta x} \quad \text{Use } f(x) = e^x.$$

$$= \lim_{\Delta x \to 0} \frac{e^x(e^{\Delta x} - 1)}{\Delta x} \quad \text{Factor numerator.}$$

$$= \lim_{\Delta x \to 0} \frac{e^x[(1 + \Delta x) - 1]}{\Delta x} \quad \text{Substitute } 1 + \Delta x \text{ for } e^{\Delta x}.$$

$$= \lim_{\Delta x \to 0} \frac{e^x(\Delta x)}{\Delta x} \quad \text{Divide out like factor.}$$

$$= \lim_{\Delta x \to 0} e^x \quad \text{Simplify.}$$

$$= e^x \quad \text{Evaluate limit.}$$

If u is a function of x, you can apply the Chain Rule to obtain the derivative of e^u with respect to x. Both formulas are summarized below.

Derivative of the Natural Exponential Function

Let u be a differentiable function of x.

1. $\dfrac{d}{dx}[e^x] = e^x$
2. $\dfrac{d}{dx}[e^u] = e^u \dfrac{du}{dx}$

TECHNOLOGY

Let $f(x) = e^x$. Use a graphing utility to evaluate $f(x)$ and the numerical derivative of $f(x)$ at each x-value. Explain the results.

a. $x = -2$ b. $x = 0$ c. $x = 2$

Example 1 Interpreting a Derivative

Find the slopes of the tangent lines to

$$f(x) = e^x \qquad \text{Original function}$$

at the points $(0, 1)$ and $(1, e)$. What conclusion can you make?

SOLUTION Because the derivative of f is

$$f'(x) = e^x \qquad \text{Derivative}$$

it follows that the slope of the tangent line to the graph of f is

$$f'(0) = e^0 = 1 \qquad \text{Slope at point } (0, 1)$$

at the point $(0, 1)$ and

$$f'(1) = e^1 = e \qquad \text{Slope at point } (1, e)$$

at the point $(1, e)$, as shown in Figure 4.11. From this pattern, you can see that the slope of the tangent line to the graph of $f(x) = e^x$ at any point (x, e^x) is equal to the y-coordinate of the point.

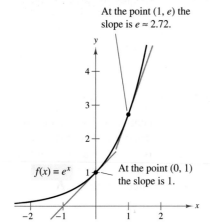

FIGURE 4.11

At the point $(1, e)$ the slope is $e \approx 2.72$.
At the point $(0, 1)$ the slope is 1.
$f(x) = e^x$

✓ CHECKPOINT 1

Find the equations of the tangent lines to $f(x) = e^x$ at the points $(0, 1)$ and $(1, e)$. ∎

Example 2 Differentiating Exponential Functions

Differentiate each function.

a. $f(x) = e^{2x}$ **b.** $f(x) = e^{-3x^2}$

c. $f(x) = 6e^{x^3}$ **d.** $f(x) = e^{-x}$

SOLUTION

a. Let $u = 2x$. Then $du/dx = 2$, and you can apply the Chain Rule.

$$f'(x) = e^u \frac{du}{dx} = e^{2x}(2) = 2e^{2x}$$

b. Let $u = -3x^2$. Then $du/dx = -6x$, and you can apply the Chain Rule.

$$f'(x) = e^u \frac{du}{dx} = e^{-3x^2}(-6x) = -6xe^{-3x^2}$$

c. Let $u = x^3$. Then $du/dx = 3x^2$, and you can apply the Chain Rule.

$$f'(x) = 6e^u \frac{du}{dx} = 6e^{x^3}(3x^2) = 18x^2 e^{x^3}$$

d. Let $u = -x$. Then $du/dx = -1$, and you can apply the Chain Rule.

$$f'(x) = e^u \frac{du}{dx} = e^{-x}(-1) = -e^{-x}$$

STUDY TIP

In Example 2, notice that when you differentiate an exponential function, the exponent does not change. For instance, the derivative of $y = e^{3x}$ is $y' = 3e^{3x}$. In both the function and its derivative, the exponent is $3x$.

✓ CHECKPOINT 2

Differentiate each function.

a. $f(x) = e^{3x}$

b. $f(x) = e^{-2x^3}$

c. $f(x) = 4e^{x^2}$

d. $f(x) = e^{-2x}$ ∎

SECTION 4.3 Derivatives of Exponential Functions

The differentiation rules that you studied in Chapter 2 can be used with exponential functions, as shown in Example 3.

Example 3 Differentiating Exponential Functions

Differentiate each function.

a. $f(x) = xe^x$ **b.** $f(x) = \dfrac{e^x - e^{-x}}{2}$

c. $f(x) = \dfrac{e^x}{x}$ **d.** $f(x) = xe^x - e^x$

SOLUTION

a. $f(x) = xe^x$ Write original function.
$f'(x) = xe^x + e^x(1)$ Product Rule
$= xe^x + e^x$ Simplify.

b. $f(x) = \dfrac{e^x - e^{-x}}{2}$ Write original function.
$= \tfrac{1}{2}(e^x - e^{-x})$ Rewrite.
$f'(x) = \tfrac{1}{2}(e^x + e^{-x})$ Constant Multiple Rule

c. $f(x) = \dfrac{e^x}{x}$ Write original function.
$f'(x) = \dfrac{xe^x - e^x(1)}{x^2}$ Quotient Rule
$= \dfrac{e^x(x - 1)}{x^2}$ Simplify.

d. $f(x) = xe^x - e^x$ Write original function.
$f'(x) = [xe^x + e^x(1)] - e^x$ Product and Difference Rules
$= xe^x + e^x - e^x$
$= xe^x$ Simplify.

✓ CHECKPOINT 3

Differentiate each function.

a. $f(x) = x^2 e^x$ **b.** $f(x) = \dfrac{e^x + e^{-x}}{2}$

c. $f(x) = \dfrac{e^x}{x^2}$ **d.** $f(x) = x^2 e^x - e^x$ ■

TECHNOLOGY

If you have access to a symbolic differentiation utility, try using it to find the derivatives of the functions in Example 3.

Applications

In Chapter 3, you learned how to use derivatives to analyze the graphs of functions. The next example applies those techniques to a function composed of exponential functions. In the example, notice that $e^a = e^b$ implies that $a = b$.

Example 4 Analyzing a Catenary

© Don Hammond/Design Pics/CORBIS

Utility wires strung between poles have the shape of a catenary.

When a telephone wire is hung between two poles, the wire forms a U-shaped curve called a **catenary**. For instance, the function

$$y = 30(e^{x/60} + e^{-x/60}), \quad -30 \leq x \leq 30$$

models the shape of a telephone wire strung between two poles that are 60 feet apart (x and y are measured in feet). Show that the lowest point on the wire is midway between the two poles. How much does the wire sag between the two poles?

SOLUTION The derivative of the function is

$$y' = 30\left[e^{x/60}\left(\tfrac{1}{60}\right) + e^{-x/60}\left(-\tfrac{1}{60}\right)\right]$$
$$= \tfrac{1}{2}(e^{x/60} - e^{-x/60}).$$

To find the critical numbers, set the derivative equal to zero.

$\tfrac{1}{2}(e^{x/60} - e^{-x/60}) = 0$	Set derivative equal to 0.
$e^{x/60} - e^{-x/60} = 0$	Multiply each side by 2.
$e^{x/60} = e^{-x/60}$	Add $e^{-x/60}$ to each side.
$\dfrac{x}{60} = -\dfrac{x}{60}$	If $e^a = e^b$, then $a = b$.
$x = -x$	Multiply each side by 60.
$2x = 0$	Add x to each side.
$x = 0$	Divide each side by 2.

Using the First-Derivative Test, you can determine that the critical number $x = 0$ yields a relative minimum of the function. From the graph in Figure 4.12, you can see that this relative minimum is actually a minimum on the interval $[-30, 30]$. To find how much the wire sags between the two poles, you can compare its height at each pole with its height at the midpoint.

$y = 30(e^{-30/60} + e^{-(-30)/60}) \approx 67.7$ feet	Height at left pole
$y = 30(e^{0/60} + e^{-(0)/60}) = 60$ feet	Height at midpoint
$y = 30(e^{30/60} + e^{-(30)/60}) \approx 67.7$ feet	Height at right pole

From this, you can see that the wire sags about 7.7 feet.

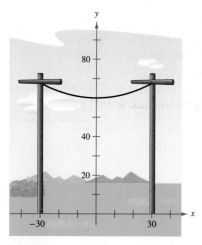

FIGURE 4.12

✓CHECKPOINT 4

Use a graphing utility to graph the function in Example 4. Verify the minimum value. Use the information in the example to choose an appropriate viewing window. ∎

Example 5 Forensics

At 8:30 A.M., a coroner was called to the home of a person who had died during the night in a room that had a constant temperature of 70°F. In order to estimate the time of death, the coroner took the person's temperature twice. From these two temperatures, the coroner was able to relate body temperature and time elapsed since death using the model

$$T = 28.6e^{-0.09996t} + 70, \quad t \geq 0$$

where T is the body temperature (in degrees Fahrenheit) and t is the time elapsed (in hours) since the person died.

a. Use the model to estimate the rate of change of the body temperature when $t = 1$.

b. Use the model to estimate the time of death. Assume the body temperature was 85.7°F at 9:00 A.M.

STUDY TIP

The formula used in Example 5 is derived from a general cooling principle called *Newton's Law of Cooling*. You will learn more about this principle later in Section 10.4.

SOLUTION

a. The derivative of the function is

$$T' = 28.6e^{-0.09996t}(-0.09996)$$
$$= -2.858856e^{-0.09996t}.$$

The rate of change of the body temperature when $t = 1$ (one hour since the person died) was

$$T' = -2.858856e^{-0.09996(1)} \approx -2.59.$$

So, the rate of change of the body temperature when $t = 1$ was -2.59°F per hour.

b. To estimate the time of death, you can use a graphical approach. After experimenting to find a reasonable viewing window, you can obtain a graph of T that is similar to that shown in Figure 4.13. Using the *trace* feature, you can determine that $T = 85.7$ when $t \approx 6$. This means that at 9:00 A.M., six hours had elapsed since the person died. So, the person died at 3:00 A.M.

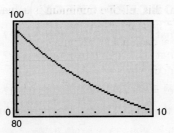

FIGURE 4.13

✓ CHECKPOINT 5

Use the model in Example 5 to estimate the rate of change of the body temperature when $t = 6$. ∎

The Normal Probability Density Function

If you take a course in statistics or biostatistics, you will spend quite a bit of time studying the characteristics and use of the **normal probability density function** given by

$$f(x) = \frac{1}{\sigma\sqrt{2\pi}} e^{-(x-\mu)^2/2\sigma^2}$$

where σ is the lowercase Greek letter sigma, and μ is the lowercase Greek letter mu. In this formula, σ represents the *standard deviation* of the probability distribution, and μ represents the *mean* of the probability distribution.

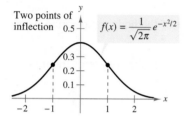

FIGURE 4.14 The graph of the normal probability density function is bell-shaped.

Example 6 Exploring a Probability Density Function

Show that the graph of the normal probability density function

$$f(x) = \frac{1}{\sqrt{2\pi}} e^{-x^2/2} \qquad \text{Original function}$$

has points of inflection at $x = \pm 1$.

SOLUTION Begin by finding the second derivative of the function.

$$f'(x) = \frac{1}{\sqrt{2\pi}}(-x)e^{-x^2/2} \qquad \text{First derivative}$$

$$f''(x) = \frac{1}{\sqrt{2\pi}}[(-x)(-x)e^{-x^2/2} + (-1)e^{-x^2/2}] \qquad \text{Second derivative}$$

$$= \frac{1}{\sqrt{2\pi}}(e^{-x^2/2})(x^2 - 1) \qquad \text{Simplify.}$$

By setting the second derivative equal to 0, you can determine that $x = \pm 1$. By testing the concavity of the graph, you can then conclude that these x-values yield points of inflection, as shown in Figure 4.14.

✓ CHECKPOINT 6

Graph the normal probability density function

$$f(x) = \frac{1}{4\sqrt{2\pi}} e^{-x^2/32}$$

and approximate the points of inflection. ∎

CONCEPT CHECK

1. What is the derivative of $f(x) = e^x$?
2. What is the derivative of $f(x) = e^u$? (Assume that u is a differentiable function of x.)
3. If $e^a = e^b$, then a is equal to what?
4. In the normal probability density function given by

$$f(x) = \frac{1}{\sigma\sqrt{2\pi}} e^{-(x-\mu)^2/2\sigma^2}$$

identify what is represented by (a) σ and (b) μ.

SECTION 4.3 Derivatives of Exponential Functions

Skills Review 4.3

The following warm-up exercises involve skills that were covered in earlier sections. You will use these skills in the exercise set for this section. For additional help, review Sections 0.4, 2.4, and 3.2.

In Exercises 1–4, factor the expression.

1. $x^2 e^x - \frac{1}{2}e^x$
2. $(xe^{-x})^{-1} + e^x$
3. $xe^x - e^{2x}$
4. $e^x - xe^{-x}$

In Exercises 5–8, find the derivative of the function.

5. $f(x) = \dfrac{3}{7x^2}$
6. $g(x) = 3x^2 - \dfrac{x}{6}$
7. $f(x) = (4x - 3)(x^2 + 9)$
8. $f(t) = \dfrac{t - 2}{\sqrt{t}}$

In Exercises 9 and 10, find the relative extrema of the function.

9. $f(x) = \frac{1}{8}x^3 - 2x$
10. $f(x) = x^4 - 2x^2 + 5$

Exercises 4.3

See www.CalcChat.com for worked-out solutions to odd-numbered exercises.

In Exercises 1–4, find the slope of the tangent line to the exponential function at the point (0, 1).

1. $y = e^{3x}$

2. $y = e^{2x}$

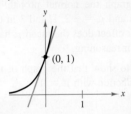

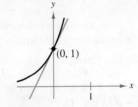

3. $y = e^{-x}$
4. $y = e^{-2x}$

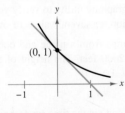

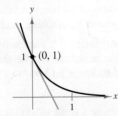

In Exercises 5–16, find the derivative of the function.

5. $y = e^{5x}$
6. $y = e^{1-x}$
7. $y = e^{-x^2}$
8. $f(x) = e^{1/x}$
9. $f(x) = e^{-1/x^2}$
10. $g(x) = e^{\sqrt{x}}$
11. $f(x) = (x^2 + 1)e^{4x}$
12. $y = 4x^3 e^{-x}$
13. $f(x) = \dfrac{2}{(e^x + e^{-x})^3}$
14. $f(x) = \dfrac{(e^x + e^{-x})^4}{2}$
15. $y = xe^x - 4e^{-x}$
16. $y = x^2 e^x - 2xe^x + 2e^x$

In Exercises 17–22, determine an equation of the tangent line to the function at the given point.

17. $y = e^{-2x + x^2}$, $(2, 1)$
18. $g(x) = e^{x^3}$, $\left(-1, \dfrac{1}{e}\right)$
19. $y = x^2 e^{-x}$, $\left(2, \dfrac{4}{e^2}\right)$
20. $y = \dfrac{x}{e^{2x}}$, $\left(1, \dfrac{1}{e^2}\right)$
21. $y = (e^{2x} + 1)^3$, $(0, 8)$
22. $y = (e^{4x} - 2)^2$, $(0, 1)$

In Exercises 23–26, find the second derivative.

23. $f(x) = 2e^{3x} + 3e^{-2x}$
24. $f(x) = (1 + 2x)e^{4x}$
25. $f(x) = 5e^{-x} - 2e^{-5x}$
26. $f(x) = (3 + 2x)e^{-3x}$

In Exercises 27–30, graph and analyze the function. Include extrema, points of inflection, and asymptotes in your analysis.

27. $f(x) = \dfrac{1}{2 - e^{-x}}$
28. $f(x) = \dfrac{e^x - e^{-x}}{2}$
29. $f(x) = x^2 e^{-x}$
30. $f(x) = xe^{-x}$

Ⓣ In Exercises 31 and 32, use a graphing utility to graph the function. Determine any asymptotes of the graph.

31. $f(x) = \dfrac{8}{1 + e^{-0.5x}}$
32. $g(x) = \dfrac{8}{1 + e^{-0.5/x}}$

In Exercises 33–36, solve the equation for x.

33. $e^{-3x} = e$

34. $e^x = 1$

35. $e^{\sqrt{x}} = e^3$

36. $e^{-1/x} = e^{1/2}$

37. **Sharks** The length l of a tiger shark (in centimeters) can be modeled by
$$l = 337 - 276e^{-0.178t}$$
where t is the shark's age (in years). Use the model to estimate the rate of change of the body length when $t = 3$.

38. **Yeast Growth** The amount Y of yeast in a culture can be modeled by
$$Y = \frac{663}{1 + 72e^{-0.547t}}, \quad 0 \le t \le 18$$
where t represents the time (in hours). Find the rates at which the amount of yeast in the culture is changing when (a) $t = 1$, (b) $t = 9$, and (c) $t = 16$.

39. **Ebbinghaus Model** The *Ebbinghaus Model* for human memory is $p = (100 - a)e^{-bt} + a$, where p is the percent retained after t weeks. (The constants a and b vary from one person to another.) If $a = 20$ and $b = 0.5$, at what rate is information being retained after 1 week? After 3 weeks?

40. **Agriculture** The yield V (in pounds per acre) for an orchard at age t (in years) is modeled by
$$V = 7955.6e^{-0.0458/t}.$$
At what rates is the yield changing when (a) $t = 5$ years, (b) $t = 10$ years, and (c) $t = 25$ years?

41. **Employment** For the years 1996 through 2005, the number y (in millions) of employed people in the United States can be modeled by
$$y = 98.020 + 6.2472t - 0.24964t^2 + 0.000002e^t$$
where t represents the year, with $t = 6$ corresponding to 1996. *(Source: U.S. Bureau of Labor Statistics)*

(a) Use a graphing utility to graph the model.

(b) Use the graph to estimate the rates of change in the number of employed people in 1996, 2000, and 2005.

(c) Confirm the results of part (b) analytically.

42. **Population Growth** A conservation organization releases 100 animals of an endangered species into a game preserve. The organization believes that the preserve has a carrying capacity of 1000 animals and that the growth p of the pack will be modeled by the logistic curve
$$p = \frac{1000}{1 + 9e^{-0.1656t}}$$
where t is measured in years.

(a) Use a graphing utility to graph the model.

(b) Use the graph to estimate when the rate of change of the number of endangered species will begin to decrease.

43. **Probability** A survey of high school seniors from a certain school district who took the SAT has determined that the mean score on the mathematics portion was 650 with a standard deviation of 12.5.

(a) Assuming the data can be modeled by a normal probability density function, find a model for these data.

(b) Use a graphing utility to graph the model. Be sure to choose an appropriate viewing window.

(c) Find the derivative of the model.

(d) Show that $f' > 0$ for $x < \mu$ and $f' < 0$ for $x > \mu$.

44. **Probability** A survey of a college freshman class has determined that the mean height of females in the class is 64 inches with a standard deviation of 3.2 inches.

(a) Assuming the data can be modeled by a normal probability density function, find a model for these data.

(b) Use a graphing utility to graph the model. Be sure to choose an appropriate viewing window.

(c) Find the derivative of the model.

(d) Show that $f' > 0$ for $x < \mu$ and $f' < 0$ for $x > \mu$.

45. Use a graphing utility to graph the normal probability density function with $\mu = 0$ and $\sigma = 2, 3,$ and 4 in the same viewing window. What effect does the standard deviation σ have on the function? Explain your reasoning.

46. Use a graphing utility to graph the normal probability density function with $\sigma = 1$ and $\mu = -2, 1,$ and 3 in the same viewing window. What effect does the mean μ have on the function? Explain your reasoning.

47. Use Example 6 as a model to show that the graph of the normal probability density function with $\mu = 0$
$$f(x) = \frac{1}{\sigma\sqrt{2\pi}}e^{-x^2/2\sigma^2}$$
has points of inflection at $x = \pm\sigma$. What is the maximum value of the function? Use a graphing utility to verify your answer by graphing the function for several values of σ.

48. **Athletics** A parachutist jumps from a plane and opens the parachute at a height of 2000 feet. The height of the parachutist is
$$h = 1950 + 50e^{-1.6t} - 20t$$
where h is the height (in feet) and t is the time (in seconds) since the parachute opened.

(a) Find dh/dt and use a graphing utility to graph dh/dt.

(b) Evaluate dh/dt for $t = 0, 1, 5, 10,$ and 15.

(c) Interpret your results for parts (a) and (b).

Mid-Chapter Quiz

See www.CalcChat.com for worked-out solutions to odd-numbered exercises.

Take this quiz as you would take a quiz in class. When you are done, check your work against the answers given in the back of the book.

In Exercises 1–4, evaluate each expression.

1. $4(4^2)$
2. $\left(\dfrac{2}{3}\right)^3$
3. $81^{1/3}$
4. $\left(\dfrac{4}{9}\right)^2$

In Exercises 5–12, use properties of exponents to simplify the expression.

5. $4^3(4^2)$
6. $\left(\dfrac{1}{6}\right)^{-3}$
7. $\dfrac{3^8}{3^5}$
8. $(5^{1/2})(3^{1/2})$
9. $(e^2)(e^5)$
10. $(e^{2/3})(e^3)$
11. $\dfrac{e^2}{e^{-4}}$
12. $(e^{-1})^{-3}$

(T) In Exercises 13–18, use a graphing utility to graph the function.

13. $f(x) = 3^x - 2$
14. $f(x) = 5^{-x} + 2$
15. $f(x) = 6^{x-3}$
16. $f(x) = e^{x+2}$
17. $f(x) = 250e^{0.15x}$
18. $f(x) = \dfrac{5}{1 + e^x}$

19. A survey has determined that the mean time a student uses a math lab is 5.4 hours per week with a standard deviation of 0.5 hour.

   (a) Assuming the data can be modeled by a normal probability density function, find a model for these data.

   (T) (b) Use a graphing utility to graph the model. Be sure to choose an appropriate viewing window.

   (c) Find the derivative of the model.

   (d) Show that $f' > 0$ for $x < \mu$ and $f' < 0$ for $x > \mu$.

In Exercises 20–23, find the derivative of the function.

20. $y = e^{5x}$
21. $y = e^{x-4}$
22. $y = 5e^{x+2}$
23. $y = 3e^x - xe^x$

24. Determine an equation of the tangent line to $y = e^{-2x}$ at the point $(0, 1)$.

25. Graph and analyze the function $f(x) = 0.5x^2 e^{-0.5x}$. Include extrema, points of inflection, and asymptotes in your analysis.

Section 4.4
Logarithmic Functions

- Sketch the graphs of natural logarithmic functions.
- Use properties of logarithms to simplify, expand, and condense logarithmic expressions.
- Use inverse properties of exponential and logarithmic functions to solve exponential and logarithmic equations.
- Use properties of natural logarithms to answer questions about real-life situations.

The Natural Logarithmic Function

From your previous algebra courses, you should be somewhat familiar with logarithms. For instance, the **common logarithm** $\log_{10} x$ is defined as

$$\log_{10} x = b \quad \text{if and only if} \quad 10^b = x.$$

The base of common logarithms is 10. In calculus, the most useful base for logarithms is the number e.

Definition of the Natural Logarithmic Function

The **natural logarithmic function,** denoted by $\ln x$, is defined as

$$\ln x = b \quad \text{if and only if} \quad e^b = x.$$

$\ln x$ is read as "el en of x" or as "the natural log of x."

This definition implies that the natural logarithmic function and the natural exponential function are inverse functions. So, every logarithmic equation can be written in an equivalent exponential form and every exponential equation can be written in logarithmic form. Here are some examples.

Logarithmic form:	*Exponential form:*
$\ln 1 = 0$	$e^0 = 1$
$\ln e = 1$	$e^1 = e$
$\ln \dfrac{1}{e} = -1$	$e^{-1} = \dfrac{1}{e}$
$\ln 2 \approx 0.693$	$e^{0.693} \approx 2$

Because the functions $f(x) = e^x$ and $g(x) = \ln x$ are inverse functions, their graphs are reflections of each other in the line $y = x$. This reflective property is illustrated in Figure 4.15. The figure also contains a summary of several properties of the graph of the natural logarithmic function.

Notice that the domain of the natural logarithmic function is the set of *positive real numbers*—be sure you see that $\ln x$ is not defined for zero or for negative numbers. You can test this on your calculator. If you try evaluating $\ln(-1)$ or $\ln 0$, your calculator should indicate that the value is not a real number.

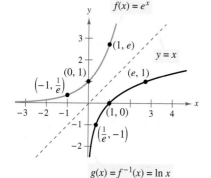

$g(x) = \ln x$
- Domain: $(0, \infty)$
- Range: $(-\infty, \infty)$
- Intercept: $(1, 0)$
- Always increasing
- $\ln x \to \infty$ as $x \to \infty$
- $\ln x \to -\infty$ as $x \to 0^+$
- Continuous
- One-to-one

FIGURE 4.15

SECTION 4.4 Logarithmic Functions

Example 1 Graphing Logarithmic Functions

Sketch the graph of each function.

a. $f(x) = \ln(x + 1)$ **b.** $f(x) = 2\ln(x - 2)$

SOLUTION

a. Because the natural logarithmic function is defined only for positive values, the domain of the function is $x + 1 > 0$, or

$x > -1.$ Domain

To sketch the graph, begin by constructing a table of values, as shown below. Then plot the points in the table and connect them with a smooth curve, as shown in Figure 4.16(a).

x	-0.5	0	0.5	1	1.5	2
$\ln(x+1)$	-0.693	0	0.405	0.693	0.916	1.099

b. The domain of this function is $x - 2 > 0$, or

$x > 2.$ Domain

A table of values for the function is shown below, and its graph is shown in Figure 4.16(b).

x	2.5	3	3.5	4	4.5	5
$2\ln(x-2)$	-1.386	0	0.811	1.386	1.833	2.197

TECHNOLOGY

What happens when you take the logarithm of a negative number? Some graphing utilities do not give an error message for $\ln(-1)$. Instead, the graphing utility displays a complex number. For the purposes of this text, however, it is assumed that the domain of the logarithmic function is the set of positive real numbers.

✓ CHECKPOINT 1

Use a graphing utility to complete the table and graph the function.

$f(x) = \ln(x + 2)$

x	-1.5	-1	-0.5
$f(x)$			

x	0	0.5	1
$f(x)$			

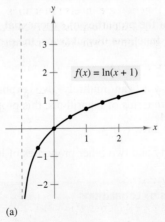
(a)

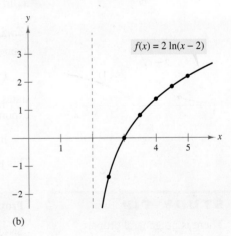
(b)

FIGURE 4.16

STUDY TIP

How does the graph of $f(x) = \ln(x + 1)$ relate to the graph of $y = \ln x$? The graph of f is a translation of the graph of $y = \ln x$ one unit to the left.

Properties of Logarithmic Functions

Recall from Section 1.4 that inverse functions have the property that

$$f(f^{-1}(x)) = x \quad \text{and} \quad f^{-1}(f(x)) = x.$$

The properties listed below follow from the fact that the natural logarithmic function and the natural exponential function are inverse functions.

Inverse Properties of Logarithms and Exponents

1. $\ln e^x = x$ 2. $e^{\ln x} = x$

Example 2 Applying Inverse Properties

Simplify each expression.

a. $\ln e^{\sqrt{2}}$ **b.** $e^{\ln 3x}$

SOLUTION

a. Because $\ln e^x = x$, it follows that
$$\ln e^{\sqrt{2}} = \sqrt{2}.$$

b. Because $e^{\ln x} = x$, it follows that
$$e^{\ln 3x} = 3x.$$

✓ CHECKPOINT 2

Simplify each expression.

a. $\ln e^3$ **b.** $e^{\ln(x+1)}$

Most of the properties of exponential functions can be rewritten in terms of logarithmic functions. For instance, the property

$$e^x e^y = e^{x+y}$$

states that you can multiply two exponential expressions by adding their exponents. In terms of logarithms, this property becomes

$$\ln xy = \ln x + \ln y.$$

This property and two other properties of logarithms are summarized below.

> **STUDY TIP**
> There is no general property that can be used to rewrite $\ln(x + y)$. Specifically, $\ln(x + y)$ is not equal to $\ln x + \ln y$.

Properties of Logarithms

1. $\ln xy = \ln x + \ln y$ 2. $\ln \dfrac{x}{y} = \ln x - \ln y$

3. $\ln x^n = n \ln x$

SECTION 4.4 Logarithmic Functions

Rewriting a logarithm of a single quantity as the sum, difference, or multiple of logarithms is called *expanding* the logarithmic expression. The reverse procedure is called *condensing* a logarithmic expression.

TECHNOLOGY

Try using a graphing utility to verify the results of Example 3(b). That is, try graphing the functions

$$y = \ln \sqrt{x^2 + 1}$$

and

$$y = \frac{1}{2}\ln(x^2 + 1).$$

Because these two functions are equivalent, their graphs should coincide.

Example 3 Expanding Logarithmic Expressions

Use the properties of logarithms to rewrite each expression as a sum, difference, or multiple of logarithms. (Assume $x > 0$ and $y > 0$.)

a. $\ln \dfrac{10}{9}$ **b.** $\ln \sqrt{x^2 + 1}$ **c.** $\ln \dfrac{xy}{5}$ **d.** $\ln[x^2(x+1)]$

SOLUTION

a. $\ln \dfrac{10}{9} = \ln 10 - \ln 9$ Property 2

b. $\ln \sqrt{x^2+1} = \ln(x^2+1)^{1/2}$ Rewrite with rational exponent.

$\quad\quad\quad\quad\quad\;\; = \tfrac{1}{2}\ln(x^2+1)$ Property 3

c. $\ln \dfrac{xy}{5} = \ln(xy) - \ln 5$ Property 2

$\quad\quad\quad\;\; = \ln x + \ln y - \ln 5$ Property 1

d. $\ln[x^2(x+1)] = \ln x^2 + \ln(x+1)$ Property 1

$\quad\quad\quad\quad\quad\;\; = 2\ln x + \ln(x+1)$ Property 3

✓ CHECKPOINT 3

Use the properties of logarithms to rewrite each expression as a sum, difference, or multiple of logarithms. (Assume $x > 0$ and $y > 0$.)

a. $\ln \dfrac{2}{5}$ **b.** $\ln \sqrt[3]{x+2}$ **c.** $\ln \dfrac{x}{5y}$ **d.** $\ln x(x+1)^2$ ■

Example 4 Condensing Logarithmic Expressions

Use the properties of logarithms to rewrite each expression as the logarithm of a single quantity. (Assume $x > 0$ and $y > 0$.)

a. $\ln x + 2\ln y$

b. $2\ln(x+2) - 3\ln x$

SOLUTION

a. $\ln x + 2\ln y = \ln x + \ln y^2$ Property 3

$\quad\quad\quad\quad\quad\;\; = \ln xy^2$ Property 1

b. $2\ln(x+2) - 3\ln x = \ln(x+2)^2 - \ln x^3$ Property 3

$\quad\quad\quad\quad\quad\quad\quad\quad\; = \ln \dfrac{(x+2)^2}{x^3}$ Property 2

✓ CHECKPOINT 4

Use the properties of logarithms to rewrite each expression as the logarithm of a single quantity. (Assume $x > 0$ and $y > 0$.)

a. $4\ln x + 3\ln y$

b. $\ln(x+1) - 2\ln(x+3)$ ■

Solving Exponential and Logarithmic Equations

The inverse properties of logarithms and exponents can be used to solve exponential and logarithmic equations, as shown in the next two examples.

STUDY TIP

In the examples on this page, note that the key step in solving an exponential equation is to take the log of each side, and the key step in solving a logarithmic equation is to exponentiate each side.

Example 5 Solving Exponential Equations

Solve each equation.

a. $e^x = 5$ **b.** $10 + e^{0.1t} = 14$

SOLUTION

a. $e^x = 5$ Write original equation.
$\ln e^x = \ln 5$ Take natural log of each side.
$x = \ln 5$ Inverse property: $\ln e^x = x$

b. $10 + e^{0.1t} = 14$ Write original equation.
$e^{0.1t} = 4$ Subtract 10 from each side.
$\ln e^{0.1t} = \ln 4$ Take natural log of each side.
$0.1t = \ln 4$ Inverse property: $\ln e^{0.1t} = 0.1t$
$t = 10 \ln 4$ Multiply each side by 10.

✓ **CHECKPOINT 5**

Solve each equation.

a. $e^x = 6$ **b.** $5 + e^{0.2t} = 10$ ∎

Example 6 Solving Logarithmic Equations

Solve each equation.

a. $\ln x = 5$ **b.** $3 + 2 \ln x = 7$

SOLUTION

a. $\ln x = 5$ Write original equation.
$e^{\ln x} = e^5$ Exponentiate each side.
$x = e^5$ Inverse property: $e^{\ln x} = x$

b. $3 + 2 \ln x = 7$ Write original equation.
$2 \ln x = 4$ Subtract 3 from each side.
$\ln x = 2$ Divide each side by 2.
$e^{\ln x} = e^2$ Exponentiate each side.
$x = e^2$ Inverse property: $e^{\ln x} = x$

✓ **CHECKPOINT 6**

Solve each equation.

a. $\ln x = 4$ **b.** $4 + 5 \ln x = 19$ ∎

SECTION 4.4 Logarithmic Functions

Application

Example 7 Human Memory Model

Students participating in a psychology experiment attended several lectures on a subject and were given an exam. Every month for a year after the exam, the students were retested to see how much of the material they remembered. The average scores for the group are given by the *human memory model*

$$f(t) = 75 - 6 \ln(t + 1), \quad 0 \le t \le 12$$

where t is the time in months. The graph of f is shown in Figure 4.17.

a. What was the average score on the original exam ($t = 0$)?

b. What was the average score at the end of $t = 2$ months?

c. What was the average score at the end of $t = 6$ months?

SOLUTION

a. The original average score was

$$\begin{aligned} f(0) &= 75 - 6 \ln(0 + 1) & &\text{Substitute 0 for } t. \\ &= 75 - 6 \ln 1 & &\text{Simplify.} \\ &= 75 - 6(0) & &\text{Property of natural logarithms} \\ &= 75. & &\text{Solution} \end{aligned}$$

b. After 2 months, the average score was

$$\begin{aligned} f(2) &= 75 - 6 \ln(2 + 1) & &\text{Substitute 2 for } t. \\ &= 75 - 6 \ln 3 & &\text{Simplify.} \\ &\approx 75 - 6(1.0986) & &\text{Use a calculator.} \\ &\approx 68.4. & &\text{Solution} \end{aligned}$$

c. After 6 months, the average score was

$$\begin{aligned} f(6) &= 75 - 6 \ln(6 + 1) & &\text{Substitute 6 for } t. \\ &= 75 - 6 \ln 7 & &\text{Simplify.} \\ &\approx 75 - 6(1.9459) & &\text{Use a calculator.} \\ &\approx 63.3. & &\text{Solution} \end{aligned}$$

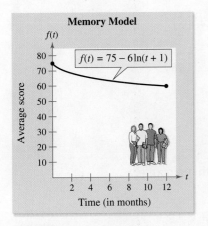

FIGURE 4.17

✓ CHECKPOINT 7

In Example 7, what was the average score at the end of $t = 12$ months?

CONCEPT CHECK

1. What are common logarithms and natural logarithms?
2. Write "logarithm of x with base 3" symbolically.
3. What are the domain and range of $f(x) = \ln x$?
4. Explain the relationship between the functions $f(x) = \ln x$ and $g(x) = e^x$.

Skills Review 4.4

The following warm-up exercises involve skills that were covered in earlier sections. You will use these skills in the exercise set for this section. For additional help, review Sections 0.1, 0.3, and 1.4.

In Exercises 1–8, use the properties of exponents to simplify the expression.

1. $(4^2)(4^{-3})$
2. $(2^3)^2$
3. $\dfrac{3^4}{3^{-2}}$
4. $\left(\dfrac{3}{2}\right)^{-3}$
5. e^0
6. $(3e)^4$
7. $\left(\dfrac{2}{e^3}\right)^{-1}$
8. $\left(\dfrac{4e^2}{25}\right)^{-3/2}$

In Exercises 9–12, solve for x.

9. $0 < x + 4$
10. $0 < x^2 + 1$
11. $0 < \sqrt{x^2 - 1}$
12. $0 < x - 5$

In Exercises 13 and 14, show that f and g are inverse functions by showing that $f(g(x)) = x$ and $g(f(x)) = x$.

13. $f(x) = \dfrac{x}{2} + 3$, $g(x) = 2x - 6$
14. $f(x) = \sqrt[3]{x - 8}$, $g(x) = x^3 + 8$

Exercises 4.4

See www.CalcChat.com for worked-out solutions to odd-numbered exercises.

In Exercises 1–8, write the logarithmic equation as an exponential equation, or vice versa.

1. $\ln 2 = 0.6931\ldots$
2. $\ln 9 = 2.1972\ldots$
3. $\ln 0.2 = -1.6094\ldots$
4. $\ln 0.05 = -2.9957\ldots$
5. $e^0 = 1$
6. $e^2 = 7.3891\ldots$
7. $e^{-3} = 0.0498\ldots$
8. $e^{0.25} = 1.2840\ldots$

In Exercises 9–12, match the function with its graph. [The graphs are labeled (a)–(d).]

9. $f(x) = 2 + \ln x$
10. $f(x) = -\ln x$
11. $f(x) = \ln(x + 2)$
12. $f(x) = -\ln(x - 1)$

In Exercises 13–18, sketch the graph of the function.

13. $y = \ln(x - 1)$
14. $y = \ln|x|$
15. $y = \ln 2x$
16. $y = 5 + \ln x$
17. $y = 3 \ln x$
18. $y = \frac{1}{4} \ln x$

(T) In Exercises 19–22, analytically show that the functions are inverse functions. Then use a graphing utility to show this graphically.

19. $f(x) = e^{2x}$
    $g(x) = \ln \sqrt{x}$
20. $f(x) = e^x - 1$
    $g(x) = \ln(x + 1)$
21. $f(x) = e^{2x-1}$
    $g(x) = \frac{1}{2} + \ln \sqrt{x}$
22. $f(x) = e^{x/3}$
    $g(x) = \ln x^3$

In Exercises 23–28, apply the inverse properties of logarithmic and exponential functions to simplify the expression.

23. $\ln e^{x^2}$
24. $\ln e^{2x-1}$
25. $e^{\ln(5x+2)}$
26. $e^{\ln \sqrt{x}}$
27. $-1 + \ln e^{2x}$
28. $-8 + e^{\ln x^3}$

(a)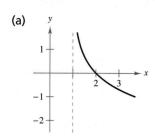

(b)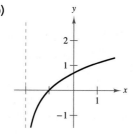

(c)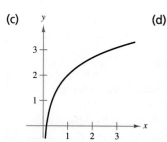

(d)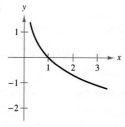

In Exercises 29 and 30, use the properties of logarithms and the fact that $\ln 2 \approx 0.6931$ and $\ln 3 \approx 1.0986$ to approximate the logarithm. Then use a calculator to confirm your approximation.

29. (a) $\ln 6$ (b) $\ln \frac{3}{2}$ (c) $\ln 81$ (d) $\ln \sqrt{3}$
30. (a) $\ln 0.25$ (b) $\ln 24$ (c) $\ln \sqrt[3]{12}$ (d) $\ln \frac{1}{72}$

In Exercises 31–40, use the properties of logarithms to write the expression as a sum, difference, or multiple of logarithms.

31. $\ln \frac{2}{3}$
32. $\ln \frac{1}{5}$
33. $\ln xyz$
34. $\ln \frac{xy}{z}$
35. $\ln \sqrt{x^2 + 1}$
36. $\ln \sqrt{\frac{x^3}{x+1}}$
37. $\ln [z(z-1)^2]$
38. $\ln(x \sqrt[3]{x^2+1})$
39. $\ln \frac{3x(x+1)}{(2x+1)^2}$
40. $\ln \frac{2x}{\sqrt{x^2-1}}$

In Exercises 41–50, write the expression as the logarithm of a single quantity.

41. $\ln(x-2) - \ln(x+2)$
42. $\ln(2x+1) + \ln(2x-1)$
43. $3 \ln x + 2 \ln y - 4 \ln z$
44. $2 \ln 3 - \frac{1}{2} \ln(x^2 + 1)$
45. $3[\ln x + \ln(x+3) - \ln(x+4)]$
46. $\frac{1}{3}[2 \ln(x+3) + \ln x - \ln(x^2 - 1)]$
47. $\frac{3}{2}[\ln x(x^2 + 1) - \ln(x+1)]$
48. $2[\ln x + \frac{1}{4} \ln(x+1)]$
49. $\frac{1}{3} \ln(x+1) - \frac{2}{3} \ln(x-1)$
50. $\frac{1}{2} \ln(x-2) + \frac{3}{2} \ln(x+2)$

In Exercises 51–74, solve for x or t.

51. $e^{\ln x} = 4$
52. $e^{\ln x^2} - 9 = 0$
53. $\ln x = 0$
54. $2 \ln x = 4$
55. $\ln 2x = 2.4$
56. $\ln 4x = 1$
57. $3 \ln 5x = 10$
58. $2 \ln 4x = 7$
59. $e^{x+1} = 4$
60. $e^{-0.5x} = 0.075$
61. $300 e^{-0.2t} = 700$
62. $400 e^{-0.0174t} = 1000$
63. $4 e^{2x-1} - 1 = 5$
64. $2 e^{-x+1} - 5 = 9$
65. $\dfrac{10}{1 + 4e^{-0.01x}} = 2.5$
66. $\dfrac{50}{1 + 12e^{-0.02x}} = 10.5$
67. $5^{2x} = 15$
68. $2^{1-x} = 6$
69. $500(1.07)^t = 1000$
70. $400(1.06)^t = 1300$
71. $\left(1 + \dfrac{0.07}{12}\right)^{12t} = 3$
72. $\left(1 + \dfrac{0.06}{12}\right)^{12t} = 5$
73. $\left(16 - \dfrac{0.878}{26}\right)^{3t} = 30$
74. $\left(4 - \dfrac{2.471}{40}\right)^{9t} = 21$

75. **Population Growth** The population P (in thousands) of Orlando, Florida for the years 1980 through 2005 can be modeled by
$$P = 131 e^{0.019t}$$
where $t = 0$ corresponds to 1980. *(Source: U.S. Census Bureau)*

(a) According to this model, what was the population of Orlando in 2005?

(b) According to this model, in what year will Orlando have a population of 300,000?

76. **Population Growth** The population P (in thousands) of Houston, Texas for the years 1980 through 2005 can be modeled by
$$P = 1576 e^{0.01t}$$
where $t = 0$ corresponds to 1980. *(Source: U.S. Census Bureau)*

(a) According to this model, what was the population of Houston in 2005?

(b) According to this model, in what year will Houston have a population of 2,500,000?

Carbon Dating In Exercises 77–80, you are given the ratio of carbon atoms in a fossil. Use the information to estimate the age of the fossil. In living organic material, the ratio of radioactive carbon isotopes to the total number of carbon atoms is about 1 to 10^{12}. (See Example 2 in Section 4.1.) When organic material dies, its radioactive carbon isotopes begin to decay, with a half-life of about 5715 years. So, the ratio R of carbon isotopes to carbon-14 atoms is modeled by $R = 10^{-12}\left(\frac{1}{2}\right)^{t/5715}$, where t is the time (in years) and $t = 0$ represents the time when the organic material died.

77. $R = 0.32 \times 10^{-12}$
78. $R = 0.27 \times 10^{-12}$
79. $R = 0.22 \times 10^{-12}$
80. $R = 0.13 \times 10^{-12}$

AP/Wide World Photos

In 1995, archeologist Johan Reinhard discovered the frozen remains of a young Incan woman atop Mt. Ampato in Peru. Carbon dating was used to estimate the age of the "Ice Maiden" at 500 years.

81. Surfing the Internet From 1999 through 2005, the number N (in millions) of people in the United States who surfed the Internet two or more times per week as a leisure activity can be modeled by

$$N = 19.257 + 10.64 \ln(t - 8)$$

where $t = 9$ corresponds to 1999. *(Source: Mediamark Research, Inc.)*

(a) Use the model to estimate the number of Internet surfers in 2003.

(b) Use the model to estimate the year during which the number of Internet surfers reached 40 million.

82. Learning Theory Students in a mathematics class were given an exam and then retested monthly with equivalent exams. The average scores S (on a 100-point scale) for the class can be modeled by $S = 80 - 14 \ln(t + 1)$, $0 \le t \le 12$, where t is the time in months.

(a) What was the average score on the original exam?

(b) What was the average score after 4 months?

(c) After how many months was the average score 46?

(T) 83. Learning Theory In a group project in learning theory, a mathematical model for the proportion P of correct responses after n trials was found to be

$$P = \frac{0.83}{1 + e^{-0.2n}}.$$

(a) Use a graphing utility to graph the function.

(b) Use the graph to determine any horizontal asymptotes of the graph of the function. Interpret the meaning of the upper asymptote in the context of the problem.

(c) After how many trials will 60% of the responses be correct?

(T) 84. Agriculture The yield V (in pounds per acre) of an orchard at age t (in years) is modeled by

$$V = 7955.6e^{-0.0458/t}.$$

(a) Use a graphing utility to graph the function.

(b) Determine the horizontal asymptote of the graph of the function. Interpret its meaning in the context of the problem.

(c) Find the time necessary to obtain a yield of 7900 pounds per acre.

(T) 85. Pressure and Altitude The atmospheric pressure decreases with increasing altitude. At sea level, the average air pressure is one atmosphere (1.033227 kilograms per square centimeter). The table shows the pressures p (in atmospheres) at selected altitudes h (in kilometers).

h	0	5	10	15	20	25
p	1	0.55	0.25	0.12	0.06	0.02

(a) Use a graphing utility to find a model of the form $p = a + b \ln h$ for the data. Explain why the result is an error message.

(b) Use a graphing utility to find the logarithmic model $h = a + b \ln p$ for the data.

(c) Use a graphing utility to plot the data and graph the model.

(d) Use the model to estimate the altitude when $p = 0.75$.

(e) Use the model to estimate the pressure when $h = 13$.

(S) 86. Demonstrate that

$$\frac{\ln x}{\ln y} \ne \ln \frac{x}{y} = \ln x - \ln y$$

by using a spreadsheet to complete the table.

x	y	$\dfrac{\ln x}{\ln y}$	$\ln \dfrac{x}{y}$	$\ln x - \ln y$
1	2			
3	4			
10	5			
4	0.5			

(S) 87. Use a spreadsheet to complete the table using $f(x) = \dfrac{\ln x}{x}$.

x	1	5	10	10^2	10^4	10^6
$f(x)$						

(a) Use the table to estimate the limit: $\lim\limits_{x \to \infty} f(x)$.

(T) (b) Use a graphing utility to estimate the relative extrema of f.

(T) In Exercises 88 and 89, use a graphing utility to verify that the functions are equivalent for $x > 0$.

88. $f(x) = \ln \dfrac{x^2}{4}$

$g(x) = 2 \ln x - \ln 4$

89. $f(x) = \ln \sqrt{x(x^2 + 1)}$

$g(x) = \tfrac{1}{2}[\ln x + \ln(x^2 + 1)]$

True or False? In Exercises 90–95, determine whether the statement is true or false given that $f(x) = \ln x$. If it is false, explain why or give an example that shows it is false.

90. $f(0) = 0$

91. $f(ax) = f(a) + f(x)$, $a > 0, x > 0$

92. $f(x - 2) = f(x) - f(2)$, $x > 2$

93. $\sqrt{f(x)} = \tfrac{1}{2}f(x)$

94. If $f(u) = 2f(v)$, then $v = u^2$.

95. If $f(x) < 0$, then $0 < x < 1$.

Section 4.5
Derivatives of Logarithmic Functions

- Find derivatives of natural logarithmic functions.
- Use calculus to analyze the graphs of functions that involve the natural logarithmic function.
- Use the definition of logarithms and the change-of-base formula to evaluate logarithmic expressions involving other bases.
- Find derivatives of exponential and logarithmic functions involving other bases.

Derivatives of Logarithmic Functions

Differentiation can be used to develop the derivative of the natural logarithmic function.

$$y = \ln x \quad \text{Natural logarithmic function}$$
$$e^y = x \quad \text{Write in exponential form.}$$
$$\frac{d}{dx}[e^y] = \frac{d}{dx}[x] \quad \text{Differentiate with respect to } x.$$
$$e^y \frac{dy}{dx} = 1 \quad \text{Chain Rule}$$
$$\frac{dy}{dx} = \frac{1}{e^y} \quad \text{Divide each side by } e^y.$$
$$\frac{dy}{dx} = \frac{1}{x} \quad \text{Substitute } x \text{ for } e^y.$$

This result and its Chain Rule version are summarized below.

Derivative of the Natural Logarithmic Function

Let u be a differentiable function of x.

1. $\dfrac{d}{dx}[\ln x] = \dfrac{1}{x}$ 2. $\dfrac{d}{dx}[\ln u] = \dfrac{1}{u}\dfrac{du}{dx}$

Example 1   Differentiating a Logarithmic Function

Find the derivative of

$$f(x) = \ln 2x.$$

SOLUTION Let $u = 2x$. Then $du/dx = 2$, and you can apply the Chain Rule as shown.

$$f'(x) = \frac{1}{u}\frac{du}{dx} = \frac{1}{2x}(2) = \frac{1}{x}$$

✓ CHECKPOINT 1

Find the derivative of $f(x) = \ln 5x$. ∎

DISCOVERY

Sketch the graph of $y = \ln x$ on a piece of paper. Draw tangent lines to the graph at various points. How do the slopes of these tangent lines change as you move to the right? Is the slope ever equal to zero? Use the formula for the derivative of the logarithmic function to confirm your conclusions.

Example 2 Differentiating Logarithmic Functions

Find the derivative of each function.

a. $f(x) = \ln(2x^2 + 4)$ **b.** $f(x) = x \ln x$ **c.** $f(x) = \dfrac{\ln x}{x}$

SOLUTION

a. Let $u = 2x^2 + 4$. Then $du/dx = 4x$, and you can apply the Chain Rule.

$$f'(x) = \frac{1}{u}\frac{du}{dx} \qquad \text{Chain Rule}$$

$$= \frac{1}{2x^2 + 4}(4x)$$

$$= \frac{2x}{x^2 + 2} \qquad \text{Simplify.}$$

b. Using the Product Rule, you can find the derivative.

$$f'(x) = x\frac{d}{dx}[\ln x] + (\ln x)\frac{d}{dx}[x] \qquad \text{Product Rule}$$

$$= x\left(\frac{1}{x}\right) + (\ln x)(1)$$

$$= 1 + \ln x \qquad \text{Simplify.}$$

c. Using the Quotient Rule, you can find the derivative.

$$f'(x) = \frac{x\dfrac{d}{dx}[\ln x] - (\ln x)\dfrac{d}{dx}[x]}{x^2} \qquad \text{Quotient Rule}$$

$$= \frac{x\left(\dfrac{1}{x}\right) - \ln x}{x^2}$$

$$= \frac{1 - \ln x}{x^2} \qquad \text{Simplify.}$$

Example 3 Rewriting Before Differentiating

Find the derivative of $f(x) = \ln\sqrt{x + 1}$.

SOLUTION

$$f(x) = \ln\sqrt{x + 1} \qquad \text{Write original function.}$$

$$= \ln(x + 1)^{1/2} \qquad \text{Rewrite with rational exponent.}$$

$$= \frac{1}{2}\ln(x + 1) \qquad \text{Property of logarithms}$$

$$f'(x) = \frac{1}{2}\left(\frac{1}{x + 1}\right) \qquad \text{Differentiate.}$$

$$= \frac{1}{2(x + 1)} \qquad \text{Simplify.}$$

STUDY TIP

When you are differentiating logarithmic functions, it is often helpful to use the properties of logarithms to rewrite the function *before* differentiating. To see the advantage of rewriting before differentiating, try using the Chain Rule to differentiate $f(x) = \ln\sqrt{x + 1}$ and compare your work with that shown in Example 3.

✓ **CHECKPOINT 2**

Find the derivative of each function.

a. $f(x) = \ln(x^2 - 4)$

b. $f(x) = x^2 \ln x$

c. $f(x) = -\dfrac{\ln x}{x^2}$ ∎

✓ **CHECKPOINT 3**

Find the derivative of $f(x) = \ln \sqrt[3]{x + 1}$. ∎

SECTION 4.5 Derivatives of Logarithmic Functions

DISCOVERY

What is the domain of the function $f(x) = \ln\sqrt{x+1}$ in Example 3? What is the domain of the function $f'(x) = 1/[2(x+1)]$? In general, you must be sure that you understand the domains of functions involving logarithms. For example, are the domains of the functions $y_1 = \ln x^2$ and $y_2 = 2\ln x$ the same? Try graphing them on your graphing utility.

The next example is an even more dramatic illustration of the benefit of rewriting a function before differentiating.

Example 4 Rewriting Before Differentiating

Find the derivative of $f(x) = \ln[x(x^2 + 1)^2]$.

SOLUTION

$$\begin{aligned}
f(x) &= \ln[x(x^2 + 1)^2] &&\text{Write original function.} \\
&= \ln x + \ln(x^2 + 1)^2 &&\text{Logarithmic properties} \\
&= \ln x + 2\ln(x^2 + 1) &&\text{Logarithmic properties} \\
f'(x) &= \frac{1}{x} + 2\left(\frac{2x}{x^2 + 1}\right) &&\text{Differentiate.} \\
&= \frac{1}{x} + \frac{4x}{x^2 + 1} &&\text{Simplify.}
\end{aligned}$$

✓ CHECKPOINT 4

Find the derivative of $f(x) = \ln[x^2\sqrt{x^2 + 1}]$. ∎

STUDY TIP

Finding the derivative of the function in Example 4 without first rewriting would be a formidable task.

$$f'(x) = \frac{1}{x(x^2 + 1)^2} \frac{d}{dx}[x(x^2 + 1)^2]$$

You might try showing that this yields the same result obtained in Example 4, but be careful—the algebra is messy.

TECHNOLOGY

A symbolic differentiation utility will not generally list the derivative of the logarithmic function in the form obtained in Example 4. Use a symbolic differentiation utility to find the derivative of the function in Example 4. Show that the two forms are equivalent by rewriting the answer obtained in Example 4.

Applications

Example 5 Analyzing a Graph

Analyze the graph of the function $f(x) = \dfrac{x^2}{2} - \ln x$.

SOLUTION From Figure 4.18, it appears that the function has a minimum at $x = 1$. To find the minimum analytically, find the critical numbers by setting the derivative of f equal to zero and solving for x.

$$f(x) = \dfrac{x^2}{2} - \ln x \quad \text{Write original function.}$$

$$f'(x) = x - \dfrac{1}{x} \quad \text{Differentiate.}$$

$$x - \dfrac{1}{x} = 0 \quad \text{Set derivative equal to 0.}$$

$$x = \dfrac{1}{x} \quad \text{Add } 1/x \text{ to each side.}$$

$$x^2 = 1 \quad \text{Multiply each side by } x.$$

$$x = \pm 1 \quad \text{Take square root of each side.}$$

Of these two possible critical numbers, only the positive one lies in the domain of f. By applying the First-Derivative Test, you can confirm that the function has a relative minimum when $x = 1$.

FIGURE 4.18

✓ CHECKPOINT 5

Determine the relative extrema of the function

$$f(x) = x - 2 \ln x. \ \blacksquare$$

Example 6 Finding a Rate of Change

A group of 200 college students was tested every 6 months over a four-year period. The group was composed of students who took Spanish during the fall semester of their freshman year and did not take subsequent Spanish courses. The average test score p (in percent) is modeled by

$$p = 91.6 - 15.6 \ln(t + 1), \quad 0 \leq t \leq 48$$

where t is the time in months, as shown in Figure 4.19. At what rate was the average score changing after 1 year?

SOLUTION The rate of change is

$$\dfrac{dp}{dt} = -\dfrac{15.6}{t + 1}.$$

When $t = 12$, $dp/dt = -1.2$, which means that the average score was decreasing at the rate of 1.2% per month.

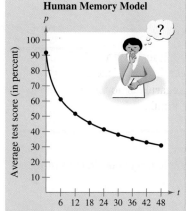

FIGURE 4.19

✓ CHECKPOINT 6

Suppose the average test score p in Example 6 was modeled by $p = 92.3 - 16.9 \ln(t + 1)$, where t is the time in months. How would the rate at which the average test score changed after 1 year compare with that of the model in Example 6? ■

Other Bases

This chapter began with a definition of a general exponential function

$$f(x) = a^x$$

where a is a positive number such that $a \neq 1$. The corresponding **logarithm to the base** a is defined by

$$\log_a x = b \quad \text{if and only if} \quad a^b = x.$$

As with the natural logarithmic function, the domain of the logarithmic function to the base a is the set of positive numbers.

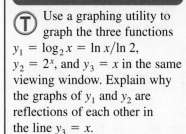

Example 7 Evaluating Logarithms

Evaluate each logarithm without using a calculator.

a. $\log_2 8$ **b.** $\log_{10} 100$ **c.** $\log_{10} \frac{1}{10}$ **d.** $\log_3 81$

SOLUTION

a. $\log_2 8 = 3$ $\qquad\qquad 2^3 = 8$

b. $\log_{10} 100 = 2$ $\qquad\quad 10^2 = 100$

c. $\log_{10} \frac{1}{10} = -1$ $\qquad 10^{-1} = \frac{1}{10}$

d. $\log_3 81 = 4$ $\qquad\qquad 3^4 = 81$

Logarithms to the base 10 are called **common logarithms.** Most calculators have only two logarithm keys—a natural logarithm key denoted by [LN] and a common logarithm key denoted by [LOG]. Logarithms to other bases can be evaluated with the following change-of-base formula.

$$\log_a x = \frac{\ln x}{\ln a} \qquad \text{Change-of-base formula}$$

Example 8 Evaluating Logarithms

Use the change-of-base formula and a calculator to evaluate each logarithm.

a. $\log_2 3$ **b.** $\log_3 6$ **c.** $\log_2(-1)$

SOLUTION In each case, use the change-of-base formula and a calculator.

a. $\log_2 3 = \frac{\ln 3}{\ln 2} \approx 1.585$ $\qquad \log_a x = \frac{\ln x}{\ln a}$

b. $\log_3 6 = \frac{\ln 6}{\ln 3} \approx 1.631$ $\qquad \log_a x = \frac{\ln x}{\ln a}$

c. $\log_2(-1)$ is not defined.

To find derivatives of exponential or logarithmic functions to bases other than e, you can either convert to base e or use the differentiation rules shown on the next page.

✓ CHECKPOINT 7

Evaluate each logarithm without using a calculator.

a. $\log_2 16$

b. $\log_{10} \frac{1}{100}$

c. $\log_2 \frac{1}{32}$

d. $\log_5 125$ ■

✓ CHECKPOINT 8

Use the change-of-base formula and a calculator to evaluate each logarithm.

a. $\log_2 5$

b. $\log_3 18$

c. $\log_4 80$

d. $\log_{16} 0.25$ ■

STUDY TIP

Remember that you can convert to base e using the formulas

$$a^x = e^{(\ln a)x}$$

and

$$\log_a x = \left(\frac{1}{\ln a}\right) \ln x.$$

Other Bases and Differentiation

Let u be a differentiable function of x.

1. $\dfrac{d}{dx}[a^x] = (\ln a)a^x$

2. $\dfrac{d}{dx}[a^u] = (\ln a)a^u \dfrac{du}{dx}$

3. $\dfrac{d}{dx}[\log_a x] = \left(\dfrac{1}{\ln a}\right)\dfrac{1}{x}$

4. $\dfrac{d}{dx}[\log_a u] = \left(\dfrac{1}{\ln a}\right)\left(\dfrac{1}{u}\right)\dfrac{du}{dx}$

PROOF By definition, $a^x = e^{(\ln a)x}$. So, you can prove the first rule by letting $u = (\ln a)x$ and differentiating with base e to obtain

$$\frac{d}{dx}[a^x] = \frac{d}{dx}[e^{(\ln a)x}] = e^u \frac{du}{dx} = e^{(\ln a)x}(\ln a) = (\ln a)a^x.$$

Example 9 Finding a Rate of Change

Radioactive carbon isotopes have a half-life of 5715 years. If 1 gram of the isotopes is present in an object now, the amount A (in grams) that will be present after t years is

$$A = \left(\frac{1}{2}\right)^{t/5715}.$$

At what rate is the amount changing when $t = 10{,}000$ years?

SOLUTION The derivative of A with respect to t is

$$\frac{dA}{dt} = \left(\ln \frac{1}{2}\right)\left(\frac{1}{2}\right)^{t/5715}\left(\frac{1}{5715}\right).$$

When $t = 10{,}000$, the rate at which the amount is changing is

$$\left(\ln \frac{1}{2}\right)\left(\frac{1}{2}\right)^{10{,}000/5715}\left(\frac{1}{5715}\right) \approx -0.000036$$

which implies that the amount of isotopes in the object is decreasing at the rate of 0.000036 gram per year.

✓ **CHECKPOINT 9**

Use a graphing utility to graph the model in Example 9. Describe the rate at which the amount is changing as time t increases. ■

CONCEPT CHECK

1. What is the derivative of $f(x) = \ln x$?
2. What is the derivative of $f(x) = \ln u$? (Assume u is a differentiable function of x.)
3. Complete the following: The change-of-base formula for base e is given by $\log_a x =$ _____.
4. Logarithms to the base e are called natural logarithms. What are logarithms to the base 10 called?

SECTION 4.5 Derivatives of Logarithmic Functions

Skills Review 4.5

The following warm-up exercises involve skills that were covered in earlier sections. You will use these skills in the exercise set for this section. For additional help, review Sections 2.1, 2.6, and 4.4.

In Exercises 1–6, expand the logarithmic expression.

1. $\ln(x+1)^2$
2. $\ln x(x+1)$
3. $\ln \dfrac{x}{x+1}$
4. $\ln \left(\dfrac{x}{x-3}\right)^3$
5. $\ln \dfrac{4x(x-7)}{x^2}$
6. $\ln x^3(x+1)$

In Exercises 7 and 8, find the slope of the tangent line to the graph of the function at the given point.

7. $f(x) = x^2 + 2x + 3$; $(0, 3)$
8. $f(x) = \dfrac{1}{\sqrt{x+1}}$; $(0, 1)$

In Exercises 9 and 10, find the second derivative of f.

9. $f(x) = x^2(x+1) - 3x^3$
10. $f(x) = -\dfrac{1}{x^2}$

Exercises 4.5

See www.CalcChat.com for worked-out solutions to odd-numbered exercises.

In Exercises 1–4, find the slope of the tangent line to the graph of the function at the point $(1, 0)$.

1. $y = \ln x^3$
2. $y = \ln x^{5/2}$
3. $y = \ln x^2$
4. $y = \ln x^{1/2}$

In Exercises 5–26, find the derivative of the function.

5. $y = \ln x^2$
6. $f(x) = \ln 2x$
7. $y = \ln(x^2 + 3)$
8. $f(x) = \ln(1 - x^2)$
9. $y = \ln \sqrt{x - 4}$
10. $y = \ln(1 - x)^{3/2}$
11. $y = (\ln x)^4$
12. $y = (\ln x^2)^2$
13. $f(x) = 2x \ln x$
14. $y = \dfrac{\ln x}{x^2}$
15. $y = \ln(x\sqrt{x^2 - 1})$
16. $y = \ln \dfrac{x}{x^2 + 1}$
17. $y = \ln \dfrac{x}{x+1}$
18. $y = \ln \dfrac{x^2}{x^2 + 1}$
19. $y = \ln \sqrt[3]{\dfrac{x-1}{x+1}}$
20. $y = \ln \sqrt{\dfrac{x+1}{x-1}}$
21. $y = \ln \dfrac{\sqrt{4 + x^2}}{x}$
22. $y = \ln(x\sqrt{4 + x^2})$
23. $g(x) = e^{-x} \ln x$
24. $f(x) = x \ln e^{x^2}$
25. $g(x) = \ln \dfrac{e^x + e^{-x}}{2}$
26. $f(x) = \ln \dfrac{1 + e^x}{1 - e^x}$

In Exercises 27–30, write the expression with base e.

27. 2^x
28. 3^x
29. $\log_4 x$
30. $\log_3 x$

In Exercises 31–38, use a calculator to evaluate the logarithm. Round to three decimal places.

31. $\log_4 7$
32. $\log_6 10$
33. $\log_2 48$
34. $\log_5 12$
35. $\log_3 \tfrac{1}{2}$
36. $\log_7 \tfrac{2}{9}$
37. $\log_{1/5} 31$
38. $\log_{2/3} 32$

In Exercises 39–48, find the derivative of the function.

39. $y = 3^x$
40. $y = \left(\frac{1}{4}\right)^x$
41. $f(x) = \log_2 x$
42. $g(x) = \log_5 x$
43. $h(x) = 4^{2x-3}$
44. $y = 6^{5x}$
45. $y = \log_{10}(x^2 + 6x)$
46. $f(x) = 10^{x^2}$
47. $y = x2^x$
48. $y = x3^{x+1}$

In Exercises 49–52, determine an equation of the tangent line to the function at the given point.

Function	Point
49. $y = x \ln x$	$(1, 0)$
50. $y = \dfrac{\ln x}{x}$	$\left(e, \dfrac{1}{e}\right)$
51. $y = \log_3 x$	$(27, 3)$
52. $g(x) = \log_{10} 2x$	$(5, 1)$

In Exercises 53–58, find the second derivative of the function.

53. $f(x) = x \ln \sqrt{x} + 2x$
54. $f(x) = 3 + 2 \ln x$
55. $f(x) = 2 + x \ln x$
56. $f(x) = \dfrac{\ln x}{x} + x$
57. $f(x) = 5^x$
58. $f(x) = \log_{10} x$

59. Sound Intensity The relationship between the number of decibels β and the intensity of a sound I in watts per square centimeter is given by

$$\beta = 10 \log_{10}\left(\dfrac{I}{10^{-16}}\right).$$

Find the rate of change of the number of decibels when the intensity is 10^{-4} watt per square centimeter.

60. Chemistry The temperatures T (°F) at which water boils at selected pressures p (pounds per square inch) can be modeled by

$$T = 87.97 + 34.96 \ln p + 7.91 \sqrt{p}.$$

Find the rate of change of the temperature when the pressure is 60 pounds per square inch.

In Exercises 61–66, find the slope of the graph at the indicated point. Then write an equation of the tangent line to the graph of the function at the given point.

61. $f(x) = 1 + 2x \ln x$, $(1, 1)$
62. $f(x) = 2 \ln x^3$, $(e, 6)$
63. $f(x) = \ln \dfrac{5(x+2)}{x}$, $(-2.5, 0)$
64. $f(x) = \ln(x\sqrt{x+3})$, $(1.2, 0.9)$
65. $f(x) = x \log_2 x$, $(1, 0)$
66. $f(x) = x^2 \log_3 x$, $(1, 0)$

In Exercises 67–72, graph and analyze the function. Include any relative extrema and points of inflection in your analysis. Use a graphing utility to verify your results.

67. $y = x - \ln x$
68. $y = \dfrac{x}{\ln x}$
69. $y = \dfrac{\ln x}{x}$
70. $y = x \ln x$
71. $y = x^2 \ln \dfrac{x}{4}$
72. $y = (\ln x)^2$

73. PCB Contamination The concentration C of polychlorinated biphenyl (PCB) contamination (in milligrams per kilogram) in a sediment sample taken from Lake Hartwell in South Carolina can be modeled by $C = 1.42 + 13.33 \ln x$, $x \geq 1$, where x is the depth of the sediment in centimeters. *(Source: American Chemical Society)*

(a) What happens to the concentration as the depth increases?

(b) Find the rate of change of C with respect to x when $x = 10$. Interpret the meaning in the context of the problem.

74. Population of France The populations P (in millions) of France from 2000 through 2005 are shown in the table.

t	0	1	2	3	4	5
P	59.4	59.7	59.9	60.2	60.4	60.7

The data can be modeled by $P = 58.98 + 0.9 \ln(t + 1)$, where $t = 0$ corresponds to 2000. *(Source: U.S. Census Bureau)*

(a) Use a graphing utility to plot the data and graph the model. How well does the model fit the data?

(b) At what rate was the population changing in 2002? in 2004?

75. Drug Concentration The concentration C of a drug in the bloodstream (in milligrams per milliliter) t hours after an injection can be modeled by

$$C = 500 - 220.3 \ln(t + 1), \quad 0 \leq t \leq 8.$$

(a) Use a graphing utility to graph the model.

(b) Use the model to approximate the concentration 4 hours after the injection.

(c) Use the model to approximate the concentration 8 hours after the injection.

(d) At what rate is the concentration changing when $t = 4$? when $t = 8$?

(e) Is the model valid for determining the concentration 9 or more hours after the injection? Explain why or why not.

76. Radioactive Decay After t years, the remaining mass y (in grams) of 1 gram of a radioactive element whose half-life is 12.3 years is given by

$$y = \left(\frac{1}{2}\right)^{t/12.3}.$$

(a) At what rate is the mass changing when $t = 10$ years?

(b) At what rate is the mass changing when $t = 15$ years?

77. Temperature An object at a temperature of 160°C was removed from a furnace and placed in a room at 20°C. The temperature T of the object was measured each hour h and recorded in the table. A model for the data is given by $T = 20[1 + 7(2^{-h})]$. The graph of this model is shown in the figure.

h	0	1	2	3	4	5
T	160°	90°	56°	38°	29°	24°

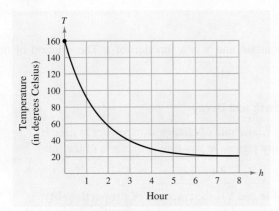

(a) Use the graph to determine when the temperature is changing most rapidly.

(b) Use the model to determine the rates of change in temperature when $h = 0$, $h = 1$, and $h = 3$.

78. Earthquake Intensity On the Richter scale, the magnitude R of an earthquake of intensity I is given by

$$R = \frac{\ln I - \ln I_0}{\ln 10}$$

where I_0 is the minimum intensity used for comparison. Assume $I_0 = 1$.

(a) Find the intensity of the 1906 San Francisco earthquake for which $R = 8.3$.

(b) Find the intensity of the May 26, 2006 earthquake in Java, Indonesia for which $R = 6.3$.

(c) Find the factor by which the intensity is increased when the value of R is doubled.

(d) Find dR/dI.

79. Learning Theory Students in a learning theory study were given an exam and then retested monthly for 6 months with an equivalent exam. The data obtained in the study are shown in the table, where t is the time in months after the initial exam and s is the average score for the class.

t	1	2	3	4	5	6
s	84.2	78.4	72.1	68.5	67.1	65.3

(a) Use these data to find a logarithmic equation that relates t and s.

(b) Use a graphing utility to plot the data and graph the model. How well does the model fit the data?

(c) Find the rate of change of s with respect to t when $t = 2$. Interpret the meaning in the context of the problem.

Life Science Capsule

Philippe Psaila/Photo Researchers, Inc.

Agricultural and food scientists play an important role in ensuring agricultural productivity and maintaining and ensuring the safety of the nation's food supply. About 1 in 4 are government-employed, and about 1 in 3 are self-employed. A bachelor's degree in agricultural science is sufficient for some jobs in applied research; a master's or Ph.D degree is required for basic research or teaching.

80. Research Project Use your school's library, the Internet, or some other reference source to collect data about a life science career (number employed over a 20-year period, for example) and find a mathematical model to represent the data.

Section 4.6

Exponential Growth and Decay

- Use exponential growth and decay to model real-life situations.

Exponential Growth and Decay

In this section, you will learn to create models of *exponential growth and decay*. Real-life situations that involve exponential growth and decay deal with a substance or population whose *rate of change at any time t is proportional to the amount of the substance present at that time*. For example, the rate of decomposition of a radioactive substance is proportional to the amount of radioactive substance at a given instant. In its simplest form, this relationship is described by the equation below.

Rate of change of y is proportional to y.

$$\frac{dy}{dt} = ky$$

In this equation, k is a constant and y is a function of t. The solution of this equation is shown below.

Law of Exponential Growth and Decay

If y is a positive quantity whose rate of change with respect to time is proportional to the quantity present at any time t, then y is of the form

$$y = Ce^{kt}$$

where C is the **initial value** and k is the **constant of proportionality.** **Exponential growth** is indicated by $k > 0$ and **exponential decay** by $k < 0$.

PROOF Because the rate of change of y is proportional to y, you can write

$$\frac{dy}{dt} = ky.$$

You can see that $y = Ce^{kt}$ is a solution of this equation by differentiating to obtain $dy/dt = kCe^{kt}$ and substituting

$$\frac{dy}{dt} = kCe^{kt} = k(Ce^{kt}) = ky.$$

DISCOVERY

Use a graphing utility to graph $y = Ce^{2t}$ for $C = 1, 2,$ and 5. How does the value of C affect the shape of the graph? Now graph $y = 2e^{kt}$ for $k = -2, -1, 0, 1,$ and 2. How does the value of k affect the shape of the graph? Which function grows faster, $y = e^x$ or $y = x^{10}$?

STUDY TIP

In the model $y = Ce^{kt}$, C is called the "initial value" because when $t = 0$

$$y = Ce^{k(0)} = C(1) = C.$$

SECTION 4.6 Exponential Growth and Decay

Applications

Much of the cost of nuclear energy is the cost of disposing of radioactive waste. Because of the long half-life of the waste, it must be stored in containers that will remain undisturbed for thousands of years.

Radioactive decay is measured in terms of **half-life,** the number of years required for half of the atoms in a sample of radioactive material to decay. The half-lives of some common radioactive isotopes are as shown.

Uranium (^{238}U)	4,470,000,000 years
Plutonium (^{239}Pu)	24,100 years
Carbon (^{14}C)	5,715 years
Radium (^{226}Ra)	1,599 years
Einsteinium (^{254}Es)	276 days
Nobelium (^{257}No)	25 seconds

Example 1 MAKE A DECISION   Modeling Radioactive Decay

A sample contains 1 gram of radium. Will more than 0.5 gram of radium remain after 1000 years?

SOLUTION Let y represent the mass (in grams) of the radium in the sample. Because the rate of decay is proportional to y, you can apply the Law of Exponential Decay to conclude that y is of the form $y = Ce^{kt}$, where t is the time in years. From the given information, you know that $y = 1$ when $t = 0$. Substituting these values into the model produces

$$1 = Ce^{k(0)} \qquad \text{Substitute 1 for } y \text{ and 0 for } t.$$

which implies that $C = 1$. Because radium has a half-life of 1599 years, you know that $y = \frac{1}{2}$ when $t = 1599$. Substituting these values into the model allows you to solve for k.

$$y = e^{kt} \qquad \text{Exponential decay model}$$
$$\tfrac{1}{2} = e^{k(1599)} \qquad \text{Substitute } \tfrac{1}{2} \text{ for } y \text{ and 1599 for } t.$$
$$\ln \tfrac{1}{2} = 1599k \qquad \text{Take natural log of each side.}$$
$$\tfrac{1}{1599} \ln \tfrac{1}{2} = k \qquad \text{Divide each side by 1599.}$$

FIGURE 4.20

So, $k \approx -0.0004335$, and the exponential decay model is $y = e^{-0.0004335t}$. To find the amount of radium remaining in the sample after 1000 years, substitute $t = 1000$ into the model. This produces

$$y = e^{-0.0004335(1000)} \approx 0.648 \text{ gram.}$$

✓ CHECKPOINT 1

Use the model in Example 1 to determine the number of years required for a one-gram sample of radium to decay to 0.4 gram. ■

Yes, more than 0.5 gram of radium will remain after 1000 years. The graph of the model is shown in Figure 4.20.

Note: Instead of approximating the value of k in Example 1, you could leave the value exact and obtain

$$y = e^{\ln[(1/2)^{(t/1599)}]} = \frac{1}{2}^{(t/1599)}.$$

This version of the model clearly shows the "half-life." When $t = 1599$, the value of y is $\frac{1}{2}$. When $t = 2(1599)$, the value of y is $\frac{1}{4}$, and so on.

300 **CHAPTER 4** Exponential and Logarithmic Functions

> **Guidelines for Modeling Exponential Growth and Decay**
>
> 1. Use the given information to write *two* sets of conditions involving y and t.
> 2. Substitute the given conditions into the model $y = Ce^{kt}$ and use the results to solve for the constants C and k. (If one of the conditions involves $t = 0$, substitute that value first to solve for C.)
> 3. Use the model $y = Ce^{kt}$ to answer the question.

Example 2 Modeling Population Growth

In a research experiment, a population of fruit flies is increasing in accordance with the exponential growth model. After 2 days, there are 100 flies, and after 4 days, there are 300 flies. How many flies will there be after 5 days?

SOLUTION Let y be the number of flies at time t. From the given information, you know that $y = 100$ when $t = 2$ and $y = 300$ when $t = 4$. Substituting this information into the model $y = Ce^{kt}$ produces

$$100 = Ce^{2k} \quad \text{and} \quad 300 = Ce^{4k}.$$

To solve for k, solve for C in the first equation and substitute the result into the second equation.

$300 = Ce^{4k}$ Second equation

$300 = \left(\dfrac{100}{e^{2k}}\right)e^{4k}$ Substitute $100/e^{2k}$ for C.

$\dfrac{300}{100} = e^{2k}$ Divide each side by 100.

$\ln 3 = 2k$ Take natural log of each side.

$\dfrac{1}{2}\ln 3 = k$ Solve for k.

Using $k = \frac{1}{2}\ln 3 \approx 0.5493$, you can determine that $C \approx 100/e^{2(0.5493)} \approx 33$. So, the exponential growth model is

$$y = 33e^{0.5493t}$$

as shown in Figure 4.21. This implies that, after 5 days, the population is

$$y = 33e^{0.5493(5)} \approx 514 \text{ flies.}$$

✓ CHECKPOINT 2

Find the exponential growth model if a population of fruit flies is 100 after 2 days and 400 after 4 days. ■

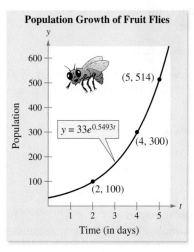

Algebra Review

For help with the algebra in Example 2, see Example 1(c) in the *Chapter 4 Algebra Review*, on page 306.

FIGURE 4.21

Example 3 Finding Doubling Time

Lemnaceae, also known as duckweed, is a family of flowering plants. A certain Lemnaceae plant is increasing in accordance with the exponential growth model. The population can be modeled by

$$y = 300e^{0.0231t} \qquad \text{Exponential growth model}$$

where y is the number of plants and t is the time in hours. How long will it take for the initial population to double?

SOLUTION The initial population ($t = 0$) is

$$y = 300e^{0.0231(0)} = 300.$$

To find when the initial population doubles, let $y = 2(300) = 600$ and solve for t.

$600 = 300e^{0.0231t}$	Substitute 600 for y.
$2 = e^{0.0231t}$	Divide each side by 300.
$\ln 2 = 0.0231t$	Take natural log of each side.
$30 \approx t$	Use a calculator.

So, it will take about 30 hours for the population to double.

✓ CHECKPOINT 3

Use the model in Example 3 to determine how long it will take for the initial population of 300 to triple. ■

Each of the examples in this section uses the exponential growth model in which the base is e. Exponential growth, however, can be modeled with *any* base. That is, the model

$$y = Ca^{bt}$$

also represents exponential growth. (To see this, note that the model can be written in the form $y = Ce^{(\ln a)bt}$.) In some real-life settings, bases other than e are more convenient. For instance, in Example 1, knowing that the half-life of radium is 1599 years, you can immediately write the exponential decay model as

$$y = \left(\frac{1}{2}\right)^{t/1599}.$$

Using this model, the amount of radium left in the sample after 1000 years is

$$y = \left(\frac{1}{2}\right)^{1000/1599} \approx 0.648 \text{ gram}$$

which is the same answer obtained in Example 1.

The smallest flowering plants in the world are from the Lemnaceae (duckweed) family. An average plant is 0.6 millimeter long and 0.3 millimeter wide. Shown above is Wolffia arrhiza, the smallest European flowering plant.

STUDY TIP

Can you see why you can immediately write the model $y = \left(\frac{1}{2}\right)^{t/1599}$ for the radioactive decay described in Example 1? Notice that when $t = 1599$, the value of y is $\frac{1}{2}$, when $t = 3198$, the value of y is $\frac{1}{4}$, and so on.

Example 4 Modeling Drug Concentration

Soon after an injection, the concentration (in milligrams per milliliter) of a drug in a patient's bloodstream is 500 milligrams per milliliter. After 6 hours, 50 milligrams per milliliter of the drug remain in the bloodstream. Assume the amount of the drug in the bloodstream follows an exponential pattern of decline. What is the concentration of the drug after 4 hours?

SOLUTION   Let y represent the concentration (in milligrams per milliliter) of the drug, let t represent the time (in hours), and consider the exponential decay model

$$y = Ce^{kt}. \qquad \text{Exponential decay model}$$

From the given information, you know that $y = 500$ when $t = 0$. Using this information, you have

$$500 = Ce^0$$

which implies that $C = 500$. To solve for k, use the fact that $y = 50$ when $t = 6$.

$y = 500e^{kt}$	Exponential decay model
$50 = 500e^{k(6)}$	Substitute 50 for y and 6 for t.
$0.1 = e^{6k}$	Divide each side by 500.
$\ln 0.1 = 6k$	Take natural log of each side.
$\frac{1}{6} \ln 0.1 = k$	Divide each side by 6.

So, $k = \frac{1}{6} \ln 0.1 \approx -0.3838$, which means that the model is

$$y = 500e^{-0.3838t}.$$

After 4 hours ($t = 4$), the concentration of the drug is

$$y = 500e^{-0.3838(4)}$$
$$\approx 108 \text{ milligrams per milliliter}$$

as shown in Figure 4.22.

Algebra Review

For help with the algebra in Example 4, see Example 1(b) in the *Chapter 4 Algebra Review*, on page 306.

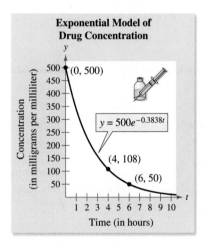

FIGURE 4.22

✓ CHECKPOINT 4

Use the model in Example 4 to determine when the amount of the drug remaining in the bloodstream is 250 milligrams per milliliter ■

TECHNOLOGY

Most graphing utilities have programs that allow you to find the least squares regression exponential model for the data. Depending on the type of graphing utility, you can fit the data to a model of the form

$$y = ab^x \qquad \text{Model with base } b$$

or

$$y = ae^{bx}. \qquad \text{Model with base } e$$

Consult your graphing utility's user manual for more information.

CONCEPT CHECK

1. Describe what the values of C and k represent in the exponential growth and decay model, $y = Ce^{kt}$.
2. For what values of k is $y = Ce^{kt}$ an exponential growth model? an exponential decay model?
3. Can the base used in an exponential growth model be a number other than e?
4. In exponential growth, is the rate of growth constant? Explain why or why not.

SECTION 4.6 Exponential Growth and Decay

Skills Review 4.6

The following warm-up exercises involve skills that were covered in earlier sections. You will use these skills in the exercise set for this section. For additional help, review Sections 4.3 and 4.4.

In Exercises 1–4, solve the equation for k.

1. $12 = 24e^{4k}$
2. $10 = 3e^{5k}$
3. $25 = 16e^{-0.01k}$
4. $22 = 32e^{-0.02k}$

In Exercises 5–8, find the derivative of the function.

5. $y = 32e^{0.23t}$
6. $y = 18e^{0.072t}$
7. $y = 24e^{-1.4t}$
8. $y = 25e^{-0.001t}$

In Exercises 9–12, simplify the expression.

9. $e^{\ln 4}$
10. $4e^{\ln 3}$
11. $e^{\ln(2x+1)}$
12. $e^{\ln(x^2+1)}$

Exercises 4.6

See www.CalcChat.com for worked-out solutions to odd-numbered exercises.

In Exercises 1–10, find the exponential function $y = Ce^{kt}$ that passes through the two given points.

1.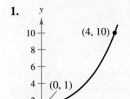
2.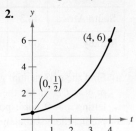
3.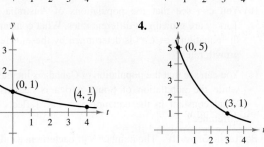
4.
5.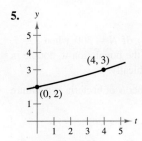
6.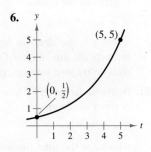
7.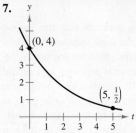
8.
9.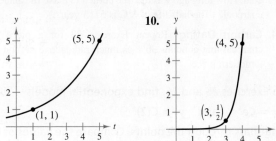
10.

In Exercises 11–14, use the given information to write an equation for y. Confirm your result analytically by showing that the function satisfies the equation $dy/dt = Cy$. Does the function represent exponential growth or exponential decay?

11. $\dfrac{dy}{dt} = 2y$, $y = 10$ when $t = 0$

12. $\dfrac{dy}{dt} = -\dfrac{2}{3}y$, $y = 20$ when $t = 0$

13. $\dfrac{dy}{dt} = -4y$, $y = 30$ when $t = 0$

14. $\dfrac{dy}{dt} = 5.2y$, $y = 18$ when $t = 0$

Radioactive Decay In Exercises 15–20, complete the table for each radioactive isotope.

Isotope	Half-life (in years)	Initial quantity	Amount after 1000 years	Amount after 10,000 years
15. ^{226}Ra	1599	10 grams		
16. ^{226}Ra	1599		1.5 grams	
17. ^{14}C	5715			2 grams
18. ^{14}C	5715	3 grams		
19. ^{239}Pu	24,100		2.1 grams	
20. ^{239}Pu	24,100			0.4 gram

21. Radioactive Decay What percent of a present amount of radioactive radium (^{226}Ra) will remain after 900 years?

22. Radioactive Decay Find the half-life of a radioactive material if after 1 year 99.57% of the initial amount remains.

23. Carbon Dating ^{14}C dating assumes that the carbon dioxide on the Earth today has the same radioactive content as it did centuries ago. If this is true, then the amount of ^{14}C absorbed by a tree that grew several centuries ago should be the same as the amount of ^{14}C absorbed by a similar tree today. A piece of ancient charcoal contains only 15% as much of the radioactive carbon as a piece of modern charcoal. How long ago was the tree burned to make the ancient charcoal? (The half-life of ^{14}C is 5715 years.)

24. Carbon Dating Repeat Exercise 23 for a piece of charcoal that contains 30% as much radioactive carbon as a modern piece.

In Exercises 25 and 26, find exponential models
$$y_1 = Ce^{k_1 t} \quad \text{and} \quad y_2 = C(2)^{k_2 t}$$
that pass through the points. Compare the values of k_1 and k_2. Briefly explain your results.

25. $(0, 5), (12, 20)$

26. $(0, 8), \left(20, \frac{1}{2}\right)$

27. Population Growth The number of a certain type of bacteria increases continuously at a rate proportional to the number present. There are 150 bacteria present at a given time and 450 present 5 hours later.

(a) How many bacteria will there be 10 hours after the initial time?

(b) How long will it take for the population to double?

(c) Does the answer to part (b) depend on the starting time? Explain your reasoning.

28. School Enrollment In 1970, the total enrollment in public universities and colleges in the United States was 5.7 million students. By 2004, enrollment had risen to 13.7 million students. Assume enrollment can be modeled by exponential growth. *(Source: U.S. Census Bureau)*

(a) Estimate the total enrollments in 1980, 1990, and 2000.

(b) Estimate the time required for the enrollment to double from the 2004 figure.

(c) By what percent is the enrollment increasing each year?

29. Population The table shows the populations (in millions) of five countries in 2005 and the projected populations (in millions) for the year 2010. *(Source: U.S. Census Bureau)*

Country	2005	2010
Australia	20.1	20.9
Canada	32.8	34.3
Philippines	87.9	95.9
South Africa	44.3	43.3
Turkey	69.7	73.3

(a) Find the exponential growth or decay model, $y = Ce^{kt}$ or $y = Ce^{-kt}$, for the population of each country by letting $t = 0$ correspond to 2000. Use the model to predict the population of each country in 2030.

(b) You can see that the populations of Australia and Turkey are growing at different rates. What constant in the equation $y = Ce^{kt}$ is determined by these different growth rates?

(c) You can see that the population of Canada is increasing while the population of South Africa is decreasing. What constant in the equation $y = Ce^{kt}$ reflects this difference?

30. Bacteria Growth The number N of bacteria in a culture is given by the model
$$N = 100e^{kt}$$
where t is the time (in hours). If $N = 300$ when $t = 5$, estimate the time required for the population to double in size. Verify your estimate graphically.

31. Bacteria Growth The number N of bacteria in a culture is given by the model
$$N = 250e^{kt}$$
where t is the time (in hours). If $N = 280$ when $t = 10$, estimate the time required for the population to double in size. Verify your estimate graphically.

32. Depreciation A piece of medical equipment that costs $32,000 new has a book value of $18,000 after 2 years.

(a) Find the linear model $V = mt + b$.

(b) Find the exponential model $V = Ce^{kt}$.

(c) Use a graphing utility to graph the two models in the same viewing window. Which model depreciates faster in the first year?

(d) Use the linear model to find the book values of the medical equipment after 1 year and after 3 years.

(e) Use the exponential model to find the book values of the medical equipment after 1 year and after 3 years.

33. Depreciation A piece of laboratory equipment that costs $2000 new has a book value of $500 after 2 years.

(a) Find the linear model $V = mt + b$.

(b) Find the exponential model $V = Ce^{kt}$.

(c) Use a graphing utility to graph the two models in the same viewing window. Which model depreciates faster in the first year?

(d) Use the linear model to find the book values of the laboratory equipment after 1 year and after 3 years.

(e) Use the exponential model to find the book values of the laboratory equipment after 1 year and after 3 years.

34. MAKE A DECISION: FARMS The number of farms (in thousands) in the United States was 2196 in 1995 and 2101 in 2005. *(Source: U.S. Department of Agriculture)*

(a) Use an exponential decay model to estimate the number of farms in the United States in 2011.

(b) Use a linear model to estimate the number of farms in the United States in 2011.

(c) Use a graphing utility to graph the models from parts (a) and (b). Which model do you think is more accurate?

35. MAKE A DECISION: TREATMENT FACILITIES The number of substance abuse treatment facilities in the United States was 10,746 in 1995 and 13,367 in 2005. *(Source: U.S. Substance Abuse and Mental Health Services Administration)*

(a) Use the *regression* feature of a graphing utility to find an exponential growth model and a linear model for the data.

(b) Use the exponential growth model to estimate the number of substance abuse treatment facilities in the United States in 2011.

(c) Use the linear model to estimate the number of substance abuse treatment facilities in the United States in 2011.

(d) Use a graphing utility to graph the models from part (a). Which model do you think is more accurate?

36. Learning Curve The management of a factory finds that the maximum number of units a worker can produce in a day is 30. The learning curve for the number of units N produced per day after a new employee has worked t days is modeled by

$$N = 30(1 - e^{kt}).$$

After 20 days on the job, a worker is producing 19 units in a day. How many days should pass before this worker is producing 25 units per day?

37. Learning Curve The management in Exercise 36 requires that a new employee be producing at least 20 units per day after 30 days on the job.

(a) Find a learning curve model that describes this minimum requirement.

(b) Find the number of days before a minimal achiever is producing 25 units per day.

38. Forestry The value V (in dollars) of a tract of timber can be modeled by

$$V = 100,000 e^{0.75\sqrt{t}}$$

where $t = 0$ corresponds to 1990. If money earns interest at a rate of 4%, compounded continuously, then the present value A of the timber at any time t is $A = Ve^{-0.04t}$. Find the year in which the timber should be harvested to maximize the present value.

39. Forestry Repeat Exercise 38 using the model

$$V = 100,000 e^{0.6\sqrt{t}}.$$

40. MAKE A DECISION: MODELING DATA The table shows the populations P (in millions) of the United States from 1960 through 2005. *(Source: U.S. Census Bureau)*

Year	1960	1970	1980	1990	2000	2005
Population, P	181	205	228	250	282	297

(a) Use the 1960 and 1970 data to find an exponential model P_1 for the data. Let $t = 0$ represent 1960.

(b) Use a graphing utility to find an exponential model P_2 for the data. Let $t = 0$ represent 1960.

(c) Use a graphing utility to plot the data and graph both models in the same viewing window. Compare the actual data with the predictions of the models. Which model is more accurate?

41. Extended Application To work an extended application analyzing the federal outlays (in billions of dollars) for general science, space, and other technology for the years 1995 through 2005, visit this text's website at college.hmco.com/info/larsonapplied. *(Data Source: Office of Management and Budget)*

Algebra Review

Solving Exponential and Logarithmic Equations

To find the extrema or points of inflection of an exponential or logarithmic function, you must know how to solve exponential and logarithmic equations. A few examples are given on page 284. Some additional examples are presented in this Algebra Review.

As with all equations, remember that your basic goal is to isolate the variable on one side of the equation. To do this, you use inverse operations. For instance, to get rid of an exponential expression such as e^{2x}, take the natural log of each side and use the property $\ln e^{2x} = 2x$. Similarly, to get rid of a logarithmic expression such as $\log_2 3x$, exponentiate each side and use the property $2^{\log_2 3x} = 3x$.

Example 1 Solving Exponential Equations

Solve each exponential equation.

a. $25 = 5e^{7t}$ **b.** $50 = 500e^{k(6)}$ **c.** $300 = \left(\dfrac{100}{e^{2k}}\right)e^{4k}$

SOLUTION

a.
$25 = 5e^{7t}$	Write original equation.
$5 = e^{7t}$	Divide each side by 5.
$\ln 5 = \ln e^{7t}$	Take natural log of each side.
$\ln 5 = 7t$	Apply the property $\ln e^a = a$.
$\frac{1}{7}\ln 5 = t$	Divide each side by 7.

b.
$50 = 500e^{k(6)}$	Example 4, page 302
$0.1 = e^{6k}$	Divide each side by 500.
$\ln 0.1 = \ln e^{6k}$	Take natural log of each side.
$\ln 0.1 = 6k$	Apply the property $\ln e^a = a$.
$\frac{1}{6}\ln 0.1 = k$	Divide each side by 6.

c.
$300 = \left(\dfrac{100}{e^{2k}}\right)e^{4k}$	Example 2, page 300
$300 = (100)\dfrac{e^{4k}}{e^{2k}}$	Rewrite product.
$300 = 100e^{4k-2k}$	To divide powers, subtract exponents.
$300 = 100e^{2k}$	Simplify.
$3 = e^{2k}$	Divide each side by 100.
$\ln 3 = \ln e^{2k}$	Take natural log of each side.
$\ln 3 = 2k$	Apply the property $\ln e^a = a$.
$\frac{1}{2}\ln 3 = k$	Divide each side by 2.

Example 2 Solving Logarithmic Equations

Solve each logarithmic equation.

a. $\ln x = 2$ **b.** $5 + 2\ln x = 4$
c. $2\ln 3x = 4$ **d.** $\ln x - \ln(x - 1) = 1$

SOLUTION

a. $\ln x = 2$ Write original equation.
$e^{\ln x} = e^2$ Exponentiate each side.
$x = e^2$ Apply the property $e^{\ln a} = a$.

b. $5 + 2\ln x = 4$ Write original equation.
$2\ln x = -1$ Subtract 5 from each side.
$\ln x = -\dfrac{1}{2}$ Divide each side by 2.
$e^{\ln x} = e^{-1/2}$ Exponentiate each side.
$x = e^{-1/2}$ Apply the property $e^{\ln a} = a$.

c. $2\ln 3x = 4$ Write original equation.
$\ln 3x = 2$ Divide each side by 2.
$e^{\ln 3x} = e^2$ Exponentiate each side.
$3x = e^2$ Apply the property $e^{\ln a} = a$.
$x = \tfrac{1}{3}e^2$ Divide each side by 3.

d. $\ln x - \ln(x - 1) = 1$ Write original equation.
$\ln \dfrac{x}{x - 1} = 1$ $\ln m - \ln n = \ln(m/n)$
$e^{\ln(x/x-1)} = e^1$ Exponentiate each side.
$\dfrac{x}{x - 1} = e^1$ Apply the property $e^{\ln a} = a$.
$x = ex - e$ Multiply each side by $x - 1$.
$x - ex = -e$ Subtract ex from each side.
$x(1 - e) = -e$ Factor.
$x = \dfrac{-e}{1 - e}$ Divide each side by $1 - e$.
$x = \dfrac{e}{e - 1}$ Simplify.

STUDY TIP

Because the domain of a logarithmic function generally does not include all real numbers, be sure to check for extraneous solutions.

Chapter Summary and Study Strategies

After studying this chapter, you should have acquired the following skills. The exercise numbers are keyed to the Review Exercises that begin on page 310. Answers to odd-numbered Review Exercises are given in the back of the text.*

Section 4.1   Review Exercises

- Use the properties of exponents to evaluate and simplify exponential expressions and functions.   1–16

$$a^0 = 1, \quad a^x a^y = a^{x+y}, \quad \frac{a^x}{a^y} = a^{x-y}, \quad (a^x)^y = a^{xy}$$

$$(ab)^x = a^x b^x, \quad \left(\frac{a}{b}\right)^x = \frac{a^x}{b^x}, \quad a^{-x} = \frac{1}{a^x}$$

- Use properties of exponents to answer questions about real life.   17, 18

Section 4.2

- Sketch and analyze the graphs of exponential functions.   19–34
- Evaluate limits of exponential functions in real life.   35
- Evaluate functions involving the natural exponential function.   37–40
- Graph logistic growth functions.   41, 42
- Answer questions involving the natural exponential function as a real-life model.   36, 43, 44

Section 4.3

- Find the derivatives of natural exponential functions.   45–52

$$\frac{d}{dx}[e^x] = e^x, \quad \frac{d}{dx}[e^u] = e^u \frac{du}{dx}$$

- Use calculus to analyze the graphs of functions that involve the natural exponential function.   53–60

Section 4.4

- Use the definition of the natural logarithmic function to write exponential equations in logarithmic form, and vice versa.   61–64

$$\ln x = b \quad \text{if and only if} \quad e^b = x.$$

- Sketch the graphs of natural logarithmic functions.   65–68
- Use properties of logarithms to expand and condense logarithmic expressions.   69–74

$$\ln xy = \ln x + \ln y, \quad \ln \frac{x}{y} = \ln x - \ln y, \quad \ln x^n = n \ln x$$

---

* Use a wide range of valuable study aids to help you master the material in this chapter. The *Student Solutions Guide* includes step-by-step solutions to all odd-numbered exercises to help you review and prepare. The student website at *college.hmco.com/info/larsonapplied* offers algebra help and a *Graphing Technology Guide*. The *Graphing Technology Guide* contains step-by-step commands and instructions for a wide variety of graphing calculators, including the most recent models.

Section 4.4 (continued)

Review Exercises

- Use inverse properties of exponential and logarithmic functions to solve exponential and logarithmic equations.

  $\ln e^x = x$, $e^{\ln x} = x$

  75–90, 119–124

- Use properties of natural logarithms to answer questions about real life.

  91, 92

Section 4.5

- Find the derivatives of natural logarithmic functions.

  $\dfrac{d}{dx}[\ln x] = \dfrac{1}{x}$, $\dfrac{d}{dx}[\ln u] = \dfrac{1}{u}\dfrac{du}{dx}$

  93–106

- Use calculus to analyze the graphs of functions that involve the natural logarithmic function.

  107–110

- Use the definition of logarithms to evaluate logarithmic expressions involving other bases.

  $\log_a x = b$ if and only if $a^b = x$

  111–114

- Use the change-of-base formula to evaluate logarithmic expressions involving other bases.

  $\log_a x = \dfrac{\ln x}{\ln a}$

  115–118

- Find the derivatives of exponential and logarithmic functions involving other bases.

  $\dfrac{d}{dx}[a^x] = (\ln a)a^x$, $\dfrac{d}{dx}[a^u] = (\ln a)a^u \dfrac{du}{dx}$

  $\dfrac{d}{dx}[\log_a x] = \left(\dfrac{1}{\ln a}\right)\dfrac{1}{x}$, $\dfrac{d}{dx}[\log_a u] = \left(\dfrac{1}{\ln a}\right)\left(\dfrac{1}{u}\right)\dfrac{du}{dx}$

  125–128

- Use calculus to answer questions about real-life rates of change.

  129

Section 4.6

- Use exponential growth and decay to model real-life situations.

  130–135

Study Strategies

- **Classifying Differentiation Rules** Differentiation rules fall into two basic classes: (1) general rules that apply to all differentiable functions; and (2) specific rules that apply to special types of functions. At this point in the course, you have studied six general rules: the Constant Rule, the Constant Multiple Rule, the Sum Rule, the Difference Rule, the Product Rule, and the Quotient Rule. Although these rules were introduced in the context of algebraic functions, remember that they can also be used with exponential and logarithmic functions. You have also studied three specific rules: the Power Rule, the derivative of the natural exponential function, and the derivative of the natural logarithmic function. Each of these rules comes in two forms: the "simple" version, such as $D_x[e^x] = e^x$, and the Chain Rule version, such as $D_x[e^u] = e^u(du/dx)$.

- **To Memorize or Not to Memorize?** When studying mathematics, you need to memorize some formulas and rules. Much of this will come from practice—the formulas that you use most often will be committed to memory. Some formulas, however, are used only infrequently. With these, it is helpful to be able to *derive* the formula from a *known* formula. For instance, knowing the Log Rule for differentiation and the change-of-base formula, $\log_a x = (\ln x)/(\ln a)$, allows you to derive the formula for the derivative of a logarithmic function to base a.

Review Exercises

In Exercises 1–4, evaluate the expression.

1. $32^{3/5}$
2. $25^{3/2}$
3. $\left(\frac{1}{16}\right)^{-3/2}$
4. $\left(\frac{27}{8}\right)^{-1/3}$

In Exercises 5–12, use the properties of exponents to simplify the expression.

5. $\left(\frac{9}{16}\right)^0$
6. $(9^{1/3})(3^{1/3})$
7. $\dfrac{6^3}{36^2}$
8. $\dfrac{1}{4}\left(\dfrac{1}{2}\right)^{-3}$
9. $(e^2)^5$
10. $\dfrac{e^6}{e^4}$
11. $(e^{-1})(e^4)$
12. $(e^{1/2})(e^3)$

(T) In Exercises 13–16, evaluate the function for the indicated value of x. If necessary, use a graphing utility, rounding your answers to three decimal places.

13. $f(x) = 2^{x+3}$, $x = 4$
14. $f(x) = 4^{x-1}$, $x = -2$
15. $f(x) = 1.02^x$, $x = 10$
16. $f(x) = 1.12^x$, $x = 1.3$

17. **Red Blood Cell Volume** The red blood cell volume V (in milliliters) for a woman with a body surface area of B square meters can be modeled by

$V = 258.82(2.317)^B$, $1.39 \le B \le 1.87$.

(Source: Journal of Nuclear Medicine)

(a) Use the model to estimate the red blood cell volumes for body surface areas of 1.4, 1.6, and 1.75 square meters.

(b) The given model is based on real-life data. Do you think the model would be valid for a body surface area less than 1.39 square meters or greater than 1.87 square meters? Explain.

18. **Depreciation** Suppose that the value of a piece of medical equipment is 25% of its original value after 8 years. If a hospital purchases the equipment for $55,000, its value t years after the date of purchase should be

$V(t) = 55,000(0.25)^{t/8}$.

Use the model to approximate the value of the equipment (a) 2 years and (b) 5 years after it is purchased.

In Exercises 19–24, match the function with its graph. [The graphs are labeled (a)–(f).]

(a)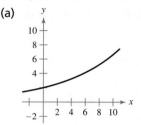
(b)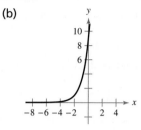
(c)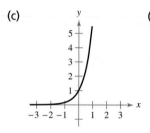
(d)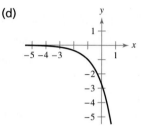
(e)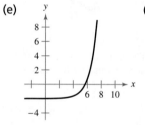
(f)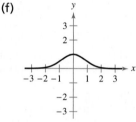

19. $f(x) = 6^x$
20. $f(x) = 3^{x+2}$
21. $f(x) = 2^{-x^2}$
22. $f(x) = -e^{x+1}$
23. $f(x) = -2 + e^{x-5}$
24. $f(x) = 2e^{0.12x}$

In Exercises 25–34, sketch the graph of the function.

25. $f(x) = 9^{x/2}$
26. $g(x) = 16^{3x/2}$
27. $f(t) = \left(\frac{1}{6}\right)^t$
28. $g(t) = \left(\frac{1}{3}\right)^{-t}$
29. $f(x) = \left(\frac{1}{2}\right)^{2x} + 4$
30. $g(x) = \left(\frac{2}{3}\right)^{2x} + 1$
31. $f(x) = e^{-x} + 1$
32. $g(x) = e^{2x} - 1$
33. $f(x) = 1 - e^x$
34. $g(x) = 2 + e^{x-1}$

35. Biology: Endangered Species Biologists consider a species of a plant or animal to be endangered if it is expected to become extinct in less than 20 years. The population y of a certain species is modeled by

$$y = 1096e^{-0.39t}$$

(see figure). Is this species endangered? Explain your reasoning.

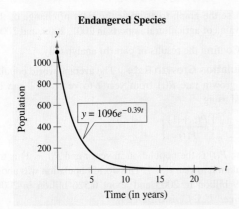

Endangered Species

$y = 1096e^{-0.39t}$

(T) 36. Population The population $P(t)$ of a town increases according to the model

$$P(t) = 2500e^{0.2197t}$$

where t is the time in years, with $t = 0$ corresponding to 2000.

(a) Use a graphing utility to graph the function.

(b) Find the populations in 2000, 2002, 2005, and 2007.

(c) Use the model to estimate when the population will exceed 3500.

In Exercises 37–40, evaluate the function at each indicated value.

37. $f(x) = 5e^{x-1}$

    (a) $x = 2$ (b) $x = \frac{1}{2}$ (c) $x = 10$

38. $f(t) = e^{4t} - 2$

    (a) $t = 0$ (b) $t = 2$ (c) $t = -\frac{3}{4}$

39. $g(t) = 6e^{-0.2t}$

    (a) $t = 17$ (b) $t = 50$ (c) $t = 100$

40. $g(x) = \dfrac{24}{1 + e^{-0.3x}}$

    (a) $x = 0$ (b) $x = 300$ (c) $x = 1000$

41. Biology A lake is stocked with 500 fish and the fish population P begins to increase according to the logistic growth model

$$P = \dfrac{10{,}000}{1 + 19e^{-t/5}}, \quad t \geq 0$$

where t is measured in months.

(T) (a) Use a graphing utility to graph the function.

(b) Estimate the number of fish in the lake after 4 months.

(c) Does the population have a limit as t increases without bound? Explain your reasoning.

(d) After how many months is the population increasing most rapidly? Explain your reasoning.

42. Medicine On a college campus of 5000 students, the spread of a flu virus through the student body is modeled by

$$P = \dfrac{5000}{1 + 4999e^{-0.8t}}, \quad t \geq 0$$

where P is the total number of infected people and t is the time, measured in days.

(T) (a) Use a graphing utility to graph the function.

(b) How many students will be infected after 5 days?

(c) According to this model, will all the students on campus become infected with the flu? Explain your reasoning.

43. Vital Statistics The population P (in millions) of people 65 years old and over in the United States from 1990 through 2005 can be modeled by

$$P = 29.7e^{0.01t}, \quad 0 \leq t \leq 15$$

where $t = 0$ corresponds to 1990. Use this model to estimate the populations of people 65 years old and over in 1990, 2000, and 2005. *(Source: U.S. Census Bureau)*

44. Higher Education The number S (in thousands) of individuals enrolled in college from 2000 through 2004 can be modeled by

$$S = 15{,}435e^{0.03t}, \quad 0 \leq t \leq 4$$

where $t = 0$ corresponds to 2000. Use this model to estimate the numbers of students enrolled in 2000 and 2003. *(Source: U.S. Center for Education Statistics)*

In Exercises 45–52, find the derivative of the function.

45. $y = 4e^{x^2}$

46. $y = 4e^{\sqrt{x}}$

47. $y = \dfrac{x}{e^{2x}}$

48. $y = x^2 e^x$

49. $y = \sqrt{4e^{4x}}$

50. $y = \sqrt[3]{2e^{3x}}$

51. $y = \dfrac{5}{1 + e^{2x}}$

52. $y = \dfrac{10}{1 - 2e^x}$

In Exercises 53–60, graph and analyze the function. Include any relative extrema, points of inflection, and asymptotes in your analysis.

53. $f(x) = 4e^{-x}$
54. $f(x) = 2e^{x^2}$
55. $f(x) = x^3 e^x$
56. $f(x) = \dfrac{e^x}{x^2}$
57. $f(x) = \dfrac{1}{xe^x}$
58. $f(x) = \dfrac{x^2}{e^x}$
59. $f(x) = xe^{2x}$
60. $f(x) = xe^{-2x}$

In Exercises 61 and 62, write the logarithmic equation as an exponential equation.

61. $\ln 12 = 2.4849\ldots$
62. $\ln 0.6 = -0.5108\ldots$

In Exercises 63 and 64, write the exponential equation as a logarithmic equation.

63. $e^{1.5} = 4.4816\ldots$
64. $e^{-4} = 0.0183\ldots$

In Exercises 65–68, sketch the graph of the function.

65. $y = \ln(4 - x)$
66. $y = 5 + \ln x$
67. $y = \ln \dfrac{x}{3}$
68. $y = -2 \ln x$

In Exercises 69–74, use the properties of logarithms to write the expression as a sum, difference, or multiple of logarithms.

69. $\ln \sqrt{x^2(x-1)}$
70. $\ln \sqrt[3]{x^2 - 1}$
71. $\ln \dfrac{x^2}{(x+1)^3}$
72. $\ln \dfrac{x^2}{x^2 + 1}$
73. $\ln \left(\dfrac{1-x}{3x}\right)^3$
74. $\ln \left(\dfrac{x-1}{x+1}\right)^2$

In Exercises 75–90, solve the equation for x.

75. $e^{\ln x} = 3$
76. $e^{\ln(x+2)} = 5$
77. $\ln x = 3e^{-1}$
78. $\ln x = 2e^5$
79. $\ln 2x - \ln(3x - 1) = 0$
80. $\ln x - \ln(x + 1) = 2$
81. $e^{2x-1} - 6 = 0$
82. $4e^{2x-3} - 5 = 0$
83. $\ln x + \ln(x - 3) = 0$
84. $2 \ln x + \ln(x - 2) = 0$
85. $e^{-1.386x} = 0.25$
86. $e^{-0.01x} - 5.25 = 0$
87. $100(1.21)^x = 110$
88. $500(1.075)^{120x} = 100{,}000$
89. $\dfrac{40}{1 - 5e^{-0.01x}} = 200$
90. $\dfrac{50}{1 - 2e^{-0.001x}} = 1000$

91. **Agricultural Imports** The total value V of agricultural imports (in billions of dollars) to the United States for the years 2000 through 2005 can be modeled by

$$V = 30.25 + 5.282t + 0.1131t^2 + 8.82065e^{-t}$$

where $t = 0$ corresponds to 2000. *(Source: U.S. Department of Agriculture)*

(a) Use a graphing utility to graph the model.

(b) Use the graph to estimate the rates of change of the value of agricultural imports in 2001, 2002, and 2004.

(c) Confirm the results of part (b) analytically.

92. **Population Growth Rate** The average world population growth rate $R(t)$ from year t to year $t + 1$ can be found using

$$R(t) = \ln\left[\dfrac{P(t+1)}{P(t)}\right]$$

where $P(t)$ is the population in year t and $P(t + 1)$ is the population in year $t + 1$. The world population was about 6.449 billion in 2005 and about 6.526 billion in 2006. *(Source: U.S. Census Bureau)*

(a) Find the growth rate from 2005 to 2006.

(b) From 2004 to 2005, the growth rate was 1.19%. Use the model to approximate the 2004 world population.

In Exercises 93–106, find the derivative of the function.

93. $f(x) = \ln 3x^2$
94. $y = \ln \sqrt{x}$
95. $y = \ln \dfrac{x(x-1)}{x-2}$
96. $y = \ln \dfrac{x^2}{x+1}$
97. $f(x) = \ln e^{2x+1}$
98. $f(x) = \ln e^{x^2}$
99. $y = \dfrac{\ln x}{x^3}$
100. $y = \dfrac{x^2}{\ln x}$
101. $y = \ln(x^2 - 2)^{2/3}$
102. $y = \ln \sqrt[3]{x^3 + 1}$
103. $f(x) = \ln \left(x^2 \sqrt{x+1}\right)$
104. $f(x) = \ln \dfrac{x}{\sqrt{x+1}}$
105. $y = \ln \dfrac{e^x}{1 + e^x}$
106. $y = \ln \left(e^{2x}\sqrt{e^{2x} - 1}\right)$

In Exercises 107–110, graph and analyze the function. Include any relative extrema and points of inflection in your analysis.

107. $y = \ln(x + 3)$
108. $y = \dfrac{8 \ln x}{x^2}$
109. $y = \ln \dfrac{10}{x+2}$
110. $y = \ln \dfrac{x^2}{9 - x^2}$

In Exercises 111–114, evaluate the logarithm.

111. $\log_7 49$
112. $\log_2 32$
113. $\log_{10} 1$
114. $\log_4 \frac{1}{64}$

In Exercises 115–118, use the change-of-base formula to evaluate the logarithm. Round the result to three decimal places.

115. $\log_5 13$
116. $\log_4 18$
117. $\log_{16} 64$
118. $\log_4 125$

pH Levels In Exercises 119–124, use the acidity model given by pH = $-\log_{10}[H^+]$, where acidity (pH) is a measure of the hydrogen ion concentration $[H^+]$ (measured in moles of hydrogen per liter) of a solution.

119. Find the pH if $[H^+] = 11.3 \times 10^{-6}$.
120. Find the pH if $[H^+] = 2.3 \times 10^{-5}$.
121. Compute $[H^+]$ for a solution in which pH = 3.2.
122. Compute $[H^+]$ for a solution in which pH = 5.8.
123. The pH of a solution is decreased by one unit. The hydrogen ion concentration is increased by what factor?
124. Apple juice has a pH of 2.9 and drinking water has a pH of 8.0. The hydrogen ion concentration of apple juice is how many times the concentration of drinking water?

In Exercises 125–128, find the derivative of the function.

125. $y = \log_3(2x - 1)$
126. $y = \log_{10} \frac{3}{x}$
127. $y = \log_2 \frac{1}{x^2}$
128. $y = \log_{16}(x^2 - 3x)$

129. **Depreciation** After t years the value V of a laboratory instrument purchased for \$695 is given by

$V = 695(0.75)^t$.

(a) Sketch a graph of the function and determine the value of the instrument 2 years after it was purchased.

(b) Find the rates of change of V with respect to t when $t = 1$ and $t = 4$.

(c) After how many years will the instrument be worth \$100?

130. **Population Growth** A population is growing continuously at the rate of $2\frac{1}{2}$% per year. Find the times necessary for the population to (a) double in size and (b) triple in size.

131. **Medical Science** A medical solution contains 500 milligrams of a drug per milliliter when the solution is prepared. After 40 days, it contains only 300 milligrams per milliliter. Assuming that the rate of decomposition is proportional to the concentration present, find an equation giving the concentration A after t days.

132. **Radioactive Decay** The half-life of cobalt-60 is 5.2 years. Find the time it would take for a sample of 0.5 gram of cobalt-60 to decay to 0.1 gram.

133. **Radioactive Decay** A sample of radioactive waste is taken from a nuclear plant. The sample contains 50 grams of strontium-90 at time $t = 0$ years and 42.031 grams after 7 years. What is the half-life of strontium-90?

134. **Unemployment Insurance** The average employer contribution rate A (in percent) for state unemployment insurance was 1.77 in 1999 and 2.92 in 2005 (see figure). Use an exponential growth model to predict the average employer contribution rate in 2008. *(Source: U.S. Employment and Training Administration)*

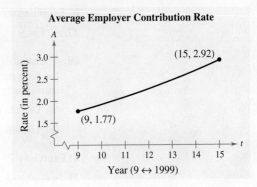

135. **Nonprofit Associations** The number N of social welfare nonprofit associations was 1885 in 1995 and 2218 in 2006 (see figure). Use an exponential growth model to predict the number of social welfare nonprofit associations in 2009. *(Source: Gale Group)*

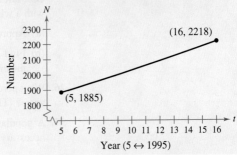

Chapter Test

See www.CalcChat.com for worked-out solutions to odd-numbered exercises.

Take this test as you would take a test in class. When you are done, check your work against the answers given in the back of the book.

In Exercises 1–4, use properties of exponents to simplify the expression.

1. $3^2(3^{-2})$
2. $\left(\dfrac{2^3}{2^{-5}}\right)^{-1}$
3. $(e^{1/2})(e^4)$
4. $(e^3)(e^{-1})$

(T) In Exercises 5–10, use a graphing utility to graph the function.

5. $f(x) = 5^{x-2}$
6. $f(x) = 4^{-x}$
7. $f(x) = 3^{x-3}$
8. $f(x) = 8 + \ln x^2$
9. $f(x) = \ln(x - 5)$
10. $f(x) = 0.5 \ln x$

In Exercises 11–13, use the properties of logarithms to write the expression as a sum, difference, or multiple of logarithms.

11. $\ln \dfrac{3}{2}$
12. $\ln \sqrt{x + y}$
13. $\ln \dfrac{x + 1}{y}$

In Exercises 14–16, condense the logarithmic expression.

14. $\ln y + \ln(x + 1)$
15. $3 \ln 2 - 2 \ln(x - 1)$
16. $2 \ln x + \ln y - \ln(z + 4)$

In Exercises 17–19, solve the equation.

17. $e^{x-1} = 9$
18. $10e^{2x+1} = 900$
19. $50(1.06)^x = 1500$

20. The number N of employees (in thousands) at general medical and surgical hospitals for the years 2000 through 2005 can be modeled by $N = 3767e^{0.02t}$, where $t = 0$ corresponds to 2000. *(Source: U.S. Bureau of Labor Statistics)*

    (a) Use the model to estimate the numbers of employees in 2000 and 2005.

    (b) Use the model to estimate when the number of employees reached 4 million.

In Exercises 21–24, find the derivative of the function.

21. $y = e^{-3x} + 5$
22. $y = 7e^{x+2} + 2x$
23. $y = \ln(3 + x^2)$
24. $y = \ln \dfrac{5x}{x + 2}$

25. The speed S (in miles per hour) of the wind near the center of a tornado and the distance d (in miles) the tornado travels are related by the model

    $S = 93 \log_{10} d + 65.$

    On March 18, 1925, a large tornado struck portions of Missouri, Illinois, and Indiana with a wind speed near the center of about 283 miles per hour.

    (a) Approximate the distance traveled by this tornado.

    (b) Find the rate of change of S with respect to d when $d = 100$. Interpret the meaning of this rate of change in the context of the problem.

26. What percent of a present amount of radioactive radium (^{226}Ra) will remain after 1200 years? (The half-life of ^{266}Ra is 1599 years.)

27. A population is growing continuously at the rate of 1.75% per year. Find the time necessary for the population to double in size.

Trigonometric Functions 5

The Bay of Fundy, seen here at Grand Manan Island, New Brunswick, Canada, has the highest tide differences in the world. This photo shows low tide in the morning. The depth of water during certain times of the day can be modeled by trigonometric functions. (See Section 5.4, Exercise 60.)

5.1 Radian Measure of Angles
5.2 The Trigonometric Functions
5.3 Graphs of Trigonometric Functions
5.4 Derivatives of Trigonometric Functions

Applications

Trigonometric functions have many real-life applications. The applications listed below represent a sample of the applications in this chapter.

- Sprinkler System, Exercise 53, page 323
- Empire State Building, Exercise 69, page 334
- Biology: Predator-Prey Cycles, Exercise 80, page 345
- Meteorology, Exercise 59, page 353
- Applying Fertilizer, Exercise 79, page 354

Section 5.1
Radian Measure of Angles

- Find coterminal angles.
- Convert from degree to radian measure and from radian to degree measure.
- Use formulas relating to triangles.

Angles and Degree Measure

As shown in Figure 5.1, an **angle** has three parts: an **initial ray**, a **terminal ray**, and a **vertex**. An angle is in **standard position** if its initial ray coincides with the positive x-axis and its vertex is at the origin.

Figure 5.2 shows the degree measures of several common angles. Note that θ (the lowercase Greek letter theta) is used to represent an angle and its measure. Angles whose measures are between 0° and 90° are **acute**, and angles whose measures are between 90° and 180° are **obtuse**. An angle whose measure is 90° is a **right** angle, and an angle whose measure is 180° is a **straight** angle.

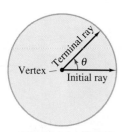

FIGURE 5.1 Standard Position of an Angle

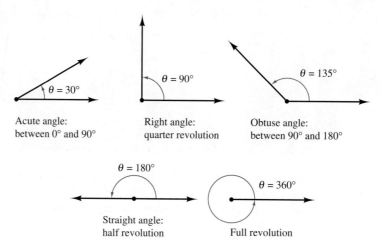

FIGURE 5.2

Positive angles are measured *counterclockwise* beginning with the initial ray. Negative angles are measured *clockwise*. For instance, Figure 5.3 shows an angle whose measure is $-45°$.

Merely knowing where an angle's initial and terminal rays are located does not allow you to assign a measure to the angle. To measure an angle, you must know how the terminal ray was revolved. For example, Figure 5.3 shows that the angle measuring $-45°$ has the same terminal ray as the angle measuring 315°. Such angles are called **coterminal**.

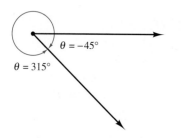

FIGURE 5.3 Coterminal Angles

SECTION 5.1 Radian Measure of Angles

Although it may seem strange to consider angle measures that are larger than 360°, such angles have very useful applications in trigonometry. An angle that is larger than 360° is one whose terminal ray has revolved more than one full revolution counterclockwise. Figure 5.4 shows two angles measuring more than 360°. In a similar way, you can generate an angle whose measure is less than −360° by revolving a terminal ray more than one full revolution clockwise.

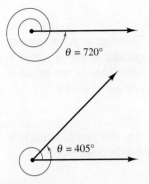

FIGURE 5.4

Example 1 Finding Coterminal Angles

For each angle, find a coterminal angle θ such that $0° \leq \theta < 360°$.

a. 450°
b. 750°
c. −160°
d. −390°

SOLUTION

a. To find an angle coterminal to 450°, subtract 360°, as shown in Figure 5.5(a).

$$\theta = 450° - 360° = 90°$$

b. To find an angle that is coterminal to 750°, subtract 2(360°), as shown in Figure 5.5(b).

$$\theta = 750° - 2(360°) = 750° - 720° = 30°$$

c. To find an angle coterminal to −160°, add 360°, as shown in Figure 5.5(c).

$$\theta = -160° + 360° = 200°$$

d. To find an angle that is coterminal to −390°, add 2(360°), as shown in Figure 5.5(d).

$$\theta = -390° + 2(360°) = -390° + 720° = 330°$$

✓CHECKPOINT 1

For each angle, find a coterminal angle θ such that $0° \leq \theta < 360°$.

a. $\theta = -210°$
b. $\theta = -330°$
c. $\theta = 495°$
d. $\theta = 390°$ ∎

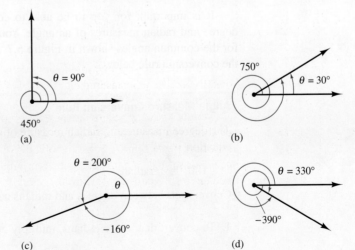

FIGURE 5.5

318 CHAPTER 5 Trigonometric Functions

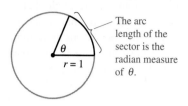

The arc length of the sector is the radian measure of θ.

FIGURE 5.6

Radian Measure

A second way to measure angles is in terms of radians. To assign a radian measure to an angle θ, consider θ to be the central angle of a circular sector of radius 1, as shown in Figure 5.6. The radian measure of θ is then defined to be the length of the arc of the sector. Recall that the circumference of a circle is given by

$$\text{Circumference} = (2\pi)(\text{radius}).$$

So, the circumference of a circle of radius 1 is simply 2π, and you can conclude that the radian measure of an angle measuring 360° is 2π. In other words

$$360° = 2\pi \text{ radians}$$

or

$$180° = \pi \text{ radians}.$$

Figure 5.7 gives the radian measures of several common angles.

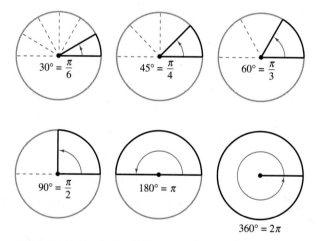

FIGURE 5.7 Radian Measures of Several Common Angles

It is important for you to be able to convert back and forth between the degree and radian measures of an angle. You should remember the conversions for the common angles shown in Figure 5.7. For other conversions, you can use the conversion rule below.

Angle Measure Conversion Rule

The degree measure and radian measure of an angle are related by the equation

$$180° = \pi \text{ radians}.$$

Conversions between degrees and radians can be done as follows.

1. To convert degrees to radians, multiply degrees by $\dfrac{\pi \text{ radians}}{180°}$.

2. To convert radians to degrees, multiply radians by $\dfrac{180°}{\pi \text{ radians}}$.

SECTION 5.1 Radian Measure of Angles

Example 2 Converting from Degrees to Radians

Convert each degree measure to radian measure.

a. 135° **b.** 40° **c.** 540° **d.** −270°

SOLUTION To convert from degree measure to radian measure, multiply the degree measure by $(\pi \text{ radians})/180°$.

a. $135° = (135 \text{ degrees})\left(\dfrac{\pi \text{ radians}}{180 \text{ degrees}}\right) = \dfrac{3\pi}{4}$ radians

b. $40° = (40 \text{ degrees})\left(\dfrac{\pi \text{ radians}}{180 \text{ degrees}}\right) = \dfrac{2\pi}{9}$ radian

c. $540° = (540 \text{ degrees})\left(\dfrac{\pi \text{ radians}}{180 \text{ degrees}}\right) = 3\pi$ radians

d. $-270° = (-270 \text{ degrees})\left(\dfrac{\pi \text{ radians}}{180 \text{ degrees}}\right) = -\dfrac{3\pi}{2}$ radians

✓ CHECKPOINT 2

Convert each degree measure to radian measure.

a. 225°
b. −45°
c. 240°
d. 150°

Although it is common to list radian measure in multiples of π, this is not necessary. For instance, if the degree measure of an angle is 79.3°, the radian measure is

$$79.3° = (79.3 \text{ degrees})\left(\dfrac{\pi \text{ radians}}{180 \text{ degrees}}\right) \approx 1.384 \text{ radians}.$$

Example 3 Converting from Radians to Degrees

Convert each radian measure to degree measure.

a. $-\dfrac{\pi}{2}$ **b.** $\dfrac{7\pi}{4}$ **c.** $\dfrac{11\pi}{6}$ **d.** $\dfrac{9\pi}{2}$

SOLUTION To convert from radian measure to degree measure, multiply the radian measure by $180°/(\pi \text{ radians})$.

a. $-\dfrac{\pi}{2}$ radians $= \left(-\dfrac{\pi}{2} \text{ radians}\right)\left(\dfrac{180 \text{ degrees}}{\pi \text{ radians}}\right) = -90°$

b. $\dfrac{7\pi}{4}$ radians $= \left(\dfrac{7\pi}{4} \text{ radians}\right)\left(\dfrac{180 \text{ degrees}}{\pi \text{ radians}}\right) = 315°$

c. $\dfrac{11\pi}{6}$ radians $= \left(\dfrac{11\pi}{6} \text{ radians}\right)\left(\dfrac{180 \text{ degrees}}{\pi \text{ radians}}\right) = 330°$

d. $\dfrac{9\pi}{2}$ radians $= \left(\dfrac{9\pi}{2} \text{ radians}\right)\left(\dfrac{180 \text{ degrees}}{\pi \text{ radians}}\right) = 810°$

TECHNOLOGY

Most calculators and graphing utilities have both degree and radian modes. You should learn how to use your calculator to convert from degrees to radians, and vice versa. Use a calculator or graphing utility to verify the results of Examples 2 and 3.*

✓ CHECKPOINT 3

Convert each radian measure to degree measure.

a. $\dfrac{5\pi}{3}$ b. $\dfrac{7\pi}{6}$
c. $\dfrac{3\pi}{2}$ d. $-\dfrac{3\pi}{4}$

*Specific calculator keystroke instructions for operations in this and other technology boxes can be found at *college.hmco.com/info/larsonapplied*.

Triangles

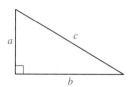

FIGURE 5.8 $a^2 + b^2 = c^2$

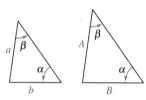

FIGURE 5.9 $\dfrac{a}{b} = \dfrac{A}{B}$

A Summary of Rules About Triangles

1. The sum of the angles of a triangle is 180°.
2. The sum of the two acute angles of a right triangle is 90°.
3. **Pythagorean Theorem** The sum of the squares of the legs of a right triangle is equal to the square of the hypotenuse, as shown in Figure 5.8.
4. **Similar Triangles** If two triangles are similar (have the same angle measures), then the ratios of the corresponding sides are equal, as shown in Figure 5.9.
5. The area of a triangle is equal to one-half the base times the height. That is, $A = \tfrac{1}{2}bh$.
6. Each angle of an equilateral triangle measures 60°.
7. Each acute angle of an isosceles right triangle measures 45°.
8. The altitude of an equilateral triangle bisects its base.

Example 4   Finding the Area of a Triangle

Find the area of an equilateral triangle with one-foot sides.

SOLUTION To use the formula $A = \tfrac{1}{2}bh$, you must first find the height of the triangle, as shown in Figure 5.10. To do this, apply the Pythagorean Theorem to the shaded portion of the triangle.

$$h^2 + \left(\frac{1}{2}\right)^2 = 1^2 \qquad \text{Pythagorean Theorem}$$

$$h^2 = \frac{3}{4} \qquad \text{Simplify.}$$

$$h = \frac{\sqrt{3}}{2} \qquad \text{Solve for } h.$$

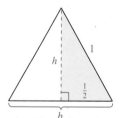

FIGURE 5.10

✓ **CHECKPOINT 4**

Find the area of an isosceles right triangle with a hypotenuse of $\sqrt{2}$ feet. ■

So, the area of the triangle is

$$A = \frac{1}{2}bh = \frac{1}{2}(1)\left(\frac{\sqrt{3}}{2}\right) = \frac{\sqrt{3}}{4} \text{ square foot.}$$

CONCEPT CHECK

1. The measure of an angle is 35°. Is the angle obtuse or acute?
2. Is the angle whose measure is 45° coterminal to an angle whose measure is 315°?
3. What is the measure of a right angle? What is the measure of a straight angle?
4. Name the three parts of an angle.

SECTION 5.1 Radian Measure of Angles

Skills Review 5.1

The following warm-up exercises involve skills that were covered in earlier sections. You will use these skills in the exercise set for this section. For additional help, review Section 1.1.

In Exercises 1 and 2, find the area of the triangle.

1. Base: 10 cm; height: 7 cm
2. Base: 4 in.; height: 6 in.

In Exercises 3–6, let a and b represent the lengths of the legs, and let c represent the length of the hypotenuse, of a right triangle. Solve for the missing side length.

3. $a = 5$, $b = 12$
4. $a = 3$, $c = 5$
5. $a = 8$, $c = 17$
6. $b = 8$, $c = 10$

In Exercises 7–10, let a, b, and c represent the side lengths of a triangle. Use the information below to determine whether the figure is a right triangle, an isosceles triangle, or an equilateral triangle.

7. $a = 4$, $b = 4$, $c = 4$
8. $a = 3$, $b = 3$, $c = 4$
9. $a = 12$, $b = 16$, $c = 20$
10. $a = 1$, $b = 1$, $c = \sqrt{2}$

Exercises 5.1

See www.CalcChat.com for worked-out solutions to odd-numbered exercises.

In Exercises 1–4, determine two coterminal angles (one positive and one negative) for each angle. Give the answers in degrees.

1. (a) 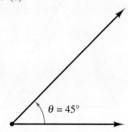 (b)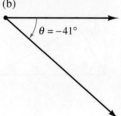

2. (a) 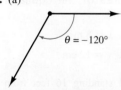 (b)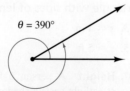

3. (a) 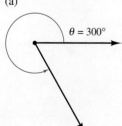 (b)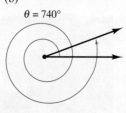

4. (a) 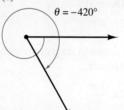 (b)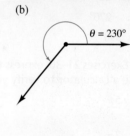

In Exercises 5–8, determine two coterminal angles (one positive and one negative) for each angle. Give the answers in radians.

5. (a) 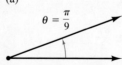 (b)

6. (a) 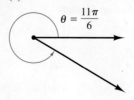 (b)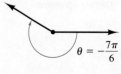

7. (a) 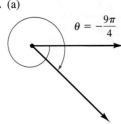 (b)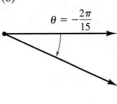

8. (a) 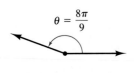 (b)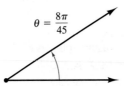

In Exercises 9–20, express the angle in radian measure as a multiple of π. Use a calculator to verify your result.

9. 30°
10. 60°
11. 270°
12. 210°
13. 315°
14. 120°
15. −20°
16. −240°
17. −270°
18. −315°
19. 330°
20. 405°

In Exercises 21–30, express the angle in degree measure. Use a calculator to verify your result.

21. $\dfrac{5\pi}{2}$
22. $\dfrac{5\pi}{4}$
23. $\dfrac{7\pi}{3}$
24. $\dfrac{\pi}{9}$
25. $-\dfrac{\pi}{12}$
26. $-\dfrac{7\pi}{4}$
27. $\dfrac{9\pi}{4}$
28. $-\dfrac{3\pi}{2}$
29. $\dfrac{19\pi}{6}$
30. $\dfrac{8\pi}{3}$

In Exercises 31–34, find the indicated measure of the angle. Express radian measure as a multiple of π.

Degree Measure	Radian Measure
31. −270°	
32.	$\dfrac{\pi}{9}$
33. 144°	
34.	$-\dfrac{7\pi}{12}$

In Exercises 35–42, solve the triangle for the indicated side and/or angle.

35.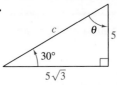
36.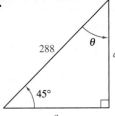
37.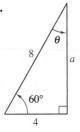
38.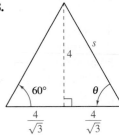
39.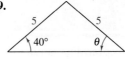
40.
41.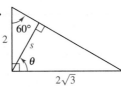
42.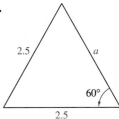

In Exercises 43–46, find the area of the equilateral triangle with sides of length s.

43. $s = 4$ in.
44. $s = 8$ m
45. $s = 5$ ft
46. $s = 12$ cm

47. Height A person 6 feet tall standing 16 feet from a streetlight casts a shadow 8 feet long (see figure). What is the height of the streetlight?

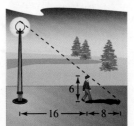

48. Length A guy wire is stretched from a broadcasting tower at a point 200 feet above the ground to an anchor 125 feet from the base (see figure). How long is the wire?

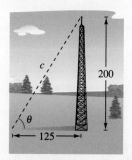

49. Arc Length Let r represent the radius of a circle, θ the central angle (measured in radians), and s the length of the arc intercepted by the angle (see figure). Use the relationship $\theta = s/r$ and a spreadsheet to complete the table.

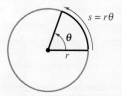

r	8 ft	15 in.	85 cm		
s	12 ft			96 in.	8642 mi
θ		1.6	$\frac{3\pi}{4}$	4	$\frac{2\pi}{3}$

50. Arc Length The minute hand on a clock is $3\frac{1}{2}$ inches long (see figure). Through what distance does the tip of the minute hand move in 25 minutes?

51. Distance A tractor tire that is 5 feet in diameter d is partially filled with a liquid ballast for additional traction. To check the air pressure, the tractor operator rotates the tire until the valve stem is at the top so that the liquid will not enter the gauge. On a given occasion, the operator notes that the tire must be rotated 80° to have the stem in the proper position (see figure).

(a) Find the radian measure of this rotation.

(b) How far must the tractor be moved to get the valve stem in the proper position?

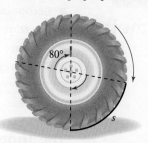

52. Speed of Revolution A centrifuge is revolving at an angular speed of 6283 radians per minute.

(a) At this angular speed, how many revolutions per minute would the centrifuge make?

(b) How long would it take the centrifuge to make 10,000 revolutions?

Area of a Sector of a Circle In Exercises 53 and 54, use the following information. A sector of a circle is the region bounded by two radii of the circle and their intercepted arc (see figure).

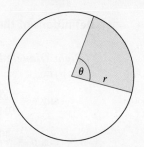

For a circle of radius r, the area A of a sector of the circle with central angle θ (measured in radians) is given by $A = \frac{1}{2}r^2\theta$.

53. Sprinkler System A sprinkler system on a farm is set to spray water over a distance of 70 feet and rotates through an angle of 120°. Find the area of the region.

54. Windshield Wiper A car's rear windshield wiper rotates 125°. The wiper mechanism has a total length of 25 inches and wipes the windshield over a distance of 14 inches. Find the area covered by the wiper.

True or False? In Exercises 55–58, determine whether the statement is true or false. If it is false, explain why or give an example that shows it is false.

55. An angle whose measure is 75° is obtuse.

56. $\theta = -35°$ is coterminal to 325°.

57. A right triangle can have one angle whose measure is 89°.

58. An angle whose measure is π radians is a straight angle.

Section 5.2

The Trigonometric Functions

- Recognize trigonometric functions.
- Use trigonometric identities.
- Evaluate trigonometric functions and solve right triangles.
- Solve trigonometric equations.

The Trigonometric Functions

There are two common approaches to the study of trigonometry. In one case the trigonometric functions are defined as ratios of two sides of a right triangle. In the other case these functions are defined in terms of a point on the terminal side of an arbitrary angle. The first approach is the one generally used in surveying, navigation, and astronomy, where a typical problem involves a triangle, three of whose six parts (sides and angles) are known and three of which are to be determined. The second approach is the one normally used in science and economics, where the periodic nature of the trigonometric functions is emphasized. In the definitions below, the six trigonometric functions are defined from both viewpoints.

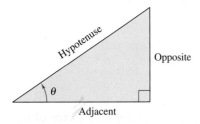

FIGURE 5.11

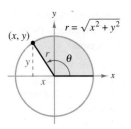

FIGURE 5.12

Definitions of the Trigonometric Functions

Right Triangle Definition: $0 < \theta < \dfrac{\pi}{2}$. (See Figure 5.11.)

$$\sin \theta = \frac{\text{opp.}}{\text{hyp.}} \qquad \csc \theta = \frac{\text{hyp.}}{\text{opp.}}$$

$$\cos \theta = \frac{\text{adj.}}{\text{hyp.}} \qquad \sec \theta = \frac{\text{hyp.}}{\text{adj.}}$$

$$\tan \theta = \frac{\text{opp.}}{\text{adj.}} \qquad \cot \theta = \frac{\text{adj.}}{\text{opp.}}$$

Circular Function Definition: θ is any angle in standard position and (x, y) is a point on the terminal ray of the angle. (See Figure 5.12.)

$$\sin \theta = \frac{y}{r} \qquad \csc \theta = \frac{r}{y}$$

$$\cos \theta = \frac{x}{r} \qquad \sec \theta = \frac{r}{x}$$

$$\tan \theta = \frac{y}{x} \qquad \cot \theta = \frac{x}{y}$$

The full names of the trigonometric functions are **sine, cosecant, cosine, secant, tangent,** and **cotangent.**

Trigonometric Identities

In the circular function definition of the six trigonometric functions, the value of r is always positive. From this, it follows that the signs of the trigonometric functions are determined from the signs of x and y, as shown in Figure 5.13.

The trigonometric reciprocal identities below are also direct consequences of the definitions.

$$\sin\theta = \frac{1}{\csc\theta} \qquad \cos\theta = \frac{1}{\sec\theta} \qquad \tan\theta = \frac{\sin\theta}{\cos\theta} = \frac{1}{\cot\theta}$$

$$\csc\theta = \frac{1}{\sin\theta} \qquad \sec\theta = \frac{1}{\cos\theta} \qquad \cot\theta = \frac{\cos\theta}{\sin\theta} = \frac{1}{\tan\theta}$$

Furthermore, because

$$\sin^2\theta + \cos^2\theta = \left(\frac{y}{r}\right)^2 + \left(\frac{x}{r}\right)^2$$

$$= \frac{x^2 + y^2}{r^2}$$

$$= \frac{r^2}{r^2}$$

$$= 1$$

you can obtain the Pythagorean Identity $\sin^2\theta + \cos^2\theta = 1$. Other trigonometric identities are listed below. In the list, ϕ is the lowercase Greek letter phi.

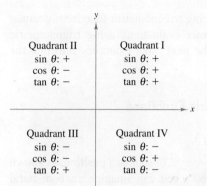

FIGURE 5.13

STUDY TIP

The symbol $\sin^2\theta$ is used to represent $(\sin\theta)^2$.

Trigonometric Identities

Pythagorean Identities

$\sin^2\theta + \cos^2\theta = 1$

$\tan^2\theta + 1 = \sec^2\theta$

$\cot^2\theta + 1 = \csc^2\theta$

Sum or Difference of Two Angles

$\sin(\theta \pm \phi) = \sin\theta\cos\phi \pm \cos\theta\sin\phi$

$\cos(\theta \pm \phi) = \cos\theta\cos\phi \mp \sin\theta\sin\phi$

$\tan(\theta \pm \phi) = \dfrac{\tan\theta \pm \tan\phi}{1 \mp \tan\theta\tan\phi}$

Double Angle

$\sin 2\theta = 2\sin\theta\cos\theta$

$\cos 2\theta = 2\cos^2\theta - 1 = 1 - 2\sin^2\theta$

Reduction Formulas

$\sin(-\theta) = -\sin\theta$

$\cos(-\theta) = \cos\theta$

$\tan(-\theta) = -\tan\theta$

$\sin\theta = -\sin(\theta - \pi)$

$\cos\theta = -\cos(\theta - \pi)$

$\tan\theta = \tan(\theta - \pi)$

Half Angle

$\sin^2\theta = \tfrac{1}{2}(1 - \cos 2\theta)$

$\cos^2\theta = \tfrac{1}{2}(1 + \cos 2\theta)$

Although an angle can be measured in either degrees or radians, radian measure is preferred in calculus. So, all angles in the remainder of this chapter are assumed to be measured in radians unless otherwise indicated. In other words, sin 3 means the sine of 3 radians, and sin 3° means the sine of 3 degrees.

Evaluating Trigonometric Functions

There are two common methods of evaluating trigonometric functions: decimal approximations using a calculator and exact evaluations using trigonometric identities and formulas from geometry. The next three examples illustrate the second method.

Example 1 Evaluating Trigonometric Functions

Evaluate the sine, cosine, and tangent of $\pi/3$.

SOLUTION Begin by drawing the angle $\theta = \pi/3$ in standard position, as shown in Figure 5.14. Because $\pi/3$ radians is $60°$, you can imagine an equilateral triangle with sides of length 1 and with θ as one of its angles. Because the altitude of the triangle bisects its base, you know that $x = \frac{1}{2}$. So, using the Pythagorean Theorem, you have

$$y = \sqrt{r^2 - x^2} = \sqrt{1^2 - \left(\frac{1}{2}\right)^2} = \sqrt{\frac{3}{4}} = \frac{\sqrt{3}}{2}.$$

Now, using $x = \frac{1}{2}$, $y = \frac{1}{2}\sqrt{3}$, and $r = 1$, you can find the values of the sine, cosine, and tangent as shown.

$$\sin\frac{\pi}{3} = \frac{y}{r} = \frac{\frac{1}{2}\sqrt{3}}{1} = \frac{\sqrt{3}}{2}$$

$$\cos\frac{\pi}{3} = \frac{x}{r} = \frac{\frac{1}{2}}{1} = \frac{1}{2}$$

$$\tan\frac{\pi}{3} = \frac{y}{x} = \frac{\frac{1}{2}\sqrt{3}}{\frac{1}{2}} = \sqrt{3}$$

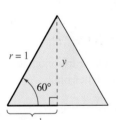

FIGURE 5.14

✓ CHECKPOINT 1

Evaluate the sine, cosine, and tangent of $\frac{\pi}{6}$. ■

The sines, cosines, and tangents of several common angles are listed in the table below. You should remember, or be able to derive, these values.

Trigonometric Values of Common Angles

θ (degrees)	0	30°	45°	60°	90°	180°	270°
θ (radians)	0	$\frac{\pi}{6}$	$\frac{\pi}{4}$	$\frac{\pi}{3}$	$\frac{\pi}{2}$	π	$\frac{3\pi}{2}$
$\sin\theta$	0	$\frac{1}{2}$	$\frac{\sqrt{2}}{2}$	$\frac{\sqrt{3}}{2}$	1	0	-1
$\cos\theta$	1	$\frac{\sqrt{3}}{2}$	$\frac{\sqrt{2}}{2}$	$\frac{1}{2}$	0	-1	0
$\tan\theta$	0	$\frac{\sqrt{3}}{3}$	1	$\sqrt{3}$	Undefined	0	Undefined

STUDY TIP

Learning the table of values at the right is worth the effort because doing so will increase both your efficiency and your confidence. Here is a pattern for the sine function that may help you remember the values.

θ	0	30°	45°	60°	90°
$\sin\theta$	$\frac{\sqrt{0}}{2}$	$\frac{\sqrt{1}}{2}$	$\frac{\sqrt{2}}{2}$	$\frac{\sqrt{3}}{2}$	$\frac{\sqrt{4}}{2}$

Reverse the order to get cosine values of the same angles.

SECTION 5.2 The Trigonometric Functions

To extend the use of the values in the table on the previous page to angles in quadrants other than the first quadrant, you can use the concept of a reference angle, as shown in Figure 5.15, together with the appropriate quadrant sign. The **reference angle θ'** for an angle θ is the smallest positive angle between the terminal side of θ and the x-axis. For instance, the reference angle for 135° is 45° and the reference angle for 210° is 30°.

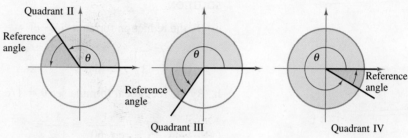

Reference angle: $\pi - \theta$ Reference angle: $\theta - \pi$ Reference angle: $2\pi - \theta$

FIGURE 5.15

To find the value of a trigonometric function of any angle θ, first determine the function value for the associated reference angle θ'. Then, depending on the quadrant in which θ lies, prefix the appropriate sign to the function value.

Example 2 Evaluating Trigonometric Functions

Evaluate each trigonometric function.

a. $\sin \dfrac{3\pi}{4}$ **b.** $\tan 330°$ **c.** $\cos \dfrac{7\pi}{6}$

SOLUTION

a. Because the reference angle for $3\pi/4$ is $\pi/4$ and the sine is positive in the second quadrant, you can write

$$\sin \frac{3\pi}{4} = \sin \frac{\pi}{4} \qquad \text{Reference angle}$$
$$= \frac{\sqrt{2}}{2}. \qquad \text{See Figure 5.16(a).}$$

b. Because the reference angle for 330° is 30° and the tangent is negative in the fourth quadrant, you can write

$$\tan 330° = -\tan 30° \qquad \text{Reference angle}$$
$$= -\frac{\sqrt{3}}{3}. \qquad \text{See Figure 5.16(b).}$$

c. Because the reference angle for $7\pi/6$ is $\pi/6$ and the cosine is negative in the third quadrant, you can write

$$\cos \frac{7\pi}{6} = -\cos \frac{\pi}{6} \qquad \text{Reference angle}$$
$$= -\frac{\sqrt{3}}{2}. \qquad \text{See Figure 5.16(c).}$$

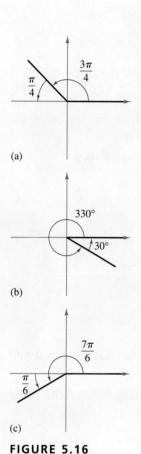

(a)
(b)
(c)
FIGURE 5.16

✓ **CHECKPOINT 2**

Evaluate each trigonometric function.

a. $\sin \dfrac{5\pi}{6}$

b. $\cos 135°$

c. $\tan \dfrac{5\pi}{3}$

CHAPTER 5 Trigonometric Functions

Example 3 Evaluating Trigonometric Functions

Evaluate each trigonometric function.

a. $\sin\left(-\dfrac{\pi}{3}\right)$ b. $\sec 60°$ c. $\cos 15°$ d. $\sin 2\pi$ e. $\cot 0°$ f. $\tan \dfrac{9\pi}{4}$

SOLUTION

a. By the reduction formula $\sin(-\theta) = -\sin\theta$,
$$\sin\left(-\dfrac{\pi}{3}\right) = -\sin\dfrac{\pi}{3} = -\dfrac{\sqrt{3}}{2}.$$

b. By the reciprocal formula $\sec\theta = 1/\cos\theta$,
$$\sec 60° = \dfrac{1}{\cos 60°} = \dfrac{1}{1/2} = 2.$$

c. By the difference formula $\cos(\theta - \phi) = \cos\theta\cos\phi + \sin\theta\sin\phi$,
$$\cos 15° = \cos(45° - 30°)$$
$$= (\cos 45°)(\cos 30°) + (\sin 45°)(\sin 30°)$$
$$= \left(\dfrac{\sqrt{2}}{2}\right)\left(\dfrac{\sqrt{3}}{2}\right) + \left(\dfrac{\sqrt{2}}{2}\right)\left(\dfrac{1}{2}\right)$$
$$= \dfrac{\sqrt{6} + \sqrt{2}}{4}.$$

d. Because the reference angle for 2π is 0,
$$\sin 2\pi = \sin 0 = 0.$$

e. Using the reciprocal formula $\cot\theta = 1/\tan\theta$ and the fact that $\tan 0° = 0$, you can conclude that $\cot 0°$ is undefined.

f. Because the reference angle for $9\pi/4$ is $\pi/4$ and the tangent is positive in the first quadrant,
$$\tan\dfrac{9\pi}{4} = \tan\dfrac{\pi}{4} = 1.$$

✓**CHECKPOINT 3**

Evaluate each trigonometric function.

a. $\sin\left(-\dfrac{\pi}{6}\right)$ b. $\csc 45°$

c. $\cos 75°$ d. $\cos 2\pi$

e. $\sec 0°$ f. $\cot\dfrac{13\pi}{4}$

TECHNOLOGY

Examples 1, 2, and 3 all involve standard angles such as $\pi/6$ and $\pi/3$. To evaluate trigonometric functions involving nonstandard angles, you should use a calculator. When doing this, remember to set the calculator to the proper mode—either *degree* mode or *radian* mode. Furthermore, most calculators have only three trigonometric functions: sine, cosine, and tangent. To evaluate the other three functions, you should combine these keys with the reciprocal key. For instance, you can evaluate the secant of $\pi/7$ as shown:

Function	Calculator Steps	Display
$\sec \dfrac{\pi}{7}$	COS π ÷ 7) x^{-1}	1.109916264

SECTION 5.2 The Trigonometric Functions

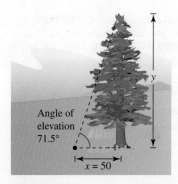

FIGURE 5.17

Example 4 Solving a Right Triangle

A surveyor is standing 50 feet from the base of a large tree. The surveyor measures the angle of elevation to the top of the tree as 71.5°. How tall is the tree?

SOLUTION Referring to Figure 5.17, you can see that

$$\tan 71.5° = \frac{y}{x}$$

where $x = 50$ and y is the height of the tree. So, you can determine the height of the tree to be

$$y = (x)(\tan 71.5°) \approx (50)(2.98868) \approx 149.4 \text{ feet.}$$

✓ CHECKPOINT 4

Find the height of a building that casts a 75-foot shadow when the angle of elevation of the sun is 35°. ∎

Example 5 Calculating Peripheral Vision

To measure the extent of your peripheral vision, stand 1 foot from the corner of a room, facing the corner. Have a friend move an object along the wall until you can just barely see it. If the object is 2 feet from the corner, as shown in Figure 5.18, what is the total angle of your peripheral vision?

SOLUTION Let α represent the total angle of your peripheral vision. As shown in Figure 5.19, you can model the physical situation with an isosceles right triangle whose legs are $\sqrt{2}$ feet and whose hypotenuse is 2 feet. In the triangle, the angle θ is given by

$$\tan \theta = \frac{y}{x} = \frac{\sqrt{2}}{\sqrt{2} - 1} \approx 3.414.$$

Using the inverse tangent function of a calculator, you can determine that $\theta \approx 73.7°$. So, $\alpha/2 \approx 180° - 73.7° = 106.3°$, which implies that $\alpha \approx 212.6°$. In other words, the total angle of your peripheral vision is about 212.6°.

Some occupations, such as that of a fighter pilot, require excellent vision, including good depth perception and good peripheral vision.

© George Hall/CORBIS

✓ CHECKPOINT 5

If the object in Example 5 is 4 feet from the corner, find the total angle of your peripheral vision. ∎

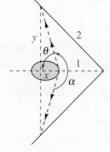

FIGURE 5.18

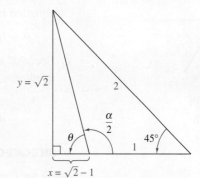

FIGURE 5.19

Solving Trigonometric Equations

> **Algebra Review**
>
> For more examples of the algebra involved in solving trigonometric equations, see the *Chapter 5 Algebra Review*, on pages 355 and 356.

An important part of the study of trigonometry is learning how to solve trigonometric equations. For example, consider the equation

$$\sin \theta = 0.$$

You know that $\theta = 0$ is one solution. Also, in Example 3(d), you saw that $\theta = 2\pi$ is another solution. But these are not the only solutions. In fact, this equation has infinitely many solutions. Any one of the values of θ shown below will work.

$$\ldots, -3\pi, -2\pi, -\pi, 0, \pi, 2\pi, 3\pi, \ldots$$

To simplify the situation, the search for solutions is usually restricted to the interval $0 \le \theta \le 2\pi$, as shown in Example 6.

Example 6 Solving Trigonometric Equations

Solve for θ in each equation. Assume $0 \le \theta \le 2\pi$.

a. $\sin \theta = -\dfrac{\sqrt{3}}{2}$ **b.** $\cos \theta = 1$ **c.** $\tan \theta = 1$

SOLUTION

a. To solve the equation $\sin \theta = -\sqrt{3}/2$, first remember that

$$\sin \frac{\pi}{3} = \frac{\sqrt{3}}{2}.$$

Because the sine is negative in the third and fourth quadrants, it follows that you are seeking values of θ in these quadrants that have a reference angle of $\pi/3$. The two angles fitting these criteria are

$$\theta = \pi + \frac{\pi}{3} = \frac{4\pi}{3} \quad \text{and} \quad \theta = 2\pi - \frac{\pi}{3} = \frac{5\pi}{3}$$

as indicated in Figure 5.20.

b. To solve $\cos \theta = 1$, remember that $\cos 0 = 1$ and note that in the interval $[0, 2\pi]$, the only angles whose reference angles are zero are zero, π, and 2π. Of these, zero and 2π have cosines of 1. (The cosine of π is -1.) So, the equation has two solutions:

$$\theta = 0 \quad \text{and} \quad \theta = 2\pi.$$

c. Because $\tan \pi/4 = 1$ and the tangent is positive in the first and third quadrants, it follows that the two solutions are

$$\theta = \frac{\pi}{4} \quad \text{and} \quad \theta = \pi + \frac{\pi}{4} = \frac{5\pi}{4}.$$

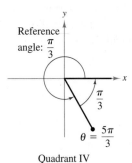

FIGURE 5.20

✓ CHECKPOINT 6

Solve for θ in each equation. Assume $0 \le \theta \le 2\pi$.

a. $\cos \theta = \dfrac{\sqrt{2}}{2}$ **b.** $\tan \theta = -\sqrt{3}$ **c.** $\sin \theta = -\dfrac{1}{2}$

Example 7 Solving a Trigonometric Equation

Solve the equation for θ.

$$\cos 2\theta = 2 - 3 \sin \theta, \quad 0 \leq \theta \leq 2\pi$$

SOLUTION You can use the double-angle identity $\cos 2\theta = 1 - 2 \sin^2 \theta$ to rewrite the original equation, as shown.

$$\cos 2\theta = 2 - 3 \sin \theta$$
$$1 - 2 \sin^2 \theta = 2 - 3 \sin \theta$$
$$0 = 2 \sin^2 \theta - 3 \sin \theta + 1$$
$$0 = (2 \sin \theta - 1)(\sin \theta - 1)$$

For $2 \sin \theta - 1 = 0$, you have $\sin \theta = \frac{1}{2}$, which has solutions of

$$\theta = \frac{\pi}{6} \quad \text{and} \quad \theta = \frac{5\pi}{6}.$$

For $\sin \theta - 1 = 0$, you have $\sin \theta = 1$, which has a solution of

$$\theta = \frac{\pi}{2}.$$

So, for $0 \leq \theta \leq 2\pi$, the three solutions are

$$\theta = \frac{\pi}{6}, \frac{\pi}{2}, \text{ and } \frac{5\pi}{6}.$$

✓ CHECKPOINT 7

Solve the equation for θ.

$$\sin 2\theta + \sin \theta = 0, \quad 0 \leq \theta \leq 2\pi \ \blacksquare$$

STUDY TIP

In Example 7, note that the expression $2 \sin^2 \theta - 3 \sin \theta + 1$ is a quadratic in $\sin \theta$, and as such can be factored. For instance, if you let $x = \sin \theta$, then the quadratic factors as

$$2x^2 - 3x + 1 = (2x - 1)(x - 1).$$

CONCEPT CHECK

1. Relative to the angle θ, name the three sides of a right triangle.
2. In the right triangle definition of trigonometric functions, $\sin \theta$ is equal to what?
3. In the circular function definition of trigonometric functions, $\cos \theta$ is equal to what?
4. The smallest positive angle between the terminal side of an angle θ and the x-axis is denoted θ'. What is the angle θ' called?

Skills Review 5.2

The following warm-up exercises involve skills that were covered in earlier sections. You will use these skills in the exercise set for this section. For additional help, review Sections 0.4 and 5.1.

In Exercises 1–8, convert the angle to radian measure.

1. $135°$
2. $315°$
3. $-210°$
4. $-300°$
5. $-120°$
6. $-225°$
7. $540°$
8. $390°$

In Exercises 9–16, solve for x.

9. $x^2 - x = 0$
10. $2x^2 + x = 0$
11. $2x^2 - x = 1$
12. $x^2 - 2x = 3$
13. $x^2 - 2x = -1$
14. $2x^2 + x = 1$
15. $x^2 - 5x = -6$
16. $x^2 + x = 2$

In Exercises 17–20, solve for t.

17. $\frac{2\pi}{24}(t - 4) = \frac{\pi}{2}$
18. $\frac{2\pi}{12}(t - 2) = \frac{\pi}{4}$
19. $\frac{2\pi}{365}(t - 10) = \frac{\pi}{4}$
20. $\frac{2\pi}{12}(t - 4) = \frac{\pi}{2}$

Exercises 5.2

See www.CalcChat.com for worked-out solutions to odd-numbered exercises.

In Exercises 1–6, determine all six trigonometric functions for the angle θ.

1.
2.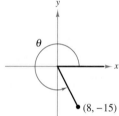
3.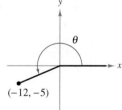
4.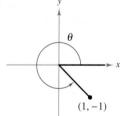
5.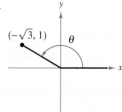
6.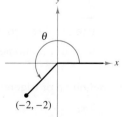

In Exercises 7–12, find the indicated trigonometric function from the given function.

7. Given $\sin \theta = \frac{1}{2}$, find $\csc \theta$.

8. Given $\sin \theta = \frac{1}{3}$, find $\tan \theta$.

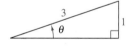

9. Given $\cos \theta = \frac{4}{5}$, find $\cot \theta$.

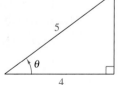

10. Given $\sec \theta = \frac{13}{5}$, find $\cot \theta$.

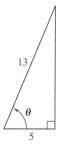

11. Given $\cot \theta = \frac{15}{8}$, find $\sec \theta$.

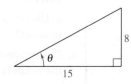

12. Given $\tan \theta = \frac{1}{2}$, find $\sin \theta$.

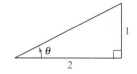

In Exercises 13–18, sketch a right triangle corresponding to the trigonometric function of the angle θ and find the other five trigonometric functions of θ.

13. $\sin \theta = \frac{1}{3}$
14. $\cot \theta = 5$
15. $\sec \theta = 2$
16. $\cos \theta = \frac{5}{7}$
17. $\tan \theta = 3$
18. $\csc \theta = 4.25$

In Exercises 19–24, determine the quadrant in which θ lies.

19. $\sin \theta < 0$, $\cos \theta > 0$
20. $\sin \theta > 0$, $\cos \theta < 0$
21. $\sin \theta > 0$, $\sec \theta > 0$
22. $\cot \theta < 0$, $\cos \theta > 0$
23. $\csc \theta > 0$, $\tan \theta < 0$
24. $\cos \theta > 0$, $\tan \theta < 0$

In Exercises 25–30, construct an appropriate triangle to complete the table. ($0 \le \theta \le 90°$, $0 \le \theta \le \pi/2$)

Function	θ (deg)	θ (rad)	Function Value
25. sin	30°		
26. cos	45°		
27. tan		$\pi/3$	
28. sec		$\pi/4$	
29. cot			1
30. tan			$\sqrt{3}/3$

In Exercises 31–38, evaluate the sines, cosines, and tangents of the angles *without* using a calculator.

31. (a) 60° (b) $-\frac{2\pi}{3}$
32. (a) $\frac{\pi}{4}$ (b) $\frac{5\pi}{4}$
33. (a) $-\frac{\pi}{6}$ (b) 150°
34. (a) $-\frac{\pi}{2}$ (b) $\frac{\pi}{2}$
35. (a) 225° (b) $-225°$
36. (a) 300° (b) 330°
37. (a) 750° (b) 510°
38. (a) $\frac{10\pi}{3}$ (b) $\frac{17\pi}{3}$

In Exercises 39–46, use a calculator to evaluate the trigonometric functions to four decimal places.

39. (a) $\sin 10°$ (b) $\csc 10°$
40. (a) $\sec 225°$ (b) $\sec 135°$
41. (a) $\tan \frac{\pi}{9}$ (b) $\tan \frac{10\pi}{9}$
42. (a) $\cot 4.5$ (b) $\tan 4.5$
43. (a) $\cos(-110°)$ (b) $\cos 250°$
44. (a) $\tan 240°$ (b) $\cot 210°$
45. (a) $\csc 2.62$ (b) $\csc 150°$
46. (a) $\sin(-0.65)$ (b) $\sin 5.63$

In Exercises 47–52, find two values of θ corresponding to each function. List the measure of θ in radians ($0 \le \theta \le 2\pi$). Do not use a calculator.

47. (a) $\sin \theta = \frac{1}{2}$ (b) $\sin \theta = -\frac{1}{2}$
48. (a) $\cos \theta = \frac{\sqrt{2}}{2}$ (b) $\cos \theta = -\frac{\sqrt{2}}{2}$
49. (a) $\csc \theta = \frac{2\sqrt{3}}{3}$ (b) $\cot \theta = -1$
50. (a) $\sec \theta = 2$ (b) $\sec \theta = -2$
51. (a) $\tan \theta = 1$ (b) $\cot \theta = -\sqrt{3}$
52. (a) $\sin \theta = \frac{\sqrt{3}}{2}$ (b) $\sin \theta = -\frac{\sqrt{3}}{2}$

In Exercises 53–62, solve the equation for θ ($0 \le \theta \le 2\pi$). For some of the equations you should use the trigonometric identities listed in this section. Use the *trace* feature of a graphing utility to verify your results.

53. $2 \sin^2 \theta = 1$
54. $\tan^2 \theta = 3$
55. $\tan^2 \theta - \tan \theta = 0$
56. $2 \cos^2 \theta - \cos \theta = 1$
57. $\sin 2\theta - \cos \theta = 0$
58. $\cos 2\theta + 3 \cos \theta + 2 = 0$
59. $\sin \theta = \cos \theta$
60. $\sec \theta \csc \theta = 2 \csc \theta$
61. $\cos^2 \theta + \sin \theta = 1$
62. $\cos \frac{\theta}{2} - \cos \theta = 1$

In Exercises 63–68, solve for x, y, or r as indicated.

63. Solve for y.

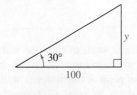

64. Solve for x.

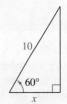

65. Solve for x.

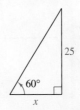

66. Solve for r.

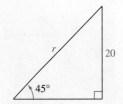

67. Solve for r.

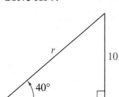

68. Solve for x.

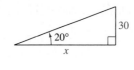

69. Empire State Building You are standing 45 meters from the base of the Empire State Building. You estimate that the angle of elevation to the top of the 86th floor is 82°. If the total height of the building is another 123 meters above the 86th floor, what is the approximate height of the building? One of your friends is on the 86th floor. What is the distance between you and your friend?

70. Height A six-foot person walks from the base of a broadcasting tower directly toward the tip of the shadow cast by the tower. When the person is 132 feet from the tower and 3 feet from the tip of the shadow, the person's shadow starts to appear beyond the tower's shadow.

(a) Draw the right triangle that gives a visual representation of the problem. Show the known quantities of the triangle and use a variable to indicate the height of the tower.

(b) Use a trigonometric function to write an equation involving the unknown quantity.

(c) What is the height of the tower?

71. Length A 20-foot ladder leaning against the side of a house makes a 75° angle with the ground (see figure). How far up the side of the house does the ladder reach?

72. Width of a River A biologist wants to know the width w of a river in order to set instruments to study the pollutants in the water. From point A the biologist walks downstream 100 feet and sights to point C. From this sighting it is determined that $\theta = 50°$ (see figure). How wide is the river?

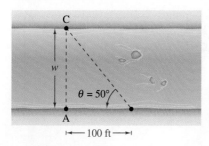

73. Height of a Mountain In traveling across flat land, you notice a mountain directly in front of you. Its angle of elevation (to the peak) is 3.5°. After you drive 13 miles closer to the mountain, the angle of elevation is 9°. Approximate the height of the mountain.

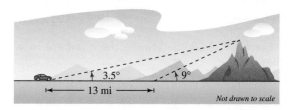

74. Distance From a 150-foot observation tower on the coast, a Coast Guard officer sights a boat in difficulty. The angle of depression of the boat is 3° (see figure). How far is the boat from the shoreline?

75. Medicine The temperature T in degrees Fahrenheit of a patient t hours after arriving at the emergency room of a hospital at 10:00 P.M. is given by

$$T(t) = 98.6 + 4 \cos \frac{\pi t}{36}, \quad 0 \le t \le 18.$$

Find the patient's temperature at each time.

(a) 10:00 P.M.   (b) 4:00 A.M.   (c) 10:00 A.M.

At what time do you expect the patient's temperature to return to normal? Explain your reasoning.

76. Machine Shop Calculations A tapered shaft has a diameter of 5 centimeters at the small end and is 15 centimeters long (see figure). The taper is 3°. Find the diameter d of the large end of the shaft.

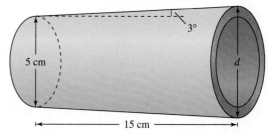

In Exercises 77 and 78, use a graphing utility or a spreadsheet to complete the table. Then graph the function.

x	0	2	4	6	8	10
$f(x)$						

77. $f(x) = \frac{2}{5}x + 2 \sin \frac{\pi x}{5}$ **78.** $f(x) = \frac{1}{2}(5 - x) + 3 \cos \frac{\pi x}{5}$

Mid-Chapter Quiz

Take this quiz as you would take a quiz in class. When you are done, check your work against the answers given in the back of the book.

In Exercises 1–4, express the angle in radian measure as a multiple of π. Use a calculator to verify your result.

1. $15°$ **2.** $105°$ **3.** $-80°$ **4.** $35°$

In Exercises 5–8, express the angle in degree measure. Use a calculator to verify your result.

5. $\dfrac{2\pi}{3}$ **6.** $\dfrac{4\pi}{15}$ **7.** $-\dfrac{4\pi}{3}$ **8.** $\dfrac{11\pi}{12}$

In Exercises 9–14, evaluate the trigonometric function *without* using a calculator.

9. $\sin\left(-\dfrac{\pi}{4}\right)$ **10.** $\cos 210°$ **11.** $\tan\dfrac{5\pi}{6}$

12. $\cot 45°$ **13.** $\sec(-60°)$ **14.** $\csc\dfrac{3\pi}{2}$

In Exercises 15–17, solve the equation for θ $(0 \leq \theta \leq 2\pi)$.

15. $\tan\theta - 1 = 0$

16. $\cos^2\theta - 2\cos\theta + 1 = 0$

17. $\sin^2\theta = 3\cos^2\theta$

In Exercises 18–20, find the indicated side and/or angle.

18. 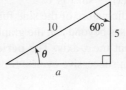 **19.** 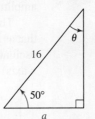 **20.**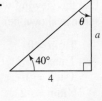

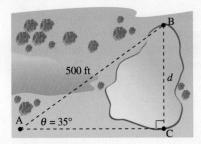

Figure for 21

21. A map maker needs to determine the distance d across a small lake. The distance from point A to point B is 500 feet and the angle θ is $35°$ (see figure). Find the distance d.

22. A ramp $17\tfrac{1}{2}$ feet in length rises to a loading platform that is $3\tfrac{1}{2}$ feet off the ground (see figure). Find the angle that the ramp makes with the gound.

Section 5.3

Graphs of Trigonometric Functions

- Sketch graphs of trigonometric functions.
- Evaluate limits of trigonometric functions.
- Use trigonometric functions to model real-life situations.

DISCOVERY

When the real number line is wrapped around the unit circle, each real number t corresponds with a point $(x, y) = (\cos t, \sin t)$ on the circle. You can visualize this graphically by setting your graphing utility to *simultaneous* mode. For instance, using *radian* and *parametric* modes as well, let

$$X_{1T} = \cos(T)$$
$$Y_{1T} = \sin(T)$$
$$X_{2T} = T$$
$$Y_{2T} = \sin(T).$$

Use the viewing window settings shown below.

Tmin = 0
Tmax = 6.3
Tstep = .1
Xmin = −2
Xmax = 7
Xscl = 1
Ymin = −3
Ymax = 3
Yscl = 1

Now graph the functions. Notice how the graphing utility traces out the unit circle and the sine function simultaneously.

Try changing Y2T to cos T or tan T.

Graphs of Trigonometric Functions

When you are sketching the graph of a trigonometric function, it is common to use x (rather than θ) as the independent variable. On the simplest level, you can sketch the graph of a function such as

$$f(x) = \sin x$$

by constructing a table of values, plotting the resulting points, and connecting them with a smooth curve, as shown in Figure 5.21. Some examples of values are shown in the table below.

x	0	$\dfrac{\pi}{6}$	$\dfrac{\pi}{4}$	$\dfrac{\pi}{3}$	$\dfrac{\pi}{2}$	$\dfrac{2\pi}{3}$	$\dfrac{3\pi}{4}$	$\dfrac{5\pi}{6}$	π
$\sin x$	0.00	0.50	0.71	0.87	1.00	0.87	0.71	0.50	0.00

In Figure 5.21, note that the maximum value of $\sin x$ is 1 and the minimum value is -1. The **amplitude** of the sine function (or the cosine function) is defined to be half of the difference between its maximum and minimum values. So, the amplitude of $f(x) = \sin x$ is 1.

The periodic nature of the sine function becomes evident when you observe that as x increases beyond 2π, the graph repeats itself over and over, continuously oscillating about the x-axis. The **period** of the function is the distance (on the x-axis) between successive cycles. So, the period of $f(x) = \sin x$ is 2π.

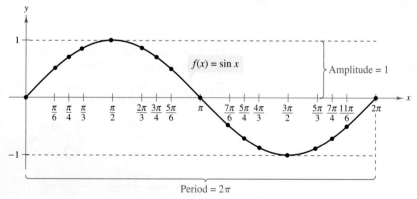

FIGURE 5.21

SECTION 5.3 Graphs of Trigonometric Functions

Figure 5.22 shows the graphs of at least one cycle of all six trigonometric functions.

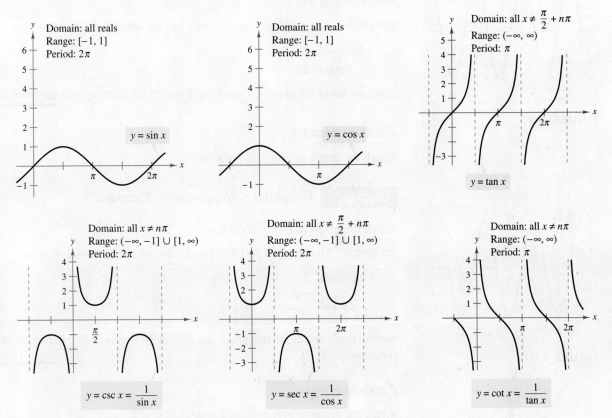

FIGURE 5.22 Graphs of the Six Trigonometric Functions

Familiarity with the graphs of the six basic trigonometric functions allows you to sketch graphs of more general functions such as

$$y = a \sin bx$$

and

$$y = a \cos bx.$$

Note that the function $y = a \sin bx$ oscillates between $-a$ and a and so has an amplitude of

$$|a|. \qquad \text{Amplitude of } y = a \sin bx$$

Furthermore, because $bx = 0$ when $x = 0$ and $bx = 2\pi$ when $x = 2\pi/b$, it follows that the function $y = a \sin bx$ has a period of

$$\frac{2\pi}{|b|}. \qquad \text{Period of } y = a \sin bx$$

338 CHAPTER 5 Trigonometric Functions

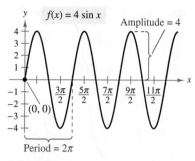

FIGURE 5.23

Example 1 Graphing a Trigonometric Function

Sketch the graph of $f(x) = 4 \sin x$.

SOLUTION The graph of $f(x) = 4 \sin x$ has the characteristics below.

Amplitude: 4

Period: 2π

Three cycles of the graph are shown in Figure 5.23, starting with the point (0, 0).

✓ CHECKPOINT 1

Sketch the graph of $g(x) = 2 \cos x$. ∎

Example 2 Graphing a Trigonometric Function

Sketch the graph of $f(x) = 3 \cos 2x$.

SOLUTION The graph of $f(x) = 3 \cos 2x$ has the characteristics below.

Amplitude: 3

Period: $\dfrac{2\pi}{2} = \pi$

Almost three cycles of the graph are shown in Figure 5.24, starting with the maximum point (0, 3).

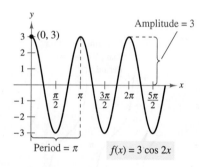

FIGURE 5.24

✓ CHECKPOINT 2

Sketch the graph of $g(x) = 2 \sin 4x$. ∎

Example 3 Graphing a Trigonometric Function

Sketch the graph of $f(x) = -2 \tan 3x$.

SOLUTION The graph of this function has a period of $\pi/3$. The vertical asymptotes of this tangent function occur at

$$x = \ldots, -\frac{\pi}{6}, \underbrace{\frac{\pi}{6}, \frac{\pi}{2}, \frac{5\pi}{6}}_{\text{Period } = \frac{\pi}{3}}, \ldots$$

Several cycles of the graph are shown in Figure 5.25, starting with the vertical asymptote $x = -\pi/6$.

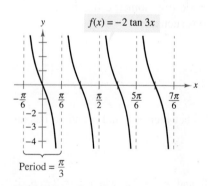

FIGURE 5.25

✓ CHECKPOINT 3

Sketch the graph of $g(x) = \tan 4x$. ∎

Limits of Trigonometric Functions

The sine and cosine functions are continuous over the entire real line. So, you can use direct substitution to evaluate a limit such as

$$\lim_{x \to 0} \sin x = \sin 0 = 0.$$

When direct substitution with a trigonometric limit yields an indeterminate form, such as 0/0, you can rely on technology to help evaluate the limit. The next example examines the limit of a function that you will encounter again in Section 5.4.

Example 4 Evaluating a Trigonometric Limit

Use a calculator to evaluate the function

$$f(x) = \frac{\sin x}{x}$$

at several x-values near $x = 0$. Then use the result to estimate

$$\lim_{x \to 0} \frac{\sin x}{x}.$$

Use a graphing utility (set in *radian* mode) to confirm your result.

SOLUTION The table shows several values of the function at x-values near zero. (Note that the function is undefined when $x = 0$.)

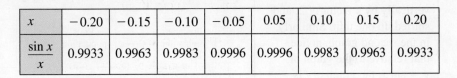

x	-0.20	-0.15	-0.10	-0.05	0.05	0.10	0.15	0.20
$\dfrac{\sin x}{x}$	0.9933	0.9963	0.9983	0.9996	0.9996	0.9983	0.9963	0.9933

From the table, it appears that the limit is 1. That is

$$\lim_{x \to 0} \frac{\sin x}{x} = 1.$$

Figure 5.26 shows the graph of $f(x) = (\sin x)/x$. From this graph, it appears that $f(x)$ gets closer and closer to 1 as x approaches zero (from either side).

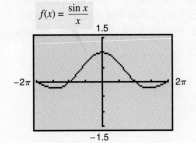

FIGURE 5.26

✓ CHECKPOINT 4

Use a calculator to evaluate the function

$$f(x) = \frac{1 - \cos x}{x}$$

at several x-values near $x = 0$. Then use the result to estimate

$$\lim_{x \to 0} \frac{1 - \cos x}{x}.$$

DISCOVERY

Try using the technique illustrated in Example 4 to evaluate

$$\lim_{x \to 0} \frac{\sin 5x}{5x}.$$

Can you hypothesize the limit of the general form

$$\lim_{x \to 0} \frac{\sin nx}{nx}$$

where n is a positive integer?

TECHNOLOGY

Graphing Trigonometric Functions

Ⓣ A graphing utility allows you to explore the effects of the constants a, b, c, and d on the graph of a function of the form

$$f(x) = a \sin[b(x - c)] + d.$$

After trying several values for these constants, you can see that a determines the amplitude, b determines the period, c determines the horizontal shift, and d determines the vertical shift. Two examples are shown below. In each case, the graph of f is compared with the graph of $y = \sin x$. For instance, in the first graph, notice that relative to the graph of $y = \sin x$, the graph of f is shifted $\pi/2$ units to the right, stretched vertically by a factor of 2, and shifted up one unit. Similarly, in the second graph, notice that relative to the graph of $y = \sin x$, the graph of f is shifted $\pi/2$ units to the right, stretched horizontally by a factor of $\frac{1}{2}$, stretched vertically by a factor of 3, and shifted down two units.

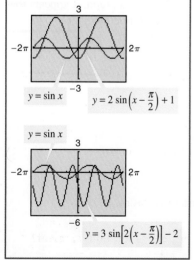

Applications

There are many examples of periodic phenomena in both business and biology. Many businesses have cyclical sales patterns, and plant growth is affected by the day-night cycle. The next example describes the cyclical pattern followed by many types of predator-prey populations, such as coyotes and rabbits.

Example 5 Modeling Predator-Prey Cycles

The population P of a predator at time t (in months) is modeled by

$$P = 10{,}000 + 3000 \sin \frac{2\pi t}{24}, \quad t \geq 0$$

and the population p of its primary food source (its prey) is modeled by

$$p = 15{,}000 + 5000 \cos \frac{2\pi t}{24}, \quad t \geq 0.$$

Graph both models on the same set of axes and explain the oscillations in the size of each population.

SOLUTION Each function has a period of 24 months. The predator population has an amplitude of 3000 and oscillates about the line $y = 10{,}000$. The prey population has an amplitude of 5000 and oscillates about the line $y = 15{,}000$. The graphs of the two models are shown in Figure 5.27. The cycles of this predator-prey population are explained by the diagram below.

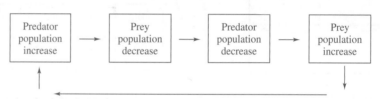

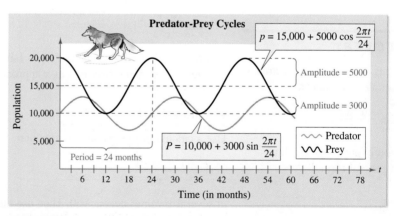

FIGURE 5.27

✓ CHECKPOINT 5

Write the keystrokes required to graph correctly the predator-prey cycle from Example 5 using a graphing utility. ∎

SECTION 5.3 Graphs of Trigonometric Functions

Example 6 Modeling Biorhythms

A popular theory that attempts to explain the ups and downs of everyday life states that each of us has three cycles, which begin at birth. These three cycles can be modeled by sine waves

Physical (23 days): $\quad P = \sin \dfrac{2\pi t}{23}, \quad t \geq 0$

Emotional (28 days): $\quad E = \sin \dfrac{2\pi t}{28}, \quad t \geq 0$

Intellectual (33 days): $\quad I = \sin \dfrac{2\pi t}{33}, \quad t \geq 0$

where t is the number of days since birth. Describe the biorhythms during the month of September 2007 for a person who was born on July 20, 1987.

SOLUTION Figure 5.28 shows the person's biorhythms during the month of September 2007. Note that September 1, 2007 was the 7348th day of the person's life.

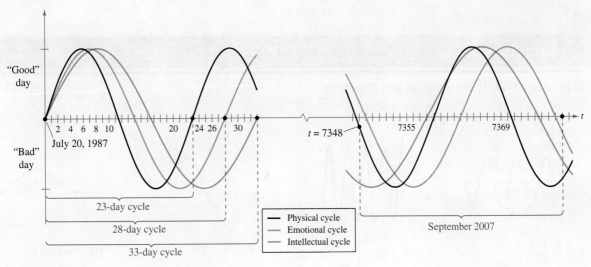

FIGURE 5.28

✓ CHECKPOINT 6

Use a graphing utility to describe the biorhythms of the person in Example 6 during the month of January 2007. Assume that January 1, 2007 is the 7105th day of the person's life. ■

CONCEPT CHECK

1. What is the amplitude of $f(x) = \sin x$?
2. What is the period of $f(x) = \cos x$?
3. What does the amplitude of a sine function or a cosine function represent?
4. What does the period of a sine function or a cosine function represent?

342 CHAPTER 5 Trigonometric Functions

Skills Review 5.3

The following warm-up exercises involve skills that were covered in earlier sections. You will use these skills in the exercise set for this section. For additional help, review Sections 1.5 and 5.2.

In Exercises 1 and 2, find the limit.

1. $\lim_{x \to 2} (x^2 + 4x + 2)$
2. $\lim_{x \to 3} (x^3 - 2x^2 + 1)$

In Exercises 3–10, evaluate the trigonometric function without using a calculator.

3. $\cos \dfrac{\pi}{2}$
4. $\sin \pi$
5. $\tan \dfrac{5\pi}{4}$
6. $\cot \dfrac{2\pi}{3}$
7. $\sin \dfrac{11\pi}{6}$
8. $\cos \dfrac{5\pi}{6}$
9. $\cos \dfrac{5\pi}{3}$
10. $\sin \dfrac{4\pi}{3}$

In Exercises 11–18, use a calculator to evaluate the trigonometric function to four decimal places.

11. $\cos 15°$
12. $\sin 220°$
13. $\sin 275°$
14. $\cos 310°$
15. $\sin 103°$
16. $\cos 72°$
17. $\tan 327°$
18. $\tan 140°$

Exercises 5.3

See www.CalcChat.com for worked-out solutions to odd-numbered exercises.

In Exercises 1–14, find the period and amplitude.

1. $y = 2 \sin 2x$
2. $y = 3 \cos 3x$
7. $y = -2 \sin x$
8. $y = -\cos \dfrac{2x}{3}$

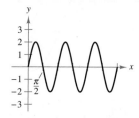
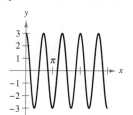
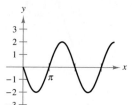
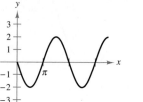

3. $y = \dfrac{3}{2} \cos \dfrac{x}{2}$
4. $y = -2 \sin \dfrac{x}{3}$

9. $y = -2 \sin 10x$
10. $y = \dfrac{1}{3} \sin 8x$
11. $y = \dfrac{1}{2} \sin \dfrac{2x}{3}$
12. $y = 5 \cos \dfrac{x}{4}$
13. $y = 3 \sin 4\pi x$
14. $y = \dfrac{2}{3} \cos \dfrac{\pi x}{10}$

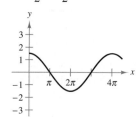
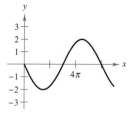

5. $y = \dfrac{1}{2} \cos \pi x$
6. $y = \dfrac{5}{2} \cos \dfrac{\pi x}{2}$

In Exercises 15–20, find the period of the function.

15. $y = 3 \tan x$
16. $y = 7 \tan 2\pi x$
17. $y = 3 \sec 5x$
18. $y = \csc 4x$
19. $y = \cot \dfrac{\pi x}{6}$
20. $y = 5 \tan \dfrac{2\pi x}{3}$

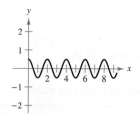
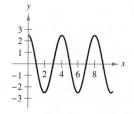

In Exercises 21–26, match the trigonometric function with the correct graph and give the period of the function. [The graphs are labeled (a)–(f).]

(a)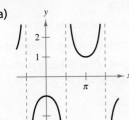

(b)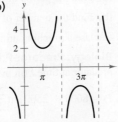

(c)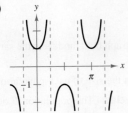

(d)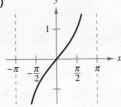

(e)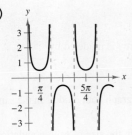

(f)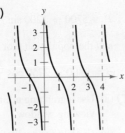

21. $y = \sec 2x$
22. $y = \frac{1}{2} \csc 2x$
23. $y = \cot \frac{\pi x}{2}$
24. $y = -\sec x$
25. $y = 2 \csc \frac{x}{2}$
26. $y = \tan \frac{x}{2}$

In Exercises 27–36, sketch the graph of the function by hand. Use a graphing utility to verify your sketch.

27. $y = \sin \frac{x}{2}$
28. $y = 4 \sin \frac{x}{3}$
29. $y = 2 \cos \frac{\pi x}{3}$
30. $y = \frac{3}{2} \cos \frac{2x}{3}$
31. $y = -2 \sin 6x$
32. $y = -3 \cos 4x$
33. $y = \cos 2\pi x$
34. $y = \frac{3}{2} \sin \frac{\pi x}{4}$
35. $y = 2 \tan x$
36. $y = 2 \cot x$

In Exercises 37–46, sketch the graph of the function.

37. $y = -\sin \frac{2\pi x}{3}$
38. $y = 10 \cos \frac{\pi x}{6}$
39. $y = \cot 2x$
40. $y = 3 \tan \pi x$
41. $y = \csc \frac{2x}{3}$
42. $y = \csc \frac{x}{4}$
43. $y = 2 \sec 2x$
44. $y = \sec \pi x$
45. $y = \csc 2\pi x$
46. $y = -\tan x$

In Exercises 47–56, complete the table (using a spreadsheet or a graphing utility set in *radian* mode) to estimate $\lim_{x \to 0} f(x)$.

x	-0.1	-0.01	-0.001	0.001	0.01	0.1
$f(x)$						

47. $f(x) = \frac{\sin 4x}{2x}$
48. $f(x) = \frac{\sin 2x}{\sin 3x}$
49. $f(x) = \frac{\sin x}{5x}$
50. $f(x) = \frac{1 - \cos 2x}{x}$
51. $f(x) = \frac{3(1 - \cos x)}{x}$
52. $f(x) = \frac{2 \sin(x/4)}{x}$
53. $f(x) = \frac{\tan 2x}{x}$
54. $f(x) = \frac{\tan 4x}{3x}$
55. $f(x) = \frac{\sin^2 x}{x}$
56. $f(x) = \frac{1 - \cos^2 x}{2x}$

In Exercises 57–60, use a graphing utility to graph the function f and find $\lim_{x \to 0} f(x)$.

57. $f(x) = \frac{\sin x}{2x}$
58. $f(x) = \frac{\sin 5x}{2x}$
59. $f(x) = \frac{\sin 5x}{\sin 2x}$
60. $f(x) = \frac{\tan 2x}{3x}$

Graphical Reasoning In Exercises 61–64, find a and d for $f(x) = a \cos x + d$ such that the graph of f matches the figure.

61.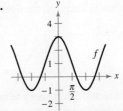

62.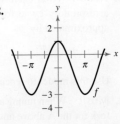

63.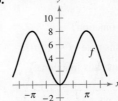

64.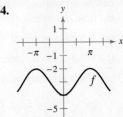

Phase Shift In Exercises 65–68, match the function with the correct graph. [The graphs are labeled (a)–(d).]

(a)
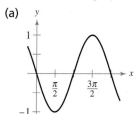

(b)
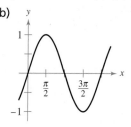

(c)
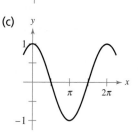

(d)
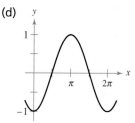

65. $y = \sin x$

66. $y = \sin\left(x - \dfrac{\pi}{2}\right)$

67. $y = \sin(x - \pi)$

68. $y = \sin\left(x - \dfrac{3\pi}{2}\right)$

69. **Health** For a person at rest, the velocity v (in liters per second) of air flow into and out of the lungs during a respiratory cycle is given by

$$v = 0.9 \sin \dfrac{\pi t}{3}$$

where t is the time in seconds. Inhalation occurs when $v > 0$, and exhalation occurs when $v < 0$.

(a) Find the time for one full respiratory cycle.

(b) Find the number of cycles per minute.

(T) (c) Use a graphing utility to graph the velocity function.

(T) 70. **Health** After exercising for a few minutes, a person has a respiratory cycle for which the velocity of air flow is approximated by

$$y = 1.75 \sin \dfrac{\pi t}{2}.$$

Use this model to repeat Exercise 69.

71. **Music** When tuning a piano, a technician strikes a tuning fork for the A above middle C and sets up wave motion that can be approximated by $y = 0.001 \sin 880\pi t$, where t is the time in seconds.

(a) What is the period p of this function?

(b) What is the frequency f of this note ($f = 1/p$)?

(T) (c) Use a graphing utility to graph this function.

72. **Health** The function $P = 100 - 20 \cos(5\pi t/3)$ approximates the blood pressure P (in millimeters of mercury) at time t in seconds for a person at rest.

(a) Find the period of the function.

(b) Find the number of heartbeats per minute.

(T) (c) Use a graphing utility to graph the pressure function.

73. **Biology: Predator-Prey Cycles** The population P of a predator at time t (in months) is modeled by

$$P = 8000 + 2500 \sin \dfrac{2\pi t}{24}$$

and the population p of its prey is modeled by

$$p = 12{,}000 + 4000 \cos \dfrac{2\pi t}{24}.$$

(T) (a) Use a graphing utility to graph both models in the same viewing window.

(b) Explain the oscillations in the size of each population.

74. **Biology: Predator-Prey Cycles** The population P of a predator at time t (in months) is modeled by

$$P = 5700 + 1200 \sin \dfrac{2\pi t}{24}$$

and the population p of its prey is modeled by

$$p = 9800 + 2750 \cos \dfrac{2\pi t}{24}.$$

(T) (a) Use a graphing utility to graph both models in the same viewing window.

(b) Explain the oscillations in the size of each population.

(T) In Exercises 75 and 76, use a graphing utility to graph the functions in the same viewing window where $0 \le x \le 2$.

75. (a) $y = \dfrac{4}{\pi} \sin \pi x$

(b) $y = \dfrac{4}{\pi}\left(\sin \pi x + \dfrac{1}{3} \sin 3\pi x\right)$

76. (a) $y = \dfrac{1}{2} - \dfrac{4}{\pi^2} \cos \pi x$

(b) $y = \dfrac{1}{2} - \dfrac{4}{\pi^2}\left(\cos \pi x + \dfrac{1}{9} \cos 3\pi x\right)$

77. **Biorhythms** For the person born on July 20, 1987, use the biorhythm cycles given in Example 6 to calculate this person's three energy levels on December 31, 2011. Assume this is the 8930th day of the person's life.

(S) 78. **Biorhythms** Use your birthday and the concept of biorhythms as given in Example 6 to calculate your three energy levels on December 31, 2011. Use a spreadsheet to calculate the number of days between your birthday and December 31, 2011.

79. MAKE A DECISION: CONSTRUCTION WORKERS The number W (in thousands) of construction workers employed in the United States during 2006 can be modeled by

$$W = 7594 + 455.2 \sin(0.41t - 1.713)$$

where t is the time in months, with $t = 1$ corresponding to January 1. *(Source: U.S. Bureau of Labor Statistics)*

(T) (a) Use a graphing utility to graph W.

(b) Did the number of construction workers exceed 8 million in 2006? If so, during which month(s)?

(B) **80. Biology: Predator-Prey Cycles** The graph below demonstrates snowshoe hare and lynx population fluctuations. The cycles of each population follow a periodic pattern. Describe several factors that could be contributing to the cyclical patterns. *(Source: Adapted from Levine/Miller, Biology: Discovering Life, Second Edition)*

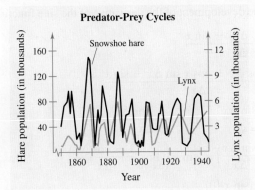

(B) **81. Physics** Use the graphs below to answer each question.

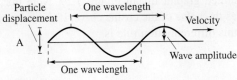

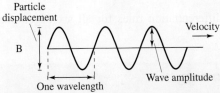

(a) Which graph (A or B) has a longer wavelength, or period?

(b) Which graph (A or B) has a greater amplitude?

(c) The frequency of a graph is the number of oscillations or cycles that occur during a given period of time. Which graph (A or B) has a greater frequency?

(d) Based on the definition of frequency, how are frequency and period related?

(Source: Adapted from Shipman/Wilson/Todd, An Introduction to Physical Science, Eleventh Edition)

82. Meteorology The normal monthly high temperatures for Erie, Pennsylvania are approximated by

$$H(t) = 56.94 - 20.86 \cos \frac{\pi t}{6} - 11.58 \sin \frac{\pi t}{6}$$

and the normal monthly low temperatures are approximated by

$$L(t) = 41.80 - 17.13 \cos \frac{\pi t}{6} - 13.39 \sin \frac{\pi t}{6}$$

where t is the time in months, with $t = 1$ corresponding to January. *(Source: National Oceanic and Atmospheric Administration)* Use the figure to answer the questions below.

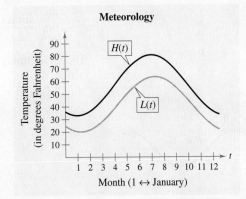

(a) During what part of the year is the difference between the normal high and low temperatures greatest? When is it smallest?

(b) The sun is the farthest north in the sky around June 21, but the graph shows the highest temperatures at a later date. Approximate the lag time of the temperatures relative to the position of the sun.

True or False? In Exercises 83–86, determine whether the statement is true or false. If it is false, explain why or give an example that shows it is false.

83. The amplitude of $f(x) = -3 \cos 2x$ is -3.

84. The period of $f(x) = 5 \cot\left(-\frac{4x}{3}\right)$ is $\frac{3\pi}{2}$.

85. $\lim_{x \to 0} \frac{\sin 5x}{3x} = \frac{5}{3}$

86. One solution of $\tan x = 1$ is $\frac{5\pi}{4}$.

87. Extended Application To work an extended application analyzing the mean monthly temperatures and precipitation in Honolulu, Hawaii, visit this text's website at *college.hmco.com/info/larsonapplied*. *(Source: National Oceanic and Atmospheric Administration)*

Section 5.4

Derivatives of Trigonometric Functions

- Find derivatives of trigonometric functions.
- Find the relative extrema of trigonometric functions.
- Use derivatives of trigonometric functions to answer questions about real-life situations.

Derivatives of Trigonometric Functions

In Example 4 and Checkpoint 4 in the preceding section, you looked at two important trigonometric limits:

$$\lim_{\Delta x \to 0} \frac{\sin \Delta x}{\Delta x} = 1 \quad \text{and} \quad \lim_{\Delta x \to 0} \frac{1 - \cos \Delta x}{\Delta x} = 0.$$

These two limits are used in the development of the derivative of the sine function.

$$\frac{d}{dx}[\sin x] = \lim_{\Delta x \to 0} \frac{\sin(x + \Delta x) - \sin x}{\Delta x}$$

$$= \lim_{\Delta x \to 0} \frac{\sin x \cos \Delta x + \cos x \sin \Delta x - \sin x}{\Delta x}$$

$$= \lim_{\Delta x \to 0} \left(\cos x \frac{\sin \Delta x}{\Delta x} - \sin x \frac{1 - \cos \Delta x}{\Delta x} \right)$$

$$= \cos x \left(\lim_{\Delta x \to 0} \frac{\sin \Delta x}{\Delta x} \right) - \sin x \left(\lim_{\Delta x \to 0} \frac{1 - \cos \Delta x}{\Delta x} \right)$$

$$= (\cos x)(1) - (\sin x)(0)$$

$$= \cos x.$$

This differentiation rule is illustrated graphically in Figure 5.29. Note that the *slope* of the sine curve determines the *value* of the cosine curve. If u is a function of x, the Chain Rule version of this differentiation rule is

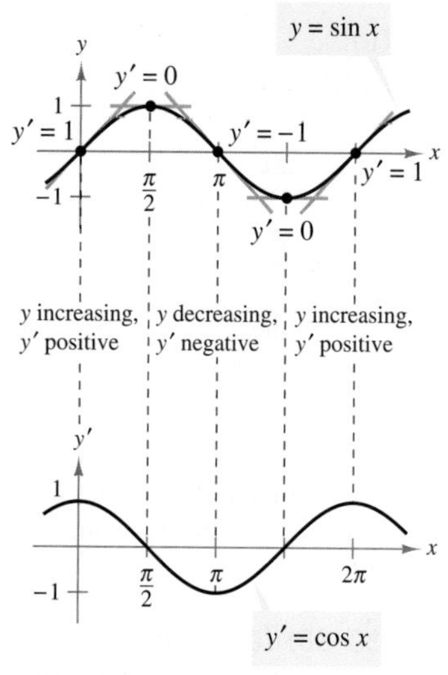

$$\frac{d}{dx}[\sin u] = \cos u \frac{du}{dx}.$$

The Chain Rule versions of the differentiation rules for all six trigonometric functions are listed below.

FIGURE 5.29

$$\frac{d}{dx}[\sin x] = \cos x$$

STUDY TIP

To help you remember these differentiation rules, note that each trigonometric function that begins with a "c" has a negative sign in its derivative.

Derivatives of Trigonometric Functions

$$\frac{d}{dx}[\sin u] = \cos u \frac{du}{dx} \qquad \frac{d}{dx}[\cos u] = -\sin u \frac{du}{dx}$$

$$\frac{d}{dx}[\tan u] = \sec^2 u \frac{du}{dx} \qquad \frac{d}{dx}[\cot u] = -\csc^2 u \frac{du}{dx}$$

$$\frac{d}{dx}[\sec u] = \sec u \tan u \frac{du}{dx} \qquad \frac{d}{dx}[\csc u] = -\csc u \cot u \frac{du}{dx}$$

SECTION 5.4 Derivatives of Trigonometric Functions

Example 1 Differentiating Trigonometric Functions

Differentiate each function.

a. $y = \sin 2x$ **b.** $y = \cos(x - 1)$ **c.** $y = \tan 3x$

SOLUTION

a. Letting $u = 2x$, you obtain

$$\frac{dy}{dx} = \cos u \frac{du}{dx} = \cos 2x \frac{d}{dx}[2x] = (\cos 2x)(2) = 2\cos 2x.$$

b. Letting $u = x - 1$, you can see that $u' = 1$. So, the derivative is simply

$$\frac{dy}{dx} = -\sin(x - 1).$$

c. Letting $u = 3x$, you have $u' = 3$, which implies that

$$\frac{dy}{dx} = 3\sec^2 3x.$$

✓ **CHECKPOINT 1**

Differentiate each function.

a. $y = \cos 4x$
b. $y = \sin(x^2 - 1)$
c. $y = \tan \dfrac{x}{2}$ ∎

Example 2 Differentiating a Trigonometric Function

Differentiate $f(x) = \cos 3x^2$.

SOLUTION Letting $u = 3x^2$, you obtain

$$f'(x) = -\sin u \frac{du}{dx} \qquad \text{Apply Cosine Differentiation Rule.}$$

$$= -\sin 3x^2 \frac{d}{dx}[3x^2] \qquad \text{Substitute } 3x^2 \text{ for } u.$$

$$= -(\sin 3x^2)(6x)$$

$$= -6x \sin 3x^2. \qquad \text{Simplify.}$$

✓ **CHECKPOINT 2**

Differentiate each function.

a. $g(x) = \sin \sqrt{x}$
b. $g(x) = 2\cos x^3$ ∎

Example 3 Differentiating a Trigonometric Function

Differentiate $f(x) = \tan^4 3x$.

SOLUTION By the Power Rule, you can write

$$\frac{d}{dx}[(\tan 3x)^4] = 4(\tan 3x)^3 \frac{d}{dx}[\tan 3x]$$

$$= 4(\tan^3 3x)(3)(\sec^2 3x)$$

$$= 12\tan^3 3x \sec^2 3x.$$

✓ **CHECKPOINT 3**

Differentiate each function.

a. $y = \sin^3 x$
b. $y = \cos^4 2x$ ∎

TECHNOLOGY

When you use a symbolic differentiation utility to differentiate trigonometric functions, you can easily obtain results that appear to be different from those you would obtain by hand. Try using a symbolic differentiation utility to differentiate the function in Example 3. How does your result compare with the given solution?

Example 4 Differentiating a Trigonometric Function

Differentiate $y = \csc \dfrac{x}{2}$.

SOLUTION

$$y = \csc \dfrac{x}{2} \qquad \text{Write original function.}$$

$$\dfrac{dy}{dx} = -\csc \dfrac{x}{2} \cot \dfrac{x}{2} \dfrac{d}{dx}\left[\dfrac{x}{2}\right] \qquad \text{Apply Cosecant Differentiation Rule.}$$

$$= -\dfrac{1}{2} \csc \dfrac{x}{2} \cot \dfrac{x}{2} \qquad \text{Simplify.}$$

✓ **CHECKPOINT 4**

Differentiate each function.

a. $y = \sec 4x$ **b.** $y = \cot x^2$

Example 5 Differentiating a Trigonometric Function

Differentiate $f(t) = \sqrt{\sin 4t}$.

SOLUTION Begin by rewriting the function in rational exponent form. Then apply the General Power Rule to find the derivative.

$$f(t) = (\sin 4t)^{1/2} \qquad \text{Rewrite with rational exponent.}$$

$$f'(t) = \left(\dfrac{1}{2}\right)(\sin 4t)^{-1/2} \dfrac{d}{dt}[\sin 4t] \qquad \text{Apply General Power Rule.}$$

$$= \left(\dfrac{1}{2}\right)(\sin 4t)^{-1/2}(4 \cos 4t)$$

$$= \dfrac{2 \cos 4t}{\sqrt{\sin 4t}} \qquad \text{Simplify.}$$

Example 6 Differentiating a Trigonometric Function

Differentiate $y = x \sin x$.

SOLUTION Using the Product Rule, you can write

$$y = x \sin x \qquad \text{Write original function.}$$

$$\dfrac{dy}{dx} = x \dfrac{d}{dx}[\sin x] + \sin x \dfrac{d}{dx}[x] \qquad \text{Apply Product Rule.}$$

$$= x \cos x + \sin x. \qquad \text{Simplify.}$$

✓ **CHECKPOINT 6**

Differentiate each function.

a. $y = x^2 \cos x$ **b.** $y = t \sin 2t$

STUDY TIP

Notice that all of the differentiation rules that you learned in earlier chapters can be applied to trigonometric functions. For instance, Example 5 uses the General Power Rule and Example 6 uses the Product Rule.

✓ **CHECKPOINT 5**

Differentiate each function.

a. $f(x) = \sqrt{\cos 2x}$

b. $f(x) = \sqrt[3]{\tan 3x}$

SECTION 5.4 Derivatives of Trigonometric Functions

Relative Extrema of Trigonometric Functions

Example 7 Finding Relative Extrema

Find the relative extrema of

$$y = \frac{x}{2} - \sin x$$

on the interval $(0, 2\pi)$.

SOLUTION To find the relative extrema of the function, begin by finding its critical numbers. The derivative of y is

$$\frac{dy}{dx} = \frac{1}{2} - \cos x.$$

By setting the derivative equal to zero, you obtain $\cos x = \frac{1}{2}$. So, in the interval $(0, 2\pi)$, the critical numbers are $x = \pi/3$ and $x = 5\pi/3$. Using the First-Derivative Test, you can conclude that $\pi/3$ yields a relative minimum and $5\pi/3$ yields a relative maximum, as shown in Figure 5.30.

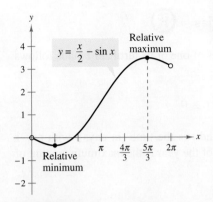

FIGURE 5.30

✓ **CHECKPOINT 7**

Find the relative extrema of

$$y = \frac{x}{2} - \cos x$$

on the interval $(0, 2\pi)$. ∎

STUDY TIP

Recall that the critical numbers of a function $y = f(x)$ are the x-values for which $f'(x) = 0$ or $f'(x)$ is undefined.

Example 8 Finding Relative Extrema

Find the relative extrema of $f(x) = 2 \sin x - \cos 2x$ on the interval $(0, 2\pi)$.

SOLUTION

$f(x) = 2 \sin x - \cos 2x$	Write original function.
$f'(x) = 2 \cos x + 2 \sin 2x$	Differentiate.
$0 = 2 \cos x + 2 \sin 2x$	Set derivative equal to 0.
$0 = 2 \cos x + 4 \cos x \sin x$	Identity: $\sin 2x = 2 \cos x \sin x$
$0 = 2(\cos x)(1 + 2 \sin x)$	Factor.

From this, you can see that the critical numbers occur when $\cos x = 0$ and when $\sin x = -\frac{1}{2}$. So, in the interval $(0, 2\pi)$, the critical numbers are

$$x = \frac{\pi}{2}, \frac{3\pi}{2}, \frac{7\pi}{6}, \text{ and } \frac{11\pi}{6}.$$

Using the First-Derivative Test, you can determine that $(\pi/2, 3)$ and $(3\pi/2, -1)$ are relative maxima, and $(7\pi/6, -\frac{3}{2})$ and $(11\pi/6, -\frac{3}{2})$ are relative minima, as shown in Figure 5.31.

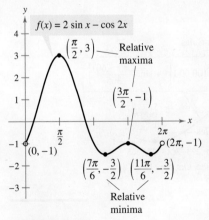

FIGURE 5.31

✓ **CHECKPOINT 8**

Find the relative extrema of $y = \frac{1}{2} \sin 2x + \cos x$ on the interval $(0, 2\pi)$. ∎

TECHNOLOGY

Graphing Trigonometric Functions

Because of the difficulty of solving some trigonometric equations, it can be difficult to find the critical numbers of a trigonometric function. For example, consider the function $f(x) = 2 \sin x - \cos 3x$. Setting the derivative of this function equal to zero produces

$f'(x) = 2 \cos x + 3 \sin 3x = 0$.

This equation is difficult to solve analytically. So, it is difficult to find the relative extrema of f analytically. With a graphing utility, however, you can easily graph the function and use the *zoom* feature to estimate the relative extrema. In the graph shown below, notice that the function has three relative minima and three relative maxima in the interval $(0, 2\pi)$.

$f(x) = 2 \sin x - \cos 3x$

Try using technology to locate the relative extrema of the function

$f(x) = 2 \sin x - \cos 4x$.

How many relative extrema does this function have in the interval $(0, 2\pi)$?

Applications

Example 9 Modeling Seasonal Sales

A fertilizer manufacturer finds that the sales of one of its fertilizer brands follows a seasonal pattern that can be modeled by

$$F = 100{,}000 \left[1 + \sin \frac{2\pi(t - 60)}{365} \right], \quad t \geq 0$$

where F is the amount sold (in pounds) and t is the time (in days), with $t = 1$ corresponding to January 1. On which day of the year is the maximum amount of fertilizer sold?

SOLUTION The derivative of the model is

$$\frac{dF}{dt} = 100{,}000 \left(\frac{2\pi}{365} \right) \cos \frac{2\pi(t - 60)}{365}.$$

Setting this derivative equal to zero produces

$$\cos \frac{2\pi(t - 60)}{365} = 0.$$

Because cosine is zero at $\pi/2$ and $3\pi/2$, you can find the critical numbers as shown.

$\dfrac{2\pi(t - 60)}{365} = \dfrac{\pi}{2}$ $\qquad\qquad$ $\dfrac{2\pi(t - 60)}{365} = \dfrac{3\pi}{2}$

$t - 60 = \dfrac{365}{4}$ $\qquad\qquad$ $t - 60 = \dfrac{3(365)}{4}$

$t = \dfrac{365}{4} + 60 \approx 151$ $\qquad\qquad$ $t = \dfrac{3(365)}{4} + 60 \approx 334$

The 151st day of the year is May 31 and the 334th day of the year is November 30. From the graph in Figure 5.32, you can see that, according to the model, the maximum sales occur on May 31.

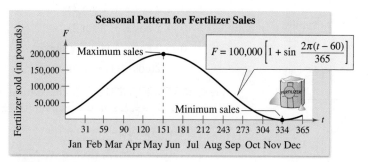

FIGURE 5.32

✓ CHECKPOINT 9

Using the model from Example 9, find the rate at which sales are changing when $t = 59$. ∎

SECTION 5.4 Derivatives of Trigonometric Functions

Example 10
MAKE A DECISION Modeling Temperature Change

The temperature T (in degrees Fahrenheit) during a given 24-hour period can be modeled by

$$T = 70 + 15 \sin \frac{\pi(t - 8)}{12}, \quad t \geq 0$$

where t is the time (in hours), with $t = 0$ corresponding to midnight, as shown in Figure 5.33. Find the rate at which the temperature is changing at 6 A.M.

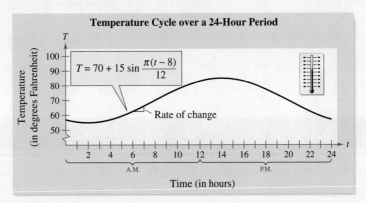

FIGURE 5.33

SOLUTION The rate of change of the temperature is given by the derivative

$$\frac{dT}{dt} = \frac{15\pi}{12} \cos \frac{\pi(t - 8)}{12}.$$

Because 6 A.M. corresponds to $t = 6$, the rate of change at 6 A.M. is

$$\frac{15\pi}{12} \cos\left(\frac{-2\pi}{12}\right) = \frac{5\pi}{4} \cos\left(-\frac{\pi}{6}\right)$$

$$= \frac{5\pi}{4}\left(\frac{\sqrt{3}}{2}\right)$$

$$\approx 3.4° \text{ per hour.}$$

✓**CHECKPOINT 10**

In Example 10, find the rate at which the temperature is changing at 8 P.M. ■

CONCEPT CHECK

1. Given $\frac{d}{dx}[\sin u] = \cos u \frac{du}{dx}$, you know that the slope of the sine curve determines the value of what curve?
2. In the differentiation rules for all six trigonometric functions, identify each trigonometric function that has a negative sign in its derivative. What do these functions have in common?
3. Can the General Power Rule and the Product Rule be applied to the differentiation of trigonometric functions?
4. Identify the trigonometric function whose derivative is $-\sin u \frac{du}{dx}$.

Skills Review 5.4

The following warm-up exercises involve skills that were covered in earlier sections. You will use these skills in the exercise set for this section. For additional help, review Sections 2.2, 2.4, 2.5, 3.2, and 5.2.

In Exercises 1–4, find the derivative of the function.

1. $f(x) = 3x^3 - 2x^2 + 4x - 7$
2. $g(x) = (x^3 + 4)^4$
3. $f(x) = (x - 1)(x^2 + 2x + 3)$
4. $g(x) = \dfrac{2x}{x^2 + 5}$

In Exercises 5 and 6, find the relative extrema of the function.

5. $f(x) = x^2 + 4x + 1$
6. $f(x) = \tfrac{1}{3}x^3 - 4x + 2$

In Exercises 7–10, solve the trigonometric equation for x where $0 \le x \le 2\pi$.

7. $\sin x = \dfrac{\sqrt{3}}{2}$
8. $\cos x = -\dfrac{1}{2}$
9. $\cos \dfrac{x}{2} = 0$
10. $\sin \dfrac{x}{2} = -\dfrac{\sqrt{2}}{2}$

Exercises 5.4

See www.CalcChat.com for worked-out solutions to odd-numbered exercises.

In Exercises 1–26, find the derivative of the function.

1. $y = \tfrac{1}{2} - 3 \sin x$
2. $y = 5 + \sin x$
3. $y = x^2 - \cos x$
4. $g(t) = \pi \cos t - \dfrac{1}{t^2}$
5. $f(x) = 4\sqrt{x} + 3 \cos x$
6. $f(x) = \sin x + \cos x$
7. $f(t) = t^2 \cos t$
8. $f(x) = (x + 1) \cos x$
9. $g(t) = \dfrac{\cos t}{t}$
10. $f(x) = \dfrac{\sin x}{x}$
11. $y = \tan x + x^2$
12. $y = x^3 - \cot x$
13. $y = e^{x^2} \sec x$
14. $y = e^{-x} \sin x$
15. $y = \cos 3x + \sin^2 x$
16. $y = \csc^2 x - \cos 2x$
17. $y = \sec \pi x$
18. $y = \tfrac{1}{2} \csc 2x$
19. $y = x \sin \dfrac{1}{x}$
20. $y = x^2 \cos \dfrac{1}{x}$
21. $y = 3 \tan 4x$
22. $y = \tan e^x$
23. $y = 2 \tan^2 4x$
24. $y = -\sin^4 2x$
25. $y = e^{2x} \sin 2x$
26. $y = e^{-x} \cos \dfrac{x}{2}$

In Exercises 27–38, find the derivative of the function and simplify your answer by using the trigonometric identities listed in Section 5.2.

27. $y = \cos^2 x$
28. $y = \tfrac{1}{4} \sin^2 2x$
29. $y = \cos^2 x - \sin^2 x$
30. $y = \dfrac{x}{2} + \dfrac{\sin 2x}{4}$
31. $y = \sin^2 x - \cos 2x$
32. $y = 3 \sin x - 2 \sin^3 x$
33. $y = \tan x - x$
34. $y = \cot x + x$
35. $y = \dfrac{\sin^3 x}{3} - \dfrac{\sin^5 x}{5}$
36. $y = \dfrac{\sec^7 x}{7} - \dfrac{\sec^5 x}{5}$
37. $y = \ln(\sin^2 x)$
38. $y = \tfrac{1}{2} \ln(\cos^2 x)$

In Exercises 39–46, find an equation of the tangent line to the graph of the function at the given point.

	Function	Point
39.	$y = \tan x$	$\left(-\dfrac{\pi}{4}, -1\right)$
40.	$y = \sec x$	$\left(\dfrac{\pi}{3}, 2\right)$
41.	$y = \sin 4x$	$(\pi, 0)$
42.	$y = \csc^2 x$	$\left(\dfrac{\pi}{2}, 1\right)$
43.	$y = \cot x$	$\left(\dfrac{3\pi}{4}, -1\right)$
44.	$y = \sin x \cos x$	$\left(\dfrac{3\pi}{2}, 0\right)$
45.	$y = \ln(\sin x + 2)$	$\left(\dfrac{3\pi}{2}, 0\right)$
46.	$y = \sqrt{\sin x}$	$\left(\dfrac{\pi}{6}, \dfrac{\sqrt{2}}{2}\right)$

In Exercises 47–52, find the slope of the tangent line to the given sine function at the origin. Compare this value with the number of complete cycles in the interval $[0, 2\pi]$.

47. $y = \sin \dfrac{5x}{4}$

48. $y = \sin \dfrac{5x}{2}$

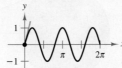

49. $y = \sin 2x$

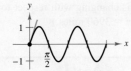

50. $y = \sin \dfrac{3x}{2}$

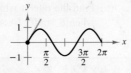

51. $y = \sin x$

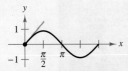

52. $y = \sin \dfrac{x}{2}$

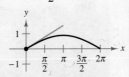

In Exercises 53–58, determine the relative extrema of the function on the interval $(0, 2\pi)$. Use a graphing utility to confirm your result.

53. $y = 2\sin x + \sin 2x$
54. $y = 2\cos x + \cos 2x$
55. $y = x - 2\sin x$
56. $y = e^{-x}\sin x$
57. $y = e^x \cos x$
58. $y = \sec \dfrac{x}{2}$

59. **Meteorology** The number of hours of daylight D in New Orleans can be modeled by

$$D = 12.13 - 1.87 \cos \dfrac{\pi(t - 0.07)}{6}, \quad 0 \le t \le 12$$

where t represents the month, with $t = 0$ corresponding to January 1. Find the month t in which New Orleans has the maximum number of daylight hours. What is this maximum number of daylight hours? (Source: U.S. Naval Observatory)

60. **Tides** Throughout the day, the depth D of water (in meters) at the end of a dock varies with the tides. The depth for one particular day can be modeled by

$$D = 3.5 + 1.5 \cos \dfrac{\pi t}{6}, \quad 0 \le t \le 24$$

where $t = 0$ represents midnight.

(a) Determine dD/dt.

(b) Evaluate dD/dt for $t = 4$ and $t = 20$ and interpret your results.

(c) Find the time(s) when the water depth is the greatest and the time(s) when the water depth is the least.

(d) What is the greatest depth? What is the least depth? Did you have to use calculus to determine these depths? Explain your reasoning.

61. **Biology** Plants do not grow at constant rates during a normal 24-hour period because their growth is affected by sunlight. Suppose that the growth of a certain plant species in a controlled environment is given by the model $h = 0.2t + 0.03 \sin 2\pi t$, where h is the height of the plant in inches and t is the time in days, with $t = 0$ corresponding to midnight of day 1 (see figure). During what time of day is the rate of growth of this plant (a) a maximum and (b) a minimum?

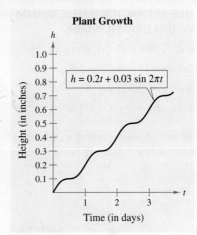

62. **Meteorology** The normal average daily temperature in degrees Fahrenheit for a city is given by

$$T = 55 - 21 \cos \dfrac{2\pi(t - 32)}{365}$$

where t is the time in days, with $t = 1$ corresponding to January 1. Find the expected dates of the warmest and coldest days.

In Exercises 63–68, use a graphing utility (a) to graph f and f' on the same coordinate axes over the specified interval, (b) to find the critical numbers of f, and (c) to find the interval(s) on which f' is positive and the interval(s) on which it is negative. Note the behavior of f in relation to the sign of f'.

Function	Interval
63. $f(t) = t^2 \sin t$	$(0, 2\pi)$
64. $f(x) = \dfrac{x}{2} + \cos \dfrac{x}{2}$	$(0, 4\pi)$
65. $f(x) = \sin x - \tfrac{1}{3}\sin 3x + \tfrac{1}{5}\sin 5x$	$(0, \pi)$
66. $f(x) = x \sin x$	$(0, \pi)$
67. $f(x) = \sqrt{2x} \sin x$	$(0, 2\pi)$
68. $f(x) = 4e^{-0.5x} \sin \pi x$	$(0, 4)$

354   **CHAPTER 5**   Trigonometric Functions

(T) In Exercises 69–74, use a graphing utility to find the relative extrema of the trigonometric function. Let $0 < x < 2\pi$.

69. $f(x) = \dfrac{x}{\sin x}$

70. $f(x) = \dfrac{x^2 - 2}{\sin x} - 5x$

71. $f(x) = \ln x \cos x$

72. $f(x) = \ln x \sin x$

73. $f(x) = \sin(0.1x^2)$

74. $f(x) = \sin \sqrt{x}$

True or False? In Exercises 75–78, determine whether the statement is true or false. If it is false, explain why or give an example that shows it is false.

75. If $y = (1 - \cos x)^{1/2}$, then $y' = \frac{1}{2}(1 - \cos x)^{-1/2}$.

76. If $f(x) = \sin^2(2x)$, then $f'(x) = 2(\sin 2x)(\cos 2x)$.

77. If $y = x \sin^2 x$, then $y' = 2x \sin x$.

78. The minimum value of $y = 3 \sin x + 2$ is -1.

79. Applying Fertilizer A tractor towing a cylindrical tank with the dimensions shown is applying liquid fertilizer to a tract of land.

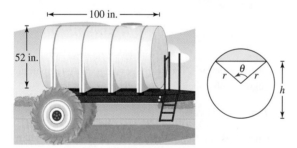

(a) The figure at the right shows a cross section of the tank. The shaded area (called a *segment*) represents the region where liquid fertilizer is *not* present. The area of a segment is given by

$$A = \frac{1}{2}r^2(\theta - \sin \theta)$$

where r is the radius of the tank and θ is in radians. Show that the volume V (in cubic inches) of fertilizer in the tank is given by

$$V = 67{,}600\pi - 33{,}800\theta + 33{,}800 \sin \theta.$$

(b) Find the value of θ when the height h of fertilizer in the tank is 40 inches.

(c) Use the results of parts (a) and (b) to find the rate at which V is changing with respect to θ when h is 40 inches.

(Submitted by R.J. Rishel, Rishel Landscaping Services)

80. Satellites When satellites observe Earth, they can scan only part of Earth's surface. Some satellites have sensors that can measure the angle θ shown in the figure. Let h represent the satellite's distance from Earth's surface and let r represent Earth's radius.

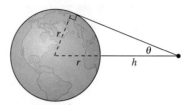

(a) Show that $h = r(\csc \theta - 1)$.

(b) Find the rate at which h is changing with respect to θ when $\theta = 30°$. (Assume $r = 3960$ miles.)

Social Science Capsule

© Keith Hunter/Arcaid/CORBIS

Everyone enjoys attractively designed residential areas, public parks, college campuses, shopping centers, golf courses, parkways, and industrial parks. Landscape architects design these areas so that they are not only functional but also pleasing to the eye and compatible with the natural environment. They plan the locations of buildings, roads, and walkways and the arrangement of flowers, shrubs, and trees. More than 26 percent of all landscape architects are self-employed. A bachelor's degree in landscape architecture is the minimum requirement for entry-level jobs.

81. Research Project Use your school's library, the Internet, or some other reference source to gather information about how trigonometry and calculus play an important role in the practice of landscape architecture. Write a short paper that summarizes your findings.

Algebra Review

Solving Trigonometric Equations

Solving a trigonometric equation requires the use of trigonometry, but it also requires the use of algebra. Some examples of solving trigonometric equations were presented on pages 330 and 331. Here are several others.

Example 1 Solving Trigonometric Equations

Solve each trigonometric equation.

a. $\sin x + \sqrt{2} = -\sin x$

b. $3 \tan^2 x = 1$

c. $\cot x \cos^2 x = 2 \cot x$

SOLUTION

a.
$$\sin x + \sqrt{2} = -\sin x \quad \text{Write original equation.}$$
$$\sin x + \sin x = -\sqrt{2} \quad \text{Add } \sin x \text{ to, and subtract } \sqrt{2} \text{ from, each side.}$$
$$2 \sin x = -\sqrt{2} \quad \text{Combine like terms.}$$
$$\sin x = -\frac{\sqrt{2}}{2} \quad \text{Divide each side by 2.}$$
$$x = \frac{5\pi}{4}, \frac{7\pi}{4}, \quad 0 \leq x \leq 2\pi$$

b.
$$3 \tan^2 x = 1 \quad \text{Write original equation.}$$
$$\tan^2 x = \frac{1}{3} \quad \text{Divide each side by 3.}$$
$$\tan x = \pm \frac{\sqrt{3}}{3} \quad \text{Extract square roots.}$$
$$x = \frac{\pi}{6}, \frac{5\pi}{6}, \frac{7\pi}{6}, \frac{11\pi}{6}, \quad 0 \leq x \leq 2\pi$$

c.
$$\cot x \cos^2 x = 2 \cot x \quad \text{Write original equation.}$$
$$\cot x \cos^2 x - 2 \cot x = 0 \quad \text{Subtract } 2 \cot x \text{ from each side.}$$
$$\cot x (\cos^2 x - 2) = 0 \quad \text{Factor.}$$

Setting each factor equal to zero, you obtain the solutions in the interval $0 \leq x \leq 2\pi$ as shown.

$$\cot x = 0 \quad \text{and} \quad \cos^2 x - 2 = 0$$
$$x = \frac{\pi}{2}, \frac{3\pi}{2} \quad \cos^2 x = 2$$
$$\cos x = \pm \sqrt{2}$$

No solution is obtained from $\cos x = \pm \sqrt{2}$ because $\pm \sqrt{2}$ are outside the range of the cosine function.

Example 2 Solving Trigonometric Equations

Solve each trigonometric equation.

a. $2\sin^2 x - \sin x - 1 = 0$

b. $2\sin^2 x + 3\cos x - 3 = 0$

c. $2\cos 3t - 1 = 0$

SOLUTION

a.
$2\sin^2 x - \sin x - 1 = 0$	Write original equation.
$(2\sin x + 1)(\sin x - 1) = 0$	Factor.

Setting each factor equal to zero, you obtain the solutions in the interval $[0, 2\pi]$ as shown.

$$2\sin x + 1 = 0 \quad \text{and} \quad \sin x - 1 = 0$$

$$\sin x = -\frac{1}{2} \qquad\qquad \sin x = 1$$

$$x = \frac{7\pi}{6}, \frac{11\pi}{6} \qquad\qquad x = \frac{\pi}{2}$$

b.
$2\sin^2 x + 3\cos x - 3 = 0$	Write original equation.
$2(1 - \cos^2 x) + 3\cos x - 3 = 0$	Pythagorean Identity
$-2\cos^2 x + 3\cos x - 1 = 0$	Combine like terms.
$2\cos^2 x - 3\cos x + 1 = 0$	Multiply each side by -1.
$(2\cos x - 1)(\cos x - 1) = 0$	Factor.

Setting each factor equal to zero, you obtain the solutions in the interval $[0, 2\pi]$ as shown.

$$2\cos x - 1 = 0 \quad \text{and} \quad \cos x - 1 = 0$$

$$2\cos x = 1 \qquad\qquad \cos x = 1$$

$$\cos x = \frac{1}{2} \qquad\qquad x = 0, 2\pi$$

$$x = \frac{\pi}{3}, \frac{5\pi}{3}$$

c.
$2\cos 3t - 1 = 0$	Write original equation.
$2\cos 3t = 1$	Add 1 to each side.
$\cos 3t = \dfrac{1}{2}$	Divide each side by 2.

$$3t = \frac{\pi}{3}, \frac{5\pi}{3}, \quad 0 \le 3t \le 2\pi$$

$$t = \frac{\pi}{9}, \frac{5\pi}{9}, \quad 0 \le t \le \frac{2}{3}\pi$$

In the interval $0 \le t \le 2\pi$, there are four other solutions.

Chapter Summary and Study Strategies

After studying this chapter, you should have acquired the following skills.
The exercise numbers are keyed to the Review Exercises that begin on page 359.
Answers to odd-numbered Review Exercises are given in the back of the text.*

Section 5.1	Review Exercises
■ Find coterminal angles.	1–8
■ Convert from degree to radian measure and from radian to degree measure. π radians $= 180°$	9–20
■ Use formulas relating to triangles.	21–24
■ Use formulas relating to triangles to solve real-life problems.	25, 26

Section 5.2

- Find the reference angles for given angles. 27–34
- Evaluate trigonometric functions exactly. 35–46

Right Triangle Definition: $0 < \theta < \dfrac{\pi}{2}$.

$$\sin \theta = \frac{\text{opp.}}{\text{hyp.}} \qquad \cos \theta = \frac{\text{adj.}}{\text{hyp.}} \qquad \tan \theta = \frac{\text{opp.}}{\text{adj.}}$$

$$\csc \theta = \frac{\text{hyp.}}{\text{opp.}} \qquad \sec \theta = \frac{\text{hyp.}}{\text{adj.}} \qquad \cot \theta = \frac{\text{adj.}}{\text{opp.}}$$

Circular Function Definition: θ is any angle in standard position and (x, y) is a point on the terminal ray of the angle.

$$\sin \theta = \frac{y}{r} \qquad \cos \theta = \frac{x}{r} \qquad \tan \theta = \frac{y}{x}$$

$$\csc \theta = \frac{r}{y} \qquad \sec \theta = \frac{r}{x} \qquad \cot \theta = \frac{x}{y}$$

- Use a calculator to approximate values of trigonometric functions. 47–54
- Solve right triangles. 55–58
- Solve trigonometric equations. 59–64
- Use right triangles to solve real-life problems. 65, 66

Section 5.3

- Sketch graphs of trigonometric functions. 67–74
- Use trigonometric functions to model real-life situations. 75, 76

---

* Use a wide range of valuable study aids to help you master the material in this chapter. The *Student Solutions Guide* includes step-by-step solutions to all odd-numbered exercises to help you review and prepare. The student website at *college.hmco.com/info/larsonapplied* offers algebra help and a *Graphing Technology Guide*. The *Graphing Technology Guide* contains step-by-step commands and instructions for a wide variety of graphing calculators, including the most recent models.

Section 5.4   Review Exercises

- Find derivatives of trigonometric functions.   77–88

$$\frac{d}{dx}[\sin u] = \cos u \frac{du}{dx} \qquad \frac{d}{dx}[\cos u] = -\sin u \frac{du}{dx}$$

$$\frac{d}{dx}[\tan u] = \sec^2 u \frac{du}{dx} \qquad \frac{d}{dx}[\cot u] = -\csc^2 u \frac{du}{dx}$$

$$\frac{d}{dx}[\sec u] = \sec u \tan u \frac{du}{dx} \qquad \frac{d}{dx}[\csc u] = -\csc u \cot u \frac{du}{dx}$$

- Find the equations of tangent lines to graphs of trigonometric functions.   89–94
- Analyze the graphs of trigonometric functions.   95–98
- Use relative extrema to solve real-life problems.   99, 100

Study Strategies

- **Degree and Radian Modes** When using a computer or calculator to evaluate or graph a trigonometric function, be sure that you use the proper mode—*radian* mode or *degree* mode.

- **Checking the Form of an Answer** Because of the abundance of trigonometric identities, solutions of problems in this chapter can take a variety of forms. For instance, the expressions $-\ln|\cot x| + C$ and $\ln|\tan x| + C$ are equivalent. So, when you are checking your solutions with those given in the back of the text, remember that your solution might be correct, even if its form doesn't agree precisely with that given in the text.

- **Using Technology** Throughout this chapter, remember that technology can help you *graph* trigonometric functions, *evaluate* trigonometric functions, *differentiate* trigonometric functions, and *integrate* trigonometric functions. Consider, for instance, the difficulty of sketching the graph of the function below *without* using a graphing utility.

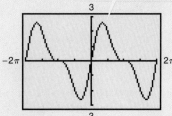

Review Exercises

See www.CalcChat.com for worked-out solutions to odd-numbered exercises.

In Exercises 1–8, determine two coterminal angles (one positive and one negative) for the angle.

1. 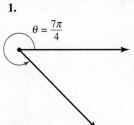 $\theta = \dfrac{7\pi}{4}$

2. 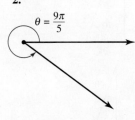 $\theta = \dfrac{9\pi}{5}$

3. 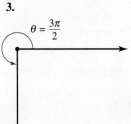 $\theta = \dfrac{3\pi}{2}$

4. 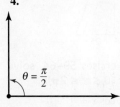 $\theta = \dfrac{\pi}{2}$

5. 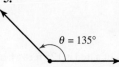 $\theta = 135°$

6. 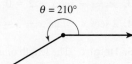 $\theta = 210°$

7. 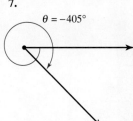 $\theta = -405°$

8. 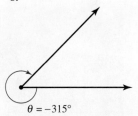 $\theta = -315°$

In Exercises 9–16, convert the degree measure to radian measure. Use a calculator to verify your results.

9. $210°$
10. $300°$
11. $-60°$
12. $-30°$
13. $-480°$
14. $-540°$
15. $110°$
16. $320°$

In Exercises 17–20, convert the radian measure to degree measure. Use a calculator to verify your results.

17. $\dfrac{4\pi}{3}$

18. $\dfrac{5\pi}{6}$

19. $-\dfrac{2\pi}{3}$

20. $-\dfrac{11\pi}{6}$

In Exercises 21–24, solve the triangle for the indicated side and/or angle.

21.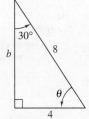

22.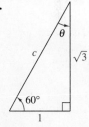

23.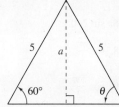

24.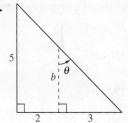

25. **Length** To stabilize a 75-foot tower for a radio antenna, a guy wire must be attached from the top of the tower to an anchor 50 feet from the base. How long is the wire?

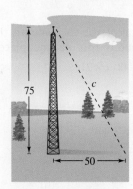

26. Height A ladder of length 16 feet leans against the side of a house. The bottom of the ladder is 4.4 feet from the house (see figure). Find the height h of the top of the ladder.

In Exercises 27–34, find the reference angle for the given angle.

27. $\dfrac{2\pi}{3}$

28. $\dfrac{9\pi}{4}$

29. $-\dfrac{5\pi}{6}$

30. $-\dfrac{5\pi}{3}$

31. 240°

32. 300°

33. 420°

34. 480°

In Exercises 35–46, evaluate the trigonometric function without using a calculator.

35. $\cos(-45°)$

36. $\sin 240°$

37. $\tan \dfrac{2\pi}{3}$

38. $\sec \dfrac{\pi}{4}$

39. $\sin \dfrac{5\pi}{3}$

40. $\cos \dfrac{5\pi}{2}$

41. $\cot\left(-\dfrac{5\pi}{6}\right)$

42. $\tan\left(-\dfrac{5\pi}{3}\right)$

43. $\sec(-180°)$

44. $\csc(-270°)$

45. $\cos\left(-\dfrac{4\pi}{3}\right)$

46. $\cot\left(-\dfrac{11\pi}{6}\right)$

(T) In Exercises 47–54, use a calculator to evaluate the trigonometric function. Round to four decimal places.

47. $\tan 33°$

48. $\cot 216°$

49. $\sec \dfrac{12\pi}{5}$

50. $\csc \dfrac{2\pi}{9}$

51. $\sin\left(-\dfrac{\pi}{9}\right)$

52. $\cos\left(-\dfrac{3\pi}{7}\right)$

53. $\cos 105°$

54. $\sin 224°$

In Exercises 55–58, solve for x, y, or r as indicated.

55.

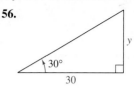

56.

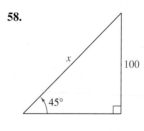

57.

58.

In Exercises 59–64, solve the trigonometric equation for x ($0 \leq x \leq 2\pi$).

59. $2\cos x + 1 = 0$

60. $2\cos^2 x = 1$

61. $2\sin^2 x + 3\sin x + 1 = 0$

62. $\cos^3 x = \cos x$

63. $\sec^2 x - \sec x - 2 = 0$

64. $2\sec^2 x + \tan^2 x - 3 = 0$

65. Height The length of a shadow of a tree is 125 feet when the angle of elevation of the sun is 33° (see figure). Approximate the height h of the tree.

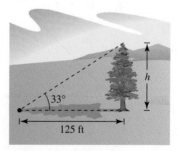

66. Distance A passenger in an airplane flying at 35,000 feet sees two towns directly to the left of the airplane. The angles of depression to the towns are 32° and 76° (see figure). How far apart are the towns?

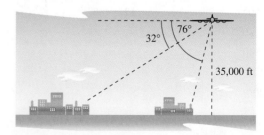

In Exercises 67–74, sketch a graph of the trigonometric function.

67. $y = 2 \cos 6x$
68. $y = \sin 2\pi x$
69. $y = \frac{1}{3} \tan x$
70. $y = \cot \frac{x}{2}$
71. $y = 3 \sin \frac{2x}{5}$
72. $y = 8 \cos\left(-\frac{x}{4}\right)$
73. $y = \sec 2\pi x$
74. $y = 3 \csc 2x$

75. **Biology: Predator-Prey Cycles** The population P of a predator at time t (in months) is modeled by

$$P = 6200 + 1700 \sin \frac{2\pi t}{24}$$

and the population p of its prey is modeled by

$$p = 9650 + 3300 \cos \frac{2\pi t}{24}.$$

(T) (a) Use a graphing utility to graph both models in the same viewing window.

(b) Explain the oscillations in the size of each population.

(T) 76. **Health** The pressure P (in millimeters of mercury) against the walls of the blood vessels of a person is modeled by

$$P = 100 - 20 \cos \frac{8\pi}{3} t$$

where t is the time (in seconds). Use a graphing utility to graph the model. One cycle is equivalent to one heartbeat. What is the person's pulse rate in heartbeats per minute?

In Exercises 77–88, find the derivative of the function.

77. $y = \sin 5\pi x$
78. $y = \tan(4x - \pi)$
79. $y = -x \tan x$
80. $y = \csc 3x + \cot 3x$
81. $y = \frac{\cos x}{x^2}$
82. $y = \frac{\cos(x - 1)}{x - 1}$
83. $y = \sin^2 x + x$
84. $y = x \cos x - \sin x$
85. $y = \csc^4 x$
86. $y = \sec^2 2x$
87. $y = e^x \cot x$
88. $y = e^{\sin x}$

In Exercises 89–94, find an equation of the tangent line to the graph of the function at the given point.

89. $y = \cos 2x$, $\left(\frac{\pi}{4}, 0\right)$
90. $y = -x \cos x$, $(0, 0)$
91. $y = \frac{1}{2} \sin^2 x$, $\left(\frac{\pi}{2}, \frac{1}{2}\right)$
92. $y = \frac{x}{\cos x}$, $(0, 0)$
93. $y = x \tan 2x$, $(0, 0)$
94. $y = \tan \pi e^x$, $(0, 0)$

In Exercises 95–98, find the relative extrema of the function on the interval $(0, 2\pi)$.

95. $f(x) = \frac{x}{2} + \cos x$
96. $f(x) = \sin x \cos x$

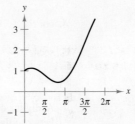
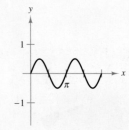

97. $f(x) = \sin^2 x + \sin x$
98. $f(x) = \frac{1}{2 + \sin x}$

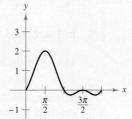
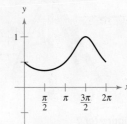

99. **Construction Workers** The number W (in thousands) of construction workers employed in the United States during 2006 can be modeled by

$$W = 7594 + 455.2 \sin(0.4t - 1.713)$$

where t is the time in months, with $t = 1$ corresponding to January 1. Approximate the month t in which the number of construction workers employed was a maximum. What was the maximum number of construction workers employed? *(Source: U.S. Bureau of Labor Statistics)*

100. **Amusement Park Workers** The number W (in thousands) of amusement park workers employed in the United States during 2006 can be modeled by

$$W = 139.8 + 37.33 \sin(0.612t - 2.66)$$

where t is the time in months, with $t = 1$ corresponding to January 1. Approximate the month t in which the number of amusement park workers employed was a maximum. What was the maximum number of amusement park workers employed? *(Source: U.S. Bureau of Labor Statistics)*

Chapter Test

See www.CalcChat.com for worked-out solutions to odd-numbered exercises.

Take this test as you would take a test in class. When you are done, check your work against the answers given in the back of the book.

In Exercises 1–6, copy and complete the table. Use a calculator if necessary.

Function	θ (deg)	θ (rad)	Function value
1. sin	67.5°		
2. cos		$\dfrac{\pi}{5}$	
3. tan	15°		
4. cot		$-\dfrac{\pi}{6}$	
5. sec	$-40°$		
6. csc		$-\dfrac{5\pi}{4}$	

7. A digital camera tripod has a height of 25 inches, and an angle of 24° is formed between the height and the leg of length ℓ (see figure). Find the length ℓ.

In Exercises 8–10, solve the equation for θ ($0 \leq \theta \leq 2\pi$).

8. $2 \sin \theta - \sqrt{2} = 0$
9. $\cos^2 \theta - \sin^2 \theta = 0$
10. $\csc \theta = \sqrt{3} \sec \theta$

In Exercises 11–13, sketch the graph of the function.

11. $y = 3 \sin 2x$
12. $y = 4 \cos 3\pi x$
13. $y = \cot \dfrac{\pi x}{5}$

In Exercises 14–16, (a) find the derivative of the function and (b) find the relative extrema of the function on the interval $(0, 2\pi)$.

14. $y = \cos x - \cos^2 x$
15. $y = \sec\left(x - \dfrac{\pi}{4}\right)$
16. $y = \dfrac{1}{3 - \sin(x + \pi)}$

17. The population P of a predator at time t (in months) is modeled by

$$P = 7300 + 2900 \sin \dfrac{2\pi t}{24}$$

and the population p of its prey is modeled by

$$p = 10{,}000 + 3700 \cos \dfrac{2\pi t}{24}.$$

(T) (a) Use a graphing utility to graph both models in the same viewing window.

(b) Explain the oscillations in the size of each population.

Figure for 7

Integration and Its Applications

6

Many people think of polluted water as being saturated with sewage or poisonous chemicals. However, some polluted lakes are filled with large amounts of algae or choked with aquatic plants. Above, a large pond on Torry Island in Lake Okeechobee, Florida, before and after suppression of waterlettuce. (For another example of cleaning a polluted lake, see Section 6.2, Example 7.)

Applications

Integration has many real-life applications. The applications listed below represent a sample of the applications in this chapter.

- Make a Decision: Vital Statistics, Exercise 70, page 374
- Women in the Labor Force, Exercise 49, page 383
- Livestock Inventory, Exercises 59 and 60, page 390
- Drug Concentration, Exercise 86, page 402
- Modeling a Body of Water, Exercise 32, page 417

6.1 Antiderivatives and Indefinite Integrals

6.2 Integration by Substitution and the General Power Rule

6.3 Exponential and Logarithmic Integrals

6.4 Area and the Fundamental Theorem of Calculus

6.5 The Area of a Region Bounded by Two Graphs

6.6 Volumes of Solids of Revolution

Section 6.1
Antiderivatives and Indefinite Integrals

- Understand the definition of antiderivative.
- Use indefinite integral notation for antiderivatives.
- Use basic integration rules to find antiderivatives.
- Use initial conditions to find particular solutions of indefinite integrals.
- Use antiderivatives to solve real-life problems.

Antiderivatives

Up to this point in the text, you have been concerned primarily with this problem: given a function, find its derivative. Many important applications of calculus involve the inverse problem: given the derivative of a function, find the function. For example, suppose you are given

$$f'(x) = 2, \quad g'(x) = 3x^2, \quad \text{and} \quad s'(t) = 4t.$$

Your goal is to determine the functions f, g, and s. By making educated guesses, you might come up with the following functions.

$f(x) = 2x$ because $\dfrac{d}{dx}[2x] = 2.$

$g(x) = x^3$ because $\dfrac{d}{dx}[x^3] = 3x^2.$

$s(t) = 2t^2$ because $\dfrac{d}{dt}[2t^2] = 4t.$

This operation of determining the original function from its derivative is the inverse operation of differentiation. It is called **antidifferentiation.**

Definition of Antiderivative

A function F is an **antiderivative** of a function f if for every x in the domain of f, it follows that $F'(x) = f(x)$.

If $F(x)$ is an antiderivative of $f(x)$, then $F(x) + C$, where C is any constant, is also an antiderivative of $f(x)$. For example,

$$F(x) = x^3, \quad G(x) = x^3 - 5, \quad \text{and} \quad H(x) = x^3 + 0.3$$

are all antiderivatives of $3x^2$ because the derivative of each is $3x^2$. As it turns out, *all* antiderivatives of $3x^2$ are of the form $x^3 + C$. So, the process of antidifferentiation does not determine a single function, but rather a *family* of functions, each differing from the others by a constant.

STUDY TIP
In this text, the phrase "$F(x)$ is an antiderivative of $f(x)$" is used synonymously with "F is an antiderivative of f."

SECTION 6.1 Antiderivatives and Indefinite Integrals

Notation for Antiderivatives and Indefinite Integrals

The antidifferentiation process is also called **integration** and is denoted by the symbol

$$\int \qquad \text{Integral sign}$$

which is called an **integral sign.** The symbol

$$\int f(x)\, dx \qquad \text{Indefinite integral}$$

is the **indefinite integral** of $f(x)$, and it denotes the family of antiderivatives of $f(x)$. That is, if $F'(x) = f(x)$ for all x, then you can write

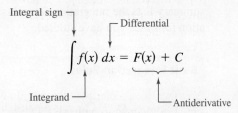

where $f(x)$ is the **integrand** and C is the **constant of integration.** The differential dx in the indefinite integral identifies the variable of integration. That is, the symbol $\int f(x)\, dx$ denotes the "antiderivative of f *with respect to x*" just as the symbol dy/dx denotes the "derivative of y *with respect to x.*"

DISCOVERY

Verify that $F_1(x) = x^2 - 2x$, $F_2(x) = x^2 - 2x - 1$, and $F_3(x) = (x - 1)^2$ are all antiderivatives of $f(x) = 2x - 2$. Use a graphing utility to graph F_1, F_2, and F_3 in the same coordinate plane. How are their graphs related? What can you say about the graph of any other antiderivative of f?

Integral Notation of Antiderivatives

The notation

$$\int f(x)\, dx = F(x) + C$$

where C is an arbitrary constant, means that F is an antiderivative of f. That is, $F'(x) = f(x)$ for all x in the domain of f.

Example 1 Notation for Antiderivatives

Using integral notation, you can write the three antiderivatives from the beginning of this section as shown.

a. $\displaystyle\int 2\, dx = 2x + C$ **b.** $\displaystyle\int 3x^2\, dx = x^3 + C$ **c.** $\displaystyle\int 4t\, dt = 2t^2 + C$

✓ **CHECKPOINT 1**

Rewrite each antiderivative using integral notation.

a. $\dfrac{d}{dx}[3x] = 3$ **b.** $\dfrac{d}{dx}[x^2] = 2x$ **c.** $\dfrac{d}{dt}[3t^3] = 9t^2$

Finding Antiderivatives

The inverse relationship between the operations of integration and differentiation can be shown symbolically, as follows.

$$\frac{d}{dx}\left[\int f(x)\,dx\right] = f(x)$$ Differentiation is the inverse of integration.

$$\int f'(x)\,dx = f(x) + C$$ Integration is the inverse of differentiation.

This inverse relationship between integration and differentiation allows you to obtain integration formulas directly from differentiation formulas. The following summary lists the integration formulas that correspond to some of the differentiation formulas you have studied.

Basic Integration Rules

1. $\int k\,dx = kx + C$, k is a constant.   Constant Rule

2. $\int k f(x)\,dx = k \int f(x)\,dx$ Constant Multiple Rule

3. $\int [f(x) + g(x)]\,dx = \int f(x)\,dx + \int g(x)\,dx$ Sum Rule

4. $\int [f(x) - g(x)]\,dx = \int f(x)\,dx - \int g(x)\,dx$ Difference Rule

5. $\int x^n\,dx = \dfrac{x^{n+1}}{n+1} + C$, $n \neq -1$ Simple Power Rule

STUDY TIP
You will study the General Power Rule for integration in Section 6.2 and the Exponential and Log Rules in Section 6.3.

Be sure you see that the Simple Power Rule has the restriction that n cannot be -1. So, you *cannot* use the Simple Power Rule to evaluate the integral

$$\int \frac{1}{x}\,dx.$$

To evaluate this integral, you need the Log Rule, which is described in Section 6.3.

STUDY TIP
In Example 2(b), the integral $\int 1\,dx$ is usually shortened to the form $\int dx$.

Example 2   Finding Indefinite Integrals

Find each indefinite integral.

a. $\displaystyle\int \frac{1}{2}\,dx$ b. $\displaystyle\int 1\,dx$ c. $\displaystyle\int -5\,dt$

SOLUTION

a. $\displaystyle\int \frac{1}{2}\,dx = \frac{1}{2}x + C$ b. $\displaystyle\int 1\,dx = x + C$ c. $\displaystyle\int -5\,dt = -5t + C$

✓**CHECKPOINT 2**

Find each indefinite integral.

a. $\displaystyle\int 5\,dx$

b. $\displaystyle\int -1\,dr$

c. $\displaystyle\int 2\,dt$

SECTION 6.1 Antiderivatives and Indefinite Integrals

TECHNOLOGY

If you have access to a symbolic integration program, try using it to find antiderivatives.

Example 3 Finding an Indefinite Integral

Find $\int 3x\, dx$.

SOLUTION

$$\begin{aligned}
\int 3x\, dx &= 3 \int x\, dx && \text{Constant Multiple Rule} \\
&= 3 \int x^1\, dx && \text{Rewrite } x \text{ as } x^1. \\
&= 3\left(\frac{x^2}{2}\right) + C && \text{Simple Power Rule with } n = 1 \\
&= \frac{3}{2}x^2 + C && \text{Simplify.}
\end{aligned}$$

✓ **CHECKPOINT 3**

Find $\int 5x\, dx$.

In finding indefinite integrals, a strict application of the basic integration rules tends to produce cumbersome constants of integration. For instance, in Example 3, you could have written

$$\int 3x\, dx = 3 \int x\, dx = 3\left(\frac{x^2}{2} + C\right) = \frac{3}{2}x^2 + 3C.$$

However, because C represents *any* constant, it is unnecessary to write $3C$ as the constant of integration. You can simply write $\frac{3}{2}x^2 + C$.

In Example 3, note that the general pattern of integration is similar to that of differentiation.

Original Integral: $\int 3x\, dx$ ⇒ Rewrite: $3\int x^1\, dx$ ⇒ Integrate: $3\left(\frac{x^2}{2}\right) + C$ ⇒ Simplify: $\frac{3}{2}x^2 + C$

STUDY TIP

Remember that you can check your answer to an antidifferentiation problem by differentiating. For instance, in Example 4(b), you can check that $\frac{2}{3}x^{3/2}$ is the correct antiderivative by differentiating to obtain

$$\frac{d}{dx}\left[\frac{2}{3}x^{3/2}\right] = \left(\frac{2}{3}\right)\left(\frac{3}{2}\right)x^{1/2}$$

$$= \sqrt{x}.$$

Example 4 Rewriting Before Integrating

Find each indefinite integral.

a. $\int \frac{1}{x^3}\, dx$

b. $\int \sqrt{x}\, dx$

SOLUTION

Original Integral	Rewrite	Integrate	Simplify
a. $\int \dfrac{1}{x^3}\, dx$	$\int x^{-3}\, dx$	$\dfrac{x^{-2}}{-2} + C$	$-\dfrac{1}{2x^2} + C$
b. $\int \sqrt{x}\, dx$	$\int x^{1/2}\, dx$	$\dfrac{x^{3/2}}{3/2} + C$	$\dfrac{2}{3}x^{3/2} + C$

✓ **CHECKPOINT 4**

Find each indefinite integral.

a. $\int \dfrac{1}{x^2}\, dx$ **b.** $\int \sqrt[3]{x}\, dx$

368   **CHAPTER 6**   Integration and Its Applications

With the five basic integration rules, you can integrate *any* polynomial function, as demonstrated in the next example.

Example 5   Integrating Polynomial Functions

Find each indefinite integral.

a. $\displaystyle\int (x + 2)\, dx$ **b.** $\displaystyle\int (3x^4 - 5x^2 + x)\, dx$

SOLUTION

a. $\displaystyle\int (x + 2)\, dx = \int x\, dx + \int 2\, dx$ Apply Sum Rule.

$\qquad\qquad\qquad\quad = \dfrac{x^2}{2} + C_1 + 2x + C_2$ Integrate.

$\qquad\qquad\qquad\quad = \dfrac{x^2}{2} + 2x + C$ $C = C_1 + C_2$

The second line in the solution is usually omitted.

b. Try to identify each basic integration rule used to evaluate this integral.

$\displaystyle\int (3x^4 - 5x^2 + x)\, dx = 3\left(\dfrac{x^5}{5}\right) - 5\left(\dfrac{x^3}{3}\right) + \dfrac{x^2}{2} + C$

$\qquad\qquad\qquad\qquad\qquad = \dfrac{3}{5}x^5 - \dfrac{5}{3}x^3 + \dfrac{1}{2}x^2 + C$

Example 6   Rewriting Before Integrating

Find $\displaystyle\int \dfrac{x + 1}{\sqrt{x}}\, dx$.

SOLUTION   Begin by rewriting the quotient in the integrand as a sum. Then rewrite each term using rational exponents.

$\displaystyle\int \dfrac{x + 1}{\sqrt{x}}\, dx = \int \left(\dfrac{x}{\sqrt{x}} + \dfrac{1}{\sqrt{x}}\right) dx$ Rewrite as a sum.

$\qquad\qquad\qquad = \displaystyle\int (x^{1/2} + x^{-1/2})\, dx$ Rewrite using rational exponents.

$\qquad\qquad\qquad = \dfrac{x^{3/2}}{3/2} + \dfrac{x^{1/2}}{1/2} + C$ Apply Power Rule.

$\qquad\qquad\qquad = \dfrac{2}{3}x^{3/2} + 2x^{1/2} + C$ Simplify.

$\qquad\qquad\qquad = \dfrac{2}{3}\sqrt{x}(x + 3) + C$ Factor.

✓ CHECKPOINT 5

Find each indefinite integral.

a. $\displaystyle\int (x + 4)\, dx$

b. $\displaystyle\int (4x^3 - 5x + 2)\, dx$

STUDY TIP

When integrating quotients, remember *not* to integrate the numerator and denominator separately. For instance, in Example 6, be sure you understand that

$\displaystyle\int \dfrac{x + 1}{\sqrt{x}}\, dx = \dfrac{2}{3}\sqrt{x}(x + 3) + C$

is not the same as

$\dfrac{\displaystyle\int (x + 1)\, dx}{\displaystyle\int \sqrt{x}\, dx} = \dfrac{\frac{1}{2}x^2 + x + C_1}{\frac{2}{3}x\sqrt{x} + C_2}.$

Algebra Review

For help with the algebra in Example 6, see Example 1(a) in the *Chapter 6 Algebra Review*, on page 418.

✓ CHECKPOINT 6

Find $\displaystyle\int \dfrac{x + 2}{\sqrt{x}}\, dx$.

Particular Solutions

You have already seen that the equation $y = \int f(x)\,dx$ has many solutions, each differing from the others by a constant. This means that the graphs of any two antiderivatives of f are vertical translations of each other. For example, Figure 6.1 shows the graphs of several antiderivatives of the form

$$y = F(x) = \int (3x^2 - 1)\,dx = x^3 - x + C$$

for various integer values of C. Each of these antiderivatives is a solution of the *differential equation* $dy/dx = 3x^2 - 1$. A **differential equation** in x and y is an equation that involves x, y, and derivatives of y. The **general solution** of $dy/dx = 3x^2 - 1$ is $F(x) = x^3 - x + C$.

In many applications of integration, you are given enough information to determine a **particular solution**. To do this, you need to know the value of $F(x)$ for only one value of x. (This information is called an **initial condition**.) For example, in Figure 6.1, there is only one curve that passes through the point $(2, 4)$. To find this curve, use the information below.

$F(x) = x^3 - x + C$ General solution
$F(2) = 4$ Initial condition

By using the initial condition in the general solution, you can determine that $F(2) = 2^3 - 2 + C = 4$, which implies that $C = -2$. So, the particular solution is

$F(x) = x^3 - x - 2.$ Particular solution

Example 7 Finding a Particular Solution

Find the general solution of

$$F'(x) = 2x - 2$$

and find the particular solution that satisfies the initial condition $F(1) = 2$.

SOLUTION Begin by integrating to find the general solution.

$F(x) = \int (2x - 2)\,dx$ Integrate $F'(x)$ to obtain $F(x)$.

$= x^2 - 2x + C$ General solution

Using the initial condition $F(1) = 2$, you can write

$$F(1) = 1^2 - 2(1) + C = 2$$

which implies that $C = 3$. So, the particular solution is

$F(x) = x^2 - 2x + 3.$ Particular solution

This solution is shown graphically in Figure 6.2. Note that each of the gray curves represents a solution of the equation $F'(x) = 2x - 2$. The black curve, however, is the only solution that passes through the point $(1, 2)$, which means that $F(x) = x^2 - 2x + 3$ is the only solution that satisfies the initial condition.

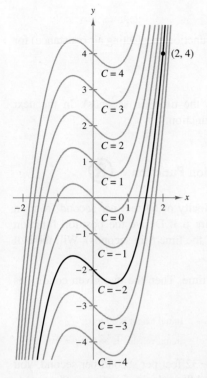

FIGURE 6.1

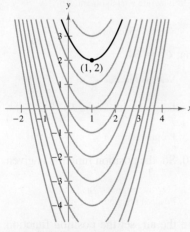

FIGURE 6.2

✓CHECKPOINT 7

Find the general solution of $F'(x) = 4x + 2$, and find the particular solution that satisfies the initial condition $F(1) = 8$. ∎

Applications

In Chapter 2, you used the general position function (neglecting air resistance) for a falling object

$$s(t) = -16t^2 + v_0 t + s_0$$

where $s(t)$ is the height (in feet) and t is the time (in seconds). In the next example, integration is used to *derive* this function.

Example 8
MAKE A DECISION Deriving a Position Function

A ball is thrown upward with an initial velocity of 64 feet per second from an initial height of 80 feet, as shown in Figure 6.3. Derive the position function giving the height s (in feet) as a function of the time t (in seconds). Will the ball be in the air for more than 5 seconds?

SOLUTION Let $t = 0$ represent the initial time. Then the two given conditions can be written as

$s(0) = 80$ Initial height is 80 feet.

$s'(0) = 64$.   Initial velocity is 64 feet per second.

Because the acceleration due to gravity is -32 feet per second per second, you can integrate the acceleration function to find the velocity function as shown.

$s''(t) = -32$ Acceleration due to gravity

$s'(t) = \int -32\, dt$ Integrate $s''(t)$ to obtain $s'(t)$.

$= -32t + C_1$ Velocity function

Using the initial velocity, you can conclude that $C_1 = 64$.

$s'(t) = -32t + 64$ Velocity function

$s(t) = \int (-32t + 64)\, dt$ Integrate $s'(t)$ to obtain $s(t)$.

$= -16t^2 + 64t + C_2$ Position function

Using the initial height, it follows that $C_2 = 80$. So, the position function is given by

$s(t) = -16t^2 + 64t + 80$.   Position function

To find the amount of time the ball will be in the air, set the position function equal to 0 and solve for t.

$-16t^2 + 64t + 80 = 0$ Set $s(t)$ equal to zero.

$-16(t + 1)(t - 5) = 0$ Factor.

$t = -1, \quad t = 5$ Solve for t.

Because the time must be positive, you can conclude that the ball will be in the air for 5 seconds. So, the ball will not be in the air for more than 5 seconds.

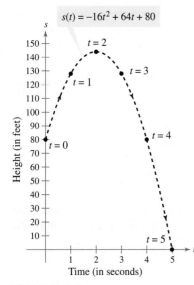

FIGURE 6.3

✓ **CHECKPOINT 8**

Derive the position function for a ball thrown upward with an initial velocity of 32 feet per second from an initial height of 48 feet. When does the ball hit the ground? With what velocity does the ball hit the ground? ■

Example 9 Finding a Population Function

Since 1990, the rate of increase in the number of divorced adults D (in millions) in the United States from 1990 through 2005 can be modeled by

$$D' = -0.004t + 0.49$$

where t is the year, with $t = 0$ corresponding to 1990. The number of divorced adults in 2005 was 22.1 million. *(Source: U.S. Census Bureau)*

a. Find the model for the number of divorced adults in the United States.

b. Use the model to predict the number of divorced adults in 2012.

SOLUTION

a. To find the model for the number of divorced adults in the United States, integrate D'.

$$D = \int (-0.004t + 0.49)\, dt \qquad \text{Integrate } D' \text{ to obtain } D.$$

$$= -0.004\left(\frac{t^2}{2}\right) + 0.49t + C \qquad \text{Apply Power Rule.}$$

$$= -0.002t^2 + 0.49t + C \qquad \text{Model for divorced adults.}$$

In 2005 ($t = 15$), the number of divorced adults was 22.1 million. To solve for C, use the initial condition that $D = 22.1$ when $t = 15$.

$$22.1 = -0.002(15)^2 + 0.49(15) + C \qquad \text{Substitute 22.1 for } D \text{ and 15 for } t.$$

$$15.2 = C \qquad \text{Solve for } C.$$

So, the model for the number of divorced adults in the United States is given by

$$D = -0.002t^2 + 0.49t + 15.2. \qquad \text{Model for divorced adults.}$$

b. According to the model, the number of divorced adults in 2012 ($t = 22$) will be

$$D = -0.002(22)^2 + 0.49(22) + 15.2$$

$$\approx 25.0 \text{ million.}$$

✓ CHECKPOINT 9

Since 1998, the rate of increase in the number of children with stay-at-home mothers H (in millions) in the United States from 1998 through 2005 can be modeled by

$$H' = -0.01706t + 0.4693$$

where t is the year, with $t = 8$ corresponding to 1998. The number of children with stay-at-home mothers in 2005 was 11.246 million. *(Source: U.S. Census Bureau)*

a. Find the model for the number of children with stay-at-home mothers in the United States.

b. Use the model to predict the number of children with stay-at-home mothers in 2012. ∎

CONCEPT CHECK

1. How can you check your answer to an antidifferentiation problem?
2. Write what is meant by the symbol $\int f(x)\, dx$ in words.
3. Given $\int (2x + 1)\, dx = x^2 + x + C$, identify (a) the integrand and (b) the antiderivative.
4. True or false: The antiderivative of a second-degree polynomial function is a third-degree polynomial function.

372 CHAPTER 6 Integration and Its Applications

Skills Review 6.1

The following warm-up exercises involve skills that were covered in earlier sections. You will use these skills in the exercise set for this section. For additional help, review Sections 0.3 and 1.2.

In Exercises 1–6, rewrite the expression using rational exponents.

1. $\dfrac{\sqrt{x}}{x}$

2. $\sqrt[3]{2x}(2x)$

3. $\sqrt{5x^3} + \sqrt{x^5}$

4. $\dfrac{1}{\sqrt{x}} + \dfrac{1}{\sqrt[3]{x^2}}$

5. $\dfrac{(x+1)^3}{\sqrt{x+1}}$

6. $\dfrac{\sqrt{x}}{\sqrt[3]{x}}$

In Exercises 7–10, let $(x, y) = (2, 2)$, and solve the equation for C.

7. $y = x^2 + 5x + C$

8. $y = 3x^3 - 6x + C$

9. $y = -16x^2 + 26x + C$

10. $y = -\dfrac{1}{4}x^4 - 2x^2 + C$

Exercises 6.1

See www.CalcChat.com for worked-out solutions to odd-numbered exercises.

In Exercises 1–8, verify the statement by showing that the derivative of the right side is equal to the integrand of the left side.

1. $\displaystyle\int \left(-\dfrac{9}{x^4}\right) dx = \dfrac{3}{x^3} + C$

2. $\displaystyle\int \dfrac{4}{\sqrt{x}} dx = 8\sqrt{x} + C$

3. $\displaystyle\int \left(4x^3 - \dfrac{1}{x^2}\right) dx = x^4 + \dfrac{1}{x} + C$

4. $\displaystyle\int \left(1 - \dfrac{1}{\sqrt[3]{x^2}}\right) dx = x - 3\sqrt[3]{x} + C$

5. $\displaystyle\int 2\sqrt{x}(x - 3)\, dx = \dfrac{4x^{3/2}(x-5)}{5} + C$

6. $\displaystyle\int 4\sqrt{x}(x^2 - 2)\, dx = \dfrac{8x^{3/2}(3x^2 - 14)}{21} + C$

7. $\displaystyle\int (x-2)(x+2)\, dx = \dfrac{1}{3}x^3 - 4x + C$

8. $\displaystyle\int \dfrac{x^2 - 1}{x^{3/2}}\, dx = \dfrac{2(x^2 + 3)}{3\sqrt{x}} + C$

In Exercises 9–20, find the indefinite integral and check your result by differentiation.

9. $\displaystyle\int 6\, dx$

10. $\displaystyle\int -4\, dx$

11. $\displaystyle\int 5t^2\, dt$

12. $\displaystyle\int 3t^4\, dt$

13. $\displaystyle\int 5x^{-3}\, dx$

14. $\displaystyle\int 4y^{-3}\, dy$

15. $\displaystyle\int du$

16. $\displaystyle\int dr$

17. $\displaystyle\int e\, dt$

18. $\displaystyle\int e^3\, dy$

19. $\displaystyle\int y^{3/2}\, dy$

20. $\displaystyle\int v^{-1/2}\, dv$

In Exercises 21–26, complete the table.

Original Integral	Rewrite	Integrate	Simplify
21. $\displaystyle\int \sqrt[3]{x}\, dx$			
22. $\displaystyle\int \dfrac{1}{x^2}\, dx$			
23. $\displaystyle\int \dfrac{1}{x\sqrt{x}}\, dx$			
24. $\displaystyle\int x(x^2 + 3)\, dx$			
25. $\displaystyle\int \dfrac{1}{2x^3}\, dx$			
26. $\displaystyle\int \dfrac{1}{(3x)^2}\, dx$			

In Exercises 27–38, find the indefinite integral and check your result by differentiation.

27. $\int (x + 3)\, dx$

28. $\int (5 - x)\, dx$

29. $\int (x^3 + 2)\, dx$

30. $\int (x^3 - 4x + 2)\, dx$

31. $\int \left(\sqrt[3]{x} - \dfrac{1}{2\sqrt[3]{x}}\right) dx$

32. $\int \left(\sqrt{x} + \dfrac{1}{2\sqrt{x}}\right) dx$

33. $\int \sqrt[3]{x^2}\, dx$

34. $\int \left(\sqrt[4]{x^3} + 1\right) dx$

35. $\int \dfrac{1}{x^4}\, dx$

36. $\int \dfrac{1}{4x^2}\, dx$

37. $\int \dfrac{2x^3 + 1}{x^3}\, dx$

38. $\int \dfrac{t^2 + 2}{t^2}\, dt$

(T) In Exercises 39–44, use a symbolic integration utility to find the indefinite integral.

39. $\int u(3u^2 + 1)\, du$

40. $\int \sqrt{x}(x + 1)\, dx$

41. $\int (x + 1)(3x - 2)\, dx$

42. $\int (2t^2 - 1)^2\, dt$

43. $\int y^2 \sqrt{y}\, dy$

44. $\int (1 + 3t)t^2\, dt$

In Exercises 45–48, the graph of the derivative of a function is given. Sketch the graphs of *two* functions that have the given derivative. (There is more than one correct answer.)

45.
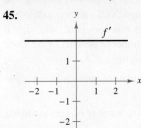

46.
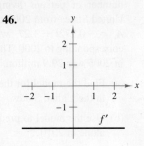

47.
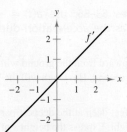

48.
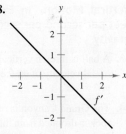

In Exercises 49–54, find the particular solution $y = f(x)$ that satisfies the differential equation and initial condition.

49. $f'(x) = 4x, \quad f(0) = 6$

50. $f'(x) = \tfrac{1}{5}x - 2, \quad f(10) = -10$

51. $f'(x) = 2(x - 1), \quad f(3) = 2$

52. $f'(x) = (2x - 3)(2x + 3), \quad f(3) = 0$

53. $f'(x) = \dfrac{2 - x}{x^3}, \quad x > 0, \quad f(2) = \dfrac{3}{4}$

54. $f'(x) = \dfrac{x^2 - 5}{x^2}, \quad x > 0, \quad f(1) = 2$

In Exercises 55 and 56, find the equation for y, given the derivative and the indicated point on the curve.

55. $\dfrac{dy}{dx} = -5x - 2$

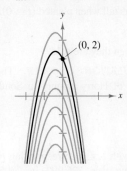

56. $\dfrac{dy}{dx} = 2(x - 1)$

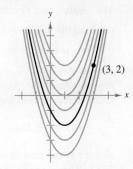

In Exercises 57 and 58, find the equation of the function f whose graph passes through the point.

	Derivative	Point
57.	$f'(x) = 2x$	$(-2, -2)$
58.	$f'(x) = 2\sqrt{x}$	$(4, 12)$

In Exercises 59–62, find a function f that satisfies the conditions.

59. $f''(x) = 2, \quad f'(2) = 5, \quad f(2) = 10$

60. $f''(x) = x^2, \quad f'(0) = 6, \quad f(0) = 3$

61. $f''(x) = x^{-2/3}, \quad f'(8) = 6, \quad f(0) = 0$

62. $f''(x) = x^{-3/2}, \quad f'(1) = 2, \quad f(9) = -4$

Vertical Motion In Exercises 63–66, use $a(t) = -32$ feet per second per second as the acceleration due to gravity.

63. A ball is thrown vertically upward from the ground with an initial velocity of 60 feet per second. How high will the ball go?

64. The Grand Canyon is 6000 feet deep at the deepest part. A rock is dropped from this height. Express the height s of the rock as a function of the time t (in seconds). How long will it take the rock to hit the canyon floor?

65. With what initial velocity must an object be thrown upward from the ground to reach the height of the Washington Monument (550 feet)?

66. A balloon rising vertically with a velocity of 16 feet per second releases a sandbag at an instant when the balloon is 64 feet above the ground.

    (a) How many seconds after its release will the bag strike the ground?

    (b) With what velocity will the bag strike the ground?

67. **Tree Growth** An evergreen nursery usually sells a certain shrub after 6 years of growth and shaping. The growth rate during those 6 years is approximated by

    $$\frac{dh}{dt} = 1.5t + 5$$

    where t is the time in years and h is the height in centimeters. The seedlings are 12 centimeters tall when planted ($t = 0$).

    (a) Find the height after t years.

    (b) How tall are the shrubs when they are sold?

68. **MAKE A DECISION: POPULATION GROWTH** The growth rate of the population of Horry County in South Carolina can be modeled by

    $$\frac{dP}{dt} = 105.46t + 2642.7$$

    where t is the time in years, with $t = 0$ corresponding to 1970. The county's population was 226,992 in 2005. *(Source: U.S. Census Bureau)*

    (a) Find the model for Horry County's population.

    (b) Use the model to predict the population in 2012. Does your answer seem reasonable? Explain your reasoning.

69. **Population Growth** The growth rate of a city's population is modeled by

    $$\frac{dP}{dt} = 500t^{1.06}$$

    where t is the time in years. The city's population is now 50,000. What will the population be in 10 years?

70. **MAKE A DECISION: VITAL STATISTICS** The rate of increase in the number of married couples M (in thousands) in the United States from 1970 through 2005 can be modeled by

    $$\frac{dM}{dt} = 1.218t^2 - 44.72t + 709.1$$

    where t is the time in years, with $t = 0$ corresponding to 1970. The number of married couples in 2005 was 59,513 thousand. *(Source: U.S. Census Bureau)*

    (a) Find the model for the number of married couples in the United States.

    (b) Use the model to predict the number of married couples in the United States in 2012. Does your answer seem reasonable? Explain your reasoning.

71. **MAKE A DECISION: INTERNET USERS** The rate of growth in the number of Internet users I (in millions) in the world from 1991 through 2004 can be modeled by

    $$\frac{dI}{dt} = -0.25t^3 + 5.319t^2 - 19.34t + 21.03$$

    where t is the time in years, with $t = 1$ corresponding to 1991. The number of Internet users in 2004 was 863 million. *(Source: International Telecommunication Union)*

    (a) Find the model for the number of Internet users in the world.

    (b) Use the model to predict the number of Internet users in the world in 2012. Does your answer seem reasonable? Explain your reasoning.

72. **School Enrollment** The rate of increase in student enrollment S (in millions) in U.S. private schools from 1995 through 2005 can be modeled by

    $$S' = 0.0024t + 0.186$$

    where t is the year, with $t = 5$ corresponding to 1995. The number of students enrolled in private schools in 2003 was 10.5 million. *(Source: U.S. National Center for Education Statistics)*

    (a) Find the model for the number of students enrolled in private schools in the United States.

    (b) Use the model to predict the number of students enrolled in private schools in 2010.

73. **Living Arrangements** The rate of increase in the number of persons living alone A (in millions) in the United States from 2000 through 2005 can be modeled by $A' = -0.272t + 1.27$, where t is the year, with $t = 0$ corresponding to 2000. The number of persons living alone in 2005 was 29.9 million. *(Source: U.S. Census Bureau)*

    (a) Find the model for the number of persons living alone in the United States.

    (b) Use the model to predict the number of persons living alone in 2010.

Section 6.2
Integration by Substitution and the General Power Rule

- Use the General Power Rule to find indefinite integrals.
- Use substitution to find indefinite integrals.
- Use the General Power Rule to solve real-life problems.

The General Power Rule

In Section 6.1, you used the Simple Power Rule

$$\int x^n \, dx = \frac{x^{n+1}}{n+1} + C, \quad n \neq -1$$

to find antiderivatives of functions expressed as powers of x alone. In this section, you will study a technique for finding antiderivatives of more complicated functions.

To begin, consider how you might find the antiderivative of $2x(x^2+1)^3$. Because you are hunting for a function whose derivative is $2x(x^2+1)^3$, you might discover the antiderivative as shown.

$$\frac{d}{dx}[(x^2+1)^4] = 4(x^2+1)^3(2x) \qquad \text{Use Chain Rule.}$$

$$\frac{d}{dx}\left[\frac{(x^2+1)^4}{4}\right] = (x^2+1)^3(2x) \qquad \text{Divide both sides by 4.}$$

$$\frac{(x^2+1)^4}{4} + C = \int 2x(x^2+1)^3 \, dx \qquad \text{Write in integral form.}$$

The key to this solution is the presence of the factor $2x$ in the integrand. In other words, this solution works because $2x$ is precisely the derivative of (x^2+1). Letting $u = x^2 + 1$, you can write

$$\int \underbrace{(x^2+1)^3}_{u^3} \underbrace{2x \, dx}_{du} = \int u^3 \, du$$

$$= \frac{u^4}{4} + C.$$

This is an example of the **General Power Rule** for integration.

General Power Rule for Integration

If u is a differentiable function of x, then

$$\int u^n \frac{du}{dx} \, dx = \int u^n \, du = \frac{u^{n+1}}{n+1} + C, \quad n \neq -1.$$

When using the General Power Rule, you must first identify a factor u of the integrand that is raised to a power. Then, you must show that its derivative du/dx is also a factor of the integrand. This is demonstrated in Example 1.

CHAPTER 6 Integration and Its Applications

Example 1 Applying the General Power Rule

Find each indefinite integral.

a. $\displaystyle\int 3(3x-1)^4\, dx$ b. $\displaystyle\int (2x+1)(x^2+x)\, dx$

c. $\displaystyle\int 3x^2\sqrt{x^3-2}\, dx$ d. $\displaystyle\int \dfrac{-4x}{(1-2x^2)^2}\, dx$

SOLUTION

a. $\displaystyle\int 3(3x-1)^4\, dx = \int \overbrace{(3x-1)^4}^{u^n}\overbrace{(3)}^{du/dx}\, dx$ Let $u = 3x-1$.

$\qquad = \dfrac{(3x-1)^5}{5} + C$ General Power Rule

b. $\displaystyle\int (2x+1)(x^2+x)\, dx = \int \overbrace{(x^2+x)}^{u^n}\overbrace{(2x+1)}^{du/dx}\, dx$ Let $u = x^2+x$.

$\qquad = \dfrac{(x^2+x)^2}{2} + C$ General Power Rule

c. $\displaystyle\int 3x^2\sqrt{x^3-2}\, dx = \int \overbrace{(x^3-2)^{1/2}}^{u^n}\overbrace{(3x^2)}^{du/dx}\, dx$ Let $u = x^3-2$.

$\qquad = \dfrac{(x^3-2)^{3/2}}{3/2} + C$ General Power Rule

$\qquad = \dfrac{2}{3}(x^3-2)^{3/2} + C$ Simplify.

d. $\displaystyle\int \dfrac{-4x}{(1-2x^2)^2}\, dx = \int \overbrace{(1-2x^2)^{-2}}^{u^n}\overbrace{(-4x)}^{du/dx}\, dx$ Let $u = 1-2x^2$.

$\qquad = \dfrac{(1-2x^2)^{-1}}{-1} + C$ General Power Rule

$\qquad = -\dfrac{1}{1-2x^2} + C$ Simplify.

✓ **CHECKPOINT 1**

Find each indefinite integral.

a. $\displaystyle\int (3x^2+6)(x^3+6x)^2\, dx$ b. $\displaystyle\int 2x\sqrt{x^2-2}\, dx$

STUDY TIP

Example 1(b) illustrates a case of the General Power Rule that is sometimes overlooked—when the power is $n = 1$. In this case, the rule takes the form

$$\int u\,\dfrac{du}{dx}\, dx = \dfrac{u^2}{2} + C.$$

STUDY TIP

Remember that you can verify the result of an indefinite integral by differentiating the function. Check the answer to Example 1(d) by differentiating the function

$$F(x) = -\dfrac{1}{1-2x^2} + C.$$

$$\dfrac{d}{dx}\left[-\dfrac{1}{1-2x^2} + C\right]$$

$$= \dfrac{-4x}{(1-2x^2)^2}$$

SECTION 6.2 Integration by Substitution and the General Power Rule

Many times, part of the derivative du/dx is missing from the integrand, and in *some* cases you can make the adjustments that allow you to apply the General Power Rule.

Algebra Review

For help with the algebra in Example 2, see Example 1(b) in the *Chapter 6 Algebra Review*, on page 418.

Example 2 Multiplying and Dividing by a Constant

Find $\int x(3 - 4x^2)^2 \, dx$.

SOLUTION Let $u = 3 - 4x^2$. To apply the General Power Rule, you need to create $du/dx = -8x$ as a factor of the integrand. You can accomplish this by multiplying and dividing by the constant -8.

$$\int x(3 - 4x^2)^2 \, dx = \int \left(-\frac{1}{8}\right) \overbrace{(3 - 4x^2)^2}^{u^n} \overbrace{(-8x)}^{du/dx} dx \qquad \text{Multiply and divide by } -8.$$

$$= -\frac{1}{8} \int (3 - 4x^2)^2 (-8x) \, dx \qquad \text{Factor } -\tfrac{1}{8} \text{ out of integrand.}$$

$$= \left(-\frac{1}{8}\right) \frac{(3 - 4x^2)^3}{3} + C \qquad \text{General Power Rule}$$

$$= -\frac{(3 - 4x^2)^3}{24} + C \qquad \text{Simplify.}$$

STUDY TIP

Try using the Chain Rule to check the result of Example 2. After differentiating $-\frac{1}{24}(3 - 4x^2)^3$ and simplifying, you should obtain the original integrand.

✓ **CHECKPOINT 2**

Find $\int x^3(3x^4 + 1)^2 \, dx$.

Example 3 A Failure of the General Power Rule

Find $\int -8(3 - 4x^2)^2 \, dx$.

SOLUTION Let $u = 3 - 4x^2$. As in Example 2, to apply the General Power Rule you must create $du/dx = -8x$ as a factor of the integrand. In Example 2, you could do this by multiplying and dividing by a constant, and then factoring that constant out of the integrand. This strategy doesn't work with variables. That is,

$$\int -8(3 - 4x^2)^2 \, dx \neq \frac{1}{x} \int (3 - 4x^2)^2(-8x) \, dx.$$

To find this indefinite integral, you can expand the integrand and use the Simple Power Rule.

$$\int -8(3 - 4x^2)^2 \, dx = \int (-72 + 192x^2 - 128x^4) \, dx$$

$$= -72x + 64x^3 - \frac{128}{5}x^5 + C$$

STUDY TIP

In Example 3, be sure you see that you cannot factor variable quantities outside the integral sign. After all, if this were permissible, then you could move the entire integrand outside the integral sign and eliminate the need for all integration rules except the rule $\int dx = x + C$.

✓ **CHECKPOINT 3**

Find $\int 2(3x^4 + 1)^2 \, dx$.

Example 4 Applying the General Power Rule

When an integrand contains an extra constant factor that is not needed as part of du/dx, you can simply move the factor outside the integral sign, as shown in the next example.

Find $\int 7x^2 \sqrt{x^3 + 1} \, dx$.

SOLUTION Let $u = x^3 + 1$. Then you need to create $du/dx = 3x^2$ by multiplying and dividing by 3. The constant factor $\frac{7}{3}$ is not needed as part of du/dx, and can be moved outside the integral sign.

$$\int 7x^2 \sqrt{x^3 + 1} \, dx = \int 7x^2(x^3 + 1)^{1/2} \, dx \qquad \text{Rewrite with rational exponent.}$$

$$= \int \frac{7}{3}(x^3 + 1)^{1/2}(3x^2) \, dx \qquad \text{Multiply and divide by 3.}$$

$$= \frac{7}{3}\int (x^3 + 1)^{1/2}(3x^2) \, dx \qquad \text{Factor } \tfrac{7}{3} \text{ outside integral.}$$

$$= \frac{7}{3} \frac{(x^3 + 1)^{3/2}}{3/2} + C \qquad \text{General Power Rule}$$

$$= \frac{14}{9}(x^3 + 1)^{3/2} + C \qquad \text{Simplify.}$$

✓ CHECKPOINT 4

Find $\int 5x\sqrt{x^2 - 1} \, dx$. ∎

Algebra Review

For help with the algebra in Example 4, see Example 1(c) in the *Chapter 6 Algebra Review*, on page 418.

TECHNOLOGY

Ⓣ If you use a symbolic integration utility to find indefinite integrals, you should be in for some surprises. This is true because integration is not nearly as straightforward as differentiation. By trying different integrands, you should be able to find several that the program cannot solve: in such situations, it may list a new indefinite integral. You should also be able to find several that have horrendous antiderivatives, some with functions that you may not recognize.

SECTION 6.2 Integration by Substitution and the General Power Rule

Substitution

The integration technique used in Examples 1, 2, and 4 depends on your ability to recognize or create an integrand of the form $u^n \, du/dx$. With more complicated integrands, it is difficult to recognize the steps needed to make the integrand fit a basic integration formula. When this occurs, an alternative procedure called **substitution** or **change of variables** can be helpful. With this procedure, you completely rewrite the integral in terms of u and du. That is, if $u = f(x)$, then $du = f'(x) \, dx$, and the General Power Rule takes the form

$$\int u^n \frac{du}{dx} dx = \int u^n \, du. \qquad \text{General Power Rule}$$

DISCOVERY

Calculate the derivative of each function. Which one is the antiderivative of $f(x) = \sqrt{1 - 3x}$?

$F(x) = (1 - 3x)^{3/2} + C$

$F(x) = \frac{2}{3}(1 - 3x)^{3/2} + C$

$F(x) = -\frac{2}{9}(1 - 3x)^{3/2} + C$

Example 5 Integrating by Substitution

Find $\int \sqrt{1 - 3x} \, dx$.

SOLUTION Begin by letting $u = 1 - 3x$. Then, $du/dx = -3$ and $du = -3 \, dx$. This implies that $dx = -\frac{1}{3} du$, and you can find the indefinite integral as shown.

$$\int \sqrt{1 - 3x} \, dx = \int (1 - 3x)^{1/2} \, dx \qquad \text{Rewrite with rational exponent.}$$

$$= \int u^{1/2} \left(-\frac{1}{3} du\right) \qquad \text{Substitute for } x \text{ and } dx.$$

$$= -\frac{1}{3} \int u^{1/2} \, du \qquad \text{Factor } -\frac{1}{3} \text{ out of integrand.}$$

$$= -\frac{1}{3} \frac{u^{3/2}}{3/2} + C \qquad \text{Apply Power Rule.}$$

$$= -\frac{2}{9} u^{3/2} + C \qquad \text{Simplify.}$$

$$= -\frac{2}{9}(1 - 3x)^{3/2} + C \qquad \text{Substitute } 1 - 3x \text{ for } u.$$

✓ CHECKPOINT 5

Find $\int \sqrt{1 - 2x} \, dx$ by the method of substitution. ■

The basic steps for integration by substitution are outlined in the guidelines below.

Guidelines for Integration by Substitution

1. Let u be a function of x (usually part of the integrand).
2. Solve for x and dx in terms of u and du.
3. Convert the entire integral to u-variable form.
4. After integrating, rewrite the antiderivative as a function of x.
5. Check your answer by differentiating.

Example 6   Integration by Substitution

Find $\int x\sqrt{x^2 - 1}\, dx$.

SOLUTION Consider the substitution $u = x^2 - 1$, which produces $du = 2x\, dx$. To create $2x\, dx$ as part of the integral, multiply and divide by 2.

$$\int x\sqrt{x^2 - 1}\, dx = \frac{1}{2} \int \overbrace{(x^2 - 1)^{1/2}}^{u^{1/n}} \overbrace{2x\, dx}^{du} \qquad \text{Multiply and divide by 2.}$$

$$= \frac{1}{2} \int u^{1/2}\, du \qquad \text{Substitute for } x \text{ and } dx.$$

$$= \frac{1}{2} \frac{u^{3/2}}{3/2} + C \qquad \text{Apply Power Rule.}$$

$$= \frac{1}{3} u^{3/2} + C \qquad \text{Simplify.}$$

$$= \frac{1}{3}(x^2 - 1)^{3/2} + C \qquad \text{Substitute for } u.$$

You can check this result by differentiating.

$$\frac{d}{dx}\left[\frac{1}{3}(x^2 - 1)^{3/2} + C\right] = \frac{1}{3}\left(\frac{3}{2}\right)(x^2 - 1)^{1/2}(2x)$$

$$= \frac{1}{2}(2x)(x^2 - 1)^{1/2}$$

$$= x\sqrt{x^2 - 1}$$

✓ CHECKPOINT 6

Find $\int x\sqrt{x^2 + 4}\, dx$ by the method of substitution. ■

To become efficient at integration, you should learn to use *both* techniques discussed in this section. For simpler integrals, you should use pattern recognition and create du/dx by multiplying and dividing by an appropriate constant. For more complicated integrals, you should use a formal change of variables, as shown in Examples 5 and 6. For the integrals in this section's exercise set, try working several of the problems twice—once with pattern recognition and once using formal substitution.

DISCOVERY

Suppose you were asked to evaluate one of the integrals below. Which one would you choose? Explain your reasoning.

$$\int \sqrt{x^2 + 1}\, dx \quad \text{or} \quad \int x\sqrt{x^2 + 1}\, dx$$

SECTION 6.2 Integration by Substitution and the General Power Rule

Application

Example 7 MAKE A DECISION Pollution

A contaminated lake is treated with a bactericide. The rate of change in harmful bacteria t days after treatment can be modeled by

$$\frac{dB}{dt} = -\frac{3000}{(1 + 0.2t)^2}, \quad t \geq 0$$

where B is the number of bacteria per milliliter of water and t is the number of days since treatment. The initial number of bacteria is 10,000 per milliliter of water. Use the model to estimate the number of bacteria after 5 days. Will the number of bacteria be more than 2000 per milliliter of water?

SOLUTION Because dB/dt is negative, the number of bacteria is decreasing. Begin by integrating dB/dt to find a model for the number of bacteria B. Use the initial condition that $B = 10,000$ when $t = 0$.

$$\frac{dB}{dt} = -\frac{3000}{(1 + 0.2t)^2} \qquad \text{Rate of change in harmful bacteria}$$

$$B = \int \frac{-3000}{(1 + 0.2t)^2}\, dt \qquad \text{Integrate to obtain } B.$$

$$= \int (-3000)(1 + 0.2t)^{-2}\, dt \qquad \text{Rewrite.}$$

$$= 15,000(1 + 0.2t)^{-1} + C \qquad \text{General Power Rule}$$

$$= \frac{15,000}{1 + 0.2t} + C \qquad \text{Rewrite.}$$

$$= \frac{15,000}{1 + 0.2t} - 5000 \qquad \text{Use initial condition to find } C.$$

Using this model, you can estimate that the number of bacteria after 5 days is 2500 per milliliter of water. Yes, after 5 days the number of bacteria will be more than 2000 per milliliter of water.

STUDY TIP

When you use the initial condition to find the value of C in Example 7, you substitute 10,000 for B and 0 for t.

$$B = \frac{15,000}{1 + 0.2t} + C$$

$$10,000 = \frac{15,000}{1 + 0.2(0)} + C$$

$$10,000 = 15,000 + C$$

$$-5000 = C$$

✓ **CHECKPOINT 7**

In Example 7, when will the number of bacteria per milliliter of water be 0? ■

CONCEPT CHECK

1. When using the General Power Rule for an integrand that contains an extra constant factor that is not needed as part of du/dx, what can you do with the factor?
2. Write the General Power Rule for integration.
3. Write the guidelines for integration by substitution.
4. Explain why the General Power Rule works for finding $\int 2x\sqrt{x^2 + 1}\, dx$, but not for finding $\int 2\sqrt{x^2 + 1}\, dx$.

Skills Review 6.2

The following warm-up exercises involve skills that were covered in earlier sections. You will use these skills in the exercise set for this section. For additional help, review Sections 0.3, 0.5, and 6.1.

In Exercises 1–10, find the indefinite integral.

1. $\int (2x^3 + 1)\, dx$
2. $\int (x^{1/2} + 3x - 4)\, dx$
3. $\int \frac{1}{x^2}\, dx$
4. $\int \frac{1}{3t^3}\, dt$
5. $\int (1 + 2t)t^{3/2}\, dt$
6. $\int \sqrt{x}(2x - 1)\, dx$
7. $\int \frac{5x^3 + 2}{x^2}\, dx$
8. $\int \frac{2x^2 - 5}{x^4}\, dx$
9. $\int (x^2 + 1)^2\, dx$
10. $\int (x^3 - 2x + 1)^2\, dx$

In Exercises 11–14, simplify the expression.

11. $\left(-\dfrac{5}{4}\right)\dfrac{(x-2)^4}{4}$
12. $\left(\dfrac{1}{6}\right)\dfrac{(x-1)^{-2}}{-2}$
13. $(6)\dfrac{(x^2+3)^{2/3}}{2/3}$
14. $\left(\dfrac{5}{2}\right)\dfrac{(1-x^3)^{-1/2}}{-1/2}$

Exercises 6.2

See www.CalcChat.com for worked-out solutions to odd-numbered exercises.

In Exercises 1–8, identify u and du/dx for the integral $\int u^n (du/dx)\, dx$.

1. $\int (5x^2 + 1)^2 (10x)\, dx$
2. $\int (3 - 4x^2)^3 (-8x)\, dx$
3. $\int \sqrt{1 - x^2}\, (-2x)\, dx$
4. $\int 3x^2 \sqrt{x^3 + 1}\, dx$
5. $\int \left(4 + \dfrac{1}{x^2}\right)^5 \left(\dfrac{-2}{x^3}\right) dx$
6. $\int \dfrac{1}{(1+2x)^2}(2)\, dx$
7. $\int (1 + \sqrt{x})^3 \left(\dfrac{1}{2\sqrt{x}}\right) dx$
8. $\int (4 - \sqrt{x})^2 \left(\dfrac{-1}{2\sqrt{x}}\right) dx$

In Exercises 9–28, find the indefinite integral and check the result by differentiation.

9. $\int (1 + 2x)^4 (2)\, dx$
10. $\int (x^2 - 1)^3 (2x)\, dx$
11. $\int \sqrt{4x^2 - 5}\,(8x)\, dx$
12. $\int \sqrt[3]{1 - 2x^2}\,(-4x)\, dx$
13. $\int (x - 1)^4\, dx$
14. $\int (x - 3)^{5/2}\, dx$
15. $\int 2x(x^2 - 1)^7\, dx$
16. $\int x(1 - 2x^2)^3\, dx$
17. $\int \dfrac{x^2}{(1 + x^3)^2}\, dx$
18. $\int \dfrac{x^2}{(x^3 - 1)^2}\, dx$
19. $\int \dfrac{x + 1}{(x^2 + 2x - 3)^2}\, dx$
20. $\int \dfrac{6x}{(1 + x^2)^3}\, dx$
21. $\int \dfrac{x - 2}{\sqrt{x^2 - 4x + 3}}\, dx$
22. $\int \dfrac{4x + 6}{(x^2 + 3x + 7)^3}\, dx$
23. $\int 5u \sqrt[3]{1 - u^2}\, du$
24. $\int u^3 \sqrt{u^4 + 2}\, du$
25. $\int \dfrac{4y}{\sqrt{1 + y^2}}\, dy$
26. $\int \dfrac{3x^2}{\sqrt{1 - x^3}}\, dx$
27. $\int \dfrac{-3}{\sqrt{2t + 3}}\, dt$
28. $\int \dfrac{t + 2t^2}{\sqrt{t}}\, dt$

(T) In Exercises 29–34, use a symbolic integration utility to find the indefinite integral.

29. $\int \dfrac{x^3}{\sqrt{1 - x^4}}\, dx$
30. $\int \dfrac{3x}{\sqrt{1 - 4x^2}}\, dx$
31. $\int \left(1 + \dfrac{4}{t^2}\right)^2 \left(\dfrac{1}{t^3}\right) dt$
32. $\int \left(1 + \dfrac{1}{t}\right)^3 \left(\dfrac{1}{t^2}\right) dt$
33. $\int (x^3 + 3x + 9)(x^2 + 1)\, dx$
34. $\int (7 - 3x - 3x^2)(2x + 1)\, dx$

SECTION 6.2 Integration by Substitution and the General Power Rule

In Exercises 35–42, use formal substitution (as illustrated in Examples 5 and 6) to find the indefinite integral.

35. $\int 12x(6x^2 - 1)^3 \, dx$

36. $\int 3x^2(1 - x^3)^2 \, dx$

37. $\int x^2(2 - 3x^3)^{3/2} \, dx$

38. $\int t\sqrt{t^2 + 1} \, dt$

39. $\int \dfrac{x}{\sqrt{x^2 + 25}} \, dx$

40. $\int \dfrac{3}{\sqrt{2x + 1}} \, dx$

41. $\int \dfrac{x^2 + 1}{\sqrt{x^3 + 3x + 4}} \, dx$

42. $\int \sqrt{x}(4 - x^{3/2})^2 \, dx$

In Exercises 43–46, (a) perform the integration in two ways: once using the Simple Power Rule and once using the General Power Rule. (b) Explain the difference in the results. (c) Which method do you prefer? Explain your reasoning.

43. $\int (x - 1)^2 \, dx$

44. $\int (3 - x)^2 \, dx$

45. $\int x(x^2 - 1)^2 \, dx$

46. $\int x(2x^2 + 1)^2 \, dx$

47. Find the equation of the function f whose graph passes through the point $\left(0, \frac{4}{3}\right)$ and whose derivative is
$f'(x) = x\sqrt{1 - x^2}$.

48. Find the equation of the function f whose graph passes through the point $\left(0, \frac{7}{3}\right)$ and whose derivative is
$f'(x) = x\sqrt{1 - x^2}$.

49. **Women in the Labor Force** The rate of increase in the number of women W (in millions) in the U.S. civilian labor force from 2000 through 2005 can be modeled by
$$\dfrac{dW}{dt} = \dfrac{1280}{(15.08 - 0.128t)^2}$$
where t is the year, with $t = 0$ corresponding to 2000. In 2000, there were about 66.3 million women in the labor force. *(Source: U.S. Bureau of Labor Statistics)*

 (a) Find a function for the number of women in the labor force.

 (b) How many women were in the labor force in 2005?

50. **Forestry** A lumber company has determined that the rate of weight loss W (in pounds) for a ponderosa pine log after t days of drying time can be approximated by
$$\dfrac{dW}{dt} = \dfrac{12}{\sqrt{16t + 9}}, \quad 0 \le t \le 100.$$
No weight loss occurs until the tree is cut ($t = 0$).

 (a) Find the weight loss function.

 (b) Find the weight loss after 100 days.

51. **Gardening** An evergreen nursery usually sells a type of shrub after 5 years of growth and shaping. The growth rate during those 5 years is approximated by
$$\dfrac{dh}{dt} = \dfrac{17.6t}{\sqrt{17.6t^2 + 1}}$$
where t is time in years and h is height in inches. The seedlings are 6 inches tall when planted ($t = 0$).

 (a) Find the height function.

 (b) How tall are the shrubs when they are sold?

52. **Cell Growth** The growth rate of a type of plant cell C (in hundreds) is modeled by
$$\dfrac{dC}{dt} = 4\sqrt{t + 1}$$
where t is the time in days. When $t = 0$, $C = 9$.

 (a) Find a function for the number of cells.

 (b) Use a graphing utility to graph dC/dt and C in the same viewing window.

 (c) Use the *zoom* and *trace* features of a graphing utility to determine the number of cells and the corresponding growth rate when $t = 5$.

 (d) Verify the results of part (c) analytically.

53. **Suburban Growth** The growth rate of the population P (in millions) in the suburbs of a southeastern U.S. city from 2000 through 2006 can be modeled by
$$\dfrac{dP}{dt} = 0.06t(0.005t^2 + 1)^2$$
where t is the year, with $t = 0$ corresponding to 2000. In 2005, the population was 3 million.

 (a) Find the population function.

 (b) Use a graphing utility to graph dP/dt and P in the same viewing window.

 (c) Use the *zoom* and *trace* features of a graphing utility to determine the population and the corresponding growth rate in the year 2003.

 (d) Verify the results of part (c) analytically.

In Exercises 54 and 55, use a symbolic integration utility to find the indefinite integral. Verify the result by differentiating.

54. $\int \dfrac{1}{\sqrt{x} + \sqrt{x + 1}} \, dx$

55. $\int \dfrac{x}{\sqrt{3x + 2}} \, dx$

Section 6.3

Exponential and Logarithmic Integrals

- Use the Exponential Rule to find indefinite integrals.
- Use the Log Rule to find indefinite integrals.

Using the Exponential Rule

Each of the differentiation rules for exponential functions has its corresponding integration rule.

Integrals of Exponential Functions

Let u be a differentiable function of x.

$$\int e^x \, dx = e^x + C \qquad \text{Simple Exponential Rule}$$

$$\int e^u \frac{du}{dx} \, dx = \int e^u \, du = e^u + C \qquad \text{General Exponential Rule}$$

Example 1 Integrating Exponential Functions

Find each indefinite integral.

a. $\displaystyle\int 2e^x \, dx$ **b.** $\displaystyle\int 2e^{2x} \, dx$ **c.** $\displaystyle\int (e^x + x) \, dx$

SOLUTION

a. $\displaystyle\int 2e^x \, dx = 2 \int e^x \, dx$ Constant Multiple Rule

$\qquad\qquad\quad = 2e^x + C$ Simple Exponential Rule

b. $\displaystyle\int 2e^{2x} \, dx = \int e^{2x}(2) \, dx$ Let $u = 2x$, then $\dfrac{du}{dx} = 2$.

$\qquad\qquad\quad = \displaystyle\int e^u \frac{du}{dx} \, dx$

$\qquad\qquad\quad = e^{2x} + C$ General Exponential Rule

c. $\displaystyle\int (e^x + x) \, dx = \int e^x \, dx + \int x \, dx$ Sum Rule

$\qquad\qquad\quad = e^x + \dfrac{x^2}{2} + C$ Simple Exponential and Power Rules

You can check each of these results by differentiating.

✓ **CHECKPOINT 1**

Find each indefinite integral.

a. $\displaystyle\int 3e^x \, dx$

b. $\displaystyle\int 5e^{5x} \, dx$

c. $\displaystyle\int (e^x - x) \, dx$

SECTION 6.3 Exponential and Logarithmic Integrals

TECHNOLOGY

If you use a symbolic integration utility to find antiderivatives of exponential or logarithmic functions, you can easily obtain results that are beyond the scope of this course. For instance, the antiderivative of e^{x^2} involves the imaginary unit i and the probability function called "ERF." In this course, you are not expected to interpret or use such results. You can simply state that the function cannot be integrated using elementary functions.

Example 2 Integrating an Exponential Function

Find $\int e^{3x+1} \, dx$.

SOLUTION Let $u = 3x + 1$, then $du/dx = 3$. You can introduce the missing factor of 3 in the integrand by multiplying and dividing by 3.

$$\int e^{3x+1} \, dx = \frac{1}{3} \int e^{3x+1}(3) \, dx \quad \text{Multiply and divide by 3.}$$
$$= \frac{1}{3} \int e^u \frac{du}{dx} \, dx \quad \text{Substitute } u \text{ and } \frac{du}{dx}.$$
$$= \frac{1}{3} e^u + C \quad \text{General Exponential Rule}$$
$$= \frac{1}{3} e^{3x+1} + C \quad \text{Substitute for } u.$$

✓ CHECKPOINT 2

Find $\int e^{2x+3} \, dx$. ■

Algebra Review

For help with the algebra in Example 3, see Example 1(d) in the *Chapter 6 Algebra Review*, on page 418.

Example 3 Integrating an Exponential Function

Find $\int 5xe^{-x^2} \, dx$.

SOLUTION Let $u = -x^2$, then $du/dx = -2x$. You can create the factor $-2x$ in the integrand by multiplying and dividing by -2.

$$\int 5xe^{-x^2} \, dx = \int \left(-\frac{5}{2}\right) e^{-x^2}(-2x) \, dx \quad \text{Multiply and divide by } -2.$$
$$= -\frac{5}{2} \int e^{-x^2}(-2x) \, dx \quad \text{Factor } -\frac{5}{2} \text{ out of the integrand.}$$
$$= -\frac{5}{2} \int e^u \frac{du}{dx} \, dx \quad \text{Substitute } u \text{ and } \frac{du}{dx}.$$
$$= -\frac{5}{2} e^u + C \quad \text{General Exponential Rule}$$
$$= -\frac{5}{2} e^{-x^2} + C \quad \text{Substitute for } u.$$

STUDY TIP

Remember that you cannot introduce a missing *variable* in the integrand. For instance, you cannot find $\int e^{x^2} \, dx$ by multiplying and dividing by $2x$ and then factoring $1/(2x)$ out of the integral. That is,

$$\int e^{x^2} \, dx \neq \frac{1}{2x} \int e^{x^2}(2x) \, dx.$$

✓ CHECKPOINT 3

Find $\int 4xe^{x^2} \, dx$. ■

DISCOVERY

The General Power Rule is not valid for $n = -1$. Can you find an antiderivative for u^{-1}?

Using the Log Rule

When the Power Rules for integration were introduced in Sections 6.1 and 6.2, you saw that they work for powers other than $n = -1$.

$$\int x^n \, dx = \frac{x^{n+1}}{n+1} + C, \quad n \neq -1 \qquad \text{Simple Power Rule}$$

$$\int u^n \frac{du}{dx} \, dx = \int u^n \, du = \frac{u^{n+1}}{n+1} + C, \quad n \neq -1 \qquad \text{General Power Rule}$$

The Log Rules for integration allow you to integrate functions of the form $\int x^{-1} \, dx$ and $\int u^{-1} \, du$.

Integrals of Logarithmic Functions

Let u be a differentiable function of x.

$$\int \frac{1}{x} \, dx = \ln|x| + C \qquad \text{Simple Logarithmic Rule}$$

$$\int \frac{du/dx}{u} \, dx = \int \frac{1}{u} \, du = \ln|u| + C \qquad \text{General Logarithmic Rule}$$

STUDY TIP

Notice the absolute values in the Log Rules. For those special cases in which u or x cannot be negative, you can omit the absolute value. For instance, in Example 4(b), it is not necessary to write the antiderivative as $\ln|x^2| + C$ because x^2 cannot be negative.

You can verify each of these rules by differentiating. For instance, to verify that $d/dx[\ln|x|] = 1/x$, notice that

$$\frac{d}{dx}[\ln x] = \frac{1}{x} \quad \text{and} \quad \frac{d}{dx}[\ln(-x)] = \frac{-1}{-x} = \frac{1}{x}.$$

Example 4 Integrating Logarithmic Functions

Find each indefinite integral.

a. $\displaystyle\int \frac{4}{x} \, dx$ **b.** $\displaystyle\int \frac{2x}{x^2} \, dx$ **c.** $\displaystyle\int \frac{3}{3x+1} \, dx$

SOLUTION

a. $\displaystyle\int \frac{4}{x} \, dx = 4 \int \frac{1}{x} \, dx$ Constant Multiple Rule

$= 4 \ln|x| + C$ Simple Logarithmic Rule

b. $\displaystyle\int \frac{2x}{x^2} \, dx = \int \frac{du/dx}{u} \, dx$ Let $u = x^2$, then $\dfrac{du}{dx} = 2x$.

$= \ln|u| + C$ General Logarithmic Rule

$= \ln x^2 + C$ Substitute for u.

c. $\displaystyle\int \frac{3}{3x+1} \, dx = \int \frac{du/dx}{u} \, dx$ Let $u = 3x+1$, then $\dfrac{du}{dx} = 3$.

$= \ln|u| + C$ General Logarithmic Rule

$= \ln|3x+1| + C$ Substitute for u.

✓ CHECKPOINT 4

Find each indefinite integral.

a. $\displaystyle\int \frac{2}{x} \, dx$

b. $\displaystyle\int \frac{3x^2}{x^3} \, dx$

c. $\displaystyle\int \frac{2}{2x+1} \, dx$

Example 5 Using the Log Rule

Find $\int \dfrac{1}{2x-1}\,dx$.

SOLUTION Let $u = 2x - 1$, then $du/dx = 2$. You can create the necessary factor of 2 in the integrand by multiplying and dividing by 2.

$$\int \dfrac{1}{2x-1}\,dx = \dfrac{1}{2}\int \dfrac{2}{2x-1}\,dx \qquad \text{Multiply and divide by 2.}$$

$$= \dfrac{1}{2}\int \dfrac{du/dx}{u}\,dx \qquad \text{Substitute } u \text{ and } \dfrac{du}{dx}.$$

$$= \dfrac{1}{2}\ln|u| + C \qquad \text{General Log Rule}$$

$$= \dfrac{1}{2}\ln|2x-1| + C \qquad \text{Substitute for } u.$$

✓ **CHECKPOINT 5**

Find $\int \dfrac{1}{4x+1}\,dx$. ■

Example 6 Using the Log Rule

Find $\int \dfrac{6x}{x^2+1}\,dx$.

SOLUTION Let $u = x^2 + 1$, then $du/dx = 2x$. You can create the necessary factor of $2x$ in the integrand by factoring a 3 out of the integrand.

$$\int \dfrac{6x}{x^2+1}\,dx = 3\int \dfrac{2x}{x^2+1}\,dx \qquad \text{Factor 3 out of integrand.}$$

$$= 3\int \dfrac{du/dx}{u}\,dx \qquad \text{Substitute } u \text{ and } \dfrac{du}{dx}.$$

$$= 3\ln|u| + C \qquad \text{General Log Rule}$$

$$= 3\ln(x^2+1) + C \qquad \text{Substitute for } u.$$

✓ **CHECKPOINT 6**

Find $\int \dfrac{3x}{x^2+4}\,dx$. ■

Algebra Review

For help with the algebra in the integral at the right, see Example 2(d) in the *Chapter 6 Algebra Review*, on page 419.

Integrals to which the Log Rule can be applied are often given in disguised form. For instance, if a rational function has a numerator of degree greater than or equal to that of the denominator, you should use long division to rewrite the integrand. Here is an example.

$$\int \dfrac{x^2+6x+1}{x^2+1}\,dx = \int \left(1 + \dfrac{6x}{x^2+1}\right)dx$$

$$= x + 3\ln(x^2+1) + C$$

> **Algebra Review**
>
> For help with the algebra in Example 7, see Example 2(a)–(c) in the *Chapter 6 Algebra Review*, on page 419.

The next example summarizes some additional situations in which it is helpful to rewrite the integrand in order to recognize the antiderivative.

Example 7 Rewriting Before Integrating

Find each indefinite integral.

a. $\displaystyle\int \frac{3x^2 + 2x - 1}{x^2}\,dx$ **b.** $\displaystyle\int \frac{1}{1 + e^{-x}}\,dx$ **c.** $\displaystyle\int \frac{x^2 + x + 1}{x - 1}\,dx$

SOLUTION

a. Begin by rewriting the integrand as the sum of three fractions.

$$\int \frac{3x^2 + 2x - 1}{x^2}\,dx = \int \left(\frac{3x^2}{x^2} + \frac{2x}{x^2} - \frac{1}{x^2}\right)dx$$

$$= \int \left(3 + \frac{2}{x} - \frac{1}{x^2}\right)dx$$

$$= 3x + 2\ln|x| + \frac{1}{x} + C$$

b. Begin by rewriting the integrand by multiplying and dividing by e^x.

$$\int \frac{1}{1 + e^{-x}}\,dx = \int \left(\frac{e^x}{e^x}\right)\frac{1}{1 + e^{-x}}\,dx$$

$$= \int \frac{e^x}{e^x + 1}\,dx$$

$$= \ln(e^x + 1) + C$$

c. Begin by dividing the numerator by the denominator.

$$\int \frac{x^2 + x + 1}{x - 1}\,dx = \int \left(x + 2 + \frac{3}{x - 1}\right)dx$$

$$= \frac{x^2}{2} + 2x + 3\ln|x - 1| + C$$

> **✓ CHECKPOINT 7**
>
> Find each indefinite integral.
>
> **a.** $\displaystyle\int \frac{4x^2 - 3x + 2}{x^2}\,dx$
>
> **b.** $\displaystyle\int \frac{2}{e^{-x} + 1}\,dx$
>
> **c.** $\displaystyle\int \frac{x^2 + 2x + 4}{x + 1}\,dx$ ∎

> **STUDY TIP**
>
> The Exponential and Log Rules are necessary to solve certain real-life problems, such as population growth. You will see such problems in the exercise set for this section.

CONCEPT CHECK

1. Write the General Exponential Rule for integration.
2. Write the General Logarithmic Rule for integration.
3. Which integration rule allows you to integrate functions of the form $\displaystyle\int e^u \frac{du}{dx}\,dx$?
4. Which integration rule allows you to integrate $\displaystyle\int x^{-1}\,dx$?

SECTION 6.3 Exponential and Logarithmic Integrals

Skills Review 6.3

The following warm-up exercises involve skills that were covered in earlier sections. You will use these skills in the exercise set for this section. For additional help, review Sections 4.4, 6.1, and 6.2.

In Exercises 1 and 2, find the domain of the function.

1. $y = \ln(2x - 5)$
2. $y = \ln(x^2 - 5x + 6)$

In Exercises 3–6, use long division to rewrite the quotient.

3. $\dfrac{x^2 + 4x + 2}{x + 2}$
4. $\dfrac{x^2 - 6x + 9}{x - 4}$
5. $\dfrac{x^3 + 4x^2 - 30x - 4}{x^2 - 4x}$
6. $\dfrac{x^4 - x^3 + x^2 + 15x + 2}{x^2 + 5}$

In Exercises 7–10, evaluate the integral.

7. $\displaystyle\int \left(x^3 + \dfrac{1}{x^2}\right) dx$
8. $\displaystyle\int \dfrac{x^2 + 2x}{x} dx$
9. $\displaystyle\int \dfrac{x^3 + 4}{x^2} dx$
10. $\displaystyle\int \dfrac{x + 3}{x^3} dx$

Exercises 6.3

See www.CalcChat.com for worked-out solutions to odd-numbered exercises.

In Exercises 1–12, use the Exponential Rule to find the indefinite integral.

1. $\displaystyle\int 2e^{2x} dx$
2. $\displaystyle\int -3e^{-3x} dx$
3. $\displaystyle\int e^{4x} dx$
4. $\displaystyle\int e^{-0.25x} dx$
5. $\displaystyle\int 9xe^{-x^2} dx$
6. $\displaystyle\int 3xe^{0.5x^2} dx$
7. $\displaystyle\int 5x^2 e^{x^3} dx$
8. $\displaystyle\int (2x + 1)e^{x^2 + x} dx$
9. $\displaystyle\int (x^2 + 2x)e^{x^3 + 3x^2 - 1} dx$
10. $\displaystyle\int 3(x - 4)e^{x^2 - 8x} dx$
11. $\displaystyle\int 5e^{2-x} dx$
12. $\displaystyle\int 3e^{-(x+1)} dx$

In Exercises 13–28, use the Log Rule to find the indefinite integral.

13. $\displaystyle\int \dfrac{1}{x + 1} dx$
14. $\displaystyle\int \dfrac{1}{x - 5} dx$
15. $\displaystyle\int \dfrac{1}{3 - 2x} dx$
16. $\displaystyle\int \dfrac{1}{6x - 5} dx$
17. $\displaystyle\int \dfrac{2}{3x + 5} dx$
18. $\displaystyle\int \dfrac{5}{2x - 1} dx$
19. $\displaystyle\int \dfrac{x}{x^2 + 1} dx$
20. $\displaystyle\int \dfrac{x^2}{3 - x^3} dx$
21. $\displaystyle\int \dfrac{x^2}{x^3 + 1} dx$
22. $\displaystyle\int \dfrac{x}{x^2 + 4} dx$
23. $\displaystyle\int \dfrac{x + 3}{x^2 + 6x + 7} dx$
24. $\displaystyle\int \dfrac{x^2 + 2x + 3}{x^3 + 3x^2 + 9x + 1} dx$
25. $\displaystyle\int \dfrac{1}{x \ln x} dx$
26. $\displaystyle\int \dfrac{1}{x(\ln x)^2} dx$
27. $\displaystyle\int \dfrac{e^{-x}}{1 - e^{-x}} dx$
28. $\displaystyle\int \dfrac{e^x}{1 + e^x} dx$

(T) In Exercises 29–38, use a symbolic integration utility to find the indefinite integral.

29. $\displaystyle\int \dfrac{1}{x^2} e^{2/x} dx$
30. $\displaystyle\int \dfrac{1}{x^3} e^{1/4x^2} dx$
31. $\displaystyle\int \dfrac{1}{\sqrt{x}} e^{\sqrt{x}} dx$
32. $\displaystyle\int \dfrac{e^{1/\sqrt{x}}}{x^{3/2}} dx$
33. $\displaystyle\int (e^x - 2)^2 dx$
34. $\displaystyle\int (e^x - e^{-x})^2 dx$
35. $\displaystyle\int \dfrac{e^{-x}}{1 + e^{-x}} dx$
36. $\displaystyle\int \dfrac{3e^x}{2 + e^x} dx$
37. $\displaystyle\int \dfrac{4e^{2x}}{5 - e^{2x}} dx$
38. $\displaystyle\int \dfrac{-e^{3x}}{2 - e^{3x}} dx$

In Exercises 39–54, use any basic integration formula or formulas to find the indefinite integral. State which integration formula(s) you used to find the integral.

39. $\int \dfrac{e^{2x} + 2e^x + 1}{e^x}\,dx$

40. $\int (6x + e^x)\sqrt{3x^2 + e^x}\,dx$

41. $\int e^x\sqrt{1 - e^x}\,dx$

42. $\int \dfrac{2(e^x - e^{-x})}{(e^x + e^{-x})^2}\,dx$

43. $\int \dfrac{1}{(x - 1)^2}\,dx$

44. $\int \dfrac{1}{\sqrt{x + 1}}\,dx$

45. $\int 4e^{2x-1}\,dx$

46. $\int (5e^{-2x} + 1)\,dx$

47. $\int \dfrac{x^3 - 8x}{2x^2}\,dx$

48. $\int \dfrac{x - 1}{4x}\,dx$

49. $\int \dfrac{2}{1 + e^{-x}}\,dx$

50. $\int \dfrac{3}{1 + e^{-3x}}\,dx$

51. $\int \dfrac{x^2 + 2x + 5}{x - 1}\,dx$

52. $\int \dfrac{x - 3}{x + 3}\,dx$

53. $\int \dfrac{1 + e^{-x}}{1 + xe^{-x}}\,dx$

54. $\int \dfrac{5}{e^{-5x} + 7}\,dx$

In Exercises 55 and 56, find the equation of the function f whose graph passes through the point.

55. $f'(x) = \dfrac{x^2 + 4x + 3}{x - 1}$, $(2, 4)$

56. $f'(x) = \dfrac{x^3 - 4x^2 + 3}{x - 3}$, $(4, -1)$

57. **Biology** A population of bacteria is growing at the rate of

$$\dfrac{dP}{dt} = \dfrac{3000}{1 + 0.25t}$$

where t is the time in days. When $t = 0$, the population is 1000.

(a) Write an equation that models the population P in terms of the time t.

(b) What is the population after 3 days?

(c) After how many days will the population be 12,000?

58. **Biology** Because of an insufficient oxygen supply, the trout population in a lake is dying. The population's rate of change can be modeled by

$$\dfrac{dP}{dt} = -125e^{-t/20}$$

where t is the time in days. When $t = 0$, the population is 2500.

(a) Write an equation that models the population P in terms of the time t.

(b) What is the population after 15 days?

(c) According to this model, how long will it take for the entire trout population to die?

59. **Livestock Inventory** From 2000 through 2005, the rate of change in the number of sheep and lambs on farms L (in millions) in the United States can be modeled by

$$\dfrac{dL}{dt} = -\dfrac{4.28}{t} + 0.14$$

where t is the year, with $t = 10$ corresponding to 2000. In 2003, the number of sheep and lambs was 6.3 million. (Source: U.S. Department of Agriculture)

(a) Find a model for the number of sheep and lambs from 2000 through 2005.

(b) Use the model to predict the number of sheep and lambs in 2010.

60. **Livestock Inventory** From 2000 through 2005, the rate of change in the number of cattle on farms C (in millions) in the United States can be modeled by

$$\dfrac{dC}{dt} = -0.62 - 0.135t^2 + 0.0439e^t$$

where t is the year, with $t = 0$ corresponding to 2000. In 2002, the number of cattle was 96.7 million. (Source: U.S. Department of Agriculture)

(a) Find a model for the number of cattle from 2000 through 2005.

(b) Use the model to predict the number of cattle in 2007.

61. **Average Salary** From 2000 through 2005, the average salary for public school nurses S (in dollars) in the United States changed at the rate of $dS/dt = 1724.1e^{-t/4.2}$, where $t = 0$ corresponds to 2000. In 2005, the average salary was $40,520. (Source: Educational Research Service)

(a) Write a model that gives the average salary per year.

(b) Use the model to find the average salary in 2002.

62. **Average Salary** From 2000 through 2006, the average salary for associate professors S (in thousands of dollars) at public universities in the United States changed at the rate of

$$\dfrac{dS}{dt} = 0.029t + \dfrac{18.35}{t}$$

where $t = 10$ corresponds to 2000. In 2006, the average salary was 66.3 thousand dollars. (Source: American Association of University Professors)

(a) Write a model that gives the average salary per year.

(b) Use the model to find the average salary in 2003.

True or False? In Exercises 63 and 64, determine whether the statement is true or false. If it is false, explain why or give an example that shows it is false.

63. $(\ln x)^{1/2} = \tfrac{1}{2}(\ln x)$

64. $\int \ln x = \left(\dfrac{1}{x}\right) + C$

Mid-Chapter Quiz

Take this quiz as you would take a quiz in class. When you are done, check your work against the answers given in the back of the book.

In Exercises 1–9, find the indefinite integral and check your result by differentiation.

1. $\int 3 \, dx$
2. $\int 10x \, dx$
3. $\int \frac{1}{x^5} \, dx$
4. $\int (x^2 - 2x + 15) \, dx$
5. $\int x(x + 4) \, dx$
6. $\int (6x + 1)^3(6) \, dx$
7. $\int (x^2 - 5x)(2x - 5) \, dx$
8. $\int \frac{3x^2}{(x^3 + 3)^3} \, dx$
9. $\int \sqrt{5x + 2} \, dx$

In Exercises 10 and 11, find the particular solution $y = f(x)$ that satisfies the differential equation and initial condition.

10. $f'(x) = 16x$, $f(0) = 1$
11. $f'(x) = 9x^2 + 4$, $f(1) = 5$

12. Find the equation of the function f whose graph passes through the point $(0, 1)$ and whose derivative is
$$f'(x) = 2x^2 + 1.$$

In Exercises 13–15, use the Exponential Rule to find the indefinite integral. Check your result by differentiation.

13. $\int 5e^{5x+4} \, dx$
14. $\int (x + 2e^{2x}) \, dx$
15. $\int 3x^2 e^{x^3} \, dx$

In Exercises 16–18, use the Log Rule to find the indefinite integral.

16. $\int \frac{2}{2x - 1} \, dx$
17. $\int \frac{-2x}{x^2 + 3} \, dx$
18. $\int \frac{3(3x^2 + 4x)}{x^3 + 2x^2} \, dx$

19. The growth rate P (in hundreds) of a type of cell is modeled by
$$\frac{dP}{dt} = 3\sqrt{t + 2}$$
where t is the time in days. When $t = 2$, $P = 25$.

(a) Find a function for the number of cells.

(b) Use a graphing utility to graph dP/dt and P in the same viewing window.

(c) Use the *zoom* and *trace* features of a graphing utility to determine the number of cells and the corresponding growth rate when $t = 7$.

(d) Verify the results of part (c) analytically.

Section 6.4
Area and the Fundamental Theorem of Calculus

- Evaluate definite integrals.
- Evaluate definite integrals using the Fundamental Theorem of Calculus.
- Use definite integrals to find the net change in a quantity.
- Find the average values of functions over closed intervals.
- Use properties of even and odd functions to help evaluate definite integrals.
- Use definite integrals to solve real-life problems.

Area and Definite Integrals

From your study of geometry, you know that area is a number that defines the size of a bounded region. For simple regions, such as rectangles, triangles, and circles, area can be found using geometric formulas.

In this section, you will learn how to use calculus to find the areas of nonstandard regions, such as the region R shown in Figure 6.4.

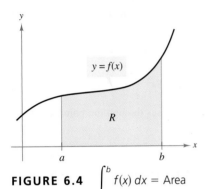

FIGURE 6.4 $\int_a^b f(x)\, dx =$ Area

Definition of a Definite Integral

Let f be nonnegative and continuous on the closed interval $[a, b]$. The area of the region bounded by the graph of f, the x-axis, and the lines $x = a$ and $x = b$ is denoted by

$$\text{Area} = \int_a^b f(x)\, dx.$$

The expression $\int_a^b f(x)\, dx$ is called the **definite integral** from a to b, where a is the **lower limit of integration** and b is the **upper limit of integration**.

Example 1   Evaluating a Definite Integral

Evaluate $\int_0^2 2x\, dx$.

SOLUTION This definite integral represents the area of the region bounded by the graph of $f(x) = 2x$, the x-axis, and the line $x = 2$, as shown in Figure 6.5. The region is triangular, with a height of four units and a base of two units.

$$\int_0^2 2x\, dx = \frac{1}{2}(\text{base})(\text{height}) \quad \text{Formula for area of triangle}$$

$$= \frac{1}{2}(2)(4) \quad \text{Substitute 2 for base and 4 for height.}$$

$$= 4 \quad \text{Simplify.}$$

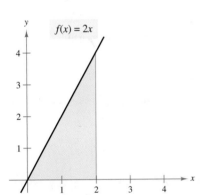

FIGURE 6.5

✓**CHECKPOINT 1**

Evaluate the definite integral using a geometric formula. Illustrate your answer with an appropriate sketch.

$$\int_0^3 4x\, dx$$ ■

SECTION 6.4 Area and the Fundamental Theorem of Calculus

The Fundamental Theorem of Calculus

Consider the function A, which denotes the area of the region shown in Figure 6.6. To discover the relationship between A and f, let x increase by an amount Δx. This increases the area by ΔA. Let $f(m)$ and $f(M)$ denote the minimum and maximum values of f on the interval $[x, x + \Delta x]$.

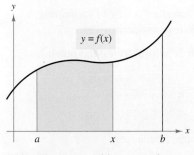

FIGURE 6.6 $A(x) =$ Area from a to x

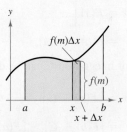 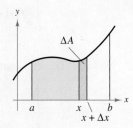 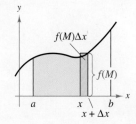

FIGURE 6.7

As indicated in Figure 6.7, you can write the inequality below.

$$f(m)\, \Delta x \leq \Delta A \leq f(M)\, \Delta x \qquad \text{See Figure 6.7.}$$

$$f(m) \leq \frac{\Delta A}{\Delta x} \leq f(M) \qquad \text{Divide each term by } \Delta x.$$

$$\lim_{\Delta x \to 0} f(m) \leq \lim_{\Delta x \to 0} \frac{\Delta A}{\Delta x} \leq \lim_{\Delta x \to 0} f(M) \qquad \text{Take limit of each term.}$$

$$f(x) \leq A'(x) \leq f(x) \qquad \text{Definition of derivative of } A(x)$$

So, $f(x) = A'(x)$, and $A(x) = F(x) + C$, where $F'(x) = f(x)$. Because $A(a) = 0$, it follows that $C = -F(a)$. So, $A(x) = F(x) - F(a)$, which implies that

$$A(b) = \int_a^b f(x)\, dx$$
$$= F(b) - F(a).$$

This equation tells you that *if you can find an antiderivative of f*, then you can use the antiderivative to evaluate the definite integral

$$\int_a^b f(x)\, dx.$$

This result is called the **Fundamental Theorem of Calculus**.

The Fundamental Theorem of Calculus

If f is nonnegative and continuous on the closed interval $[a, b]$, then

$$\int_a^b f(x)\, dx = F(b) - F(a)$$

where F is any function such that

$$F'(x) = f(x)$$

for all x in $[a, b]$.

Guidelines for Using the Fundamental Theorem of Calculus

1. The Fundamental Theorem of Calculus describes a way of *evaluating* a definite integral, not a procedure for finding antiderivatives.

2. In applying the Fundamental Theorem, it is helpful to use the notation
$$\int_a^b f(x)\, dx = F(x)\Big]_a^b$$
$$= F(b) - F(a).$$

3. The constant of integration C can be dropped because
$$\int_a^b f(x)\, dx = \Big[F(x) + C\Big]_a^b$$
$$= [F(b) + C] - [F(a) + C]$$
$$= F(b) - F(a) + C - C$$
$$= F(b) - F(a).$$

In the development of the Fundamental Theorem of Calculus, f was assumed to be nonnegative on the closed interval $[a, b]$. As such, the definite integral was defined as an area. Now, with the Fundamental Theorem, the definition can be extended to include functions that are negative on all or part of the closed interval $[a, b]$. Specifically, if f is *any* function that is continuous on a closed interval $[a, b]$, then the **definite integral** of $f(x)$ from a to b is defined as

$$\int_a^b f(x)\, dx = F(b) - F(a)$$

where F is an antiderivative of f. Remember that definite integrals do not necessarily represent areas and can be negative, zero, or positive.

STUDY TIP

Be sure you see the distinction between indefinite and definite integrals. The *indefinite integral*

$$\int f(x)\, dx$$

denotes a family of *functions*, each of which is an antiderivative of f, whereas the *definite integral*

$$\int_a^b f(x)\, dx$$

is a *number*.

Properties of Definite Integrals

Let f and g be continuous on the closed interval $[a, b]$.

1. $\displaystyle\int_a^b kf(x)\, dx = k\int_a^b f(x)\, dx$, k is a constant.

2. $\displaystyle\int_a^b [f(x) \pm g(x)]\, dx = \int_a^b f(x)\, dx \pm \int_a^b g(x)\, dx$

3. $\displaystyle\int_a^b f(x)\, dx = \int_a^c f(x)\, dx + \int_c^b f(x)\, dx$, $a < c < b$

4. $\displaystyle\int_a^a f(x)\, dx = 0$

5. $\displaystyle\int_a^b f(x)\, dx = -\int_b^a f(x)\, dx$

SECTION 6.4 Area and the Fundamental Theorem of Calculus

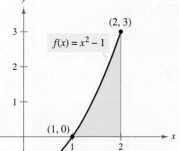

FIGURE 6.8 Area = $\int_1^2 (x^2 - 1)\, dx$

✓ CHECKPOINT 2

Find the area of the region bounded by the x-axis and the graph of $f(x) = x^2 + 1$, $2 \le x \le 3$. ∎

Example 2 Finding Area by the Fundamental Theorem

Find the area of the region bounded by the x-axis and the graph of
$$f(x) = x^2 - 1, \quad 1 \le x \le 2.$$

SOLUTION Note that $f(x) \ge 0$ on the interval $1 \le x \le 2$, as shown in Figure 6.8. So, you can represent the area of the region by a definite integral. To find the area, use the Fundamental Theorem of Calculus.

$$\begin{aligned}
\text{Area} &= \int_1^2 (x^2 - 1)\, dx && \text{Definition of definite integral} \\
&= \left(\frac{x^3}{3} - x\right)\Big|_1^2 && \text{Find antiderivative.} \\
&= \left(\frac{2^3}{3} - 2\right) - \left(\frac{1^3}{3} - 1\right) && \text{Apply Fundamental Theorem.} \\
&= \frac{4}{3} && \text{Simplify.}
\end{aligned}$$

So, the area of the region is $\frac{4}{3}$ square units.

STUDY TIP

It is easy to make errors in signs when evaluating definite integrals. To avoid such errors, enclose the values of the antiderivative at the upper and lower limits of integration in separate sets of parentheses, as shown above.

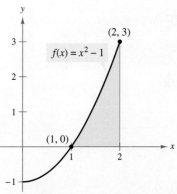

FIGURE 6.9

✓ CHECKPOINT 3

Evaluate $\int_0^1 (2t + 3)^3\, dt$. ∎

Example 3 Evaluating a Definite Integral

Evaluate the definite integral
$$\int_0^1 (4t + 1)^2\, dt$$
and sketch the region whose area is represented by the integral.

SOLUTION

$$\begin{aligned}
\int_0^1 (4t + 1)^2\, dt &= \frac{1}{4}\int_0^1 (4t + 1)^2 (4)\, dt && \text{Multiply and divide by 4.} \\
&= \frac{1}{4}\left[\frac{(4t + 1)^3}{3}\right]_0^1 && \text{Find antiderivative.} \\
&= \frac{1}{4}\left[\left(\frac{5^3}{3}\right) - \left(\frac{1}{3}\right)\right] && \text{Apply Fundamental Theorem.} \\
&= \frac{31}{3} && \text{Simplify.}
\end{aligned}$$

The region is shown in Figure 6.9.

Example 4 Evaluating Definite Integrals

Evaluate each definite integral.

a. $\displaystyle\int_0^3 e^{2x}\,dx$ **b.** $\displaystyle\int_1^2 \frac{1}{x}\,dx$ **c.** $\displaystyle\int_1^4 -3\sqrt{x}\,dx$

SOLUTION

a. $\displaystyle\int_0^3 e^{2x}\,dx = \frac{1}{2}e^{2x}\Big]_0^3 = \frac{1}{2}(e^6 - e^0) \approx 201.21$

b. $\displaystyle\int_1^2 \frac{1}{x}\,dx = \ln x\Big]_1^2 = \ln 2 - \ln 1 = \ln 2 \approx 0.69$

c. $\displaystyle\int_1^4 -3\sqrt{x}\,dx = -3\int_1^4 x^{1/2}\,dx$ Rewrite with rational exponent.

$\qquad = -3\left[\dfrac{x^{3/2}}{3/2}\right]_1^4$ Find antiderivative.

$\qquad = -2x^{3/2}\Big]_1^4$

$\qquad = -2(4^{3/2} - 1^{3/2})$ Apply Fundamental Theorem.

$\qquad = -2(8 - 1)$

$\qquad = -14$ Simplify.

✓ CHECKPOINT 4

Evaluate each definite integral.

a. $\displaystyle\int_0^1 e^{4x}\,dx$

b. $\displaystyle\int_2^5 -\frac{1}{x}\,dx$

STUDY TIP

In Example 4(c), note that the value of a definite integral can be negative.

Example 5 Interpreting Absolute Value

Evaluate $\displaystyle\int_0^2 |2x - 1|\,dx$.

SOLUTION The region represented by the definite integral is shown in Figure 6.10. From the definition of absolute value, you can write

$$|2x - 1| = \begin{cases} -(2x - 1), & x < \frac{1}{2} \\ 2x - 1, & x \geq \frac{1}{2} \end{cases}.$$

Using Property 3 of definite integrals, you can rewrite the integral as two definite integrals.

$\displaystyle\int_0^2 |2x - 1|\,dx = \int_0^{1/2} -(2x - 1)\,dx + \int_{1/2}^2 (2x - 1)\,dx$

$\qquad = \left[-x^2 + x\right]_0^{1/2} + \left[x^2 - x\right]_{1/2}^2$

$\qquad = \left(-\dfrac{1}{4} + \dfrac{1}{2}\right) - (0 + 0) + (4 - 2) - \left(\dfrac{1}{4} - \dfrac{1}{2}\right) = \dfrac{5}{2}$

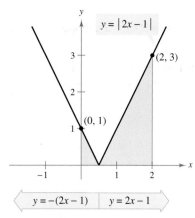

FIGURE 6.10

✓ CHECKPOINT 5

Evaluate $\displaystyle\int_0^5 |x - 2|\,dx$.

Definite Integrals and Net Change

Consider the height s of a free-falling object at time t on the closed interval $[a, b]$. The net change in the height of the object from $t = a$ to $t = b$ is given by

$$s(b) - s(a). \qquad \text{Net change in position}$$

Because s has a continuous derivative s' on $[a, b]$, you can use the Fundamental Theorem of Calculus to obtain

$$s(b) - s(a) = \int_a^b s'(t)\, dt. \qquad \text{Net change using a definite integral}$$

So, when you only know the velocity of the object at a time t, you can still determine the net change in the object's position. In general, if you know the rate of change of a quantity, then you can determine the net change in the quantity by using the **Net Change Formula.**

Net Change Formula

The net change in a function f on the interval $[a, b]$ is given by

$$f(b) - f(a) = \int_a^b f'(x)\, dx$$

where f' is continuous on $[a, b]$.

Example 6 Finding a Net Change

Neglecting air resistance, the velocity (in feet per second) of an object after it is dropped from a hovering helicopter can be modeled by

$$v(t) = -32t.$$

Find the net change in the object's position 2 seconds after it is dropped from the helicopter.

SOLUTION Because $v(t) = s'(t)$ is continuous on $[0, 2]$, you can integrate the velocity function to find the net change in the object's position as follows.

$$\begin{aligned}
s(b) - s(a) &= \int_a^b s'(t)\, dt & &\text{Net Change Formula} \\
&= \int_0^2 (-32t)\, dt & &\text{Substitute 0 for } a, 2 \text{ for } b, \text{ and } v(t) \text{ for } s'(t). \\
&= -16t^2 \Big]_0^2 & &\text{Find antiderivative.} \\
&= -16(4 - 0) & &\text{Apply Fundamental Theorem.} \\
&= -64 & &\text{Simplify.}
\end{aligned}$$

After 2 seconds, the net change in the object's position is -64 feet. This means the object is 64 feet below where it was dropped from the helicopter.

✓ CHECKPOINT 6

Find the net change in the object's position 3 seconds after it is dropped from the helicopter. ■

Average Value

The *average value* of a function on a closed interval is defined below.

> **Definition of the Average Value of a Function**
>
> If f is continuous on $[a, b]$, then the **average value** of f on $[a, b]$ is
>
> $$\text{Average value of } f \text{ on } [a, b] = \frac{1}{b-a} \int_a^b f(x)\, dx.$$

Example 7 — MAKE A DECISION: Finding an Average Value

The concentration y of a drug (in milligrams per milliliter) in a patient's bloodstream after t hours can be modeled by

$$y = 500e^{-0.4t}.$$

Determine the average concentration of the drug during the first 5 hours after the drug is administered. Will the average concentration be greater than 200 milligrams per milliliter?

SOLUTION The average concentration can be found by integrating y over the interval $[0, 5]$.

$$\begin{aligned}
\text{Average concentration} &= \frac{1}{5} \int_0^5 500e^{-0.4t}\, dt \\
&= 100 \int_0^5 e^{-0.4t}\, dt \\
&= 100 \left[-2.5 e^{-0.4t} \right]_0^5 \\
&= 216.2 \text{ milligrams per milliliter} \qquad \text{(See Figure 6.11)}
\end{aligned}$$

Yes, the average concentration of the drug in the patient's bloodstream will be greater than 200 milligrams per milliliter.

TECHNOLOGY

Symbolic integration utilities can be used to evaluate definite integrals as well as indefinite integrals. If you have access to such a program, try using it to evaluate several of the definite integrals in this section.

✓ CHECKPOINT 7

Use the model in Example 7 to determine the average concentration of the drug in the patient's bloodstream during the first 8 hours after the drug is administered. ■

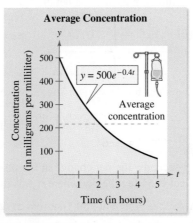

FIGURE 6.11

Even and Odd Functions

Several common functions have graphs that are symmetric with respect to the y-axis or the origin, as shown in Figure 6.12. If the graph of f is symmetric with respect to the y-axis, as in Figure 6.12(a), then

$$f(-x) = f(x) \qquad \text{Even function}$$

and f is called an **even** function. If the graph of f is symmetric with respect to the origin, as in Figure 6.12(b), then

$$f(-x) = -f(x) \qquad \text{Odd function}$$

and f is called an **odd** function.

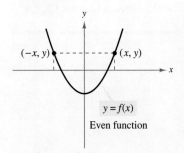

(a) y-axis symmetry

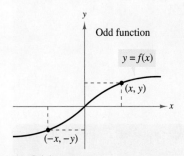

(b) Origin symmetry

FIGURE 6.12

Integration of Even and Odd Functions

1. If f is an *even* function, then $\int_{-a}^{a} f(x)\,dx = 2\int_{0}^{a} f(x)\,dx$.

2. If f is an *odd* function, then $\int_{-a}^{a} f(x)\,dx = 0$.

Example 8 Integrating Even and Odd Functions

a. Evaluate $\int_{-2}^{2} x^2\,dx$. **b.** Evaluate $\int_{-2}^{2} x^3\,dx$.

SOLUTION

a. Because $f(x) = x^2$ is even,

$$\int_{-2}^{2} x^2\,dx = 2\int_{0}^{2} x^2\,dx = 2\left[\frac{x^3}{3}\right]_{0}^{2} = 2\left(\frac{8}{3} - 0\right) = \frac{16}{3}.$$

b. Because $f(x) = x^3$ is odd,

$$\int_{-2}^{2} x^3\,dx = 0.$$

✓ CHECKPOINT 8

Evaluate each definite integral.

a. $\int_{-1}^{1} x^4\,dx$

b. $\int_{-1}^{1} x^5\,dx$

CONCEPT CHECK

1. Complete the following: The indefinite integral $\int f(x)\,dx$ denotes a family of _____, each of which is a(n) _____ of f, whereas the definite integral $\int_{a}^{b} f(x)\,dx$ is a _____.

2. If f is an odd function, then $\int_{-a}^{a} f(x)\,dx$ equals what?

3. State the Fundamental Theorem of Calculus.

4. Write the formula for finding the average value of a function f on [a, b] given that f is continuous on [a, b].

400 CHAPTER 6 Integration and Its Applications

Skills Review 6.4

The following warm-up exercises involve skills that were covered in earlier sections. You will use these skills in the exercise set for this section. For additional help, review Sections 1.2, 1.4, and 6.1–6.3.

In Exercises 1–4, sketch the graph of the function.

1. $f(x) = x - \dfrac{1}{2}$

2. $f(x) = x^2 + 3$

3. $f(x) = |x + 3|$

4. $f(x) = \dfrac{1}{\sqrt{x^2 + 1}}$

In Exercises 5–8, find the indefinite integral.

5. $\displaystyle\int (3x + 7)\, dx$

6. $\displaystyle\int \left(x^{3/2} + 2\sqrt{x}\right) dx$

7. $\displaystyle\int \dfrac{1}{5x}\, dx$

8. $\displaystyle\int e^{-6x}\, dx$

In Exercises 9 and 10, evaluate the expression when $a = 5$ and $b = 3$.

9. $\left(\dfrac{a}{5} - a\right) - \left(\dfrac{b}{5} - b\right)$

10. $\left(6a - \dfrac{a^3}{3}\right) - \left(6b - \dfrac{b^3}{3}\right)$

Exercises 6.4

See www.CalcChat.com for worked-out solutions to odd-numbered exercises.

 In Exercises 1 and 2, use a graphing utility to graph the integrand. Use the graph to determine whether the definite integral is positive, negative, or zero.

1. $\displaystyle\int_0^3 \dfrac{5x}{x^2 + 1}\, dx$

2. $\displaystyle\int_{-2}^{2} x\sqrt{x^2 + 1}\, dx$

In Exercises 3–12, sketch the region whose area is represented by the definite integral. Then use a geometric formula to evaluate the integral.

3. $\displaystyle\int_0^2 3\, dx$

4. $\displaystyle\int_0^3 4\, dx$

5. $\displaystyle\int_0^4 x\, dx$

6. $\displaystyle\int_0^4 \dfrac{x}{2}\, dx$

7. $\displaystyle\int_0^5 (x + 1)\, dx$

8. $\displaystyle\int_0^3 (2x + 1)\, dx$

9. $\displaystyle\int_{-2}^{3} |x - 1|\, dx$

10. $\displaystyle\int_{-1}^{4} |x - 2|\, dx$

11. $\displaystyle\int_{-3}^{3} \sqrt{9 - x^2}\, dx$

12. $\displaystyle\int_0^2 \sqrt{4 - x^2}\, dx$

In Exercises 13 and 14, use the values $\int_0^5 f(x)\, dx = 6$ and $\int_0^5 g(x)\, dx = 2$ to evaluate the definite integral.

13. (a) $\displaystyle\int_0^5 [f(x) + g(x)]\, dx$
    (b) $\displaystyle\int_0^5 [f(x) - g(x)]\, dx$
    (c) $\displaystyle\int_0^5 -4f(x)\, dx$
    (d) $\displaystyle\int_0^5 [f(x) - 3g(x)]\, dx$

14. (a) $\displaystyle\int_0^5 2g(x)\, dx$
    (b) $\displaystyle\int_5^0 f(x)\, dx$
    (c) $\displaystyle\int_5^5 f(x)\, dx$
    (d) $\displaystyle\int_0^5 [f(x) - f(x)]\, dx$

In Exercises 15–22, find the area of the region.

15. $y = x - x^2$

16. $y = 1 - x^4$

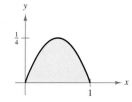
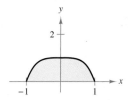

17. $y = \dfrac{1}{x^2}$

18. $y = \dfrac{2}{\sqrt{x}}$

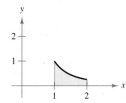
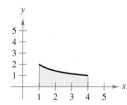

19. $y = 3e^{-x/2}$

20. $y = 2e^{x/2}$

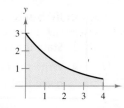
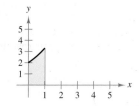

SECTION 6.4 Area and the Fundamental Theorem of Calculus

21. $y = \dfrac{x^2 + 4}{x}$ **22.** $y = \dfrac{x - 2}{x}$

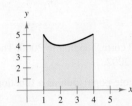
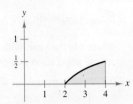

In Exercises 23–46, evaluate the definite integral.

23. $\displaystyle\int_0^1 2x\, dx$ **24.** $\displaystyle\int_2^7 3v\, dv$

25. $\displaystyle\int_{-1}^0 (x - 2)\, dx$ **26.** $\displaystyle\int_2^5 (-3x + 4)\, dx$

27. $\displaystyle\int_{-1}^1 (2t - 1)^2\, dt$ **28.** $\displaystyle\int_0^1 (1 - 2x)^2\, dx$

29. $\displaystyle\int_0^3 (x - 2)^3\, dx$ **30.** $\displaystyle\int_2^2 (x - 3)^4\, dx$

31. $\displaystyle\int_{-1}^1 (\sqrt[3]{t} - 2)\, dt$ **32.** $\displaystyle\int_1^4 \sqrt{\dfrac{2}{x}}\, dx$

33. $\displaystyle\int_1^4 \dfrac{u - 2}{\sqrt{u}}\, du$ **34.** $\displaystyle\int_0^1 \dfrac{x - \sqrt{x}}{3}\, dx$

35. $\displaystyle\int_{-1}^0 (t^{1/3} - t^{2/3})\, dt$ **36.** $\displaystyle\int_0^4 (x^{1/2} + x^{1/4})\, dx$

37. $\displaystyle\int_0^4 \dfrac{1}{\sqrt{2x + 1}}\, dx$ **38.** $\displaystyle\int_0^2 \dfrac{x}{\sqrt{1 + 2x^2}}\, dx$

39. $\displaystyle\int_0^1 e^{-2x}\, dx$ **40.** $\displaystyle\int_1^2 e^{1-x}\, dx$

41. $\displaystyle\int_1^3 \dfrac{e^{3/x}}{x^2}\, dx$ **42.** $\displaystyle\int_{-1}^1 (e^x - e^{-x})\, dx$

43. $\displaystyle\int_0^1 e^{2x}\sqrt{e^{2x} + 1}\, dx$ **44.** $\displaystyle\int_0^1 \dfrac{e^{-x}}{\sqrt{e^{-x} + 1}}\, dx$

45. $\displaystyle\int_0^2 \dfrac{x}{1 + 4x^2}\, dx$ **46.** $\displaystyle\int_0^1 \dfrac{e^{2x}}{e^{2x} + 1}\, dx$

In Exercises 47–50, evaluate the definite integral by the most convenient method. Explain your approach.

47. $\displaystyle\int_{-1}^1 |4x|\, dx$ **48.** $\displaystyle\int_0^3 |2x - 3|\, dx$

49. $\displaystyle\int_0^4 (2 - |x - 2|)\, dx$ **50.** $\displaystyle\int_{-4}^4 (4 - |x|)\, dx$

 In Exercises 51–54, evaluate the definite integral by hand. Then use a symbolic integration utility to evaluate the definite integral. Briefly explain any differences in your results.

51. $\displaystyle\int_{-1}^2 \dfrac{x}{x^2 - 9}\, dx$ **52.** $\displaystyle\int_2^3 \dfrac{x + 1}{x^2 + 2x - 3}\, dx$

53. $\displaystyle\int_0^3 \dfrac{2e^x}{2 + e^x}\, dx$ **54.** $\displaystyle\int_1^2 \dfrac{(2 + \ln x)^3}{x}\, dx$

 In Exercises 55–60, evaluate the definite integral by hand. Then use a graphing utility to graph the region whose area is represented by the integral.

55. $\displaystyle\int_1^3 (4x - 3)\, dx$ **56.** $\displaystyle\int_0^2 (x + 4)\, dx$

57. $\displaystyle\int_0^1 (x - x^3)\, dx$ **58.** $\displaystyle\int_0^2 (2 - x)\sqrt{x}\, dx$

59. $\displaystyle\int_2^4 \dfrac{3x^2}{x^3 - 1}\, dx$ **60.** $\displaystyle\int_0^{\ln 6} \dfrac{e^x}{2}\, dx$

In Exercises 61–64, find the area of the region bounded by the graphs of the equations. Use a graphing utility to verify your results.

61. $y = 3x^2 + 1,\quad y = 0,\quad x = 0,\quad$ and $\quad x = 2$
62. $y = 1 + \sqrt{x},\quad y = 0,\quad x = 0,\quad$ and $\quad x = 4$
63. $y = 4/x\quad y = 0,\quad x = 1,\quad$ and $\quad x = 3$
64. $y = e^x,\quad y = 0,\quad x = 0,\quad$ and $\quad x = 2$

 In Exercises 65–72, use a graphing utility to graph the function over the interval. Find the average value of the function over the interval. Then find all x-values in the interval for which the function is equal to its average value.

Function	Interval
65. $f(x) = 4 - x^2$	$[-2, 2]$
66. $f(x) = x - 2\sqrt{x}$	$[0, 4]$
67. $f(x) = 2e^x$	$[-1, 1]$
68. $f(x) = e^{x/4}$	$[0, 4]$
69. $f(x) = x\sqrt{4 - x^2}$	$[0, 2]$
70. $f(x) = \dfrac{1}{(x - 3)^2}$	$[0, 2]$
71. $f(x) = \dfrac{6x}{x^2 + 1}$	$[0, 7]$
72. $f(x) = \dfrac{4x}{x^2 + 1}$	$[0, 1]$

In Exercises 73–76, state whether the function is even, odd, or neither.

73. $f(x) = 3x^4$ **74.** $g(x) = x^3 - 2x$

75. $g(t) = 2t^5 - 3t^2$ **76.** $f(t) = 5t^4 + 1$

77. Use the value $\displaystyle\int_0^1 x^2\, dx = \tfrac{1}{3}$ to evaluate each definite integral. Explain your reasoning.

(a) $\displaystyle\int_{-1}^0 x^2\, dx$ (b) $\displaystyle\int_{-1}^1 x^2\, dx$ (c) $\displaystyle\int_0^1 -x^2\, dx$

78. Use the value $\int_0^2 x^3\,dx = 4$ to evaluate each definite integral. Explain your reasoning.

(a) $\displaystyle\int_{-2}^0 x^3\,dx$ (b) $\displaystyle\int_{-2}^2 x^3\,dx$ (c) $\displaystyle\int_0^2 3x^3\,dx$

79. Number of Farms The rate of change in the number of farms F (in thousands) in the United States from 2000 through 2005 can be modeled by $dF/dt = 1.08t - 15.4$, where t is the year, with $t = 0$ corresponding to 2000. Find the net change in the number of farms from 2000 to 2005. *(Source: U.S. Department of Agriculture)*

80. Volunteers The rate of change in the number of people V (in thousands) who did volunteer work in the United States from 2000 through 2006 can be modeled by

$$\frac{dV}{dt} = 119.85t^2 - 30.0e^t + 37{,}261e^{-t}$$

where t is the year, with $t = 2$ corresponding to 2002. Find the net change in the number of volunteers from 2002 to 2006. *(Source: U.S. Bureau of Labor Statistics)*

81. Medical Equipment The total cost of purchasing and maintaining a piece of medical equipment for x years can be modeled by

$$C = 5000\left(25 + 3\int_0^x t^{1/4}\,dt\right).$$

Find the total cost after (a) 1 year, (b) 5 years, and (c) 10 years.

82. Depreciation A hospital purchases a new medical instrument for which the rate of depreciation can be modeled by

$$\frac{dV}{dt} = 10{,}000(t - 6), \quad 0 \le t \le 5$$

where V is the value of the instrument after t years. Set up and evaluate the definite integral that yields the total loss of value of the instrument over the first 3 years.

83. Blood Flow The velocity v of the flow of blood at a distance r from the center of an artery of radius R can be modeled by $v = k(R^2 - r^2)$, $k > 0$, where k is a constant. Find the average velocity along a radius of the artery. (Use 0 and R as the limits of integration.)

84. Biology In the North Sea, codfish are in danger of becoming extinct because a large proportion of the catch is being taken before the cod can reach breeding age. The fishing quotas set in the United Kingdom for the years 1999 through 2006 can be approximated by the equation

$$y = -0.7020t^3 + 29.802t^2 - 422.77t + 2032.9$$

where y is the total catch weight (in thousands of kilograms) and t is the year, with $t = 9$ corresponding to 1999. Determine the average recommended quota during the years 1995 through 2006. *(Source: International Council for Exploration of the Sea)*

85. Coyote Population The rate of change in the coyote population C in a wilderness area can be modeled by

$$\frac{dC}{dt} = 147e^{-0.42t}$$

where t is the year, with $t = 0$ corresponding to 2000. The coyote population was 500 in 2002.

(a) Write a model for the coyote population as a function of t.

(b) Find the average coyote population from 2000 through 2005.

86. Drug Concentration The graph shows the concentration C (in milligrams per milliliter) of a drug in a patient's bloodstream t hours after the drug is administered orally. The part of the graph that represents the drug entering the bloodstream is called the *absorption step*. During the absorption step, the concentration increases to a maximum value.

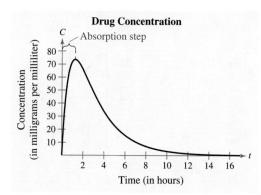

A model for the concentration is

$$C = -170(e^{-1.4t} - e^{-0.4t}).$$

(Source: Bioanalytical Systems, Inc.)

(a) Find the duration of the absorption step and the maximum concentration of the drug.

(b) Estimate the area of the region bounded by the graph of C and the t-axis by evaluating the definite integral

$$C = \int_0^{16} -170(e^{-1.4t} - e^{-0.4t})\,dt.$$

(c) Discuss how the results of parts (a) and (b) might be helpful to pharmaceutical companies or people working in the medical profession.

In Exercises 87–90, use a symbolic integration utility to evaluate the definite integral.

87. $\displaystyle\int_3^6 \frac{x}{3\sqrt{x^2-8}}\,dx$ **88.** $\displaystyle\int_{1/2}^1 (x+1)\sqrt{1-x}\,dx$

89. $\displaystyle\int_2^5 \left(\frac{1}{x^2} - \frac{1}{x^3}\right)dx$ **90.** $\displaystyle\int_0^1 x^3(x^3+1)^3\,dx$

Section 6.5
The Area of a Region Bounded by Two Graphs

- Find the areas of regions bounded by two graphs.
- Use the areas of regions bounded by two graphs to solve real-life problems.

Area of a Region Bounded by Two Graphs

With a few modifications, you can extend the use of definite integrals from finding the area of a region *under a graph* to finding the area of a region *bounded by two graphs*. To see how this is done, consider the region bounded by the graphs of f, g, $x = a$, and $x = b$, as shown in Figure 6.13. If the graphs of both f and g lie above the x-axis, then you can interpret the area of the region between the graphs as the area of the region under the graph of g subtracted from the area of the region under the graph of f, as shown in Figure 6.13.

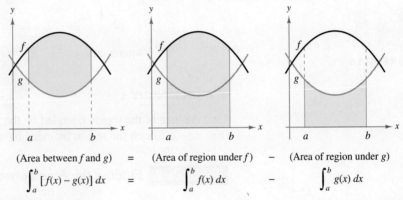

(Area between f and g) = (Area of region under f) − (Area of region under g)

$$\int_a^b [f(x) - g(x)]\, dx = \int_a^b f(x)\, dx - \int_a^b g(x)\, dx$$

FIGURE 6.13

Although Figure 6.13 depicts the graphs of f and g lying above the x-axis, this is not necessary, and the same integrand $[f(x) - g(x)]$ can be used as long as both functions are continuous and $g(x) \leq f(x)$ on the interval $[a, b]$.

Area of a Region Bounded by Two Graphs

If f and g are continuous on $[a, b]$ and $g(x) \leq f(x)$ for all x in the interval, then the area of the region bounded by the graphs of f, g, $x = a$, and $x = b$ is given by

$$A = \int_a^b [f(x) - g(x)]\, dx.$$

DISCOVERY

Sketch the graph of $f(x) = x^3 - 4x$ and shade in the regions bounded by the graph of f and the x-axis. Write the appropriate integral(s) for this area.

Example 1 Finding the Area Bounded by Two Graphs

Find the area of the region bounded by the graphs of

$$y = x^2 + 2 \quad \text{and} \quad y = x$$

for $0 \leq x \leq 1$.

SOLUTION Begin by sketching the graphs of both functions, as shown in Figure 6.14. From the figure, you can see that $x \leq x^2 + 2$ for all x in $[0, 1]$. So, you can let $f(x) = x^2 + 2$ and $g(x) = x$. Then compute the area as shown.

$$\text{Area} = \int_a^b [f(x) - g(x)] \, dx \qquad \text{Area between } f \text{ and } g$$

$$= \int_0^1 [(x^2 + 2) - (x)] \, dx \qquad \text{Substitute for } a, b, f, \text{ and } g.$$

$$= \int_0^1 (x^2 - x + 2) \, dx$$

$$= \left[\frac{x^3}{3} - \frac{x^2}{2} + 2x \right]_0^1 \qquad \text{Find antiderivative.}$$

$$= \frac{11}{6} \text{ square units} \qquad \text{Apply Fundamental Theorem.}$$

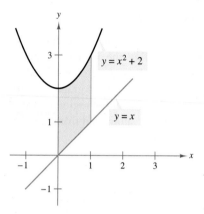

FIGURE 6.14

✓ CHECKPOINT 1

Find the area of the region bounded by the graphs of $y = x^2 + 1$ and $y = x$ for $0 \leq x \leq 2$. Sketch the region bounded by the graphs. ■

Example 2 Finding the Area Between Intersecting Graphs

Find the area of the region bounded by the graphs of

$$y = 2 - x^2 \quad \text{and} \quad y = x.$$

SOLUTION In this problem, the values of a and b are not given and you must compute them by finding the points of intersection of the two graphs. To do this, equate the two functions and solve for x. When you do this, you will obtain $x = -2$ and $x = 1$. In Figure 6.15, you can see that the graph of $f(x) = 2 - x^2$ lies above the graph of $g(x) = x$ for all x in the interval $[-2, 1]$.

$$\text{Area} = \int_a^b [f(x) - g(x)] \, dx \qquad \text{Area between } f \text{ and } g$$

$$= \int_{-2}^1 [(2 - x^2) - (x)] \, dx \qquad \text{Substitute for } a, b, f, \text{ and } g.$$

$$= \int_{-2}^1 (-x^2 - x + 2) \, dx$$

$$= \left[-\frac{x^3}{3} - \frac{x^2}{2} + 2x \right]_{-2}^1 \qquad \text{Find antiderivative.}$$

$$= \frac{9}{2} \text{ square units} \qquad \text{Apply Fundamental Theorem.}$$

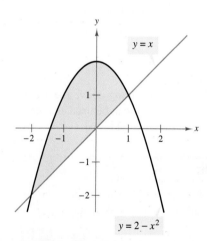

FIGURE 6.15

✓ CHECKPOINT 2

Find the area of the region bounded by the graphs of $y = 3 - x^2$ and $y = 2x$. ■

SECTION 6.5 The Area of a Region Bounded by Two Graphs

Example 3 Finding an Area Below the x-Axis

Find the area of the region bounded by the graph of

$$y = x^2 - 3x - 4$$

and the x-axis.

SOLUTION Begin by finding the x-intercepts of the graph. To do this, set the function equal to zero and solve for x.

$x^2 - 3x - 4 = 0$ Set function equal to 0.
$(x - 4)(x + 1) = 0$ Factor.
$x = 4, x = -1$ Solve for x.

From Figure 6.16, you can see that $x^2 - 3x - 4 \leq 0$ for all x in the interval $[-1, 4]$. So, you can let $f(x) = 0$ and $g(x) = x^2 - 3x - 4$, and compute the area as shown.

$$\text{Area} = \int_a^b [f(x) - g(x)]\, dx \quad \text{Area between } f \text{ and } g$$

$$= \int_{-1}^4 [(0) - (x^2 - 3x - 4)]\, dx \quad \text{Substitute for } a, b, f, \text{ and } g.$$

$$= \int_{-1}^4 (-x^2 + 3x + 4)\, dx$$

$$= \left[-\frac{x^3}{3} + \frac{3x^2}{2} + 4x \right]_{-1}^4 \quad \text{Find antiderivative.}$$

$$= \frac{125}{6} \text{ square units} \quad \text{Apply Fundamental Theorem.}$$

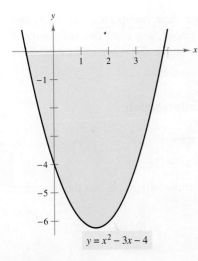

FIGURE 6.16

STUDY TIP

When finding the area of a region bounded by two graphs, be sure to use the integrand $[f(x) - g(x)]$. Be sure you realize that you cannot interchange $f(x)$ and $g(x)$. For instance, when solving Example 3, if you subtract $f(x)$ from $g(x)$, you will obtain an answer of $-\frac{125}{6}$, which is not correct.

✓ CHECKPOINT 3

Find the area of the region bounded by the graph of $y = x^2 - x - 2$ and the x-axis. ■

TECHNOLOGY

Most graphing utilities can display regions that are bounded by two graphs. For instance, to graph the region in Example 3, set the viewing window to $-1 \leq x \leq 4$ and $-7 \leq y \leq 1$. Consult your user's manual for specific keystrokes that are used to shade the graph. You should obtain the graph shown at the right.*

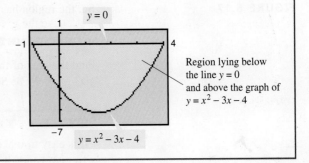

Region lying below the line $y = 0$ and above the graph of $y = x^2 - 3x - 4$

---

*Specific calculator keystroke instructions for operations in this and other technology boxes can be found at *college.hmco.com/info/larsonapplied*.

Sometimes two graphs intersect at more than two points. To determine the area of the region bounded by two such graphs, you must find *all* points of intersection and check to see which graph is above the other in each interval determined by the points.

Example 4 Using Multiple Points of Intersection

Find the area of the region bounded by the graphs of

$$f(x) = 3x^3 - x^2 - 10x \quad \text{and} \quad g(x) = -x^2 + 2x.$$

SOLUTION To find the points of intersection of the two graphs, set the functions equal to each other and solve for x.

$f(x) = g(x)$	Set $f(x)$ equal to $g(x)$.
$3x^3 - x^2 - 10x = -x^2 + 2x$	Substitute for $f(x)$ and $g(x)$.
$3x^3 - 12x = 0$	Write in general form.
$3x(x^2 - 4) = 0$	
$3x(x - 2)(x + 2) = 0$	Factor.
$x = 0, x = 2, x = -2$	Solve for x.

These three points of intersection determine two intervals of integration: $[-2, 0]$ and $[0, 2]$. In Figure 6.17, you can see that $g(x) \le f(x)$ in the interval $[-2, 0]$, and that $f(x) \le g(x)$ in the interval $[0, 2]$. So, you must use two integrals to determine the area of the region bounded by the graphs of f and g: one for the interval $[-2, 0]$ and one for the interval $[0, 2]$.

$$\begin{aligned}
\text{Area} &= \int_{-2}^{0} [f(x) - g(x)]\, dx + \int_{0}^{2} [g(x) - f(x)]\, dx \\
&= \int_{-2}^{0} (3x^3 - 12x)\, dx + \int_{0}^{2} (-3x^3 + 12x)\, dx \\
&= \left[\frac{3x^4}{4} - 6x^2 \right]_{-2}^{0} + \left[-\frac{3x^4}{4} + 6x^2 \right]_{0}^{2} \\
&= (0 - 0) - (12 - 24) + (-12 + 24) - (0 + 0) \\
&= 24
\end{aligned}$$

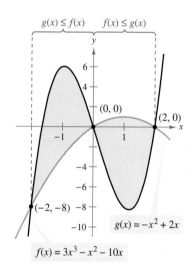

FIGURE 6.17

So, the region has an area of 24 square units.

✓ CHECKPOINT 4

Find the area of the region bounded by the graphs of $f(x) = x^3 + 2x^2 - 3x$ and $g(x) = x^2 + 3x$. Sketch a graph of the region. ∎

STUDY TIP

It is easy to make an error when calculating areas such as that in Example 4. To give yourself some idea about the reasonableness of your solution, you could make a careful sketch of the region on graph paper and then use the grid on the graph paper to approximate the area. Try doing this with the graph shown in Figure 6.17. Is your approximation close to 24 square units?

SECTION 6.5 The Area of a Region Bounded by Two Graphs

Application

There are many types of applications involving the area of a region bounded by two graphs. Example 5 shows one of these applications.

Example 5 Number of Households

For the years 1995 through 2005, the number of households N (in thousands) in the United States can be modeled by

$$N_1 = 18.32t^2 + 1178.3t + 92{,}099, \quad 5 \le t \le 15$$

where t is the year, with $t = 5$ corresponding to 1995. Before 1995, the U.S. Census Bureau used different models to project the number of households in the United States. One series of projections was given by the model

$$N_2 = 1.35t^2 + 1078.4t + 92{,}323, \quad 5 \le t \le 15.$$

This projection by the U.S. Census Bureau underestimated the number of households from 1995 through 2005. Find the number of households the projection underestimated. *(Source: U.S. Census Bureau)*

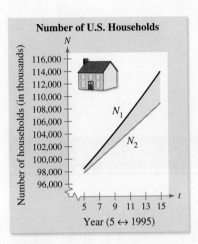

FIGURE 6.18

SOLUTION The difference in the actual number of households and the projected number of households can be represented as the area of the region between the graphs of N_1 and N_2, as shown in Figure 6.18.

$$\begin{aligned}
\text{Difference in households} &= \int_5^{15} (N_1 - N_2)\, dt \\
&= \int_5^{15} (16.97t^2 + 99.9t - 224)\, dt \\
&= \left[\frac{16.97}{3} t^3 + \frac{99.9}{2} t^2 - 224t \right]_5^{15} \\
&\approx 26{,}134.
\end{aligned}$$

The projection underestimated the number of households by about 26,134 thousand, or 26,134,000.

✓ CHECKPOINT 5

Repeat Example 5 for a different projection of the number of households using the model

$$N_2 = 0.32t^2 + 1021.3t + 92{,}486, \quad 5 \le t \le 15. \ \blacksquare$$

CONCEPT CHECK

1. When finding the area of a region bounded by two graphs, you use the integrand $[f(x) - g(x)]$. Identify what f and g represent.
2. Complete the following: To determine the area of the region bounded by two graphs that intersect at two or more points, begin by finding _____ points of intersection. Then check to see which graph is _____ the other in each interval determined by the points.
3. Consider the functions f and g, where f and g are continuous on $[a, b]$ and $g(x) \le f(x)$ for all x in the interval. How can you find the area of the region bounded by the graphs of f, g, $x = a$, and $x = b$?
4. The graphs of f and g lie below the x-axis, both functions are continuous, and $g(x) \le f(x)$ on $[a, b]$. For the area of the region bounded by the graphs of f and g, can you use the integrand $[f(x) - g(x)]$?

408 CHAPTER 6 Integration and Its Applications

Skills Review 6.5

The following warm-up exercises involve skills that were covered in earlier sections. You will use these skills in the exercise set for this section. For additional help, review Sections 1.1 and 1.2.

In Exercises 1–4, simplify the expression.

1. $(-x^2 + 4x + 3) - (x + 1)$
2. $(-2x^2 + 3x + 9) - (-x + 5)$
3. $(-x^3 + 3x^2 - 1) - (x^2 - 4x + 4)$
4. $(3x + 1) - (-x^3 + 9x + 2)$

In Exercises 5–10, find the points of intersection of the graphs.

5. $f(x) = x^2 - 4x + 4$, $g(x) = 4$
6. $f(x) = -3x^2$, $g(x) = 6 - 9x$
7. $f(x) = x^2$, $g(x) = -x + 6$
8. $f(x) = \frac{1}{2}x^3$, $g(x) = 2x$
9. $f(x) = x^2 - 3x$, $g(x) = 3x - 5$
10. $f(x) = e^x$, $g(x) = e$

Exercises 6.5

See www.CalcChat.com for worked-out solutions to odd-numbered exercises.

In Exercises 1–8, find the area of the region.

1. $f(x) = x^2 - 6x$
   $g(x) = 0$

2. $f(x) = x^2 + 2x + 1$
   $g(x) = 2x + 5$

7. $f(x) = e^x - 1$
   $g(x) = 0$

8. $f(x) = -x + 3$
   $g(x) = 2x^{-1}$

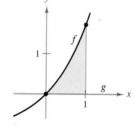

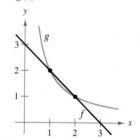

3. $f(x) = x^2 - 4x + 3$
   $g(x) = -x^2 + 2x + 3$

4. $f(x) = x^2$
   $g(x) = x^3$

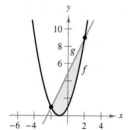

In Exercises 9–14, the integrand of the definite integral is a difference of two functions. Sketch the graph of each function and shade the region whose area is represented by the integral.

9. $\int_0^4 [(x + 1) - \frac{1}{2}x]\, dx$

10. $\int_{-1}^1 [(1 - x^2) - (x^2 - 1)]\, dx$

11. $\int_{-2}^2 [2x^2 - (x^4 - 2x^2)]\, dx$

12. $\int_{-4}^0 [(x - 6) - (x^2 + 5x - 6)]\, dx$

13. $\int_{-1}^2 [(y^2 + 2) - 1]\, dy$

14. $\int_{-2}^3 [(y + 6) - y^2]\, dy$

5. $f(x) = 3(x^3 - x)$
   $g(x) = 0$

6. $f(x) = (x - 1)^3$
   $g(x) = x - 1$

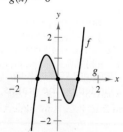

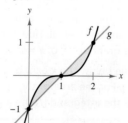

SECTION 6.5 The Area of a Region Bounded by Two Graphs

Think About It In Exercises 15 and 16, determine which value best approximates the area of the region bounded by the graphs of f and g. (Make your selection on the basis of a sketch of the region and not by performing any calculations.)

15. $f(x) = x + 1$, $g(x) = (x - 1)^2$
    (a) -2 (b) 2 (c) 10 (d) 4 (e) 8

16. $f(x) = 2 - \frac{1}{2}x$, $g(x) = 2 - \sqrt{x}$
    (a) 1 (b) 6 (c) -3 (d) 3 (e) 4

In Exercises 17–32, sketch the region bounded by the graphs of the functions and find the area of the region.

17. $y = \frac{1}{x^2}$, $y = 0$, $x = 1$, $x = 5$
18. $y = x^3 - 2x + 1$, $y = -2x$, $x = 1$
19. $f(x) = \sqrt[3]{x}$, $g(x) = x$
20. $f(x) = \sqrt{3x} + 1$, $g(x) = x + 1$
21. $y = x^2 - 4x + 3$, $y = 3 + 4x - x^2$
22. $y = 4 - x^2$, $y = x^2$
23. $y = xe^{-x^2}$, $y = 0$, $x = 0$, $x = 1$
24. $y = \frac{e^{1/x}}{x^2}$, $y = 0$, $x = 1$, $x = 3$
25. $y = \frac{8}{x}$, $y = x^2$, $y = 0$, $x = 1$, $x = 4$
26. $y = \frac{1}{x}$, $y = x^3$, $x = \frac{1}{2}$, $x = 1$
27. $f(x) = e^{0.5x}$, $g(x) = -\frac{1}{x}$, $x = 1$, $x = 2$
28. $f(x) = \frac{1}{x}$, $g(x) = -e^x$, $x = \frac{1}{2}$, $x = 1$
29. $f(y) = y^2$, $g(y) = y + 2$
30. $f(y) = y(2 - y)$, $g(y) = -y$
31. $f(y) = \sqrt{y}$, $y = 9$, $x = 0$
32. $f(y) = y^2 + 1$, $g(y) = 4 - 2y$

In Exercises 33–36, use a graphing utility to graph the region bounded by the graphs of the functions. Write the definite integral that represents the area of the region. (*Hint:* Multiple integrals may be necessary.)

33. $f(x) = 2x$, $g(x) = 4 - 2x$, $h(x) = 0$
34. $f(x) = x(x^2 - 3x + 3)$, $g(x) = x^2$
35. $y = \frac{4}{x}$, $y = x$, $x = 1$, $x = 4$
36. $y = x^3 - 4x^2 + 1$, $y = x - 3$

In Exercises 37–40, use a graphing utility to graph the region bounded by the graphs of the functions, and find the area of the region.

37. $f(x) = x^2 - 4x$, $g(x) = 0$
38. $f(x) = 3 - 2x - x^2$, $g(x) = 0$
39. $f(x) = x^2 + 2x + 1$, $g(x) = x + 1$
40. $f(x) = -x^2 + 4x + 2$, $g(x) = x + 2$

In Exercises 41 and 42, use integration to find the area of the triangular region having the given vertices.

41. $(0, 0)$, $(4, 0)$, $(4, 4)$ 42. $(0, 0)$, $(4, 0)$, $(6, 4)$

43. **MAKE A DECISION: JOB OFFERS** A college graduate has two job offers. The starting salary for each is $32,000, and after 8 years of service each will pay $54,000. The salary increase for each offer is shown in the figure. From a strictly monetary viewpoint, which is the better offer? Explain.

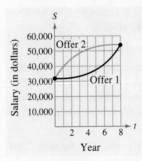
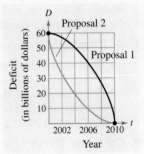

Figure for 43   Figure for 44

44. **MAKE A DECISION: BUDGET DEFICITS** A state legislature is debating two proposals for eliminating the annual budget deficits by the year 2010. The rate of decrease of the deficits for each proposal is shown in the figure. From the viewpoint of minimizing the cumulative state deficit, which is the better proposal? Explain.

45. **Fuel Cost** The projected fuel cost C (in millions of dollars per year) for an airline company from 2007 through 2013 is $C_1 = 568.5 + 7.15t$, where $t = 7$ corresponds to 2007. If the company purchases more efficient airplane engines, fuel cost is expected to decrease and to follow the model $C_2 = 525.6 + 6.43t$. How much can the company save with the more efficient engines? Explain your reasoning.

46. **Health** An epidemic was spreading such that t weeks after its outbreak it had infected

$$N_1(t) = 0.1t^2 + 0.5t + 150, \quad 0 \le t \le 50$$

people. Twenty-five weeks after the outbreak, a vaccine was developed and administered to the public. At that point, the number of people infected was governed by the model $N_2(t) = -0.2t^2 + 6t + 200$. Approximate the number of people that the vaccine prevented from becoming ill during the epidemic.

47. National Health Expenditures For the years 2000 through 2004, the total national health expenditures N (in billions of dollars) in the United States can be modeled by

$$N_1 = 3.07t^2 + 118.2t + 1357, \quad 0 \le t \le 4$$

where t is the year, with $t = 0$ corresponding to 2000. In 1999, the projected total national health expenditures from 2000 through 2004 were given by the model

$$N_2 = 2t^2 + 112t + 1311, \quad 0 \le t \le 4.$$

This projection underestimated the total national health expenditures from 2000 through 2004. Find the amount the projection underestimated. *(Source: U.S. Centers for Medicare and Medicaid Services)*

48. Consumer Trends For the years 1996 through 2004, the per capita consumption of fresh pineapples (in pounds per year) in the United States can be modeled by

$$C(t) = \begin{cases} -0.046t^2 + 1.07t - 2.9, & 6 \le t \le 10 \\ -0.164t^2 + 4.53t - 26.8, & 10 < t \le 14 \end{cases}$$

where t is the year, with $t = 6$ corresponding to 1996. *(Source: U.S. Department of Agriculture)*

(T) (a) Use a graphing utility to graph this model.

(b) Suppose the fresh pineapple consumption from 2001 through 2004 had continued to follow the model for 1996 through 2000. How many more or fewer pounds of fresh pineapples would have been consumed from 2001 through 2004?

49. Consumer Trends For the years 1995 through 2004, the production of beef (in billions of pounds) in the United States can be modeled by

$$B(t) = \begin{cases} -0.02708t^4 + 0.8153t^3 - 8.96t^2 + 42.85t - 50, \\ \quad 5 \le t < 10 \\ 0.2t^4 - 9.8t^3 + 178.85t^2 - 1440.95t + 4351.4, \\ \quad 10 \le t \le 14 \end{cases}$$

where t is the year, with $t = 5$ corresponding to 1995. *(Source: U.S. Department of Agriculture)*

(T) (a) Use a graphing utility to graph this model.

(b) Suppose the production of beef from 2000 through 2001 had continued to follow the model for 1995 through 1999. How many more or fewer pounds of beef would have been produced from 2000 through 2001?

50. Lorenz Curve Economists use *Lorenz curves* to illustrate the distribution of income in a country. Letting x represent the percent of families in a country and y the percent of total income, the model $y = x$ would represent a country in which each family had the same income. The Lorenz curve, $y = f(x)$, represents the actual income distribution. The area between these two models indicates the "income inequality" of a country.

In 2005, the Lorenz curve for the United States could be modeled by

$$y = (0.00061x^2 + 0.0218x + 1.723)^2, \quad 0 \le x \le 100$$

where x is measured from the poorest to the wealthiest families. Find the income inequality for the United States in 2005. *(Source: U.S. Census Bureau)*

51. Income Distribution Using the Lorenz curve in Exercise 50 and a spreadsheet, complete the table, which lists the percent of total income earned by each quintile in the United States in 2005.

Quintile	Lowest	2nd	3rd	4th	Highest
Percent					

52. Extended Application To work an extended application analyzing the receipts and expenditures for the Old-Age and Survivors Insurance Trust Fund (Social Security Trust Fund) from 1990 through 2005, visit this text's website at *college.hmco.com/info/larsonapplied*. *(Data Source: Social Security Administration)*

Social Science Capsule

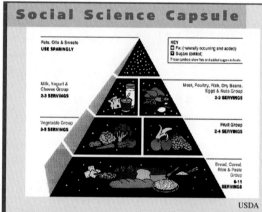

The United States Department of Agriculture (USDA) promotes agricultural trade and production and works to ensure food safety in America and abroad. The Economic Research Service (ERS) provides the USDA with information regarding but not limited to farming practices, food safety, nutrition programs, natural resources, and the rural economy.

53. Research Project Use your school's library, (T) the Internet, or some other reference source to research projected and actual statistical data published by the USDA over a period of time. Use a graphing utility to graph a scatter plot of the data. Fit models to the data. Determine whether the projection underestimated or overestimated the actual data for that period of time. Find the amount the projection underestimated or overestimated.

Section 6.6
Volumes of Solids of Revolution

- Use the Disk Method to find volumes of solids of revolution.
- Use the Washer Method to find volumes of solids of revolution with holes.
- Use solids of revolution to solve real-life problems.

The Disk Method

Another important application of the definite integral is its use in finding the volume of a three-dimensional solid. In this section, you will study a particular type of three-dimensional solid—one whose cross sections are similar. You will begin with solids of revolution. Some samples of these solids are axles, funnels, pills, bottles, and pistons.

As shown in Figure 6.19, a **solid of revolution** is formed by revolving a plane region about a line. The line is called the **axis of revolution.**

To develop a formula for finding the volume of a solid of revolution, consider a continuous function f that is nonnegative on the interval $[a, b]$. Suppose that the area of the region is approximated by n rectangles, each of width Δx, as shown in Figure 6.20. By revolving the rectangles about the x-axis, you obtain n circular disks, each with a volume of $\pi[f(x_i)]^2 \Delta x$. The volume of the solid formed by revolving the region about the x-axis is approximately equal to the sum of the volumes of the n disks. Moreover, by taking the limit as n approaches infinity, you can see that the exact volume is given by a definite integral. This result is called the **Disk Method.**

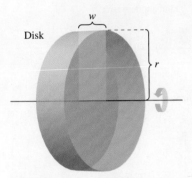

FIGURE 6.19 Volume: $\pi r^2 w$

The Disk Method

The volume of the solid formed by revolving the region bounded by the graph of f and the x-axis ($a \leq x \leq b$) about the x-axis is

$$\text{Volume} = \pi \int_a^b [f(x)]^2 \, dx.$$

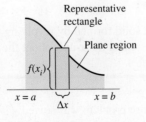
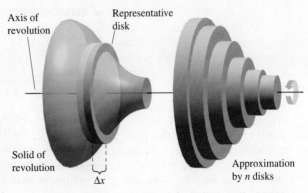

FIGURE 6.20 Disk Method

Example 1 Finding the Volume of a Solid of Revolution

Find the volume of the solid formed by revolving the region bounded by the graph of $f(x) = -x^2 + x$ and the x-axis about the x-axis.

SOLUTION Begin by sketching the region bounded by the graph of f and the x-axis. As shown in Figure 6.21(a), sketch a representative rectangle whose height is $f(x)$ and whose width is Δx. From this rectangle, you can see that the radius of the solid is

$$\text{Radius} = f(x) = -x^2 + x.$$

Using the Disk Method, you can find the volume of the solid of revolution.

$$\begin{aligned}
\text{Volume} &= \pi \int_0^1 [f(x)]^2 \, dx && \text{Disk Method} \\
&= \pi \int_0^1 (-x^2 + x)^2 \, dx && \text{Substitute for } f(x). \\
&= \pi \int_0^1 (x^4 - 2x^3 + x^2) \, dx && \text{Expand integrand.} \\
&= \pi \left[\frac{x^5}{5} - \frac{x^4}{2} + \frac{x^3}{3} \right]_0^1 && \text{Find antiderivative.} \\
&= \frac{\pi}{30} && \text{Apply Fundamental Theorem.} \\
&\approx 0.105 && \text{Round to three decimal places.}
\end{aligned}$$

So, the volume of the solid is about 0.105 cubic unit.

TECHNOLOGY

Try using the integration capabilities of a graphing utility to verify the solution in Example 1. Consult your user's manual for specific keystrokes.

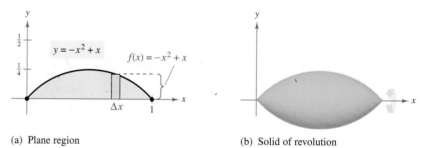

(a) Plane region

(b) Solid of revolution

FIGURE 6.21

✓ CHECKPOINT 1

Find the volume of the solid formed by revolving the region bounded by the graph of $f(x) = -x^2 + 4$ and the x-axis about the x-axis. ■

STUDY TIP

In Example 1, the entire problem was solved *without* referring to the three-dimensional sketch given in Figure 6.21(b). In general, to set up the integral for calculating the volume of a solid of revolution, a sketch of the plane region is more useful than a sketch of the solid, because the radius is more readily visualized in the plane region.

The Washer Method

You can extend the Disk Method to find the volume of a solid of revolution with a *hole*. Consider a region that is bounded by the graphs of f and g, as shown in Figure 6.22(a). If the region is revolved about the x-axis, then the volume of the resulting solid can be found by applying the Disk Method to f and g and subtracting the results.

$$\text{Volume} = \pi \int_a^b [f(x)]^2 \, dx - \pi \int_a^b [g(x)]^2 \, dx$$

Writing this as a single integral produces the **Washer Method.**

The Washer Method

Let f and g be continuous and nonnegative on the closed interval $[a, b]$, as shown in Figure 6.22(a). If $g(x) \leq f(x)$ for all x in the interval, then the volume of the solid formed by revolving the region bounded by the graphs of f and g ($a \leq x \leq b$) about the x-axis is

$$\text{Volume} = \pi \int_a^b \{[f(x)]^2 - [g(x)]^2\} \, dx.$$

$f(x)$ is the **outer radius** and $g(x)$ is the **inner radius.**

In Figure 6.22(b), note that the solid of revolution has a hole. Moreover, the radius of the hole is $g(x)$, the inner radius.

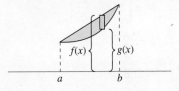

(a) Plane region

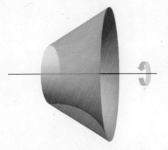

(b) Solid of revolution with hole

FIGURE 6.22 Washer Method

TECHNOLOGY

Some graphing utilities have the capability of generating (or have built-in software capable of generating) a solid of revolution. If you have access to such a utility, use it to graph some of the solids of revolution described in this section. For instance, the solid in Example 1 might be similar in appearance to the one shown in the figure at the right.

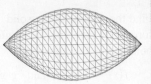

Generated by Mathematica

Example 2 Using the Washer Method

Find the volume of the solid formed by revolving the region bounded by the graphs of

$$f(x) = \sqrt{25 - x^2} \quad \text{and} \quad g(x) = 3$$

about the x-axis (see Figure 6.23).

SOLUTION First find the points of intersection of f and g by setting $f(x)$ equal to $g(x)$ and solving for x.

$f(x) = g(x)$	Set $f(x)$ equal to $g(x)$.
$\sqrt{25 - x^2} = 3$	Substitute for $f(x)$ and $g(x)$.
$25 - x^2 = 9$	Square each side.
$16 = x^2$	
$\pm 4 = x$	Solve for x.

Using $f(x)$ as the outer radius and $g(x)$ as the inner radius, you can find the volume of the solid as shown.

$$\text{Volume} = \pi \int_{-4}^{4} \{[f(x)]^2 - [g(x)]^2\} \, dx \qquad \text{Washer Method}$$

$$= \pi \int_{-4}^{4} \left[(\sqrt{25 - x^2})^2 - (3)^2\right] dx \qquad \text{Substitute for } f(x) \text{ and } g(x).$$

$$= \pi \int_{-4}^{4} (16 - x^2) \, dx \qquad \text{Simplify.}$$

$$= \pi \left[16x - \frac{x^3}{3}\right]_{-4}^{4} \qquad \text{Find antiderivative.}$$

$$= \frac{256\pi}{3} \qquad \text{Apply Fundamental Theorem.}$$

$$\approx 268.08 \qquad \text{Round to two decimal places.}$$

So, the volume of the solid is about 268.08 cubic inches.

✓ CHECKPOINT 2

Find the volume of the solid formed by revolving the region bounded by the graphs of

$$f(x) = 5 - x^2 \quad \text{and} \quad g(x) = 1$$

about the x-axis. ∎

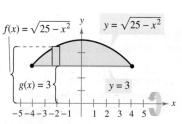
(a) Plane region

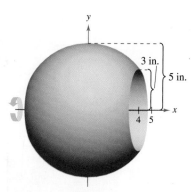
(b) Solid of revolution

FIGURE 6.23

SECTION 6.6 Volumes of Solids of Revolution

Example 3 Estimating an Egg's Volume

Scientific studies use the volume of an egg to determine if egg size is a good predictor of egg constituents and neonate characteristics, such as body composition.

A goose egg can be modeled as a solid of revolution formed by revolving the graph of

$$y = \frac{1}{30}\sqrt{7569 - 400x^2}, \quad -4.35 \le x \le 4.35$$

about the x-axis, as shown in Figure 6.24. Use this model to estimate the volume of the egg. (In the model, x and y are measured in centimeters.)

SOLUTION To find the volume of the solid of revolution, use the Disk Method.

$$\begin{aligned}
\text{Volume} &= \pi \int_{-4.35}^{4.35} [f(x)]^2 \, dx && \text{Disk Method} \\
&= \pi \int_{-4.35}^{4.35} \left(\frac{1}{30}\sqrt{7569 - 400x^2}\right)^2 dx && \text{Substitute for } f(x). \\
&= \frac{\pi}{900} \int_{-4.35}^{4.35} (7569 - 400x^2) \, dx && \text{Simplify.} \\
&\approx 153 \text{ cubic centimeters} && \text{Volume.}
\end{aligned}$$

So, the volume of the egg is about 153 cubic centimeters.

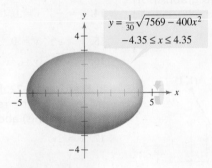

FIGURE 6.24

✓ CHECKPOINT 3

An egg can be modeled as a solid of revolution formed by revolving the graph of

$$f(x) = \frac{3}{4}\sqrt{16 - x^2}, \quad -4 \le x \le 4$$

about the x-axis. Use this model, where x and y are measured in centimeters, to find the volume of the egg. ∎

CONCEPT CHECK

1. How is a solid of revolution formed?
2. If a solid of revolution has a hole, should you find its volume using the Disk Method or the Washer Method?
3. Write the formula for the Disk Method.
4. Write the formula for the Washer Method.

Skills Review 6.6

The following warm-up exercises involve skills that were covered in earlier sections. You will use these skills in the exercise set for this section. For additional help, review Sections 1.2 and 6.4.

In Exercises 1–6, solve for x.

1. $x^2 = 2x$
2. $-x^2 + 4x = x^2$
3. $x = -x^3 + 5x$
4. $x^2 + 1 = x + 3$
5. $-x + 4 = \sqrt{4x - x^2}$
6. $\sqrt{x - 1} = \frac{1}{2}(x - 1)$

In Exercises 7–10, evaluate the integral.

7. $\int_0^2 2e^{2x}\, dx$
8. $\int_{-1}^3 \frac{2x + 1}{x^2 + x + 2}\, dx$
9. $\int_0^2 x\sqrt{x^2 + 1}\, dx$
10. $\int_1^5 \frac{(\ln x)^2}{x}\, dx$

Exercises 6.6

See www.CalcChat.com for worked-out solutions to odd-numbered exercises.

In Exercises 1–16, find the volume of the solid formed by revolving the region bounded by the graph(s) of the equation(s) about the x-axis.

1. $y = \sqrt{4 - x^2}$
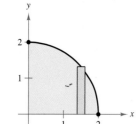

2. $y = x^2$
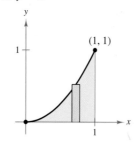

3. $y = \sqrt{x}$
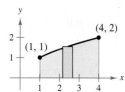

4. $y = \sqrt{4 - x^2}$
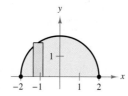

5. $y = 4 - x^2$, $y = 0$
6. $y = x$, $y = 0$, $x = 4$
7. $y = 1 - \frac{1}{4}x^2$, $y = 0$
8. $y = x^2 + 1$, $y = 5$
9. $y = -x + 1$, $y = 0$, $x = 0$
10. $y = x$, $y = e^{x-1}$, $x = 0$
11. $y = \sqrt{x} + 1$, $y = 0$, $x = 0$, $x = 9$
12. $y = \sqrt{x}$, $y = 0$, $x = 4$

13. $y = 2x^2$, $y = 0$, $x = 2$
14. $y = \dfrac{1}{x}$, $y = 0$, $x = 1$, $x = 3$
15. $y = e^x$, $y = 0$, $x = 0$, $x = 1$
16. $y = x^2$, $y = 4x - x^2$

In Exercises 17–22, find the volume of the solid formed by revolving the region bounded by the graph(s) of the equation(s) about the y-axis.

17. $x = 2y$
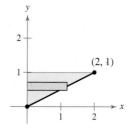

18. $x = 2 - y$
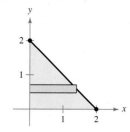

19. $x = y^3$
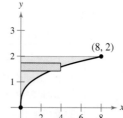

20. $x = \sqrt[3]{y}$
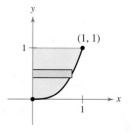

21. $y = x^2$

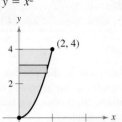

22. $y = \sqrt{1 - x^2}$

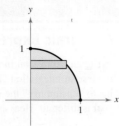

23. Volume The line segment from $(0, 0)$ to $(6, 3)$ is revolved about the x-axis to form a cone. What is the volume of the cone?

24. Volume The line segment from $(0, 0)$ to $(4, 8)$ is revolved about the y-axis to form a cone. What is the volume of the cone?

25. Volume Use the Disk Method to verify that the volume of a right circular cone is $\frac{1}{3}\pi r^2 h$, where r is the radius of the base and h is the height.

26. Volume Use the Disk Method to verify that the volume of a sphere of radius r is $\frac{4}{3}\pi r^3$.

27. Volume The right half of the ellipse

$$x^2 + 4y^2 = 4$$

is revolved about the y-axis to form an oblate spheroid (shaped like an M&M'S® candy). Find the volume of the spheroid.

28. Volume The upper half of the ellipse

$$9x^2 + 16y^2 = 144$$

is revolved about the x-axis to form a prolate spheroid (shaped like a football). Find the volume of the spheroid.

29. Volume A tank on the wing of a jet airplane is modeled by revolving the region bounded by the graph of $y = \frac{1}{8}x^2\sqrt{2 - x}$ and the x-axis about the x-axis, where x and y are measured in meters (see figure). Find the volume of the tank.

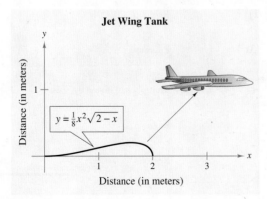

30. Volume A soup bowl can be modeled as a solid of revolution formed by revolving the graph of

$$y = \sqrt{\frac{x}{2} + 1}, \quad 0 \le x \le 4$$

about the x-axis. Use this model, where x and y are measured in inches, to find the volume of the soup bowl.

31. Biology A pond is to be stocked with a species of fish. The food supply in 500 cubic feet of pond water can adequately support one fish. The pond is nearly circular, is 20 feet deep at its center, and has a radius of 200 feet. The bottom of the pond can be modeled by

$$y = 20[(0.005x)^2 - 1].$$

(a) How much water is in the pond?

(b) How many fish can the pond support?

32. Modeling a Body of Water A pond is approximately circular, with a diameter of 400 feet. Starting at the center, the depth of the water is measured every 25 feet (see figure) and recorded in the table.

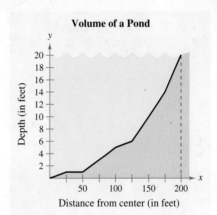

x	0	25	50	75	100
Depth	20	19	19	17	15

x	125	150	175	200
Depth	14	10	6	0

(a) Use a graphing utility to plot the depths and graph the model of the pond's depth, $y = 20 - 0.00045x^2$.

(b) Use the model in part (a) to find the pond's volume.

(c) Use the result of part (b) to approximate the number of gallons of water in the pond (1 ft³ ≈ 7.48 gal).

Algebra Review

"Unsimplifying" an Algebraic Expression

In algebra it is often helpful to write an expression in simplest form. In this chapter, you have seen that the reverse is often true in integration. That is, to make an integrand fit an integration formula, it often helps to "unsimplify" the expression. To do this, you use the same algebraic rules, but your goal is different. Here are some examples.

Example 1 Rewriting Algebraic Expressions

Rewrite each algebraic expression as indicated in the example.

a. $\dfrac{x+1}{\sqrt{x}}$ Example 6, page 368

b. $x(3-4x^2)^2$ Example 2, page 377

c. $7x^2\sqrt{x^3+1}$ Example 4, page 378

d. $5xe^{-x^2}$ Example 3, page 385

SOLUTION

a.
$$\dfrac{x+1}{\sqrt{x}} = \dfrac{x}{\sqrt{x}} + \dfrac{1}{\sqrt{x}} \qquad \text{Rewrite as two fractions.}$$
$$= \dfrac{x^1}{x^{1/2}} + \dfrac{1}{x^{1/2}} \qquad \text{Rewrite with rational exponents.}$$
$$= x^{1-1/2} + x^{-1/2} \qquad \text{Properties of exponents}$$
$$= x^{1/2} + x^{-1/2} \qquad \text{Simplify exponent.}$$

b.
$$x(3-4x^2)^2 = \dfrac{-8}{-8}x(3-4x^2)^2 \qquad \text{Multiply and divide by } -8.$$
$$= \left(-\dfrac{1}{8}\right)(-8)x(3-4x^2)^2 \qquad \text{Regroup.}$$
$$= \left(-\dfrac{1}{8}\right)(3-4x^2)^2(-8x) \qquad \text{Regroup.}$$

c.
$$7x^2\sqrt{x^3+1} = 7x^2(x^3+1)^{1/2} \qquad \text{Rewrite with rational exponent.}$$
$$= \dfrac{3}{3}(7x^2)(x^3+1)^{1/2} \qquad \text{Multiply and divide by 3.}$$
$$= \dfrac{7}{3}(3x^2)(x^3+1)^{1/2} \qquad \text{Regroup.}$$
$$= \dfrac{7}{3}(x^3+1)^{1/2}(3x^2) \qquad \text{Regroup.}$$

d.
$$5xe^{-x^2} = \dfrac{-2}{-2}(5x)e^{-x^2} \qquad \text{Multiply and divide by } -2.$$
$$= \left(-\dfrac{5}{2}\right)(-2x)e^{-x^2} \qquad \text{Regroup.}$$
$$= \left(-\dfrac{5}{2}\right)e^{-x^2}(-2x) \qquad \text{Regroup.}$$

Algebra Review

Example 2 Rewriting Algebraic Expressions

Rewrite each algebraic expression.

a. $\dfrac{3x^2 + 2x - 1}{x^2}$ Example 7(a), page 388

b. $\dfrac{1}{1 + e^{-x}}$ Example 7(b), page 388

c. $\dfrac{x^2 + x + 1}{x - 1}$ Example 7(c), page 388

d. $\dfrac{x^2 + 6x + 1}{x^2 + 1}$ Bottom of page 387

SOLUTION

a. $\dfrac{3x^2 + 2x - 1}{x^2} = \dfrac{3x^2}{x^2} + \dfrac{2x}{x^2} - \dfrac{1}{x^2}$ Rewrite as separate fractions

$= 3 + \dfrac{2}{x} - x^{-2}$ Properties of exponents.

$= 3 + 2\left(\dfrac{1}{x}\right) - x^{-2}$ Regroup.

b. $\dfrac{1}{1 + e^{-x}} = \left(\dfrac{e^x}{e^x}\right)\dfrac{1}{1 + e^{-x}}$ Multiply and divide by e^x.

$= \dfrac{e^x}{e^x + e^x(e^{-x})}$ Multiply.

$= \dfrac{e^x}{e^x + e^{x-x}}$ Property of exponents

$= \dfrac{e^x}{e^x + e^0}$ Simplify exponent.

$= \dfrac{e^x}{e^x + 1}$ $e^0 = 1$

c. $\dfrac{x^2 + x + 1}{x - 1} = x + 2 + \dfrac{3}{x - 1}$ Use long division as shown below.

$$\begin{array}{r} x + 2 \\ x - 1 \overline{\smash{)}\, x^2 + x + 1} \\ \underline{x^2 - x} \\ 2x + 1 \\ \underline{2x - 2} \\ 3 \end{array}$$

d. $\dfrac{x^2 + 6x + 1}{x^2 + 1} = 1 + \dfrac{6x}{x^2 + 1}$ Use long division as shown below.

$$\begin{array}{r} 1 \\ x^2 + 1 \overline{\smash{)}\, x^2 + 6x + 1} \\ \underline{x^2 + 1} \\ 6x \end{array}$$

Chapter Summary and Study Strategies

After studying this chapter, you should have acquired the following skills.
The exercise numbers are keyed to the Review Exercises that begin on page 422.
Answers to odd-numbered Review Exercises are given in the back of the text.*

Section 6.1   Review Exercises

- Use basic integration rules to find indefinite integrals.   1–10

$$\int k \, dx = kx + C \qquad \int [f(x) - g(x)] \, dx = \int f(x) \, dx - \int g(x) \, dx$$

$$\int kf(x) \, dx = k \int f(x) \, dx \qquad \int x^n \, dx = \frac{x^{n+1}}{n+1} + C, \quad n \neq -1$$

$$\int [f(x) + g(x)] \, dx = \int f(x) \, dx + \int g(x) \, dx$$

- Use initial conditions to find particular solutions of indefinite integrals.   11–14
- Use antiderivatives to solve real-life problems.   15, 16, 71

Section 6.2

- Use the General Power Rule or integration by substitution to find indefinite integrals.   17–24

$$\int u^n \frac{du}{dx} \, dx = \int u^n \, du = \frac{u^{n+1}}{n+1} + C, \quad n \neq -1$$

- Use the General Power Rule or integration by substitution to solve real-life problems.   25, 26

Section 6.3

- Use the Exponential and Log Rules to find indefinite integrals.   27–32

$$\int e^x \, dx = e^x + C \qquad \int \frac{1}{x} \, dx = \ln|x| + C$$

$$\int e^u \frac{du}{dx} \, dx = \int e^u \, du = e^u + C \qquad \int \frac{du/dx}{u} \, dx = \int \frac{1}{u} \, du = \ln|u| + C$$

- Use a symbolic integration utility to find indefinite integrals.   33, 34
- Use Exponential and Log Rules to solve real-life problems.   70

Section 6.4

- Find the areas of regions using a geometric formula.   35, 36
- Find the areas of regions bounded by the graph of a function and the x-axis.   37–44
- Use properties of definite integrals.   45, 46
- Use the Fundamental Theorem of Calculus to evaluate definite integrals.   47–64

$$\int_a^b f(x) \, dx = F(x) \Big]_a^b = F(b) - F(a), \quad \text{where} \quad F'(x) = f(x)$$

---

* Use a wide range of valuable study aids to help you master the material in this chapter. The *Student Solutions Guide* includes step-by-step solutions to all odd-numbered exercises to help you review and prepare. The student website at *college.hmco.com/info/larsonapplied* offers algebra help and a *Graphing Technology Guide*. The *Graphing Technology Guide* contains step-by-step commands and instructions for a wide variety of graphing calculators, including the most recent models.

Section 6.4 (continued)

Review Exercises

- Use definite integrals to find the net change in a quantity. 15 (c), 16 (c)

$$\text{Net change} = f(b) - f(a) = \int_a^b f'(t)\, dt$$

- Find average values of functions over closed intervals. 65–68

$$\text{Average value} = \frac{1}{b-a}\int_a^b f(x)\, dx$$

- Use average values to solve real-life problems. 69, 72
- Use properties of even and odd functions to help evaluate definite integrals. 73–76

  Even function: $f(-x) = f(x)$ Odd function: $f(-x) = -f(x)$

  If f is an *even* function, then $\int_{-a}^a f(x)\, dx = 2\int_0^a f(x)\, dx$.

  If f is an *odd* function, then $\int_{-a}^a f(x)\, dx = 0$.

Section 6.5

- Find the areas of regions bounded by two (or more) graphs. 77–86

$$A = \int_a^b [f(x) - g(x)]\, dx$$

- Use the areas of regions bounded by two graphs to solve real-life problems. 87, 88

Section 6.6

- Use the Disk Method to find volumes of solids of revolution. 89–92

$$\text{Volume} = \pi \int_a^b [f(x)]^2\, dx$$

- Use the Washer Method to find volumes of solids of revolution with holes. 93–96

$$\text{Volume} = \pi \int_a^b \{[f(x)]^2 - [g(x)]^2\}\, dx$$

- Use solids of revolution to solve real-life problems. 97, 98

Study Strategies

- **Indefinite and Definite Integrals** When evaluating integrals, remember that an indefinite integral is a *family of antiderivatives*, each differing by a constant C, whereas a definite integral is a *number*.
- **Checking Antiderivatives by Differentiating** When finding an antiderivative, remember that you can check your result by differentiating. For example, you can check that the antiderivative

$$\int (3x^3 - 4x)\, dx = \frac{3}{4}x^4 - 2x^2 + C \quad \text{is correct by differentiating to obtain} \quad \frac{d}{dx}\left[\frac{3}{4}x^4 - 2x^2 + C\right] = 3x^3 - 4x.$$

  Because the derivative is equal to the original integrand, you know that the antiderivative is correct.

- **Grouping Symbols and the Fundamental Theorem** When using the Fundamental Theorem of Calculus to evaluate a definite integral, you can avoid sign errors by using grouping symbols. Here is an example.

$$\int_1^3 (x^3 - 9x)\, dx = \left[\frac{x^4}{4} - \frac{9x^2}{2}\right]_1^3 = \left[\frac{3^4}{4} - \frac{9(3^2)}{2}\right] - \left[\frac{1^4}{4} - \frac{9(1^2)}{2}\right] = \frac{81}{4} - \frac{81}{2} - \frac{1}{4} + \frac{9}{2} = -16$$

Review Exercises

In Exercises 1–10, find the indefinite integral.

1. $\int 16\, dx$
2. $\int \frac{3}{5}x\, dx$
3. $\int (2x^2 + 5x)\, dx$
4. $\int (5 - 6x^2)\, dx$
5. $\int \frac{2}{3\sqrt[3]{x}}\, dx$
6. $\int 6x^2\sqrt{x}\, dx$
7. $\int \left(\sqrt[3]{x^4} + 3x\right) dx$
8. $\int \left(\frac{4}{\sqrt{x}} + \sqrt{x}\right) dx$
9. $\int \frac{2x^4 - 1}{\sqrt{x}}\, dx$
10. $\int \frac{1 - 3x}{x^2}\, dx$

In Exercises 11–14, find the particular solution, $y = f(x)$, that satisfies the conditions.

11. $f'(x) = 3x + 1$, $f(2) = 6$
12. $f'(x) = x^{-1/3} - 1$, $f(8) = 4$
13. $f''(x) = 2x^2$, $f'(3) = 10$, $f(3) = 6$
14. $f''(x) = \frac{6}{\sqrt{x}} + 3$, $f'(1) = 12$, $f(4) = 56$

15. **Vertical Motion** An object is projected upward from the ground with an initial velocity of 80 feet per second.
    (a) How long does it take the object to rise to its maximum height?
    (b) What is the maximum height?
    (c) Find the net change in the object's position 1 second after it leaves the ground.

16. **Yellow Bullhead Fish** The rate of change in the mass M (in grams) with respect to the total length L (in millimeters) of a yellow bullhead fish can be modeled by

    $$\frac{dM}{dL} = 0.000116 L^{1.9004}.$$

    A fish that is 200 millimeters long has a mass of 100 grams. *(Source: U.S. Geological Survey)*

    (a) Find a model for the mass function.
    (b) Find the mass of a fish that is 250 millimeters long.
    (c) Find the net change in the mass of a fish when the length increases from 200 to 300 millimeters.

In Exercises 17–24, find the indefinite integral.

17. $\int (1 + 5x)^2\, dx$
18. $\int (x - 6)^{4/3}\, dx$
19. $\int \frac{1}{\sqrt{5x - 1}}\, dx$
20. $\int \frac{4x}{\sqrt{1 - 3x^2}}\, dx$
21. $\int x(1 - 4x^2)\, dx$
22. $\int \frac{x^2}{(x^3 - 4)^2}\, dx$
23. $\int (x^4 - 2x)(2x^3 - 1)\, dx$
24. $\int \frac{\sqrt{x}}{(1 - x^{3/2})^3}\, dx$

25. **Meteorology** The amount A of snow accumulation (in centimeters) from a storm changes according to the model

    $$\frac{dA}{dt} = 2t(0.001t^2 + 0.5)^{1/4}, \quad 0 \le t \le 6$$

    where t is measured in hours. Find the amounts of accumulation after (a) 3 hours and (b) 5 hours.

26. **Mold Spores** The number N of mold spores (in thousands) that have accumulated in a damp basement changes according to the model

    $$\frac{dN}{dt} = \frac{25t}{\sqrt{5t^2 + 1000}}, \quad 0 \le t \le 8$$

    where t is measured in weeks. When $t = 0$, $N = 150$. Find the numbers of spores after (a) 2 weeks and (b) 6 weeks.

In Exercises 27–32, find the indefinite integral.

27. $\int 3e^{-3x}\, dx$
28. $\int (2t - 1)e^{t^2 - t}\, dt$
29. $\int (x - 1)e^{x^2 - 2x}\, dx$
30. $\int \frac{4}{6x - 1}\, dx$
31. $\int \frac{x^2}{1 - x^3}\, dx$
32. $\int \frac{x - 4}{x^2 - 8x}\, dx$

(T) In Exercises 33 and 34, use a symbolic integration utility to find the indefinite integral.

33. $\int \frac{(\sqrt{x} + 1)^2}{\sqrt{x}}\, dx$
34. $\int \frac{e^{5x}}{5 + e^{5x}}\, dx$

In Exercises 35 and 36, sketch the region whose area is given by the definite integral. Then use a geometric formula to evaluate the integral.

35. $\int_0^5 (5 - |x - 5|)\, dx$
36. $\int_{-4}^{4} \sqrt{16 - x^2}\, dx$

In Exercises 37–44, find the area of the region.

37. $f(x) = 4 - 2x$	38. $f(x) = 3x + 6$

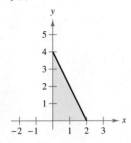 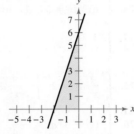

39. $f(x) = 4 - x^2$	40. $f(x) = 9 - x^2$

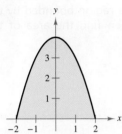 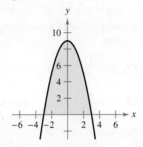

41. $f(y) = (y - 2)^2$	42. $f(x) = \sqrt{9 - x^2}$

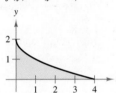 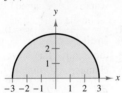

43. $f(x) = \dfrac{2}{x + 1}$	44. $f(x) = 2xe^{x^2-4}$

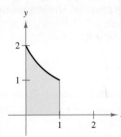 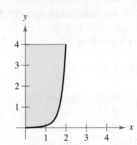

45. Given $\displaystyle\int_2^6 f(x)\,dx = 10$ and $\displaystyle\int_2^6 g(x)\,dx = 3$, evaluate each definite integral.

(a) $\displaystyle\int_2^6 [f(x) + g(x)]\,dx$	(b) $\displaystyle\int_2^6 [f(x) - g(x)]\,dx$

(c) $\displaystyle\int_2^6 [2f(x) - 3g(x)]\,dx$	(d) $\displaystyle\int_2^6 5f(x)\,dx$

46. Given $\displaystyle\int_0^3 f(x)\,dx = 4$ and $\displaystyle\int_3^6 f(x)\,dx = -1$, evaluate each definite integral.

(a) $\displaystyle\int_0^6 f(x)\,dx$	(b) $\displaystyle\int_6^3 f(x)\,dx$

(c) $\displaystyle\int_4^4 f(x)\,dx$	(d) $\displaystyle\int_3^6 -10f(x)\,dx$

In Exercises 47–60, use the Fundamental Theorem of Calculus to evaluate the definite integral.

47. $\displaystyle\int_0^4 (2 + x)\,dx$	48. $\displaystyle\int_{-1}^1 (t^2 + 2)\,dt$

49. $\displaystyle\int_4^9 x\sqrt{x}\,dx$	50. $\displaystyle\int_1^4 2x\sqrt{x}\,dx$

51. $\displaystyle\int_{-1}^1 (4t^3 - 2t)\,dt$	52. $\displaystyle\int_{-2}^2 (x^4 + 2x^2 - 5)\,dx$

53. $\displaystyle\int_0^3 \dfrac{1}{\sqrt{1+x}}\,dx$	54. $\displaystyle\int_3^6 \dfrac{x}{3\sqrt{x^2 - 8}}\,dx$

55. $\displaystyle\int_1^2 \left(\dfrac{1}{x^2} - \dfrac{1}{x^3}\right) dx$	56. $\displaystyle\int_0^1 x^2(x^3 + 1)^3\,dx$

57. $\displaystyle\int_1^3 \dfrac{(3 + \ln x)}{x}\,dx$

58. $\displaystyle\int_0^{\ln 5} e^{x/5}\,dx$

59. $\displaystyle\int_{-1}^1 3xe^{x^2-1}\,dx$

60. $\displaystyle\int_1^3 \dfrac{1}{x(\ln x + 2)^2}\,dx$

In Exercises 61–64, sketch the graph of the region whose area is given by the integral, and find the area.

61. $\displaystyle\int_1^3 (2x - 1)\,dx$

62. $\displaystyle\int_0^2 (x + 4)\,dx$

63. $\displaystyle\int_3^4 (x^2 - 9)\,dx$

64. $\displaystyle\int_{-1}^2 (-x^2 + x + 2)\,dx$

In Exercises 65–68, find the average value of the function on the closed interval. Then find all x-values in the interval for which the function is equal to its average value.

65. $f(x) = \dfrac{1}{\sqrt{x}}$, $[4, 9]$	66. $f(x) = \dfrac{20 \ln x}{x}$, $[2, 10]$

67. $f(x) = e^{5-x}$, $[2, 5]$	68. $f(x) = x^3$, $[0, 2]$

69. Consumer Awareness Suppose the price p of gasoline can be modeled by

$$p = 0.0782t^2 - 0.352t + 1.75$$

where $t = 1$ corresponds to January 1, 2001. Find the cost of gasoline for an automobile that is driven 15,000 miles per year and gets 33 miles per gallon from 2001 through 2006. *(Source: U.S. Department of Energy)*

70. Swarm of Ants The rate of change in the number of ants A swarming on a dropped ice cream cone can be modeled by

$$\frac{dA}{dt} = \frac{120}{1 + 0.075t}$$

where t is the time in hours. When $t = 0$, one ant is on the ice cream cone.

(a) Write an equation that models the number of ants A in terms of the time t.

(b) How many ants will be on the ice cream cone after 3 hours?

(c) After how many hours will there be 1000 ants on the ice cream cone?

71. Consumer Trends The rate of change in the price of lean and extra lean beef B (in dollars per pound) in the United States from 1999 through 2006 can be modeled by

$$\frac{dB}{dt} = -0.0391t + 0.6108$$

where t is the year, with $t = 9$ corresponding to 1999. The price of 1 pound of lean and extra lean beef in 2006 was $2.95. *(Source: U.S. Bureau of Labor Statistics)*

(a) Find the price function in terms of the year.

(b) If the price of beef per pound continues to change at this rate, in what year does the model predict the price per pound of lean and extra lean beef will surpass $3.25? Explain your reasoning.

72. Medical Science The volume V (in liters) of air in the lungs during a five-second respiratory cycle is approximated by the model

$$V = 0.1729t + 0.1522t^2 - 0.0374t^3$$

where t is time in seconds.

(T) (a) Use a graphing utility to graph the equation on the interval $[0, 5]$.

(b) Determine the intervals on which the function is increasing and decreasing.

(c) Determine the maximum volume during the respiratory cycle.

(d) Determine the average volume of air in the lungs during one cycle.

(e) Briefly explain your results for parts (a) through (d).

In Exercises 73–76, explain how the given value can be used to evaluate the second integral.

73. $\int_0^2 6x^5 \, dx = 64, \quad \int_{-2}^2 6x^5 \, dx$

74. $\int_0^3 (x^4 + x^2) \, dx = 57.6, \quad \int_{-3}^3 (x^4 + x^2) \, dx$

75. $\int_1^2 \frac{4}{x^2} \, dx = 2, \quad \int_{-2}^{-1} \frac{4}{x^2} \, dx$

76. $\int_0^1 (x^3 - x) \, dx = -\frac{1}{4}, \quad \int_{-1}^0 (x^3 - x) \, dx$

In Exercises 77–84, sketch the region bounded by the graphs of the equations. Then find the area of the region.

77. $y = \frac{1}{x^2}, \quad y = 0, \quad x = 1, \quad x = 5$

78. $y = \frac{1}{x^2}, \quad y = 4, \quad x = 5$

79. $y = x, \quad y = x^3$

80. $y = 1 - \frac{1}{2}x, \quad y = x - 2, \quad y = 1$

81. $y = \frac{4}{\sqrt{x+1}}, \quad y = 0, \quad x = 0, \quad x = 8$

82. $y = \sqrt{x}(x - 1), \quad y = 0$

83. $y = (x - 3)^2, \quad y = 8 - (x - 3)^2$

84. $y = 4 - x, \quad y = x^2 - 5x + 8, \quad x = 0$

(T) In Exercises 85 and 86, use a graphing utility to graph the region bounded by the graphs of the equations. Then find the area of the region.

85. $y = x, \quad y = 2 - x^2$

86. $y = x, \quad y = x^5$

87. Population of the United States From 2000 through 2005, the population P (in millions) of the United States can be modeled by

$$P_1 = 282.52 e^{0.0098t}, \quad 0 \le t \le 5$$

where t is the year, with $t = 0$ corresponding to 2000. Before 2000, one model that was used to project the population from 2000 through 2005 was

$$P_2 = 275.69 e^{0.0117t}, \quad 0 \le t \le 5.$$

This projection underestimated the population from 2000 through 2005. Find the amount the projection underestimated. *(Source: U.S. Census Bureau)*

88. Psychology: Sleep Patterns The graph shows three areas, representing awake time, REM (rapid eye movement) sleep time, and non-REM sleep time, over a typical individual's lifetime. Make generalizations about the amounts of total sleep, non-REM sleep, and REM sleep an individual gets as he or she gets older. If you wanted to estimate mathematically the amount of non-REM sleep an individual gets between birth and age 50, how would you do so? How would you mathematically estimate the amount of REM sleep an individual gets during this interval? *(Source: Adapted from Bernstein/Clarke-Stewart/Roy/Wickens,* Psychology, *Seventh Edition)*

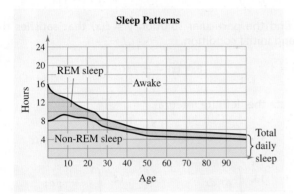

In Exercises 89–92, use the Disk Method to find the volume of the solid of revolution formed by revolving the region about the x-axis.

89. $y = \dfrac{1}{\sqrt{x}}$

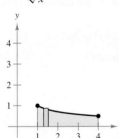

90. $y = \sqrt{16 - x}$

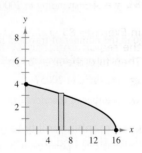

91. $y = e^{1-x}$

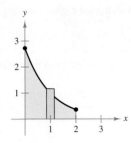

92. $y = \dfrac{1}{x}$

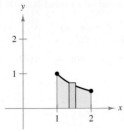

In Exercises 93–96, find the volume of the solid of revolution formed by revolving the region about the x-axis.

93. $y = 2x + 1,\ y = 1,\ x = 2$

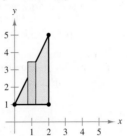

94. $y = \sqrt{x},\ y = 2,\ x = 0$

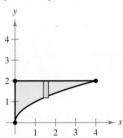

95. $y = x^2,\ y = x^3$

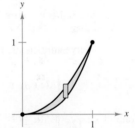

96. $y = 4 - \tfrac{1}{2}x^2,\ y = 2$

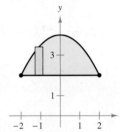

97. Egg Volume A chicken egg can be modeled as a solid of revolution formed by revolving the graph of $y = \tfrac{2}{3}\sqrt{9 - x^2},\ -3 \le x \le 3$, about the x-axis (see figure). Use this model, where x and y are measured in centimeters, to find the volume of the egg.

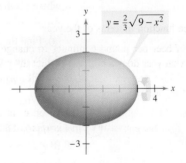

98. Volume A hardened plastic valve part is in the shape of a sphere with a hole drilled through it. The radius of the sphere is 1 inch and the radius of the hole is 0.25 inch (see figure). What is the volume of the part?

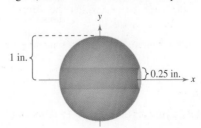

Chapter Test

See www.CalcChat.com for worked-out solutions to odd-numbered exercises.

Take this test as you would take a test in class. When you are done, check your work against the answers given in the back of the book.

In Exercises 1–6, find the indefinite integral.

1. $\displaystyle\int (9x^2 - 4x + 13)\, dx$
2. $\displaystyle\int (x + 1)^2\, dx$
3. $\displaystyle\int 4x^3 \sqrt{x^4 - 7}\, dx$
4. $\displaystyle\int \frac{5x - 6}{\sqrt{x}}\, dx$
5. $\displaystyle\int 15e^{3x}\, dx$
6. $\displaystyle\int \frac{3x^2 - 11}{x^3 - 11x}\, dx$

In Exercises 7 and 8, find the particular solution $y = f(x)$ that satisfies the differential equation and initial condition.

7. $f'(x) = e^x + 1,\quad f(0) = 1$
8. $f'(x) = \dfrac{1}{x},\quad f(-1) = 2$

In Exercises 9–14, evaluate the definite integral.

9. $\displaystyle\int_0^1 16x\, dx$
10. $\displaystyle\int_{-3}^{3} (3 - 2x)\, dx$
11. $\displaystyle\int_{-1}^{1} (x^3 + x^2)\, dx$
12. $\displaystyle\int_{-1}^{2} \frac{2x}{\sqrt{x^2 + 1}}\, dx$
13. $\displaystyle\int_0^3 e^{4x}\, dx$
14. $\displaystyle\int_{-2}^{3} \frac{1}{x + 3}\, dx$

15. The rate of change in the number N of overseas international travelers (in millions) to the United States from 2000 through 2005 can be modeled by

$$\frac{dN}{dt} = 1.714t - 5.06$$

where t is the year, with $t = 0$ corresponding to 2000. In 2000, the number of travelers was about 26 million. *(Source: U.S. Department of Commerce)*

(a) Find the net change in the number of travelers from 2000 to 2005.

(b) Write a model for the number of travelers as a function of t.

(c) How many travelers were there in 2005?

(d) What was the average number of travelers from 2000 through 2005?

Ⓣ In Exercises 16 and 17, use a graphing utility to graph the region bounded by the graphs of the functions. Then find the area of the region.

16. $f(x) = 6,\quad g(x) = x^2 - x - 6$
17. $f(x) = \sqrt[3]{x},\quad g(x) = x^2$

18. A computer-generated model of a funnel is made by revolving the region bounded by the lines

$$y = 3 - x,\quad y = 0,\quad \text{and}\quad x = 0$$

about the x-axis, where x and y are measured in inches (see figure). Find the volume of the funnel.

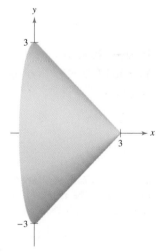

Figure for 18

Techniques of Integration

7

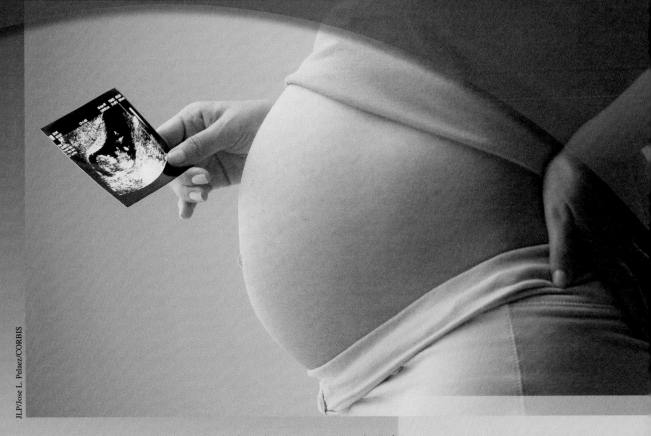

Integration is used to estimate the weight of a fetus. (See Section 7.2, Exercise 51.)

Applications

Integration has many real-life applications. The applications listed below represent a sample of the applications in this chapter.

- Hispanic Enrollment in Schools, Exercise 67, page 435
- Biology, Exercises 53 and 54, page 444
- Meteorology, Exercises 57 and 58, pages 453–454
- Zoology, Exercise 37, page 461
- Women's Heights, Exercise 43, page 480

7.1 Integration by Parts
7.2 Partial Fractions and Logistic Growth
7.3 Integrals of Trigonometric Functions
7.4 The Definite Integral as the Limit of a Sum
7.5 Numerical Integration
7.6 Improper Integrals

Section 7.1
Integration by Parts

- Use integration by parts to find indefinite and definite integrals.
- Use integration by parts to solve real-life problems.

Integration by Parts

In this section, you will study an integration technique called **integration by parts**. This technique is particularly useful for integrands involving the products of algebraic and exponential or logarithmic functions, such as

$$\int x^2 e^x \, dx \quad \text{and} \quad \int x \ln x \, dx.$$

Integration by parts is based on the Product Rule for differentiation.

$$\frac{d}{dx}[uv] = u\frac{dv}{dx} + v\frac{du}{dx} \qquad \text{Product Rule}$$

$$uv = \int u\frac{dv}{dx}\,dx + \int v\frac{du}{dx}\,dx \qquad \text{Integrate each side.}$$

$$uv = \int u\,dv + \int v\,du \qquad \text{Write in differential form.}$$

$$\int u\,dv = uv - \int v\,du \qquad \text{Rewrite.}$$

Integration by Parts

Let u and v be differentiable functions of x.

$$\int u\,dv = uv - \int v\,du$$

STUDY TIP

When using integration by parts, note that you can first choose dv or first choose u. After you choose, however, the choice of the other factor is determined—it must be the remaining portion of the integrand. Also note that dv must contain the differential dx of the original integral.

Note that the formula for integration by parts expresses the original integral in terms of another integral. Depending on the choices of u and dv, it may be easier to evaluate the second integral than the original one.

Guidelines for Integration by Parts

1. Let dv be the most complicated portion of the integrand that fits a basic integration formula. Let u be the remaining factor.
2. Let u be the portion of the integrand whose derivative is a function simpler than u. Let dv be the remaining factor.

Example 1 Integration by Parts

Find $\int xe^x \, dx$.

SOLUTION To apply integration by parts, you must rewrite the original integral in the form $\int u \, dv$. That is, you must break $xe^x \, dx$ into two factors—one "part" representing u and the other "part" representing dv. There are several ways to do this.

$$\int \underbrace{(x)}_{u}\underbrace{(e^x \, dx)}_{dv} \qquad \int \underbrace{(e^x)}_{u}\underbrace{(x \, dx)}_{dv} \qquad \int \underbrace{(1)}_{u}\underbrace{(xe^x \, dx)}_{dv} \qquad \int \underbrace{(xe^x)}_{u}\underbrace{(dx)}_{dv}$$

Following the guidelines, you should choose the first option because $dv = e^x \, dx$ is the most complicated portion of the integrand that fits a basic integration formula *and* because the derivative of $u = x$ is simpler than x.

$$dv = e^x \, dx \quad \Longrightarrow \quad v = \int dv = \int e^x \, dx = e^x$$

$$u = x \quad \Longrightarrow \quad du = dx$$

With these substitutions, you can apply the integration by parts formula as shown.

$$\int xe^x \, dx = xe^x - \int e^x \, dx \qquad \int u \, dv = uv - \int v \, du$$

$$= xe^x - e^x + C \qquad \text{Integrate } \int e^x \, dx.$$

✓ CHECKPOINT 1

Find $\int xe^{2x} \, dx$.

STUDY TIP

In Example 1, notice that you do not need to include a constant of integration when solving $v = \int e^x \, dx = e^x$. To see why this is true, try replacing e^x by $e^x + C_1$ in the solution.

$$\int xe^x \, dx = x(e^x + C_1) - \int (e^x + C_1) \, dx$$

After integrating, you can see that the terms involving C_1 subtract out.

TECHNOLOGY

If you have access to a symbolic integration utility, try using it to solve several of the exercises in this section. Note that the form of the integral may be slightly different from what you obtain when solving the exercise by hand.

STUDY TIP

To remember the integration by parts formula, you might like to use the "Z" pattern below. The top row represents the original integral, the diagonal row represents uv, and the bottom row represents the new integral.

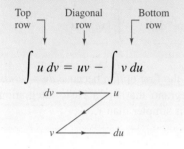

Example 2 Integration by Parts

Find $\int x^2 \ln x \, dx$.

SOLUTION For this integral, x^2 is more easily integrated than $\ln x$. Furthermore, the derivative of $\ln x$ is simpler than $\ln x$. So, you should choose $dv = x^2 \, dx$.

$$dv = x^2 \, dx \quad \Longrightarrow \quad v = \int dv = \int x^2 \, dx = \frac{x^3}{3}$$

$$u = \ln x \quad \Longrightarrow \quad du = \frac{1}{x} \, dx$$

Using these substitutions, apply the integration by parts formula as shown.

$$\int x^2 \ln x \, dx = \frac{x^3}{3} \ln x - \int \left(\frac{x^3}{3}\right)\left(\frac{1}{x}\right) dx \qquad \int u \, dv = uv - \int v \, du$$

$$= \frac{x^3}{3} \ln x - \frac{1}{3} \int x^2 \, dx \qquad \text{Simplify.}$$

$$= \frac{x^3}{3} \ln x - \frac{x^3}{9} + C \qquad \text{Integrate.}$$

✓ **CHECKPOINT 2**

Find $\int x \ln x \, dx$.

Example 3 Integration by Parts with a Single Factor

Find $\int \ln x \, dx$.

SOLUTION This integral is unusual because it has only one factor. In such cases, you should choose $dv = dx$ and choose u to be the single factor.

$$dv = dx \quad \Longrightarrow \quad v = \int dv = \int dx = x$$

$$u = \ln x \quad \Longrightarrow \quad du = \frac{1}{x} \, dx$$

Using these substitutions, apply the integration by parts formula as shown.

$$\int \ln x \, dx = x \ln x - \int (x)\left(\frac{1}{x}\right) dx \qquad \int u \, dv = uv - \int v \, du$$

$$= x \ln x - \int dx \qquad \text{Simplify.}$$

$$= x \ln x - x + C \qquad \text{Integrate.}$$

✓ **CHECKPOINT 3**

Differentiate $y = x \ln x - x + C$ to show that it is the antiderivative of $\ln x$. ∎

Example 4 Using Integration by Parts Repeatedly

Find $\int x^2 e^x \, dx$.

SOLUTION Using the guidelines, notice that the derivative of x^2 becomes simpler, whereas the derivative of e^x does not. So, you should let $u = x^2$ and let $dv = e^x \, dx$.

$$dv = e^x \, dx \implies v = \int dv = \int e^x \, dx = e^x$$

$$u = x^2 \implies du = 2x \, dx$$

Using these substitutions, apply the integration by parts formula as shown.

$$\int x^2 e^x \, dx = x^2 e^x - \int 2x e^x \, dx \quad \text{First application of integration by parts}$$

To evaluate the new integral on the right, apply integration by parts a second time, using the substitutions below.

$$dv = e^x \, dx \implies v = \int dv = \int e^x \, dx = e^x$$

$$u = 2x \implies du = 2 \, dx$$

Using these substitutions, apply the integration by parts formula as shown.

$$\int x^2 e^x \, dx = x^2 e^x - \int 2x e^x \, dx \quad \text{First application of integration by parts}$$

$$= x^2 e^x - \left(2x e^x - \int 2 e^x \, dx\right) \quad \text{Second application of integration by parts}$$

$$= x^2 e^x - 2x e^x + 2 e^x + C \quad \text{Integrate.}$$

$$= e^x(x^2 - 2x + 2) + C \quad \text{Simplify.}$$

You can confirm this result by differentiating.

✓ CHECKPOINT 4

Find $\int x^3 e^x \, dx$.

When making repeated applications of integration by parts, be careful not to interchange the substitutions in successive applications. For instance, in Example 4, the first substitutions were $dv = e^x \, dx$ and $u = x^2$. If in the second application you had switched to $dv = 2x \, dx$ and $u = e^x$, you would have reversed the previous integration and returned to the *original* integral.

$$\int x^2 e^x \, dx = x^2 e^x - \left(x^2 e^x - \int x^2 e^x \, dx\right)$$

$$= \int x^2 e^x \, dx$$

STUDY TIP

Remember that you can check an indefinite integral by differentiating. For instance, in Example 4, try differentiating the antiderivative

$$e^x(x^2 - 2x + 2) + C$$

to check that you obtain the original integrand, $x^2 e^x$.

Example 5 Evaluating a Definite Integral

Evaluate $\int_1^e \ln x \, dx$.

SOLUTION Integration by parts was used to find the antiderivative of $\ln x$ in Example 3. Using this result, you can evaluate the definite integral as shown.

$$\int_1^e \ln x \, dx = \Big[x \ln x - x \Big]_1^e \qquad \text{Use result of Example 3.}$$
$$= (e \ln e - e) - (1 \ln 1 - 1) \qquad \text{Apply Fundamental Theorem.}$$
$$= (e - e) - (0 - 1)$$
$$= 1 \qquad \text{Simplify.}$$

The area represented by this definite integral is shown in Figure 7.1.

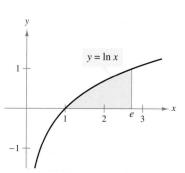

FIGURE 7.1

✓ CHECKPOINT 5

Evaluate $\int_0^1 x^2 e^x \, dx$.

Before starting the exercises in this section, remember that it is not enough to know *how* to use the various integration techniques. You also must know *when* to use them. Integration is first and foremost a problem of recognition—recognizing which formula or technique to apply to obtain an antiderivative. Often, a slight alteration of an integrand will necessitate the use of a different integration technique. Here are some examples.

Integral	Technique	Antiderivative
$\int x \ln x \, dx$	Integration by parts	$\dfrac{x^2}{2} \ln x - \dfrac{x^2}{4} + C$
$\int \dfrac{\ln x}{x} \, dx$	Power Rule: $\int u^n \dfrac{du}{dx} dx$	$\dfrac{(\ln x)^2}{2} + C$
$\int \dfrac{1}{x \ln x} \, dx$	Log Rule: $\int \dfrac{1}{u} \dfrac{du}{dx} dx$	$\ln|\ln x| + C$

As you gain experience with integration by parts, your skill in determining u and dv will improve. The summary below gives suggestions for choosing u and dv.

Summary of Common Uses of Integration by Parts

1. $\int x^n e^{ax} \, dx$ Let $u = x^n$ and $dv = e^{ax} \, dx$. (Examples 1 and 4)

2. $\int x^n \ln x \, dx$ Let $u = \ln x$ and $dv = x^n \, dx$. (Examples 2 and 3)

SECTION 7.1 Integration by Parts

Researchers such as psychologists use definite integrals to represent the probability that an event will occur. For instance, a probability of 0.5 means that an event will occur about 50% of the time.

Application

Integration can be used to find the probability that an event will occur. In such an application, the real-life situation is modeled by a *probability density function f*, and the probability that x will lie between a and b is represented by

$$P(a \leq x \leq b) = \int_a^b f(x)\, dx.$$

The probability $P(a \leq x \leq b)$ must be a number between 0 and 1.

Example 6 Finding a Probability

A psychologist finds that the probability that a participant in a memory experiment will recall between a and b percent (in decimal form) of the material is

$$P(a \leq x \leq b) = \int_a^b \frac{-64}{9e^{-8} - 1} xe^{-8x}\, dx, \quad 0 \leq a \leq b \leq 1.$$

Find the probability that a randomly chosen participant will recall between 0% and 60% of the material.

SOLUTION You can use the Constant Multiple Rule to rewrite the integral as

$$\frac{-64}{9e^{-8} - 1} \int_a^b xe^{-8x}\, dx.$$

Using integration by parts, let $dv = e^{-8x}\, dx$.

$$dv = e^{-8x}\, dx \quad \Longrightarrow \quad v = \int dv = \int e^{-8x}\, dx = -\frac{1}{8} e^{-8x}$$

$$u = x \quad \Longrightarrow \quad du = dx$$

This implies that

$$\int xe^{-8x}\, dx = x\left(-\frac{1}{8} e^{-8x}\right) + \frac{1}{8} \int e^{-8x}\, dx$$

$$= -\frac{1}{8} xe^{-8x} + \frac{1}{8}\left(-\frac{1}{8} e^{-8x}\right) = -\frac{1}{8} e^{-8x}\left(x + \frac{1}{8}\right).$$

So, the probability with $a = 0$ and $b = 0.6$ is

$$\frac{-64}{9e^{-8} - 1} \int_0^{0.6} xe^{-8x}\, dx = \left(\frac{-64}{9e^{-8} - 1}\right)\left[-\frac{1}{8} e^{-8x}\left(x + \frac{1}{8}\right)\right]_0^{0.6} \approx 0.955.$$

The probability is about 95.5%, as indicated in Figure 7.2.

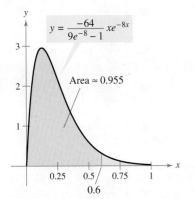

FIGURE 7.2

✓ CHECKPOINT 6

Use Example 6 to find the probability that a participant will recall between 0% and 40% of the material. ∎

CONCEPT CHECK

1. Integration by parts is based on what differentiation rule?
2. Write the formula for integration by parts.
3. State the guidelines for integration by parts.
4. Without integrating, which formula or technique of integration would you use to find $\int xe^{4x}\, dx$? Explain your reasoning.

Skills Review 7.1

The following warm-up exercises involve skills that were covered in earlier sections. You will use these skills in the exercise set for this section. For additional help, review Sections 4.3, 4.5, and 6.5.

In Exercises 1–6, find $f'(x)$.

1. $f(x) = \ln(x + 1)$
2. $f(x) = \ln(x^2 - 1)$
3. $f(x) = e^{x^3}$
4. $f(x) = e^{-x^2}$
5. $f(x) = x^2 e^x$
6. $f(x) = xe^{-2x}$

In Exercises 7–10, find the area between the graphs of f and g.

7. $f(x) = -x^2 + 4$, $g(x) = x^2 - 4$
8. $f(x) = -x^2 + 2$, $g(x) = 1$
9. $f(x) = 4x$, $g(x) = x^2 - 5$
10. $f(x) = x^3 - 3x^2 + 2$, $g(x) = x - 1$

Exercises 7.1

See www.CalcChat.com for worked-out solutions to odd-numbered exercises.

In Exercises 1–4, identify u and dv for finding the integral using integration by parts. (Do not evaluate the integral.)

1. $\int xe^{3x}\, dx$
2. $\int x^2 e^{3x}\, dx$
3. $\int x \ln 2x\, dx$
4. $\int \ln 4x\, dx$

In Exercises 5–10, use integration by parts to find the indefinite integral.

5. $\int xe^{3x}\, dx$
6. $\int xe^{-x}\, dx$
7. $\int x^2 e^{-x}\, dx$
8. $\int x^2 e^{2x}\, dx$
9. $\int \ln 2x\, dx$
10. $\int \ln x^2\, dx$

In Exercises 11–36, find the indefinite integral. (Hint: Integration by parts is not required for all the integrals.)

11. $\int e^{4x}\, dx$
12. $\int e^{-2x}\, dx$
13. $\int xe^{4x}\, dx$
14. $\int xe^{-2x}\, dx$
15. $\int xe^{x^2}\, dx$
16. $\int x^2 e^{x^3}\, dx$
17. $\int \dfrac{x}{e^x}\, dx$
18. $\int \dfrac{2x}{e^x}\, dx$
19. $\int 2x^2 e^x\, dx$
20. $\int \dfrac{1}{2}x^3 e^x\, dx$
21. $\int t \ln(t + 1)\, dt$
22. $\int x^3 \ln x\, dx$
23. $\int (x - 1)e^x\, dx$
24. $\int x^4 \ln x\, dx$
25. $\int \dfrac{e^{1/t}}{t^2}\, dt$
26. $\int \dfrac{1}{x(\ln x)^3}\, dx$
27. $\int x(\ln x)^2\, dx$
28. $\int \ln 3x\, dx$
29. $\int \dfrac{(\ln x)^2}{x}\, dx$
30. $\int \dfrac{1}{x \ln x}\, dx$
31. $\int \dfrac{\ln x}{x^2}\, dx$
32. $\int \dfrac{\ln 2x}{x^2}\, dx$
33. $\int x\sqrt{x - 1}\, dx$
34. $\int \dfrac{x}{\sqrt{x - 1}}\, dx$
35. $\int x(x + 1)^2\, dx$
36. $\int \dfrac{x}{\sqrt{2 + 3x}}\, dx$

In Exercises 37–44, evaluate the definite integral.

37. $\int_1^2 x^2 e^x\, dx$
38. $\int_0^2 \dfrac{x^2}{e^x}\, dx$
39. $\int_0^4 \dfrac{x}{e^{x/2}}\, dx$
40. $\int_1^2 x^2 \ln x\, dx$
41. $\int_1^e x^5 \ln x\, dx$
42. $\int_1^e 2x \ln x\, dx$
43. $\int_{-1}^0 \ln(x + 2)\, dx$
44. $\int_0^1 \ln(1 + 2x)\, dx$

In Exercises 45–48, find the area of the region bounded by the graphs of the equations. Then use a graphing utility to graph the region and verify your answer.

45. $y = x^3 e^x$, $y = 0$, $x = 0$, $x = 2$
46. $y = (x^2 - 1)e^x$, $y = 0$, $x = -1$, $x = 1$
47. $y = x^2 \ln x$, $y = 0$, $x = 1$, $x = e$
48. $y = \dfrac{\ln x}{x^2}$, $y = 0$, $x = 1$, $x = e$

In Exercises 49 and 50, use integration by parts to verify the formula.

49. $\displaystyle\int x^n \ln x \, dx = \dfrac{x^{n+1}}{(n+1)^2}[-1 + (n+1)\ln x] + C$, $n \ne -1$

50. $\displaystyle\int x^n e^{ax} \, dx = \dfrac{x^n e^{ax}}{a} - \dfrac{n}{a}\int x^{n-1} e^{ax} \, dx$, $n > 0$

In Exercises 51–54, use the results of Exercises 49 and 50 to find the indefinite integral.

51. $\displaystyle\int x^2 e^{5x} \, dx$
52. $\displaystyle\int x e^{-3x} \, dx$
53. $\displaystyle\int x^{-2} \ln x \, dx$
54. $\displaystyle\int x^{1/2} \ln x \, dx$

In Exercises 55–58, find the area of the region bounded by the graphs of the given equations.

55. $y = xe^{-x}$, $y = 0$, $x = 4$

56. $y = \tfrac{1}{9}xe^{-x/3}$, $y = 0$, $x = 0$, $x = 3$

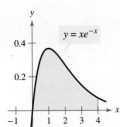
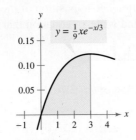

57. $y = x \ln x$, $y = 0$, $x = e$

58. $y = x^{-3} \ln x$, $y = 0$, $x = e$

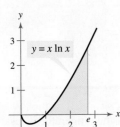

In Exercises 59–62, use a symbolic integration utility to evaluate the integral.

59. $\displaystyle\int_0^2 t^3 e^{-4t} \, dt$
60. $\displaystyle\int_1^4 \ln x (x^2 + 4) \, dx$
61. $\displaystyle\int_0^5 x^4 (25 - x^2)^{3/2} \, dx$
62. $\displaystyle\int_1^e x^9 \ln x \, dx$

63. **Probability** The probability of recall in a memory experiment is modeled by
$$P(a \le x \le b) = \int_a^b \dfrac{e}{e-2} xe^{-x} \, dx, \quad 0 \le a \le b \le 1$$
where x is the percent of recall (in decimal form). What is the probability of recalling (a) between 40% and 80% and (b) between 0% and 50%?

64. **Ore Samples** The probability of finding between a and b percent iron in ore samples is modeled by
$$P(a \le x \le b) = \int_a^b \dfrac{75}{14}\left(\dfrac{x}{\sqrt{4 + 5x}}\right) dx, \quad 0 \le a \le b \le 1$$
where x is the percent of iron in a sample (in decimal form). What is the probability of finding (a) between 0% and 25% iron and (b) between 50% and 100% iron in a sample?

65. **Memory Model** A model for the ability M of a child to memorize, measured on a scale from 0 to 10, is
$$M = 1 + 1.6t \ln t, \quad 0 < t \le 4$$
where t is the child's age in years. Find the average values of this model (a) between the child's first and second birthdays and (b) between the child's third and fourth birthdays.

66. **Growth Rate** The rate of change in the length L (in centimeters) of a species of vine can be modeled by
$$\dfrac{dL}{dt} = 0.18te^{0.01t}, \quad 0 \le t \le 30$$
where t is the number of days after germination has taken place. Find the net change in the length of the vine between 7 and 21 days after germination.

67. **Hispanic Enrollment in Schools** The percent P of Hispanic enrollment in U.S. public schools from 2001 through 2005 can be modeled by
$$P = 17.3 + 0.70\sqrt{t} \ln t, \quad 1 \le t \le 5$$
where t is the year, with $t = 1$ corresponding to 2001. Find the average percent of Hispanic enrollment from 2001 through 2005. *(Source: National Center for Education Statistics)*

68. **Extended Application** To work an extended application analyzing the numbers of children served in federally supported programs for the disabled from 1995 through 2006, visit this text's website at *college.hmco.com/info/larsonapplied*. *(Data Source: National Center for Education Statistics)*

Section 7.2

Partial Fractions and Logistic Growth

- Use partial fractions to find indefinite integrals.
- Use logistic growth functions to model real-life situations.

Partial Fractions

In Sections 6.2 and 7.1, you studied integration by substitution and by parts. In this section you will study a third technique called **partial fractions.** This technique involves the decomposition of a rational function into the sum of two or more simple rational functions. For instance, suppose you know that

$$\frac{x+7}{x^2-x-6} = \frac{2}{x-3} - \frac{1}{x+2}.$$

Knowing the "partial fractions" on the right side will allow you to integrate the left side as shown.

$$\int \frac{x+7}{x^2-x-6}\,dx = \int \left(\frac{2}{x-3} - \frac{1}{x+2}\right)dx$$

$$= 2\int \frac{1}{x-3}\,dx - \int \frac{1}{x+2}\,dx$$

$$= 2\ln|x-3| - \ln|x+2| + C$$

This method depends on your ability to factor the denominator of the original rational function *and* to find the partial fraction decomposition of the function.

STUDY TIP

Recall that finding the partial fraction decomposition of a rational function is a *precalculus* topic. Explain how you could verify that

$$\frac{1}{x-1} + \frac{2}{x+2}$$

is the partial fraction decomposition of

$$\frac{3x}{x^2+x-2}.$$

Partial Fractions

To find the partial fraction decomposition of the *proper* rational function $p(x)/q(x)$, factor $q(x)$ and write an equation that has the form

$$\frac{p(x)}{q(x)} = (\text{sum of partial fractions}).$$

For each *distinct* linear factor $ax + b$, the right side should include a term of the form

$$\frac{A}{ax+b}.$$

For each *repeated* linear factor $(ax + b)^n$, the right side should include n terms of the form

$$\frac{A_1}{ax+b} + \frac{A_2}{(ax+b)^2} + \cdots + \frac{A_n}{(ax+b)^n}.$$

STUDY TIP

A rational function $p(x)/q(x)$ is *proper* if the degree of the numerator is less than the degree of the denominator.

SECTION 7.2 Partial Fractions and Logistic Growth

Example 1 Finding a Partial Fraction Decomposition

Write the partial fraction decomposition for

$$\frac{x+7}{x^2-x-6}.$$

SOLUTION Begin by factoring the denominator as $x^2 - x - 6 = (x-3)(x+2)$. Then, write the partial fraction decomposition as

$$\frac{x+7}{x^2-x-6} = \frac{A}{x-3} + \frac{B}{x+2}.$$

To solve this equation for A and B, multiply each side of the equation by the least common denominator $(x-3)(x+2)$. This produces the **basic equation** as shown.

$x + 7 = A(x+2) + B(x-3)$ Basic equation

Because this equation is true for all x, you can substitute any convenient values of x into the equation. The x-values that are especially convenient are the ones that make particular factors equal to 0.

> **Algebra Review**
> You can check the result in Example 1 by subtracting the partial fractions to obtain the original fraction, as shown in Example 1(a) in the *Chapter 7 Algebra Review*, on page 481.

To solve for B, substitute $x = -2$:

$x + 7 = A(x+2) + B(x-3)$	Write basic equation.
$-2 + 7 = A(-2+2) + B(-2-3)$	Substitute -2 for x.
$5 = A(0) + B(-5)$	Simplify.
$-1 = B$	Solve for B.

To solve for A, substitute $x = 3$:

$x + 7 = A(x+2) + B(x-3)$	Write basic equation.
$3 + 7 = A(3+2) + B(3-3)$	Substitute 3 for x.
$10 = A(5) + B(0)$	Simplify.
$2 = A$	Solve for A.

Now that you have solved the basic equation for A and B, you can write the partial fraction decomposition as

$$\frac{x+7}{x^2-x-6} = \frac{2}{x-3} - \frac{1}{x+2}$$

as indicated at the beginning of this section.

✓**CHECKPOINT 1**

Write the partial fraction decomposition for $\dfrac{x+8}{x^2+7x+12}$. ∎

STUDY TIP
Be sure you see that the substitutions for x in Example 1 are chosen for their convenience in solving for A and B. The value $x = -2$ is selected because it eliminates the term $A(x+2)$, and the value $x = 3$ is chosen because it eliminates the term $B(x-3)$.

TECHNOLOGY

The use of partial fractions depends on your ability to factor the denominator. If this cannot be easily done, then partial fractions should not be used. For instance, consider the integral

$$\int \frac{5x^2 + 20x + 6}{x^3 + 2x^2 + x + 1} dx.$$

This integral is only slightly different from that in Example 2, yet it is immensely more difficult to solve. A symbolic integration utility was unable to solve this integral. Of course, if the integral is a definite integral (as is true in many applied problems), then you can use an approximation technique such as the Midpoint Rule (see Section 7.4).

Algebra Review

You can check the partial fraction decomposition in Example 2 by combining the partial fractions to obtain the original fraction, as shown in Example 1(b) in the *Chapter 7 Algebra Review*, on page 481. Also, for help with the algebra used to simplify the answer, see Example 1(c) on page 481.

Example 2 Integrating with Repeated Factors

Find $\int \dfrac{5x^2 + 20x + 6}{x^3 + 2x^2 + x} dx$.

SOLUTION Begin by factoring the denominator as $x(x+1)^2$. Then, write the partial fraction decomposition as

$$\frac{5x^2 + 20x + 6}{x(x+1)^2} = \frac{A}{x} + \frac{B}{x+1} + \frac{C}{(x+1)^2}.$$

To solve this equation for A, B, and C, multiply each side of the equation by the least common denominator $x(x+1)^2$.

$$5x^2 + 20x + 6 = A(x+1)^2 + Bx(x+1) + Cx \qquad \text{Basic equation}$$

Now, solve for A and C by substituting $x = -1$ and $x = 0$ into the basic equation.

Substitute $x = -1$:

$$5(-1)^2 + 20(-1) + 6 = A(-1+1)^2 + B(-1)(-1+1) + C(-1)$$
$$-9 = A(0) + B(0) - C$$
$$9 = C \qquad \text{Solve for } C.$$

Substitute $x = 0$:

$$5(0)^2 + 20(0) + 6 = A(0+1)^2 + B(0)(0+1) + C(0)$$
$$6 = A(1) + B(0) + C(0)$$
$$6 = A \qquad \text{Solve for } A.$$

At this point, you have exhausted the convenient choices for x and have yet to solve for B. When this happens, you can use *any* other x-value along with the known values of A and C.

Substitute $x = 1$, $A = 6$, and $C = 9$:

$$5(1)^2 + 20(1) + 6 = (6)(1+1)^2 + B(1)(1+1) + (9)(1)$$
$$31 = 6(4) + B(2) + 9(1)$$
$$-1 = B \qquad \text{Solve for } B.$$

Now that you have solved for A, B, and C, you can use the partial fraction decomposition to integrate.

$$\int \frac{5x^2 + 20x + 6}{x^3 + 2x^2 + x} dx = \int \left(\frac{6}{x} - \frac{1}{x+1} + \frac{9}{(x+1)^2} \right) dx$$

$$= 6 \ln|x| - \ln|x+1| + 9 \frac{(x+1)^{-1}}{-1} + C$$

$$= \ln \left| \frac{x^6}{x+1} \right| - \frac{9}{x+1} + C$$

✓ CHECKPOINT 2

Find $\int \dfrac{3x^2 + 7x + 4}{x^3 + 4x^2 + 4x} dx$. ■

SECTION 7.2 Partial Fractions and Logistic Growth

You can use the partial fraction decomposition technique outlined in Examples 1 and 2 only with a *proper* rational function—that is, a rational function whose numerator is of lower degree than its denominator. If the numerator is of equal or greater degree, you must divide first. For instance, the rational function

$$\frac{x^3}{x^2+1}$$

is improper because the degree of the numerator is greater than the degree of the denominator. Before applying partial fractions to this function, you should divide the denominator into the numerator to obtain

$$\frac{x^3}{x^2+1} = x - \frac{x}{x^2+1}.$$

Example 3 Integrating an Improper Rational Function

Find $\int \frac{x^5+x-1}{x^4-x^3} \, dx$.

SOLUTION This rational function is improper—its numerator has a degree greater than that of its denominator. So, you should begin by dividing the denominator into the numerator to obtain

$$\frac{x^5+x-1}{x^4-x^3} = x + 1 + \frac{x^3+x-1}{x^4-x^3}.$$

Now, applying partial fraction decomposition produces

$$\frac{x^3+x-1}{x^3(x-1)} = \frac{A}{x} + \frac{B}{x^2} + \frac{C}{x^3} + \frac{D}{x-1}.$$

Multiplying both sides by the least common denominator $x^3(x-1)$ produces the basic equation.

$$x^3+x-1 = Ax^2(x-1) + Bx(x-1) + C(x-1) + Dx^3 \quad \text{Basic equation}$$

Using techniques similar to those in the first two examples, you can solve for A, B, C, and D to obtain

$$A = 0, \quad B = 0, \quad C = 1, \quad \text{and} \quad D = 1.$$

So, you can integrate as shown.

$$\int \frac{x^5+x-1}{x^4-x^3} \, dx = \int \left(x + 1 + \frac{x^3+x-1}{x^4-x^3} \right) dx$$

$$= \int \left(x + 1 + \frac{1}{x^3} + \frac{1}{x-1} \right) dx$$

$$= \frac{x^2}{2} + x - \frac{1}{2x^2} + \ln|x-1| + C$$

> **Algebra Review**
> You can check the partial fraction decomposition in Example 3 by combining the partial fractions to obtain the original fraction, as shown in Example 2(a) in the *Chapter 7 Algebra Review*, on page 482.

✓ CHECKPOINT 3

Find $\int \frac{x^4-x^3+2x^2+x+1}{x^3+x^2} \, dx$.

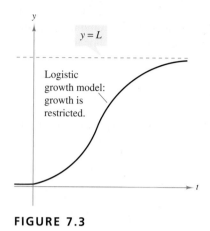

FIGURE 7.3

Logistic Growth Function

In Section 4.6, you saw that exponential growth occurs in situations for which the rate of growth is proportional to the quantity present at any given time. That is, if y is the quantity at time t, then $dy/dt = ky$. The general solution of this differential equation is $y = Ce^{kt}$. Exponential growth is unlimited. As long as C and k are positive, the value of Ce^{kt} can be made arbitrarily large by choosing sufficiently large values of t.

In many real-life situations, however, the growth of a quantity is limited and cannot increase beyond a certain size L, as shown in Figure 7.3. This upper limit L is called the **carrying capacity,** which is the maximum population $y(t)$ that can be sustained or supported as time t increases. A model that is often used for this type of growth is the **logistic differential equation**

$$\frac{dy}{dt} = ky\left(1 - \frac{y}{L}\right) \qquad \text{Logistic differential equation}$$

where k and L are positive constants. A population that satisfies this equation does not grow without bound, but approaches L as t increases. The general solution of this differential equation is called the *logistic growth model* and is derived in Example 4.

STUDY TIP
The graph of
$$y = \frac{L}{1 + be^{-kt}}$$
is called the *logistic curve,* as shown in Figure 7.3.

Algebra Review
For help with the algebra used to solve for y in Example 4, see Example 2(b) in the *Chapter 7 Algebra Review,* on page 482.

✓ CHECKPOINT 4
Show that if
$$y = \frac{1}{1 + be^{-kt}}, \text{ then }$$
$$\frac{dy}{dt} = ky(1 - y).$$

[*Hint:* First find $ky(1 - y)$ in terms of t, then find dy/dt and show that they are equivalent.] ■

Example 4 Deriving the Logistic Growth Model

Solve the equation $\dfrac{dy}{dt} = ky\left(1 - \dfrac{y}{L}\right)$.

SOLUTION

$$\frac{dy}{dt} = ky\left(1 - \frac{y}{L}\right) \qquad \text{Write differential equation.}$$

$$\frac{1}{y(1 - y/L)}\, dy = k\, dt \qquad \text{Write in differential form.}$$

$$\int \frac{1}{y(1 - y/L)}\, dy = \int k\, dt \qquad \text{Integrate each side.}$$

$$\int \left(\frac{1}{y} + \frac{1}{L - y}\right) dy = \int k\, dt \qquad \text{Rewrite left side using partial fractions.}$$

$$\ln|y| - \ln|L - y| = kt + C \qquad \text{Find antiderivative of each side.}$$

$$\ln\left|\frac{L - y}{y}\right| = -kt - C \qquad \text{Multiply each side by } -1 \text{ and simplify.}$$

$$\left|\frac{L - y}{y}\right| = e^{-kt - C} = e^{-C}e^{-kt} \qquad \text{Exponentiate each side.}$$

$$\frac{L - y}{y} = be^{-kt} \qquad \text{Let } \pm e^{-C} = b.$$

Solving this equation for y produces the **logistic growth model** $y = \dfrac{L}{1 + be^{-kt}}$.

SECTION 7.2 Partial Fractions and Logistic Growth

Example 5 Comparing Logistic Growth Functions

Use a graphing utility to investigate the effects of the values of L, b, and k on the graph of

$$y = \frac{L}{1 + be^{-kt}}.$$ Logistic growth function ($L > 0, b > 0, k > 0$)

SOLUTION The value of L determines the horizontal asymptote of the graph to the right. In other words, as t increases without bound, the graph approaches a limit of L (see Figure 7.4).

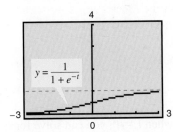 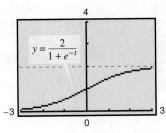 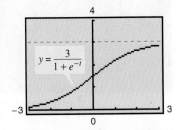

FIGURE 7.4

The value of b determines the point of inflection of the graph. If $0 < b < 1$, the point of inflection is to the left of the y-axis. If $b = 1$, the point of inflection occurs at $t = 0$. If $b > 1$, the point of inflection is to the right of the y-axis. (See Figure 7.5.)

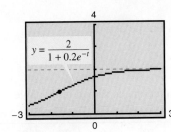 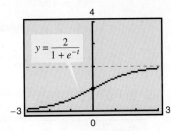 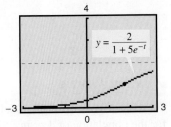

FIGURE 7.5

The value of k determines the rate of growth of the graph. For fixed values of b and L, larger values of k correspond to higher rates of growth (see Figure 7.6).

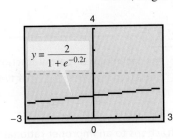 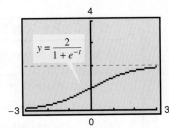 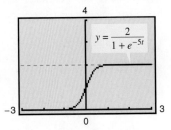

FIGURE 7.6

✓ CHECKPOINT 5

Find the horizontal asymptote of the graph of $y = \dfrac{4}{1 + 5e^{-6t}}.$

442 CHAPTER 7 Techniques of Integration

Daniel J. Cox/Getty Images

Example 6 Modeling a Population

The state game commission releases 100 deer into a game preserve. During the first 5 years, the population increases to 432 deer. The commission believes that the population can be modeled by logistic growth with a limit of 2000 deer. Write a logistic growth model for this population. Then use the model to create a table showing the size of the deer population over the next 30 years.

SOLUTION Let y represent the number of deer in year t. Assuming a logistic growth model means that the rate of change in the population is proportional to both y and $(1 - y/2000)$. That is

$$\frac{dy}{dt} = ky\left(1 - \frac{y}{2000}\right), \quad 100 \leq y \leq 2000.$$

The solution of this equation is

$$y = \frac{2000}{1 + be^{-kt}}.$$

Using the fact that $y = 100$ when $t = 0$, you can solve for b.

$$100 = \frac{2000}{1 + be^{-k(0)}} \quad \Longrightarrow \quad b = 19$$

Then, using the fact that $y = 432$ when $t = 5$, you can solve for k.

$$432 = \frac{2000}{1 + 19e^{-k(5)}} \quad \Longrightarrow \quad k \approx 0.33106$$

So, the logistic growth model for the population is

$$y = \frac{2000}{1 + 19e^{-0.33106t}}. \qquad \text{Logistic growth model}$$

The population, in five-year intervals, is shown in the table.

Time, t	0	5	10	15	20	25	30
Population, y	100	432	1181	1766	1951	1990	1998

✓ CHECKPOINT 6

Write the logistic growth model for the population of deer in Example 6 if the game preserve could contain a limit of 4000 deer.

CONCEPT CHECK

1. Complete the following: The technique of partial fractions involves the decomposition of a _____ function into the _____ of two or more simple _____ functions.

2. What is a proper rational function?

3. Before applying partial fractions to an improper rational function, what should you do?

4. Describe what the value of L represents in the logistic growth function $$y = \frac{L}{1 + be^{-kt}}.$$

SECTION 7.2 Partial Fractions and Logistic Growth

Skills Review 7.2

The following warm-up exercises involve skills that were covered in earlier sections. You will use these skills in the exercise set for this section. For additional help, review Sections 0.4 and 0.5.

In Exercises 1–8, factor the expression.

1. $x^2 - 16$
2. $x^2 - 25$
3. $x^2 - x - 12$
4. $x^2 + x - 6$
5. $x^3 - x^2 - 2x$
6. $x^3 - 4x^2 + 4x$
7. $x^3 - 4x^2 + 5x - 2$
8. $x^3 - 5x^2 + 7x - 3$

In Exercises 9–14, rewrite the improper rational expression as the sum of a proper rational expression and a polynomial.

9. $\dfrac{x^2 - 2x + 1}{x - 2}$
10. $\dfrac{2x^2 - 4x + 1}{x - 1}$
11. $\dfrac{x^3 - 3x^2 + 2}{x - 2}$
12. $\dfrac{x^3 + 2x - 1}{x + 1}$
13. $\dfrac{x^3 + 4x^2 + 5x + 2}{x^2 - 1}$
14. $\dfrac{x^3 + 3x^2 - 4}{x^2 - 1}$

Exercises 7.2

See www.CalcChat.com for worked-out solutions to odd-numbered exercises.

In Exercises 1–12, write the partial fraction decomposition for the expression.

1. $\dfrac{2(x + 20)}{x^2 - 25}$
2. $\dfrac{3x + 11}{x^2 - 2x - 3}$
3. $\dfrac{8x + 3}{x^2 - 3x}$
4. $\dfrac{10x + 3}{x^2 + x}$
5. $\dfrac{4x - 13}{x^2 - 3x - 10}$
6. $\dfrac{7x + 5}{6(2x^2 + 3x + 1)}$
7. $\dfrac{3x^2 - 2x - 5}{x^3 + x^2}$
8. $\dfrac{3x^2 - x + 1}{x(x + 1)^2}$
9. $\dfrac{x + 1}{3(x - 2)^2}$
10. $\dfrac{3x - 4}{(x - 5)^2}$
11. $\dfrac{8x^2 + 15x + 9}{(x + 1)^3}$
12. $\dfrac{6x^2 - 5x}{(x + 2)^3}$

In Exercises 13–32, use partial fractions to find the indefinite integral.

13. $\displaystyle\int \dfrac{1}{x^2 - 1}\,dx$
14. $\displaystyle\int \dfrac{4}{x^2 - 4}\,dx$
15. $\displaystyle\int \dfrac{-2}{x^2 - 16}\,dx$
16. $\displaystyle\int \dfrac{-4}{x^2 - 4}\,dx$
17. $\displaystyle\int \dfrac{1}{2x^2 - x}\,dx$
18. $\displaystyle\int \dfrac{2}{x^2 - 2x}\,dx$
19. $\displaystyle\int \dfrac{10}{x^2 - 10x}\,dx$
20. $\displaystyle\int \dfrac{5}{x^2 + x - 6}\,dx$
21. $\displaystyle\int \dfrac{3}{x^2 + x - 2}\,dx$
22. $\displaystyle\int \dfrac{1}{4x^2 - 9}\,dx$
23. $\displaystyle\int \dfrac{5 - x}{2x^2 + x - 1}\,dx$
24. $\displaystyle\int \dfrac{x + 1}{x^2 + 4x + 3}\,dx$
25. $\displaystyle\int \dfrac{x^2 - 4x - 4}{x^3 - 4x}\,dx$
26. $\displaystyle\int \dfrac{x^2 + 12x + 12}{x^3 - 4x}\,dx$
27. $\displaystyle\int \dfrac{x + 2}{x^2 - 4x}\,dx$
28. $\displaystyle\int \dfrac{4x^2 + 2x - 1}{x^3 + x^2}\,dx$
29. $\displaystyle\int \dfrac{2x - 3}{(x - 1)^2}\,dx$
30. $\displaystyle\int \dfrac{x^4}{(x - 1)^3}\,dx$
31. $\displaystyle\int \dfrac{3x^2 + 3x + 1}{x(x^2 + 2x + 1)}\,dx$
32. $\displaystyle\int \dfrac{3x}{x^2 - 6x + 9}\,dx$

In Exercises 33–40, evaluate the definite integral.

33. $\displaystyle\int_4^5 \dfrac{1}{9 - x^2}\,dx$
34. $\displaystyle\int_0^1 \dfrac{3}{2x^2 + 5x + 2}\,dx$
35. $\displaystyle\int_1^5 \dfrac{x - 1}{x^2(x + 1)}\,dx$
36. $\displaystyle\int_0^1 \dfrac{x^2 - x}{x^2 + x + 1}\,dx$
37. $\displaystyle\int_0^1 \dfrac{x^3}{x^2 - 2}\,dx$
38. $\displaystyle\int_0^1 \dfrac{x^3 - 1}{x^2 - 4}\,dx$
39. $\displaystyle\int_1^2 \dfrac{x^3 - 4x^2 - 3x + 3}{x^2 - 3x}\,dx$
40. $\displaystyle\int_2^4 \dfrac{x^4 - 4}{x^2 - 1}\,dx$

In Exercises 41–44, find the area of the shaded region.

41. $y = \dfrac{14}{16 - x^2}$

42. $y = \dfrac{-4}{x^2 - x - 6}$

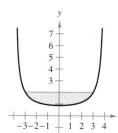
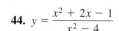

43. $y = \dfrac{x + 1}{x^2 - x}$

44. $y = \dfrac{x^2 + 2x - 1}{x^2 - 4}$

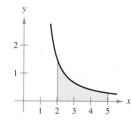
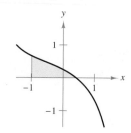

In Exercises 45 and 46, find the area of the region bounded by the graphs of the given equations.

45. $y = \dfrac{12}{x^2 + 5x + 6}, \quad y = 0, \quad x = 0, \quad x = 1$

46. $y = \dfrac{-24}{x^2 - 16}, \quad y = 0, \quad x = 1, \quad x = 3$

In Exercises 47–50, write the partial fraction decomposition for the rational expression. Check your result algebraically. Then assign a value to the constant a and use a graphing utility to check the result graphically.

47. $\dfrac{1}{a^2 - x^2}$

48. $\dfrac{1}{x(x + a)}$

49. $\dfrac{1}{x(a - x)}$

50. $\dfrac{1}{(x + 1)(a - x)}$

51. Prenatal Development The weight of a human fetus is about 0.04 ounce after 8 gestational weeks. During the next 35 weeks, the weight of this fetus increases at a rate of

$$\dfrac{dB}{dt} = \dfrac{24{,}361e^{-0.193t}}{(1 + 784e^{-0.193t})^2}, \quad 8 \le t \le 43$$

where B is the weight of the fetus in ounces and t is the time in weeks. Use the fact that $B = 0.04$ ounce when $t = 8$ to find the weight (in ounces) of the fetus after 25 weeks.

52. Medicine On a college campus, 50 students return from semester break with a contagious flu virus. The virus has a history of spreading at a rate of

$$\dfrac{dN}{dt} = \dfrac{100e^{-0.1t}}{(1 + 4e^{-0.1t})^2}$$

where N is the number of students infected after t days.

(a) Find the model giving the number of students infected with the virus in terms of the number of days since returning from semester break.

(b) If nothing is done to stop the virus from spreading, will the virus spread to infect half the student population of 1000 students? Explain your answer.

53. Biology A conservation organization releases 100 animals of an endangered species into a game preserve. The organization believes the population of the species will increase at a rate of

$$\dfrac{dN}{dt} = \dfrac{125e^{-0.125t}}{(1 + 9e^{-0.125t})^2}$$

where N is the population and t is the time in months.

(a) Use the fact that $N = 100$ when $t = 0$ to find the population after 2 years.

(b) Find the limiting size of the population as time increases without bound.

T 54. Biology A conservation organization releases 100 animals of an endangered species into a game preserve. During the first 2 years, the population increases to 134 animals. The organization believes that the preserve has a capacity of 1000 animals and that the herd will grow according to a logistic growth model. That is, the size y of the herd will follow the equation

$$\int \dfrac{1}{y(1 - y/1000)}\, dy = \int k\, dt$$

where t is measured in years. Find this logistic curve. (To solve for the constant of integration C and the proportionality constant k, assume $y = 100$ when $t = 0$ and $y = 134$ when $t = 2$.) Use a graphing utility to graph your solution.

55. Health: Epidemic A single infected individual enters a community of 500 individuals susceptible to the disease. The disease spreads at a rate proportional to the product of the total number infected and the number of susceptible individuals not yet infected. A model for the time it takes for the disease to spread to x individuals is

$$t = 5010 \int \dfrac{1}{(x + 1)(500 - x)}\, dx$$

where t is the time in hours.

(a) Find the time it takes for 75% of the population to become infected (when $t = 0$, $x = 1$).

(b) Find the number of people infected after 100 hours.

56. Biology One gram of a bacterial culture is present at time $t = 0$, and 10 grams is the upper limit of the culture's weight. The time required for the culture to grow to y grams is modeled by

$$kt = \int \frac{1}{y(1 - y/10)}\, dy$$

where y is the weight of the culture (in grams) and t is the time in hours.

(a) Verify that the weight of the culture at time t is modeled by

$$y = \frac{10}{1 + 9e^{-kt}}.$$

Use the fact that $y = 1$ when $t = 0$.

(b) Use the graph to determine the constant k.

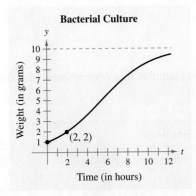

Bacterial Culture

57. Biology: Population Growth The graph shows the logistic growth curves for two species of the single-celled *Paramecium* in a laboratory culture. During which time intervals is the rate of growth of each species increasing? During which time intervals is the rate of growth of each species decreasing? Which species has a higher limiting population under these conditions? *(Source: Adapted from Levine/Miller, Biology: Discovering Life, Second Edition)*

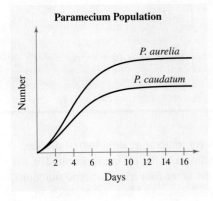

Paramecium Population

58. Environment The predicted costs C (in hundreds of thousands of dollars) for a company to remove $p\%$ of a chemical from its waste water are shown in the table.

p	0	10	20	30	40
C	0	0.7	1.0	1.3	1.7

p	50	60	70	80	90
C	2.0	2.7	3.6	5.5	11.2

A model for the data is given by

$$C = \frac{124p}{(10 + p)(100 - p)}, \quad 0 \leq p < 100.$$

Use the model to find the average cost of removing between 75% and 80% of the chemical.

59. Population Growth The population of the United States was 76 million people in 1900 and reached 300 million people in 2006. From 1900 through 2006, assume the population of the United States can be modeled by logistic growth with a limit of 839.1 million people. *(Source: U.S. Census Bureau)*

(a) Write a differential equation of the form

$$\frac{dy}{dt} = ky\left(1 - \frac{y}{L}\right)$$

where y represents the population of the United States (in millions of people) and t represents the number of years since 1900.

(b) Find the logistic growth model $y = \dfrac{L}{1 + be^{-kt}}$ for this population.

(c) Use a graphing utility to graph the model from part (b). Then estimate the year in which the population of the United States will reach 400 million people.

60. Writing What is the first step you would perform in integrating $\displaystyle\int \frac{x^2}{x - 5}\, dx$? Explain. (Do not integrate.)

61. Writing State the method you would use to evaluate each integral. Explain why you chose that method. (Do not integrate.)

(a) $\displaystyle\int \frac{2x + 1}{x^2 + x - 8}\, dx$

(b) $\displaystyle\int \frac{7x + 4}{x^2 + 2x - 8}\, dx$

Section 7.3
Integrals of Trigonometric Functions

- Find the six basic trigonometric integrals.
- Solve trigonometric integrals.
- Use trigonometric integrals to solve real-life problems.

The Six Basic Trigonometric Integrals

For each trigonometric differentiation rule, there is a corresponding integration rule. For instance, corresponding to the differentiation rule

$$\frac{d}{dx}[\cos u] = -\sin u \frac{du}{dx}$$

is the integration rule

$$\int \sin u \, du = -\cos u + C.$$

The list below contains the integration formulas that correspond to the six basic trigonometric differentiation rules.

Integrals Involving Trigonometric Functions

Differentiation Rule   *Integration Rule*

$$\frac{d}{dx}[\sin u] = \cos u \frac{du}{dx} \qquad \int \cos u \, du = \sin u + C$$

$$\frac{d}{dx}[\cos u] = -\sin u \frac{du}{dx} \qquad \int \sin u \, du = -\cos u + C$$

$$\frac{d}{dx}[\tan u] = \sec^2 u \frac{du}{dx} \qquad \int \sec^2 u \, du = \tan u + C$$

$$\frac{d}{dx}[\sec u] = \sec u \tan u \frac{du}{dx} \qquad \int \sec u \tan u \, du = \sec u + C$$

$$\frac{d}{dx}[\cot u] = -\csc^2 u \frac{du}{dx} \qquad \int \csc^2 u \, du = -\cot u + C$$

$$\frac{d}{dx}[\csc u] = -\csc u \cot u \frac{du}{dx} \qquad \int \csc u \cot u \, du = -\csc u + C$$

STUDY TIP

Note that this list gives you formulas for integrating only two of the six trigonometric functions: the sine function and the cosine function. The list does not show you how to integrate the other four trigonometric functions. Rules for integrating those functions are discussed later in this section.

SECTION 7.3 Integrals of Trigonometric Functions

Example 1 Integrating a Trigonometric Function

Find $\int 2 \cos x \, dx$.

SOLUTION Let $u = x$. Then $du = dx$.

$\int 2 \cos x \, dx = 2 \int \cos x \, dx$ Apply Constant Multiple Rule.

$= 2 \int \cos u \, du$ Substitute for x and dx.

$= 2 \sin u + C$ Integrate.

$= 2 \sin x + C$ Substitute for u.

✓ **CHECKPOINT 1**

Find $\int 5 \sin x \, dx$.

Example 2 Integrating a Trigonometric Function

Find $\int 3x^2 \sin x^3 \, dx$.

SOLUTION Let $u = x^3$. Then $du = 3x^2 \, dx$.

$\int 3x^2 \sin x^3 \, dx = \int (\sin x^3) 3x^2 \, dx$ Rewrite integrand.

$= \int \sin u \, du$ Substitute for x^3 and $3x^2 \, dx$.

$= -\cos u + C$ Integrate.

$= -\cos x^3 + C$ Substitute for u.

✓ **CHECKPOINT 2**

Find $\int 4x^3 \cos x^4 \, dx$.

Example 3 Integrating a Trigonometric Function

Find $\int \sec 3x \tan 3x \, dx$.

SOLUTION Let $u = 3x$. Then $du = 3 \, dx$.

$\int \sec 3x \tan 3x \, dx = \frac{1}{3} \int (\sec 3x \tan 3x) 3 \, dx$ Multiply and divide by 3.

$= \frac{1}{3} \int \sec u \tan u \, du$ Substitute for $3x$ and $3 \, dx$.

$= \frac{1}{3} \sec u + C$ Integrate.

$= \frac{1}{3} \sec 3x + C$ Substitute for u.

TECHNOLOGY

If you have access to a symbolic integration utility, try using it to integrate the functions in Examples 1, 2, and 3. Does your utility give the same results that are given in the examples?

✓ **CHECKPOINT 3**

Find $\int \sec^2 5x \, dx$.

CHAPTER 7 Techniques of Integration

Example 4 Integrating a Trigonometric Function

Find $\int e^x \sec^2 e^x \, dx$.

SOLUTION Let $u = e^x$. Then $du = e^x \, dx$.

$$\int e^x \sec^2 e^x \, dx = \int (\sec^2 e^x) e^x \, dx \qquad \text{Rewrite integrand.}$$

$$= \int \sec^2 u \, du \qquad \text{Substitute for } e^x \text{ and } e^x \, dx.$$

$$= \tan u + C \qquad \text{Integrate.}$$

$$= \tan e^x + C \qquad \text{Substitute for } u.$$

✓ **CHECKPOINT 4**

Find $\int 2 \csc 2x \cot 2x \, dx$.

The next two examples use the General Power Rule for integration and the General Log Rule for integration. Recall from Chapter 6 that these rules are

$$\int u^n \frac{du}{dx} \, dx = \frac{u^{n+1}}{n+1} + C, \quad n \neq -1 \qquad \text{General Power Rule}$$

and

$$\int \frac{du/dx}{u} \, dx = \ln|u| + C. \qquad \text{General Log Rule}$$

The key to using these two rules is identifying the proper substitution for u. For instance, in the next example, the proper choice for u is $\sin 4x$.

STUDY TIP

It is a good idea to check your answers to integration problems by differentiating. In Example 5, for instance, try differentiating the answer

$$y = \frac{1}{12} \sin^3 4x + C.$$

You should obtain the original integrand, as shown.

$$y' = \frac{1}{12} 3(\sin 4x)^2 (\cos 4x) 4$$

$$= \sin^2 4x \cos 4x$$

Example 5 Using the General Power Rule

Find $\int \sin^2 4x \cos 4x \, dx$.

SOLUTION Let $u = \sin 4x$. Then $du/dx = 4 \cos 4x$.

$$\int \sin^2 4x \cos 4x \, dx = \frac{1}{4} \int \overbrace{(\sin 4x)^2}^{u^2} \overbrace{(4 \cos 4x)}^{du/dx} \, dx \qquad \text{Rewrite integrand.}$$

$$= \frac{1}{4} \int u^2 \, du \qquad \text{Substitute for } \sin 4x \text{ and } 4 \cos 4x \, dx.$$

$$= \frac{1}{4} \frac{u^3}{3} + C \qquad \text{Integrate.}$$

$$= \frac{1}{4} \frac{(\sin 4x)^3}{3} + C \qquad \text{Substitute for } u.$$

$$= \frac{1}{12} \sin^3 4x + C \qquad \text{Simplify.}$$

✓ **CHECKPOINT 5**

Find $\int \cos^3 2x \sin 2x \, dx$.

Example 6 Using the Log Rule

Find $\int \dfrac{\sin x}{\cos x}\, dx$.

SOLUTION Let $u = \cos x$. Then $du/dx = -\sin x$.

$$\int \frac{\sin x}{\cos x}\, dx = -\int \frac{-\sin x}{\cos x}\, dx \qquad \text{Rewrite integrand.}$$

$$= -\int \frac{du/dx}{u}\, dx \qquad \text{Substitute for } \cos x \text{ and } -\sin x.$$

$$= -\ln|u| + C \qquad \text{Apply Log Rule.}$$

$$= -\ln|\cos x| + C \qquad \text{Substitute for } u.$$

✓ CHECKPOINT 6

Find $\int \dfrac{\cos x}{\sin x}\, dx$.

Example 7 Evaluating a Definite Integral

Evaluate $\displaystyle\int_0^{\pi/4} \cos 2x\, dx$.

SOLUTION

$$\int_0^{\pi/4} \cos 2x\, dx = \left[\frac{1}{2}\sin 2x\right]_0^{\pi/4}$$

$$= \frac{1}{2} - 0 = \frac{1}{2}$$

✓ CHECKPOINT 7

Find $\displaystyle\int_0^{\pi/2} \sin 2x\, dx$.

Example 8 Finding Area by Integration

Find the area of the region bounded by the x-axis and one arc of the graph of $y = \sin x$.

SOLUTION As indicated in Figure 7.7, this area is given by

$$\text{Area} = \int_0^{\pi} \sin x\, dx$$

$$= \Big[-\cos x\Big]_0^{\pi}$$

$$= -(-1) - (-1)$$

$$= 2.$$

So, the region has an area of 2 square units.

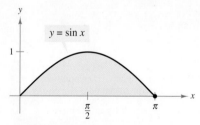

FIGURE 7.7

✓ CHECKPOINT 8

Find the area of the region bounded by the graphs of $y = \cos x$ and $y = 0$ for $0 \le x \le \dfrac{\pi}{2}$.

Other Trigonometric Integrals

At the beginning of this section, the integration rules for the sine and cosine functions were listed. Now, using the result of Example 6, you have an integration rule for the tangent function. That rule is

$$\int \tan x \, dx = \int \frac{\sin x}{\cos x} dx = -\ln|\cos x| + C.$$

Integration formulas for the other three trigonometric functions can be developed in a similar way. For instance, to obtain the integration formula for the secant function, you can integrate as shown.

$$\int \sec x \, dx = \int \frac{\sec x(\sec x + \tan x)}{\sec x + \tan x} dx$$

$$= \int \frac{\sec^2 x + \sec x \tan x}{\sec x + \tan x} dx \qquad \text{Use substitution with } u = \sec x + \tan x.$$

$$= \ln|\sec x + \tan x| + C$$

These formulas, and integration formulas for the other two trigonometric functions, are summarized below.

Integrals of Trigonometric Functions

$$\int \tan u \, du = -\ln|\cos u| + C \qquad \int \sec u \, du = \ln|\sec u + \tan u| + C$$

$$\int \cot u \, du = \ln|\sin u| + C \qquad \int \csc u \, du = \ln|\csc u - \cot u| + C$$

Example 9 Integrating a Trigonometric Function

Find $\int \tan 4x \, dx$.

SOLUTION Let $u = 4x$. Then $du = 4 \, dx$.

$$\int \tan 4x \, dx = \frac{1}{4} \int (\tan 4x) 4 \, dx \qquad \text{Rewrite integrand.}$$

$$= \frac{1}{4} \int \tan u \, du \qquad \text{Substitute for } 4x \text{ and } 4 \, dx.$$

$$= -\frac{1}{4} \ln|\cos u| + C \qquad \text{Integrate.}$$

$$= -\frac{1}{4} \ln|\cos 4x| + C \qquad \text{Substitute for } u.$$

✓ **CHECKPOINT 9**

Find $\int \sec 2x \, dx$.

Application

In the next example, recall from Section 6.4 that the average value of a function f over an interval $[a, b]$ is given by

$$\frac{1}{b-a}\int_a^b f(x)\,dx.$$

Example 10 — MAKE A DECISION: Finding an Average Temperature

The temperature T (in degrees Fahrenheit) during a 24-hour period can be modeled by

$$T = 72 + 18 \sin \frac{\pi(t-8)}{12}$$

where t is the time (in hours), with $t = 0$ corresponding to midnight. Will the average temperature during the four-hour period from noon to 4 P.M. be greater than 90°?

SOLUTION To find the average temperature A, use the formula for the average value of a function over an interval.

$$A = \frac{1}{4}\int_{12}^{16}\left[72 + 18\sin\frac{\pi(t-8)}{12}\right]dt$$

$$= \frac{1}{4}\left[72t + 18\left(\frac{12}{\pi}\right)\left(-\cos\frac{\pi(t-8)}{12}\right)\right]_{12}^{16}$$

$$= \frac{1}{4}\left[72(16) + 18\left(\frac{12}{\pi}\right)\left(\frac{1}{2}\right) - 72(12) + 18\left(\frac{12}{\pi}\right)\left(\frac{1}{2}\right)\right]$$

$$= \frac{1}{4}\left(288 + \frac{216}{\pi}\right)$$

$$= 72 + \frac{54}{\pi} \approx 89.2°$$

So, the average temperature is 89.2°, as indicated in Figure 7.8. No, the average temperature from noon to 4 P.M. will not be greater than 90°.

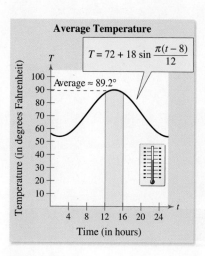

FIGURE 7.8

✓ CHECKPOINT 10

Use the function in Example 10 to find the average temperature from 9 A.M. to noon. ■

CONCEPT CHECK

1. For each trigonometric differentiation rule, is there a corresponding integration rule?

2. For the differentiation rule $\frac{d}{dx}[\sin u] = \cos u \frac{du}{dx}$, what is the corresponding integration rule?

3. For the differentiation rule $\frac{d}{dx}[\cos u] = -\sin u \frac{du}{dx}$, what is the corresponding integration rule?

4. For the integration rule $\int \sec^2 u\,du = \tan u + C$, what is the corresponding differentiation rule?

452 CHAPTER 7 Techniques of Integration

Skills Review 7.3

The following warm-up exercises involve skills that were covered in earlier sections. You will use these skills in the exercise set for this section. For additional help, review Sections 5.2 and 6.4.

In Exercises 1–8, evaluate the trigonometric function.

1. $\cos \dfrac{5\pi}{4}$
2. $\sin \dfrac{7\pi}{6}$
3. $\sin\left(-\dfrac{\pi}{3}\right)$
4. $\cos\left(-\dfrac{\pi}{6}\right)$
5. $\tan \dfrac{5\pi}{6}$
6. $\cot \dfrac{5\pi}{3}$
7. $\sec \pi$
8. $\cos \dfrac{\pi}{2}$

In Exercises 9–16, simplify the expression using the trigonometric identities.

9. $\sin x \sec x$
10. $\csc x \cos x$
11. $\cos^2 x(\sec^2 x - 1)$
12. $\sin^2 x(\csc^2 x - 1)$
13. $\sec x \sin\left(\dfrac{\pi}{2} - x\right)$
14. $\cot x \cos\left(\dfrac{\pi}{2} - x\right)$
15. $\cot x \sec x$
16. $\cot x(\sin^2 x)$

In Exercises 17–20, evaluate the definite integral.

17. $\displaystyle\int_0^4 (x^2 + 3x - 4)\, dx$
18. $\displaystyle\int_{-1}^1 (1 - x^2)\, dx$
19. $\displaystyle\int_0^2 x(4 - x^2)\, dx$
20. $\displaystyle\int_0^1 x(9 - x^2)\, dx$

Exercises 7.3

See www.CalcChat.com for worked-out solutions to odd-numbered exercises.

In Exercises 1–34, find the indefinite integral.

1. $\displaystyle\int (2\sin x + 3\cos x)\, dx$
2. $\displaystyle\int (t^2 - \sin t)\, dt$
3. $\displaystyle\int (1 - \csc t \cot t)\, dt$
4. $\displaystyle\int (\theta^2 + \sec^2 \theta)\, d\theta$
5. $\displaystyle\int (\csc^2 \theta - \cos \theta)\, d\theta$
6. $\displaystyle\int (\sec y \tan y - \sec^2 y)\, dy$
7. $\displaystyle\int \sin 2x\, dx$
8. $\displaystyle\int \cos 6x\, dx$
9. $\displaystyle\int 2x \cos x^2\, dx$
10. $\displaystyle\int 2x \sin x^2\, dx$
11. $\displaystyle\int \sec^2 \dfrac{x}{2}\, dx$
12. $\displaystyle\int \csc^2 4x\, dx$
13. $\displaystyle\int \tan 3x\, dx$
14. $\displaystyle\int \csc \dfrac{x}{3} \cot \dfrac{x}{3}\, dx$
15. $\displaystyle\int \tan^3 x \sec^2 x\, dx$
16. $\displaystyle\int \sqrt{\cot x}\, \csc^2 x\, dx$
17. $\displaystyle\int \cot \pi x\, dx$
18. $\displaystyle\int \tan 5x\, dx$
19. $\displaystyle\int \csc 2x\, dx$
20. $\displaystyle\int \sec \dfrac{x}{2}\, dx$
21. $\displaystyle\int \dfrac{\sec^2 x}{\tan x}\, dx$
22. $\displaystyle\int \dfrac{\sin x}{\cos^2 x}\, dx$
23. $\displaystyle\int \dfrac{\sec x \tan x}{\sec x - 1}\, dx$
24. $\displaystyle\int \dfrac{\cos t}{1 + \sin t}\, dt$
25. $\displaystyle\int \dfrac{\sin x}{1 + \cos x}\, dx$
26. $\displaystyle\int \dfrac{\sin \sqrt{x}}{\sqrt{x}}\, dx$
27. $\displaystyle\int \dfrac{\csc^2 x}{\cot^3 x}\, dx$
28. $\displaystyle\int \dfrac{1 - \cos \theta}{\theta - \sin \theta}\, d\theta$
29. $\displaystyle\int e^x \sin e^x\, dx$
30. $\displaystyle\int e^{-x} \tan e^{-x}\, dx$
31. $\displaystyle\int e^{\sin x} \cos x\, dx$
32. $\displaystyle\int e^{\sec x} \sec x \tan x\, dx$
33. $\displaystyle\int (\sin x + \cos x)^2\, dx$
34. $\displaystyle\int (1 + \tan \theta)^2\, d\theta$

In Exercises 35–38, use integration by parts to find the indefinite integral.

35. $\int x \cos x \, dx$

36. $\int x \sin x \, dx$

37. $\int x \sec^2 x \, dx$

38. $\int \theta \sec \theta \tan \theta \, d\theta$

In Exercises 39–46, evaluate the definite integral. Use a symbolic integration utility to verify your results.

39. $\int_0^{\pi/4} \cos \frac{4x}{3} \, dx$

40. $\int_0^{\pi/6} \sin 6x \, dx$

41. $\int_{\pi/2}^{2\pi/3} \sec^2 \frac{x}{2} \, dx$

42. $\int_0^{\pi/2} (x + \cos x) \, dx$

43. $\int_{\pi/12}^{\pi/4} \csc 2x \cot 2x \, dx$

44. $\int_0^{\pi/8} \sin 2x \cos 2x \, dx$

45. $\int_0^1 \tan(1 - x) \, dx$

46. $\int_0^{\pi/4} \sec x \tan x \, dx$

In Exercises 47–52, determine the area of the region.

47. $y = \cos \frac{x}{4}$

48. $y = \tan x$

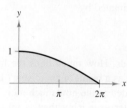
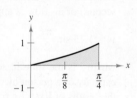

49. $y = x + \sin x$

50. $y = \frac{x}{2} + \cos x$

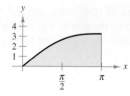
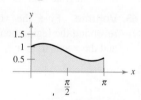

51. $y = \sin x + \cos 2x$

52. $y = 2 \sin x + \sin 2x$

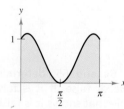
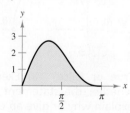

53. **Consumer Trends** Energy consumption in the United States is seasonal. For instance, primary residential energy consumption can be approximated by the model

$$Q = 588 + 390 \cos(0.46t - 0.25), \quad 0 \leq t \leq 12$$

where Q is the monthly consumption (in trillion Btu) and t is the time in months, with $t = 1$ corresponding to January. Find the average rate of domestic energy consumption during a year. *(Source: Energy Information Administration)*

54. **Inventory** The stockpile level of liquefied petroleum gases in the United States in 2006 can be approximated by the model

$$Q = 109 + 32 \cos \frac{\pi(t + 3)}{6}$$

where Q is measured in millions of barrels and t is the time in months, with $t = 1$ corresponding to January. Find the average levels given by this model during

(a) the first quarter ($0 \leq t \leq 3$).

(b) the second quarter ($3 \leq t \leq 6$).

(c) the entire year ($0 \leq t \leq 12$).

(Source: Energy Information Administration)

55. **Construction Workers** The number W (in thousands) of construction workers employed in the United States during 2006 can be modeled by

$$W = 7594 + 455.2 \sin(0.41t - 1.713)$$

where t is the time in months, with $t = 1$ corresponding to January. Use a graphing utility to estimate the average numbers of construction workers during (a) the first quarter ($0 \leq t \leq 3$), (b) the second quarter ($3 \leq t \leq 6$), and (c) the entire year ($0 \leq t \leq 12$). *(Source: U.S. Bureau of Labor Statistics)*

56. **Buffon's Needle Experiment** A piece of paper is ruled with parallel lines 2 inches apart. A two-inch needle is tossed randomly onto the piece of paper (see figure). The probability that the needle will touch a line is

$$P = \frac{2}{\pi} \int_0^{\pi/2} \sin \theta \, d\theta$$

where θ is the acute angle between the needle and any one of the parallel lines. Find this probability.

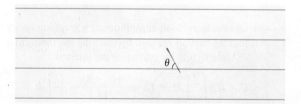

57. **Meteorology** The average monthly precipitation P in inches, including rain, snow, and ice, in Sacramento, California can be modeled by

$$P = 2.47 \sin(0.40t + 1.80) + 2.08, \quad 0 \leq t \leq 12$$

where t is the time in months, with $t = 1$ corresponding to January. Find the total annual precipitation in Sacramento. *(Source: National Oceanic and Atmospheric Administration)*

58. Meteorology The average monthly precipitation P in inches, including rain, snow, and ice, in Bismarck, North Dakota can be modeled by

$$P = 1.07 \sin(0.59t + 3.94) + 1.52, \quad 0 \le t \le 12$$

where t is the time in months, with $t = 1$ corresponding to January. *(Source: National Oceanic and Atmospheric Administration)*

(a) Find the maximum and minimum precipitation and the month in which each occurs.

(b) Determine the average monthly precipitation for the year.

(c) Find the total annual precipitation in Bismarck.

T 59. Temperature Suppose that the temperature in degrees Fahrenheit is given by

$$T = 72 + 12 \sin \frac{\pi(t - 8)}{12}$$

where t is the time in hours, with $t = 0$ corresponding to midnight. Furthermore, suppose that it costs \$0.30 to cool a particular house 1° for 1 hour.

(a) Use the integration capabilities of a graphing utility to find the cost C of cooling this house between 8 A.M. and 8 P.M., if the thermostat is set at 72° (see figure) and the cost is given by

$$C = 0.3 \int_{8}^{20} \left[72 + 12 \sin \frac{\pi(t - 8)}{12} - 72 \right] dt.$$

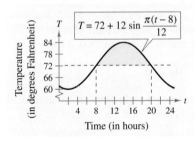

(b) Use the integration capabilities of a graphing utility to find the savings realized by resetting the thermostat to 78° (see figure) by evaluating the integral

$$C = 0.3 \int_{10}^{18} \left[72 + 12 \sin \frac{\pi(t - 8)}{12} - 78 \right] dt.$$

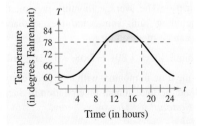

60. Water Supply A model for the flow rate of water at a pumping station on a given day is

$$R = 53 + 7 \sin\left(\frac{\pi t}{6} + 3.6\right) + 9 \cos\left(\frac{\pi t}{12} + 8.9\right)$$

where $0 \le t \le 24$. R is the flow rate in thousands of gallons per hour, and t is the time in hours.

T (a) Use a graphing utility to graph the rate function and approximate the maximum flow rate at the pumping station.

(b) Approximate the total volume of water pumped in 1 day.

61. Health For a person at rest, the velocity v (in liters per second) of air flow into and out of the lungs during a respiratory cycle is approximated by

$$v = 0.9 \sin \frac{\pi t}{3}$$

where t is the time in seconds. Find the volume in liters of air inhaled during one cycle by integrating this function over the interval $[0, 3]$.

62. Health After exercising for a few minutes, a person has a respiratory cycle for which the velocity v (in liters per second) of air flow is approximated by

$$v = 1.75 \sin \frac{\pi t}{2}$$

where t is the time in seconds. How much does the lung capacity of a person increase as a result of exercising? Use the results of Exercise 61 to determine how much more air is inhaled during a cycle after exercising than is inhaled during a cycle at rest. (Note that the cycle is shorter and you must integrate over the interval $[0, 2]$.)

63. Volume Find the volume of the solid formed by revolving the region bounded by the graph of $y = \sqrt{\sin x}$ and the x-axis $(0 \le x \le \pi)$ about the x-axis (see figure).

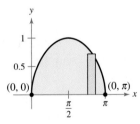

True or False? In Exercises 64 and 65, determine whether the statement is true or false. If it is false, explain why or give an example that shows it is false.

64. $\displaystyle\int_{a}^{b} \sin x \, dx = \int_{a}^{b+2\pi} \sin x \, dx$

65. $4 \displaystyle\int \sin x \cos x \, dx = 0$

Mid-Chapter Quiz

Take this quiz as you would take a quiz in class. When you are done, check your work against the answers given in the back of the book.

In Exercises 1–9, find the indefinite integral.

1. $\displaystyle\int xe^{5x}\,dx$

2. $\displaystyle\int \ln x^3\,dx$

3. $\displaystyle\int (x+1)\ln x\,dx$

4. $\displaystyle\int \frac{10}{x^2-25}\,dx$

5. $\displaystyle\int \frac{x-14}{x^2+2x-8}\,dx$

6. $\displaystyle\int \frac{5x-1}{(x+1)^2}\,dx$

7. $\displaystyle\int \sin 5x\,dx$

8. $\displaystyle\int x\csc x^2\,dx$

9. $\displaystyle\int \frac{\cos\sqrt{x}}{2\sqrt{x}}\,dx$

In Exercises 10–18, evaluate the definite integral.

10. $\displaystyle\int_{-2}^{0} xe^{x/2}\,dx$

11. $\displaystyle\int_{1}^{e} (\ln x)^2\,dx$

12. $\displaystyle\int_{0}^{1} \frac{x}{\sqrt{1+2x}}\,dx$

13. $\displaystyle\int_{1}^{4} \frac{3x+1}{x(x+1)}\,dx$

14. $\displaystyle\int_{4}^{5} \frac{120}{(x-3)(x+5)}\,dx$

15. $\displaystyle\int_{1}^{2} \frac{1}{x(0.1+0.2x)}\,dx$

16. $\displaystyle\int_{0}^{\pi} \sec^2\frac{x}{3}\tan\frac{x}{3}\,dx$

17. $\displaystyle\int_{1/4}^{1/2} \cos \pi x\,dx$

18. $\displaystyle\int_{\pi/4}^{\pi/2} \frac{e^{\cot x}}{\sin^2 x}\,dx$

19. The probability of finding between a and b percent aluminum in bauxite (aluminum ore) samples is modeled by

$$P(a \le x \le b) = \int_{a}^{b} 2x^3 e^{x^2}\,dx,\quad 0 \le a \le b \le 1$$

where x is the percent of aluminum in a sample (in decimal form).

(a) What is the probability of finding between 0% and 40% aluminum in a sample?

(b) What is the probability of finding between 80% and 100% aluminum in a sample?

20. The population of a colony of bees can be modeled by logistic growth. The capacity of the colony's hive is 100,000 bees. One day in the early spring, there are 25,000 bees in the hive. Thirteen days later, the population of the hive has increased to 28,000 bees. Write a logistic growth model for the colony.

(T) 21. The number W (in thousands) of amusement park workers employed in the United States during 2006 can be modeled by

$$W = 139.8 + 37.33 \sin(0.612t - 2.66)$$

where t is the time in months, with $t = 1$ corresponding to January. Use a graphing utility to estimate the average number of amusement park workers during

(a) the first quarter ($0 \le t \le 3$).

(b) the second quarter ($3 \le t \le 6$).

(c) the entire year ($0 \le t \le 12$).

(Source: U.S. Bureau of Labor Statistics)

Section 7.4

The Definite Integral as the Limit of a Sum

- Use the Midpoint Rule to approximate definite integrals.
- Use a symbolic integration utility to approximate definite integrals.

The Midpoint Rule

In Section 6.4, you learned that you cannot use the Fundamental Theorem of Calculus to evaluate a definite integral unless you can find an antiderivative of the integrand. In cases where this cannot be done, you can approximate the value of the integral using an approximation technique. One such technique is called the **Midpoint Rule.** (Two other techniques are discussed in Section 7.5.)

Example 1 Approximating the Area of a Plane Region

Use the five rectangles in Figure 7.9 to approximate the area of the region bounded by the graph of $f(x) = -x^2 + 5$, the x-axis, and the lines $x = 0$ and $x = 2$.

SOLUTION You can find the heights of the five rectangles by evaluating f at the midpoint of each of the following intervals.

$$\left[0, \frac{2}{5}\right], \quad \left[\frac{2}{5}, \frac{4}{5}\right], \quad \left[\frac{4}{5}, \frac{6}{5}\right], \quad \left[\frac{6}{5}, \frac{8}{5}\right], \quad \left[\frac{8}{5}, \frac{10}{5}\right]$$

Evaluate f at the midpoints of these intervals.

The width of each rectangle is $\frac{2}{5}$. So, the sum of the five areas is

$$\text{Area} \approx \frac{2}{5}f\left(\frac{1}{5}\right) + \frac{2}{5}f\left(\frac{3}{5}\right) + \frac{2}{5}f\left(\frac{5}{5}\right) + \frac{2}{5}f\left(\frac{7}{5}\right) + \frac{2}{5}f\left(\frac{9}{5}\right)$$

$$= \frac{2}{5}\left[f\left(\frac{1}{5}\right) + f\left(\frac{3}{5}\right) + f\left(\frac{5}{5}\right) + f\left(\frac{7}{5}\right) + f\left(\frac{9}{5}\right)\right]$$

$$= \frac{2}{5}\left(\frac{124}{25} + \frac{116}{25} + \frac{100}{25} + \frac{76}{25} + \frac{44}{25}\right)$$

$$= \frac{920}{125}$$

$$= 7.36.$$

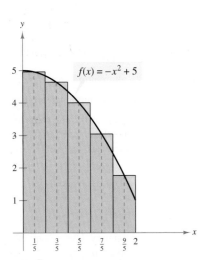

FIGURE 7.9

✓ **CHECKPOINT 1**

Use four rectangles to approximate the area of the region bounded by the graph of $f(x) = x^2 + 1$, the x-axis, $x = 0$ and $x = 2$. ■

For the region in Example 1, you can find the exact area with a definite integral. That is,

$$\text{Area} = \int_0^2 (-x^2 + 5)\, dx = \frac{22}{3} \approx 7.33.$$

The approximation procedure used in Example 1 is the **Midpoint Rule.** You can use the Midpoint Rule to approximate *any* definite integral—not just those representing area. The basic steps are summarized below.

Guidelines for Using the Midpoint Rule

To approximate the definite integral $\int_a^b f(x)\, dx$ with the Midpoint Rule, use the steps below.

1. Divide the interval $[a, b]$ into n subintervals, each of width
$$\Delta x = \frac{b-a}{n}.$$

2. Find the midpoint of each subinterval.
$$\text{Midpoints} = \{x_1, x_2, x_3, \ldots, x_n\}$$

3. Evaluate f at each midpoint and form the sum as shown.
$$\int_a^b f(x)\, dx \approx \frac{b-a}{n}[f(x_1) + f(x_2) + f(x_3) + \cdots + f(x_n)]$$

An important characteristic of the Midpoint Rule is that the approximation tends to improve as n increases. The table below shows the approximations for the area of the region described in Example 1 for various values of n. For example, for $n = 10$, the Midpoint Rule yields

$$\int_0^2 (-x^2 + 5)\, dx \approx \frac{2}{10}\left[f\left(\frac{1}{10}\right) + f\left(\frac{3}{10}\right) + \cdots + f\left(\frac{19}{10}\right)\right]$$
$$= 7.34.$$

n	5	10	15	20	25	30
Approximation	7.3600	7.3400	7.3363	7.3350	7.3344	7.3341

Note that as n increases, the approximation gets closer and closer to the exact value of the integral, which was found to be

$$\frac{22}{3} \approx 7.3333.$$

STUDY TIP

In Example 1, the Midpoint Rule is used to approximate an integral whose exact value can be found with the Fundamental Theorem of Calculus. This was done to illustrate the accuracy of the rule. In practice, of course, you would use the Midpoint Rule to approximate the values of definite integrals for which you cannot find an antiderivative. Examples 2 and 3 illustrate such integrals.

TECHNOLOGY

Programming the Midpoint Rule

The easiest way to use the Midpoint Rule to approximate the definite integral $\int_a^b f(x)\, dx$ is to program it into a computer or programmable calculator. For instance, the pseudocode below will help you write a program to evaluate the Midpoint Rule. (Appendix E lists this program for several models of graphing utilities.)

Program

- *Prompt for value of a.*
- *Input value of a.*
- *Prompt for value of b.*
- *Input value of b.*
- *Prompt for value of n.*
- *Input value of n.*
- *Initialize sum of areas.*
- *Calculate width of subinterval.*
- *Initialize counter.*
- *Begin loop.*
- *Calculate left endpoint.*
- *Calculate right endpoint.*
- *Calculate midpoint of subinterval.*
- *Add area to sum.*
- *Test counter.*
- *End loop.*
- *Display approximation.*

Before executing the program, enter the function. When the program is executed, you will be prompted to enter the lower and upper limits of integration and the number of subintervals you want to use.

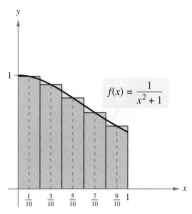

FIGURE 7.10

Example 2 Using the Midpoint Rule

Use the Midpoint Rule with $n = 5$ to approximate $\int_0^1 \frac{1}{x^2 + 1} \, dx$.

SOLUTION With $n = 5$, the interval $[0, 1]$ is divided into five subintervals.

$$\left[0, \frac{1}{5}\right], \quad \left[\frac{1}{5}, \frac{2}{5}\right], \quad \left[\frac{2}{5}, \frac{3}{5}\right], \quad \left[\frac{3}{5}, \frac{4}{5}\right], \quad \left[\frac{4}{5}, 1\right]$$

The midpoints of these intervals are $\frac{1}{10}, \frac{3}{10}, \frac{5}{10}, \frac{7}{10}$, and $\frac{9}{10}$. Because each subinterval has a width of $\Delta x = (1 - 0)/5 = \frac{1}{5}$, you can approximate the value of the definite integral as shown.

$$\int_0^1 \frac{1}{x^2 + 1} \, dx \approx \frac{1}{5}\left(\frac{1}{1.01} + \frac{1}{1.09} + \frac{1}{1.25} + \frac{1}{1.49} + \frac{1}{1.81}\right)$$
$$\approx 0.786$$

The region whose area is represented by the definite integral is shown in Figure 7.10. The actual area of this region is $\pi/4 \approx 0.785$. So, the approximation is off by only 0.001.

✓ CHECKPOINT 2

Use the Midpoint Rule with $n = 4$ to approximate the area of the region bounded by the graph of $f(x) = 1/(x^2 + 2)$, the x-axis, and the lines $x = 0$ and $x = 1$. ∎

Example 3 Using the Midpoint Rule

Use the Midpoint Rule with $n = 10$ to approximate $\int_1^3 \sqrt{x^2 + 1} \, dx$.

SOLUTION Begin by dividing the interval $[1, 3]$ into 10 subintervals. The midpoints of these intervals are

$$\frac{11}{10}, \quad \frac{13}{10}, \quad \frac{3}{2}, \quad \frac{17}{10}, \quad \frac{19}{10}, \quad \frac{21}{10}, \quad \frac{23}{10}, \quad \frac{5}{2}, \quad \frac{27}{10}, \quad \text{and} \quad \frac{29}{10}.$$

Because each subinterval has a width of $\Delta x = (3 - 1)/10 = \frac{1}{5}$, you can approximate the value of the definite integral as shown.

$$\int_1^3 \sqrt{x^2 + 1} \, dx \approx \frac{1}{5}\left[\sqrt{(1.1)^2 + 1} + \sqrt{(1.3)^2 + 1} + \cdots + \sqrt{(2.9)^2 + 1}\right]$$
$$\approx 4.504$$

The region whose area is represented by the definite integral is shown in Figure 7.11. Using techniques that are not within the scope of this course, it can be shown that the actual area is

$$\tfrac{1}{2}\left[3\sqrt{10} + \ln(3 + \sqrt{10}) - \sqrt{2} - \ln(1 + \sqrt{2})\right] \approx 4.505.$$

So, the approximation is off by only 0.001.

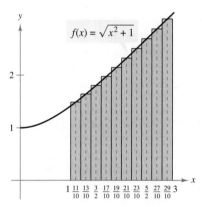

FIGURE 7.11

STUDY TIP

The Midpoint Rule is necessary for solving certain real-life problems, such as measuring irregular areas like bodies of water (see Exercise 37).

✓ CHECKPOINT 3

Use the Midpoint Rule with $n = 4$ to approximate the area of the region bounded by the graph of $f(x) = \sqrt{x^2 - 1}$, the x-axis, and the lines $x = 2$ and $x = 4$. ∎

The Definite Integral as the Limit of a Sum

Consider the closed interval $[a, b]$, divided into n subintervals whose midpoints are x_i and whose widths are $\Delta x = (b - a)/n$. In this section, you have seen that the midpoint approximation

$$\int_a^b f(x)\,dx \approx f(x_1)\,\Delta x + f(x_2)\,\Delta x + f(x_3)\,\Delta x + \cdots + f(x_n)\,\Delta x$$

$$= [f(x_1) + f(x_2) + f(x_3) + \cdots + f(x_n)]\,\Delta x$$

becomes better and better as n increases. In fact, the limit of this sum as n approaches infinity is exactly equal to the definite integral. That is,

$$\int_a^b f(x)\,dx = \lim_{n \to \infty} [f(x_1) + f(x_2) + f(x_3) + \cdots + f(x_n)]\,\Delta x.$$

It can be shown that this limit is valid as long as x_i is *any* point in the *i*th interval.

Example 4 Approximating a Definite Integral

Use a computer, a programmable calculator, or a symbolic integration utility to approximate the definite integral

$$\int_0^1 e^{-x^2}\,dx.$$

SOLUTION Using the program on page 457, with $n = 10, 20, 30, 40$, and 50, it appears that the value of the integral is approximately 0.7468. If you have access to a computer or calculator with a built-in program for approximating definite integrals, try using it to approximate this integral. When a computer with such a built-in program was used to approximate the integral, it returned a value of 0.746824.

✓ CHECKPOINT 4

Use a computer, a programmable calculator, or a symbolic integration utility to approximate the definite integral

$$\int_0^1 e^{x^2}\,dx.$$ ■

CONCEPT CHECK

1. Complete the following: In cases where the Fundamental Theorem of Calculus cannot be used to evaluate a definite integral, you can approximate the value of the integral using the _____ _____.
2. True or false: The Midpoint Rule can be used to approximate any definite integral.
3. When the Midpoint Rule is used, as the number of subintervals *n* increases, does the approximation of a definite integral become better or worse?
4. State the guidelines for using the Midpoint Rule.

460 CHAPTER 7 Techniques of Integration

Skills Review 7.4

The following warm-up exercises involve skills that were covered in earlier sections. You will use these skills in the exercise set for this section. For additional help, review Sections 0.2 and 1.5.

In Exercises 1–6, find the midpoint of the interval.

1. $[0, \frac{1}{3}]$
2. $[\frac{1}{10}, \frac{2}{10}]$
3. $[\frac{3}{20}, \frac{4}{20}]$
4. $[1, \frac{7}{6}]$
5. $[2, \frac{31}{15}]$
6. $[\frac{26}{9}, 3]$

In Exercises 7–10, find the limit.

7. $\lim_{x \to \infty} \dfrac{2x^2 + 4x - 1}{3x^2 - 2x}$
8. $\lim_{x \to \infty} \dfrac{4x + 5}{7x - 5}$
9. $\lim_{x \to \infty} \dfrac{x - 7}{x^2 + 1}$
10. $\lim_{x \to \infty} \dfrac{5x^3 + 1}{x^3 + x^2 + 4}$

Exercises 7.4

See www.CalcChat.com for worked-out solutions to odd-numbered exercises.

In Exercises 1–4, use the Midpoint Rule with $n = 4$ to approximate the area of the region. Compare your result with the exact area obtained with a definite integral.

1. $f(x) = -2x + 3$, $[0, 1]$
2. $f(x) = \dfrac{1}{x}$, $[1, 5]$

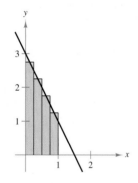
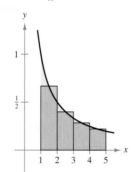

3. $f(x) = \sqrt{x}$, $[0, 1]$
4. $f(x) = 1 - x^2$, $[-1, 1]$

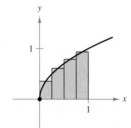
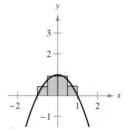

In Exercises 5–16, use the Midpoint Rule with $n = 4$ to approximate the area of the region bounded by the graph of f and the x-axis over the interval. Compare your result with the exact area. Sketch the region.

Function	Interval
5. $f(x) = 4 - x^2$	$[0, 2]$
6. $f(x) = 4x^2$	$[0, 2]$
7. $f(x) = x^2 + 3$	$[-1, 1]$
8. $f(x) = 4 - x^2$	$[-2, 2]$
9. $f(x) = 2x^2$	$[1, 3]$
10. $f(x) = 3x^2 + 1$	$[-1, 3]$
11. $f(x) = 2x - x^3$	$[0, 1]$
12. $f(x) = x^2 - x^3$	$[0, 1]$
13. $f(x) = x^2 - x^3$	$[-1, 0]$
14. $f(x) = x(1 - x)^2$	$[0, 1]$
15. $f(x) = x^2(3 - x)$	$[0, 3]$
16. $f(x) = x^2 + 4x$	$[0, 4]$

(T) In Exercises 17–22, use a program similar to the one described on page 457 to approximate the area of the region. How large must n be to obtain an approximation that is correct to within 0.01?

17. $\displaystyle\int_0^4 (2x^2 + 3)\, dx$

18. $\displaystyle\int_0^4 (2x^3 + 3)\, dx$

19. $\displaystyle\int_1^2 (2x^2 - x + 1)\, dx$

20. $\int_{1}^{2} (x^3 - 1)\, dx$

21. $\int_{1}^{4} \dfrac{1}{x+1}\, dx$

22. $\int_{1}^{2} \sqrt{x+2}\, dx$

In Exercises 23–26, use the Midpoint Rule with $n = 4$ to approximate the area of the region. Compare your result with the exact area obtained with a definite integral.

23. $f(y) = \tfrac{1}{4}y$, $[2, 4]$

24. $f(y) = 2y$, $[0, 2]$

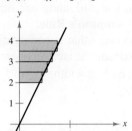
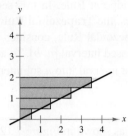

25. $f(y) = y^2 + 1$, $[0, 4]$

26. $f(y) = 4y - y^2$, $[0, 4]$

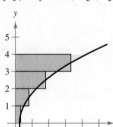
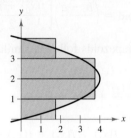

In Exercises 27–30, use the Midpoint Rule with $n = 4$ to approximate the definite integral.

27. $\int_{0}^{2} \dfrac{1}{x+1}\, dx$

28. $\int_{0}^{4} \sqrt{1 + x^2}\, dx$

29. $\int_{-1}^{1} \dfrac{1}{x^2 + 1}\, dx$

30. $\int_{1}^{5} \dfrac{\sqrt{x-1}}{x}\, dx$

(T) In Exercises 31 and 32, use a computer or a programmable calculator to approximate the definite integral using the Midpoint Rule with $n = 4, 8, 12, 16,$ and 20.

31. $\int_{0}^{4} \sqrt{2 + 3x^2}\, dx$

32. $\int_{0}^{2} \dfrac{5}{x^3 + 1}\, dx$

In Exercises 33 and 34, use the Midpoint Rule with $n = 10$ to approximate the area of the region bounded by the graphs of the equations.

33. $y = \sqrt{\dfrac{x^3}{4-x}}$, $y = 0$, $x = 3$

34. $y = x\sqrt{\dfrac{4-x}{4+x}}$, $y = 0$, $x = 4$

35. Velocity and Acceleration The following table lists the velocities v (in feet per second) of an accelerating car over a 20-second interval. Use the Midpoint Rule to approximate the distance in feet that the car travels during the 20 seconds.

Time, t	0	5	10	15	20
Velocity, v	0.0	29.3	51.3	66.0	73.3

36. Agronomy To estimate the surface area of a soil testing site, an agronomist takes several measurements, as shown in the figure. Estimate the surface area of the testing site using the Midpoint Rule.

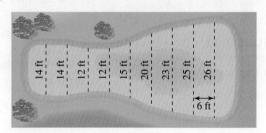

37. Zoology To estimate the surface area of a pond, a zoologist takes several measurements, as shown in the figure. Estimate the surface area of the pond using the Midpoint Rule.

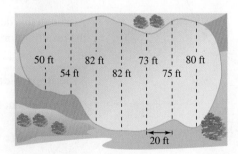

(T) **38. Numerical Approximation** Use the Midpoint Rule with $n = 4$ to approximate π where

$$\pi = \int_{0}^{1} \dfrac{4}{1 + x^2}\, dx.$$

Then use a graphing utility to evaluate the definite integral. Compare your results.

Section 7.5

Numerical Integration

- Use the Trapezoidal Rule to approximate definite integrals.
- Use Simpson's Rule to approximate definite integrals.
- Analyze the sizes of the errors when approximating definite integrals with the Trapezoidal Rule and Simpson's Rule.

Trapezoidal Rule

In Section 7.4, you studied one technique for approximating the value of a *definite* integral—the Midpoint Rule. In this section, you will study two other approximation techniques: the **Trapezoidal Rule** and **Simpson's Rule.**

To develop the Trapezoidal Rule, consider a function f that is nonnegative and continuous on the closed interval $[a, b]$. To approximate the area represented by $\int_a^b f(x)\,dx$, partition the interval into n subintervals, each of width

$$\Delta x = \frac{b - a}{n}. \quad \text{Width of each subinterval}$$

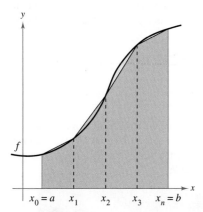

FIGURE 7.12 The area of the region can be approximated using four trapezoids.

Next, form n trapezoids, as shown in Figure 7.12. As you can see in Figure 7.13, the area of the first trapezoid is

$$\text{Area of first trapezoid} = \left(\frac{b - a}{n}\right)\left[\frac{f(x_0) + f(x_1)}{2}\right].$$

The areas of the other trapezoids follow a similar pattern, and the sum of the n areas is

$$\left(\frac{b - a}{n}\right)\left[\frac{f(x_0) + f(x_1)}{2} + \frac{f(x_1) + f(x_2)}{2} + \cdots + \frac{f(x_{n-1}) + f(x_n)}{2}\right]$$

$$= \left(\frac{b - a}{2n}\right)[f(x_0) + f(x_1) + f(x_1) + f(x_2) + \cdots + f(x_{n-1}) + f(x_n)]$$

$$= \left(\frac{b - a}{2n}\right)[f(x_0) + 2f(x_1) + 2f(x_2) + \cdots + 2f(x_{n-1}) + f(x_n)].$$

Although this development assumes f to be continuous *and* nonnegative on $[a, b]$, the resulting formula is valid as long as f is continuous on $[a, b]$.

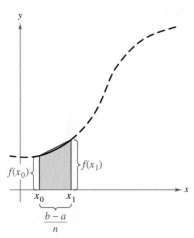

FIGURE 7.13

The Trapezoidal Rule

If f is continuous on $[a, b]$, then

$$\int_a^b f(x)\,dx \approx \left(\frac{b - a}{2n}\right)[f(x_0) + 2f(x_1) + \cdots + 2f(x_{n-1}) + f(x_n)].$$

STUDY TIP

The coefficients in the Trapezoidal Rule have the pattern

1 2 2 2 ... 2 2 1.

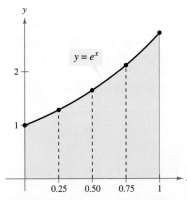

FIGURE 7.14 Four Subintervals

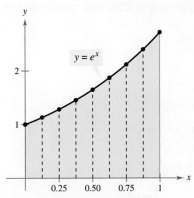

FIGURE 7.15 Eight Subintervals

✓ **CHECKPOINT 1**

Use the Trapezoidal Rule with $n = 4$ to approximate

$$\int_0^1 e^{2x}\, dx.$$ ■

TECHNOLOGY

Ⓣ A graphing utility can also evaluate a definite integral that does not have an elementary function as an antiderivative. Use the integration capabilities of a graphing utility to approximate the integral $\int_0^1 e^{x^2}\, dx$.*

Example 1 Using the Trapezoidal Rule

Use the Trapezoidal Rule to approximate $\int_0^1 e^x\, dx$. Compare the results for $n = 4$ and $n = 8$.

SOLUTION When $n = 4$, the width of each subinterval is

$$\frac{1-0}{4} = \frac{1}{4}$$

and the endpoints of the subintervals are

$$x_0 = 0, \quad x_1 = \frac{1}{4}, \quad x_2 = \frac{1}{2}, \quad x_3 = \frac{3}{4}, \quad \text{and} \quad x_4 = 1$$

as indicated in Figure 7.14. So, by the Trapezoidal Rule

$$\int_0^1 e^x\, dx = \frac{1}{8}(e^0 + 2e^{0.25} + 2e^{0.5} + 2e^{0.75} + e^1)$$

$$\approx 1.7272. \qquad \text{Approximation using } n = 4$$

When $n = 8$, the width of each subinterval is

$$\frac{1-0}{8} = \frac{1}{8}$$

and the endpoints of the subintervals are

$$x_0 = 0, \quad x_1 = \frac{1}{8}, \quad x_2 = \frac{1}{4}, \quad x_3 = \frac{3}{8}, \quad x_4 = \frac{1}{2}$$

$$x_5 = \frac{5}{8}, \quad x_6 = \frac{3}{4}, \quad x_7 = \frac{7}{8}, \quad \text{and} \quad x_8 = 1$$

as indicated in Figure 7.15. So, by the Trapezoidal Rule

$$\int_0^1 e^x\, dx = \frac{1}{16}(e^0 + 2e^{0.125} + 2e^{0.25} + \cdots + 2e^{0.875} + e^1)$$

$$\approx 1.7205. \qquad \text{Approximation using } n = 8$$

Of course, for *this particular* integral, you could have found an antiderivative and used the Fundamental Theorem of Calculus to find the exact value of the definite integral. The exact value is

$$\int_0^1 e^x\, dx = e - 1 \approx 1.718282. \qquad \text{Exact value}$$

There are two important points that should be made concerning the Trapezoidal Rule. First, the approximation tends to become more accurate as n increases. For instance, in Example 1, if $n = 16$, the Trapezoidal Rule yields an approximation of 1.7188. Second, although you could have used the Fundamental Theorem of Calculus to evaluate the integral in Example 1, this theorem cannot be used to evaluate an integral as simple as $\int_0^1 e^{x^2}\, dx$, because e^{x^2} has no elementary function as an antiderivative. Yet the Trapezoidal Rule can be easily applied to this integral.

---

*Specific calculator keystroke instructions for operations in this and other technology boxes can be found at *college.hmco.com/info/larsonapplied.*

Simpson's Rule

One way to view the Trapezoidal Rule is to say that on each subinterval, f is approximated by a first-degree polynomial. In Simpson's Rule, f is approximated by a second-degree polynomial on each subinterval.

To develop Simpson's Rule, partition the interval $[a, b]$ into an *even number* n of subintervals, each of width

$$\Delta x = \frac{b - a}{n}.$$

On the subinterval $[x_0, x_2]$, approximate the function f by the second-degree polynomial $p(x)$ that passes through the points

$$(x_0, f(x_0)), \quad (x_1, f(x_1)), \quad \text{and} \quad (x_2, f(x_2))$$

as shown in Figure 7.16. The Fundamental Theorem of Calculus can be used to show that

$$\int_{x_0}^{x_2} f(x)\,dx \approx \int_{x_0}^{x_2} p(x)\,dx$$

$$= \left(\frac{x_2 - x_0}{6}\right)\left[p(x_0) + 4p\left(\frac{x_0 + x_2}{2}\right) + p(x_2)\right]$$

$$= \frac{2[(b - a)/n]}{6}[p(x_0) + 4p(x_1) + p(x_2)]$$

$$= \left(\frac{b - a}{3n}\right)[f(x_0) + 4f(x_1) + f(x_2)].$$

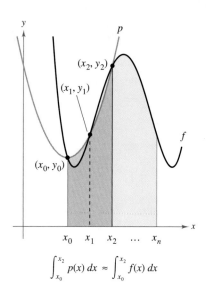

$$\int_{x_0}^{x_2} p(x)\,dx \approx \int_{x_0}^{x_2} f(x)\,dx$$

FIGURE 7.16

Repeating this process on the subintervals $[x_{i-2}, x_i]$ produces

$$\int_a^b f(x)\,dx \approx \left(\frac{b - a}{3n}\right)[f(x_0) + 4f(x_1) + f(x_2) + f(x_2) + 4f(x_3) + f(x_4) + \cdots + f(x_{n-2}) + 4f(x_{n-1}) + f(x_n)].$$

By grouping like terms, you can obtain the approximation shown below, which is known as Simpson's Rule. This rule is named after the English mathematician Thomas Simpson (1710–1761).

> **Simpson's Rule (n Is Even)**
>
> If f is continuous on $[a, b]$, then
>
> $$\int_a^b f(x)\,dx \approx \left(\frac{b - a}{3n}\right)[f(x_0) + 4f(x_1) + 2f(x_2) + 4f(x_3) + \cdots + 4f(x_{n-1}) + f(x_n)].$$

> **STUDY TIP**
>
> The Trapezoidal Rule and Simpson's Rule are necessary for solving certain real-life problems, such as approximating the number of square feet of land in a lot. You will see such problems in the exercise set for this section.

> **STUDY TIP**
>
> The coefficients in Simpson's Rule have the pattern
>
> 1 4 2 4 2 4 ... 4 2 4 1.

SECTION 7.5 Numerical Integration

In Example 1, the Trapezoidal Rule was used to estimate the value of

$$\int_0^1 e^x \, dx.$$

The next example uses Simpson's Rule to approximate the same integral.

Example 2 Using Simpson's Rule

Use Simpson's Rule to approximate

$$\int_0^1 e^x \, dx.$$

Compare the results for $n = 4$ and $n = 8$.

SOLUTION When $n = 4$, the width of each subinterval is $(1 - 0)/4 = \frac{1}{4}$ and the endpoints of the subintervals are

$$x_0 = 0, \quad x_1 = \frac{1}{4}, \quad x_2 = \frac{1}{2}, \quad x_3 = \frac{3}{4}, \quad \text{and} \quad x_4 = 1$$

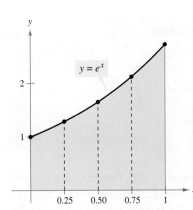

FIGURE 7.17 Four Subintervals

as indicated in Figure 7.17. So, by Simpson's Rule

$$\int_0^1 e^x \, dx = \frac{1}{12}(e^0 + 4e^{0.25} + 2e^{0.5} + 4e^{0.75} + e^1)$$

$$\approx 1.718319. \qquad \text{Approximation using } n = 4$$

When $n = 8$, the width of each subinterval is $(1 - 0)/8 = \frac{1}{8}$ and the endpoints of the subintervals are

$$x_0 = 0, \quad x_1 = \frac{1}{8}, \quad x_2 = \frac{1}{4}, \quad x_3 = \frac{3}{8}, \quad x_4 = \frac{1}{2}$$

$$x_5 = \frac{5}{8}, \quad x_6 = \frac{3}{4}, \quad x_7 = \frac{7}{8}, \quad \text{and} \quad x_8 = 1$$

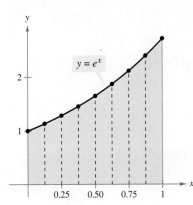

FIGURE 7.18 Eight Subintervals

as indicated in Figure 7.18. So, by Simpson's Rule

$$\int_0^1 e^x \, dx = \frac{1}{24}(e^0 + 4e^{0.125} + 2e^{0.25} + \cdots + 4e^{0.875} + e^1)$$

$$\approx 1.718284. \qquad \text{Approximation using } n = 8$$

Recall that the exact value of this integral is

$$\int_0^1 e^x \, dx = e - 1 \approx 1.718282. \qquad \text{Exact value}$$

So, with only eight subintervals, you obtained an approximation that is correct to the nearest 0.000002—an impressive result.

STUDY TIP

Comparing the results of Examples 1 and 2, you can see that for a given value of n, Simpson's Rule tends to be more accurate than the Trapezoidal Rule.

✓ CHECKPOINT 2

Use Simpson's Rule with $n = 4$ to approximate $\int_0^1 e^{2x} \, dx$.

TECHNOLOGY

Programming Simpson's Rule

In Section 7.4, you saw how to program the Midpoint Rule into a computer or programmable calculator. The pseudocode below can be used to write a program that will evaluate Simpson's Rule. (Appendix E lists this program for several models of graphing utilities.)

Program

- Prompt for value of a.
- Input value of a.
- Prompt for value of b.
- Input value of b.
- Prompt for value of n/2.
- Input value of n/2.
- Initialize sum of areas.
- Calculate width of subinterval.
- Initialize counter.
- Begin loop.
- Calculate left endpoint.
- Calculate right endpoint.
- Calculate midpoint of subinterval.
- Store left endpoint.
- Evaluate f(x) at left endpoint.
- Store midpoint of subinterval.
- Evaluate f(x) at midpoint.
- Store right endpoint.
- Evaluate f(x) at right endpoint.
- Store Simpson's Rule.
- Check value of index.
- End loop.
- Display approximation.

Before executing the program, enter the function. When the program is executed, you will be prompted to enter the lower and upper limits of integration, and *half* the number of subintervals you want to use.

Error Analysis

In Examples 1 and 2, you were able to calculate the exact value of the integral and compare that value with the approximations to see how good they were. In practice, you need to have a different way of telling how good an approximation is: such a way is provided in the next result.

Errors in the Trapezoidal Rule and Simpson's Rule

The errors E in approximating $\int_a^b f(x)\,dx$ are as shown.

Trapezoidal Rule: $|E| \leq \dfrac{(b-a)^3}{12n^2}\big[\max|f''(x)|\big]$, $a \leq x \leq b$

Simpson's Rule: $|E| \leq \dfrac{(b-a)^5}{180n^4}\big[\max|f^{(4)}(x)|\big]$, $a \leq x \leq b$

This result indicates that the errors generated by the Trapezoidal Rule and Simpson's Rule have upper bounds dependent on the extreme values of $f''(x)$ and $f^{(4)}(x)$ in the interval $[a, b]$. Furthermore, the bounds for the errors can be made arbitrarily small by *increasing* n. To determine what value of n to choose, consider the following steps.

Trapezoidal Rule

1. Find $f''(x)$.
2. Find the maximum of $|f''(x)|$ on the interval $[a, b]$.
3. Set up the inequality
$$|E| \leq \frac{(b-a)^3}{12n^2}\big[\max|f''(x)|\big].$$
4. For an error less than ϵ, solve for n in the inequality
$$\frac{(b-a)^3}{12n^2}\big[\max|f''(x)|\big] < \epsilon.$$
5. Partition $[a, b]$ into n subintervals and apply the Trapezoidal Rule.

Simpson's Rule

1. Find $f^{(4)}(x)$.
2. Find the maximum of $|f^{(4)}(x)|$ on the interval $[a, b]$.
3. Set up the inequality
$$|E| \leq \frac{(b-a)^5}{180n^4}\big[\max|f^{(4)}(x)|\big].$$
4. For an error less than ϵ, solve for n in the inequality
$$\frac{(b-a)^5}{180n^4}\big[\max|f^{(4)}(x)|\big] < \epsilon.$$
5. Partition $[a, b]$ into n subintervals and apply Simpson's Rule.

SECTION 7.5 Numerical Integration

Example 3 Using the Trapezoidal Rule

Use the Trapezoidal Rule to estimate the value of $\int_0^1 e^{-x^2}\,dx$ such that the approximation error is less than 0.01.

SOLUTION

1. Begin by finding the second derivative of $f(x) = e^{-x^2}$.

$$f(x) = e^{-x^2}$$
$$f'(x) = -2xe^{-x^2}$$
$$f''(x) = 4x^2 e^{-x^2} - 2e^{-x^2}$$
$$= 2e^{-x^2}(2x^2 - 1)$$

2. f'' has only one critical number in the interval $[0, 1]$, and the maximum value of $|f''(x)|$ on this interval is $|f''(0)| = 2$.

3. The error E using the Trapezoidal Rule is bounded by

$$|E| \leq \frac{(b-a)^3}{12n^2}(2) = \frac{1}{12n^2}(2) = \frac{1}{6n^2}.$$

4. To ensure that the approximation has an error of less than 0.01, you should choose n such that

$$\frac{1}{6n^2} < 0.01.$$

Solving for n, you can determine that n must be 5 or more.

5. Partition $[0, 1]$ into five subintervals, as shown in Figure 7.19. Then apply the Trapezoidal Rule to obtain

$$\int_0^1 e^{-x^2}\,dx = \frac{1}{10}\left(\frac{1}{e^0} + \frac{2}{e^{0.04}} + \frac{2}{e^{0.16}} + \frac{2}{e^{0.36}} + \frac{2}{e^{0.64}} + \frac{1}{e^1}\right)$$
$$\approx 0.744.$$

So, with an error no larger than 0.01, you know that

$$0.734 \leq \int_0^1 e^{-x^2}\,dx \leq 0.754.$$

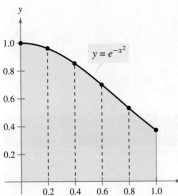

FIGURE 7.19

✓ CHECKPOINT 3

Use the Trapezoidal Rule to estimate the value of

$$\int_0^1 \sqrt{1+x^2}\,dx$$

such that the approximation error is less than 0.01. ■

CONCEPT CHECK

1. For the Trapezoidal Rule, the number of subintervals n can be odd or even. For Simpson's Rule, n must be what?
2. As the number of subintervals n increases, does an approximation given by the Trapezoidal Rule or Simpson's Rule tend to become less accurate or more accurate?
3. Write the formulas for (a) the Trapezoidal Rule and (b) Simpson's Rule.
4. The Trapezoidal Rule and Simpson's Rule yield approximations of a definite integral $\int_a^b f(x)\,dx$ based on polynomial approximations of f. The polynomial used for each should be of what degree?

468 CHAPTER 7 Techniques of Integration

Skills Review 7.5

The following warm-up exercises involve skills that were covered in earlier sections. You will use these skills in the exercise set for this section. For additional help, review Sections 0.1, 2.2, 2.6, 3.2, 4.3, and 4.5.

In Exercises 1–6, find the indicated derivative.

1. $f(x) = \dfrac{1}{x}$, $f''(x)$
2. $f(x) = \ln(2x + 1)$, $f^{(4)}(x)$
3. $f(x) = 2\ln x$, $f^{(4)}(x)$

4. $f(x) = x^3 - 2x^2 + 7x - 12$, $f''(x)$
5. $f(x) = e^{2x}$, $f^{(4)}(x)$
6. $f(x) = e^{x^2}$, $f''(x)$

In Exercises 7 and 8, find the absolute maximum of f on the interval.

7. $f(x) = -x^2 + 6x + 9$, $[0, 4]$
8. $f(x) = \dfrac{8}{x^3}$, $[1, 2]$

In Exercises 9 and 10, solve for n.

9. $\dfrac{1}{4n^2} < 0.001$
10. $\dfrac{1}{16n^4} < 0.0001$

Exercises 7.5

See www.CalcChat.com for worked-out solutions to odd-numbered exercises.

In Exercises 1–14, use the Trapezoidal Rule and Simpson's Rule to approximate the value of the definite integral for the indicated value of n. Compare these results with the exact value of the definite integral. Round your answers to four decimal places.

1. $\displaystyle\int_0^2 x^2\, dx$, $n = 4$
2. $\displaystyle\int_0^1 \left(\dfrac{x^2}{2} + 1\right) dx$, $n = 4$

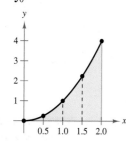
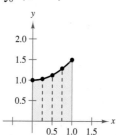

3. $\displaystyle\int_0^2 (x^4 + 1)\, dx$, $n = 4$
4. $\displaystyle\int_1^2 \dfrac{1}{x}\, dx$, $n = 4$

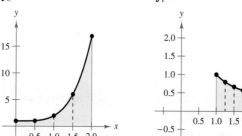

5. $\displaystyle\int_0^2 x^3\, dx$, $n = 8$
6. $\displaystyle\int_1^3 (4 - x^2)\, dx$, $n = 4$

7. $\displaystyle\int_1^2 \dfrac{1}{x}\, dx$, $n = 8$
8. $\displaystyle\int_1^2 \dfrac{1}{x^2}\, dx$, $n = 4$

9. $\displaystyle\int_0^4 \sqrt{x}\, dx$, $n = 8$
10. $\displaystyle\int_0^2 \sqrt{1 + x}\, dx$, $n = 4$

11. $\displaystyle\int_4^9 \sqrt{x}\, dx$, $n = 8$
12. $\displaystyle\int_0^8 \sqrt[3]{x}\, dx$, $n = 8$

13. $\displaystyle\int_0^1 \dfrac{1}{1 + x}\, dx$, $n = 4$
14. $\displaystyle\int_0^2 x\sqrt{x^2 + 1}\, dx$, $n = 4$

In Exercises 15–26, approximate the integral using (a) the Trapezoidal Rule and (b) Simpson's Rule for the indicated value of n. (Round your answers to three significant digits.)

15. $\displaystyle\int_0^1 \dfrac{1}{1 + x^2}\, dx$, $n = 4$
16. $\displaystyle\int_0^2 \dfrac{1}{\sqrt{1 + x^3}}\, dx$, $n = 4$

17. $\displaystyle\int_0^2 \sqrt{1 + x^3}\, dx$, $n = 4$
18. $\displaystyle\int_0^1 \sqrt{1 - x}\, dx$, $n = 4$

19. $\displaystyle\int_0^1 \sqrt{1 - x^2}\, dx$, $n = 4$
20. $\displaystyle\int_0^1 \sqrt{1 - x^2}\, dx$, $n = 8$

21. $\displaystyle\int_0^2 e^{-x^2}\, dx$, $n = 2$
22. $\displaystyle\int_0^2 e^{-x^2}\, dx$, $n = 4$

23. $\displaystyle\int_0^3 \dfrac{1}{2 - 2x + x^2}\, dx$, $n = 6$

24. $\displaystyle\int_0^3 \dfrac{x}{2 + x + x^2}\, dx$, $n = 6$

25. $\int_0^{\pi/4} x \tan x \, dx$, $n = 4$ **26.** $\int_0^{\pi/2} \sqrt{1 + \cos^2 x} \, dx$, $n = 4$

27. Approximate the area of the shaded region using (a) the Trapezoidal Rule and (b) Simpson's Rule with $n = 4$.

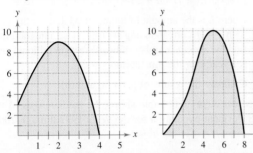

Figure for 27   Figure for 28

28. Approximate the area of the shaded region using (a) the Trapezoidal Rule and (b) Simpson's Rule with $n = 8$.

Probability In Exercises 29–32, use a program similar to the Simpson's Rule program on page 466 with $n = 6$ to approximate the indicated normal probability. The standard normal probability density function is $f(x) = (1/\sqrt{2\pi})e^{-x^2/2}$. If x is chosen at random from a population with this density, then the probability that x lies in the interval $[a, b]$ is $P(a \le x \le b) = \int_a^b f(x) \, dx$.

29. $P(0 \le x \le 1)$ **30.** $P(0 \le x \le 2)$

31. $P(0 \le x \le 4)$ **32.** $P(0 \le x \le 1.5)$

Surveying In Exercises 33 and 34, use a program similar to the Simpson's Rule program on page 466 to estimate the number of square feet of land in the lot, where x and y are measured in feet, as shown in the figures. In each case, the land is bounded by a stream and two straight roads.

33.

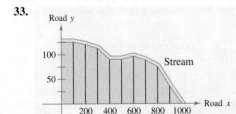

x	0	100	200	300	400	500
y	125	125	120	112	90	90

x	600	700	800	900	1000
y	95	88	75	35	0

34.

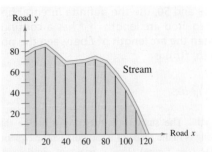

x	0	10	20	30	40	50	60
y	75	81	84	76	67	68	69

x	70	80	90	100	110	120
y	72	68	56	42	23	0

In Exercises 35–38, use the error formulas to find bounds for the errors in approximating the integral using (a) the Trapezoidal Rule and (b) Simpson's Rule. (Let $n = 4$.)

35. $\int_0^2 x^3 \, dx$ **36.** $\int_0^1 \dfrac{1}{x+1} \, dx$

37. $\int_0^1 e^{x^3} \, dx$ **38.** $\int_0^1 e^{-x^2} \, dx$

In Exercises 39–42, use the error formulas to find n such that the errors in the approximation of the definite integral are less than 0.0001 using (a) the Trapezoidal Rule and (b) Simpson's Rule.

39. $\int_0^1 x^3 \, dx$ **40.** $\int_1^3 \dfrac{1}{x} \, dx$

41. $\int_1^3 e^{2x} \, dx$ **42.** $\int_3^5 \ln x \, dx$

In Exercises 43–46, use a program similar to the Simpson's Rule program on page 466 to approximate the integral. Use $n = 100$.

43. $\int_1^4 x\sqrt{x+4} \, dx$ **44.** $\int_1^4 x^2\sqrt{x+4} \, dx$

45. $\int_2^5 10xe^{-x} \, dx$ **46.** $\int_2^5 10x^2 e^{-x} \, dx$

In Exercises 47 and 48, use a program similar to the Simpson's Rule program on page 466 with $n = 4$ to find the area of the region bounded by the graphs of the equations.

47. $y = x\sqrt[3]{x+4}$, $y = 0$, $x = 1$, $x = 5$

48. $y = \sqrt{2 + 3x^2}$, $y = 0$, $x = 1$, $x = 3$

In Exercises 49 and 50, use the definite integral below to find the required arc length. If f has a continuous derivative, then the arc length of f between the points $(a, f(a))$ and $(b, f(b))$ is

$$\int_b^a \sqrt{1 + [f'(x)]^2}\, dx.$$

49. **Arc Length** The suspension cable on a bridge that is 400 feet long is in the shape of a parabola whose equation is $y = x^2/800$ (see figure). Use a program similar to the Simpson's Rule program on page 466 with $n = 12$ to approximate the length of the cable.

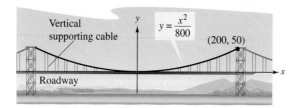

50. **Arc Length** A fleeing hare leaves its burrow $(0, 0)$ and moves due north (up the y-axis). At the same time, a pursuing lynx leaves from 1 yard east of the burrow $(1, 0)$ and always moves toward the fleeing hare (see figure). If the lynx's speed is twice that of the hare's, the equation of the lynx's path is

$$y = \frac{1}{3}(x^{3/2} - 3x^{1/2} + 2).$$

Find the distance traveled by the lynx by integrating over the interval $[0, 1]$.

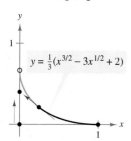

51. **Lumber Use** The table shows the amounts of lumber used for residential upkeep and improvements (in billions of board-feet per year) for the years 1997 through 2005. *(Source: U.S. Forest Service)*

Year	1997	1998	1999	2000	2001
Amount	15.1	14.7	15.1	16.4	17.0

Year	2002	2003	2004	2005
Amount	17.8	18.3	20.0	20.6

(a) Use Simpson's Rule to estimate the average number of board-feet (in billions) used per year over the time period.

(b) A model for the data is

$$L = 6.613 + 0.93t + 2095.7e^{-t}, \quad 7 \le t \le 15$$

where L is the amount of lumber used and t is the year, with $t = 7$ corresponding to 1997. Use integration to find the average number of board-feet (in billions) used per year over the time period.

(c) Compare the results of parts (a) and (b).

52. **Median Age** The table shows the median ages of the U.S. resident population for the years 1997 through 2005. *(Source: U.S. Census Bureau)*

Year	1997	1998	1999	2000	2001
Median age	34.7	34.9	35.2	35.3	35.6

Year	2002	2003	2004	2005
Median age	35.7	35.9	36.0	36.2

(a) Use Simpson's Rule to estimate the average age over the time period.

(b) A model for the data is $A = 31.5 + 1.21\sqrt{t}$, $7 \le t \le 15$, where A is the median age and t is the year, with $t = 7$ corresponding to 1997. Use integration to find the average age over the time period.

(c) Compare the results of parts (a) and (b).

53. **Medicine** The rate at which a body assimilates a 12-hour cold tablet can be modeled by $dC/dt = 8 - \ln(t^2 - 2t + 4)$, $0 \le t \le 12$, where dC/dt is measured in milligrams per hour and t is the time in hours. Use Simpson's Rule with $n = 8$ to estimate the total amount of the drug absorbed into the body during the 12 hours.

54. **Medicine** The concentration M (in grams per liter) of a six-hour allergy medicine in a body is modeled by $M = 12 - 4\ln(t^2 - 4t + 6)$, $0 \le t \le 6$, where t is the time in hours since the allergy medication was taken. Use Simpson's Rule with $n = 6$ to estimate the average level of concentration in the body over the six-hour period.

55. **Consumer Trends** The rate of change S in the number of subscribers to a newly introduced magazine is modeled by $dS/dt = 1000t^2 e^{-t}$, $0 \le t \le 6$, where t is the time in years. Use Simpson's Rule with $n = 12$ to estimate the total increase in the number of subscribers during the first 6 years.

56. Prove that Simpson's Rule is exact when used to approximate the integral of a cubic polynomial function, and demonstrate the result for $\int_0^1 x^3\, dx$, $n = 2$.

Section 7.6
Improper Integrals

- Recognize improper integrals.
- Evaluate improper integrals with infinite limits of integration.
- Evaluate improper integrals with infinite integrands.
- Use improper integrals to solve real-life problems.

Improper Integrals

The definition of the definite integral

$$\int_a^b f(x)\, dx$$

includes the requirements that the interval $[a, b]$ be finite and that f be continuous on $[a, b]$. In this section, you will study integrals that do not satisfy these requirements because of one of the conditions below.

1. One or both of the limits of integration are infinite.
2. f has an infinite discontinuity in the interval $[a, b]$.

Integrals having either of these characteristics are called **improper integrals**. For instance, the integrals

$$\int_0^\infty e^{-x}\, dx \quad \text{and} \quad \int_{-\infty}^\infty \frac{1}{x^2 + 1}\, dx$$

are improper because one or both limits of integration are infinite, as indicated in Figure 7.20. Similarly, the integrals

$$\int_1^5 \frac{1}{\sqrt{x-1}}\, dx \quad \text{and} \quad \int_{-2}^2 \frac{1}{(x+1)^2}\, dx$$

are improper because their integrands have an **infinite discontinuity**—that is, they approach infinity somewhere in the interval of integration, as indicated in Figure 7.21.

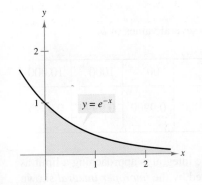
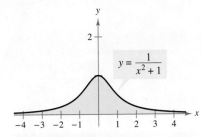

FIGURE 7.20

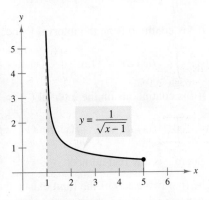
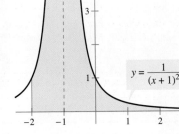

FIGURE 7.21

DISCOVERY

Use a graphing utility to calculate the definite integral $\int_0^b e^{-x}\, dx$ for $b = 10$ and for $b = 20$. What is the area of the region bounded by the graph of $y = e^{-x}$ and the two coordinate axes?

472 CHAPTER 7 Techniques of Integration

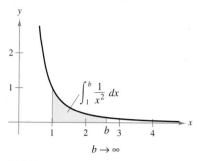

FIGURE 7.22

Integrals with Infinite Limits of Integration

To see how to evaluate an improper integral, consider the integral shown in Figure 7.22. As long as b is a real number that is greater than 1 (no matter how large), this is a definite integral whose value is

$$\int_1^b \frac{1}{x^2}\, dx = \left[-\frac{1}{x}\right]_1^b$$

$$= -\frac{1}{b} + 1$$

$$= 1 - \frac{1}{b}.$$

The table shows the values of this integral for several values of b.

b	2	5	10	100	1000	10,000
$\int_1^b \frac{1}{x^2}\, dx = 1 - \frac{1}{b}$	0.5000	0.8000	0.9000	0.9900	0.9990	0.9999

From this table, it appears that the value of the integral is approaching a limit as b increases without bound. This limit is denoted by the *improper integral* shown below.

$$\int_1^\infty \frac{1}{x^2}\, dx = \lim_{b\to\infty} \int_1^b \frac{1}{x^2}\, dx$$

$$= \lim_{b\to\infty} \left(1 - \frac{1}{b}\right)$$

$$= 1$$

Improper Integrals (Infinite Limits of Integration)

1. If f is continuous on the interval $[a, \infty)$, then

$$\int_a^\infty f(x)\, dx = \lim_{b\to\infty} \int_a^b f(x)\, dx.$$

2. If f is continuous on the interval $(-\infty, b]$, then

$$\int_{-\infty}^b f(x)\, dx = \lim_{a\to-\infty} \int_a^b f(x)\, dx.$$

3. If f is continuous on the interval $(-\infty, \infty)$, then

$$\int_{-\infty}^\infty f(x)\, dx = \int_{-\infty}^c f(x)\, dx + \int_c^\infty f(x)\, dx$$

where c is any real number.

In the first two cases, if the limit exists, then the improper integral **converges**; otherwise, the improper integral **diverges**. In the third case, the integral on the left will diverge if either one of the integrals on the right diverges.

Example 1 Evaluating an Improper Integral

Determine the convergence or divergence of $\int_{1}^{\infty} \frac{1}{x} \, dx$.

SOLUTION Begin by applying the definition of an improper integral.

$$\int_{1}^{\infty} \frac{1}{x} \, dx = \lim_{b \to \infty} \int_{1}^{b} \frac{1}{x} \, dx \quad \text{Definition of improper integral}$$

$$= \lim_{b \to \infty} \Big[\ln x\Big]_{1}^{b} \quad \text{Find antiderivative.}$$

$$= \lim_{b \to \infty} (\ln b - 0) \quad \text{Apply Fundamental Theorem.}$$

$$= \infty \quad \text{Evaluate limit.}$$

Because the limit is infinite, the improper integral diverges.

> **TECHNOLOGY**
>
> Symbolic integration utilities evaluate improper integrals in much the same way that they evaluate definite integrals. Use a symbolic integration utility to evaluate
>
> $$\int_{-\infty}^{-1} \frac{1}{x^2} \, dx.$$

✓ CHECKPOINT 1

Determine the convergence or divergence of each improper integral.

a. $\int_{1}^{\infty} \frac{1}{x^3} \, dx$ **b.** $\int_{1}^{\infty} \frac{1}{\sqrt{x}} \, dx$

As you begin to work with improper integrals, you will find that integrals that appear to be similar can have very different values. For instance, consider the two improper integrals

$$\int_{1}^{\infty} \frac{1}{x} \, dx = \infty \quad \text{Divergent integral}$$

and

$$\int_{1}^{\infty} \frac{1}{x^2} \, dx = 1. \quad \text{Convergent integral}$$

The first integral diverges and the second converges to 1. Graphically, this means that the areas shown in Figure 7.23 are very different. The region lying between the graph of $y = 1/x$ and the x-axis (for $x \geq 1$) has an *infinite* area, and the region lying between the graph of $y = 1/x^2$ and the x-axis (for $x \geq 1$) has a *finite* area.

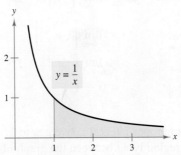

Diverges (infinite area)

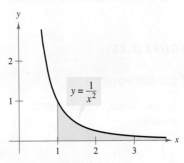

Converges (finite area)

FIGURE 7.23

474 CHAPTER 7 Techniques of Integration

Example 2 Evaluating an Improper Integral

Evaluate the improper integral.

$$\int_{-\infty}^{0} \frac{1}{(1-2x)^{3/2}} \, dx$$

SOLUTION Begin by applying the definition of an improper integral.

$$\int_{-\infty}^{0} \frac{1}{(1-2x)^{3/2}} \, dx = \lim_{a \to -\infty} \int_{a}^{0} \frac{1}{(1-2x)^{3/2}} \, dx \quad \text{Definition of improper integral}$$

$$= \lim_{a \to -\infty} \left[\frac{1}{\sqrt{1-2x}} \right]_{a}^{0} \quad \text{Find antiderivative.}$$

$$= \lim_{a \to -\infty} \left(1 - \frac{1}{\sqrt{1-2a}} \right) \quad \text{Apply Fundamental Theorem.}$$

$$= 1 - 0 \quad \text{Evaluate limit.}$$

$$= 1 \quad \text{Simplify.}$$

So, the improper integral converges to 1. As shown in Figure 7.24, this implies that the region lying between the graph of $y = 1/(1-2x)^{3/2}$ and the x-axis (for $x \leq 0$) has an area of 1 square unit.

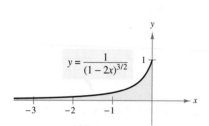

FIGURE 7.24

✓ CHECKPOINT 2

Evaluate the improper integral, if possible.

$$\int_{-\infty}^{0} \frac{1}{(x-1)^2} \, dx$$

Example 3 Evaluating an Improper Integral

Evaluate the improper integral.

$$\int_{0}^{\infty} 2xe^{-x^2} \, dx$$

SOLUTION Begin by applying the definition of an improper integral.

$$\int_{0}^{\infty} 2xe^{-x^2} \, dx = \lim_{b \to \infty} \int_{0}^{b} 2xe^{-x^2} \, dx \quad \text{Definition of improper integral}$$

$$= \lim_{b \to \infty} \left[-e^{-x^2} \right]_{0}^{b} \quad \text{Find antiderivative.}$$

$$= \lim_{b \to \infty} \left(-e^{-b^2} + 1 \right) \quad \text{Apply Fundamental Theorem.}$$

$$= 0 + 1 \quad \text{Evaluate limit.}$$

$$= 1 \quad \text{Simplify.}$$

So, the improper integral converges to 1. As shown in Figure 7.25, this implies that the region lying between the graph of $y = 2xe^{-x^2}$ and the x-axis (for $x \geq 0$) has an area of 1 square unit.

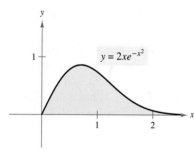

FIGURE 7.25

✓ CHECKPOINT 3

Evaluate the improper integral, if possible.

$$\int_{-\infty}^{0} e^{2x} \, dx$$

Integrals with Infinite Integrands

Improper Integrals (Infinite Integrands)

1. If f is continuous on the interval $[a, b)$ and approaches infinity at b, then
$$\int_a^b f(x)\, dx = \lim_{c \to b^-} \int_a^c f(x)\, dx.$$

2. If f is continuous on the interval $(a, b]$ and approaches infinity at a, then
$$\int_a^b f(x)\, dx = \lim_{c \to a^+} \int_c^b f(x)\, dx.$$

3. If f is continuous on the interval $[a, b]$, except for some c in (a, b) at which f approaches infinity, then
$$\int_a^b f(x)\, dx = \int_a^c f(x)\, dx + \int_c^b f(x)\, dx.$$

In the first two cases, if the limit exists, then the improper integral **converges**; otherwise, the improper integral **diverges**. In the third case, the improper integral on the left diverges if either of the improper integrals on the right diverges.

Example 4 Evaluating an Improper Integral

Evaluate $\int_1^2 \dfrac{1}{\sqrt[3]{x-1}}\, dx.$

SOLUTION

$$\int_1^2 \frac{1}{\sqrt[3]{x-1}}\, dx = \lim_{c \to 1^+} \int_c^2 \frac{1}{\sqrt[3]{x-1}}\, dx \qquad \text{Definition of improper integral}$$

$$= \lim_{c \to 1^+} \left[\frac{3}{2}(x-1)^{2/3} \right]_c^2 \qquad \text{Find antiderivative.}$$

$$= \lim_{c \to 1^+} \left[\frac{3}{2} - \frac{3}{2}(c-1)^{2/3} \right] \qquad \text{Apply Fundamental Theorem.}$$

$$= \frac{3}{2} - 0 \qquad \text{Evaluate limit.}$$

$$= \frac{3}{2} \qquad \text{Simplify.}$$

So, the integral converges to $\frac{3}{2}$. This implies that the region shown in Figure 7.26 has an area of $\frac{3}{2}$ square units.

✓ CHECKPOINT 4

Evaluate $\int_1^2 \dfrac{1}{\sqrt{x-1}}\, dx.$ ■

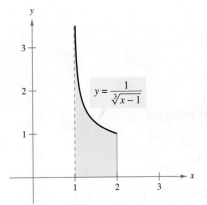

FIGURE 7.26

TECHNOLOGY

Use a graphing utility to verify the result of Example 4 by calculating each definite integral.

$$\int_{1.01}^2 \frac{1}{\sqrt[3]{x-1}}\, dx$$

$$\int_{1.001}^2 \frac{1}{\sqrt[3]{x-1}}\, dx$$

$$\int_{1.0001}^2 \frac{1}{\sqrt[3]{x-1}}\, dx$$

Example 5 Evaluating an Improper Integral

Evaluate $\int_{1}^{2} \dfrac{2}{x^2 - 2x}\, dx$.

SOLUTION

$$\int_{1}^{2} \dfrac{2}{x^2 - 2x}\, dx = \int_{1}^{2} \left(\dfrac{1}{x-2} - \dfrac{1}{x}\right) dx \quad \text{Use partial fractions.}$$

$$= \lim_{c \to 2^-} \int_{1}^{c} \left(\dfrac{1}{x-2} - \dfrac{1}{x}\right) dx \quad \text{Definition of improper integral}$$

$$= \lim_{c \to 2^-} \Big[\ln|x-2| - \ln|x|\Big]_{1}^{c} \quad \text{Find antiderivative.}$$

$$= -\infty \quad \text{Evaluate limit.}$$

So, you can conclude that the integral diverges. This implies that the region shown in Figure 7.27 has an infinite area.

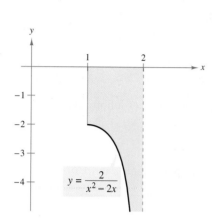

FIGURE 7.27

✓ CHECKPOINT 5

Evaluate $\int_{1}^{3} \dfrac{3}{x^2 - 3x}\, dx$.

Example 6 Evaluating an Improper Integral

Evaluate $\int_{-1}^{2} \dfrac{1}{x^3}\, dx$.

SOLUTION This integral is improper because the integrand has an infinite discontinuity at the interior value $x = 0$, as shown in Figure 7.28. So, you can write

$$\int_{-1}^{2} \dfrac{1}{x^3}\, dx = \int_{-1}^{0} \dfrac{1}{x^3}\, dx + \int_{0}^{2} \dfrac{1}{x^3}\, dx.$$

By applying the definition of an improper integral, you can show that each of these integrals diverges. So, the original improper integral also diverges.

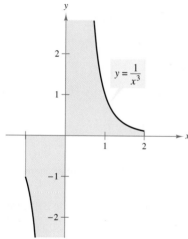

FIGURE 7.28

✓ CHECKPOINT 6

Evaluate $\int_{-1}^{1} \dfrac{1}{x^2}\, dx$.

STUDY TIP

Had you not recognized that the integral in Example 6 was improper, you would have obtained the incorrect result

$$\int_{-1}^{2} \dfrac{1}{x^3}\, dx = \left[-\dfrac{1}{2x^2}\right]_{-1}^{2} = -\dfrac{1}{8} + \dfrac{1}{2} = \dfrac{3}{8}. \quad \text{Incorrect}$$

Improper integrals in which the integrands have infinite discontinuities *between* the limits of integration are often overlooked, so keep alert for such possibilities. Even symbolic integrators can have trouble with this type of integral and can give the same incorrect result.

Applications

In Section 4.3, you studied the graph of the *normal probability density function*

$$f(x) = \frac{1}{\sigma\sqrt{2\pi}} e^{-(x-\mu)^2/2\sigma^2}.$$

This function is used in statistics to represent a population that is normally distributed with a mean of μ and a standard deviation of σ. Specifically, if an outcome x is chosen at random from the population, the probability that x will have a value between a and b is

$$P(a \leq x \leq b) = \int_a^b \frac{1}{\sigma\sqrt{2\pi}} e^{-(x-\mu)^2/2\sigma^2}\, dx.$$

As shown in Figure 7.29, the probability $P(-\infty < x < \infty)$ is

$$P(-\infty < x < \infty) = \int_{-\infty}^{\infty} \frac{1}{\sigma\sqrt{2\pi}} e^{-(x-\mu)^2/2\sigma^2}\, dx = 1.$$

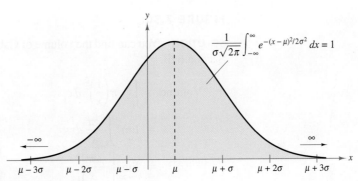

FIGURE 7.29

Example 7 Finding a Probability

The mean height of American men (from 20 to 29 years old) is 70 inches, and the standard deviation is 3 inches. A 20- to 29-year-old man is chosen at random from the population. What is the probability that he is 6 feet tall or taller? *(Source: U.S. National Center for Health Statistics)*

SOLUTION Using a mean of $\mu = 70$ and a standard deviation of $\sigma = 3$, the probability $P(72 \leq x < \infty)$ is given by the improper integral

$$P(72 \leq x < \infty) = \int_{72}^{\infty} \frac{1}{3\sqrt{2\pi}} e^{-(x-70)^2/18}\, dx.$$

Using a symbolic integration utility, you can approximate the value of this integral to be 0.252. So, the probability that the man is 6 feet tall or taller is about 25.2%.

✓CHECKPOINT 7

Use Example 7 to find the probability that a 20- to 29-year-old man chosen at random from the population is 6 feet 6 inches tall or taller. ■

Victor Baldizon/NBAE via Getty Images

Many professional basketball players are over $6\frac{1}{2}$ feet tall. If a man is chosen at random from the population, the probability that he is $6\frac{1}{2}$ feet tall or taller is less than half of one percent.

Example 8 Finding Volume

The solid formed by revolving the graph of $f(x) = \dfrac{1}{x}$ for $1 \leq x < \infty$ about the x-axis is called **Gabriel's horn** (see Figure 7.30). Find the volume of Gabriel's horn.

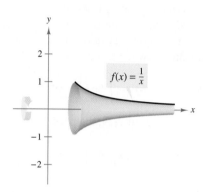

FIGURE 7.30

SOLUTION You can find the volume of Gabriel's horn using the Disk Method (see Section 6.6).

$$\begin{aligned}
\text{Volume} &= \int_1^\infty \pi\left(\frac{1}{x}\right)^2 dx \\
&= \lim_{b \to \infty} \int_1^b \frac{\pi}{x^2} dx \\
&= \lim_{b \to \infty} \left[-\frac{\pi}{x}\right]_1^b \\
&= \lim_{b \to \infty} \left(\pi - \frac{\pi}{b}\right) = \pi
\end{aligned}$$

So, Gabriel's horn has a volume of π cubic units.

✓ CHECKPOINT 8

Use a symbolic integration utility to help you decide which of the improper integrals converge. What can you conclude?

(a) $\displaystyle\int_1^\infty \frac{1}{x^{0.9}} dx$ (b) $\displaystyle\int_1^\infty \frac{1}{x^{1.0}} dx$ (c) $\displaystyle\int_1^\infty \frac{1}{x^{1.1}} dx$

CONCEPT CHECK

1. Integrals are improper integrals if they have either of what two characteristics?
2. Describe the different types of improper integrals.
3. Define the term *converges* as it applies to improper integrals.
4. Define the term *diverges* as it applies to improper integrals.

SECTION 7.6 Improper Integrals

Skills Review 7.6

The following warm-up exercises involve skills that were covered in earlier sections. You will use these skills in the exercise set for this section. For additional help, review Sections 1.5, 4.1, and 4.4.

In Exercises 1–6, find the limit.

1. $\lim\limits_{x \to 2} (2x + 5)$
2. $\lim\limits_{x \to 1} \left(\dfrac{1}{x} + 2x^2 \right)$
3. $\lim\limits_{x \to -4} \dfrac{x + 4}{x^2 - 16}$
4. $\lim\limits_{x \to 0} \dfrac{x^2 - 2x}{x^3 + 3x^2}$
5. $\lim\limits_{x \to 1} \dfrac{1}{\sqrt{x - 1}}$
6. $\lim\limits_{x \to -3} \dfrac{x^2 + 2x - 3}{x + 3}$

In Exercises 7–10, evaluate the expression (a) when $x = b$ and (b) when $x = 0$.

7. $\dfrac{4}{3}(2x - 1)^3$
8. $\dfrac{1}{x - 5} + \dfrac{3}{(x - 2)^2}$
9. $\ln(5 - 3x^2) - \ln(x + 1)$
10. $e^{3x^2} + e^{-3x^2}$

Exercises 7.6

See www.CalcChat.com for worked-out solutions to odd-numbered exercises.

In Exercises 1–4, decide whether the integral is improper. Explain your reasoning.

1. $\displaystyle\int_0^1 \dfrac{dx}{3x - 2}$
2. $\displaystyle\int_1^3 \dfrac{dx}{x^2}$
3. $\displaystyle\int_0^1 \dfrac{2x - 5}{x^2 - 5x + 6}\,dx$
4. $\displaystyle\int_1^\infty x^2\,dx$

In Exercises 5–10, explain why the integral is improper and determine whether it diverges or converges. Evaluate the integral if it converges.

5. $\displaystyle\int_0^4 \dfrac{1}{\sqrt{x}}\,dx$

6. $\displaystyle\int_3^4 \dfrac{1}{\sqrt{x - 3}}\,dx$

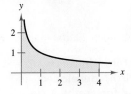 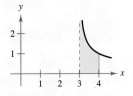

7. $\displaystyle\int_0^2 \dfrac{1}{(x - 1)^{2/3}}\,dx$
8. $\displaystyle\int_0^2 \dfrac{1}{(x - 1)^2}\,dx$

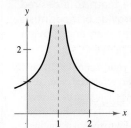 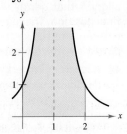

9. $\displaystyle\int_0^\infty e^{-x}\,dx$
10. $\displaystyle\int_{-\infty}^0 e^{2x}\,dx$

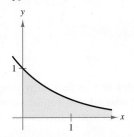 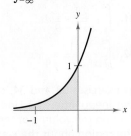

In Exercises 11–22, determine whether the improper integral diverges or converges. Evaluate the integral if it converges.

11. $\displaystyle\int_1^\infty \dfrac{1}{x^2}\,dx$
12. $\displaystyle\int_1^\infty \dfrac{1}{\sqrt[3]{x}}\,dx$
13. $\displaystyle\int_0^\infty e^{x/3}\,dx$
14. $\displaystyle\int_0^\infty \dfrac{5}{e^{2x}}\,dx$
15. $\displaystyle\int_5^\infty \dfrac{x}{\sqrt{x^2 - 16}}\,dx$
16. $\displaystyle\int_{1/2}^\infty \dfrac{1}{\sqrt{2x - 1}}\,dx$
17. $\displaystyle\int_{-\infty}^0 e^{-x}\,dx$
18. $\displaystyle\int_{-\infty}^{-1} \dfrac{1}{x^2}\,dx$
19. $\displaystyle\int_1^\infty \dfrac{e^{\sqrt{x}}}{\sqrt{x}}\,dx$
20. $\displaystyle\int_{-\infty}^0 \dfrac{x}{x^2 + 1}\,dx$
21. $\displaystyle\int_{-\infty}^\infty 2xe^{-3x^2}\,dx$
22. $\displaystyle\int_{-\infty}^\infty x^2 e^{-x^3}\,dx$

In Exercises 23–32, determine whether the improper integral diverges or converges. Evaluate the integral if it converges, and check your results with the results obtained by using the integration capabilities of a graphing utility.

23. $\int_0^1 \dfrac{1}{1-x}\,dx$

24. $\int_0^{27} \dfrac{5}{\sqrt[3]{x}}\,dx$

25. $\int_0^9 \dfrac{1}{\sqrt{9-x}}\,dx$

26. $\int_0^2 \dfrac{x}{\sqrt{4-x^2}}\,dx$

27. $\int_0^1 \dfrac{1}{x^2}\,dx$

28. $\int_0^1 \dfrac{1}{x}\,dx$

29. $\int_0^2 \dfrac{1}{\sqrt[3]{x-1}}\,dx$

30. $\int_0^2 \dfrac{1}{(x-1)^{4/3}}\,dx$

31. $\int_3^4 \dfrac{1}{\sqrt{x^2-9}}\,dx$

32. $\int_3^5 \dfrac{1}{x^2\sqrt{x^2-9}}\,dx$

In Exercises 33 and 34, consider the region satisfying the inequalities. (a) Find the area of the region, and (b) find the volume of the solid generated by revolving the region about the x-axis.

33. $y \le \dfrac{1}{x^2}$, $y \ge 0$, $x \ge 1$

34. $y \le e^{-x}$, $y \ge 0$, $x \ge 0$

Ⓢ In Exercises 35–38, use a spreadsheet to complete the table for the specified values of a and n to demonstrate that

$$\lim_{x\to\infty} x^n e^{-ax} = 0, \quad a > 0, \quad n > 0.$$

x	1	10	25	50
$x^n e^{-ax}$				

35. $a = 1$, $n = 1$
36. $a = 2$, $n = 4$
37. $a = \frac{1}{2}$, $n = 2$
38. $a = \frac{1}{2}$, $n = 5$

In Exercises 39–42, use the results of Exercises 35–38 to evaluate the improper integral.

39. $\int_0^\infty x^2 e^{-x}\,dx$

40. $\int_0^\infty (x-1)e^{-x}\,dx$

41. $\int_0^\infty x e^{-2x}\,dx$

42. $\int_0^\infty x e^{-x}\,dx$

43. **Women's Heights** The mean height of American women between the ages of 30 and 39 is 64.5 inches, and the standard deviation is 2.7 inches. Find the probability that a 30- to 39-year-old woman chosen at random is

   (a) between 5 and 6 feet tall.

   (b) 5 feet 8 inches or taller.

   (c) 6 feet or taller.

   (Source: U.S. National Center for Health Statistics)

44. **Quality Control** A company manufactures wooden yardsticks. The lengths of the yardsticks are normally distributed with a mean of 36 inches and a standard deviation of 0.2 inch. Find the probability that a yardstick is

   (a) longer than 35.5 inches.

   (b) longer than 35.9 inches.

Life Science Capsule

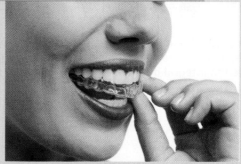

© Align Technology, Inc.

Two graduates of Stanford University, Zia Christi and Kelsey Wirth, developed an alternative to metal braces for teeth: a clear, removable aligner. They used 3-D graphics to mold a person's teeth and to prepare a series of aligners that gently shift the teeth to a desired final position. The aligners provide orthodontists with an alternative treatment plan and provide patients with convenience.

45. **Research Project** Use your school's library, the Internet, or some other reference source to research a company that uses mathematical algorithms and imaging software in the medical field. Write a short paper that summarizes your findings.

Algebra Review

Algebra and Integration Techniques

Integration techniques involve many different algebraic skills. Study the examples in this Algebra Review. Be sure that you understand the algebra used in each step.

Example 1 Algebra and Integration Techniques

Perform each operation and simplify.

a. $\dfrac{2}{x-3} - \dfrac{1}{x+2}$ — Example 1, page 437

$= \dfrac{2(x+2)}{(x-3)(x+2)} - \dfrac{(x-3)}{(x-3)(x+2)}$ — Rewrite with common denominator.

$= \dfrac{2(x+2) - (x-3)}{(x-3)(x+2)}$ — Rewrite as single fraction.

$= \dfrac{2x + 4 - x + 3}{x^2 - x - 6}$ — Multiply factors.

$= \dfrac{x+7}{x^2 - x - 6}$ — Combine like terms.

b. $\dfrac{6}{x} - \dfrac{1}{x+1} + \dfrac{9}{(x+1)^2}$ — Example 2, page 438

$= \dfrac{6(x+1)^2}{x(x+1)^2} - \dfrac{x(x+1)}{x(x+1)^2} + \dfrac{9x}{x(x+1)^2}$ — Rewrite with common denominator.

$= \dfrac{6(x+1)^2 - x(x+1) + 9x}{x(x+1)^2}$ — Rewrite as single fraction.

$= \dfrac{6x^2 + 12x + 6 - x^2 - x + 9x}{x^3 + 2x^2 + x}$ — Multiply factors.

$= \dfrac{5x^2 + 20x + 6}{x^3 + 2x^2 + x}$ — Combine like terms.

c. $6\ln|x| - \ln|x+1| + 9\dfrac{(x+1)^{-1}}{-1}$ — Example 2, page 438

$= \ln|x|^6 - \ln|x+1| + 9\dfrac{(x+1)^{-1}}{-1}$ — $m \ln n = \ln n^m$

$= \ln|x^6| - \ln|x+1| + 9\dfrac{(x+1)^{-1}}{-1}$ — Property of absolute value

$= \ln\dfrac{|x^6|}{|x+1|} + 9\dfrac{(x+1)^{-1}}{-1}$ — $\ln m - \ln n = \ln\dfrac{m}{n}$

$= \ln\left|\dfrac{x^6}{x+1}\right| + 9\dfrac{(x+1)^{-1}}{-1}$ — $\dfrac{|a|}{|b|} = \left|\dfrac{a}{b}\right|$

$= \ln\left|\dfrac{x^6}{x+1}\right| - 9(x+1)^{-1}$ — Rewrite sum as difference.

$= \ln\left|\dfrac{x^6}{x+1}\right| - \dfrac{9}{x+1}$ — Rewrite with positive exponent.

Example 2 Algebra and Integration Techniques

Perform each operation and simplify.

a. $x + 1 + \dfrac{1}{x^3} + \dfrac{1}{x-1}$

b. Solve for y: $\ln|y| - \ln|L - y| = kt + C$

SOLUTION

a. $x + 1 + \dfrac{1}{x^3} + \dfrac{1}{x-1}$ *Example 3, page 439*

$= \dfrac{(x+1)(x^3)(x-1)}{x^3(x-1)} + \dfrac{x-1}{x^3(x-1)} + \dfrac{x^3}{x^3(x-1)}$ Rewrite with common denominator.

$= \dfrac{(x+1)(x^3)(x-1) + (x-1) + x^3}{x^3(x-1)}$ Rewrite as single fraction.

$= \dfrac{(x^2-1)(x^3) + x - 1 + x^3}{x^3(x-1)}$ $(x+1)(x-1) = x^2 - 1$

$= \dfrac{x^5 - x^3 + x - 1 + x^3}{x^4 - x^3}$ Multiply factors.

$= \dfrac{x^5 + x - 1}{x^4 - x^3}$ Combine like terms.

b. $\ln|y| - \ln|L - y| = kt + C$ *Example 4, page 440*

$-\ln|y| + \ln|L - y| = -kt - C$ Multiply each side by -1.

$\ln\left|\dfrac{L-y}{y}\right| = -kt - C$ $\ln x - \ln y = \ln \dfrac{x}{y}$

$\left|\dfrac{L-y}{y}\right| = e^{-kt-C}$ Exponentiate each side.

$\left|\dfrac{L-y}{y}\right| = e^{-C}e^{-kt}$ $x^{n+m} = x^n x^m$

$\dfrac{L-y}{y} = \pm e^{-C}e^{-kt}$ Property of absolute value

$L - y = be^{-kt}y$ Let $\pm e^{-C} = b$ and multiply each side by y.

$L = y + be^{-kt}y$ Add y to each side.

$L = y(1 + be^{-kt})$ Factor.

$\dfrac{L}{1 + be^{-kt}} = y$ Divide.

Chapter Summary and Study Strategies

After studying this chapter, you should have acquired the following skills. The exercise numbers are keyed to the Review Exercises that begin on page 485. Answers to odd-numbered Review Exercises are given in the back of the text.*

Section 7.1 — Review Exercises
- Use integration by parts to find indefinite integrals. — 1–6
$$\int u\, dv = uv - \int v\, du$$
- Use integration by parts repeatedly to find indefinite integrals. — 7, 8
- Use integration by parts to solve real-life problems. — 9, 10

Section 7.2
- Use partial fractions to find indefinite integrals. — 11–18
- Use logistic growth functions to model real-life situations. — 19, 20
$$y = \frac{L}{1 + be^{-kt}}$$

Section 7.3
- Solve trigonometric integrals. — 21–32

$$\int \cos u\, du = \sin u + C \qquad \int \sin u\, du = -\cos u + C$$

$$\int \sec^2 u\, du = \tan u + C \qquad \int \sec u \tan u\, du = \sec u + C$$

$$\int \csc^2 u\, du = -\cot u + C \qquad \int \csc u \cot u\, du = -\csc u + C$$

$$\int \tan u\, du = -\ln|\cos u| + C \qquad \int \sec u\, du = \ln|\sec u + \tan u| + C$$

$$\int \cot u\, du = \ln|\sin u| + C \qquad \int \csc u\, du = \ln|\csc u - \cot u| + C$$

- Find the areas of regions in the plane. — 33–36
- Use trigonometric integrals to solve real-life problems. — 37

Section 7.4
- Use the Midpoint Rule to approximate definite integrals. — 39–42
$$\int_a^b f(x)\, dx \approx \frac{b-a}{n}[f(x_1) + f(x_2) + f(x_3) + \cdots + f(x_n)]$$
- Use the Midpoint Rule to solve real-life problems. — 38

---

* Use a wide range of valuable study aids to help you master the material in this chapter. The *Student Solutions Guide* includes step-by-step solutions to all odd-numbered exercises to help you review and prepare. The student website at *college.hmco.com/info/larsonapplied* offers algebra help and a *Graphing Technology Guide*. The *Graphing Technology Guide* contains step-by-step commands and instructions for a wide variety of graphing calculators, including the most recent models.

Section 7.5

	Review Exercises
■ Use the Trapezoidal Rule to approximate definite integrals.	43–46

$$\int_a^b f(x)\,dx \approx \left(\frac{b-a}{2n}\right)[f(x_0) + 2f(x_1) + \cdots + 2f(x_{n-1}) + f(x_n)]$$

■ Use Simpson's Rule to approximate definite integrals. 47–50

$$\int_a^b f(x)\,dx \approx \left(\frac{b-a}{3n}\right)[f(x_0) + 4f(x_1) + 2f(x_2) + 4f(x_3) + \cdots + 4f(x_{n-1}) + f(x_n)]$$

■ Analyze the sizes of the errors when approximating definite integrals with the Trapezoidal Rule. 51, 52

$$|E| \leq \frac{(b-a)^3}{12n^2}[\max|f''(x)|], \quad a \leq x \leq b$$

■ Analyze the sizes of the errors when approximating definite integrals with Simpson's Rule. 53, 54

$$|E| \leq \frac{(b-a)^5}{180n^4}[\max|f^{(4)}(x)|], \quad a \leq x \leq b$$

Section 7.6

■ Evaluate improper integrals with infinite limits of integration. 55–58

$$\int_a^\infty f(x)\,dx = \lim_{b\to\infty}\int_a^b f(x)\,dx, \quad \int_{-\infty}^b f(x)\,dx = \lim_{a\to-\infty}\int_a^b f(x)\,dx,$$

$$\int_{-\infty}^\infty f(x)\,dx = \int_{-\infty}^c f(x)\,dx + \int_c^\infty f(x)\,dx$$

■ Evaluate improper integrals with infinite integrands. 59–62

$$\int_a^b f(x)\,dx = \lim_{c\to b^-}\int_a^c f(x)\,dx, \quad \int_a^b f(x)\,dx = \lim_{c\to a^+}\int_c^b f(x)\,dx,$$

$$\int_a^b f(x)\,dx = \int_a^c f(x)\,dx + \int_c^b f(x)\,dx$$

■ Use improper integrals to solve real-life problems. 63, 64

Study Strategies

■ **Use a Variety of Approaches** To be efficient at finding antiderivatives, you need to use a variety of approaches.

   1. Check to see whether the integral fits one of the basic integration formulas—you should have these formulas memorized.
   2. Try an integration technique such as substitution, integration by parts, or partial fractions to rewrite the integral in a form that fits one of the basic integration formulas.
   3. Use a symbolic integration utility.

■ **Use Numerical Integration** When solving a definite integral, remember that you cannot apply the Fundamental Theorem of Calculus unless you can find an antiderivative of the integrand. This is not always possible—even with a symbolic integration utility. In such cases, you can use a numerical technique such as the Midpoint Rule, the Trapezoidal Rule, or Simpson's Rule to approximate the value of the integral.

■ **Improper Integrals** When solving integration problems, remember that the symbols used to denote definite integrals are the same as those used to denote improper integrals. Evaluating an improper integral as a definite integral can lead to an incorrect value. For instance, if you evaluated the integral

$$\int_{-2}^1 \frac{1}{x^2}\,dx$$

as though it were a definite integral, you would obtain a value of $-\frac{3}{2}$. This is not, however, correct. This integral is actually a divergent improper integral.

Review Exercises

See www.CalcChat.com for worked-out solutions to odd-numbered exercises.

In Exercises 1–6, use integration by parts to find the indefinite integral.

1. $\displaystyle\int \frac{\ln x}{\sqrt{x}}\,dx$
2. $\displaystyle\int \sqrt{x}\ln x\,dx$
3. $\displaystyle\int (x+1)e^x\,dx$
4. $\displaystyle\int \ln\!\left(\frac{x}{x+1}\right)dx$
5. $\displaystyle\int \frac{x}{\sqrt{x+1}}\,dx$
6. $\displaystyle\int \frac{x}{\sqrt{3+2x}}\,dx$

In Exercises 7 and 8, use integration by parts repeatedly to find the indefinite integral. Use a symbolic integration utility to verify your answer.

7. $\displaystyle\int 2x^2 e^{2x}\,dx$
8. $\displaystyle\int (\ln x)^3\,dx$

9. **Probability** The probability of recall in an experiment is found to be

$$P(a\le x\le b)=\int_a^b \frac{96}{11}\!\left(\frac{x}{\sqrt{9+16x}}\right)dx,\ 0\le a\le b\le 1$$

where x represents the percent of recall (see figure). Find the probability that a randomly chosen individual will recall (a) between 0% and 80% of the material and (b) between 0% and 50% of the material.

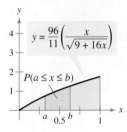

10. **Autistic Children** The number N of autistic children (in thousands), aged 3 to 21 years old, served by U.S. federally supported programs for the disabled for the school years ending in 2001 through 2006 can be modeled by

$N=96+11.8t\ln t$, $1\le t\le 6$

where t is the school year, with $t=1$ corresponding to the school year ending in 2001. Find the average number of autistic children served from 2001 through 2006. (Source: National Center for Education Statistics)

In Exercises 11–18, use partial fractions to find the indefinite integral.

11. $\displaystyle\int \frac{1}{x(x+5)}\,dx$
12. $\displaystyle\int \frac{4x-2}{3(x-1)^2}\,dx$
13. $\displaystyle\int \frac{x-28}{x^2-x-6}\,dx$
14. $\displaystyle\int \frac{4x^2-x-5}{x^2(x+5)}\,dx$
15. $\displaystyle\int \frac{x^2}{x^2+2x-15}\,dx$
16. $\displaystyle\int \frac{x^2+2x-12}{x(x+3)}\,dx$
17. $\displaystyle\int \frac{x}{(2+3x)^2}\,dx$
18. $\displaystyle\int \frac{1}{x^2-4}\,dx$

19. **Orthodontics** A new orthodontic product initially sells 1250 units per week. After 24 weeks, the number of units sold increases to 6500. The sales can be modeled by logistic growth with a limit of 10,000 units per week.

(a) Find a logistic growth model for the number of units.

(b) Use the model to complete the table.

Time, t	0	3	6	12	24
Sales, y					

(c) Use the graph shown below to approximate the time t when sales will be 7500.

20. **Biology** A conservation society has introduced a population of 300 ring-necked pheasants into a new area. After 5 years, the population has increased to 966. The population can be modeled by logistic growth with a limit of 2700 pheasants.

(a) Find a logistic growth model for the population of ring-necked pheasants.

(b) How many pheasants were present after 4 years?

(c) How long will it take to establish a population of 1750 pheasants?

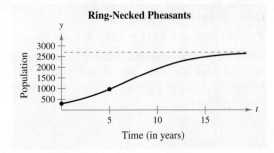

In Exercises 21–32, find or evaluate the integral.

21. $\displaystyle\int (3\sin x - 2\cos x)\,dx$

22. $\displaystyle\int \csc 5x \cot 5x\,dx$

23. $\displaystyle\int \sin^3 x \cos x\,dx$

24. $\displaystyle\int 2x \sec^2 x^2\,dx$

25. $\displaystyle\int_0^\pi (1 + \sin x)\,dx$

26. $\displaystyle\int_{-\pi}^\pi \frac{1}{2}(1 + \cos 2x)\,dx$

27. $\displaystyle\int_{-\pi/6}^{\pi/6} \sec^2 x\,dx$

28. $\displaystyle\int_{\pi/6}^{\pi/2} \csc^2 x\,dx$

29. $\displaystyle\int_{-\pi/3}^{\pi/3} 4\sec x \tan x\,dx$

30. $\displaystyle\int_{\pi/6}^{\pi/3} \csc x \cot x\,dx$

31. $\displaystyle\int_{-\pi/2}^{\pi/2} (2x + \cos x)\,dx$

32. $\displaystyle\int_0^\pi 2x \sin x^2\,dx$

In Exercises 33–36, find the area of the region.

33. $y = \sin 3x$

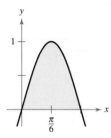

34. $y = \cot x$

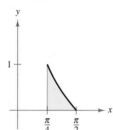

35. $y = 2\sin x + \cos 3x$

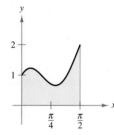

36. $y = 2\cos x + \cos 2x$

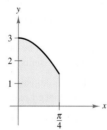

37. **Meteorology** The average monthly precipitation P (in inches), including rain, snow, and ice, in San Francisco, California can be modeled by

$$P = 2.91\sin(0.4t + 1.81) + 2.38, \quad 0 \le t \le 12$$

where t is the time in months, with $t = 1$ corresponding to January. Find the total annual precipitation in San Francisco. *(Source: National Oceanic and Atmospheric Administration)*

38. **Surface Area** Use the Midpoint Rule to estimate the surface area of the oil spill shown in the figure.

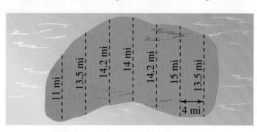

(T) In Exercises 39–42, use the Midpoint Rule with $n = 4$ to approximate the definite integral. Then use a programmable calculator or computer to approximate the definite integral with $n = 20$. Compare the two approximations.

39. $\displaystyle\int_0^2 (x^2 + 1)^2\,dx$

40. $\displaystyle\int_{-1}^1 \sqrt{1 - x^2}\,dx$

41. $\displaystyle\int_0^1 \frac{1}{x^2 + 1}\,dx$

42. $\displaystyle\int_{-1}^1 e^{3 - x^2}\,dx$

In Exercises 43–46, use the Trapezoidal Rule to approximate the definite integral.

43. $\displaystyle\int_1^3 \frac{1}{x^2}\,dx,\ n = 4$

44. $\displaystyle\int_0^2 (x^2 + 1)\,dx,\ n = 4$

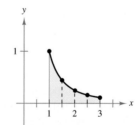
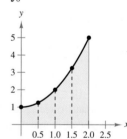

45. $\displaystyle\int_1^2 \frac{1}{1 + \ln x}\,dx,\ n = 4$

46. $\displaystyle\int_0^2 \frac{1}{\sqrt{1 + x^3}}\,dx,\ n = 8$

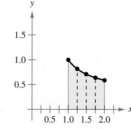
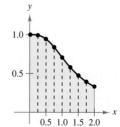

In Exercises 47–50, use Simpson's Rule to approximate the definite integral.

47. $\int_{1}^{2} \frac{1}{x^3}\, dx$, $n = 4$

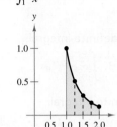

48. $\int_{1}^{2} x^3\, dx$, $n = 4$

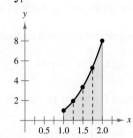

49. $\int_{0}^{1} \frac{x^{3/2}}{2 - x^2}\, dx$, $n = 4$

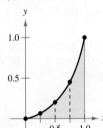

50. $\int_{0}^{1} e^{x^2}\, dx$, $n = 6$

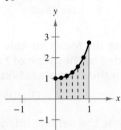

In Exercises 51 and 52, use the error formula to find bounds for the error in approximating the integral using the Trapezoidal Rule.

51. $\int_{0}^{2} e^{2x}\, dx$, $n = 4$

52. $\int_{0}^{2} e^{2x}\, dx$, $n = 8$

In Exercises 53 and 54, use the error formula to find bounds for the error in approximating the integral using Simpson's Rule.

53. $\int_{2}^{4} \frac{1}{x - 1}\, dx$, $n = 4$

54. $\int_{2}^{4} \frac{1}{x - 1}\, dx$, $n = 8$

In Exercises 55–62, determine whether the improper integral diverges or converges. Evaluate the integral if it converges.

55. $\int_{0}^{\infty} 4xe^{-2x^2}\, dx$

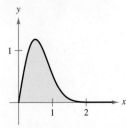

56. $\int_{-\infty}^{0} \frac{3}{(1 - 3x)^{2/3}}\, dx$

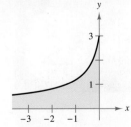

57. $\int_{-\infty}^{0} \frac{1}{3x^2}\, dx$

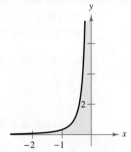

58. $\int_{0}^{\infty} 2x^2 e^{-x^3}\, dx$

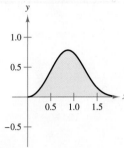

59. $\int_{0}^{4} \frac{1}{\sqrt{4x}}\, dx$

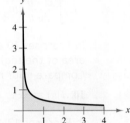

60. $\int_{1}^{2} \frac{x}{16(x - 1)^2}\, dx$

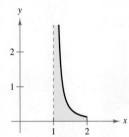

61. $\int_{2}^{3} \frac{1}{\sqrt{x - 2}}\, dx$

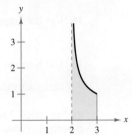

62. $\int_{0}^{2} \frac{x + 2}{(x - 1)^2}\, dx$

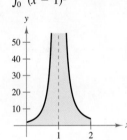

63. SAT Scores In 2006, the Scholastic Aptitude Test (SAT) math scores for college-bound seniors roughly followed a normal distribution

$$y = 0.0035e^{-(x - 518)^2/26{,}450}, \quad 200 \leq x \leq 800$$

where x is the SAT score for mathematics. Find the probability that a senior chosen at random had an SAT score (a) between 500 and 650, (b) 650 or better, and (c) 750 or better. *(Source: College Board)*

64. ACT Scores In 2006, the ACT composite scores for college-bound seniors followed a normal distribution

$$y = 0.0831e^{-(x - 21.1)^2/46.08}, \quad 1 \leq x \leq 36$$

where x is the composite ACT score. Find the probability that a senior chosen at random had an ACT score (a) between 16.3 and 25.9, (b) 25.9 or better, and (c) 30.7 or better. *(Source: ACT, Inc.)*

CHAPTER 7 Techniques of Integration

Chapter Test

See www.CalcChat.com for worked-out solutions to odd-numbered exercises.

Take this test as you would take a test in class. When you are done, check your work against the answers given in the back of the book.

In Exercises 1–3, use integration by parts to find the indefinite integral.

1. $\displaystyle\int xe^{x+1}\,dx$

2. $\displaystyle\int 9x^2 \ln x\,dx$

3. $\displaystyle\int x^2 e^{-x/3}\,dx$

In Exercises 4–6, use partial fractions to find the indefinite integral.

4. $\displaystyle\int \frac{18}{x^2 - 81}\,dx$

5. $\displaystyle\int \frac{3x}{(3x+1)^2}\,dx$

6. $\displaystyle\int \frac{x+4}{x^2+2x}\,dx$

In Exercises 7–9, find the indefinite integral.

7. $\displaystyle\int \frac{\cos\theta}{\sin\theta}\,d\theta$

8. $\displaystyle\int \tan 5\theta\,d\theta$

9. $\displaystyle\int \csc 2\theta\,d\theta$

In Exercises 10 and 11, use the Midpoint Rule with $n = 4$ to approximate the area of the region bounded by the graph of f and the x-axis over the interval. Compare your result with the exact area. Sketch the region.

10. $f(x) = 3x^2$, $[0, 1]$

11. $f(x) = x^2 + 1$, $[-1, 1]$

In Exercises 12–14, evaluate the definite integral.

12. $\displaystyle\int_0^1 \ln(3-2x)\,dx$

13. $\displaystyle\int_5^{10} \frac{28}{x^2 - x - 12}\,dx$

14. $\displaystyle\int_0^{\pi/2} \cos 2x\,dx$

15. Use the Trapezoidal Rule with $n = 4$ to approximate $\displaystyle\int_1^2 \frac{1}{x^2\sqrt{x^2+4}}\,dx$.

16. Use Simpson's Rule with $n = 4$ to approximate $\displaystyle\int_0^1 9xe^{3x}\,dx$. Compare your result with the exact value of the definite integral.

In Exercises 17–19, determine whether the improper integral converges or diverges. Evaluate the integral if it converges.

17. $\displaystyle\int_0^\infty e^{-3x}\,dx$

18. $\displaystyle\int_0^9 \frac{2}{\sqrt{x}}\,dx$

19. $\displaystyle\int_{-\infty}^0 \frac{1}{(4x-1)^{2/3}}\,dx$

20. To estimate the surface area of a swamp, a botanist takes several measurements, as shown in the figure. Estimate the surface area of the swamp using the Midpoint Rule.

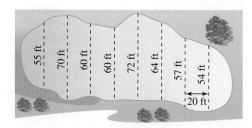

Matrices

8

Matrices can be used to solve systems of linear equations to determine the amounts of solutions needed for an acid mixture. (See Section 8.2, Exercises 49 and 50.)

Applications

Matrices have many real-life applications. The applications listed below represent a sample of the applications in this chapter.

- Nutrition, Exercises 49 and 50, page 497
- Agriculture, Exercises 43 and 44, page 505
- Breeding Facility, Exercise 67, page 516
- Leslie Matrix, Exercises 51–54, page 528
- Child Support, Exercise 67, page 537

8.1 Systems of Equations in Two Variables

8.2 Systems of Linear Equations in More than Two Variables

8.3 Matrices and Systems of Equations

8.4 Operations with Matrices

8.5 The Inverse of a Square Matrix

Section 8.1

Systems of Equations in Two Variables

- Use the method of substitution to solve systems of linear equations in two variables.
- Use the method of substitution to solve systems of nonlinear equations in two variables.
- Use the method of elimination to solve systems of linear equations in two variables.
- Interpret graphically the numbers of solutions of systems of linear equations in two variables.
- Use systems of linear equations in two variables to model and solve real-life problems.

The Method of Substitution

Many problems in science, business, and engineering involve two or more equations in two or more variables. To solve such problems, you need to find solutions of a **system of equations**. Here is an example of a system of two equations in two unknowns.

$$\begin{cases} 2x + y = 5 & \text{Equation 1} \\ 3x - 2y = 4 & \text{Equation 2} \end{cases}$$

A **solution** of this system is an ordered pair that satisfies each equation in the system. Finding the set of all solutions is called **solving the system of equations.** For instance, the ordered pair $(2, 1)$ is a solution of this system. To check this, you can substitute 2 for x and 1 for y in *each* equation.

Check $(2, 1)$ in Equation 1	Check $(2, 1)$ in Equation 2
$2x + y = 5$	$3x - 2y = 4$
$2(2) + 1 \stackrel{?}{=} 5$	$3(2) - 2(1) \stackrel{?}{=} 4$
$4 + 1 = 5$	$6 - 2 = 4$
$5 = 5$	$4 = 4$

In this section, you will study two ways to solve a system of equations, beginning with the **method of substitution.**

STUDY TIP

The method of substitution is equivalent to finding the point(s) of intersection of two graphs (see Section 1.2).

Method of Substitution

1. *Solve* one of the equations for one variable in terms of the other.
2. *Substitute* the expression found in Step 1 into the other equation to obtain an equation in one variable.
3. *Solve* the equation obtained in Step 2.
4. *Back-substitute* the value obtained in Step 3 into the expression obtained in Step 1 to find the value of the other variable.
5. *Check* that the solution satisfies *each* of the original equations.

SECTION 8.1 Systems of Equations in Two Variables

TECHNOLOGY

Use a graphing utility to graph $y_1 = 4 - x$ and $y_2 = x - 2$ in the same viewing window. Use the *zoom* and *trace* features to find the coordinates of the point of intersection. What is the relationship between the point of intersection and the solution found in Example 1?*

Algebra Review

For help in solving equations for specific variables, see Example 1 in the *Chapter 8 Algebra Review*, on page 538. For help in solving single-variable equations, see Example 2 in the *Chapter 8 Algebra Review*, on page 538.

STUDY TIP

Because many steps are required to solve a system of equations, it is very easy to make errors in arithmetic. So, you should always check your solution by substituting it into *each* equation in the original system.

Example 1 Solving a System of Equations by Substitution

Solve the system of equations.

$$\begin{cases} x + y = 4 & \text{Equation 1} \\ x - y = 2 & \text{Equation 2} \end{cases}$$

SOLUTION Begin by solving for y in Equation 1.

$y = 4 - x$ Solve for y in Equation 1.

Next, substitute this expression for y into Equation 2 and solve the resulting single-variable equation for x.

$x - y = 2$	Write Equation 2.
$x - (4 - x) = 2$	Substitute $4 - x$ for y.
$x - 4 + x = 2$	Distributive Property
$2x = 6$	Combine like terms.
$x = 3$	Divide each side by 2.

Finally, you can solve for y by *back-substituting* $x = 3$ into the equation $y = 4 - x$, to obtain

$y = 4 - x$	Write revised Equation 1.
$y = 4 - 3$	Substitute 3 for x.
$y = 1$.	Solve for y.

The solution is the ordered pair $(3, 1)$. You can check this solution as follows.

Check $(3, 1)$ in Equation 1
$x + y = 4$
$3 + 1 \stackrel{?}{=} 4$
$4 = 4$

Check $(3, 1)$ in Equation 2
$x - y = 2$
$3 - 1 \stackrel{?}{=} 2$
$2 = 2$

Because $(3, 1)$ satisfies both equations in the system, it is a solution of the system of equations.

✓CHECKPOINT 1

Solve the system of equations.

$$\begin{cases} 2x + y = 6 \\ -x + y = 0 \end{cases}$$ ■

The term *back-substitution* implies that you work *backwards*. First you solve for one of the variables, and then you substitute that value *back* into one of the equations in the system to find the value of the other variable.

---

*Specific calculator keystroke instructions for operations in this and other technology boxes can be found at *college.hmco.com/info/larsonapplied.*

Nonlinear Systems of Equations

The equations in Example 1 are linear. The method of substitution can also be used to solve systems in which one or both of the equations are nonlinear.

Example 2 Substitution: Two-Solution Case

Solve the system of equations.

$$\begin{cases} x^2 + 4x - y = 7 & \text{Equation 1} \\ 2x - y = -1 & \text{Equation 2} \end{cases}$$

SOLUTION Begin by solving for y in Equation 2 to obtain $y = 2x + 1$. Next, substitute this expression for y into Equation 1 and solve for x.

$$x^2 + 4x - (2x + 1) = 7 \qquad \text{Substitute } 2x + 1 \text{ for } y \text{ in Equation 1.}$$
$$x^2 + 2x - 1 = 7 \qquad \text{Simplify.}$$
$$x^2 + 2x - 8 = 0 \qquad \text{Write in general form.}$$
$$(x + 4)(x - 2) = 0 \qquad \text{Factor.}$$
$$x = -4, 2 \qquad \text{Solve for } x.$$

Back-substituting these values of x to solve for the corresponding values of y produces the solutions $(-4, -7)$ and $(2, 5)$. Check these in the original system.

✓ **CHECKPOINT 2**

Solve the system of equations.

$$\begin{cases} x^2 - y = 0 \\ 2x + y = 0 \end{cases}$$ ■

The Method of Elimination

A second method for solving a system of equations is the **method of elimination**. The key step in this method is to obtain, for one of the variables, coefficients that differ only in sign so that *adding* the equations eliminates the variable.

$$\begin{array}{rl} 3x + 5y = 7 & \text{Equation 1} \\ \underline{-3x - 2y = -1} & \text{Equation 2} \\ 3y = 6 & \text{Add equations.} \end{array}$$

Note that by adding the two equations, you eliminate the x-terms and obtain a single equation in y. Solving this equation for y produces $y = 2$, which you can then back-substitute into one of the original equations to solve for x.

> **Method of Elimination**
>
> To use the **method of elimination** to solve a system of two linear equations in x and y, perform the following steps.
>
> 1. *Obtain coefficients* for x (or y) that differ only in sign by multiplying all terms of one or both equations by suitably chosen constants.
>
> 2. *Add* the equations to eliminate one variable, and solve the resulting equation.
>
> 3. *Back-substitute* the value obtained in Step 2 into either of the original equations and solve for the other variable.
>
> 4. *Check* your solution in both of the original equations.

SECTION 8.1 Systems of Equations in Two Variables

Example 3 Solving a System of Equations by Elimination

Solve the system of linear equations.

$$\begin{cases} 2x - 3y = -7 & \text{Equation 1} \\ 3x + y = -5 & \text{Equation 2} \end{cases}$$

SOLUTION For this system, you can obtain coefficients that differ only in sign by multiplying Equation 2 by 3.

$$2x - 3y = -7 \quad \Longrightarrow \quad 2x - 3y = -7 \quad \text{Write Equation 1.}$$
$$3x + y = -5 \quad \Longrightarrow \quad 9x + 3y = -15 \quad \text{Multiply Equation 2 by 3.}$$
$$\overline{} \qquad \overline{11x = -22} \quad \text{Add equations.}$$

So, $x = -2$. By back-substituting this value of x into Equation 1, you can solve for y.

$$2x - 3y = -7 \quad \text{Write Equation 1.}$$
$$2(-2) - 3y = -7 \quad \text{Substitute } -2 \text{ for } x.$$
$$-3y = -3 \quad \text{Combine like terms.}$$
$$y = 1 \quad \text{Solve for } y.$$

The solution is $(-2, 1)$, as shown in Figure 8.1.

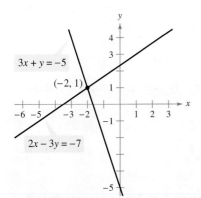

FIGURE 8.1

✓ CHECKPOINT 3

Solve the system of linear equations.

$$\begin{cases} 2x + 3y = 18 \\ 5x - y = 11 \end{cases}$$

Graphical Interpretations of Solutions

It is possible for a *general* system of equations to have exactly one solution, two or more solutions, or no solution. If a system of *linear* equations has two different solutions, it must have an *infinite* number of solutions.

Graphical Interpretations of Solutions

For a system of two linear equations in two variables, the number of solutions is one of the following.

Number of Solutions	*Graphical Interpretation*
1. Exactly one solution	The two lines intersect at one point. (See Example 3.)
2. Infinitely many solutions	The two lines coincide (are identical). (See Example 5.)
3. No solution	The two lines are parallel. (See Example 4.)

A system of linear equations is **consistent** if it has at least one solution. A consistent system with exactly one solution is *independent*, whereas a consistent system with infinitely many solutions is *dependent*. A system is **inconsistent** if it has no solution.

494   CHAPTER 8   Matrices

Example 4   No-Solution Case: Method of Elimination

Solve the system of linear equations.

$$\begin{cases} x - 2y = 3 & \text{Equation 1} \\ -2x + 4y = 1 & \text{Equation 2} \end{cases}$$

SOLUTION   To obtain coefficients that differ only in sign, multiply Equation 1 by 2.

$x - 2y = 3$	$\Rightarrow$	$2x - 4y = 6$ — Multiply Equation 1 by 2.
$-2x + 4y = 1$	$\Rightarrow$	$-2x + 4y = 1$ — Write Equation 2.
		$0 = 7$ — False statement

Because there are no values of x and y for which $0 = 7$, you can conclude that the system is inconsistent and has no solution. The lines corresponding to the two equations in this system are shown in Figure 8.2. Note that the two lines are parallel and therefore have no point of intersection.

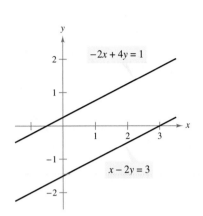

FIGURE 8.2

✓ CHECKPOINT 4

Solve the system of linear equations.

$$\begin{cases} 9x + 6y = 20 \\ 3x + 2y = 1 \end{cases}$$

In Example 4, note that the occurrence of a false statement, such as $0 = 7$, indicates that the system has no solution. In the next example, note that the occurrence of a statement that is true for all values of the variables, such as $0 = 0$, indicates that the system has infinitely many solutions.

Example 5   Many-Solution Case: Method of Elimination

Solve the system of linear equations.

$$\begin{cases} 2x - y = 1 & \text{Equation 1} \\ 4x - 2y = 2 & \text{Equation 2} \end{cases}$$

SOLUTION   To obtain coefficients that differ only in sign, multiply Equation 2 by $-\frac{1}{2}$.

$2x - y = 1$	$\Rightarrow$	$2x - y = 1$ — Write Equation 1.
$4x - 2y = 2$	$\Rightarrow$	$-2x + y = -1$ — Multiply Equation 2 by $-\frac{1}{2}$.
		$0 = 0$ — Add equations.

Because the two equations turn out to be equivalent (have the same solution set), you can conclude that the system has infinitely many solutions. The solution set consists of all points (x, y) lying on the line $2x - y = 1$, as shown in Figure 8.3. Letting $x = a$, where a is any real number, you can see that the solutions to the system are $(a, 2a - 1)$.

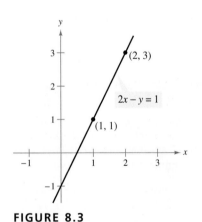

FIGURE 8.3

✓ CHECKPOINT 5

Solve the system of linear equations.

$$\begin{cases} -5x + 6y = -3 \\ 20x - 24y = 12 \end{cases}$$

Application

Example 6
MAKE A DECISION Medicine

A nurse has two iodine solutions. One is a 10% solution and the other is a 2% solution. How many milliliters of each solution should the nurse mix to obtain 100 milliliters of a 5% iodine solution? Will the nurse need more than 60 milliliters of the 2% iodine solution?

SOLUTION Let x equal the number of milliliters of the 10% solution, and let y equal the number of milliliters of the 2% solution. The total amount of the final mixture is

$$x + y = 100 \qquad \text{Equation 1}$$

and the amount of iodine in the final mixture is

$$0.10x + 0.02y = 0.05(100) = 5. \qquad \text{Equation 2}$$

The system of equations to solve is

$$\begin{cases} x + y = 100 \\ 0.10x + 0.02y = 5 \end{cases}$$

Begin by solving for x in Equation 1.

$$x = 100 - y \qquad \text{Solve for } x \text{ in Equation 1.}$$

Next, substitute this expression for x into Equation 2 and solve the resulting equation for y.

$$0.10(100 - y) + 0.02y = 5 \qquad \text{Substitute for } x \text{ in Equation 2.}$$
$$10 - 0.10y + 0.02y = 5 \qquad \text{Distributive Property}$$
$$-0.08y = -5 \qquad \text{Simplify.}$$
$$y = 62.5 \qquad \text{Divide each side by } -0.08.$$

So, $y = 62.5$ and $x = 100 - (62.5) = 37.5$. The nurse will need to mix 37.5 milliliters of the 10% solution and 62.5 milliliters of the 2% solution to obtain 100 milliliters of a 5% solution. Yes, the nurse will need more than 60 milliliters of the 2% solution.

✓**CHECKPOINT 6**

Use the method of elimination to solve the system in Example 6. Which method is easier? ■

CONCEPT CHECK

1. Complete the following: A set of two or more equations in two or more variables is called a _____ of _____.
2. List the steps used to solve a system of equations by the method of substitution.
3. List the steps used to solve a system of equations by the method of elimination.
4. When solving a system of equations by substitution, how do you recognize that the system has no solution?

496 CHAPTER 8 Matrices

Skills Review 8.1

The following warm-up exercises involve skills that were covered in earlier sections. You will use these skills in the exercise set for this section. For additional help, review Section 1.3.

In Exercises 1 and 2, solve the equation.

1. $3x + (x - 5) = 15 + 4$

2. $y^2 + (y - 2)^2 = 2$

In Exercises 3 and 4, determine whether the lines represented by the pair of equations are parallel, perpendicular, or neither.

3. $2x - 3y = -10$
$3x + 2y = 11$

4. $4x - 12y = 5$
$-2x + 6y = 3$

Exercises 8.1

See www.CalcChat.com for worked-out solutions to odd-numbered exercises.

In Exercises 1–4, determine whether each ordered pair is a solution of the system of equations.

1. $\begin{cases} 4x - y = 1 \\ 6x + y = -6 \end{cases}$
(a) $(0, -3)$ (b) $(-1, -4)$
(c) $\left(-\frac{3}{2}, -2\right)$ (d) $\left(-\frac{1}{2}, -3\right)$

2. $\begin{cases} 4x^2 + y = 3 \\ -x - y = 11 \end{cases}$
(a) $(2, -13)$ (b) $(2, -9)$
(c) $\left(-\frac{3}{2}, -\frac{31}{3}\right)$ (d) $\left(-\frac{7}{4}, -\frac{37}{4}\right)$

3. $\begin{cases} y = -2e^x \\ 3x - y = 2 \end{cases}$
(a) $(-2, 0)$ (b) $(0, -2)$
(c) $(0, -3)$ (d) $(-1, 2)$

4. $\begin{cases} -\log x + 3 = y \\ \frac{1}{9}x + y = \frac{28}{9} \end{cases}$
(a) $\left(9, \frac{37}{9}\right)$ (b) $(10, 2)$
(c) $(1, 3)$ (d) $(2, 4)$

In Exercises 5–18, solve the system by the method of substitution.

5. $\begin{cases} x - y = 0 \\ 5x - 3y = 10 \end{cases}$

6. $\begin{cases} x + 2y = 1 \\ 5x - 4y = -23 \end{cases}$

7. $\begin{cases} 2x - y + 2 = 0 \\ 4x + y - 5 = 0 \end{cases}$

8. $\begin{cases} 6x - 3y - 4 = 0 \\ x + 2y - 4 = 0 \end{cases}$

9. $\begin{cases} 1.5x + 0.8y = 2.3 \\ 0.3x - 0.2y = 0.1 \end{cases}$

10. $\begin{cases} 0.5x + 3.2y = 9.0 \\ 0.2x - 1.6y = -3.6 \end{cases}$

11. $\begin{cases} \frac{1}{5}x + \frac{1}{2}y = 8 \\ x + y = 20 \end{cases}$

12. $\begin{cases} \frac{1}{2}x + \frac{3}{4}y = 10 \\ \frac{3}{4}x - y = 4 \end{cases}$

13. $\begin{cases} 6x + 5y = -3 \\ -x - \frac{5}{6}y = -7 \end{cases}$

14. $\begin{cases} -\frac{2}{3}x + y = 2 \\ 2x - 3y = 6 \end{cases}$

15. $\begin{cases} x^2 - y = 0 \\ 2x + y = 0 \end{cases}$

16. $\begin{cases} x - 2y = 0 \\ 3x - y^2 = 0 \end{cases}$

17. $\begin{cases} x - y = -1 \\ x^2 - y = -4 \end{cases}$

18. $\begin{cases} y = -x \\ y = x^3 + 3x^2 + 2x \end{cases}$

In Exercises 19–38, solve the system by the method of elimination.

19. $\begin{cases} x + 2y = 4 \\ x - 2y = 1 \end{cases}$

20. $\begin{cases} 3x - 5y = 2 \\ 2x + 5y = 13 \end{cases}$

21. $\begin{cases} -4x + 5y = -12 \\ 2x - y = 3 \end{cases}$

22. $\begin{cases} x + 7y = 12 \\ 3x - 5y = 10 \end{cases}$

23. $\begin{cases} 3x + 2y = 10 \\ 2x + 5y = 3 \end{cases}$

24. $\begin{cases} 2r + 4s = 5 \\ 16r + 50s = 55 \end{cases}$

25. $\begin{cases} 5u + 6v = 24 \\ 3u + 5v = 18 \end{cases}$

26. $\begin{cases} 3x + 11y = 4 \\ -2x - 5y = 9 \end{cases}$

27. $\begin{cases} \frac{9}{5}x + \frac{6}{5}y = 4 \\ 9x + 6y = 3 \end{cases}$

28. $\begin{cases} \frac{3}{4}x + y = \frac{1}{8} \\ \frac{9}{4}x + 3y = \frac{3}{8} \end{cases}$

29. $\begin{cases} \dfrac{x}{4} + \dfrac{y}{6} = 1 \\ x - y = 3 \end{cases}$

30. $\begin{cases} \dfrac{2}{3}x + \dfrac{1}{6}y = \dfrac{2}{3} \\ 4x + y = 4 \end{cases}$

31. $\begin{cases} -5x + 6y = -3 \\ 20x - 24y = 12 \end{cases}$

32. $\begin{cases} 7x + 8y = 6 \\ -14x - 16y = -12 \end{cases}$

33. $\begin{cases} 0.05x - 0.03y = 0.21 \\ 0.07x + 0.02y = 0.16 \end{cases}$

34. $\begin{cases} 0.2x - 0.5y = -27.8 \\ 0.3x + 0.4y = 68.7 \end{cases}$

35. $\begin{cases} 4b + 3m = 3 \\ 3b + 11m = 13 \end{cases}$

36. $\begin{cases} 2x + 5y = 8 \\ 5x + 8y = 10 \end{cases}$

37. $\begin{cases} \dfrac{x + 3}{4} + \dfrac{y - 1}{3} = 1 \\ 2x - y = 12 \end{cases}$

38. $\begin{cases} \dfrac{x - 1}{2} + \dfrac{y + 2}{3} = 4 \\ x - 2y = 5 \end{cases}$

In Exercises 39–46, use any method to solve the system.

39. $\begin{cases} 3x - 5y = 7 \\ 2x + y = 9 \end{cases}$

40. $\begin{cases} -x + 3y = 17 \\ 4x + 3y = 7 \end{cases}$

41. $\begin{cases} y = 2x - 5 \\ y = 5x - 11 \end{cases}$

42. $\begin{cases} 7x + 3y = 16 \\ y = x + 2 \end{cases}$

43. $\begin{cases} x - 5y = 21 \\ 6x + 5y = 21 \end{cases}$
44. $\begin{cases} y = -3x - 8 \\ y = 15 - 2x \end{cases}$
45. $\begin{cases} -2x + 8y = 19 \\ y = x - 3 \end{cases}$
46. $\begin{cases} 4x - 3y = 6 \\ -5x + 7y = -1 \end{cases}$

47. **Airplane Speed** An airplane flying into a headwind travels the 1800-mile flying distance between Los Angeles, California and South Bend, Indiana in 3 hours and 36 minutes. On the return flight, the distance is traveled in 3 hours. Find the airspeed of the plane and the speed of the wind, assuming that both remain constant.

48. **Airplane Speed** Two planes start from Los Angeles International Airport and fly in opposite directions. The second plane starts $\frac{1}{2}$ hour after the first plane, but its speed is 80 kilometers per hour faster. Find the airspeed of each plane if 2 hours after the first plane departs the planes are 3200 kilometers apart.

49. **Nutrition** Two cheeseburgers and one small order of french fries from a fast-food restaurant contain a total of 850 calories. Three cheeseburgers and two small orders of french fries contain a total of 1390 calories. Find the caloric content of each item.

50. **Nutrition** One eight-ounce glass of apple juice and one eight-ounce glass of orange juice contain a total of 185 milligrams of vitamin C. Two eight-ounce glasses of apple juice and three eight-ounce glasses of orange juice contain a total of 452 milligrams of vitamin C. How much vitamin C is in an eight-ounce glass of each type of juice?

51. **Body Mass Index** Body Mass Index (BMI) is a measure of body fat based on height and weight. The 75th percentile BMI for females, ages 9 to 20, is growing more slowly than that of males of the same age range. Models that represent the 75th percentile BMI for males and females, ages 9 to 20, are given by

    $B = 0.73a + 11$ Males
    $B = 0.61a + 12.8$ Females

    where B is the BMI (kg/m^2) and a represents the age, with $a = 9$ corresponding to 9 years of age. Use a graphing utility to determine whether the BMI for males will ever exceed the BMI for females. *(Source: National Center for Health Statistics)*

52. **Financial Aid** The average awards for Federal Pell Grants and Federal Perkins Loans from 1995 through 2005 can be approximated by the models

    $A = -2.051t^3 + 56.87t^2 - 376.7t + 2238$ Federal Pell Grant
    $A = -1.810t^3 + 56.64t^2 - 476.4t + 2711$ Federal Perkins Loan

    where A is the award (in dollars) and t represents the year, with $t = 5$ corresponding to 1995. Use a graphing utility to determine whether Federal Perkins Loan awards will ever exceed Federal Pell Grant awards. Do you think these models will continue to be accurate? Explain your reasoning. *(Source: U.S. Department of Education)*

53. **SAT or ACT?** The numbers of participants in SAT and ACT testing from 1995 through 2005 can be approximated by the models

    $y = 0.68t^2 + 28.1t + 903$ SAT
    $y = -0.485t^3 + 14.88t^2 - 115.1t + 1201$ ACT

    where y is the number of participants (in thousands) and t represents the year, with $t = 5$ corresponding to 1995. Use a graphing utility to determine whether the number of participants in ACT testing will exceed the number of participants in SAT testing. Do you think these models will continue to be accurate? Explain your reasoning. *(Source: College Board; ACT, Inc.)*

54. **Prescriptions** The numbers of prescriptions P (in thousands) filled at two pharmacies from 2002 through 2008 are shown in the table.

Year	Pharmacy A	Pharmacy B
2002	18.1	19.5
2003	18.6	19.9
2004	19.2	20.4
2005	19.6	20.8
2006	20.0	21.1
2007	20.4	21.4
2008	21.3	22.0

    (a) Use a graphing utility or a spreadsheet to create a scatter plot of the data for pharmacy A and use the *regression* feature to find a linear model. Let x represent the year, with $x = 2$ corresponding to 2002. Repeat the procedure for pharmacy B.

    (b) Assuming the numbers for the given seven years are representative of future years, will the number of prescriptions filled at pharmacy A ever exceed the number of prescriptions filled at pharmacy B?

55. **Extended Application** To work an extended application analyzing the dormitory charges (in thousands of dollars) at private institutions in the United States for the years 1990 through 2005, visit this text's website at *college.hmco.com/info/larsonapplied*. *(Data Source: U.S. National Center for Education Statistics)*

Section 8.2
Systems of Linear Equations in More than Two Variables

- Use back-substitution to solve linear systems in row-echelon form.
- Use Gaussian elimination to solve systems of linear equations.
- Solve nonsquare systems of linear equations.
- Use systems of linear equations in three or more variables to model and solve real-life problems.

Row-Echelon Form and Back-Substitution

The method of elimination can be applied to a system of linear equations in more than two variables. In fact, this method easily adapts to computer use for solving linear systems with dozens of variables.

When elimination is used to solve a system of linear equations, the goal is to rewrite the system in a form to which back-substitution can be applied. To see how this works, consider the following two systems of linear equations.

System of Three Linear Equations in Three Variables: (See Example 2.)

$$\begin{cases} x - 2y + 3z = 9 \\ -x + 3y = -4 \\ 2x - 5y + 5z = 17 \end{cases}$$

Equivalent System in Row-Echelon Form: (See Example 1.)

$$\begin{cases} x - 2y + 3z = 9 \\ y + 3z = 5 \\ z = 2 \end{cases}$$

The second system is said to be in **row-echelon form,** which means that it has a "stair-step" pattern with leading coefficients of 1. After comparing the two systems, it should be clear that it is easier to solve the system in row-echelon form, using back-substitution.

Algebra Review

For help in using back-substitution to solve systems of equations, see Example 3 in the *Chapter 8 Algebra Review*, on page 539.

Example 1 Using Back-Substitution to Solve a System in Row-Echelon Form

Solve the system of linear equations.

$$\begin{cases} x - 2y + 3z = 9 & \text{Equation 1} \\ y + 3z = 5 & \text{Equation 2} \\ z = 2 & \text{Equation 3} \end{cases}$$

SOLUTION From Equation 3, you know the value of z. To solve for y, substitute $z = 2$ into Equation 2 to obtain

$$y + 3(2) = 5 \quad \text{Substitute 2 for } z.$$
$$y = -1. \quad \text{Solve for } y.$$

Finally, substitute $y = -1$ and $z = 2$ into Equation 1 to obtain

$$x - 2(-1) + 3(2) = 9 \quad \text{Substitute } -1 \text{ for } y \text{ and 2 for } z.$$
$$x = 1. \quad \text{Solve for } x.$$

The solution is $x = 1$, $y = -1$, and $z = 2$, which can be written as the **ordered triple** $(1, -1, 2)$. Check this in the original system of equations.

✓CHECKPOINT 1

Use back-substitution to solve the system of linear equations.

$$\begin{cases} 2x - y + 5z = 24 \\ y + 2z = 6 \\ z = 4 \end{cases}$$

Gaussian Elimination

Two systems of equations are *equivalent* if they have the same solution set. To solve a system that is not in row-echelon form, first convert it to an *equivalent* system that is in row-echelon form by using the following operations.

> **Operations That Produce Equivalent Systems**
>
> Each of the following **row operations** on a system of linear equations produces an *equivalent* system of linear equations.
>
> 1. Interchange two equations.
> 2. Multiply one of the equations by a nonzero constant.
> 3. Add a multiple of one of the equations to another equation to replace the latter equation.

Example 2 Using Gaussian Elimination to Solve a System

Solve the system of linear equations.

$$\begin{cases} x - 2y + 3z = 9 & \text{Equation 1} \\ -x + 3y = -4 & \text{Equation 2} \\ 2x - 5y + 5z = 17 & \text{Equation 3} \end{cases}$$

SOLUTION Because the leading coefficient of the first equation is 1, you can begin by saving the x at the upper left and eliminating the other x-terms from the first column.

$$\begin{cases} x - 2y + 3z = 9 \\ y + 3z = 5 \\ 2x - 5y + 5z = 17 \end{cases}$$
Adding the first equation to the second equation produces a new second equation.

$$\begin{cases} x - 2y + 3z = 9 \\ y + 3z = 5 \\ -y - z = -1 \end{cases}$$
Adding -2 times the first equation to the third equation produces a new third equation.

Now that all but the first x have been eliminated from the first column, go to work on the second column. (You need to eliminate y from the third equation.)

$$\begin{cases} x - 2y + 3z = 9 \\ y + 3z = 5 \\ 2z = 4 \end{cases}$$
Adding the second equation to the third equation produces a new third equation.

Finally, you need a coefficient of 1 for z in the third equation.

$$\begin{cases} x - 2y + 3z = 9 \\ y + 3z = 5 \\ z = 2 \end{cases}$$
Multiplying the third equation by $\frac{1}{2}$ produces a new third equation.

This is the same system that was solved in Example 1, and, as in that example, you can conclude that the solution is $x = 1$, $y = -1$, and $z = 2$.

STUDY TIP

Arithmetic errors are often made when performing elementary row operations. You should note the operation performed in each step so that you can go back and check your work.

✓ **CHECKPOINT 2**

Solve the system of linear equations.

$$\begin{cases} x + y + z = 6 \\ 2x - y + z = 3 \\ 3x - z = 0 \end{cases}$$

As shown in Example 2, rewriting a system of linear equations in row-echelon form usually involves a chain of equivalent systems, each of which is obtained by using one of the three basic row operations listed on the previous page. This process is called **Gaussian elimination,** after the German mathematician Carl Friedrich Gauss (1777–1855).

The next example involves an inconsistent system—one that has no solution. The key to recognizing an inconsistent system is that at some stage in the elimination process you obtain a false statement such as $0 = -2$.

Example 3 An Inconsistent System

Solve the system of linear equations.

$$\begin{cases} x - 3y + z = 1 & \text{Equation 1} \\ 2x - y - 2z = 2 & \text{Equation 2} \\ x + 2y - 3z = -1 & \text{Equation 3} \end{cases}$$

SOLUTION

$$\begin{cases} x - 3y + z = 1 \\ 5y - 4z = 0 \\ x + 2y - 3z = -1 \end{cases}$$

Adding -2 times the first equation to the second equation produces a new second equation.

$$\begin{cases} x - 3y + z = 1 \\ 5y - 4z = 0 \\ 5y - 4z = -2 \end{cases}$$

Adding -1 times the first equation to the third equation produces a new third equation.

$$\begin{cases} x - 3y + z = 1 \\ 5y - 4z = 0 \\ 0 = -2 \end{cases}$$

Adding -1 times the second equation to the third equation produces a new third equation.

Because $0 = -2$ is a false statement, you can conclude that this system is inconsistent and so has no solution. Moreover, because this system is equivalent to the original system, you can conclude that the original system also has no solution.

✓ **CHECKPOINT 3**

Solve the system of linear equations.

$$\begin{cases} 2x + y - z = 7 \\ x - 2y + 2z = -9 \\ 3x - y + z = 5 \end{cases}$$

As with a system of linear equations in two variables, the solution(s) of a system of linear equations in more than two variables must fall into one of three categories.

The Number of Solutions of a Linear System

For a system of linear equations, exactly one of the following is true.

1. There is exactly one solution.
2. There are infinitely many solutions.
3. There is no solution.

SECTION 8.2 Systems of Linear Equations in More than Two Variables

Example 4 A System with Infinitely Many Solutions

Solve the system of linear equations.

$$\begin{cases} x + y - 3z = -1 & \text{Equation 1} \\ y - z = 0 & \text{Equation 2} \\ -x + 2y = 1 & \text{Equation 3} \end{cases}$$

SOLUTION

$$\begin{cases} x + y - 3z = -1 \\ y - z = 0 \\ 3y - 3z = 0 \end{cases}$$

Adding the first equation to the third equation produces a new third equation.

$$\begin{cases} x + y - 3z = -1 \\ y - z = 0 \\ 0 = 0 \end{cases}$$

Adding -3 times the second equation to the third equation produces a new third equation.

This result means that Equation 3 depends on Equations 1 and 2 in the sense that it gives no additional information about the variables. Because $0 = 0$ is a true statement, you can conclude that this system will have infinitely many solutions. However, it is incorrect to say simply that the solution is "infinite." You must also specify the correct form of the solution. So, the original system is equivalent to the system

$$\begin{cases} x + y - 3z = -1 \\ y - z = 0 \end{cases}.$$

In the last equation, solve for y in terms of z to obtain $y = z$. Back-substituting for y in the first equation produces $x = 2z - 1$. Finally, letting $z = a$, where a is a real number, the solutions to the given system are all of the form $x = 2a - 1$, $y = a$, and $z = a$. So, every ordered triple of the form

$$(2a - 1, a, a), \quad a \text{ is a real number}$$

is a solution of the system.

In Example 4, there are other ways to write the same infinite set of solutions. For instance, letting $x = b$, the solutions could have been written as

$$\left(b, \tfrac{1}{2}(b + 1), \tfrac{1}{2}(b + 1)\right), \quad b \text{ is a real number.}$$

To convince yourself that this description produces the same set of solutions, consider the following.

Substitution	Solution	
$a = 0$	$(2(0) - 1, 0, 0) = (-1, 0, 0)$	Same solution
$b = -1$	$\left(-1, \tfrac{1}{2}(-1 + 1), \tfrac{1}{2}(-1 + 1)\right) = (-1, 0, 0)$	
$a = 1$	$(2(1) - 1, 1, 1) = (1, 1, 1)$	Same solution
$b = 1$	$\left(1, \tfrac{1}{2}(1 + 1), \tfrac{1}{2}(1 + 1)\right) = (1, 1, 1)$	
$a = 2$	$(2(2) - 1, 2, 2) = (3, 2, 2)$	Same solution
$b = 3$	$\left(3, \tfrac{1}{2}(3 + 1), \tfrac{1}{2}(3 + 1)\right) = (3, 2, 2)$	

STUDY TIP

In Example 4, x and y are solved in terms of the third variable z. To write the correct form of the solution to the system that does not use any of the three variables of the system, let a represent any real number and let $z = a$. Then solve for x and y. The solution can then be written in terms of a, which is not one of the variables of the system.

✓ CHECKPOINT 4

Solve the system of linear equations.

$$\begin{cases} x + 2y - 7z = -4 \\ 2x + y + z = 13 \\ 3x + 9y - 36z = -33 \end{cases}$$

STUDY TIP

When comparing descriptions of an infinite solution set, keep in mind that there is more than one way to describe the set.

Nonsquare Systems

So far, each system of linear equations you have looked at has been *square*, which means that the number of equations is equal to the number of variables. In a **nonsquare** system, the number of equations differs from the number of variables. A system of linear equations cannot have a unique solution unless there are at least as many equations as there are variables in the system.

Example 5   A System with Fewer Equations than Variables

Solve the system of linear equations.

$$\begin{cases} x - 2y + z = 2 & \text{Equation 1} \\ 2x - y - z = 1 & \text{Equation 2} \end{cases}$$

SOLUTION   Begin by rewriting the system in row-echelon form.

$$\begin{cases} x - 2y + z = 2 \\ \phantom{x - {}} 3y - 3z = -3 \end{cases}$$
Adding -2 times the first equation to the second equation produces a new second equation.

$$\begin{cases} x - 2y + z = 2 \\ y - z = -1 \end{cases}$$
Multiplying the second equation by $\frac{1}{3}$ produces a new second equation.

Solve for y in terms of z, to obtain

$$y = z - 1.$$

By back-substituting into Equation 1, you can solve for x, as follows.

$x - 2y + z = 2$ Write Equation 1.
$x - 2(z - 1) + z = 2$ Substitute for y in Equation 1.
$x - 2z + 2 + z = 2$ Distributive Property
$x = z$ Solve for x.

Finally, by letting $z = a$, where a is a real number, you have the solution

$$x = a, \quad y = a - 1, \quad \text{and} \quad z = a.$$

So, every ordered triple of the form

$(a, a - 1, a)$, a is a real number

is a solution of the system. Because there were originally three variables and only two equations, the system cannot have a unique solution.

✓ **CHECKPOINT 5**

Solve the system of linear equations.

$$\begin{cases} x - 2y + 5z = 2 \\ 4x - z = 0 \end{cases}$$

In Example 5, try choosing some values of a to obtain different solutions of the system, such as $(1, 0, 1)$, $(2, 1, 2)$, and $(3, 2, 3)$. Then check each of the solutions in the original system to verify that they are solutions of the original system.

SECTION 8.2 Systems of Linear Equations in More than Two Variables

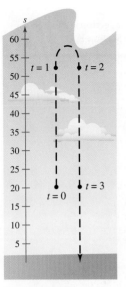

FIGURE 8.4

Application

Example 6 Vertical Motion

The height at time t of an object that is moving in a (vertical) line with constant acceleration a is given by the **position equation**

$$s = \tfrac{1}{2}at^2 + v_0 t + s_0.$$

The height s is measured in feet, the acceleration a is measured in feet per second squared, t is measured in seconds, v_0 is the initial velocity (at $t = 0$), and s_0 is the initial height. Find the values of a, v_0, and s_0 if $s = 52$ at $t = 1$, $s = 52$ at $t = 2$, and $s = 20$ at $t = 3$, and interpret the result. (See Figure 8.4.)

SOLUTION By substituting the three values of t and s into the position equation, you can obtain three linear equations in a, v_0, and s_0.

When $t = 1$: $\tfrac{1}{2}a(1)^2 + v_0(1) + s_0 = 52$ ⟹ $a + 2v_0 + 2s_0 = 104$

When $t = 2$: $\tfrac{1}{2}a(2)^2 + v_0(2) + s_0 = 52$ ⟹ $2a + 2v_0 + s_0 = 52$

When $t = 3$: $\tfrac{1}{2}a(3)^2 + v_0(3) + s_0 = 20$ ⟹ $9a + 6v_0 + 2s_0 = 40$

This produces the following system of linear equations.

$$\begin{cases} a + 2v_0 + 2s_0 = 104 \\ 2a + 2v_0 + s_0 = 52 \\ 9a + 6v_0 + 2s_0 = 40 \end{cases}$$

After rewriting the system in row-echelon form

$$\begin{cases} a + 2v_0 + 2s_0 = 104 \\ v_0 + \tfrac{3}{2}s_0 = 78 \\ s_0 = 20 \end{cases}$$

you can use back-substitution to find $a = -32$, $v_0 = 48$, and $s_0 = 20$. This solution results in a position equation of $s = -16t^2 + 48t + 20$ and implies that the object was thrown upward at a velocity of 48 feet per second from a height of 20 feet.

✓ **CHECKPOINT 6**

Find the position equation

$$s = \frac{1}{2}at^2 + v_0 t + s_0$$

for an object moving vertically at the given heights at the specified times.

At $t = 1$ second, $s = 128$ feet
At $t = 2$ seconds, $s = 80$ feet
At $t = 3$ seconds, $s = 0$ feet

CONCEPT CHECK

1. When a system of linear equations is written in row-echelon form, what are the leading coefficients?
2. Complete the following: The process used to write a system of linear equations in row-echelon form is called _____ elimination.
3. Name the three operations that can be used on a system of linear equations to produce an equivalent system.
4. When using Gaussian elimination to solve a system of linear equations, how can you recognize that the system has no solution?

Skills Review 8.2

The following warm-up exercises involve skills that were covered in earlier sections. You will use these skills in the exercise set for this section. For additional help, review Section 8.1.

In Exercises 1–4, solve the system of linear equations.

1. $\begin{cases} x + y = 25 \\ \phantom{x +{}} y = 10 \end{cases}$
2. $\begin{cases} 2x - 3y = 4 \\ 6x \phantom{{}- 3y} = -12 \end{cases}$
3. $\begin{cases} x + y = 32 \\ x - y = 24 \end{cases}$
4. $\begin{cases} 2r - s = 5 \\ r + 2s = 10 \end{cases}$

In Exercises 5–8, determine whether the ordered triple is a solution of the equation.

5. $5x - 3y + 4z = 2$
   $(-1, -2, 1)$
6. $x - 2y + 12z = 9$
   $(6, 3, 2)$
7. $2x - 5y + 3z = -9$
   $(a - 2, a + 1, a)$
8. $-5x + y + z = 21$
   $(a - 4, 4a + 1, a)$

In Exercises 9 and 10, solve for x in terms of a.

9. $x + 2y - 3z = 4$
   $y = 1 - a, \quad z = a$
10. $x - 3y + 5z = 4$
    $y = 2a + 3, \quad z = a$

Exercises 8.2

See www.CalcChat.com for worked-out solutions to odd-numbered exercises.

In Exercises 1–4, determine whether each ordered triple is a solution of the system of equations.

1. $\begin{cases} 3x - y + z = 1 \\ 2x \phantom{{}- y} - 3z = -14 \\ \phantom{2x -{}} 5y + 2z = 8 \end{cases}$
   (a) $(2, 0, -3)$ (b) $(-2, 0, 8)$
   (c) $(0, -1, 3)$ (d) $(-1, 0, 4)$

2. $\begin{cases} 3x + 4y - z = 17 \\ 5x - y + 2z = -2 \\ 2x - 3y + 7z = -21 \end{cases}$
   (a) $(3, -1, 2)$ (b) $(1, 3, -2)$
   (c) $(4, 1, -3)$ (d) $(1, -2, 2)$

3. $\begin{cases} 4x + y - z = 0 \\ -8x - 6y + z = -\frac{7}{4} \\ 3x - y \phantom{{}+ z} = -\frac{9}{4} \end{cases}$
   (a) $\left(\frac{1}{2}, -\frac{3}{4}, -\frac{7}{4}\right)$ (b) $\left(-\frac{3}{2}, \frac{5}{4}, -\frac{5}{4}\right)$
   (c) $\left(-\frac{1}{2}, \frac{3}{4}, -\frac{5}{4}\right)$ (d) $\left(-\frac{1}{2}, \frac{1}{6}, -\frac{3}{4}\right)$

4. $\begin{cases} -4x - y - 8z = -6 \\ \phantom{-4x -{}} y + z = 0 \\ 4x - 7y \phantom{{}+ z} = 6 \end{cases}$
   (a) $(-2, -2, 2)$ (b) $\left(-\frac{33}{2}, -10, 10\right)$
   (c) $\left(\frac{1}{8}, -\frac{1}{2}, \frac{1}{2}\right)$ (d) $\left(-\frac{11}{2}, -4, 4\right)$

In Exercises 5–10, use back-substitution to solve the system of linear equations.

5. $\begin{cases} 2x - y + 5z = 24 \\ \phantom{2x -{}} y + 2z = 6 \\ \phantom{2x - y +{}} z = 4 \end{cases}$
6. $\begin{cases} 4x - 3y - 2z = 21 \\ \phantom{4x -{}} 6y - 5z = -8 \\ \phantom{4x - 3y -{}} z = -2 \end{cases}$
7. $\begin{cases} 2x + y - 3z = 10 \\ \phantom{2x +{}} y + z = 12 \\ \phantom{2x + y +{}} z = 2 \end{cases}$
8. $\begin{cases} x - y + 2z = 22 \\ 3y - 8z = -9 \\ z = -3 \end{cases}$
9. $\begin{cases} 4x - 2y + z = 8 \\ -y + z = 4 \\ z = 2 \end{cases}$
10. $\begin{cases} 5x \phantom{{}- 2y} - 8z = 22 \\ 3y - 5z = 10 \\ z = -4 \end{cases}$

In Exercises 11 and 12, perform the row operation and write the equivalent system.

11. Add Equation 1 to Equation 2.

$\begin{cases} x - 2y + 3z = 5 & \text{Equation 1} \\ -x + 3y - 5z = 4 & \text{Equation 2} \\ 2x \phantom{{}- 2y} - 3z = 0 & \text{Equation 3} \end{cases}$

What did this operation accomplish?

12. Add -2 times Equation 1 to Equation 3.

$$\begin{cases} x - 2y + 3z = 5 & \text{Equation 1} \\ -x + 3y - 5z = 4 & \text{Equation 2} \\ 2x - 3z = 0 & \text{Equation 3} \end{cases}$$

What did this operation accomplish?

In Exercises 13–32, solve the system of linear equations and check any solution algebraically.

13. $\begin{cases} x + y + z = 3 \\ x - 2y + 4z = 5 \\ 3y + 4z = 5 \end{cases}$

14. $\begin{cases} 2x + 2z = 2 \\ 5x + 3y = 4 \\ 3y - 4z = 4 \end{cases}$

15. $\begin{cases} 2x + 4y + z = 1 \\ x - 2y - 3z = 2 \\ x + y - z = -1 \end{cases}$

16. $\begin{cases} 6y + 4z = -12 \\ 3x + 3y = 9 \\ 2x - 3z = 10 \end{cases}$

17. $\begin{cases} 2x + 4y - z = 7 \\ 2x - 4y + 2z = -6 \\ x + 4y + z = 0 \end{cases}$

18. $\begin{cases} 5x - 3y + 2z = 3 \\ 2x + 4y - z = 7 \\ x - 11y + 4z = 3 \end{cases}$

19. $\begin{cases} 2x + y - 3z = 4 \\ 4x + 2z = 10 \\ -2x + 3y - 13z = -8 \end{cases}$

20. $\begin{cases} 3x - 3y + 6z = 6 \\ x + 2y - z = 5 \\ 5x - 8y + 13z = 7 \end{cases}$

21. $\begin{cases} x + 2z = 5 \\ 3x - y - z = 1 \\ 6x - y + 5z = 16 \end{cases}$

22. $\begin{cases} x - 3y + 2z = 18 \\ 5x - 13y + 12z = 80 \end{cases}$

23. $\begin{cases} 2x - 3y + z = -2 \\ -4x + 9y = 7 \end{cases}$

24. $\begin{cases} 2x + 3y + 3z = 7 \\ 4x + 18y + 15z = 44 \end{cases}$

25. $\begin{cases} x + 3w = 4 \\ 2y - z - w = 0 \\ 3y - 2w = 1 \\ 2x - y + 4z = 5 \end{cases}$

26. $\begin{cases} x + y + z + w = 6 \\ 2x + 3y - w = 0 \\ -3x + 4y + z + 2w = 4 \\ x + 2y - z + w = 0 \end{cases}$

27. $\begin{cases} x + 4z = 1 \\ x + y + 10z = 10 \\ 2x - y + 2z = -5 \end{cases}$

28. $\begin{cases} 2x - 2y - 6z = -4 \\ -3x + 2y + 6z = 1 \\ x - y - 5z = -3 \end{cases}$

29. $\begin{cases} 2x + 3y = 0 \\ 4x + 3y - z = 0 \\ 8x + 3y + 3z = 0 \end{cases}$

30. $\begin{cases} 4x + 3y + 17z = 0 \\ 5x + 4y + 22z = 0 \\ 4x + 2y + 19z = 0 \end{cases}$

31. $\begin{cases} 12x + 5y + z = 0 \\ 23x + 4y - z = 0 \end{cases}$

32. $\begin{cases} 2x - y - z = 0 \\ -2x + 6y + 4z = 2 \end{cases}$

Vertical Motion In Exercises 33 and 34, an object moving vertically is at the given heights at the specified times. Find the position equation $s = \frac{1}{2}at^2 + v_0 t + s_0$ for the object.

33. At $t = 1$ second, $s = 452$ feet

At $t = 2$ seconds, $s = 372$ feet

At $t = 3$ seconds, $s = 260$ feet

34. At $t = 1$ second, $s = 132$ feet

At $t = 2$ seconds, $s = 100$ feet

At $t = 3$ seconds, $s = 36$ feet

In Exercises 35–38, find the equation of the parabola

$y = ax^2 + bx + c$

that passes through the points. To verify your result, use a graphing utility to plot the points and graph the parabola.

35. $(0, 0), (2, -2), (4, 0)$ **36.** $(0, 3), (1, 4), (2, 3)$

37. $(2, 0), (3, -1), (4, 0)$ **38.** $(1, 3), (2, 2), (3, -3)$

In Exercises 39–42, find the equation of the circle

$x^2 + y^2 + Dx + Ey + F = 0$

that passes through the points. To verify your result, use a graphing utility to plot the points and graph the circle.

39. $(0, 0), (2, 2), (4, 0)$ **40.** $(0, 0), (0, 6), (3, 3)$

41. $(-3, -1), (2, 4), (-6, 8)$ **42.** $(0, 0), (0, -2), (3, 0)$

43. Agriculture A mixture of 5 pounds of fertilizer A, 13 pounds of fertilizer B, and 4 pounds of fertilizer C provides the optimal nutrients for a plant. Commercial brand X contains equal parts of fertilizer B and fertilizer C. Commercial brand Y contains 1 part of fertilizer A and 2 parts of fertilizer B. Commercial brand Z contains 2 parts of fertilizer A, 5 parts of fertilizer B, and 2 parts of fertilizer C. How much of each fertilizer brand is needed to obtain the desired mixture?

44. Agriculture A mixture of 12 liters of chemical A, 16 liters of chemical B, and 26 liters of chemical C is required to kill a destructive crop insect. Commercial spray X contains 1, 2, and 2 parts, respectively, of these chemicals. Commercial spray Y contains only chemical C. Commercial spray Z contains only chemicals A and B in equal amounts. How much of each type of commercial spray is needed to get the desired mixture?

45. Coffee Mixture A coffee manufacturer sells a 10-pound package of coffee that consists of three flavors of coffee. Vanilla-flavored coffee costs $2 per pound, hazelnut-flavored coffee costs $2.50 per pound, and mocha-flavored coffee costs $3 per pound. The package contains the same amount of hazelnut coffee as mocha coffee. The cost of the 10-pound package is $26. How many pounds of each type of coffee are in the package?

46. Floral Arrangements A florist is creating 10 centerpieces for a wedding. The florist can use roses that cost $2.50 each, lilies that cost $4 each, and irises that cost $2 each to make the bouquets. The customer has a budget of $300 and wants each bouquet to contain 12 flowers, with twice as many roses used as the other two types of flowers combined. How many of each type of flower should be in each centerpiece?

47. Advertising A health insurance company advertises on television, on radio, and in the local newspaper. The marketing department has an advertising budget of $42,000 per month. A television ad costs $1000, a radio ad costs $200, and a newspaper ad costs $500. The department wants to run 60 ads per month, and have as many television ads as radio and newspaper ads combined. How many of each type of ad can the department run each month?

48. Radio You work as a disc jockey at your college radio station. You are supposed to play 32 songs within two hours. You are to choose the songs from the latest rock, dance, and pop albums. You want to play twice as many rock songs as pop songs and four more pop songs than dance songs. How many of each type of song will you play?

49. Acid Mixture A chemist needs 10 liters of a 25% acid solution. The solution is to be mixed from three solutions whose concentrations are 10%, 20%, and 50%. How many liters of each solution will satisfy each condition?

(a) Use 2 liters of the 50% solution.

(b) Use as little as possible of the 50% solution.

(c) Use as much as possible of the 50% solution.

50. Acid Mixture A chemist needs 12 gallons of a 20% acid solution. The solution is to be mixed from three solutions whose concentrations are 10%, 15%, and 25%. How many gallons of each solution will satisfy each condition?

(a) Use 4 gallons of the 25% solution.

(b) Use as little as possible of the 25% solution.

(c) Use as much as possible of the 25% solution.

In Exercises 51–54, find two systems of linear equations that have the ordered triple as a solution. (There are many correct answers.)

51. $(4, -1, 2)$

52. $(-5, -2, 1)$

53. $\left(3, -\frac{1}{2}, \frac{7}{4}\right)$

54. $\left(-\frac{3}{2}, 4, -7\right)$

True or False? In Exercises 55 and 56, determine whether the statement is true or false. Justify your answer.

55. The system
$$\begin{cases} x + 3y - 6z = -16 \\ 2y - z = -1 \\ z = 3 \end{cases}$$
is in row-echelon form.

56. If a system of three linear equations is inconsistent, then its graph has no points common to all three equations.

Life Science Capsule

Scott Bauer/USDA ARS Archive

Chemists and materials scientists search for and use new knowledge about chemicals. Chemical research has led to the discovery and development of new and improved synthetic fibers, paints, adhesives, drugs, cosmetics, electronic components, lubricants, and thousands of other products. Research on the chemistry of living things spurs advances in medicine, agriculture, food processing, and other fields. Often, chemists use systems of equations to analyze the results of experiments. A bachelor's degree in chemistry or a related discipline is the minimum educational requirement for chemists and materials scientists; however, many research jobs require a master's degree, or more often a Ph.D.

57. Research Project Use your school's library, the Internet, or some other reference source to gather information about a specific product or process that was developed through the work of chemists or materials scientists. Write a short paper that summarizes your findings.

Section 8.3
Matrices and Systems of Equations

- Write matrices and identify their orders.
- Perform elementary row operations on matrices.
- Use matrices and Gaussian elimination to solve systems of linear equations.
- Use matrices and Gauss-Jordan elimination to solve systems of linear equations.

Matrices

In this section, you will study a streamlined technique for solving systems of linear equations. This technique involves the use of a rectangular array of real numbers called a **matrix**. The plural of matrix is *matrices*.

> **Definition of Matrix**
>
> If m and n are positive integers, an $m \times n$ (read "m by n") matrix is a rectangular array
>
> $$\begin{array}{c} \phantom{\text{Row 1}} \\ \text{Row 1} \\ \text{Row 2} \\ \text{Row 3} \\ \vdots \\ \text{Row } m \end{array} \begin{bmatrix} \overset{\text{Column 1}}{a_{11}} & \overset{\text{Column 2}}{a_{12}} & \overset{\text{Column 3}}{a_{13}} & \cdots & \overset{\text{Column } n}{a_{1n}} \\ a_{21} & a_{22} & a_{23} & \cdots & a_{2n} \\ a_{31} & a_{32} & a_{33} & \cdots & a_{3n} \\ \vdots & \vdots & \vdots & & \vdots \\ a_{m1} & a_{m2} & a_{m3} & \cdots & a_{mn} \end{bmatrix}$$
>
> in which each **entry**, a_{ij}, of the matrix is a number. An $m \times n$ matrix has m rows and n columns. Matrices are usually denoted by capital letters.

The entry in the ith row and jth column is denoted by the *double subscript* notation a_{ij}. For instance, a_{23} refers to the entry in the second row, third column. A matrix having m rows and n columns is said to be of **order** $m \times n$. If $m = n$, the matrix is **square** of order n. For a square matrix, the entries $a_{11}, a_{22}, a_{33}, \ldots$ are the **main diagonal** entries. A matrix that has only one row is called a **row matrix**, and a matrix that has only one column is called a **column matrix**.

Example 1 Orders of Matrices

Determine the order of each matrix.

a. $[2]$ **b.** $\begin{bmatrix} 1 & -3 & 0 & \frac{1}{2} \end{bmatrix}$ **c.** $\begin{bmatrix} 0 & 0 \\ 0 & 0 \end{bmatrix}$ **d.** $\begin{bmatrix} 5 & 0 \\ 2 & -2 \\ -7 & 4 \end{bmatrix}$

SOLUTION

a. This matrix has *one* row and *one* column. The order of the matrix is 1×1.
b. This matrix has *one* row and *four* columns. The order of the matrix is 1×4.
c. This matrix has *two* rows and *two* columns. The order of the matrix is 2×2.
d. This matrix has *three* rows and *two* columns. The order of the matrix is 3×2.

✓**CHECKPOINT 1**

Determine the order of the matrix.

$[7 \ 0]$ ■

A matrix derived from a system of linear equations (each written in standard form with the constant term on the right) is the **augmented matrix** of the system. Moreover, the matrix derived from the coefficients of the system (but not including the constant terms) is the **coefficient matrix** of the system.

System   *Augmented Matrix*   *Coefficient Matrix*

$$\begin{cases} x - 4y + 3z = 5 \\ -x + 3y - z = -3 \\ 2x - 4z = 6 \end{cases} \quad \left[\begin{array}{rrr:r} 1 & -4 & 3 & 5 \\ -1 & 3 & -1 & -3 \\ 2 & 0 & -4 & 6 \end{array}\right] \quad \left[\begin{array}{rrr} 1 & -4 & 3 \\ -1 & 3 & -1 \\ 2 & 0 & -4 \end{array}\right]$$

> **STUDY TIP**
> The vertical dots in an augmented matrix separate the coefficients of the linear system from the constant terms.

Note the use of 0 for the coefficient of the missing y-variable in the third equation, and also note the fourth column of constant terms in the augmented matrix.

When forming either the coefficient matrix or the augmented matrix of a system, you should begin by vertically aligning the variables in the equations and using zeros for the coefficients of the missing variables.

Given System   *Line Up Variables*   *Form Augmented Matrix*

$$\begin{cases} x + 3y = 9 \\ -y + 4z = -2 \\ x - 5z = 0 \end{cases} \quad \begin{cases} x + 3y = 9 \\ -y + 4z = -2 \\ x - 5z = 0 \end{cases} \quad \begin{matrix} R_1 \\ R_2 \\ R_3 \end{matrix} \left[\begin{array}{rrr:r} 1 & 3 & 0 & 9 \\ 0 & -1 & 4 & -2 \\ 1 & 0 & -5 & 0 \end{array}\right]$$

> **STUDY TIP**
> The augmented matrix at the far right has three rows and four columns, so it is a 3×4 matrix. The notation R_n is used to designate the nth row in the matrix. For example, Row 1 is represented by R_1.

Elementary Row Operations

In the preceding section, you studied three operations that can be used on a system of linear equations to produce an equivalent system.

1. Interchange two equations.
2. Multiply an equation by a nonzero constant.
3. Add a multiple of an equation to another equation.

In matrix terminology, these three operations correspond to **elementary row operations.** An elementary row operation on an augmented matrix of a given system of linear equations produces a new augmented matrix corresponding to a new (but equivalent) system of linear equations. Two matrices are **row-equivalent** if one can be obtained from the other by a sequence of elementary row operations.

> **TECHNOLOGY**
> Most graphing utilities can perform elementary row operations on matrices. Consult the user's guide for your graphing utility for specific keystrokes.
> After performing a row operation, the new row-equivalent matrix that is displayed on your graphing utility is stored in the *answer* variable. You should use the *answer* variable and not the original matrix for subsequent row operations.

> **Elementary Row Operations**
>
> 1. Interchange two rows.
> 2. Multiply a row by a nonzero constant.
> 3. Add a multiple of a row to another row.

Although elementary row operations are simple to perform, they involve a lot of arithmetic. Because it is easy to make a mistake, you should get in the habit of noting the elementary row operations performed in each step so that you can go back and check your work.

SECTION 8.3 Matrices and Systems of Equations

Example 2 Elementary Row Operations

a. Interchange the first and second rows of the original matrix.

Original Matrix
$$\begin{bmatrix} 0 & 1 & 3 & 4 \\ -1 & 2 & 0 & 3 \\ 2 & -3 & 4 & 1 \end{bmatrix}$$

New Row-Equivalent Matrix
$$\begin{matrix} R_2 \\ R_1 \end{matrix} \begin{bmatrix} -1 & 2 & 0 & 3 \\ 0 & 1 & 3 & 4 \\ 2 & -3 & 4 & 1 \end{bmatrix}$$

b. Multiply the first row of the original matrix by $\frac{1}{2}$.

Original Matrix
$$\begin{bmatrix} 2 & -4 & 6 & -2 \\ 1 & 3 & -3 & 0 \\ 5 & -2 & 1 & 2 \end{bmatrix}$$

New Row-Equivalent Matrix
$$\frac{1}{2}R_1 \rightarrow \begin{bmatrix} 1 & -2 & 3 & -1 \\ 1 & 3 & -3 & 0 \\ 5 & -2 & 1 & 2 \end{bmatrix}$$

c. Add -2 times the first row of the original matrix to the third row.

Original Matrix
$$\begin{bmatrix} 1 & 2 & -4 & 3 \\ 0 & 3 & -2 & -1 \\ 2 & 1 & 5 & -2 \end{bmatrix}$$

New Row-Equivalent Matrix
$$-2R_1 + R_3 \rightarrow \begin{bmatrix} 1 & 2 & -4 & 3 \\ 0 & 3 & -2 & -1 \\ 0 & -3 & 13 & -8 \end{bmatrix}$$

Note that the elementary row operation is written beside the row that is *changed*.

✓ CHECKPOINT 2

Identify the elementary row operation(s) being performed to obtain the new row-equivalent matrix.

Original Matrix
$$\begin{bmatrix} -2 & 6 & 4 \\ 3 & -1 & 8 \end{bmatrix}$$

New Row-Equivalent Matrix
$$\begin{bmatrix} 1 & -3 & -2 \\ 3 & -1 & 8 \end{bmatrix}$$

The matrix

$$\begin{bmatrix} 1 & -2 & 3 & \vdots & 9 \\ 0 & 1 & 3 & \vdots & 5 \\ 0 & 0 & 1 & \vdots & 2 \end{bmatrix}$$

is said to be in **row-echelon form**. The term *echelon* refers to the stair-step pattern formed by the nonzero elements of the matrix. To be in this form, a matrix must have the following properties.

Row-Echelon Form and Reduced Row-Echelon Form

A matrix in **row-echelon form** has the following properties.

1. Any rows consisting entirely of zeros occur at the bottom of the matrix.

2. For each row that does not consist entirely of zeros, the first nonzero entry is 1 (called a **leading 1**).

3. For two successive (nonzero) rows, the leading 1 in the higher row is farther to the left than the leading 1 in the lower row.

A matrix in *row-echelon* form is in **reduced row-echelon form** if every column that has a leading 1 has zeros in every position above and below its leading 1. (See Example 5.)

Gaussian Elimination with Back-Substitution

Gaussian elimination with back-substitution works well for solving systems of linear equations by hand or with a computer. For this algorithm, the order in which the elementary row operations are performed is important. You should operate from left to right by columns, using elementary row operations to obtain zeros in all entries directly below the leading 1's.

Example 3 Gaussian Elimination with Back-Substitution

Solve the system $\begin{cases} x - 2y + 3z = 9 \\ -x + 3y = -4 \\ 2x - 5y + 5z = 17 \end{cases}$

STUDY TIP

In Example 2 in Section 8.2, you used Gaussian elimination with back-substitution to solve a system of linear equations. The example at the right demonstrates the matrix version of Gaussian elimination for the same system. The two methods are essentially the same. The basic difference is that with matrices you do not need to keep writing the variables.

SOLUTION

$$\begin{bmatrix} 1 & -2 & 3 & \vdots & 9 \\ -1 & 3 & 0 & \vdots & -4 \\ 2 & -5 & 5 & \vdots & 17 \end{bmatrix}$$ Write augmented matrix.

$$R_1 + R_2 \rightarrow \begin{bmatrix} 1 & -2 & 3 & \vdots & 9 \\ 0 & 1 & 3 & \vdots & 5 \\ 2 & -5 & 5 & \vdots & 17 \end{bmatrix}$$ Add the first row to the second row.

$$-2R_1 + R_3 \rightarrow \begin{bmatrix} 1 & -2 & 3 & \vdots & 9 \\ 0 & 1 & 3 & \vdots & 5 \\ 0 & -1 & -1 & \vdots & -1 \end{bmatrix}$$ Add -2 times the first row to the third row.

$$R_2 + R_3 \rightarrow \begin{bmatrix} 1 & -2 & 3 & \vdots & 9 \\ 0 & 1 & 3 & \vdots & 5 \\ 0 & 0 & 2 & \vdots & 4 \end{bmatrix}$$ Add the second row to the third row.

$$\tfrac{1}{2}R_3 \rightarrow \begin{bmatrix} 1 & -2 & 3 & \vdots & 9 \\ 0 & 1 & 3 & \vdots & 5 \\ 0 & 0 & 1 & \vdots & 2 \end{bmatrix}$$ Multiply the third row by $\tfrac{1}{2}$.

The matrix is now in row-echelon form, and the corresponding system is

$$\begin{cases} x - 2y + 3z = 9 \\ y + 3z = 5 \\ z = 2 \end{cases}$$

At this point, you can use back-substitution to find x and y.

$y + 3(2) = 5$ Substitute 2 for z.
$y = -1$ Solve for y.
$x - 2(-1) + 3(2) = 9$ Substitute -1 for y and 2 for z.
$x = 1$ Solve for x.

The solution is $x = 1$, $y = -1$, and $z = 2$.

✓ CHECKPOINT 3

Use matrices to solve the system of equations.

$\begin{cases} x + 2y = 7 \\ 2x + y = 8 \end{cases}$

SECTION 8.3 Matrices and Systems of Equations

The procedure for using Gaussian elimination with back-substitution is summarized below.

> **Gaussian Elimination with Back-Substitution**
>
> 1. Write the augmented matrix of the system of linear equations.
> 2. Use elementary row operations to rewrite the augmented matrix in row-echelon form.
> 3. Write the system of linear equations corresponding to the matrix in row-echelon form, and use back-substitution to find the solution.

When solving a system of linear equations, remember that it is possible for the system to have no solution. If, in the elimination process, you obtain a row with zeros except for the last entry, it is unnecessary to continue the elimination process. You can simply conclude that the system has no solution, or is *inconsistent*.

Example 4 A System with No Solution

Solve the system $\begin{cases} x - y + 2z = 4 \\ x \phantom{{}-y} + z = 6. \\ 2x - 3y + 5z = 4 \end{cases}$

SOLUTION

$$\begin{bmatrix} 1 & -1 & 2 & \vdots & 4 \\ 1 & 0 & 1 & \vdots & 6 \\ 2 & -3 & 5 & \vdots & 4 \end{bmatrix}$$ Write augmented matrix.

$$\begin{matrix} \\ -R_1 + R_2 \rightarrow \\ -2R_1 + R_3 \rightarrow \end{matrix} \begin{bmatrix} 1 & -1 & 2 & \vdots & 4 \\ 0 & 1 & -1 & \vdots & 2 \\ 0 & -1 & 1 & \vdots & -4 \end{bmatrix}$$ Perform row operations.

$$\begin{matrix} \\ \\ R_2 + R_3 \rightarrow \end{matrix} \begin{bmatrix} 1 & -1 & 2 & \vdots & 4 \\ 0 & 1 & -1 & \vdots & 2 \\ 0 & 0 & 0 & \vdots & -2 \end{bmatrix}$$ Perform row operations.

Note that the third row of this matrix consists of zeros except for the last entry. This means that the original system of linear equations is inconsistent. You can see why this is true by converting back to a system of linear equations.

$$\begin{cases} x - y + 2z = 4 \\ \phantom{x -{}} y - z = 2 \\ 0 = -2 \end{cases}$$

Because the third equation is not possible, the system has no solution.

✓ **CHECKPOINT 4**

Use matrices to solve the system of equations.

$$\begin{cases} -x + 2y = 1.5 \\ 2x - 4y = 3 \end{cases}$$ ∎

Gauss-Jordan Elimination

With Gaussian elimination, elementary row operations are applied to a matrix to obtain a (row-equivalent) row-echelon form of the matrix. A second method of elimination, called **Gauss-Jordan elimination,** after Carl Friedrich Gauss and Wilhelm Jordan (1842–1899), continues the reduction process until a *reduced* row-echelon form is obtained. This procedure is demonstrated in Example 5.

Example 5 Gauss-Jordan Elimination

Use Gauss-Jordan elimination to solve the system $\begin{cases} x - 2y + 3z = 9 \\ -x + 3y = -4 \\ 2x - 5y + 5z = 17 \end{cases}$

STUDY TIP

The advantage of using Gauss-Jordan elimination to solve a system of linear equations is that the solution of the system is easily found without using back-substitution, as illustrated in Example 5.

SOLUTION In Example 3, Gaussian elimination was used to obtain the following row-echelon form of the linear system above.

$$\begin{bmatrix} 1 & -2 & 3 & \vdots & 9 \\ 0 & 1 & 3 & \vdots & 5 \\ 0 & 0 & 1 & \vdots & 2 \end{bmatrix}$$

Now, apply elementary row operations until you obtain zeros above each of the leading 1's, as follows.

$2R_2 + R_1 \rightarrow \begin{bmatrix} 1 & 0 & 9 & \vdots & 19 \\ 0 & 1 & 3 & \vdots & 5 \\ 0 & 0 & 1 & \vdots & 2 \end{bmatrix}$ Perform operations on R_1 so second column has a zero above its leading 1.

$\begin{matrix} -9R_3 + R_1 \rightarrow \\ -3R_3 + R_2 \rightarrow \end{matrix} \begin{bmatrix} 1 & 0 & 0 & \vdots & 1 \\ 0 & 1 & 0 & \vdots & -1 \\ 0 & 0 & 1 & \vdots & 2 \end{bmatrix}$ Perform operations on R_1 and R_2 so third column has zeros above its leading 1.

The matrix is now in reduced row-echelon form. Converting back to a system of linear equations, you have

$\begin{cases} x = 1 \\ y = -1 \\ z = 2 \end{cases}$

Now you can simply read the solution, $x = 1$, $y = -1$, and $z = 2$, which can be written as the ordered triple $(1, -1, 2)$.

✓ CHECKPOINT 5

Use matrices to solve the system of equations.

$\begin{cases} x - 3z = -2 \\ 3x + y - 2z = 5 \\ 2x + 2y + z = 4 \end{cases}$ ■

The elimination procedures described in this section sometimes result in fractional coefficients. For instance, in the elimination procedure for the system

$\begin{cases} 2x - 5y + 5z = 17 \\ 3x - 2y + 3z = 11 \\ -3x + 3y = -6 \end{cases}$

you may be inclined to multiply the first row by $\frac{1}{2}$ to produce a leading 1, which will require that you work with fractional coefficients. You can sometimes avoid fractions by judiciously choosing the order in which you apply elementary row operations.

SECTION 8.3 Matrices and Systems of Equations

Recall that when there are fewer equations than variables in a system of equations, then the system has either no solution or infinitely many solutions.

Example 6 A System with an Infinite Number of Solutions

Solve the system $\begin{cases} 2x + 4y - 2z = 0 \\ 3x + 5y = 1 \end{cases}$.

SOLUTION

$$\begin{bmatrix} 2 & 4 & -2 & \vdots & 0 \\ 3 & 5 & 0 & \vdots & 1 \end{bmatrix}$$

$\frac{1}{2}R_1 \rightarrow \begin{bmatrix} 1 & 2 & -1 & \vdots & 0 \\ 3 & 5 & 0 & \vdots & 1 \end{bmatrix}$

$-3R_1 + R_2 \rightarrow \begin{bmatrix} 1 & 2 & -1 & \vdots & 0 \\ 0 & -1 & 3 & \vdots & 1 \end{bmatrix}$

$-R_2 \rightarrow \begin{bmatrix} 1 & 2 & -1 & \vdots & 0 \\ 0 & 1 & -3 & \vdots & -1 \end{bmatrix}$

$-2R_2 + R_1 \rightarrow \begin{bmatrix} 1 & 0 & 5 & \vdots & 2 \\ 0 & 1 & -3 & \vdots & -1 \end{bmatrix}$

The corresponding system of equations is

$$\begin{cases} x + 5z = 2 \\ y - 3z = -1 \end{cases}.$$

Solving for x and y in terms of z, you have $x = -5z + 2$ and $y = 3z - 1$. To write a solution of the system that does not use any of the three variables of the system, let a represent any real number and let $z = a$. Now substitute a for z in the equations for x and y.

$$x = -5z + 2 = -5a + 2$$
$$y = 3z - 1 = 3a - 1$$

So, the solution set can be written as an ordered triple of the form

$$(-5a + 2, 3a - 1, a)$$

where a is any real number. Remember that a solution set of this form represents an infinite number of solutions. Try substituting values of a to obtain a few solutions. Then check each solution in the original equation.

✓ **CHECKPOINT 6**

Use matrices to solve the system of equations.

$$\begin{cases} x + y - 5z = 3 \\ x - 2z = 1 \\ 2x - y - z = 0 \end{cases}$$

CONCEPT CHECK

1. What is the order of a matrix that has two rows and one column?
2. What is the name of a matrix that has only one row?
3. What is a coefficient matrix?
4. What is the relationship between the three elementary row operations performed on an augmented matrix and the operations that lead to equivalent systems of linear equations?

Skills Review 8.3

The following warm-up exercises involve skills that were covered in earlier sections. You will use these skills in the exercise set for this section. For additional help, review Sections 8.1 and 8.2.

In Exercises 1 and 2, determine whether $x = 1$, $y = 3$, and $z = -1$ is a solution of the system.

1. $\begin{cases} 4x - 2y + 3z = -5 \\ x + 3y - z = 11 \\ -x + 2y = 5 \end{cases}$

2. $\begin{cases} -x + 2y + z = 4 \\ 2x - 3z = 5 \\ 3x + 5y - 2z = 21 \end{cases}$

In Exercises 3–6, use back-substitution to solve the system of linear equations.

3. $\begin{cases} 2x - 3y = 4 \\ y = 2 \end{cases}$

4. $\begin{cases} 5x + 4y = 0 \\ y = -3 \end{cases}$

5. $\begin{cases} x - 3y + z = 0 \\ y - 3z = 8 \\ z = 2 \end{cases}$

6. $\begin{cases} 2x - 5y + 3z = -2 \\ y - 4z = 0 \\ z = 1 \end{cases}$

Exercises 8.3

See www.CalcChat.com for worked-out solutions to odd-numbered exercises.

In Exercises 1–6, determine the order of the matrix.

1. $\begin{bmatrix} 1 & 9 \end{bmatrix}$

2. $\begin{bmatrix} 5 & -3 & 8 & 7 \end{bmatrix}$

3. $\begin{bmatrix} 2 \\ 36 \\ 3 \end{bmatrix}$

4. $\begin{bmatrix} -3 & 7 & 15 & 0 \\ 0 & 0 & 3 & 3 \\ 1 & 1 & 6 & 7 \end{bmatrix}$

5. $\begin{bmatrix} 33 & 45 \\ -9 & 20 \end{bmatrix}$

6. $\begin{bmatrix} -7 & 6 & 4 \\ 0 & -5 & 1 \end{bmatrix}$

In Exercises 7–12, write the augmented matrix for the system of linear equations.

7. $\begin{cases} 4x - 3y = -5 \\ -x + 3y = 12 \end{cases}$

8. $\begin{cases} 7x + 4y = 22 \\ 5x - 9y = 15 \end{cases}$

9. $\begin{cases} x + 10y - 2z = 2 \\ 5x - 3y + 4z = 0 \\ 2x + y = 6 \end{cases}$

10. $\begin{cases} -x - 8y + 5z = 8 \\ -7x - 15z = -38 \\ 3x - y + 8z = 20 \end{cases}$

11. $\begin{cases} 7x - 5y + z = 13 \\ 19x - 8z = 10 \end{cases}$

12. $\begin{cases} 9x + 2y - 3z = 20 \\ -25y + 11z = -5 \end{cases}$

In Exercises 13–18, write the system of linear equations represented by the augmented matrix. (Use variables x, y, z, and w, if applicable.)

13. $\begin{bmatrix} 1 & 2 & \vdots & 7 \\ 2 & -3 & \vdots & 4 \end{bmatrix}$

14. $\begin{bmatrix} 7 & -5 & \vdots & 0 \\ 8 & 3 & \vdots & -2 \end{bmatrix}$

15. $\begin{bmatrix} 2 & 0 & 5 & \vdots & -12 \\ 0 & 1 & -2 & \vdots & 7 \\ 6 & 3 & 0 & \vdots & 2 \end{bmatrix}$

16. $\begin{bmatrix} 4 & -5 & -1 & \vdots & 18 \\ -11 & 0 & 6 & \vdots & 25 \\ 3 & 8 & 0 & \vdots & -29 \end{bmatrix}$

17. $\begin{bmatrix} 9 & 12 & 3 & 0 & \vdots & 0 \\ -2 & 18 & 5 & 2 & \vdots & 10 \\ 1 & 7 & -8 & 0 & \vdots & -4 \\ 3 & 0 & 2 & 0 & \vdots & -10 \end{bmatrix}$

18. $\begin{bmatrix} 6 & 2 & -1 & -5 & \vdots & -25 \\ -1 & 0 & 7 & 3 & \vdots & 7 \\ 4 & -1 & -10 & 6 & \vdots & 23 \\ 0 & 8 & 1 & -11 & \vdots & -21 \end{bmatrix}$

In Exercises 19–22, fill in the blank(s) using elementary row operations to form a row-equivalent matrix.

19. $\begin{bmatrix} 1 & 4 & 3 \\ 2 & 10 & 5 \end{bmatrix}$

$\begin{bmatrix} 1 & 4 & 3 \\ 0 & \square & -1 \end{bmatrix}$

20. $\begin{bmatrix} 3 & 6 & 8 \\ 4 & -3 & 6 \end{bmatrix}$

$\begin{bmatrix} 1 & \square & \frac{8}{3} \\ 4 & -3 & 6 \end{bmatrix}$

21. $\begin{bmatrix} 1 & 1 & 4 & -1 \\ 3 & 8 & 10 & 3 \\ -2 & 1 & 12 & 6 \end{bmatrix}$

$\begin{bmatrix} 1 & 1 & 4 & -1 \\ 0 & 5 & \square & \square \\ 0 & 3 & \square & \square \end{bmatrix}$

$\begin{bmatrix} 1 & 1 & 4 & -1 \\ 0 & 1 & -\frac{2}{5} & \frac{6}{5} \\ 0 & 3 & \square & \square \end{bmatrix}$

22. $\begin{bmatrix} 2 & 4 & 8 & 3 \\ 1 & -1 & -3 & 2 \\ 2 & 6 & 4 & 9 \end{bmatrix}$

$\begin{bmatrix} 1 & \square & \square & \square \\ 1 & -1 & -3 & 2 \\ 2 & 6 & 4 & 9 \end{bmatrix}$

$\begin{bmatrix} 1 & 2 & 4 & \frac{3}{2} \\ 0 & \square & -7 & \frac{1}{2} \\ 0 & 2 & \square & \square \end{bmatrix}$

In Exercises 23–26, identify the elementary row operation(s) being performed to obtain the new row-equivalent matrix.

23. Original Matrix $\begin{bmatrix} -2 & 5 & 1 \\ 3 & -1 & -8 \end{bmatrix}$ New Row-Equivalent Matrix $\begin{bmatrix} 13 & 0 & -39 \\ 3 & -1 & -8 \end{bmatrix}$

24. Original Matrix $\begin{bmatrix} 3 & -1 & -4 \\ -4 & 3 & 7 \end{bmatrix}$ New Row-Equivalent Matrix $\begin{bmatrix} 3 & -1 & -4 \\ 5 & 0 & -5 \end{bmatrix}$

25. Original Matrix $\begin{bmatrix} 0 & -1 & -5 & 5 \\ -1 & 3 & -7 & 6 \\ 4 & -5 & 1 & 3 \end{bmatrix}$ New Row-Equivalent Matrix $\begin{bmatrix} -1 & 3 & -7 & 6 \\ 0 & -1 & -5 & 5 \\ 0 & 7 & -27 & 27 \end{bmatrix}$

26. Original Matrix $\begin{bmatrix} -1 & -2 & 3 & -2 \\ 2 & -5 & 1 & -7 \\ 5 & 4 & -7 & 6 \end{bmatrix}$ New Row-Equivalent Matrix $\begin{bmatrix} -1 & -2 & 3 & -2 \\ 0 & -9 & 7 & -11 \\ 0 & -6 & 8 & -4 \end{bmatrix}$

27. Perform the sequence of row operations on the matrix. What did the operations accomplish?
$$\begin{bmatrix} 1 & 2 & 3 \\ 2 & -1 & -4 \\ 3 & 1 & -1 \end{bmatrix}$$
(a) Add -2 times R_1 to R_2.
(b) Add -3 times R_1 to R_3.
(c) Add -1 times R_2 to R_3.
(d) Multiply R_2 by $-\frac{1}{5}$.
(e) Add -2 times R_2 to R_1.

28. Perform the sequence of row operations on the matrix. What did the operations accomplish?
$$\begin{bmatrix} 7 & 1 \\ 0 & 2 \\ -3 & 4 \\ 4 & 1 \end{bmatrix}$$
(a) Add R_3 to R_4.
(b) Interchange R_1 and R_4.
(c) Add 3 times R_1 to R_3.
(d) Add -7 times R_1 to R_4.
(e) Multiply R_2 by $\frac{1}{2}$.
(f) Add the appropriate multiples of R_2 to R_1, R_3, and R_4.

In Exercises 29–32, determine whether the matrix is in row-echelon form. If it is, determine if it is also in reduced row-echelon form.

29. $\begin{bmatrix} 1 & 0 & 0 & 0 \\ 0 & 1 & 1 & 5 \\ 0 & 0 & 0 & 0 \end{bmatrix}$

30. $\begin{bmatrix} 1 & 3 & 0 & 0 \\ 0 & 0 & 1 & 8 \\ 0 & 0 & 0 & 0 \end{bmatrix}$

31. $\begin{bmatrix} 2 & 0 & 4 & 0 \\ 0 & -1 & 3 & 6 \\ 0 & 0 & 1 & 5 \end{bmatrix}$

32. $\begin{bmatrix} 1 & 0 & 2 & 1 \\ 0 & 1 & -3 & 10 \\ 0 & 0 & 1 & 0 \end{bmatrix}$

In Exercises 33–36, write the matrix in row-echelon form. (Remember that the row-echelon form of a matrix is not unique.)

33. $\begin{bmatrix} 1 & 1 & 0 & 5 \\ -2 & -1 & 2 & -10 \\ 3 & 6 & 7 & 14 \end{bmatrix}$

34. $\begin{bmatrix} 1 & 2 & -1 & 3 \\ 3 & 7 & -5 & 14 \\ -2 & -1 & -3 & 8 \end{bmatrix}$

35. $\begin{bmatrix} 1 & -1 & -1 & 1 \\ 5 & -4 & 1 & 8 \\ -6 & 8 & 18 & 0 \end{bmatrix}$

36. $\begin{bmatrix} 1 & -3 & 0 & -7 \\ -3 & 10 & 1 & 23 \\ 4 & -10 & 2 & -24 \end{bmatrix}$

(T) In Exercises 37–42, use the matrix capabilities of a graphing utility to write the matrix in *reduced* row-echelon form.

37. $\begin{bmatrix} 3 & 3 & 3 \\ -1 & 0 & -4 \\ 2 & 4 & -2 \end{bmatrix}$

38. $\begin{bmatrix} 1 & 3 & 2 \\ 5 & 15 & 9 \\ 2 & 6 & 10 \end{bmatrix}$

39. $\begin{bmatrix} 1 & 2 & 3 & -5 \\ 1 & 2 & 4 & -9 \\ -2 & -4 & -4 & 3 \\ 4 & 8 & 11 & -14 \end{bmatrix}$

40. $\begin{bmatrix} -2 & 3 & -1 & -2 \\ 4 & -2 & 5 & 8 \\ 1 & 5 & -2 & 0 \\ 3 & 8 & -10 & -30 \end{bmatrix}$

41. $\begin{bmatrix} -3 & 5 & 1 & 12 \\ 1 & -1 & 1 & 4 \end{bmatrix}$

42. $\begin{bmatrix} 5 & 1 & 2 & 4 \\ -1 & 5 & 10 & -32 \end{bmatrix}$

In Exercises 43–46, write the system of linear equations represented by the augmented matrix. Then use back-substitution to solve. (Use variables x, y, and z, if applicable.)

43. $\begin{bmatrix} 1 & -2 & \vdots & 4 \\ 0 & 1 & \vdots & -3 \end{bmatrix}$

44. $\begin{bmatrix} 1 & 5 & \vdots & 0 \\ 0 & 1 & \vdots & -1 \end{bmatrix}$

45. $\begin{bmatrix} 1 & -1 & 2 & \vdots & 4 \\ 0 & 1 & -1 & \vdots & 2 \\ 0 & 0 & 1 & \vdots & -2 \end{bmatrix}$

46. $\begin{bmatrix} 1 & 2 & -2 & \vdots & -1 \\ 0 & 1 & 1 & \vdots & 9 \\ 0 & 0 & 1 & \vdots & -3 \end{bmatrix}$

In Exercises 47–62, use matrices to solve the system of equations (if possible). Use Gaussian elimination with back-substitution or Gauss-Jordan elimination.

47. $\begin{cases} 2x + 6y = 16 \\ 2x + 3y = 7 \end{cases}$

48. $\begin{cases} 3x - 2y = -27 \\ x + 3y = 13 \end{cases}$

49. $\begin{cases} -x + y = 4 \\ 2x - 4y = -34 \end{cases}$

50. $\begin{cases} -2x + 6y = -22 \\ x + 2y = -9 \end{cases}$

51. $\begin{cases} 5x - 5y = -5 \\ -2x - 3y = 7 \end{cases}$

52. $\begin{cases} x - 3y = 5 \\ -2x + 6y = -10 \end{cases}$

53. $\begin{cases} 2x - y + 3z = 24 \\ 2y - z = 14 \\ 7x - 5y = 6 \end{cases}$

54. $\begin{cases} -x + y - z = -14 \\ 2x - y + z = 21 \\ 3x + 2y + z = 19 \end{cases}$

55. $\begin{cases} 2x + 2y - z = 2 \\ x - 3y + z = -28 \\ -x + y = 14 \end{cases}$

56. $\begin{cases} x + 2y - 3z = -28 \\ 4y + 2z = 0 \\ -x + y - z = -5 \end{cases}$

57. $\begin{cases} 3x - 2y + z = 15 \\ -x + y + 2z = -10 \\ x - y - 4z = 14 \end{cases}$

58. $\begin{cases} 2x + 3z = 3 \\ 4x - 3y + 7z = 5 \\ 8x - 9y + 15z = 9 \end{cases}$

59. $\begin{cases} x + 2y + z + 2w = 8 \\ 3x + 7y + 6z + 9w = 26 \end{cases}$

60. $\begin{cases} 4x + 12y - 7z - 20w = 22 \\ 3x + 9y - 5z - 28w = 30 \end{cases}$

61. $\begin{cases} -x + y = -22 \\ 3x + 4y = 4 \\ 4x - 8y = 32 \end{cases}$

62. $\begin{cases} x + 2y = 0 \\ x + y = 6 \\ 3x - 2y = 8 \end{cases}$

In Exercises 63–66, determine whether the two systems of linear equations yield the same solution. If so, find the solution using matrices.

63. (a) $\begin{cases} x - 2y + z = -6 \\ y - 5z = 16 \\ z = -3 \end{cases}$ (b) $\begin{cases} x + y - 2z = 6 \\ y + 3z = -8 \\ z = -3 \end{cases}$

64. (a) $\begin{cases} x - 3y + 4z = -11 \\ y - z = -4 \\ z = 2 \end{cases}$ (b) $\begin{cases} x + 4y = -11 \\ y + 3z = 4 \\ z = 2 \end{cases}$

65. (a) $\begin{cases} x - 4y + 5z = 27 \\ y - 7z = -54 \\ z = 8 \end{cases}$ (b) $\begin{cases} x - 6y + z = 15 \\ y + 5z = 42 \\ z = 8 \end{cases}$

66. (a) $\begin{cases} x + 3y - z = 19 \\ y + 6z = -18 \\ z = -4 \end{cases}$ (b) $\begin{cases} x - y + 3z = -15 \\ y - 2z = 14 \\ z = -4 \end{cases}$

67. **Breeding Facility** A city zoo borrowed $2,200,000 at simple annual interest to construct a breeding facility. Some of the money was borrowed at 8%, some at 9%, and some at 12%. Use a system of equations to determine how much was borrowed at each rate if the total annual interest was $204,000 and the amount borrowed at 8% was twice the amount borrowed at 12%. Solve the system using matrices.

68. **Museum** A natural history museum borrowed $2,000,000 at simple annual interest to purchase new exhibits. Some of the money was borrowed at 7%, some at 8.5%, and some at 9.5%. Use a system of equations to determine how much was borrowed at each rate if the total annual interest was $169,750 and the amount borrowed at 8.5% was four times the amount borrowed at 9.5%. Solve the system using matrices.

69. **Health and Wellness** For the years 1997 through 2008, the number of new cases of waterborne disease in a small city increased in a pattern that was approximately linear (see figure). Find the least squares regression line $y = at + b$ for the data shown in the figure by solving the following system using matrices. Let t represent the year, with $t = 0$ corresponding to 1997.

$\begin{cases} 12b + 66a = 831 \\ 66b + 506a = 5643 \end{cases}$

Use the result to estimate the number of new cases of the waterborne disease in 2011. Is the estimate reasonable? Explain.

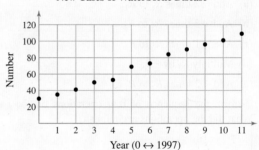

True or False? In Exercises 70–72, determine whether the statement is true or false. Justify your answer.

70. $\begin{bmatrix} 5 & 0 & -2 & 7 \\ -1 & 3 & -6 & 0 \end{bmatrix}$ is a 4×2 matrix.

71. The matrix

$\begin{bmatrix} 0 & 0 & 0 & 0 \\ 0 & 0 & 1 & -4 \\ 0 & 1 & 0 & 2 \\ 1 & 0 & 0 & 5 \end{bmatrix}$

is in reduced row-echelon form.

72. The method of Gaussian elimination reduces a matrix until a reduced row-echelon form is obtained.

Mid-Chapter Quiz

Take this quiz as you would take a quiz in class. When you are done, check your work against the answers given in the back of the book.

In Exercises 1 and 2, solve the system by substitution or elimination.

1. $\begin{cases} 2.5x - y = 6 \\ 3x + 4y = 2 \end{cases}$
2. $\begin{cases} \frac{1}{2}x + \frac{1}{3}y = 1 \\ x - 2y = -2 \end{cases}$

In Exercises 3–6, solve the system of equations.

3. $\begin{cases} 2x + 3y - z = -7 \\ x + 3z = 10 \\ 2y + z = -1 \end{cases}$
4. $\begin{cases} x + y - 2z = 12 \\ 2x - y - z = 6 \\ y - z = 6 \end{cases}$
5. $\begin{cases} 3x + 2y + z = -17 \\ -x + y + z = 4 \\ x - y - z = 3 \end{cases}$
6. $\begin{cases} 2x + y + 3z = 1 \\ 2x + 6y + 8z = 3 \\ 6x + 8y + 18z = 5 \end{cases}$

In Exercises 7 and 8, write a matrix of the given order.

7. 4×3
8. 3×1

In Exercises 9 and 10, write the augmented matrix for the system of equations.

9. $\begin{cases} 3x + 2y = -2 \\ 5x - y = 19 \end{cases}$
10. $\begin{cases} x + 3z = -5 \\ x + 2y - z = 3 \\ 3x + 4z = 0 \end{cases}$

11. Use Gaussian elimination with back-substitution to solve the augmented matrix found in Exercise 9.

12. Use Gauss-Jordan elimination to solve the augmented matrix found in Exercise 10.

In Exercises 13 and 14, fill in the blank to form a new row-equivalent matrix.

13. $\begin{bmatrix} 1 & 1 & 1 \\ 5 & -2 & 4 \end{bmatrix}$

    $\begin{bmatrix} 1 & 1 & 1 \\ 0 & & -1 \end{bmatrix}$

14. $\begin{bmatrix} -3 & 3 & 12 \\ 18 & -8 & 4 \end{bmatrix}$

    $\begin{bmatrix} 1 & -1 & \\ 18 & -8 & 4 \end{bmatrix}$

15. A mixture of 15 liters of chemical A, 11 liters of chemical B, and 20 liters of chemical C is required to kill a destructive crop insect. Commercial spray X contains only chemicals A and B in equal amounts. Commercial spray Y contains 3, 2, and 1 parts, respectively, of these chemicals. Commercial spray Z contains only chemical C. How much of each type of commercial spray is needed to get the desired mixture?

Section 8.4

Operations with Matrices

- Decide whether two matrices are equal.
- Add and subtract matrices and multiply matrices by scalars.
- Multiply two matrices.
- Use matrix multiplication to represent and solve systems of linear equations.

Equality of Matrices

In Section 8.3, you used matrices to solve systems of linear equations. There is a rich mathematical theory of matrices, and its applications are numerous. This section and the next introduce some fundamentals of matrix theory. It is standard mathematical convention to represent matrices in any of the following three ways.

Representation of Matrices

1. A matrix can be denoted by an uppercase letter such as A, B, or C.

2. A matrix can be denoted by a representative element enclosed in brackets, such as $[a_{ij}]$, $[b_{ij}]$, or $[c_{ij}]$.

3. A matrix can be denoted by a rectangular array of numbers such as

$$A = [a_{ij}] = \begin{bmatrix} a_{11} & a_{12} & a_{13} & \cdots & a_{1n} \\ a_{21} & a_{22} & a_{23} & \cdots & a_{2n} \\ a_{31} & a_{32} & a_{33} & \cdots & a_{3n} \\ \vdots & \vdots & \vdots & & \vdots \\ a_{m1} & a_{m2} & a_{m3} & \cdots & a_{mn} \end{bmatrix}.$$

Two matrices $A = [a_{ij}]$ and $B = [b_{ij}]$ are **equal** if they have the same order ($m \times n$) and $a_{ij} = b_{ij}$ for $1 \leq i \leq m$ and $1 \leq j \leq n$. In other words, two matrices are equal if their corresponding entries are equal.

Example 1 Equality of Matrices

Solve for a_{11}, a_{12}, a_{21}, and a_{22} in the following matrix equation.

$$\begin{bmatrix} a_{11} & a_{12} \\ a_{21} & a_{22} \end{bmatrix} = \begin{bmatrix} 2 & -1 \\ -3 & 0 \end{bmatrix}$$

SOLUTION Because two matrices are equal only if their corresponding entries are equal, you can conclude that $a_{11} = 2$, $a_{12} = -1$, $a_{21} = -3$, and $a_{22} = 0$.

Be sure you see that for two matrices to be equal, they must have the same order *and* their corresponding entries must be equal. For instance,

$$\begin{bmatrix} 2 & -1 \\ \sqrt{4} & \frac{1}{2} \end{bmatrix} = \begin{bmatrix} 2 & -1 \\ 2 & 0.5 \end{bmatrix} \quad \text{but} \quad \begin{bmatrix} 2 & -1 \\ 3 & 4 \\ 0 & 0 \end{bmatrix} \neq \begin{bmatrix} 2 & -1 \\ 3 & 4 \end{bmatrix}.$$

✓**CHECKPOINT 1**

Solve for x and y in the following matrix equation.

$$\begin{bmatrix} x & -2 \\ 7 & y \end{bmatrix} = \begin{bmatrix} -4 & -2 \\ 7 & 22 \end{bmatrix}$$ ■

Matrix Addition and Scalar Multiplication

In this section, three basic matrix operations will be covered. The first two are matrix addition and scalar multiplication. In matrix addition, you can add two matrices (of the same order) by adding their corresponding entries.

TECHNOLOGY

Most graphing utilities have the capability of performing matrix operations. Consult the user's guide for your graphing utility for specific keystrokes. Try using a graphing utility to find the sum of the matrices

$$A = \begin{bmatrix} 2 & -3 \\ -1 & 0 \end{bmatrix} \text{ and}$$

$$B = \begin{bmatrix} -1 & 4 \\ 2 & -5 \end{bmatrix}.$$

Definition of Matrix Addition

If $A = [a_{ij}]$ and $B = [b_{ij}]$ are matrices of order $m \times n$, their sum is the $m \times n$ matrix given by

$$A + B = [a_{ij} + b_{ij}].$$

The sum of two matrices of different orders is undefined.

Example 2 Addition of Matrices

a. $\begin{bmatrix} -1 & 2 \\ 0 & 1 \end{bmatrix} + \begin{bmatrix} 1 & 3 \\ -1 & 2 \end{bmatrix} = \begin{bmatrix} -1+1 & 2+3 \\ 0+(-1) & 1+2 \end{bmatrix} = \begin{bmatrix} 0 & 5 \\ -1 & 3 \end{bmatrix}$

b. $\begin{bmatrix} 0 & 1 & -2 \\ 1 & 2 & 3 \end{bmatrix} + \begin{bmatrix} 0 & 0 & 0 \\ 0 & 0 & 0 \end{bmatrix} = \begin{bmatrix} 0 & 1 & -2 \\ 1 & 2 & 3 \end{bmatrix}$

c. The sum of

$$A = \begin{bmatrix} 2 & 1 & 0 \\ 4 & 0 & -1 \\ 3 & -2 & 2 \end{bmatrix} \text{ and } B = \begin{bmatrix} 0 & 1 \\ -1 & 3 \\ 2 & 4 \end{bmatrix}$$

is undefined because A is of order 3×3 and B is of order 3×2.

✓ CHECKPOINT 2

Evaluate the expression.

$$\begin{bmatrix} 1 \\ -3 \\ -2 \end{bmatrix} + \begin{bmatrix} -1 \\ 3 \\ 2 \end{bmatrix}. \quad \blacksquare$$

In operations with matrices, numbers are usually referred to as **scalars.** In this text, scalars will always be real numbers. You can multiply a matrix A by a scalar c by multiplying each entry in A by c.

Definition of Scalar Multiplication

If $A = [a_{ij}]$ is an $m \times n$ matrix and c is a scalar, the **scalar multiple** of A by c is the $m \times n$ matrix given by

$$cA = [ca_{ij}].$$

The symbol $-A$ represents the negation of A, which is the scalar product $(-1)A$. Moreover, if A and B are of the same order, then $A - B$ represents the sum of A and $(-1)B$. That is,

$$A - B = A + (-1)B. \qquad \text{Subtraction of matrices}$$

The order of operations for matrix expressions is similar to that for real numbers. In particular, you perform scalar multiplication before matrix addition and subtraction, as shown in Example 3(c) on the next page.

DISCOVERY

Consider the matrices A, B, and C below. Perform the indicated operations and compare the results.

$$A = \begin{bmatrix} 3 & -1 \\ 4 & 7 \end{bmatrix}, B = \begin{bmatrix} -2 & 0 \\ 8 & 1 \end{bmatrix},$$

$$C = \begin{bmatrix} 5 & 2 \\ 2 & -6 \end{bmatrix}$$

a. Find $A + B$ and $B + A$.

b. Find $A + B$, then add C to the resulting matrix. Find $B + C$, then add A to the resulting matrix.

c. Find $2A$ and $2B$, then add the two resulting matrices. Find $A + B$, then multiply the resulting matrix by 2.

✓ CHECKPOINT 3

For the following matrices, find (a) $2A$, (b) $-A$, and (c) $3A - 2B$.

$$A = \begin{bmatrix} 6 & -1 \\ 2 & 4 \\ -3 & 5 \end{bmatrix}, B = \begin{bmatrix} 1 & 4 \\ -1 & 5 \\ 1 & 10 \end{bmatrix}$$

STUDY TIP

It is often convenient to rewrite the scalar multiple cA by factoring c out of every entry in the matrix. For instance, in the following example, the scalar has been factored out of the matrix.

$$\begin{bmatrix} \tfrac{1}{2} & -\tfrac{3}{2} \\ \tfrac{5}{2} & \tfrac{1}{2} \end{bmatrix} = \begin{bmatrix} \tfrac{1}{2}(1) & \tfrac{1}{2}(-3) \\ \tfrac{1}{2}(5) & \tfrac{1}{2}(1) \end{bmatrix}$$

$$= \tfrac{1}{2} \begin{bmatrix} 1 & -3 \\ 5 & 1 \end{bmatrix}$$

Example 3 Scalar Multiplication and Matrix Subtraction

For the following matrices, find (a) $3A$, (b) $-B$, and (c) $3A - B$.

$$A = \begin{bmatrix} 2 & 2 & 4 \\ -3 & 0 & -1 \\ 2 & 1 & 2 \end{bmatrix} \text{ and } B = \begin{bmatrix} 2 & 0 & 0 \\ 1 & -4 & 3 \\ -1 & 3 & 2 \end{bmatrix}$$

SOLUTION

a. $3A = 3\begin{bmatrix} 2 & 2 & 4 \\ -3 & 0 & -1 \\ 2 & 1 & 2 \end{bmatrix}$ Scalar multiplication

$= \begin{bmatrix} 3(2) & 3(2) & 3(4) \\ 3(-3) & 3(0) & 3(-1) \\ 3(2) & 3(1) & 3(2) \end{bmatrix}$ Multiply each entry by 3.

$= \begin{bmatrix} 6 & 6 & 12 \\ -9 & 0 & -3 \\ 6 & 3 & 6 \end{bmatrix}$ Simplify.

b. $-B = (-1)\begin{bmatrix} 2 & 0 & 0 \\ 1 & -4 & 3 \\ -1 & 3 & 2 \end{bmatrix}$ Definition of negation

$= \begin{bmatrix} -2 & 0 & 0 \\ -1 & 4 & -3 \\ 1 & -3 & -2 \end{bmatrix}$ Multiply each entry by -1.

c. $3A - B = \begin{bmatrix} 6 & 6 & 12 \\ -9 & 0 & -3 \\ 6 & 3 & 6 \end{bmatrix} - \begin{bmatrix} 2 & 0 & 0 \\ 1 & -4 & 3 \\ -1 & 3 & 2 \end{bmatrix}$ Matrix subtraction

$= \begin{bmatrix} 4 & 6 & 12 \\ -10 & 4 & -6 \\ 7 & 0 & 4 \end{bmatrix}$ Subtract corresponding entries.

The properties of matrix addition and scalar multiplication are similar to those of addition and multiplication of real numbers.

Properties of Matrix Addition and Scalar Multiplication

Let A, B, and C be $m \times n$ matrices and let c and d be scalars.

1. $A + B = B + A$ Commutative Property of Matrix Addition
2. $A + (B + C) = (A + B) + C$ Associative Property of Matrix Addition
3. $(cd)A = c(dA)$ Associative Property of Scalar Multiplication
4. $1A = A$ Scalar Identity Property
5. $c(A + B) = cA + cB$ Distributive Property
6. $(c + d)A = cA + dA$ Distributive Property

SECTION 8.4 Operations with Matrices

Note that the Associative Property of Matrix Addition allows you to write expressions such as $A + B + C$ without ambiguity because the same sum occurs no matter how the matrices are grouped. This same reasoning applies to sums of four or more matrices.

Example 4 Addition of More than Two Matrices

By adding corresponding entries, you obtain the following sum of four matrices.

$$\begin{bmatrix} 1 \\ 2 \\ -3 \end{bmatrix} + \begin{bmatrix} -1 \\ -1 \\ 2 \end{bmatrix} + \begin{bmatrix} 0 \\ 1 \\ 4 \end{bmatrix} + \begin{bmatrix} 2 \\ -3 \\ -2 \end{bmatrix} = \begin{bmatrix} 2 \\ -1 \\ 1 \end{bmatrix}$$

Example 5 Using the Distributive Property

$$3\left(\begin{bmatrix} -2 & 0 \\ 4 & 1 \end{bmatrix} + \begin{bmatrix} 4 & -2 \\ 3 & 7 \end{bmatrix}\right) = 3\begin{bmatrix} -2 & 0 \\ 4 & 1 \end{bmatrix} + 3\begin{bmatrix} 4 & -2 \\ 3 & 7 \end{bmatrix}$$

$$= \begin{bmatrix} -6 & 0 \\ 12 & 3 \end{bmatrix} + \begin{bmatrix} 12 & -6 \\ 9 & 21 \end{bmatrix}$$

$$= \begin{bmatrix} 6 & -6 \\ 21 & 24 \end{bmatrix}$$

✓ **CHECKPOINT 4**

Evaluate the expression.

$$\begin{bmatrix} -5 \\ 3 \end{bmatrix} + \begin{bmatrix} 7 \\ 0 \end{bmatrix} + \begin{bmatrix} -1 \\ 5 \end{bmatrix}$$

✓ **CHECKPOINT 5**

Evaluate the expression.

$$2\left(\begin{bmatrix} -1 \\ -3 \end{bmatrix} + \begin{bmatrix} 5 \\ 0 \end{bmatrix}\right)$$

STUDY TIP

In Example 5, you could add the two matrices first and then multiply the matrix by 3, as follows. Notice that you obtain the same result.

$$3\left(\begin{bmatrix} -2 & 0 \\ 4 & 1 \end{bmatrix} + \begin{bmatrix} 4 & -2 \\ 3 & 7 \end{bmatrix}\right)$$

$$= 3\begin{bmatrix} 2 & -2 \\ 7 & 8 \end{bmatrix} = \begin{bmatrix} 6 & -6 \\ 21 & 24 \end{bmatrix}$$

One important property of addition of real numbers is that the number 0 is the additive identity. That is, $c + 0 = c$ for any real number c. For matrices, a similar property holds. That is, if A is an $m \times n$ matrix and O is the $m \times n$ **zero matrix** consisting entirely of zeros, then $A + O = A$.

In other words, O is the **additive identity** for the set of all $m \times n$ matrices. For example, the following matrices are the additive identities for the set of all 2×3 and 2×2 matrices.

$$O = \begin{bmatrix} 0 & 0 & 0 \\ 0 & 0 & 0 \end{bmatrix} \quad \text{and} \quad O = \begin{bmatrix} 0 & 0 \\ 0 & 0 \end{bmatrix}$$

$\underbrace{}_{2 \times 3 \text{ zero matrix}}$ $\underbrace{}_{2 \times 2 \text{ zero matrix}}$

The algebra of real numbers and the algebra of matrices have many similarities. For example, compare the following solutions.

Real Numbers (Solve for x.)	$m \times n$ Matrices (Solve for X.)
$x + a = b$	$X + A = B$
$x + a + (-a) = b + (-a)$	$X + A + (-A) = B + (-A)$
$x + 0 = b - a$	$X + O = B - A$
$x = b - a$	$X = B - A$

STUDY TIP

Remember that matrices are denoted by capital letters. So, when you solve for X, you are solving for a *matrix* that makes the matrix equation true.

The algebra of real numbers and the algebra of matrices also have important differences, which will be discussed later.

Example 6 Solving a Matrix Equation

Solve for X in the Equation $3X + A = B$, where

$$A = \begin{bmatrix} 1 & -2 \\ 0 & 3 \end{bmatrix} \text{ and } B = \begin{bmatrix} -3 & 4 \\ 2 & 1 \end{bmatrix}.$$

SOLUTION Begin by solving the equation for X to obtain

$$3X = B - A \implies X = \frac{1}{3}(B - A).$$

Now, using the matrices A and B, you have

$$X = \frac{1}{3}\left(\begin{bmatrix} -3 & 4 \\ 2 & 1 \end{bmatrix} - \begin{bmatrix} 1 & -2 \\ 0 & 3 \end{bmatrix}\right) \quad \text{Substitute the matrices.}$$

$$= \frac{1}{3}\begin{bmatrix} -4 & 6 \\ 2 & -2 \end{bmatrix} \quad \text{Subtract matrix } A \text{ from matrix } B.$$

$$= \begin{bmatrix} -\frac{4}{3} & 2 \\ \frac{2}{3} & -\frac{2}{3} \end{bmatrix}. \quad \text{Multiply the matrix by } \frac{1}{3}.$$

✓ **CHECKPOINT 6**

Solve for X in the equation $2X + A = B$, where

$$A = \begin{bmatrix} 1 & 5 \\ -2 & 0 \end{bmatrix} \text{ and }$$

$$B = \begin{bmatrix} -3 & 1 \\ 0 & 4 \end{bmatrix}.$$

Matrix Multiplication

The third basic matrix operation is **matrix multiplication.** At first glance, the definition may seem unusual. You will see later, however, that this definition of the product of two matrices has many practical applications.

> **Definition of Matrix Multiplication**
>
> If $A = [a_{ij}]$ is an $m \times n$ matrix and $B = [b_{ij}]$ is an $n \times p$ matrix, the product AB is an $m \times p$ matrix
>
> $$AB = [c_{ij}]$$
>
> where $c_{ij} = a_{i1}b_{1j} + a_{i2}b_{2j} + a_{i3}b_{3j} + \cdots + a_{in}b_{nj}.$

The definition of matrix multiplication indicates a *row-by-column* multiplication, where the entry in the ith row and jth column of the product AB is obtained by multiplying the entries in the ith row of A by the corresponding entries in the jth column of B and then adding the results. The general pattern for matrix multiplication is as follows.

$$\begin{bmatrix} a_{11} & a_{12} & a_{13} & \cdots & a_{1n} \\ a_{21} & a_{22} & a_{23} & \cdots & a_{2n} \\ a_{31} & a_{32} & a_{33} & \cdots & a_{3n} \\ \vdots & \vdots & \vdots & & \vdots \\ a_{i1} & a_{i2} & a_{i3} & \cdots & a_{in} \\ \vdots & \vdots & \vdots & & \vdots \\ a_{m1} & a_{m2} & a_{m3} & \cdots & a_{mn} \end{bmatrix} \begin{bmatrix} b_{11} & b_{12} & \cdots & b_{1j} & \cdots & b_{1p} \\ b_{21} & b_{22} & \cdots & b_{2j} & \cdots & b_{2p} \\ b_{31} & b_{32} & \cdots & b_{3j} & \cdots & b_{3p} \\ \vdots & \vdots & & \vdots & & \vdots \\ b_{n1} & b_{n2} & \cdots & b_{nj} & \cdots & b_{np} \end{bmatrix} = \begin{bmatrix} c_{11} & c_{12} & \cdots & c_{1j} & \cdots & c_{1p} \\ c_{21} & c_{22} & \cdots & c_{2j} & \cdots & c_{2p} \\ \vdots & \vdots & & \vdots & & \vdots \\ c_{i1} & c_{i2} & \cdots & c_{ij} & \cdots & c_{ip} \\ \vdots & \vdots & & \vdots & & \vdots \\ c_{m1} & c_{m2} & \cdots & c_{mj} & \cdots & c_{mp} \end{bmatrix}$$

$$a_{i1}b_{1j} + a_{i2}b_{2j} + a_{i3}b_{3j} + \cdots + a_{in}b_{nj} = c_{ij}$$

Example 7 Finding the Product of Two Matrices

The product AB shown below is defined because the number of columns of A is equal to the number of rows of B. Moreover, the product AB has order 3×2. To find the entries of the product, multiply each row of A by each column of B, as follows.

$$AB = \begin{bmatrix} -1 & 3 \\ 4 & -2 \\ 5 & 0 \end{bmatrix} \begin{bmatrix} -3 & 2 \\ -4 & 1 \end{bmatrix}$$

$$= \begin{bmatrix} (-1)(-3) + (3)(-4) & (-1)(2) + (3)(1) \\ (4)(-3) + (-2)(-4) & (4)(2) + (-2)(1) \\ (5)(-3) + (0)(-4) & (5)(2) + (0)(1) \end{bmatrix}$$

$$= \begin{bmatrix} -9 & 1 \\ -4 & 6 \\ -15 & 10 \end{bmatrix}$$

✓ **CHECKPOINT 7**

Find the product AB, where

$$A = \begin{bmatrix} -1 & 3 \\ 4 & -5 \\ 0 & 2 \end{bmatrix} \text{ and }$$

$$B = \begin{bmatrix} 1 & 2 \\ 0 & 7 \end{bmatrix}.$$

Be sure you understand that for the product of two matrices to be defined, the number of *columns* of the first matrix must equal the number of *rows* of the second matrix. That is, the middle two indices must be the same. The outside two indices give the order of the product, as shown below.

$$\underset{m \times n}{A} \times \underset{n \times p}{B} = \underset{m \times p}{AB}$$

Equal — Order of AB

DISCOVERY

Use the following matrices to find AB, BA, $(AB)C$, and $A(BC)$. What do your results tell you about matrix multiplication, commutativity, and associativity?

$$A = \begin{bmatrix} 1 & 2 \\ 3 & 4 \end{bmatrix},$$

$$B = \begin{bmatrix} 0 & 1 \\ 2 & 3 \end{bmatrix},$$

$$C = \begin{bmatrix} 3 & 0 \\ 0 & 1 \end{bmatrix}$$

Example 8 Patterns in Matrix Multiplication

a. $\begin{bmatrix} 3 & 4 \\ -2 & 5 \end{bmatrix} \begin{bmatrix} 1 & 0 \\ 0 & 1 \end{bmatrix} = \begin{bmatrix} 3 & 4 \\ -2 & 5 \end{bmatrix}$

$\quad 2 \times 2 \qquad\quad 2 \times 2 \qquad\quad 2 \times 2$

b. $\begin{bmatrix} 6 & 2 & 0 \\ 3 & -1 & 2 \\ 1 & 4 & 6 \end{bmatrix} \begin{bmatrix} 1 \\ 2 \\ -3 \end{bmatrix} = \begin{bmatrix} 10 \\ -5 \\ -9 \end{bmatrix}$

$\quad 3 \times 3 \qquad\qquad 3 \times 1 \qquad 3 \times 1$

✓ **CHECKPOINT 8**

Find the product AB, where

$$A = \begin{bmatrix} -1 & 3 \\ 1 & 2 \end{bmatrix} \text{ and }$$

$$B = \begin{bmatrix} 3 & 0 \\ 5 & -2 \end{bmatrix}.$$

c. The product AB for the following matrices is not defined.

$$A = \begin{bmatrix} -2 & 1 \\ 1 & -3 \\ 1 & 4 \end{bmatrix} \text{ and } B = \begin{bmatrix} -2 & 3 & 1 & 4 \\ 0 & 1 & -1 & 2 \\ 2 & -1 & 0 & 1 \end{bmatrix}$$

$\qquad\quad 3 \times 2 \qquad\qquad\qquad\qquad 3 \times 4$

✓ CHECKPOINT 9

Find (a) AB and (b) BA, where

$$A = \begin{bmatrix} 7 \\ 8 \\ -1 \end{bmatrix} \text{ and } B = \begin{bmatrix} 1 & 1 & 2 \end{bmatrix}.$$

Example 9 Patterns in Matrix Multiplication

a. $\begin{bmatrix} 1 & -2 & -3 \end{bmatrix} \begin{bmatrix} 2 \\ -1 \\ 1 \end{bmatrix} = \begin{bmatrix} 1 \end{bmatrix}$

$\quad 1 \times 3 \quad\quad 3 \times 1 \quad 1 \times 1$

b. $\begin{bmatrix} 2 \\ -1 \\ 1 \end{bmatrix} \begin{bmatrix} 1 & -2 & -3 \end{bmatrix} = \begin{bmatrix} 2 & -4 & -6 \\ -1 & 2 & 3 \\ 1 & -2 & -3 \end{bmatrix}$

$\quad 3 \times 1 \quad\quad 1 \times 3 \quad\quad\quad 3 \times 3$

In Example 9, note that the two products are different. Even if AB and BA are defined, matrix multiplication is not, in general, commutative. That is, for most matrices, $AB \neq BA$. This is one way in which the algebra of real numbers and the algebra of matrices differ.

Properties of Matrix Multiplication

Let A, B, and C be matrices and let c be a scalar.

1. $A(BC) = (AB)C$ Associative Property of Multiplication
2. $A(B + C) = AB + AC$ Distributive Property
3. $(A + B)C = AC + BC$ Distributive Property
4. $c(AB) = (cA)B = A(cB)$ Associative Property of Scalar Multiplication

---

Definition of Identity Matrix

The $n \times n$ matrix that consists of 1's on its main diagonal and 0's elsewhere is called the **identity matrix of order n** and is denoted by

$$I_n = \begin{bmatrix} 1 & 0 & 0 & \cdots & 0 \\ 0 & 1 & 0 & \cdots & 0 \\ 0 & 0 & 1 & \cdots & 0 \\ \vdots & \vdots & \vdots & & \vdots \\ 0 & 0 & 0 & \cdots & 1 \end{bmatrix}. \quad \text{Identity matrix}$$

Note that an identity matrix must be *square*. When the order is understood to be n, you can denote I_n simply by I.

If A is an $n \times n$ matrix, the identity matrix has the property that $AI_n = A$ and $I_n A = A$. For example,

$$\begin{bmatrix} 3 & -2 & 5 \\ 1 & 0 & 4 \\ -1 & 2 & -3 \end{bmatrix} \begin{bmatrix} 1 & 0 & 0 \\ 0 & 1 & 0 \\ 0 & 0 & 1 \end{bmatrix} = \begin{bmatrix} 3 & -2 & 5 \\ 1 & 0 & 4 \\ -1 & 2 & -3 \end{bmatrix} \quad AI = A$$

and

$$\begin{bmatrix} 1 & 0 & 0 \\ 0 & 1 & 0 \\ 0 & 0 & 1 \end{bmatrix} \begin{bmatrix} 3 & -2 & 5 \\ 1 & 0 & 4 \\ -1 & 2 & -3 \end{bmatrix} = \begin{bmatrix} 3 & -2 & 5 \\ 1 & 0 & 4 \\ -1 & 2 & -3 \end{bmatrix}. \quad IA = A$$

Application

Matrix multiplication can be used to represent a system of linear equations. Note how the system below can be written as the matrix equation $AX = B$, where A is the *coefficient matrix* of the system, and X and B are column matrices.

System of Equations

$$\begin{cases} a_{11}x_1 + a_{12}x_2 + a_{13}x_3 = b_1 \\ a_{21}x_1 + a_{22}x_2 + a_{23}x_3 = b_2 \\ a_{31}x_1 + a_{32}x_2 + a_{33}x_3 = b_3 \end{cases}$$

Matrix Equation

$$\begin{bmatrix} a_{11} & a_{12} & a_{13} \\ a_{21} & a_{22} & a_{23} \\ a_{31} & a_{32} & a_{33} \end{bmatrix} \begin{bmatrix} x_1 \\ x_2 \\ x_3 \end{bmatrix} = \begin{bmatrix} b_1 \\ b_2 \\ b_3 \end{bmatrix}$$

$$A \quad \times \quad X \,=\, B$$

Example 10 Solving a System of Linear Equations

Consider the system $\begin{cases} x_1 - 2x_2 + x_3 = -4 \\ x_2 + 2x_3 = 4 \\ 2x_1 + 3x_2 - 2x_3 = 2 \end{cases}$.

Write this system as a matrix equation, $AX = B$. Then use Gauss-Jordan elimination on the augmented matrix $[A \vdots B]$ to solve for the matrix X.

SOLUTION In matrix form, $AX = B$, the system can be written as follows.

$$\begin{bmatrix} 1 & -2 & 1 \\ 0 & 1 & 2 \\ 2 & 3 & -2 \end{bmatrix} \begin{bmatrix} x_1 \\ x_2 \\ x_3 \end{bmatrix} = \begin{bmatrix} -4 \\ 4 \\ 2 \end{bmatrix}$$

The augmented matrix is formed by adjoining matrix B to matrix A.

$$[A \vdots B] = \begin{bmatrix} 1 & -2 & 1 & \vdots & -4 \\ 0 & 1 & 2 & \vdots & 4 \\ 2 & 3 & -2 & \vdots & 2 \end{bmatrix}$$

Using Gauss-Jordan elimination, you can rewrite this equation as

$$[I \vdots X] = \begin{bmatrix} 1 & 0 & 0 & \vdots & -1 \\ 0 & 1 & 0 & \vdots & 2 \\ 0 & 0 & 1 & \vdots & 1 \end{bmatrix}.$$

So, the solution of the system of linear equations is $x_1 = -1$, $x_2 = 2$, and $x_3 = 1$, and the solution of the matrix equation is

$$X = \begin{bmatrix} x_1 \\ x_2 \\ x_3 \end{bmatrix} = \begin{bmatrix} -1 \\ 2 \\ 1 \end{bmatrix}.$$

STUDY TIP

The notation $[A \vdots B]$ represents the augmented matrix formed when matrix B is adjoined to matrix A. The notation $[I \vdots X]$ represents the reduced row-echelon form of the augmented matrix that yields the *solution* of the system.

✓ CHECKPOINT 10

Repeat Example 10 for the following system of linear equations.

$$\begin{cases} x_1 - 2x_2 + 3x_3 = 9 \\ -x_1 + 3x_2 - x_3 = -6 \\ 2x_1 - 5x_2 + 5x_3 = 17 \end{cases}$$ ∎

CONCEPT CHECK

1. When are two matrices equal?
2. Is it possible to add two matrices of different orders?
3. What is meant by $-A$?
4. For the product of two matrices to be defined, the number of columns of the first matrix must equal what?

Skills Review 8.4

The following warm-up exercises involve skills that were covered in earlier sections. You will use these skills in the exercise set for this section. For additional help, review Section 8.3.

In Exercises 1 and 2, determine whether the matrix is in *reduced row-echelon form*.

1. $\begin{bmatrix} 0 & 1 & 0 & -5 \\ 1 & 0 & 3 & 2 \\ 0 & 0 & 1 & 0 \end{bmatrix}$

2. $\begin{bmatrix} 1 & 0 & 0 & 2 & 3 \\ 0 & 0 & 0 & 0 & 0 \\ 0 & 1 & 1 & 3 & 10 \end{bmatrix}$

In Exercises 3 and 4, write the augmented matrix for the system of linear equations.

3. $\begin{cases} -5x + 10y = 12 \\ 7x - 3y = 0 \end{cases}$

4. $\begin{cases} 10x + 15y - 9z = 42 \\ 6x - 5y = 0 \end{cases}$

In Exercises 5–7, solve the system of linear equations represented by the augmented matrix.

5. $\begin{bmatrix} 1 & 0 & \vdots & 0 \\ 0 & 1 & \vdots & 2 \end{bmatrix}$

6. $\begin{bmatrix} 1 & 0 & -1 & \vdots & 2 \\ 0 & 1 & 1 & \vdots & 3 \end{bmatrix}$

7. $\begin{bmatrix} 1 & -1 & 0 & \vdots & 3 \\ 0 & 1 & -2 & \vdots & 1 \\ 0 & 0 & 1 & \vdots & -1 \end{bmatrix}$

Exercises 8.4

See www.CalcChat.com for worked-out solutions to odd-numbered exercises.

In Exercises 1–4, find x and y.

1. $\begin{bmatrix} x & 5 \\ 0 & y \end{bmatrix} = \begin{bmatrix} -2 & 5 \\ 0 & 4 \end{bmatrix}$

2. $\begin{bmatrix} -5 & x \\ y & 8 \end{bmatrix} = \begin{bmatrix} -5 & 13 \\ 12 & 8 \end{bmatrix}$

3. $\begin{bmatrix} 16 & 4 & 5 & 4 \\ -3 & 13 & 15 & 6 \\ 0 & 2 & 4 & 0 \end{bmatrix} = \begin{bmatrix} 16 & 4 & 2x+1 & 4 \\ -3 & 13 & & 15 & 3x \\ 0 & 2 & 3y-5 & 0 \end{bmatrix}$

4. $\begin{bmatrix} x+2 & 8 & -3 \\ 1 & 2y & 2x \\ 7 & -2 & y+2 \end{bmatrix} = \begin{bmatrix} 2x+6 & 8 & -3 \\ 1 & 18 & -8 \\ 7 & -2 & 11 \end{bmatrix}$

In Exercises 5–10, if possible, find (a) $A + B$, (b) $A - B$, (c) $3A$, and (d) $3A - 2B$.

5. $A = \begin{bmatrix} 1 & -1 \\ 2 & -1 \end{bmatrix}$, $B = \begin{bmatrix} 2 & -1 \\ -1 & 8 \end{bmatrix}$

6. $A = \begin{bmatrix} 1 & 2 \\ 2 & 1 \end{bmatrix}$, $B = \begin{bmatrix} -3 & -2 \\ 4 & 2 \end{bmatrix}$

7. $A = \begin{bmatrix} 1 & -1 \\ -3 & 1 \\ 4 & 1 \end{bmatrix}$, $B = \begin{bmatrix} 4 & -2 \\ 0 & 1 \\ 2 & 3 \end{bmatrix}$

8. $A = \begin{bmatrix} 2 & 1 & 1 \\ -1 & -1 & 4 \end{bmatrix}$, $B = \begin{bmatrix} 2 & -3 & 4 \\ -3 & 1 & -2 \end{bmatrix}$

9. $A = \begin{bmatrix} 6 & 0 & 3 \\ -1 & -4 & 0 \end{bmatrix}$, $B = \begin{bmatrix} 8 & -1 \\ 4 & -3 \end{bmatrix}$

10. $A = \begin{bmatrix} 3 \\ 2 \\ -1 \end{bmatrix}$, $B = \begin{bmatrix} -4 & 6 & 2 \end{bmatrix}$

In Exercises 11–14, evaluate the expression.

11. $\begin{bmatrix} -5 & 0 \\ 3 & -6 \end{bmatrix} + \begin{bmatrix} 7 & 1 \\ -2 & -1 \end{bmatrix} + \begin{bmatrix} -10 & -8 \\ 14 & 6 \end{bmatrix}$

12. $\begin{bmatrix} 6 & 8 \\ -1 & 0 \end{bmatrix} + \begin{bmatrix} 0 & 5 \\ -3 & -1 \end{bmatrix} + \begin{bmatrix} -11 & -7 \\ 2 & -1 \end{bmatrix}$

13. $\frac{1}{2}([5 \quad -2 \quad 4 \quad 0] + [14 \quad 6 \quad -18 \quad 9])$

14. $-3\left(\begin{bmatrix} 0 & -3 \\ 7 & 2 \end{bmatrix} + \begin{bmatrix} -6 & 3 \\ 8 & 1 \end{bmatrix}\right) - 2\begin{bmatrix} 4 & -4 \\ 7 & -9 \end{bmatrix}$

In Exercises 15–18, use the matrix capabilities of a graphing utility to evaluate the expression. Round your results to three decimal places, if necessary.

15. $\frac{3}{7}\begin{bmatrix} 2 & 5 \\ -1 & -4 \end{bmatrix} + 6\begin{bmatrix} -3 & 0 \\ 2 & 2 \end{bmatrix}$

16. $55\left(\begin{bmatrix} 14 & -11 \\ -22 & 19 \end{bmatrix} + \begin{bmatrix} -22 & 20 \\ 13 & 6 \end{bmatrix}\right)$

17. $-\begin{bmatrix} 3.211 & 6.829 \\ -1.004 & 4.914 \\ 0.055 & -3.889 \end{bmatrix} - \begin{bmatrix} -1.630 & -3.090 \\ 5.256 & 8.335 \\ -9.768 & 4.251 \end{bmatrix}$

18. $-12\left(\begin{bmatrix} 6 & 20 \\ 1 & -9 \\ -2 & 5 \end{bmatrix} + \begin{bmatrix} 14 & -15 \\ -8 & -6 \\ 7 & 0 \end{bmatrix} + \begin{bmatrix} -31 & -19 \\ 16 & 10 \\ 24 & -10 \end{bmatrix}\right)$

In Exercises 19–22, solve for X in the equation, given

$A = \begin{bmatrix} -2 & -1 \\ 1 & 0 \\ 3 & -4 \end{bmatrix}$ and $B = \begin{bmatrix} 0 & 3 \\ 2 & 0 \\ -4 & -1 \end{bmatrix}$.

19. $X = 3A - 2B$
20. $2X = 2A - B$
21. $2X + 3A = B$
22. $2A + 4B = -2X$

In Exercises 23–28, if possible, find AB and state the order of the result.

23. $A = \begin{bmatrix} 2 & 1 \\ -3 & 4 \\ 1 & 6 \end{bmatrix}$, $B = \begin{bmatrix} 0 & -1 & 0 \\ 4 & 0 & 2 \\ 8 & -1 & 7 \end{bmatrix}$

24. $A = \begin{bmatrix} 1 & 0 & 3 & -2 \\ 6 & 13 & 8 & -17 \end{bmatrix}$, $B = \begin{bmatrix} 1 & 6 \\ 4 & 2 \end{bmatrix}$

25. $A = \begin{bmatrix} 0 & -1 & 0 \\ 4 & 0 & 2 \\ 8 & -1 & 7 \end{bmatrix}$, $B = \begin{bmatrix} 2 & 1 \\ -3 & 4 \\ 1 & 6 \end{bmatrix}$

26. $A = \begin{bmatrix} 5 & 0 & 0 \\ 0 & -8 & 0 \\ 0 & 0 & 7 \end{bmatrix}$, $B = \begin{bmatrix} \frac{1}{5} & 0 & 0 \\ 0 & -\frac{1}{8} & 0 \\ 0 & 0 & \frac{1}{2} \end{bmatrix}$

27. $A = \begin{bmatrix} 0 & 0 & 5 \\ 0 & 0 & -3 \\ 0 & 0 & 4 \end{bmatrix}$, $B = \begin{bmatrix} 6 & -11 & 4 \\ 8 & 16 & 4 \\ 0 & 0 & 0 \end{bmatrix}$

28. $A = \begin{bmatrix} 10 \\ 12 \end{bmatrix}$, $B = \begin{bmatrix} 6 & -2 & 1 & 6 \end{bmatrix}$

In Exercises 29–32, use the matrix capabilities of a graphing utility to find AB, if possible.

29. $A = \begin{bmatrix} 5 & 6 & -3 \\ -2 & 5 & 1 \\ 10 & -5 & 5 \end{bmatrix}$, $B = \begin{bmatrix} 1 & -1 & 2 \\ 8 & 1 & 4 \\ 4 & -2 & 9 \end{bmatrix}$

30. $A = \begin{bmatrix} 11 & -12 & 4 \\ 14 & 10 & 12 \\ 6 & -2 & 9 \end{bmatrix}$, $B = \begin{bmatrix} 12 & 10 \\ -5 & 12 \\ 15 & 16 \end{bmatrix}$

31. $A = \begin{bmatrix} -3 & 8 & -6 & 8 \\ -12 & 15 & 9 & 6 \\ 5 & -1 & 1 & 5 \end{bmatrix}$, $B = \begin{bmatrix} 3 & 1 & 6 \\ 24 & 15 & 14 \\ 16 & 10 & 21 \\ 8 & -4 & 10 \end{bmatrix}$

32. $A = \begin{bmatrix} -2 & 4 & 8 \\ 21 & 5 & 6 \\ 13 & 2 & 6 \end{bmatrix}$, $B = \begin{bmatrix} 2 & 0 \\ -7 & 15 \\ 32 & 14 \\ 0.5 & 1.6 \end{bmatrix}$

In Exercises 33–36, if possible, find (a) AB, (b) BA, and (c) A^2. (Note: $A^2 = AA$.)

33. $A = \begin{bmatrix} 1 & 2 \\ 4 & 2 \end{bmatrix}$, $B = \begin{bmatrix} 2 & -1 \\ -1 & 8 \end{bmatrix}$

34. $A = \begin{bmatrix} 2 & -1 \\ 1 & 4 \end{bmatrix}$, $B = \begin{bmatrix} 0 & 0 \\ 3 & -3 \end{bmatrix}$

35. $A = \begin{bmatrix} 6 \\ 9 \\ -2 \end{bmatrix}$, $B = \begin{bmatrix} 2 & 2 & 1 \end{bmatrix}$

36. $A = \begin{bmatrix} 3 & 2 & 1 \end{bmatrix}$, $B = \begin{bmatrix} 2 \\ 3 \\ 0 \end{bmatrix}$

In Exercises 37–40, evaluate the expression. Use the matrix capabilities of a graphing utility to verify your answer.

37. $\begin{bmatrix} 3 & 1 \\ 0 & -2 \end{bmatrix} \begin{bmatrix} 1 & 0 \\ -2 & 2 \end{bmatrix} \begin{bmatrix} 1 & 0 \\ 2 & 4 \end{bmatrix}$

38. $-3\left(\begin{bmatrix} 6 & 5 & -1 \\ 1 & -2 & 0 \end{bmatrix} \begin{bmatrix} 0 & 3 \\ -1 & -3 \\ 4 & 1 \end{bmatrix}\right)$

39. $\begin{bmatrix} 0 & 2 & -2 \\ 4 & 1 & 2 \end{bmatrix}\left(\begin{bmatrix} 4 & 0 \\ 0 & -1 \\ -1 & 2 \end{bmatrix} + \begin{bmatrix} -2 & 3 \\ -3 & 5 \\ 0 & -3 \end{bmatrix}\right)$

40. $\begin{bmatrix} 3 \\ -1 \\ 5 \\ 7 \end{bmatrix}([5 \quad -6] + [7 \quad -1] + [-8 \quad 9])$

In Exercises 41–46, (a) write the system of linear equations as a matrix equation, $AX = B$, and (b) use Gauss-Jordan elimination on the augmented matrix $[A \vdots B]$ to solve for the matrix X.

41. $\begin{cases} -x_1 + x_2 = 4 \\ -2x_1 + x_2 = 0 \end{cases}$

42. $\begin{cases} 2x_1 + 3x_2 = 5 \\ x_1 + 4x_2 = 10 \end{cases}$

43. $\begin{cases} -2x_1 - 3x_2 = -4 \\ 6x_1 + x_2 = -36 \end{cases}$

44. $\begin{cases} -4x_1 + 9x_2 = -13 \\ x_1 - 3x_2 = 12 \end{cases}$

45. $\begin{cases} x_1 - 2x_2 + 3x_3 = 9 \\ -x_1 + 3x_2 - x_3 = -6 \\ 2x_1 - 5x_2 + 5x_3 = 17 \end{cases}$

46. $\begin{cases} x_1 - 5x_2 + 2x_3 = -20 \\ -3x_1 + x_2 - x_3 = 8 \\ -2x_2 + 5x_3 = -16 \end{cases}$

47. Voting Preferences The matrix

$$P = \begin{bmatrix} 0.6 & 0.1 & 0.1 \\ 0.2 & 0.7 & 0.1 \\ 0.2 & 0.2 & 0.8 \end{bmatrix} \begin{matrix} R \\ D \\ I \end{matrix}$$

with columns labeled R, D, I (From) and rows labeled R, D, I (To), is called a *stochastic matrix*. Each entry p_{ij} ($i \neq j$) represents the proportion of the voting population that changes from party i to party j, and p_{ii} represents the proportion that remains loyal to the party from one election to the next. Compute and interpret P^2.

(T) 48. Voting Preferences Use a graphing utility to find P^3, P^4, P^5, P^6, P^7, and P^8 for the matrix given in Exercise 47. Can you detect a pattern as P is raised to higher powers?

49. Exercise The numbers of calories burned by individuals of different body weights performing different types of aerobic exercises for 20-minute time periods are shown in matrix A.

$$A = \begin{bmatrix} 109 & 136 \\ 127 & 159 \\ 64 & 79 \end{bmatrix} \begin{matrix} \text{Bicycling} \\ \text{Jogging} \\ \text{Walking} \end{matrix}$$

with columns for 120-lb person and 150-lb person (Calories burned).

(a) A 120-pound person and a 150-pound person bicycled for 40 minutes, jogged for 10 minutes, and walked for 60 minutes. Organize the time spent exercising in a matrix B.

(b) Compute BA and interpret the result.

50. Agriculture A fruit grower raises two crops: apples and peaches. Each of these crops is sent to three different outlets for sale. These outlets are The Farmer's Market, The Fruit Stand, and The Fruit Farm. The numbers of bushels of apples sent to the three outlets are 125, 100, and 75, respectively. The numbers of bushels of peaches sent to the three outlets are 100, 175, and 125, respectively. The profit per bushel for apples is $3.50 and the profit per bushel for peaches is $6.00.

(a) Write a matrix A that represents the number of bushels of each crop i that are shipped to each outlet j. State what each entry a_{ij} of the matrix represents.

(b) Write a matrix B that represents the profit per bushel of each fruit. State what each entry b_{ij} of the matrix represents.

(c) Find the product BA and state what each entry of the matrix represents.

Leslie Matrix In Exercises 51–54, use the following information. The Leslie Matrix L can be used to determine the growth and age distributions of a population over a period of time. A population is grouped into n age classes of equal duration. If p_i is the probability that a member of the ith age class will survive to become a member of the $(i + 1)$th age class, and b_i is the average number of offspring produced by a member of the ith age class, then the Leslie Matrix for the population is given by

$$L = \begin{bmatrix} b_1 & b_2 & b_3 & \cdots & b_{n-1} & b_n \\ p_1 & 0 & 0 & \cdots & 0 & 0 \\ 0 & p_2 & 0 & \cdots & 0 & 0 \\ \vdots & \vdots & \vdots & & \vdots & \vdots \\ 0 & 0 & 0 & \cdots & p_{n-1} & 0 \end{bmatrix}.$$

If the number of population members in each age class in a given time period is represented by the age distribution matrix X_i, where

$$X_i = \begin{bmatrix} x_1 \\ x_2 \\ \vdots \\ x_n \end{bmatrix} \begin{matrix} \text{Number in first age class} \\ \text{Number in second age class} \\ \vdots \\ \text{Number in } n\text{th age class} \end{matrix}$$

then the age distribution matrix for the next time period can be found by $X_{i+1} = LX_i$. Use the Leslie Matrix L and age distribution matrix X_1 to find the age distribution matrices X_2 and X_3.

51. $L = \begin{bmatrix} 0 & 2 \\ \frac{1}{2} & 0 \end{bmatrix}$, $X_1 = \begin{bmatrix} 10 \\ 10 \end{bmatrix}$

52. $L = \begin{bmatrix} 0 & 4 \\ \frac{1}{16} & 0 \end{bmatrix}$, $X_1 = \begin{bmatrix} 160 \\ 80 \end{bmatrix}$

53. $L = \begin{bmatrix} 0 & 3 & 4 \\ 1 & 0 & 0 \\ 0 & \frac{1}{2} & 0 \end{bmatrix}$, $X_1 = \begin{bmatrix} 12 \\ 10 \\ 12 \end{bmatrix}$

54. Population Growth A population of rabbits raised in a research laboratory has the following characteristics.

(a) Half of the rabbits survive their first year. Of those, half survive their second year. The maximum life span is 3 years. During the first year, the rabbits produce no offspring. The average number of offspring is 6 during the second year and 8 during the third year.

(b) The laboratory population now consists of 24 rabbits in the first age class, 24 in the second, and 20 in the third.

How many rabbits will be in each age class in 1 year? in 2 years?

Section 8.5
The Inverse of a Square Matrix

- Verify that two matrices are inverses of each other.
- Use Gauss-Jordan elimination to find the inverses of matrices.
- Use a formula to find the inverses of 2 × 2 matrices.
- Use inverse matrices to solve systems of linear equations.

The Inverse of a Matrix

This section further develops the algebra of matrices. To begin, consider the real number equation $ax = b$. To solve this equation for x, multiply each side of the equation by a^{-1} (provided that $a \neq 0$).

$$ax = b$$
$$(a^{-1}a)x = a^{-1}b$$
$$(1)x = a^{-1}b$$
$$x = a^{-1}b$$

The number a^{-1} is called the *multiplicative inverse of a* because $a^{-1}a = 1$. The definition of the multiplicative **inverse of a matrix** is similar.

Definition of the Inverse of a Square Matrix

Let A be an $n \times n$ matrix and let I_n be the $n \times n$ identity matrix. If there exists a matrix A^{-1} such that

$$AA^{-1} = I_n = A^{-1}A$$

then A^{-1} is called the **inverse** of A. The symbol A^{-1} is read "A inverse."

Example 1 The Inverse of a Matrix

Show that B is the inverse of A, where

$$A = \begin{bmatrix} -1 & 2 \\ -1 & 1 \end{bmatrix} \quad \text{and} \quad B = \begin{bmatrix} 1 & -2 \\ 1 & -1 \end{bmatrix}.$$

SOLUTION To show that B is the inverse of A, show that $AB = I = BA$, as follows.

$$AB = \begin{bmatrix} -1 & 2 \\ -1 & 1 \end{bmatrix}\begin{bmatrix} 1 & -2 \\ 1 & -1 \end{bmatrix} = \begin{bmatrix} -1+2 & 2-2 \\ -1+1 & 2-1 \end{bmatrix} = \begin{bmatrix} 1 & 0 \\ 0 & 1 \end{bmatrix}$$

$$BA = \begin{bmatrix} 1 & -2 \\ 1 & -1 \end{bmatrix}\begin{bmatrix} -1 & 2 \\ -1 & 1 \end{bmatrix} = \begin{bmatrix} -1+2 & 2-2 \\ -1+1 & 2-1 \end{bmatrix} = \begin{bmatrix} 1 & 0 \\ 0 & 1 \end{bmatrix}$$

As you can see, $AB = I = BA$. This is an example of a square matrix that has an inverse. Note that not all square matrices have an inverse.

Recall that it is not always true that $AB = BA$, even if both products are defined. However, if A and B are both square matrices and $AB = I_n$, it can be shown that $BA = I_n$. So, in Example 1, you need only to check that $AB = I_2$.

✓ **CHECKPOINT 1**

Show that B is the inverse of A, where

$$A = \begin{bmatrix} 2 & 1 \\ 5 & 3 \end{bmatrix} \quad \text{and}$$

$$B = \begin{bmatrix} 3 & -1 \\ -5 & 2 \end{bmatrix}.$$

Finding Inverse Matrices

If a matrix A has an inverse, A is called **invertible** (or **nonsingular**); otherwise, A is called **singular**. A nonsquare matrix cannot have an inverse. To see this, note that if A is of order $m \times n$ and B is of order $n \times m$ (where $m \neq n$), the products AB and BA are of different orders and so cannot be equal to each other. Not all square matrices have inverses (see the matrix at the bottom of page 532). If, however, a matrix does have an inverse, that inverse is unique. Example 2 shows how to use a system of equations to find the inverse of a matrix.

Example 2 Finding the Inverse of a Matrix

Find the inverse of

$$A = \begin{bmatrix} 1 & 4 \\ -1 & -3 \end{bmatrix}.$$

SOLUTION To find the inverse of A, try to solve the matrix equation $AX = I$ for X.

$$\overset{A}{\begin{bmatrix} 1 & 4 \\ -1 & -3 \end{bmatrix}} \overset{X}{\begin{bmatrix} x_{11} & x_{12} \\ x_{21} & x_{22} \end{bmatrix}} = \overset{I}{\begin{bmatrix} 1 & 0 \\ 0 & 1 \end{bmatrix}}$$

$$\begin{bmatrix} x_{11} + 4x_{21} & x_{12} + 4x_{22} \\ -x_{11} - 3x_{21} & -x_{12} - 3x_{22} \end{bmatrix} = \begin{bmatrix} 1 & 0 \\ 0 & 1 \end{bmatrix}$$

Equating corresponding entries, you obtain two systems of linear equations.

$$\begin{cases} x_{11} + 4x_{21} = 1 \\ -x_{11} - 3x_{21} = 0 \end{cases} \quad \text{Linear system with two variables, } x_{11} \text{ and } x_{21}.$$

$$\begin{cases} x_{12} + 4x_{22} = 0 \\ -x_{12} - 3x_{22} = 1 \end{cases} \quad \text{Linear system with two variables, } x_{12} \text{ and } x_{22}.$$

Solve the first system using elementary row operations to determine that $x_{11} = -3$ and $x_{21} = 1$. From the second system you can determine that $x_{12} = -4$ and $x_{22} = 1$. Therefore, the inverse of A is

$$X = A^{-1}$$
$$= \begin{bmatrix} -3 & -4 \\ 1 & 1 \end{bmatrix}.$$

You can use matrix multiplication to check this result.

$$AA^{-1} = \begin{bmatrix} 1 & 4 \\ -1 & -3 \end{bmatrix} \begin{bmatrix} -3 & -4 \\ 1 & 1 \end{bmatrix} = \begin{bmatrix} 1 & 0 \\ 0 & 1 \end{bmatrix} \checkmark$$

$$A^{-1}A = \begin{bmatrix} -3 & -4 \\ 1 & 1 \end{bmatrix} \begin{bmatrix} 1 & 4 \\ -1 & -3 \end{bmatrix} = \begin{bmatrix} 1 & 0 \\ 0 & 1 \end{bmatrix} \checkmark$$

✓ CHECKPOINT 2

Find the inverse of $A = \begin{bmatrix} -1 & 2 \\ -2 & 5 \end{bmatrix}$. ∎

In Example 2, note that the two systems of linear equations have the *same coefficient matrix A*. Rather than solve the two systems represented by

$$\begin{bmatrix} 1 & 4 & \vdots & 1 \\ -1 & -3 & \vdots & 0 \end{bmatrix}$$

and

$$\begin{bmatrix} 1 & 4 & \vdots & 0 \\ -1 & -3 & \vdots & 1 \end{bmatrix}$$

separately, you can solve them *simultaneously* by *adjoining* the identity matrix to the coefficient matrix to obtain

$$\begin{matrix} A & & I \end{matrix}$$
$$\begin{bmatrix} 1 & 4 & \vdots & 1 & 0 \\ -1 & -3 & \vdots & 0 & 1 \end{bmatrix}.$$

This "doubly augmented" matrix can be represented as $[A \vdots I]$. By applying Gauss-Jordan elimination to this matrix, you can solve *both* systems with a single elimination process.

$$\begin{bmatrix} 1 & 4 & \vdots & 1 & 0 \\ -1 & -3 & \vdots & 0 & 1 \end{bmatrix}$$

$$R_1 + R_2 \to \begin{bmatrix} 1 & 4 & \vdots & 1 & 0 \\ 0 & 1 & \vdots & 1 & 1 \end{bmatrix}$$

$$-4R_2 + R_1 \to \begin{bmatrix} 1 & 0 & \vdots & -3 & -4 \\ 0 & 1 & \vdots & 1 & 1 \end{bmatrix}$$

So, from the "doubly augmented" matrix $[A \vdots I]$, you obtain the matrix $[I \vdots A^{-1}]$.

$$\begin{matrix} A & & I & & & I & & A^{-1} \end{matrix}$$
$$\begin{bmatrix} 1 & 4 & \vdots & 1 & 0 \\ -1 & -3 & \vdots & 0 & 1 \end{bmatrix} \implies \begin{bmatrix} 1 & 0 & \vdots & -3 & -4 \\ 0 & 1 & \vdots & 1 & 1 \end{bmatrix}$$

This procedure (or algorithm) works for any square matrix that has an inverse.

TECHNOLOGY

Most graphing utilities can find the inverse of a square matrix. To do so, you may have to use the inverse key $\boxed{x^{-1}}$. Consult the user's guide for your graphing utility for specific keystrokes.

Finding an Inverse Matrix

Let A be a square matrix of order n.

1. Write the $n \times 2n$ matrix that consists of the given matrix A on the left and the $n \times n$ identity matrix I on the right to obtain $[A \vdots I]$.

2. If possible, row reduce A to I using elementary row operations on the *entire* matrix $[A \vdots I]$. The result will be the matrix $[I \vdots A^{-1}]$. If this is not possible, A is not invertible.

3. Check your work by multiplying to see that $AA^{-1} = I = A^{-1}A$.

Example 3 Finding the Inverse of a Matrix

Find the inverse of $A = \begin{bmatrix} 1 & -1 & 0 \\ 1 & 0 & -1 \\ 6 & -2 & -3 \end{bmatrix}$.

SOLUTION Begin by adjoining the identity matrix to A to form the matrix

$$[A \vdots I] = \begin{bmatrix} 1 & -1 & 0 & \vdots & 1 & 0 & 0 \\ 1 & 0 & -1 & \vdots & 0 & 1 & 0 \\ 6 & -2 & -3 & \vdots & 0 & 0 & 1 \end{bmatrix}.$$

Use elementary row operations to obtain the form $[I \vdots A^{-1}]$, as follows.

$$\begin{matrix} \\ -R_1 + R_2 \to \\ -6R_1 + R_3 \to \end{matrix} \begin{bmatrix} 1 & -1 & 0 & \vdots & 1 & 0 & 0 \\ 0 & 1 & -1 & \vdots & -1 & 1 & 0 \\ 0 & 4 & -3 & \vdots & -6 & 0 & 1 \end{bmatrix}$$

$$\begin{matrix} R_2 + R_1 \to \\ \\ -4R_2 + R_3 \to \end{matrix} \begin{bmatrix} 1 & 0 & -1 & \vdots & 0 & 1 & 0 \\ 0 & 1 & -1 & \vdots & -1 & 1 & 0 \\ 0 & 0 & 1 & \vdots & -2 & -4 & 1 \end{bmatrix}$$

$$\begin{matrix} R_3 + R_1 \to \\ R_3 + R_2 \to \\ \end{matrix} \begin{bmatrix} 1 & 0 & 0 & \vdots & -2 & -3 & 1 \\ 0 & 1 & 0 & \vdots & -3 & -3 & 1 \\ 0 & 0 & 1 & \vdots & -2 & -4 & 1 \end{bmatrix} = [I \vdots A^{-1}]$$

So, the matrix A is invertible and its inverse is

$$A^{-1} = \begin{bmatrix} -2 & -3 & 1 \\ -3 & -3 & 1 \\ -2 & -4 & 1 \end{bmatrix}.$$

Confirm this result by multiplying A and A^{-1} to obtain I, as follows.

$$AA^{-1} = \begin{bmatrix} 1 & -1 & 0 \\ 1 & 0 & -1 \\ 6 & -2 & -3 \end{bmatrix} \begin{bmatrix} -2 & -3 & 1 \\ -3 & -3 & 1 \\ -2 & -4 & 1 \end{bmatrix} = \begin{bmatrix} 1 & 0 & 0 \\ 0 & 1 & 0 \\ 0 & 0 & 1 \end{bmatrix} = I$$

STUDY TIP

Be sure to check your solutions because it is easy to make algebraic errors when using elementary row operations.

✓**CHECKPOINT 3**

Find the inverse of

$$A = \begin{bmatrix} 1 & 1 & 1 \\ 2 & 4 & 3 \\ 2 & 5 & 4 \end{bmatrix}.$$ ∎

The process shown in Example 3 applies to any $n \times n$ matrix A. When using this algorithm, if the matrix A does not reduce to the identity matrix, then A does not have an inverse. For instance, the following matrix has no inverse.

$$A = \begin{bmatrix} 1 & 2 & 0 \\ 3 & -1 & 2 \\ -2 & 3 & -2 \end{bmatrix}$$

To confirm that matrix A above has no inverse, adjoin the identity matrix to A to form $[A \vdots I]$ and perform elementary row operations on the matrix. After doing so, you will see that it is impossible to obtain the identity matrix I on the left. Therefore, A is not invertible.

The Inverse of a 2 × 2 Matrix

Using Gauss-Jordan elimination to find the inverse of a matrix works well (even as a computer technique) for matrices of order 3×3 or greater. For 2×2 matrices, however, many people prefer to use a formula for the inverse rather than Gauss-Jordan elimination. This simple formula, which works *only* for 2×2 matrices, is explained as follows. If A is a 2×2 matrix given by

$$A = \begin{bmatrix} a & b \\ c & d \end{bmatrix}$$

then A is invertible if and only if $ad - bc \neq 0$. Moreover, if $ad - bc \neq 0$, the inverse is given by

$$A^{-1} = \frac{1}{ad - bc} \begin{bmatrix} d & -b \\ -c & a \end{bmatrix}.$$ Formula for inverse of matrix A

The denominator $ad - bc$ is called the **determinant** of the 2×2 matrix A.

> **DISCOVERY**
>
> Use a graphing utility with matrix capabilities to find the inverse of the matrix
>
> $$A = \begin{bmatrix} 1 & -3 \\ -2 & 6 \end{bmatrix}.$$
>
> What message appears on the screen? Why does the graphing utility display this message?

Example 4 Finding the Inverses of 2 × 2 Matrices

If possible, find the inverse of each matrix.

a. $A = \begin{bmatrix} 3 & -1 \\ -2 & 2 \end{bmatrix}$

b. $B = \begin{bmatrix} 3 & -1 \\ -6 & 2 \end{bmatrix}$

SOLUTION

a. For the matrix A, apply the formula for the inverse of a 2×2 matrix to obtain
$$ad - bc = (3)(2) - (-1)(-2)$$
$$= 4.$$

Because this quantity is not zero, the inverse is formed by interchanging the entries on the main diagonal, changing the signs of the other two entries, and multiplying by the scalar $\frac{1}{4}$, as follows.

$$A^{-1} = \frac{1}{4} \begin{bmatrix} 2 & 1 \\ 2 & 3 \end{bmatrix}$$ Substitute for a, b, c, d, and the determinant.

$$= \begin{bmatrix} \frac{1}{2} & \frac{1}{4} \\ \frac{1}{2} & \frac{3}{4} \end{bmatrix}$$ Multiply by the scalar $\frac{1}{4}$.

b. For the matrix B, you have
$$ad - bc = (3)(2) - (-1)(-6)$$
$$= 0$$

which means that B is not invertible.

✓**CHECKPOINT 4**

Find the inverse of $A = \begin{bmatrix} 5 & -2 \\ 2 & 3 \end{bmatrix}$. ∎

534 CHAPTER 8 Matrices

Systems of Linear Equations

You know that a system of linear equations can have exactly one solution, infinitely many solutions, or no solution. If the coefficient matrix A of a *square* system (a system that has the same number of equations as variables) is invertible, the system has a unique solution, which is defined as follows.

> **A System of Equations with a Unique Solution**
>
> If A is an invertible matrix, the system of linear equations represented by $AX = B$ has a unique solution given by
>
> $$X = A^{-1}B.$$

Example 5 Solving a System Using an Inverse Matrix

Use an inverse matrix to solve the system.

$$\begin{cases} 2x + 3y + z = 4 \\ 3x + 3y + z = 8 \\ 2x + 4y + z = 5 \end{cases}$$

SOLUTION Begin by writing the system in the matrix form $AX = B$.

$$\begin{bmatrix} 2 & 3 & 1 \\ 3 & 3 & 1 \\ 2 & 4 & 1 \end{bmatrix} \begin{bmatrix} x \\ y \\ z \end{bmatrix} = \begin{bmatrix} 4 \\ 8 \\ 5 \end{bmatrix}$$

Next, use Gauss-Jordan elimination to find A^{-1}.

$$A^{-1} = \begin{bmatrix} -1 & 1 & 0 \\ -1 & 0 & 1 \\ 6 & -2 & -3 \end{bmatrix}$$

Finally, multiply B by A^{-1} on the left to obtain the solution.

$$X = A^{-1}B = \begin{bmatrix} -1 & 1 & 0 \\ -1 & 0 & 1 \\ 6 & -2 & -3 \end{bmatrix} \begin{bmatrix} 4 \\ 8 \\ 5 \end{bmatrix} = \begin{bmatrix} 4 \\ 1 \\ -7 \end{bmatrix}$$

So, the solution is $x = 4$, $y = 1$, and $z = -7$.

TECHNOLOGY

To solve a system of equations with a graphing utility, enter the matrices A and B in the matrix editor. Then, using the inverse key, solve for X.

A $\boxed{x^{-1}}$ B $\boxed{\text{ENTER}}$

The screen will display the solution, matrix X.

✓ **CHECKPOINT 5**

Use an inverse matrix to solve the system.

$$\begin{cases} -x + y + z = 4 \\ 2x - y - 3z = -7 \\ -2x + 3y + 2z = 10 \end{cases}$$

CONCEPT CHECK

1. What is the product of a square matrix of order n and its inverse?
2. Matrix A is a singular matrix of order n. Does a matrix B exist such that $AB = I$? Explain.
3. Consider the matrix $A = \begin{bmatrix} x_{11} & x_{12} \\ x_{21} & x_{22} \end{bmatrix}$, where $x_{11} \cdot x_{22} = x_{12} \cdot x_{21}$. Is A invertible? Explain.
4. Matrix A is nonsingular. Can a system of linear equations represented by $AX = B$ have infinitely many solutions? Explain.

SECTION 8.5 The Inverse of a Square Matrix

Skills Review 8.5

The following warm-up exercises involve skills that were covered in earlier sections. You will use these skills in the exercise set for this section. For additional help, review Sections 8.3 and 8.4.

In Exercises 1–6, perform the indicated matrix operations.

1. $4\begin{bmatrix} 1 & 6 \\ 0 & -4 \\ 12 & 2 \end{bmatrix}$

2. $\dfrac{1}{2}\begin{bmatrix} 11 & 10 & 48 \\ 1 & 0 & 16 \\ 0 & 2 & 8 \end{bmatrix}$

3. $\begin{bmatrix} 5 & 20 \\ -7 & 15 \end{bmatrix} - 3\begin{bmatrix} 6 & 3 \\ 4 & -2 \end{bmatrix}$

4. $\begin{bmatrix} 1 & 0 \\ 0 & 1 \end{bmatrix}\begin{bmatrix} 6 & 5 \\ 3 & -2 \end{bmatrix}$

5. $\begin{bmatrix} 2 & 0 & 0 \\ 0 & -1 & 0 \\ 0 & 0 & 3 \end{bmatrix}\begin{bmatrix} \tfrac{1}{2} & 0 & 0 \\ 0 & -1 & 0 \\ 0 & 0 & \tfrac{1}{3} \end{bmatrix}$

6. $\begin{bmatrix} 1 & -1 & 0 \\ 1 & 0 & -1 \\ 6 & -2 & -3 \end{bmatrix}\begin{bmatrix} -2 & -3 & 1 \\ -3 & -3 & 1 \\ -2 & -4 & 1 \end{bmatrix}$

In Exercises 7 and 8, rewrite the matrix in reduced row-echelon form.

7. $\begin{bmatrix} 3 & -2 & 1 & 0 \\ 4 & -3 & 0 & 1 \end{bmatrix}$

8. $\begin{bmatrix} 1 & 1 & 2 & 1 & 0 & 0 \\ -1 & 0 & 3 & 0 & 1 & 0 \\ 1 & 2 & 8 & 0 & 0 & 1 \end{bmatrix}$

Exercises 8.5

See www.CalcChat.com for worked-out solutions to odd-numbered exercises.

In Exercises 1–8, show that B is the inverse of A.

1. $A = \begin{bmatrix} 5 & 3 \\ 3 & 2 \end{bmatrix}$, $B = \begin{bmatrix} 2 & -3 \\ -3 & 5 \end{bmatrix}$

2. $A = \begin{bmatrix} 1 & -1 \\ -1 & 2 \end{bmatrix}$, $B = \begin{bmatrix} 2 & 1 \\ 1 & 1 \end{bmatrix}$

3. $A = \begin{bmatrix} 1 & 2 \\ 3 & 4 \end{bmatrix}$, $B = \begin{bmatrix} -2 & 1 \\ \tfrac{3}{2} & -\tfrac{1}{2} \end{bmatrix}$

4. $A = \begin{bmatrix} 1 & -1 \\ 2 & 3 \end{bmatrix}$, $B = \begin{bmatrix} \tfrac{3}{5} & \tfrac{1}{5} \\ -\tfrac{2}{5} & \tfrac{1}{5} \end{bmatrix}$

5. $A = \begin{bmatrix} 2 & -17 & 11 \\ -1 & 11 & -7 \\ 0 & 3 & -2 \end{bmatrix}$, $B = \begin{bmatrix} 1 & 1 & 2 \\ 2 & 4 & -3 \\ 3 & 6 & -5 \end{bmatrix}$

6. $A = \begin{bmatrix} -4 & 1 & 5 \\ -1 & 2 & 4 \\ 0 & -1 & -1 \end{bmatrix}$, $B = \begin{bmatrix} -\tfrac{1}{2} & 1 & \tfrac{3}{2} \\ \tfrac{1}{4} & -1 & -\tfrac{11}{4} \\ -\tfrac{1}{4} & 1 & \tfrac{7}{4} \end{bmatrix}$

7. $A = \begin{bmatrix} 2 & 0 & 1 & 1 \\ 3 & 0 & 0 & 1 \\ -1 & 1 & -2 & 1 \\ 4 & -1 & 1 & 0 \end{bmatrix}$, $B = \begin{bmatrix} -1 & 2 & -1 & -1 \\ -4 & 9 & -5 & -6 \\ 0 & 1 & -1 & -1 \\ 3 & -5 & 3 & 3 \end{bmatrix}$

8. $A = \begin{bmatrix} -2 & 0 & 1 & 0 \\ 1 & -1 & -3 & 0 \\ -2 & -1 & 0 & -2 \\ 0 & 1 & 3 & -1 \end{bmatrix}$, $B = \begin{bmatrix} -3 & -3 & 1 & -2 \\ 12 & 14 & -5 & 10 \\ -5 & -6 & 2 & -4 \\ -3 & -4 & 1 & -3 \end{bmatrix}$

In Exercises 9–24, find the inverse of the matrix (if it exists).

9. $\begin{bmatrix} 2 & 0 \\ 0 & 3 \end{bmatrix}$

10. $\begin{bmatrix} 1 & 2 \\ 3 & 7 \end{bmatrix}$

11. $\begin{bmatrix} 1 & -2 \\ 2 & -3 \end{bmatrix}$

12. $\begin{bmatrix} -7 & 33 \\ 4 & -19 \end{bmatrix}$

13. $\begin{bmatrix} -1 & 1 \\ -2 & 1 \end{bmatrix}$

14. $\begin{bmatrix} 11 & 1 \\ -1 & 0 \end{bmatrix}$

15. $\begin{bmatrix} 2 & 4 \\ 4 & 8 \end{bmatrix}$

16. $\begin{bmatrix} 2 & 3 \\ 1 & 4 \end{bmatrix}$

17. $\begin{bmatrix} 2 & 7 & 1 \\ -3 & -9 & 2 \end{bmatrix}$

18. $\begin{bmatrix} -2 & 5 \\ 6 & -15 \\ 0 & 1 \end{bmatrix}$

19. $\begin{bmatrix} 1 & 1 & 1 \\ 3 & 5 & 4 \\ 3 & 6 & 5 \end{bmatrix}$

20. $\begin{bmatrix} 1 & 2 & 2 \\ 3 & 7 & 9 \\ -1 & -4 & -7 \end{bmatrix}$

21. $\begin{bmatrix} 1 & 0 & 0 \\ 3 & 4 & 0 \\ 2 & 5 & 5 \end{bmatrix}$

22. $\begin{bmatrix} 1 & 0 & 0 \\ 3 & 0 & 0 \\ 2 & 5 & 5 \end{bmatrix}$

23. $\begin{bmatrix} -8 & 0 & 0 & 0 \\ 0 & 1 & 0 & 0 \\ 0 & 0 & 4 & 0 \\ 0 & 0 & 0 & -5 \end{bmatrix}$

24. $\begin{bmatrix} 1 & 3 & -2 & 0 \\ 0 & 2 & 4 & 6 \\ 0 & 0 & -2 & 1 \\ 0 & 0 & 0 & 5 \end{bmatrix}$

In Exercises 25–34, use the matrix capabilities of a graphing utility to find the inverse of the matrix (if it exists).

25. $\begin{bmatrix} 1 & 2 & -1 \\ 3 & 7 & -10 \\ -5 & -7 & -15 \end{bmatrix}$

26. $\begin{bmatrix} 10 & 5 & -7 \\ -5 & 1 & 4 \\ 3 & 2 & -2 \end{bmatrix}$

27. $\begin{bmatrix} 1 & 1 & 2 \\ 3 & 1 & 0 \\ -2 & 0 & 3 \end{bmatrix}$

28. $\begin{bmatrix} 3 & 2 & 2 \\ 2 & 2 & 2 \\ -4 & 4 & 3 \end{bmatrix}$

29. $\begin{bmatrix} -\frac{1}{2} & \frac{3}{4} & \frac{1}{4} \\ 1 & 0 & -\frac{3}{2} \\ 0 & -1 & \frac{1}{2} \end{bmatrix}$

30. $\begin{bmatrix} -\frac{5}{6} & \frac{1}{3} & \frac{11}{6} \\ 0 & \frac{2}{3} & 2 \\ 1 & -\frac{1}{2} & -\frac{5}{2} \end{bmatrix}$

31. $\begin{bmatrix} 0.1 & 0.2 & 0.3 \\ -0.3 & 0.2 & 0.2 \\ 0.5 & 0.4 & 0.4 \end{bmatrix}$

32. $\begin{bmatrix} 0.6 & 0 & -0.3 \\ 0.7 & -1 & 0.2 \\ 1 & 0 & -0.9 \end{bmatrix}$

33. $\begin{bmatrix} -1 & 0 & 1 & 0 \\ 0 & 2 & 0 & -1 \\ 2 & 0 & -1 & 0 \\ 0 & -1 & 0 & 1 \end{bmatrix}$

34. $\begin{bmatrix} 1 & -2 & -1 & -2 \\ 3 & -5 & -2 & -3 \\ 2 & -5 & -2 & -5 \\ -1 & 4 & 4 & 11 \end{bmatrix}$

In Exercises 35–40, use the formula on page 533 to find the inverse of the 2 × 2 matrix (if it exists).

35. $\begin{bmatrix} 1 & -3 \\ 5 & 4 \end{bmatrix}$

36. $\begin{bmatrix} 7 & 12 \\ -8 & -5 \end{bmatrix}$

37. $\begin{bmatrix} -4 & -6 \\ 2 & 3 \end{bmatrix}$

38. $\begin{bmatrix} -12 & 3 \\ 5 & -2 \end{bmatrix}$

39. $\begin{bmatrix} \frac{7}{2} & -\frac{3}{4} \\ \frac{1}{5} & \frac{4}{5} \end{bmatrix}$

40. $\begin{bmatrix} -\frac{1}{4} & \frac{9}{4} \\ \frac{5}{3} & \frac{8}{9} \end{bmatrix}$

In Exercises 41–44, use the inverse matrix found in Exercise 11 to solve the system of linear equations.

41. $\begin{cases} x - 2y = 5 \\ 2x - 3y = 10 \end{cases}$

42. $\begin{cases} x - 2y = 0 \\ 2x - 3y = 3 \end{cases}$

43. $\begin{cases} x - 2y = 4 \\ 2x - 3y = 2 \end{cases}$

44. $\begin{cases} x - 2y = 1 \\ 2x - 3y = -2 \end{cases}$

In Exercises 45 and 46, use the inverse matrix found in Exercise 19 to solve the system of linear equations.

45. $\begin{cases} x + y + z = 0 \\ 3x + 5y + 4z = 5 \\ 3x + 6y + 5z = 2 \end{cases}$

46. $\begin{cases} x + y + z = -1 \\ 3x + 5y + 4z = 2 \\ 3x + 6y + 5z = 0 \end{cases}$

In Exercises 47 and 48, use the inverse matrix found in Exercise 34 to solve the system of linear equations.

47. $\begin{cases} x_1 - 2x_2 - x_3 - 2x_4 = 0 \\ 3x_1 - 5x_2 - 2x_3 - 3x_4 = 1 \\ 2x_1 - 5x_2 - 2x_3 - 5x_4 = -1 \\ -x_1 + 4x_2 + 4x_3 + 11x_4 = 2 \end{cases}$

48. $\begin{cases} x_1 - 2x_2 - x_3 - 2x_4 = 1 \\ 3x_1 - 5x_2 - 2x_3 - 3x_4 = -2 \\ 2x_1 - 5x_2 - 2x_3 - 5x_4 = 0 \\ -x_1 + 4x_2 + 4x_3 + 11x_4 = -3 \end{cases}$

In Exercises 49–56, use an inverse matrix to solve (if possible) the system of linear equations.

49. $\begin{cases} 3x + 4y = -2 \\ 5x + 3y = 4 \end{cases}$

50. $\begin{cases} 18x + 12y = 13 \\ 30x + 24y = 23 \end{cases}$

51. $\begin{cases} -0.4x + 0.8y = 1.6 \\ 2x - 4y = 5 \end{cases}$

52. $\begin{cases} 0.2x - 0.6y = 2.4 \\ -x + 1.4y = -8.8 \end{cases}$

53. $\begin{cases} -\frac{1}{4}x + \frac{3}{8}y = -2 \\ \frac{3}{2}x + \frac{3}{4}y = -12 \end{cases}$

54. $\begin{cases} \frac{5}{6}x - y = -20 \\ \frac{4}{3}x - \frac{7}{2}y = -51 \end{cases}$

55. $\begin{cases} 4x - y + z = -5 \\ 2x + 2y + 3z = 10 \\ 5x - 2y + 6z = 1 \end{cases}$

56. $\begin{cases} 4x - 2y + 3z = -2 \\ 2x + 2y + 5z = 16 \\ 8x - 5y - 2z = 4 \end{cases}$

In Exercises 57–62, use the matrix capabilities of a graphing utility to solve (if possible) the system of linear equations.

57. $\begin{cases} 5x - 3y + 2z = 2 \\ 2x + 2y - 3z = 3 \\ x - 7y + 8z = -4 \end{cases}$

58. $\begin{cases} 2x + 3y + 5z = 4 \\ 3x + 5y + 9z = 7 \\ 5x + 9y + 17z = 13 \end{cases}$

59. $\begin{cases} 3x - 2y + z = -29 \\ -4x + y - 3z = 37 \\ x - 5y + z = -24 \end{cases}$

60. $\begin{cases} -8x + 7y - 10z = -151 \\ 12x + 3y - 5z = 86 \\ 15x - 9y + 2z = 187 \end{cases}$

61. $\begin{cases} 7x - 3y + 2w = 41 \\ -2x + y - w = -13 \\ 4x + z - 2w = 12 \\ -x + y - w = -8 \end{cases}$

62. $\begin{cases} 2x + 5y + w = 11 \\ x + 4y + 2z - 2w = -7 \\ 2x - 2y + 5z + w = 3 \\ x - 3w = -1 \end{cases}$

Raw Materials In Exercises 63–66, consider a company that specializes in potting soil. Each bag of potting soil for seedlings requires 2 units of sand, 1 unit of loam, and 1 unit of peat moss. Each bag of potting soil for general potting requires 1 unit of sand, 2 units of loam, and 1 unit of peat moss. Each bag of potting soil for hardwood plants requires 2 units of sand, 2 units of loam, and 2 units of peat moss. Find the numbers of bags of the three types of potting soil that the company can produce with the given amounts of raw materials.

63. 500 units of sand
    500 units of loam
    400 units of peat moss

64. 500 units of sand
    750 units of loam
    450 units of peat moss

65. 350 units of sand
445 units of loam
345 units of peat moss

66. 975 units of sand
1050 units of loam
725 units of peat moss

67. Child Support The total amounts y of child support collections (in billions of dollars) for the years 2000 through 2005 are shown in the table. The least squares regression line $y = at + b$ for these data is found by solving the system

$$\begin{cases} 6b + 3a = 123.1 \\ 3b + 19a = 79.2 \end{cases}.$$

Let t represent the year, with $t = 0$ corresponding to 2002.
(Source: U.S. Department of Health and Human Services)

Year, t	2000	2001	2002
Collections, t	17.9	19.0	20.1

Year, t	2003	2004	2005
Collections, t	21.2	21.9	23.0

(T) (a) Use a graphing utility to find an inverse matrix that can be used to solve this system.

(b) Find the equation of the least squares regression line $y = at + b$.

(c) Use the result of part (b) to estimate the amount of child support collections in 2007.

68. Assistance for Needy Families The numbers y of needy families receiving temporary assistance (in millions) for the years 2000 through 2005 are shown in the table. The least squares regression line $y = at + b$ for these data is found by solving the system

$$\begin{cases} 6b + 9a = 30.4 \\ 9b + 31a = 41.2 \end{cases}.$$

Let t represent the year, with $t = 0$ corresponding to 2001.
(Source: U.S. Administration for Children and Families)

Year, t	2000	2001	2002
Collections, t	5.8	5.4	5.1

Year, t	2003	2004	2005
Collections, t	4.9	4.7	4.5

(T) (a) Use a graphing utility to find an inverse matrix that can be used to solve this system.

(b) Find the equation of the least squares regression line $y = at + b$.

(c) Use the result of part (b) to estimate the number of needy families receiving temporary assistance in 2007.

69. Circuit Analysis Consider the circuit shown in the figure. The currents I_1, I_2, and I_3, in amperes, are the solution of the system of linear equations

$$\begin{cases} 2I_1 + 4I_3 = E_1 \\ I_2 + 4I_3 = E_2 \\ I_1 + I_2 - I_3 = 0 \end{cases}$$

where E_1 and E_2 are voltages. Use the inverse of the coefficient matrix of this system to find the unknown currents for the voltages.

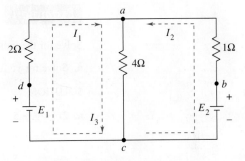

(a) $E_1 = 14$ volts, $E_2 = 28$ volts

(b) $E_1 = 24$ volts, $E_2 = 23$ volts

True or False? In Exercises 70 and 71, determine whether the statement is true or false. Justify your answer.

70. Multiplication of an invertible matrix and its inverse is commutative.

71. If you multiply two square matrices and obtain the identity matrix, you can assume that the matrices are inverses of one another.

72. If A is a 2×2 matrix $A = \begin{bmatrix} a & b \\ c & d \end{bmatrix}$, then A is invertible if and only if $ad - bc \neq 0$. If $ad - bc \neq 0$, verify that the inverse is

$$A^{-1} = \frac{1}{ad - bc} \begin{bmatrix} d & -b \\ -c & a \end{bmatrix}.$$

73. Consider matrices of the form

$$A = \begin{bmatrix} a_{11} & 0 & 0 & 0 & \cdots & 0 \\ 0 & a_{22} & 0 & 0 & \cdots & 0 \\ 0 & 0 & a_{33} & 0 & \cdots & 0 \\ \vdots & \vdots & \vdots & \vdots & & \vdots \\ 0 & 0 & 0 & 0 & \cdots & a_{nn} \end{bmatrix}.$$

(a) Write a 2×2 matrix and a 3×3 matrix in the form of A. Find the inverse of each.

(b) Use the result of part (a) to make a conjecture about the inverses of matrices in the form of A.

Algebra Review

Solving Equations

When using the method of substitution to solve a linear system, you need the algebraic skill of solving an equation in one variable. Recall that to solve a linear equation, you can add or subtract the same quantity from each side of the equation. You can also multiply or divide each side of the equation by the same *nonzero* quantity.

Example 1 Solving Equations

Solve for the indicated variable.

a. Solve for y: $2x - y - 5 = 0$ **b.** Solve for x: $x - 3y + 7 = 0$

SOLUTION

a.
$2x - y - 5 = 0$	Write original equation.
$2x - y = 5$	Add 5 to each side.
$-y = -2x + 5$	Subtract $2x$ from each side.
$y = 2x - 5$	Multiply each side by -1.

b.
$x - 3y + 7 = 0$	Write original equation.
$x - 3y = -7$	Subtract 7 from each side.
$x = 3y - 7$	Add $3y$ to each side.

Example 2 Solving Equations

Solve each equation.

a. $y - 3(4y - 2) = 1$ **b.** $-2x + 6(2x - 3) = 4$

SOLUTION

a.
$y - 3(4y - 2) = 1$	Write original equation.
$y - 12y + 6 = 1$	Distributive Property
$-11y + 6 = 1$	Combine like terms.
$-11y = -5$	Subtract 6 from each side.
$y = \dfrac{5}{11}$	Divide each side by -11.

b.
$-2x + 6(2x - 3) = 4$	Write original equation.
$-2x + 12x - 18 = 4$	Distributive Property
$10x - 18 = 4$	Combine like terms.
$10x = 22$	Add 18 to each side.
$x = \dfrac{11}{5}$	Divide each side by 10.

Example 3 Using Back-Substitution

Solve for x, y, and z.

a. $\begin{cases} x - 2y + z = 0 & \text{Equation 1} \\ y - 2z = 0 & \text{Equation 2} \\ z = 3 & \text{Equation 3} \end{cases}$

b. $\begin{cases} x + 3y - z = 0 & \text{Equation 1} \\ y + 3z = 0 & \text{Equation 2} \\ z = \frac{1}{3} & \text{Equation 3} \end{cases}$

SOLUTION

a. From Equation 3, you already know the value of z. To solve for y, substitute $z = 3$ into Equation 2 and solve for y.

$y - 2(3) = 0$ Substitute 3 for z.

$y - 6 = 0$ Multiply.

$y = 6$ Add 6 to each side.

Finally, substitute $y = 6$ and $z = 3$ into Equation 1 and solve for x.

$x - 2(6) + 3 = 0$ Substitute 6 for y and 3 for z.

$x - 12 + 3 = 0$ Multiply.

$x - 9 = 0$ Combine like terms.

$x = 9$ Add 9 to each side.

b. From Equation 3, you already know the value of z. To solve for y, substitute $z = \frac{1}{3}$ into Equation 2 and solve for y.

$y + 3\left(\dfrac{1}{3}\right) = 0$ Substitute $\frac{1}{3}$ for z.

$y + 1 = 0$ Multiply.

$y = -1$ Subtract 1 from each side.

Finally, substitute $y = -1$ and $z = \frac{1}{3}$ into Equation 1 and solve for x.

$x + 3(-1) - \dfrac{1}{3} = 0$ Substitute -1 for y and $\frac{1}{3}$ for z.

$x - 3 - \dfrac{1}{3} = 0$ Multiply.

$x - \dfrac{10}{3} = 0$ Combine like terms.

$x = \dfrac{10}{3}$ Add $\frac{10}{3}$ to each side.

Chapter Summary and Study Strategies

After studying this chapter, you should have acquired the following skills. The exercise numbers are keyed to the Review Exercises that begin on page 542. Answers to odd-numbered Review Exercises are given in the back of the text.*

Section 8.1   Review Exercises

- Solve a system of linear equations by the method of substitution.   1–4
- Solve a system of nonlinear equations by the method of substitution.   5, 6
- Solve a system of linear equations by the method of elimination.   7–14
- Interpret the solution of a linear system graphically.   15, 16
- Construct and use a system of equations to solve a real-life problem.   17–20

Section 8.2

- Solve a linear system in row-echelon form using back-substitution.   21, 22
- Solve a linear system using Gaussian elimination.   23–26
- Solve a nonsquare linear system.   27, 28
- Construct and use a linear system in three or more variables to solve a real-life problem.   29, 30
- Find the equation of a parabola or a circle using a linear system in three variables.   31–34

Section 8.3

- Determine the order of a matrix.   35, 36

$$\begin{array}{c} \phantom{\text{Row 1}} \quad \text{Column 1} \quad \text{Column 2} \quad \text{Column 3} \quad \cdots \quad \text{Column } n \\ \begin{array}{c} \text{Row 1} \\ \text{Row 2} \\ \text{Row 3} \\ \vdots \\ \text{Row } m \end{array} \begin{bmatrix} a_{11} & a_{12} & a_{13} & \cdots & a_{1n} \\ a_{21} & a_{22} & a_{23} & \cdots & a_{2n} \\ a_{31} & a_{32} & a_{33} & \cdots & a_{3n} \\ \vdots & \vdots & \vdots & & \vdots \\ a_{m1} & a_{m2} & a_{m3} & \cdots & a_{mn} \end{bmatrix} \end{array}$$

A matrix having m rows and n columns is said to be of order $m \times n$.

- Perform elementary row operations on a matrix in order to write the matrix in row-echelon form or reduced row-echelon form.   37–40
    - Interchange two rows.
    - Multiply a row by a nonzero constant.
    - Add a multiple of a row to another row.
- Solve a system of linear equations using Gaussian elimination or Gauss-Jordan elimination.   41–48

---

* Use a wide range of valuable study aids to help you master the material in this chapter. The *Student Solutions Guide* includes step-by-step solutions to all odd-numbered exercises to help you review and prepare. The student website at *college.hmco.com/info/larsonapplied* offers algebra help and a *Graphing Technology Guide*. The *Graphing Technology Guide* contains step-by-step commands and instructions for a wide variety of graphing calculators, including the most recent models.

Section 8.4

	Review Exercises
■ Add or subtract two matrices and multiply a matrix by a scalar.	49–52

If $A = [a_{ij}]$ and $B = [b_{ij}]$ are $m \times n$ matrices and c is a scalar, then

$$A + B = [a_{ij} + b_{ij}] \quad \text{and} \quad cA = [ca_{ij}].$$

■ Find the product of two matrices.   53–62

If $A = [a_{ij}]$ is an $m \times n$ matrix and $B = [b_{ij}]$ is an $n \times p$ matrix, then AB is an $m \times p$ matrix.

$$AB = [c_{ij}] \quad \text{where} \quad c_{ij} = a_{i1}b_{1j} + a_{i2}b_{2j} + a_{i3}b_{3j} + \cdots + a_{in}b_{nj}.$$

■ Solve a matrix equation.   63–66

■ Use matrix multiplication to represent and solve a system of linear equations.   67–70

Section 8.5

■ Verify that a matrix B is the inverse of a given matrix A by showing   71–74
$$AB = I \quad \text{or} \quad BA = I$$

■ Find the inverse of a matrix.   75, 76

■ Find the inverse of a 2×2 matrix using a formula.   77, 78

$$A^{-1} = \frac{1}{ad - bc} \begin{bmatrix} d & -b \\ -c & a \end{bmatrix}.$$

■ Use an inverse matrix to solve a system of linear equations.   79–90

If A is an invertible matrix, then $AX = B$ has a unique solution given by $X = A^{-1}B$.

Study Strategies

■ **Units of Variables in Applied Problems** When using systems of equations to solve real-life applications, be sure to keep track of the unit(s) assigned to each variable. This will allow you to write each equation of the system correctly based on the constraints given in the application.

■ **Using Technology** Performing operations with matrices can be tedious. You can use a graphing utility to accomplish the following.
- Perform elementary row operations on matrices.
- Reduce matrices to row-echelon form and reduced row-echelon form.
- Add and subtract matrices.
- Multiply matrices.
- Multiply matrices by scalars.
- Find inverses of matrices.
- Solve systems of equations using matrices.

Review Exercises

See www.CalcChat.com for worked-out solutions to odd-numbered exercises.

In Exercises 1–6, solve the system by the method of substitution.

1. $\begin{cases} x + 3y = 10 \\ 4x - 5y = -28 \end{cases}$
2. $\begin{cases} 3x - y - 13 = 0 \\ 4x + 3y - 26 = 0 \end{cases}$
3. $\begin{cases} \frac{1}{2}x + \frac{3}{5}y = -2 \\ 2x + y = 6 \end{cases}$
4. $\begin{cases} 1.3x + 0.9y = 7.5 \\ 0.4x - 0.5y = -0.8 \end{cases}$
5. $\begin{cases} x^2 + y^2 = 100 \\ x + 2y = 20 \end{cases}$
6. $\begin{cases} y = x^3 - 2x^2 - 2x - 3 \\ y = -x^2 + 4x - 3 \end{cases}$

In Exercises 7–14, solve the system by the method of elimination.

7. $\begin{cases} 2x - 3y = 21 \\ 3x + y = 4 \end{cases}$
8. $\begin{cases} 3u + 5v = 9 \\ 12u + 10v = 22 \end{cases}$
9. $\begin{cases} 4x - 3y = 10 \\ 8x - 6y = 20 \end{cases}$
10. $\begin{cases} 3x + 4y = 18 \\ 6x + 8y = 18 \end{cases}$
11. $\begin{cases} 1.25x - 2y = 3.5 \\ 5x - 8y = 14 \end{cases}$
12. $\begin{cases} 1.5x + 2.5y = 8.5 \\ 6x + 10y = 24 \end{cases}$
13. $\begin{cases} \frac{x-2}{3} + \frac{y+3}{4} = 5 \\ 2x - y = 7 \end{cases}$
14. $\begin{cases} \frac{3}{5}x + \frac{2}{7}y = 10 \\ x + 2y = 38 \end{cases}$

In Exercises 15 and 16, describe the graph of the solution of the linear system.

15. $\begin{cases} 2x + y = -1 \\ 3x - 2y = -5 \end{cases}$
16. $\begin{cases} x - 2y = -1 \\ -2x + 4y = 2 \end{cases}$

17. **Choice of Newscasts** Television Stations A and B are competing for the 6 P.M. newscast audience. Station A is implementing a new newscast format for the 6 P.M. audience. Models that represent the numbers of 6 P.M. viewers each month for the two stations are given by

$\begin{cases} y = 950x + 10{,}000 & \text{Station A (new format)} \\ y = -875x + 18{,}000 & \text{Station B} \end{cases}$

where y is the number of viewers and x represents the month, with $x = 1$ corresponding to the first month of the new format. Use the models to estimate when the number of viewers for Station A's 6 P.M. newscast will exceed the number of viewers for Station B's 6 P.M. newscast.

(T) 18. **Comparing Populations** For the years 2000 through 2005, the population of Vermont grew more slowly than that of Alaska. Two models that approximate the populations of the two states are

$\begin{cases} P = 7.7t + 626 & \text{Alaska} \\ P = 2.8t + 610 & \text{Vermont} \end{cases}$

where P is the population (in thousands) and t represents the year, with $t = 0$ corresponding to 2000. Use a graphing utility to determine whether the population of Vermont will exceed the population of Alaska. *(Source: U.S. Census Bureau)*

19. **Acid Mixture** Twelve gallons of a 25% acid solution is obtained by mixing a 10% solution with a 50% solution.

   (T) (a) Write a system of equations that represents the problem and use a graphing utility to graph the equations in the same viewing window.

   (b) How much of each solution is required to obtain the specified concentration of the final mixture?

20. **Acid Mixture** Twenty gallons of a 30% acid solution is obtained by mixing a 12% solution with a 60% solution.

   (T) (a) Write a system of equations that represents the problem and use a graphing utility to graph the equations in the same viewing window.

   (b) How much of each solution is required to obtain the specified concentration of the final mixture?

In Exercises 21–28, solve the system of equations and check your solution algebraically.

21. $\begin{cases} 4x - 3y + 2z = 1 \\ 2y - 4z = 2 \\ z = 2 \end{cases}$
22. $\begin{cases} 2x + y - 4z = 6 \\ 3y + z = 2 \\ z = -4 \end{cases}$
23. $\begin{cases} 2x + y + z = 6 \\ x - 4y - z = 3 \\ x + y + z = 4 \end{cases}$
24. $\begin{cases} x + 3y - z = 13 \\ 2x - 5z = 23 \\ 4x - y - 2z = 4 \end{cases}$
25. $\begin{cases} 2x + 6y - z = 1 \\ x - 3y + z = 2 \\ \frac{3}{2}x + \frac{3}{2}y = 6 \end{cases}$
26. $\begin{cases} x + y + z + w = 8 \\ 4y + 5z - 2w = 3 \\ 2x + 3y - z = -2 \\ 3x + 2y - 4w = -20 \end{cases}$
27. $\begin{cases} x + y + z = 10 \\ -2x + 3y + 4z = 22 \end{cases}$
28. $\begin{cases} 5x - 12y + 7z = 16 \\ 3x - 7y + 4z = 9 \end{cases}$

29. Agriculture A mixture of 6 gallons of chemical A, 8 gallons of chemical B, and 13 gallons of chemical C is required to kill a destructive crop insect. Commercial spray X contains 1, 2, and 2 parts, respectively, of these chemicals. Commercial spray Y contains only chemical C. Commercial spray Z contains chemicals A, B, and C in equal amounts. How much of each type of commercial spray is needed to get the desired mixture?

30. Vertical Motion An object moving vertically is at the given heights at the specified times. Find the position equation $s = \frac{1}{2}at^2 + v_0 t + s_0$ for the object.

(a) At $t = 1$ second, $s = 134$ feet
At $t = 2$ seconds, $s = 86$ feet
At $t = 3$ seconds, $s = 6$ feet

(b) At $t = 1$ second, $s = 184$ feet
At $t = 2$ seconds, $s = 116$ feet
At $t = 3$ seconds, $s = 16$ feet

In Exercises 31 and 32, find the equation of the parabola $y = ax^2 + bx + c$ that passes through the points. To verify your result, use a graphing utility to plot the points and graph the parabola.

31. 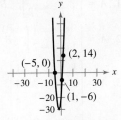 32.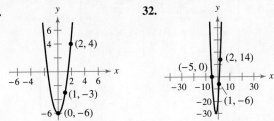

In Exercises 33 and 34, find the equation of the circle $x^2 + y^2 + Dx + Ey + F = 0$ that passes through the points. To verify your result, use a graphing utility to plot the points and graph the circle.

33. 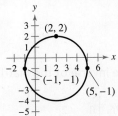 34.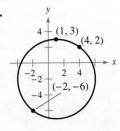

In Exercises 35 and 36, determine the order of the matrix.

35. $\begin{bmatrix} 3 & 7 & 4 & -2 \\ 1 & 8 & 6 & 1 \end{bmatrix}$

36. $\begin{bmatrix} 5 \\ -1 \\ 2 \\ 4 \end{bmatrix}$

In Exercises 37 and 38, write the matrix in row-echelon form.

37. $\begin{bmatrix} 1 & 3 & 0 & 2 \\ 3 & 10 & 1 & 8 \\ 2 & 3 & 3 & 10 \end{bmatrix}$

38. $\begin{bmatrix} 1 & 2 & -1 & 0 \\ -2 & -3 & 3 & 4 \\ 4 & 0 & 1 & 3 \end{bmatrix}$

In Exercises 39 and 40, write the matrix in *reduced* row-echelon form.

39. $\begin{bmatrix} 1 & 2 & 3 \\ -2 & 0 & 2 \\ 2 & 1 & 2 \end{bmatrix}$

40. $\begin{bmatrix} 2 & 3 & 1 & -5 \\ 1 & 0 & 5 & 2 \\ -1 & 4 & 3 & 6 \\ 0 & -2 & 6 & -8 \end{bmatrix}$

In Exercises 41–48, use matrices to solve the system of equations (if possible). Use Gaussian elimination with back-substitution or Gauss-Jordan elimination.

41. $\begin{cases} 4x - 3y = 18 \\ x + y = 1 \end{cases}$

42. $\begin{cases} 2x + 4y = 16 \\ -x + 3y = 17 \end{cases}$

43. $\begin{cases} 2x + 3y - z = 13 \\ 3x + z = 8 \\ x - 2y + 3z = -4 \end{cases}$

44. $\begin{cases} 3x + 4y + 2z = 5 \\ 2x + 3y = 7 \\ 2y - 3z = 12 \end{cases}$

45. $\begin{cases} x + 2y + 2z = 10 \\ 2x + 3y + 5z = 20 \end{cases}$

46. $\begin{cases} 3x + 10y + 4z = 20 \\ x + 3y - 2z = 8 \end{cases}$

47. $\begin{cases} 2x + y - 3z = 4 \\ x + 2y + 2z = 10 \\ x - 2z = 12 \\ x + y + z = 6 \end{cases}$

48. $\begin{cases} 2x + 4y + 2z = 10 \\ x + 3z = 9 \\ 3x - 2y = 4 \\ x + y + z = 8 \end{cases}$

In Exercises 49–52, find (a) $A + B$, (b) $A - B$, (c) $4A$, and (d) $4A - 3B$.

49. $A = \begin{bmatrix} -1 & 5 \\ 2 & 1 \end{bmatrix}$, $B = \begin{bmatrix} 4 & 2 \\ -6 & 3 \end{bmatrix}$

50. $A = \begin{bmatrix} 1 & 0 & 2 \\ -1 & 3 & 5 \\ 2 & -2 & 3 \end{bmatrix}$, $B = \begin{bmatrix} 2 & 0 & 1 \\ 3 & -4 & 6 \\ 1 & 2 & -3 \end{bmatrix}$

51. $A = \begin{bmatrix} 1 & 3 & -2 & 6 \\ 0 & 1 & 3 & 2 \end{bmatrix}$,
$B = \begin{bmatrix} 2 & 1 & 4 & -5 \\ 3 & -6 & 3 & -2 \end{bmatrix}$

52. $A = \begin{bmatrix} 3 \\ -2 \\ 3 \end{bmatrix}$, $B = \begin{bmatrix} -1 \\ 4 \\ 5 \end{bmatrix}$

In Exercises 53–58, find AB, if possible.

53. $A = \begin{bmatrix} 1 & 4 \\ -2 & -1 \\ 3 & 2 \end{bmatrix}$, $B = \begin{bmatrix} -4 \\ 3 \end{bmatrix}$

54. $A = \begin{bmatrix} 3 \\ 2 \\ 4 \\ 6 \end{bmatrix}$, $B = \begin{bmatrix} 2 & 0 & -1 \end{bmatrix}$

55. $A = \begin{bmatrix} 4 & 0 & 0 \\ 0 & 3 & 0 \\ 0 & 0 & -2 \end{bmatrix}$, $B = \begin{bmatrix} \frac{1}{4} & 0 & 0 \\ 0 & \frac{1}{3} & 0 \\ 0 & 0 & -\frac{1}{2} \end{bmatrix}$

56. $A = \begin{bmatrix} 3 & 1 \\ 4 & 7 \\ 1 & 1 \end{bmatrix}$, $B = \begin{bmatrix} 1 & 2 & -2 \\ 3 & 4 & 0 \\ 0 & 1 & 0 \end{bmatrix}$

57. $A = \begin{bmatrix} 1 & 2 & 3 & 6 & -1 \\ 2 & 8 & 0 & 0 & 2 \end{bmatrix}$, $B = \begin{bmatrix} 3 & 2 \\ 4 & -1 \end{bmatrix}$

58. $A = \begin{bmatrix} 0 & 0 & 2 \\ 1 & 0 & 6 \\ 0 & 2 & 2 \end{bmatrix}$, $B = \begin{bmatrix} 3 & 4 & 0 & 1 \\ 2 & 1 & 0 & 0 \\ 0 & 0 & 1 & 1 \end{bmatrix}$

In Exercises 59 and 60, if possible, find (a) AB, (b) BA, and (c) A^2. (Note: $A^2 = AA$.)

59. $A = \begin{bmatrix} 1 & -3 & 4 \end{bmatrix}$, $B = \begin{bmatrix} 2 \\ -2 \\ -1 \end{bmatrix}$

60. $A = \begin{bmatrix} 1 & 0 & 2 \\ 3 & 1 & -2 \\ 1 & 1 & 1 \end{bmatrix}$, $B = \begin{bmatrix} 2 & 0 & 0 \\ 1 & -2 & 1 \\ 5 & 4 & -2 \end{bmatrix}$

61. **Leslie Matrix** Use the Leslie Matrix L and age distribution matrix X_1 to find the age distribution matrices X_2 and X_3.

$L = \begin{bmatrix} 0 & 2 & 2 & 0 \\ \frac{1}{4} & 0 & 0 & 0 \\ 0 & 1 & 0 & 0 \\ 0 & 0 & \frac{1}{2} & 0 \end{bmatrix}$, $X_1 = \begin{bmatrix} 100 \\ 100 \\ 100 \\ 100 \end{bmatrix}$

62. **Population Growth** A population has the following characteristics.

 (a) A total of 75% of the population survives its first year. Of that 75%, 25% survives the second year. The maximum life span is 3 years. The average number of offspring for each member of the population is 2 the first year, 4 the second year, and 2 the third year.

 (b) The population now consists of 120 members in each of the three age classes.

How many members will there be in each age class in 1 year? in 2 years?

In Exercises 63–66, solve for X when

$A = \begin{bmatrix} 1 & -2 \\ 0 & 1 \\ 2 & 3 \end{bmatrix}$ and $B = \begin{bmatrix} 0 & 1 \\ 1 & 1 \\ 3 & 5 \end{bmatrix}$.

63. $X = 4A - 3B$

64. $X = 5B + 2A$

65. $2X - 3A = B$

66. $4X - 8B = 4A$

In Exercises 67–70, (a) write the system of linear equations as a matrix equation, $AX = B$, and (b) use Gauss-Jordan elimination on the augmented matrix $[A \vdots B]$ to solve for the matrix X.

67. $\begin{cases} 5x_1 + 4x_2 = 2 \\ -x_1 + x_2 = -22 \end{cases}$

68. $\begin{cases} 2x_1 - 5x_2 = 2 \\ 3x_1 - 7x_2 = 1 \end{cases}$

69. $\begin{cases} 2x_1 + 3x_2 + x_3 = 10 \\ 2x_1 - 3x_2 - 3x_3 = 22 \\ 4x_1 - 2x_2 + 3x_3 = -2 \end{cases}$

70. $\begin{cases} 2x_1 + 3x_2 + 3x_3 = 3 \\ 6x_1 + 6x_2 + 12x_3 = 13 \\ 12x_1 + 9x_2 - x_3 = 2 \end{cases}$

In Exercises 71–74, show that B is the inverse of A.

71. $A = \begin{bmatrix} -4 & -1 \\ 7 & 2 \end{bmatrix}$, $B = \begin{bmatrix} -2 & -1 \\ 7 & 4 \end{bmatrix}$

72. $A = \begin{bmatrix} 5 & -1 \\ 11 & -2 \end{bmatrix}$, $B = \begin{bmatrix} -2 & 1 \\ -11 & 5 \end{bmatrix}$

73. $A = \begin{bmatrix} 1 & 2 & 1 \\ 3 & 6 & 4 \\ 0 & 1 & 3 \end{bmatrix}$, $B = \begin{bmatrix} -14 & 5 & -2 \\ 9 & -3 & 1 \\ -3 & 1 & 0 \end{bmatrix}$

74. $A = \begin{bmatrix} 2 & 0 & 1 & 2 \\ 3 & 0 & 0 & 1 \\ -1 & 1 & 2 & 0 \\ 0 & -1 & 2 & 2 \end{bmatrix}$,

$B = \frac{1}{9} \begin{bmatrix} -4 & 6 & 1 & 1 \\ 10 & -6 & 2 & -7 \\ -7 & 6 & 4 & 4 \\ 12 & -9 & -3 & -3 \end{bmatrix}$

In Exercises 75 and 76, find the inverse of the matrix.

75. $\begin{bmatrix} -1 & 0 & 0 \\ 0 & 2 & 0 \\ 0 & 0 & 4 \end{bmatrix}$

76. $\begin{bmatrix} 3 & 2 & 2 \\ 0 & 2 & 1 \\ 1 & 0 & 1 \end{bmatrix}$

In Exercises 77 and 78, use the formula on page 533 to find the inverse of the matrix.

77. $\begin{bmatrix} 1 & 3 \\ 2 & 5 \end{bmatrix}$

78. $\begin{bmatrix} -2 & 1 \\ 4 & 3 \end{bmatrix}$

In Exercises 79 and 80, use the inverse matrix found in Exercise 77 to solve the system of linear equations.

79. $\begin{cases} x + 3y = 15 \\ 2x + 5y = 26 \end{cases}$
80. $\begin{cases} x + 3y = 7 \\ 2x + 5y = 11 \end{cases}$

In Exercises 81 and 82, use the inverse matrix found in Exercise 76 to solve the system of linear equations.

81. $\begin{cases} 3x + 2y + 2z = 13 \\ 2y + z = 4 \\ x + z = 5 \end{cases}$

82. $\begin{cases} 3x + 2y + 2z = 12 \\ 2y + z = 13 \\ x + z = 3 \end{cases}$

In Exercises 83–86, use an inverse matrix to solve the system of linear equations.

83. $\begin{cases} -3x + 10y = 8 \\ 5x - 17y = -13 \end{cases}$

84. $\begin{cases} 5x - y = 13 \\ -9x + 2y = -24 \end{cases}$

85. $\begin{cases} 3x + 2y - z = 6 \\ x - y + 2z = -1 \\ 5x + y + z = 7 \end{cases}$

86. $\begin{cases} -x + 4y - 2z = 12 \\ 2x - 9y + 5z = -25 \\ -x + 5y - 4z = 10 \end{cases}$

Raw Materials In Exercises 87 and 88, you are making three types of windshield washer fluid in chemistry class. Fluid A requires 9 cups of water, 1 cup of isopropyl alcohol, and 1 tablespoon of detergent. Fluid B requires 10 cups of water, 3 cups of isopropyl alcohol, and 1 tablespoon of detergent. Fluid C requires 14 cups of water, 2 cups of isopropyl alcohol, and 2 tablespoons of detergent. A system of linear equations (where x, y, and z represent fluids A, B, and C, respectively) is as follows.

$\begin{cases} 9x + 10y + 14z = \text{(cups of water)} \\ x + 3y + 2z = \text{(cups of isopropyl alcohol)} \\ x + y + 2z = \text{(tablespoons of detergent)} \end{cases}$

Use the inverse of the coefficient matrix of this system to find the numbers of units of fluids A, B, and C that you can produce with the given amounts of ingredients.

87. 240 cups of water
44 cups of isopropyl alcohol
28 tablespoons of detergent

88. 235 cups of water
41 cups of isopropyl alcohol
29 tablespoons of detergent

89. Lead Production The amounts y of lead (in thousands of metric tons) produced in the years 2000 through 2005 are shown in the table. The least squares regression line $y = at + b$ for the data is found by solving the system

$\begin{cases} 6b + 3a = 2691 \\ 3b + 19a = 1188 \end{cases}.$

Let t represent the year, with $t = 0$ corresponding to 2002.
(Source: U.S. Geological Survey)

Year, t	2000	2001	2002
Production, y	468	466	451

Year, t	2003	2004	2005
Production, y	449	430	427

(T) (a) Use a graphing utility to find an inverse matrix to solve this system.

(b) Find an equation of the least squares regression line $y = at + b$.

(c) Use the result of part (b) to estimate the amount of lead produced in 2007.

90. Medical Index The values of the Consumer Price Index (CPI) y for physician services (in dollars) for the years 2000 through 2005 are shown in the table. The least squares regression line $y = at + b$ for the data is found by solving the system

$\begin{cases} 6b + 9a = 1592.4 \\ 9b + 31a = 2536.2 \end{cases}.$

Let t represent the year, with $t = 0$ corresponding to 2001.
(Source: U.S. Bureau of Labor Statistics)

Year, t	2000	2001	2002
CPI, y	244.7	253.6	260.6

Year, t	2003	2004	2005
CPI, y	267.7	278.3	287.5

(T) (a) Use a graphing utility to find an inverse matrix to solve this system.

(b) Find an equation of the least squares regression line $y = at + b$.

(c) Use the result of part (b) to estimate the Consumer Price Index for physician services in 2007.

Chapter Test

Take this test as you would take a test in class. When you are done, check your work against the answers given in the back of the book.

In Exercises 1–6, solve the system of equations using the indicated method.

1. *Substitution*
$$\begin{cases} 5x - 7y = -18 \\ 4x + 3y = 20 \end{cases}$$

2. *Substitution*
$$\begin{cases} x + y = 3 \\ x^2 + y = 9 \end{cases}$$

3. *Substitution*
$$\begin{cases} 3x - 2y = 6 \\ 2x^2 + 2y = 8 \end{cases}$$

4. *Elimination*
$$\begin{cases} 1.5x - 2y = 8 \\ 2.5x + 2y = 5.75 \end{cases}$$

5. *Elimination*
$$\begin{cases} 2x - 4y + z = 11 \\ x + 2y + 3z = 9 \\ 3y + 5z = 12 \end{cases}$$

6. *Elimination*
$$\begin{cases} 3x - 2y + z = 16 \\ 5x - z = 6 \\ 2x - y - z = 3 \end{cases}$$

In Exercises 7 and 8, write the augmented matrix for the system of linear equations.

7. $\begin{cases} 2x + y + 4z = 2 \\ x + 4y - z = 0 \\ -x + 3y + 3z = -1 \end{cases}$

8. $\begin{cases} 3x + 4y + 2z = 4 \\ 2x + 3y = -2 \\ 2y - 3z = -13 \end{cases}$

In Exercises 9–11, use matrices to solve the system of equations.

9. $\begin{cases} x + 2y + 3z = 16 \\ 5x + 4y - z = 22 \end{cases}$

10. $\begin{cases} x - 2y + z = 14 \\ y - 3z = 2 \\ z = -6 \end{cases}$

11. $\begin{cases} 2x - 3y + z = 14 \\ x + 2y = -4 \\ y - z = -4 \end{cases}$

In Exercises 12–15, use the matrices to find the indicated matrix.

$A = \begin{bmatrix} 1 & 3 \\ 2 & 4 \end{bmatrix}$, $B = \begin{bmatrix} 2 & -1 & 3 \\ 4 & 0 & 1 \end{bmatrix}$, $C = \begin{bmatrix} 0 & -2 \\ 3 & 5 \end{bmatrix}$, $D = \begin{bmatrix} 3 \\ 2 \\ -1 \end{bmatrix}$

12. $2A + C$ **13.** CA **14.** BD **15.** A^2

In Exercises 16–18, find the inverse of the matrix.

16. $A = \begin{bmatrix} 2 & -1 \\ -3 & 4 \end{bmatrix}$

17. $A = \begin{bmatrix} 1 & 0 \\ 0 & 1 \end{bmatrix}$

18. $A = \begin{bmatrix} 3 & 4 & 2 \\ 2 & 3 & 0 \\ 0 & 2 & -3 \end{bmatrix}$

19. One hundred liters of a 50% solution is obtained by mixing a 60% solution with a 20% solution. How many liters of each solution must be used to obtained the desired mixture?

20. In the 2004 presidential election, approximately 120.884 million voters divided their votes among three presidential candidates. George W. Bush received approximately 2.978 million votes more than John Kerry. Ralph Nader received approximately 0.096% of the votes. Write and solve a system of equations to find the total number of votes cast for each candidate. Let B represent the total votes cast for Bush, K the total votes cast for Kerry, and N the total votes cast for Nader. *(Source: U.S. House of Representatives)*

Functions of Several Variables

9

You can use least squares regression analysis to model the number of students served by an autism disability program in the United States. (See Section 9.6, Example 4.)

Applications

Functions of several variables have many real-life applications. The applications listed below represent a sample of the applications in this chapter.

- Geology, Exercises 59 and 60, page 555
- Infant Development, Exercise 45, page 572
- Shannon Diversity Index, Exercise 46, page 593
- Make a Decision: Optometry, Exercise 36, page 603
- Population Density, Exercises 31 and 32, page 619

- **9.1** The Three-Dimensional Coordinate System
- **9.2** Surfaces in Space
- **9.3** Functions of Several Variables
- **9.4** Partial Derivatives
- **9.5** Extrema of Functions of Two Variables
- **9.6** Least Squares Regression Analysis
- **9.7** Double Integrals and Area in the Plane
- **9.8** Applications of Double Integrals

Section 9.1

The Three-Dimensional Coordinate System

- Plot points in space.
- Find distances between points in space and find midpoints of line segments in space.
- Write the standard forms of the equations of spheres and find the centers and radii of spheres.
- Sketch the coordinate plane traces of surfaces.

The Three-Dimensional Coordinate System

Recall from Section 1.1 that the Cartesian plane is determined by two perpendicular number lines called the x-axis and the y-axis. These axes together with their point of intersection (the origin) allow you to develop a two-dimensional coordinate system for identifying points in a plane. To identify a point in space, you must introduce a third dimension to the model. The geometry of this three-dimensional model is called **solid analytic geometry.**

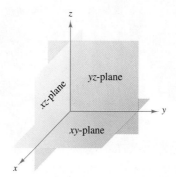

FIGURE 9.1

DISCOVERY

Describe the location of a point (x, y, z) if $x = 0$. Describe the location of a point (x, y, z) if $x = 0$ and $y = 0$. What can you conclude about the ordered triple (x, y, z) if the point is located on the y-axis? What can you conclude about the ordered triple (x, y, z) if the point is located in the xz-plane?

You can construct a **three-dimensional coordinate system** by passing a z-axis perpendicular to both the x- and y-axes at the origin. Figure 9.1 shows the positive portion of each coordinate axis. Taken as pairs, the axes determine three **coordinate planes:** the ***xy*-plane,** the ***xz*-plane,** and the ***yz*-plane.** These three coordinate planes separate the three-dimensional coordinate system into eight **octants.** The first octant is the one for which all three coordinates are positive. In this three-dimensional system, a point P in space is determined by an ordered triple (x, y, z), where x, y, and z are as follows.

x = directed distance from yz-plane to P
y = directed distance from xz-plane to P
z = directed distance from xy-plane to P

A three-dimensional coordinate system can have either a **left-handed** or a **right-handed** orientation. To determine the orientation of a system, imagine that you are standing at the origin, with your arms pointing in the directions of the positive x- and y-axes, and with the z-axis pointing up, as shown in Figure 9.2. The system is right-handed or left-handed depending on which hand points along the x-axis. In this text, you will work exclusively with the right-handed system.

Right-handed system   Left-handed system

FIGURE 9.2

SECTION 9.1 The Three-Dimensional Coordinate System 549

Example 1 Plotting Points in Space

Plot each point in space.

a. $(2, -3, 3)$

b. $(-2, 6, 2)$

c. $(1, 4, 0)$

d. $(2, 2, -3)$

SOLUTION To plot the point $(2, -3, 3)$, notice that $x = 2$, $y = -3$, and $z = 3$. To help visualize the point (see Figure 9.3), locate the point $(2, -3)$ in the xy-plane (denoted by a cross). The point $(2, -3, 3)$ lies three units above the cross. The other three points are also shown in the figure.

✓ CHECKPOINT 1

Plot each point on the three-dimensional coordinate system.

a. $(2, 5, 1)$

b. $(-2, -4, 3)$

c. $(4, 0, -5)$

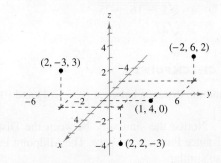

FIGURE 9.3

The Distance and Midpoint Formulas

Many of the formulas established for the two-dimensional coordinate system can be extended to three dimensions. For example, to find the distance between two points in space, you can use the Pythagorean Theorem twice, as shown in Figure 9.4. By doing this, you will obtain the formula for the distance between two points in space.

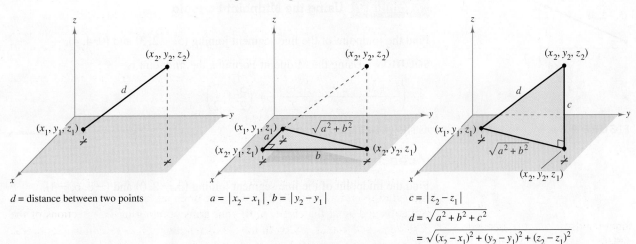

FIGURE 9.4

Distance Formula in Space

The distance between the points (x_1, y_1, z_1) and (x_2, y_2, z_2) is
$$d = \sqrt{(x_2 - x_1)^2 + (y_2 - y_1)^2 + (z_2 - z_1)^2}.$$

Example 2 Finding the Distance Between Two Points

Find the distance between $(1, 0, 2)$ and $(2, 4, -3)$.

SOLUTION

$$\begin{aligned}
d &= \sqrt{(x_2 - x_1)^2 + (y_2 - y_1)^2 + (z_2 - z_1)^2} && \text{Write Distance Formula.} \\
&= \sqrt{(2 - 1)^2 + (4 - 0)^2 + (-3 - 2)^2} && \text{Substitute.} \\
&= \sqrt{1 + 16 + 25} && \text{Simplify.} \\
&= \sqrt{42} && \text{Simplify.}
\end{aligned}$$

✓ CHECKPOINT 2

Find the distance between $(2, 3, -1)$ and $(0, 5, 3)$. ■

Notice the similarity between the Distance Formula in the plane and the Distance Formula in space. The Midpoint Formulas in the plane and in space are also similar.

Midpoint Formula in Space

The midpoint of the line segment joining the points (x_1, y_1, z_1) and (x_2, y_2, z_2) is
$$\text{Midpoint} = \left(\frac{x_1 + x_2}{2}, \frac{y_1 + y_2}{2}, \frac{z_1 + z_2}{2}\right).$$

Example 3 Using the Midpoint Formula

Find the midpoint of the line segment joining $(5, -2, 3)$ and $(0, 4, 4)$.

SOLUTION Using the Midpoint Formula, the midpoint is
$$\left(\frac{5 + 0}{2}, \frac{-2 + 4}{2}, \frac{3 + 4}{2}\right) = \left(\frac{5}{2}, 1, \frac{7}{2}\right)$$

as shown in Figure 9.5.

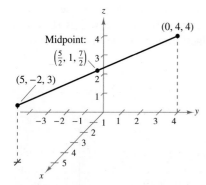

FIGURE 9.5

✓ CHECKPOINT 3

Find the midpoint of the line segment joining $(3, -2, 0)$ and $(-8, 6, -4)$. ■

SECTION 9.1 The Three-Dimensional Coordinate System

The Equation of a Sphere

A **sphere** with center at (h, k, l) and radius r is defined to be the set of all points (x, y, z) such that the distance between (x, y, z) and (h, k, l) is r, as shown in Figure 9.6. Using the Distance Formula, this condition can be written as

$$\sqrt{(x - h)^2 + (y - k)^2 + (z - l)^2} = r.$$

By squaring both sides of this equation, you obtain the standard equation of a sphere.

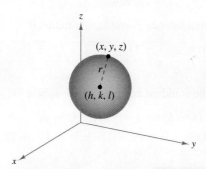

FIGURE 9.6 Sphere: Radius r, Center (h, k, l)

Standard Equation of a Sphere

The **standard equation of a sphere** whose center is (h, k, l) and whose radius is r is

$$(x - h)^2 + (y - k)^2 + (z - l)^2 = r^2.$$

Example 4 Finding the Equation of a Sphere

Find the standard equation of the sphere whose center is $(2, 4, 3)$ and whose radius is 3. Does this sphere intersect the xy-plane?

SOLUTION

$(x - h)^2 + (y - k)^2 + (z - l)^2 = r^2$	Write standard equation.
$(x - 2)^2 + (y - 4)^2 + (z - 3)^2 = 3^2$	Substitute.
$(x - 2)^2 + (y - 4)^2 + (z - 3)^2 = 9$	Simplify.

From the graph shown in Figure 9.7, you can see that the center of the sphere lies three units above the xy-plane. Because the sphere has a radius of 3, you can conclude that it does intersect the xy-plane—at the point $(2, 4, 0)$.

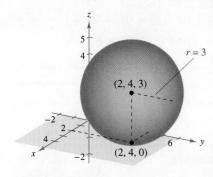

FIGURE 9.7

✓ CHECKPOINT 4

Find the standard equation of the sphere whose center is $(4, 3, 2)$ and whose radius is 5. ■

Example 5 Finding the Equation of a Sphere

Find the equation of the sphere that has the points $(3, -2, 6)$ and $(-1, 4, 2)$ as endpoints of a diameter.

SOLUTION By the Midpoint Formula, the center of the sphere is

$$(h, k, l) = \left(\frac{3 + (-1)}{2}, \frac{-2 + 4}{2}, \frac{6 + 2}{2}\right) \quad \text{Apply Midpoint Formula.}$$
$$= (1, 1, 4). \quad \text{Simplify.}$$

By the Distance Formula, the radius is

$$r = \sqrt{(3 - 1)^2 + (-2 - 1)^2 + (6 - 4)^2}$$
$$= \sqrt{17}. \quad \text{Simplify.}$$

So, the standard equation of the sphere is

$$(x - h)^2 + (y - k)^2 + (z - l)^2 = r^2 \quad \text{Write formula for a sphere.}$$
$$(x - 1)^2 + (y - 1)^2 + (z - 4)^2 = 17. \quad \text{Substitute.}$$

✓ **CHECKPOINT 5**

Find the equation of the sphere that has the points $(-2, 5, 7)$ and $(4, 1, -3)$ as endpoints of a diameter. ■

Example 6 Finding the Center and Radius of a Sphere

Find the center and radius of the sphere whose equation is

$$x^2 + y^2 + z^2 - 2x + 4y - 6z + 8 = 0.$$

SOLUTION You can obtain the standard equation of the sphere by completing the square. To do this, begin by grouping terms with the same variable. Then add "the square of half the coefficient of each linear term" to each side of the equation. For instance, to complete the square of $(x^2 - 2x)$, add $\left[\frac{1}{2}(-2)\right]^2 = 1$ to each side.

$$x^2 + y^2 + z^2 - 2x + 4y - 6z + 8 = 0$$
$$(x^2 - 2x + \quad) + (y^2 + 4y + \quad) + (z^2 - 6z + \quad) = -8$$
$$(x^2 - 2x + 1) + (y^2 + 4y + 4) + (z^2 - 6z + 9) = -8 + 1 + 4 + 9$$
$$(x - 1)^2 + (y + 2)^2 + (z - 3)^2 = 6$$

So, the center of the sphere is $(1, -2, 3)$, and its radius is $\sqrt{6}$, as shown in Figure 9.8.

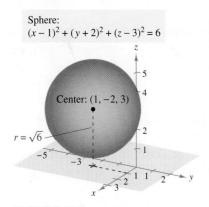

FIGURE 9.8

✓ **CHECKPOINT 6**

Find the center and radius of the sphere whose equation is

$$x^2 + y^2 + z^2 + 6x - 8y + 2z - 10 = 0. \quad ■$$

Note in Example 6 that the points satisfying the equation of the sphere are "surface points," not "interior points." In general, the collection of points satisfying an equation involving x, y, and z is called a **surface in space**.

Traces of Surfaces

Finding the intersection of a surface with one of the three coordinate planes (or with a plane parallel to one of the three coordinate planes) helps visualize the surface. Such an intersection is called a **trace** of the surface. For example, the xy-trace of a surface consists of all points that are common to both the surface *and* the xy-plane. Similarly, the xz-trace of a surface consists of all points that are common to both the surface and the xz-plane.

Example 7 Finding a Trace of a Surface

Sketch the xy-trace of the sphere whose equation is

$$(x - 3)^2 + (y - 2)^2 + (z + 4)^2 = 5^2.$$

SOLUTION To find the xy-trace of this surface, use the fact that every point in the xy-plane has a z-coordinate of zero. This means that if you substitute $z = 0$ into the original equation, the resulting equation will represent the intersection of the surface with the xy-plane.

$(x - 3)^2 + (y - 2)^2 + (z + 4)^2 = 5^2$ Write original equation.
$(x - 3)^2 + (y - 2)^2 + (0 + 4)^2 = 25$ Let $z = 0$ to find xy-trace.
$(x - 3)^2 + (y - 2)^2 + 16 = 25$
$(x - 3)^2 + (y - 2)^2 = 9$
$(x - 3)^2 + (y - 2)^2 = 3^2$ Equation of circle

From this equation, you can see that the xy-trace is a circle of radius 3, as shown in Figure 9.9.

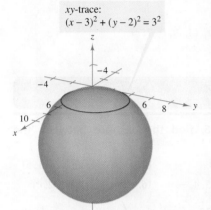

xy-trace:
$(x - 3)^2 + (y - 2)^2 = 3^2$

Sphere:
$(x - 3)^2 + (y - 2)^2 + (z + 4)^2 = 5^2$

FIGURE 9.9

✓ CHECKPOINT 7

Find the equation of the xy-trace of the sphere whose equation is

$$(x + 1)^2 + (y - 2)^2 + (z + 3)^2 = 5^2.$$

CONCEPT CHECK

1. Name the three coordinate planes of a three-dimensional coordinate system formed by passing a z-axis perpendicular to both the x- and y-axes at the origin.
2. A point in the three-dimensional coordinate system has coordinates (x_1, y_1, z_1). Describe what each coordinate measures.
3. Give the formula for the distance between the points (x_1, y_1, z_1) and (x_2, y_2, z_2).
4. Give the standard equation of a sphere of radius r centered at (h, k, l).

554 CHAPTER 9 Functions of Several Variables

Skills Review 9.1

The following warm-up exercises involve skills that were covered in earlier sections. You will use these skills in the exercise set for this section. For additional help, review Sections 1.1 and 1.2.

In Exercises 1–4, find the distance between the points.

1. $(5, 1), (3, 5)$ **2.** $(2, 3), (-1, -1)$ **3.** $(-5, 4), (-5, -4)$ **4.** $(-3, 6), (-3, -2)$

In Exercises 5–8, find the midpoint of the line segment connecting the points.

5. $(2, 5), (6, 9)$ **6.** $(-1, -2), (3, 2)$ **7.** $(-6, 0), (6, 6)$ **8.** $(-4, 3), (2, -1)$

In Exercises 9 and 10, write the standard form of the equation of the circle.

9. Center: $(2, 3)$; radius: 2

10. Endpoints of a diameter: $(4, 0), (-2, 8)$

Exercises 9.1

See www.CalcChat.com for worked-out solutions to odd-numbered exercises.

In Exercises 1–4, plot the points on the same three-dimensional coordinate system.

1. (a) $(2, 1, 3)$
 (b) $(-1, 2, 1)$

2. (a) $(3, -2, 5)$
 (b) $\left(\frac{3}{2}, 4, -2\right)$

3. (a) $(5, -2, 2)$
 (b) $(5, -2, -2)$

4. (a) $(0, 4, -5)$
 (b) $(4, 0, 5)$

In Exercises 5 and 6, approximate the coordinates of the points.

5.
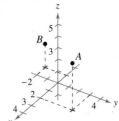

6.
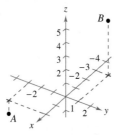

In Exercises 7–10, find the coordinates of the point.

7. The point is located three units behind the yz-plane, four units to the right of the xz-plane, and five units above the xy-plane.

8. The point is located seven units in front of the yz-plane, two units to the left of the xz-plane, and one unit below the xy-plane.

9. The point is located on the x-axis, 10 units in front of the yz-plane.

10. The point is located in the yz-plane, three units to the right of the xz-plane, and two units above the xy-plane.

11. Think About It What is the z-coordinate of any point in the xy-plane?

12. Think About It What is the x-coordinate of any point in the yz-plane?

In Exercises 13–16, find the distance between the two points.

13. $(4, 1, 5), (8, 2, 6)$ **14.** $(-4, -1, 1), (2, -1, 5)$

15. $(-1, -5, 7), (-3, 4, -4)$ **16.** $(8, -2, 2), (8, -2, 4)$

In Exercises 17–20, find the coordinates of the midpoint of the line segment joining the two points.

17. $(6, -9, 1), (-2, -1, 5)$ **18.** $(4, 0, -6), (8, 8, 20)$

19. $(-5, -2, 5), (6, 3, -7)$ **20.** $(0, -2, 5), (4, 2, 7)$

In Exercises 21–24, find (x, y, z).

21.
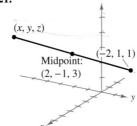

22.
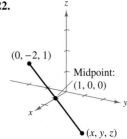

23.

24.
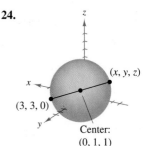

In Exercises 25–28, find the lengths of the sides of the triangle with the given vertices, and determine whether the triangle is a right triangle, an isosceles triangle, or neither of these.

25. $(0, 0, 0), (2, 2, 1), (2, -4, 4)$
26. $(5, 3, 4), (7, 1, 3), (3, 5, 3)$
27. $(-2, 2, 4), (-2, 2, 6), (-2, 4, 8)$
28. $(5, 0, 0), (0, 2, 0), (0, 0, -3)$

29. **Think About It** The triangle in Exercise 25 is translated five units upward along the z-axis. Determine the coordinates of the translated triangle.

30. **Think About It** The triangle in Exercise 26 is translated three units to the right along the y-axis. Determine the coordinates of the translated triangle.

In Exercises 31–40, find the standard equation of the sphere.

31.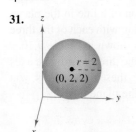
32.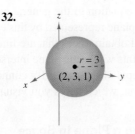
33.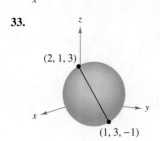
34.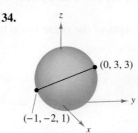

35. Center: $(1, 1, 5)$; radius: 3
36. Center: $(4, -1, 1)$; radius: 5
37. Endpoints of a diameter: $(2, 0, 0), (0, 6, 0)$
38. Endpoints of a diameter: $(1, 0, 0), (0, 5, 0)$
39. Center: $(-2, 1, 1)$; tangent to the xy-plane
40. Center: $(1, 2, 0)$; tangent to the yz-plane

In Exercises 41–46, find the sphere's center and radius.

41. $x^2 + y^2 + z^2 - 5x = 0$
42. $x^2 + y^2 + z^2 - 8y = 0$
43. $x^2 + y^2 + z^2 - 2x + 6y + 8z + 1 = 0$
44. $x^2 + y^2 + z^2 - 4y + 6z + 4 = 0$
45. $2x^2 + 2y^2 + 2z^2 - 4x - 12y - 8z + 3 = 0$
46. $4x^2 + 4y^2 + 4z^2 - 8x + 16y + 11 = 0$

In Exercises 47–50, sketch the xy-trace of the sphere.

47. $(x - 1)^2 + (y - 3)^2 + (z - 2)^2 = 25$
48. $(x + 1)^2 + (y + 2)^2 + (z - 2)^2 = 16$
49. $x^2 + y^2 + z^2 - 6x - 10y + 6z + 30 = 0$
50. $x^2 + y^2 + z^2 - 4y + 2z - 60 = 0$

In Exercises 51–54, sketch the yz-trace of the sphere.

51. $x^2 + (y + 3)^2 + z^2 = 25$
52. $(x + 2)^2 + (y - 3)^2 + z^2 = 9$
53. $x^2 + y^2 + z^2 - 4x - 4y - 6z - 12 = 0$
54. $x^2 + y^2 + z^2 - 6x - 10y + 6z + 30 = 0$

In Exercises 55–58, sketch the trace of the intersection of each plane with the given sphere.

55. $x^2 + y^2 + z^2 = 25$
   (a) $z = 3$ (b) $x = 4$
56. $x^2 + y^2 + z^2 = 169$
   (a) $x = 5$ (b) $y = 12$
57. $x^2 + y^2 + z^2 - 4x - 6y + 9 = 0$
   (a) $x = 2$ (b) $y = 3$
58. $x^2 + y^2 + z^2 - 8x - 6z + 16 = 0$
   (a) $x = 4$ (b) $z = 3$

59. **Geology** Crystals are classified according to their symmetry. Crystals shaped like cubes are classified as isometric. The vertices of an isometric crystal mapped onto a three-dimensional coordinate system are shown in the figure. Determine (x, y, z).

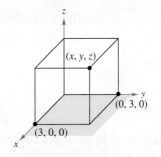
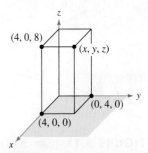

Figure for 59   Figure for 60

60. **Geology** Crystals are classified according to their symmetry. Crystals shaped like rectangular prisms are classified as tetragonal. The vertices of a tetragonal crystal mapped onto a three-dimensional coordinate system are shown in the figure. Determine (x, y, z).

61. **Architecture** A spherical building has a diameter of 165 feet. The center of the building is placed at the origin of a three-dimensional coordinate system. What is the equation of the sphere?

Section 9.2
Surfaces in Space

- Sketch planes in space.
- Draw planes in space with different numbers of intercepts.
- Classify quadric surfaces in space.

Equations of Planes in Space

In Section 9.1, you studied one type of surface in space—a sphere. In this section, you will study a second type—a plane in space. The **general equation of a plane** in space is

$$ax + by + cz = d. \qquad \text{General equation of a plane}$$

Note the similarity of this equation to the general equation of a line in the plane. In fact, if you intersect the plane represented by this equation with each of the three coordinate planes, you will obtain traces that are lines, as shown in Figure 9.10.

In Figure 9.10, the points where the plane intersects the three coordinate axes are the x-, y-, and z-intercepts of the plane. By connecting these three points, you can form a triangular region, which helps you visualize the plane in space.

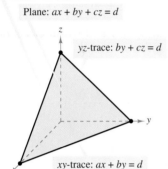

FIGURE 9.10

Example 1 Sketching a Plane in Space

Find the x-, y-, and z-intercepts of the plane given by

$$3x + 2y + 4z = 12.$$

Then sketch the plane.

SOLUTION To find the x-intercept, let both y and z be zero.

$$3x + 2(0) + 4(0) = 12 \qquad \text{Substitute 0 for } y \text{ and } z.$$
$$3x = 12 \qquad \text{Simplify.}$$
$$x = 4 \qquad \text{Solve for } x.$$

So, the x-intercept is $(4, 0, 0)$. To find the y-intercept, let x and z be zero and conclude that $y = 6$. So, the y-intercept is $(0, 6, 0)$. Similarly, by letting x and y be zero, you can determine that $z = 3$ and that the z-intercept is $(0, 0, 3)$. Figure 9.11 shows the triangular portion of the plane formed by connecting the three intercepts.

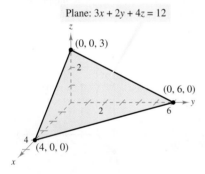

FIGURE 9.11 Sketch Made by Connecting Intercepts: $(4, 0, 0), (0, 6, 0), (0, 0, 3)$

✓ CHECKPOINT 1

Find the x-, y-, and z-intercepts of the plane given by

$$2x + 4y + z = 8.$$

Then sketch the plane. ■

Drawing Planes in Space

The planes shown in Figures 9.10 and 9.11 have three intercepts. When this occurs, we suggest that you draw the plane by sketching the triangular region formed by connecting the three intercepts.

It is possible for a plane in space to have fewer than three intercepts. This occurs when one or more of the coefficients in the equation $ax + by + cz = d$ is zero. Figure 9.12 shows some planes in space that have only one intercept, and Figure 9.13 shows some that have only two intercepts. In each figure, note the use of dashed lines and shading to give the illusion of three dimensions.

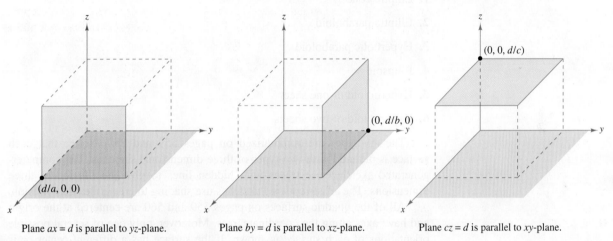

Plane $ax = d$ is parallel to yz-plane.   Plane $by = d$ is parallel to xz-plane.   Plane $cz = d$ is parallel to xy-plane.

FIGURE 9.12 Planes Parallel to Coordinate Planes

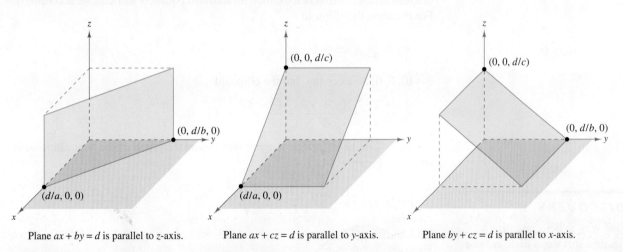

Plane $ax + by = d$ is parallel to z-axis.   Plane $ax + cz = d$ is parallel to y-axis.   Plane $by + cz = d$ is parallel to x-axis.

FIGURE 9.13 Planes Parallel to Coordinate Axes

DISCOVERY

What is the equation of each plane?

a. xy-plane   **b.** xz-plane   **c.** yz-plane

Quadric Surfaces

A third common type of surface in space is a **quadric surface.** Every quadric surface has an equation of the form

$$Ax^2 + By^2 + Cz^2 + Dx + Ey + Fz + G = 0. \qquad \text{Second-degree equation}$$

There are six basic types of quadric surfaces.

1. Elliptic cone
2. Elliptic paraboloid
3. Hyperbolic paraboloid
4. Ellipsoid
5. Hyperboloid of one sheet
6. Hyperboloid of two sheets

The six types are summarized on pages 559 and 560. Notice that each surface is pictured with two types of three-dimensional sketches. The computer-generated sketches use traces with hidden lines to give the illusion of three dimensions. The artist-rendered sketches use shading to create the same illusion.

All of the quadric surfaces on pages 559 and 560 are centered at the origin and have axes along the coordinate axes. Moreover, only one of several possible orientations of each surface is shown. If the surface has a different center or is oriented along a different axis, then its standard equation will change accordingly. For instance, the ellipsoid

$$\frac{x^2}{1^2} + \frac{y^2}{3^2} + \frac{z^2}{2^2} = 1$$

has $(0, 0, 0)$ as its center, but the ellipsoid

$$\frac{(x-2)^2}{1^2} + \frac{(y+1)^2}{3^2} + \frac{(z-4)^2}{2^2} = 1$$

has $(2, -1, 4)$ as its center. A computer-generated graph of the first ellipsoid is shown in Figure 9.14.

DISCOVERY

One way to help visualize a quadric surface is to determine the intercepts of the surface with the coordinate axes. What are the intercepts of the ellipsoid in Figure 9.14?

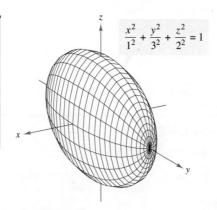

FIGURE 9.14

SECTION 9.2 Surfaces in Space 559

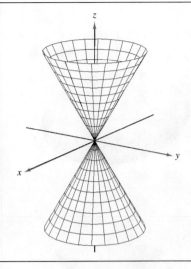

Elliptic Cone

$$\frac{x^2}{a^2} + \frac{y^2}{b^2} - \frac{z^2}{c^2} = 0$$

Trace	Plane
Ellipse	Parallel to xy-plane
Hyperbola	Parallel to xz-plane
Hyperbola	Parallel to yz-plane

The axis of the cone corresponds to the variable whose coefficient is negative. The traces in the coordinate planes parallel to this axis are intersecting lines.

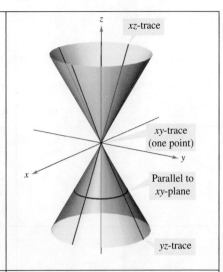

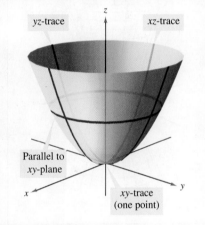

Elliptic Paraboloid

$$z = \frac{x^2}{a^2} + \frac{y^2}{b^2}$$

Trace	Plane
Ellipse	Parallel to xy-plane
Parabola	Parallel to xz-plane
Parabola	Parallel to yz-plane

The axis of the paraboloid corresponds to the variable raised to the first power.

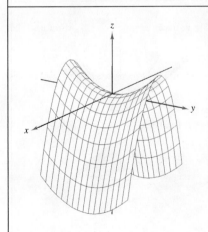

Hyperbolic Paraboloid

$$z = \frac{y^2}{b^2} - \frac{x^2}{a^2}$$

Trace	Plane
Hyperbola	Parallel to xy-plane
Parabola	Parallel to xz-plane
Parabola	Parallel to yz-plane

The axis of the paraboloid corresponds to the variable raised to the first power.

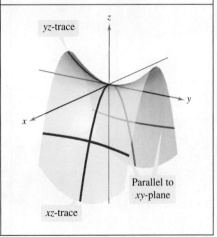

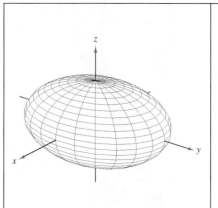

Ellipsoid

$$\frac{x^2}{a^2} + \frac{y^2}{b^2} + \frac{z^2}{c^2} = 1$$

Trace	Plane
Ellipse	Parallel to xy-plane
Ellipse	Parallel to xz-plane
Ellipse	Parallel to yz-plane

The surface is a sphere if the coefficients a, b, and c are equal and nonzero.

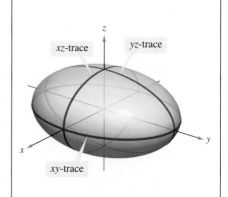

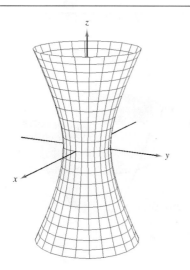

Hyperboloid of One Sheet

$$\frac{x^2}{a^2} + \frac{y^2}{b^2} - \frac{z^2}{c^2} = 1$$

Trace	Plane
Ellipse	Parallel to xy-plane
Hyperbola	Parallel to xz-plane
Hyperbola	Parallel to yz-plane

The axis of the hyperboloid corresponds to the variable whose coefficient is negative.

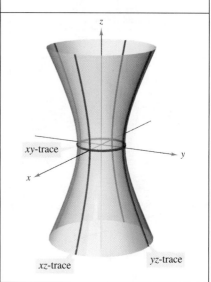

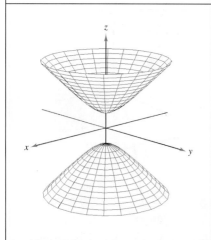

Hyperboloid of Two Sheets

$$\frac{z^2}{c^2} - \frac{x^2}{a^2} - \frac{y^2}{b^2} = 1$$

Trace	Plane
Ellipse	Parallel to xy-plane
Hyperbola	Parallel to xz-plane
Hyperbola	Parallel to yz-plane

The axis of the hyperboloid corresponds to the variable whose coefficient is positive. There is no trace in the coordinate plane perpendicular to this axis.

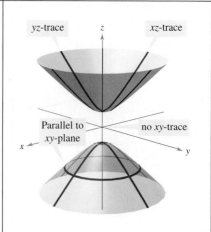

SECTION 9.2 Surfaces in Space

When classifying quadric surfaces, note that the two types of paraboloids have one variable raised to the first power. The other four types of quadric surfaces have equations that are of second degree in *all* three variables.

Example 2 Classifying a Quadric Surface

Classify the surface given by $x - y^2 - z^2 = 0$. Describe the traces of the surface in the xy-plane, the xz-plane, and the plane given by $x = 1$.

SOLUTION Because x is raised only to the first power, the surface is a paraboloid whose axis is the x-axis, as shown in Figure 9.15. In standard form, the equation is

$$x = y^2 + z^2.$$

The traces in the xy-plane, the xz-plane, and the plane given by $x = 1$ are as shown.

Trace in xy-plane ($z = 0$):	$x = y^2$	Parabola
Trace in xz-plane ($y = 0$):	$x = z^2$	Parabola
Trace in plane $x = 1$:	$y^2 + z^2 = 1$	Circle

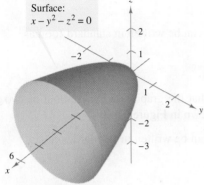

Surface: $x - y^2 - z^2 = 0$

FIGURE 9.15 Elliptic Paraboloid

These three traces are shown in Figure 9.16. From the traces, you can see that the surface is an elliptic (or circular) paraboloid. If you have access to a three-dimensional graphing utility, try using it to graph this surface. If you do this, you will discover that sketching surfaces in space is not a simple task—even with a graphing utility.

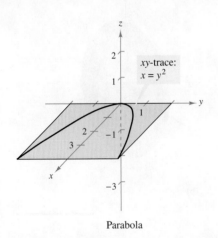
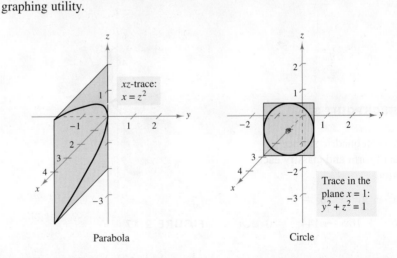

FIGURE 9.16

✓CHECKPOINT 2

Classify the surface given by $x^2 + y^2 - z^2 = 1$. Describe the traces of the surface in the xy-plane, the yz-plane, the xz-plane, and the plane given by $z = 3$. ∎

Example 3 Classifying Quadric Surfaces

Classify the surface given by each equation.

a. $x^2 - 4y^2 - 4z^2 - 4 = 0$

b. $x^2 + 4y^2 + z^2 - 4 = 0$

SOLUTION

a. The equation $x^2 - 4y^2 - 4z^2 - 4 = 0$ can be written in standard form as

$$\frac{x^2}{4} - y^2 - z^2 = 1. \quad \text{Standard form}$$

From the standard form, you can see that the graph is a hyperboloid of two sheets, with the x-axis as its axis, as shown in Figure 9.17(a).

b. The equation $x^2 + 4y^2 + z^2 - 4 = 0$ can be written in standard form as

$$\frac{x^2}{4} + y^2 + \frac{z^2}{4} = 1. \quad \text{Standard form}$$

From the standard form, you can see that the graph is an ellipsoid, as shown in Figure 9.17(b).

✓ CHECKPOINT 3

Write each quadric surface in standard form and classify each equation.

a. $4x^2 + 9y^2 - 36z = 0$

b. $36x^2 + 16y^2 - 144z^2 = 0$

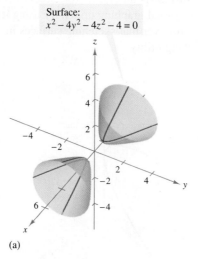

Surface:
$x^2 - 4y^2 - 4z^2 - 4 = 0$
(a)

Surface:
$x^2 + 4y^2 + z^2 - 4 = 0$
(b)

FIGURE 9.17

CONCEPT CHECK

1. Give the general equation of a plane in space.
2. List the six basic types of quadric surfaces.
3. Which types of quadric surfaces have equations that are of second degree in *all* three variables? Which types of quadric surfaces have equations that have one variable raised to the first power?
4. Is it possible for a plane in space to have fewer than three intercepts? If so, when does this occur?

Skills Review 9.2

The following warm-up exercises involve skills that were covered in earlier sections. You will use these skills in the exercise set for this section. For additional help, review Sections 1.2 and 9.1.

In Exercises 1–4, find the x- and y-intercepts of the function.

1. $3x + 4y = 12$
2. $6x + y = -8$
3. $-2x + y = -2$
4. $-x - y = 5$

In Exercises 5–8, rewrite the expression by completing the square.

5. $x^2 + y^2 + z^2 - 2x - 4y - 6z + 15 = 0$
6. $x^2 + y^2 - z^2 - 8x + 4y - 6z + 11 = 0$
7. $z - 2 = x^2 + y^2 + 2x - 2y$
8. $x^2 + y^2 + z^2 - 6x + 10y + 26z = -202$

In Exercises 9 and 10, write the equation of the sphere in standard form.

9. $16x^2 + 16y^2 + 16z^2 = 4$
10. $9x^2 + 9y^2 + 9z^2 = 36$

Exercises 9.2

See www.CalcChat.com for worked-out solutions to odd-numbered exercises.

In Exercises 1–12, find the intercepts and sketch the graph of the plane.

1. $4x + 2y + 6z = 12$
2. $3x + 6y + 2z = 6$
3. $3x + 3y + 5z = 15$
4. $x + y + z = 3$
5. $2x - y + 3z = 4$
6. $2x - y + z = 4$
7. $z = 8$
8. $x = 5$
9. $y + z = 5$
10. $x + 2y = 4$
11. $x + y - z = 0$
12. $x - 3z = 3$

In Exercises 13–20, find the distance between the point and the plane (see figure). The distance D between a point (x_0, y_0, z_0) and the plane $ax + by + cz + d = 0$ is

$$D = \frac{|ax_0 + by_0 + cz_0 + d|}{\sqrt{a^2 + b^2 + c^2}}$$

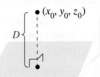

13. $(0, 0, 0)$, $2x + 3y + z = 12$
14. $(0, 0, 0)$, $8x - 4y + z = 8$
15. $(1, 5, -4)$, $3x - y + 2z = 6$
16. $(3, 2, 1)$, $x - y + 2z = 4$
17. $(1, 0, -1)$, $2x - 4y + 3z = 12$
18. $(2, -1, 0)$, $3x + 3y + 2z = 6$
19. $(3, 2, -1)$, $2x - 3y + 4z = 24$
20. $(-2, 1, 0)$, $2x + 5y - z = 20$

In Exercises 21–30, determine whether the planes $a_1x + b_1y + c_1z = d_1$ and $a_2x + b_2y + c_2z = d_2$ are parallel, perpendicular, or neither. The planes are parallel if there exists a nonzero constant k such that $a_1 = ka_2$, $b_1 = kb_2$, and $c_1 = kc_2$, and are perpendicular if $a_1a_2 + b_1b_2 + c_1c_2 = 0$.

21. $5x - 3y + z = 4$, $x + 4y + 7z = 1$
22. $3x + y - 4z = 3$, $-9x - 3y + 12z = 4$
23. $x - 5y - z = 1$, $5x - 25y - 5z = -3$
24. $x + 3y + 2z = 6$, $4x - 12y + 8z = 24$
25. $x + 2y = 3$, $4x + 8y = 5$
26. $x + 3y + z = 7$, $x - 5z = 0$
27. $2x + y = 3$, $3x - 5z = 0$
28. $2x - z = 1$, $4x + y + 8z = 10$
29. $x = 6$, $y = -1$
30. $x = -2$, $y = 4$

In Exercises 31–34, describe the traces of the surface in the given planes.

Surface	Planes
31. $x^2 - y - z^2 = 0$	xy-plane, $y = 1$, yz-plane
32. $y = x^2 + z^2$	xy-plane, $y = 1$, yz-plane
33. $\dfrac{x^2}{4} + y^2 + z^2 = 1$	xy-plane, xz-plane, yz-plane
34. $y^2 + z^2 - x^2 = 1$	xy-plane, xz-plane, yz-plane

In Exercises 35–40, match the equation with its graph. [The graphs are labeled (a)–(f).]

(a)
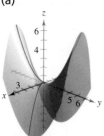

(b)
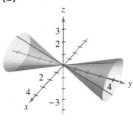

(c)
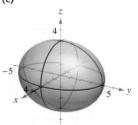

(d)
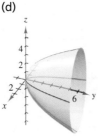

(e)
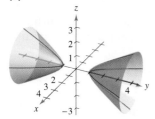

(f)
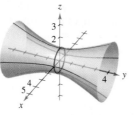

35. $\dfrac{x^2}{9} + \dfrac{y^2}{16} + \dfrac{z^2}{9} = 1$

36. $15x^2 - 4y^2 + 15z^2 = -4$

37. $4x^2 - y^2 + 4z^2 = 4$

38. $y^2 = 4x^2 + 9z^2$

39. $4x^2 - 4y + z^2 = 0$

40. $4x^2 - y^2 + 4z = 0$

In Exercises 41–52, identify the quadric surface.

41. $x^2 + \dfrac{y^2}{4} + z^2 = 1$

42. $\dfrac{x^2}{9} + \dfrac{y^2}{16} + \dfrac{z^2}{16} = 1$

43. $25x^2 + 25y^2 - z^2 = 5$

44. $9x^2 + 4y^2 - 8z^2 = 72$

45. $x^2 - y + z^2 = 0$

46. $z = 4x^2 + y^2$

47. $x^2 - y^2 + z = 0$

48. $z^2 - x^2 - \dfrac{y^2}{4} = 1$

49. $2x^2 - y^2 + 2z^2 = -4$

50. $4y = x^2 + z^2$

51. $z^2 = 9x^2 + y^2$

52. $z^2 = 2x^2 + 2y^2$

Think About It In Exercises 53–56, each figure is a graph of the quadric surface $z = x^2 + y^2$. Match each of the four graphs with the point in space from which the paraboloid is viewed. The four points are (0, 0, 20), (0, 20, 0), (20, 0, 0), and (10, 10, 20).

53.
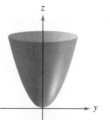

54.

55.
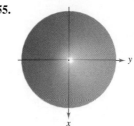

56.
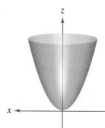

57. Modeling Data Per capita consumptions (in gallons) of different types of plain milk in the United States from 1999 through 2004 are shown in the table. Consumptions of reduced-fat (1%) and skim milks, reduced-fat milk (2%), and whole milk are represented by the variables x, y, and z, respectively. *(Source: U.S. Department of Agriculture)*

Year	1999	2000	2001	2002	2003	2004
x	6.2	6.1	5.9	5.8	5.6	5.5
y	7.3	7.1	7.0	7.0	6.9	6.9
z	7.8	7.7	7.4	7.3	7.2	6.9

A model for the data in the table is given by $-1.25x + 0.125y + z = 0.95$.

(a) Complete a fourth row of the table using the model to approximate z for the given values of x and y. Compare the approximations with the actual values of z.

(b) According to this model, increases in consumption of milk types y and z would correspond to what kind of change in consumption of milk type x?

58. Physical Science Because of the forces caused by its rotation, Earth is actually an oblate ellipsoid rather than a sphere. The equatorial radius is 3963 miles and the polar radius is 3950 miles (see figure). Find an equation of the ellipsoid. Assume that the center of Earth is at the origin and the xy-trace ($z = 0$) corresponds to the equator.

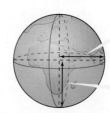

Equatorial radius = 3963 mi

Polar radius = 3950 mi

Section 9.3
Functions of Several Variables

- Evaluate functions of several variables.
- Find the domains and ranges of functions of several variables.
- Read contour maps and sketch level curves of functions of two variables.
- Use functions of several variables to answer questions about real-life situations.

Functions of Several Variables

So far in this text, you have studied functions of a single independent variable. Many quantities in science, business, and technology, however, are functions not of one, but of two or more variables. For instance, the wind chill (the temperature it "feels like" outside during winter weather) is a function of two variables, air temperature *and* wind speed (see Example 5). The notation for a function of two or more variables is similar to that for a function of a single variable. Here are two examples.

$$z = f(\underbrace{x, y}_{\text{2 variables}}) = x^2 + xy \qquad \text{Function of two variables}$$

and

$$w = f(\underbrace{x, y, z}_{\text{3 variables}}) = x + 2y - 3z \qquad \text{Function of three variables}$$

Definition of a Function of Two Variables

Let D be a set of ordered pairs of real numbers. If to each ordered pair (x, y) in D there corresponds a unique real number $f(x, y)$, then f is called a **function of x and y**. The set D is the **domain** of f, and the corresponding set of z-values is the **range** of f. Functions of three, four, or more variables are defined similarly.

Example 1 Evaluating Functions of Several Variables

a. For $f(x, y) = 2x^2 - y^2$, you can evaluate $f(2, 3)$ as shown.

$$f(2, 3) = 2(2)^2 - (3)^2$$
$$= 8 - 9$$
$$= -1$$

b. For $f(x, y, z) = e^x(y + z)$, you can evaluate $f(0, -1, 4)$ as shown.

$$f(0, -1, 4) = e^0(-1 + 4)$$
$$= (1)(3)$$
$$= 3$$

✓ CHECKPOINT 1

Find the function values of $f(x, y)$.

a. For $f(x, y) = x^2 + 2xy$, find $f(2, -1)$.

b. For $f(x, y, z) = \dfrac{2x^2 z}{y^3}$, find $f(-3, 2, 1)$. ■

The Graph of a Function of Two Variables

A function of two variables can be represented graphically as a surface in space by letting $z = f(x, y)$. When sketching the graph of a function of x and y, remember that even though the graph is three-dimensional, the domain of the function is two-dimensional—it consists of the points in the xy-plane for which the function is defined. As with functions of a single variable, unless specifically restricted, the domain of a function of two variables is assumed to be the set of all points (x, y) for which the defining equation has meaning. In other words, to each point (x, y) in the domain of f there corresponds a point (x, y, z) on the surface, and conversely, to each point (x, y, z) on the surface there corresponds a point (x, y) in the domain of f.

Example 2 Finding the Domain and Range of a Function

Find the domain and range of the function

$$f(x, y) = \sqrt{64 - x^2 - y^2}.$$

SOLUTION Because no restrictions are given, the domain is assumed to be the set of all points for which the defining equation makes sense.

$64 - x^2 - y^2 \geq 0$ Quantity inside radical must be nonnegative.

$x^2 + y^2 \leq 64$ Domain of the function

So, the domain is the set of all points that lie on or inside the circle given by $x^2 + y^2 = 8^2$. The range of f is the set

$0 \leq z \leq 8.$ Range of the function

As shown in Figure 9.18, the graph of the function is a hemisphere.

Hemisphere:
$f(x, y) = \sqrt{64 - x^2 - y^2}$

Domain: $x^2 + y^2 \leq 64$
Range: $0 \leq z \leq 8$

FIGURE 9.18

✓**CHECKPOINT 2**

Find the domain and range of the function

$f(x, y) = \sqrt{9 - x^2 - y^2}.$ ■

TECHNOLOGY

Some three-dimensional graphing utilities can graph equations in x, y, and z. Others are programmed to graph only functions of x and y. A surface in space represents the graph of a function of x and y only if each vertical line intersects the surface at most once. For instance, the surface shown in Figure 9.18 passes this vertical line test, but the surface at the right (drawn by *Mathematica*) does not represent the graph of a function of x and y.

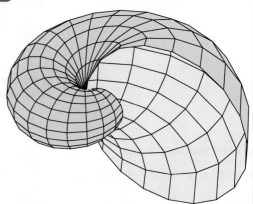

Some vertical lines intersect this surface more than once. So, the surface does not pass the Vertical Line Test and is not a function of x and y.

Contour Maps and Level Curves

A **contour map** of a surface is created by *projecting* traces, taken in evenly spaced planes that are parallel to the *xy*-plane, onto the *xy*-plane. Each projection is a **level curve** of the surface.

Contour maps are used to create weather maps, topographical maps, and population density maps. For instance, Figure 9.19(a) shows a graph of a "mountain and valley" surface given by $z = f(x, y)$. Each of the level curves in Figure 9.19(b) represents the intersection of the surface $z = f(x, y)$ with a plane $z = c$, where $c = 828, 830, \ldots, 854$.

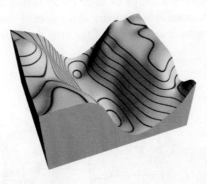

(a) Surface (b) Contour map

FIGURE 9.19

Example 3 Reading a Contour Map

The "contour map" in Figure 9.20 was computer generated using data collected by satellite instrumentation. Color is used to show the "ozone hole" in Earth's atmosphere. The purple and blue areas represent the lowest levels of ozone and the green areas represent the highest level. Describe the areas that have the lowest levels of ozone. *(Source: National Aeronautics and Space Administration)*

SOLUTION The lowest levels of ozone are over Antarctica and the Antarctic Ocean. The ozone layer acts to protect life on Earth by blocking harmful ultraviolet rays from the sun. The "ozone hole" in the polar region of the Southern Hemisphere is an area in which there is a severe depletion of the ozone levels in the atmosphere. It is primarily caused by compounds that release chlorine and bromine gases into the atmosphere.

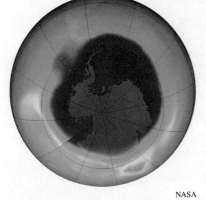

FIGURE 9.20

✓ CHECKPOINT 3

When the level curves of a contour map are close together, is the surface represented by the contour map steep or nearly level? When the level curves of a contour map are far apart, is the surface represented by the contour map steep or nearly level? ■

Example 4 Reading a Contour Map

The contour map shown in Figure 9.21 represents the economy of the United States. Discuss the use of color to represent the level curves. *(Source: U.S. Census Bureau)*

SOLUTION You can see from the key that the light yellow regions are mainly used in crop production. The gray areas represent regions that are unproductive. Manufacturing centers are denoted by large red dots and mineral deposits are denoted by small black dots.

One advantage of such a map is that it allows you to "see" the components of the country's economy at a glance. From the map it is clear that the Midwest is responsible for most of the crop production in the United States.

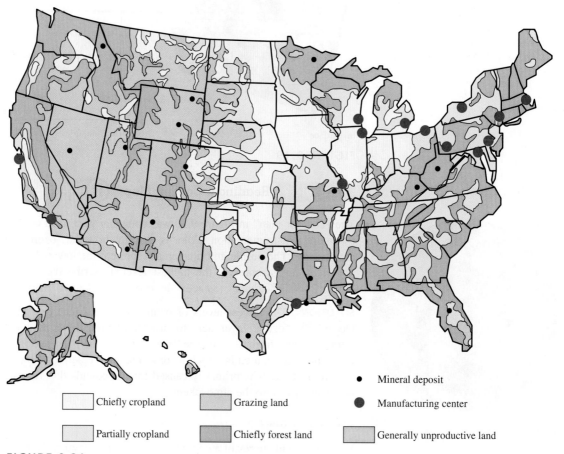

FIGURE 9.21

✓ CHECKPOINT 4

Use Figure 9.21 to describe how Alaska contributes to the U.S. economy. Does Alaska contain any manufacturing centers? Does Alaska contain any mineral deposits? ■

Application

Example 5 Finding Wind Chill

The wind chill C (in degrees Fahrenheit) is given by

$$C = f(T, v) = 35.74 + 0.6215T - 35.75v^{0.16} + 0.4275Tv^{0.16}$$

where v is the wind speed (in miles per hour) and T is the temperature (in degrees Fahrenheit).

a. Find the wind chill when the temperature is 0°F and the wind speed is 10 miles per hour.

b. Find the wind chill when the temperature is 5°F and the wind speed is 30 miles per hour.

SOLUTION

a. When the temperature is 0°F and the wind speed is 10 miles per hour, the wind chill is

$$C = f(0, 10)$$
$$= 35.74 + 0.6215(0) - 35.75(10)^{0.16} + 0.4275(0)(10)^{0.16}$$
$$\approx -16°F.$$

b. When the temperature is 5°F and the wind speed is 30 miles per hour, the wind chill is

$$C = f(5, 30)$$
$$= 35.74 + 0.6215(5) - 35.75(30)^{0.16} + 0.4275(5)(30)^{0.16}$$
$$\approx -19°F.$$

In 2001, the U.S. National Weather Service implemented the current wind chill formula. The new formula is based on heat loss from exposed skin and was tested on human volunteers. When the temperature is 5° and the wind speed is 30 miles per hour, exposed human skin can freeze in 30 minutes. The previous formula underestimated the time required for human flesh to freeze.

© Michael Dwyer/Alamy

✓ CHECKPOINT 5

Use the wind chill formula in Example 5 to answer the following.

a. Find the wind chill when the temperature is 0°F and the wind speed is 20 miles per hour.

b. Find the wind chill when the temperature is 10°F and the wind speed is 5 miles per hour. ∎

CONCEPT CHECK

1. The function $f(x, y) = x + y$ is a function of how many variables?
2. What is a graph of a function of two variables?
3. Give a description of the domain of a function of two variables.
4. How is a contour map created? What is a level curve?

CHAPTER 9 Functions of Several Variables

Skills Review 9.3

The following warm-up exercises involve skills that were covered in earlier sections. You will use these skills in the exercise set for this section. For additional help, review Sections 0.3 and 1.4.

In Exercises 1–4, evaluate the function when $x = -3$.

1. $f(x) = 5 - 2x$ **2.** $f(x) = -x^2 + 4x + 5$ **3.** $y = \sqrt{4x^2 - 3x + 4}$ **4.** $y = \sqrt[3]{34 - 4x + 2x^2}$

In Exercises 5–8, find the domain of the function.

5. $f(x) = 5x^2 + 3x - 2$ **6.** $g(x) = \dfrac{1}{2x} - \dfrac{2}{x+3}$ **7.** $h(y) = \sqrt{y - 5}$ **8.** $f(y) = \sqrt{y^2 - 5}$

In Exercises 9 and 10, evaluate the expression.

9. $(476)^{0.65}$ **10.** $(251)^{0.35}$

Exercises 9.3

See www.CalcChat.com for worked-out solutions to odd-numbered exercises.

In Exercises 1–14, find the function values.

1. $f(x, y) = \dfrac{x}{y}$
(a) $f(3, 2)$ (b) $f(-1, 4)$ (c) $f(30, 5)$
(d) $f(5, y)$ (e) $f(x, 2)$ (f) $f(5, t)$

2. $f(x, y) = 4 - x^2 - 4y^2$
(a) $f(0, 0)$ (b) $f(0, 1)$ (c) $f(2, 3)$
(d) $f(1, y)$ (e) $f(x, 0)$ (f) $f(t, 1)$

3. $f(x, y) = xe^y$
(a) $f(5, 0)$ (b) $f(3, 2)$ (c) $f(2, -1)$
(d) $f(5, y)$ (e) $f(x, 2)$ (f) $f(t, t)$

4. $g(x, y) = \ln|x + y|$
(a) $g(2, 3)$ (b) $g(5, 6)$ (c) $g(e, 0)$
(d) $g(0, 1)$ (e) $g(2, -3)$ (f) $g(e, e)$

5. $h(x, y, z) = \dfrac{xy}{z}$
(a) $h(2, 3, 9)$ (b) $h(1, 0, 1)$

6. $f(x, y, z) = \sqrt{x + y + z}$
(a) $f(0, 5, 4)$ (b) $f(6, 8, -3)$

7. $V(r, h) = \pi r^2 h$
(a) $V(3, 10)$ (b) $V(5, 2)$

8. $F(r, N) = 500\left(1 + \dfrac{r}{12}\right)^N$
(a) $F(0.09, 60)$ (b) $F(0.14, 240)$

9. $A(P, r, t) = P\left[\left(1 + \dfrac{r}{12}\right)^{12t} - 1\right]\left(1 + \dfrac{12}{r}\right)$
(a) $A(100, 0.10, 10)$ (b) $A(275, 0.0925, 40)$

10. $A(P, r, t) = Pe^{rt}$
(a) $A(500, 0.10, 5)$ (b) $A(1500, 0.12, 20)$

11. $f(x, y) = \displaystyle\int_x^y (2t - 3)\, dt$
(a) $f(1, 2)$ (b) $f(1, 4)$

12. $g(x, y) = \displaystyle\int_x^y \dfrac{1}{t}\, dt$
(a) $g(4, 1)$ (b) $g(6, 3)$

13. $f(x, y) = x^2 - 2y$
(a) $f(x + \Delta x, y)$ (b) $\dfrac{f(x, y + \Delta y) - f(x, y)}{\Delta y}$

14. $f(x, y) = 3xy + y^2$
(a) $f(x + \Delta x, y)$ (b) $\dfrac{f(x, y + \Delta y) - f(x, y)}{\Delta y}$

In Exercises 15–18, describe the region R in the xy-plane that corresponds to the domain of the function, and find the range of the function.

15. $f(x, y) = \sqrt{16 - x^2 - y^2}$
16. $f(x, y) = x^2 + y^2 - 1$
17. $f(x, y) = e^{x/y}$
18. $f(x, y) = \ln(x + y)$

In Exercises 19–28, describe the region R in the xy-plane that corresponds to the domain of the function.

19. $z = \sqrt{4 - x^2 - y^2}$ **20.** $z = \sqrt{4 - x^2 - 4y^2}$
21. $f(x, y) = x^2 + y^2$ **22.** $f(x, y) = \dfrac{x}{y}$

23. $f(x, y) = \dfrac{1}{xy}$

24. $g(x, y) = \dfrac{1}{x - y}$

25. $h(x, y) = x\sqrt{y}$

26. $f(x, y) = \sqrt{xy}$

27. $g(x, y) = \ln(4 - x - y)$

28. $f(x, y) = ye^{1/x}$

In Exercises 29–32, match the graph of the surface with one of the contour maps. [The contour maps are labeled (a)–(d).]

(a)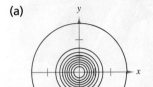

(b)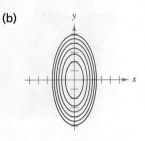

(c)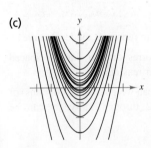

(d)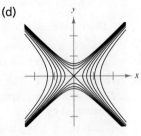

29. $f(x, y) = x^2 + \dfrac{y^2}{4}$

30. $f(x, y) = e^{1 - x^2 + y^2}$

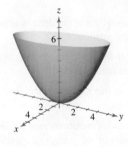

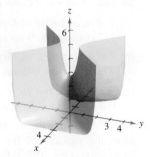

31. $f(x, y) = e^{1 - x^2 - y^2}$

32. $f(x, y) = \ln|y - x^2|$

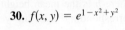

In Exercises 33–40, describe the level curves of the function. Sketch the level curves for the given c-values.

Function	c-Values
33. $z = x + y$	$c = -1, 0, 2, 4$
34. $z = 6 - 2x - 3y$	$c = 0, 2, 4, 6, 8, 10$
35. $z = \sqrt{25 - x^2 - y^2}$	$c = 0, 1, 2, 3, 4, 5$
36. $f(x, y) = x^2 + y^2$	$c = 0, 2, 4, 6, 8$
37. $f(x, y) = xy$	$c = \pm 1, \pm 2, \ldots, \pm 6$
38. $z = e^{xy}$	$c = 1, 2, 3, 4, \tfrac{1}{2}, \tfrac{1}{3}, \tfrac{1}{4}$
39. $f(x, y) = \dfrac{x}{x^2 + y^2}$	$c = \pm\tfrac{1}{2}, \pm 1, \pm\tfrac{3}{2}, \pm 2$
40. $f(x, y) = \ln(x - y)$	$c = 0, \pm\tfrac{1}{2}, \pm 1, \pm\tfrac{3}{2}, \pm 2$

41. **Volume** An oxygen tank is constructed by welding hemispheres to the ends of a right circular cylinder. Write the volume V of the tank as a function of r and l, where r is the radius of the cylinder and hemispheres, and l is the length of the cylinder.

42. **Ideal Gas Law** According to the Ideal Gas Law, $PV = kT$, where P is pressure, V is volume, T is temperature (in Kelvins), and k is a constant of proportionality. A tank contains 2600 cubic inches of nitrogen at a pressure of 20 pounds per square inch and a temperature of 300 K.

    (a) Determine k.

    (b) Write P as a function of V and T and describe the level curves.

43. **Forestry** *The Doyle Log Rule* is one of several methods used to determine the lumber yield of a log in board-feet in terms of its diameter d in inches and its length L in feet. The number of board-feet is given by

$$N(d, L) = \left(\dfrac{d - 4}{4}\right)^2 L.$$

    (a) Find the number of board-feet of lumber in a log with a diameter of 22 inches and a length of 12 feet.

    (b) Find $N(30, 12)$.

44. **Queuing Model** The average amount of time that a customer waits in line for service is given by

$$W(x, y) = \dfrac{1}{x - y}, \quad y < x$$

where y is the average arrival rate and x is the average service rate (x and y are measured in the number of customers per hour). Evaluate W at each point.

   (a) (15, 10)   (b) (12, 9)   (c) (12, 6)   (d) (4, 2)

45. Infant Development The mean head circumference c (in centimeters) of U.S. female infants ages 2 through 12 months can be approximated by

$$c = 0.58l - 0.36a + 6.70$$

where l is the mean recumbent length of the infant (in centimeters) and a is the corresponding age (in months). Use this function of two variables and a spreadsheet to complete the table. *(Source: Centers for Disease Control and Prevention)*

a	2	4	6	8	10	12
l	57.40	62.57	66.91	69.01	72.00	75.83
c						

46. Dew Point The dew point temperature is the temperature to which air must be cooled for dew to form. If T is the measured temperature (in degrees Celsius) and H is the percent relative humidity (in decimal form), the dew point temperature D (in degrees Celsius) can be approximated by

$$D = \frac{237.7x}{17.27 - x}, \text{ where } x = \frac{17.27T}{237.7 + T} + \ln H.$$

Use these functions and a spreadsheet to complete the table.

T	20°	30°	20°	30°	20°	30°
H	0.50	0.50	0.75	0.75	0.90	0.90
x						
D						

47. Meteorology Meteorologists measure the atmospheric pressure in millibars. From these observations they create weather maps on which the curves of equal atmospheric pressure (isobars) are drawn (see figure). On the map, the closer the isobars the higher the wind speed. Match points A, B, and C with (a) highest pressure, (b) lowest pressure, and (c) highest wind velocity.

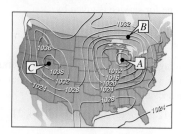

48. Geology The contour map below represents color-coded seismic amplitudes of a fault horizon and a projected contour map, which is used in earthquake studies. *(Source: Adapted from Shipman/Wilson/Todd, An Introduction to Physical Science, Tenth Edition)*

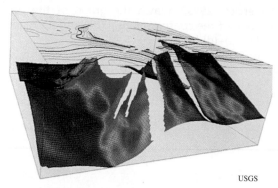

USGS

Shipman, An Introduction to Physical Science 10/e, 2003, Houghton Mifflin Company

(a) Discuss the use of color to represent the level curves.

(b) Do the level curves correspond to equally spaced amplitudes? Explain your reasoning.

49. Average Weight The mean weight w (in kilograms) of U.S. male children ages 2 through 18 years can be approximated by

$$w = -0.59h + 7.25a + 47.37$$

where h is the mean height (in centimeters) and a is the corresponding age (in years). *(Source: Centers for Disease Control and Prevention)*

(a) The mean height of 6-year-old male children is about 117 centimeters. What is the mean weight?

(b) The mean weight of 13-year-old male children is about 47.9 kilograms. What is the mean height?

(c) Which of the two variables in this model has the greater influence on weight? Explain.

50. Basal Metabolic Rate The *basal metabolic rate* (BMR) gives the measure of the amount of energy expended by a person at rest. The BMR (in calories per day) for males and females can be approximated by

$$B_m(w, h, a) = 66 + 13.7w + 5h - 6.8a \qquad \text{Males}$$
$$B_f(w, h, a) = 655 + 9.6w + 1.8h - 4.7a \qquad \text{Females}$$

where w is the person's weight (in kilograms), h is the person's height (in centimeters), and a is the person's age (in years).

(a) Evaluate $B_m(72, 173, 30)$ and $B_f(72, 173, 30)$.

(b) Which of the three variables in the models has the greatest influence on BMR? Explain.

Section 9.4
Partial Derivatives

- Find the first partial derivatives of functions of two variables.
- Find the slopes of surfaces in the *x*- and *y*-directions and use partial derivatives to answer questions about real-life situations.
- Find the partial derivatives of functions of several variables.
- Find higher-order partial derivatives.

Functions of Two Variables

Real-life applications of functions of several variables are often concerned with how changes in one of the variables will affect the values of the functions. For instance, a biologist studying the net assimilation of carbon dioxide by a plant might conduct several experiments while keeping light intensity constant and letting temperature vary.

You can follow a similar procedure to find the rate of change of a function f with respect to one of its independent variables. That is, you find the derivative of f with respect to one independent variable while holding the other variable(s) constant. This process is called **partial differentiation,** and each derivative is called a **partial derivative.** A function of several variables has as many partial derivatives as it has independent variables.

STUDY TIP

Note that this definition indicates that partial derivatives of a function of two variables are determined by temporarily considering one variable to be fixed. For instance, if $z = f(x, y)$, then to find $\partial z/\partial x$, you consider y to be constant and differentiate with respect to x. Similarly, to find $\partial z/\partial y$, you consider x to be constant and differentiate with respect to y.

Partial Derivatives of a Function of Two Variables

If $z = f(x, y)$, then the **first partial derivatives of f with respect to x and y** are the functions $\partial z/\partial x$ and $\partial z/\partial y$, defined as shown.

$$\frac{\partial z}{\partial x} = \lim_{\Delta x \to 0} \frac{f(x + \Delta x, y) - f(x, y)}{\Delta x} \qquad y \text{ is held constant.}$$

$$\frac{\partial z}{\partial y} = \lim_{\Delta y \to 0} \frac{f(x, y + \Delta y) - f(x, y)}{\Delta y} \qquad x \text{ is held constant.}$$

Example 1 Finding Partial Derivatives

Find $\partial z/\partial x$ and $\partial z/\partial y$ for the function $z = 3x - x^2y^2 + 2x^3y$.

SOLUTION

$$\frac{\partial z}{\partial x} = 3 - 2xy^2 + 6x^2y \qquad \text{Hold } y \text{ constant and differentiate with respect to } x.$$

$$\frac{\partial z}{\partial y} = -2x^2y + 2x^3 \qquad \text{Hold } x \text{ constant and differentiate with respect to } y.$$

✓CHECKPOINT 1

Find $\dfrac{\partial z}{\partial x}$ and $\dfrac{\partial z}{\partial y}$ for $z = 2x^2 - 4x^2y^3 + y^4$. ■

Notation for First Partial Derivatives

The first partial derivatives of $z = f(x, y)$ are denoted by

$$\frac{\partial z}{\partial x} = f_x(x, y) = z_x = \frac{\partial}{\partial x}[f(x, y)]$$

and

$$\frac{\partial z}{\partial y} = f_y(x, y) = z_y = \frac{\partial}{\partial y}[f(x, y)].$$

The values of the first partial derivatives at the point (a, b) are denoted by

$$\frac{\partial z}{\partial x}\bigg|_{(a, b)} = f_x(a, b) \quad \text{and} \quad \frac{\partial z}{\partial y}\bigg|_{(a, b)} = f_y(a, b).$$

TECHNOLOGY

Symbolic differentiation utilities can be used to find partial derivatives of a function of two variables. Try using a symbolic differentiation utility to find the first partial derivatives of the function in Example 2.

Example 2 Finding and Evaluating Partial Derivatives

Find the first partial derivatives of $f(x, y) = xe^{x^2 y}$ and evaluate each at the point $(1, \ln 2)$.

SOLUTION To find the first partial derivative with respect to x, hold y constant and differentiate using the Product Rule.

$$f_x(x, y) = x\frac{\partial}{\partial x}[e^{x^2 y}] + e^{x^2 y}\frac{\partial}{\partial x}[x] \qquad \text{Apply Product Rule.}$$
$$= x(2xy)e^{x^2 y} + e^{x^2 y} \qquad \text{y is held constant.}$$
$$= e^{x^2 y}(2x^2 y + 1) \qquad \text{Simplify.}$$

At the point $(1, \ln 2)$, the value of this derivative is

$$f_x(1, \ln 2) = e^{(1)^2(\ln 2)}[2(1)^2(\ln 2) + 1] \qquad \text{Substitute for x and y.}$$
$$= 2(2 \ln 2 + 1) \qquad \text{Simplify.}$$
$$\approx 4.773. \qquad \text{Use a calculator.}$$

To find the first partial derivative with respect to y, hold x constant and differentiate to obtain

$$f_y(x, y) = x(x^2)e^{x^2 y} \qquad \text{Apply Constant Multiple Rule.}$$
$$= x^3 e^{x^2 y}. \qquad \text{Simplify.}$$

At the point $(1, \ln 2)$, the value of this derivative is

$$f_y(1, \ln 2) = (1)^3 e^{(1)^2(\ln 2)} \qquad \text{Substitute for x and y.}$$
$$= 2. \qquad \text{Simplify.}$$

✓CHECKPOINT 2

Find the first partial derivatives of $f(x, y) = x^2 y^3$ and evaluate each at the point $(1, 2)$. ■

Graphical Interpretation of Partial Derivatives

At the beginning of this course, you studied graphical interpretations of the derivative of a function of a single variable. There, you found that $f'(x_0)$ represents the slope of the tangent line to the graph of $y = f(x)$ at the point (x_0, y_0). The partial derivatives of a function of two variables also have useful graphical interpretations. Consider the function

$$z = f(x, y). \qquad \text{Function of two variables}$$

As shown in Figure 9.22(a), the graph of this function is a surface in space. If the variable y is fixed, say at $y = y_0$, then

$$z = f(x, y_0) \qquad \text{Function of one variable}$$

is a function of one variable. The graph of this function is the curve that is the intersection of the plane $y = y_0$ and the surface $z = f(x, y)$. On this curve, the partial derivative

$$f_x(x, y_0) \qquad \text{Slope in } x\text{-direction}$$

represents the slope in the plane $y = y_0$, as shown in Figure 9.22(a). In a similar way, if the variable x is fixed, say at $x = x_0$, then

$$z = f(x_0, y) \qquad \text{Function of one variable}$$

is a function of one variable. Its graph is the intersection of the plane $x = x_0$ and the surface $z = f(x, y)$. On this curve, the partial derivative

$$f_y(x_0, y) \qquad \text{Slope in } y\text{-direction}$$

represents the slope in the plane $x = x_0$, as shown in Figure 9.22(b).

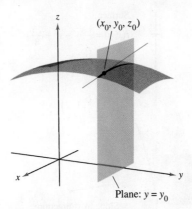

(a) $f_x(x, y_0)$ = slope in x-direction

(b) $f_y(x_0, y)$ = slope in y-direction

FIGURE 9.22

DISCOVERY

How can partial derivatives be used to find *relative extrema* of graphs of functions of two variables?

576   CHAPTER 9   Functions of Several Variables

Example 3   Finding Slopes in the *x*- and *y*-Directions

Find the slopes of the surface given by

$$f(x, y) = -\frac{x^2}{2} - y^2 + \frac{25}{8}$$

at the point $\left(\frac{1}{2}, 1, 2\right)$ in (a) the *x*-direction and (b) the *y*-direction.

SOLUTION

a. To find the slope in the *x*-direction, hold *y* constant and differentiate with respect to *x* to obtain

$$f_x(x, y) = -x. \qquad \text{Partial derivative with respect to } x$$

At the point $\left(\frac{1}{2}, 1, 2\right)$, the slope in the *x*-direction is

$$f_x\left(\tfrac{1}{2}, 1\right) = -\tfrac{1}{2} \qquad \text{Slope in } x\text{-direction}$$

as shown in Figure 9.23(a).

b. To find the slope in the *y*-direction, hold *x* constant and differentiate with respect to *y* to obtain

$$f_y(x, y) = -2y. \qquad \text{Partial derivative with respect to } y$$

At the point $\left(\frac{1}{2}, 1, 2\right)$, the slope in the *y*-direction is

$$f_y\left(\tfrac{1}{2}, 1\right) = -2 \qquad \text{Slope in } y\text{-direction}$$

as shown in Figure 9.23(b).

✓ **CHECKPOINT 3**

Find the slopes of the surface given by

$$f(x, y) = 4x^2 + 9y^2 + 36$$

at the point $(1, -1, 49)$ in the *x*-direction and the *y*-direction. ■

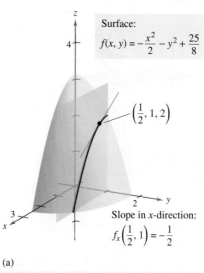
(a)

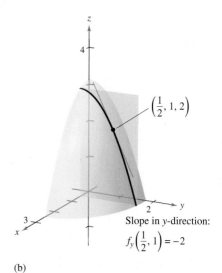
(b)

FIGURE 9.23

DISCOVERY

Find the partial derivatives f_x and f_y at $(0, 0)$ for the function in Example 3. What are the slopes of *f* in the *x*- and *y*-directions at $(0, 0)$? Describe the shape of the graph of *f* at this point.

Example 4 Body Mass Index

Body Mass Index (BMI) is a measure of body fat based on weight and height of an adult person. BMI can be found using the formula

$$B = \frac{703w}{h^2}$$

where w is the weight (in pounds) and h is height (in inches).

a. Find the rate of change of BMI with respect to height when $w = 180$ pounds and $h = 70$ inches.

b. Find the rate of change of BMI with respect to weight when $w = 180$ pounds and $h = 70$ inches.

SOLUTION

a. To find the rate of change of B with respect to h, hold w constant and differentiate with respect to h.

$$B = \frac{703w}{h^2} \quad \text{Write original function.}$$

$$\frac{\partial B}{\partial h} = -\frac{1406w}{h^3} \quad \text{Partial derivative with respect to } h$$

When $w = 180$ pounds and $h = 70$ inches, the rate of change of B with respect to h is

$$\frac{\partial B}{\partial h} = -\frac{1406(180)}{70^3} \quad \text{Substitute 180 for } w \text{ and 70 for } h$$

$$\approx -0.74 \quad \text{Simplify.}$$

So, when weight is constant at 180 pounds, BMI will decrease by about 0.74 for every inch that height increases.

b. To find the rate of change of B with respect to w, hold h constant and differentiate with respect to w.

$$B = \frac{703w}{h^2} \quad \text{Write original function.}$$

$$\frac{\partial B}{\partial w} = \frac{703}{h^2} \quad \text{Partial derivative with respect to } w$$

When $w = 180$ pounds and $h = 70$ inches, the rate of change of B with respect to w is

$$\frac{\partial B}{\partial w} = \frac{703}{70^2} \quad \text{Substitute 70 for } h.$$

$$\approx 0.14 \quad \text{Simplify.}$$

So, when height is constant at 70 inches, BMI will increase by about 0.14 for every pound that weight increases.

✓**CHECKPOINT 4**

Refer to Example 4.

(a) Find the rate of change of BMI with respect to height when $w = 150$ pounds and $h = 64$ inches.

(b) Find the rate of change of BMI with respect to weight when $w = 150$ pounds and $h = 64$ inches. ■

Functions of Three Variables

The concept of a partial derivative can be extended naturally to functions of three or more variables. For instance, the function $w = f(x, y, z)$ has three partial derivatives, each of which is formed by considering two of the variables to be constant. That is, to define the partial derivative of w with respect to x, consider y and z to be constant and write

$$\frac{\partial w}{\partial x} = f_x(x, y, z) = \lim_{\Delta x \to 0} \frac{f(x + \Delta x, y, z) - f(x, y, z)}{\Delta x}.$$

To define the partial derivative of w with respect to y, consider x and z to be constant and write

$$\frac{\partial w}{\partial y} = f_y(x, y, z) = \lim_{\Delta y \to 0} \frac{f(x, y + \Delta y, z) - f(x, y, z)}{\Delta y}.$$

To define the partial derivative of w with respect to z, consider x and y to be constant and write

$$\frac{\partial w}{\partial z} = f_z(x, y, z) = \lim_{\Delta z \to 0} \frac{f(x, y, z + \Delta z) - f(x, y, z)}{\Delta z}.$$

Example 5 Finding Partial Derivatives of a Function

Find the three partial derivatives of the function

$$w = xe^{xy+2z}.$$

SOLUTION Holding y and z constant, you obtain

$$\frac{\partial w}{\partial x} = x\frac{\partial}{\partial x}[e^{xy+2z}] + e^{xy+2z}\frac{\partial}{\partial x}[x] \quad \text{Apply Product Rule.}$$
$$= x(ye^{xy+2z}) + e^{xy+2z}(1) \quad \text{Hold } y \text{ and } z \text{ constant.}$$
$$= (xy + 1)e^{xy+2z}. \quad \text{Simplify.}$$

Holding x and z constant, you obtain

$$\frac{\partial w}{\partial y} = x(x)e^{xy+2z} \quad \text{Hold } x \text{ and } z \text{ constant.}$$
$$= x^2 e^{xy+2z}. \quad \text{Simplify.}$$

Holding x and y constant, you obtain

$$\frac{\partial w}{\partial z} = x(2)e^{xy+2z} \quad \text{Hold } x \text{ and } y \text{ constant.}$$
$$= 2xe^{xy+2z}. \quad \text{Simplify.}$$

✓ CHECKPOINT 5

Find the three partial derivatives of the function

$$w = x^2 y \ln(xz). \quad \blacksquare$$

TECHNOLOGY

A symbolic differentiation utility can be used to find the partial derivatives of a function of three or more variables. Try using a symbolic differentiation utility to find the partial derivative $f_y(x, y, z)$ for the function in Example 5.

STUDY TIP

Note that in Example 5 the Product Rule was used only when finding the partial derivative with respect to x. Can you see why?

Higher-Order Partial Derivatives

As with ordinary derivatives, it is possible to take second, third, and higher-order partial derivatives of a function of several variables, provided such derivatives exist. Higher-order derivatives are denoted by the order in which the differentiation occurs. For instance, there are four different ways to find a second partial derivative of $z = f(x, y)$.

$\dfrac{\partial}{\partial x}\left(\dfrac{\partial f}{\partial x}\right) = \dfrac{\partial^2 f}{\partial x^2} = f_{xx}$ Differentiate twice with respect to x.

$\dfrac{\partial}{\partial y}\left(\dfrac{\partial f}{\partial y}\right) = \dfrac{\partial^2 f}{\partial y^2} = f_{yy}$ Differentiate twice with respect to y.

$\dfrac{\partial}{\partial y}\left(\dfrac{\partial f}{\partial x}\right) = \dfrac{\partial^2 f}{\partial y \partial x} = f_{xy}$ Differentiate first with respect to x and then with respect to y.

$\dfrac{\partial}{\partial x}\left(\dfrac{\partial f}{\partial y}\right) = \dfrac{\partial^2 f}{\partial x \partial y} = f_{yx}$ Differentiate first with respect to y and then with respect to x.

The third and fourth cases are **mixed partial derivatives.** Notice that with the two types of notation for mixed partials, different conventions are used for indicating the order of differentiation. For instance, the partial derivative

$\dfrac{\partial}{\partial y}\left(\dfrac{\partial f}{\partial x}\right) = \dfrac{\partial^2 f}{\partial y \partial x}$ Right-to-left order

indicates differentiation with respect to x first, but the partial derivative

$(f_y)_x = f_{yx}$ Left-to-right order

indicates differentiation with respect to y first. To remember this, note that in each case you differentiate first with respect to the variable "nearest" f.

Example 6   Finding Second Partial Derivatives

Find the second partial derivatives of

$f(x, y) = 3xy^2 - 2y + 5x^2y^2$

and determine the value of $f_{xy}(-1, 2)$.

SOLUTION   Begin by finding the first partial derivatives.

$f_x(x, y) = 3y^2 + 10xy^2$ $f_y(x, y) = 6xy - 2 + 10x^2y$

Then, differentiating with respect to x and y produces

$f_{xx}(x, y) = 10y^2,$ $f_{yy}(x, y) = 6x + 10x^2$
$f_{xy}(x, y) = 6y + 20xy,$ $f_{yx}(x, y) = 6y + 20xy.$

Finally, the value of $f_{xy}(x, y)$ at the point $(-1, 2)$ is

$f_{xy}(-1, 2) = 6(2) + 20(-1)(2) = 12 - 40 = -28.$

STUDY TIP

Notice in Example 6 that the two mixed partials are equal. This is often the case. In fact, it can be shown that if a function has continuous second partial derivatives, then the order in which the partial derivatives are taken is irrelevant.

✓ CHECKPOINT 6

Find the second partial derivatives of

$f(x, y) = 4x^2y^2 + 2x + 4y^2.$ ■

A function of two variables has two first partial derivatives and four second partial derivatives. For a function of three variables, there are three first partials

$$f_x, f_y, \text{ and } f_z$$

and nine second partials

$$f_{xx}, f_{xy}, f_{xz}, f_{yx}, f_{yy}, f_{yz}, f_{zx}, f_{zy}, \text{ and } f_{zz}$$

of which six are mixed partials. To find partial derivatives of order three and higher, follow the same pattern used to find second partial derivatives. For instance, if $z = f(x, y)$, then

$$z_{xxx} = \frac{\partial}{\partial x}\left(\frac{\partial^2 f}{\partial x^2}\right) = \frac{\partial^3 f}{\partial x^3} \quad \text{and} \quad z_{xxy} = \frac{\partial}{\partial y}\left(\frac{\partial^2 f}{\partial x^2}\right) = \frac{\partial^3 f}{\partial y \partial x^2}.$$

Example 7 Finding Second Partial Derivatives

Find the second partial derivatives of

$$f(x, y, z) = ye^x + x \ln z.$$

SOLUTION Begin by finding the first partial derivatives.

$$f_x(x, y, z) = ye^x + \ln z, \quad f_y(x, y, z) = e^x, \quad f_z(x, y, z) = \frac{x}{z}$$

Then, differentiate with respect to x, y, and z to find the nine second partial derivatives.

$$f_{xx}(x, y, z) = ye^x, \quad f_{xy}(x, y, z) = e^x, \quad f_{xz}(x, y, z) = \frac{1}{z}$$

$$f_{yx}(x, y, z) = e^x, \quad f_{yy}(x, y, z) = 0, \quad f_{yz}(x, y, z) = 0$$

$$f_{zx}(x, y, z) = \frac{1}{z}, \quad f_{zy}(x, y, z) = 0, \quad f_{zz}(x, y, z) = -\frac{x}{z^2}$$

✓ CHECKPOINT 7

Find the second partial derivatives of $f(x, y, z) = xe^y + 2xz + y^2$. ■

CONCEPT CHECK

1. Write the notation that denotes the first partial derivative of $z = f(x, y)$ with respect to x.
2. Write the notation that denotes the first partial derivative of $z = f(x, y)$ with respect to y.
3. Let f be a function of two variables x and y. Describe the procedure for finding the first partial derivatives.
4. Define the first partial derivatives of a function f of two variables x and y.

SECTION 9.4 Partial Derivatives

Skills Review 9.4

The following warm-up exercises involve skills that were covered in earlier sections. You will use these skills in the exercise set for this section. For additional help, review Sections 2.2, 2.4, 2.5, 4.3, and 4.5.

In Exercises 1–8, find the derivative of the function.

1. $f(x) = \sqrt{x^2 + 3}$
2. $g(x) = (3 - x^2)^3$
3. $g(t) = te^{2t+1}$
4. $f(x) = e^{2x}\sqrt{1 - e^{2x}}$
5. $f(x) = \ln(3 - 2x)$
6. $u(t) = \ln\sqrt{t^3 - 6t}$
7. $g(x) = \dfrac{5x^2}{(4x - 1)^2}$
8. $f(x) = \dfrac{(x + 2)^3}{(x^2 - 9)^2}$

In Exercises 9 and 10, evaluate the derivative at the point (2, 4).

9. $f(x) = x^2 e^{x-2}$
10. $g(x) = x\sqrt{x^2 - x + 2}$

Exercises 9.4

See www.CalcChat.com for worked-out solutions to odd-numbered exercises.

In Exercises 1–14, find the first partial derivatives with respect to x and with respect to y.

1. $z = 3x + 5y - 1$
2. $z = x^2 - 2y$
3. $f(x, y) = 3x - 6y^2$
4. $f(x, y) = x + 4y^{3/2}$
5. $f(x, y) = \dfrac{x}{y}$
6. $z = x\sqrt{y}$
7. $f(x, y) = \sqrt{x^2 + y^2}$
8. $f(x, y) = \dfrac{xy}{x^2 + y^2}$
9. $z = x^2 e^{2y}$
10. $z = xe^{x+y}$
11. $h(x, y) = e^{-(x^2+y^2)}$
12. $g(x, y) = e^{x/y}$
13. $z = \ln\dfrac{x + y}{x - y}$
14. $g(x, y) = \ln(x^2 + y^2)$

In Exercises 15–20, let $f(x, y) = 3x^2ye^{x-y}$ and $g(x, y) = 3xy^2e^{y-x}$. Find each of the following.

15. $f_x(x, y)$
16. $f_y(x, y)$
17. $g_x(x, y)$
18. $g_y(x, y)$
19. $f_x(1, 1)$
20. $g_x(-2, -2)$

In Exercises 21–28, evaluate f_x and f_y at the point.

Function	Point
21. $f(x, y) = 3x^2 + xy - y^2$	(2, 1)
22. $f(x, y) = x^2 - 3xy + y^2$	(1, −1)
23. $f(x, y) = e^{3xy}$	(0, 4)
24. $f(x, y) = e^x y^2$	(0, 2)
25. $f(x, y) = \dfrac{xy}{x - y}$	(2, −2)
26. $f(x, y) = \dfrac{4xy}{\sqrt{x^2 + y^2}}$	(1, 0)
27. $f(x, y) = \ln(x^2 + y^2)$	(1, 0)
28. $f(x, y) = \ln\sqrt{xy}$	(−1, −1)

In Exercises 29–32, find the first partial derivatives with respect to x, y, and z.

29. $w = xyz$
30. $w = x^2 - 3xy + 4yz + z^3$
31. $w = \dfrac{2z}{x + y}$
32. $w = \sqrt{x^2 + y^2 + z^2}$

In Exercises 33–38, evaluate w_x, w_y, and w_z at the point.

Function	Point
33. $w = \sqrt{x^2 + y^2 + z^2}$	$(2, -1, 2)$
34. $w = \dfrac{xy}{x + y + z}$	$(1, 2, 0)$
35. $w = \ln\sqrt{x^2 + y^2 + z^2}$	$(3, 0, 4)$
36. $w = \dfrac{1}{\sqrt{1 - x^2 - y^2 - z^2}}$	$(0, 0, 0)$
37. $w = 2xz^2 + 3xyz - 6y^2z$	$(1, -1, 2)$
38. $w = xye^{z^2}$	$(2, 1, 0)$

In Exercises 39–42, find values of x and y such that $f_x(x, y) = 0$ and $f_y(x, y) = 0$ simultaneously.

39. $f(x, y) = x^2 + 4xy + y^2 - 4x + 16y + 3$
40. $f(x, y) = 3x^3 - 12xy + y^3$
41. $f(x, y) = \dfrac{1}{x} + \dfrac{1}{y} + xy$
42. $f(x, y) = \ln(x^2 + y^2 + 1)$

In Exercises 43–48, find the slope of the surface at the given point in (a) the x-direction and (b) the y-direction.

43. $z = xy$
    $(1, 2, 2)$

44. $z = \sqrt{25 - x^2 - y^2}$
    $(3, 0, 4)$

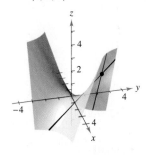
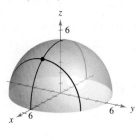

45. $z = 4 - x^2 - y^2$
    $(1, 1, 2)$

46. $z = x^2 - y^2$
    $(-2, 1, 3)$

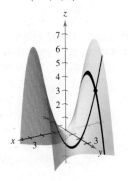

47. $z = e^{-x}\cos y$
    $(0, 0, 1)$

48. $z = \cos(2x - y)$
    $\left(\dfrac{\pi}{4}, \dfrac{\pi}{3}, \dfrac{\sqrt{3}}{2}\right)$

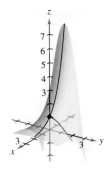
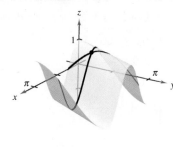

In Exercises 49–58, find the four second partial derivatives. Observe that the second mixed partials are equal.

49. $z = x^2 - 2xy + 3y^2$
50. $z = y^3 - 4xy^2 - 1$
51. $z = \dfrac{e^{2xy}}{4x}$
52. $z = \dfrac{x^2 - y^2}{2xy}$
53. $z = x^3 - 4y^2$
54. $z = \sqrt{9 - x^2 - y^2}$
55. $z = \dfrac{1}{x - y}$
56. $z = \dfrac{x}{x + y}$
57. $z = xe^{-y^2}$
58. $z = xe^y + ye^x$

In Exercises 59–62, evaluate the second partial derivatives f_{xx}, f_{xy}, f_{yy}, and f_{yx} at the point.

Function	Point
59. $f(x, y) = x^4 - 3x^2y^2 + y^2$	$(1, 0)$
60. $f(x, y) = \sqrt{x^2 + y^2}$	$(0, 2)$
61. $f(x, y) = \ln(x - y)$	$(2, 1)$
62. $f(x, y) = x^2e^y$	$(-1, 0)$

63. Milk Consumption Per capita consumptions (in gallons) of different types of plain milk in the United States from 1999 through 2004 are shown in the table. Consumptions of reduced-fat milk (1%) and skim milks, reduced-fat milk (2%), and whole milk are represented by the variables x, y, and z, respectively. *(Source: U.S. Department of Agriculture)*

Year	1999	2000	2001	2002	2003	2004
x	6.2	6.1	5.9	5.8	5.6	5.5
y	7.3	7.1	7.0	7.0	6.9	6.9
z	7.8	7.7	7.4	7.3	7.2	6.9

A model for the data is given by
$$z = 1.25x - 0.125y + 0.95.$$

(a) Find $\dfrac{\partial z}{\partial x}$ and $\dfrac{\partial z}{\partial y}$.

(b) Interpret the partial derivatives in the context of the problem.

64. Health Care The table shows the amounts of public expenditures (in billions of dollars) for workers' compensation x, public assistance y, and Medicare z for selected years. *(Source: Centers for Medicare and Medicaid Services)*

Year	1999	2000	2001	2002	2003	2004
x	22.2	24.9	28.1	30.1	31.5	33.5
y	189.1	207.5	233.2	258.2	281.2	303.7
z	213.2	225.2	248.3	266.3	283.8	309.0

A model for the data is given by
$$z = -1.28x + 0.94y + 62.3.$$

(a) Find $\dfrac{\partial z}{\partial x}$ and $\dfrac{\partial z}{\partial y}$.

(b) Interpret the partial derivatives in the context of the problem.

65. Chemistry The temperature at any point (x, y) in a steel plate is given by
$$T = 500 - 0.6x^2 - 1.5y^2$$
where x and y are measured in meters. At the point $(2, 3)$, find the rates of change of the temperature with respect to the distances moved along the plate in the directions of the x- and y-axes.

66. Chemistry A measure of what hot weather feels like to an average person is the Apparent Temperature Index. A model for this index is
$$A = 0.885t - 78.7h + 1.20th + 2.70$$

where A is the apparent temperature, t is the air temperature, and h is the relative humidity in decimal form. *(Source: The UMAP Journal)*

(a) Find $\partial A/\partial t$ and $\partial A/\partial h$ when $t = 90°F$ and $h = 0.80$.

(b) Which has a greater effect on A, air temperature or humidity? Explain your reasoning.

67. Think About It Let N be the number of applicants to a university, p the charge for food and housing at the university, and t the tuition. Suppose that N is a function of p and t such that $\partial N/\partial p < 0$ and $\partial N/\partial t < 0$. How would you interpret the fact that both partials are negative?

68. Psychology Early in the twentieth century, an intelligence test called the *Stanford-Binet Test* (more commonly known as the *IQ test*) was developed. In this test, an individual's mental age M is divided by the individual's chronological age C and the quotient is multiplied by 100. The result is the individual's *IQ*.

$$IQ(M, C) = \frac{M}{C} \times 100$$

Find the partial derivatives of *IQ* with respect to M and with respect to C. Evaluate the partial derivatives at the point $(12, 10)$ and interpret the result. *(Source: Adapted from Bernstein/Clark-Stewart/Roy/Wickens, Psychology, Fourth Edition)*

Social Science Capsule

Multiple-variable experiments are designed, created, and used quite often in the field of psychology. Empirisoft created the cost-effective software packages MediaLab and DirectRT, which allow users to create PC-based multimedia psychology experiments.

69. Research Project Use your school's library, the Internet, or some other reference source to research a company that uses software to create user-friendly multiple-variable experiments. Write a short paper that summarizes your findings.

Mid-Chapter Quiz

Take this quiz as you would take a quiz in class. When you are done, check your work against the answers given in the back of the book.

In Exercises 1–3, (a) plot the points on a three-dimensional coordinate system, (b) find the distance between the points, and (c) find the coordinates of the midpoint of the line segment joining the points.

1. $(1, 3, 2), (-1, 2, 0)$
2. $(-1, 4, 3), (5, 1, -6)$
3. $(0, -3, 3), (3, 0, -3)$

In Exercises 4 and 5, find the standard equation of the sphere.

4. Center: $(2, -1, 3)$; radius: 4
5. Endpoints of a diameter: $(0, 3, 1), (2, 5, -5)$

6. Find the center and radius of the sphere whose equation is
$$x^2 + y^2 + z^2 - 8x - 2y - 6z - 23 = 0.$$

In Exercises 7–9, find the intercepts and sketch the graph of the plane.

7. $2x + 3y + z = 6$
8. $x - 2z = 4$
9. $z = -5$

In Exercises 10–12, identify the quadric surface.

10. $\dfrac{x^2}{4} + \dfrac{y^2}{9} + \dfrac{z^2}{16} = 1$
11. $z^2 - x^2 - y^2 = 25$
12. $81z - 9x^2 - y^2 = 0$

In Exercises 13–15, find $f(1, 0)$ and $f(4, -1)$.

13. $f(x, y) = x - 9y^2$
14. $f(x, y) = \sqrt{4x^2 + y}$
15. $f(x, y) = \ln(x + 3y)$

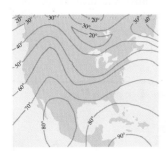

Figure for 16

16. The contour map shows level curves of equal temperature (isotherms), measured in degrees Fahrenheit, across North America on a spring day. Use the map to find the approximate range of temperatures in (a) the Great Lakes region, (b) the United States, and (c) Mexico.

In Exercises 17 and 18, find f_x and f_y and evaluate each at the point $(-2, 3)$.

17. $f(x, y) = x^2 + 2y^2 - 3x - y + 1$
18. $f(x, y) = \dfrac{3x - y^2}{x + y}$

19. Assume that Earth is a sphere with a radius of 3963 miles. If the center of Earth is placed at the origin of a three-dimensional coordinate system, what is the equation of the sphere? Lines of longitude that run north-south could be represented by what trace(s)? What shape would each of these traces form? Why? Lines of latitude that run east-west could be represented by what trace(s)? Why? What shape would each of these traces form? Why?

Section 9.5
Extrema of Functions of Two Variables

- Understand the relative extrema of functions of two variables.
- Use the First-Partials Test to find the relative extrema of functions of two variables.
- Use the Second-Partials Test to find the relative extrema of functions of two variables.
- Use relative extrema to answer questions about real-life situations.

Relative Extrema

Earlier in the text, you learned how to use derivatives to find the relative minimum and relative maximum values of a function of a single variable. In this section, you will learn how to use partial derivatives to find the relative minimum and relative maximum values of a function of two variables.

> **Relative Extrema of a Function of Two Variables**
>
> Let f be a function defined on a region containing (x_0, y_0). The function f has a **relative maximum** at (x_0, y_0) if there is a circular region R centered at (x_0, y_0) such that
>
> $$f(x, y) \leq f(x_0, y_0) \qquad \text{f has a relative maximum at (x_0, y_0).}$$
>
> for all (x, y) in R. The function f has a **relative minimum** at (x_0, y_0) if there is a circular region R centered at (x_0, y_0) such that
>
> $$f(x, y) \geq f(x_0, y_0) \qquad \text{f has a relative minimum at (x_0, y_0).}$$
>
> for all (x, y) in R.

To say that f has a relative maximum at (x_0, y_0) means that the point (x_0, y_0, z_0) is at least as high as all nearby points on the graph of $z = f(x, y)$. Similarly, f has a relative minimum at (x_0, y_0) if (x_0, y_0, z_0) is at least as low as all nearby points on the graph. (See Figure 9.24.)

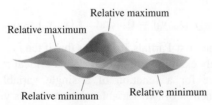

FIGURE 9.24 Relative Extrema

As in single-variable calculus, you need to distinguish between relative extrema and absolute extrema of a function of two variables. The number $f(x_0, y_0)$ is an absolute maximum of f in the region R if it is greater than or equal to all other function values in the region. For instance, the function $f(x, y) = -(x^2 + y^2)$ graphs as a paraboloid, opening downward, with vertex at $(0, 0, 0)$. (See Figure 9.25.) The number $f(0, 0) = 0$ is an absolute maximum of the function over the entire xy-plane. An absolute minimum of f in a region is defined similarly.

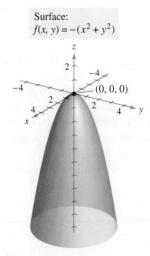

FIGURE 9.25 f has an absolute maximum at $(0, 0, 0)$.

The First-Partials Test for Relative Extrema

To locate the relative extrema of a function of two variables, you can use a procedure that is similar to the First-Derivative Test used for functions of a single variable.

First-Partials Test for Relative Extrema

If f has a relative extremum at (x_0, y_0) on an open region R in the xy-plane, and the first partial derivatives of f exist in R, then

$$f_x(x_0, y_0) = 0$$

and

$$f_y(x_0, y_0) = 0$$

as shown in Figure 9.26.

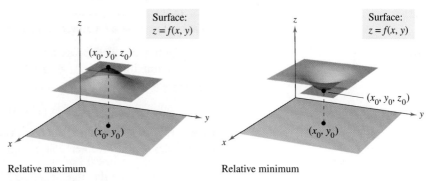

Relative maximum Relative minimum

FIGURE 9.26

An *open* region in the xy-plane is similar to an open interval on the real number line. For instance, the region R consisting of the interior of the circle $x^2 + y^2 = 1$ is an open region. If the region R consists of the interior of the circle *and* the points on the circle, then it is a *closed* region.

A point (x_0, y_0) is a **critical point** of f if $f_x(x_0, y_0)$ or $f_y(x_0, y_0)$ is undefined or if

$$f_x(x_0, y_0) = 0 \quad \text{and} \quad f_y(x_0, y_0) = 0. \qquad \text{Critical point}$$

The First-Partials Test states that if the first partial derivatives exist, then you need only examine values of $f(x, y)$ at critical points to find the relative extrema. As is true for a function of a single variable, however, the critical points of a function of two variables do not always yield relative extrema. For instance, the point $(0, 0)$ is a critical point of the surface shown in Figure 9.27, but $f(0, 0)$ is not a relative extremum of the function. Such points are called **saddle points** of the function.

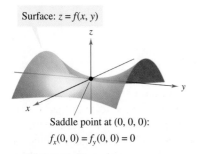

Saddle point at $(0, 0, 0)$:
$f_x(0, 0) = f_y(0, 0) = 0$

FIGURE 9.27

Example 1 Finding Relative Extrema

Find the relative extrema of

$$f(x, y) = 2x^2 + y^2 + 8x - 6y + 20.$$

SOLUTION Begin by finding the first partial derivatives of f.

$$f_x(x, y) = 4x + 8 \quad \text{and} \quad f_y(x, y) = 2y - 6$$

Because these partial derivatives are defined for all points in the xy-plane, the only critical points are those for which both first partial derivatives are zero. To locate these points, set $f_x(x, y)$ and $f_y(x, y)$ equal to 0, and solve the resulting system of equations.

$4x + 8 = 0$ Set $f_x(x, y)$ equal to 0.

$2y - 6 = 0$ Set $f_y(x, y)$ equal to 0.

The solution of this system is $x = -2$ and $y = 3$. So, the point $(-2, 3)$ is the only critical number of f. From the graph of the function, shown in Figure 9.28, you can see that this critical point yields a relative minimum of the function. So, the function has only one relative extremum, which is

$f(-2, 3) = 3.$ Relative minimum

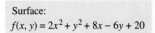

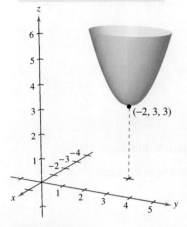

FIGURE 9.28

✓ CHECKPOINT 1

Find the relative extrema of $f(x, y) = x^2 + 2y^2 + 16x - 8y + 8$. ■

Example 1 shows a relative minimum occurring at one type of critical point—the type for which both $f_x(x, y)$ and $f_y(x, y)$ are zero. The next example shows a relative maximum that occurs at the other type of critical point—the type for which either $f_x(x, y)$ or $f_y(x, y)$ is undefined.

Example 2 Finding Relative Extrema

Find the relative extrema of

$$f(x, y) = 1 - (x^2 + y^2)^{1/3}.$$

SOLUTION Begin by finding the first partial derivatives of f.

$$f_x(x, y) = -\frac{2x}{3(x^2 + y^2)^{2/3}} \quad \text{and} \quad f_y(x, y) = -\frac{2y}{3(x^2 + y^2)^{2/3}}$$

These partial derivatives are defined for all points in the xy-plane *except* the point $(0, 0)$. So, $(0, 0)$ is a critical point of f. Moreover, this is the only critical point, because there are no other values of x and y for which either partial is undefined or for which both partials are zero. From the graph of the function, shown in Figure 9.29, you can see that this critical point yields a relative maximum of the function. So, the function has only one relative extremum, which is

$f(0, 0) = 1.$ Relative maximum

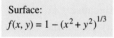

FIGURE 9.29 $f_x(x, y)$ and $f_y(x, y)$ are undefined at $(0, 0)$.

✓ CHECKPOINT 2

Find the relative extrema of

$$f(x, y) = \sqrt{1 - \frac{x^2}{16} - \frac{y^2}{4}}.$$

■

STUDY TIP

Note in the Second-Partials Test that if $d > 0$, then $f_{xx}(a, b)$ and $f_{yy}(a, b)$ must have the same sign. So, you can replace $f_{xx}(a, b)$ with $f_{yy}(a, b)$ in the first two parts of the test.

Algebra Review

For help in solving the system of equations

$$y - x^3 = 0$$
$$x - y^3 = 0$$

in Example 3, see Example 1(a) in the *Chapter 9 Algebra Review*, on page 620.

The Second-Partials Test for Relative Extrema

For functions such as those in Examples 1 and 2, you can determine the *types* of extrema at the critical points by sketching the graph of the function. For more complicated functions, a graphical approach is not so easy to use. The **Second-Partials Test** is an analytical test that can be used to determine whether a critical number yields a relative minimum, a relative maximum, or neither.

Second-Partials Test for Relative Extrema

Let f have continuous second partial derivatives on an open region containing (a, b) for which $f_x(a, b) = 0$ and $f_y(a, b) = 0$. To test for relative extrema of f, consider the quantity

$$d = f_{xx}(a, b)f_{yy}(a, b) - [f_{xy}(a, b)]^2.$$

1. If $d > 0$ and $f_{xx}(a, b) > 0$, then f has a **relative minimum** at (a, b).
2. If $d > 0$ and $f_{xx}(a, b) < 0$, then f has a **relative maximum** at (a, b).
3. If $d < 0$, then $(a, b, f(a, b))$ is a **saddle point.**
4. The test gives no information if $d = 0$.

Example 3 Applying the Second-Partials Test

Find the relative extrema and saddle points of $f(x, y) = xy - \frac{1}{4}x^4 - \frac{1}{4}y^4$.

SOLUTION Begin by finding the critical points of f. Because $f_x(x, y) = y - x^3$ and $f_y(x, y) = x - y^3$ are defined for all points in the xy-plane, the only critical points are those for which both first partial derivatives are zero. By solving the equations $y - x^3 = 0$ and $x - y^3 = 0$ simultaneously, you can determine that the critical points are $(1, 1)$, $(-1, -1)$, and $(0, 0)$. Furthermore, because

$$f_{xx}(x, y) = -3x^2, \quad f_{yy}(x, y) = -3y^2, \quad \text{and} \quad f_{xy}(x, y) = 1$$

you can use the quantity $d = f_{xx}(a, b)f_{yy}(a, b) - [f_{xy}(a, b)]^2$ to classify the critical points as shown.

Critical Point	d	$f_{xx}(x, y)$	Conclusion
$(1, 1)$	$(-3)(-3) - 1 = 8$	-3	Relative maximum
$(-1, -1)$	$(-3)(-3) - 1 = 8$	-3	Relative maximum
$(0, 0)$	$(0)(0) - 1 = -1$	0	Saddle point

The graph of f is shown in Figure 9.30.

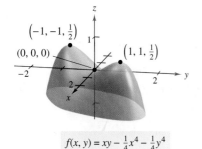

$f(x, y) = xy - \frac{1}{4}x^4 - \frac{1}{4}y^4$

FIGURE 9.30

✓**CHECKPOINT 3**

Find the relative extrema and saddle points of $f(x, y) = \dfrac{y^2}{16} - \dfrac{x^2}{4}$. ■

Applications of Extrema

Example 4 Finding a Maximum

A lake is to be stocked with smallmouth and largemouth bass. Let x represent the number of smallmouth bass and let y represent the number of largemouth bass in the lake. The weight of each fish is dependent on the population densities. After a six-month period, the weight of a single smallmouth bass is given by

$$W_1 = 3 - 0.002x - 0.001y$$

and the weight of a single largemouth bass is given by

$$W_2 = 4.5 - 0.004x - 0.005y.$$

Assuming that no fish die during the six-month period, how many smallmouth and largemouth bass should be stocked in the lake so that the total weight T of bass in the lake is a maximum?

SOLUTION The weight of the smallmouth bass population is xW_1, and the weight of the largemouth bass population is yW_2. So, the total weight function is

$$\begin{aligned}T &= xW_1 + yW_2 &&\text{Write total weight function.}\\ &= x(3 - 0.002x - 0.001y) + y(4.5 - 0.004x - 0.005y) &&\text{Substitute.}\\ &= 3x - 0.002x^2 + 4.5y - 0.005y^2 - 0.005xy. &&\text{Simplify.}\end{aligned}$$

The maximum total weight occurs when the two first partial derivatives are zero.

$$T_x = 3 - 0.004x - 0.005y = 0$$
$$T_y = 4.5 - 0.005x - 0.01y = 0$$

By solving this system simultaneously, you can conclude that the solution is $x = 500$ and $y = 200$. The second partial derivatives of T are

$$T_{xx} = -0.004, \quad T_{yy} = -0.01, \quad \text{and} \quad T_{xy} = -0.005.$$

Because $T_{xx} < 0$ and

$$d = T_{xx}(500, 200)T_{yy}(500, 200) - [T_{xy}(500, 200)]^2 = 0.000015 > 0$$

you can conclude by the Second-Partials Test that $(500, 200)$ yields a relative maximum. The lake should be stocked with 500 smallmouth bass and 200 largemouth bass, and the maximum total weight is

$$T(500, 200) = 1200 \text{ pounds}.$$

✓ CHECKPOINT 4

Repeat Example 4 assuming the weight of a single smallmouth bass is given by

$$W_1 = 2.5 - 0.002x - 0.0025y$$

and the weight of a single largemouth bass is given by

$$W_2 = 3 - 0.002x - 0.003y. \ ∎$$

Steve Maslowski/Photo Researchers, Inc.

Bass can help keep a pond healthy. In sufficient numbers, they keep other fish populations in check, thereby balancing the food chain.

Algebra Review

For help in solving the system of equations in Example 4, see Example 1(b) in the *Chapter 9 Algebra Review*, on page 620.

Algebra Review

For help in solving the system of equations

$$y(24 - 12x - 4y) = 0$$
$$x(24 - 6x - 8y) = 0$$

in Example 5, see Example 2(a) in the *Chapter 9 Algebra Review*, on page 621.

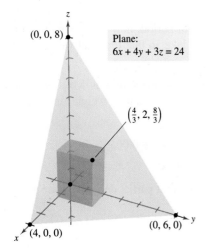

FIGURE 9.31

Example 5 Finding a Maximum Volume

Consider all possible rectangular boxes that are resting on the *xy*-plane with one vertex at the origin and the opposite vertex in the plane $6x + 4y + 3z = 24$, as shown in Figure 9.31. Of all such boxes, which has the greatest volume?

SOLUTION Because one vertex of the box lies in the plane given by $6x + 4y + 3z = 24$ or $z = \frac{1}{3}(24 - 6x - 4y)$, you can write the volume of the box as

$$V = xyz \quad\quad \text{Volume} = (\text{width})(\text{length})(\text{height})$$
$$= xy\left(\tfrac{1}{3}\right)(24 - 6x - 4y) \quad\quad \text{Substitute for } z.$$
$$= \tfrac{1}{3}(24xy - 6x^2y - 4xy^2). \quad\quad \text{Simplify.}$$

To find the critical numbers, set the first partial derivatives equal to zero.

$$V_x = \tfrac{1}{3}(24y - 12xy - 4y^2) \quad\quad \text{Partial with respect to } x$$
$$= \tfrac{1}{3}y(24 - 12x - 4y) = 0 \quad\quad \text{Factor and set equal to 0.}$$
$$V_y = \tfrac{1}{3}(24x - 6x^2 - 8xy) \quad\quad \text{Partial with respect to } y$$
$$= \tfrac{1}{3}x(24 - 6x - 8y) = 0 \quad\quad \text{Factor and set equal to 0.}$$

The four solutions of this system are $(0, 0)$, $(0, 6)$, $(4, 0)$, and $\left(\tfrac{4}{3}, 2\right)$. Using the Second-Partials Test, you can determine that the maximum volume occurs when the width is $x = \tfrac{4}{3}$ and the length is $y = 2$. For these values, the height of the box is

$$z = \tfrac{1}{3}\left[24 - 6\left(\tfrac{4}{3}\right) - 4(2)\right] = \tfrac{8}{3}.$$

So, the maximum volume is

$$V = xyz = \left(\tfrac{4}{3}\right)(2)\left(\tfrac{8}{3}\right) = \tfrac{64}{9} \text{ cubic units.}$$

✓ CHECKPOINT 5

Find the maximum volume of a box that is resting on the *xy*-plane with one vertex at the origin and the opposite vertex in the plane $2x + 4y + z = 8$. ■

CONCEPT CHECK

1. Given a function of two variables *f*, state how you can determine whether (x_0, y_0) is a critical point of *f*.
2. The point $(a, b, f(a, b))$ is a saddle point if what is true?
3. If $d > 0$ and $f_{xx}(a, b) > 0$, then what does *f* have at (a, b): a relative minimum or a relative maximum?
4. If $d > 0$ and $f_{xx}(a, b) < 0$, then what does *f* have at (a, b): a relative minimum or a relative maximum?

SECTION 9.5 Extrema of Functions of Two Variables

Skills Review 9.5

The following warm-up exercises involve skills that were covered in earlier sections. You will use these skills in the exercise set for this section. For additional help, review Sections 8.1 and 9.4.

In Exercises 1–8, solve the system of equations.

1. $\begin{cases} 5x = 15 \\ 3x - 2y = 5 \end{cases}$
2. $\begin{cases} \frac{1}{2}y = 3 \\ -x + 5y = 19 \end{cases}$
3. $\begin{cases} x + y = 5 \\ x - y = -3 \end{cases}$
4. $\begin{cases} x + y = 8 \\ 2x - y = 4 \end{cases}$
5. $\begin{cases} 2x - y = 8 \\ 3x - 4y = 7 \end{cases}$
6. $\begin{cases} 2x - 4y = 14 \\ 3x + y = 7 \end{cases}$
7. $\begin{cases} x^2 + x = 0 \\ 2yx + y = 0 \end{cases}$
8. $\begin{cases} 3y^2 + 6y = 0 \\ xy + x + 2 = 0 \end{cases}$

In Exercises 9–14, find all first and second partial derivatives of the function.

9. $z = 4x^3 - 3y^2$
10. $z = 2x^5 - y^3$
11. $z = x^4 - \sqrt{xy} + 2y$
12. $z = 2x^2 - 3xy + y^2$
13. $z = ye^{xy^2}$
14. $z = xe^{xy}$

Exercises 9.5

See www.CalcChat.com for worked-out solutions to odd-numbered exercises.

In Exercises 1–4, find any critical points and relative extrema of the function.

1. $f(x, y) = x^2 - y^2 + 4x - 8y - 11$
2. $f(x, y) = x^2 + y^2 + 2x - 6y + 6$
3. $f(x, y) = \sqrt{x^2 + y^2 + 1}$
4. $f(x, y) = \sqrt{25 - (x - 2)^2 - y^2}$

In Exercises 5–20, examine the function for relative extrema and saddle points.

5. $f(x, y) = (x - 1)^2 + (y - 3)^2$
6. $f(x, y) = 9 - (x - 3)^2 - (y + 2)^2$
7. $f(x, y) = 2x^2 + 2xy + y^2 + 2x - 3$
8. $f(x, y) = -x^2 - 5y^2 + 8x - 10y - 13$
9. $f(x, y) = -5x^2 + 4xy - y^2 + 16x + 10$
10. $f(x, y) = x^2 + 6xy + 10y^2 - 4y + 4$
11. $f(x, y) = 3x^2 + 2y^2 - 12x - 4y + 7$
12. $f(x, y) = -3x^2 - 2y^2 + 3x - 4y + 5$
13. $f(x, y) = x^2 - y^2 + 4x - 4y - 8$
14. $f(x, y) = x^2 - 3xy - y^2$
15. $f(x, y) = \frac{1}{2}xy$

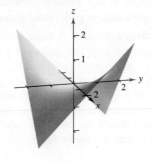

16. $f(x, y) = x + y + 2xy - x^2 - y^2$

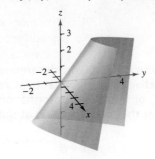

17. $f(x, y) = (x + y)e^{1 - x^2 - y^2}$

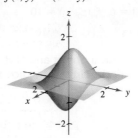

18. $f(x, y) = 3e^{-(x^2 + y^2)}$

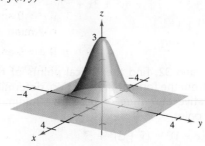

19. $f(x, y) = 4e^{xy}$

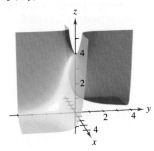

20. $f(x, y) = -\dfrac{3}{x^2 + y^2 + 1}$

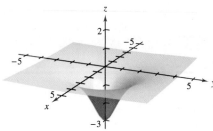

Think About It In Exercises 21–24, determine whether there is a relative maximum, a relative minimum, a saddle point, or insufficient information to determine the nature of the function $f(x, y)$ at the critical point (x_0, y_0).

21. $f_{xx}(x_0, y_0) = 9$, $f_{yy}(x_0, y_0) = 4$, $f_{xy}(x_0, y_0) = 6$

22. $f_{xx}(x_0, y_0) = -3$, $f_{yy}(x_0, y_0) = -8$, $f_{xy}(x_0, y_0) = 2$

23. $f_{xx}(x_0, y_0) = -9$, $f_{yy}(x_0, y_0) = 6$, $f_{xy}(x_0, y_0) = 10$

24. $f_{xx}(x_0, y_0) = 25$, $f_{yy}(x_0, y_0) = 8$, $f_{xy}(x_0, y_0) = 10$

In Exercises 25–30, find the critical points and test for relative extrema. List the critical points for which the Second-Partials Test fails.

25. $f(x, y) = (xy)^2$

26. $f(x, y) = \sqrt{x^2 + y^2}$

27. $f(x, y) = x^3 + y^3$

28. $f(x, y) = x^3 + y^3 - 3x^2 + 6y^2 + 3x + 12y + 7$

29. $f(x, y) = x^{2/3} + y^{2/3}$

30. $f(x, y) = (x^2 + y^2)^{2/3}$

In Exercises 31 and 32, find the critical points of the function and, from the form of the function, determine whether a relative maximum or a relative minimum occurs at each point.

31. $f(x, y, z) = (x - 1)^2 + (y + 3)^2 + z^2$

32. $f(x, y, z) = 6 - [x(y + 2)(z - 1)]^2$

In Exercises 33–36, find three positive numbers x, y, and z that satisfy the given conditions.

33. The sum is 30 and the product is a maximum.

34. The sum is 32 and $P = xy^2z$ is a maximum.

35. The sum is 30 and the sum of the squares is a minimum.

36. The sum is 1 and the sum of the squares is a minimum.

37. Medicine In order to treat a certain bacterial infection, a combination of two drugs is being tested. Studies have shown that the duration of the infection in laboratory tests can be modeled by

$$D(x, y) = x^2 + 2y^2 - 18x - 24y + 2xy + 120$$

where x is the dosage in hundreds of milligrams of the first drug and y is the dosage in hundreds of milligrams of the second drug. Determine the partial derivatives of D with respect to x and with respect to y. Find the amount of each drug necessary to minimize the duration of the infection.

38. Animal Shelter An animal shelter buys two different brands of dog food. The number of dogs that can be fed from x pounds of the first brand and y pounds of the second brand is given by the model

$$D(x, y) = -x^2 + 52x - y^2 + 44y - 1128.$$

(a) Find the number of pounds of each brand of dog food that should be ordered so that the maximum number of dogs can be fed.

(b) What is the maximum number of dogs that can be fed?

39. Volume Find the dimensions of a rectangular package of maximum volume that may be sent by a shipping company assuming that the sum of the length and the girth (perimeter of a cross section; see figure) cannot exceed 96 inches.

40. Volume Repeat Exercise 39 assuming that the sum of the length and the girth cannot exceed 144 inches.

41. Painting a Room A home improvement contractor is painting the walls and ceiling of a rectangular room. The volume of the room is 4608 cubic feet. The walls will be given three coats of paint and the ceiling will be given two coats of paint. Find the room dimensions that result in the minimum amount of paint being used. How many gallons of paint should the contractor buy? (Assume 1 gallon of paint covers 500 square feet.)

42. Weight A wooden storage crate has an open top (see figure). The volume of the crate is 8 cubic feet. The material for the base of the crate weighs twice as much per square foot as the material for the sides. Find the dimensions that minimize the weight of the crate.

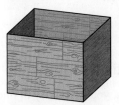

43. Gardening A home gardener plans to enclose two rectangular gardens with fencing. The dimensions of the gardens, shown in the figure, are in meters.

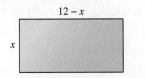
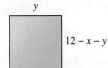

(a) Find the values of x and y that maximize the total area enclosed.

(b) What is the maximum total area enclosed?

(c) How many meters of fencing are needed?

44. Volume Find the values of x and y that maximize the total volume of the two rectangular solids in the figure. What is the maximum total volume?

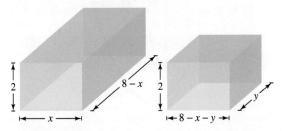

45. Hardy-Weinberg Law Common blood types are determined genetically by the three alleles A, B, and O. (An allele is any one of a group of possible mutational forms of a gene.) A person whose blood type is AA, BB, or OO is homozygous. A person whose blood type is AB, AO, or BO is heterozygous. The Hardy-Weinberg Law states that the proportion P of heterozygous individuals in any given population is modeled by

$$P(p, q, r) = 2pq + 2pr + 2qr$$

where p represents the percent of allele A in the population, q represents the percent of allele B in the population, and r represents the percent of allele O in the population. Use the fact that $p + q + r = 1$ (the sum of the three must equal 100%) to show that the maximum proportion of heterozygous individuals in any population is $\frac{2}{3}$.

46. Shannon Diversity Index One way to measure species diversity is to use the Shannon diversity index H. If a habitat consists of three species, A, B, and C its Shannon diversity index is

$$H = -x \ln x - y \ln y - z \ln z$$

where x is the percent of species A in the habitat, y is the percent of species B in the habitat, and z is the percent of species C in the habitat.

(a) Use the fact that $x + y + z = 1$ (the sum of the three must equal 100%) to show that the maximum value of H occurs when $x = y = z = \frac{1}{3}$.

(b) Use the results of part (a) to show that the maximum value of H in this habitat is $\ln 3$.

47. Biology A microbiologist must prepare a culture medium in which to grow a certain type of bacteria. The percent of salt contained in this medium is given by

$$S = 12xyz$$

where x, y, and z are the amounts of nutrient solutions (in liters) to be mixed in the medium. For the bacteria to grow, the medium must be 13% salt. Nutrient solutions x, y, and z cost \$1, \$2, and \$3 per liter, respectively. How much of each nutrient solution should be used to minimize the cost of the culture medium?

48. Biology Repeat Exercise 47 for a salt-content model of

$$S = 0.01x^2y^2z.$$

49. Heat Loss A heated storage room is shaped like a rectangular box and has a volume of 1000 cubic feet. Because warm air rises, the heat loss per unit of area through the ceiling is five times as great as the heat loss through the floor. The heat loss through the four walls is three times as great as the heat loss through the floor. Determine the room dimensions that will minimize heat loss and therefore minimize heating costs.

50. Volume A storage tank is in the shape of a cylinder with hemispherical ends (see figure). The tank must hold 1000 liters of fluid. Determine the radius r and length h that minimize the amount of material used in the construction of the tank.

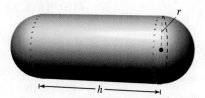

True or False? In Exercises 51 and 52, determine whether the statement is true or false. If it is false, explain why or give an example that shows it is false.

51. A saddle point always occurs at a critical point.

52. If $f(x, y)$ has a relative maximum at (x_0, y_0, z_0), then $f_x(x_0, y_0) = f_y(x_0, y_0) = 0$.

Section 9.6
Least Squares Regression Analysis

- Find the sums of the squared errors for mathematical models.
- Find the least squares regression lines for data.
- Find the least squares regression quadratics for data.

Measuring the Accuracy of a Mathematical Model

When seeking a mathematical model to fit real-life data, you should try to find a model that is both as *simple* and as *accurate* as possible. For instance, a simple linear model for the points shown in Figure 9.32(a) is

$$f(x) = 1.8566x - 5.0246. \qquad \text{Linear model}$$

However, Figure 9.32(b) shows that by choosing a slightly more complicated quadratic model

$$g(x) = 0.1996x^2 - 0.7281x + 1.3749 \qquad \text{Quadratic model}$$

you can obtain significantly greater accuracy.

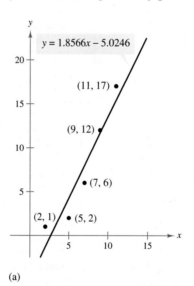
(a)

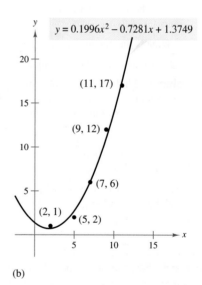
(b)

FIGURE 9.32

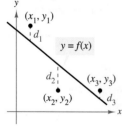

Sum of the squared errors:
$S = d_1^2 + d_2^2 + d_3^2$

FIGURE 9.33

To measure how well the model $y = f(x)$ fits a collection of points, sum the squares of the differences between the actual y-values and the model's y-values. This sum is called the **sum of the squared errors** and is denoted by S. Graphically, S can be interpreted as the sum of the squares of the vertical distances between the graph of f and the given points in the plane, as shown in Figure 9.33. If the model is a perfect fit, then $S = 0$. However, when a perfect fit is not feasible, you should use a model that minimizes S.

SECTION 9.6 Least Squares Regression Analysis

Definition of the Sum of the Squared Errors

The **sum of the squared errors** for the model $y = f(x)$ with respect to the points $(x_1, y_1), (x_2, y_2), \ldots, (x_n, y_n)$ is given by

$$S = [f(x_1) - y_1]^2 + [f(x_2) - y_2]^2 + \cdots + [f(x_n) - y_n]^2.$$

Example 1 Finding the Sums of the Squared Errors

Find the sums of the squared errors for the linear and quadratic models

$f(x) = 1.8566x - 5.0246$ Linear model

$g(x) = 0.1996x^2 - 0.7281x + 1.3749$ Quadratic model

(see Figure 9.32) with respect to the points

$(2, 1), (5, 2), (7, 6), (9, 12), (11, 17).$

SOLUTION Begin by evaluating each model at the given x-values, as shown in the table.

x	2	5	7	9	11
Actual y-value	1	2	6	12	17
Linear model, $f(x)$	-1.3114	4.2584	7.9716	11.6848	15.3980
Quadratic model, $g(x)$	0.7171	2.7244	6.0586	10.9896	17.5174

For the linear model f, the sum of the squared errors is

$S = (-1.3114 - 1)^2 + (4.2584 - 2)^2 + (7.9716 - 6)^2$
$\quad + (11.6848 - 12)^2 + (15.3980 - 17)^2$
$\approx 16.9959.$

Similarly, the sum of the squared errors for the quadratic model g is

$S = (0.7171 - 1)^2 + (2.7244 - 2)^2 + (6.0586 - 6)^2$
$\quad + (10.9896 - 12)^2 + (17.5174 - 17)^2$
$\approx 1.8968.$

STUDY TIP

In Example 1, note that the sum of the squared errors for the quadratic model is less than the sum of the squared errors for the linear model, which confirms that the quadratic model is a better fit.

✓ CHECKPOINT 1

Find the sums of the squared errors for the linear and quadratic models

$f(x) = 2.85x - 6.1$

$g(x) = 0.1964x^2 + 0.4929x - 0.6$

with respect to the points $(2, 1), (4, 5), (6, 9), (8, 16), (10, 24)$. Then decide which model is a better fit. ■

Least Squares Regression Line

The sum of the squared errors can be used to determine which of several models is the best fit for a collection of data. In general, if the sum of the squared errors of f is less than the sum of the squared errors of g, then f is said to be a better fit for the data than g. In regression analysis, you consider all possible models of a certain type. The one that is defined to be the best-fitting model is the one with the least sum of the squared errors. Example 2 shows how to use the optimization techniques described in Section 9.5 to find the best-fitting linear model for a collection of data.

Example 2 Finding the Best Linear Model

Algebra Review

For help in solving the system of equations in Example 2, see Example 2(b) in the *Chapter 9 Algebra Review*, on page 621.

Find the values of a and b such that the linear model

$$f(x) = ax + b$$

has a minimum sum of the squared errors for the points

$$(-3, 0), (-1, 1), (0, 2), (2, 3).$$

SOLUTION The sum of the squared errors is

$$\begin{aligned} S &= [f(x_1) - y_1]^2 + [f(x_2) - y_2]^2 + [f(x_3) - y_3]^2 + [f(x_4) - y_4]^2 \\ &= (-3a + b - 0)^2 + (-a + b - 1)^2 + (b - 2)^2 + (2a + b - 3)^2 \\ &= 14a^2 - 4ab + 4b^2 - 10a - 12b + 14. \end{aligned}$$

To find the values of a and b for which S is a minimum, you can use the techniques described in Section 9.5. That is, find the partial derivatives of S.

$$\frac{\partial S}{\partial a} = 28a - 4b - 10 \qquad \text{Differentiate with respect to } a.$$

$$\frac{\partial S}{\partial b} = -4a + 8b - 12 \qquad \text{Differentiate with respect to } b.$$

Next, set each partial derivative equal to zero.

$$28a - 4b - 10 = 0 \qquad \text{Set } \partial S/\partial a \text{ equal to 0.}$$

$$-4a + 8b - 12 = 0 \qquad \text{Set } \partial S/\partial b \text{ equal to 0.}$$

The solution of this system of linear equations is

$$a = \frac{8}{13} \quad \text{and} \quad b = \frac{47}{26}.$$

So, the best-fitting linear model for the given points is

$$f(x) = \frac{8}{13}x + \frac{47}{26}.$$

The graph of this model is shown in Figure 9.34.

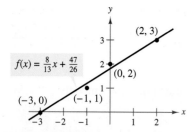

FIGURE 9.34

✓ CHECKPOINT 2

Find the values of a and b such that the linear model $f(x) = ax + b$ has a minimum sum of the squared errors for the points $(-2, 0), (0, 2), (2, 5), (4, 7)$. ∎

SECTION 9.6 Least Squares Regression Analysis

The line in Example 2 is called the **least squares regression line** for the given data. The solution shown in Example 2 can be generalized to find a formula for the least squares regression line, as shown below. Consider the linear model

$$f(x) = ax + b$$

and the points $(x_1, y_1), (x_2, y_2), \ldots, (x_n, y_n)$. The sum of the squared errors is

$$S = \sum_{i=1}^{n} [f(x_i) - y_i]^2 = \sum_{i=1}^{n} (ax_i + b - y_i)^2.$$

To minimize S, set the partial derivatives $\partial S/\partial a$ and $\partial S/\partial b$ equal to zero and solve for a and b. The results are summarized below.

The Least Squares Regression Line

The **least squares regression line** for the points

$$(x_1, y_1), (x_2, y_2), \ldots, (x_n, y_n)$$

is $y = ax + b$, where

$$a = \frac{n\sum_{i=1}^{n} x_i y_i - \sum_{i=1}^{n} x_i \sum_{i=1}^{n} y_i}{n\sum_{i=1}^{n} x_i^2 - \left(\sum_{i=1}^{n} x_i\right)^2} \quad \text{and} \quad b = \frac{1}{n}\left(\sum_{i=1}^{n} y_i - a\sum_{i=1}^{n} x_i\right).$$

In the formula for the least squares regression line, note that if the x-values are symmetrically spaced about zero, then

$$\sum_{i=1}^{n} x_i = 0$$

and the formulas for a and b simplify to

$$a = \frac{n\sum_{i=1}^{n} x_i y_i}{n\sum_{i=1}^{n} x_i^2} \quad \text{and} \quad b = \frac{1}{n}\sum_{i=1}^{n} y_i.$$

Note also that only the *development* of the least squares regression line involves partial derivatives. The *application* of this formula is simply a matter of computing the values of a and b—a task that is performed much more simply on a calculator or a computer than by hand.

DISCOVERY

Graph the three points $(2, 2)$, $(2, 1)$, and $(2.1, 1.5)$ and visually estimate the least squares regression line for these data. Now use the formulas on this page or a graphing utility to show that the equation of the line is actually $y = 1.5$. In general, the least squares regression line for "nearly vertical" data can be quite unusual. Show that by interchanging the roles of x and y, you can obtain a better linear approximation.

Example 3 Modeling Temperature

The table shows the chirp rates n (in number of chirps per minute) of snowy tree crickets at various air temperatures T (in degrees Fahrenheit). Find the least squares regression line for the data and use the result to estimate the temperature at which the snowy tree cricket chirps 172 times per minute.

n	43	68	80	108	131	152
T	50	55	60	65	70	75

SOLUTION You need to find the linear model that best fits the points

(43, 50), (68, 55), (80, 60), (108, 65), (131, 70), (152, 75).

Using a calculator with a built-in least squares regression program, you can determine that the best-fitting line is $T = 40.4 + 0.23n$. With this model, you can estimate the temperature, using $n = 172$, to be

$$T = 40.4 + 0.23(172) \approx 80°F.$$

This result is shown graphically in Figure 9.35.

✓ CHECKPOINT 3

The weights w (in grams) of infants and their body surface areas B (in square centimeters) are shown in the table. Find the least squares regression line for the data and use the result to estimate the body surface area of an infant that weighs 9000 grams.

w	5000	5980	6200	6850	7420	8170	9340	10,180
B	3050	3365	3440	3650	3860	4120	4530	4825

TECHNOLOGY

Most graphing utilities and spreadsheet software programs have a built-in linear regression program. When you run such a program, the "r-value" gives a measure of how well the model fits the data. The closer the value of $|r|$ is to 1, the better the fit. For the data in Example 3, $r \approx 0.997$, which implies that the model is a very good fit. Use a graphing utility or a spreadsheet software program to find the least squares regression line and compare your results with those in Example 3. (Consult the user's manual of a graphing utility or a spreadsheet software program for specific instructions.)*

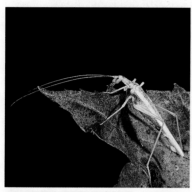

© Peter Arnold, Inc./Alamy

The chirp rate of a snowy tree cricket and the air temperature have a linear relationship. A formula published in 1897 by Amos Dolbear gave a linear relationship between the chirp rate of crickets and temperature. His formula can be simplified to counting the number of chirps in 15 seconds and adding 40 to estimate the temperature in degrees Fahrenheit.

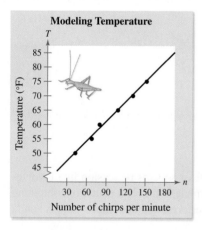

FIGURE 9.35

---

*Specific calculator keystroke instructions for operations in this and other technology boxes can be found at *college.hmco.com/info/larsonapplied*.

Least Squares Regression Quadratic

When using regression analysis to model data, remember that the least squares regression line provides only the best *linear* model for a set of data. It does not necessarily provide the best possible model. For instance, in Example 1, you saw that the quadratic model was a better fit than the linear model.

Regression analysis can be performed with many different types of models, such as exponential or logarithmic models. The following development shows how to find the best-fitting quadratic model for a collection of data points. Consider a quadratic model of the form

$$f(x) = ax^2 + bx + c.$$

The sum of the squared errors for this model is

$$S = \sum_{i=1}^{n}[f(x_i) - y_i]^2 = \sum_{i=1}^{n}(ax_i^2 + bx_i + c - y_i)^2.$$

To find the values of a, b, and c that minimize S, set the three partial derivatives, $\partial S/\partial a$, $\partial S/\partial b$, and $\partial S/\partial c$, equal to zero.

$$\frac{\partial S}{\partial a} = \sum_{i=1}^{n} 2x_i^2(ax_i^2 + bx_i + c - y_i) = 0$$

$$\frac{\partial S}{\partial b} = \sum_{i=1}^{n} 2x_i(ax_i^2 + bx_i + c - y_i) = 0$$

$$\frac{\partial S}{\partial c} = \sum_{i=1}^{n} 2(ax_i^2 + bx_i + c - y_i) = 0$$

By expanding this system, you obtain the result given in the summary below.

Least Squares Regression Quadratic

The **least squares regression quadratic** for the points

$$(x_1, y_1), (x_2, y_2), \ldots, (x_n, y_n)$$

is $y = ax^2 + bx + c$, where a, b, and c are the solutions of the system of equations below.

$$a\sum_{i=1}^{n} x_i^4 + b\sum_{i=1}^{n} x_i^3 + c\sum_{i=1}^{n} x_i^2 = \sum_{i=1}^{n} x_i^2 y_i$$

$$a\sum_{i=1}^{n} x_i^3 + b\sum_{i=1}^{n} x_i^2 + c\sum_{i=1}^{n} x_i = \sum_{i=1}^{n} x_i y_i$$

$$a\sum_{i=1}^{n} x_i^2 + b\sum_{i=1}^{n} x_i + cn = \sum_{i=1}^{n} y_i$$

TECHNOLOGY

Most graphing utilities have a built-in program for finding the least squares regression quadratic. This program works just like the program for the least squares regression line. You should use this program to verify your solutions to the exercises.

Example 4 Modeling Autism Cases

The numbers y (in thousands) of students ages 12 to 17 served by an autism disability program in the United States from 1997 through 2005 are shown in the table. Find the least squares regression quadratic for the data and use the result to estimate the number of students in 2010. *(Source: U.S. Department of Education)*

Year	1997	1998	1999	2000	2001	2002	2003	2004	2005
y	12	15	19	22	29	37	47	60	72

SOLUTION Let t represent the year, with $t = 7$ corresponding to 1997. Then you need to find the quadratic model that best fits the points

$(7, 12)$, $(8, 15)$, $(9, 19)$, $(10, 22)$, $(11, 29)$, $(12, 37)$,
$(13, 47)$, $(14, 60)$, $(15, 72)$.

Using a calculator with a built-in least squares regression program, you can determine that the best-fitting quadratic is $y = 0.83t^2 - 10.8t + 48$. With this model, you can estimate the number of students ages 12 to 17 served by an autism disability program in 2010, using $t = 20$, to be

$$y = 0.83(20)^2 - 10.8(20) + 48 = 164 \text{ thousand}.$$

This result is shown graphically in Figure 9.36.

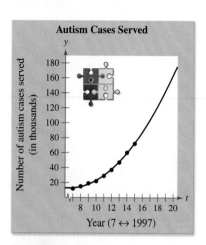

FIGURE 9.36

✓ CHECKPOINT 4

The numbers y (in thousands) of students ages 12 to 17 served by any disability program in the United States from 1997 through 2005 are shown in the table. Find the least squares regression quadratic for the data and use the result to estimate the number of students in 2010. *(Source: U.S. Department of Education)*

Year	1997	1998	1999	2000	2001	2002	2003	2004	2005
y	2412	2499	2596	2690	2797	2904	2970	3017	3018

CONCEPT CHECK

1. What are the two main goals when seeking a mathematical model to fit real-life data?
2. What does S, the sum of the squared errors, measure?
3. Describe how to find the least squares regression line for a given set of data.
4. Describe how to find the least squares regression quadratic for a given set of data.

Skills Review 9.6

The following warm-up exercises involve skills that were covered in earlier sections. You will use these skills in the exercise set for this section. For additional help, review Sections 0.3 and 9.4.

In Exercises 1 and 2, evaluate the expression.

1. $(2.5 - 1)^2 + (3.25 - 2)^2 + (4.1 - 3)^2$

2. $(1.1 - 1)^2 + (2.08 - 2)^2 + (2.95 - 3)^2$

In Exercises 3 and 4, find the partial derivatives of S.

3. $S = a^2 + 6b^2 - 4a - 8b - 4ab + 6$

4. $S = 4a^2 + 9b^2 - 6a - 4b - 2ab + 8$

In Exercises 5–10, evaluate the sum.

5. $\sum_{i=1}^{5} i$

6. $\sum_{i=1}^{6} 2i$

7. $\sum_{i=1}^{4} \frac{1}{i}$

8. $\sum_{i=1}^{3} i^2$

9. $\sum_{i=1}^{6} (2 - i)^2$

10. $\sum_{i=1}^{5} (30 - i^2)$

Exercises 9.6

See www.CalcChat.com for worked-out solutions to odd-numbered exercises.

In Exercises 1–6, (a) find the least squares regression line and (b) calculate S, the sum of the squared errors. Use the regression capabilities of a graphing utility or a spreadsheet to verify your results.

1.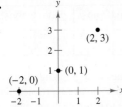

2.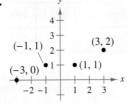

3.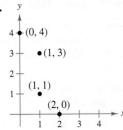

4.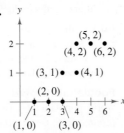

5.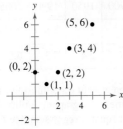

6.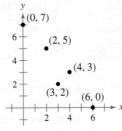

In Exercises 7–10, find the least squares regression line for the points. Use the regression capabilities of a graphing utility or a spreadsheet to verify your results. Then plot the points and graph the regression line.

7. $(-2, -1), (0, 0), (2, 3)$

8. $(-3, 0), (-1, 1), (1, 1), (3, 2)$

9. $(-2, 4), (-1, 1), (0, -1), (1, -3)$

10. $(-5, -3), (-4, -2), (-2, -1), (-1, 1)$

In Exercises 11–20, use the regression capabilities of a graphing utility or a spreadsheet to find the least squares regression line for the given points.

11. $(-2, 0), (-1, 1), (0, 1), (1, 2), (2, 3)$

12. $(-4, -1), (-2, 0), (2, 4), (4, 5)$

13. $(-2, 2), (2, 6), (3, 7)$

14. $(-5, 1), (1, 3), (2, 3), (2, 5)$

15. $(-3, 4), (-1, 2), (1, 1), (3, 0)$

16. $(-10, 10), (-5, 8), (3, 6), (7, 4), (5, 0)$

17. $(0, 0), (1, 1), (3, 4), (4, 2), (5, 5)$

18. $(1, 0), (3, 3), (5, 6)$

19. $(0, 6), (4, 3), (5, 0), (8, -4), (10, -5)$

20. $(6, 4), (1, 2), (3, 3), (8, 6), (11, 8), (13, 8)$

In Exercises 21–24, use the regression capabilities of a graphing utility or a spreadsheet to find the least squares regression quadratic for the given points. Then plot the points and graph the least squares regression quadratic.

21. $(-2, 0), (-1, 0), (0, 1), (1, 2), (2, 5)$
22. $(-4, 5), (-2, 6), (2, 6), (4, 2)$
23. $(0, 0), (2, 2), (3, 6), (4, 12)$
24. $(0, 10), (1, 9), (2, 6), (3, 0)$

In Exercises 25–28, use the regression capabilities of a graphing utility or a spreadsheet to find linear and quadratic models for the data. State which model best fits the data.

25. $(-4, 1), (-3, 2), (-2, 2), (-1, 4), (0, 6), (1, 8), (2, 9)$
26. $(-1, -4), (0, -3), (1, -3), (2, 0), (4, 5), (6, 9), (9, 3)$
27. $(0, 769), (1, 677), (2, 601), (3, 543), (4, 489), (5, 411)$
28. $(1, 10.3), (2, 14.2), (3, 18.9), (4, 23.7), (5, 29.1), (6, 35)$

29. **Baseball Participation** The numbers y of 45- to 54-year-olds (in thousands) who participated in baseball as a leisure activity from 1999 through 2004, where $t = 9$ corresponds to 1999, are given in the table. Use the regression capabilities of a graphing utility or a spreadsheet to find the least squares regression line for the data. Estimate the number of 45- to 54-year-olds who participated in baseball in 2005. *(Source: National Sporting Goods Association)*

Year, t	9	10	11	12	13	14
Participants, y	702	737	838	847	971	1058

30. **Agriculture** An agronomist used four test plots to determine the relationship between the wheat yield y (in bushels per acre) and the amount of fertilizer x (in hundreds of pounds per acre). The results are shown in the table.

Fertilizer, x	1.0	1.5	2.0	2.5
Yield, y	35	44	50	56

(a) Use the regression capabilities of a graphing utility or a spreadsheet to find the least squares regression line for the data.

(b) Estimate the yield for a fertilizer application of 160 pounds per acre.

31. **Finance: Median Income** The table shows the median income levels for various age levels in the United States. Use the regression capabilities of a graphing utility or a spreadsheet to find the least squares regression quadratic for the data and use the resulting model to estimate the median income for someone who is 28 years old. *(Source: U.S. Census Bureau)*

Age level, x	20	30	40
Median income, y	28,800	47,400	58,100

Age level, x	50	60	70
Median income, y	62,400	52,300	26,000

Table for 31

32. **Engineering** After the development of a new turbocharger for an automobile engine, the experimental data listed in the table were obtained for speed in miles per hour at two-second intervals.

Time, x	0	2	4	6	8	10
Speed, y	0	20	40	60	75	80

(a) Use the regression capabilities of a graphing utility or a spreadsheet to find the least squares regression quadratic for the data.

(b) Use the model to estimate the speed after 5 seconds.

33. **Infant Mortality** To study the numbers y of infant deaths per 1000 live births in the United States, a medical researcher obtains the data listed in the table. *(Source: U.S. National Center for Health Statistics)*

Year	1980	1985	1990	1995	2000	2005
Deaths, y	12.6	10.6	9.2	7.6	6.9	6.8

(a) Use the regression capabilities of a graphing utility or a spreadsheet to find the least squares regression line for the data and use this line to estimate the number of infant deaths in 2010. Let $t = 0$ represent 1980.

(b) Use the regression capabilities of a graphing utility or a spreadsheet to find the least squares regression quadratic for the data and use the model to estimate the number of infant deaths in 2010.

34. **Population Growth** The table gives the approximate world populations y (in billions) for six different years. *(Source: U.S. Census Bureau)*

Year	1800	1850	1900	1950	1990	2005
Time, t	-2	-1	0	1	1.8	2.1
Population, y	0.8	1.1	1.6	2.4	5.3	6.5

(a) During the 1800s, population growth was almost linear. Use the regression capabilities of a graphing utility or a spreadsheet to find a least squares regression line for those years and use the line to estimate the population in 1875.

(b) Use the regression capabilities of a graphing utility or a spreadsheet to find a least squares regression quadratic for the data from 1850 through 2005 and use the model to estimate the population in the year 2010.

(c) Even though the rate of growth of the population has begun to decline, most demographers believe the population size will pass the 8 billion mark sometime in the next 25 years. What do you think?

35. MAKE A DECISION: METEOROLOGY A meteorologist measures the atmospheric pressure P (in kilograms per square meter) at altitude h (in kilometers). The data are shown below.

Altitude, h	0	5	10	15	20
Pressure, P	10,332	5583	2376	1240	517

(a) Use a graphing utility or a spreadsheet to create a scatter plot of the data.

(b) Use the regression capabilities of a graphing utility or a spreadsheet to find an appropriate model for the data.

(c) Explain why you chose the type of model that you created in part (b).

36. MAKE A DECISION: OPTOMETRY The endpoints of the interval over which distinct vision is possible are called the near point and far point of the eye. With increasing age, these points normally change. The table shows the approximate near points y (in inches) for various ages x (in years).

Age, x	16	32	44	50	60
Near point, y	3.0	4.7	9.8	19.7	39.4

(a) Use a graphing utility or a spreadsheet to create a scatter plot of the data.

(b) Use the regression capabilities of a graphing utility or a spreadsheet to find an appropriate model for the data.

(c) Explain why you chose the type of model that you created in part (b).

In Exercises 37–40, use the regression capabilities of a graphing utility or a spreadsheet to find any model that best fits the data points.

37. (1, 13), (2, 16.5), (4, 24), (5, 28), (8, 39), (11, 50.25), (17, 72), (20, 85)

38. (1, 5.5), (3, 7.75), (6, 15.2), (8, 23.5), (11, 46), (15, 110)

39. (1, 1.5), (2.5, 8.5), (5, 13.5), (8, 16.7), (9, 18), (20, 22)

40. (0, 0.5), (1, 7.6), (3, 60), (4.2, 117), (5, 170), (7.9, 380)

In Exercises 41–46, plot the points and determine whether the data have positive, negative, or no linear correlation (see figures below). Then use a graphing utility to find the value of r and confirm your result. The number r is called the *correlation coefficient*. It is a measure of how well the model fits the data. Correlation coefficients vary between -1 and 1, and the closer $|r|$ is to 1, the better the model.

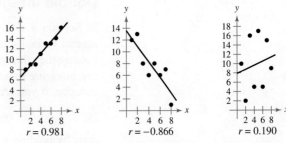

Positive correlation   Negative correlation   No correlation

41. (1, 4), (2, 6), (3, 8), (4, 11), (5, 13), (6, 15)

42. (1, 7.5), (2, 7), (3, 7), (4, 6), (5, 5), (6, 4.9)

43. (1, 3), (2, 6), (3, 2), (4, 3), (5, 9), (6, 1)

44. (0.5, 2), (0.75, 1.75), (1, 3), (1.5, 3.2), (2, 3.7), (2.6, 4)

45. (1, 36), (2, 10), (3, 0), (4, 4), (5, 16), (6, 36)

46. (0.5, 9), (1, 8.5), (1.5, 7), (2, 5.5), (2.5, 5), (3, 3.5)

True or False? In Exercises 47–52, determine whether the statement is true or false. If it is false, explain why or give an example that shows it is false.

47. Data that are modeled by $y = 3.29x - 4.17$ have a negative correlation.

48. Data that are modeled by $y = -0.238x + 25$ have a negative correlation.

49. If the correlation coefficient is $r \approx -0.98781$, the model is a good fit.

50. A correlation coefficient of $r \approx 0.201$ implies that the data have no correlation.

51. A linear regression model with a positive correlation will have a slope that is greater than 0.

52. If the correlation coefficient for a linear regression model is close to -1, the regression line cannot be used to describe the data.

53. Extended Application To work an extended application analyzing the number of dentist office employees and nursing care facility employees in the United States from 1999 through 2005, visit this text's website at *college.hmco.com/info/larsonapplied*. (Data Source: U.S. Bureau of Labor Statistics)

Section 9.7
Double Integrals and Area in the Plane

- Evaluate double integrals.
- Use double integrals to find the areas of regions.

Double Integrals

In Section 9.4, you learned that it is meaningful to differentiate functions of several variables by differentiating with respect to one variable at a time while holding the other variable(s) constant. It should not be surprising to learn that you can *integrate* functions of two or more variables using a similar procedure. For instance, if you are given the partial derivative

$$f_x(x, y) = 2xy \qquad \text{Partial with respect to } x$$

then, by holding y constant, you can integrate with respect to x to obtain

$$\int f_x(x, y)\, dx = f(x, y) + C(y)$$
$$= x^2 y + C(y).$$

This procedure is called **partial integration with respect to x.** Note that the "constant of integration" $C(y)$ is assumed to be a function of y, because y is fixed during integration with respect to x. Similarly, if you are given the partial derivative

$$f_y(x, y) = x^2 + 2 \qquad \text{Partial with respect to } y$$

then, by holding x constant, you can integrate with respect to y to obtain

$$\int f_y(x, y)\, dy = f(x, y) + C(x)$$
$$= x^2 y + 2y + C(x).$$

In this case, the "constant of integration" $C(x)$ is assumed to be a function of x, because x is fixed during integration with respect to y.

To evaluate a definite integral of a function of two or more variables, you can apply the Fundamental Theorem of Calculus to one variable while holding the other variable(s) constant, as shown.

$$\int_1^{2y} 2xy\, dx = x^2 y \Big]_1^{2y} = (2y)^2 y - (1)^2 y$$

x is the variable of integration and y is fixed.

Replace x by the limits of integration.

$$= 4y^3 - y.$$

The result is a function of y.

Note that you omit the constant of integration, just as you do for a definite integral of a function of one variable.

SECTION 9.7 Double Integrals and Area in the Plane

Example 1 Finding Partial Integrals

Find each partial integral.

a. $\int_1^x (2x^2 y^{-2} + 2y) \, dy$ **b.** $\int_y^{5y} \sqrt{x-y} \, dx$

SOLUTION

a. $\int_1^x (2x^2 y^{-2} + 2y) \, dy = \left[\dfrac{-2x^2}{y} + y^2 \right]_1^x$ Hold x constant.

$= \left(\dfrac{-2x^2}{x} + x^2 \right) - \left(\dfrac{-2x^2}{1} + 1 \right)$

$= 3x^2 - 2x - 1$

b. $\int_y^{5y} \sqrt{x-y} \, dx = \left[\dfrac{2}{3}(x-y)^{3/2} \right]_y^{5y}$ Hold y constant.

$= \dfrac{2}{3}[(5y-y)^{3/2} - (y-y)^{3/2}] = \dfrac{16}{3} y^{3/2}$

✓ CHECKPOINT 1

Find each partial integral.

a. $\int_1^x (4xy + y^3) \, dy$

b. $\int_y^{y^2} \dfrac{1}{x+y} \, dx$

In Example 1(a), note that the definite integral defines a function of x and can *itself* be integrated. An "integral of an integral" is called a **double integral**. With a function of two variables, there are two types of double integrals.

$$\int_a^b \int_{g_1(x)}^{g_2(x)} f(x,y) \, dy \, dx = \int_a^b \left[\int_{g_1(x)}^{g_2(x)} f(x,y) \, dy \right] dx$$

$$\int_a^b \int_{g_1(y)}^{g_2(y)} f(x,y) \, dx \, dy = \int_a^b \left[\int_{g_1(y)}^{g_2(y)} f(x,y) \, dx \right] dy$$

STUDY TIP

Notice that the difference between the two types of double integrals is the order in which the integration is performed, $dy \, dx$ or $dx \, dy$.

TECHNOLOGY

A symbolic integration utility can be used to evaluate double integrals. To do this, you need to enter the integrand, then integrate twice—once with respect to one of the variables and then with respect to the other variable. Use a symbolic integration utility to evaluate the double integral in Example 2.

Example 2 Evaluating a Double Integral

Evaluate $\int_1^2 \int_0^x (2xy + 3) \, dy \, dx$.

SOLUTION

$\int_1^2 \int_0^x (2xy + 3) \, dy \, dx = \int_1^2 \left[\int_0^x (2xy + 3) \, dy \right] dx$

$= \int_1^2 \left[xy^2 + 3y \right]_0^x dx$

$= \int_1^2 (x^3 + 3x) \, dx$

$= \left[\dfrac{x^4}{4} + \dfrac{3x^2}{2} \right]_1^2$

$= \left(\dfrac{2^4}{4} + \dfrac{3(2^2)}{2} \right) - \left(\dfrac{1^4}{4} + \dfrac{3(1^2)}{2} \right) = \dfrac{33}{4}$

✓ CHECKPOINT 2

Evaluate the double integral.

$\int_1^2 \int_0^x (5x^2 y - 2) \, dy \, dx$

Finding Area with a Double Integral

One of the simplest applications of a double integral is finding the area of a plane region. For instance, consider the region R that is bounded by

$$a \leq x \leq b \quad \text{and} \quad g_1(x) \leq y \leq g_2(x).$$

Using the techniques described in Section 6.5, you know that the area of R is

$$\int_a^b [g_2(x) - g_1(x)]\, dx.$$

This same area is also given by the double integral

$$\int_a^b \int_{g_1(x)}^{g_2(x)} dy\, dx$$

because

$$\int_a^b \int_{g_1(x)}^{g_2(x)} dy\, dx = \int_a^b \left[y \right]_{g_1(x)}^{g_2(x)} dx = \int_a^b [g_2(x) - g_1(x)]\, dx.$$

Figure 9.37 shows the two basic types of plane regions whose areas can be determined by double integrals.

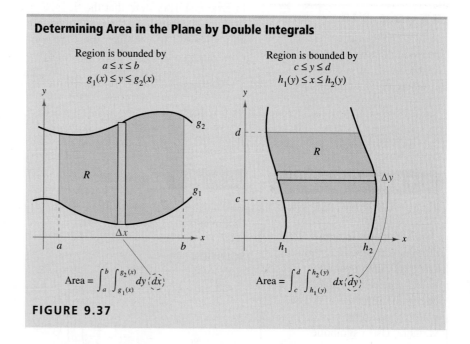

FIGURE 9.37

STUDY TIP

In Figure 9.37, note that the horizontal or vertical orientation of the narrow rectangle indicates the order of integration. The "outer" variable of integration always corresponds to the width of the rectangle. Notice also that the outer limits of integration for a double integral are constant, whereas the inner limits may be functions of the outer variable.

SECTION 9.7 Double Integrals and Area in the Plane

Example 3 Finding Area with a Double Integral

Use a double integral to find the area of the rectangular region shown in Figure 9.38.

SOLUTION The bounds for x are $1 \leq x \leq 5$ and the bounds for y are $2 \leq y \leq 4$. So, the area of the region is

$$\int_1^5 \int_2^4 dy\, dx = \int_1^5 \Big[y\Big]_2^4 dx \qquad \text{Integrate with respect to } y.$$

$$= \int_1^5 (4 - 2)\, dx \qquad \text{Apply Fundamental Theorem of Calculus.}$$

$$= \int_1^5 2\, dx \qquad \text{Simplify.}$$

$$= \Big[2x\Big]_1^5 \qquad \text{Integrate with respect to } x.$$

$$= 10 - 2 \qquad \text{Apply Fundamental Theorem of Calculus.}$$

$$= 8 \text{ square units.} \qquad \text{Simplify.}$$

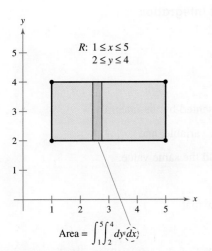

FIGURE 9.38

You can confirm this by noting that the rectangle measures two units by four units.

✓ CHECKPOINT 3

Use a double integral to find the area of the rectangular region shown in Example 3 by integrating with respect to x and then with respect to y. ■

Example 4 Finding Area with a Double Integral

Use a double integral to find the area of the region bounded by the graphs of $y = x^2$ and $y = x^3$.

SOLUTION As shown in Figure 9.39, the two graphs intersect when $x = 0$ and $x = 1$. Choosing x to be the outer variable, the bounds for x are $0 \leq x \leq 1$ and the bounds for y are $x^3 \leq y \leq x^2$. This implies that the area of the region is

$$\int_0^1 \int_{x^3}^{x^2} dy\, dx = \int_0^1 \Big[y\Big]_{x^3}^{x^2} dx \qquad \text{Integrate with respect to } y.$$

$$= \int_0^1 (x^2 - x^3)\, dx \qquad \text{Apply Fundamental Theorem of Calculus.}$$

$$= \left[\frac{x^3}{3} - \frac{x^4}{4}\right]_0^1 \qquad \text{Integrate with respect to } x.$$

$$= \frac{1}{3} - \frac{1}{4} \qquad \text{Apply Fundamental Theorem of Calculus.}$$

$$= \frac{1}{12} \text{ square unit.} \qquad \text{Simplify.}$$

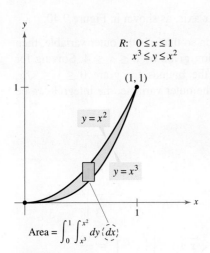

FIGURE 9.39

✓ CHECKPOINT 4

Use a double integral to find the area of the region bounded by the graphs of $y = 2x$ and $y = x^2$. ■

In setting up double integrals, the most difficult task is likely to be determining the correct limits of integration. This can be simplified by making a sketch of the region R and identifying the appropriate bounds for x and y.

Example 5 Changing the Order of Integration

For the double integral

$$\int_0^2 \int_{y^2}^4 dx\, dy$$

a. sketch the region R whose area is represented by the integral,

b. rewrite the integral so that x is the outer variable, and

c. show that both orders of integration yield the same value.

SOLUTION

a. From the limits of integration, you know that

$$y^2 \leq x \leq 4 \qquad \text{Variable bounds for } x$$

which means that the region R is bounded on the left by the parabola $x = y^2$ and on the right by the line $x = 4$. Furthermore, because

$$0 \leq y \leq 2 \qquad \text{Constant bounds for } y$$

you know that the region lies above the x-axis, as shown in Figure 9.40.

b. If you interchange the order of integration so that x is the outer variable, then x will have constant bounds of integration given by $0 \leq x \leq 4$. Solving for y in the equation $x = y^2$ implies that the bounds for y are $0 \leq y \leq \sqrt{x}$, as shown in Figure 9.41. So, with x as the outer variable, the integral can be written as

$$\int_0^4 \int_0^{\sqrt{x}} dy\, dx.$$

c. Both integrals yield the same value.

$$\int_0^2 \int_{y^2}^4 dx\, dy = \int_0^2 \Big[x\Big]_{y^2}^4 dy = \int_0^2 (4 - y^2)\, dy = \left[4y - \frac{y^3}{3}\right]_0^2 = \frac{16}{3}$$

$$\int_0^4 \int_0^{\sqrt{x}} dy\, dx = \int_0^4 \Big[y\Big]_0^{\sqrt{x}} dx = \int_0^4 \sqrt{x}\, dx = \left[\frac{2}{3}x^{3/2}\right]_0^4 = \frac{16}{3}$$

✓ CHECKPOINT 5

For the double integral $\int_0^2 \int_{2y}^4 dx\, dy$,

a. sketch the region R whose area is represented by the integral,

b. rewrite the integral so that x is the outer variable, and

c. show that both orders of integration yield the same value. ■

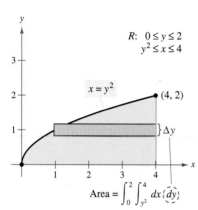

FIGURE 9.40

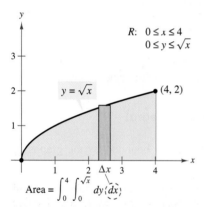

FIGURE 9.41

STUDY TIP

To designate a double integral or an area of a region without specifying a particular order of integration, you can use the symbol

$$\iint_R dA$$

where $dA = dx\, dy$ or $dA = dy\, dx$.

SECTION 9.7 Double Integrals and Area in the Plane

Example 6 Finding Area with a Double Integral

Use a double integral to calculate the area denoted by

$$\int_R \int dA$$

where R is the region bounded by $y = x$ and $y = x^2 - x$.

SOLUTION Begin by sketching the region R, as shown in Figure 9.42. From the sketch, you can see that vertical rectangles of width dx are more convenient than horizontal ones. So, x is the outer variable of integration and its constant bounds are $0 \le x \le 2$. This implies that the bounds for y are $x^2 - x \le y \le x$, and the area is given by

$$\int_R \int dA = \int_0^2 \int_{x^2-x}^x dy\, dx \qquad \text{Substitute bounds for region.}$$

$$= \int_0^2 \Big[y \Big]_{x^2-x}^x dx \qquad \text{Integrate with respect to } y.$$

$$= \int_0^2 [x - (x^2 - x)]\, dx \qquad \text{Apply Fundamental Theorem of Calculus.}$$

$$= \int_0^2 (2x - x^2)\, dx \qquad \text{Simplify.}$$

$$= \left[x^2 - \frac{x^3}{3} \right]_0^2 \qquad \text{Integrate with respect to } x.$$

$$= 4 - \frac{8}{3} \qquad \text{Apply Fundamental Theorem of Calculus.}$$

$$= \frac{4}{3} \text{ square units.} \qquad \text{Simplify.}$$

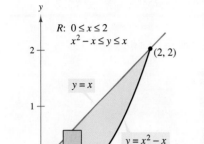

FIGURE 9.42

✓ **CHECKPOINT 6**

Use a double integral to calculate the area denoted by $\int_R \int dA$ where R is the region bounded by $y = 2x + 3$ and $y = x^2$. ■

As you are working the exercises for this section, you should be aware that the primary uses of double integrals will be discussed in Section 9.8. Double integrals by way of areas in the plane have been introduced so that you can gain practice in finding the limits of integration. When setting up a double integral, remember that your first step should be to sketch the region R. After doing this, you have two choices of integration orders: $dx\, dy$ or $dy\, dx$.

CONCEPT CHECK

1. What is an "integral of an integral" called?
2. In the double integral $\int_0^2 \int_0^1 dy\, dx$, in what order is the integration performed? (Do not integrate.)
3. True or false: Changing the order of integration will sometimes change the value of a double integral.
4. To designate a double integral or an area of a region without specifying a particular order of integration, what symbol can you use?

Skills Review 9.7

The following warm-up exercises involve skills that were covered in earlier sections. You will use these skills in the exercise set for this section. For additional help, review Sections 6.2–6.5.

In Exercises 1–12, evaluate the definite integral.

1. $\int_0^1 dx$
2. $\int_0^2 3\, dy$
3. $\int_1^4 2x^2\, dx$
4. $\int_0^1 2x^3\, dx$
5. $\int_1^2 (x^3 - 2x + 4)\, dx$
6. $\int_0^2 (4 - y^2)\, dy$
7. $\int_1^2 \dfrac{2}{7x^2}\, dx$
8. $\int_1^4 \dfrac{2}{\sqrt{x}}\, dx$
9. $\int_0^2 \dfrac{2x}{x^2+1}\, dx$
10. $\int_2^e \dfrac{1}{y-1}\, dy$
11. $\int_0^2 xe^{x^2+1}\, dx$
12. $\int_0^1 e^{-2y}\, dy$

In Exercises 13–16, sketch the region bounded by the graphs of the equations.

13. $y = x$, $y = 0$, $x = 3$
14. $y = x$, $y - 3$, $x = 0$
15. $y = 4 - x^2$, $y = 0$, $x = 0$
16. $y = x^2$, $y = 4x$

Exercises 9.7

See www.CalcChat.com for worked-out solutions to odd-numbered exercises.

In Exercises 1–10, evaluate the partial integral.

1. $\int_0^x (2x - y)\, dy$
2. $\int_x^{x^2} \dfrac{y}{x}\, dy$
3. $\int_1^{2y} \dfrac{y}{x}\, dx$
4. $\int_0^{e^y} y\, dx$
5. $\int_0^{\sqrt{4-x^2}} x^2 y\, dy$
6. $\int_{x^2}^{\sqrt{x}} (x^2 + y^2)\, dy$
7. $\int_1^{e^y} \dfrac{y \ln x}{x}\, dx$
8. $\int_{-\sqrt{1-y^2}}^{\sqrt{1-y^2}} (x^2 + y^2)\, dx$
9. $\int_0^x y e^{xy}\, dy$
10. $\int_y^3 \dfrac{xy}{\sqrt{x^2+1}}\, dx$

In Exercises 11–24, evaluate the double integral.

11. $\int_0^1 \int_0^2 (x + y)\, dy\, dx$
12. $\int_0^2 \int_0^2 (6 - x^2)\, dy\, dx$
13. $\int_0^4 \int_0^3 xy\, dy\, dx$
14. $\int_0^1 \int_0^x \sqrt{1 - x^2}\, dy\, dx$
15. $\int_0^1 \int_0^y (x + y)\, dx\, dy$
16. $\int_0^2 \int_{3y^2-6y}^{2y-y^2} 3y\, dx\, dy$
17. $\int_1^2 \int_0^4 (3x^2 - 2y^2 + 1)\, dx\, dy$
18. $\int_0^1 \int_y^{2y} (1 + 2x^2 + 2y^2)\, dx\, dy$
19. $\int_0^2 \int_0^{\sqrt{1-y^2}} -5xy\, dx\, dy$
20. $\int_0^4 \int_0^x \dfrac{2}{x^2+1}\, dy\, dx$
21. $\int_0^2 \int_0^{6x^2} x^3\, dy\, dx$
22. $\int_{-1}^1 \int_{-2}^2 (x^2 - y^2)\, dy\, dx$
23. $\int_0^\infty \int_0^\infty e^{-(x+y)/2}\, dy\, dx$
24. $\int_0^\infty \int_0^\infty xye^{-(x^2+y^2)}\, dx\, dy$

In Exercises 25–32, sketch the region R whose area is given by the double integral. Then change the order of integration and show that both orders yield the same area.

25. $\int_0^1 \int_0^2 dy\, dx$

26. $\int_1^2 \int_2^4 dx\, dy$

27. $\int_0^1 \int_{2y}^2 dx\, dy$

28. $\int_0^4 \int_0^{\sqrt{x}} dy\, dx$

29. $\int_0^2 \int_{x/2}^1 dy\, dx$

30. $\int_0^4 \int_{\sqrt{x}}^2 dy\, dx$

31. $\int_0^1 \int_{y^2}^{\sqrt[3]{y}} dx\, dy$

32. $\int_{-2}^2 \int_0^{4-y^2} dx\, dy$

In Exercises 33 and 34, evaluate the double integral. Note that it is necessary to change the order of integration.

33. $\int_0^3 \int_y^3 e^{x^2} dx\, dy$

34. $\int_0^2 \int_x^2 e^{-y^2} dy\, dx$

In Exercises 35–40, use a double integral to find the area of the specified region.

35.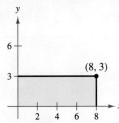

36.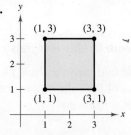

37.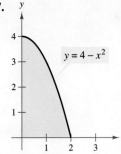

38.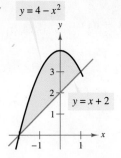

39.

40.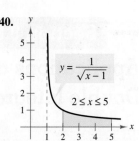

In Exercises 41–46, use a double integral to find the area of the region bounded by the graphs of the equations.

41. $y = 9 - x^2$, $y = 0$

42. $y = x^{3/2}$, $y = x$

43. $2x - 3y = 0$, $x + y = 5$, $y = 0$

44. $xy = 9$, $y = x$, $y = 0$, $x = 9$

45. $y = x$, $y = 2x$, $x = 2$

46. $y = x^2 + 2x + 1$, $y = 3(x + 1)$

(T) In Exercises 47–54, use a symbolic integration utility to evaluate the double integral.

47. $\int_0^1 \int_0^2 e^{-x^2 - y^2} dx\, dy$

48. $\int_0^2 \int_{x^2}^{2x} (x^3 + 3y^2) dy\, dx$

49. $\int_1^2 \int_0^x e^{xy} dy\, dx$

50. $\int_1^2 \int_y^{2y} \ln(x + y) dx\, dy$

51. $\int_0^1 \int_x^1 \sqrt{1 - x^2} dy\, dx$

52. $\int_0^3 \int_0^{x^2} \sqrt{x}\sqrt{1 + x} dy\, dx$

53. $\int_0^2 \int_{\sqrt{4-x^2}}^{4-x^2/4} \frac{xy}{x^2 + y^2 + 1} dy\, dx$

54. $\int_0^4 \int_0^y \frac{2}{(x + 1)(y + 1)} dx\, dy$

True or False? In Exercises 55 and 56, determine whether the statement is true or false. If it is false, explain why or give an example that shows it is false.

55. $\int_{-1}^1 \int_{-2}^2 y\, dy\, dx = \int_{-1}^1 \int_{-2}^2 y\, dx\, dy$

56. $\int_2^5 \int_1^6 x\, dy\, dx = \int_1^6 \int_2^5 x\, dx\, dy$

Section 9.8
Applications of Double Integrals

- Use double integrals to find the volumes of solids.
- Use double integrals to find the average values of real-life models.

Volume of a Solid Region

In Section 9.7, you used double integrals as an alternative way to find the area of a plane region. In this section, you will study the primary uses of double integrals: to find the volume of a solid region and to find the average value of a function.

Consider a function $z = f(x, y)$ that is continuous and nonnegative over a region R. Let S be the solid region that lies between the xy-plane and the surface

$$z = f(x, y) \qquad \text{Surface lying above the } xy\text{-plane}$$

directly above the region R, as shown in Figure 9.43. You can find the volume of S by integrating $f(x, y)$ over the region R.

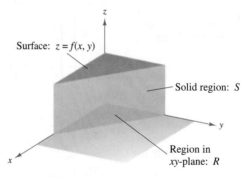

FIGURE 9.43

Determining Volume with Double Integrals

If R is a bounded region in the xy-plane and f is continuous and nonnegative over R, then the **volume of the solid** region between the surface $z = f(x, y)$ and R is given by the double integral

$$\int_R \int f(x, y)\, dA$$

where $dA = dx\, dy$ or $dA = dy\, dx$.

SECTION 9.8 Applications of Double Integrals

Example 1 Finding the Volume of a Solid

Find the volume of the solid region bounded in the first octant by the plane

$$z = 2 - x - 2y.$$

SOLUTION

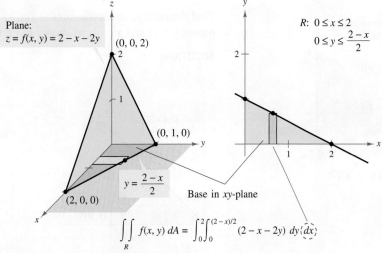

FIGURE 9.44

To set up the double integral for the volume, it is helpful to sketch both the solid region and the plane region R in the xy-plane. In Figure 9.44, you can see that the region R is bounded by the lines $x = 0$, $y = 0$, and $y = \frac{1}{2}(2 - x)$. One way to set up the double integral is to choose x as the outer variable. With that choice, the constant bounds for x are $0 \leq x \leq 2$ and the variable bounds for y are $0 \leq y \leq \frac{1}{2}(2 - x)$. So, the volume of the solid region is

$$\begin{aligned}
V &= \int_0^2 \int_0^{(2-x)/2} (2 - x - 2y)\, dy\, dx \\
&= \int_0^2 \left[(2 - x)y - y^2 \right]_0^{(2-x)/2} dx \\
&= \int_0^2 \left\{ (2 - x)\left(\frac{1}{2}\right)(2 - x) - \left[\frac{1}{2}(2 - x)\right]^2 \right\} dx \\
&= \frac{1}{4} \int_0^2 (2 - x)^2\, dx \\
&= \left[-\frac{1}{12}(2 - x)^3 \right]_0^2 \\
&= \frac{2}{3} \text{ cubic unit.}
\end{aligned}$$

STUDY TIP

Example 1 uses $dy\, dx$ as the order of integration. Try using the other order, $dx\, dy$, as indicated in Figure 9.45, to find the volume of the region. Do you get the same result as in Example 1?

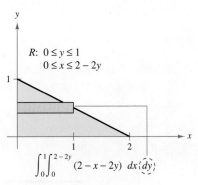

FIGURE 9.45

✓ **CHECKPOINT 1**

Find the volume of the solid region bounded in the first octant by the plane $z = 4 - 2x - y$. ■

614　CHAPTER 9　Functions of Several Variables

In Example 1, the order of integration was arbitrary. Although the example used x as the outer variable, you could just as easily have used y as the outer variable. The next example describes a situation in which one order of integration is more convenient than the other.

Example 2　Comparing Different Orders of Integration

Find the volume under the surface $f(x, y) = e^{-x^2}$ bounded by the xz-plane and the planes $y = x$ and $x = 1$, as shown in Figure 9.46.

SOLUTION

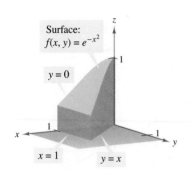

FIGURE 9.46

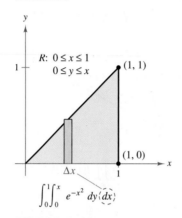 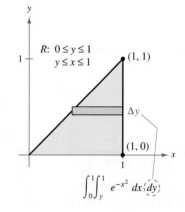

FIGURE 9.47

In the xy-plane, the bounds of region R are the lines $y = 0$, $x = 1$, and $y = x$. The two possible orders of integration are indicated in Figure 9.47. If you attempt to evaluate the two double integrals shown in the figure, you will discover that the one on the right involves finding the antiderivative of e^{-x^2}, which you know is not an elementary function. The integral on the left, however, can be evaluated more easily, as shown.

$$
\begin{aligned}
V &= \int_0^1 \int_0^x e^{-x^2}\, dy\, dx \\
&= \int_0^1 \left[e^{-x^2} y \right]_0^x dx \\
&= \int_0^1 x e^{-x^2}\, dx \\
&= \left[-\frac{1}{2} e^{-x^2} \right]_0^1 \\
&= -\frac{1}{2}\left(\frac{1}{e} - 1 \right) \approx 0.316 \text{ cubic unit}
\end{aligned}
$$

TECHNOLOGY

Use a symbolic integration utility to evaluate the double integral in Example 2.

✓ CHECKPOINT 2

Find the volume under the surface $f(x, y) = e^{x^2}$ bounded by the xz-plane and the planes $y = 2x$ and $x = 1$. ■

SECTION 9.8 Applications of Double Integrals

Guidelines for Finding the Volume of a Solid

1. Write the equation of the surface in the form $z = f(x, y)$ and sketch the solid region.
2. Sketch the region R in the xy-plane and determine the order and limits of integration.
3. Evaluate the double integral

$$\int_R \int f(x, y) \, dA$$

using the order and limits determined in the second step.

The first step above suggests that you sketch the three-dimensional solid region. This is a good suggestion, but it is not always feasible and is not as important as making a sketch of the two-dimensional region R.

Example 3 Finding the Volume of a Solid

Find the volume of the solid bounded above by the surface

$$f(x, y) = 6x^2 - 2xy$$

and below by the plane region R shown in Figure 9.48.

SOLUTION Because the region R is bounded by the parabola $y = 3x - x^2$ and the line $y = x$, the limits for y are $x \leq y \leq 3x - x^2$. The limits for x are $0 \leq x \leq 2$, and the volume of the solid is

$$V = \int_0^2 \int_x^{3x-x^2} (6x^2 - 2xy) \, dy \, dx$$

$$= \int_0^2 \left[6x^2 y - xy^2 \right]_x^{3x-x^2} dx$$

$$= \int_0^2 \left[(18x^3 - 6x^4 - 9x^3 + 6x^4 - x^5) - (6x^3 - x^3) \right] dx$$

$$= \int_0^2 (4x^3 - x^5) \, dx$$

$$= \left[x^4 - \frac{x^6}{6} \right]_0^2$$

$$= \frac{16}{3} \text{ cubic units.}$$

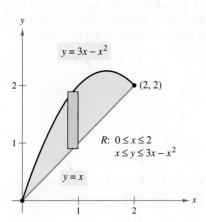

FIGURE 9.48

✓ **CHECKPOINT 3**

Find the volume of the solid bounded above by the surface $f(x, y) = 4x^2 + 2xy$ and below by the plane region bounded by $y = x^2$ and $y = 2x$. ∎

A *population density function* $p = f(x, y)$ is a model that describes the density (in people per square unit) of a region. To find the population of a region R, evaluate the double integral

$$\int_R \int f(x, y) \, dA.$$

Example 4
MAKE A DECISION Finding the Population of a Region

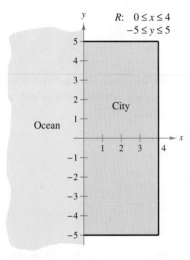

FIGURE 9.49

The population density (in people per square mile) of the city shown in Figure 9.49 can be modeled by

$$f(x, y) = \frac{50,000}{x + |y| + 1}$$

where x and y are measured in miles. Approximate the city's population. Is the city's average population density less than 10,000 people per square mile?

SOLUTION Because the model involves the absolute value of y, it follows that the population density is symmetrical about the x-axis. So, the population in the first quadrant is equal to the population in the fourth quadrant. This means that you can find the total population by doubling the population in the first quadrant.

$$\begin{aligned}
\text{Population} &= 2 \int_0^4 \int_0^5 \frac{50,000}{x + y + 1} \, dy \, dx \\
&= 100,000 \int_0^4 \Big[\ln(x + y + 1) \Big]_0^5 dx \\
&= 100,000 \int_0^4 [\ln(x + 6) - \ln(x + 1)] \, dx \\
&= 100,000 \Big[(x + 6) \ln(x + 6) - (x + 6) - \\
&\quad\quad (x + 1) \ln(x + 1) + (x + 1) \Big]_0^4 \\
&= 100,000 \Big[(x + 6) \ln(x + 6) - (x + 1) \ln(x + 1) - 5 \Big]_0^4 \\
&= 100,000 [10 \ln(10) - 5 \ln(5) - 5 - 6 \ln(6) + 5] \\
&\approx 422,810 \text{ people}
\end{aligned}$$

So, the city's population is about 422,810. Because the city covers a region 4 miles wide and 10 miles long, its area is 40 square miles. So, the average population density is

$$\text{Average population density} = \frac{422,810}{40}$$

$$\approx 10,570 \text{ people per square mile.}$$

No, the city's average population density is not less than 10,000 people per square mile.

✓ **CHECKPOINT 4**

In Example 4, what integration technique was used to integrate

$$\int [\ln(x + 6) - \ln(x + 1)] \, dx?$$

Average Value of a Function over a Region

> **Average Value of a Function over a Region**
>
> If f is integrable over the plane region R with area A, then its **average value** over R is
>
> $$\text{Average value} = \frac{1}{A}\int\int_R f(x, y)\, dA.$$

Example 5 Finding Average Profit

A medical supply company determines that the profit for selling x units of one product and y units of a second product is modeled by

$$P = -(x - 200)^2 - (y - 100)^2 + 5000.$$

The weekly sales of product 1 vary between 150 and 200 units, and the weekly sales of product 2 vary between 80 and 100 units. Estimate the average weekly profit for the two products.

SOLUTION Because $150 \leq x \leq 200$ and $80 \leq y \leq 100$, you can estimate the weekly profit to be the average of the profit function over the rectangular region shown in Figure 9.50. Because the area of this rectangular region is $(50)(20) = 1000$, it follows that the average profit V is

$$\begin{aligned}
V &= \frac{1}{1000}\int_{150}^{200}\int_{80}^{100}[-(x-200)^2 - (y-100)^2 + 5000]\, dy\, dx\\
&= \frac{1}{1000}\int_{150}^{200}\left[-(x-200)^2 y - \frac{(y-100)^3}{3} + 5000y\right]_{80}^{100} dx\\
&= \frac{1}{1000}\int_{150}^{200}\left[-20(x-200)^2 - \frac{292{,}000}{3}\right] dx\\
&= \frac{1}{3000}\left[-20(x-200)^3 + 292{,}000x\right]_{150}^{200}\\
&\approx \$4033.
\end{aligned}$$

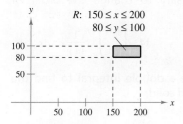

FIGURE 9.50

✓ CHECKPOINT 5

Find the average value of $f(x, y) = 4 - \frac{1}{2}x - \frac{1}{2}y$ over the region $0 \leq x \leq 2$ and $0 \leq y \leq 2$. ∎

CONCEPT CHECK

1. Complete the following: The double integral $\int_R \int f(x, y)\, dA$ gives the _____ of the solid region between the surface $z = f(x, y)$ and the bounded region in the xy-plane R.
2. State the guidelines for finding the volume of a solid.
3. What does a population density function describe?
4. What is the average value of $f(x, y)$ over the plane region R?

618 CHAPTER 9 Functions of Several Variables

Skills Review 9.8

The following warm-up exercises involve skills that were covered in earlier sections. You will use these skills in the exercise set for this section. For additional help, review Sections 6.4 and 9.7.

In Exercises 1–4, sketch the region that is described.

1. $0 \le x \le 2,\ 0 \le y \le 1$
2. $1 \le x \le 3,\ 2 \le y \le 3$
3. $0 \le x \le 4,\ 0 \le y \le 2x - 1$
4. $0 \le x \le 2,\ 0 \le y \le x^2$

In Exercises 5–10, evaluate the double integral.

5. $\int_0^1 \int_1^2 dy\, dx$
6. $\int_0^3 \int_0^3 dx\, dy$
7. $\int_0^1 \int_0^x x\, dy\, dx$
8. $\int_0^4 \int_1^y y\, dx\, dy$
9. $\int_1^3 \int_x^{x^2} 2\, dy\, dx$
10. $\int_0^1 \int_x^{-x^2+2} dy\, dx$

Exercises 9.8

See www.CalcChat.com for worked-out solutions to odd-numbered exercises.

In Exercises 1–8, sketch the region of integration and evaluate the double integral.

1. $\int_0^2 \int_0^1 (3x + 4y)\, dy\, dx$
2. $\int_0^3 \int_0^1 (2x + 6y)\, dy\, dx$
3. $\int_0^1 \int_y^{\sqrt{y}} x^2 y^2\, dx\, dy$
4. $\int_0^6 \int_{y/2}^3 (x + y)\, dx\, dy$
5. $\int_0^1 \int_0^{\sqrt{1-x^2}} y\, dy\, dx$
6. $\int_0^2 \int_0^{4-x^2} xy^2\, dy\, dx$
7. $\int_{-a}^a \int_{-\sqrt{a^2-x^2}}^{\sqrt{a^2-x^2}} dy\, dx$
8. $\int_0^a \int_0^{\sqrt{a^2-x^2}} dy\, dx$

In Exercises 9–12, set up the integral for both orders of integration and use the more convenient order to evaluate the integral over the region R.

9. $\iint_R xy\, dA$
   R: rectangle with vertices at $(0, 0), (0, 5), (3, 5), (3, 0)$

10. $\iint_R x\, dA$
    R: semicircle bounded by $y = \sqrt{25 - x^2}$ and $y = 0$

11. $\iint_R \frac{y}{x^2 + y^2}\, dA$
    R: triangle bounded by $y = x, y = 2x, x = 2$

12. $\iint_R \frac{y}{1 + x^2}\, dA$
    R: region bounded by $y = 0, y = \sqrt{x}, x = 4$

In Exercises 13 and 14, evaluate the double integral. Note that it is necessary to change the order of integration.

13. $\int_0^1 \int_{y/2}^{1/2} e^{-x^2}\, dx\, dy$
14. $\int_0^{\ln 10} \int_{e^x}^{10} \frac{1}{\ln y}\, dy\, dx$

In Exercises 15–26, use a double integral to find the volume of the specified solid.

15.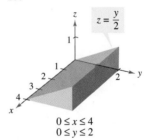

16.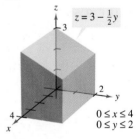

17.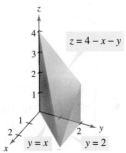

18.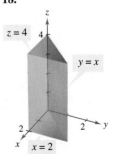

19.

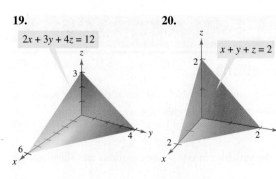

21.

23.

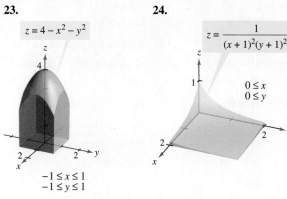

25.

20.

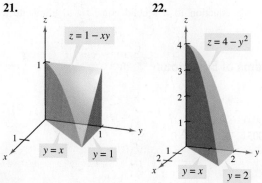

22.

24.

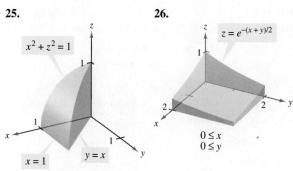

26.

In Exercises 27–30, use a double integral to find the volume of the solid bounded by the graphs of the equations.

27. $z = xy$, $z = 0$, $y = 0$, $y = 4$, $x = 0$, $x = 1$

28. $z = x$, $z = 0$, $y = x$, $y = 0$, $x = 0$, $x = 4$

29. $z = x^2$, $z = 0$, $x = 0$, $x = 2$, $y = 0$, $y = 4$

30. $z = x + y$, $x^2 + y^2 = 4$ (first octant)

31. Population Density The population density (in people per square mile) for a coastal town can be modeled by

$$f(x, y) = \frac{120{,}000}{(2 + x + y)^3}$$

where x and y are measured in miles. What is the population inside the rectangular area defined by the vertices $(0, 0)$, $(2, 0)$, $(0, 2)$, and $(2, 2)$?

32. Population Density The population density (in people per square mile) for a coastal town on an island can be modeled by

$$f(x, y) = \frac{5000xe^y}{1 + 2x^2}$$

where x and y are measured in miles. What is the population inside the rectangular area defined by the vertices $(0, 0)$, $(4, 0)$, $(0, -2)$, and $(4, -2)$?

In Exercises 33–36, find the average value of $f(x, y)$ over the region R.

33. $f(x, y) = x$
R: rectangle with vertices $(0, 0)$, $(4, 0)$, $(4, 2)$, $(0, 2)$

34. $f(x, y) = xy$
R: rectangle with vertices $(0, 0)$, $(4, 0)$, $(4, 2)$, $(0, 2)$

35. $f(x, y) = x^2 + y^2$
R: square with vertices $(0, 0)$, $(2, 0)$, $(2, 2)$, $(0, 2)$

36. $f(x, y) = e^{x+y}$
R: triangle with vertices $(0, 0)$, $(0, 1)$, $(1, 1)$

37. Average Weekly Profit A biomedical supply firm's weekly profit from marketing two products is given by

$$P = 192x_1 + 576x_2 - x_1^2 - 5x_2^2 - 2x_1x_2 - 5000$$

where x_1 and x_2 represent the numbers of units of each product sold weekly. Estimate the average weekly profit if x_1 varies between 40 and 50 units and x_2 varies between 45 and 50 units.

38. Average Weekly Profit After a change in marketing, the weekly profit of the firm in Exercise 37 is given by

$$P = 200x_1 + 580x_2 - x_1^2 - 5x_2^2 - 2x_1x_2 - 7500.$$

Estimate the average weekly profit if x_1 varies between 55 and 65 units and x_2 varies between 50 and 60 units.

Algebra Review

Solving Systems of Equations

Nonlinear System in Two Variables

$$\begin{cases} 4x + 3y = 6 \\ x^2 - y = 4 \end{cases}$$

Linear System in Three Variables

$$\begin{cases} -x + 2y + 4z = 2 \\ 2x - y + z = 0 \\ 6x + 2z = 3 \end{cases}$$

Two of the sections in this chapter (9.5, and 9.6) involve solutions of systems of equations. These systems can be linear or nonlinear, as shown at the left.

There are many techniques for solving a system of linear equations. Two of the more common ones are listed here.

1. *Substitution*: Solve for one of the variables in one of the equations and substitute the value into another equation.

2. *Elimination*: Add multiples of one equation to a second equation to eliminate a variable in the second equation.

Example 1 Solving Systems of Equations

Solve each system of equations.

a. $\begin{cases} y - x^3 = 0 \\ x - y^3 = 0 \end{cases}$ **b.** $\begin{cases} 3 - 0.004x - 0.005y = 0 \\ 4.5 - 0.005x - 0.01y = 0 \end{cases}$

SOLUTION

a. Example 3, page 588

$\begin{cases} y - x^3 = 0 \\ x - y^3 = 0 \end{cases}$	Equation 1 Equation 2
$y = x^3$	Solve for y in Equation 1.
$x - (x^3)^3 = 0$	Substitute x^3 for y in Equation 2.
$x - x^9 = 0$	$(x^m)^n = x^{mn}$
$x(x-1)(x+1)(x^2+1)(x^4+1) = 0$	Factor.
$x = 0$	Set factors equal to zero.
$x = 1$	Set factors equal to zero.
$x = -1$	Set factors equal to zero.

b. Example 4, page 589

$\begin{cases} 3 - 0.004x - 0.005y = 0 \\ 4.5 - 0.005x - 0.01y = 0 \end{cases}$	Multiply Equation 1 by 1000. Multiply Equation 2 by 100.
$\begin{cases} 3000 - 4x - 5y = 0 \\ 450 - 0.5x - y = 0 \end{cases}$	Equation 1 Equation 2
$y = 450 - 0.5x$	Solve for y in Equation 2.
$3000 - 4x - 5(450 - 0.5x) = 0$	Substitute for y in Equation 1.
$3000 - 4x - 2250 + 2.5x = 0$	Multiply.
$-1.5x = -750$	Combine like terms.
$x = 500$	Divide each side by -1.5.
$y = 450 - 0.5(500)$	Find y by substituting x.
$y = 200$	Solve for y.

Algebra Review

Example 2 — Solving Systems of Equations

Solve each system of equations.

a. $\begin{cases} y(24 - 12x - 4y) = 0 \\ x(24 - 6x - 8y) = 0 \end{cases}$ **b.** $\begin{cases} 28a - 4b = 10 \\ -4a + 8b = 12 \end{cases}$

SOLUTION

a. Example 5, page 590

Before solving this system of equations, factor 4 out of the first equation and factor 2 out of the second equation.

$\begin{cases} y(24 - 12x - 4y) = 0 \\ x(24 - 6x - 8y) = 0 \end{cases}$ Original Equation 1
Original Equation 2

$\begin{cases} y(4)(6 - 3x - y) = 0 \\ x(2)(12 - 3x - 4y) = 0 \end{cases}$ Factor 4 out of Equation 1.
Factor 2 out of Equation 2.

$\begin{cases} y(6 - 3x - y) = 0 \\ x(12 - 3x - 4y) = 0 \end{cases}$ Equation 1
Equation 2

In each equation, either factor can be 0, so you obtain four different linear systems. For the first system, substitute $y = 0$ into the second equation to obtain $x = 4$.

$\begin{cases} y = 0 \\ 12 - 3x - 4y = 0 \end{cases}$ $(4, 0)$ is a solution.

You can solve the second system by the method of elimination.

$\begin{cases} 6 - 3x - y = 0 \\ 12 - 3x - 4y = 0 \end{cases}$ $\left(\frac{4}{3}, 2\right)$ is a solution.

The third system is already solved.

$\begin{cases} y = 0 \\ x = 0 \end{cases}$ $(0, 0)$ is a solution.

You can solve the last system by substituting $x = 0$ into the first equation to obtain $y = 6$.

$\begin{cases} 6 - 3x - y = 0 \\ x = 0 \end{cases}$ $(0, 6)$ is a solution.

b. Example 2, page 596

$\begin{cases} 28a - 4b = 10 \\ -4a + 8b = 12 \end{cases}$ Equation 1
Equation 2

$-2a + 4b = 6$ Divide Equation 2 by 2.

$26a = 16$ Add new equation to Equation 1.

$a = \frac{8}{13}$ Divide each side by 26.

$28\left(\frac{8}{13}\right) - 4b = 10$ Substitute for a in Equation 1.

$b = \frac{47}{26}$ Solve for b.

Chapter Summary and Study Strategies

After studying this chapter, you should have acquired the following skills. The exercise numbers are keyed to the Review Exercises that begin on page 624. Answers to odd-numbered Review Exercises are given in the back of the text.*

Section 9.1 — Review Exercises
- Plot points in space. — 1, 2
- Find the distance between two points in space. — 3, 4
  $$d = \sqrt{(x_2 - x_1)^2 + (y_2 - y_1)^2 + (z_2 - z_1)^2}$$
- Find the midpoints of line segments in space. — 5, 6
  $$\text{Midpoint} = \left(\frac{x_1 + x_2}{2}, \frac{y_1 + y_2}{2}, \frac{z_1 + z_2}{2}\right)$$
- Write the standard forms of the equations of spheres. — 7–10, 16
  $$(x - h)^2 + (y - k)^2 + (z - l)^2 = r^2$$
- Find the centers and radii of spheres. — 11, 12
- Sketch the coordinate plane traces of spheres. — 13, 14
- Find a point in space. — 15

Section 9.2
- Sketch planes in space. — 17–20
- Classify quadric surfaces in space. — 21–28

Section 9.3
- Evaluate functions of several variables. — 29, 30, 62
- Find the domains and ranges of functions of several variables. — 31, 32
- Sketch the level curves of functions of two variables. — 33–36
- Use functions of several variables to answer questions about real-life situations. — 37–42

Section 9.4
- Find the first partial derivatives of functions of several variables. — 43–52
  $$\frac{\partial z}{\partial x} = \lim_{\Delta x \to 0} \frac{f(x + \Delta x, y) - f(x, y)}{\Delta x} \qquad \frac{\partial z}{\partial y} = \lim_{\Delta y \to 0} \frac{f(x, y + \Delta y) - f(x, y)}{\Delta y}$$
- Find the slopes of surfaces in the x- and y-directions. — 53–56
- Find the second partial derivatives of functions of several variables. — 57–60
- Use partial derivatives to answer questions about real-life situations. — 61

Section 9.5
- Find the relative extrema of functions of two variables. — 63–70

---

* Use a wide range of valuable study aids to help you master the material in this chapter. The *Student Solutions Guide* includes step-by-step solutions to all odd-numbered exercises to help you review and prepare. The student website at *college.hmco.com/info/larsonapplied* offers algebra help and a *Graphing Technology Guide*. The *Graphing Technology Guide* contains step-by-step commands and instructions for a wide variety of graphing calculators, including the most recent models.

Section 9.6

- Find the least squares regression line, $y = ax + b$, for data and calculate the sum of the squared errors for data. — 71, 72

$$a = \frac{n\sum_{i=1}^{n} x_i y_i - \sum_{i=1}^{n} x_i \sum_{i=1}^{n} y_i}{n\sum_{i=1}^{n} x_i^2 - \left(\sum_{i=1}^{n} x_i\right)^2} \quad \text{and} \quad b = \frac{1}{n}\left(\sum_{i=1}^{n} y_i - a\sum_{i=1}^{n} x_i\right)$$

- Use least squares regression quadratics to model real-life data. — 73, 75
- Use least squares regression lines to model real-life data. — 74, 76
- Find the least squares regression quadratics for data. — 77, 78

Section 9.7

- Evaluate double integrals. — 79–82
- Use double integrals to find the areas of regions. — 83–86

Section 9.8

- Use double integrals to find the volumes of solids. — 87, 88

$$\text{Volume} = \int\int_R f(x, y)\, dA$$

- Use double integrals to find the average values of real-life models. — 89

$$\text{Average value} = \frac{1}{A}\int\int_R f(x, y)\, dA$$

Study Strategies

- **Comparing Two Dimensions with Three Dimensions** Many of the formulas and techniques in this chapter are generalizations of formulas and techniques used in earlier chapters in the text. Here are several examples.

Two-Dimensional Coordinate System	Three-Dimensional Coordinate System
Distance Formula $d = \sqrt{(x_2 - x_1)^2 + (y_2 - y_1)^2}$	*Distance Formula* $d = \sqrt{(x_2 - x_1)^2 + (y_2 - y_1)^2 + (z_2 - z_1)^2}$
Midpoint Formula $\text{Midpoint} = \left(\frac{x_1 + x_2}{2}, \frac{y_1 + y_2}{2}\right)$	*Midpoint Formula* $\text{Midpoint} = \left(\frac{x_1 + x_2}{2}, \frac{y_1 + y_2}{2}, \frac{z_1 + z_2}{2}\right)$
Equation of Circle $(x - h)^2 + (y - k)^2 = r^2$	*Equation of Sphere* $(x - h)^2 + (y - k)^2 + (z - l)^2 = r^2$
Equation of Line $ax + by = c$	*Equation of Plane* $ax + by + cz = d$
Derivative of $y = f(x)$ $\frac{dy}{dx} = \lim_{\Delta x \to 0} \frac{f(x + \Delta x) - f(x)}{\Delta x}$	*Partial Derivative of* $z = f(x, y)$ $\frac{\partial z}{\partial x} = \lim_{\Delta x \to 0} \frac{f(x + \Delta x, y) - f(x, y)}{\Delta x}$
Area of Region $A = \int_a^b f(x)\, dx$	*Volume of Region* $V = \int\int_R f(x, y)\, dA$

Review Exercises

In Exercises 1 and 2, plot the points.

1. $(2, -1, 4), (-1, 3, -3)$
2. $(1, -2, -3), (-4, -3, 5)$

In Exercises 3 and 4, find the distance between the two points.

3. $(0, 0, 0), (2, 5, 9)$
4. $(-4, 1, 5), (1, 3, 7)$

In Exercises 5 and 6, find the midpoint of the line segment joining the two points.

5. $(2, 6, 4), (-4, 2, 8)$
6. $(5, 0, 7), (-1, -2, 9)$

In Exercises 7–10, find the standard form of the equation of the sphere.

7.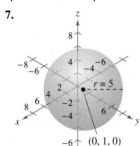
8.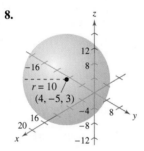
9.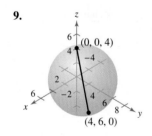
10.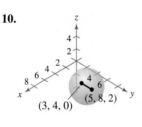

In Exercises 11 and 12, find the center and radius of the sphere.

11. $x^2 + y^2 + z^2 + 4x - 2y - 8z + 5 = 0$
12. $x^2 + y^2 + z^2 + 4y - 10z - 7 = 0$

In Exercises 13 and 14, sketch the *xy*-trace of the sphere.

13. $(x + 2)^2 + (y - 1)^2 + (z - 3)^2 = 25$
14. $(x - 1)^2 + (y + 3)^2 + (z - 6)^2 = 72$

15. **Geology** Crystals shaped like rectangular prisms are classified as tetragonal. The vertices of a tetragonal crystal mapped onto a three-dimensional coordinate system are shown in the figure. Determine (x, y, z).

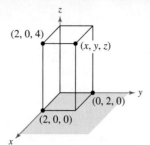

Figure for 15

16. **Health** A spherical exercise ball has a diameter of 8 inches. The center of the ball is placed at the origin of a three-dimensional coordinate system. What is the equation of the sphere?

In Exercises 17–20, find the intercepts and sketch the graph of the plane.

17. $x + 2y + 3z = 6$
18. $2y + z = 4$
19. $3x - 6z = 12$
20. $4x - y + 2z = 8$

In Exercises 21–28, identify the surface.

21. $x^2 + y^2 + z^2 - 2x + 4y - 6z + 5 = 0$
22. $16x^2 + 16y^2 - 9z^2 = 0$
23. $x^2 + \dfrac{y^2}{16} + \dfrac{z^2}{9} = 1$
24. $x^2 - \dfrac{y^2}{16} - \dfrac{z^2}{9} = 1$
25. $z = \dfrac{x^2}{9} + y^2$
26. $-4x^2 + y^2 + z^2 = 4$
27. $z = \sqrt{x^2 + y^2}$
28. $z = 9x + 3y - 5$

In Exercises 29 and 30, find the function values.

29. $f(x, y) = xy^2$
    (a) $f(2, 3)$ (b) $f(0, 1)$
    (c) $f(-5, 7)$ (d) $f(-2, -4)$

30. $f(x, y) = \dfrac{x^2}{y}$
    (a) $f(6, 9)$ (b) $f(8, 4)$
    (c) $f(t, 2)$ (d) $f(r, r)$

In Exercises 31 and 32, describe the region R in the xy-plane that corresponds to the domain of the function. Then find the range of the function.

31. $f(x, y) = \sqrt{1 - x^2 - y^2}$

32. $f(x, y) = \dfrac{1}{x + y}$

In Exercises 33–36, describe the level curves of the function. Sketch the level curves for the given c-values.

33. $z = 10 - 2x - 5y$, $c = 0, 2, 4, 5, 10$

34. $z = \sqrt{9 - x^2 - y^2}$, $c = 0, 1, 2, 3$

35. $z = (xy)^2$, $c = 1, 4, 9, 12, 16$

36. $z = y - x^2$, $c = 0, \pm 1, \pm 2$

37. Meteorology The contour map shown below represents the average yearly precipitation for Iowa. *(Source: U.S. National Oceanic and Atmospheric Administration)*

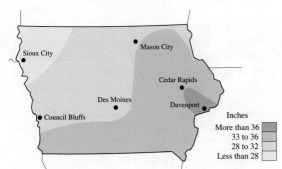

(a) Discuss the use of color to represent the level curves.

(b) Which part of Iowa receives the most precipitation?

(c) Which part of Iowa receives the least precipitation?

38. Population Density The contour map represents the population density of New York (see figure). *(Source: U.S. Census Bureau)*

(a) Discuss the use of color to represent the level curves.

(b) Do the level curves correspond to equally spaced population densities?

(c) Describe how to obtain a more detailed contour map.

39. Chemistry The acidity of rainwater is measured in units called pH, and smaller pH values are increasingly acidic. The map shows the curves of equal pH and gives evidence that downwind of heavily industrialized areas the acidity has been increasing. Using the level curves on the map, determine the direction of the prevailing winds in the northeastern United States.

40. Infant Development In the United States the mean head circumference c (in centimeters) of male infants ages 2 through 12 months can be approximated by $c = 0.36l - 0.028a + 19.35$, where l is the mean recumbent length of the infant (in centimeters) and a is the corresponding age (in months). Use this function of two variables and a spreadsheet to complete the table. *(Source: Centers for Disease Control and Prevention)*

a	2	4	6	8	10	12
l	58.73	64.15	68.49	71.69	73.18	77.81
c						

41. Average Weight The mean weight w (in kilograms) of U.S. female children ages 2 through 18 years can be approximated by $w = 0.06h + 3.09a - 2.32$, where h is the mean height (in centimeters) and a is the corresponding age (in years). *(Source: Centers for Disease Control and Prevention)*

(a) The mean height of 4-year-old female children is about 103 centimeters. What is the mean weight?

(b) The mean weight of 11-and-a-half-year-old female children is about 42.4 kilograms. What is the mean height?

(c) Which of the two variables in this model has the greater influence on weight? Explain.

42. Biomechanics The Froude number F, defined as

$$F = \dfrac{v^2}{gl}$$

where v represents velocity, g represents gravitational acceleration, and l represents stride length, is an example of a "similarity criterion." Find the Froude number of a rabbit for which velocity is 2 meters per second, gravitational acceleration is 3 meters per second squared, and stride length is 0.75 meter.

In Exercises 43–52, find all first partial derivatives.

43. $f(x, y) = x^2y + 3xy + 2x - 5y$
44. $f(x, y) = 4xy + xy^2 - 3x^2y$
45. $z = \dfrac{x^2}{y^2}$
46. $z = (xy + 2x + 4y)^2$
47. $f(x, y) = \ln(2x + 3y)$
48. $f(x, y) = \ln\sqrt{2x + 3y}$
49. $f(x, y) = xe^y + ye^x$
50. $f(x, y) = x^2e^{-2y}$
51. $w = xyz^2$
52. $w = 3xy - 5xz + 2yz$

In Exercises 53–56, find the slope of the surface at the indicated point in (a) the *x*-direction and (b) the *y*-direction.

53. $z = 3x - 4y + 9$, $(3, 2, 10)$
54. $z = 4x^2 - y^2$, $(2, 4, 0)$
55. $z = 8 - x^2 - y^2$, $(1, 2, 3)$
56. $z = x^2 - y^2$, $(5, -4, 9)$

In Exercises 57–60, find all second partial derivatives.

57. $f(x, y) = 3x^2 - xy + 2y^3$
58. $f(x, y) = \dfrac{y}{x + y}$
59. $f(x, y) = \sqrt{1 + x + y}$
60. $f(x, y) = x^2e^{-y^2}$

61. **Medical Science** The surface area *A* of an average human body in square centimeters can be approximated by the model

    $A(w, h) = 101.4w^{0.425}h^{0.725}$

    where *w* is the weight in pounds and *h* is the height in inches.

    (a) Determine the partial derivatives of *A* with respect to *w* and with respect to *h*.

    (b) Evaluate $\partial A/\partial w$ at $(180, 70)$. Explain your result.

62. **Medicine** In order to treat a certain bacterial infection, a combination of two drugs is being tested. Studies have shown that the duration *D* (in hours) of the infection in laboratory tests can be modeled by

    $D(x, y) = x^2 + 2y^2 - 18x - 24y + 2xy + 120$

    where *x* is the dosage in hundreds of milligrams of the first drug and *y* is the dosage in hundreds of milligrams of the second drug. Evaluate $D(5, 2.5)$ and $D(7.5, 8)$ and interpret your results.

In Exercises 63–70, find any critical points and relative extrema of the function.

63. $f(x, y) = x^2 + 2y^2$
64. $f(x, y) = x^3 - 3xy + y^2$
65. $f(x, y) = 1 - (x + 2)^2 + (y - 3)^2$
66. $f(x, y) = e^x - x + y^2$
67. $f(x, y) = x^3 + y^2 - xy$
68. $f(x, y) = y^2 + xy + 3y - 2x + 5$
69. $f(x, y) = x^3 + y^3 - 3x - 3y + 2$
70. $f(x, y) = -x^2 - y^2$

In Exercises 71 and 72, (a) use the method of least squares to find the least squares regression line and (b) calculate the sum of the squared errors.

71. $(-2, -3), (-1, -1), (1, 2), (3, 2)$
72. $(-3, -1), (-2, -1), (0, 0), (1, 1), (2, 1)$

73. **Lumber Consumption** The amounts *y* of softwoods (in billions of board-feet) consumed in the United States from 1999 through 2005, where *t* = 9 corresponds to 1999, are given in the table. *(Source: U.S. Forest Service)*

Year, *t*	9	10	11	12
Softwoods, *y*	54.5	54.0	53.7	56.4

Year, *t*	13	14	15
Softwoods, *y*	56.5	62.0	64.4

(a) Use the regression capabilities of a graphing utility or a spreadsheet to find the least squares regression quadratic for the data.

(b) Estimate the amount of softwoods (in billions of board-feet) consumed in 2008.

74. **Shrimp** The numbers *y* of shrimp (in millions of pounds) supplied in the United States from 1999 through 2004, where *t* = 9 corresponds to 1999, are given in the table. *(Source: U.S. National Oceanic and Atmospheric Administration)*

Year, *t*	9	10	11	12	13	14
Shrimp *y*	1084	1172	1312	1430	1608	1669

(a) Use the regression capabilities of a graphing utility or a spreadsheet to find the least squares regression line for the data.

(b) Estimate the number of shrimp (in millions of pounds) supplied in 2008.

75. Mineral Industries The numbers y of nonmetallic mineral production workers (in thousands) in the United States from 2000 through 2005, where $t = 0$ corresponds to 2000, are given in the table. *(Source: U.S. Bureau of Labor Statistics)*

Year, t	0	1	2	3	4	5
Workers, y	87	83	80	78	81	83

(T)(S) (a) Use the regression capabilities of a graphing utility or a spreadsheet to find the least squares regression quadratic for the data.

(b) Estimate the number of nonmetallic mineral production workers in 2008.

76. Work Force The table gives the percents x and numbers y (in millions) of women in the work force for selected years. *(Source: U.S. Bureau of Labor Statistics)*

Year	1970	1975	1980	1985
Percent, x	43.3	46.3	51.5	54.5
Number, y	31.5	37.5	45.5	51.1

Year	1990	1995	2000	2005
Percent, x	57.5	58.9	59.9	59.3
Number, y	56.8	60.9	66.3	69.3

(T)(S) (a) Use the regression capabilities of a graphing utility or a spreadsheet to find the least squares regression line for the data.

(b) According to this model, approximately how many women enter the labor force for each one-point increase in the percent of women in the labor force?

(T)(S) In Exercises 77 and 78, use the regression capabilities of a graphing utility or a spreadsheet to find the least squares regression quadratic for the given points. Plot the points and graph the least squares regression quadratic.

77. $(-1, 9), (0, 7), (1, 5), (2, 6), (4, 23)$

78. $(0, 10), (2, 9), (3, 7), (4, 4), (5, 0)$

In Exercises 79–82, evaluate the double integral.

79. $\int_0^1 \int_0^{1+x} (4x - 2y) \, dy \, dx$

80. $\int_{-3}^3 \int_0^4 (x - y^2) \, dx \, dy$

81. $\int_1^2 \int_1^{2y} \frac{x}{y^2} \, dx \, dy$

82. $\int_0^4 \int_0^{\sqrt{16-x^2}} 2x \, dy \, dx$

In Exercises 83–86, use a double integral to find the area of the region.

83.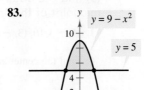

84.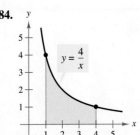

85.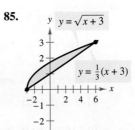

86.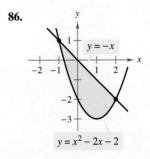

87. Find the volume of the solid bounded by the graphs of $z = (xy)^2$, $z = 0$, $y = 0$, $y = 4$, $x = 0$, and $x = 4$.

88. Find the volume of the solid bounded by the graphs of $z = x + y$, $z = 0$, $x = 0$, $x = 3$, $y = x$, and $y = 0$.

89. Average Elevation In a triangular coastal area, the elevation in miles above sea level at the point (x, y) is modeled by

$$f(x, y) = 0.25 - 0.025x - 0.01y$$

where x and y are measured in miles (see figure). Find the average elevation of the triangular area.

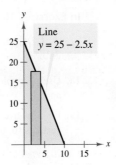

Chapter Test

See www.CalcChat.com for worked-out solutions to odd-numbered exercises.

Take this test as you would take a test in class. When you are done, check your work against the answers given in the back of the book.

In Exercises 1–3, (a) plot the points on a three-dimensional coordinate system, (b) find the distance between the points, and (c) find the coordinates of the midpoint of the line segment joining the points.

1. $(1, -3, 0), (3, -1, 0)$ **2.** $(-2, 2, 3), (-4, 0, 2)$ **3.** $(3, -7, 2), (5, 11, -6)$

4. Find the center and radius of the sphere whose equation is
$$x^2 + y^2 + z^2 - 20x + 10y - 10z + 125 = 0.$$

In Exercise 5–7, identify the surface.

5. $3x - y - z = 0$ **6.** $36x^2 + 9y^2 - 4z^2 = 0$ **7.** $4x^2 - y^2 - 16z = 0$

In Exercises 8–10, find $f(3, 3)$ and $f(1, 1)$.

8. $f(x, y) = x^2 + xy + 1$ **9.** $f(x, y) = \dfrac{x + 2y}{3x - y}$ **10.** $f(x, y) = xy \ln \dfrac{x}{y}$

In Exercises 11 and 12, find f_x and f_y and evaluate each at the point $(10, -1)$.

11. $f(x, y) = 3x^2 + 9xy^2 - 2$ **12.** $f(x, y) = x\sqrt{x + y}$

In Exercises 13 and 14, find any critical points, relative extrema, and saddle points of the function.

13. $f(x, y) = 3x^2 + 4y^2 - 6x + 16y - 4$
14. $f(x, y) = 4xy - x^4 - y^4$

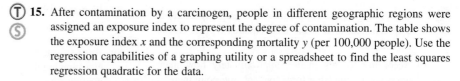

15. After contamination by a carcinogen, people in different geographic regions were assigned an exposure index to represent the degree of contamination. The table shows the exposure index x and the corresponding mortality y (per 100,000 people). Use the regression capabilities of a graphing utility or a spreadsheet to find the least squares regression quadratic for the data.

Exposure, x	1.35	2.67	3.93	5.14	7.43
Mortality, y	118.5	135.2	167.3	197.6	204.7

In Exercises 16 and 17, evaluate the double integral.

16. $\displaystyle\int_0^1 \int_x^1 (30x^2 y - 1)\, dy\, dx$ **17.** $\displaystyle\int_0^{\sqrt{e-1}} \int_0^{2y} \dfrac{1}{y^2 + 1}\, dx\, dy$

18. Use a double integral to find the area of the region bounded by the graphs of $y = 3$ and $y = x^2 - 2x + 3$ (see figure).

19. Find the average value of $f(x, y) = x^2 + y$ over the region defined by a rectangle with vertices $(0, 0), (1, 0), (1, 3),$ and $(0, 3)$.

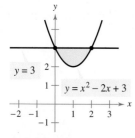

Figure for 18

Differential Equations 10

You can use differential equations to model how quickly a characteristic will pass from one generation to the next in a population of beetles. (See Section 10.4, Example 4.)

10.1 Solutions of Differential Equations

10.2 Separation of Variables

10.3 First-Order Linear Differential Equations

10.4 Applications of Differential Equations

Applications

Differential equations have many real-life applications. The applications listed below represent a sample of the applications in this chapter.

- Biology, Exercise 51, page 635
- Radioactive Decay, Exercises 53 and 54, page 642
- Learning Curve, Exercise 39, page 648
- Newton's Law of Cooling, Exercises 21–24, page 655
- Medical Science, Exercises 39–42, page 656

Section 10.1

Solutions of Differential Equations

- Find general solutions of differential equations.
- Find particular solutions of differential equations.

General Solution of a Differential Equation

A **differential equation** is an equation involving a differentiable function and one or more of its derivatives. For instance,

$$y' + 2y = 0 \qquad \text{Differential equation}$$

is a differential equation. A function $y = f(x)$ is a **solution** of a differential equation if the equation is satisfied when y and its derivatives are replaced by $f(x)$ and its derivatives. For example,

$$y = e^{-2x} \qquad \text{Solution of differential equation}$$

is a solution of the differential equation shown above. To see this, substitute for y and $y' = -2e^{-2x}$ in the original equation.

$$y' + 2y = -2e^{-2x} + 2(e^{-2x}) \qquad \text{Substitute for } y \text{ and } y'.$$
$$= 0$$

In the same way, you can show that $y = 2e^{-2x}$, $y = -3e^{-2x}$, and $y = \frac{1}{2}e^{-2x}$ are also solutions of the differential equation. In fact, each function given by

$$y = Ce^{-2x} \qquad \text{General solution}$$

where C is a real number, is a solution of the equation. This family of solutions is called the **general solution** of the differential equation.

Example 1 Verifying Solutions

Determine whether the function is a solution of the differential equation $y'' - y = 0$.

a. $y = Ce^x$ **b.** $y = Ce^{-x}$

SOLUTION

a. Because $y' = Ce^x$ and $y'' = Ce^x$, it follows that

$$y'' - y = Ce^x - Ce^x$$
$$= 0.$$

So, $y = Ce^x$ is a solution.

b. Because $y' = -Ce^{-x}$ and $y'' = Ce^{-x}$, it follows that

$$y'' - y = Ce^{-x} - Ce^{-x}$$
$$= 0.$$

So, $y = Ce^{-x}$ is also a solution.

✓ **CHECKPOINT 1**

Determine whether $y = Ce^{4x}$ is a solution of the differential equation $y' = y$. ■

Particular Solutions and Initial Conditions

A **particular solution** of a differential equation is any solution that is obtained by assigning specific values to the arbitrary constant(s) in the general solution.*

Geometrically, the general solution of a differential equation represents a family of curves known as **solution curves.** For instance, the general solution of the differential equation $xy' - 2y = 0$ is

$$y = Cx^2. \qquad \text{General solution}$$

Figure 10.1 shows several solution curves corresponding to different values of C.

Particular solutions of a differential equation are obtained from **initial conditions** placed on the unknown function and its derivatives. For instance, in Figure 10.1, suppose you want to find the particular solution whose graph passes through the point $(1, 3)$. This initial condition can be written as

$$y = 3 \quad \text{when} \quad x = 1. \qquad \text{Initial condition}$$

Substituting these values into the general solution produces $3 = C(1)^2$, which implies that $C = 3$. So, the particular solution is

$$y = 3x^2. \qquad \text{Particular solution}$$

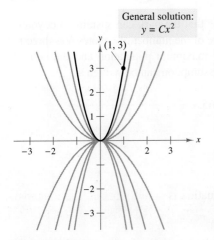

FIGURE 10.1 Solution Curves for $xy' - 2y = 0$

Example 2 Finding a Particular Solution

For the differential equation $xy' - 3y = 0$, verify that $y = Cx^3$ is a solution, and find the particular solution determined by the initial condition $y = 2$ when $x = -3$.

SOLUTION You know that $y = Cx^3$ is a solution because $y' = 3Cx^2$ and

$$xy' - 3y = x(3Cx^2) - 3(Cx^3)$$
$$= 0.$$

Furthermore, the initial condition $y = 2$ when $x = -3$ yields

$$y = Cx^3 \qquad \text{General solution}$$
$$2 = C(-3)^3 \qquad \text{Substitute initial condition.}$$
$$-\frac{2}{27} = C \qquad \text{Solve for } C.$$

and you can conclude that the particular solution is

$$y = -\frac{2x^3}{27}. \qquad \text{Particular solution}$$

Try checking this solution by substituting for y and y' in the original differential equation.

STUDY TIP

To determine a particular solution, the number of initial conditions must match the number of constants in the general solution.

✓ CHECKPOINT 2

For the differential equation $xy' - 2y = 0$, verify that $y = Cx^2$ is a solution, and find the particular solution determined by the initial condition $y = 1$ when $x = 4$. ■

---

*Some differential equations have solutions other than those given by their general solutions. These are called **singular solutions.** In this brief discussion of differential equations, singular solutions will not be discussed.

Example 3 Finding a Particular Solution

The manager of a national park determines that the park can sustain 65 coyotes. The manager assumes that the rate of change of the number of coyotes N is directly proportional to the difference between the maximum population and the current population. As a differential equation, this assumption can be written as

$$\frac{dN}{dt} = k(65 - N), \quad 0 \leq N \leq 65.$$

$\underbrace{\frac{dN}{dt}}_{\text{Rate of change of }N} = \underbrace{k}_{\substack{\text{is propor-}\\\text{tional to}}} \underbrace{(65 - N)}_{\substack{\text{the difference}\\\text{between}\\\text{65 and }N.}}$

The general solution of this differential equation is

$$N = 65 - Ce^{-kt} \qquad \text{General solution}$$

where t is the time in years. When $t = 0$, the population is 30, and when $t = 2$, the population has increased to 50. Sketch the graph of the population function when $0 \leq t \leq 5$.

SOLUTION Substituting the initial condition $N = 30$ when $t = 0$ into the general solution, you can conclude that $C = 35$.

$N = 65 - Ce^{-kt}$	General solution
$30 = 65 - Ce^{-k(0)}$	Substitute 30 for N and 0 for t.
$35 = C$	Solve for C.

So, $N = 65 - 35e^{-kt}$. Next, use $N = 50$ when $t = 2$ to find k.

$50 = 65 - 35e^{-k(2)}$	Substitute 50 for N and 2 for t.
$\frac{3}{7} = e^{-2k}$	Simplify.
$-\frac{1}{2} \ln \frac{3}{7} = k.$	Solve for k.

So, $k \approx 0.4236$ and the particular solution is

$$N = 65 - 35e^{-0.4236t}. \qquad \text{Particular solution}$$

The table shows the populations when $0 \leq t \leq 5$, and the graph of the solution is shown in Figure 10.2.

t	0	1	2	3	4	5
N	30	42	50	55	59	61

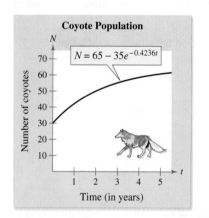

FIGURE 10.2

✓ CHECKPOINT 3

Repeat Example 3 using the initial conditions $N = 25$ when $t = 0$ and $N = 40$ when $t = 2$. ∎

In the first three examples in this section, each solution was given in explicit form, such as $y = f(x)$. Sometimes you will encounter solutions for which it is more convenient to write the solution in implicit form, as shown in Example 4.

Example 4 Sketching Graphs of Solutions

Given that

$$2y^2 - x^2 = C \qquad \text{General solution}$$

is the general solution of the differential equation

$$2yy' - x = 0$$

sketch the particular solutions represented by $C = 0$, $C = \pm 1$, and $C = \pm 4$.

SOLUTION The particular solutions represented by $C = 0$, $C = \pm 1$, and $C = \pm 4$ are shown in Figure 10.3.

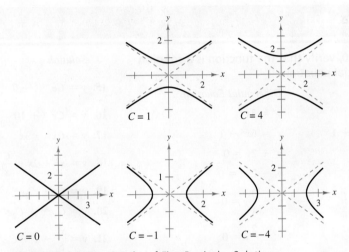

FIGURE 10.3 Graphs of Five Particular Solutions

✓ CHECKPOINT 4

Given that $y = Cx^2$ is the general solution of $xy' - 2y = 0$, sketch the particular solutions represented by $C = 1$, $C = 2$, and $C = 4$. ∎

CONCEPT CHECK

1. Complete the following: A _____ equation is an equation involving a differentiable function and one or more of its derivatives.
2. Complete the following: Because each function given by $y = Ce^{-2x}$ is a solution of $y' + 2y = 0$, $y = Ce^{-2x}$ is the _____ solution of $y' + 2y = 0$.
3. Explain why $y' - 3y = 0$ is a differential equation.
4. In general, describe in words a particular solution of a differential equation.

634 CHAPTER 10 Differential Equations

Skills Review 10.1

The following warm-up exercises involve skills that were covered in earlier sections. You will use these skills in the exercise set for this section. For additional help, review Sections 2.2, 2.6, 4.3, and 4.4.

In Exercises 1–4, find the first and second derivatives of the function.

1. $y = 3x^2 + 2x + 1$
2. $y = -2x^3 - 8x + 4$
3. $y = -3e^{2x}$
4. $y = -3e^{x^2}$

In Exercises 5 and 6, solve for k.

5. $0.5 = 9 - 9e^{-k}$
6. $14.75 = 25 - 25e^{-2k}$

Exercises 10.1

See www.CalcChat.com for worked-out solutions to odd-numbered exercises.

In Exercises 1–10, verify that the function is a solution of the differential equation.

Solution	Differential Equation
1. $y = x^3 + 5$	$y' = 3x^2$
2. $y = 2x^3 - x + 1$	$y' = 6x^2 - 1$
3. $y = e^{-2x}$	$y' + 2y = 0$
4. $y = 3e^{x^2}$	$y' - 2xy = 0$
5. $y = 2x^3$	$y' - \dfrac{3}{x}y = 0$
6. $y = 4x^2$	$y' - \dfrac{2}{x}y = 0$
7. $y = x^2$	$x^2 y'' - 2y = 0$
8. $y = \dfrac{1}{x}$	$xy'' + 2y' = 0$
9. $y = 2e^{2x}$	$y'' - y' - 2y = 0$
10. $y = e^{x^3}$	$y'' - 3x^2 y' - 6xy = 0$

In Exercises 11–24, verify that the function is a solution of the differential equation for any value of C.

Solution	Differential Equation
11. $y = \dfrac{1}{x} + C$	$\dfrac{dy}{dx} = -\dfrac{1}{x^2}$
12. $y = \sqrt{4 - x^2} + C$	$\dfrac{dy}{dx} = -\dfrac{x}{\sqrt{4 - x^2}}$
13. $y = Ce^{4x}$	$\dfrac{dy}{dx} = 4y$
14. $y = Ce^{-4x}$	$\dfrac{dy}{dx} = -4y$
15. $y = Ce^{-t/3} + 7$	$3\dfrac{dy}{dt} + y - 7 = 0$
16. $y = Ce^{-t} + 10$	$y' + y - 10 = 0$
17. $y = Cx^2 - 3x$	$xy' - 3x - 2y = 0$
18. $y = x^2 + 2x + \dfrac{C}{x}$	$xy' + y = x(3x + 4)$
19. $y = C_1 + C_2 e^x$	$y'' - y' = 0$
20. $y = C_1 e^{4x} + C_2 e^{-x}$	$y'' - 3y' - 4y = 0$
21. $y = \dfrac{x^3}{5} - x + C\sqrt{x}$	$2xy' - y = x^3 - x$
22. $y = Ce^{x-x^2}$	$y' + (2x - 1)y = 0$
23. $y = x \ln x + Cx + 4$	$x(y' - 1) - (y - 4) = 0$
24. $y = x(\ln x + C)$	$x + y - xy' = 0$

In Exercises 25–28, determine whether the function is a solution of the differential equation $y^{(4)} - 16y = 0$.

25. $y = e^{-2x}$
26. $y = 5 \ln x$
27. $y = \dfrac{4}{x}$
28. $y = 4e^{2x}$

In Exercises 29–32, determine whether the function is a solution of the differential equation $y''' - 3y' + 2y = 0$.

29. $y = \dfrac{2}{9}xe^{-2x}$
30. $y = 4e^x + \dfrac{2}{9}xe^{-2x}$
31. $y = xe^x$
32. $y = x \ln x$

In Exercises 33–36, verify that the general solution satisfies the differential equation. Then find the particular solution that satisfies the initial condition.

33. General solution: $y = Ce^{-2x}$
    Differential equation: $y' + 2y = 0$
    Initial condition: $y = 3$ when $x = 0$

34. General solution: $y = C_1 + C_2 \ln x$
    Differential equation: $xy'' + y' = 0$
    Initial condition: $y = 5$ and $y' = 0.5$ when $x = 1$

35. General solution: $y = C_1 e^{4x} + C_2 e^{-3x}$
    Differential equation: $y'' - y' - 12y = 0$
    Initial condition: $y = 5$ and $y' = 6$ when $x = 0$

36. General solution: $y = Ce^{x-x^2}$
    Differential equation: $y' + (2x - 1)y = 0$
    Initial condition: $y = 2$ when $x = 1$

(T) In Exercises 37 and 38, the general solution of the differential equation is given. Use a graphing utility to graph the particular solutions that correspond to the indicated values of C.

General Solution	Differential Equation	C-Values
37. $y = C(x + 2)^2$	$(x + 2)y' - 2y = 0$	$0, \pm 1, \pm 2$
38. $y = Ce^{-x}$	$y' + y = 0$	$0, \pm 1, \pm 2$

In Exercises 39–46, use integration to find the general solution of the differential equation.

39. $\dfrac{dy}{dx} = 3x^2$

40. $\dfrac{dy}{dx} = \dfrac{1}{1 + x}$

41. $\dfrac{dy}{dx} = \dfrac{x + 3}{x}$

42. $\dfrac{dy}{dx} = \dfrac{x - 2}{x}$

43. $\dfrac{dy}{dx} = \dfrac{1}{x^2 - 1}$

44. $\dfrac{dy}{dx} = \dfrac{x}{1 + x^2}$

45. $\dfrac{dy}{dx} = \cos 4x$

46. $\dfrac{dy}{dx} = 4 \sin x$

In Exercises 47–50, some of the curves corresponding to different values of C in the general solution of the differential equation are shown in the figure. Find the particular solution that passes through the point plotted on the graph.

47. $y^2 = Cx^3$
    $2xy' - 3y = 0$

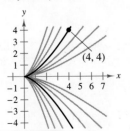

48. $2x^2 - y^2 = C$
    $yy' - 2x = 0$

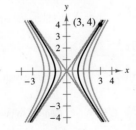

49. $y = Ce^x$
    $y' - y = 0$

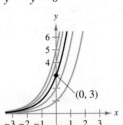

50. $y^2 = 2Cx$
    $2xy' - y = 0$

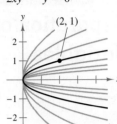

51. **Biology** The limiting capacity of the habitat of a wildlife herd is 750. The growth rate dN/dt of the herd is proportional to the unutilized opportunity for growth, as described by the differential equation

$$\dfrac{dN}{dt} = k(750 - N).$$

The general solution of this differential equation is $N = 750 - Ce^{-kt}$. When $t = 0$, the population of the herd is 100. After 2 years, the population has grown to 160.

(a) Write the population function N as a function of t.

(T) (b) Use a graphing utility to graph the population function.

(c) What is the population of the herd after 4 years?

52. **Safety** Assume that the rate of change per hour in the number of miles s of road cleared by a snowplow is inversely proportional to the depth h of the snow. This rate of change is described by the differential equation

$$\dfrac{ds}{dh} = \dfrac{k}{h}.$$ Show that $s = 25 - \dfrac{13}{\ln 3} \ln \dfrac{h}{2}$ is a solution of this differential equation.

53. Show that $y = a + Ce^{k(1-b)t}$ is a solution of the differential equation

$$y = a + b(y - a) + \left(\dfrac{1}{k}\right)\left(\dfrac{dy}{dt}\right)$$

where k is a constant.

54. The function $y = Ce^{kx}$ is a solution of the differential equation

$$\dfrac{dy}{dx} = 0.07y.$$

Is it possible to determine C or k from the information given? If so, find its value.

True or False? In Exercises 55 and 56, determine whether the statement is true or false. If it is false, explain why or give an example that shows it is false.

55. A differential equation can have more than one solution.

56. If $y = f(x)$ is a solution of a differential equation, then $y = f(x) + C$ is also a solution.

Section 10.2
Separation of Variables

- Use separation of variables to solve differential equations.
- Use differential equations to model and solve real-life problems.

Separation of Variables

The simplest type of differential equation is one of the form $y' = f(x)$. You know that this type of equation can be solved by integration to obtain

$$y = \int f(x)\, dx.$$

In this section, you will learn how to use integration to solve another important family of differential equations—those in which the variables can be separated. This technique is called **separation of variables.**

TECHNOLOGY

You can use a symbolic integration utility to solve a differential equation that has separable variables. Use a symbolic integration utility to solve the differential equation

$$y' = \frac{x}{y^2 + 1}.$$

Separation of Variables

If f and g are continuous functions, then the differential equation

$$\frac{dy}{dx} = f(x)g(y)$$

has a general solution of

$$\int \frac{1}{g(y)}\, dy = \int f(x)\, dx + C.$$

Essentially, the technique of separation of variables is just what its name implies. For a differential equation involving x and y, you separate the variables by grouping the x variables on one side and the y variables on the other. After separating variables, integrate each side to obtain the general solution.

Example 1 Solving a Differential Equation

Find the general solution of

$$\frac{dy}{dx} = \frac{x}{y^2 + 1}.$$

SOLUTION Begin by separating variables, then integrate each side.

$$\frac{dy}{dx} = \frac{x}{y^2 + 1} \qquad \text{Differential equation}$$

$$(y^2 + 1)\, dy = x\, dx \qquad \text{Separate variables.}$$

$$\int (y^2 + 1)\, dy = \int x\, dx \qquad \text{Integrate each side.}$$

$$\frac{y^3}{3} + y = \frac{x^2}{2} + C \qquad \text{General solution}$$

✓**CHECKPOINT 1**

Find the general solution of

$$\frac{dy}{dx} = \frac{x^2}{y}.$$
■

Example 2 Solving a Differential Equation

Find the general solution of

$$\frac{dy}{dx} = \frac{x}{y}.$$

SOLUTION Begin by separating variables, then integrate each side.

$\dfrac{dy}{dx} = \dfrac{x}{y}$	Differential equation
$y\, dy = x\, dx$	Separate variables.
$\displaystyle\int y\, dy = \int x\, dx$	Integrate each side.
$\dfrac{y^2}{2} = \dfrac{x^2}{2} + C_1$	Find antiderivative of each side.
$y^2 = x^2 + C$	Multiply each side by 2.

So, the general solution is $y^2 = x^2 + C$. Note that C_1 is used as a temporary constant of integration in anticipation of multiplying each side of the equation by 2 to produce the constant C.

✓ **CHECKPOINT 2**

Find the general solution of

$$\frac{dy}{dx} = \frac{x+1}{y}.$$ ∎

Example 3 Solving a Differential Equation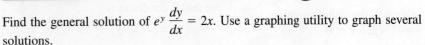

Find the general solution of $e^y \dfrac{dy}{dx} = 2x$. Use a graphing utility to graph several solutions.

SOLUTION Begin by separating variables, then integrate each side.

$e^y \dfrac{dy}{dx} = 2x$	Differential equation
$e^y\, dy = 2x\, dx$	Separate variables.
$\displaystyle\int e^y\, dy = \int 2x\, dx$	Integrate each side.
$e^y = x^2 + C$	Find antiderivative of each side.

By taking the natural logarithm of each side, you can write the general solution as

$$y = \ln(x^2 + C). \qquad \text{General solution}$$

The graphs of the particular solutions given by $C = 0$, $C = 5$, $C = 10$, and $C = 15$ are shown in Figure 10.4.

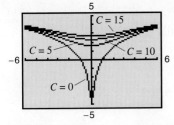

FIGURE 10.4

✓ **CHECKPOINT 3**

Find the general solution of

$$2y \frac{dy}{dx} = -2x.$$

Use a graphing utility to graph the particular solutions given by $C = 1$, $C = 2$, and $C = 4$. ∎

Example 4 Finding a Particular Solution

Solve the differential equation

$$xe^{x^2} + yy' = 0$$

subject to the initial condition $y = 1$ when $x = 0$.

SOLUTION

$$xe^{x^2} + yy' = 0 \qquad \text{Differential equation}$$

$$y\frac{dy}{dx} = -xe^{x^2} \qquad \text{Subtract } xe^{x^2} \text{ from each side.}$$

$$y\, dy = -xe^{x^2}\, dx \qquad \text{Separate variables.}$$

$$\int y\, dy = \int -xe^{x^2}\, dx \qquad \text{Integrate each side.}$$

$$\frac{y^2}{2} = -\frac{1}{2}e^{x^2} + C_1 \qquad \text{Find antiderivative of each side.}$$

$$y^2 = -e^{x^2} + C \qquad \text{Multiply each side by 2.}$$

To find the particular solution, substitute the initial condition values to obtain

$$(1)^2 = -e^{(0)^2} + C.$$

This implies that $1 = -1 + C$, or $C = 2$. So, the particular solution that satisfies the initial condition is

$$y^2 = -e^{x^2} + 2. \qquad \text{Particular solution}$$

✓ **CHECKPOINT 4**

Solve the differential equation

$$e^x + yy' = 0$$

subject to the initial condition $y = 2$ when $x = 0$. ∎

Example 5 Solving a Differential Equation

Example 3 in Section 10.1 uses the differential equation

$$\frac{dN}{dt} = k(65 - N), \quad 0 \le N \le 65$$

to model a coyote population. Solve this differential equation.

SOLUTION

$$\frac{dN}{dt} = k(65 - N) \qquad \text{Differential equation}$$

$$\frac{1}{65 - N}\, dN = k\, dt \qquad \text{Separate variables.}$$

$$\int \frac{1}{65 - N}\, dN = \int k\, dt \qquad \text{Integrate each side.}$$

$$-\ln(65 - N) = kt + C_1 \qquad \text{Find antiderivative of each side.}$$

$$\ln(65 - N) = -kt - C_1 \qquad \text{Multiply each side by } -1.$$

$$65 - N = Ce^{-kt} \qquad \text{Exponentiate and let } e^{-C_1} = C.$$

$$N = 65 - Ce^{-kt} \qquad \text{Solve for } N.$$

STUDY TIP

In Example 5, the context of the original model indicates that $(65 - N)$ is positive. So, when you integrate $1/(65 - N)$, you can write $-\ln(65 - N)$, rather than $-\ln|65 - N|$.

Also note in Example 5 that the solution agrees with the one that was given in Example 3 in Section 10.1.

✓ **CHECKPOINT 5**

Solve the differential equation

$$\frac{dy}{dx} = k(10 - y),$$

$$0 \le y \le 10. \quad ∎$$

SECTION 10.2 Separation of Variables

Example 6 Using Graphical Information

Find the equation of the graph that has the characteristics listed below.

1. At each point (x, y) on the graph, the slope is $-x/(2y)$.
2. The graph passes through the point $(2, 1)$.

SOLUTION Using the information about the slope of the graph, you can write the differential equation

$$\frac{dy}{dx} = -\frac{x}{2y}.$$

Using the point on the graph, you can determine the initial condition $y = 1$ when $x = 2$.

$$\frac{dy}{dx} = -\frac{x}{2y} \quad \text{Differential equation}$$

$$2y \, dy = -x \, dx \quad \text{Separate variables.}$$

$$\int 2y \, dy = \int -x \, dx \quad \text{Integrate each side.}$$

$$y^2 = -\frac{x^2}{2} + C_1 \quad \text{Find antiderivative of each side.}$$

$$2y^2 = -x^2 + C \quad \text{Multiply each side by 2.}$$

$$x^2 + 2y^2 = C \quad \text{Simplify.}$$

Applying the initial condition yields

$$(2)^2 + 2(1)^2 = C$$

which implies that $C = 6$. So, the equation that satisfies the two given conditions is

$$x^2 + 2y^2 = 6. \quad \text{Particular solution}$$

As shown in Figure 10.5, the graph of this equation is an ellipse.

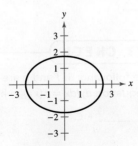

FIGURE 10.5

✓ CHECKPOINT 6

Find the equation of the graph that has the characteristics listed below.

1. At each point (x, y) on the graph, the slope is $2x/y$.
2. The graph passes through the point $(2, 4)$. ■

Application

Example 7 Modeling Advertising Awareness

A new drug is introduced through an advertising campaign to a population of 1 million potential customers. The rate at which the population hears about the drug is assumed to be proportional to the number of people who are not yet aware of the drug. By the end of 1 year, half of the population has heard of the drug. How many will have heard of it by the end of 2 years?

SOLUTION Let y be the number of people (in millions) at time t who have heard of the drug. This means that $(1 - y)$ is the number of people who have not heard of it, and dy/dt is the rate at which the population hears about the drug. From the given assumption, you can write the differential equation as shown.

$$\frac{dy}{dt} = k(1 - y) \qquad \text{Rate of change of } y \text{ is proportional to the difference between 1 and } y.$$

You can solve this equation using separation of variables.

$$dy = k(1 - y)\,dt \qquad \text{Differential form}$$

$$\frac{dy}{1 - y} = k\,dt \qquad \text{Separate variables.}$$

$$\ln|1 - y| = -kt - C_1 \qquad \text{Integrate and multiply each side by } -1.$$

$$1 - y = e^{-kt - C_1} \qquad \text{Assume } y < 1.$$

$$y = 1 - Ce^{-kt} \qquad \text{General solution}$$

To solve for the constants C and k, use the initial conditions. That is, because $y = 0$ when $t = 0$, you can determine that $C = 1$. Similarly, because $y = 0.5$ when $t = 1$, it follows that $0.5 = 1 - e^{-k}$, which implies that $k = \ln 2 \approx 0.693$. So, the particular solution is $y = 1 - e^{-0.693t}$. This model is shown in Figure 10.6. Using the model, you can determine that the number of people who have heard of the product after 2 years is

$$y = 1 - e^{-0.693(2)} \approx 0.75 \text{ or } 750{,}000 \text{ people.}$$

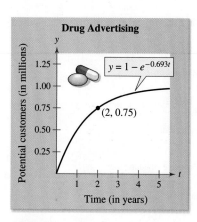

FIGURE 10.6

✓ CHECKPOINT 7

Repeat Example 7 assuming that there are 2 million potential customers and that one fourth of the population has heard of the drug by the end of 1 year. ■

CONCEPT CHECK

1. Complete the following: If f and g are continuous functions, then the differential equation $dy/dx = f(x)g(y)$ has a general solution of
$$\int \frac{1}{g(y)}\,dy = \underline{\qquad} + C.$$

2. True or false: The differential equation $\dfrac{dy}{dx} = \dfrac{3x}{y}$ can be written in separated variables form.

3. True or false: The differential equation $\dfrac{dy}{dx} = \dfrac{3x}{y} + 1$ can be written in separated variables form.

4. In your own words, describe how to solve differential equations that can be solved by separation of variables.

SECTION 10.2 Separation of Variables

Skills Review 10.2

The following warm-up exercises involve skills that were covered in earlier sections. You will use these skills in the exercise set for this section. For additional help, review Sections 4.4, 6.2, and 6.3.

In Exercises 1–6, find the indefinite integral and check your result by differentiating.

1. $\int x^{3/2}\, dx$

2. $\int (t^3 - t^{1/3})\, dt$

3. $\int \dfrac{2}{x-5}\, dx$

4. $\int \dfrac{y}{2y^2+1}\, dy$

5. $\int e^{2y}\, dy$

6. $\int xe^{1-x^2}\, dx$

In Exercises 7–10, solve the equation for C or k.

7. $(3)^2 - 6(3) = 1 + C$

8. $(-1)^2 + (-2)^2 = C$

9. $10 = 2e^{2k}$

10. $(6)^2 - 3(6) = e^{-k}$

Exercises 10.2

See www.CalcChat.com for worked-out solutions to odd-numbered exercises.

In Exercises 1–6, decide whether the variables in the differential equation can be separated.

1. $\dfrac{dy}{dx} = \dfrac{x}{y+3}$

2. $\dfrac{dy}{dx} = \dfrac{x+1}{x}$

3. $\dfrac{dy}{dx} = \dfrac{1}{x} + 1$

4. $\dfrac{dy}{dx} = \dfrac{x}{x+y}$

5. $\dfrac{dy}{dx} = x - y$

6. $x\dfrac{dy}{dx} = \dfrac{1}{y}$

In Exercises 7–32, use separation of variables to find the general solution of the differential equation.

7. $\dfrac{dy}{dx} = 2x$

8. $\dfrac{dy}{dx} = \dfrac{1}{x}$

9. $\dfrac{dr}{ds} = 0.05r$

10. $\dfrac{dr}{ds} = 0.05s$

11. $\dfrac{dy}{dx} = \dfrac{3x}{2y}$

12. $\dfrac{dy}{dx} = \dfrac{3y}{2x}$

13. $3y^2\dfrac{dy}{dx} = 1$

14. $\dfrac{dy}{dx} = x^2 y$

15. $(y+1)\dfrac{dy}{dx} = 2x$

16. $(1+y)\dfrac{dy}{dx} - 4x = 0$

17. $y' - xy = 0$

18. $y' - y = 5$

19. $\dfrac{dy}{dt} = \dfrac{e^t}{4y}$

20. $e^y \dfrac{dy}{dt} = 3t^2 + 1$

21. $\dfrac{dy}{dx} = \sqrt{1-y}$

22. $\dfrac{dy}{dx} = \sqrt{\dfrac{x}{y}}$

23. $(2+x)y' = 2y$

24. $y' = (2x-1)(y+3)$

25. $xy' = y$

26. $y' - y(x+1) = 0$

27. $y' = \dfrac{x}{y} - \dfrac{x}{1+y}$

28. $\dfrac{dy}{dx} = \dfrac{x^2+2}{3y^2}$

29. $e^x(y'+1) = 1$

30. $yy' - 2xe^x = 0$

31. $y\dfrac{dy}{dx} = \sin x$

32. $y\dfrac{dy}{dx} = 6\cos(\pi x)$

(T) In Exercises 33–36, (a) find the general solution of the differential equation and (b) use a graphing utility to graph the particular solutions given by $C = 1$, $C = 2$, and $C = 4$.

33. $\dfrac{dy}{dx} = x$

34. $\dfrac{dy}{dx} = -\dfrac{2x}{y}$

35. $\dfrac{dy}{dx} = 4 - y$

36. $\dfrac{dy}{dx} = 0.25x(4-y)$

In Exercises 37–44, use the initial condition to find the particular solution of the differential equation.

Differential Equation	Initial Condition
37. $yy' - e^x = 0$	$y = 4$ when $x = 0$
38. $\sqrt{x} + \sqrt{y}\,y' = 0$	$y = 4$ when $x = 1$
39. $x(y+4) + y' = 0$	$y = -5$ when $x = 0$
40. $\dfrac{dy}{dx} = x^2(1+y)$	$y = 3$ when $x = 0$
41. $dP - 6P\,dt = 0$	$P = 5$ when $t = 0$
42. $dT + k(T-70)\,dt = 0$	$T = 140$ when $t = 0$
43. $\dfrac{dy}{dx} = y\cos x$	$y = 1$ when $x = 0$
44. $\dfrac{dy}{dx} = 2xy\sin x^2$	$y = 1$ when $x = 0$

In Exercises 45 and 46, find an equation of the graph that passes through the point and has the specified slope. Then graph the equation.

45. Point: $(-1, 1)$

Slope: $y' = \dfrac{6x}{5y}$

46. Point: $(8, 2)$

Slope: $y' = \dfrac{2y}{3x}$

Velocity In Exercises 47 and 48, solve the differential equation to find velocity v as a function of time t if $v = 0$ when $t = 0$. The differential equation models the motion of two people on a toboggan after consideration of the force of gravity, friction, and air resistance.

47. $12.5 \dfrac{dv}{dt} = 43.2 - 1.25v$ **48.** $12.5 \dfrac{dv}{dt} = 43.2 - 1.75v$

49. Biology: Cell Growth The growth rate of a spherical cell with volume V is proportional to its surface area S. For a sphere, the surface area and volume are related by $S = kV^{2/3}$. So, a model for the cell's growth is

$$\dfrac{dV}{dt} = kV^{2/3}.$$

Solve this differential equation.

50. Weight Gain A calf that weighed w_0 pounds at birth gains weight at the rate $dw/dt = 1200 - w$, where w is weight (in pounds) and t is time (in years). Solve the differential equation.

51. Weight Gain A calf that weighed 60 pounds at birth gains weight at the rate $dw/dt = k(1200 - w)$, where w is weight (in pounds) and t is time (in years). Solve the differential equation.

(T) (a) Use a graphing utility to graph the solutions for $k = 0.8, 0.9,$ and 1.

(b) If the animal is sold when its weight reaches 800 pounds, find the time of sale for each of the models in part (a).

(c) What is the maximum weight of the animal for each of the models?

52. Learning Theory The management of a medical claims office has found that one employee can process at most 1000 claims per day. The number M of claims processed per day by a new employee will increase at a rate proportional to the difference between 1000 and M. This is described by the differential equation

$$\dfrac{dM}{dt} = k(1000 - M)$$

where t is the time in days and $M(0) = 0$. Solve this differential equation.

(T) (a) Use a graphing utility to graph the solutions for $k = 0.8, 0.9,$ and 1.

(b) For each of the models in part (a), find the amount of time it takes for an employee to be able to process 500 claims per day.

(c) For each of the models in part (a), find the amount of time it takes for an employee to be able to process 750 claims per day.

53. Radioactive Decay The rate of decomposition of radioactive radium is proportional to the amount present at any time. The half-life of radioactive radium is 1599 years. What percent of a present amount will remain after 25 years?

54. Radioactive Decay The rate of decomposition of radioactive einsteinium is proportional to the amount present at any time. The half-life of radioactive einsteinium is 276 days. If 0.5 gram remains after 100 days, what was the initial amount?

Life Science Capsule

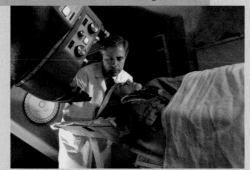
© Yoav Levy/Phototake

Radiation oncologists are physicians who specialize in treating cancer through radiation therapies and methods. They investigate the use of high-energy x-rays, electron beams, gamma rays, and radioactive isotopes to kill cancer cells without exceeding the maximum doses that are safe for normal tissues. Radiation oncologists frequently use differential equations in their research. Becoming a radiation oncologist requires many years of education. Most start with a Bachelor of Science degree. Subsequently, four years of medical school and a year of internship are completed before entering residency graduate education, which usually takes about four more years.

55. Research Project Use your school's library, the Internet, or some other reference source to investigate the educational requirements of a life science or social science career such as the one described above. Are any specialized residencies, internships, or certifications required? Write a short paper that summarizes your findings.

Mid-Chapter Quiz

In Exercises 1 and 2, verify that the function is a solution of the differential equation for any value of C.

Solution	Differential Equation
1. $y = Ce^{-x/2}$	$2y' + y = 0$
2. $y = C_1 \cos x + C_2 \sin x$	$y'' + y = 0$

In Exercises 3 and 4, verify that the general solution satisfies the differential equation. Then find the particular solution that satisfies the initial condition.

3. General solution: $y = C_1 \sin 3x + C_2 \cos 3x$
   Differential equation: $y'' + 9y = 0$
   Initial condition: $y = 2$ and $y' = 1$ when $x = \pi/6$

4. General solution: $y = C_1 x + C_2 x^3$
   Differential equation: $x^2 y'' - 3xy' + 3y = 0$
   Initial condition: $y = 0$ and $y' = 4$ when $x = 2$

In Exercises 5–7, use separation of variables to find the general solution of the differential equation.

5. $\dfrac{dy}{dx} = -4x + 4$

6. $y' = (x + 2)(y - 1)$

7. $y \dfrac{dy}{dx} = \dfrac{1}{x^2 - 1}$

(T) In Exercises 8 and 9, (a) find the general solution of the differential equation and (b) use a graphing utility to graph the particular solutions given by $C = 0$ and $C = \pm 1$.

8. $\dfrac{dy}{dx} = \dfrac{x^2 + 1}{2y}$

9. $\dfrac{dy}{dx} = \dfrac{y}{x - 3}$

In Exercises 10 and 11, use the initial condition to find the particular solution of the differential equation.

Differential Equation	Initial Condition
10. $y' + 2y - 1 = 0$	$y = 1$ when $x = 0$
11. $\dfrac{dy}{dx} = y \sin \pi x$	$y = -3$ when $x = \dfrac{1}{2}$

(T) 12. Find an equation of the graph that passes through the point $(0, 2)$ and has a slope of $3x^2 y$ at any point (x, y). Then use a graphing utility to graph the equation.

13. Ignoring resistance, a sailboat starting from rest accelerates at a rate proportional to the difference between the velocities of the wind and the boat. With a 20 knot wind, this acceleration is described by the differential equation

$$\frac{dv}{dt} = k(20 - v)$$

where v is the velocity of the boat and t is the time in hours. After half an hour, the boat is moving at 10 knots. Write the velocity as a function of time.

Section 10.3

First-Order Linear Differential Equations

- Solve first-order linear differential equations.
- Use first-order linear differential equations to model and solve real-life problems.

First-Order Linear Differential Equations

Definition of a First-Order Linear Differential Equation

A **first-order linear differential equation** is an equation of the form

$$y' + P(x)y = Q(x)$$

where P and Q are functions of x. An equation that is written in this form is said to be in **standard form.**

STUDY TIP

The term "first-order" refers to the fact that the highest-order derivative of y in the equation is the first derivative.

To solve a linear differential equation, write it in standard form to identify the functions $P(x)$ and $Q(x)$. Then integrate $P(x)$ and form the expression

$$u(x) = e^{\int P(x)\,dx} \qquad \text{Integrating factor}$$

which is called an **integrating factor.** The general solution of the equation is

$$y = \frac{1}{u(x)} \int Q(x)u(x)\,dx. \qquad \text{General solution}$$

Example 1 Solving a Linear Differential Equation

Find the general solution of $y' + y = e^x$.

SOLUTION For this equation, $P(x) = 1$ and $Q(x) = e^x$. So, the integrating factor is

$$u(x) = e^{\int P(x)\,dx}$$
$$= e^{\int dx} = e^x.$$

This implies that the general solution is

$$y = \frac{1}{u(x)} \int Q(x)u(x)\,dx$$
$$= \frac{1}{e^x} \int e^x(e^x)\,dx$$
$$= e^{-x}\left(\frac{1}{2}e^{2x} + C\right) = \frac{1}{2}e^x + Ce^{-x}. \qquad \text{General solution}$$

✓ CHECKPOINT 1

Find the general solution of

$$y' - y = 10. \blacksquare$$

In Example 1, the differential equation was given in standard form. For equations that are not written in standard form, you should first convert to standard form so that you can identify the functions $P(x)$ and $Q(x)$.

SECTION 10.3 First-Order Linear Differential Equations

DISCOVERY

Solve for y' in the differential equation in Example 2. Use this equation for y' to determine the slopes of y at the points $(1, 0)$ and $(e^{-1/2}, -1/2e)$. Now graph the particular solution $y = x^2 \ln x$ and estimate the slopes at $x = 1$ and $x = e^{-1/2}$. What happens to the slope of y as x approaches zero?

Example 2 Solving a Linear Differential Equation

Find the general solution of

$$xy' - 2y = x^2.$$

Assume $x > 0$.

SOLUTION Begin by writing the equation in standard form.

$$y' - \left(\frac{2}{x}\right)y = x \qquad \text{Standard form, } y' + P(x)y = Q(x)$$

In this form, you can see that $P(x) = -2/x$ and $Q(x) = x$. So,

$$\int P(x)\,dx = -\int \frac{2}{x}\,dx$$
$$= -2\ln x$$
$$= -\ln x^2$$

which implies that the integrating factor is

$$u(x) = e^{\int P(x)\,dx}$$
$$= e^{-\ln x^2}$$
$$= \frac{1}{x^2}. \qquad \text{Integrating factor}$$

This implies that the general solution is

$$y = \frac{1}{u(x)}\int Q(x)u(x)\,dx \qquad \text{Form of general solution}$$
$$= \frac{1}{1/x^2}\int x\left(\frac{1}{x^2}\right)dx \qquad \text{Substitute.}$$
$$= x^2 \int \frac{1}{x}\,dx \qquad \text{Simplify.}$$
$$= x^2(\ln x + C). \qquad \text{General solution}$$

✓ CHECKPOINT 2

Find the general solution of

$$xy' - y = x.$$

Assume $x > 0$. ∎

TECHNOLOGY

From Example 2, you can see that it can be difficult to solve a linear differential equation. Fortunately, the task is greatly simplified by symbolic integration utilities. Use a symbolic integration utility to find the particular solution of the differential equation in Example 2, given the initial condition $y = 1$ when $x = 1$.

Guidelines for Solving a Linear Differential Equation

1. Write the equation in standard form
$$y' + P(x)y = Q(x).$$

2. Find the integrating factor
$$u(x) = e^{\int P(x)\,dx}.$$

3. Evaluate the integral below to find the general solution.
$$y = \frac{1}{u(x)}\int Q(x)u(x)\,dx$$

Application

Example 3 Intravenous Feeding

Glucose is added intravenously to the bloodstream at the rate of q units per minute, and the body removes glucose from the bloodstream at a rate proportional to the amount present. Assume that A is the amount of glucose in the bloodstream at time t and that the rate of change of the amount of glucose is

$$\frac{dA}{dt} = q - kA$$

where k is a constant. Find the general solution of the differential equation.

SOLUTION In standard form, this linear differential equation is

$$\frac{dA}{dt} + kA = q \qquad \text{Standard form}$$

which implies that $P(t) = k$ and $Q(t) = q$. So, the integrating factor is

$$\begin{aligned} u(t) &= e^{\int P(t)\,dt} \\ &= e^{\int k\,dt} \\ &= e^{kt} \end{aligned} \qquad \text{Integrating factor}$$

and the general solution is

$$\begin{aligned} A &= e^{-kt}\int q e^{kt}\,dt \\ &= e^{-kt}\left(\frac{q}{k}e^{kt} + C\right) \\ &= \frac{q}{k} + Ce^{-kt}. \qquad \text{General solution} \end{aligned}$$

✓ CHECKPOINT 3

Use the result of Example 3 to find the particular solution determined by the initial condition $A = 0$ when $t = 0$. (Assume $k = 0.05$ and $q = 0.05$.) ∎

CONCEPT CHECK

1. Given a first-order linear differential equation, what does the term "first-order" refer to?
2. True or false: $y' + \frac{1}{x}y = x + 1$ is a first-order linear differential equation.
3. Give the standard form of a first-order linear differential equation. What is its integrating factor?
4. Give the guidelines for solving a first-order linear differential equation.

SECTION 10.3 First-Order Linear Differential Equations

Skills Review 10.3

The following warm-up exercises involve skills that were covered in earlier sections. You will use these skills in the exercise set for this section. For additional help, review Sections 4.2, 4.4, and 6.1–6.3.

In Exercises 1–4, simplify the expression.

1. $e^{-x}(e^{2x} + e^x)$

2. $\dfrac{1}{e^{-x}}(e^{-x} + e^{2x})$

3. $e^{-\ln x^3}$

4. $e^{2\ln x + x}$

In Exercises 5–10, find the indefinite integral.

5. $\displaystyle\int 4e^{2x}\, dx$

6. $\displaystyle\int xe^{3x^2}\, dx$

7. $\displaystyle\int \dfrac{1}{2x+5}\, dx$

8. $\displaystyle\int \dfrac{x+1}{x^2+2x+3}\, dx$

9. $\displaystyle\int (4x-3)^2\, dx$

10. $\displaystyle\int x(1-x^2)^2\, dx$

Exercises 10.3

See www.CalcChat.com for worked-out solutions to odd-numbered exercises.

In Exercises 1–6, write the linear differential equation in standard form.

1. $x^3 - 2x^2 y' + 3y = 0$
2. $y' - 5(2x - y) = 0$
3. $xy' + y = xe^x$
4. $xy' + y = x^3 y$
5. $y + 1 = (x - 1)y'$
6. $x = x^2(y' + y)$

In Exercises 7–18, solve the differential equation.

7. $\dfrac{dy}{dx} + 3y = 6$

8. $\dfrac{dy}{dx} + 5y = 15$

9. $\dfrac{dy}{dx} + y = e^{-x}$

10. $\dfrac{dy}{dx} + 3y = e^{-3x}$

11. $\dfrac{dy}{dx} = \dfrac{x^2 + 3}{x}$

12. $\dfrac{dy}{dx} = \dfrac{e^{-2x}}{1 + e^{-2x}}$

13. $y' + 5xy = x$

14. $y' + 5y = e^{5x}$

15. $(x - 1)y' + y = x^2 - 1$

16. $xy' + y = x^2 + 1$

17. $x^3 y' + 2y = e^{1/x^2}$

18. $xy' + y = x^2 \ln x$

In Exercises 19–22, solve for y in two ways.

19. $y' + y = 4$

20. $y' + 10y = 5$

21. $y' - 2xy = 2x$

22. $y' + 4xy = x$

In Exercises 23–26, match the differential equation with its solution.

Differential Equation	Solution
23. $y' - 2x = 0$	(a) $y = Ce^{x^2}$
24. $y' - 2y = 0$	(b) $y = -\tfrac{1}{2} + Ce^{x^2}$
25. $y' - 2xy = 0$	(c) $y = x^2 + C$
26. $y' - 2xy = x$	(d) $y = Ce^{2x}$

In Exercises 27–34, find the particular solution that satisfies the initial condition.

Differential Equation	Initial Condition
27. $y' + y = 6e^x$	$y = 3$ when $x = 0$
28. $y' + 2y = e^{-2x}$	$y = 4$ when $x = 1$
29. $xy' + y = 0$	$y = 2$ when $x = 2$
30. $y' + y = x$	$y = 4$ when $x = 0$
31. $y' + 3x^2 y = 3x^2$	$y = 6$ when $x = 0$
32. $y' + (2x - 1)y = 0$	$y = 2$ when $x = 1$
33. $xy' - 2y = -x^2$	$y = 5$ when $x = 1$
34. $x^2 y' - 4xy = 10$	$y = 10$ when $x = 1$

35. Sales The rate of change (in thousands of units) in sales S of a biomedical syringe is modeled by

$$\frac{dS}{dt} = 0.2(100 - S) + 0.2t$$

where t is the time in years. Solve this differential equation and use the result to complete the table.

t	0	1	2	3	4	5	6	7	8	9	10
S	0										

36. Sales The rate of change in sales S of pharmaceutical equipment is modeled by

$$\frac{dS}{dt} = k_1(L - S) + k_2 t$$

where t is the time in years and $S = 0$ when $t = 0$. Solve this differential equation for S as a function of t.

37. Vertical Motion A falling object encounters air resistance that is proportional to its velocity v. The acceleration due to gravity is -9.8 meters per second per second. The rate of change in velocity is

$$\frac{dv}{dt} = kv - 9.8.$$

Solve this differential equation to find v as a function of time t.

38. Velocity A booster rocket carrying an observation satellite is launched into space. The rocket and satellite have mass m and are subject to air resistance proportional to the velocity v at any time t. A differential equation that models the velocity of the rocket and satellite is

$$m\frac{dv}{dt} = -mg - kv$$

where g is the acceleration due to gravity. Solve the differential equation for v as a function of t.

39. Learning Curve The management at a medical supply factory has found that the maximum number of units an employee can produce in a day is 40. The rate of increase in the number of units N produced with respect to time t (in days) by a new employee is proportional to $40 - N$. This rate of change of performance with respect to time can be modeled by

$$\frac{dN}{dt} = k(40 - N).$$

(a) Solve this differential equation.

(b) Find the particular solution for a new employee who produced 10 units on the first day at the factory and 19 units on the twentieth day.

40. Investment Let $A(t)$ be the amount in a fund earning interest at the annual rate of r, compounded continuously. If a continuous cash flow of P dollars per year is withdrawn from the fund, then the rate of decrease of A is given by the differential equation

$$\frac{dA}{dt} = rA - P$$

where $A = A_0$ when $t = 0$.

(a) Solve this equation for A as a function of t.

(b) Use the result of part (a) to find A when $A_0 =$ $\$2,000,000$, $r = 7\%$, $P = \$250,000$, and $t = 5$ years.

(c) Find A_0 if a retired person wants a continuous cash flow of $\$40,000$ per year for 20 years. Assume that the person's investment will earn 8%, compounded continuously.

41. Health An infectious disease spreads through a large population according to the model

$$\frac{dy}{dt} = \frac{1-y}{4}$$

where y is the percent of the population exposed to the disease, and t is the time in years.

(a) Solve this differential equation, assuming $y(0) = 0$.

(b) Find the number of years it takes for half of the population to have been exposed to the disease.

(c) Find the percent of the population that has been exposed to the disease after 4 years.

42. Extended Application To work an extended application analyzing a person's weight loss, visit this text's website at *college.hmco/info/larsonapplied.com*. (Data Source: The College Mathematics Journal)

Section 10.4
Applications of Differential Equations

- Use differential equations to model and solve real-life problems.

Applications of Differential Equations

Example 1 Modeling a Chemical Reaction

During a chemical reaction, substance A is converted into substance B at a rate that is proportional to the square of the amount of substance A. When $t = 0$, 60 grams of A are present, and after 1 hour $(t = 1)$, only 10 grams of A remain unconverted. How much of A is present after 2 hours?

SOLUTION Let y be the amount of unconverted substance A at any time t. From the given assumption about the conversion rate, you can write the differential equation as shown.

$$\underbrace{\frac{dy}{dt}}_{\text{Rate of change of } y} = \underbrace{k}_{\text{is proportional to}} \underbrace{y^2}_{\text{the square of } y.}$$

Using separation of variables *or* a symbolic integration utility, you can find the general solution to be

$$y = \frac{-1}{kt + C}. \qquad \text{General solution}$$

To solve for the constants C and k, use the initial conditions. That is, because $y = 60$ when $t = 0$, you can determine that $C = -\frac{1}{60}$. Similarly, because $y = 10$ when $t = 1$, it follows that

$$10 = \frac{-1}{k - (1/60)}$$

which implies that $k = -\frac{1}{12}$. So, the particular solution is

$$y = \frac{-1}{(-1/12)t - (1/60)} \qquad \text{Substitute for } k \text{ and } C.$$

$$= \frac{60}{5t + 1}. \qquad \text{Particular solution}$$

Using the model, you can determine that the amount of unconverted substance A after 2 hours is

$$y = \frac{60}{5(2) + 1}$$

$$\approx 5.45 \text{ grams}.$$

In Figure 10.7, note that the chemical conversion is occurring rapidly during the first hour. Then, as more and more of substance A is converted, the conversion rate slows down.

DISCOVERY

In Example 1, the rate of conversion was assumed to be proportional to the *square* of the amount of unconverted substance A. How would the result change if the rate of conversion were assumed to be proportional to the amount of unconverted substance A?

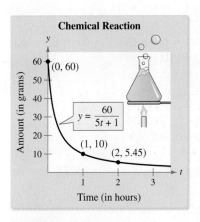

FIGURE 10.7

✓ CHECKPOINT 1

Use the chemical reaction model in Example 1 to find the amount y (in grams) as a function of t (in hours) given that $y = 40$ grams when $t = 0$ and $y = 5$ grams when $t = 2$. ■

The next example concerns **Newton's Law of Cooling,** which states that the rate of change in the temperature of an object is proportional to the difference between the object's temperature and the temperature of the surrounding medium.

Example 2 Newton's Law of Cooling

Let y represent the temperature (in °F) of an object in a room whose temperature is kept at a constant 60°. If the object cools from 100° to 90° in 10 minutes, how much longer will it take for its temperature to decrease to 80°?

SOLUTION From Newton's Law of Cooling, you know that the rate of change in y is proportional to the difference between y and 60. This can be written as

$$y' = k(y - 60), \quad 80 \leq y \leq 100.$$

To solve this differential equation, use separation of variables, as shown.

$$\frac{dy}{dt} = k(y - 60) \qquad \text{Differential equation}$$

$$\left(\frac{1}{y - 60}\right) dy = k \, dt \qquad \text{Separate variables.}$$

$$\int \frac{1}{y - 60} \, dy = \int k \, dt \qquad \text{Integrate each side.}$$

$$\ln|y - 60| = kt + C_1 \qquad \text{Find antiderivative of each side.}$$

Because $y > 60$, $|y - 60| = y - 60$, and you can omit the absolute value signs. Using exponential notation, you have

$$y - 60 = e^{kt + C_1} \quad \Longrightarrow \quad y = 60 + Ce^{kt}. \qquad C = e^{C_1}$$

Using $y = 100$ when $t = 0$, you obtain $100 = 60 + Ce^{k(0)} = 60 + C$, which implies that $C = 40$. Because $y = 90$ when $t = 10$,

$$90 = 60 + 40e^{k(10)}$$

$$30 = 40e^{10k}$$

$$k = \tfrac{1}{10} \ln \tfrac{3}{4} \approx -0.02877.$$

So, the model is

$$y = 60 + 40e^{-0.02877t} \qquad \text{Cooling model}$$

and finally, when $y = 80$, you obtain

$$80 = 60 + 40e^{-0.02877t}$$

$$20 = 40e^{-0.02877t}$$

$$\tfrac{1}{2} = e^{-0.02877t}$$

$$\ln \tfrac{1}{2} = -0.02877t$$

$$t \approx 24.09 \text{ minutes.}$$

So, it will require about 14.09 *more* minutes for the object to cool to a temperature of 80° (see Figure 10.8).

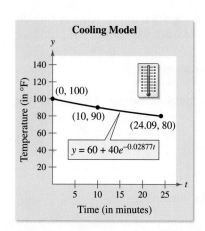

FIGURE 10.8

✓ CHECKPOINT 2

Let y represent the temperature (in °C) of an object in a room whose temperature is kept at a constant 20°. If the object cools from 40° to 35° in 10 minutes, how much longer will it take for its temperature to decrease to 30°? ■

SECTION 10.4 Applications of Differential Equations

Earlier in the text, you studied two models for population growth: *exponential growth*, which assumes that the rate of change of y is proportional to y, and *logistic growth*, which assumes that the rate of change of y is proportional to y and $1 - y/L$, where L is the population limit.

The next example describes a third type of growth model called a **Gompertz growth model**. This model assumes that the rate of change of y is proportional to y and the natural log of L/y, where L is the population limit.

Example 3 Modeling Population Growth

A population of 20 wolves has been introduced into a national park. The forest service estimates that the maximum population the park can sustain is 200 wolves. After 3 years, the population is estimated to be 40 wolves. If the population follows a Gompertz growth model, how many wolves will there be 10 years after their introduction?

SOLUTION Let y be the number of wolves at any time t. From the given assumption about the rate of growth of the population, you can write the differential equation as shown.

$$\frac{dy}{dt} = ky \ln \frac{200}{y}$$

Rate of change of y is proportional to the product of y and the log of the ratio of 200 and y.

Using separation of variables *or* a symbolic integration utility, you can find the general solution to be

$$y = 200e^{-Ce^{-kt}}. \qquad \text{General solution}$$

To solve for the constants C and k, use the initial conditions. That is, because $y = 20$ when $t = 0$, you can determine that

$$C = \ln 10$$
$$\approx 2.3026.$$

Similarly, because $y = 40$ when $t = 3$, it follows that

$$40 = 200e^{-2.3026e^{-k(3)}}$$

which implies that $k \approx 0.1194$. So, the particular solution is

$$y = 200e^{-2.3026e^{-0.1194t}}. \qquad \text{Particular solution}$$

Using the model, you can estimate the wolf population after 10 years to be

$$y = 200e^{-2.3026e^{-0.1194(10)}}$$
$$\approx 100 \text{ wolves}.$$

In Figure 10.9, note that after 10 years the population has reached about half of the estimated maximum population. Try checking the growth model to see that it yields $y = 20$ when $t = 0$ and $y = 40$ when $t = 3$.

Algebra Review

For help with the algebra in solving for C in Example 3, see Example 1 in the *Chapter 10 Algebra Review*, on page 657. For help with the algebra in solving for k in Example 3, see Example 3 in the *Chapter 10 Algebra Review*, on page 658.

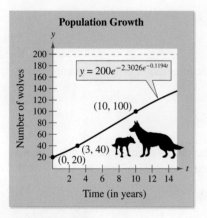

FIGURE 10.9

✓ CHECKPOINT 3

A population of 10 wolves has been introduced into a national park. The forest service estimates that the maximum population the park can sustain is 150 wolves. After 3 years, the population is estimated to be 25 wolves. If the population follows a Gompertz growth model, how many wolves will there be 10 years after their introduction? ∎

In genetics, a commonly used hybrid selection model is based on the differential equation

$$\frac{dy}{dt} = ky(1-y)(a-by).$$

In this model, y represents the portion of the population that has a certain characteristic and t represents the time (measured in generations). The numbers a, b, and k are constants that depend on the genetic characteristic that is being studied.

Algebra Review

For help with the algebra in solving for C in Example 4, see Example 2 in the *Chapter 10 Algebra Review,* on page 657. For help with the algebra in solving for k in Example 4, see Example 4 in the *Chapter 10 Algebra Review,* on page 658.

Example 4 Modeling Hybrid Selection

You are studying a population of beetles to determine how quickly characteristic D will pass from one generation to the next. At the beginning of your study ($t = 0$), you find that half the population has characteristic D. After four generations ($t = 4$), you find that 80% of the population has characteristic D. Use the hybrid selection model above with $a = 2$ and $b = 1$ to find the percent of the population that will have characteristic D after 10 generations.

SOLUTION Using $a = 2$ and $b = 1$, the differential equation for the hybrid selection model is

$$\frac{dy}{dt} = ky(1-y)(2-y).$$

Using separation of variables *or* a symbolic integration utility, you can find the general solution to be

$$\frac{y(2-y)}{(1-y)^2} = Ce^{2kt}. \quad \text{General solution}$$

To solve for the constants C and k, use the initial conditions. That is, because $y = 0.5$ when $t = 0$, you can determine that $C = 3$. Similarly, because $y = 0.8$ when $t = 4$, it follows that

$$\frac{0.8(1.2)}{(0.2)^2} = 3e^{8k}$$

which implies that

$$k = \frac{1}{8} \ln 8 \approx 0.2599.$$

So, the particular solution is

$$\frac{y(2-y)}{(1-y)^2} = 3e^{0.5199t}. \quad \text{Particular solution}$$

Using the model, you can estimate the percent of the population that will have characteristic D after 10 generations to be given by

$$\frac{y(2-y)}{(1-y)^2} = 3e^{0.5199(10)}.$$

Using a symbolic algebra utility, you can solve this equation for y to obtain $y \approx 0.96$. The graph of the model is shown in Figure 10.10.

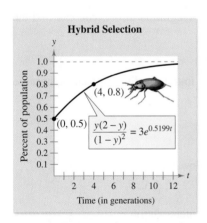

FIGURE 10.10

✓ CHECKPOINT 4

Repeat Example 4 given that only 25% of the population has characteristic D when $t = 0$ and 50% of the population has characteristic D when $t = 4$. ■

SECTION 10.4 Applications of Differential Equations

Example 5
MAKE A DECISION Modeling a Chemical Mixture

A tank contains 40 gallons of a solution composed of 90% water and 10% alcohol. A second solution containing half water and half alcohol is added to the tank at the rate of 4 gallons per minute. At the same time, the tank is being drained at the rate of 4 gallons per minute, as shown in Figure 10.11. Assuming that the solution is stirred constantly, will there be at least 14 gallons of alcohol in the tank after 10 minutes?

SOLUTION Let y be the number of gallons of alcohol in the tank at any time t. The *percent* of alcohol in the 40-gallon tank at any time is $y/40$. Moreover, because 4 gallons of solution are being drained each minute, the rate of change of y is

$$\frac{dy}{dt} = -4\left(\frac{y}{40}\right) + 2$$

Rate of change of y is equal to the amount of alcohol draining out plus the amount of alcohol entering.

where 2 represents the number of gallons of alcohol entering each minute in the 50% solution. In standard form, this linear differential equation is

$$y' + \frac{1}{10}y = 2. \qquad \text{Standard form}$$

Using an integrating factor *or* a symbolic integration utility, you can find the general solution to be

$$y = 20 + Ce^{-t/10}. \qquad \text{General solution}$$

Because $y = 4$ when $t = 0$, you can conclude that $C = -16$. So, the particular solution is

$$y = 20 - 16e^{-t/10}. \qquad \text{Particular solution}$$

Using this model, you can determine that the amount of alcohol in the tank when $t = 10$ is

$$y = 20 - 16e^{-(10)/10} \approx 14.1 \text{ gallons.}$$

Yes, there will be at least 14 gallons of alcohol in the tank after 10 minutes.

FIGURE 10.11

✓ **CHECKPOINT 5**

A tank contains 50 gallons of a solution composed of 90% water and 10% alcohol. A second solution containing half water and half alcohol is added to the tank at the rate of 5 gallons per minute. At the same time, the tank is being drained at the rate of 5 gallons per minute. Assuming that the solution is stirred constantly, how much alcohol will be in the tank after 10 minutes? ■

CONCEPT CHECK

1. What does the exponential growth model assume about the rate of change of y?
2. What does the logistic growth model assume about the rate of change of y?
3. What does the Gompertz growth model assume about the rate of change of y?
4. In the logistic and Gompertz growth models, what does L represent?

CHAPTER 10 Differential Equations

Skills Review 10.4

The following warm-up exercises involve skills that were covered in earlier sections. You will use these skills in the exercise set for this section. For additional help, review Sections 4.6, 10.2, and 10.3.

In Exercises 1–4, use separation of variables to find the general solution of the differential equation.

1. $\dfrac{dy}{dx} = 3x$ **2.** $2y\dfrac{dy}{dx} = 3$ **3.** $\dfrac{dy}{dx} = 2xy$ **4.** $\dfrac{dy}{dx} = \dfrac{x-4}{4y^3}$

In Exercises 5–8, use an integrating factor to solve the first-order linear differential equation.

5. $y' + 2y = 4$ **6.** $y' + 2y = e^{-2x}$ **7.** $y' + xy = x$ **8.** $xy' + 2y = x^2$

In Exercises 9 and 10, write the equation that models the statement.

9. The rate of change of y with respect to x is proportional to the square of x.

10. The rate of change of x with respect to t is proportional to the difference of x and t.

Exercises 10.4

See www.CalcChat.com for worked-out solutions to odd-numbered exercises.

In Exercises 1–6, assume that the rate of change in y is proportional to y. Solve the resulting differential equation $dy/dx = ky$ and find the particular solution that passes through the points.

1. (0, 1), (3, 2)
2. (0, 4), (1, 6)
3. (0, 4), (4, 1)
4. (0, 60), (5, 30)
5. (2, 2), (3, 4)
6. (1, 4), (2, 1)

7. Population Growth The rate of change of the population of a city is proportional to the population P at any time t (in years). In 1998, the population was 400,000, and the constant of proportionality was 0.015. Estimate the population of the city in the year 2005.

8. Fruit Flies The rate of change of an experimental population of fruit flies is proportional to the population P at any time t (in days). There were 100 flies after the second day of the experiment and 300 flies after the fourth day. Approximately how many flies were in the original population?

In Exercises 9–12, the rate of change of y is proportional to the product of y and the difference of L and y. Solve the resulting differential equation $dy/dx = ky(L - y)$ and find the particular solution that passes through the points for the given value of L.

9. $L = 20$; (0, 1), (5, 10)
10. $L = 100$; (0, 10), (5, 30)
11. $L = 5000$; (0, 250), (25, 2000)
12. $L = 1000$; (0, 100), (4, 750)

13. Biology At any time t (in years), the rate of growth of the population N of deer in a state park is proportional to the product of N and $L - N$, where $L = 500$ is the maximum number of deer the park can maintain. When $t = 0$, $N = 100$, and when $t = 4$, $N = 200$.

(a) Write N as a function of t.

(b) Find N when $t = 1$.

(c) Find t when $N = 350$.

14. Biology At any time t (in years), the rate of growth of the population N of fish in a pond is proportional to the product of N and $L - N$, where $L = 1000$ is the maximum number of fish the pond can maintain. The constant of proportionality is 0.0007. When $t = 0$, $N = 200$.

(a) Write N as a function of t.

(b) Find N when $t = 1$.

(c) Find t when $N = 600$.

Learning Theory In Exercises 15 and 16, assume that the rate of change in the proportion P of correct responses after n trials is proportional to the product of P and $L - P$, where L is the limiting proportion of correct responses.

15. Write and solve the differential equation for this learning theory model.

(T) **16.** Use the solution of Exercise 15 to write P as a function of n, and then use a graphing utility to graph the solution.

(a) $L = 1.00$
   $P = 0.50$ when $n = 0$
   $P = 0.85$ when $n = 4$

(b) $L = 0.80$
   $P = 0.25$ when $n = 0$
   $P = 0.60$ when $n = 10$

Chemical Reaction In Exercises 17 and 18, use the chemical reaction model described in Example 1 to find the amount y (in grams) as a function of time t (in hours), and use a graphing utility to graph the function.

17. $y = 45$ grams when $t = 0$; $y = 4$ grams when $t = 2$

18. $y = 75$ grams when $t = 0$; $y = 12$ grams when $t = 1$

Chemical Reaction In Exercises 19 and 20, assume that during a chemical reaction, a compound changes into another compound at a rate proportional to the unchanged amount y.

19. Write the differential equation for the chemical reaction model. Find the particular solution if the initial amount of the original compound is 20 grams and the amount remaining after 1 hour is 16 grams.

20. Using the result of Exercise 19, when will 75% of the compound have been changed? When will 95% of the compound have been changed?

Newton's Law of Cooling In Exercises 21 and 22, use the Newton's Law of Cooling model described in Example 2 to find the temperature y of an object (in °F) as a function of time t. In each exercise, y_0 is the temperature of the surrounding medium.

21. $y_0 = 70°F$; $y = 350°F$ when $t = 0$; $y = 150°F$ when $t = 45$ minutes

22. $y_0 = 0°F$; $y = 70°F$ when $t = 0$; $y = 48°F$ when $t = 1$ hour

23. **Newton's Law of Cooling** A room is kept at a constant temperature of 70°F. An object placed in the room cools from 350°F to 150°F in 45 minutes. Use the result of Exercise 21 to determine how much longer it will take for the object to cool to a temperature of 80°F.

24. **Newton's Law of Cooling** Food at a temperature of 70°F is placed in a freezer that is set at 0°F. After 1 hour, the temperature of the food is 48°F.

    (a) Use the result of Exercise 22 to find the temperature of the food after it has been in the freezer for 6 hours.

    (b) How long will it take the food to cool to a temperature of 10°F?

In Exercises 25 and 26, use the Gompertz growth model described in Example 3 to find the growth function, and sketch its graph. In each exercise, L is the population limit.

25. $L = 500$; $y = 100$ when $t = 0$; $y = 150$ when $t = 2$

26. $L = 5000$; $y = 500$ when $t = 0$; $y = 625$ when $t = 1$

27. **Biology** A population of eight beavers has been introduced into a new wetlands area. Biologists estimate that the maximum population the wetlands can sustain is 60 beavers. After 3 years, the population is 15 beavers. If the population follows a Gompertz growth model, how many beavers will be in the wetlands after 10 years?

28. **Biology** A population of 30 rabbits has been introduced into a new region. It is estimated that the maximum population the region can sustain is 400 rabbits. After 1 year, the population is estimated to be 90 rabbits. If the population follows a Gompertz growth model, how many rabbits will be present after 3 years?

Biology In Exercises 29 and 30, use the hybrid selection model described in Example 4 to find the percent of the population that has the indicated characteristic.

29. You are studying a population of mayflies to determine how quickly characteristic A will pass from one generation to the next. At the start of the study, half the population has characteristic A. After four generations, 75% of the population has characteristic A. Find the percent of the population that will have characteristic A after 10 generations. (Assume $a = 2$ and $b = 1$.)

30. A research team is studying a population of snails to determine how quickly characteristic B will pass from one generation to the next. At the start of the study, 40% of the snails have characteristic B. After five generations, 80% of the population has characteristic B. Find the percent of the population that will have characteristic B after eight generations. (Assume $a = 2$ and $b = 1$.)

31. **Chemical Mixture** A 100-gallon tank is full of a solution containing 25 pounds of a concentrate. Starting at time $t = 0$, distilled water is admitted to the tank at the rate of 5 gallons per minute, and the well-stirred solution is withdrawn at the same rate.

    (a) Find the amount Q of the concentrate in the solution as a function of t. (*Hint:* Solve $Q' = -5(Q/100)$.)

    (b) Find the time required for the amount of concentrate in the tank to reach 15 pounds.

32. **Chemical Mixture** A 200-gallon tank is half full of distilled water. At time $t = 0$, a solution containing 0.5 pound of concentrate per gallon enters the tank at the rate of 5 gallons per minute, and the well-stirred mixture is withdrawn at the same rate.

    (a) Find the amount Q of the concentrate in the solution as a function of t. (*Hint:* Solve $Q' = -5(Q/100) + 5/2$.)

    (b) Find the amount of concentrate in the tank after 30 minutes.

33. Chemistry A wet towel hung from a clothesline to dry loses moisture through evaporation at a rate proportional to its moisture content. If after 1 hour the towel has lost 40% of its original moisture content, after how long will it have lost 80%?

34. Biology Let x and y be the sizes of two internal organs of a particular mammal at time t. Empirical data indicate that the relative growth rates of these two organs are equal, and can be modeled by

$$\frac{1}{x}\frac{dx}{dt} = \frac{1}{y}\frac{dy}{dt}.$$

Use this differential equation to write y as a function of x.

35. Population Growth When predicting population growth, demographers must consider birth and death rates as well as the net change caused by the difference between the rates of immigration and emigration. Let P be the population at time t and let N be the net increase per unit time due to the difference between immigration and emigration. So, the rate of growth of the population is given by

$$\frac{dP}{dt} = kP + N, \quad N \text{ is constant.}$$

Solve this differential equation to find P as a function of time.

36. Meteorology The barometric pressure y (in inches of mercury) at an altitude of x miles above sea level decreases at a rate proportional to the current pressure according to the model

$$\frac{dy}{dx} = -0.2y$$

where $y = 29.92$ inches when $x = 0$. Find the barometric pressure (a) at the top of Mt. St. Helens (8364 feet) and (b) at the top of Mt. McKinley (20,320 feet).

37. Investment A hospital starts at time $t = 0$ to invest part of its profit at a rate of P dollars per year in a fund for future expansion. Assume that the fund earns r percent interest per year compounded continuously. So, the rate of growth of the amount A in the fund is given by

$$\frac{dA}{dt} = rA + P$$

where $A = 0$ when $t = 0$, and r is in decimal form. Solve this differential equation for A as a function of t.

38. Investment Use the result of Exercise 37 for each situation.

(a) Find A when $P = \$100{,}000$, $r = 12\%$, and $t = 5$ years.

(b) Find P when $A = \$120{,}000{,}000$, $r = 16\frac{1}{4}\%$, and $t = 8$ years.

(3) Find t when $P = \$75{,}000$, $A = \$800{,}000$, and $r = 13\%$.

Medical Science In Exercises 39–42, a medical researcher wants to determine the concentration C (in moles per liter) of a tracer drug injected into a moving fluid. Solve this problem by considering a single-compartment dilution model (see figure). Assume that the fluid is continuously mixed and that the volume of fluid in the compartment is constant.

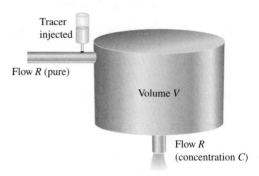

39. If the tracer is injected instantaneously at time $t = 0$, then the concentration of the fluid in the compartment begins diluting according to the differential equation

$$\frac{dC}{dt} = \left(-\frac{R}{V}\right)C, \quad C = C_0 \text{ when } t = 0.$$

(a) Solve this differential equation to find the concentration as a function of time.

(b) Find the limit of C as $t \to \infty$.

T 40. Use the solution of the differential equation in Exercise 39 to find the concentration as a function of time, and use a graphing utility to graph the function.

(a) $V = 2$ liters, $R = 0.5$ L/min, and $C_0 = 0.6$ mol/L

(b) $V = 2$ liters, $R = 1.5$ L/min, and $C_0 = 0.6$ mol/L

41. In Exercises 39 and 40, it was assumed that there was a single initial injection of the tracer drug into the compartment. Now consider the case in which the tracer is continuously injected (beginning at $t = 0$) at a constant rate of Q mol/min. The concentration of the fluid in the compartment begins diluting according to the differential equation

$$\frac{dC}{dt} = \frac{Q}{V} - \left(\frac{R}{V}\right)C, \quad C = 0 \text{ when } t = 0.$$

(a) Solve this differential equation to find the concentration as a function of time.

(b) Find the limit of C as $t \to \infty$.

T 42. Use the solution of the differential equation in Exercise 41 to find the concentration as a function of time, and use a graphing utility to graph the function.

(a) $Q = 2$ mol/min, $V = 2$ liters, and $R = 0.5$ L/min

(b) $Q = 1$ mol/min, $V = 2$ liters, and $R = 1.0$ L/min

Algebra Review

Solving Equations

To solve for the constants in the general solution of a differential equation, you will need to use algebraic skills. For instance, to solve for C in the differential equation

$$y = 200e^{-Ce^{-kt}}$$

you must know how to solve an exponential equation, as shown in Example 1.

Example 1 Solving for C

Solve $y = 200e^{-Ce^{-kt}}$ for C when $y = 20$ and $t = 0$.

SOLUTION

$y = 200e^{-Ce^{-kt}}$	Example 3, page 651
$20 = 200e^{-Ce^{-k(0)}}$	Substitute 20 for y and 0 for t.
$\dfrac{1}{10} = e^{-C}$	Divide each side by 200.
$\dfrac{1}{10} = \dfrac{1}{e^C}$	Definition of negative exponent.
$e^C = 10$	Cross-multiply.
$C = \ln 10$	Take natural logarithm of each side.
$C \approx 2.3026$	Approximate.

Example 2 Solving for C

Solve $\dfrac{y(2-y)}{(1-y)^2} = Ce^{2kt}$ for C when $y = 0.5$ and $t = 0$.

SOLUTION

$\dfrac{y(2-y)}{(1-y)^2} = Ce^{2kt}$	Example 4, page 652
$\dfrac{0.5(2-0.5)}{(1-0.5)^2} = Ce^{2k(0)}$	Substitute 0.5 for y and 0 for t.
$\dfrac{0.75}{0.5^2} = C$	Simplify.
$\dfrac{0.75}{0.25} = C$	Evaluate power.
$3 = C$	Divide.

Example 3 Solving for k

Solve $y = 200e^{-2.3026e^{-kt}}$ for k when $y = 40$ and $t = 3$.

SOLUTION

$$y = 200e^{-2.3026e^{-kt}}$$ Example 3, page 651

$$40 = 200e^{-2.3026e^{-k(3)}}$$ Substitute 40 for y and 3 for t.

$$\frac{1}{5} = e^{-2.3026e^{-3k}}$$ Divide each side by 200.

$$\ln \frac{1}{5} = -2.3026e^{-3k}$$ Take natural logarithm of each side.

$$\frac{\ln(1/5)}{-2.3026} = e^{-3k}$$ Divide each side by -2.3026.

$$\ln\left(\frac{\ln(1/5)}{-2.3026}\right) = -3k$$ Take natural logarithm of each side.

$$-\frac{1}{3}\ln\left(\frac{\ln(1/5)}{-2.3026}\right) = k$$ Multiply each side by $-\frac{1}{3}$.

$$0.1194 \approx k$$ Approximate.

Example 4 Solving for k

Solve $\dfrac{y(2-y)}{(1-y)^2} = 3e^{2kt}$ for k when $y = 0.8$ and $t = 4$.

SOLUTION

$$\frac{y(2-y)}{(1-y)^2} = 3e^{2kt}$$ Example 4, page 652

$$\frac{0.8(2-0.8)}{(1-0.8)^2} = 3e^{2k(4)}$$ Substitute 0.8 for y and 4 for t.

$$\frac{0.8(1.2)}{0.2^2} = 3e^{8k}$$ Simplify.

$$24 = 3e^{8k}$$ Evaluate fraction.

$$8 = e^{8k}$$ Divide each side by 3.

$$\ln 8 = 8k$$ Take natural logarithm of each side.

$$\frac{1}{8}\ln 8 = k$$ Multiply each side by $\frac{1}{8}$.

$$0.2599 \approx k$$ Approximate.

Chapter Summary and Study Strategies

After studying this chapter, you should have acquired the following skills. The exercise numbers are keyed to the Review Exercises that begin on page 660. Answers to odd-numbered Review Exercises are given in the back of the text.*

Section 10.1 Review Exercises
- Find general solutions of differential equations. 1–10
- Find particular solutions of differential equations. 11–24

Section 10.2
- Decide whether the variables in differential equations can be separated. 25–28
- Solve differential equations using separation of variables. 29–48

    If f and g are continuous functions, then the differential equation

    $$\frac{dy}{dx} = f(x)g(y)$$

    has a general solution of

    $$\int \frac{1}{g(y)}\, dy = \int f(x)\, dx + C.$$

- Use differential equations to model real-life problems and separation of variables to solve them. 49, 50

Section 10.3
- Write linear differential equations in standard form. 51–54
- Solve first-order linear differential equations. 55–66

    A first-order linear differential equation is an equation of the form

    $$y' + P(x)y = Q(x)$$

    where P and Q are functions of x. An equation that is written in this form is said to be in standard form.

Section 10.4
- Use differential equations to model and solve real-life problems. 67–80

Study Strategies

- **Using Technology** Throughout this chapter, remember that technology can help you *solve* a differential equation and *graph* a particular solution.

---

* Use a wide range of valuable study aids to help you master the material in this chapter. The *Student Solutions Guide* includes step-by-step solutions to all odd-numbered exercises to help you review and prepare. The student website at *college.hmco.com/info/larsonapplied* offers algebra help and a *Graphing Technology Guide*. The *Graphing Technology Guide* contains step-by-step commands and instructions for a wide variety of graphing calculators, including the most recent models.

Review Exercises

In Exercises 1–4, verify that the function is a solution of the differential equation for any value of C.

Solution	Differential Equation
1. $y = Ce^{x/2}$	$2\dfrac{dy}{dx} = y$
2. $y = \dfrac{1}{x} - \dfrac{1}{x^2} + C$	$\dfrac{dy}{dx} = \dfrac{2-x}{x^3}$
3. $y = C_1 e^x + C_2 e^{-x}$	$y'' - y = 0$
4. $y = C_1 e^{x/2} + C_2 e^{-2x}$	$2y'' + 3y' - 2y = 0$

In Exercises 5–10, use integration to find a general solution of the differential equation.

5. $\dfrac{dy}{dx} = 2x^2 + 5$
6. $\dfrac{dy}{dx} = x^3 - 2x$
7. $\dfrac{dy}{dx} = \cos 2x$
8. $\dfrac{dy}{dx} = 2 \sin x$
9. $\dfrac{dy}{dx} = 2x\sqrt{x-7}$
10. $\dfrac{dy}{dx} = 3e^{-x/3}$

In Exercises 11–14, verify that the function is a solution of the differential equation.

Solution	Differential Equation
11. $y = e^{4x}$	$y' = 4y$
12. $y = e^{-x}$	$3y' + 4y = e^{-x}$
13. $y = x^2$	$\dfrac{1}{2}y' - \dfrac{1}{x}y = 0$
14. $y = \cos x + 3 \sin x$	$y'' + y = 0$

In Exercises 15–18, determine whether the function is a solution of the differential equation $xy' - 2y = x^3 e^x$.

15. $y = x^2$
16. $y = \ln x$
17. $y = x^2 e^x$
18. $y = x^2 e^x - 5x^2$

In Exercises 19 and 20, verify that the general solution satisfies the differential equation. Then find the particular solution that satisfies the initial condition.

19. General solution: $y = Ce^{-5x}$
    Differential equation: $y' + 5y = 0$
    Initial condition: $y = 1$ when $x = 0$

20. General solution: $y = Ce^{-x^2}$
    Differential equation: $y' + 2xy = 0$
    Initial condition: $y = -2$ when $x = 0$

(T) In Exercises 21 and 22, the general solution of the differential equation is given. Use a graphing utility to graph the particular solutions that correspond to the indicated values of C.

General Solution	Differential Equation	C-Values
21. $y = Ce^{3/x}$	$x^2 y' + 3y = 0$	1, 2, 4
22. $y = \dfrac{C}{x+1}$	$(x+1)y' + y = 0$	$0, \pm 1, \pm 2$

In Exercises 23 and 24, some of the curves corresponding to different values of C in the general solution of the differential equation are given. Find the particular solution that passes through the point shown on the graph.

23. $y = C(x^2 + 1)$
    $(x^2 + 1)y' = 2xy$

24. $y = Ce^x - 3$
    $y' - y = 3$

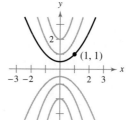
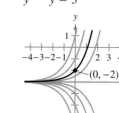

In Exercises 25–28, decide whether the variables in the differential equation can be separated.

25. $\dfrac{dy}{dx} = \dfrac{y}{x+3}$
26. $\dfrac{dy}{dx} = \dfrac{y}{x+y}$
27. $\dfrac{dy}{dx} = \dfrac{x^2 + y}{y}$
28. $\dfrac{dy}{dx} = \dfrac{1}{y} - 1$

In Exercises 29–40, use separation of variables to find the general solution of the differential equation.

29. $\dfrac{dy}{dx} = 4x$
30. $\dfrac{dy}{dx} = \dfrac{2}{x}$
31. $4y^3 \dfrac{dy}{dx} = 5$
32. $\dfrac{dy}{dx} = 3x^2 y$
33. $y' + 2xy^2 = 0$
34. $y' - y = 12$
35. $y' = (x+1)(y+1)$
36. $(3 + 2y)\dfrac{dy}{dx} = 2x$
37. $\dfrac{dy}{dx} = -\dfrac{y+2}{2x^3}$
38. $\dfrac{dy}{dx} = \dfrac{y}{x} - \dfrac{y}{x+1}$
39. $\dfrac{dy}{dx} = \dfrac{\cos x}{y}$
40. $y' = \dfrac{\sin x}{y^2}$

In Exercises 41 and 42, (a) find the general solution of the differential equation and (b) use a graphing utility to graph the particular solutions given by $C = 0$, $C = \pm 1$, and $C = \pm 2$.

41. $\dfrac{dy}{dx} = 3x^2$

42. $\dfrac{dy}{dx} = 0.5x(2 - y)$

In Exercises 43–46, use the initial condition to find the particular solution of the differential equation.

Differential Equation	Initial Condition
43. $yy' + e^x = 0$	$y = 2$ when $x = 0$
44. $\dfrac{dy}{dx} = 2xy \cos x^2$	$y = 1$ when $x = 0$
45. $2xy' - \ln x^2 = 0$	$y = 2$ when $x = 1$
46. $\dfrac{dr}{ds} = e^{r-2s}$	$r = 0$ when $s = 0$

In Exercises 47 and 48, find an equation of the graph that passes through the point and has the specified slope. Then graph the equation.

47. Point: $(1, 1)$

Slope: $y' = -\dfrac{9x}{16y}$

48. Point: $(-2, -1)$

Slope: $y' = y^2 x$

49. **Fuel Economy** An automobile gets 28 miles per gallon of gasoline for speeds up to 50 miles per hour. Over 50 miles per hour, the number of miles per gallon drops at the rate of 12 percent for each 10 miles per hour.

(a) Find the number of miles per gallon y as a function of the speed s by solving the differential equation

$$\dfrac{dy}{ds} = -0.012y, \quad s > 50.$$

(b) Use the function in part (a) to complete the table.

Speed	50	55	60	65	70
Miles per gallon					

50. **Advertising Awareness** A newly opened chiropractor's office is introduced through an advertising campaign to a population of 120,000 potential customers. The rate at which the population hears about the new office is assumed to be proportional to the number of people who are not yet aware of it. This is described by the differential equation

$$\dfrac{dy}{dt} = k(120 - y)$$

where y is the number of people (in thousands) who have heard of the new office, and t is the time (in months). By the end of 6 months, half of the population has heard of the new office. How many will have heard of it by the end of 12 months?

In Exercises 51–54, write the first-order linear differential equation in standard form.

51. $x^4 + 4x^2 y' - 4y = 0$

52. $y' + 10(2x + y) = 0$

53. $x = 2x^3(y' - y)$

54. $y - 2 = (x + 2)y'$

In Exercises 55–62, solve the first-order linear differential equation.

55. $\dfrac{dy}{dx} - 4y = 8$

56. $\dfrac{dy}{dx} - 10y = 20$

57. $\dfrac{dy}{dx} - \dfrac{y}{x} = 2x - 3$

58. $\dfrac{dy}{dx} - \dfrac{4y}{x} = 3x + 2$

59. $y' - y = 9$

60. $y' + 2xy = 4x$

61. $\dfrac{dy}{dx} + \dfrac{y}{x} = 3x + 4$

62. $\dfrac{dy}{dx} + \dfrac{2y}{x} = 3x + 1$

In Exercises 63–66, solve for y in two ways.

63. $y' + y = 6$

64. $y' + 4y = 2$

65. $y' + 2xy = x$

66. $y' - 3xy = 3x$

67. **Newton's Law of Cooling** A steel ingot whose temperature is 1500°F is placed in a room whose temperature is a constant 90°F. One hour later, the temperature of the ingot is 1120°F. What is the ingot's temperature 5 hours after it is placed in the room?

68. **Newton's Law of Cooling** Food at a temperature of 160°F is removed from an oven and placed in a room that is kept at a constant temperature of 68°F. After 1 hour, the temperature of the food is 100°F.

(a) Find the temperature of the food after it has been in the room 2 hours.

(b) How long will it take the food to cool to a temperature of 70°F?

69. **Economics: Pareto's Law** According to the economist Vilfredo Pareto (1848–1923), the rate of decrease of the number of people y in a stable economy having an income of at least x dollars is directly proportional to the number of such people and inversely proportional to their income x. This is modeled by the differential equation

$$\dfrac{dy}{dx} = -k\dfrac{y}{x}.$$

Solve this differential equation.

70. Economics: Pareto's Law In 2005, 19.9 million people in the United States earned at least $75,000 and 101.7 million people earned at least $25,000 (see figure). Assume that Pareto's Law holds and use the result of Exercise 69 to determine the number of people (in millions) who earned (a) at least $20,000 and (b) at least $100,000. *(Source: U.S. Census Bureau)*

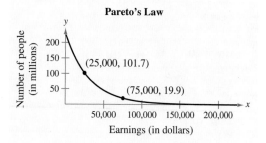

71. Meteorology The barometric pressure y (in inches of mercury) at an altitude of x miles above sea level decreases at a rate proportional to the current pressure according to the model $dy/dx = -0.2y$, where $y = 29.92$ inches when $x = 0$.

(a) Find the barometric pressure at the top of Mt. Kosciuszko in Australia, which is 7310 feet above sea level.

(b) Find the barometric pressure at the top of Mt. Kilimanjaro in Tanzania, which is 19,340 feet above sea level.

72. Chemistry A wet shirt hung from a clothesline to dry loses moisture through evaporation at a rate proportional to its moisture content. If after 1 hour the shirt has lost 60% of its original moisture content, after how long will it have lost 90%?

73. Radioactive Decay The rate of decomposition of radioactive carbon is proportional to the amount present at any time. The half-life of radioactive carbon is 5715 years. What percent of a present amount will remain after 1000 years?

74. Radioactive Decay The rate of decomposition of radioactive plutonium is proportional to the amount present at any time. The half-life of radioactive plutonium is 24,100 years. If 1.5 grams remain after 10,000 years, what was the initial amount?

Chemical Reaction In Exercises 75 and 76, assume that during a chemical reaction, a compound changes into another compound at a rate proportional to the cube root of the unchanged amount y.

75. Write the differential equation for the chemical reaction model. Find the particular solution if the initial amount of the original compound is 27 grams and the amount remaining after 1 hour is 8 grams.

76. Using the result of Exercise 75, when will 80% of the compound have been changed? When will 99% of the compound have been changed?

77. Chemical Mixture A tank contains 30 gallons of a solution composed of 80% water and 20% alcohol. A second solution containing half water and half alcohol is added to the tank at the rate of 6 gallons per minute. At the same time, the tank is being drained at the rate of 6 gallons per minute, as shown in the figure. Assuming that the solution is stirred constantly, how much alcohol will be in the tank after 10 minutes?

78. Chemical Mixture A tank contains 20 gallons of a solution composed of 90% water and 10% alcohol. A second solution containing half water and half alcohol is added to the tank at the rate of 2 gallons per minute. At the same time, the tank is being drained at the rate of 2 gallons per minute, as shown in the figure. Assuming that the solution is stirred constantly, how much alcohol will be in the tank after 10 minutes?

79. Biology A population of 12 pelicans has been introduced in a new wetlands area. Biologists estimate that the maximum population the wetlands can sustain is 64 pelicans. After 3 years, the population is 28 pelicans. If the population follows a Gompertz growth model, how many pelicans will be in the wetlands after eight years?

80. Biology You are studying a population of ladybugs to determine how quickly characteristic C will pass from one generation to the next. At the start of the study, 25% of the population has characteristic C. After three generations, 60% of the population has characteristic C. Use the hybrid selection model discussed in Section 10.4, Example 4, to find the percent of the population that will have characteristic C after seven generations. (Assume $a = 2$ and $b = 1$.)

Chapter Test

See www.CalcChat.com for worked-out solutions to odd-numbered exercises.

Take this test as you would take a test in class. When you are done, check your work against the answers given in the back of the book.

In Exercises 1 and 2, verify that the function is a solution of the differential equation.

Solution	Differential Equation
1. $y = e^{-2x}$	$3y' + 2y = -4e^{-2x}$
2. $y = \dfrac{1}{x+1}$	$y'' - \dfrac{2y}{(x+1)^2} = 0$

In Exercises 3–6, use separation of variables to find the general solution of the differential equation.

3. $yy' = x$

4. $\dfrac{dy}{dx} = \dfrac{2x}{y}$

5. $\dfrac{dy}{dx} = \dfrac{\cos \pi x}{3y^2}$

6. $y' = 2(xy - x)$

In Exercises 7–10, solve the first-order linear differential equation.

7. $y' - 2y = e^{2x}$

8. $y' - x = y$

9. $y' = \dfrac{x^2 - y}{x}$

10. $\dfrac{1}{x}\dfrac{dy}{dx} = xy + x$

In Exercises 11–13, use the initial condition to find the particular solution of the differential equation.

Differential Equation	Initial Condition
11. $y' + x^2y - x^2 = 0$	$y = 0$ when $x = 0$
12. $y'e^{-x^2} = 2xy$	$y = e$ when $x = 0$
13. $x\dfrac{dy}{dx} = \dfrac{\ln x}{7}$	$y = -2$ when $x = 1$

14. A lamb that weighed 10 pounds at birth gains weight at the rate

$$\dfrac{dw}{dt} = k(200 - w)$$

where w is weight (in pounds) and t is time (in years). Solve the differential equation.

(T) (a) Use a graphing utility to graph the solutions for $k = 0.8, 0.9,$ and 1.

(b) If the animal is sold when its weight reaches 150 pounds, find the time of sale for each of the models in part (a).

(c) What is the maximum weight of the animal for each of the models?

15. A room is kept at a constant temperature of $72°F$. An object placed in the room cools from $400°F$ to $160°F$ in 40 minutes. Determine how much longer it will take for the object to cool to a temperature of $100°F$.

11 Probability and Calculus

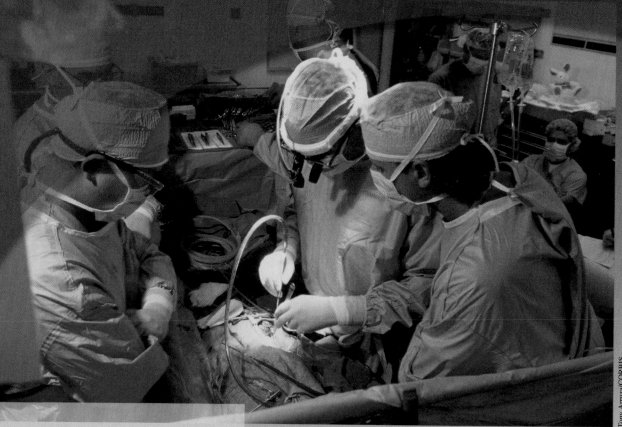

Calculus and probability theory can be used to determine the expected time for a patient to recover after a certain medical procedure. (See Section 11.3, Exercise 36.)

11.1 Discrete Probability
11.2 Continuous Random Variables
11.3 Expected Value and Variance

Applications

Probability has many real-life applications. The applications listed below represent a sample of the applications in this chapter.

- Biology, Exercise 23, page 672
- Learning Theory, Exercise 30, page 680
- Shelf Life, Exercise 34, page 680
- Make a Decision: Useful Life, Exercise 47, page 690
- Medical Science, Exercise 49, page 690

Section 11.1
Discrete Probability

- Describe sample spaces of experiments.
- Assign values to, and form frequency distributions for, discrete random variables.
- Find the probabilities of events for discrete random variables.
- Find the expected values or means of discrete random variables.
- Find the variances and standard deviations of discrete random variables.

Sample Spaces

When assigning measurements to the uncertainties of everyday life, people often use ambiguous terminology such as "fairly certain," "probable," and "highly unlikely." Probability theory allows you to remove this ambiguity by assigning a number to the likelihood of the occurrence of an event. This number is called the **probability** that the event will occur. For example, if you toss a fair coin, the probability that it will land heads up is one-half or 0.5.

In probability theory, any happening whose result is uncertain is called an **experiment**. The possible results of the experiment are **outcomes**, the set of all possible outcomes of the experiment is the **sample space** of the experiment, and any subcollection of a sample space is an **event**.

For instance, consider an experiment in which a coin is tossed. The sample space of this experiment consists of two outcomes: either the coin will land heads up (denoted by H) or it will land tails up (denoted by T). So, the sample space S is

$$S = \{H, T\}. \quad \text{Sample space}$$

In this text, all outcomes of a sample space are assumed to be equally likely. For instance, when a coin is tossed, H and T are assumed to be equally likely.

Thomas Wiewandt/Getty Images

When a weather forecaster states that there is a 50% chance of thunderstorms, it means that thunderstorms have occurred on half of all days that have had similar weather conditions.

Example 1 Finding a Sample Space

An experiment consists of tossing a six-sided die.

a. What is the sample space?

b. Describe the event corresponding to a number greater than 2 turning up.

SOLUTION

a. The sample space S consists of six outcomes, which can be represented by the numbers 1 through 6. That is,

$$S = \{1, 2, 3, 4, 5, 6\}. \quad \text{Sample space}$$

Note that each of the outcomes in the sample space is equally likely.

b. The event E corresponding to a number greater than 2 turning up is a subset of S. That is

$$E = \{3, 4, 5, 6\}. \quad \text{Event}$$

✓ CHECKPOINT 1

An experiment consists of tossing two six-sided dice.

a. What is the sample space?

b. Describe the event corresponding to a sum greater than or equal to seven points when the dice are tossed. ■

Discrete Random Variables

A function that assigns a numerical value to each of the outcomes in a sample space is called a **random variable.** For instance, in the sample space $S = \{HH, HT, TH, TT\}$, the outcomes could be assigned the numbers 2, 1, and 0, depending on the number of heads in the outcome.

Algebra Review

For examples of how to count the number of ways an event can happen, see the *Chapter 11 Algebra Review* on pages 691 and 692.

Definition of Discrete Random Variable

Let S be a sample space. A **random variable** is a function x that assigns a numerical value to each outcome in S. If the set of values taken on by the random variable is finite, then the random variable is **discrete.** The number of times a specific value of x occurs is the **frequency** of x and is denoted by $n(x)$.

Example 2 Finding Frequencies

Three coins are tossed. A random variable assigns the number 0, 1, 2, or 3 to each possible outcome, depending on the number of heads that turn up.

$$S = \{HHH, HHT, HTH, HTT, THH, THT, TTH, TTT\}$$
$$\phantom{S = \{}\downarrow\downarrow\downarrow\downarrow\downarrow\downarrow\downarrow\downarrow$$
$$\phantom{S = \{}32212110$$

Find the frequencies of 0, 1, 2, and 3. Then use a bar graph to represent the result.

SOLUTION To find the frequencies, simply count the number of occurrences of each value of the random variable, as shown in the table.

Random variable, x	0	1	2	3
Frequency of x, $n(x)$	1	3	3	1

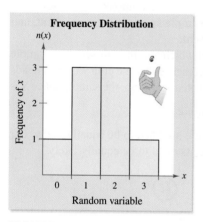

FIGURE 11.1

This table is called a **frequency distribution** of the random variable. The result is shown graphically by the bar graph in Figure 11.1.

✓CHECKPOINT 2

Use a graphing utility to create a bar graph similar to the one shown in Figure 11.1, representing the frequency for tossing two six-sided dice. Let the random variable be the sum of the points when the dice are tossed. ■

STUDY TIP

In Example 2, note that the sample space consists of *eight* outcomes, each of which is *equally likely*. The sample space does *not* consist of the outcomes "zero heads," "one head," "two heads," and "three heads." You cannot consider these events to be outcomes because they are not equally likely.

Discrete Probability

The probability of a random variable x is

$$P(x) = \frac{\text{Frequency of } x}{\text{Number of outcomes in } S}$$

$$= \frac{n(x)}{n(S)}$$

Probability

where $n(S)$ is the number of equally likely outcomes in the sample space. By this definition, it follows that the probability of an event must be a number between 0 and 1. That is, $0 \leq P(x) \leq 1$.

The collection of probabilities corresponding to the values of the random variable is called the **probability distribution** of the random variable. If the range of a discrete random variable consists of m different values $\{x_1, x_2, x_3, \ldots, x_m\}$, then the sum of the probabilities of x_i is 1. This can be written as

$$P(x_1) + P(x_2) + P(x_3) + \cdots + P(x_m) = 1.$$

Example 3 Finding a Probability Distribution

Five coins are tossed. Graph the probability distribution for the random variable giving the number of heads that turn up.

SOLUTION

x	Event	$n(x)$
0	TTTTT	1
1	HTTTT, THTTT, TTHTT, TTTHT, TTTTH	5
2	HHTTT, HTHTT, HTTHT, HTTTH, THHTT, THTHT, THTTH, TTHHT, TTHTH, TTTHH	10
3	HHHTT, HHTHT, HHTTH, HTHHT, HTHTH, HTTHH, THHHT, THHTH, THTHH, TTHHH	10
4	HHHHT, HHHTH, HHTHH, HTHHH, THHHH	5
5	HHHHH	1

The number of outcomes in the sample space is $n(S) = 32$. The probability of each value of the random variable is shown in the table.

Random variable, x	0	1	2	3	4	5
Probability, $P(x)$	$\frac{1}{32}$	$\frac{5}{32}$	$\frac{10}{32}$	$\frac{10}{32}$	$\frac{5}{32}$	$\frac{1}{32}$

A graph of this probability distribution is shown in Figure 11.2. Note that values of the random variable are represented by intervals on the x-axis. Observe that the sum of the probabilities is 1.

✓ CHECKPOINT 3

Two six-sided dice are tossed. Graph the probability distribution for the random variable giving the sum of the points when the dice are tossed. ∎

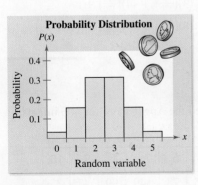

FIGURE 11.2

Expected Value

Suppose you repeated the coin-tossing experiment in Example 3 several times. On the average, how many heads would you expect to turn up? From Figure 11.2, it seems reasonable that the average number of heads would be $2\frac{1}{2}$. This "average" is the **expected value** of the random variable.

STUDY TIP

Although the expected value of x is denoted by $E(x)$, the mean of x is usually denoted by the lowercase Greek letter μ (pronounced "mu"). Because the mean often occurs near the center of the values in the range of the random variable, it is called a **measure of central tendency**.

Definition of Expected Value

If the range of a discrete random variable consists of m different values $\{x_1, x_2, x_3, \ldots, x_m\}$, then the **expected value** of the random variable is

$$E(x) = x_1 P(x_1) + x_2 P(x_2) + x_3 P(x_3) + \cdots + x_m P(x_m).$$

The expected value is also called the **mean** of the random variable.

✓ CHECKPOINT 4

Two six-sided dice are tossed. Find the expected value for the sum of the points. ■

Example 4 Finding an Expected Value

Five coins are tossed. Find the expected value of the number of heads that will turn up.

SOLUTION Using the results of Example 3, you obtain the expected value as shown.

$$E(x) = \overbrace{(0)\left(\tfrac{1}{32}\right)}^{\text{0 Heads}} + \overbrace{(1)\left(\tfrac{5}{32}\right)}^{\text{1 Head}} + \overbrace{(2)\left(\tfrac{10}{32}\right)}^{\text{2 Heads}} + \overbrace{(3)\left(\tfrac{10}{32}\right)}^{\text{3 Heads}} + \overbrace{(4)\left(\tfrac{5}{32}\right)}^{\text{4 Heads}} + \overbrace{(5)\left(\tfrac{1}{32}\right)}^{\text{5 Heads}}$$

$$= \tfrac{80}{32} = 2.5$$

Example 5 Finding an Expected Value

Just after birth, a hospital rates each infant on the Apgar score. The possible scores are $0, 1, \ldots, 9, 10$, where 10 is the best possible score. During the past year, the hospital recorded the scores for 370 infants, as shown in the table at the left. From these data, what is the average Apgar score the hospital should expect for an infant?

SOLUTION To answer this question, calculate the expected value of the score.

$$E(x) = (0)\left(\tfrac{1}{370}\right) + (1)\left(\tfrac{2}{370}\right) + (2)\left(\tfrac{1}{370}\right) + (3)\left(\tfrac{3}{370}\right) + (4)\left(\tfrac{7}{370}\right) +$$
$$(5)\left(\tfrac{16}{370}\right) + (6)\left(\tfrac{66}{370}\right) + (7)\left(\tfrac{135}{370}\right) + (8)\left(\tfrac{91}{370}\right) + (9)\left(\tfrac{44}{370}\right) + (10)\left(\tfrac{4}{370}\right)$$

$$= \tfrac{2626}{370} \approx 7.1$$

Apgar score	0	1	2
Number of infants	1	2	1

Apgar score	3	4	5
Number of infants	3	7	16

Apgar score	6	7
Number of infants	66	135

Apgar score	8	9	10
Number of infants	91	44	4

The Apgar score is based on color, heart rate, reflex irritability, muscle tone, and respiratory effort. The score was developed by Virginia Apgar in 1952 as a method of quickly assessing the health of an infant immediately after childbirth.

✓ CHECKPOINT 5

Repeat Example 5 for a hospital that recorded the Apgar score for 400 infants as shown in the table below.

Apgar score	0	1	2	3	4	5	6	7	8	9	10
Number of infants	1	1	2	2	8	15	73	149	99	47	3

■

Variance and Standard Deviation

The expected value or mean gives a measure of the average value assigned by a random variable. But the mean does not tell the whole story. For instance, all three of the distributions shown below have a mean of 2.

Distribution 1

Random variable, x	0	1	2	3	4
Frequency of x, $n(x)$	2	2	2	2	2

Distribution 2

Random variable, x	0	1	2	3	4
Frequency of x, $n(x)$	0	3	4	3	0

Distribution 3

Random variable, x	0	1	2	3	4
Frequency of x, $n(x)$	5	0	0	0	5

Even though each distribution has the same mean, the patterns of the distributions are quite different. In the first distribution, each value has the same frequency. In the second, the values are clustered about the mean. In the third distribution, the values are far from the mean. To measure how much the distribution varies from the mean, you can use the concepts of variance and standard deviation.

Definitions of Variance and Standard Deviation

Consider a random variable whose range is $\{x_1, x_2, x_3, \ldots, x_m\}$ with a mean of μ. The **variance** of the random variable is

$$V(x) = (x_1 - \mu)^2 P(x_1) + (x_2 - \mu)^2 P(x_2) + \cdots + (x_m - \mu)^2 P(x_m).$$

The **standard deviation** of the random variable is

$$\sigma = \sqrt{V(x)}$$

(σ is the lowercase Greek letter sigma).

DISCOVERY

The average grade on the calculus final in a class of 20 students was 80 out of 100 possible points. Describe a distribution of grades for which 10 students scored above 95 points. Describe another distribution of grades for which only one student scored above 85. In general, how does the standard deviation influence the grade distribution in a course?

A small standard deviation indicates that most of the values of the random variable are clustered near the mean. As the standard deviation becomes larger and larger, the distribution becomes more and more spread out. For instance, in the three distributions above, you would expect the second to have the smallest standard deviation and the third to have the largest. This is confirmed in Example 6.

670   CHAPTER 11   Probability and Calculus

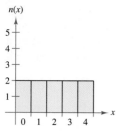

(a) Mean = 2; standard deviation ≈ 1.41

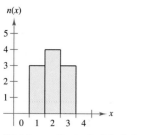

(b) Mean = 2; standard deviation ≈ 0.77

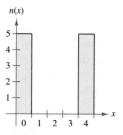

(c) Mean = 2; standard deviation = 2

FIGURE 11.3

Example 6 Finding Variance and Standard Deviation

Find the variance and standard deviation of each of the three distributions shown on the previous page.

SOLUTION

a. For distribution 1, the mean is $\mu = 2$, the variance is

$$V(x) = (0-2)^2\left(\tfrac{2}{10}\right) + (1-2)^2\left(\tfrac{2}{10}\right) + (2-2)^2\left(\tfrac{2}{10}\right) +$$
$$(3-2)^2\left(\tfrac{2}{10}\right) + (4-2)^2\left(\tfrac{2}{10}\right)$$
$$= 2 \qquad \text{Variance}$$

and the standard deviation is $\sigma = \sqrt{2} \approx 1.41$.

b. For distribution 2, the mean is $\mu = 2$, the variance is

$$V(x) = (0-2)^2\left(\tfrac{0}{10}\right) + (1-2)^2\left(\tfrac{3}{10}\right) + (2-2)^2\left(\tfrac{4}{10}\right) +$$
$$(3-2)^2\left(\tfrac{3}{10}\right) + (4-2)^2\left(\tfrac{0}{10}\right)$$
$$= 0.6 \qquad \text{Variance}$$

and the standard deviation is $\sigma = \sqrt{0.6} \approx 0.77$.

c. For distribution 3, the mean is $\mu = 2$, the variance is

$$V(x) = (0-2)^2\left(\tfrac{5}{10}\right) + (1-2)^2\left(\tfrac{0}{10}\right) + (2-2)^2\left(\tfrac{0}{10}\right) +$$
$$(3-2)^2\left(\tfrac{0}{10}\right) + (4-2)^2\left(\tfrac{5}{10}\right)$$
$$= 4 \qquad \text{Variance}$$

and the standard deviation is $\sigma = \sqrt{4} = 2$.

As you can see in Figure 11.3, the second distribution has the smallest standard deviation and the third distribution has the largest.

✓ CHECKPOINT 6

Find the variance and standard deviation of the distribution shown in the table. Then graph the distribution.

Random variable, x	0	1	2	3	4
Frequency of x, $n(x)$	1	2	4	2	1

CONCEPT CHECK

1. What is an experiment?
2. What is a sample space?
3. Complete the following: The expected value of a random variable is also called the _____ of the random variable.
4. What is a probability distribution?

SECTION 11.1 Discrete Probability

Skills Review 11.1

The following warm-up exercises involve skills that were covered in earlier sections. You will use these skills in the exercise set for this section. For additional help, review Sections 0.3 and 0.5.

In Exercises 1 and 2, solve for x.

1. $\frac{1}{9} + \frac{2}{3} + \frac{2}{9} = x$

2. $\frac{1}{3} + \frac{5}{12} + \frac{1}{8} + \frac{1}{12} + \frac{x}{24} = 1$

In Exercises 3–6, evaluate the expression.

3. $0\left(\frac{1}{16}\right) + 1\left(\frac{3}{16}\right) + 2\left(\frac{8}{16}\right) + 3\left(\frac{3}{16}\right) + 4\left(\frac{1}{16}\right)$
4. $0\left(\frac{1}{12}\right) + 1\left(\frac{2}{12}\right) + 2\left(\frac{6}{12}\right) + 3\left(\frac{2}{12}\right) + 4\left(\frac{1}{12}\right)$
5. $(0-1)^2\left(\frac{1}{4}\right) + (1-1)^2\left(\frac{1}{2}\right) + (2-1)^2\left(\frac{1}{4}\right)$
6. $(0-2)^2\left(\frac{1}{12}\right) + (1-2)^2\left(\frac{2}{12}\right) + (2-2)^2\left(\frac{6}{12}\right) + (3-2)^2\left(\frac{2}{12}\right) + (4-2)^2\left(\frac{1}{12}\right)$

In Exercises 7–10, write the fraction as a percent. Round your answers to 2 decimal places, if necessary.

7. $\frac{3}{8}$
8. $\frac{9}{11}$
9. $\frac{13}{24}$
10. $\frac{112}{256}$

Exercises 11.1

See www.CalcChat.com for worked-out solutions to odd-numbered exercises.

In Exercises 1–4, list or describe the elements in the specified set.

1. **Coin Toss** A coin is tossed three times.
   (a) The sample space S
   (b) The event A that at least two heads occur
   (c) The event B that no more than one head occurs

2. **Coin Toss** A coin is tossed. If a head occurs, the coin is tossed again; otherwise, a die is tossed.
   (a) The sample space S
   (b) The event A that 4, 5, or 6 occurs on the die
   (c) The event B that two heads occur

3. **Poll** Three people are asked their opinions on a political issue. They can answer "In favor" (I), "Opposed" (O), or "Undecided" (U).
   (a) The sample space S
   (b) The event A that at least two people are in favor
   (c) The event B that no more than one person is opposed

4. **Health Insurance Fraud** Four cases of health insurance fraud are examined. The method of fraud is "stolen insurance card" (S), "counterfeit insurance card" (C), "mail order" (M), or "other" (O).
   (a) The sample space S
   (b) The event A that at least three cases are mail order fraud
   (c) The event B that no more than one case is counterfeit insurance card fraud

5. **Coin Toss** Two coins are tossed. A random variable assigns the number 0, 1, or 2 to each possible outcome, depending on the number of heads that turn up. Find the frequencies of 0, 1, and 2.

6. **Coin Toss** Four coins are tossed. A random variable assigns the number 0, 1, 2, 3, or 4 to each possible outcome, depending on the number of heads that turn up. Find the frequencies of 0, 1, 2, 3, and 4.

7. **Exam** Three students answer a true-false question on an examination. A random variable assigns the number 0, 1, 2, or 3 to each outcome, depending on the number of answers of *true* among the three students. Find the frequencies of 0, 1, 2, and 3.

8. **Exam** Four students answer a true-false question on an examination. A random variable assigns the number 0, 1, 2, 3, or 4 to each outcome, depending on the number of answers of *true* among the four students. Find the frequencies of 0, 1, 2, 3, and 4.

9. Poll Three people have been nominated for president of a college class. From a small poll it is estimated that Jane has a probability of 0.29 of winning and Larry has a probability of 0.47. What is the probability of the third candidate winning the election?

10. Random Selection In a class of 72 students, 44 are girls and, of these, 12 are going to college. Of the 28 boys in the class, 9 are going to college. If a student is selected at random from the class, what is the probability that the person chosen is (a) going to college, (b) not going to college, and (c) a girl who is not going to college?

11. Quality Control A component of a spacecraft has both a main system and a backup system. The probability of at least one of the systems performing satisfactorily throughout the duration of the flight is 0.9855. What is the probability of both of them failing?

12. Random Selection A card is chosen at random from a standard 52-card deck of playing cards. What is the probability that the card will be black and a face card?

In Exercises 13 and 14, find the missing value of the probability distribution.

13.

x	0	1	2	3	4
$P(x)$	0.20	0.35	0.15	?	0.05

14.

x	0	1	2	3	4	5
$P(x)$	0.05	?	0.25	0.30	0.15	0.10

In Exercises 15–18, determine whether the table represents a probability distribution. If it is a probability distribution, sketch its graph. If it is not a probability distribution, state any properties that are not satisfied.

15.

x	0	1	2	3
$P(x)$	0.10	0.45	0.30	0.15

16.

x	0	1	2	3	4	5
$P(x)$	0.05	0.30	0.10	0.40	0.15	0.20

17.

x	0	1	2	3	4
$P(x)$	$\frac{12}{50}$	$\frac{20}{50}$	$\frac{8}{50}$	$\frac{10}{50}$	$-\frac{5}{50}$

18.

x	0	1	2	3	4	5
$P(x)$	$\frac{8}{30}$	$\frac{2}{30}$	$\frac{6}{30}$	$\frac{3}{30}$	$\frac{4}{30}$	$\frac{7}{30}$

In Exercises 19–22, sketch a graph of the probability distribution and find the required probabilities.

19.

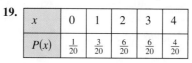

(a) $P(1 \leq x \leq 3)$

(b) $P(x \geq 2)$

20.

(a) $P(x \leq 2)$

(b) $P(x > 2)$

21.

x	0	1	2	3	4	5
$P(x)$	0.041	0.189	0.247	0.326	0.159	0.038

(a) $P(x \leq 3)$

(b) $P(x > 3)$

22.

x	0	1	2	3
$P(x)$	0.027	0.189	0.441	0.343

(a) $P(1 \leq x \leq 2)$

(b) $P(x < 2)$

23. Biology Consider a couple who have four children. Assume that it is equally likely that each child is a girl or a boy.

(a) Complete the set to form the sample space consisting of 16 elements.

$S = \{gggg, gggb, ggbg, \ldots\}$

(b) Complete the table, in which the random variable x is the number of girls in the family.

x	0	1	2	3	4
$P(x)$					

(c) Use the table in part (b) to sketch the graph of the probability distribution.

(d) Use the table in part (b) to find the probability that at least one of the children is a boy.

SECTION 11.1 Discrete Probability 673

24. Die Toss Consider the experiment of tossing a 12-sided die twice.

(a) Complete the set to form the sample space of 144 elements. Note that each element is an ordered pair in which the entries are the numbers of points on the first and second tosses, respectively.
$S = \{(1, 1), (1, 2), \ldots, (2, 1), (2, 2), \ldots\}$

(b) Complete the table, in which the random variable x is the sum of the number of points.

x	2	3	4	5	6	7	8	9
$P(x)$								

x	10	11	12	13	14	15	16	17
$P(x)$								

x	18	19	20	21	22	23	24
$P(x)$							

(c) Use the table in part (b) to sketch the graph of the probability distribution.

(d) Use the table in part (b) to find $P(15 \leq x \leq 19)$.

In Exercises 25 and 26, find $E(x)$, $V(x)$, and σ for the given probability distribution.

25.

x	1	2	3	4	5
$P(x)$	$\frac{1}{16}$	$\frac{3}{16}$	$\frac{8}{16}$	$\frac{3}{16}$	$\frac{1}{16}$

26.

x	−5000	−2500	300
$P(x)$	0.008	0.052	0.940

27. Health The table shows the probability distribution of the numbers of AIDS cases diagnosed in the United States in 2005 by age group. *(Source: Centers for Disease Control and Prevention)*

Age, a	14 and under	15–24	25–34	35–44
$P(a)$	0.003	0.056	0.212	0.380

Age, a	45–54	55–64	65 and over
$P(a)$	0.254	0.075	0.020

(a) Sketch the probability distribution.

(b) Find the probability that an individual diagnosed with AIDS was from 15 to 44 years of age.

(c) Find the probability that an individual diagnosed with AIDS was at least 35 years of age.

28. Education The table gives the probability distribution of the educational attainments of people in the United States in 2005, ages 25 years old and over, where $x = 0$ represents no high school diploma, $x = 1$ represents a high school diploma, $x = 2$ represents some college, $x = 3$ represents an associate's degree, $x = 4$ represents a bachelor's degree, and $x = 5$ represents an advanced degree. *(Source: U.S. Census Bureau)*

x	0	1	2	3	4	5
$P(x)$	0.148	0.322	0.168	0.086	0.181	0.095

(a) Sketch the probability distribution.

(b) Determine $E(x)$, $V(x)$, and σ. Explain the meanings of these values.

In Exercises 29 and 30, find the mean and variance of the discrete random variable x.

29. Die Toss x is (a) the number of points when a four-sided die is tossed once and (b) the sum of the points when the four-sided die is tossed twice.

30. Coin Toss x is the number of heads when a coin is tossed four times.

31. Athletics A baseball fan examined the record of a favorite baseball player's performance during his last 50 games. The numbers of games in which the player had zero, one, two, three, and four hits are recorded in the table.

Number of hits	0	1	2	3	4
Frequency	14	26	7	2	1

(a) Complete the table, where x is the number of hits.

x	0	1	2	3	4
$P(x)$					

(b) Use the table in part (a) to sketch the graph of the probability distribution.

(c) Use the table in part (a) to find $P(1 \leq x \leq 3)$.

(d) Determine $E(x)$, $V(x)$, and σ. Explain your results.

32. Extended Application To work an extended application analyzing the health insurance coverage status of people in the United States by age, visit this text's website at *college.hmco.com/info/larsonapplied*. *(Data Source: U.S. Census Bureau)*

Section 11.2

Continuous Random Variables

- Verify continuous probability density functions and use continuous probability density functions to find probabilities.
- Use continuous probability density functions to answer questions about real-life situations.

Continuous Random Variables

In many applications of probability, it is useful to consider a random variable whose range is an interval on the real number line. Such a random variable is called **continuous.** For instance, the random variable that measures the height of a person in a population is continuous.

To define the probability of an event involving a continuous random variable, you cannot simply count the number of ways the event can occur (as you can with a discrete random variable). Rather, you need to define a function f called a **probability density function.**

Definition of Probability Density Function

Consider a function f of a continuous random variable x whose range is the interval $[a, b]$. The function is a **probability density function** if it is nonnegative and continuous on the interval $[a, b]$ and if

$$\int_a^b f(x)\, dx = 1.$$

The probability that x lies in the interval $[c, d]$ is

$$P(c \le x \le d) = \int_c^d f(x)\, dx$$

as shown in Figure 11.4. If the range of the continuous random variable is an infinite interval, then the integrals are improper integrals.

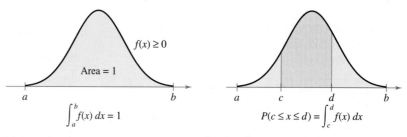

FIGURE 11.4 Probability Density Function

SECTION 11.2 Continuous Random Variables

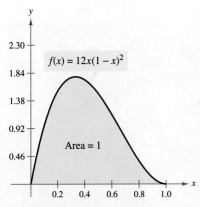

FIGURE 11.5

Example 1 Verifying a Probability Density Function

Show that

$$f(x) = 12x(1-x)^2$$

is a probability density function over the interval $[0, 1]$.

SOLUTION Begin by observing that f is continuous and nonnegative on the interval $[0, 1]$.

$$f(x) = 12x(1-x)^2 \geq 0, \quad 0 \leq x \leq 1 \qquad f(x) \text{ is nonnegative on } [0,1].$$

Next, evaluate the integral below.

$$\int_0^1 12x(1-x)^2 \, dx = 12\int_0^1 (x^3 - 2x^2 + x) \, dx \qquad \text{Expand polynomial.}$$

$$= 12\left[\frac{x^4}{4} - \frac{2x^3}{3} + \frac{x^2}{2}\right]_0^1 \qquad \text{Integrate.}$$

$$= 12\left(\frac{1}{4} - \frac{2}{3} + \frac{1}{2}\right) \qquad \text{Apply Fundamental Theorem of Calculus.}$$

$$= 1 \qquad \text{Simplify.}$$

Because this value is 1, you can conclude that f is a probability density function over the interval $[0, 1]$. The graph of f is shown in Figure 11.5.

✓ **CHECKPOINT 1**

Show that $f(x) = \frac{1}{2}x$ is a probability density function over the interval $[0, 2]$. ■

The next example deals with an infinite interval and its corresponding improper integral.

Example 2 Verifying a Probability Density Function

Show that

$$f(t) = 0.1e^{-0.1t}$$

is a probability density function over the infinite interval $[0, \infty)$.

SOLUTION Begin by observing that f is continuous and nonnegative on the interval $[0, \infty)$.

$$f(t) = 0.1e^{-0.1t} \geq 0, \quad 0 \leq t \qquad f(t) \text{ is nonnegative on } [0, \infty).$$

Next, evaluate the integral below.

$$\int_0^\infty 0.1e^{-0.1t}\, dt = \lim_{b \to \infty} \left[-e^{-0.1t}\right]_0^b \qquad \text{Improper integral}$$

$$= \lim_{b \to \infty} (-e^{-0.1b} + 1) \qquad \text{Evaluate limit.}$$

$$= 1$$

Because this value is 1, you can conclude that f is a probability density function over the interval $[0, \infty)$. The graph of f is shown in Figure 11.6.

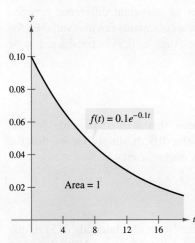

FIGURE 11.6

✓ **CHECKPOINT 2**

Show that $f(x) = 2e^{-2x}$ is a probability density function over the interval $[0, \infty)$. ■

Example 3 Finding a Probability

For the probability density function in Example 1

$$f(x) = 12x(1-x)^2$$

find the probability that x lies in the interval $\frac{1}{2} \le x \le \frac{3}{4}$.

SOLUTION

$$P\left(\tfrac{1}{2} \le x \le \tfrac{3}{4}\right) = 12 \int_{1/2}^{3/4} x(1-x)^2 \, dx \qquad \text{Integrate } f(x) \text{ over } \left[\tfrac{1}{2}, \tfrac{3}{4}\right].$$

$$= 12 \int_{1/2}^{3/4} (x^3 - 2x^2 + x) \, dx \qquad \text{Expand polynomial.}$$

$$= 12 \left[\frac{x^4}{4} - \frac{2x^3}{3} + \frac{x^2}{2} \right]_{1/2}^{3/4} \qquad \text{Integrate.}$$

$$= 12 \left[\frac{\left(\tfrac{3}{4}\right)^4}{4} - \frac{2\left(\tfrac{3}{4}\right)^3}{3} + \frac{\left(\tfrac{3}{4}\right)^2}{2} - \frac{\left(\tfrac{1}{2}\right)^4}{4} + \frac{2\left(\tfrac{1}{2}\right)^3}{3} - \frac{\left(\tfrac{1}{2}\right)^2}{2} \right]$$

$$\approx 0.262 \qquad \text{Simplify.}$$

So, the probability that x lies in the interval $\left[\tfrac{1}{2}, \tfrac{3}{4}\right]$ is approximately 0.262 or 26.2%, as indicated in Figure 11.7.

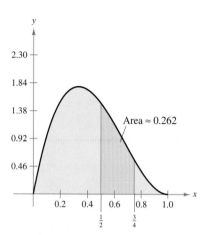

FIGURE 11.7

✓ CHECKPOINT 3

Find the probability that x lies in the interval $\tfrac{1}{2} \le x \le 1$ for the probability density function in Checkpoint 1. ∎

In Example 3, note that if you had been asked to find the probability that x lies in any of the intervals $\tfrac{1}{2} < x < \tfrac{3}{4}$, $\tfrac{1}{2} \le x < \tfrac{3}{4}$, or $\tfrac{1}{2} < x \le \tfrac{3}{4}$, you would have obtained the same solution. In other words, the inclusion of either endpoint adds nothing to the probability. This demonstrates an important difference between discrete and continuous random variables. For a continuous random variable, the probability that x will be precisely one value (such as 0.5) is considered to be zero, because

$$P(0.5 \le x \le 0.5) = \int_{0.5}^{0.5} f(x) \, dx = 0.$$

You should not interpret this result to mean that it is impossible for the continuous random variable x to have the value 0.5. It simply means that the probability that x will have this *exact* value is insignificant.

Example 4 Finding a Probability

Consider a probability density function defined over the interval $[0, 5]$. If the probability that x lies in the interval $[0, 2]$ is 0.7, what is the probability that x lies in the interval $[2, 5]$?

SOLUTION Because the probability that x lies in the interval $[0, 5]$ is 1, you can conclude that the probability that x lies in the interval $[2, 5]$ is $1 - 0.7 = 0.3$.

✓ CHECKPOINT 4

A probability density function is defined over the interval $[0, 4]$. The probability that x lies in $[0, 1]$ is 0.6. What is the probability that x lies in $[1, 4]$? ∎

Applications

Example 5 Modeling the Lifetime of a Product

The useful lifetime (in years) of a medical product is modeled by the probability density function $f(t) = 0.1e^{-0.1t}$ for $0 \leq t < \infty$. Find the probability that a randomly selected unit will have a lifetime falling in each interval.

a. No more than 2 years

b. More than 2 years, but no more than 4 years

c. More than 4 years

SOLUTION

a. The probability that the unit will last no more than 2 years is

$$P(0 \leq t \leq 2) = 0.1 \int_0^2 e^{-0.1t}\, dt \quad \text{Integrate } f(t) \text{ over } [0, 2].$$
$$= \left[-e^{-0.1t}\right]_0^2 \quad \text{Find antiderivative.}$$
$$= -e^{-0.2} + 1 \quad \text{Apply Fundamental Theorem of Calculus.}$$
$$\approx 0.181. \quad \text{Approximate.}$$

b. The probability that the unit will last more than 2 years, but no more than 4 years, is

$$P(2 < t \leq 4) = 0.1 \int_2^4 e^{-0.1t}\, dt \quad \text{Integrate } f(t) \text{ over } [2, 4].$$
$$= \left[-e^{-0.1t}\right]_2^4 \quad \text{Find antiderivative.}$$
$$= -e^{-0.4} + e^{-0.2} \quad \text{Apply Fundamental Theorem of Calculus.}$$
$$\approx 0.148. \quad \text{Approximate.}$$

c. The probability that the unit will last more than 4 years is

$$P(4 < t < \infty) = 0.1 \int_4^\infty e^{-0.1t}\, dt \quad \text{Integrate } f(t) \text{ over } [4, \infty).$$
$$= \lim_{b \to \infty} \left[-e^{-0.1t}\right]_4^b \quad \text{Improper integral}$$
$$= \lim_{b \to \infty} \left(-e^{-0.1b} + e^{-0.4}\right) \quad \text{Evaluate limit.}$$
$$= e^{-0.4}$$
$$\approx 0.670. \quad \text{Approximate.}$$

These three probabilities are illustrated graphically in Figure 11.8. Note that the sum of the three probabilities is 1.

✓ CHECKPOINT 5

For the product in Example 5, find the probability that a randomly selected unit will have a lifetime of more than 10 years. ■

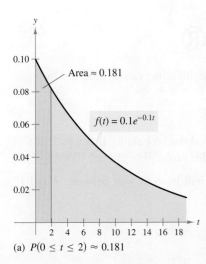

(a) $P(0 \leq t \leq 2) \approx 0.181$

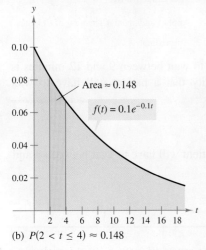

(b) $P(2 < t \leq 4) \approx 0.148$

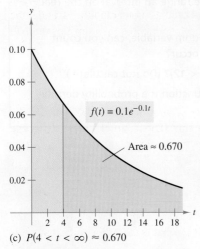

(c) $P(4 < t < \infty) \approx 0.670$

FIGURE 11.8

Example 6
MAKE A DECISION Waiting Time

The waiting time for patients arriving at a health clinic can be modeled by

$$f(t) = \frac{1}{12}e^{-t/12}$$

where t is the waiting time (in minutes). Find the probability that a patient will have to wait between 9 and 12 minutes. Is this probability at least 10%?

SOLUTION The probability that a patient will have to wait between 9 and 12 minutes is

$$P(9 \leq t \leq 12) = \frac{1}{12}\int_9^{12} e^{-t/12}\, dt \quad \text{Integrate } f(t) \text{ over } [9, 12].$$

$$= \left[-e^{-t/12}\right]_9^{12} \quad \text{Find antiderivative.}$$

$$= -e^{-1} + e^{-3/4} \quad \text{Apply Fundamental Theorem of Calculus.}$$

$$\approx 0.104. \quad \text{Approximate.}$$

So, the probability that a patient will have to wait between 9 and 12 minutes is about 0.104, or 10.4%. Yes, the probability that a patient will have to wait between 9 and 12 minutes is at least 10%.

✓ CHECKPOINT 6

In Example 6, find the probability that a patient will have to wait between 8 and 16 minutes. ■

CONCEPT CHECK

1. For which type of random variable is the range an interval on the real number line?
2. For an event involving a continuous random variable, can you count the number of ways that the event can occur?
3. In Example 6, is $P(9 \leq x \leq 12) = P(9 < x < 12)$? (Do not calculate.)
4. List the conditions that determine if a function is a probability density function.

SECTION 11.2 Continuous Random Variables

Skills Review 11.2

The following warm-up exercises involve skills that were covered in earlier sections. You will use these skills in the exercise set for this section. For additional help, review Sections 1.6, 6.4, and 7.6.

In Exercises 1–4, determine whether f is continuous and nonnegative on the given interval.

1. $f(x) = \dfrac{1}{x}$, $[1, 4]$
2. $f(x) = x^2 - 1$, $[0, 1]$
3. $f(x) = 3 - x$, $[1, 5]$
4. $f(x) = e^{-x}$, $[0, 1]$

In Exercises 5–10, evaluate the definite integral.

5. $\displaystyle\int_0^4 \dfrac{1}{4}\, dx$
6. $\displaystyle\int_1^3 \dfrac{1}{4}\, dx$
7. $\displaystyle\int_0^2 \dfrac{2-x}{2}\, dx$
8. $\displaystyle\int_1^2 \dfrac{2-x}{2}\, dx$
9. $\displaystyle\int_0^\infty 0.4 e^{-0.4t}\, dt$
10. $\displaystyle\int_0^\infty 3 e^{-3t}\, dt$

Exercises 11.2

See www.CalcChat.com for worked-out solutions to odd-numbered exercises.

In Exercises 1–14, use a graphing utility to graph the function. Then determine whether the function f represents a probability density function over the given interval. If f is not a probability density function, identify the condition(s) that is (are) not satisfied.

1. $f(x) = \tfrac{1}{8}$, $[0, 8]$
2. $f(x) = \tfrac{1}{5}$, $[0, 4]$
3. $f(x) = \dfrac{4-x}{8}$, $[0, 4]$
4. $f(x) = \dfrac{x}{18}$, $[0, 6]$
5. $f(x) = 6x(1 - 2x)$, $[0, 1]$
6. $f(x) = \tfrac{1}{36} x(6 - x)$, $[0, 6]$
7. $f(x) = \tfrac{1}{5} e^{-x/5}$, $[0, 5]$
8. $f(x) = \tfrac{1}{6} e^{-x/6}$, $[0, \infty)$
9. $f(x) = 2\sqrt{4 - x}$, $[0, 2]$
10. $f(x) = x^2(1 - x)$, $[0, 1]$
11. $f(x) = \tfrac{4}{27} x^2 (3 - x)$, $[0, 3]$
12. $f(x) = \tfrac{2}{9} x(3 - x)$, $[0, 3]$
13. $f(x) = \tfrac{1}{3} e^{-x/3}$, $[0, \infty)$
14. $f(x) = \tfrac{1}{4}$, $[8, 12]$

In Exercises 15–22, sketch the graph of the probability density function over the indicated interval and find the indicated probabilities.

15. $f(x) = \tfrac{1}{5}$, $[0, 5]$
    (a) $P(0 < x < 3)$
    (b) $P(1 < x < 3)$
    (c) $P(3 < x < 5)$
    (d) $P(x \geq 1)$

16. $f(x) = \tfrac{1}{10}$, $[0, 10]$
    (a) $P(0 < x < 6)$
    (b) $P(4 < x < 6)$
    (c) $P(8 < x < 10)$
    (d) $P(x \geq 2)$

17. $f(x) = \dfrac{x}{50}$, $[0, 10]$
    (a) $P(0 < x < 6)$
    (b) $P(4 < x < 6)$
    (c) $P(8 < x < 10)$
    (d) $P(x \geq 2)$

18. $f(x) = \dfrac{2x}{25}$, $[0, 5]$
    (a) $P(0 < x < 3)$
    (b) $P(1 < x < 3)$
    (c) $P(3 < x < 5)$
    (d) $P(x \geq 1)$

19. $f(x) = \tfrac{3}{16} \sqrt{x}$, $[0, 4]$
    (a) $P(0 < x < 2)$
    (b) $P(2 < x < 4)$
    (c) $P(1 < x < 3)$
    (d) $P(x \leq 3)$

20. $f(x) = \dfrac{5}{4(x + 1)^2}$, $[0, 4]$
    (a) $P(0 < x < 2)$
    (b) $P(2 < x < 4)$
    (c) $P(1 < x < 3)$
    (d) $P(x \leq 3)$

21. $f(t) = \tfrac{1}{3} e^{-t/3}$, $[0, \infty)$
    (a) $P(t < 2)$
    (b) $P(t \geq 2)$
    (c) $P(1 < t < 4)$
    (d) $P(t = 3)$

22. $f(t) = \tfrac{3}{256}(16 - t^2)$, $[-4, 4]$
    (a) $P(t < -2)$
    (b) $P(t > 2)$
    (c) $P(-1 < t < 1)$
    (d) $P(t > -2)$

In Exercises 23–28, find the constant k such that the function f is a probability density function over the given interval.

23. $f(x) = kx$, $[1, 4]$

24. $f(x) = kx^3$, $[0, 4]$

25. $f(x) = \dfrac{k}{b-a}$, $[a, b]$

26. $f(x) = k(4 - x^2)$, $[0, 1]$

27. $f(x) = k\sqrt{x}(1 - x)$, $[0, 1]$

28. $f(x) = ke^{-x/2}$, $[0, \infty)$

29. **Demand** The daily demand for gasoline x (in millions of gallons) in a city is described by the probability density function
$$f(x) = 0.41 - 0.08x, \quad [0, 4].$$
Find the probabilities that the daily demand for gasoline will be (a) no more than 3 million gallons and (b) at least 2 million gallons.

30. **Learning Theory** The time t (in hours) required for a new employee to learn the procedures for hazardous material handling in his or her work area is described by the probability density function
$$f(t) = \tfrac{5}{324} t \sqrt{9 - t}, \quad [0, 9].$$
Find the probabilities that a new employee will learn the procedures (a) in less than 3 hours and (b) in more than 4 hours but less than 8 hours.

(T) In Exercises 31–34, use a symbolic integration utility to find the required probabilities using the *exponential density function*
$$f(t) = \tfrac{1}{\lambda} e^{-t/\lambda}, \quad [0, \infty).$$

31. **Waiting Time** The waiting times (in minutes) for service at the checkout at a grocery store are exponentially distributed with $\lambda = 3$. Find the probabilities of waiting (a) less than 2 minutes, (b) more than 2 minutes but less than 4 minutes, and (c) at least 2 minutes.

32. **Unloading Time** The lengths of time (in hours) required to unload Red Cross supply trucks at the site of a natural disaster are exponentially distributed with $\lambda = \tfrac{3}{4}$. What proportion of the trucks can be unloaded in less than 1 hour?

33. **Useful Life** The lifetimes (in years) of hearing aid batteries are exponentially distributed with $\lambda = 5$. Find the probabilities that the lifetime of a given hearing aid battery will be (a) less than 6 years, (b) more than 2 years but less than 6 years, and (c) more than 8 years.

34. **Shelf Life** The shelf life (in years) of a drug is exponentially distributed with $\lambda = 3.5$. A pharmaceutical company has a large quantity of this drug in stock and plans to replace the supply during regularly scheduled inventory replenishment periods. How much time should elapse between inventory replenishment periods if at least 90% of the supply is to remain within the shelf life throughout the period?

35. **Meteorology** A meteorologist predicts that the amount of rainfall (in inches) expected for a certain coastal community during a hurricane has the probability density function
$$f(x) = \dfrac{\pi}{30} \sin \dfrac{\pi x}{15}, \quad 0 \le x \le 15.$$
Find and interpret the probabilities.

(a) $P(0 \le x \le 10)$ (b) $P(10 \le x \le 15)$

(c) $P(0 \le x < 5)$ (d) $P(12 \le x \le 15)$

(T) 36. **Coin Toss** The probability of obtaining 49, 50, or 51 heads when a fair coin is tossed 100 times is
$$P(49 \le x \le 51) \approx \int_{48.5}^{51.5} \dfrac{1}{5\sqrt{2\pi}} e^{-(x-50)^2/50} \, dx.$$
Use a computer or graphing utility and Simpson's Rule (with $n = 12$) to approximate this integral.

Life Science Capsule

Donna Miles/OSHA News Photo

Occupational health and safety specialists inspect places of employment for unsafe equipment and working conditions. They also design programs to correct safety and health problems in the workplace and assist employers in complying with Occupational Safety and Health Administration (OSHA) regulations and standards. Many employers, including the federal government, require a bachelor's degree in occupational health, safety, or a related field.

37. **Research Project** Use your school's library, the Internet, or some other reference source to gather information about how occupational health and safety specialists can use probability and calculus to design workplace health or safety programs. Write a short paper that summarizes your findings.

Section 11.3
Expected Value and Variance

- Find the expected values or means of continuous probability density functions.
- Find the variances and standard deviations of continuous probability density functions.
- Find the medians of continuous probability density functions.
- Use special probability density functions to answer questions about real-life situations.

Expected Value

In Section 11.1, you studied the concepts of expected value (or mean), variance, and standard deviation of *discrete* random variables. In this section, you will extend these concepts to *continuous* random variables.

Definition of Expected Value

If f is a probability density function of a continuous random variable x over the interval $[a, b]$, then the **expected value** or **mean** of x is

$$\mu = E(x) = \int_a^b xf(x)\, dx.$$

Example 1 Finding Expected Value

Find the expected value of x for the probability density function

$$f(x) = \frac{1}{36}(-x^2 + 6x), \quad 0 \le x \le 6.$$

SOLUTION

$$\mu = E(x) = \frac{1}{36}\int_0^6 x(-x^2 + 6x)\, dx$$

$$= \frac{1}{36}\int_0^6 (-x^3 + 6x^2)\, dx$$

$$= \frac{1}{36}\left[\frac{-x^4}{4} + 2x^3\right]_0^6$$

$$= 3$$

In Figure 11.9, you can see that an expected value of 3 seems reasonable because the region is symmetric about the line $x = 3$.

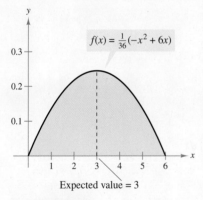

FIGURE 11.9

✓ CHECKPOINT 1

Find the expected value of the probability density function $f(x) = \frac{1}{32}3x(4 - x)$ on the interval $[0, 4]$. ∎

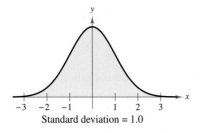
Standard deviation = 1.0

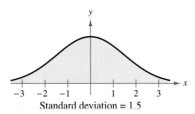
Standard deviation = 1.5

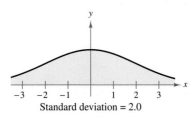
Standard deviation = 2.0

FIGURE 11.10

Variance and Standard Deviation

Definitions of Variance and Standard Deviation

If f is a probability density function of a continuous random variable x over the interval $[a, b]$, then the **variance** of x is

$$V(x) = \int_a^b (x - \mu)^2 f(x)\, dx$$

where μ is the mean of x. The **standard deviation** of x is

$$\sigma = \sqrt{V(x)}.$$

Recall from Section 11.1 that distributions that are clustered about the mean tend to have smaller standard deviations than distributions that are more dispersed. For instance, all three of the probability density distributions shown in Figure 11.10 have a mean of $\mu = 0$, but they have different standard deviations. Because the first distribution is clustered more closely about the mean, its standard deviation is the smallest of the three.

Example 2 Finding Variance and Standard Deviation

Find the variance and standard deviation of the probability density function

$$f(x) = 2 - 2x, \quad 0 \leq x \leq 1.$$

SOLUTION Begin by finding the mean.

$$\mu = \int_0^1 x(2 - 2x)\, dx = \frac{1}{3} \qquad \text{Mean}$$

Next, apply the formula for variance.

$$\begin{aligned}
V(x) &= \int_0^1 \left(x - \frac{1}{3}\right)^2 (2 - 2x)\, dx \\
&= \int_0^1 \left(-2x^3 + \frac{10x^2}{3} - \frac{14x}{9} + \frac{2}{9}\right) dx \\
&= \left[-\frac{x^4}{2} + \frac{10x^3}{9} - \frac{7x^2}{9} + \frac{2x}{9}\right]_0^1 \\
&= \frac{1}{18} \qquad \text{Variance}
\end{aligned}$$

Finally, you can conclude that the standard deviation is

$$\sigma = \sqrt{\frac{1}{18}} \approx 0.236. \qquad \text{Standard deviation}$$

✓CHECKPOINT 2

Find the variance and standard deviation of the probability density function in Checkpoint 1. ∎

The integral for variance can be difficult to evaluate. The following alternative formula is often simpler.

Alternative Formula for Variance

If f is a probability density function of a continuous random variable x over the interval $[a, b]$, then the **variance** of x is

$$V(x) = \int_a^b x^2 f(x)\, dx - \mu^2$$

where μ is the mean of x.

Example 3 Using the Alternative Formula

Find the standard deviation of the probability density function

$$f(x) = \frac{2}{\pi(x^2 - 2x + 2)}, \quad 0 \leq x \leq 2.$$

What percent of the distribution lies within one standard deviation of the mean?

SOLUTION Begin by using a symbolic integration utility to find the mean.

$$\mu = \int_0^2 \left[\frac{2}{\pi(x^2 - 2x + 2)} \right] (x)\, dx$$
$$= 1 \qquad \text{Mean}$$

Next, use a symbolic integration utility to find the variance.

$$V(x) = \int_0^2 \left[\frac{2}{\pi(x^2 - 2x + 2)} \right] (x^2)\, dx - 1^2$$
$$\approx 0.273 \qquad \text{Variance}$$

This implies that the standard deviation is

$$\sigma \approx \sqrt{0.273}$$
$$\approx 0.522. \qquad \text{Standard deviation}$$

To find the percent of the distribution that lies within one standard deviation of the mean, integrate the probability density function between $\mu - \sigma = 0.478$ and $\mu + \sigma = 1.522$.

$$\int_{0.478}^{1.522} \frac{2}{\pi(x^2 - 2x + 2)}\, dx \approx 0.613$$

So, about 61.3% of the distribution lies within one standard deviation of the mean. This result is illustrated in Figure 11.11.

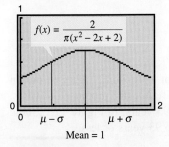

FIGURE 11.11

✓CHECKPOINT 3

Use a symbolic integration utility to find the percent of the distribution in Example 3 that lies within 1.5 standard deviations of the mean. ■

Median

The mean of a probability density function is an example of a **measure of central tendency.** Another useful measure of central tendency is the **median.**

Definition of Median

If f is a probability density function of a continuous random variable x over the interval $[a, b]$, then the **median** of x is the number m such that

$$\int_a^m f(x)\, dx = 0.5.$$

Example 4 Comparing Mean and Median

In Example 5 in Section 11.2, the probability density function

$$f(t) = 0.1e^{-0.1t}, \quad 0 \le t < \infty$$

was used to model the useful lifetime of a medical product. Find the mean and median useful lifetimes.

SOLUTION Using integration by parts or a symbolic integration utility, you can find the mean to be

$$\mu = \int_0^\infty 0.1 t e^{-0.1t}\, dt$$
$$= 10 \text{ years.} \qquad \text{Mean}$$

The median is given by

$$\int_0^m 0.1 e^{-0.1t}\, dt = 0.5$$
$$-e^{-0.1m} + 1 = 0.5$$
$$e^{-0.1m} = 0.5$$
$$-0.1m = \ln 0.5$$
$$m = -10 \ln 0.5$$
$$m \approx 6.93 \text{ years.} \qquad \text{Median}$$

From this, you can see that the mean and median of a probability distribution can be quite different. Using the mean, the "average" lifetime of a product is 10 years, but using the median, the "average" lifetime is 6.93 years. In Figure 11.12, note that half of the products have usable lifetimes of 6.93 years or less.

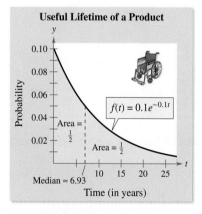

FIGURE 11.12

✓CHECKPOINT 4

Find the mean and median of the probability density function $f(x) = 2e^{-2x}$, $0 \le x < \infty$. ∎

Special Probability Density Functions

The remainder of this section describes three common types of probability density functions: uniform, exponential, and normal. The **uniform probability density function** is defined as

$$f(x) = \frac{1}{b-a}, \quad a \leq x \leq b.$$ Uniform probability density function

This probability density function represents a continuous random variable for which each outcome is equally likely.

Example 5 Analyzing a Probability Density Function

Find the expected value and standard deviation of the uniform probability density function

$$f(x) = \frac{1}{8}, \quad 0 \leq x \leq 8.$$

SOLUTION The expected value (or mean) is

$$\mu = \int_0^8 \frac{1}{8} x \, dx$$

$$= \left[\frac{x^2}{16}\right]_0^8$$

$$= 4. \qquad \text{Expected value}$$

The variance is

$$V(x) = \int_0^8 \frac{1}{8} x^2 \, dx - 4^2$$

$$= \left[\frac{x^3}{24}\right]_0^8 - 16$$

$$\approx 5.333. \qquad \text{Variance}$$

The standard deviation is

$$\sigma \approx \sqrt{5.333}$$

$$\approx 2.309. \qquad \text{Standard deviation}$$

The graph of f is shown in Figure 11.13.

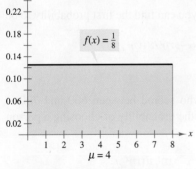

FIGURE 11.13 Uniform Probability Density Function

✓CHECKPOINT 5

Find the expected value and standard deviation of the uniform probability density function $f(x) = \frac{1}{2}, 0 \leq x \leq 2$. ∎

STUDY TIP

Try showing that the mean and the variance of the general uniform probability density function $f(x) = 1/(b - a)$ are $\mu = \frac{1}{2}(a + b)$ and $V(x) = \frac{1}{12}(b - a)^2$.

The second special type of probability density function is the **exponential probability density function** and has the form

$$f(x) = ae^{-ax}, \quad 0 \le x < \infty.$$

Exponential probability density function, $a > 0$

The probability density function in Example 4 is of this type. Try showing that this function has a mean of $1/a$ and a variance of $1/a^2$.

The third special type of probability density function (and the most widely used) is the **normal probability density function** given by

$$f(x) = \frac{1}{\sigma\sqrt{2\pi}} e^{-(x-\mu)^2/2\sigma^2}, \quad -\infty < x < \infty.$$

Normal probability density function

The expected value of this function is μ, and the standard deviation is σ. Figure 11.14 shows the graph of a typical normal probability density function.

A normal probability density function for which $\mu = 0$ and $\sigma = 1$ is called a **standard normal probability density function**.

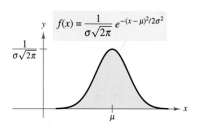

FIGURE 11.14 Normal Probability Density Function

Example 6 Finding a Probability

In 2006, the scores for the *Graduate Management Admission Test* (GMAT) could be modeled by a normal probability density function with a mean of $\mu = 527$ and a standard deviation of $\sigma = 117$. If you select a person who took the GMAT in 2006, what is the probability that the person scored between 600 and 700? What is the probability that the person scored between 700 and 800? *(Source: Graduate Management Admission Council)*

SOLUTION Using a calculator or computer, you can find the first probability to be

$$P(600 \le x \le 700) = \int_{600}^{700} \frac{1}{117\sqrt{2\pi}} e^{-(x-527)^2/2(117)^2} dx$$
$$\approx 0.197.$$

So, the probability of choosing a person who scored between 600 and 700 is about 19.7%. In a similar way, you can find the probability of choosing a person who scored between 700 and 800 to be

$$P(700 \le x \le 800) = \int_{700}^{800} \frac{1}{117\sqrt{2\pi}} e^{-(x-527)^2/2(117)^2} dx$$
$$\approx 0.060$$

or about 6.0%.

TECHNOLOGY

The normal probability density function does not have an antiderivative that is an elementary function. However, it does have an antiderivative. Finding its antiderivative analytically is beyond the scope of this text. You can use a graphing utility, a symbolic integration utility, or a spreadsheet to evaluate this function.* Use one of these tools to evaluate the function in Example 6.

✓ CHECKPOINT 6

From Example 6, find the probability that a person selected at random scored between 400 and 600. ■

STUDY TIP

In Example 6, note that probabilities connected with normal distributions must be evaluated with a table of values, with a symbolic integration utility, or with a spreadsheet.

---

*Specific calculator keystroke instructions for operations in this and other technology boxes can be found at *college.hmco.com/info/larsonapplied.*

SECTION 11.3 Expected Value and Variance

Example 7 Modeling Body Weight

Assume that the weights of adult male rhesus monkeys are normally distributed with a mean of 15 pounds and a standard deviation of 3 pounds. In a typical population of adult male rhesus monkeys, what percent of the monkeys would have weights within one standard deviation of the mean?

SOLUTION For this population, the normal probability density function is

$$f(x) = \frac{1}{3\sqrt{2\pi}} e^{-(x-15)^2/18}.$$

The probability that a randomly chosen adult male monkey will weigh between 12 and 18 pounds (that is, within 3 pounds of 15 pounds) is

$$P(12 \leq x \leq 18) = \int_{12}^{18} \frac{1}{3\sqrt{2\pi}} e^{-(x-15)^2/18}\, dx$$
$$\approx 0.683.$$

So, about 68% of the adult male rhesus monkeys have weights that lie within one standard deviation of the mean, as shown in Figure 11.15.

© Shay Fogelman/Alamy

In addition to being popular zoo animals, rhesus monkeys are commonly used in medical and behavioral research. Research on rhesus monkeys led to the discovery of the Rh factor in human red blood cells.

✓ **CHECKPOINT 7**

Use the results of Example 7 to find the probability that an adult male rhesus monkey chosen at random will weigh more than 18 pounds. ■

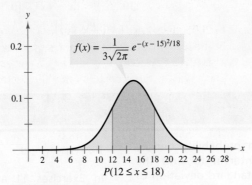

FIGURE 11.15

The result described in Example 7 can be generalized to all normal distributions. That is, in any normal distribution, the probability that x lies within one standard deviation of the mean is about 68%. For normal distributions, 95.4% of the x-values lie within two standard deviations of the mean, and almost all (99.7%) of the x-values lie within three standard deviations of the mean.

CONCEPT CHECK

1. Complete the following: The mean and median are both measures of _____ _____.
2. Which probability density function represents a continuous random variable for which each outcome is equally likely?
3. What is the mean of the standard normal probability density function?
4. What is the standard deviation of the standard normal probability density function?

Skills Review 11.3

The following warm-up exercises involve skills that were covered in earlier sections. You will use these skills in the exercise set for this section. For additional help, review Sections 6.4 and 11.2.

In Exercises 1–4, solve for m.

1. $\int_0^m \dfrac{1}{10}\,dx = 0.5$
2. $\int_0^m \dfrac{1}{16}\,dx = 0.5$
3. $\int_0^m \dfrac{1}{3} e^{-t/3}\,dt = 0.5$
4. $\int_0^m \dfrac{1}{9} e^{-t/9}\,dt = 0.5$

In Exercises 5–8, evaluate the definite integral.

5. $\int_0^2 \dfrac{x^2}{2}\,dx$
6. $\int_1^2 x(4 - 2x)\,dx$
7. $\int_2^5 x^2\left(\dfrac{1}{3}\right)dx - \left(\dfrac{7}{2}\right)^2$
8. $\int_2^4 x^2\left(\dfrac{4-x}{2}\right)dx - \left(\dfrac{8}{3}\right)^2$

In Exercises 9 and 10, find the indicated probability using the given probability density function.

9. $f(x) = \tfrac{1}{8}, \ [0, 8]$
  (a) $P(x \le 2)$
  (b) $P(3 < x < 7)$

10. $f(x) = 6x - 6x^2, \ [0, 1]$
  (a) $P(x \le \tfrac{1}{2})$
  (b) $P(\tfrac{1}{4} \le x \le \tfrac{3}{4})$

Exercises 11.3

See www.CalcChat.com for worked-out solutions to odd-numbered exercises.

In Exercises 1–6, use the given probability density function over the indicated interval to find the (a) mean, (b) variance, and (c) standard deviation of the random variable. Sketch the graph of the density function and locate the mean on the graph.

1. $f(x) = \tfrac{1}{3}, \ [0, 3]$
2. $f(x) = \tfrac{1}{4}, \ [0, 4]$
3. $f(t) = \dfrac{t}{18}, \ [0, 6]$
4. $f(x) = \dfrac{4}{3x^2}, \ [1, 4]$
5. $f(x) = \tfrac{5}{2} x^{3/2}, \ [0, 1]$
6. $f(x) = \tfrac{3}{16}\sqrt{4 - x}, \ [0, 4]$

(T) In Exercises 7–10, use a graphing utility to graph the function and approximate the mean. Then find the mean analytically. Compare your results.

7. $f(x) = 6x(1 - x), \ [0, 1]$
8. $f(x) = \tfrac{3}{32} x(4 - x), \ [0, 4]$
9. $f(x) = \dfrac{4}{3(x+1)^2}, \ [0, 3]$
10. $f(x) = \tfrac{1}{18}\sqrt{9 - x}, \ [0, 9]$

In Exercises 11 and 12, find the median of the exponential probability density function.

11. $f(t) = \tfrac{1}{9} e^{-t/9}, \ [0, \infty)$
12. $f(t) = \tfrac{2}{5} e^{-2t/5}, \ [0, \infty)$

In Exercises 13–18, identify the probability density function. Then find the mean, variance, and standard deviation without integrating.

13. $f(x) = \tfrac{1}{10}, \ [0, 10]$
14. $f(x) = 0.2, \ [0, 5]$
15. $f(x) = \tfrac{1}{8} e^{-x/8}, \ [0, \infty)$
16. $f(x) = \tfrac{5}{3} e^{-5x/3}, \ [0, \infty)$
17. $f(x) = \dfrac{1}{11\sqrt{2\pi}} e^{-(x-100)^2/242}, \ (-\infty, \infty)$
18. $f(x) = \dfrac{1}{6\sqrt{2\pi}} e^{-(x-30)^2/72}, \ (-\infty, \infty)$

In Exercises 19–24, use a symbolic integration utility to find the mean, standard deviation, and given probability.

Function	Probability
19. $f(x) = \frac{1}{\sqrt{2\pi}} e^{-x^2/2}$	$P(0 \le x \le 0.85)$
20. $f(x) = \frac{1}{\sqrt{2\pi}} e^{-x^2/2}$	$P(-1.21 \le x \le 1.21)$
21. $f(x) = \frac{1}{6} e^{-x/6}$	$P(x \ge 2.23)$
22. $f(x) = \frac{3}{4} e^{-3x/4}$	$P(x \ge 0.27)$
23. $f(x) = \frac{1}{2\sqrt{2\pi}} e^{-(x-8)^2/8}$	$P(3 \le x \le 13)$
24. $f(x) = \frac{1}{1.5\sqrt{2\pi}} e^{-(x-2)^2/4.5}$	$P(-2.5 \le x \le 2.5)$

In Exercises 25 and 26, let x be a random variable that is normally distributed with the given mean and standard deviation. Find the indicated probabilities using a symbolic integration utility.

25. $\mu = 50$, $\sigma = 10$
    (a) $P(x > 55)$
    (b) $P(x > 60)$
    (c) $P(x < 60)$
    (d) $P(30 < x < 55)$

26. $\mu = 70$, $\sigma = 14$
    (a) $P(x > 65)$
    (b) $P(x < 98)$
    (c) $P(x < 49)$
    (d) $P(56 < x < 75)$

27. **Transportation** The arrival times t of a bus at a bus stop are uniformly distributed between 10:00 A.M. and 10:10 A.M.
    (a) Find the mean and standard deviation of the random variable t.
    (b) What is the probability that you will miss the bus if you arrive at the bus stop at 10:03 A.M.?

28. **Transportation** Repeat Exercise 27 for a bus that arrives between 10:00 A.M. and 10:05 A.M.

29. **Useful Life** The times t until failure of an appliance are exponentially distributed with a mean of 2 years.
    (a) Find the probability density function of the random variable t.
    (b) Find the probability that the appliance will fail in less than 1 year.

30. **Useful Life** The lifetimes of batteries are normally distributed with a mean of 400 hours and a standard deviation of 24 hours. You purchased one of the batteries, and its useful life was 340 hours.
    (a) How far, in standard deviations, did the useful life of your battery fall short of the expected life?
    (b) What percent of all other batteries of this type have useful lives that exceed yours?

31. **Waiting Time** The waiting times t for service in a store are exponentially distributed with a mean of 5 minutes.
    (a) Find the probability density function of the random variable t.
    (b) Find the probability that t is within one standard deviation of the mean.

32. **License Renewal** The lengths of time t spent at a driver's license renewal center are exponentially distributed with a mean of 15 minutes.
    (a) Find the probability density function of the random variable t.
    (b) Find the probability that t is within one standard deviation of the mean.

33. **Education** The scores on a national exam are normally distributed with a mean of 150 and a standard deviation of 16. You scored 174 on the exam.
    (a) How far, in standard deviations, did your score exceed the national mean?
    (b) What percent of those who took the exam had scores lower than yours?

34. **Education** The scores on a qualifying exam for entrance into a post-secondary school are normally distributed with a mean of 120 and a standard deviation of 10.5. To qualify for admittance, the candidates must score in the top 10%. Find the lowest possible qualifying score.

35. **Metallurgy** The percent of iron x in samples of ore is a random variable with the probability density function $f(x) = \frac{1155}{32} x^3 (1-x)^{3/2}$, $[0, 1]$. Determine the expected percent of iron in each ore sample.

36. **Medicine** The time t (in days) until recovery after a certain medical procedure is a random variable with the probability density function
$$f(t) = \frac{1}{2\sqrt{t-2}}, \quad [3, 6].$$
    (a) Find the probability that a patient selected at random will take more than 4 days to recover.
    (b) Determine the expected time for recovery.

In Exercises 37–42, find the mean and median.

37. $f(x) = \frac{1}{11}$, $[0, 11]$
38. $f(x) = 0.05$, $[0, 20]$
39. $f(x) = 4(1 - 2x)$, $[0, \frac{1}{2}]$
40. $f(x) = \frac{4}{3} - \frac{2}{3}x$, $[0, 1]$
41. $f(x) = \frac{1}{5} e^{-x/5}$, $[0, \infty)$
42. $f(x) = \frac{2}{3} e^{-2x/3}$, $[0, \infty)$

43. Cost The daily cost (in dollars) of electricity x in a city is a random variable with the probability density function $f(x) = 0.28e^{-0.28x}$, $0 \leq x < \infty$. Find the median daily cost of electricity.

44. Consumer Trends The number of coupons used by a customer in a grocery store is a random variable with the probability density function

$$f(x) = \frac{2x+1}{12}, \quad 0 \leq x \leq 3.$$

Find the expected number of coupons a customer will use.

45. Useful Life The lifetimes of tires are normally distributed with a mean of 50,000 miles and a standard deviation of 3000 miles. For how many miles should this tire be guaranteed if the manufacturer does not want to replace more than 10% of the tires under the terms of the guarantee?

46. Manufacturing An automatic filling machine fills cans so that the weights are normally distributed with a mean of μ and a standard deviation of σ. The value of μ can be controlled by settings on the machine, but σ depends on the precision and design of the machine. For a particular substance, $\sigma = 0.15$ ounce. If 12-ounce cans are being filled, determine the setting for μ such that no more than 5% of the cans weigh less than the stated weight.

47. MAKE A DECISION: USEFUL LIFE A storage battery has an expected lifetime of 4.5 years with a standard deviation of 0.5 year. Assume that the useful lives of these batteries are normally distributed.

(a) Use a computer or graphing utility and Simpson's Rule (with $n = 12$) to approximate the probability that a given battery will last for 4 to 5 years.

(b) Will 10% of the batteries last less than 3 years?

48. MAKE A DECISION: WAGES The employees of a large corporation are paid an average wage of $14.50 per hour with a standard deviation of $1.50. Assume that these wages are normally distributed.

(a) Use a computer or graphing utility and Simpson's Rule (with $n = 10$) to approximate the percent of employees that earn hourly wages of $11.00 to $14.00.

(b) Will 20% of the employees be paid more than $16.00 per hour?

49. Medical Science A medical research team has determined that for a group of 500 females, the length of pregnancy from conception to birth varies according to an approximately normal distribution with a mean of 266 days and a standard deviation of 16 days.

(a) Use a graphing utility to graph the distribution.

(b) Use a symbolic integration utility to approximate the probability that a pregnancy will last from 240 days to 280 days.

(c) Use a symbolic integration utility to approximate the probability that a pregnancy will last more than 280 days.

50. Education In 2006, the scores for the ACT Test could be modeled by a normal probability density function with a mean of 21.1 and a standard deviation of 4.8. *(Source: ACT, Inc.)*

(a) Use a graphing utility to graph the distribution.

(b) Use a symbolic integration utility to approximate the probability that a person who took the ACT scored between 24 and 36.

(c) Use a symbolic integration utility to approximate the probability that a person who took the ACT scored more than 26.

51. Fuel Mileage Assume that the fuel mileages of all 2007 model vehicles weighing less than 8500 pounds are normally distributed with a mean of 20.6 miles per gallon and a standard deviation of 4.9 miles per gallon. *(Source: U.S. Environmental Protection Agency)*

(a) Use a graphing utility to graph the distribution.

(b) Use a symbolic integration utility to approximate the probability that a vehicle's fuel mileage is between 25 and 30 miles per gallon.

(c) Use a symbolic integration utility to approximate the probability that a vehicle's fuel mileage is less than 18 miles per gallon.

52. Mr. Average During the 2006–2007 professional basketball season, Jason Richardson of the Golden State Warriors was deemed "Mr. Average," based on his closeness to the league averages in the four statistical categories shown below. *(Source: National Basketball Association)*

	Jason Richardson	Entire NBA
Height	6 ft 6 in.	6 ft 6.93 in.
Weight	225 lbs	221.55 lbs
Age	26	26.67
Years in League	5	4.42

(a) The heights h of adult males (ages 20–29) in the United States are normally distributed with a mean of 5 feet 10 inches and a standard deviation of 3 inches. Find the probability density function for the random variable h and find the percent of adult males in the United States who are taller than Jason Richardson.

(b) The shortest person to play in the NBA during the 2006–2007 season was Earl Boykins (5 feet 5 inches) of the Denver Nuggets. What percent of adult males in the United States are taller than Earl Boykins?

Algebra Review

Using Counting Principles

In discrete probability, one of the basic skills is being able to count the number of ways an event can happen. To do this, the strategies below can be helpful.

STUDY TIP

If n is a positive integer, n *factorial* is defined as

$$n! = 1 \cdot 2 \cdot 3 \cdot 4 \cdots (n-1) \cdot n.$$

1. *The Fundamental Counting Principle:* The number of ways that two or more events can occur is the product of the numbers of ways each event can occur by itself. These ways can be listed graphically using a tree diagram.

2. *Permutations:* The number of permutations of n elements is $n!$.

3. *Combinations:* The number of combinations of n elements taken r at a time is

$$_nC_r = \frac{n!}{(n-r)!r!}.$$

Example 1 Counting the Ways an Event Can Happen

a. In how many ways can you form a five-letter password if no letter is used more than once?

b. Your class is divided into five work groups containing three, four, four, three, and five people. In how many ways can you poll one person from each group?

c. In how many orders can seven runners finish a race if there are no ties?

d. You have 12 phone calls to return. In how many orders can you return them?

SOLUTION

a. For the first letter of the password, you have 26 choices. For the second letter, you have 25 choices. For the third letter, you have 24 choices, and so on.

Number of ways $= 26 \cdot 25 \cdot 24 \cdot 23 \cdot 22$ Counting Principle

$= 7{,}893{,}600$ Multiply.

b. Number of ways $= 3 \cdot 4 \cdot 4 \cdot 3 \cdot 5$ Counting Principle

$= 720$ Multiply.

c. The solution is given by the number of permutations of the seven runners.

Number of ways $= 7! = 5040$ Multiply.

d. The solution is given by the number of permutations of the 12 phone calls.

Number of ways $= 12!$ Use a calculator.

$= 479{,}001{,}600$

 TECHNOLOGY

Most graphing utilities have a factorial key. Consult your user's manual for specific keystrokes for your graphing utility.

Example 2 **Counting the Ways an Event Can Happen**

In how many different ways can you choose a three-person group from a class of 20 people? From a class of 40 people?

SOLUTION The number of ways to choose a three-person group from a class of 20 people is given by the number of combinations of 20 elements taken three at a time.

$$
\begin{aligned}
\text{Number of ways} &= {}_{20}C_3 && \text{Combination} \\
&= \frac{20!}{17!\,3!} && \text{Formula for combination} \\
&= \frac{20 \cdot 19 \cdot 18 \cdot 17!}{17! \cdot 3!} \\
&= \frac{20 \cdot 19 \cdot 18}{3 \cdot 2 \cdot 1} && \text{Divide out like factors.} \\
&= 20 \cdot 19 \cdot 3 && \text{Divide out like factors.} \\
&= 1140 && \text{Multiply.}
\end{aligned}
$$

The number of ways to choose a three-person group from a class of 40 people is given by ${}_{40}C_3$, which is 9880.

Example 3 **Counting the Ways an Event Can Happen**

To test for defective units, you are to choose a sample of 10 from a manufacturing production run of 2000 units. How many different samples of 10 are possible?

SOLUTION The solution is given by the number of combinations of 2000 elements taken 10 at a time.

$$
\begin{aligned}
\text{Number of ways} &= {}_{2000}C_{10} && \text{Combination} \\
&= \frac{2000!}{1990!\,10!} && \text{Formula for combination} \\
&\approx 2.76 \times 10^{26} && \text{Use a graphing utility.} \\
&= 276{,}000{,}000{,}000{,}000{,}000{,}000{,}000{,}000
\end{aligned}
$$

From these examples, you can see that combinations and permutations can be very large numbers.

TECHNOLOGY

 Most graphing utilities have a combination key. Consult your user's manual for specific keystrokes for your graphing utility.

Chapter Summary and Study Strategies

After studying this chapter, you should have acquired the following skills. The exercise numbers are keyed to the Review Exercises that begin on page 694. Answers to odd-numbered Review Exercises are given in the back of the text.*

Section 11.1 | Review Exercises

- Describe sample spaces of experiments. 1–4
- Assign values to discrete random variables. 5, 6
- Form frequency distributions for discrete random variables. 7, 8
- Find the probabilities of events for discrete random variables. 9–12

$$P(x) = \frac{\text{Frequency of } x}{\text{Number of outcomes in } S} = \frac{n(x)}{n(S)}$$

- Find the expected values or means of discrete random variables. 13–15

$$\mu = E(x) = x_1 P(x_1) + x_2 P(x_2) + x_3 P(x_3) + \cdots + x_m P(x_m)$$

- Find the variances and standard deviations of discrete random variables. 16–20

$$V(x) = (x_1 - \mu)^2 P(x_1) + \cdots + (x_m - \mu)^2 P(x_m), \quad \sigma = \sqrt{V(x)}$$

Section 11.2

- Verify continuous probability density functions. 21–26
- Use continuous probability density functions to find probabilities. 27–30

$$P(c \le x \le d) = \int_c^d f(x)\, dx$$

- Use continuous probability density functions to answer questions about real-life situations. 31, 32

Section 11.3

- Find the means of continuous probability density functions. 33–36

$$\mu = E(x) = \int_a^b x f(x)\, dx$$

- Find the variances and standard deviations of continuous probability density functions. 37–40

$$V(x) = \int_a^b (x - \mu)^2 f(x)\, dx, \quad \sigma = \sqrt{V(x)}$$

- Find the medians of continuous probability density functions. 41–44

$$\int_a^m f(x)\, dx = 0.5$$

- Use special probability density functions to answer questions about real-life situations. 45–52

Study Strategies

- **Using Technology** Integrals that are used for continuous probability density functions tend to be difficult to evaluate by hand. When evaluating such integrals, we suggest that you use a symbolic integration utility or that you use a numerical integration technique such as Simpson's Rule with a programmable calculator.

---

* Use a wide range of valuable study aids to help you master the material in this chapter. The *Student Solutions Guide* includes step-by-step solutions to all odd-numbered exercises to help you review and prepare. The student website at *college.hmco.com/info/larsonapplied* offers algebra help and a *Graphing Technology Guide*. The *Graphing Technology Guide* contains step-by-step commands and instructions for a wide variety of graphing calculators, including the most recent models.

Review Exercises

In Exercises 1–4, describe the sample space of the experiment.

1. A month of the year is chosen for vacation.
2. A letter from the word *calculus* is selected.
3. A student must answer three questions from a selection of four essay questions.
4. A winner in a game show must choose two out of five prizes.
5. **Lottery** Three numbers are drawn in a lottery. Each number is a digit from 0 to 9. Find the sample space giving the number of 7's drawn.
6. **Quality Control** As cans of energy drink are filled on the production line, four are randomly selected and labeled with an "S" if the weight is satisfactory or with a "U" if the weight is unsatisfactory. Find the sample space giving the satisfactory/unsatisfactory classification of the four cans in the selected group.

In Exercises 7 and 8, complete the table to form the frequency distribution of the random variable x. Then construct a bar graph to represent the result.

7. A computer randomly selects a three-digit bar code. Each digit can be 0 or 1, and x is the number of 1's in the bar code.

x	0	1	2	3
$n(x)$				

8. A cat has a litter of four kittens. Let x represent the number of male kittens.

x	0	1	2	3	4
$n(x)$					

In Exercises 9 and 10, sketch a graph of the given probability distribution and find the required probabilities.

9.

x	1	2	3	4	5
$P(x)$	$\frac{1}{18}$	$\frac{7}{18}$	$\frac{5}{18}$	$\frac{3}{18}$	$\frac{2}{18}$

(a) $P(2 \leq x \leq 4)$
(b) $P(x \geq 3)$

10.

x	-2	-1	1	3	5
$P(x)$	$\frac{1}{11}$	$\frac{2}{11}$	$\frac{4}{11}$	$\frac{3}{11}$	$\frac{1}{11}$

(a) $P(x < 0)$
(b) $P(x > 1)$

11. **Dice Toss** Consider an experiment in which two six-sided dice are tossed. Find the indicated probabilities.

(a) The probability that the total is 8
(b) The probability that the total is greater than 4
(c) The probability that doubles are thrown
(d) The probability of getting double 6's

12. **Random Selection** Consider an experiment in which one card is randomly selected from a standard deck of 52 playing cards. Find the probabilities of

(a) selecting a face card.
(b) selecting a card that is not a face card.
(c) selecting a black card that is not a face card.
(d) selecting a card whose value is 6 or less.

13. **Education** An instructor gave a 25-point quiz to 52 students. Use the frequency distribution shown below to find the mean quiz score.

Score	9	10	11	12	13	14	15	16	17
Frequency	1	0	1	0	0	0	3	4	7

Score	18	19	20	21	22	23	24	25
Frequency	3	0	9	11	6	3	0	4

14. **Cost Increases** A pharmaceutical company uses three different chemicals, A, B, and C, to create a nutritional supplement. The table shown below gives the cost and the percent increase of the cost of each of the three chemicals. Find the mean percent increase of the three chemicals.

Chemical	Percent Increase	Cost of Materials
A	8%	$650
B	23%	$375
C	16%	$800

15. **Games of Chance** A service organization is selling $5 raffle tickets as part of a fundraising program. The first and second prizes are $3000 and $1000, respectively. In addition to the first and second prizes, there are 50 $20 gift certificates to be awarded. The number of tickets sold is 2000. Find the expected net gain to the player when one ticket is purchased.

16. **Dental Hygiene** A pharmacy sells five different models of electronic toothbrushes. During one month the sales for the five models were as shown.

    Model 1   4 sold at $125 each
    Model 2   8 sold at $80 each
    Model 3   12 sold at $65 each
    Model 4   21 sold at $38 each
    Model 5   30 sold at $12 each

    Find the variance and standard deviation of the prices.

17. **Health** A discount health product retailer stocks multiple brands of heating pads. The quantities and prices per heating pad are shown below.

    Brand 1   30 pads at $5 each
    Brand 2   25 pads at $12 each
    Brand 3   20 pads at $30 each
    Brand 4   18 pads at $49 each
    Brand 5   12 pads at $65 each

    Find the variance and standard deviation of the prices.

18. **Consumer Trends** A random survey of households recorded the number of cars per household. The results of the survey are shown in the table.

x	0	1	2	3	4	5
$P(x)$	0.10	0.28	0.39	0.17	0.04	0.02

    Find the variance and standard deviation of x.

19. **Vital Statistics** The probability distribution for the numbers of children in a sample of families is shown in the table.

x	0	1	2	3	4
$P(x)$	0.12	0.31	0.43	0.12	0.02

    Find the variance and standard deviation of x.

20. **Freshman Classes** A random survey of college students recorded the numbers of classes they took during the first semester of their freshman year. The results of the survey are shown in the table.

x	3	4	5	6	7
$P(x)$	0.18	0.29	0.41	0.10	0.02

    Find the variance and standard deviation of x.

In Exercises 21–26, use a graphing utility to graph the function. Then determine whether the function f represents a probability density function over the given interval. If f is not a probability density function, identify the condition(s) that is (are) not satisfied.

21. $f(x) = \frac{1}{12}$, $[0, 12]$
22. $f(x) = \frac{1}{8}$, $[1, 8]$
23. $f(x) = \frac{1}{4}(3 - x)$, $[0, 4]$
24. $f(x) = \frac{3}{4}x^2(2 - x)$, $[0, 2]$
25. $f(x) = \frac{1}{4\sqrt{x}}$, $[1, 9]$
26. $f(x) = 8.75x^{3/2}(1 - x)$, $[0, 2]$

In Exercises 27–30, find the indicated probability for the probability density function.

27. $f(x) = \frac{1}{50}(10 - x)$, $[0, 10]$
    $P(0 < x < 2)$

28. $f(x) = \frac{1}{36}(9 - x^2)$, $[-3, 3]$
    $P(-1 < x < 2)$

29. $f(x) = \frac{2}{(x+1)^2}$, $[0, 1]$
    $P\left(0 < x < \frac{1}{2}\right)$

30. $f(x) = \frac{3}{128}\sqrt{x}$, $[0, 16]$
    $P(4 < x < 9)$

31. **Waiting Time** Buses arrive and depart from a college every 20 minutes. The probability density function of the waiting time t (in minutes) for a person arriving at the bus stop is
    $f(t) = \frac{1}{20}$, $[0, 20]$.
    Find the probabilities that the person will wait (a) no more than 10 minutes and (b) at least 15 minutes.

32. **Medicine** The time t (in days) until recovery after a certain medical procedure is a random variable with the probability density function
    $f(t) = \frac{1}{4\sqrt{t-4}}$, $[5, 13]$.
    Find the probability that a patient selected at random will take more than 8 days to recover.

In Exercises 33–36, find the mean of the probability density function.

33. $f(x) = \frac{1}{7}$, $[0, 7]$

34. $f(x) = \frac{8-x}{32}$, $[0, 8]$

35. $f(x) = \frac{1}{6}e^{-x/6}$, $[0, \infty)$

36. $f(x) = 0.3e^{-0.3x}$, $[0, \infty)$

In Exercises 37–40, find the variance and standard deviation of the probability density function.

37. $f(x) = \frac{2}{9}x(3-x)$, $[0, 3]$

38. $f(x) = \frac{3}{16}\sqrt{x}$, $[0, 4]$

39. $f(x) = \frac{1}{2}e^{-x/2}$, $[0, \infty)$

40. $f(x) = 0.8e^{-0.8x}$, $[0, \infty)$

In Exercises 41–44, find the median of the probability density function.

41. $f(x) = 6x(1-x)$, $[0, 1]$

42. $f(x) = 12x^2(1-x)$, $[0, 1]$

43. $f(x) = 0.25e^{-x/4}$, $[0, \infty)$

44. $f(x) = \frac{5}{6}e^{-5x/6}$, $[0, \infty)$

45. Waiting Time The waiting times t (in minutes) for patients arriving at a dentist's office are exponentially distributed with the probability density function

$$f(t) = \frac{1}{15}e^{-t/15}, \quad [0, \infty).$$

Find the probabilities of waiting (a) less than 10 minutes and (b) more than 10 minutes but less than 20 minutes.

46. Useful Life The lifetime t (in hours) of a laboratory equipment component is exponentially distributed with the density function

$$f(t) = \frac{1}{350}e^{-t/350}, \quad [0, \infty).$$

Find the probability that a given component chosen at random will perform satisfactorily for more than 400 hours.

47. Botany In a botany experiment, plants are grown in a nutrient solution. The heights of the plants are found to be normally distributed with a mean of 42 centimeters and a standard deviation of 3 centimeters. Find the probability that a plant in the experiment is at least 50 centimeters tall.

48. Wages The hourly wages for the workers at a nursing care facility are normally distributed with a mean of $14.50 and a standard deviation of $1.40.

(a) What percent of the workers receive hourly wages from $13 to $15, inclusive?

(b) The highest 10% of the hourly wages are greater than what amount?

49. Heart Transplants Assume that the waiting times for heart transplants are normally distributed with a mean of 130 days and a standard deviation of 25 days. *(Source: Organ Procurement and Transplant Network)*

(a) Use a graphing utility to graph the distribution.

(b) Use a symbolic integration utility to approximate the probability that a waiting time is between 70 and 105 days.

(c) Use a symbolic integration utility to approximate the probability that a waiting time is more than 120 days.

50. Health Assume that the heights of American men (from 20 to 29 years old) are normally distributed with a mean of 70 inches and a standard deviation of 3 inches. *(Source: U.S. National Center for Health Statistics)*

(a) Use a graphing utility to graph the distribution.

(b) Use a symbolic integration utility to approximate the probability that a man's height is between 72 and 75 inches.

(c) Use a symbolic integration utility to approximate the probability that a man's height is less than 68 inches.

51. Meteorology The monthly rainfall x in a certain state is normally distributed with a mean of 3.75 inches and a standard deviation of 0.5 inch. Use a computer or a graphing utility and Simpson's Rule (with $n = 12$) to approximate the probability that in a randomly selected month the rainfall is between 3.5 and 4 inches.

52. Chemistry: Hydrogen Orbitals In chemistry, the probability of finding an electron at a particular position is greatest close to the nucleus and drops off rapidly as the distance from the nucleus increases. The graph displays the probability of finding the electron at points along a line drawn from the nucleus outward in any direction for the hydrogen 1s orbital. Make a sketch of this graph, and add to your sketch an indication of where you think the median might be. *(Source: Adapted from Zumdahl, Chemistry, Seventh Edition)*

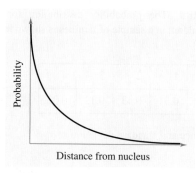

Chapter Test

See www.CalcChat.com for worked-out solutions to odd-numbered exercises.

Take this test as you would take a test in class. When you are done, check your work against the answers given in the back of the book.

1. A coin is tossed four times.
   (a) Write the sample space and frequency distribution for the possible outcomes.
   (b) What is the probability that at least two heads occur?

2. A card is chosen at random from a standard 52-card deck of playing cards. What is the probability that the card will be red and not a face card?

In Exercises 3 and 4, sketch a graph of the probability distribution and find the indicated probabilities.

3.

x	1	2	3	4
$P(x)$	$\frac{3}{16}$	$\frac{7}{16}$	$\frac{1}{16}$	$\frac{5}{16}$

(a) $P(x < 3)$ (b) $P(x \geq 3)$

4.

x	7	8	9	10	11
$P(x)$	0.21	0.13	0.19	0.42	0.05

(a) $P(7 \leq x \leq 10)$ (b) $P(x > 8)$

In Exercises 5 and 6, find $E(x)$, $V(x)$, and σ for the given probability distribution.

5.

x	0	1	2	3
$P(x)$	$\frac{2}{10}$	$\frac{1}{10}$	$\frac{4}{10}$	$\frac{3}{10}$

6.

x	-2	-1	0	1	2
$P(x)$	0.141	0.305	0.257	0.063	0.234

(T) In Exercises 7–9, use a graphing utility to graph the function. Then determine whether the function represents a probability density function over the given interval. If f is not a probability density function, identify the condition(s) that is (are) not satisfied.

7. $f(x) = \frac{\pi}{2}\sin \pi x$, $[0, 1]$ 8. $f(x) = \frac{3-x}{6}$, $[-1, 1]$ 9. $f(x) = \frac{2x}{x^2+1}$, $[0, \infty)$

In Exercises 10–12, find the indicated probabilities for the probability density function.

10. $f(x) = \frac{x}{32}$, $[0, 8]$ (a) $P(1 \leq x \leq 4)$ (b) $P(3 \leq x \leq 6)$

11. $f(x) = 4(x - x^3)$, $[0, 1]$ (a) $P(0 < x < 0.5)$ (b) $P(0.25 \leq x < 1)$

12. $f(x) = 2xe^{-x^2}$, $[0, \infty)$ (a) $P(x < 1)$ (b) $P(x \geq 1)$

In Exercises 13–15, find the mean, variance, and standard deviation of the probability density function.

13. $f(x) = \frac{1}{14}$, $[0, 14]$ 14. $f(x) = 3x - \frac{3}{2}x^2$, $[0, 1]$

15. $f(x) = e^{-x}$, $[0, \infty)$

(T) 16. An *intelligence quotient* or IQ is a number that is meant to measure intelligence. The IQs of students in a school are normally distributed with a mean of 110 and a standard deviation of 10. Use a symbolic integration utility to find the probability that a student selected at random will have an IQ within one standard deviation of the mean.

Appendices

A Differentiation and Integration Formulas

B Additional Topics in Differentiation
- **B.1 Implicit Differentiation**   A3
  Explicit and Implicit Functions • Implicit Differentiation
- **B.2 Related Rates**   A10
  Related Variables • Solving Related-Rate Problems

C Probability and Probability Distributions*
- **C.1 Probability**
  Sample Space of an Experiment • Probability of an Event
- **C.2 Probability Computations**
  Mutually Exclusive Events • Complement of an Event • Odds
- **C.3 Conditional Probability**
  Conditional Probability • Independent Events
- **C.4 Tree Diagrams and Bayes' Theorem**
  Tree Diagrams • Bayes' Theorem
- **C.5 Probability Distributions**
  Frequency Distributions • Probability Distributions
- **C.6 Normal Distributions**
  Continuous Random Variables • Normal Distributions • Standard Normal Distribution • z-Scores for Nonstandard Normal Distributions
- **C.7 Binomial Distributions**
  Binomial Experiments • Binomial Distributions • Approximating Binomial Distributions

D Properties and Measurement*
- **D.1 Review of Algebra, Geometry, and Trigonometry**
  Algebra • Properties of Logarithms • Geometry • Plane Analytic Geometry • Solid Analytic Geometry • Trigonometry • Library of Functions
- **D.2 Units of Measurements**
  Units of Measurement of Length • Units of Measurement of Area • Units of Measurement of Volume • Units of Measurement of Mass and Force • Units of Measurement of Temperature • Miscellaneous Units and Number Constants

E Graphing Utility Programs*

*Appendices C, D and E are on the website that accompanies this text (college.hmco.com/info/larsonapplied).

A Differentiation and Integration Formulas

Differentiation Formulas

1. $\dfrac{d}{dx}[cu] = cu'$

2. $\dfrac{d}{dx}[u \pm v] = u' \pm v'$

3. $\dfrac{d}{dx}[uv] = uv' + vu'$

4. $\dfrac{d}{dx}\left[\dfrac{u}{v}\right] = \dfrac{vu' - uv'}{v^2}$

5. $\dfrac{d}{dx}[c] = 0$

6. $\dfrac{d}{dx}[u^n] = nu^{n-1}u'$

7. $\dfrac{d}{dx}[x] = 1$

8. $\dfrac{d}{dx}[\ln u] = \dfrac{u'}{u}$

9. $\dfrac{d}{dx}[e^u] = e^u u'$

10. $\dfrac{d}{dx}[\sin u] = (\cos u)u'$

11. $\dfrac{d}{dx}[\cos u] = -(\sin u)u'$

12. $\dfrac{d}{dx}[\tan u] = (\sec^2 u)u'$

13. $\dfrac{d}{dx}[\cot u] = -(\csc^2 u)u'$

14. $\dfrac{d}{dx}[\sec u] = (\sec u \tan u)u'$

15. $\dfrac{d}{dx}[\csc u] = -(\csc u \cot u)u'$

Integration Formulas

1. $\displaystyle\int kf(u)\,du = k\int f(u)\,du$

2. $\displaystyle\int [f(u) \pm g(u)]\,du = \int f(u)\,du \pm \int g(u)\,du$

3. $\displaystyle\int du = u + C$

4. $\displaystyle\int k\,du = ku + C$

5. $\displaystyle\int u^n\,du = \dfrac{u^{n+1}}{n+1} + C,\quad n \neq -1$

6. $\displaystyle\int \dfrac{1}{u}\,du = \ln|u| + C$

7. $\displaystyle\int e^u\,du = e^u + C$

8. $\displaystyle\int \sin u\,du = -\cos u + C$

9. $\displaystyle\int \cos u\,du = \sin u + C$

10. $\displaystyle\int \tan u\,du = -\ln|\cos u| + C$

11. $\displaystyle\int \cot u\,du = \ln|\sin u| + C$

12. $\displaystyle\int \sec u\,du = \ln|\sec u + \tan u| + C$

13. $\displaystyle\int \csc u\,du = \ln|\csc u - \cot u| + C$

14. $\displaystyle\int \sec^2 u\,du = \tan u + C$

15. $\displaystyle\int \csc^2 u\,du = -\cot u + C$

16. $\displaystyle\int \sec u \tan u\,du = \sec u + C$

17. $\displaystyle\int \csc u \cot u\,du = -\csc u + C$

B Additional Topics in Differentiation

B.1 Implicit Differentiation

- Find derivatives explicitly.
- Find derivatives implicitly.

Explicit and Implicit Functions

In this text, most functions involving two variables have been expressed in the **explicit form** $y = f(x)$. That is, one of the two variables has been explicitly given in terms of the other. For example, in the equation

$$y = 3x - 5 \qquad \text{Explicit form}$$

the variable y is explicitly written as a function of x. Some functions, however, are not given explicitly and are only implied by a given equation, as shown in Example 1.

Example 1 Finding a Derivative Explicitly

Find dy/dx for the equation

$$xy = 1.$$

SOLUTION In this equation, y is **implicitly** defined as a function of x. One way to find dy/dx is first to solve the equation for y, then differentiate as usual.

$$xy = 1 \qquad \text{Write original equation.}$$
$$y = \frac{1}{x} \qquad \text{Solve for } y.$$
$$= x^{-1} \qquad \text{Rewrite.}$$
$$\frac{dy}{dx} = -x^{-2} \qquad \text{Differentiate with respect to } x.$$
$$= -\frac{1}{x^2} \qquad \text{Simplify.}$$

✓ CHECKPOINT 1

Find dy/dx for the equation $x^2 y = 1$. ■

The procedure shown in Example 1 works well whenever you can easily write the given function explicitly. You cannot, however, use this procedure when you are unable to solve for y as a function of x. For instance, how would you find dy/dx for the equation

$$x^2 - 2y^3 + 4y = 2$$

where it is very difficult to express y as a function of x explicitly? To do this, you can use a procedure called **implicit differentiation**.

Implicit Differentiation

To understand how to find dy/dx implicitly, you must realize that the differentiation is taking place *with respect to x*. This means that when you differentiate terms involving x alone, you can differentiate as usual. *But* when you differentiate terms involving y, you must apply the Chain Rule because you are assuming that y is defined implicitly as a differentiable function of x. Study the next example carefully. Note in particular how the Chain Rule is used to introduce the dy/dx factors in Examples 2(b) and 2(d).

Example 2 Applying the Chain Rule

Differentiate each expression with respect to x.

a. $3x^2$ **b.** $2y^3$ **c.** $x + 3y$ **d.** xy^2

SOLUTION

a. The only variable in this expression is x. So, to differentiate with respect to x, you can use the Simple Power Rule and the Constant Multiple Rule to obtain

$$\frac{d}{dx}[3x^2] = 6x.$$

b. This case is different. The variable in the expression is y, and yet you are asked to differentiate with respect to x. To do this, assume that y is a differentiable function of x and use the Chain Rule.

$$\frac{d}{dx}[2y^3] = \overbrace{2}^{c}\ \overbrace{(3)}^{n}\ \overbrace{y^2}^{u^{n-1}}\ \overbrace{\frac{dy}{dx}}^{u'} \quad \text{Chain Rule}$$

$$= 6y^2 \frac{dy}{dx}$$

c. This expression involves both x and y. By the Sum Rule and the Constant Multiple Rule, you can write

$$\frac{d}{dx}[x + 3y] = 1 + 3\frac{dy}{dx}.$$

d. By the Product Rule and the Chain Rule, you can write

$$\frac{d}{dx}[xy^2] = x\frac{d}{dx}[y^2] + y^2\frac{d}{dx}[x] \quad \text{Product Rule}$$

$$= x\left(2y\frac{dy}{dx}\right) + y^2(1) \quad \text{Chain Rule}$$

$$= 2xy\frac{dy}{dx} + y^2.$$

✓ **CHECKPOINT 2**

Differentiate each expression with respect to x.

a. $4x^3$ **b.** $3y^2$ **c.** $x + 5y$ **d.** xy^3 ∎

Implicit Differentiation

Consider an equation involving x and y in which y is a differentiable function of x. You can use the steps below to find dy/dx.

1. Differentiate both sides of the equation *with respect to x*.
2. Write the result so that all terms involving dy/dx are on the left side of the equation and all other terms are on the right side of the equation.
3. Factor dy/dx out of the terms on the left side of the equation.
4. Solve for dy/dx by dividing both sides of the equation by the left-hand factor that does not contain dy/dx.

In Example 3, note that implicit differentiation can produce an expression for dy/dx that contains both x and y.

Example 3 Finding the Slope of a Graph Implicitly

Find the slope of the tangent line to the ellipse given by $x^2 + 4y^2 = 4$ at the point $(\sqrt{2}, -1/\sqrt{2})$, as shown in Figure B.1.

SOLUTION

$$x^2 + 4y^2 = 4 \quad \text{Write original equation.}$$

$$\frac{d}{dx}[x^2 + 4y^2] = \frac{d}{dx}[4] \quad \text{Differentiate with respect to } x.$$

$$2x + 8y\left(\frac{dy}{dx}\right) = 0 \quad \text{Implicit differentiation}$$

$$8y\left(\frac{dy}{dx}\right) = -2x \quad \text{Subtract } 2x \text{ from each side.}$$

$$\frac{dy}{dx} = \frac{-2x}{8y} \quad \text{Divide each side by } 8y.$$

$$\frac{dy}{dx} = -\frac{x}{4y} \quad \text{Simplify.}$$

To find the slope at the given point, substitute $x = \sqrt{2}$ and $y = -1/\sqrt{2}$ into the derivative, as shown below.

$$-\frac{\sqrt{2}}{4(-1/\sqrt{2})} = \frac{1}{2}$$

STUDY TIP

To see the benefit of implicit differentiation, try reworking Example 3 using the explicit function

$$y = -\frac{1}{2}\sqrt{4 - x^2}.$$

The graph of this function is the lower half of the ellipse.

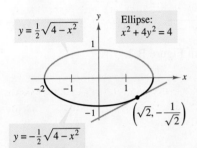

FIGURE B.1 Slope of tangent line is $\frac{1}{2}$.

✓ CHECKPOINT 3

Find the slope of the tangent line to the circle $x^2 + y^2 = 25$ at the point $(3, -4)$.

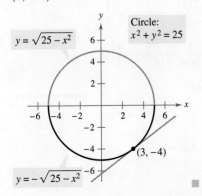

Example 4 Using Implicit Differentiation

Find dy/dx for the equation $y^3 + y^2 - 5y - x^2 = -4$. What are the slopes of the graph at the points $(1, -3)$, $(2, 0)$, and $(1, 1)$?

SOLUTION

$$y^3 + y^2 - 5y - x^2 = -4 \qquad \text{Write original equation.}$$

$$\frac{d}{dx}[y^3 + y^2 - 5y - x^2] = \frac{d}{dx}[-4] \qquad \text{Differentiate with respect to } x.$$

$$3y^2\frac{dy}{dx} + 2y\frac{dy}{dx} - 5\frac{dy}{dx} - 2x = 0 \qquad \text{Implicit differentiation}$$

$$3y^2\frac{dy}{dx} + 2y\frac{dy}{dx} - 5\frac{dy}{dx} = 2x \qquad \text{Collect } dy/dx \text{ terms.}$$

$$\frac{dy}{dx}(3y^2 + 2y - 5) = 2x \qquad \text{Factor.}$$

$$\frac{dy}{dx} = \frac{2x}{3y^2 + 2y - 5}$$

The graph of the original equation is shown in Figure B.2.

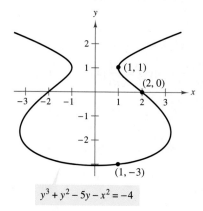

$y^3 + y^2 - 5y - x^2 = -4$

FIGURE B.2

The slopes of the graph at the points $(1, -3)$, $(2, 0)$, and $(1, 1)$ are shown below.

Point on Graph	Slope of Graph
$(1, -3)$	$\dfrac{2(1)}{3(-3)^2 + 2(-3) - 5} = \dfrac{1}{8}$
$(2, 0)$	$\dfrac{2(2)}{3(0)^2 + 2(0) - 5} = -\dfrac{4}{5}$
$(1, 1)$	Undefined

✓ CHECKPOINT 4

Find dy/dx for the equation $y^2 + x^2 - 2y - 4x = 4$. ■

Example 5 Finding the Slope of a Graph Implicitly

Find the slope of the graph of $2x^2 - y^2 = 1$ at the point $(1, 1)$.

SOLUTION Begin by finding dy/dx implicitly.

$$2x^2 - y^2 = 1 \qquad \text{Write original equation.}$$

$$4x - 2y\left(\frac{dy}{dx}\right) = 0 \qquad \text{Differentiate with respect to } x.$$

$$-2y\left(\frac{dy}{dx}\right) = -4x \qquad \text{Subtract } 4x \text{ from each side.}$$

$$\frac{dy}{dx} = \frac{2x}{y} \qquad \text{Divide each side by } -2y.$$

At the point $(1, 1)$, the slope of the graph is

$$\frac{2(1)}{1} = 2$$

as shown in Figure B.3. The graph is called a **hyperbola**.

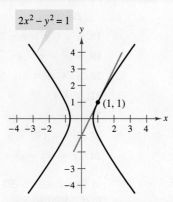

FIGURE B.3 Hyperbola

✓ CHECKPOINT 5

Find the slope of the graph of $x^2 - 9y^2 = 16$ at the point $(5, 1)$. ∎

CONCEPT CHECK

1. Complete the following: The equation $x + y = 1$ is written in _____ form and the equation $y = 1 - x$ is written in _____ form.
2. Complete the following: When you are asked to find dy/dt, you are being asked to find the derivative of _____ with respect to _____.
3. Describe the difference between the explicit form of a function and an implicit equation. Give an example of each.
4. In your own words, state the guidelines for implicit differentiation.

APPENDIX B Additional Topics in Differentiation

Skills Review B.1

The following warm-up exercises involve skills that were covered in earlier sections. You will use these skills in the exercise set for this section. For additional help, review Section 0.3.

In Exercises 1–6, solve the equation for y.

1. $x - \dfrac{y}{x} = 2$
2. $\dfrac{4}{x-3} = \dfrac{1}{y}$
3. $xy - x + 6y = 6$
4. $12 + 3y = 4x^2 + x^2 y$
5. $x^2 + y^2 = 5$
6. $x = \pm\sqrt{6 - y^2}$

In Exercises 7–10, evaluate the expression at the given point.

7. $\dfrac{3x^2 - 4}{3y^2}$, $(2, 1)$
8. $\dfrac{x^2 - 2}{1 - y}$, $(0, -3)$
9. $\dfrac{5x}{3y^2 - 12y + 5}$, $(-1, 2)$
10. $\dfrac{1}{y^2 - 2xy + x^2}$, $(4, 3)$

Exercises B.1

See www.CalcChat.com for worked-out solutions to odd-numbered exercises.

In Exercises 1–12, find dy/dx.

1. $xy = 4$
2. $3x^2 - y = 8x$
3. $y^2 = 1 - x^2$, $0 \le x \le 1$
4. $4x^2 y - \dfrac{3}{y} = 0$
5. $x^2 y^2 - 2x = 3$
6. $xy^2 + 4xy = 10$
7. $4y^2 - xy = 2$
8. $2xy^3 - x^2 y = 2$
9. $\dfrac{2y - x}{y^2 - 3} = 5$
10. $\dfrac{xy - y^2}{y - x} = 1$
11. $\dfrac{x + y}{2x - y} = 1$
12. $\dfrac{2x + y}{x - 5y} = 1$

In Exercises 13–24, find dy/dx by implicit differentiation and evaluate the derivative at the given point.

Equation	Point
13. $x^2 + y^2 = 16$	$(0, 4)$
14. $x^2 - y^2 = 25$	$(5, 0)$
15. $y + xy = 4$	$(-5, -1)$
16. $x^3 - y^2 = 0$	$(1, 1)$
17. $x^3 - xy + y^2 = 4$	$(0, -2)$
18. $x^2 y + y^2 x = -2$	$(2, -1)$
19. $x^3 y^3 - y = x$	$(0, 0)$
20. $x^3 + y^3 = 2xy$	$(1, 1)$
21. $x^{1/2} + y^{1/2} = 9$	$(16, 25)$
22. $\sqrt{xy} = x - 2y$	$(4, 1)$
23. $x^{2/3} + y^{2/3} = 5$	$(8, 1)$
24. $(x + y)^3 = x^3 + y^3$	$(-1, 1)$

In Exercises 25–30, find the slope of the graph at the given point.

25. $3x^2 - 2y + 5 = 0$

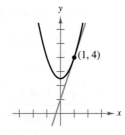

26. $4x^2 + 2y - 1 = 0$

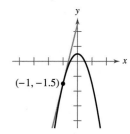

27. $x^2 + y^2 = 4$

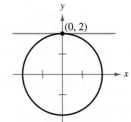

28. $4x^2 + y^2 = 4$

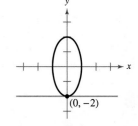

29. $4x^2 + 9y^2 = 36$

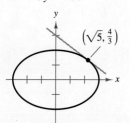

30. $x^2 - y^3 = 0$

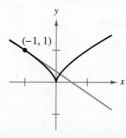

In Exercises 31–34, find dy/dx implicitly and explicitly (the explicit functions are shown on the graph) and show that the results are equivalent. Use the graph to estimate the slope of the tangent line at the labeled point. Then verify your result analytically by evaluating dy/dx at the point.

31. $x^2 + y^2 = 25$

32. $9x^2 + 16y^2 = 144$

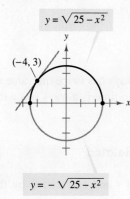

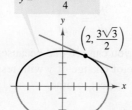

33. $x - y^2 - 1 = 0$

34. $4y^2 - x^2 = 7$

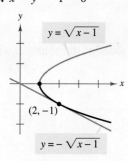

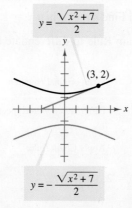

(T) In Exercises 35–40, find equations of the tangent lines to the graph at the given points. Use a graphing utility to graph the equation and the tangent lines in the same viewing window.

Equation	Points
35. $x^2 + y^2 = 100$	$(8, 6)$ and $(-6, 8)$
36. $x^2 + y^2 = 9$	$(0, 3)$ and $(2, \sqrt{5})$
37. $y^2 = 5x^3$	$(1, \sqrt{5})$ and $(1, -\sqrt{5})$
38. $4xy + x^2 = 5$	$(1, 1)$ and $(5, -1)$
39. $x^3 + y^3 = 8$	$(0, 2)$ and $(2, 0)$
40. $y^2 = \dfrac{x^3}{4 - x}$	$(2, 2)$ and $(2, -2)$

In Exercises 41–44, find the rate of change of x with respect to y.

41. $y = \dfrac{2}{0.00001x^3 + 0.1x}, \quad x > 0$

42. $y = \dfrac{4}{0.000001x^2 + 0.05x + 1}, \quad x \geq 0$

43. $y = \sqrt{\dfrac{200 - x}{2x}}, \quad 0 < x \leq 200$

44. $y = \sqrt{\dfrac{500 - x}{2x}}, \quad 0 < x \leq 500$

45. Biological Equipment Production Let x represent the units of labor and y the endowment funds invested in the production process. When 135,540 units are produced, the relationship between labor and funds can be modeled by

$$100x^{0.75}y^{0.25} = 135{,}540.$$

Find the rate of change of y with respect to x when $x = 1500$ and $y = 1000$.

46. Health: U.S. HIV/AIDS Epidemic The numbers y of cases (in thousands) of HIV/AIDS reported in the years 2001 through 2005 can be modeled by

$$y^2 - 1141.6 = 24.9099t^3 - 183.045t^2 + 452.79t$$

where t represents the year, with $t = 1$ corresponding to 2001. (*Source: U.S. Centers for Disease Control and Prevention*)

(T) (a) Use a graphing utility to graph the model and describe the results.

(b) Use the graph to estimate the year during which the number of reported cases was increasing at the greatest rate.

(c) Complete the table to estimate the year during which the number of reported cases was increasing at the greatest rate. Compare this estimate with your answer in part (b).

t	1	2	3	4	5
y					
y'					

B.2 Related Rates

- Examine related variables.
- Solve related-rate problems.

Related Variables

In this section, you will study problems involving variables that are changing with respect to time. If two or more such variables are related to each other, then their rates of change with respect to time are also related.

For instance, suppose that x and y are related by the equation $y = 2x$. If both variables are changing with respect to time, then their rates of change will also be related.

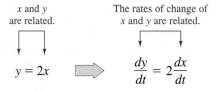

In this simple example, you can see that because y always has twice the value of x, it follows that the rate of change of y with respect to time is always twice the rate of change of x with respect to time.

Example 1 Examining Two Rates That Are Related

The variables x and y are differentiable functions of t and are related by the equation

$$y = x^2 + 3.$$

When $x = 1$, $dx/dt = 2$. Find dy/dt when $x = 1$.

SOLUTION Use the Chain Rule to differentiate both sides of the equation with respect to t.

$y = x^2 + 3$ \qquad Write original equation.

$\dfrac{d}{dt}[y] = \dfrac{d}{dt}[x^2 + 3]$ \qquad Differentiate with respect to t.

$\dfrac{dy}{dt} = 2x\dfrac{dx}{dt}$ \qquad Apply Chain Rule.

When $x = 1$ and $dx/dt = 2$, you have

$\dfrac{dy}{dt} = 2(1)(2)$

$= 4.$

✓ CHECKPOINT 1

When $x = 1$, $dx/dt = 3$. Find dy/dt when $x = 1$ if $y = x^3 + 2$.

Solving Related-Rate Problems

In Example 1, you were *given* the mathematical model.

Given equation: $y = x^2 + 3$

Given rate: $\dfrac{dx}{dt} = 2$ when $x = 1$

Find: $\dfrac{dy}{dt}$ when $x = 1$

In the next example, you are asked to *create* a similar mathematical model.

Example 2 Changing Area

A pebble is dropped into a calm pool of water, causing ripples in the form of concentric circles, as shown in the photo. The radius r of the outer ripple is increasing at a constant rate of 1 foot per second. When the radius is 4 feet, at what rate is the total area A of the disturbed water changing?

SOLUTION The variables r and A are related by the equation for the area of a circle, $A = \pi r^2$. To solve this problem, use the fact that the rate of change of the radius is given by dr/dt.

Equation: $A = \pi r^2$

Given rate: $\dfrac{dr}{dt} = 1$ when $r = 4$

Find: $\dfrac{dA}{dt}$ when $r = 4$

Using this model, you can proceed as in Example 1.

$A = \pi r^2$ Write original equation.

$\dfrac{d}{dt}[A] = \dfrac{d}{dt}[\pi r^2]$ Differentiate with respect to t.

$\dfrac{dA}{dt} = 2\pi r \dfrac{dr}{dt}$ Apply Chain Rule.

When $r = 4$ and $dr/dt = 1$, you have

$\dfrac{dA}{dt} = 2\pi(4)(1) = 8\pi$ Substitute 4 for r and 1 for dr/dt.

When the radius is 4 feet, the area is changing at a rate of 8π square feet per second.

Total area increases as the outer radius increases.

✓ **CHECKPOINT 2**

If the radius r of the outer ripple in Example 2 is increasing at a rate of 2 feet per second, at what rate is the total area changing when the radius is 3 feet? ■

STUDY TIP

In Example 2, note that the radius changes at a *constant* rate ($dr/dt = 1$ for all t), but the area changes at a *nonconstant* rate.

When $r = 1$ ft	When $r = 2$ ft	When $r = 3$ ft	When $r = 4$ ft
$\dfrac{dA}{dt} = 2\pi$ ft²/sec	$\dfrac{dA}{dt} = 4\pi$ ft²/sec	$\dfrac{dA}{dt} = 6\pi$ ft²/sec	$\dfrac{dA}{dt} = 8\pi$ ft²/sec

The solution shown in Example 2 illustrates the steps for solving a related-rate problem.

Guidelines for Solving a Related-Rate Problem

1. Identify all *given* quantities and all quantities *to be determined*. If possible, make a sketch and label the quantities.

2. Write an equation that relates all variables whose rates of change are either given or to be determined.

3. Use the Chain Rule to differentiate both sides of the equation *with respect to time*.

4. Substitute into the resulting equation all known values of the variables and their rates of change. Then solve for the required rate of change.

STUDY TIP

Be sure you notice the order of Steps 3 and 4 in the guidelines. Do not substitute the known values for the variables until after you have differentiated.

In Step 2 of the guidelines, note that you must write an equation that relates the given variables. To help you with this step, reference tables that summarize many common formulas are included in the appendices. For instance, the volume of a sphere of radius r is given by the formula

$$V = \frac{4}{3}\pi r^3$$

as listed in Appendix D.

The table below shows the mathematical models for some common rates of change that can be used in the first step of the solution of a related-rate problem.

Verbal statement	Mathematical model
The velocity of a car after traveling for 1 hour is 50 miles per hour.	x = distance traveled $\frac{dx}{dt} = 50$ when $t = 1$
Water is being pumped into a swimming pool at the rate of 10 cubic feet per minute.	V = volume of water in pool $\frac{dV}{dt} = 10$ ft³/min
A population of bacteria is increasing at the rate of 2000 per hour.	x = number in population $\frac{dx}{dt} = 2000$ bacteria per hour
Revenue is increasing at the rate of $4000 per month.	R = revenue $\frac{dR}{dt} = 4000$ dollars per month

APPENDIX B.2 Related Rates A13

Example 3 Changing Volume

Air is being pumped into a spherical balloon at the rate of 4.5 cubic inches per minute. See Figure B.4. Find the rate of change of the radius when the radius is 2 inches.

SOLUTION Let V represent the volume of the balloon and let r represent the radius. Because the volume is increasing at the rate of 4.5 cubic inches per minute, you know that $dV/dt = 4.5$. An equation that relates V and r is $V = \frac{4}{3}\pi r^3$. So, the problem can be represented by the model shown below.

Equation: $V = \dfrac{4}{3}\pi r^3$

Given rate: $\dfrac{dV}{dt} = 4.5$

Find: $\dfrac{dr}{dt}$ when $r = 2$

By differentiating the equation, you obtain

$V = \dfrac{4}{3}\pi r^3$ Write original equation.

$\dfrac{d}{dt}[V] = \dfrac{d}{dt}\left[\dfrac{4}{3}\pi r^3\right]$ Differentiate with respect to t.

$\dfrac{dV}{dt} = \dfrac{4}{3}\pi(3r^2)\dfrac{dr}{dt}$ Apply Chain Rule.

$\dfrac{1}{4\pi r^2}\dfrac{dV}{dt} = \dfrac{dr}{dt}.$ Solve for dr/dt.

When $r = 2$ and $dV/dt = 4.5$, the rate of change of the radius is

$$\dfrac{dr}{dt} = \dfrac{1}{4\pi(2^2)}(4.5) \approx 0.09 \text{ inch per minute}$$

In Example 3, note that the volume is increasing at a *constant rate* but the radius is increasing at a *variable* rate. In this particular example, the radius is increasing more and more slowly as t increases.

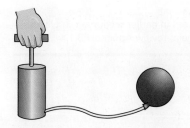

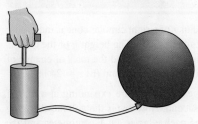

FIGURE B.4 Expanding Balloon

✓ CHECKPOINT 3

If the radius of a spherical balloon increases at a rate of 1.5 inches per minute, find the rate at which the surface area changes when the radius is 6 inches. (Formula for surface area of a sphere: $S = 4\pi r^2$) ∎

CONCEPT CHECK

1. Complete the following. Two variables x and y are changing with respect to _____. If x and y are related to each other, then their rates of change with respect to time are also _____.
2. The volume V of an object is a differentiable function of time t. Describe what dV/dt represents.
3. The area A of an object is a differentiable function of time t. Describe what dA/dt represents.
4. In your own words, state the guidelines for solving related-rate problems.

APPENDIX B Additional Topics in Differentiation

Skills Review B.2

The following warm-up exercises involve skills that were covered in earlier sections. You will use these skills in the exercise set for this section. For additional help, review Appendix B.1.

In Exercises 1–6, write a formula for the given quantity.

1. Area of a circle
2. Volume of a sphere
3. Surface area of a cube
4. Volume of a cube
5. Volume of a cone
6. Area of a triangle

In Exercises 7–10, find dy/dx by implicit differentiation.

7. $x^2 + y^2 = 9$
8. $3xy - x^2 = 6$
9. $x^2 + 2y + xy = 12$
10. $x + xy^2 - y^2 = xy$

Exercises B.2

See www.CalcChat.com for worked-out solutions to odd-numbered exercises.

In Exercises 1–4, use the given values to find dy/dt and dx/dt.

Equation	Find	Given
1. $y = \sqrt{x}$	(a) $\dfrac{dy}{dt}$	$x = 4, \dfrac{dx}{dt} = 3$
	(b) $\dfrac{dx}{dt}$	$x = 25, \dfrac{dy}{dt} = 2$
2. $y = 2(x^2 - 3x)$	(a) $\dfrac{dy}{dt}$	$x = 3, \dfrac{dx}{dt} = 2$
	(b) $\dfrac{dx}{dt}$	$x = 1, \dfrac{dy}{dt} = 5$
3. $xy = 4$	(a) $\dfrac{dy}{dt}$	$x = 8, \dfrac{dx}{dt} = 10$
	(b) $\dfrac{dx}{dt}$	$x = 1, \dfrac{dy}{dt} = -6$
4. $x^2 + y^2 = 25$	(a) $\dfrac{dy}{dt}$	$x = 3, y = 4, \dfrac{dx}{dt} = 8$
	(b) $\dfrac{dx}{dt}$	$x = 4, y = 3, \dfrac{dy}{dt} = -2$

5. **Area** Let A be the area of a circle of radius r that is changing with respect to time. If dr/dt is constant, is dA/dt constant? Explain your reasoning.

6. **Volume** Let V be the volume of a sphere of radius r that is changing with respect to time. If dr/dt is constant, is dV/dt constant? Explain your reasoning.

7. **Volume** A spherical balloon is inflated with gas at a rate of 10 cubic feet per minute. How fast is the radius of the balloon changing at the instant the radius is (a) 1 foot and (b) 2 feet?

8. **Volume** The radius r of a right circular cone is increasing at a rate of 2 inches per minute. The height h of the cone is related to the radius by $h = 3r$. Find the rates of change of the volume when (a) $r = 6$ inches and (b) $r = 24$ inches.

9. **Volume** All edges of a cube are expanding at a rate of 3 centimeters per second. How fast is the volume changing when the length of each edge is (a) 1 centimeter and (b) 10 centimeters?

10. **Surface Area** All edges of a cube are expanding at a rate of 3 centimeters per second. How fast is the surface area changing when the length of each edge is (a) 1 centimeter and (b) 10 centimeters?

11. **Moving Ladder** A 25-foot ladder is leaning against a house (see figure). The base of the ladder is pulled away from the house at a rate of 2 feet per second. How fast is the top of the ladder moving down the wall when the base is (a) 7 feet, (b) 15 feet, and (c) 24 feet from the house?

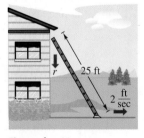
Figure for 11

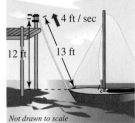
Figure for 12

12. **Boating** A boat is pulled by a winch on a dock, and the winch is 12 feet above the deck of the boat (see figure). The winch pulls the rope at a rate of 4 feet per second. Find the speed of the boat when 13 feet of rope is out. What happens to the speed of the boat as it gets closer to the dock?

13. Environment An accident at an oil drilling platform is causing a circular oil slick. The slick is 0.08 foot thick, and when the radius of the slick is 150 feet, the radius is increasing at the rate of 0.5 foot per minute. At what rate (in cubic feet per minute) is oil flowing from the site of the accident?

14. Baseball A (square) baseball diamond has sides that are 90 feet long (see figure). A player 26 feet from third base is running at a speed of 30 feet per second. At what rate is the player's distance from home plate changing?

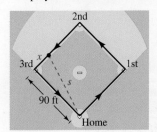

Figure for 14

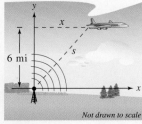

Figure for 15

15. Air Traffic Control An airplane flying at an altitude of 6 miles passes directly over a radar antenna (see figure). When the airplane is 10 miles away ($s = 10$), the radar detects that the distance s is changing at a rate of 240 miles per hour. What is the speed of the airplane?

16. Air Traffic Control An air traffic controller spots two airplanes at the same altitude converging to a point as they fly at right angles to each other. One airplane is 150 miles from the point and has a speed of 450 miles per hour. The other is 200 miles from the point and has a speed of 600 miles per hour.

(a) At what rate is the distance between the planes changing?

(b) How much time does the controller have to get one of the airplanes on a different flight path?

17. Roadway Design Cars on a certain roadway travel on a circular arc of radius r. In order not to rely on friction alone to overcome the centrifugal force, the road is banked at an angle of magnitude θ from the horizontal (see figure). The banking angle must satisfy the equation $rg \tan \theta = v^2$, where v is the velocity of the cars and $g = 32$ feet per second per second is the acceleration due to gravity. Find the relationship between the related rates dv/dt and $d\theta/dt$.

18. Security Camera A security camera is centered 50 feet above a 100-foot hallway (see figure). It is easiest to design the camera with a constant angular rate of rotation, but this results in a variable rate at which the images of the surveillance area are recorded. So, it is desirable to design a system with a variable rate of rotation and a constant rate of movement of the scanning beam along the hallway. Find a model for the variable rate of rotation if $|dx/dt| = 2$ feet per second.

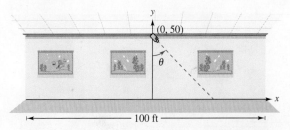

19. Depth A swimming pool is 12 meters long, 6 meters wide, 1 meter deep at the shallow end, and 3 meters deep at the deep end (see figure). Water is being pumped into the pool at $\frac{1}{4}$ cubic meter per minute, and there is 1 meter of water at the deep end.

(a) What percent of the pool is filled?

(b) At what rate is the water level rising?

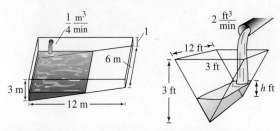

Figure for 19

Figure for 20

20. Depth A trough is 12 feet long and 3 feet across the top (see figure). Its ends are isosceles triangles with altitudes of 3 feet. If water is being pumped into the trough at 2 cubic feet per minute, how fast is the water level rising when the depth of the water h is 1 foot?

21. Adiabatic Expansion When a certain polyatomic gas undergoes adiabatic expansion, its pressure p and volume V satisfy the equation $pV^{1.3} = k$, where k is a constant. Find the relationship between the related rates dp/dt and dV/dt.

22. Electricity The combined electrical resistance R of R_1 and R_2, connected in parallel, is given by

$$\frac{1}{R} = \frac{1}{R_1} + \frac{1}{R_2}$$

where R, R_1, and R_2 are measured in ohms. R_1 and R_2 are increasing at rates of 1 and 1.5 ohms per second, respectively. At what rate is R changing when $R_1 = 50$ ohms and $R_2 = 75$ ohms?

Answers to Selected Exercises

CHAPTER 0

SECTION 0.1 (page 7)

1. Rational   3. Irrational   5. Rational
7. Rational   9. Irrational
11. (a) Yes (b) No (c) Yes
13. (a) Yes (b) No (c) No
15. $x \geq 12$ 17. $x < -\frac{1}{2}$

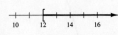

19. $x > 1$ 21. $-\frac{1}{2} < x < \frac{7}{2}$

23. $-\frac{3}{4} < x < -\frac{1}{4}$ 25. $x > 6$

27. $-\frac{3}{2} < x < 2$ 29. $2.39 \leq g \leq 3.25$

31. $w \geq 18$ 33.

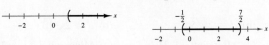

35. No, $18.3°C = 64.94°F$ which is below $68°F$.
37. (a) False (b) True (c) True (d) False

SECTION 0.2 (page 12)

1. (a) -51 (b) 51 (c) 51
3. (a) -14.99 (b) 14.99 (c) 14.99
5. (a) $-\frac{128}{75}$ (b) $\frac{128}{75}$ (c) $\frac{128}{75}$ 7. $|x| \leq 2$
9. $|x| > 2$ 11. $|x - 5| \leq 3$ 13. $|x - 2| > 2$
15. $|x - 5| < 3$ 17. $|y - a| \leq 2$
19. $-4 < x < 4$ 21. $x < -6$ or $x > 6$

23. $3 < x < 7$ 25. $x \leq -7$ or $x \geq 13$

27. $x < 6$ or $x > 14$ 29. $4 < x < 5$

31. $a - b \leq x \leq a + b$ 33. $\dfrac{a - 8b}{3} < x < \dfrac{a + 8b}{3}$

35. 16   37. 1.25   39. $\frac{1}{8}$ 41. $|M - 1083.4| < 0.2$
43. $65.8 \leq h \leq 71.2$ 45. $750 \leq n \leq 950$
47. (a) $0.45 - 0.03 \leq X \leq 0.45 + 0.03$; $0.42 \leq X \leq 0.48$
   (b) $33{,}600 \leq X \leq 38{,}400$; According to the poll, Candidate X can expect at least 33,600 votes and at most 38,400 votes.

SECTION 0.3 (page 18)

1. -54 3. $\frac{1}{2}$ 5. 4   7. 44   9. 5   11. 9
13. $\frac{1}{2}$ 15. $\frac{1}{4}$ 17. 908.3483   19. -5.3601
21. $\dfrac{3}{4y^{14}}$ 23. $10x^4$ 25. $7x^5$
27. $\frac{5}{2}(x + y)^5, x \neq -y$ 29. $3x, x > 0$
31. $2\sqrt{2}$ 33. $3x\sqrt[3]{2x^2}$ 35. $\dfrac{2x^3z}{y}\sqrt[3]{\dfrac{18z^2}{y}}$
37. $3x(x + 2)(x - 2)$ 39. $(2x^3 + 1)/x^{1/2}$
41. $3(x + 1)^{1/2}(x + 2)(x - 1)$ 43. $\dfrac{2(x - 1)^2}{(x + 1)^2}$
45. $x \geq 4$ 47. $(-\infty, \infty)$ 49. $(-\infty, 4) \cup (4, \infty)$
51. $x \neq 1, x \geq -2$ 53. 1.4 m^2 55. 2 m^2
57. About 2,251,018   59. Answers will vary.

SECTION 0.4 (page 24)

1. $\frac{1}{6}, 1$ 3. $\frac{3}{2}$ 5. $-2 \pm \sqrt{3}$ 7. $\dfrac{-3 \pm \sqrt{41}}{4}$
9. $(x - 2)^2$ 11. $(2x + 1)^2$ 13. $(3x - 1)(x - 1)$
15. $(3x - 2)(x - 1)$ 17. $(x - 2y)^2$
19. $(3 + y)(3 - y)(9 + y^2)$ 21. $(x - 2)(x^2 + 2x + 4)$
23. $(y + 4)(y^2 - 4y + 16)$ 25. $(x - y)(x^2 + xy + y^2)$
27. $(x - 4)(x - 1)(x + 1)$ 29. $(2x - 3)(x^2 + 2)$
31. $(x - 2)(2x^2 - 1)$ 33. $(x + 4)(x - 4)(x^2 + 1)$
35. 0, 5   37. ± 3 39. $\pm\sqrt{3}$ 41. 0, 6
43. $-2, 1$ 45. $-1, 6$ 47. $-1, -\frac{2}{3}$ 49. -4
51. ± 2 53. $1, \pm 2$ 55. $(-\infty, -2] \cup [2, \infty)$

A18 Answers to Selected Exercises

57. $(-\infty, 3] \cup [4, \infty)$ **59.** $(-\infty, -1] \cup \left[-\frac{1}{5}, \infty\right)$
61. $(x + 1)(x^2 - 4x - 2)$ **63.** $(x + 1)(2x^2 - 3x + 1)$
65. $-2, -1, 4$ **67.** $1, 2, 3$ **69.** $-\frac{2}{3}, -\frac{1}{2}, 3$
71. 4 **73.** $-2, -1, \frac{1}{4}$
75. Base: 2 ft by 2 ft; height: 3 ft
77. 3.4×10^{-5}

SECTION 0.5 (page 32)

1. $\dfrac{x + 3}{x - 2}$ **3.** $\dfrac{5x - 1}{x^2 + 2}$ **5.** $-\dfrac{x}{x^2 - 4}$ **7.** $\dfrac{2}{x - 3}$

9. $\dfrac{(A + C)x^2 + (A + B - 2C)x + (-2A + 2B + C)}{(x - 1)^2(x + 2)}$

11. $\dfrac{(A + B)x^2 - (6B - C)x + 3(A - 2C)}{(x - 6)(x^2 + 3)}$

13. $\dfrac{-2x^2 + x - 4}{x(x^2 + 2)}$ **15.** $-\dfrac{x^2 + 3}{(x + 1)(x - 2)(x - 3)}$

17. $\dfrac{x + 2}{(x + 1)^{3/2}}$ **19.** $-\dfrac{3t}{2\sqrt{1 + t}}$ **21.** $\dfrac{x(x^2 + 2)}{(x^2 + 1)^{3/2}}$

23. $\dfrac{2}{x^2\sqrt{x^2 + 2}}$ **25.** $\dfrac{3x(x + 2)}{(2x + 3)^{3/2}}$ **27.** $\dfrac{\sqrt{10}}{5}$

29. $\dfrac{4x\sqrt{x - 1}}{x - 1}$ **31.** $\dfrac{49\sqrt{x^2 - 9}}{x + 3}$ **33.** $\dfrac{\sqrt{14} + 2}{2}$

35. $\dfrac{x(5 + \sqrt{3})}{11}$ **37.** $\sqrt{6} - \sqrt{5}$ **39.** $\sqrt{x} - \sqrt{x - 2}$

41. $\dfrac{1}{\sqrt{x + 2} + \sqrt{2}}$ **43.** $\dfrac{4 - 3x^2}{x^4(4 - x^2)^{3/2}}$

45.

n	1	2	3	4	5
P	0.6	0.78	0.85	0.89	0.91

n	6	7	8	9	10
P	0.92	0.93	0.94	0.95	0.95

CHAPTER 1

SECTION 1.1 (page 40)

Skills Review (page 40)
1. $3\sqrt{5}$ **2.** $2\sqrt{5}$ **3.** $\frac{1}{2}$ **4.** -2 **5.** $5\sqrt{3}$
6. $-\sqrt{2}$ **7.** $x = -3, x = 9$
8. $y = -8, y = 4$ **9.** $x = 19$ **10.** $y = 1$

1.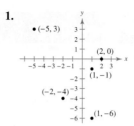

3. (a)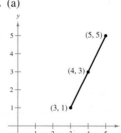

(b) $d = 2\sqrt{5}$
(c) Midpoint: $(4, 3)$

5. (a)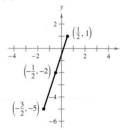

(b) $d = 2\sqrt{10}$
(c) Midpoint: $\left(-\frac{1}{2}, -2\right)$

7. (a)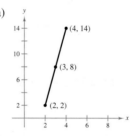

(b) $d = 2\sqrt{37}$ (c) Midpoint: $(3, 8)$

9. (a)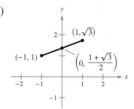

(b) $d = \sqrt{8 - 2\sqrt{3}}$ (c) Midpoint: $\left(0, \dfrac{1 + \sqrt{3}}{2}\right)$

11. (a)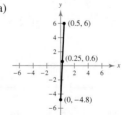

(b) $d = \sqrt{116.89}$ (c) Midpoint: $(0.25, 0.6)$

13. (a) $a = 4, b = 3, c = 5$
(b) $4^2 + 3^2 = 5^2$

15. (a) $a = 10, b = 3, c = \sqrt{109}$
(b) $10^2 + 3^2 = (\sqrt{109})^2$

17. $d_1 = \sqrt{45}, d_2 = \sqrt{20},$
$d_3 = \sqrt{65}$
$d_1^2 + d_2^2 = d_3^2$

19. $d_1 = d_2 = d_3 = d_4 = \sqrt{5}$

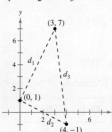
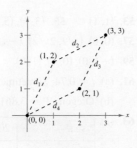

21. $x = 4, -2$ **23.** $y = \pm\sqrt{55}$

25. (a) 16.76 ft (b) 1341.04 ft²

27. Answers will vary. Sample answer:

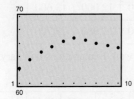

The number of subscribers appears to be increasing from 1996 to 2001 and decreasing from 2001 to 2005.

29. (a) 26.6 million (b) 24.1 million
(c) 21.8 million (d) 19.8 million

31. (a) 22.6 million (b) 27.5 million
(c) 22.9 million (d) 25.7 million

33. (a) $1164.5 million (b) Actual 2003: $1190 million
(c) No, the increase in value from 2001 to 2003 is greater than the increase in value from 2003 to 2005.

35. (a) $3006.5 million (b) Actual 2003: $3268 million
(c) No, the increase in value from 2001 to 2003 is greater than the increase in value from 2003 to 2005.

37. (a)
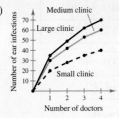

(b) The larger the clinic, the more patients a doctor can treat.

39. (a) $(-1, 2), (1, 1), (2, 3)$
(b)
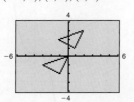

41. $\left(\dfrac{3x_1 + x_2}{4}, \dfrac{3y_1 + y_2}{4}\right), \left(\dfrac{x_1 + x_2}{2}, \dfrac{y_1 + y_2}{2}\right),$
$\left(\dfrac{x_1 + 3x_2}{4}, \dfrac{y_1 + 3y_2}{4}\right)$

43. (a) $\left(\tfrac{7}{4}, -\tfrac{7}{4}\right), \left(\tfrac{5}{2}, -\tfrac{3}{2}\right), \left(\tfrac{13}{4}, -\tfrac{5}{4}\right)$
(b) $\left(-\tfrac{3}{2}, -\tfrac{9}{4}\right), \left(-1, -\tfrac{3}{2}\right), \left(-\tfrac{1}{2}, -\tfrac{3}{4}\right)$

SECTION 1.2 (page 53)

Skills Review (page 53)

1. $y = \tfrac{1}{5}(x + 12)$ **2.** $y = x - 15$
3. $y = \dfrac{1}{x^3 + 2}$
4. $y = \pm\sqrt{x^2 + x - 6} = \pm\sqrt{(x + 3)(x - 2)}$
5. $y = -1 \pm \sqrt{9 - (x - 2)^2}$
6. $y = 5 \pm \sqrt{81 - (x + 6)^2}$ **7.** $x^2 - 4x + 4$
8. $x^2 + 6x + 9$ **9.** $x^2 - 5x + \tfrac{25}{4}$
10. $x^2 + 3x + \tfrac{9}{4}$ **11.** $(x - 2)(x - 1)$
12. $(x + 3)(x + 2)$ **13.** $\left(y - \tfrac{3}{2}\right)^2$ **14.** $\left(y - \tfrac{7}{2}\right)^2$

1. (a) Not a solution point (b) Solution point
(c) Solution point

3. (a) Solution point (b) Not a solution point
(c) Not a solution point

5. e **6.** b **7.** c **8.** f **9.** a **10.** d

11. $(0, -3), \left(\tfrac{3}{2}, 0\right)$ **13.** $(0, -2), (-2, 0), (1, 0)$

15. $(-2, 0), (0, 2), (2, 0)$ **17.** $(-2, 0), (0, 2)$

19. $(0, 0)$

21.
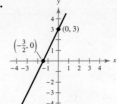

23.
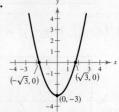

25. 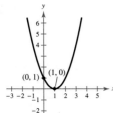 **27.**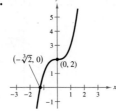

29. 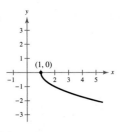 **31.**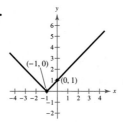

33. 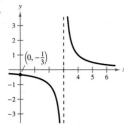 **35.**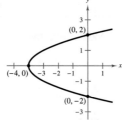

37. $x^2 + y^2 - 16 = 0$
39. $x^2 + y^2 - 4x + 2y - 4 = 0$
41. $x^2 + y^2 + 2x - 4y = 0$ **43.** $x^2 + y^2 - 6x - 8y = 0$
45. $(x - 1)^2 + (y + 3)^2 = 4$

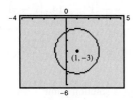

47. $(x - 2)^2 + (y - 1)^2 = 4$

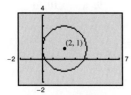

49. $\left(x - \frac{1}{2}\right)^2 + \left(y - \frac{1}{2}\right)^2 = 2$

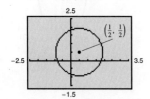

51. $(x + 1.5)^2 + (y - 3)^2 = 1$

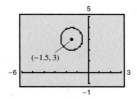

53. $(1, 1)$ **55.** $(3, 4), (5, 0)$
57. $(0, 0), \left(\sqrt{2}, 2\sqrt{2}\right), \left(-\sqrt{2}, -2\sqrt{2}\right)$
59. $(-1, 0), (0, 1), (1, 0)$
61. (a) $t \approx 0.74$; Sometime during 2000.
(b) Oregon: $P = 3,861,000$; Oklahoma: $P = 3,649,000$

63. (a)

Year	2000	2001	2002
Consumption	7.600	8.270	8.753
Model	7.500	8.260	9.260

Year	2003	2004	2005
Consumption	11.000	12.000	13.600
Model	10.500	11.980	13.700

The model fits the data well. Explanations will vary.

(b) 36.1 trillion Btu

65. (a)

Year	2000	2001	2002	2003	2004	2005
Outlay	20.6	22.9	25.9	31.5	34.4	38.3
Model	20.2	23.3	26.7	30.4	34.4	38.6

The model fits the data well. Explanations will vary.

(b) $75.8 billion

67. (a)

Year	2001	2002	2003
Transplants	14,264.0	14,737.1	15,271.3

Year	2004	2005
Transplants	15,866.7	16,523.3

(b)

Year	2001	2002	2003
Actual	14,265	14,780	15,136

Year	2004	2005
Actual	16,004	16,477

The model fits the data well.

(c) 21,746; Answers will vary.

69.

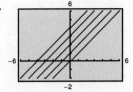

The constant term c is the y-intercept of the graph of the equation.

71.

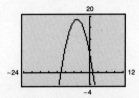

$(-10.5896, 0)$, $(1.0539, 0)$, $(0, 6.25)$

73.

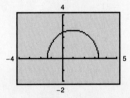

$(-1.3917, 0)$, $(3.3256, 0)$, $(0, 2.3664)$

75.

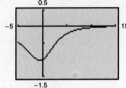

$(13.25, 0)$, $(0, -1)$

SECTION 1.3 (page 65)

Skills Review (page 65)

1. -1 **2.** 1 **3.** $\frac{1}{3}$ **4.** $-\frac{7}{6}$
5. $y = 4x + 7$ **6.** $y = 3x - 7$
7. $y = 3x - 10$ **8.** $y = -x - 7$
9. $y = 7x - 17$ **10.** $y = \frac{2}{3}x + \frac{5}{3}$

1. 1 **3.** 0
5.

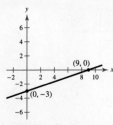

$m = \frac{1}{3}$

7.

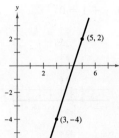

$m = 3$

9.

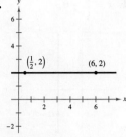

$m = 0$

11.

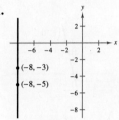

m is undefined.

13.

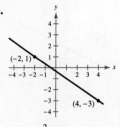

$m = -\frac{2}{3}$

15.

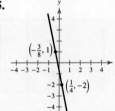

$m = -\frac{24}{5}$

17.

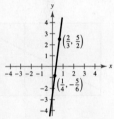

$m = 8$

19. $(0, 1), (1, 1), (3, 1)$ **21.** $(3, -6), (9, -2), (12, 0)$
23. $(0, 10), (2, 4), (3, 1)$ **25.** $(-8, 0), (-8, 2), (-8, 3)$
27. $m = -\frac{1}{5}, (0, 4)$ **29.** $m = -\frac{7}{6}, (0, 5)$
31. $m = 3, (0, -15)$ **33.** m is undefined; no y-intercept.
35. $m = 0, (0, 4)$
37. $y = 2x - 5$ **39.** $3x + y = 0$

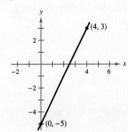

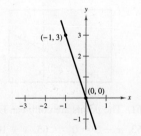

A22 Answers to Selected Exercises

41. $x - 2 = 0$

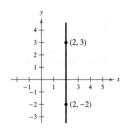

43. $y + 1 = 0$

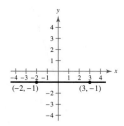

45. $3x - 6y + 7 = 0$

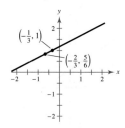

47. $4x - y + 6 = 0$

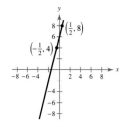

49. $3x - 4y + 12 = 0$

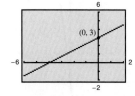

51. $x + 1 = 0$

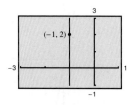

53. $y - 7 = 0$

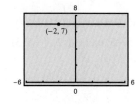

55. $4x + y + 2 = 0$

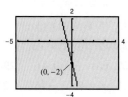

57. $9x - 12y + 8 = 0$

59. The points are not collinear. Explanations will vary.

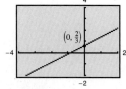

61. The points are collinear. Explanations will vary.

63. $x - 3 = 0$ **65.** $y + 10 = 0$

67. (a) $x + y + 1 = 0$ (b) $x - y + 5 = 0$

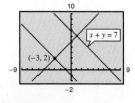

69. (a) $6x + 8y - 3 = 0$ (b) $96x - 72y + 127 = 0$

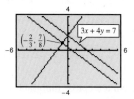

71. (a) $y = 0$ (b) $x + 1 = 0$

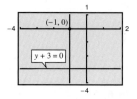

73. (a) $x - 1 = 0$ (b) $y - 1 = 0$

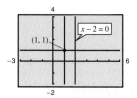

75.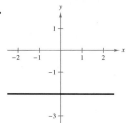

77.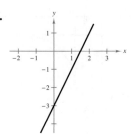

79.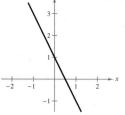

81.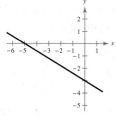

83.

85. Slope $= \frac{3}{32} \approx 0.09375$; So, the slope is steeper than recommended.

87. (a) The slope $m = 60$ tells you that the deer population increases 60 each year. The y-intercept $(0, 1300)$ tells you the deer population was 1300 in 2000.

(b) 1600 (c) 2020

89. (a) $y = 297.4t + 34{,}728$ (b) 38,296.8   (c) 40,081.2

91. (a) $y = 46.2t + 4024$; The slope $m = 46.2$ tells you that the population increases by 46.2 thousand each year.

(b) 4116.4 thousand (4,116,400)

(c) 4208.8 thousand (4,208,800)

(d) Answers will vary. Sample answer:

2002: 4103 thousand (4,103,000)

2004: 4198 thousand (4,198,000)

The estimates were close to the actual populations.

(e) The model could possibly be used to predict the population in 2009 if the population continues to grow at the same linear rate.

93. $F = \frac{9}{5}C + 32$ or $C = \frac{5}{9}F - \frac{160}{9}$

95. (a) $y = -9000t + 45{,}000$

(b)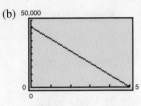

(c) $18,000.00 (d) $t = 1.89$ yr

97. (a) 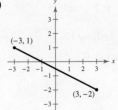   (b) $y = 129.2x + 1343$

(c) $129.2 billion/yr

(d) $2893.4 billion; This prediction seems reasonable. Explanations will vary.

MID-CHAPTER QUIZ   *(page 68)*

1. (a) 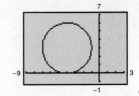   (b) $d = 3\sqrt{5}$

(c) Midpoint: $(0, -0.5)$

2. (a) 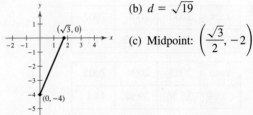   (b) $d = \sqrt{12.3125}$

(c) Midpoint: $\left(\frac{3}{8}, \frac{1}{4}\right)$

3. (a) 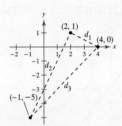   (b) $d = \sqrt{19}$

(c) Midpoint: $\left(\frac{\sqrt{3}}{2}, -2\right)$

4. $d_1 = \sqrt{5}$

$d_2 = \sqrt{45}$

$d_3 = \sqrt{50}$

$d_1^2 + d_2^2 = d_3^2$

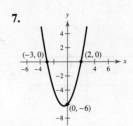

5. 5759.5 thousand

6. 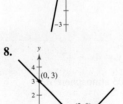   **7.**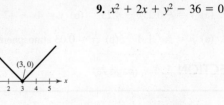

8.

9. $x^2 + 2x + y^2 - 36 = 0$

10. $x^2 - 4x + y^2 + 4y - 17 = 0$

11. $(x + 4)^2 + (y - 3)^2 = 9$

A24 Answers to Selected Exercises

12. $(x - 1)^2 + (y + 0.5)^2 = 4$

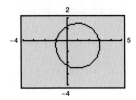

13.

Year	2000	2001	2002
Sales	22.3	26.72	31.14

Year	2003	2004	2005
Sales	35.56	39.98	44.4

14. $y = -1.2x + 0.2$ **15.** $x = -2$

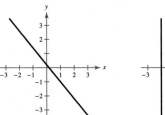
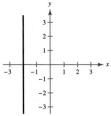

16. $y = 2$

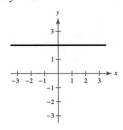

17. (a) $y = -0.25x - 4.25$ (b) $y = 4x - 17$

18. (a) $p = \frac{1}{33}d + 1$ (b) $\frac{1}{33} \approx 0.03$ atmosphere/ft

SECTION 1.4 (page 78)

Skills Review (page 78)

1. 20   2. 10   3. $x^2 + x - 6$
4. $x^3 + 9x^2 + 26x + 30$ 5. $\frac{1}{x}$ 6. $\frac{2x - 1}{x}$
7. $y = -2x + 17$ 8. $y = \frac{6}{5}x^2 + \frac{1}{5}$
9. $y = 3 \pm \sqrt{5 + (x + 1)^2}$ 10. $y = \pm\sqrt{4x^2 + 2}$
11. $y = 2x + \frac{1}{2}$ 12. $y = \frac{x^3}{2} + \frac{1}{2}$

1. y is not a function of x.   **3.** y is a function of x.
5. y is a function of x.   **7.** y is a function of x.

9.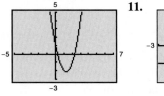

Domain: $(-\infty, \infty)$
Range: $[-2.125, \infty)$

11.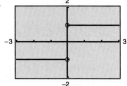

Domain: $(-\infty, 0) \cup (0, \infty)$
Range: $y = -1$ or $y = 1$

13.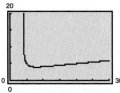

Domain: $(4, \infty)$
Range: $[4, \infty)$

15.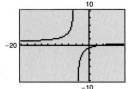

Domain: $(-\infty, -4) \cup (-4, \infty)$
Range: $(-\infty, 1) \cup (1, \infty)$

17. Domain: $(-\infty, \infty)$ **19.** Domain: $(-\infty, \infty)$
Range: $(-\infty, \infty)$ Range: $(-\infty, 4]$

21. (a) -2 (b) $3x - 5$ (c) $3x + 3\Delta x - 2$

23. (a) 4   (b) $\dfrac{1}{x + 4}$ (c) $-\dfrac{\Delta x}{x(x + \Delta x)}$

25. $\Delta x + 2x - 5$, $\Delta x \neq 0$

27. $\dfrac{1}{\sqrt{x + \Delta x + 1} + \sqrt{x + 1}}$, $\Delta x \neq 0$

29. $-\dfrac{1}{(x + \Delta x - 2)(x - 2)}$, $\Delta x \neq 0$

31. y is not a function of x.   **33.** y is a function of x.

35. (a) $2x$ (b) $10x - 25$ (c) $\dfrac{2x - 5}{5}$ (d) 5   (e) 5

37. (a) $x^2 + x$ (b) $(x^2 + 1)(x - 1) = x^3 - x^2 + x - 1$
(c) $\dfrac{x^2 + 1}{x - 1}$ (d) $x^2 - 2x + 2$ (e) x^2

39. (a) 0   (b) 0   (c) -1 (d) $\sqrt{15}$
(e) $\sqrt{x^2 - 1}$ (f) $x - 1$, $x \geq 0$

41. $f(g(x)) = 5\left(\dfrac{x-1}{5}\right) + 1 = x$

$g(f(x)) = \dfrac{5x + 1 - 1}{5} = x$

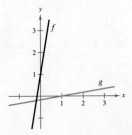

43. $f(g(x)) = 9 - \left(\sqrt{9-x}\right)^2 = 9 - (9-x) = x$

$g(f(x)) = \sqrt{9 - (9 - x^2)} = \sqrt{x^2} = x$

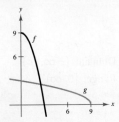

45. $f(x) = 2x - 3,\ f^{-1}(x) = \dfrac{x+3}{2}$

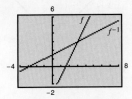

47. $f(x) = x^5,\ f^{-1}(x) = \sqrt[5]{x}$

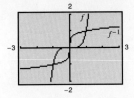

49. $f(x) = \sqrt{9 - x^2},\ 0 \le x \le 3$

$f^{-1}(x) = \sqrt{9 - x^2},\ 0 \le x \le 3$

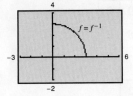

51. $f(x) = x^{2/3},\ x \ge 0$

$f^{-1}(x) = x^{3/2},\ x \ge 0$

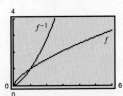

53.

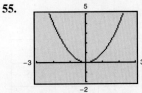

$f(x)$ is one-to-one. $f^{-1}(x) = \dfrac{3-x}{7}$

55. 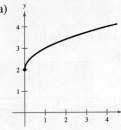 **57.**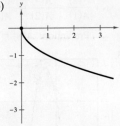

$f(x)$ is not one-to-one. $f(x)$ is not one-to-one.

59. (a) 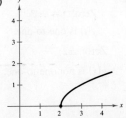 (b)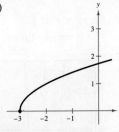

(c) 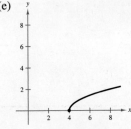 (d)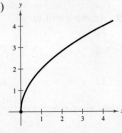

(e) 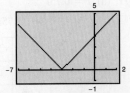 (f)

61. (a) $y = (x+3)^2$ (b) $y = -(x+6)^2 - 3$

63. (a)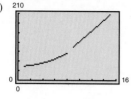

(b) 1997: $76.22 billion
2000: $122 billion
2004: $188.8 billion

65. $D = 1100 + 35r$

67. (a) Domain: $[15, 24]$; Range: $[57.95, 75.5]$
(b) 59.9 in. $\approx$ 5 ft   (c) 21.18 in.

69. $(N \circ T)(t) = 32t^2 + 36t + 204$; $(N \circ T)(t)$ is the number of bacteria in terms of time.

71. (a) $r = 0.75t$
(b) $A = \pi r^2 = \pi(0.75t)^2$

Time, t	1	2	3	4	5
Radius, r	0.75	1.5	2.25	3	3.75
Area, A	1.77	7.07	15.90	28.27	44.18

(c) $\dfrac{A(2)}{A(1)} \approx \dfrac{A(4)}{A(2)} \approx \dfrac{A(2t)}{A(t)} \approx 4$;

Prediction: $A(8) = 4 \cdot A(4) = 113.08$ ft^2

73.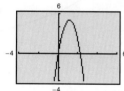

Zeros: $x = 0, \frac{9}{4}$
$f(x)$ is not one-to-one.

75.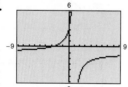

Zero: $t = -3$
$g(t)$ is one-to-one.

Zero: ± 2
$g(x)$ is not one-to-one.

77.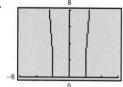

79. Answers will vary.

SECTION 1.5   (page 91)

Skills Review   (page 91)

1. (a) 7   (b) $c^2 - 3c + 3$
(c) $x^2 + 2xh + h^2 - 3x - 3h + 3$

2. (a) -4 (b) 10   (c) $3t^2 + 4$ **3.** h **4.** 4

5. Domain: $(-\infty, 0) \cup (0, \infty)$
Range: $(-\infty, 0) \cup (0, \infty)$

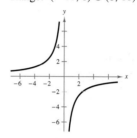

6. Domain: $[-5, 5]$
Range: $[0, 5]$

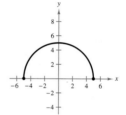

7. Domain: $(-\infty, \infty)$
Range: $[0, \infty)$

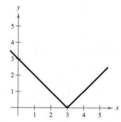

8. Domain: $(-\infty, 0) \cup (0, \infty)$
Range: $-1, 1$

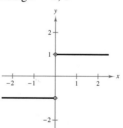

9. y is not a function of x.   **10.** y is a function of x.

1.

x	1.9	1.99	1.999	2
$f(x)$	8.8	8.98	8.998	?

x	2.001	2.01	2.1
$f(x)$	9.002	9.02	9.2

$\lim\limits_{x \to 2}(2x + 5) = 9$

3.

x	1.9	1.99	1.999	2
$f(x)$	0.2564	0.2506	0.2501	?

x	2.001	2.01	2.1
$f(x)$	0.2499	0.2494	0.2439

$$\lim_{x \to 2} \frac{x-2}{x^2-4} = \frac{1}{4}$$

5.

x	-0.1	-0.01	-0.001	0
$f(x)$	0.5132	0.5013	0.5001	?

x	0.001	0.01	0.1
$f(x)$	0.4999	0.4988	0.4881

$$\lim_{x \to 0} \frac{\sqrt{x+1}-1}{x} = 0.5$$

7.

x	-0.5	-0.1	-0.01
$f(x)$	-0.0714	-0.0641	-0.0627

x	-0.001	0
$f(x)$	-0.0625	?

$$\lim_{x \to 0^-} \frac{\frac{1}{x+4} - \frac{1}{4}}{x} = -\frac{1}{16}$$

9. (a) 1   (b) 3   **11.** (a) 1   (b) 3
13. (a) 12   (b) 27   (c) $\frac{1}{3}$
15. (a) 4   (b) 48   (c) 256
17. (a) 1   (b) 1   (c) 1
19. (a) 0   (b) 0   (c) 0
21. (a) 3   (b) -3 (c) Limit does not exist.
23. 4   **25.** -1 **27.** 0   **29.** 3   **31.** -2
33. $-\frac{3}{4}$ **35.** $\frac{35}{9}$ **37.** $\frac{1}{3}$ **39.** $-\frac{1}{20}$ **41.** 2
43. Limit does not exist.   **45.** Limit does not exist.
47. 12   **49.** Limit does not exist.
51. 2   **53.** -1 **55.** 2
57. $\dfrac{1}{2\sqrt{x+2}}$ **59.** $2t - 5$

61.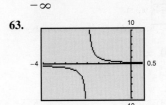

x	0	0.5	0.9	0.99
$f(x)$	-2	-2.67	-10.53	-100.5

x	0.999	0.9999	1
$f(x)$	-1000.5	$-10{,}000.5$	Undefined

$-\infty$

63.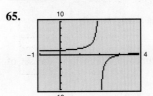

x	-3	-2.5	-2.1	-2.01
$f(x)$	-1	-2	-10	-100

x	-2.001	-2.0001	-2
$f(x)$	-1000	$-10{,}000$	Undefined

$-\infty$

65.

Limit does not exist.

67.

$-\frac{17}{9} \approx -1.8889$

69. $\lim\limits_{t \to 4}\left(98 + \dfrac{3}{t+1}\right) = 98 + \dfrac{3}{4+1} = 98.6;$

It takes 4 hours for the fever-reducing drug to reduce the body temperature to normal.

71. Yes; $\lim_{r \to 0.06} 5000(1+r)^{10} \approx 895$

73. (a)

x	-0.01	-0.001	-0.0001	0
$f(x)$	2.732	2.720	2.718	Undefined

x	0.0001	0.001	0.01
$f(x)$	2.718	2.717	2.705

$\lim_{x \to 0}(1+x)^{1/x} \approx 2.718$

(b)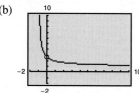

(c) Domain: $(-1, 0) \cup (0, \infty)$
Range: $(1, e) \cup (e, \infty)$

SECTION 1.6 (page 101)

Skills Review *(page 101)*

1. $\dfrac{x+4}{x-8}$ **2.** $\dfrac{x+1}{x-3}$ **3.** $\dfrac{x+2}{2(x-3)}$ **4.** $\dfrac{x-4}{x-2}$

5. $x = 0, -7$ **6.** $x = -5, 1$ **7.** $x = -\tfrac{2}{3}, -2$

8. $x = 0, 3, -8$ **9.** 13 **10.** -1

1. Continuous; The function is a polynomial.

3. Not continuous ($x \neq \pm 2$)

5. Continuous; The rational function's domain is the set of real numbers.

7. Not continuous ($x \neq 3$ and $x \neq 5$)

9. Not continuous ($x \neq \pm 2$)

11. $(-\infty, 0)$ and $(0, \infty)$; Explanations will vary. There is a discontinuity at $x = 0$, because $f(0)$ is not defined.

13. $(-\infty, -1)$ and $(-1, \infty)$; Explanations will vary. There is a discontinuity at $x = -1$, because $f(-1)$ is not defined.

15. $(-\infty, \infty)$; Explanations will vary.

17. $(-\infty, -1)$, $(-1, 1)$, and $(1, \infty)$; Explanations will vary. There are discontinuities at $x = \pm 1$, because $f(\pm 1)$ is not defined.

19. $(-\infty, \infty)$; Explanations will vary.

21. $(-\infty, 4)$, $(4, 5)$, and $(5, \infty)$; Explanations will vary. There are discontinuities at $x = 4$ and $x = 5$, because $f(4)$ and $f(5)$ are not defined.

23. Continuous on all intervals $\left(\dfrac{c}{2}, \dfrac{c}{2}+\dfrac{1}{2}\right)$, where c is an integer. Explanations will vary. There are discontinuities at $x = \dfrac{c}{2}$ where c is an integer, because $\lim_{x \to c} f\left(\dfrac{c}{2}\right)$ does not exist.

25. $(-\infty, \infty)$; Explanations will vary.

27. $(-\infty, 2]$ and $(2, \infty)$; Explanations will vary. There is a discontinuity at $x = 2$, because $\lim_{x \to 2} f(2)$ does not exist.

29. $(-\infty, -1)$ and $(-1, \infty)$; Explanations will vary. There is a discontinuity at $x = -1$, because $f(-1)$ is not defined.

31. Continuous on all intervals $(c, c+1)$, where c is an integer. Explanations will vary. There are discontinuities at $x = c$ where c is an integer, because $\lim_{x \to c} f(c)$ does not exist.

33. $(1, \infty)$; Explanations will vary. **35.** Continuous

37. Nonremovable discontinuity at $x = 2$

39.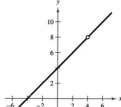

Continuous on $(-\infty, 4)$ and $(4, \infty)$

41.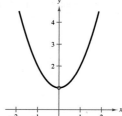

Continuous on $(-\infty, 0)$ and $(0, \infty)$

43.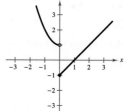

Continuous on $(-\infty, 0)$ and $(0, \infty)$

45. $a = 2$

47.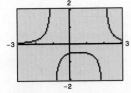

Not continuous at $x = 2$ and $x = -1$, because $f(-1)$ and $f(2)$ are not defined.

49.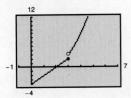

Not continuous at $x = 3$, because $\lim_{x \to 3} f(3)$ does not exist.

51.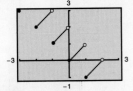

Not continuous at all integers c, because $\lim_{x \to c} f(c)$ does not exist.

53. $(-\infty, \infty)$

55. Continuous on all intervals $\left(\dfrac{c}{2}, \dfrac{c+1}{2}\right)$, where c is an integer.

57.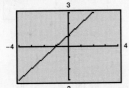

The graph of $f(x) = \dfrac{x^2 + x}{x}$ appears to be continuous on $[-4, 4]$, but f is not continuous at $x = 0$.

59. (a)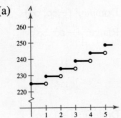

No; A is not continuous at $t = 1, 2, \ldots$.

(b) $229.50

61. $C = 3.50 - 1.90[\![1 - x]\!]$

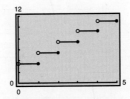

C is not continuous at $x = 1, 2, 3, \ldots$.

63. (a)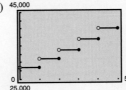

S is not continuous at $t = 1, 2, \ldots, 5$.

(b) $43,850.78

65. The model is continuous. The actual growth of the population probably would not be continuous because the population is usually recorded over larger units of time. In these cases, the population may jump between different units of time.

REVIEW EXERCISES FOR CHAPTER 1
(page 108)

1. 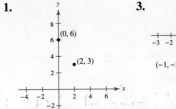 **3.**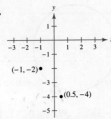

5. a   **6.** c   **7.** b   **8.** d   **9.** $\sqrt{29}$ **11.** $3\sqrt{2}$

13. $(7, 4)$ **15.** $(-8, 6)$

17. (a) 752,000   (b) 826,000   (c) 912,000
(d) 907,000

19. $(4, 7), (5, 8), (8, 10)$

21.    **23.**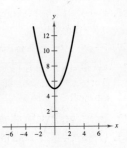

Answers to Selected Exercises

25.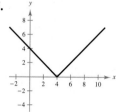
27.

29.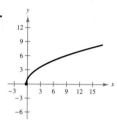

31. y-intercept: $x = 0 \Rightarrow y = -3 \Rightarrow (0, -3)$
x-intercept: $y = 0 \Rightarrow x = -\frac{3}{4} \Rightarrow \left(-\frac{3}{4}, 0\right)$

33. $x^2 + y^2 = 3^2$

35. $(x - 3)^2 + (y + 4)^2 = 25$
Center: $(3, -4)$
Radius: 5

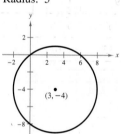

37. $(2, -3)$ **39.** $(0, 0), (1, 1), (-1, -1)$

41. (a) About mid 2001
(b) Kentucky: 4,303,000 people
South Carolina: 4,495,000 people

43. Slope: -3
y-intercept: $(0, -2)$

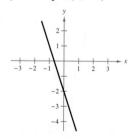

45. Slope: 0 (horizontal line)
y-intercept: $\left(0, -\frac{5}{3}\right)$

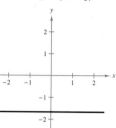

47. Slope: $-\frac{2}{5}$
y-intercept: $(0, -1)$

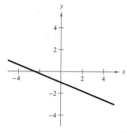

49. $\frac{6}{7}$ **51.** $\frac{20}{21}$

53. $y = -2x + 5$ **55.** $y = -4$

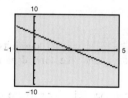 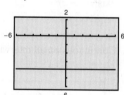

57. (a) $7x - 8y + 69 = 0$ (b) $2x + y = 0$
(c) $2x + y = 0$ (d) $2x + 3y - 12 = 0$

59. (a) $L = 45t + 204$ (b) 474 residents
(c) 789 residents

61. y is a function of x.   **63.** y is not a function of x.

65. (a) 7   (b) $3x + 7$ (c) $10 + 3\Delta x$

67. 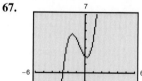   Domain: $(-\infty, \infty)$
Range: $(-\infty, \infty)$

69.    Domain: $[-1, \infty)$
Range: $[0, \infty)$

71. 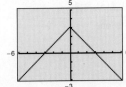 Domain: $(-\infty, \infty)$
Range: $(-\infty, 3]$

73. (a) $x^2 + 2x$ (b) $x^2 - 2x + 2$
(c) $2x^3 - x^2 + 2x - 1$ (d) $\dfrac{1 + x^2}{2x - 1}$
(e) $4x^2 - 4x + 2$ (f) $2x^2 + 1$

75. $f^{-1}(x) = \tfrac{2}{3}x$

77. $f(x)$ does not have an inverse function.

79. 7   **81.** 49   **83.** $\tfrac{10}{3}$ **85.** -2

87. $-\tfrac{1}{4}$ **89.** $-\infty$ **91.** Limit does not exist.

93. $-\tfrac{1}{16}$ **95.** $3x^2 - 1$ **97.** 0.5774

99. False, limit does not exist.

101. False, limit does not exist.

103. False, limit does not exist.

105. $(-\infty, -4)$ and $(-4, \infty)$; Explanations will vary. There is a discontinuity at $x = -4$ because $f(-4)$ is not defined.

107. $(-\infty, -1)$ and $(-1, \infty)$; Explanations will vary. There is a discontinuity at $x = -1$ because $f(-1)$ is not defined.

109. Continuous on all intervals $(c, c + 1)$, where c is an integer; Explanations will vary. There is a discontinuity at each integer because at each integer the limit does not exist.

111. $(-\infty, 0)$ and $(0, \infty)$; Explanations will vary. There is a discontinuity at $x = 0$ because the limit does not exist.

113. $a = 2$

115. (a) Explanations will vary. The function is defined for all values of x greater than zero. The function is discontinuous when $x = 5, x = 10,$ and $x = 15$.

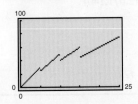

(b) $49.90

117. $C = 5 - 4[\![1 - x]\!]$

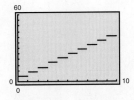

Explanations will vary. There is a discontinuity at each integer.

119. (a)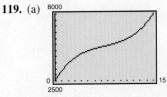

(b)

t	2	3	4	5	6
D	4001.8	4351.0	4643.3	4920.6	5181.5
Model	3937.0	4391.3	4727.4	4971.0	5147.8

t	7	8	9	10	11
D	5369.2	5478.2	5605.5	5628.7	5769.9
Model	5283.6	5404.0	5534.7	5701.5	5930.1

t	12	13	14	15
D	6198.4	6760.0	7354.7	7905.3
Model	6246.1	6675.3	7243.3	7976.0

(c) $15,007.9 billion

CHAPTER TEST   (page 112)

1. (a) $d = 5\sqrt{2}$ (b) Midpoint: $(-1.5, 1.5)$
(c) $m = -1$

2. (a) $d = 2.5$ (b) Midpoint: $(1.25, 2)$ (c) $m = 0$

3. (a) $d = 3$ (b) Midpoint: $(2\sqrt{2}, 1.5)$ (c) $m = \dfrac{\sqrt{2}}{4}$

4.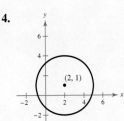

5. $t \approx 21.4$; Sometime during 2021.

6. $m = \frac{1}{5}$; $(0, -2)$

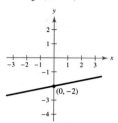

7. m is undefined; no y-intercept

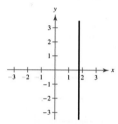

8. $m = -2.5$; $(0, 6.25)$

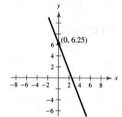

9. (a)

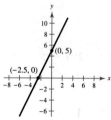

(b) Domain: $(-\infty, \infty)$

Range: $(-\infty, \infty)$

(c) $f(-3) = -1; f(-2) = 1; f(3) = 11$

(d) The function is one-to-one.

10. (a)

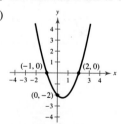

(b) Domain: $(-\infty, \infty)$

Range: $(-2.25, \infty)$

(c) $f(-3) = 10; f(-2) = 4; f(3) = 4$

(d) The function is not one-to-one.

11. (a)

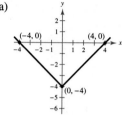

(b) Domain: $(-\infty, \infty)$

Range: $(-4, \infty)$

(c) $f(-3) = -1; f(-2) = -2; f(3) = -1$

(d) The function is not one-to-one.

12. $f^{-1}(x) = \frac{1}{4}x - \frac{3}{2}$

$f(f^{-1}(x)) = 4\left(\frac{1}{4}x - \frac{3}{2}\right) + 6 = x - 6 + 6 = x$

$f^{-1}(f(x)) = \frac{1}{4}(4x + 6) - \frac{3}{2} = x + \frac{3}{2} - \frac{3}{2} = x$

13. $f^{-1}(x) = -\frac{1}{3}x^3 + \frac{8}{3}$

$f(f^{-1}(x)) = \sqrt[3]{8 - 3\left(-\frac{1}{3}x^3 + \frac{8}{3}\right)}$

$= \sqrt[3]{8 + x^3 - 8}$

$= \sqrt[3]{x^3} = x$

$f^{-1}(f(x)) = -\frac{1}{3}\left(\sqrt[3]{8 - 3x}\right)^3 + \frac{8}{3}$

$= -\frac{1}{3}(8 - 3x) + \frac{8}{3}$

$= -\frac{8}{3} + x + \frac{8}{3} = x$

14. -1 **15.** Limit does not exist. **16.** 2 **17.** $\frac{1}{6}$

18. $(-\infty, 4)$ and $(4, \infty)$; Explanations will vary. There is a discontinuity at $x = 4$, because $f(4)$ is not defined.

19. $(-\infty, 5]$; Explanations will vary.

20. $(-\infty, \infty)$; Explanations will vary.

21. (a) The model fits the data well. Explanations will vary.

(b) 2071.14 thousand (2,071,140)

CHAPTER 2

SECTION 2.1 (page 122)

> **Skills Review** (page 122)
>
> **1.** $x = 2$ **2.** $y = 2$ **3.** $y = -x + 2$
>
> **4.** $2x$ **5.** $3x^2$ **6.** $\dfrac{1}{x^2}$ **7.** $2x$
>
> **8.** $(-\infty, 1) \cup (1, \infty)$ **9.** $(-\infty, \infty)$
>
> **10.** $(-\infty, 0) \cup (0, \infty)$

1. 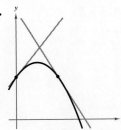 **3.**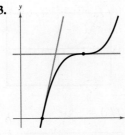

5. $m = 1$ **7.** $m = 0$ **9.** $m = -\frac{1}{3}$

11. $t = 2$: $m \approx 300$
$\quad\;$ $t = 8$: $m \approx 0$
$\quad\;$ $t = 11$: $m \approx -600$

13. 2002: $m \approx 450$
$\quad\;\;\,$ 2004: $m \approx 500$

15. $f'(x) = -2$
$\quad\;\;\,$ $f'(2) = -2$

17. $f'(x) = 0$
$\quad\;\;\,$ $f'(0) = 0$

19. $f'(x) = 2x$
$\quad\;\;\,$ $f'(2) = 4$

21. $f'(x) = 3x^2 - 1$
$\quad\;\;\,$ $f'(2) = 11$

23. $f'(x) = \dfrac{1}{\sqrt{x}}$
$\quad\;\;\,$ $f'(4) = \dfrac{1}{2}$

25. $f(x) = 3$
$\quad\;\;\,$ $f(x + \Delta x) = 3$
$\quad\;\;\,$ $f(x + \Delta x) - f(x) = 0$
$\quad\;\;\,$ $\dfrac{f(x + \Delta x) - f(x)}{\Delta x} = 0$
$\quad\;\;\,$ $\lim_{\Delta x \to 0} \dfrac{f(x + \Delta x) - f(x)}{\Delta x} = 0$

27. $f(x) = -5x$
$\quad\;\;\,$ $f(x + \Delta x) = -5x - 5\Delta x$
$\quad\;\;\,$ $f(x + \Delta x) - f(x) = -5\Delta x$
$\quad\;\;\,$ $\dfrac{f(x + \Delta x) - f(x)}{\Delta x} = -5$
$\quad\;\;\,$ $\lim_{\Delta x \to 0} \dfrac{f(x + \Delta x) - f(x)}{\Delta x} = -5$

29. $g(s) = \dfrac{1}{3}s + 2$
$\quad\;\;\,$ $g(s + \Delta s) = \dfrac{1}{3}s + \dfrac{1}{3}\Delta s + 2$
$\quad\;\;\,$ $g(s + \Delta s) - g(s) = \dfrac{1}{3}\Delta s$
$\quad\;\;\,$ $\dfrac{g(s + \Delta s) - g(s)}{\Delta s} = \dfrac{1}{3}$
$\quad\;\;\,$ $\lim_{\Delta s \to 0} \dfrac{g(s + \Delta s) - g(s)}{\Delta s} = \dfrac{1}{3}$

31. $f(x) = x^2 - 4$
$\quad\;\;\,$ $f(x + \Delta x) = x^2 + 2x\Delta x + (\Delta x)^2 - 4$
$\quad\;\;\,$ $f(x + \Delta x) - f(x) = 2x\Delta x + (\Delta x)^2$
$\quad\;\;\,$ $\dfrac{f(x + \Delta x) - f(x)}{\Delta x} = 2x + \Delta x$
$\quad\;\;\,$ $\lim_{\Delta x \to 0} \dfrac{f(x + \Delta x) - f(x)}{\Delta x} = 2x$

33. $h(t) = \sqrt{t - 1}$
$\quad\;\;\,$ $h(t + \Delta t) = \sqrt{t + \Delta t - 1}$
$\quad\;\;\,$ $h(t + \Delta t) - h(t) = \sqrt{t + \Delta t - 1} - \sqrt{t - 1}$
$\quad\;\;\,$ $\dfrac{h(t + \Delta t) - h(t)}{\Delta t} = \dfrac{1}{\sqrt{t + \Delta t - 1} + \sqrt{t - 1}}$
$\quad\;\;\,$ $\lim_{\Delta t \to 0} \dfrac{h(t + \Delta t) - h(t)}{\Delta t} = \dfrac{1}{2\sqrt{t - 1}}$

35. $f(t) = t^3 - 12t$
$\quad\;\;\,$ $f(t + \Delta t) = t^3 + 3t^2\Delta t + 3t(\Delta t)^2$
$\quad\;\;\,\;\;\;\;\;\;\;\;\;\;\;\;\;\;\; + (\Delta t)^3 - 12t - 12\Delta t$
$\quad\;\;\,$ $f(t + \Delta t) - f(t) = 3t^2\Delta t + 3t(\Delta t)^2 + (\Delta t)^3 - 12\Delta t$
$\quad\;\;\,$ $\dfrac{f(t + \Delta t) - f(t)}{\Delta t} = 3t^2 + 3t\Delta t + (\Delta t)^2 - 12$
$\quad\;\;\,$ $\lim_{\Delta t \to 0} \dfrac{f(t + \Delta t) - f(t)}{\Delta t} = 3t^2 - 12$

37. $f(x) = \dfrac{1}{x + 2}$
$\quad\;\;\,$ $f(x + \Delta x) = \dfrac{1}{x + \Delta x + 2}$
$\quad\;\;\,$ $f(x + \Delta x) - f(x) = \dfrac{-\Delta x}{(x + \Delta x + 2)(x + 2)}$
$\quad\;\;\,$ $\dfrac{f(x + \Delta x) - f(x)}{\Delta x} = \dfrac{-1}{(x + \Delta x + 2)(x + 2)}$
$\quad\;\;\,$ $\lim_{\Delta x \to 0} \dfrac{f(x + \Delta x) - f(x)}{\Delta x} = -\dfrac{1}{(x + 2)^2}$

39. $y = 2x - 2$ **41.** $y = -6x - 3$

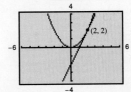 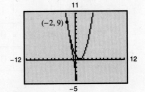

43. $y = \dfrac{x}{4} + 2$ **45.** $y = -x + 2$

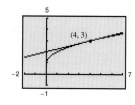 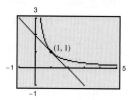

47. $y = -x + 1$
49. $y = -6x + 8$
    $y = -6x - 8$
51. $x \neq -3$ (node)   **53.** $x \neq 3$ (cusp)   **55.** $x > 1$
57. $x \neq 0$ (nonremovable discontinuity)
59. $x \neq 1$
61. $f(x) = -3x + 2$

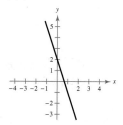

63.

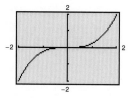

$f'(x) = \tfrac{3}{4}x^2$

x	-2	$-\tfrac{3}{2}$	-1	$-\tfrac{1}{2}$
$f(x)$	-2	-0.8438	-0.25	-0.0313
$f'(x)$	3	1.6875	0.75	0.1875

x	0	$\tfrac{1}{2}$	1	$\tfrac{3}{2}$	2
$f(x)$	0	0.0313	0.25	0.8438	2
$f'(x)$	0	0.1875	0.75	1.6875	3

65.

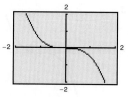

$f'(x) = -\tfrac{3}{2}x^2$

x	-2	$-\tfrac{3}{2}$	-1	$-\tfrac{1}{2}$
$f(x)$	4	1.6875	0.5	0.0625
$f'(x)$	-6	-3.375	-1.5	-0.375

x	0	$\tfrac{1}{2}$	1	$\tfrac{3}{2}$	2
$f(x)$	0	-0.0625	-0.5	-1.6875	-4
$f'(x)$	0	-0.375	-1.5	-3.375	-6

67. $f'(x) = 2x - 4$

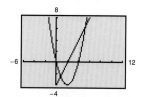

The x-intercept of the derivative indicates a point of horizontal tangency for f.

69. $f'(x) = 3x^2 - 3$

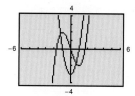

The x-intercepts of the derivative indicate points of horizontal tangency for f.

71. True   **73.** True
75.

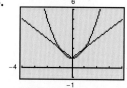

The graph of f is smooth at $(0, 1)$, but the graph of g has a sharp point at $(0, 1)$. The function g is not differentiable at $x = 0$.

SECTION 2.2 (page 134)

Skills Review (page 134)

1. (a) 8 (b) 16 (c) $\frac{1}{2}$
2. (a) $\frac{1}{36}$ (b) $\frac{1}{32}$ (c) $\frac{1}{64}$
3. $4x(3x^2 + 1)$ 4. $\frac{3}{2}x^{1/2}(x^{3/2} - 1)$ 5. $\frac{1}{4x^{3/4}}$
6. $x^2 - \frac{1}{x^{1/2}} + \frac{1}{3x^{2/3}}$ 7. $0, -\frac{2}{3}$
8. $0, \pm 1$ 9. $-10, 2$ 10. $-2, 12$

1. (a) 2 (b) $\frac{1}{2}$ 3. (a) -1 (b) $-\frac{1}{3}$ 5. 0
7. $4x^3$ 9. 4 11. $2x + 5$ 13. $-6t + 2$
15. $3t^2 - 2$ 17. $\frac{16}{3}t^{1/3}$ 19. $\frac{2}{\sqrt{x}}$ 21. $-\frac{8}{x^3} + 4x$

23. Function: $y = \frac{1}{x^3}$
    Rewrite: $y = x^{-3}$
    Differentiate: $y' = -3x^{-4}$
    Simplify: $y' = -\frac{3}{x^4}$

25. Function: $y = \frac{1}{(4x)^3}$
    Rewrite: $y = \frac{1}{64}x^{-3}$
    Differentiate: $y' = -\frac{3}{64}x^{-4}$
    Simplify: $y' = -\frac{3}{64x^4}$

27. Function: $y = \frac{\sqrt{x}}{x}$
    Rewrite: $y = x^{-1/2}$
    Differentiate: $y' = -\frac{1}{2}x^{-3/2}$
    Simplify: $y' = -\frac{1}{2x^{3/2}}$

29. -1 31. -2 33. 4 35. $2x + \frac{4}{x^2} + \frac{6}{x^3}$
37. $2x - 2 + \frac{8}{x^5}$ 39. $3x^2 + 1$ 41. $6x^2 + 16x - 1$
43. $\frac{2x^3 - 6}{x^3}$ 45. $\frac{4x^3 - 2x - 10}{x^3}$ 47. $\frac{4}{5x^{1/5}} + 1$

49. (a) $y = 2x - 2$
    (b) and (c)
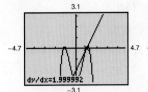

51. (a) $y = \frac{8}{15}x + \frac{22}{15}$
    (b) and (c)
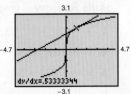

53. $(0, -1), \left(-\frac{\sqrt{6}}{2}, \frac{5}{4}\right), \left(\frac{\sqrt{6}}{2}, \frac{5}{4}\right)$ 55. $(-5, -12.5)$

57. (a)
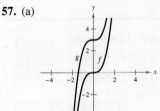
    (b) $f'(1) = g'(1) = 3$

    (c)
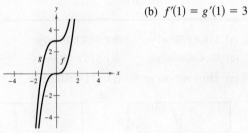
    (d) $f' = g' = 3x^2$ for every value of x.

59. (a) 3 (b) 6 (c) -3 (d) 6
61. (a) 2001: $m = -386.095$
        2004: $m = 375.14$
    (b) The results are similar.
    (c) Thousands of lb/yr/yr
63. (a) $T'(1) = -0.0754$
        $T'(8) = 0.4944$
        $T'(12) = -4.6976$
    (b) °F/mo

65.
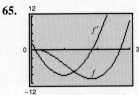
    $(0.11, 0.14), (1.84, -10.49)$

67. False. Let $f(x) = x$ and $g(x) = x + 1$.

SECTION 2.3 (page 143)

Skills Review (page 143)

1. 3 2. −7 3. $y' = 8x - 2$
4. $y' = -9t^2 + 4t$ 5. $s' = -32t + 24$
6. $y' = -32x + 54$ 7. $A' = -\frac{3}{5}r^2 + \frac{3}{5}r + \frac{1}{2}$
8. $y' = 2x^2 - 4x + 7$ 9. $y' = 12 - \dfrac{x}{2500}$
10. $y' = 74 - \dfrac{3x^2}{10{,}000}$

1. (a) $10.4 billion/yr (b) $7.4 billion/yr
   (c) $6.4 billion/yr (d) $16.6 billion/yr
   (e) $10.4 billion/yr (f) $11.4 billion/yr

3.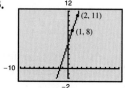

Average rate: 3
Instantaneous rates:
$f'(1) = f'(2) = 3$

5.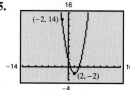

Average rate: −4
Instantaneous rates:
$h'(-2) = -8, h'(2) = 0$

7.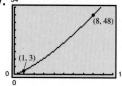

Average rate: $\frac{45}{7}$
Instantaneous rates:
$f'(1) = 4, f'(8) = 8$

9.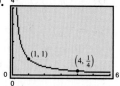

Average rate: $-\frac{1}{4}$
Instantaneous rates:
$f'(1) = -1, f'(4) = -\frac{1}{16}$

11.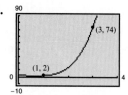

Average rate: 36
Instantaneous rates: $g'(1) = 2, g'(3) = 102$

13. (a) [5, 6]
    (b) Answers will vary. Sample answer: [3, 5]; Explanations will vary.

15. (a)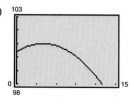

(b) For $t < 4$, positive; for $t > 4$, negative; This shows when the fever is going up and down.

(c) $T(0) = 100.4°F$
    $T(4) = 101°F$
    $T(8) = 100.4°F$
    $T(12) = 98.6°F$

(d) $T'(t) = -0.075t + 0.3$;
    the rate of change of temperature with respect to time

(e) $T'(0) = 0.3°F/h$
    $T'(4) = 0°F/h$
    $T'(8) = -0.3°F/h$
    $T'(12) = -0.6°F/h$

17. (a) $\dfrac{dH}{dv} = 33\left(\dfrac{5}{\sqrt{r}} - 1\right)$;
    Rate of change of heat loss with respect to velocity

(b) $H'(2) \approx 0.023$ kilocalorie/m³
    $H'(5) \approx 0.11$ kilocalorie/m³

19. (a) 37.5 m/sec (b) 33.$\overline{3}$ m/sec

21.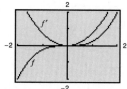

f has a horizontal tangent at $x = 0$.

23. (a) $P(0) = 117{,}216$ $P(10) = 123{,}600$
    $P(15) = 125{,}688.75$ $P(20) = 127{,}042$
    $P(25) = 127{,}659.75$
    Japan's population in 1980, 1990, 1995, 2000, and 2005.

(b) $\dfrac{dP}{dt} = -29.42t + 785.5$

(c) $P'(0) = 785.5$ $P'(10) = 491.3$ $P'(15) = 344.2$
    $P'(20) = 197.1$ $P'(25) = 50$
    Japan's population growth rate is decreasing over time.

25. $C = \dfrac{44{,}250}{x}$; Explanations will vary.

x	10	15	20	25
C	4425.00	2950.00	2212.50	1770.00
dC/dx	-442.5	-196.67	-110.63	-70.80

x	30	35	40
C	1475.00	1264.29	1106.25
dC/dx	-49.17	-36.12	-27.66

27. (a)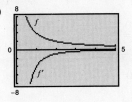

(b)

x	$\frac{1}{8}$	$\frac{1}{4}$	$\frac{1}{2}$	1
$f(x)$	32	16	8	4
$f'(x)$	-256	-64	-16	-4

x	2	3	4	5
$f(x)$	2	$\frac{4}{3}$	1	$\frac{4}{5}$
$f'(x)$	-1	$-\frac{4}{9}$	$-\frac{1}{4}$	$-\frac{4}{25}$

(c) $\left[\frac{1}{8}, \frac{1}{4}\right]$: -128; $\left[\frac{1}{4}, \frac{1}{2}\right]$: -32; $\left[\frac{1}{2}, 1\right]$: -8
  $[1, 2]$: -2; $[2, 3]$: $-\frac{2}{3}$; $[3, 4]$: $-\frac{1}{3}$; $[4, 5]$: $-\frac{1}{5}$

29. (a) 1.5 (b) $\frac{1}{15}$ (c) $\frac{1}{15}$ (d) $-\frac{1}{15}$

MID-CHAPTER QUIZ (page 146)

1. $f(x) = -x + 2$
$f(x + \Delta x) = -x - \Delta x + 2$
$f(x + \Delta x) - f(x) = -\Delta x$
$\dfrac{f(x + \Delta x) - f(x)}{\Delta x} = -1$
$\lim\limits_{\Delta x \to 0} \dfrac{f(x + \Delta x) - f(x)}{\Delta x} = -1$
$f'(x) = -1$
$f'(2) = -1$

2. $f(x) = \sqrt{x + 3}$
$f(x + \Delta x) = \sqrt{x + \Delta x + 3}$
$f(x + \Delta x) - f(x) = \sqrt{x + \Delta x + 3} - \sqrt{x + 3}$
$\dfrac{f(x + \Delta x) - f(x)}{\Delta x} = \dfrac{1}{\sqrt{x + \Delta x + 3} + \sqrt{x + 3}}$
$\lim\limits_{\Delta x \to 0} \dfrac{f(x + \Delta x) - f(x)}{\Delta x} = \dfrac{1}{2\sqrt{x + 3}}$
$f'(x) = \dfrac{1}{2\sqrt{x + 3}}$
$f'(1) = \dfrac{1}{4}$

3. $f(x) = \dfrac{4}{x}$
$f(x + \Delta x) = \dfrac{4}{x + \Delta x}$
$f(x + \Delta x) - f(x) = -\dfrac{4\Delta x}{x(x + \Delta x)}$
$\dfrac{f(x + \Delta x) - f(x)}{\Delta x} = -\dfrac{4}{x(x + \Delta x)}$
$\lim\limits_{\Delta x \to 0} \dfrac{f(x + \Delta x) - f(x)}{\Delta x} = -\dfrac{4}{x^2}$
$f'(x) = -\dfrac{4}{x^2}$
$f'(1) = -4$

4. $f'(x) = 0$ **5.** $f'(x) = 19$ **6.** $f'(x) = -6x$

7. $f'(x) = \dfrac{3}{x^{3/4}}$ **8.** $f'(x) = -\dfrac{8}{x^3}$ **9.** $f'(x) = \dfrac{1}{\sqrt{x}}$

10.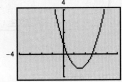

Average rate: 0
Instantaneous rates: $f'(0) = -3, f'(3) = 3$

11.

Average rate: 1
Instantaneous rates: $f'(-1) = 3, f'(1) = 7$

12.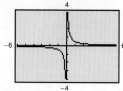

Average rate: $-\frac{1}{20}$

Instantaneous rates: $f'(2) = -\frac{1}{8}, f'(5) = -\frac{1}{50}$

13.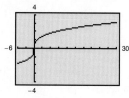

Average rate: $\frac{1}{19}$

Instantaneous rates: $f'(8) = \frac{1}{12}, f'(27) = \frac{1}{27}$

14. 2002: $m \approx 800$

2004: $m \approx 0$

15. $y = -4x - 6$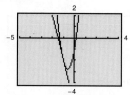

16. $y = -1$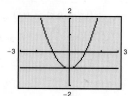

17. (a) 1.8 sec

(b) -42.6 ft/sec

SECTION 2.4 *(page 155)*

Skills Review *(page 155)*

1. $2(3x^2 + 7x + 1)$ 2. $4x^2(6 - 5x^2)$
3. $8x^2(x^2 + 2)^3 + (x^2 + 4)$
4. $(2x)(2x + 1)[2x + (2x + 1)^3]$
5. $\dfrac{23}{(2x + 7)^2}$ 6. $-\dfrac{x^2 + 8x + 4}{(x^2 - 4)^2}$
7. $-\dfrac{2(x^2 + x - 1)}{(x^2 + 1)^2}$ 8. $\dfrac{4(3x^4 - x^3 + 1)}{(1 - x^4)^2}$
9. $\dfrac{4x^3 - 3x^2 + 3}{x^2}$ 10. $\dfrac{x^2 - 2x + 4}{(x - 1)^2}$
11. 11 12. 0 13. $-\frac{1}{4}$ 14. $\frac{17}{4}$

In Exercises 1–15, the differentiation rule(s) used may vary. A sample answer is provided.

1. $f'(2) = 15$; Product Rule
3. $f'(1) = 13$; Product Rule
5. $f'(0) = 0$; Constant Multiple Rule
7. $g'(4) = 11$; Product Rule
9. $h'(6) = -5$; Quotient Rule
11. $f'(3) = \frac{3}{4}$; Quotient Rule
13. $g'(6) = -11$; Quotient Rule
15. $f'(1) = \frac{2}{5}$; Quotient Rule
17. Function: $y = \dfrac{x^2 + 2x}{x}$

Rewrite: $y = x + 2, \ x \neq 0$

Differentiate: $y' = 1, \ x \neq 0$

Simplify: $y' = 1, \ x \neq 0$

19. Function: $y = \dfrac{7}{3x^3}$

Rewrite: $y = \dfrac{7}{3}x^{-3}$

Differentiate: $y' = -7x^{-4}$

Simplify: $y' = -\dfrac{7}{x^4}$

21. Function: $y = \dfrac{4x^2 - 3x}{8\sqrt{x}}$

Rewrite: $y = \dfrac{1}{2}x^{3/2} - \dfrac{3}{8}x^{1/2}$

Differentiate: $y' = \dfrac{3}{4}x^{1/2} - \dfrac{3}{16}x^{-1/2}$

Simplify: $y' = \dfrac{3}{4}\sqrt{x} - \dfrac{3}{16\sqrt{x}}$

23. Function: $y = \dfrac{x^2 - 4x + 3}{x - 1}$

Rewrite: $y = x - 3, \ x \neq 1$

Differentiate: $y' = 1, \ x \neq 1$

Simplify: $y' = 1, \ x \neq 1$

25. $10x^4 + 12x^3 - 3x^2 - 18x - 15$; Product Rule
27. $12t^2(2t^3 - 1)$; Product Rule
29. $\dfrac{5}{6x^{1/6}} + \dfrac{1}{x^{2/3}}$; Product Rule
31. $-\dfrac{5}{(2x - 3)^2}$; Quotient Rule
33. $\dfrac{2}{(x + 1)^2}, \ x \neq 1$; Quotient Rule

35. $\dfrac{x^2 + 2x - 1}{(x + 1)^2}$; Quotient Rule

37. $\dfrac{3s^2 - 2s - 5}{2s^{3/2}}$; Quotient Rule

39. $\dfrac{2x^3 + 11x^2 - 8x - 17}{(x + 4)^2}$; Quotient Rule

41. $y = 5x - 2$ **43.** $y = \tfrac{3}{4}x - \tfrac{5}{4}$

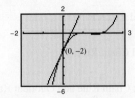 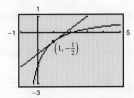

45. $y = -16x - 5$

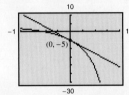

47. $(0, 0), (2, 4)$ **49.** $(0, 0), (\sqrt[3]{-4}, -2.117)$

51. **53.**

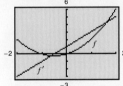 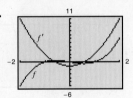

55. (a) -0.480/wk (b) 0.120/wk (c) 0.015/wk

57. 31.55 bacteria/h

59. -0.41 in./mo; The average monthly precipitation was decreasing at a rate of 0.41 in./mo in June.

61. (a) $T'(0.5) \approx 1.37$

$T'(4) = -\tfrac{1}{6}$

(b)

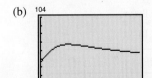

(c) After about 2.8 hours

(d) After about 11.3 hours; Explanations will vary.

63. (a) $C'(25) \approx 0.029$

$C'(75) \approx 0.182$

(b)

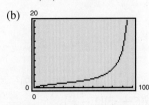

(c) C and C' approach ∞.

(d) Answers will vary.

65. (a) $M'(t) = \dfrac{300(1 - t^2)}{(t^2 + 1)^2}$

(b) $M(3) = 98$ memberships

$M'(3) = -24$ memberships/mo

(c) $M(24) \approx 20$ memberships

$M'(24) = -0.52$ memberships/mo

67. $f'(2) = 2$ **69.** $f'(2) = -10$

SECTION 2.5   (page 165)

Skills Review   (page 165)

1. $(1 - 5x)^{2/5}$ **2.** $(2x - 1)^{3/4}$

3. $(4x^2 + 1)^{-1/2}$ **4.** $(x - 6)^{-1/3}$

5. $x^{1/2}(1 - 2x)^{-1/3}$ **6.** $(2x)^{-1}(3 - 7x)^{3/2}$

7. $(x - 2)(3x^2 + 5)$ **8.** $(x - 1)(5\sqrt{x} - 1)$

9. $(x^2 + 1)^2(4 - x - x^3)$

10. $(3 - x^2)(x - 1)(x^2 + x + 1)$

	$y = f(g(x))$	$u = g(x)$	$y = f(u)$
1.	$y = (6x - 5)^4$	$u = 6x - 5$	$y = u^4$
3.	$y = (4 - x^2)^{-1}$	$u = 4 - x^2$	$y = u^{-1}$
5.	$y = \sqrt{5x - 2}$	$u = 5x - 2$	$y = \sqrt{u}$
7.	$y = (3x + 1)^{-1}$	$u = 3x + 1$	$y = u^{-1}$

9. $\dfrac{dy}{du} = 2u$ **11.** $\dfrac{dy}{du} = \dfrac{1}{2\sqrt{u}}$

$\dfrac{du}{dx} = 4$ $\dfrac{du}{dx} = -2x$

$\dfrac{dy}{dx} = 32x + 56$ $\dfrac{dy}{dx} = -\dfrac{x}{\sqrt{3 - x^2}}$

13. $\dfrac{dy}{du} = \dfrac{2}{3u^{1/3}}$

$\dfrac{du}{dx} = 20x^3 - 2$

$\dfrac{dy}{dx} = \dfrac{40x^3 - 4}{3\sqrt[3]{5x^4 - 2x}}$

15. c 17. b 19. a 21. c 23. $6(2x - 7)^2$

25. $-6(4 - 2x)^2$ 27. $6x(6 - x^2)(2 - x^2)$

29. $\dfrac{4x}{3(x^2 - 9)^{1/3}}$ 31. $\dfrac{1}{2\sqrt{t+1}}$ 33. $\dfrac{4t + 5}{2\sqrt{2t^2 + 5t + 2}}$

35. $\dfrac{6x}{(9x^2 + 4)^{2/3}}$ 37. $\dfrac{27}{4(2 - 9x)^{3/4}}$ 39. $\dfrac{4x^2}{(4 - x^3)^{7/3}}$

41. $y = 216x - 378$ 43. $y = \dfrac{8}{3}x - \dfrac{7}{3}$

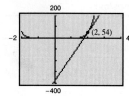
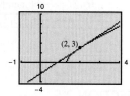

45. $y = x - 1$

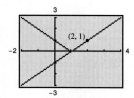

47. $f'(x) = \dfrac{1 - 3x^2 - 4x^{3/2}}{2\sqrt{x}(x^2 + 1)^2}$

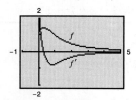

The zero of $f'(x)$ corresponds to the point on the graph of $f(x)$ where the tangent line is horizontal.

49. $f'(x) = -\dfrac{\sqrt{(x + 1)/x}}{2x(x + 1)}$

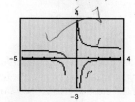

$f'(x)$ has no zeros.

In Exercises 51–65, the differentiation rule(s) used may vary. A sample answer is provided.

51. $-\dfrac{1}{(x - 2)^2}$; Chain Rule 53. $\dfrac{8}{(t + 2)^3}$; Chain Rule

55. $-\dfrac{2(2x - 3)}{(x^2 - 3x)^3}$; Chain Rule 57. $-\dfrac{2t}{(t^2 - 2)^2}$; Chain Rule

59. $27(x - 3)^2(4x - 3)$; Product Rule and Chain Rule

61. $\dfrac{3(x + 1)}{\sqrt{2x + 3}}$; Product Rule and Chain Rule

63. $\dfrac{t(5t - 8)}{2\sqrt{t - 2}}$; Product Rule and Chain Rule

65. $\dfrac{2(6 - 5x)(5x^2 - 12x + 5)}{(x^2 - 1)^3}$; Chain Rule and Quotient Rule

67. $y = \dfrac{8}{3}t + 4$

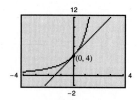

69. $y = -6t - 14$

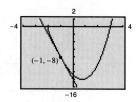

71. $y = -2x + 7$

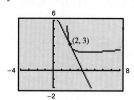

73. (a) $A'(2) \approx 0.607$ billion lb/yr
    (b) $A'(4) \approx 0.886$ billion lb/yr

75.
t	0	1	2	3	4
$\dfrac{dN}{dt}$	0	177.78	44.44	10.82	3.29

The rate of growth of N is decreasing.

77. (a) $V = \dfrac{10,000}{\sqrt[3]{t + 1}}$

(b) $-\$1322.83/\text{yr}$

(c) $-\$524.97/\text{yr}$

79. False. $y' = \frac{1}{2}(1-x)^{-1/2}(-1) = -\frac{1}{2}(1-x)^{-1/2}$

81. (a) 15 (b) -10

SECTION 2.6 (page 172)

Skills Review (page 172)

1. $t = 0, \frac{3}{2}$ **2.** $t = -2, 7$ **3.** $t = -2, 10$

4. $t = \dfrac{9 \pm 3\sqrt{10,249}}{32}$ **5.** $\dfrac{dy}{dx} = 6x^2 + 14x$

6. $\dfrac{dy}{dx} = 8x^3 + 18x^2 - 10x - 15$

7. $\dfrac{dy}{dx} = \dfrac{2x(x+7)}{(2x+7)^2}$ **8.** $\dfrac{dy}{dx} = -\dfrac{6x^2 + 10x + 15}{(2x^2 - 5)^2}$

9. Domain: $(-\infty, \infty)$ **10.** Domain: $[7, \infty)$
Range: $[-4, \infty)$ Range: $[0, \infty)$

1. 0   **3.** 2   **5.** $2t - 8$ **7.** $\dfrac{9}{2t^4}$

9. $18(2 - x^2)(5x^2 - 2)$

11. $12(x^3 - 2x)^2(11x^4 - 16x^2 + 4)$ **13.** $\dfrac{4}{(x-1)^3}$

15. $12x^2 + 24x + 16$ **17.** $60x^2 - 72x$

19. $120x + 360$ **21.** $-\dfrac{9}{2x^5}$ **23.** 260   **25.** $-\dfrac{1}{648}$

27. -126 **29.** $4x$ **31.** $\dfrac{1}{x^2}$ **33.** $12x^2 + 4$

35. $f''(x) = 6(x - 3) = 0$ when $x = 3$.

37. $f''(x) = 2(3x + 4) = 0$ when $x = -\frac{4}{3}$.

39. $f''(x) = \dfrac{x(2x^2 - 3)}{(x^2 - 1)^{3/2}} = 0$ when $x = \pm\dfrac{\sqrt{6}}{2}$.

41. $f''(x) = \dfrac{2x(x+3)(x-3)}{(x^2+3)^3}$
$= 0$ when $x = 0$ or $x = \pm 3$.

43. (a) $s(t) = -16t^2 + 144t$
$v(t) = -32t + 144$
$a(t) = -32$

(b) 4.5 sec; 324 ft

(c) $v(9) = -144$ ft/sec, which is the same speed as the initial velocity

45.

t	0	10	20	30	40	50	60
$\dfrac{ds}{dt}$	0	45	60	67.5	72	75	77.1
$\dfrac{d^2s}{dt^2}$	9	2.25	1	0.56	0.36	0.25	0.18

As time increases, velocity increases and acceleration decreases.

47. $f(x) = x^2 - 6x + 6$
$f'(x) = 2x - 6$
$f''(x) = 2$

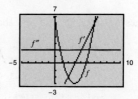

With each successive derivative, the degree decreases by 1.

49.

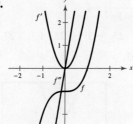

We know that with each successive derivative, the degree decreases by 1.

51. (a) $y = -41.333t^3 + 226.54t^2 - 299.9t + 8374$

(b)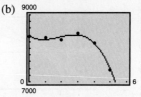

The model fits the data well.

(c) $y'(t) = -123.999t^2 + 453.08t - 299.9$
$y''(t) = -247.998t + 453.08$

(d) $y'(t) < 0$ on $[3, 5]$

(e) 2001 $(t \approx 1.83)$

(f) The first derivative is used to show that the utilized production is decreasing in part (d), and the utilized production increased at the greatest rate at the zero of the second derivative, as shown in part (e).

53. False. The product rule is
$[f(x)g(x)]' = f(x)g'(x) + g(x)f'(x)$.
55. True   **57.** $[xf(x)]^{(n)} = xf^{(n)}(x) + nf^{(n-1)}(x)$

REVIEW EXERCISES FOR CHAPTER 2
(page 178)

1. -2 **3.** 0

5. Answers will vary. Sample answer:

$t = 1$: $m \approx -3$; Launches were decreasing by about 3/yr in 2000.

$t = 2$: $m \approx 5$; Launches were increasing by about 5/yr in 2002.

$t = 4$: $m \approx -10$; Launches were decreasing by about 10/yr in 2005.

7. $t = 1$: slope ≈ 300
$t = 4$: slope ≈ -70
$t = 5$: slope ≈ -350

9. -3; -3 **11.** $2x - 4$; -2

13. $\dfrac{1}{2\sqrt{x+9}}$; $\dfrac{1}{4}$ **15.** $-\dfrac{1}{(x-5)^2}$; -1

17. -3 **19.** 0   **21.** $\dfrac{1}{6}$ **23.** -5 **25.** 1   **27.** 0

29. $y = -\dfrac{4}{3}t + 2$ **31.** $y = 2x + 2$

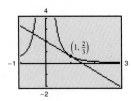

33. $y = -34x - 27$ **35.** $y = x - 1$

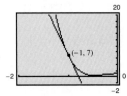 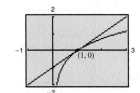

37. $y = -2x + 6$

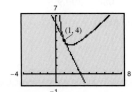

39. Average rate of change: 4
Instantaneous rate of change when $x = 0$: 3
Instantaneous rate of change when $x = 1$: 5

41. (a) About -6 launches/yr
(b) 2000: about -53 launches/yr
2005: about 15 launches/yr
(c) The number of successful space launches was decreasing in 2000 and increasing in 2005, and decreased at an average rate of 6 launches/yr over the period 2000–2005.

43. (a) $N' = \dfrac{1}{\sqrt{t}} + 1$

(b)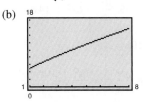

Answers will vary. Sample answer: 2001 ($t = 1$)

(c)

t	1	2	3	4
$N'(t)$	2	1.71	1.58	1.5

t	5	6	7	8
$N'(t)$	1.45	1.41	1.38	1.35

2001

(d) Answers will vary.

45. (a) [graph from 0 to 10, 4500 to 9000]

(b) $\dfrac{dP}{dt} = 2(0.035t^2 + 2.04t + 70.7)(0.07t + 2.04)$

$P'(0) = 288.456$

$P'(5) = 390.885$

$P'(10) = 518.408$

(c) Yes; Yes; The mosquito's population growth rate is increasing during the 10-day period. Explanations will vary.

47. (a) $s(t) = -16t^2 + 276$ (b) -32 ft/sec
(c) $t = 2$: -64 ft/sec
$t = 3$: -96 ft/sec
(d) About 4.15 sec   (e) About -132.8 ft/sec

In Exercises 49–67, the differentiation rule(s) used may vary. A sample answer is provided.

49. $15x^2(1 - x^2)$; Product Rule

51. $16x^3 - 33x^2 + 12x$; Product Rule
53. $\dfrac{2(3 + 5x - 3x^2)}{(x^2 + 1)^2}$; Quotient Rule
55. $30x(5x^2 + 2)^2$; General Power Rule
57. $-\dfrac{1}{(x + 1)^{3/2}}$; General Power Rule
59. $\dfrac{2x^2 + 1}{\sqrt{x^2 + 1}}$; Product Rule
61. $80x^4 - 24x^2 + 1$; Product Rule
63. $18x^5(x + 1)(2x + 3)^2$; Chain Rule
65. $x(x - 1)^4(7x - 2)$; Product Rule
67. $\dfrac{3(9t + 5)}{2\sqrt{3t + 1}(1 - 3t)^3}$; Quotient Rule
69. (a) $P'(2) = 22.31$
    $P'(4) = 9.18$
    $P'(6) = 0.83$
    $P'(20) = -2.23$
    (b)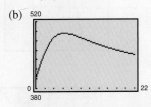

The bacteria's growth rate is increasing for the first 6 hours; then the bacteria's growth rate decreases.

71. (a) 6 board ft/in.
    (b) 18 board-ft/in.
    (c) 30 board-ft/in.
    (d) 48 board-ft/in.
73. 6   75. $-\dfrac{120}{x^6}$ 77. $\dfrac{35x^{3/2}}{2}$ 79. $\dfrac{2}{x^{2/3}}$
81. (a) $s(t) = -16t^2 + 5t + 30$ (b) About 1.534 sec
    (c) About -44.09 ft/sec   (d) -32 ft/sec^2
83. (a) $n(t) = 0.2602t^3 - 2.914t^2 + 0.75t + 612.4$
    (b)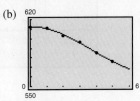

The model fits the data well.
    (c) $n'(t) = 0.7806t^2 - 5.828t + 0.75$
    $n''(t) = 1.5612t - 5.828$
    (d) $n'(t) < 0$ on $[3, 5]$

(e) 2003 ($t \approx 3.73$)
(f) The first derivative is used to show that the number of students enrolled is decreasing in part (d), and the number of students enrolled decreased at the greatest rate at the zero of the second derivative, as shown in part (e).

CHAPTER TEST (page 182)

1. $f(x) = x^2 + 1$
   $f(x + \Delta x) = x^2 + 2x\Delta x + \Delta x^2 + 1$
   $f(x + \Delta x) - f(x) = 2x\Delta x + \Delta x^2$
   $\dfrac{f(x + \Delta x) - f(x)}{\Delta x} = 2x + \Delta x$
   $\lim\limits_{\Delta x \to 0} \dfrac{f(x + \Delta x) - f(x)}{\Delta x} = 2x$
   $f'(x) = 2x$
   $f'(2) = 4$

2. $f(x) = \sqrt{x} - 2$
   $f(x + \Delta x) = \sqrt{x + \Delta x} - 2$
   $f(x + \Delta x) - f(x) = \sqrt{x + \Delta x} - \sqrt{x}$
   $\dfrac{f(x + \Delta x) - f(x)}{\Delta x} = \dfrac{1}{\sqrt{x + \Delta x} + \sqrt{x}}$
   $\lim\limits_{\Delta x \to 0} \dfrac{f(x + \Delta x) - f(x)}{\Delta x} = \dfrac{1}{2\sqrt{x}}$
   $f'(x) = \dfrac{1}{2\sqrt{x}}$
   $f'(4) = \dfrac{1}{4}$

3. $f'(t) = 3t^2 + 2$ 4. $f'(x) = 8x - 8$
5. $f'(x) = \dfrac{3\sqrt{x}}{2}$ 6. $f'(x) = 2x$ 7. $f'(x) = \dfrac{9}{x^4}$
8. $f'(x) = \dfrac{5 + x}{2\sqrt{x}} + \sqrt{x}$ 9. $f'(x) = 36x^3 + 48x$
10. $f'(x) = -\dfrac{1}{\sqrt{1 - 2x}}$
11. $f'(x) = \dfrac{(10x + 1)(5x - 1)^2}{x^2} = 250x - 75 + \dfrac{1}{x^2}$
12. $y = 2x - 2$

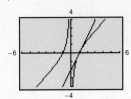

13. (a) 175.05 thousand/yr

(b) 2001: 136.2 thousand/yr

2004: 213.9 thousand/yr

(c) The number of bone graft procedures was increasing in 2001 and 2004, and increased at an average rate of about 175 thousand procedures per year.

14. 0 **15.** $-\dfrac{3}{8(3-x)^{5/2}}$ **16.** $-\dfrac{96}{(2x-1)^4}$

17. (a)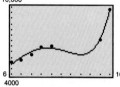

The model fits the data well.

(b) $N'(t) = 0.01578t^5 - 4.14t^3 + 249t$

$N''(t) = 0.0789t^4 - 12.42t^2 + 249$

(c) $N'(10) < 0$ and $N'(12) < 0$

(d) 2001

CHAPTER 3

SECTION 3.1 (page 191)

Skills Review (page 191)

1. $x = 0, x = 8$ 2. $x = 0, x = 24$ 3. $x = \pm 5$
4. $x = 0$ 5. $(-\infty, 3) \cup (3, \infty)$ 6. $(-\infty, 1)$
7. $(-\infty, -2) \cup (-2, 5) \cup (5, \infty)$ 8. $\left(-\sqrt{3}, \sqrt{3}\right)$
9. $x = -2$: -6 10. $x = -2$: 60
   $x = 0$: 2 $x = 0$: -4
   $x = 2$: -6 $x = 2$: 60
11. $x = -2$: $-\frac{1}{3}$ 12. $x = -2$: $\frac{1}{18}$
    $x = 0$: 1 $x = 0$: $-\frac{1}{8}$
    $x = 2$: 5 $x = 2$: $-\frac{3}{2}$

1. $f'(-1) = -\frac{8}{25}$ **3.** $f'(-3) = -\frac{2}{3}$
   $f'(0) = 0$ $f'(-2)$ is undefined.
   $f'(1) = \frac{8}{25}$ $f'(-1) = \frac{2}{3}$

5. Increasing on $(-\infty, -1)$
   Decreasing on $(-1, \infty)$

7. Increasing on $(-1, 0)$ and $(1, \infty)$
   Decreasing on $(-\infty, -1)$ and $(0, 1)$

9. No critical numbers
Increasing on $(-\infty, \infty)$

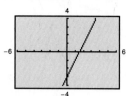

11. Critical number: $x = 1$
Increasing on $(-\infty, 1)$
Decreasing on $(1, \infty)$

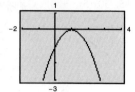

13. Critical number: $x = 3$
Decreasing on $(-\infty, 3)$
Increasing on $(3, \infty)$

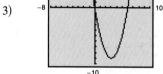

15. Critical numbers:
$x = 0, x = 4$
Increasing on $(-\infty, 0)$ and $(4, \infty)$
Decreasing on $(0, 4)$

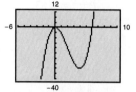

17. Critical numbers:
$x = -1, x = 1$
Decreasing on $(-\infty, -1)$
Increasing on $(1, \infty)$

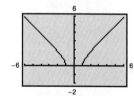

19. No critical numbers
Increasing on $(-\infty, \infty)$

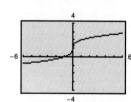

21. No critical numbers
Increasing on $(-\infty, \infty)$

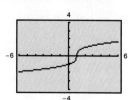

23. Critical number: $x = 1$
Increasing on $(-\infty, 1)$
Decreasing on $(1, \infty)$

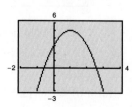

25. Critical numbers:
$x = -1, x = -\frac{5}{3}$
Increasing on $\left(-\infty, -\frac{5}{3}\right)$ and $(-1, \infty)$
Decreasing on $\left(-\frac{5}{3}, -1\right)$

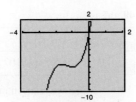

27. Critical numbers:
$x = -1, x = -\frac{2}{3}$
Decreasing on $\left(-1, -\frac{2}{3}\right)$
Increasing on $\left(-\frac{2}{3}, \infty\right)$

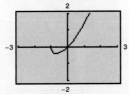

29. Critical numbers:
$x = 0, x = \frac{3}{2}$
Decreasing on $\left(-\infty, \frac{3}{2}\right)$
Increasing on $\left(\frac{3}{2}, \infty\right)$

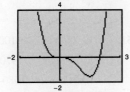

31. Critical numbers:
$x = 2, x = -2$
Decreasing on $(-\infty, -2)$ and $(2, \infty)$
Increasing on $(-2, 2)$

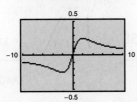

33. No critical numbers
Discontinuities: $x = \pm 4$
Increasing on $(-\infty, -4)$, $(-4, 4)$, and $(4, \infty)$

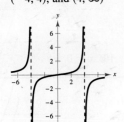

35. Critical number: $x = 0$
Discontinuity: $x = 0$
Increasing on $(-\infty, 0)$
Decreasing on $(0, \infty)$

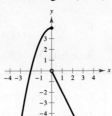

37. Critical number: $x = 1$
No discontinuity, but the function is not differentiable at $x = 1$.
Increasing on $(-\infty, 1)$
Decreasing on $(1, \infty)$

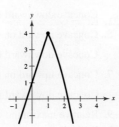

39. (a) Decreasing on $(1, 4.10)$
Increasing on $(4.10, 31)$

(b)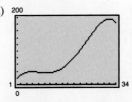

(c) About 4 days.

41. Increasing on $(1.4, 6)$

43. (a)

Increasing from 1971 to mid-1975 and from late 1980 to mid-2001

Decreasing from mid-1975 to late 1980 and from mid-2001 to 2004

(b) Answers will vary. Sample answer: By using a graphing utility, $y'(t) = 0$ when $t \approx 5.7$, $t \approx 10.8$, and $t \approx 31.8$.

SECTION 3.2 (page 201)

Skills Review (page 201)

1. $0, \pm\frac{1}{2}$ **2.** $-2, 5$ **3.** 1 **4.** $0, 125$
5. $-4 \pm \sqrt{17}$ **6.** $1 \pm \sqrt{5}$
7. Negative   **8.** Positive   **9.** Positive
10. Negative   **11.** Increasing   **12.** Decreasing

1. Relative maximum: $(1, 5)$
3. Relative minimum: $(3, -9)$
5. Relative maximum: $\left(\frac{2}{3}, \frac{28}{9}\right)$
Relative minimum: $(1, 3)$
7. No relative extrema
9. Relative maximum: $(0, 15)$
Relative minimum: $(4, -17)$
11. Relative minima: $(-0.366, 0.75)$, $(1.37, 0.75)$
Relative maximum: $\left(\frac{1}{2}, \frac{21}{16}\right)$

13.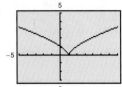

Relative minimum: $(1, 0)$

15.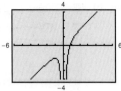

Relative maximum: $\left(-1, -\frac{3}{2}\right)$

17.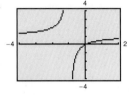

No relative extrema

19. Minimum: $(2, 2)$
Maximum: $(-1, 8)$

21. Maximum: $(0, 5)$
Minimum: $(3, -13)$

23. Minima: $(-1, -4), (2, -4)$
Maxima: $(0, 0), (3, 0)$

25. Maximum: $(2, 1)$
Minimum: $\left(0, \frac{1}{3}\right)$

27. Maximum: $(-1, 5)$
Minimum: $(0, 0)$

29. Maximum: $(-7, 4)$
Minimum: $(1, 0)$

31. 2, absolute maximum

33. Maximum: $(5, 7)$
Minimum: $(2.69, -5.55)$

35. Maximum: $(2, 2.\overline{6})$
Minima: $(0, 0), (3, 0)$

37. Minimum: $(0, 0)$
Maximum: $(1, 2)$

39. Maximum: $\left(2, \frac{1}{2}\right)$
Minimum: $(0, 0)$

41. Maximum: $\left|f''\left(\sqrt[3]{-10 + \sqrt{108}}\right)\right| \approx 1.47$

43. Maximum: $\left|f^{(4)}(0)\right| = \dfrac{56}{81}$

45. Answers will vary. Sample answer:

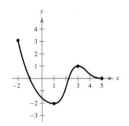

47. Mid-1972

49. (a)

t	0	0.5	1	1.5	2	2.5	3
$C(t)$	0	0.055	0.107	0.148	0.171	0.176	0.167

Maximum concentration occurred about 2.5 hours after injection.

(b)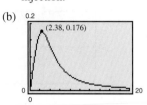

(c) $C'(t) = 0$ when $t \approx 2.38$

51. (a) 1970: 2500 per 1000 women

(b) Most rapidly: 1970–1975
Most slowly: 1980–1985

(c) Answers will vary.

SECTION 3.3 *(page 210)*

> **Skills Review** *(PAGE 210)*
>
> **1.** $f''(x) = 48x^2 - 54x$
>
> **2.** $g''(s) = 12s^2 - 18s + 2$
>
> **3.** $g''(x) = 56x^6 + 120x^4 + 72x^2 + 8$
>
> **4.** $f''(x) = \dfrac{4}{9(x - 3)^{2/3}}$ **5.** $h''(x) = \dfrac{190}{(5x - 1)^3}$
>
> **6.** $f''(x) = -\dfrac{42}{(3x + 2)^3}$ **7.** $x = \pm\dfrac{\sqrt{3}}{3}$
>
> **8.** $x = 0, 3$ **9.** $t = \pm 4$ **10.** $x = 0, \pm 5$

1. Concave upward on $(-\infty, \infty)$

3. Concave upward on $\left(-\infty, -\frac{1}{2}\right)$
Concave downward on $\left(-\frac{1}{2}, \infty\right)$

5. Concave upward on $(-\infty, -2)$ and $(2, \infty)$
Concave downward on $(-2, 2)$

7. Concave upward on $(-\infty, 2)$
Concave downward on $(2, \infty)$

9. Relative maximum: $(3, 9)$

11. Relative maximum: $(1, 3)$
Relative minimum: $\left(\frac{7}{3}, \frac{49}{27}\right)$

13. Relative minimum: $(0, -3)$

15. Relative minimum: $(0, 1)$

Answers to Selected Exercises A47

17. Relative minima: $(-3, 0)$, $(3, 0)$
Relative maximum: $(0, 3)$
19. Relative maximum: $(0, 4)$
21. No relative extrema
23. Relative maximum: $(0, 0)$
Relative minima: $(-0.5, -0.052)$, $(1, -0.\overline{3})$
25. Relative maximum: $(2, 9)$
Relative minimum: $(0, 5)$
27. Sign of $f'(x)$ on $(0, 2)$ is positive.
Sign of $f''(x)$ on $(0, 2)$ is positive.
29. Sign of $f'(x)$ on $(0, 2)$ is negative.
Sign of $f''(x)$ on $(0, 2)$ is negative.
31. $(3, 0)$ **33.** $(1, 0)$, $(3, -16)$
35. No inflection points   **37.** $\left(\frac{3}{2}, -\frac{1}{16}\right)$, $(2, 0)$

39.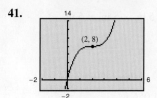
Relative maximum: $(-2, 16)$
Relative minimum: $(2, -16)$
Point of inflection: $(0, 0)$

41.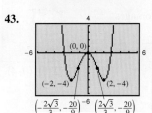
No relative extrema
Point of inflection: $(2, 8)$

43.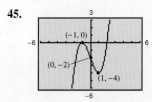
Relative maximum: $(0, 0)$
Relative minima: $(\pm 2, -4)$
Points of inflection:
$\left(\pm\dfrac{2\sqrt{3}}{3}, -\dfrac{20}{9}\right)$

45.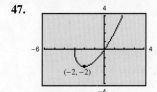
Relative maximum: $(-1, 0)$
Relative minimum: $(1, -4)$
Point of inflection: $(0, -2)$

47.
Relative minimum: $(-2, -2)$
No inflection points

49.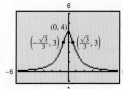
Relative maximum: $(0, 4)$
Points of inflection:
$\left(\pm\dfrac{\sqrt{3}}{3}, 3\right)$

51.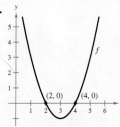

53.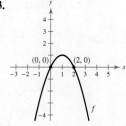

55.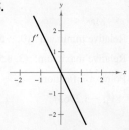

(a) f': Positive on $(-\infty, 0)$
   f: Increasing on $(-\infty, 0)$
(b) f': Negative on $(0, \infty)$
   f: Decreasing on $(0, \infty)$
(c) f': Not increasing
   f: Not concave upward
(d) f': Decreasing on $(-\infty, \infty)$
   f: Concave downward on $(-\infty, \infty)$

57. (a) f': Increasing on $(-\infty, \infty)$
(b) f: Concave upward on $(-\infty, \infty)$
(c) Relative minimum: $(-2.5, -6.25)$
No inflection points
(d)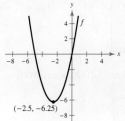

59. (a) f': Increasing on $(-\infty, 1)$
Decreasing on $(1, \infty)$
(b) f: Concave upward on $(-\infty, 1)$
Concave downward on $(1, \infty)$
(c) No relative extrema
Point of inflection: $\left(1, -\frac{1}{3}\right)$
(d)

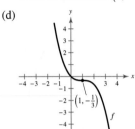

61. 8:30 P.M.

63.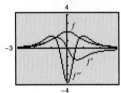

Relative minimum: $(0, -5)$
Relative maximum: $(3, 8.5)$
Point of inflection: $\left(\frac{2}{3}, -3.2963\right)$

When f' is positive, f is increasing. When f' is negative, f is decreasing. When f'' is positive, f is concave upward. When f'' is negative, f is concave downward.

65.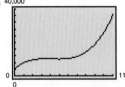

Relative maximum: $(0, 2)$
Points of inflection: $(0.58, 1.5), (-0.58, 1.5)$

When f' is positive, f is increasing. When f' is negative, f is decreasing. When f'' is positive, f is concave upward. When f'' is negative, f is concave downward.

67. (a)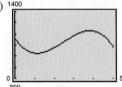

(b) November (c) October (d) October; April

69. (a)

(b) $d''(t) = 0$ when $t \approx 2.5$: Concave upward on $(0, 2.5)$
Concave downward on $(2.5, 5)$
(c) Point of inflection: $(2.5, 1128.1)$
(d) In 2001, the number of dropouts not in the labor force began to increase, but since the middle of 2002 the rate of this increase began to slow down.

71. Answers will vary.

SECTION 3.4 *(page 219)*

Skills Review *(page 219)*

1. $x + \frac{1}{2}y = 12$ **2.** $2xy = 24$ **3.** $xy = 24$
4. $\sqrt{(x_2 - x_1)^2 + (y_2 - y_1)^2} = 10$
5. $x = -3$ **6.** $x = -\frac{2}{3}, 1$ **7.** $x = \pm 5$
8. $x = 4$ **9.** $x = \pm 1$ **10.** $x = \pm 3$

1. 60, 60   **3.** 18, 9   **5.** $\sqrt{192}, \sqrt{192}$
7. $l = w = 25$ m   **9.** $l = w = 8$ ft
11. $x = 25$ ft, $y = \frac{100}{3}$ ft
13. (a) Proof
(b) $V_1 = 99$ in.^{3} (c) 5 in. × 5 in. × 5 in.
$V_2 = 125$ in.^{3}
$V_3 = 117$ in.^{3}
15. (a) 20 in. × 20 in. × 20 in.   (b) 2400 in.^{2}
17. $x = 5$ m, $y = 3\frac{1}{3}$ m   **19.** 1.056 ft^3
21. (a) 59 trees   (b) 87,025 oranges
23. (a) $L = \sqrt{x^2 + 4 + \dfrac{8}{x-1} + \dfrac{4}{(x-1)^2}}$, $x > 1$
(b)

(c) $(0, 0), (2, 0), (0, 4)$

Minimum when $x \approx 2.587$

25. $\sqrt{2}r \times \dfrac{\sqrt{2}}{2}r$

27. (a) Radius: about 1.66 in.
Height: about 3.34 in.
(b) About 52.15 in.^{2}
The height is about twice the radius.

29. $(1, 4)$ **31.** $(8, 0)$ **33.** About 1.32 in.

35. Radius of circle: $\dfrac{8}{\pi + 4}$

Side of square: $\dfrac{16}{\pi + 4}$

37. (a) $x = 1$ mi

39. Harvesting near the end of the 4th week will yield 135 bushels for a maximum value of $3645.

41. (a) $A(x) = x^2 + \dfrac{(2 - 2x)^2}{\pi}$ (b) Domain: $0 \leq x \leq 1$

(c)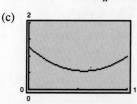

(d) Least total area: $x \approx 0.56$ and $r \approx 0.28$; About 2.24 ft for the square and 1.76 ft for the circle.

Greatest total area: $x = 0$ and $r = \dfrac{2}{\pi}$; All the wire is used for the circle.

MID-CHAPTER QUIZ *(page 222)*

1. Critical number: $x = 3$

Increasing on $(3, \infty)$

Decreasing on $(-\infty, 3)$

2. Critical numbers: $x = -4, x = 0$

Increasing on $(-\infty, -4)$ and $(0, \infty)$

Decreasing on $(-4, 0)$

3. Critical number: $x = 0$

Increasing on $(-\infty, 0)$

Decreasing on $(0, \infty)$

4. Relative minimum: $(0, -5)$

Relative maximum: $(-2, -1)$

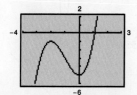

5. Relative minima: $(2, -13), (-2, -13)$

Relative maximum: $(0, 3)$

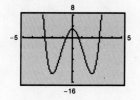

6. Relative minimum: $(0, 0)$

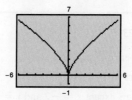

7. Minimum: $(-1, -9)$

Maximum: $(1, -5)$

8. Minimum: $(3, -54)$

Maximum: $(-3, 54)$

9. Minimum: $(0, 0)$

Maximum: $(1, 0.5)$

10. Point of inflection: $(2, -2)$

Concave downward on $(-\infty, 2)$

Concave upward on $(2, \infty)$

11. Points of inflection: $(-2, -80)$ and $(2, -80)$

Concave downward on $(-2, 2)$

Concave upward on $(-\infty, -2)$ and $(2, \infty)$

12. Relative minimum: $(1, 9)$

Relative maximum: $(-2, 36)$

13. Relative minimum: $(1, 2)$

Relative maximum: $(-1, -2)$

14. (a)

(b) $C''(t) = 0$ when $t \approx 2.03$: Concave upward on $(0, 2.03)$, concave downward on $(2.03, 5)$

(c) $(2.03, 43{,}767)$; In 2001, the number of cases began to increase, but since 2002 the rate of this increase began to slow down.

15. 50 ft by 100 ft

16. (a) Late 1999; 2005

(b) Increasing from early 1999 to late 1999; Decreasing from late 1999 to 2005

A50 Answers to Selected Exercises

SECTION 3.5 (page 230)

Skills Review (page 230)

1. Domain: all real numbers except $x = 5$
   Range: all real numbers except $y = 0$
2. Domain: all real numbers except $x = -2$ and $x = 2$
   Range: all real numbers except $y = 0$
3. Domain: $x \geq -3$ 4. Domain: $x > 3$
   Range: $y \geq 0$ Range: $y > 0$
5. 3   6. 1   7. -11 8. 4   9. $-\frac{1}{4}$
10. -2 11. 0   12. 1

---

1. Vertical asymptote: $x = 0$
   Horizontal asymptote: $y = 1$
3. Vertical asymptotes: $x = -1, x = 2$
   Horizontal asymptote: $y = 1$
5. Vertical asymptote: none
   Horizontal asymptote: $y = \frac{3}{2}$
7. Vertical asymptotes: $x = \pm 2$
   Horizontal asymptote: $y = \frac{1}{2}$
9. d   10. b   11. a   12. c
13. ∞ 15. $-\infty$ 17. $-\infty$ 19. $-\infty$

21.

x	10^0	10^1	10^2	10^3
$f(x)$	2.000	0.348	0.101	0.032

x	10^4	10^5	10^6
$f(x)$	0.010	0.003	0.001

$$\lim_{x \to \infty} \frac{x + 1}{x\sqrt{x}} = 0$$

23.

x	10^0	10^1	10^2	10^3
$f(x)$	0	49.5	49.995	49.99995

x	10^4	10^5	10^6
$f(x)$	50.0	50.0	50.0

$$\lim_{x \to \infty} \frac{x^2 - 1}{0.02x^2} = 50$$

25.

x	-10^6	-10^4	-10^2	10^0
$f(x)$	-2	-2	-1.9996	0.8944

x	10^2	10^4	10^6
$f(x)$	1.9996	2	2

$$\lim_{x \to -\infty} \frac{2x}{\sqrt{x^2 + 4}} = -2, \lim_{x \to \infty} \frac{2x}{\sqrt{x^2 + 4}} = 2$$

27. (a) ∞ (b) 5   (c) 0   29. (a) 0   (b) 1   (c) ∞
31. 2   33. 0   35. $-\infty$ 37. ∞ 39. 5

41. 43.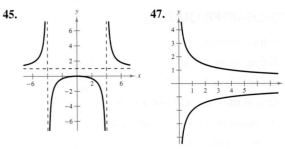

45. 47.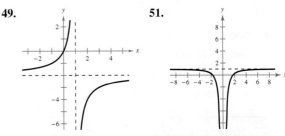

49. 51.

53. 55.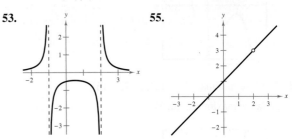

57.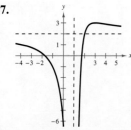

59. (a) 425; The temperature of the oven.
(b) 72; The temperature of the room.

61. (a) $C(15) = \$14{,}118$; $C(50) = \$80{,}000$; $C(90) = \$720{,}000$
(b) ∞; The limit does not exist. The cost increases without bound as the percent of air pollutants removed approaches 100%.

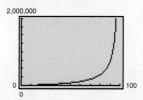

63. (a)

n	1	2	3	4	5
P	0.5	0.74	0.82	0.86	0.89

n	6	7	8	9	10
P	0.91	0.92	0.93	0.94	0.95

(b) 1
(c)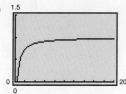

The percent of correct responses approaches 100% as the number of times the task is performed increases.

65. (a) 5 yr: 153 elk; 10 yr: 215 elk; 25 yr: 294 elk
(b) 400 elk

SECTION 3.6 *(page 240)*

Skills Review *(page 240)*

1. Vertical asymptote: $x = 0$
   Horizontal asymptote: $y = 0$
2. Vertical asymptote: $x = 2$
   Horizontal asymptote: $y = 0$
3. Vertical asymptote: $x = -3$
   Horizontal asymptote: $y = 40$
4. Vertical asymptotes: $x = 1, x = 3$
   Horizontal asymptote: $y = 1$
5. Decreasing on $(-\infty, -2)$
   Increasing on $(-2, \infty)$
6. Increasing on $(-\infty, -4)$
   Decreasing on $(-4, \infty)$
7. Increasing on $(-\infty, -1)$ and $(1, \infty)$
   Decreasing on $(-1, 1)$
8. Decreasing on $(-\infty, 0)$ and $(\sqrt[3]{2}, \infty)$
   Increasing on $(0, \sqrt[3]{2})$
9. Increasing on $(-\infty, 1)$ and $(1, \infty)$
10. Decreasing on $(-\infty, -3)$ and $(\frac{1}{3}, \infty)$
    Increasing on $(-3, \frac{1}{3})$

1.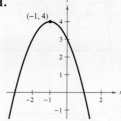

3.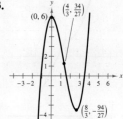

5.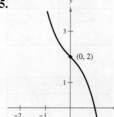

7.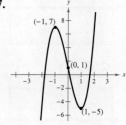

A52 Answers to Selected Exercises

9.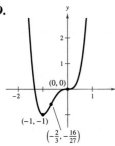

11.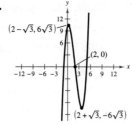

13.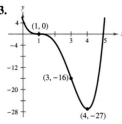

15.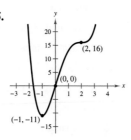

17.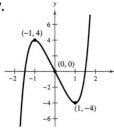

19.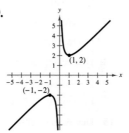

21.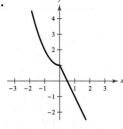

23.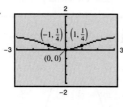

25.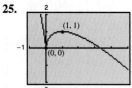

27.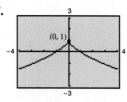

29.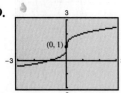

31.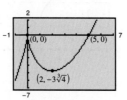

33.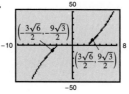

35.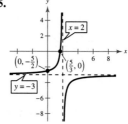

Domain: $(-\infty, 2) \cup (2, \infty)$

37.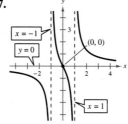

Domain: $(-\infty, -1) \cup (-1, 1) \cup (1, \infty)$

39.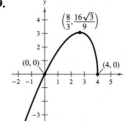

Domain: $(-\infty, 4]$

41.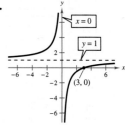

Domain: $(-\infty, 0) \cup (0, \infty)$

43.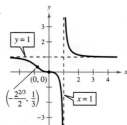

Domain: $(-\infty, 1) \cup (1, \infty)$

45. Answers will vary.

Sample answer: $f(x) = -x^3 + x^2 + x + 1$

47. Answers will vary. Sample answer:

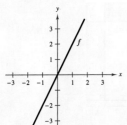

49. Answers will vary. Sample answer:

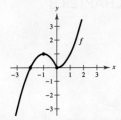

51. Answers will vary. Sample answer:

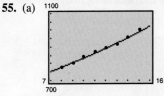

53. Answers will vary. Sample answer: $y = \dfrac{1}{x-5}$

55. (a)

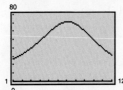

The model fits the data well.

(b) $1099.31

(c) No, because the benefits increase without bound as time approaches the year 2040 ($x = 50$), and the benefits are negative for the years past 2040.

57.

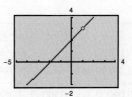

Absolute minimum: $(1, 26.71)$

Absolute maximum: $(7.11, 71.17)$

January is the coldest month and July is the warmest month.

59.

The rational function has the common factor $(x - 1)$ in the numerator and denominator. At $x = 1$, there is a hole in the graph, not a vertical asymptote.

SECTION 3.7 (page 247)

Skills Review (page 247)

1. $\dfrac{dC}{dx} = 0.18x$ 2. $\dfrac{dC}{dx} = 0.15$

3. $\dfrac{dR}{dx} = 1.25 + 0.03\sqrt{x}$ 4. $\dfrac{dR}{dx} = 15.5 - 3.1x$

5. $\dfrac{dP}{dx} = -\dfrac{0.01}{\sqrt[3]{x^2}} + 1.4$ 6. $\dfrac{dP}{dx} = -0.04x + 25$

7. $\dfrac{dA}{dx} = \dfrac{\sqrt{3}}{2}x$ 8. $\dfrac{dA}{dx} = 12x$ 9. $\dfrac{dC}{dr} = 2\pi$

10. $\dfrac{dP}{dw} = 4$ 11. $\dfrac{dS}{dr} = 8\pi r$ 12. $\dfrac{dP}{dx} = 2 + \sqrt{2}$

13. $A = \pi r^2$ 14. $A = x^2$

15. $V = x^3$ 16. $V = \tfrac{4}{3}\pi r^3$

1. $dy = 6x\, dx$ 3. $dy = 12(4x - 1)^2\, dx$

5. $dy = -\dfrac{x}{\sqrt{9 - x^2}}\, dx$ 7. 0.1005 9. -0.013245

11. $dy = 0.6$ 13. $dy = -0.04$
    $\Delta y = 0.6305$ $\Delta y \approx -0.0394$

15.

$dx = \Delta x$	dy	Δy	$\Delta y - dy$	$\dfrac{dy}{\Delta y}$
1.000	4.000	5.000	1.0000	0.8000
0.500	2.000	2.2500	0.2500	0.8889
0.100	0.400	0.4100	0.0100	0.9756
0.010	0.040	0.0401	0.0001	0.9975
0.001	0.004	0.0040	0.0000	1.0000

17.

$dx = \Delta x$	dy	Δy	$\Delta y - dy$	$\dfrac{dy}{\Delta y}$
1.000	−0.25000	−0.13889	0.11111	1.79999
0.500	−0.12500	−0.09000	0.03500	1.38889
0.100	−0.02500	−0.02324	0.00176	1.07573
0.010	−0.00250	−0.00248	0.00002	1.00806
0.001	−0.00025	−0.00025	0.00000	1.00000

19.

$dx = \Delta x$	dy	Δy	$\Delta y - dy$	$\dfrac{dy}{\Delta y}$
1.000	0.14865	0.12687	−0.02178	1.17167
0.500	0.07433	0.06823	−0.00610	1.08940
0.100	0.01487	0.01459	−0.00028	1.01919
0.010	0.00149	0.00148	−0.00001	1.00676
0.001	0.00015	0.00015	0.00000	1.00000

21. $y = 28x + 37$

For $\Delta x = -0.01, f(x + \Delta x) = -19.281302$ and $y(x + \Delta x) = -19.28$

For $\Delta x = 0.01, f(x + \Delta x) = -18.721298$ and $y(x + \Delta x) = -18.72$

23. $y = x$

For $\Delta x = -0.01, f(x + \Delta x) = -0.009999$ and $y(x + \Delta x) = -0.01$

For $\Delta x = 0.01, f(x + \Delta x) = 0.009999$ and $y(x + \Delta x) = 0.01$

25. $dN = 3216$

27. (a) About 2.091% change from 2001 to 2002.
(b) About -0.504% change from 2004 to 2005.

29. About 0.0478 mg/mL

31. (a) $dP = -0.3787$
(b)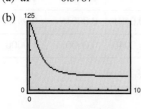

About -0.31895

(c) The differential dP is slightly smaller than the actual change in pressure from $t = 8$ to $t = 9$.

33. (a) $dA = 2x\Delta x$, $\Delta A = 2x\Delta x + (\Delta x)^2$
(b) and (c)

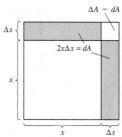

35. $\pm \frac{5}{2}\pi$ in.^{2}; $\pm \frac{1}{40}$ **37.** $\pm 2.88\pi$ in.^{3}; ± 0.01 **39.** True

REVIEW EXERCISES FOR CHAPTER 3
(page 253)

1. $x = 1$ **3.** $x = 0, x = 1$

5. Increasing on $\left(-\frac{1}{2}, \infty\right)$
Decreasing on $\left(-\infty, -\frac{1}{2}\right)$

7. Increasing on $(-\infty, 3)$ and $(3, \infty)$

9. (a) (1.38, 7.24)   (b) (1, 1.38), (7.24, 12)

(c) Normal monthly temperature is rising from early January to early July and decreasing from early July to early January.

(d)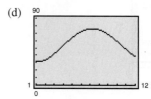

11. Relative maximum: $(0, -2)$
Relative minimum: $(1, -4)$

13. Relative minimum: $(8, -52)$

15. Relative maxima: $(-1, 1), (1, 1)$
Relative minimum: $(0, 0)$

17. Relative maximum: $(0, 6)$

19. Relative maximum: $(0, 0)$
Relative minimum: $(4, 8)$

21. Maximum: $(0, 6)$ **23.** Maxima: $(-2, 17), (4, 17)$
Minimum: $\left(-\frac{5}{2}, -\frac{1}{4}\right)$ Minima: $(-4, -15), (2, -15)$

25. Maximum: $(1, 3)$
Minimum: $(3, 4\sqrt{3} - 9)$

27. Maximum: $\left(2, \dfrac{2\sqrt{5}}{5}\right)$ **29.** Maximum: $(1, 1)$
Minimum: $(0, 0)$ Minimum: $(-1, -1)$

31.

Minimum surface area when $r \approx 0.7937$ in.

33. Concave upward on $(2, \infty)$

Concave downward on $(-\infty, 2)$

35. Concave upward on $\left(-\frac{2\sqrt{3}}{3}, \frac{2\sqrt{3}}{3}\right)$

Concave downward on $\left(-\infty, -\frac{2\sqrt{3}}{3}\right)$ and $\left(\frac{2\sqrt{3}}{3}, \infty\right)$

37. $(0, 0)$, $(4, -128)$

39. $(0, 0)$, $(1.0652, 4.5244)$, $(2.5348, 3.5246)$

41. Relative maximum: $\left(-\sqrt{3}, 6\sqrt{3}\right)$

Relative minimum: $\left(\sqrt{3}, -6\sqrt{3}\right)$

43. Relative maximum: $\left(-\frac{\sqrt{2}}{2}, \frac{1}{2}\right)$, $\left(\frac{\sqrt{2}}{2}, \frac{1}{2}\right)$

Relative minimum: $(0, 0)$

45. (a)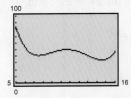

(b) Concave upward on $(5, 8.9)$ and $(13.0, 16)$

Concave downward on $(8.9, 13.0)$

(c) $(8.9, 45.0)$, $(13.0, 40.6)$

(d) In 1998, the number of fatalities due to lightning was increasing but by the end of 1998 the rate of this increase began to slow down. In 2002, the number of fatalities due to lightning was decreasing but by the beginning of 2003 the rate of this decrease began to slow down.

47. About 12.7 ft

49. (a) 70 ft by 70 ft

(b) 100 ft by 100 ft

51. (a) $\frac{dN}{dt}$ is decreasing on the entire interval $(0, 5)$.

(b) $\lim\limits_{t \to \infty} N = -\infty$

(c) Answers will vary.

53. Vertical asymptote: $x = 4$

Horizontal asymptote: $y = 2$

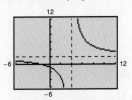

55. Vertical asymptote: $x = 0$

Horizontal asymptotes: $y = \pm 3$

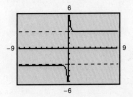

57. No vertical asymptote

Horizontal asymptote: $y = 0$

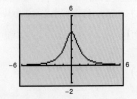

59. $-\infty$ **61.** ∞ **63.** $\frac{2}{3}$ **65.** 0

67. (a)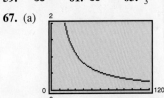

(b) $\lim\limits_{s \to \infty} T = -0.03$

69.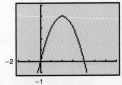

Intercepts: $(0, 0)$, $(4, 0)$

Relative maximum: $(2, 4)$

Relative minimum: none

Point of inflection: none

Asymptotes: none

Domain: $(-\infty, \infty)$

71.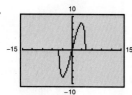

Intercepts: $(-4, 0), (0, 0), (4, 0)$
Relative maximum: $(2\sqrt{2}, 8)$
Relative minimum: $(-2\sqrt{2}, -8)$
Point of inflection: $(0, 0)$
Asymptotes: none
Domain: $[-4, 4]$

73.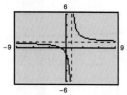

Intercepts: $(-1, 0), (0, -1)$
Horizontal asymptote: $y = 1$
Vertical asymptote: $x = 1$
Domain: $(-\infty, 1) \cup (1, \infty)$

75.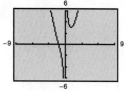

Intercept: $\left(-\sqrt[3]{2}, 0\right)$
Relative minimum: $(1, 3)$
Point of inflection: $\left(-\sqrt[3]{2}, 0\right)$
Vertical asymptote: $x = 0$
Domain: $(-\infty, 0) \cup (0, \infty)$

77. (a)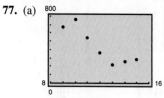

(b) Answers will vary.

(c)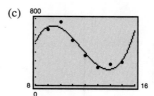

(d) Increasing: 1999 to 2000 and 2003 to 2005
Decreasing: 2000 to 2003
(e) Increasing: mid-2001 to 2005
Decreasing: 1999 to mid-2001

79. $dy = 1 - 2x \, dx$

81. $dy = -\dfrac{x}{\sqrt{36 - x^2}} \, dx$

83. (a) 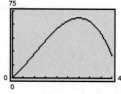 (b) 67.5

85. (a) $\Delta p \approx dp = -0.493$ (b) $\Delta p \approx dp = -0.704$

87. (a) $dS = \pm 1.8\pi$ in.^{2} (b) $dV = \pm 8.1\pi$ in.^{3}

CHAPTER TEST (page 257)

1. Critical number: $x = 0$
   Increasing on $(0, \infty)$
   Decreasing on $(-\infty, 0)$
2. Critical numbers: $x = -2, x = 2$
   Increasing on $(-\infty, -2)$ and $(2, \infty)$
   Decreasing on $(-2, 2)$
3. Critical number: $x = 5$
   Increasing on $(5, \infty)$
   Decreasing on $(-\infty, 5)$
4. Relative minimum: $(3, -14)$
   Relative maximum: $(-3, 22)$
5. Relative minima: $(-1, -7)$ and $(1, -7)$
   Relative maximum: $(0, -5)$
6. Relative maximum: $(0, 2.5)$
7. Minimum: $(-3, -1)$
   Maximum: $(0, 8)$
8. Minimum: $(0, 0)$
   Maximum: $(2.25, 9)$

9. Minimum: $(2\sqrt{3}, 2\sqrt{3})$
   Maximum: $(1, 6.5)$

10. Concave upward: $\left(\dfrac{\sqrt[3]{50}}{5}, \infty\right)$
    Concave downward: $\left(-\infty, \dfrac{\sqrt[3]{50}}{5}\right)$

11. Concave upward: $\left(-\infty, -\dfrac{2\sqrt{2}}{3}\right)$ and $\left(\dfrac{2\sqrt{2}}{3}, \infty\right)$
    Concave downward: $\left(-\dfrac{2\sqrt{2}}{3}, \dfrac{2\sqrt{2}}{3}\right)$

12. No point of inflection

13. $\left(\sqrt[3]{2}, -\dfrac{18\sqrt[3]{4}}{5}\right)$

14. Relative minimum: $(5.46, -135.14)$
    Relative maximum: $(-1.46, 31.14)$

15. Relative minimum: $(3, -97.2)$
    Relative maximum: $(-3, 97.2)$

16. Vertical asymptote: $x = 5$
    Horizontal asymptote: $y = 3$

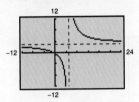

17. Horizontal asymptote: $y = 2$

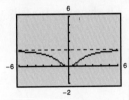

18. Vertical asymptote: $x = 1$

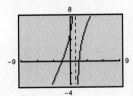

19. 1   20. ∞ 21. 3   22. $dy = 10x\,dx$

23. $dy = -\dfrac{4}{(x+3)^2}dx$ 24. $dy = 3(x+4)^2\,dx$

25. The percent was the highest in 2001 and the lowest in 2004.

CHAPTER 4

SECTION 4.1 (page 263)

Skills Review (page 263)

1. Horizontal shift to the left two units
2. Reflection about the x-axis
3. Vertical shift down one unit
4. Reflection about the y-axis
5. Horizontal shift to the right one unit
6. Vertical shift up two units
7. Nonremovable discontinuity at $x = -4$
8. Continuous on $(-\infty, \infty)$
9. Discontinuous at $x = \pm 1$
10. Continuous on $(-\infty, \infty)$
11. 5   12. $\tfrac{4}{3}$ 13. $-9, 1$ 14. $2 \pm 2\sqrt{2}$
15. $1, -5$ 16. $\tfrac{1}{2}, 1$

1. (a) 625   (b) 9   (c) $16\sqrt{2}$
   (d) 9   (e) 125   (f) 4

3. (a) 3125   (b) $\tfrac{1}{5}$ (c) 625   (d) $\tfrac{1}{125}$

5. (a) $\tfrac{1}{5}$ (b) 27   (c) 5   (d) 4096

7. (a) 4   (b) $\dfrac{\sqrt{2}}{2} \approx 0.707$ (c) $\dfrac{1}{8}$ (d) $\dfrac{\sqrt{2}}{8} \approx 0.177$

9. (a) 0.907   (b) 348.912   (c) 1.796   (d) 1.308

11. (a) About 1457   (b) About 1217   13. e   14. c

15. a   16. f   17. d   18. b

19.   21.

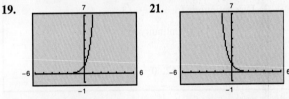

23.   25.

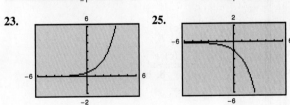

A58 Answers to Selected Exercises

27. 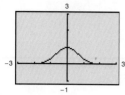 **29.**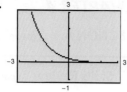

31. (a)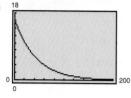

(b) 0.5 g (c) About 120 yr

33. (a) $V(t) = 30,500\left(\frac{7}{8}\right)^t$ (b) About $17,878.54

35. (a)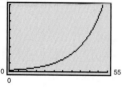

(b) About 3003 pairs

(c) About 6364 pairs

(d) Middle of 2011

(e) Explanations will vary.

SECTION 4.2 (page 269)

Skills Review (page 269)

1. Continuous on $(-\infty, \infty)$
2. Discontinuous for $x = \pm 2$
3. Discontinuous for $x = \pm\sqrt{3}$
4. Removable discontinuity at $x = 4$
5. 0 6. 0 7. 4 8. $\frac{1}{2}$ 9. $\frac{3}{2}$
10. 6 11. 0 12. 0

1. (a) e^7 (b) e^{12} (c) $\dfrac{1}{e^6}$ (d) 1

3. (a) e^5 (b) $e^{5/2}$ (c) e^6 (d) e^7

5. f **6.** e **7.** d **8.** b **9.** c **10.** a

11. 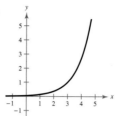 **13.**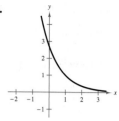

15. 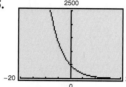 **17.**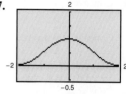

19. 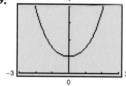 **21.**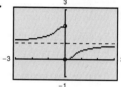

No horizontal asymptotes

Continuous on the entire real number line

Horizontal asymptote: $y = 1$

Discontinuous at $x = 0$

23. (a)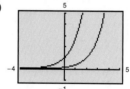

The graph of $g(x) = e^{x-2}$ is shifted horizontally two units to the right.

(b)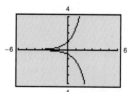

The graph of $h(x) = -\frac{1}{2}e^x$ decreases at a slower rate than e^x increases.

(c)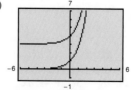

The graph of $q(x) = e^x + 3$ is shifted vertically three units upward.

25. (a) 0.1535 (b) 0.4866 (c) 0.8111

27. (a) 1960: 68,400 people
  1970: About 109,110 people
  1980: About 174,060 people
  1990: About 277,650 people
  2000: About 442,910 people
  2005: About 559,400 people

(b) The population rate does not increase at a constant rate.

(c) 2015

29.

Time	0	0.5	1	1.5	2	2.5
Cells	1	2	4	8	16	32

Time	3	3.5	4	4.5	5
Cells	64	128	256	512	1024

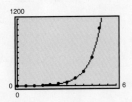

Model: $y = 4^t$ or $y = e^{t \ln 4}$

31. (a) About 16.66%   (b) About 38,800

SECTION 4.3   (page 277)

Skills Review   (page 277)

1. $\frac{1}{2}e^x(2x^2 - 1)$ **2.** $\frac{e^x(x+1)}{x}$ **3.** $e^x(x - e^x)$

4. $e^{-x}(e^{2x} - x)$ **5.** $-\frac{6}{7x^3}$ **6.** $6x - \frac{1}{6}$

7. $6(2x^2 - x + 6)$ **8.** $\frac{t+2}{2t^{3/2}}$

9. Relative maximum: $\left(-\frac{4\sqrt{3}}{3}, \frac{16\sqrt{3}}{9}\right)$

  Relative minimum: $\left(\frac{4\sqrt{3}}{3}, -\frac{16\sqrt{3}}{9}\right)$

10. Relative maximum: $(0, 5)$
  Relative minima: $(-1, 4), (1, 4)$

---

1. 3   **3.** -1 **5.** $5e^{5x}$ **7.** $-2xe^{-x^2}$

9. $\frac{2}{x^3}e^{-1/x^2}$ **11.** $e^{4x}(4x^2 + 2x + 4)$

13. $-\frac{6(e^x - e^{-x})}{(e^x + e^{-x})^4}$ **15.** $xe^x + e^x + 4e^{-x}$

17. $y = 2x - 3$ **19.** $y = \frac{4}{e^2}$ **21.** $y = 24x + 8$

23. $6(3e^{3x} + 2e^{-2x})$ **25.** $5(e^{-x} - 10e^{-5x})$

27.

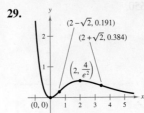

No relative extrema
No points of inflection
Horizontal asymptote to the right: $y = \frac{1}{2}$
Horizontal asymptote to the left: $y = 0$
Vertical asymptote: $x \approx -0.693$

29.

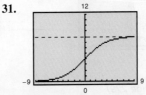

Relative minimum: $(0, 0)$
Relative maximum: $\left(2, \frac{4}{e^2}\right)$
Points of inflection: $(2 - \sqrt{2}, 0.191), (2 + \sqrt{2}, 0.384)$
Horizontal asymptote to the right: $y = 0$

31.

Horizontal asymptotes: $y = 0, y = 8$

33. $x = -\frac{1}{3}$ **35.** $x = 9$

37. $t = 3$: About 28.8 cm/yr

39. $t = 1$: -24.3%/week
  $t = 3$: -8.9%/week

41. (a)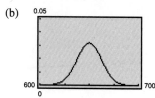

(b and c) 1996: 3.25 million people/yr
2000: 1.30 million people/yr
2005: 5.30 million people/yr

43. (a) $f(x) = \dfrac{1}{12.5\sqrt{2\pi}} e^{-(x-650)^2/312.5}$

(b)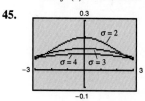

(c) $f'(x) = \dfrac{-4\sqrt{2}(x-650)e^{-2(x-650)^2/625}}{15{,}625\sqrt{\pi}}$

(d) $f'(x) = 0$ when $x = 650$,
$f'(x) > 0$ when $x < 650$,
and $f'(x) < 0$ when $x > 650$.

45.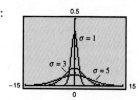

As σ increases, the graph becomes flatter.

47. Proof; Maximum: $\left(0, \dfrac{1}{\sigma\sqrt{2\pi}}\right)$; Answers will vary.

Sample answer:

MID-CHAPTER QUIZ (page 279)

1. 64   **2.** $\frac{8}{27}$ **3.** $3\sqrt[3]{3}$ **4.** $\frac{16}{81}$ **5.** 1024
6. 216   **7.** 27   **8.** $\sqrt{15}$ **9.** e^7 **10.** $e^{11/3}$
11. e^6 **12.** e^3

13. 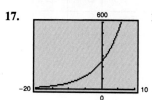 **14.**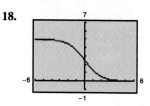

15. **16.**

17. **18.**

19. (a) $f(x) = \dfrac{2}{\sqrt{2\pi}} e^{-2(x-5.4)^2}$

(b)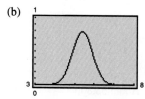

(c) $f'(x) = -\dfrac{8(x-5.4)}{\sqrt{2\pi}} e^{-2(x-5.4)^2}$

(d) $f'(x) = 0$ when $x = 5.4$, $f'(x) > 0$ when $x < 5.4$, and $f'(x) < 0$ when $x > 5.4$.

20. $5e^{5x}$ **21.** e^{x-4} **22.** $5e^{x+2}$
23. $e^x(2-x)$ **24.** $y = -2x + 1$
25.

Relative maximum: $(4, 8e^{-2})$
Relative minimum: $(0, 0)$
Points of inflection: $(4 - 2\sqrt{2}, 0.382)$, $(4 + 2\sqrt{2}, 0.767)$
Horizontal asymptote to the right: $y = 0$

SECTION 4.4 (page 286)

Skills Review (page 286)

1. $\frac{1}{4}$ 2. 64   3. 729   4. $\frac{8}{27}$ 5. 1
6. $81e^4$ 7. $\frac{e^3}{2}$ 8. $\frac{125}{8e^3}$ 9. $x > -4$
10. Any real number x 11. $x < -1$ or $x > 1$
12. $x > 5$
13. $f(g(x)) = \frac{2x - 6}{2} + 3 = x$
    $g(f(x)) = 2\left(\frac{x}{2} + 3\right) - 6 = x$
14. $f(g(x)) = \sqrt[3]{x^3 + 8} - 8 = x$
    $g(f(x)) = \left(\sqrt[3]{x - 8}\right)^3 + 8 = x$

1. $e^{0.6931\ldots} = 2$ 3. $e^{-1.6094\ldots} = 0.2$ 5. $\ln 1 = 0$
7. $\ln(0.0498\ldots) = -3$ 9. c   10. d
11. b   12. a

13. 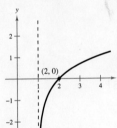   15.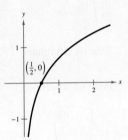

17.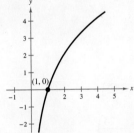

19. Answers will vary.   21. Answers will vary.

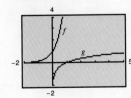

23. x^2 25. $5x + 2$ 27. $2x - 1$
29. (a) 1.7917   (b) 0.4055   (c) 4.3944   (d) 0.5493

31. $\ln 2 - \ln 3$ 33. $\ln x + \ln y + \ln z$
35. $\frac{1}{2}\ln(x^2 + 1)$ 37. $\ln z + 2\ln(z - 1)$
39. $\ln 3 + \ln x + \ln(x + 1) - 2\ln(2x + 1)$
41. $\ln\frac{x-2}{x+2}$ 43. $\ln\frac{x^3 y^2}{z^4}$ 45. $\ln\left[\frac{x(x+3)}{x+4}\right]^3$
47. $\ln\left[\frac{x(x^2+1)}{x+1}\right]^{3/2}$ 49. $\ln\frac{(x+1)^{1/3}}{(x-1)^{2/3}}$
51. $x = 4$ 53. $x = 1$
55. $x = \frac{e^{2.4}}{2} \approx 5.5116$ 57. $x = \frac{e^{10/3}}{5} \approx 5.6063$
59. $x = \ln 4 - 1 \approx 0.3863$
61. $t = \frac{\ln 7 - \ln 3}{-0.2} \approx -4.2365$
63. $x = \frac{1}{2}\left(1 + \ln\frac{3}{2}\right) \approx 0.7027$
65. $x = -100\ln\frac{3}{4} \approx 28.7682$
67. $x = \frac{\ln 15}{2\ln 5} \approx 0.8413$ 69. $t = \frac{\ln 2}{\ln 1.07} \approx 10.2448$
71. $t = \frac{\ln 3}{12\ln[1 + (0.07/12)]} \approx 15.7402$
73. $t = \frac{\ln 30}{3\ln[16 - (0.878/26)]} \approx 0.4092$
75. (a) $P(25) \approx 210{,}650$ (b) 2023
77. 9395 yr   79. 12,484 yr
81. (a) About 36.38 million people   (b) $t = 15$ or 2005
83. (a)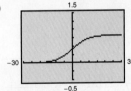

(b) Horizontal asymptotes: $y = 0$, $y = 0.83$

As the number of trials increases, the proportion of correct responses will never exceed 83%.

(c) About 4.79 trials

85. (a) $h = 0$ is not in the domain of the natural logarithmic function.
(b) $h = 0.86 - 6.447\ln p$
(c)

(d) About 2.71 km   (e) About 0.15 atmosphere

87.

x	1	5	10	10^2	10^4	10^6
f(x)	0	0.3219	0.2303	0.0461	0.0009	0.00001

(a) $\lim_{x \to \infty} \frac{\ln x}{x} = 0$

(b) Relative maximum: (2.718, 0.368)

No relative minima

89. 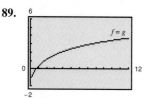 The graphs appear to be identical.

91. True

93. False; $\frac{1}{2} f(x) = \frac{1}{2} \ln x = \ln x^{1/2} \neq \sqrt{\ln x}$

95. True

SECTION 4.5 (page 295)

Skills Review (page 295)

1. $2 \ln(x + 1)$ **2.** $\ln x + \ln(x + 1)$
3. $\ln x - \ln(x + 1)$ **4.** $3[\ln x - \ln(x - 3)]$
5. $\ln 4 + \ln x + \ln(x - 7) - 2 \ln x$
6. $3 \ln x + \ln(x + 1)$
7. 2 **8.** $-\frac{1}{2}$ **9.** $-12x + 2$ **10.** $-\frac{6}{x^4}$

1. 3 **3.** 2 **5.** $\frac{2}{x}$ **7.** $\frac{2x}{x^2 + 3}$ **9.** $\frac{1}{2(x - 4)}$
11. $\frac{4}{x}(\ln x)^3$ **13.** $2 \ln x + 2$ **15.** $\frac{2x^2 - 1}{x(x^2 - 1)}$
17. $\frac{1}{x(x + 1)}$ **19.** $\frac{2}{3(x^2 - 1)}$ **21.** $-\frac{4}{x(4 + x^2)}$
23. $e^{-x}\left(\frac{1}{x} - \ln x\right)$ **25.** $\frac{e^x - e^{-x}}{e^x + e^{-x}}$ **27.** $e^{x(\ln 2)}$
29. $\frac{1}{\ln 4} \ln x$ **31.** 1.404 **33.** 5.585 **35.** -0.631
37. -2.134 **39.** $(\ln 3)3^x$ **41.** $\frac{1}{x \ln 2}$
43. $(2 \ln 4)4^{2x-3}$ **45.** $\frac{2x + 6}{(x^2 + 6x) \ln 10}$
47. $2^x(1 + x \ln 2)$ **49.** $y = x - 1$
51. $y = \frac{1}{27 \ln 3} x - \frac{1}{\ln 3} + 3$

53. $\frac{1}{2x}$ **55.** $\frac{1}{x}$ **57.** $(\ln 5)^2 5^x$
59. $\frac{d\beta}{dI} = \frac{10}{(\ln 10)I}$; About 43,429.4 db/W/cm^2
61. 2, $y = 2x - 1$ **63.** $-\frac{8}{5}, y = -\frac{8}{5}x - 4$
65. $\frac{1}{\ln 2}, y = \frac{1}{\ln 2} x - \frac{1}{\ln 2}$

67.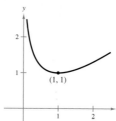

Relative minimum: (1, 1)

69.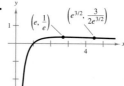

Relative maximum: $\left(e, \frac{1}{e}\right)$

Point of inflection: $\left(e^{3/2}, \frac{3}{2e^{3/2}}\right)$

71.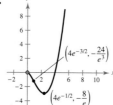

Relative minimum: $\left(4e^{-1/2}, \frac{-8}{e}\right)$

Point of inflection: $\left(4e^{-3/2}, \frac{-24}{e^3}\right)$

73. (a) The concentration increases.

(b) 1.333 mg/kg/cm

The concentration increases by 1.333 mg/kg/cm.

75. (a)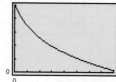

(b) About 145.44 mg/mL

(c) About 15.95 mg/mL

(d) $t = 4$: -44.06 mg/mL/hr

$t = 8$: -24.48 mg/mL/hr

(e) No, because the model yields values of C that are less than zero when $t \geq 9$.

77. (a) $h = 0$

(b) $h = 0$: $-97.04°C/hr$

$h = 1$: $-48.52°C/hr$

$h = 3$: $-12.13°C/hr$

79. (a) $s = 84.66 - 11.0 \ln t$

(b)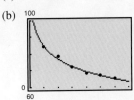

The model fits the data well.

(c) $t = 2$: $s'(2) = -5.5$; The average score is decreasing at a rate of 5.5%/mo at 2 months.

SECTION 4.6 (page 303)

Skills Review (page 303)

1. $-\frac{1}{4} \ln 2$ **2.** $\frac{1}{5} \ln \frac{10}{3}$ **3.** $-\frac{\ln(25/16)}{0.01}$

4. $-\frac{\ln(11/16)}{0.02}$ **5.** $7.36e^{0.23t}$ **6.** $1.296e^{0.072t}$

7. $-33.6e^{-1.4t}$ **8.** $-0.025e^{-0.001t}$ **9.** 4

10. 12   **11.** $2x + 1$ **12.** $x^2 + 1$

1. $y = e^{0.5756t}$ **3.** $y = e^{-0.3466t}$

5. $y = 2e^{0.1014t}$ **7.** $y = 4e^{-0.4159t}$

9. $y = 0.6687e^{0.4024t}$ **11.** $y = 10e^{2t}$, exponential growth

13. $y = 30e^{-4t}$, exponential decay

15. *Amount after 1000 years:* 6.48 g

Amount after 10,000 years: 0.13 g

17. *Initial quantity:* 6.73 g

Amount after 1000 years: 5.96 g

19. *Initial quantity:* 2.16 g

Amount after 10,000 years: 1.62 g

21. 68%   **23.** 15,642 yr

25. $k_1 = \frac{\ln 4}{12} \approx 0.1155$, so $y_1 = 5e^{0.1155t}$.

$k_2 = \frac{1}{6}$, so $y_2 = 5(2)^{t/6}$.

Explanations will vary.

27. (a) 1350   (b) $\frac{5 \ln 2}{\ln 3} \approx 3.15$ hr

(c) No; Answers will vary.

29. (a) Australia: $y = 19.3e^{0.0078t}$; About 24.4 million

Canada: $y = 31.4e^{0.0089t}$; About 41.0 million

Philippines: $y = 80.6e^{0.0174t}$; About 135.8 million

South Africa: $y = 45.3e^{-0.0046t}$; About 39.5 million

Turkey: $y = 66.3e^{0.0101t}$; About 89.8 million

(b) k (c) k

31. About 61.2 hr

33. (a) $V = -750t + 2000$

(b) $V = 2000e^{(-\ln 2)t} \approx 2000e^{-0.6931t}$

(c)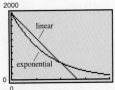

The exponential model depreciates faster in the first year.

(d) Linear: $V(1) = 1250$; $V(3) = -250$

(e) Exponential: $V(1) = 1000$; $V(3) = 250$

35. (a) $F(t) = 262.1t + 9436$

$F(t) = 9635(1.022)^t \approx 9635e^{0.0218t}$

(b) $F(21) \approx 15{,}229$

(c) $F(21) = 14{,}940$

(d)

Answers will vary.

37. (a) $N = 30(1 - e^{-0.0366t})$

(b) About 49 days

39. 2046   **41.** Answers will vary.

REVIEW EXERCISES FOR CHAPTER 4
(page 310)

1. 8 **3.** 64 **5.** 1 **7.** $\frac{1}{6}$ **9.** e^{10}
11. e^3 **13.** $f(4) = 128$ **15.** $f(10) \approx 1.219$
17. (a) $B = 1.4$: About 839.256 mL
$B = 1.6$: About 992.841 mL
$B = 1.75$: About 1026.207 mL
(b) Answers will vary.
19. c **20.** b **21.** f **22.** d **23.** e **24.** a
25.
27.

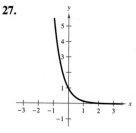
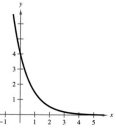

29.
31.

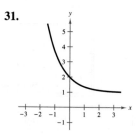

33.

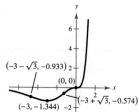

35. Yes; Because if $t = 20$, $y = 0.449 < 1$, which indicates that the species is endangered.
37. (a) $5e \approx 13.59$ (b) $5e^{-1/2} \approx 3.03$
(c) $5e^9 \approx 40{,}515.42$
39. (a) $6e^{-3.4} \approx 0.2002$ (b) $6e^{-10} \approx 0.0003$
(c) $6e^{-20} \approx 1.2367 \times 10^{-8}$
41. (a)

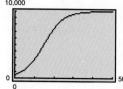

(b) $P \approx 1049$ fish
(c) Yes, P approaches 10,000 fish as t approaches ∞.
(d) The population is increasing most rapidly at the inflection point, which occurs around $t = 15$ months.
43. 1990: $P(0) = 29.7$ million
2000: $P(10) \approx 32.8$ million
2005: $P(15) \approx 34.5$ million
45. $8xe^{x^2}$ **47.** $\dfrac{1 - 2x}{e^{2x}}$ **49.** $4e^{2x}$ **51.** $-\dfrac{10e^{2x}}{(1 + e^{2x})^2}$
53.

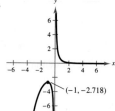

No relative extrema
No points of inflection
Horizontal asymptote: $y = 0$
55.

Relative minimum: $(-3, -1.344)$
Inflection points: $(0, 0)$, $(-3 + \sqrt{3}, -0.574)$, and $(-3 - \sqrt{3}, -0.933)$
Horizontal asymptote: $y = 0$
57.

Relative maximum: $(-1, -2.718)$
Horizontal asymptote: $y = 0$
Vertical asymptote: $x = 0$

59.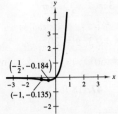

Relative minimum: $\left(-\frac{1}{2}, -0.184\right)$

Inflection point: $(-1, -0.135)$

Horizontal asymptote: $y = 0$

61. $e^{2.4849} \approx 12$ **63.** $\ln 4.4816 \approx 1.5$

65. **67.**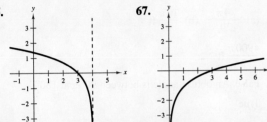

69. $\ln x + \frac{1}{2}\ln(x-1)$ **71.** $2\ln x - 3\ln(x+1)$

73. $3[\ln(1-x) - \ln 3 - \ln x]$ **75.** 3

77. $e^{3e^{-1}} \approx 3.0151$ **79.** 1   **81.** $\frac{1}{2}(\ln 6 + 1) \approx 1.3959$

83. $\frac{3+\sqrt{13}}{2} \approx 3.3028$ **85.** $-\frac{\ln(0.25)}{1.386} \approx 1.0002$

87. $\frac{\ln 1.1}{\ln 1.21} = 0.5$ **89.** $100 \ln\left(\frac{25}{4}\right) \approx 183.2581$

91. (a)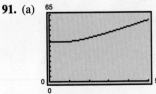

(b) Answers will vary.

(c) 2001: $2.3 billion/yr

2002: $4.5 billion/yr

2004: $6.0 billion/yr

93. $\frac{2}{x}$ **95.** $\frac{1}{x} + \frac{1}{x-1} - \frac{1}{x-2} = \frac{x^2 - 4x + 2}{x(x-2)(x-1)}$

97. 2   **99.** $\frac{1 - 3\ln x}{x^4}$ **101.** $\frac{4x}{3(x^2-2)}$

103. $\frac{2}{x} + \frac{1}{2(x+1)}$ **105.** $\frac{1}{1+e^x}$

107. 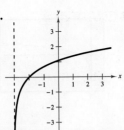   **109.**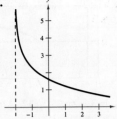

No relative extrema   No relative extrema

No points of inflection   No points of inflection

111. 2   **113.** 0   **115.** 1.594   **117.** 1.500

119. pH $\approx$ 4.95   **121.** H$^+ \approx 6.3 \times 10^{-4}$ **123.** 10

125. $\frac{2}{(2x-1)\ln 3}$ **127.** $-\frac{2}{x \ln 2}$

129. (a)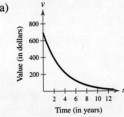

$t = 2$: $V = \$390.94$

(b) $t = 1$: $-\$149.95$/yr

$t = 4$: $-\$63.26$/yr

(c) About 6.7 years

131. $A = 500e^{-0.01277t}$ **133.** 27.9 yr

135. 2319 associations

CHAPTER TEST   (page 314)

1. 1   **2.** $\frac{1}{256}$ **3.** $e^{9/2}$ **4.** e^2

5. 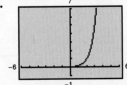   **6.**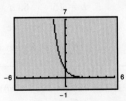

7. 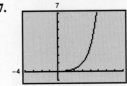   **8.**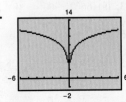

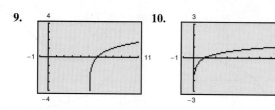

9. (graph) **10.** (graph)

11. $\ln 3 - \ln 2$ **12.** $\frac{1}{2}\ln(x+y)$
13. $\ln(x+1) - \ln y$ **14.** $\ln[y(x+1)]$
15. $\ln \frac{8}{(x-1)^2}$ **16.** $\ln \frac{x^2 y}{z+4}$
17. $x \approx 3.197$ **18.** $x \approx 1.750$ **19.** $x \approx 58.371$
20. (a) 2000: 3767 thousand employees
      2005: About 4163 thousand employees
  (b) 2003
21. $-3e^{-3x}$ **22.** $7e^{x+2} + 2$
23. $\frac{2x}{3+x^2}$ **24.** $\frac{2}{x(x+2)}$
25. (a) About 220.8 mi
  (b) $d = 100$: $S'(100) \approx 0.404$; The speed of the wind is increasing at a rate of 0.404 mi/hr/mi when the tornado has traveled 100 miles.
26. 59.4% **27.** 39.61 yr

CHAPTER 5

SECTION 5.1 (page 321)

Skills Review (page 321)

1. 35 cm² **2.** 12 in.² **3.** $c = 13$ **4.** $b = 4$
5. $b = 15$ **6.** $a = 6$ **7.** Equilateral triangle
8. Isosceles triangle **9.** Right triangle
10. Isosceles triangle and right triangle

1. (a) $405°, -315°$ (b) $319°, -401°$
3. (a) $660°, -60°$ (b) $20°, -340°$
5. (a) $\frac{19\pi}{9}, -\frac{17\pi}{9}$ (b) $\frac{8\pi}{3}, -\frac{4\pi}{3}$
7. (a) $\frac{7\pi}{4}, -\frac{\pi}{4}$ (b) $\frac{28\pi}{15}, -\frac{32\pi}{15}$
9. $\frac{\pi}{6}$ **11.** $\frac{3\pi}{2}$ **13.** $\frac{7\pi}{4}$ **15.** $-\frac{\pi}{9}$ **17.** $-\frac{3\pi}{2}$
19. $\frac{11\pi}{6}$ **21.** $450°$ **23.** $420°$ **25.** $-15°$
27. $405°$ **29.** $570°$ **31.** $-\frac{3\pi}{2}$ **33.** $\frac{4\pi}{5}$
35. $c = 10, \theta = 60°$ **37.** $a = 4\sqrt{3}, \theta = 30°$
39. $\theta = 40°$ **41.** $s = \sqrt{3}, \theta = 60°$
43. $4\sqrt{3}$ in.² **45.** $\frac{25\sqrt{3}}{4}$ ft² **47.** 18 ft

49.

r	8 ft	15 in.	85 cm	24 in.	$\frac{12{,}963}{\pi}$ mi
s	12 ft	24 in.	200.28 cm	96 in.	8642 mi
θ	1.5	1.6	$\frac{3\pi}{4}$	4	$\frac{2\pi}{3}$

51. (a) $-\frac{4\pi}{9}$ (b) $\frac{10\pi}{9}$ ft
53. $\frac{4900\pi}{3}$ ft²
55. False. An obtuse angle is between 90° and 180°.
57. True

SECTION 5.2 (page 332)

Skills Review (page 332)

1. $\frac{3\pi}{4}$ **2.** $\frac{7\pi}{4}$ **3.** $-\frac{7\pi}{6}$ **4.** $-\frac{5\pi}{3}$
5. $-\frac{2\pi}{3}$ **6.** $-\frac{5\pi}{4}$ **7.** 3π **8.** $\frac{13\pi}{6}$
9. $x = 0, 1$ **10.** $x = 0, -\frac{1}{2}$
11. $x = -\frac{1}{2}, 1$ **12.** $x = -1, 3$ **13.** $x = 1$
14. $x = -1, \frac{1}{2}$ **15.** $x = 2, 3$ **16.** $x = -2, 1$
17. $t = 10$ **18.** $t = \frac{7}{2}$ **19.** $t = \frac{445}{8}$ **20.** $t = 7$

1. $\sin\theta = \frac{4}{5},$ $\csc\theta = \frac{5}{4}$
   $\cos\theta = \frac{3}{5},$ $\sec\theta = \frac{5}{3}$
   $\tan\theta = \frac{4}{3},$ $\cot\theta = \frac{3}{4}$
3. $\sin\theta = -\frac{5}{13},$ $\csc\theta = -\frac{13}{5}$
   $\cos\theta = -\frac{12}{13},$ $\sec\theta = -\frac{13}{12}$
   $\tan\theta = \frac{5}{12},$ $\cot\theta = \frac{12}{5}$
5. $\sin\theta = \frac{1}{2},$ $\csc\theta = 2$
   $\cos\theta = -\frac{\sqrt{3}}{2},$ $\sec\theta = -\frac{2\sqrt{3}}{3}$
   $\tan\theta = -\frac{\sqrt{3}}{3},$ $\cot\theta = -\sqrt{3}$

7. $\csc\theta = 2$ **9.** $\cot\theta = \frac{4}{3}$ **11.** $\sec\theta = \frac{17}{15}$

13.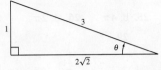

$\cos\theta = \frac{2\sqrt{2}}{3}$, $\csc\theta = 3$

$\tan\theta = \frac{\sqrt{2}}{4}$, $\sec\theta = \frac{3\sqrt{2}}{4}$

$\cot\theta = 2\sqrt{2}$

15.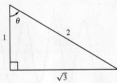

$\sin\theta = \frac{\sqrt{3}}{2}$, $\csc\theta = \frac{2\sqrt{3}}{3}$

$\cos\theta = \frac{1}{2}$

$\tan\theta = \sqrt{3}$, $\cot\theta = \frac{\sqrt{3}}{3}$

17.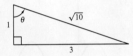

$\sin\theta = \frac{3\sqrt{10}}{10}$, $\csc\theta = \frac{\sqrt{10}}{3}$

$\cos\theta = \frac{\sqrt{10}}{10}$, $\sec\theta = \sqrt{10}$

$\cot\theta = \frac{1}{3}$

19. Quadrant IV   **21.** Quadrant I   **23.** Quadrant II

	Function	θ (deg)	θ (rad)	Function Value
25.	sin	30°	$\frac{\pi}{6}$	$\frac{1}{2}$
27.	tan	60°	$\frac{\pi}{3}$	$\sqrt{3}$
29.	cot	45°	$\frac{\pi}{4}$	1

31. (a) $\sin 60° = \frac{\sqrt{3}}{2}$ (b) $\sin\left(-\frac{2\pi}{3}\right) = -\frac{\sqrt{3}}{2}$

$\cos 60° = \frac{1}{2}$ $\cos\left(-\frac{2\pi}{3}\right) = -\frac{1}{2}$

$\tan 60° = \sqrt{3}$ $\tan\left(-\frac{2\pi}{3}\right) = \sqrt{3}$

33. (a) $\sin\left(-\frac{\pi}{6}\right) = -\frac{1}{2}$ (b) $\sin 150° = \frac{1}{2}$

$\cos\left(-\frac{\pi}{6}\right) = \frac{\sqrt{3}}{2}$ $\cos 150° = -\frac{\sqrt{3}}{2}$

$\tan\left(-\frac{\pi}{6}\right) = -\frac{\sqrt{3}}{3}$ $\tan 150° = -\frac{\sqrt{3}}{3}$

35. (a) $\sin 225° = -\frac{\sqrt{2}}{2}$ (b) $\sin(-225°) = \frac{\sqrt{2}}{2}$

$\cos 225° = -\frac{\sqrt{2}}{2}$ $\cos(-225°) = -\frac{\sqrt{2}}{2}$

$\tan 225° = 1$ $\tan(-225°) = -1$

37. (a) $\sin 750° = \frac{1}{2}$ (b) $\sin 510° = \frac{1}{2}$

$\cos 750° = \frac{\sqrt{3}}{2}$ $\cos 510° = -\frac{\sqrt{3}}{2}$

$\tan 750° = \frac{\sqrt{3}}{3}$ $\tan 510° = -\frac{\sqrt{3}}{3}$

39. (a) 0.1736   (b) 5.7588
41. (a) 0.3640   (b) 0.3640
43. (a) −0.3420   (b) −0.3420
45. (a) 2.0070   (b) 2.0000
47. (a) $\frac{\pi}{6}, \frac{5\pi}{6}$ (b) $\frac{7\pi}{6}, \frac{11\pi}{6}$
49. (a) $\frac{\pi}{3}, \frac{2\pi}{3}$ (b) $\frac{3\pi}{4}, \frac{7\pi}{4}$
51. (a) $\frac{\pi}{4}, \frac{5\pi}{4}$ (b) $\frac{5\pi}{6}, \frac{11\pi}{6}$ **53.** $\frac{\pi}{4}, \frac{3\pi}{4}, \frac{5\pi}{4}, \frac{7\pi}{4}$
55. $0, \frac{\pi}{4}, \pi, \frac{5\pi}{4}, 2\pi$ **57.** $\frac{\pi}{6}, \frac{\pi}{2}, \frac{5\pi}{6}, \frac{3\pi}{2}$ **59.** $\frac{\pi}{4}, \frac{5\pi}{4}$
61. $0, \frac{\pi}{2}, \pi, 2\pi$ **63.** $\frac{100\sqrt{3}}{3}$ **65.** $\frac{25\sqrt{3}}{3}$
67. 15.5572   **69.** About 443.2 m; about 323.3 m
71. About 19.3 ft   **73.** About 1.3 mi
75. (a) 102.6°F   (b) 102.1°F   (c) 100.6°F

At 4 P.M. the following afternoon, the patient's temperature should return to normal. This is determined by setting the function equal to 98.6 and solving for t.

77.

x	0	2	4	6	8	10
$f(x)$	0	2.7021	2.7756	1.2244	1.2979	4

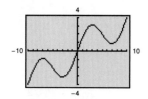

MID-CHAPTER QUIZ (page 335)

1. $\dfrac{\pi}{12}$ 2. $\dfrac{7\pi}{12}$ 3. $-\dfrac{4\pi}{9}$ 4. $\dfrac{7\pi}{36}$ 5. 120°

6. 48°   7. $-240°$ 8. 165°   9. $-\dfrac{\sqrt{2}}{2}$

10. $-\dfrac{\sqrt{3}}{2}$ 11. $-\dfrac{\sqrt{3}}{3}$ 12. 1   13. 2   14. -1

15. $\dfrac{\pi}{4}, \dfrac{5\pi}{4}$ 16. $0, 2\pi$ 17. $\dfrac{\pi}{3}, \dfrac{2\pi}{3}, \dfrac{4\pi}{3}, \dfrac{5\pi}{3}$

18. $\theta = 30°; a = 5\sqrt{3}$ 19. $\theta = 40°; a \approx 10.285$

20. $\theta = 50°; a \approx 3.356$ 21. $d \approx 286.8$ ft

22. About 11.5°

SECTION 5.3 (page 342)

> **Skills Review** (page 342)
>
> 1. 14   2. 10   3. 0   4. 0   5. 1
>
> 6. $-\dfrac{\sqrt{3}}{3}$ 7. $-\dfrac{1}{2}$ 8. $-\dfrac{\sqrt{3}}{2}$ 9. $\dfrac{1}{2}$
>
> 10. $-\dfrac{\sqrt{3}}{2}$ 11. 0.9659   12. -0.6428
>
> 13. -0.9962 14. 0.6428   15. 0.9744
>
> 16. 0.3090   17. -0.6494 18. -0.8391

1. Period: π
   Amplitude: 2
3. Period: 4π
   Amplitude: $\dfrac{3}{2}$
5. Period: 2
   Amplitude: $\dfrac{1}{2}$
7. Period: 2π
   Amplitude: 2
9. Period: $\dfrac{\pi}{5}$
   Amplitude: 2
11. Period: 3π
    Amplitude: $\dfrac{1}{2}$
13. Period: $\dfrac{1}{2}$
    Amplitude: 3

15. π 17. $\dfrac{2\pi}{5}$ 19. 6   21. c; π 22. e; π

23. f; 2   24. a; 2π 25. b; 4π 26. d; 2π

27. 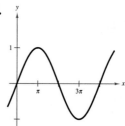   **29.**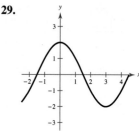

31. 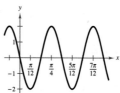   **33.**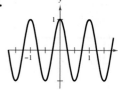

35. 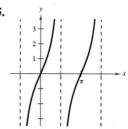   **37.**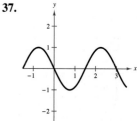

39. 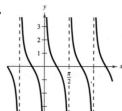   **41.**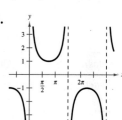

43. 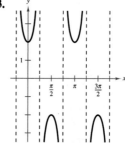   **45.**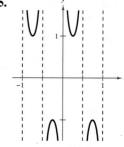

47.

x	-0.1	-0.01	-0.001
$f(x)$	1.9471	1.9995	2.0000

x	0.001	0.01	0.1
$f(x)$	2.0000	1.9995	1.9471

$\lim_{x \to 0} \dfrac{\sin 4x}{2x} = 2$

49.

x	-0.1	-0.01	-0.001
$f(x)$	0.1997	0.2000	0.2000

x	0.001	0.01	0.1
$f(x)$	0.2000	0.2000	0.1997

$\lim_{x \to 0} \dfrac{\sin x}{5x} = \dfrac{1}{5}$

51.

x	-0.1	-0.01	-0.001
$f(x)$	-0.1499	-0.0150	-0.0015

x	0.001	0.01	0.1
$f(x)$	0.0015	0.0150	0.1499

$\lim_{x \to 0} \dfrac{3(1 - \cos x)}{x} = 0$

53.

x	-0.1	-0.01	-0.001
$f(x)$	2.0271	2.0003	2.0000

x	0.001	0.01	0.1
$f(x)$	2.0000	2.0003	2.0271

$\lim_{x \to 0} \dfrac{\tan 2x}{x} = 2$

55.

x	-0.1	-0.01	-0.001
$f(x)$	-0.0997	-0.0100	-0.0010

x	0.001	0.01	0.1
$f(x)$	0.0010	0.0100	0.0997

$\lim_{x \to 0} \dfrac{\sin^2 x}{x} = 0$

57.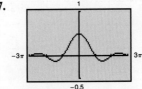

$\lim_{x \to 0} \dfrac{\sin x}{2x} = \dfrac{1}{2}$

59.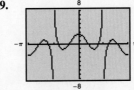

$\lim_{x \to 0} \dfrac{\sin 5x}{\sin 2x} = \dfrac{5}{2}$

61. $a = 2, d = 1$ **63.** $a = -4, d = 4$
65. b **66.** d **67.** a **68.** c
69. (a) 6 sec (b) 10
(c)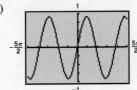

71. (a) $\dfrac{1}{440}$ (b) 440
(c)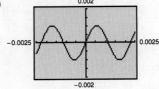

73. (a)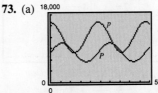

(b) As the population of the prey increases, the population of the predator increases as well. At some point, the predator eliminates the prey faster than the prey can reproduce, and the prey population decreases rapidly. As the prey becomes scarce, the predator population decreases, releasing the prey from predator pressure, and the cycle begins again.

75.

77. $P(8930) \approx 0.9977$
$E(8930) \approx -0.4339$
$I(8930) \approx -0.6182$

79. (a)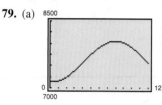

(b) Yes; June, July, August, and September

81. (a) A (b) B (c) B
(d) The frequency is the inverse of the period.

83. False. The amplitude is $|-3| = 3$. **85.** True

87. Answers will vary.

SECTION 5.4 (page 352)

Skills Review (page 352)

1. $f'(x) = 9x^2 - 4x + 4$ **2.** $g'(x) = 12x^2(x^3 + 4)^3$

3. $f'(x) = 3x^2 + 2x + 1$ **4.** $g'(x) = \dfrac{2(5 - x^2)}{(x^2 + 5)^2}$

5. Relative minimum: $(-2, -3)$

6. Relative maximum: $\left(-2, \frac{22}{3}\right)$
Relative minimum: $\left(2, -\frac{10}{3}\right)$

7. $x = \dfrac{\pi}{3}, x = \dfrac{2\pi}{3}$ **8.** $x = \dfrac{2\pi}{3}, x = \dfrac{4\pi}{3}$

9. $x = \pi$ **10.** No solution

---

1. $-3 \cos x$ **3.** $2x + \sin x$ **5.** $\dfrac{2}{\sqrt{x}} - 3 \sin x$

7. $-t^2 \sin t + 2t \cos t$ **9.** $-\dfrac{t \sin t + \cos t}{t^2}$

11. $\sec^2 x + 2x$ **13.** $e^{x^2} \sec x (\tan x + 2x)$

15. $-3 \sin 3x + 2 \sin x \cos x$ **17.** $\pi \tan \pi x \sec \pi x$

19. $\sin \dfrac{1}{x} - \dfrac{1}{x} \cos \dfrac{1}{x}$ **21.** $12 \sec^2 4x$

23. $16 \sec^2 4x \tan 4x$ **25.** $2e^{2x}(\cos 2x + \sin 2x)$

27. $-2 \cos x \sin x = -\sin 2x$

29. $-4 \cos x \sin x = -2 \sin 2x$

31. $2 \sin 2x + 2 \sin x \cos x = 3 \sin 2x$

33. $\sec^2 x - 1 = \tan^2 x$

35. $\sin^2 x \cos x - \sin^4 x \cos x = \sin^2 x \cos^3 x$

37. $\dfrac{2 \cos x}{\sin x} = 2 \cot x$ **39.** $y = 2x + \dfrac{\pi}{2} - 1$

41. $y = 4x - 4\pi$ **43.** $y = -2x + \dfrac{3}{2}\pi - 1$

45. $y = 0$

47. $\frac{5}{4}$; one complete cycle **49.** 2; two complete cycles

51. 1; one complete cycle

53. Relative maximum: $\left(\dfrac{\pi}{3}, \dfrac{3\sqrt{3}}{2}\right)$

Relative minimum: $\left(\dfrac{5\pi}{3}, -\dfrac{3\sqrt{3}}{2}\right)$

55. Relative maximum: $\left(\dfrac{5\pi}{3}, \dfrac{5\pi}{3} + \sqrt{3}\right)$

Relative minimum: $\left(\dfrac{\pi}{3}, \dfrac{\pi}{3} - \sqrt{3}\right)$

57. Relative maximum: $\left(\dfrac{\pi}{4}, 1.5509\right)$

Relative minimum: $\left(\dfrac{5\pi}{4}, -35.8885\right)$

59. July; 14 hr

61. (a) $h'(t)$ is a maximum when $t = 0$, or at midnight.
(b) $h'(t)$ is a minimum when $t = \frac{1}{2}$, or at noon.

63. (a)

(b) 0, 2.2889, 5.0870
(c) $f' > 0$ on $(0, 2.2889)$, $(5.0870, 2\pi)$
$f' < 0$ on $(2.2889, 5.0870)$

65. (a)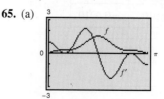

(b) 0.5236, $\pi/2$, 2.6180

(c) $f' > 0$ on $(0, 0.5236)$, $(0.5236, \pi/2)$
$f' < 0$ on $(\pi/2, 2.6180)$, $(2.6180, \pi)$

67. (a) 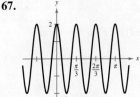 (b) 1.8366, 4.8158

(c) $f' > 0$ on $(0, 1.8366)$, $(4.8158, 2\pi)$
$f' < 0$ on $(1.8366, 4.8158)$

69. Relative maximum: $(4.49, -4.60)$

71. Relative maximum: $(1.27, 0.07)$
Relative minimum: $(3.38, -1.18)$

73. Relative maximum: $(3.96, 1)$

75. False. $y' = \frac{1}{2}\sin x(1 - \cos x)^{-1/2}$

77. False. $y' = x \sin 2x + \sin^2 x$

79. (a) Answers will vary.
(b) $\theta \approx 2$
(c) About $-48{,}000$ in.^{3}/radian

81. Answers will vary.

REVIEW EXERCISES FOR CHAPTER 5
(page 359)

1. $\dfrac{15\pi}{4}, -\dfrac{\pi}{4}$ **3.** $\dfrac{7\pi}{2}, -\dfrac{\pi}{2}$

5. $495°, -225°$ **7.** $315°, -45°$

9. $\dfrac{7\pi}{6}$ **11.** $-\dfrac{\pi}{3}$ **13.** $-\dfrac{8\pi}{3}$ **15.** $\dfrac{11\pi}{18}$

17. $240°$ **19.** $-120°$ **21.** $b = 4\sqrt{3}, \theta = 60°$

23. $a = \dfrac{5\sqrt{3}}{2}, c = 5, \theta = 60°$ **25.** About 90.14 ft

27. $\dfrac{\pi}{3}$ **29.** $\dfrac{\pi}{6}$ **31.** $60°$ **33.** $60°$ **35.** $\dfrac{\sqrt{2}}{2}$

37. $-\sqrt{3}$ **39.** $-\dfrac{\sqrt{3}}{2}$ **41.** $\sqrt{3}$ **43.** -1

45. $-\dfrac{1}{2}$ **47.** 0.6494   **49.** 3.2361   **51.** -0.3420

53. -0.2588 **55.** $r \approx 146.19$ **57.** $x \approx 68.69$

59. $\dfrac{2\pi}{3}, \dfrac{4\pi}{3}$ **61.** $\dfrac{7\pi}{6}, \dfrac{3\pi}{2}, \dfrac{11\pi}{6}$ **63.** $\dfrac{\pi}{3}, \pi, \dfrac{5\pi}{3}$

65. About 81.18 ft

67.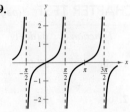

69.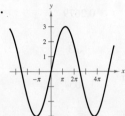

71.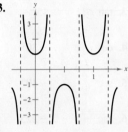

73.

75. (a)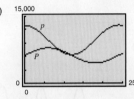

(b) Each function has a period of 24 months. The predator population has an amplitude of 1700 and oscillates about the line $y = 6200$. The prey population has an amplitude of 3300 and oscillates about the line $y = 9650$. As the predator population increases, the prey population decreases. As the predator population decreases, the prey population increases.

77. $5\pi \cos 5\pi x$ **79.** $-x \sec^2 x - \tan x$

81. $-\dfrac{x \sin x + 2 \cos x}{x^3}$

83. $2 \sin x \cos x + 1 = \sin 2x + 1$

85. $-4 \cot x \csc^4 x$ **87.** $e^x(\cot x - \csc^2 x)$

89. $y = -2x + \dfrac{\pi}{2}$ **91.** $y = \dfrac{1}{2}$ **93.** $y = 0$

95. Relative maximum: $\left(\dfrac{\pi}{6}, \dfrac{\pi + 6\sqrt{3}}{12}\right)$

Relative minimum: $\left(\dfrac{5\pi}{6}, \dfrac{5\pi - 6\sqrt{3}}{12}\right)$

97. Relative maxima: $\left(\dfrac{\pi}{2}, 2\right), \left(\dfrac{3\pi}{2}, 0\right)$

Relative minima: $\left(\dfrac{7\pi}{6}, -\dfrac{1}{4}\right), \left(\dfrac{11\pi}{6}, -\dfrac{1}{4}\right)$

99. August; About 8049 thousand workers

Answers to Selected Exercises

CHAPTER TEST (page 362)

	Function	θ (deg)	θ (rad)	Function value
1.	sin	67.5°	$\frac{3\pi}{8}$	0.9239
2.	cos	36°	$\frac{\pi}{5}$	0.8090
3.	tan	15°	$\frac{\pi}{12}$	0.2679
4.	cot	−30°	$-\frac{\pi}{6}$	$-\sqrt{3}$
5.	sec	−40°	$-\frac{2\pi}{9}$	1.3054
6.	csc	−225°	$-\frac{5\pi}{4}$	$\sqrt{2}$

7. About 27.37 in.

8. $\frac{\pi}{4}, \frac{3\pi}{4}$ 9. $\frac{\pi}{4}, \frac{3\pi}{4}, \frac{5\pi}{4}, \frac{7\pi}{4}$ 10. $\frac{\pi}{6}, \frac{7\pi}{6}$

11. 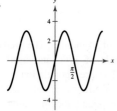   12.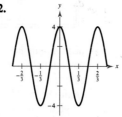

13.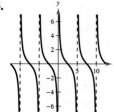

14. (a) $y' = \sin 2x - \sin x$

(b) Relative maxima: $\left(\frac{\pi}{3}, \frac{1}{4}\right), \left(\frac{5\pi}{3}, \frac{1}{4}\right)$

Relative minima: $(\pi, -2)$

15. (a) $y' = \sec\left(x - \frac{\pi}{4}\right) \tan\left(x - \frac{\pi}{4}\right)$

(b) Relative maximum: $\left(\frac{5\pi}{4}, -1\right)$

Relative minimum: $\left(\frac{\pi}{4}, 1\right)$

16. (a) $y' = \dfrac{\cos(x + \pi)}{[3 - \sin(x + \pi)]^2}$

(b) Relative maximum: $\left(\frac{3\pi}{2}, \frac{1}{2}\right)$

Relative minimum: $\left(\frac{\pi}{2}, \frac{1}{4}\right)$

17. (a)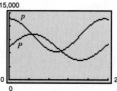

(b) Each function has a period of 24 months. The predator population has an amplitude of 2900 and oscillates about the line $y = 7300$. The prey population has an amplitude of 3700 and oscillates about the line $y = 10{,}000$. As the predator population increases, the prey population decreases. As the predator population decreases, the prey population increases.

CHAPTER 6

SECTION 6.1 (page 372)

> **Skills Review** (page 372)
>
> 1. $x^{-1/2}$ 2. $(2x)^{4/3}$ 3. $5^{1/2}x^{3/2} + x^{5/2}$
> 4. $x^{-1/2} + x^{-2/3}$ 5. $(x+1)^{5/2}$ 6. $x^{1/6}$
> 7. −12   8. −10   9. 14   10. 14

1–7. Answers will vary.   9. $6x + C$

11. $\frac{5}{3}t^3 + C$ 13. $-\dfrac{5}{2x^2} + C$ 15. $u + C$

17. $et + C$ 19. $\frac{2}{5}y^{5/2} + C$

	Rewrite	Integrate	Simplify
21.	$\int x^{1/3}\, dx$	$\dfrac{x^{4/3}}{4/3} + C$	$\dfrac{3}{4}x^{4/3} + C$
23.	$\int x^{-3/2}\, dx$	$\dfrac{x^{-1/2}}{-1/2} + C$	$-\dfrac{2}{\sqrt{x}} + C$
25.	$\dfrac{1}{2}\int x^{-3}\, dx$	$\dfrac{1}{2}\left(\dfrac{x^{-2}}{-2}\right) + C$	$-\dfrac{1}{4x^2} + C$

27. $\frac{x^2}{2} + 3x + C$ 29. $\frac{1}{4}x^4 + 2x + C$

31. $\frac{3}{4}x^{4/3} - \frac{3}{4}x^{2/3} + C$ 33. $\frac{3}{5}x^{5/3} + C$

35. $-\frac{1}{3x^3} + C$ 37. $2x - \frac{1}{2x^2} + C$

39. $\frac{3}{4}u^4 + \frac{1}{2}u^2 + C$ 41. $x^3 + \frac{x^2}{2} - 2x + C$

43. $\frac{2}{7}y^{7/2} + C$

45.   47.

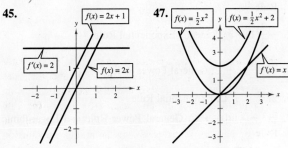

49. $f(x) = 2x^2 + 6$ 51. $f(x) = x^2 - 2x - 1$

53. $f(x) = -\frac{1}{x^2} + \frac{1}{x} + \frac{1}{2}$ 55. $y = -\frac{5}{2}x^2 - 2x + 2$

57. $f(x) = x^2 - 6$ 59. $f(x) = x^2 + x + 4$

61. $f(x) = \frac{9}{4}x^{4/3}$ 63. 56.25 ft

65. $v_0 = 40\sqrt{22} \approx 187.62$ ft/sec

67. (a) $h(t) = 0.75t^2 + 5t + 12$
    (b) 69 cm

69. About 77,868

71. (a) $I(t) = -0.0625t^4 + 1.773t^3 - 9.67t^2 + 21.03t - 0.212$ (in millions)

    (b) 20.072 million; No, this does not seem reasonable. Explanations will vary. Sample answer: A sharp decline from 863 million users to about 20 million users from the year 2004 to the year 2012 does not seem to follow the trend over the past few years, which is always increasing.

73. (a) $A = -0.136t^2 + 1.27t + 26.95$
    (b) 26.05 million persons

SECTION 6.2   (page 382)

Skills Review   (page 382)

1. $\frac{1}{2}x^4 + x + C$ 2. $\frac{3}{2}x^2 + \frac{2}{3}x^{3/2} - 4x + C$

3. $-\frac{1}{x} + C$ 4. $-\frac{1}{6t^2} + C$

5. $\frac{4}{7}t^{7/2} + \frac{2}{5}t^{5/2} + C$ 6. $\frac{4}{5}x^{5/2} - \frac{2}{3}x^{3/2} + C$

7. $\frac{5x^3 - 4}{2x} + C$ 8. $\frac{-6x^2 + 5}{3x^3} + C$

9. $\frac{1}{5}x^5 + \frac{2}{3}x^3 + x + C$

10. $\frac{1}{7}x^7 - \frac{4}{5}x^5 + \frac{1}{2}x^4 + \frac{4}{3}x^3 - 2x^2 + x + C$

11. $-\frac{5(x-2)^4}{16}$ 12. $-\frac{1}{12(x-1)^2}$

13. $9(x^2 + 3)^{2/3}$ 14. $-\frac{5}{(1-x^3)^{1/2}}$

	$\int u^n \frac{du}{dx} dx$	u	$\frac{du}{dx}$
1.	$\int (5x^2 + 1)^2 (10x)\, dx$	$5x^2 + 1$	$10x$
3.	$\int \sqrt{1-x^2}\,(-2x)\, dx$	$1 - x^2$	$-2x$
5.	$\int \left(4 + \frac{1}{x^2}\right)^5 \left(\frac{-2}{x^3}\right) dx$	$4 + \frac{1}{x^2}$	$-\frac{2}{x^3}$
7.	$\int (1 + \sqrt{x})^3 \left(\frac{1}{2\sqrt{x}}\right) dx$	$1 + \sqrt{x}$	$\frac{1}{2\sqrt{x}}$

9. $\frac{1}{5}(1 + 2x)^5 + C$ 11. $\frac{2}{3}(4x^2 - 5)^{3/2} + C$

13. $\frac{1}{5}(x-1)^5 + C$ 15. $\frac{(x^2 - 1)^8}{8} + C$

17. $-\frac{1}{3(1+x^3)} + C$ 19. $-\frac{1}{2(x^2 + 2x - 3)} + C$

21. $\sqrt{x^2 - 4x + 3} + C$ 23. $-\frac{15}{8}(1 - u^2)^{4/3} + C$

25. $4\sqrt{1 + y^2} + C$ 27. $-3\sqrt{2t + 3} + C$

29. $-\frac{1}{2}\sqrt{1 - x^4} + C$ 31. $-\frac{1}{24}\left(1 + \frac{4}{t^2}\right)^3 + C$

33. $\frac{(x^3 + 3x + 9)^2}{6} + C$ 35. $\frac{1}{4}(6x^2 - 1)^4 + C$

37. $-\frac{2}{45}(2 - 3x^3)^{5/2} + C$ 39. $\sqrt{x^2 + 25} + C$

41. $\frac{2}{3}\sqrt{x^3 + 3x + 4} + C$

43. (a) $\frac{1}{3}x^3 - x^2 + x + C_1 = \frac{1}{3}(x-1)^3 + C_2$

(b) Answers differ by a constant: $C_1 = C_2 - \frac{1}{3}$

(c) Answers will vary.

45. (a) $\frac{1}{6}x^6 - \frac{1}{2}x^4 + \frac{1}{2}x^2 + C_1 = \frac{(x^2-1)^3}{6} + C_2$

(b) Answers differ by a constant: $C_1 = C_2 - \frac{1}{6}$

(c) Answers will vary.

47. $f(x) = \frac{1}{3}[5 - (1-x^2)^{3/2}]$

49. (a) $W = \dfrac{1000}{15.08 - 0.128t} - 596.83$

(b) About 95.7 million women

51. (a) $h(t) = \sqrt{17.6t^2 + 1} + 5$ (b) 26 in.

53. (a) $P = 2(0.005t^2 + 1)^3 + 0.15$

(b)

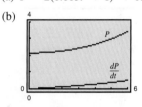

(c) $P(3) \approx 2.43$ million people; $P'(3) \approx 0.197$ million people/yr

(d) $P(3) = 2.43$ million people; $P'(3) = 0.197$ million people/yr

55. $\displaystyle\int \frac{x}{\sqrt{3x+2}}\,dx = \frac{2}{27}(3x+2)^{3/2} - \frac{4}{9}\sqrt{3x+2} + C$

SECTION 6.3 (page 389)

Skills Review (page 389)

1. $\left(\frac{5}{2}, \infty\right)$ **2.** $(-\infty, 2) \cup (3, \infty)$

3. $x + 2 - \dfrac{2}{x+2}$ **4.** $x - 2 + \dfrac{1}{x-4}$

5. $x + 8 + \dfrac{2x-4}{x^2-4x}$ **6.** $x^2 - x - 4 + \dfrac{20x+22}{x^2+5}$

7. $\frac{1}{4}x^4 - \frac{1}{x} + C$ **8.** $\frac{1}{2}x^2 + 2x + C$

9. $\frac{1}{2}x^2 - \frac{4}{x} + C$ **10.** $-\frac{1}{x} - \dfrac{3}{2x^2} + C$

1. $e^{2x} + C$ **3.** $\frac{1}{4}e^{4x} + C$ **5.** $-\frac{9}{2}e^{-x^2} + C$

7. $\frac{5}{3}e^{x^3} + C$ **9.** $\frac{1}{3}e^{x^3+3x^2-1} + C$ **11.** $-5e^{2-x} + C$

13. $\ln|x+1| + C$ **15.** $-\frac{1}{2}\ln|3-2x| + C$

17. $\frac{2}{3}\ln|3x+5| + C$

19. $\ln\sqrt{x^2+1} + C$ **21.** $\frac{1}{3}\ln|x^3+1| + C$

23. $\frac{1}{2}\ln|x^2+6x+7| + C$ **25.** $\ln|\ln x| + C$

27. $\ln|1-e^{-x}| + C$ **29.** $-\frac{1}{2}e^{2/x} + C$ **31.** $2e^{\sqrt{x}} + C$

33. $\frac{1}{2}e^{2x} - 4e^x + 4x + C$ **35.** $-\ln(1+e^{-x}) + C$

37. $-2\ln|5-e^{2x}| + C$

39. $e^x + 2x - e^{-x} + C$; Exponential Rule and General Power Rule

41. $-\frac{2}{3}(1-e^x)^{3/2} + C$; Exponential Rule

43. $-\dfrac{1}{x-1} + C$; General Power Rule

45. $2e^{2x-1} + C$; Exponential Rule

47. $\frac{1}{4}x^2 - 4\ln|x| + C$; General Power Rule and Logarithmic Rule

49. $2\ln(e^x + 1) + C$; Logarithmic Rule

51. $\frac{1}{2}x^2 + 3x + 8\ln|x-1| + C$; General Power Rule and Logarithmic Rule

53. $\ln|e^x + x| + C$; Logarithmic Rule

55. $f(x) = \frac{1}{2}x^2 + 5x + 8\ln|x-1| - 8$

57. (a) $P(t) = 1000[1 + \ln(1 + 0.25t)^{12}]$

(b) $P(3) \approx 7715$ bacteria   (c) $t \approx 6$ days

59. (a) $L(t) = -4.28\ln|t| + 0.14t + 15.46$

(b) $L(20) \approx 5.4$ million sheep and lambs

61. (a) $S(t) = -7241.22e^{-t/4.2} + 42{,}721.88$ (in dollars)

(b) $\$38{,}224.03$

63. False. $(\ln e)^{1/2} \neq \frac{1}{2}(\ln e)$

MID-CHAPTER QUIZ (page 391)

1. $3x + C$ **2.** $5x^2 + C$ **3.** $-\dfrac{1}{4x^4} + C$

4. $\dfrac{x^3}{3} - x^2 + 15x + C$ **5.** $\dfrac{x^3}{3} + 2x^2 + C$

6. $\dfrac{(6x+1)^4}{4} + C$ **7.** $\dfrac{(x^2-5x)^2}{2} + C$

8. $-\dfrac{1}{2(x^3+3)^2} + C$ **9.** $\dfrac{2}{15}(5x+2)^{3/2} + C$

10. $f(x) = 8x^2 + 1$ **11.** $f(x) = 3x^3 + 4x - 2$

12. $f(x) = \frac{2}{3}x^3 + x + 1$ **13.** $e^{5x+4} + C$

14. $\dfrac{x^2}{2} + e^{2x} + C$ **15.** $e^{x^3} + C$ **16.** $\ln|2x-1| + C$

17. $-\ln|x^2 + 3| + C$ **18.** $3\ln|x^3 + 2x^2| + C$

19. (a) $P(t) = 2(t + 2)^{3/2} + 9$

(b)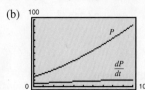

(c) $P(7) \approx 63$, or 6300 cells; $P'(7) \approx 9$, or 900 cells/day

(d) $P(7) = 63$, or 6300 cells; $P'(7) = 9$, or 900 cells/day

SECTION 6.4 (page 400)

Skills Review (page 400)

1. 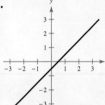   **2.**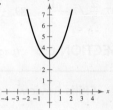

3. 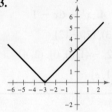   **4.**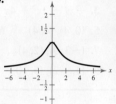

5. $\frac{3}{2}x^2 + 7x + C$ **6.** $\frac{2}{5}x^{5/2} + \frac{4}{3}x^{3/2} + C$

7. $\frac{1}{5}\ln|x| + C$ **8.** $-\frac{1}{6e^{6x}} + C$ **9.** $-\frac{8}{5}$

10. $-\frac{62}{3}$

1.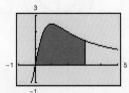

Positive

3.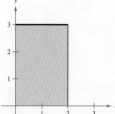

Area = 6

5.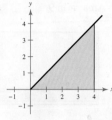

Area = 8

7.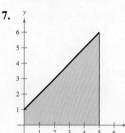

Area = $\frac{35}{2}$

9.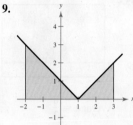

Area = $\frac{13}{2}$

11.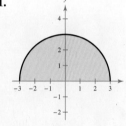

Area = $\frac{9\pi}{2}$

13. (a) 8   (b) 4   (c) -24 (d) 0

15. $\frac{1}{6}$ **17.** $\frac{1}{2}$ **19.** $6\left(1 - \frac{1}{e^2}\right)$ **21.** $8\ln 2 + \frac{15}{2}$

23. 1   **25.** $-\frac{5}{2}$ **27.** $\frac{14}{3}$ **29.** $-\frac{15}{4}$ **31.** -4

33. $\frac{2}{3}$ **35.** $-\frac{27}{20}$ **37.** 2   **39.** $\frac{1}{2}(1 - e^{-2}) \approx 0.432$

41. $\frac{e^3 - e}{3} \approx 5.789$ **43.** $\frac{1}{3}\left[(e^2 + 1)^{3/2} - 2\sqrt{2}\right] \approx 7.157$

45. $\frac{1}{8}\ln 17 \approx 0.354$ **47.** 4   **49.** 4

51. $\frac{1}{2}\ln 5 - \frac{1}{2}\ln 8 \approx -0.235$

53. $2\ln(2 + e^3) - 2\ln 3 \approx 3.993$

55. Area = 10   **57.** Area = $\frac{1}{4}$

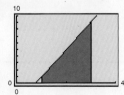   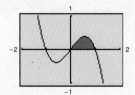

59. Area = ln 9

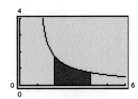

61. 10   **63.** 4 ln 3 ≈ 4.394

65.

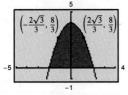

Average = $\frac{8}{3}$

$x = \pm \frac{2\sqrt{3}}{3} \approx \pm 1.155$

67.

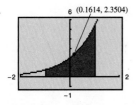

Average = $e - e^{-1} \approx 2.3504$

$x = \ln\left(\frac{e - e^{-1}}{2}\right) \approx 0.1614$

69.

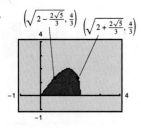

Average = $\frac{4}{3}$

$x = \sqrt{2 + \frac{2\sqrt{5}}{3}} \approx 1.868$

$x = \sqrt{2 - \frac{2\sqrt{5}}{3}} \approx 0.714$

71.

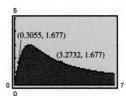

Average = $\frac{3}{7}$ ln 50 ≈ 1.677

$x \approx 0.3055$

$x \approx 3.2732$

73. Even   **75.** Neither odd nor even

77. (a) $\frac{1}{3}$ (b) $\frac{2}{3}$ (c) $-\frac{1}{3}$

Explanations will vary.

79. −63,500 farms

81. (a) $137,000   (b) $214,720.93   (c) $338,393.53

83. $\frac{2kR^3}{3}$

85. (a) $C = -350e^{-0.42t} + 651$ (b) About 505 coyotes

87. $\frac{2}{3}\sqrt{7} - \frac{1}{3}$ **89.** $\frac{39}{200}$

SECTION 6.5   (page 408)

Skills Review   (page 408)

1. $-x^2 + 3x + 2$ 2. $-2x^2 + 4x + 4$
3. $-x^3 + 2x^2 + 4x - 5$ 4. $x^3 - 6x - 1$
5. (0, 4), (4, 4)   6. (1, −3), (2, −12)
7. (−3, 9), (2, 4)   8. (−2, −4), (0, 0), (2, 4)
9. (1, −2), (5, 10)   10. (1, e)

1. 36   **3.** 9   **5.** $\frac{3}{2}$ **7.** $e - 2$

9.

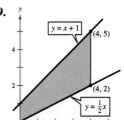

11.

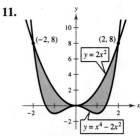

13.

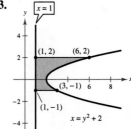

15. d

17.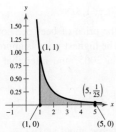

Area = $\frac{4}{5}$

19.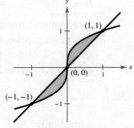

Area = $\frac{1}{2}$

21.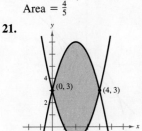

Area = $\frac{64}{3}$

23.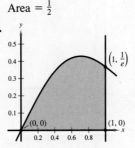

Area = $-\frac{1}{2}e^{-1} + \frac{1}{2}$

25.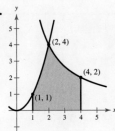

Area = $\frac{7}{3} + 8\ln 2$

27.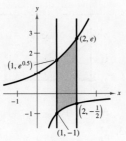

Area = $(2e + \ln 2) - 2e^{1/2}$

29.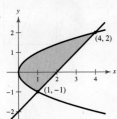

Area = $\frac{9}{2}$

31.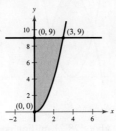

Area = 18

33.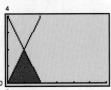

Area = $\int_0^1 2x\, dx + \int_1^2 (4 - 2x)\, dx$

35.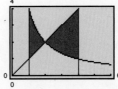

Area = $\int_1^2 \left(\frac{4}{x} - x\right) dx + \int_2^4 \left(x - \frac{4}{x}\right) dx$

37.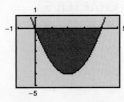

Area = $\frac{32}{3}$

39.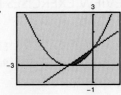

Area = $\frac{1}{6}$

41. 8

43. Offer 2 is better because the cumulative salary (area under the curve) is greater.

45. $300.6 million; Explanations will vary.

47. $256.43 billion

49. (a)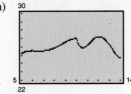

(b) About 0.5 billion lb more

51.

Quintile	Lowest	2nd	3rd	4th	Highest
Percent	2.81	6.98	14.57	27.01	45.73

53. Answers will vary.

SECTION 6.6 (page 416)

Skills Review (page 416)

1. 0, 2   **2.** 0, 2   **3.** 0, 2, −2   **4.** −1, 2
5. 2, 4   **6.** 1, 5   **7.** $e^4 - 1$ **8.** ln 7
9. $\frac{5\sqrt{5}}{3} - \frac{1}{3}$ **10.** $\frac{(\ln 5)^3}{3}$

1. $\frac{16\pi}{3}$ **3.** $\frac{15\pi}{2}$ **5.** $\frac{512\pi}{15}$ **7.** $\frac{32\pi}{15}$ **9.** $\frac{\pi}{3}$

11. $\frac{171\pi}{2}$ **13.** $\frac{128\pi}{5}$ **15.** $\frac{\pi}{2}(e^2 - 1)$ **17.** $\frac{4\pi}{3}$

19. $\dfrac{128\pi}{7}$ **21.** 8π **23.** 18π

25. $V = \pi \int_0^h \left(\dfrac{r}{h}x\right)^2 dx = \dfrac{1}{3}\pi r^2 h$

27. $\dfrac{16\pi}{3}$ **29.** $\dfrac{\pi}{30} m^3$

31. (a) $1{,}256{,}637\ \text{ft}^3$ (b) 2513 fish

REVIEW EXERCISES FOR CHAPTER 6
(page 422)

1. $16x + C$ **3.** $\dfrac{2}{3}x^3 + \dfrac{5}{2}x^2 + C$ **5.** $x^{2/3} + C$

7. $\dfrac{3}{7}x^{7/3} + \dfrac{3}{2}x^2 + C$ **9.** $\dfrac{4}{9}x^{9/2} - 2\sqrt{x} + C$

11. $f(x) = \dfrac{3}{2}x^2 + x - 2$ **13.** $f(x) = \dfrac{1}{6}x^4 - 8x + \dfrac{33}{2}$

15. (a) 2.5 sec (b) 100 ft (c) 64 ft

17. $x + 5x^2 + \dfrac{25}{3}x^3 + C$ or $\dfrac{1}{15}(1 + 5x)^3 + C_1$

19. $\dfrac{2}{5}\sqrt{5x - 1} + C$ **21.** $\dfrac{1}{2}x^2 - x^4 + C$

23. $\dfrac{1}{4}(x^4 - 2x)^2 + C$

25. (a) About 7.6 cm (b) About 21.2 cm

27. $-e^{-3x} + C$ **29.** $\dfrac{1}{2}e^{x^2 - 2x} + C$

31. $-\dfrac{1}{3}\ln|1 - x^3| + C$ **33.** $\dfrac{2}{3}x^{3/2} + 2x + 2x^{1/2} + C$

35.

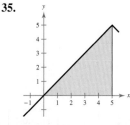

Area $= \dfrac{25}{2}$

37. $A = 4$ **39.** $A = \dfrac{32}{3}$ **41.** $A = \dfrac{8}{3}$ **43.** $A = 2\ln 2$

45. (a) 13 (b) 7 (c) 11 (d) 50

47. 16 **49.** $\dfrac{422}{5}$ **51.** 0 **53.** 2

55. $\dfrac{1}{8}$ **57.** 3.899 **59.** 0

61. **63.**

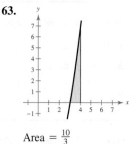

Area $= 6$ Area $= \dfrac{10}{3}$

65. Average value: $\dfrac{2}{5}$; $x = \dfrac{25}{4}$

67. Average value: $\dfrac{1}{3}(-1 + e^3) \approx 6.362$; $x \approx 3.150$

69. $\$3724.70$

71. (a) $B = -0.01955t^2 + 0.6108t - 1.818$

(b) According to the model, the price of beef per pound will never surpass $\$3.25$. The highest price is approximately $\$2.95$ per pound in 2005, and after that the prices decrease.

73. $\int_{-2}^{2} 6x^5\ dx = 0$ **75.** $\int_{-2}^{-1} \dfrac{4}{x^2}\ dx = \int_{1}^{2} \dfrac{4}{x^2}\ dx = 2$

(Odd function) (Symmetric about y-axis)

77. **79.**

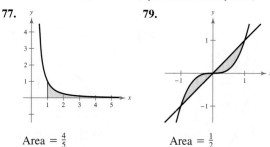

Area $= \dfrac{4}{5}$ Area $= \dfrac{1}{2}$

81.

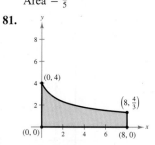

Area $= 16$

83. **85.**

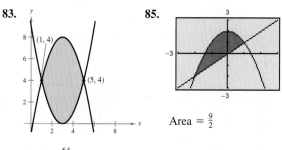

Area $= \dfrac{64}{3}$ Area $= \dfrac{9}{2}$

87. About 28.21 million people **89.** $\pi \ln 4 \approx 4.355$

91. $\dfrac{\pi}{2}(e^2 - e^{-2}) \approx 11.394$ **93.** $\dfrac{56\pi}{3}$ **95.** $\dfrac{2\pi}{35}$

97. $16\pi \approx 50.27\ \text{cm}^3$

CHAPTER TEST *(page 426)*

1. $3x^3 - 2x^2 + 13x + C$ **2.** $\dfrac{(x + 1)^3}{3} + C$

3. $\dfrac{2(x^4 - 7)^{3/2}}{3} + C$ **4.** $\dfrac{10x^{3/2}}{3} - 12x^{1/2} + C$

5. $5e^{3x} + C$ **6.** $\ln|x^3 - 11x| + C$

7. $f(x) = e^x + x$ **8.** $f(x) = \ln|x| + 2$
9. 8   **10.** 18   **11.** $\frac{2}{3}$ **12.** $2\sqrt{5} - 2\sqrt{2} \approx 1.644$
13. $\frac{1}{4}(e^{12} - 1) \approx 40{,}688.4$ **14.** $\ln 6 \approx 1.792$
15. (a) -3.875 million travelers
  (b) $N = 0.857t^2 - 5.06t + 26$
  (c) 22.125 million travelers
  (d) 20.492 million travelers

45. Area $= 2e^2 + 6 \approx 20.778$

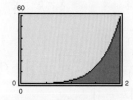

47. Area $= \frac{1}{9}(2e^3 + 1) \approx 4.575$

16. 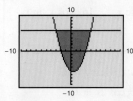 **17.**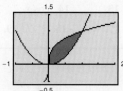

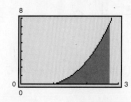

Area $= \frac{343}{6} \approx 57.167$ Area $= \frac{5}{12}$

49. Proof   **51.** $\dfrac{e^{5x}}{125}(25x^2 - 10x + 2) + C$

18. 9π in.^{3}

53. $-\dfrac{1}{x}(1 + \ln x) + C$ **55.** $1 - 5e^{-4} \approx 0.908$

CHAPTER 7

SECTION 7.1   (page 434)

57. $\frac{1}{4}(e^2 + 1) \approx 2.097$ **59.** $\frac{3}{128} - \frac{379}{128}e^{-8} \approx 0.022$

61. $\dfrac{1{,}171{,}875}{256}\pi \approx 14{,}381.070$

Skills Review   (page 434)

63. (a) 49%   (b) 34%

1. $\dfrac{1}{x + 1}$ **2.** $\dfrac{2x}{x^2 - 1}$ **3.** $3x^2 e^{x^3}$

65. (a) $3.2 \ln 2 - 0.2 \approx 2.018$
  (b) $12.8 \ln 4 - 7.2 \ln 3 - 1.8 \approx 8.035$

4. $-2xe^{-x^2}$ **5.** $e^x(x^2 + 2x)$ **6.** $e^{-2x}(1 - 2x)$

67. 18.6%

7. $\frac{64}{3}$ **8.** $\frac{4}{3}$ **9.** 36   **10.** 8

SECTION 7.2   (page 443)

1. $u = x; dv = e^{3x}\, dx$ **3.** $u = \ln 2x; dv = x\, dx$
5. $\frac{1}{3}xe^{3x} - \frac{1}{9}e^{3x} + C$ **7.** $-x^2 e^{-x} - 2xe^{-x} - 2e^{-x} + C$
9. $x \ln 2x - x + C$ **11.** $\frac{1}{4}e^{4x} + C$
13. $\frac{1}{4}xe^{4x} - \frac{1}{16}e^{4x} + C$ **15.** $\frac{1}{2}e^{x^2} + C$
17. $-xe^{-x} - e^{-x} + C$ **19.** $2x^2 e^x - 4e^x x + 4e^x + C$
21. $\frac{1}{2}t^2 \ln(t+1) - \frac{1}{2}\ln|t+1| - \frac{1}{4}t^2 + \frac{1}{2}t + C$
23. $xe^x - 2e^x + C$ **25.** $-e^{1/t} + C$
27. $\frac{1}{2}x^2 (\ln x)^2 - \frac{1}{2}x^2 \ln x + \frac{1}{4}x^2 + C$
29. $\frac{1}{3}(\ln x)^3 + C$ **31.** $-\dfrac{1}{x}(\ln x + 1) + C$
33. $\frac{2}{3}x(x - 1)^{3/2} - \frac{4}{15}(x - 1)^{5/2} + C$
35. $\frac{1}{4}x^4 + \frac{2}{3}x^3 + \frac{1}{2}x^2 + C$ **37.** $e(2e - 1) \approx 12.06$
39. $-12e^{-2} + 4 \approx 2.376$
41. $\frac{5}{36}e^6 + \frac{1}{36} \approx 56.060$ **43.** $2 \ln 2 - 1 \approx 0.386$

Skills Review   (page 443)

1. $(x - 4)(x + 4)$ **2.** $(x - 5)(x + 5)$
3. $(x - 4)(x + 3)$ **4.** $(x - 2)(x + 3)$
5. $x(x - 2)(x + 1)$ **6.** $x(x - 2)^2$
7. $(x - 2)(x - 1)^2$ **8.** $(x - 3)(x - 1)^2$
9. $x + \dfrac{1}{x - 2}$ **10.** $2x - 2 - \dfrac{1}{1 - x}$
11. $x^2 - x - 2 - \dfrac{2}{x - 2}$
12. $x^2 - x + 3 - \dfrac{4}{x + 1}$
13. $x + 4 + \dfrac{6}{x - 1}$, $x \neq -1$
14. $x + 3 + \dfrac{1}{x + 1}$, $x \neq 1$

Answers to Selected Exercises

1. $\dfrac{5}{x-5} - \dfrac{3}{x+5}$
3. $\dfrac{9}{x-3} - \dfrac{1}{x}$
5. $\dfrac{1}{x-5} + \dfrac{3}{x+2}$
7. $\dfrac{3}{x} - \dfrac{5}{x^2}$
9. $\dfrac{1}{3(x-2)} + \dfrac{1}{(x-2)^2}$
11. $\dfrac{8}{x+1} - \dfrac{1}{(x+1)^2} + \dfrac{2}{(x+1)^3}$
13. $\dfrac{1}{2}\ln\left|\dfrac{x-1}{x+1}\right| + C$
15. $\dfrac{1}{4}\ln\left|\dfrac{x+4}{x-4}\right| + C$
17. $\ln\left|\dfrac{2x-1}{x}\right| + C$
19. $\ln\left|\dfrac{x-10}{x}\right| + C$
21. $\ln\left|\dfrac{x-1}{x+2}\right| + C$
23. $\dfrac{3}{2}\ln|2x-1| - 2\ln|x+1| + C$
25. $\ln\left|\dfrac{x(x+2)}{x-2}\right| + C$
27. $\dfrac{1}{2}(3\ln|x-4| - \ln|x|) + C$
29. $2\ln|x-1| + \dfrac{1}{x-1} + C$
31. $\ln|x| + 2\ln|x+1| + \dfrac{1}{x+1} + C$
33. $\dfrac{1}{6}\ln\dfrac{4}{7} \approx -0.093$
35. $-\dfrac{4}{5} + 2\ln\dfrac{5}{3} \approx 0.222$
37. $\dfrac{1}{2} - \ln 2 \approx -0.193$
39. $4\ln 2 + \dfrac{1}{2} \approx 3.273$
41. $12 - \dfrac{7}{2}\ln 7 \approx 5.189$
43. $5\ln 2 - \ln 5 \approx 1.856$
45. $24\ln 3 - 36\ln 2 \approx 1.413$
47. $\dfrac{1}{2a}\left(\dfrac{1}{a+x} + \dfrac{1}{a-x}\right)$
49. $\dfrac{1}{a}\left(\dfrac{1}{x} + \dfrac{1}{a-x}\right)$
51. 21.16 oz
53. (a) About 166 animals (b) 200 animals
55. (a) About 66.2 hr (b) About 494 individuals
57. The rate of growth is increasing on [0, 3] for *P. aurelia* and on [0, 2] for *P. caudatum*; the rate of growth is decreasing on [3, ∞) for *P. aurelia* and on [2, ∞) for *P. caudatum*; *P. aurelia* has a higher limiting population.
59. (a) $\dfrac{dy}{dt} = ky\left(1 - \dfrac{y}{839.1}\right)$ (b) $y = \dfrac{839.1}{1 + 10e^{-0.016193t}}$
(c)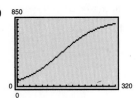
2036 ($t \approx 136$)
61. (a) Log Rule; If $u = x^2 + x - 8$, then $du = (2x+1)\,dx$.
(b) Partial fractions; $\dfrac{7x+4}{x^2+2x-8} = \dfrac{A}{x+4} + \dfrac{B}{x-2}$

SECTION 7.3 (page 452)

Skills Review (page 452)

1. $-\dfrac{\sqrt{2}}{2}$
2. $-\dfrac{1}{2}$
3. $-\dfrac{\sqrt{3}}{2}$
4. $\dfrac{\sqrt{3}}{2}$
5. $-\dfrac{\sqrt{3}}{3}$
6. $-\dfrac{\sqrt{3}}{3}$
7. -1
8. 0
9. $\tan x$
10. $\cot x$
11. $\sin^2 x$
12. $\cos^2 x$
13. 1
14. $\cos x$
15. $\csc x$
16. $\cos x \sin x$
17. $\dfrac{88}{3}$
18. $\dfrac{4}{3}$
19. 4
20. $\dfrac{17}{4}$

1. $-2\cos x + 3\sin x + C$
3. $t + \csc t + C$
5. $-\cot\theta - \sin\theta + C$
7. $-\dfrac{1}{2}\cos 2x + C$
9. $\sin x^2 + C$
11. $2\tan\dfrac{x}{2} + C$
13. $-\dfrac{1}{3}\ln|\cos 3x| + C$
15. $\dfrac{1}{4}\tan^4 x + C$
17. $\dfrac{1}{\pi}\ln|\sin \pi x| + C$
19. $\dfrac{1}{2}\ln|\csc 2x - \cot 2x| + C$
21. $\ln|\tan x| + C$
23. $\ln|\sec x - 1| + C$
25. $-\ln|1 + \cos x| + C$
27. $\dfrac{1}{2}\tan^2 x + C$
29. $-\cos e^x + C$
31. $e^{\sin x} + C$
33. $-\cos^2 x + x + C$ or $\sin^2 x + x + C$
35. $x\sin x + \cos x + C$
37. $x\tan x + \ln|\cos x| + C$
39. $\dfrac{3\sqrt{3}}{8} \approx 0.6495$
41. $2(\sqrt{3} - 1) \approx 1.4641$
43. $\dfrac{1}{2}$
45. $\ln(\cos 0) - \ln(\cos 1) \approx 0.6156$
47. 4
49. $\dfrac{\pi^2}{2} + 2 \approx 6.9348$
51. 2
53. 545.53 trillion Btu
55. (a) About 7,213,800 workers
   (b) About 7,650,200 workers
   (c) About 7,673,200 workers
57. 17.69 in.
59. (a) $C \approx \$27.50$ (b) $C \approx \$18.08$
61. 1.719 L
63. 2π
65. False, because $4\displaystyle\int \sin x \cos x\, dx = 2\sin^2 x + C$

MID-CHAPTER QUIZ (page 455)

1. $\frac{1}{5}xe^{5x} - \frac{1}{25}e^{5x} + C$ 2. $3x \ln x - 3x + C$
3. $\frac{1}{2}x^2 \ln x + x \ln x - \frac{1}{4}x^2 - x + C$
4. $\ln\left|\frac{x-5}{x+5}\right| + C$ 5. $3 \ln|x+4| - 2\ln|x-2| + C$
6. $5 \ln|x+1| + \frac{6}{x+1} + C$ 7. $-\frac{1}{5} \cos 5x + C$
8. $\frac{1}{2} \ln|\csc x^2 - \cot x^2| + C$ 9. $\sin\sqrt{x} + C$
10. $\frac{8}{e} - 4 \approx -1.0570$ 11. $e - 2 \approx 0.7183$
12. $\frac{1}{3}$ 13. $\ln 4 + 2 \ln 5 - 2 \ln 2 \approx 3.2189$
14. $15(\ln 9 - \ln 5) \approx 8.8168$
15. $10(\ln 6 - \ln 5) \approx 1.823$
16. $\frac{9}{2}$ 17. $\frac{2 - \sqrt{2}}{2\pi} \approx 0.0932$ 18. $e - 1 \approx 1.7183$
19. (a) 0.01425 or 1.43%   (b) 0.6827 or 68.3%
20. $y = \dfrac{100{,}000}{1 + 3e^{-0.01186t}}$
21. (a) About 108,000 workers
    (b) About 142,800 workers
    (c) About 135,400 workers

SECTION 7.4 (page 460)

Skills Review *(page 460)*

1. $\frac{1}{6}$ 2. $\frac{3}{20}$ 3. $\frac{7}{40}$ 4. $\frac{13}{12}$ 5. $\frac{61}{30}$ 6. $\frac{53}{18}$
7. $\frac{2}{3}$ 8. $\frac{4}{7}$ 9. 0   10. 5

1. Midpoint Rule: 2   3. Midpoint Rule: 0.6730
   Exact area: 2   Exact area: $\frac{2}{3} \approx 0.6667$
5. Midpoint Rule: 5.375   7. Midpoint Rule: 6.625
   Exact area: $\frac{16}{3} \approx 5.333$ Exact area: $\frac{20}{3} \approx 6.667$

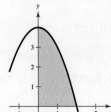
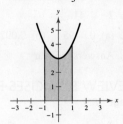

9. Midpoint Rule: 17.25   11. Midpoint Rule: 0.7578
   Exact area: $\frac{52}{3} \approx 17.33$ Exact area: 0.75

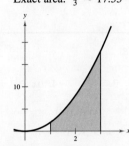
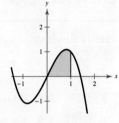

13. Midpoint Rule: 0.5703
    Exact area: $\frac{7}{12} \approx 0.5833$

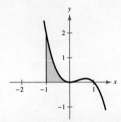

15. Midpoint Rule: 6.9609
    Exact area: 6.75

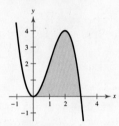

17. Area $\approx$ 54.6667,   19. Area $\approx$ 4.16,
    $n = 31$ $n = 5$
21. Area $\approx$ 0.9163,   23. Midpoint Rule: 1.5
    $n = 5$ Exact area: 1.5
25. Midpoint Rule: 25
    Exact area: $\frac{76}{3} \approx 25.33$
27. 1.0898   29. 1.5812
31.

n	Midpoint Rule
4	15.3965
8	15.4480
12	15.4578
16	15.4613
20	15.4628

33. 4.7532   **35.** 916.25 ft²   **37.** 9920 ft²

SECTION 7.5 (page 468)

> **Skills Review** (page 468)
>
> **1.** $\dfrac{2}{x^3}$ **2.** $-\dfrac{96}{(2x+1)^4}$ **3.** $-\dfrac{12}{x^4}$ **4.** $6x - 4$
> **5.** $16e^{2x}$ **6.** $e^{x^2}(4x^2 + 2)$ **7.** $(3, 18)$
> **8.** $(1, 8)$ **9.** $n < -5\sqrt{10},\ n > 5\sqrt{10}$
> **10.** $n < -5,\ n > 5$

	Exact Value	Trapezoidal Rule	Simpson's Rule
1.	2.6667	2.7500	2.6667
3.	8.4000	9.0625	8.4167
5.	4.0000	4.0625	4.0000
7.	0.6931	0.6941	0.6932
9.	5.3333	5.2650	5.3046
11.	12.6667	12.6640	12.6667
13.	0.6931	0.6970	0.6933

15. (a) 0.783   (b) 0.785
17. (a) 3.283   (b) 3.240
19. (a) 0.749   (b) 0.771
21. (a) 0.877   (b) 0.830
23. (a) 1.879   (b) 1.888
25. (a) 0.194   (b) 0.186
27. (a) 24.5   (b) 25.7
29. $0.3413 = 34.13\%$
31. $0.4999 = 49.99\%$ **33.** 89,500 ft²
35. (a) $|E| \leq 0.5$ (b) $|E| = 0$
37. (a) $|E| \leq \dfrac{5e}{64} \approx 0.212$ (b) $|E| \leq \dfrac{13e}{1024} \approx 0.035$
39. (a) $n = 71$ (b) $n = 1$
41. (a) $n = 3280$ (b) $n = 60$ **43.** 19.5215
45. 3.6558   **47.** 23.375   **49.** 416.1 ft
51. (a) 17.171 billion board-feet/yr
    (b) 17.082 billion board-feet/yr
    (c) The results are approximately equal.
53. 58.912 mg   **55.** 1878 subscribers

SECTION 7.6 (page 479)

> **Skills Review** (page 479)
>
> **1.** 9   **2.** 3   **3.** $-\tfrac{1}{8}$ **4.** Limit does not exist.
> **5.** Limit does not exist.   **6.** -4
> **7.** (a) $\tfrac{32}{3}b^3 - 16b^2 + 8b - \tfrac{4}{3}$ (b) $-\tfrac{4}{3}$
> **8.** (a) $\dfrac{b^2 - b - 11}{(b-2)^2(b-5)}$ (b) $\dfrac{11}{20}$
> **9.** (a) $\ln\left(\dfrac{5 - 3b^2}{b+1}\right)$ (b) $\ln 5 \approx 1.609$
> **10.** (a) $e^{-3b^2}(e^{6b^2} + 1)$ (b) 2

1. Improper; The integrand has an infinite discontinuity when $x = \tfrac{2}{3}$ and $0 \leq \tfrac{2}{3} \leq 1$.
3. Not improper; continuous on $[0, 1]$
5. Improper because the integrand has an infinite discontinuity when $x = 0$ and $0 \leq 0 \leq 4$; converges; 4
7. Improper because the integrand has an infinite discontinuity when $x = 1$ and $0 \leq 1 \leq 2$; converges; 6
9. Improper because the upper limit of integration is infinite; converges; 1
11. Converges; 1   **13.** Diverges
15. Diverges   **17.** Diverges   **19.** Diverges
21. Converges; 0   **23.** Diverges   **25.** Converges; 6
27. Diverges   **29.** Converges; 0
31. Converges; $\ln\left(\dfrac{4 + \sqrt{7}}{3}\right) \approx 0.7954$ **33.** (a) 1   (b) $\dfrac{\pi}{3}$

35.

x	1	10	25	50
xe^{-x}	0.3679	0.0005	0.0000	0.0000

37.

x	1	10	25	50
$x^2 e^{-(1/2)x}$	0.6065	0.6738	0.0023	0.0000

39. 2   **41.** $\tfrac{1}{4}$
43. (a) 0.9495   (b) 0.0974   (c) 0.0027
45. Answers will vary.

REVIEW EXERCISES FOR CHAPTER 7
(page 485)

1. $2\sqrt{x} \ln x - 4\sqrt{x} + C$ **3.** $xe^x + C$
5. $\tfrac{2}{3}(x - 2)\sqrt{1 + x} + C$ **7.** $x^2 e^{2x} - xe^{2x} + \tfrac{1}{2}e^{2x} + C$

9. (a) 0.675 (b) 0.290

11. $\frac{1}{5} \ln \left| \frac{x}{x+5} \right| + C$ **13.** $6 \ln|x+2| - 5 \ln|x-3| + C$

15. $x - \frac{25}{8} \ln|x+5| + \frac{9}{8} \ln|x-3| + C$

17. $\frac{1}{9} \left(\ln|2+3x| + \frac{2}{2+3x} \right) + C$

19. (a) $y = \dfrac{10{,}000}{1 + 7e^{-0.106873t}}$

(b)
Time, t	0	3	6	12	24
Sales, y	1250	1645	2134	3400	6500

(c) $t \approx 28$ weeks

21. $-3 \cos x - 2 \sin x + C$ **23.** $\frac{1}{4} \sin^4 x + C$

25. $\pi + 2$ **27.** $\frac{2\sqrt{3}}{3}$ **29.** 0   **31.** 2

33. $\frac{2}{3}$ **35.** $\frac{5}{3}$ **37.** About 19.95 in.

39. $n = 4$: 13.3203   **41.** $n = 4$: 0.7867
$n = 20$: 13.7167 $n = 20$: 0.7855

43. 0.705   **45.** 0.741   **47.** 0.376   **49.** 0.289

51. 9.0997   **53.** 0.017   **55.** Converges; 1

57. Diverges   **59.** Converges; 2   **61.** Converges; 2

63. (a) 0.441   (b) 0.119   (c) 0.015

CHAPTER TEST   (page 488)

1. $xe^{x+1} - e^{x+1} + C$ **2.** $3x^3 \ln x - x^3 + C$

3. $-3x^2 e^{-x/3} - 18xe^{-x/3} - 54e^{-x/3} + C$

4. $\ln \left| \dfrac{x-9}{x+9} \right| + C$

5. $\frac{1}{3} \ln|3x+1| + \dfrac{1}{3(3x+1)} + C$

6. $2 \ln|x| - \ln|x+2| + C$

7. $\ln|\sin \theta| + C$

8. $-\frac{1}{5} \ln|\cos 5\theta| + C$

9. $\frac{1}{2} \ln|\csc 2\theta - \cot 2\theta| + C$

10. Midpoint Rule: $\frac{63}{64} \approx 0.9844$
Exact area: 1

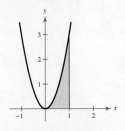

11. Midpoint Rule: $\frac{21}{8} = 2.625$
Exact area: $\frac{8}{3} = 2.\overline{6}$

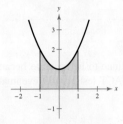

12. $-1 + \frac{3}{2} \ln 3 \approx 0.6479$ **13.** $4 \ln\left(\frac{48}{13}\right) \approx 5.2250$

14. 0   **15.** 0.2100

16. Simpson's Rule: 41.3606; Exact value: 41.1711

17. Converges; $\frac{1}{3}$ **18.** Converges; 12   **19.** Diverges

20. 9840 ft^2

CHAPTER 8

SECTION 8.1   (page 496)

Skills Review   (page 496)

1. 6   **2.** 1   **3.** Perpendicular   **4.** Parallel

1. (a) No   (b) No   (c) No   (d) Yes

3. (a) No   (b) Yes   (c) No   (d) No

5. (5, 5)   **7.** $\left(\frac{1}{2}, 3\right)$ **9.** (1, 1)   **11.** $\left(\frac{20}{3}, \frac{40}{3}\right)$

13. No solution   **15.** $(-2, 4), (0, 0)$ **17.** No solution

19. $\left(\frac{5}{2}, \frac{3}{4}\right)$ **21.** $\left(\frac{1}{2}, -2\right)$ **23.** $(4, -1)$ **25.** $\left(\frac{12}{7}, \frac{18}{7}\right)$

27. No solution   **29.** $\left(\frac{18}{5}, \frac{3}{5}\right)$

31. Infinitely many solutions; $\left(a, -\frac{1}{2} + \frac{5}{6}a\right)$

33. $\left(\frac{90}{31}, -\frac{67}{31}\right)$ **35.** $\left(-\frac{6}{35}, \frac{43}{35}\right)$ **37.** $(5, -2)$

39. (4, 1)   **41.** (2, -1)   **43.** (6, -3)   **45.** $\left(\frac{43}{6}, \frac{25}{6}\right)$

47. Speed of plane: 550 miles per hour
Speed of wind: 50 miles per hour

49. Cheeseburger: 310 calories; fries: 230 calories

51.

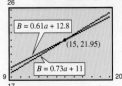

Yes, when $a = 15$.

53.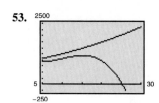

No; The models will not continue to be accurate because the model for participants in ACT testing will eventually produce negative values.

55. Answers will vary.

SECTION 8.2 (page 504)

> **Skills Review** (page 504)
>
> **1.** $(15, 10)$ **2.** $\left(-2, -\frac{8}{3}\right)$ **3.** $(28, 4)$
> **4.** $(4, 3)$ **5.** Not a solution **6.** Not a solution
> **7.** Solution **8.** Solution **9.** $x = 5a + 2$
> **10.** $x = a + 13$

1. (a) No (b) No (c) No (d) Yes
3. (a) No (b) No (c) Yes (d) No
5. $(1, -2, 4)$ **7.** $(3, 10, 2)$ **9.** $\left(\frac{1}{2}, -2, 2\right)$
11. $\begin{cases} x - 2y + 3z = 5 \\ y - 2z = 9 \\ 2x - 3z = 0 \end{cases}$

Answers will vary. Sample answer: This operation eliminated the x-variable from Equation 2.

13. $\left(\frac{5}{3}, \frac{1}{3}, 1\right)$ **15.** $(5, -3, 3)$ **17.** $\left(1, \frac{1}{2}, -3\right)$
19. $\left(-\frac{1}{2}a + \frac{5}{2}, 4a - 1, a\right)$ **21.** $(-2a + 5, -7a + 14, a)$
23. $\left(-\frac{3}{2}a + \frac{1}{2}, -\frac{2}{3}a + 1, a\right)$ **25.** $(1, 1, 1, 1)$
27. No solution **29.** $(0, 0, 0)$ **31.** $(9a, -35a, 67a)$
33. $s = -16t^2 - 32t + 500$ **35.** $y = \frac{1}{2}x^2 - 2x$
37. $y = x^2 - 6x + 8$ **39.** $x^2 + y^2 - 4x = 0$
41. $x^2 + y^2 + 6x - 8y = 0$
43. Brand X = 4 lb   **45.** Vanilla = 2 lb
Brand Y = 9 lb   Hazelnut = 4 lb
Brand Z = 9 lb   Mocha = 4 lb
47. Television = 30 ads
Radio = 10 ads
Newspaper = 20 ads

49. (a) 1 liter of 10%, 7 liters of 20%, 2 liters of 50%
(b) No liters of 10%, $8\frac{1}{3}$ liters of 20%, $1\frac{2}{3}$ liters of 50%
(c) $6\frac{1}{4}$ liters of 10%, no liters of 20%, $3\frac{3}{4}$ liters of 50%

51. Answers will vary. Sample answer:
$\begin{cases} 3x + y - z = 9 \\ x + 2y - z = 0 \\ -x + y + 3z = 1 \end{cases}$ $\begin{cases} x + y + z = 5 \\ x - 2z = 0 \\ 2y + z = 0 \end{cases}$

53. Answers will vary. Sample answer:
$\begin{cases} x + 2y - 4z = -5 \\ -x - 4y + 8z = 13 \\ x + 6y + 4z = 7 \end{cases}$ $\begin{cases} x + 2y + 4z = 9 \\ y + 2z = 3 \\ x - 4z = -4 \end{cases}$

55. False; The leading coefficient of Equation 2 is not 1.
57. Answers will vary.

SECTION 8.3 (page 514)

> **Skills Review** (page 514)
>
> **1.** Yes **2.** No **3.** $(5, 2)$ **4.** $\left(\frac{12}{5}, -3\right)$
> **5.** $(40, 14, 2)$ **6.** $\left(\frac{15}{2}, 4, 1\right)$

1. 1×2 **3.** 3×1 **5.** 2×2
7. $\begin{bmatrix} 4 & -3 & \vdots & -5 \\ -1 & 3 & \vdots & 12 \end{bmatrix}$ **9.** $\begin{bmatrix} 1 & 10 & -2 & \vdots & 2 \\ 5 & -3 & 4 & \vdots & 0 \\ 2 & 1 & 0 & \vdots & 6 \end{bmatrix}$

11. $\begin{bmatrix} 7 & -5 & 1 & \vdots & 13 \\ 19 & 0 & -8 & \vdots & 10 \end{bmatrix}$

13. $\begin{cases} x + 2y = 7 \\ 2x - 3y = 4 \end{cases}$ **15.** $\begin{cases} 2x + 5z = -12 \\ y - 2z = 7 \\ 6x + 3y = 2 \end{cases}$

17. $\begin{cases} 9x + 12y + 3z = 0 \\ -2x + 18y + 5z + 2w = 10 \\ x + 7y - 8z = -4 \\ 3x + 2z = -10 \end{cases}$

19. $\begin{bmatrix} 1 & 4 & 3 \\ 0 & 2 & -1 \end{bmatrix}$

21. $\begin{bmatrix} 1 & 1 & 4 & -1 \\ 0 & 5 & -2 & 6 \\ 0 & 3 & 20 & 4 \end{bmatrix}, \begin{bmatrix} 1 & 1 & 4 & -1 \\ 0 & 1 & -\frac{2}{5} & \frac{6}{5} \\ 0 & 3 & 20 & 4 \end{bmatrix}$

23. Add 5 times Row 2 to Row 1.
25. Interchange Row 1 and Row 2.
Add 4 times new Row 1 to Row 3.

27. (a) $\begin{bmatrix} 1 & 2 & 3 \\ 0 & -5 & -10 \\ 3 & 1 & -1 \end{bmatrix}$ (b) $\begin{bmatrix} 1 & 2 & 3 \\ 0 & -5 & -10 \\ 0 & -5 & -10 \end{bmatrix}$

(c) $\begin{bmatrix} 1 & 2 & 3 \\ 0 & -5 & -10 \\ 0 & 0 & 0 \end{bmatrix}$ (d) $\begin{bmatrix} 1 & 2 & 3 \\ 0 & 1 & 2 \\ 0 & 0 & 0 \end{bmatrix}$

(e) $\begin{bmatrix} 1 & 0 & -1 \\ 0 & 1 & 2 \\ 0 & 0 & 0 \end{bmatrix}$

The matrix is in reduced row-echelon form.

29. Reduced row-echelon form

31. Not in row-echelon form

33. $\begin{bmatrix} 1 & 1 & 0 & 5 \\ 0 & 1 & 2 & 0 \\ 0 & 0 & 1 & -1 \end{bmatrix}$ **35.** $\begin{bmatrix} 1 & -1 & -1 & 1 \\ 0 & 1 & 6 & 3 \\ 0 & 0 & 0 & 0 \end{bmatrix}$

37. $\begin{bmatrix} 1 & 0 & 0 \\ 0 & 1 & 0 \\ 0 & 0 & 1 \end{bmatrix}$ **39.** $\begin{bmatrix} 1 & 2 & 0 & 0 \\ 0 & 0 & 1 & 0 \\ 0 & 0 & 0 & 1 \\ 0 & 0 & 0 & 0 \end{bmatrix}$

41. $\begin{bmatrix} 1 & 0 & 3 & 16 \\ 0 & 1 & 2 & 12 \end{bmatrix}$

43. $\begin{cases} x - 2y = 4 \\ y = -3 \end{cases}$ **45.** $\begin{cases} x - y + 2z = 4 \\ y - z = 2 \\ z = -2 \end{cases}$

$(-2, -3)$

$(8, 0, -2)$

47. $(-1, 3)$ **49.** $(9, 13)$ **51.** $(-2, -1)$

53. $(8, 10, 6)$ **55.** $(-6, 8, 2)$ **57.** $(5, -1, -2)$

59. $(4 + 5b + 4a, 2 - 3b - 3a, b, a)$ **61.** Inconsistent

63. Yes; $(-1, 1, -3)$ **65.** No

67. \$1,200,000 at 8%, \$400,000 at 9%, \$600,000 at 12%

69. $y = 7.5t + 28$; 133 cases; The estimate seems reasonable because the data increased each year from 1997 to 2008.

71. False; Answers will vary. Sample answer: The row with all zeros is not the last row of the matrix.

MID-CHAPTER QUIZ (page 517)

1. $(2, -1)$ **2.** $\left(1, \frac{3}{2}\right)$ **3.** $(1, -2, 3)$

4. $(a + 6, a + 6, a)$ **5.** No solution **6.** $\left(\frac{3}{10}, \frac{2}{5}, 0\right)$

7. Answers will vary. **8.** Answers will vary.
Sample answer: Sample answer:

$\begin{bmatrix} 0 & 3 & 6 \\ 1 & 5 & -2 \\ 8 & 1 & 0 \\ -2 & 0 & 0 \end{bmatrix}$ $\begin{bmatrix} 0 \\ 6 \\ 6 \end{bmatrix}$

9. $\begin{bmatrix} 3 & 2 & \vdots & -2 \\ 5 & -1 & \vdots & 19 \end{bmatrix}$ **10.** $\begin{bmatrix} 1 & 0 & 3 & \vdots & -5 \\ 1 & 2 & -1 & \vdots & 3 \\ 3 & 0 & 4 & \vdots & 0 \end{bmatrix}$

11. $\left(\frac{36}{13}, -\frac{67}{13}\right)$ **12.** $(4, -2, -3)$

13. $\begin{bmatrix} 1 & 1 & 1 \\ 0 & -7 & -1 \end{bmatrix}$ **14.** $\begin{bmatrix} 1 & -1 & -4 \\ 18 & -8 & 4 \end{bmatrix}$

15. Commercial spray X: 6 liters
Commercial spray Y: 24 liters
Commercial spray Z: 16 liters

SECTION 8.4 (page 526)

Skills Review (page 526)

1. Not in reduced row-echelon form
2. Not in reduced row-echelon form
3. $\begin{bmatrix} -5 & 10 & \vdots & 12 \\ 7 & -3 & \vdots & 0 \end{bmatrix}$ **4.** $\begin{bmatrix} 10 & 15 & -9 & \vdots & 42 \\ 6 & -5 & 0 & \vdots & 0 \end{bmatrix}$
5. $(0, 2)$ **6.** $(a + 2, -a + 3, a)$ **7.** $(2, -1, -1)$

1. $x = -2, y = 4$ **3.** $x = 2, y = 3$

5. (a) $\begin{bmatrix} 3 & -2 \\ 1 & 7 \end{bmatrix}$ (b) $\begin{bmatrix} -1 & 0 \\ 3 & -9 \end{bmatrix}$ (c) $\begin{bmatrix} 3 & -3 \\ 6 & -3 \end{bmatrix}$

(d) $\begin{bmatrix} -1 & -1 \\ 8 & -19 \end{bmatrix}$

7. (a) $\begin{bmatrix} 5 & -3 \\ -3 & 2 \\ 6 & 4 \end{bmatrix}$ (b) $\begin{bmatrix} -3 & 1 \\ -3 & 0 \\ 2 & -2 \end{bmatrix}$ (c) $\begin{bmatrix} 3 & -3 \\ -9 & 3 \\ 12 & 3 \end{bmatrix}$

(d) $\begin{bmatrix} -5 & 1 \\ -9 & 1 \\ 8 & -3 \end{bmatrix}$

9. (a), (b), and (d) not possible (c) $\begin{bmatrix} 18 & 0 & 9 \\ -3 & -12 & 0 \end{bmatrix}$

11. $\begin{bmatrix} -8 & -7 \\ 15 & -1 \end{bmatrix}$ **13.** $\begin{bmatrix} \frac{19}{2} & 2 & -7 & \frac{9}{2} \end{bmatrix}$

15. $\begin{bmatrix} -17.143 & 2.143 \\ 11.571 & 10.286 \end{bmatrix}$ **17.** $\begin{bmatrix} -1.581 & -3.739 \\ -4.252 & -13.249 \\ 9.713 & -0.362 \end{bmatrix}$

19. $\begin{bmatrix} -6 & -9 \\ -1 & 0 \\ 17 & -10 \end{bmatrix}$ **21.** $\begin{bmatrix} 3 & 3 \\ -\frac{1}{2} & 0 \\ -\frac{13}{2} & \frac{11}{2} \end{bmatrix}$ **23.** Not possible

25. $\begin{bmatrix} 3 & -4 \\ 10 & 16 \\ 26 & 46 \end{bmatrix}$ **27.** $\begin{bmatrix} 0 & 0 & 0 \\ 0 & 0 & 0 \\ 0 & 0 & 0 \end{bmatrix}$

Order: 3×2 Order: 3×3

29. $\begin{bmatrix} 41 & 7 & 7 \\ 42 & 5 & 25 \\ -10 & -25 & 45 \end{bmatrix}$ **31.** $\begin{bmatrix} 151 & 25 & 48 \\ 516 & 279 & 387 \\ 47 & -20 & 87 \end{bmatrix}$

33. (a) $\begin{bmatrix} 0 & 15 \\ 6 & 12 \end{bmatrix}$ (b) $\begin{bmatrix} -2 & 2 \\ 31 & 14 \end{bmatrix}$ (c) $\begin{bmatrix} 9 & 6 \\ 12 & 12 \end{bmatrix}$

35. (a) $\begin{bmatrix} 12 & 12 & 6 \\ 18 & 18 & 9 \\ -4 & -4 & -2 \end{bmatrix}$ (b) $[28]$ (c) Not possible

37. $\begin{bmatrix} 5 & 8 \\ -4 & -16 \end{bmatrix}$ **39.** $\begin{bmatrix} -4 & 10 \\ 3 & 14 \end{bmatrix}$

41. (a) $\begin{bmatrix} -1 & 1 \\ -2 & 1 \end{bmatrix} \begin{bmatrix} x_1 \\ x_2 \end{bmatrix} = \begin{bmatrix} 4 \\ 0 \end{bmatrix}$ (b) $\begin{bmatrix} 4 \\ 8 \end{bmatrix}$

43. (a) $\begin{bmatrix} -2 & -3 \\ 6 & 1 \end{bmatrix} \begin{bmatrix} x_1 \\ x_2 \end{bmatrix} = \begin{bmatrix} -4 \\ -36 \end{bmatrix}$ (b) $\begin{bmatrix} -7 \\ 6 \end{bmatrix}$

45. (a) $\begin{bmatrix} 1 & -2 & 3 \\ -1 & 3 & -1 \\ 2 & -5 & 5 \end{bmatrix} \begin{bmatrix} x_1 \\ x_2 \\ x_3 \end{bmatrix} = \begin{bmatrix} 9 \\ -6 \\ 17 \end{bmatrix}$ (b) $\begin{bmatrix} 1 \\ -1 \\ 2 \end{bmatrix}$

47. $P^2 = \begin{bmatrix} 0.40 & 0.15 & 0.15 \\ 0.28 & 0.53 & 0.17 \\ 0.32 & 0.32 & 0.68 \end{bmatrix}$

The matrix gives the proportions of the voting population that changed parties or remained loyal to their parties from the first election to the third.

49. (a) $B = [2 \quad 0.5 \quad 3]$

(b) 120 lb 150 lb

$BA = [473.5 \quad 588.5]$

The entries represent the total calories burned.

51. $X_2 = \begin{bmatrix} 20 \\ 5 \end{bmatrix}, X_3 = \begin{bmatrix} 10 \\ 10 \end{bmatrix}$ **53.** $X_2 = \begin{bmatrix} 78 \\ 12 \\ 5 \end{bmatrix}, X_3 = \begin{bmatrix} 56 \\ 78 \\ 6 \end{bmatrix}$

SECTION 8.5 *(page 535)*

Skills Review *(page 535)*

1. $\begin{bmatrix} 4 & 24 \\ 0 & -16 \\ 48 & 8 \end{bmatrix}$ **2.** $\begin{bmatrix} \frac{11}{2} & 5 & 24 \\ \frac{1}{2} & 0 & 8 \\ 0 & 1 & 4 \end{bmatrix}$ **3.** $\begin{bmatrix} -13 & 11 \\ -19 & 21 \end{bmatrix}$

4. $\begin{bmatrix} 6 & 5 \\ 3 & -2 \end{bmatrix}$ **5.** $\begin{bmatrix} 1 & 0 & 0 \\ 0 & 1 & 0 \\ 0 & 0 & 1 \end{bmatrix}$ **6.** $\begin{bmatrix} 1 & 0 & 0 \\ 0 & 1 & 0 \\ 0 & 0 & 1 \end{bmatrix}$

7. $\begin{bmatrix} 1 & 0 & 3 & -2 \\ 0 & 1 & 4 & -3 \end{bmatrix}$

8. $\begin{bmatrix} 1 & 0 & 0 & -6 & -4 & 3 \\ 0 & 1 & 0 & 11 & 6 & -5 \\ 0 & 0 & 1 & -2 & -1 & 1 \end{bmatrix}$

1–7. $AB = I$ and $BA = I$

9. $\begin{bmatrix} \frac{1}{2} & 0 \\ 0 & \frac{1}{3} \end{bmatrix}$ **11.** $\begin{bmatrix} -3 & 2 \\ -2 & 1 \end{bmatrix}$ **13.** $\begin{bmatrix} 1 & -1 \\ 2 & -1 \end{bmatrix}$

15. Does not exist **17.** Does not exist

19. $\begin{bmatrix} 1 & 1 & -1 \\ -3 & 2 & -1 \\ 3 & -3 & 2 \end{bmatrix}$ **21.** $\begin{bmatrix} 1 & 0 & 0 \\ -\frac{3}{4} & \frac{1}{4} & 0 \\ \frac{7}{20} & -\frac{1}{4} & \frac{1}{5} \end{bmatrix}$

23. $\begin{bmatrix} -\frac{1}{8} & 0 & 0 & 0 \\ 0 & 1 & 0 & 0 \\ 0 & 0 & \frac{1}{4} & 0 \\ 0 & 0 & 0 & -\frac{1}{5} \end{bmatrix}$ **25.** $\begin{bmatrix} -175 & 37 & -13 \\ 95 & -20 & 7 \\ 14 & -3 & 1 \end{bmatrix}$

27. $\begin{bmatrix} -1.5 & 1.5 & 1 \\ 4.5 & -3.5 & -3 \\ -1 & 1 & 1 \end{bmatrix}$ **29.** $\begin{bmatrix} -12 & -5 & -9 \\ -4 & -2 & -4 \\ -8 & -4 & -6 \end{bmatrix}$

31. $\begin{bmatrix} 0 & -1.\overline{81} & 0.\overline{90} \\ -10 & 5 & 5 \\ 10 & -2.\overline{72} & -3.\overline{63} \end{bmatrix}$ **33.** $\begin{bmatrix} 1 & 0 & 1 & 0 \\ 0 & 1 & 0 & 1 \\ 2 & 0 & 1 & 0 \\ 0 & 1 & 0 & 2 \end{bmatrix}$

35. $\begin{bmatrix} \frac{4}{19} & \frac{3}{19} \\ -\frac{5}{19} & \frac{1}{19} \end{bmatrix}$ **37.** Does not exist

39. $\begin{bmatrix} \frac{16}{59} & \frac{15}{59} \\ -\frac{4}{59} & \frac{70}{59} \end{bmatrix}$ **41.** $(5, 0)$ **43.** $(-8, -6)$

45. $(3, 8, -11)$ **47.** $(2, 1, 0, 0)$ **49.** $(2, -2)$

51. No solution **53.** $(-4, -8)$

55. $(-1, 3, 2)$ **57.** $\left(\frac{5}{16}a + \frac{13}{16}, \frac{19}{16}a + \frac{11}{16}, a\right)$

59. $(-7, 3, -2)$ **61.** $(5, 0, -2, 3)$

63. Seedlings: 100 bags

General potting: 100 bags

Hardwood plants: 100 bags

65. Seedlings: 5 bags

General potting: 100 bags

Hardwood plants: 120 bags

67. (a) $A^{-1} = \begin{bmatrix} \frac{19}{105} & -\frac{1}{35} \\ -\frac{1}{35} & \frac{2}{35} \end{bmatrix}$

(b) $y = t + 20$ (c) $25 billion

69. (a) $I_1 = -3$ amperes (b) $I_1 = 2$ amperes

$I_2 = 8$ amperes $I_2 = 3$ amperes

$I_3 = 5$ amperes $I_3 = 5$ amperes

71. True; If A and B are both square matrices and $AB = I_n$, it can be shown that $BA = I_n$.

73. (a) Answers will vary.

(b) $A^{-1} = \begin{bmatrix} \frac{1}{a_{11}} & 0 & 0 & \cdots & 0 \\ 0 & \frac{1}{a_{22}} & 0 & \cdots & 0 \\ 0 & 0 & \frac{1}{a_{33}} & \cdots & 0 \\ \vdots & \vdots & \vdots & & \vdots \\ 0 & 0 & 0 & \cdots & \frac{1}{a_{nn}} \end{bmatrix}$

REVIEW EXERCISES FOR CHAPTER 8
(page 542)

1. $(-2, 4)$ **3.** $(8, -10)$ **5.** $(8, 6), (0, 10)$

7. $(3, -5)$ **9.** $\left(\frac{3}{4}a + \frac{5}{2}, a\right)$ **11.** $\left(\frac{8}{5}a + \frac{14}{5}, a\right)$ **13.** $(8, 9)$

15. The two lines intersect at one point. Solution: $(-1, 1)$

17. During the fourth month of the new format

19. (a) $\begin{cases} 0.1x + 0.5y = 0.25(12) \\ x + y = 12 \end{cases}$

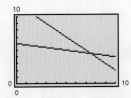

(b) 7.5 gallons of 10% solution
4.5 gallons of 50% solution

21. $(3, 5, 2)$ **23.** $(2, -1, 3)$ **25.** Inconsistent

27. $\left(\frac{1}{5}a + \frac{8}{5}, -\frac{6}{5}a + \frac{42}{5}, a\right)$

29. Commercial spray X: 10 gallons
Commercial spray Y: 5 gallons
Commercial spray Z: 12 gallons

31. $y = 2x^2 + x - 6$ **33.** $x^2 + y^2 - 4x + 2y - 4 = 0$

35. 2×4 **37.** Answers will vary. $\begin{bmatrix} 1 & 3 & 0 & 2 \\ 0 & 1 & 1 & 2 \\ 0 & 0 & 1 & 2 \end{bmatrix}$

39. $\begin{bmatrix} 1 & 0 & 0 \\ 0 & 1 & 0 \\ 0 & 0 & 1 \end{bmatrix}$ **41.** $(3, -2)$ **43.** $(3, 2, -1)$

45. $(10 - 4a, a, a)$ **47.** Inconsistent

49. (a) $\begin{bmatrix} 3 & 7 \\ -4 & 4 \end{bmatrix}$ (b) $\begin{bmatrix} -5 & 3 \\ 8 & -2 \end{bmatrix}$

(c) $\begin{bmatrix} -4 & 20 \\ 8 & 4 \end{bmatrix}$ (d) $\begin{bmatrix} -16 & 14 \\ 26 & -5 \end{bmatrix}$

51. (a) $\begin{bmatrix} 3 & 4 & 2 & 1 \\ 3 & -5 & 6 & 0 \end{bmatrix}$ (b) $\begin{bmatrix} -1 & 2 & -6 & 11 \\ -3 & 7 & 0 & 4 \end{bmatrix}$

(c) $\begin{bmatrix} 4 & 12 & -8 & 24 \\ 0 & 4 & 12 & 8 \end{bmatrix}$ (d) $\begin{bmatrix} -2 & 9 & -20 & 39 \\ -9 & 22 & 3 & 14 \end{bmatrix}$

53. $\begin{bmatrix} 8 \\ 5 \\ -6 \end{bmatrix}$ **55.** $\begin{bmatrix} 1 & 0 & 0 \\ 0 & 1 & 0 \\ 0 & 0 & 1 \end{bmatrix}$ **57.** Not possible

59. (a) $[4]$ (b) $\begin{bmatrix} 2 & -6 & 8 \\ -2 & 6 & -8 \\ -1 & 3 & -4 \end{bmatrix}$ (c) Not possible

61. $X_2 = \begin{bmatrix} 400 \\ 25 \\ 100 \\ 50 \end{bmatrix}, X_3 = \begin{bmatrix} 250 \\ 100 \\ 25 \\ 50 \end{bmatrix}$

63. $\begin{bmatrix} 4 & -11 \\ -3 & 1 \\ -1 & -3 \end{bmatrix}$ **65.** $\frac{1}{2}\begin{bmatrix} 3 & -5 \\ 1 & 4 \\ 9 & 14 \end{bmatrix}$

67. (a) $\begin{bmatrix} 5 & 4 \\ -1 & 1 \end{bmatrix} \begin{bmatrix} x_1 \\ x_2 \end{bmatrix} = \begin{bmatrix} 2 \\ -22 \end{bmatrix}$ (b) $X = \begin{bmatrix} 10 \\ -12 \end{bmatrix}$

69. (a) $\begin{bmatrix} 2 & 3 & 1 \\ 2 & -3 & -3 \\ 4 & -2 & 3 \end{bmatrix} \begin{bmatrix} x_1 \\ x_2 \\ x_3 \end{bmatrix} = \begin{bmatrix} 10 \\ 22 \\ -2 \end{bmatrix}$ (b) $X = \begin{bmatrix} 5 \\ 2 \\ -6 \end{bmatrix}$

71. $AB = I$ and $BA = I$ **73.** $AB = I$ and $BA = I$

75. $\begin{bmatrix} -1 & 0 & 0 \\ 0 & \frac{1}{2} & 0 \\ 0 & 0 & \frac{1}{4} \end{bmatrix}$ **77.** $\begin{bmatrix} -5 & 3 \\ 2 & -1 \end{bmatrix}$ **79.** $(3, 4)$

81. $\left(2, \frac{1}{2}, 3\right)$ **83.** $(-6, -1)$ **85.** $(2, -1, -2)$

87. 10 units of fluid A
8 units of fluid B
5 units of fluid C

89. (a) $\begin{bmatrix} \frac{19}{105} & -\frac{1}{35} \\ -\frac{1}{35} & \frac{2}{35} \end{bmatrix}$ (b) $y = -9t + 453$

(c) 408,000 metric tons

CHAPTER TEST (page 546)

1. $(2, 4)$ **2.** $(-2, 5), (3, 0)$ **3.** $(2, 0), (-3.5, -8.25)$

4. $\left(\frac{55}{16}, -\frac{91}{64}\right)$ **5.** $(2, -1, 3)$ **6.** $(2, -3, 4)$

7. $\begin{bmatrix} 2 & 1 & 4 & \vdots & 2 \\ 1 & 4 & -1 & \vdots & 0 \\ -1 & 3 & 3 & \vdots & -1 \end{bmatrix}$

8. $\begin{bmatrix} 3 & 4 & 2 & \vdots & 4 \\ 2 & 3 & 0 & \vdots & -2 \\ 0 & 2 & -3 & \vdots & -13 \end{bmatrix}$ **9.** $\left(\frac{7}{3}a - \frac{10}{3}, -\frac{8}{3}a + \frac{29}{3}, a\right)$

10. $(-12, -16, -6)$ **11.** $(2, -3, 1)$

12. $\begin{bmatrix} 2 & 4 \\ 7 & 13 \end{bmatrix}$ **13.** $\begin{bmatrix} -4 & -8 \\ 13 & 29 \end{bmatrix}$ **14.** $\begin{bmatrix} 1 \\ 11 \end{bmatrix}$

15. $\begin{bmatrix} 7 & 15 \\ 10 & 22 \end{bmatrix}$ **16.** $\frac{1}{5}\begin{bmatrix} 4 & 1 \\ 3 & 2 \end{bmatrix}$ **17.** $\begin{bmatrix} 1 & 0 \\ 0 & 1 \end{bmatrix}$

18. $\frac{1}{5}\begin{bmatrix} -9 & 16 & -6 \\ 6 & -9 & 4 \\ 4 & -6 & 1 \end{bmatrix}$ **19.** 75 liters of 60% solution
25 liters of 20% solution

20. $\begin{cases} B + K + N = 120{,}884{,}000 \\ B - K = 2{,}978{,}000 \\ N = 116{,}048.64 \end{cases}$

Bush: About 61,872,976 votes

Kerry: About 58,894,976 votes

Nader: About 116,049 votes

CHAPTER 9

SECTION 9.1 (page 554)

Skills Review (page 554)
1. $2\sqrt{5}$ 2. 5   3. 8   4. 8   5. (4, 7)
6. (1, 0)   7. (0, 3)   8. (−1, 1)
9. $(x - 2)^2 + (y - 3)^2 = 4$
10. $(x - 1)^2 + (y - 4)^2 = 25$

1.   **3.**

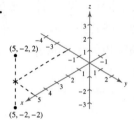

5. $A(2, 3, 4), B(-1, -2, 2)$ 7. $(-3, 4, 5)$
9. $(10, 0, 0)$ 11. 0   13. $3\sqrt{2}$ 15. $\sqrt{206}$
17. $(2, -5, 3)$ 19. $\left(\frac{1}{2}, \frac{1}{2}, -1\right)$ 21. $(6, -3, 5)$
23. $(1, 2, 1)$ 25. $3, 3\sqrt{5}, 6$; right triangle
27. $2, 2\sqrt{5}, 2\sqrt{2}$; neither right nor isosceles
29. $(0, 0, 5), (2, 2, 6), (2, -4, 9)$
31. $x^2 + (y - 2)^2 + (z - 2)^2 = 4$
33. $\left(x - \frac{3}{2}\right)^2 + (y - 2)^2 + (z - 1)^2 = \frac{21}{4}$
35. $(x - 1)^2 + (y - 1)^2 + (z - 5)^2 = 9$
37. $(x - 1)^2 + (y - 3)^2 + z^2 = 10$
39. $(x + 2)^2 + (y - 1)^2 + (z - 1)^2 = 1$
41. Center: $\left(\frac{5}{2}, 0, 0\right)$ 43. Center: $(1, -3, -4)$
Radius: $\frac{5}{2}$ Radius: 5

45. Center: $(1, 3, 2)$

Radius: $\frac{5\sqrt{2}}{2}$

47. 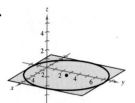   **49.**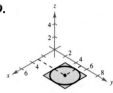

51. 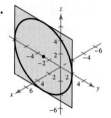   **53.**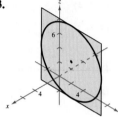

55. (a) 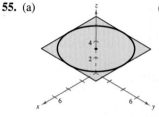 (b)

57. (a) 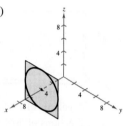 (b)

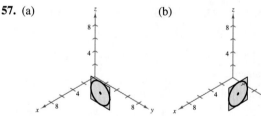

59. $(3, 3, 3)$ 61. $x^2 + y^2 + z^2 = 6806.25$

SECTION 9.2 (page 563)

Skills Review (page 563)
1. $(4, 0), (0, 3)$ 2. $\left(-\frac{4}{3}, 0\right), (0, -8)$
3. $(1, 0), (0, -2)$ 4. $(-5, 0), (0, -5)$
5. $(x - 1)^2 + (y - 2)^2 + (z - 3)^2 = -1$
6. $(x - 4)^2 + (y + 2)^2 - (z + 3)^2 = 0$
7. $(x + 1)^2 + (y - 1)^2 - z = 0$
8. $(x - 3)^2 + (y + 5)^2 + (z + 13)^2 = 1$
9. $x^2 + y^2 + z^2 = \frac{1}{4}$ 10. $x^2 + y^2 + z^2 = 4$

1. 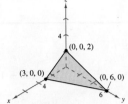 **3.**

5. 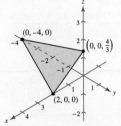 **7.**

9. 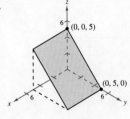 **11.**

13. $\dfrac{6\sqrt{14}}{7}$ **15.** $\dfrac{8\sqrt{14}}{7}$ **17.** $\dfrac{13\sqrt{29}}{29}$ **19.** $\dfrac{28\sqrt{29}}{29}$

21. Perpendicular **23.** Parallel **25.** Parallel
27. Neither parallel nor perpendicular **29.** Perpendicular
31. Trace in xy-plane ($z = 0$): $y = x^2$ (parabola)
Trace in plane $y = 1$: $x^2 - z^2 = 1$ (hyperbola)
Trace in yz-plane ($x = 0$): $y = -z^2$ (parabola)
33. Trace in xy-plane ($z = 0$): $\dfrac{x^2}{4} + y^2 = 1$ (ellipse)
Trace in xz-plane ($y = 0$): $\dfrac{x^2}{4} + z^2 = 1$ (ellipse)
Trace in yz-plane ($x = 0$): $y^2 + z^2 = 1$ (circle)
35. c **36.** e **37.** f **38.** b **39.** d **40.** a
41. Ellipsoid **43.** Hyperboloid of one sheet
45. Elliptic paraboloid **47.** Hyperbolic paraboloid
49. Hyperboloid of two sheets **51.** Elliptic cone
53. $(20, 0, 0)$ **55.** $(0, 0, 20)$

57. (a)

Year	1999	2000	2001
x	6.2	6.1	5.9
y	7.3	7.1	7.0
z (actual)	7.8	7.7	7.4
z (approximated)	7.8	7.7	7.5

Year	2002	2003	2004
x	5.8	5.6	5.5
y	7.0	6.9	6.9
z (actual)	7.3	7.2	6.9
z (approximated)	7.3	7.1	7.0

The approximated values of z are very close to the actual values.

(b) According to the model, increases in consumption of milk types y and z would correspond to an increase in consumption of milk type x.

SECTION 9.3 *(page 570)*

> **Skills Review** *(page 570)*
>
> **1.** 11 **2.** -16 **3.** 7
> **4.** 4 **5.** $(-\infty, \infty)$
> **6.** $(-\infty, -3) \cup (-3, 0) \cup (0, \infty)$
> **7.** $[5, \infty)$ **8.** $(-\infty, -\sqrt{5}] \cup [\sqrt{5}, \infty)$
> **9.** 55.0104 **10.** 6.9165

1. (a) $\dfrac{3}{2}$ (b) $-\dfrac{1}{4}$ (c) 6 (d) $\dfrac{5}{y}$ (e) $\dfrac{x}{2}$ (f) $\dfrac{5}{t}$
3. (a) 5 (b) $3e^2$ (c) $2e^{-1}$
   (d) $5e^y$ (e) xe^2 (f) te^t
5. (a) $\dfrac{2}{3}$ (b) 0 **7.** (a) 90π (b) 50π
9. (a) $\$20{,}655.20$ (b) $\$1{,}397{,}672.67$
11. (a) 0 (b) 6
13. (a) $x^2 + 2x\,\Delta x + (\Delta x)^2 - 2y$ (b) -2, $\Delta y \neq 0$
15. Domain: all points (x, y) inside and on the circle $x^2 + y^2 = 16$
Range: $[0, 4]$
17. Domain: all points (x, y) such that $y \neq 0$
Range: $(0, \infty)$

19. All points inside and on the circle $x^2 + y^2 = 4$
21. All points (x, y)
23. All points (x, y) such that $x \neq 0$ and $y \neq 0$
25. All points (x, y) such that $y \geq 0$
27. The half-plane below the line $y = -x + 4$
29. b **30.** d **31.** a **32.** c
33. The level curves are parallel lines.
35. The level curves are circles.

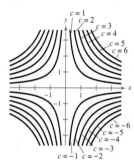
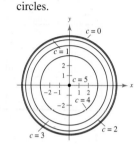

37. The level curves are hyperbolas.
39. The level curves are circles.

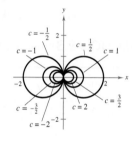

41. $V = \pi r^2 \left(l + \frac{4}{3}r\right)$

43. (a) $N(22, 12) = \left(\dfrac{22 - 4}{4}\right)^2 (12) = 243$ board-feet

(b) $N(30, 12) = \left(\dfrac{30 - 4}{4}\right)^2 (12) = 507$ board-feet

45.

a	2	4	6	8	10	12
l	57.40	62.57	66.91	69.01	72.00	75.83
c	39.27	41.55	43.35	43.85	44.86	46.36

47. (a) C (b) A (c) B
49. (a) 21.84 kg (b) About 159 cm
(c) a; Explanations will vary. Sample answer: The a-variable has a greater influence on weight because the absolute value of its coefficient is larger than the absolute value of the coefficient of the h-term.

SECTION 9.4 *(page 581)*

Skills Review *(page 581)*

1. $\dfrac{x}{\sqrt{x^2 + 3}}$ **2.** $-6x(3 - x^2)^2$ **3.** $e^{2t+1}(2t + 1)$

4. $\dfrac{e^{2x}(2 - 3e^{2x})}{\sqrt{1 - e^{2x}}}$ **5.** $-\dfrac{2}{3 - 2x}$ **6.** $\dfrac{3(t^2 - 2)}{2t(t^2 - 6)}$

7. $-\dfrac{10x}{(4x - 1)^3}$ **8.** $-\dfrac{(x + 2)^2(x^2 + 8x + 27)}{(x^2 - 9)^3}$

9. $f'(2) = 8$ **10.** $g'(2) = \dfrac{7}{2}$

1. $\dfrac{\partial z}{\partial x} = 3;\ \dfrac{\partial z}{\partial y} = 5$

3. $f_x(x, y) = 3;\ f_y(x, y) = -12y$

5. $f_x(x, y) = \dfrac{1}{y};\ f_y(x, y) = -\dfrac{x}{y^2}$

7. $f_x(x, y) = \dfrac{x}{\sqrt{x^2 + y^2}};\ f_y(x, y) = \dfrac{y}{\sqrt{x^2 + y^2}}$

9. $\dfrac{\partial z}{\partial x} = 2xe^{2y};\ \dfrac{\partial z}{\partial y} = 2x^2 e^{2y}$

11. $h_x(x, y) = -2xe^{-(x^2+y^2)};\ h_y(x, y) = -2ye^{-(x^2+y^2)}$

13. $\dfrac{\partial z}{\partial x} = -\dfrac{2y}{x^2 - y^2};\ \dfrac{\partial z}{\partial y} = \dfrac{2x}{x^2 - y^2}$

15. $f_x(x, y) = 3xye^{x-y}(2 + x)$

17. $g_x(x, y) = 3y^2 e^{y-x}(1 - x)$ **19.** 9

21. $f_x(x, y) = 6x + y, 13;\ f_y(x, y) = x - 2y, 0$

23. $f_x(x, y) = 3ye^{3xy}, 12;\ f_y(x, y) = 3xe^{3xy}, 0$

25. $f_x(x, y) = -\dfrac{y^2}{(x - y)^2}, -\dfrac{1}{4};\ f_y(x, y) = \dfrac{x^2}{(x - y)^2}, \dfrac{1}{4}$

27. $f_x(x, y) = \dfrac{2x}{x^2 + y^2}, 2;\ f_y(x, y) = \dfrac{2y}{x^2 + y^2}, 0$

29. $w_x = yz$
$w_y = xz$
$w_z = xy$

31. $w_x = -\dfrac{2z}{(x + y)^2}$
$w_y = -\dfrac{2z}{(x + y)^2}$
$w_z = \dfrac{2}{x + y}$

33. $w_x = \dfrac{x}{\sqrt{x^2 + y^2 + z^2}}, \dfrac{2}{3}$
$w_y = \dfrac{y}{\sqrt{x^2 + y^2 + z^2}}, -\dfrac{1}{3}$
$w_z = \dfrac{z}{\sqrt{x^2 + y^2 + z^2}}, \dfrac{2}{3}$

35. $w_x = \dfrac{x}{x^2 + y^2 + z^2}, \dfrac{3}{25}$

$w_y = \dfrac{y}{x^2 + y^2 + z^2}, 0$

$w_z = \dfrac{z}{x^2 + y^2 + z^2}, \dfrac{4}{25}$

37. $w_x = 2z^2 + 3yz, 2$

$w_y = 3xz - 12yz, 30$

$w_z = 4xz + 3xy - 6y^2, -1$

39. $(-6, 4)$ 41. $(1, 1)$ 43. (a) 2   (b) 1

45. (a) -2 (b) -2 47. (a) -1 (b) 0

49. $\dfrac{\partial^2 z}{\partial x^2} = 2$ 51. $\dfrac{\partial^2 z}{\partial x^2} = \dfrac{e^{2xy}(2x^2y^2 - 2xy + 1)}{2x^3}$

$\dfrac{\partial^2 z}{\partial x \partial y} = \dfrac{\partial^2 z}{\partial y \partial x} = -2$ $\dfrac{\partial^2 z}{\partial x \partial y} = \dfrac{\partial^2 z}{\partial y \partial x} = ye^{2xy}$

$\dfrac{\partial^2 z}{\partial y^2} = 6$ $\dfrac{\partial^2 z}{\partial y^2} = xe^{2xy}$

53. $\dfrac{\partial^2 z}{\partial x^2} = 6x$ 55. $\dfrac{\partial^2 z}{\partial x^2} = \dfrac{2}{(x - y)^3}$

$\dfrac{\partial^2 z}{\partial y^2} = -8$ $\dfrac{\partial^2 z}{\partial x \partial y} = \dfrac{\partial^2 z}{\partial y \partial x} = \dfrac{-2}{(x - y)^3}$

$\dfrac{\partial^2 z}{\partial y \partial x} = \dfrac{\partial^2 z}{\partial x \partial y} = 0$ $\dfrac{\partial^2 z}{\partial y^2} = \dfrac{2}{(x - y)^3}$

57. $\dfrac{\partial^2 z}{\partial x^2} = 0$

$\dfrac{\partial^2 z}{\partial y^2} = 2xe^{-y^2}(2y^2 - 1)$

$\dfrac{\partial^2 z}{\partial x \partial y} = \dfrac{\partial^2 z}{\partial y \partial x} = -2ye^{-y^2}$

59. $f_{xx}(x, y) = 12x^2 - 6y^2, 12$

$f_{xy}(x, y) = -12xy, 0$

$f_{yy}(x, y) = -6x^2 + 2, -4$

$f_{yx}(x, y) = -12xy, 0$

61. $f_{xx}(x, y) = -\dfrac{1}{(x - y)^2}, -1$

$f_{xy}(x, y) = \dfrac{1}{(x - y)^2}, 1$

$f_{yy}(x, y) = -\dfrac{1}{(x - y)^2}, -1$

$f_{yx}(x, y) = \dfrac{1}{(x - y)^2}, 1$

63. (a) $\dfrac{\partial z}{\partial x} = 1.25, \dfrac{\partial z}{\partial y} = -0.125$

(b) An increase in consumption of milk type x corresponds to an increase in consumption of milk type z. An increase in consumption of milk type y corresponds to a decrease in consumption of milk type z.

65. At $(2, 3)$: $\dfrac{dT}{dx} = -2.4$ degrees/m

At $(2, 3)$: $\dfrac{dT}{dy} = -9$ degrees/m

67. An increase in either price will cause a decrease in the number of applicants.

69. Answers will vary.

MID-CHAPTER QUIZ (page 584)

1. (a) 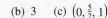   (b) 3 (c) $\left(0, \dfrac{5}{2}, 1\right)$

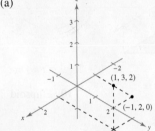

2. (a)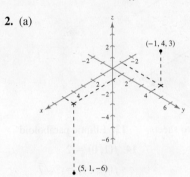

(b) $3\sqrt{14}$ (c) $\left(2, \dfrac{5}{2}, -\dfrac{3}{2}\right)$

3. (a)

(b) $3\sqrt{6}$ (c) $\left(\dfrac{3}{2}, -\dfrac{3}{2}, 0\right)$

4. $(x - 2)^2 + (y + 1)^2 + (z - 3)^2 = 16$

5. $(x - 1)^2 + (y - 4)^2 + (z + 2)^2 = 11$

6. Center: $(4, 1, 3)$; radius: 7

7.
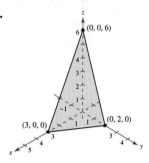

8.
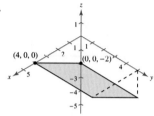

9.
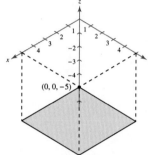

10. Ellipsoid

11. Hyperboloid of two sheets **12.** Elliptic paraboloid

13. $f(1, 0) = 1$
$f(4, -1) = -5$

14. $f(1, 0) = 2$
$f(4, -1) = 3\sqrt{7}$

15. $f(1, 0) = 0$
$f(4, -1) = 0$

16. (a) Between 30° and 50° (b) Between 40° and 80° (c) Between 70° and 90°

17. $f_x = 2x - 3;\ f_x(-2, 3) = -7$
$f_y = 4y - 1;\ f_y(-2, 3) = 11$

18. $f_x = \dfrac{y(3 + y)}{(x + y)^2};\ f_x(-2, 3) = 18$
$f_y = \dfrac{-2xy - y^2 - 3x}{(x + y)^2};\ f_y(-2, 3) = 9$

19. $x^2 + y^2 + z^2 = 3963^2$

Lines of longitude would be traces in planes passing through the z-axis. Each trace is a circle. Lines of latitude would be traces in planes parallel to the equator. They are circles.

SECTION 9.5 (page 591)

Skills Review (page 591)

1. $(3, 2)$ **2.** $(11, 6)$ **3.** $(1, 4)$ **4.** $(4, 4)$
5. $(5, 2)$ **6.** $(3, -2)$ **7.** $(0, 0), (-1, 0)$
8. $(-2, 0), (2, -2)$

9. $\dfrac{\partial z}{\partial x} = 12x^2$ $\dfrac{\partial^2 z}{\partial y^2} = -6$

$\dfrac{\partial z}{\partial y} = -6y$ $\dfrac{\partial^2 z}{\partial x \partial y} = 0$

$\dfrac{\partial^2 z}{\partial x^2} = 24x$ $\dfrac{\partial^2 z}{\partial y \partial x} = 0$

10. $\dfrac{\partial z}{\partial x} = 10x^4$ $\dfrac{\partial^2 z}{\partial y^2} = -6y$

$\dfrac{\partial z}{\partial y} = -3y^2$ $\dfrac{\partial^2 z}{\partial x \partial y} = 0$

$\dfrac{\partial^2 z}{\partial x^2} = 40x^3$ $\dfrac{\partial^2 z}{\partial y \partial x} = 0$

11. $\dfrac{\partial z}{\partial x} = 4x^3 - \dfrac{\sqrt{xy}}{2x}$ $\dfrac{\partial^2 z}{\partial y^2} = \dfrac{\sqrt{xy}}{4y^2}$

$\dfrac{\partial z}{\partial y} = -\dfrac{\sqrt{xy}}{2y} + 2$ $\dfrac{\partial^2 z}{\partial x \partial y} = -\dfrac{\sqrt{xy}}{4xy}$

$\dfrac{\partial^2 z}{\partial x^2} = 12x^2 + \dfrac{\sqrt{xy}}{4x^2}$ $\dfrac{\partial^2 z}{\partial y \partial x} = -\dfrac{\sqrt{xy}}{4xy}$

12. $\dfrac{\partial z}{\partial x} = 4x - 3y$ $\dfrac{\partial^2 z}{\partial y^2} = 2$

$\dfrac{\partial z}{\partial y} = 2y - 3x$ $\dfrac{\partial^2 z}{\partial x \partial y} = -3$

$\dfrac{\partial^2 z}{\partial x^2} = 4$ $\dfrac{\partial^2 z}{\partial y \partial x} = -3$

13. $\dfrac{\partial z}{\partial x} = y^3 e^{xy^2}$ $\dfrac{\partial^2 z}{\partial y^2} = 4x^2 y^3 e^{xy^2} + 6xy e^{xy^2}$

$\dfrac{\partial z}{\partial y} = 2xy^2 e^{xy^2} + e^{xy^2}$ $\dfrac{\partial^2 z}{\partial x \partial y} = 2xy^4 e^{xy^2} + 3y^2 e^{xy^2}$

$\dfrac{\partial^2 z}{\partial x^2} = y^5 e^{xy^2}$ $\dfrac{\partial^2 z}{\partial y \partial x} = 2xy^4 e^{xy^2} + 3y^2 e^{xy^2}$

14. $\dfrac{\partial z}{\partial x} = e^{xy}(xy + 1)$ $\dfrac{\partial^2 z}{\partial y^2} = x^3 e^{xy}$

$\dfrac{\partial z}{\partial y} = x^2 e^{xy}$ $\dfrac{\partial^2 z}{\partial x \partial y} = xe^{xy}(xy + 2)$

$\dfrac{\partial^2 z}{\partial x^2} = ye^{xy}(xy + 2)$ $\dfrac{\partial^2 z}{\partial y \partial x} = xe^{xy}(xy + 2)$

1. Critical point: $(-2, -4)$
   No relative extrema
   $(-2, -4, 1)$ is a saddle point.
3. Critical point: $(0, 0)$
   Relative minimum: $(0, 0, 1)$
5. Relative minimum: $(1, 3, 0)$
7. Relative minimum: $(-1, 1, -4)$
9. Relative maximum: $(8, 16, 74)$
11. Relative minimum: $(2, 1, -7)$
13. Saddle point: $(-2, -2, -8)$
15. Saddle point: $(0, 0, 0)$
17. Relative maximum: $\left(\frac{1}{2}, \frac{1}{2}, e^{1/2}\right)$
    Relative minimum: $\left(-\frac{1}{2}, -\frac{1}{2}, -e^{1/2}\right)$
19. Saddle point: $(0, 0, 4)$
21. Insufficient information
23. $f(x_0, y_0)$ is a saddle point.
25. Relative minima: $(a, 0, 0), (0, b, 0)$
    Second-Partials Test fails at $(a, 0)$ and $(0, b)$.
27. Saddle point: $(0, 0, 0)$
    Second-Partials Test fails at $(0, 0)$.
29. Relative minimum: $(0, 0, 0)$
    Second-Partials Test fails at $(0, 0)$.
31. Relative minimum: $(1, -3, 0)$
33. 10, 10, 10   35. 10, 10, 10
37. $D_x(x, y) = 2x - 18 + 2y$
    $D_y(x, y) = 4y - 24 + 2x$
    To minimize the duration of the infection, 600 mg of the first drug and 300 mg of the second drug are necessary.
39. 32 in. × 16 in. × 16 in.
41. Base dimensions: 24 ft × 24 ft
    Height: 8 ft
    Gallons: 7
43. (a) $x = 4, y = 4$
    (b) 48 m²
    (c) 40 m

45. $P(p, q, r) = 2pq + 2pr + 2qr$
    $p + q + r = 1$ implies that $r = 1 - p - q$.
    $P(p, q) = 2pq + 2p(1 - p - q) + 2q(1 - p - q)$
    $ = 2pq + 2p - 2p^2 - 2pq + 2q - 2pq - 2q^2$
    $ = -2pq + 2p + 2q - 2p^2 - 2q^2$
    $\frac{\partial P}{\partial p} = -2q + 2 - 4p; \quad \frac{\partial P}{\partial q} = -2p + 2 - 4q$
    Solving $\frac{\partial P}{\partial p} = \frac{\partial P}{\partial q} = 0$ gives
    $q + 2p = 1$
    $p + 2q = 1$
    so $p = q = \frac{1}{3}$ and
    $P\left(\frac{1}{3}, \frac{1}{3}\right) = -2\left(\frac{1}{9}\right) + 2\left(\frac{1}{3}\right) + 2\left(\frac{1}{3}\right) - 2\left(\frac{1}{9}\right) - 2\left(\frac{1}{9}\right)$
    $\phantom{P\left(\frac{1}{3}, \frac{1}{3}\right)} = \frac{6}{9} = \frac{2}{3}.$
47. $x = \sqrt[3]{0.065} \approx 0.402$ L
    $y = \frac{1}{2}\sqrt[3]{0.065} \approx 0.201$ L
    $z = \frac{1}{3}\sqrt[3]{0.065} \approx 0.134$ L
49. Base dimensions: 10 ft × 10 ft
    Height: 10 ft
51. True

SECTION 9.6 (page 601)

Skills Review (page 601)

1. 5.0225   2. 0.0189
3. $S_a = 2a - 4 - 4b$ 4. $S_a = 8a - 6 - 2b$
   $S_b = 12b - 8 - 4a$ $S_b = 18b - 4 - 2a$
5. 15   6. 42   7. $\frac{25}{12}$
8. 14   9. 31   10. 95

1. (a) $y = \frac{3}{4}x + \frac{4}{3}$ (b) $S = \frac{1}{6}$
3. (a) $y = -2x + 4$ (b) $S = 2$
5. (a) $y = \frac{35}{37}x + \frac{34}{37}$ (b) $S \approx 2.7568$
7. $y = x + \frac{2}{3}$ 9. $y = -2.3x - 0.9$

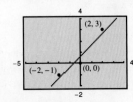
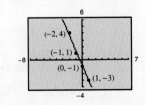

11. $y = 0.7x + 1.4$ **13.** $y = x + 4$

15. $y = -0.65x + 1.75$ **17.** $y = 0.8605x + 0.163$

19. $y = -1.1824x + 6.385$

21. $y = 0.4286x^2 + 1.2x + 0.74$ **23.** $y = x^2 - x$

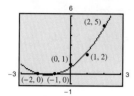 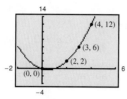

25. Linear: $y = 1.4x + 6$

Quadratic: $y = 0.12x^2 + 1.7x + 6$

The quadratic model is a better fit.

27. Linear: $y = -68.9x + 754$

Quadratic: $y = 2.82x^2 - 83.0x + 763$

The quadratic model is a better fit.

29. $y = 71.2x + 40$; 1,108,000 participants

31. $y = -54.95x^2 + 4959.5x - 50{,}050$; $45,735

33. (a) $y = -0.238t + 11.93$;

In 2010, $y \approx 4.8$ deaths per 1000 live births

(b) $y = 0.0088t^2 - 0.458t + 12.66$;

In 2010, $y \approx 6.8$ deaths per 1000 live births

35. (a)

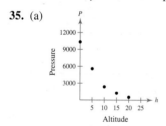

(b) $P = 28.92h^2 - 1057.9h + 10{,}250$

(c) Answers will vary. Sample answer: The quadratic model has an "r^2-value" of about 0.997 and the linear model has an "r^2-value" of about 0.884. Because $0.997 > 0.884$, the quadratic model is a better fit for the data.

37. Linear: $y = 3.757x + 9.03$

Quadratic: $y = 0.006x^2 + 3.63x + 9.4$

Either model is a good fit for the data.

39. Quadratic: $y = -0.087x^2 + 2.82x + 0.4$

41.

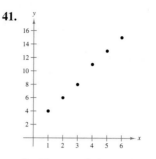

Positive correlation, $r \approx 0.9981$

43.

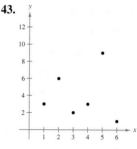

No correlation, $r = 0$

45.

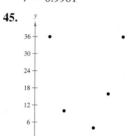

No correlation, $r \approx 0.0750$

47. False; The data modeled by $y = 3.29x - 4.17$ have a positive correlation because the slope of the line is positive.

49. True   **51.** True   **53.** Answers will vary.

SECTION 9.7   (page 610)

Skills Review (page 610)

1. 1   **2.** 6   **3.** 42   **4.** $\frac{1}{2}$ **5.** $\frac{19}{4}$

6. $\frac{16}{3}$ **7.** $\frac{1}{7}$ **8.** 4   **9.** $\ln 5$ **10.** $\ln(e - 1)$

11. $\frac{e}{2}(e^4 - 1)$ **12.** $\frac{1}{2}\left(1 - \frac{1}{e^2}\right)$

13. 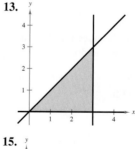   **14.**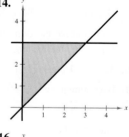

15. 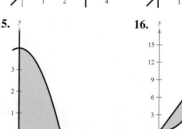   **16.**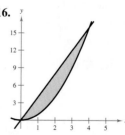

1. $\dfrac{3x^2}{2}$ **3.** $y\ln|2y|$ **5.** $x^2\left(2-\dfrac{1}{2}x^2\right)$ **7.** $\dfrac{y^3}{2}$

9. $e^{x^2}-\dfrac{e^{x^2}}{x^2}+\dfrac{1}{x^2}$ **11.** 3 **13.** 36 **15.** $\dfrac{1}{2}$

17. $\dfrac{148}{3}$ **19.** 5 **21.** 64 **23.** 4

25.

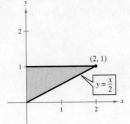

$\displaystyle\int_0^1\int_0^2 dy\,dx = \int_0^2\int_0^1 dx\,dy = 2$

27.

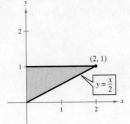

$\displaystyle\int_0^1\int_{2y}^2 dx\,dy = \int_0^2\int_0^{x/2} dy\,dx = 1$

29.

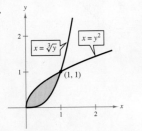

$\displaystyle\int_0^2\int_{x/2}^1 dy\,dx = \int_0^1\int_0^{2y} dx\,dy = 1$

31.

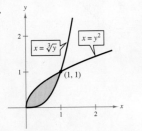

$\displaystyle\int_0^1\int_{y^2}^{\sqrt[3]{y}} dx\,dy = \int_0^1\int_{x^3}^{\sqrt{x}} dy\,dx = \dfrac{5}{12}$

33. $\dfrac{1}{2}(e^9-1)\approx 4051.042$ **35.** 24 **37.** $\dfrac{16}{3}$

39. $\dfrac{8}{3}$ **41.** 36 **43.** 5 **45.** 2 **47.** 0.6588

49. 8.1747 **51.** 0.4521 **53.** 1.1190 **55.** True

SECTION 9.8 (page 618)

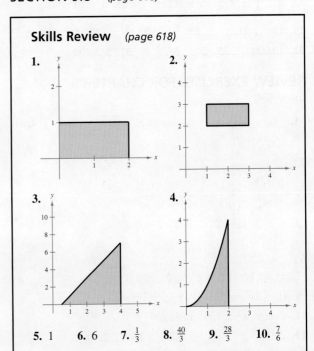

Skills Review (page 618)

5. 1 **6.** 6 **7.** $\dfrac{1}{3}$ **8.** $\dfrac{40}{3}$ **9.** $\dfrac{28}{3}$ **10.** $\dfrac{7}{6}$

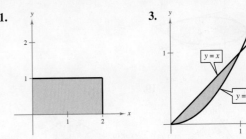

10

$\dfrac{1}{54}$

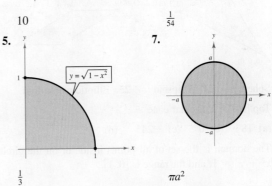

$\dfrac{1}{3}$

πa^2

9. $\displaystyle\int_0^3\int_0^5 xy\,dy\,dx = \int_0^5\int_0^3 xy\,dx\,dy = \dfrac{225}{4}$

11. $\int_0^2 \int_x^{2x} \dfrac{y}{x^2+y^2}\, dy\, dx = \int_0^2 \int_{y/2}^{y} \dfrac{y}{x^2+y^2}\, dx\, dy$
$+ \int_2^4 \int_{y/2}^{2} \dfrac{y}{x^2+y^2}\, dx\, dy = \ln \dfrac{5}{2}$

13. $\int_0^{1/2} \int_0^{2x} e^{-x^2}\, dy\, dx = 0.2212$ **15.** 4 **17.** 4

19. 12 **21.** $\dfrac{3}{8}$ **23.** $\dfrac{40}{3}$ **25.** $\dfrac{1}{3}$ **27.** 4 **29.** $\dfrac{32}{3}$

31. 10,000 **33.** 2 **35.** $\dfrac{8}{3}$ **37.** $13,400

REVIEW EXERCISES FOR CHAPTER 9
(page 624)

1. 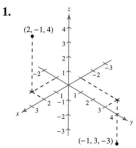   3. $\sqrt{110}$ 5. $(-1, 4, 6)$

7. $x^2 + (y-1)^2 + z^2 = 25$
9. $(x-2)^2 + (y-3)^2 + (z-2)^2 = 17$
11. Center: $(-2, 1, 4)$; radius: 4
13. 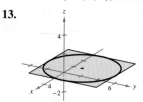   15. $(2, 2, 4)$

17. 19.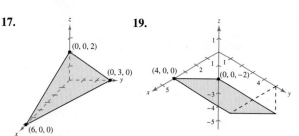

21. Sphere   23. Ellipsoid   25. Elliptic paraboloid
27. Top half of a circular cone
29. (a) 18   (b) 0   (c) -245 (d) -32
31. The domain is the set of all points inside or on the circle $x^2 + y^2 = 1$, and the range is $[0, 1]$.

33. The level curves are lines of slope $-\dfrac{2}{5}$.

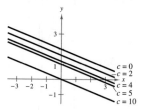

35. The level curves are hyperbolas.

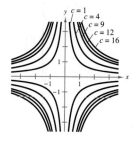

37. (a) As the color darkens from light green to dark green, the average yearly precipitation increases.
   (b) The small eastern portion containing Davenport
   (c) The northwestern portion containing Sioux City
39. Southwest to northeast
41. (a) 16.22 kg   (b) About 153 cm
   (c) a; Explanations will vary. Sample answer: The a-variable has a greater influence on weight because the absolute value of its coefficient is larger than the absolute value of the coefficient of the h-term.
43. $f_x = 2xy + 3y + 2$
   $f_y = x^2 + 3x - 5$
45. $z_x = \dfrac{2x}{y^2}$ 47. $f_x = \dfrac{2}{2x+3y}$
   $z_y = -\dfrac{2x^2}{y^3}$ $f_y = \dfrac{3}{2x+3y}$
49. $f_x = ye^x + e^y$
   $f_y = xe^y + e^x$
51. $w_x = yz^2$ 53. (a) $z_x = 3$
   $w_y = xz^2$ (b) $z_y = -4$
   $w_z = 2xyz$
55. (a) $z_x = -2x$
   At $(1, 2, 3)$, $z_x = -2$.
   (b) $z_y = -2y$
   At $(1, 2, 3)$, $z_y = -4$.

57. $f_{xx} = 6$
$f_{yy} = 12y$
$f_{xy} = f_{yx} = -1$

59. $f_{xx} = f_{yy} = f_{xy} = f_{yx} = \dfrac{-1}{4(1+x+y)^{3/2}}$

61. (a) $A_w = 43.095w^{-0.575}h^{0.725}$
$A_h = 73.515w^{0.425}h^{-0.275}$
(b) ≈ 47.35;
The surface area of an average human body increases approximately 47.35 square centimeters per pound for a human who weighs 180 pounds and is 70 inches tall.

63. Critical point: $(0, 0)$
Relative minimum: $(0, 0, 0)$

65. Critical point: $(-2, 3)$
Saddle point: $(-2, 3, 1)$

67. Critical points: $(0, 0)$, $\left(\frac{1}{6}, \frac{1}{12}\right)$
Relative minimum: $\left(\frac{1}{6}, \frac{1}{12}, -\frac{1}{432}\right)$
Saddle point: $(0, 0, 0)$

69. Critical points: $(1, 1)$, $(-1, -1)$, $(1, -1)$, $(-1, 1)$
Relative minimum: $(1, 1, -2)$
Relative maximum: $(-1, -1, 6)$
Saddle points: $(1, -1, 2)$, $(-1, 1, 2)$

71. (a) $y = \frac{60}{59}x - \frac{15}{59}$ (b) 2.746

73. (a) $y = 0.456t^2 - 9.21t + 100.4$
(b) About 82.4 billion board-feet

75. (a) $y = 0.96t^2 - 5.6t + 87$
(b) 103,640 workers

77. $y = 1.71x^2 - 2.57x + 5.56$

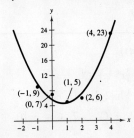

79. 1   **81.** $\frac{7}{4}$

83. $\int_{-2}^{2}\int_{5}^{9-x^2} dy\, dx = \int_{5}^{9}\int_{-\sqrt{9-y}}^{\sqrt{9-y}} dx\, dy = \frac{32}{3}$

85. $\int_{-3}^{6}\int_{1/3(x+3)}^{\sqrt{x+3}} dy\, dx = \int_{0}^{3}\int_{3y-3}^{y^2-3} dx\, dy = \frac{9}{2}$

87. $\frac{4096}{9}$ **89.** 0.0833 mi

CHAPTER TEST (page 628)

1. (a) 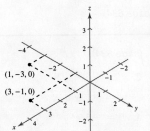 (b) $2\sqrt{2}$ (c) $(2, -2, 0)$

2. (a) 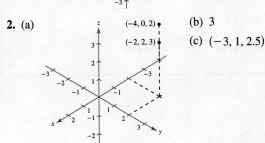 (b) 3 (c) $(-3, 1, 2.5)$

3. (a) 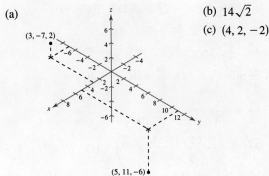 (b) $14\sqrt{2}$ (c) $(4, 2, -2)$

4. Center: $(10, -5, 5)$; radius: 5   **5.** Plane

6. Elliptic cone   **7.** Hyperbolic paraboloid

8. $f(3, 3) = 19$ **9.** $f(3, 3) = \frac{3}{2}$ **10.** $f(3, 3) = 0$
$f(1, 1) = 3$ $f(1, 1) = \frac{3}{2}$ $f(1, 1) = 0$

11. $f_x = 6x + 9y^2$; $f_x(10, -1) = 69$
$f_y = 18xy$; $f_y(10, -1) = -180$

12. $f_x = (x+y)^{1/2} + \dfrac{x}{2(x+y)^{1/2}}$; $f_x(10, -1) = \dfrac{14}{3}$
$f_y = \dfrac{x}{2(x+y)^{1/2}}$; $f_y(10, -1) = \dfrac{5}{3}$

13. Critical point: $(1, -2)$; Relative minimum: $(1, -2, -23)$

14. Critical points: $(0, 0)$, $(1, 1)$, $(-1, -1)$
Saddle point: $(0, 0, 0)$
Relative maxima: $(1, 1, 2)$, $(-1, -1, 2)$

15. $y = -1.839x^2 + 31.70x + 73.6$ **16.** $\frac{3}{2}$

17. 1   **18.** $\frac{4}{3}$ units^2 **19.** $\frac{11}{6}$

CHAPTER 10

SECTION 10.1 (page 634)

> **Skills Review** (page 634)
>
> 1. $y' = 6x + 2$
> $y'' = 6$
> 2. $y' = -6x^2 - 8$
> $y'' = -12x$
> 3. $y' = -6e^{2x}$
> $y'' = -12e^{2x}$
> 4. $y' = -6xe^{x^2}$
> $y'' = -6e^{x^2}(2x^2 + 1)$
> 5. $k = 2\ln 3 - \ln \frac{17}{2} \approx 0.0572$
> 6. $k = \ln 10 - \dfrac{\ln 41}{2} \approx 0.4458$

1. $y' = 3x^2$
3. $y' = -2e^{-2x}$ and $y' + 2y = -2e^{-2x} + 2(e^{-2x}) = 0$
5. $y' = 6x^2$ and $y' - \dfrac{3}{x}y = 6x^2 - \dfrac{3}{x}(2x^3) = 0$
7. $y'' = 2$ and $x^2 y'' - 2y = x^2(2) - 2(x^2) = 0$
9. $y' = 4e^{2x}$, $y'' = 8e^{2x}$, and
   $y'' - y' - 2y = 8e^{2x} - 4e^{2x} - 2(2e^{2x}) = 0$
11. $\dfrac{dy}{dx} = -\dfrac{1}{x^2}$
13. $\dfrac{dy}{dx} = 4Ce^{4x} = 4y$
15. $\dfrac{dy}{dt} = -\dfrac{1}{3}Ce^{-t/3}$ and
    $3\dfrac{dy}{dt} + y - 7 = 3\left(-\dfrac{1}{3}Ce^{-t/3}\right) + (Ce^{-t/3} + 7) - 7 = 0$
17. $y' = 2Cx - 3$ and
    $xy' - 3x - 2y = x(2Cx - 3) - 3x - 2(Cx^2 - 3x) = 0$
19. $y' = C_2 e^x$, $y'' = C_2 e^x$, and
    $y'' - y' = C_2 e^x - C_2 e^x = 0$
21. $y' = \dfrac{3}{5}x^2 - 1 + \dfrac{C}{2\sqrt{x}}$ and
    $2xy' - y = 2x\left(\dfrac{3}{5}x^2 - 1 + \dfrac{C}{2\sqrt{x}}\right) - \left(\dfrac{x^3}{5} - x + C\sqrt{x}\right)$
    $= x^3 - x$
23. $y' = \ln x + 1 + C$ and
    $x(y' - 1) - (y - 4) = x(\ln x + 1 + C - 1)$
    $\qquad - (x \ln x + Cx + 4 - 4) = 0$
25. Solution   27. Not a solution   29. Not a solution
31. Solution   33. $y = 3e^{-2x}$ 35. $y = 3e^{4x} + 2e^{-3x}$

37.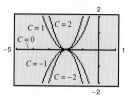

39. $y = x^3 + C$ 41. $y = x + 3\ln|x| + C$
43. $y = \dfrac{1}{2}\ln\left|\dfrac{x-1}{x+1}\right| + C$ 45. $y = \dfrac{1}{4}\sin 4x + C$
47. $y^2 = \dfrac{1}{4}x^3$ 49. $y = 3e^x$
51. (a) $N = 750 - 650e^{-0.0484t}$

(b)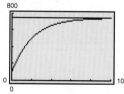

(c) $N \approx 214$

53. $y = a + Ce^{k(1-b)t}$

$\dfrac{dy}{dt} = Ck(1-b)e^{k(1-b)t}$

$a + b(y - a) + \dfrac{1}{k}\dfrac{dy}{dt} = a + b[(a + Ce^{k(1-b)t}) - a]$

$\qquad + \dfrac{1}{k}[Ck(1-b)e^{k(1-b)t}]$

$= a + bCe^{k(1-b)t} + C(1-b)e^{k(1-b)t}$

$= a + Ce^{k(1-b)t}[b + (1-b)]$

$= a + Ce^{k(1-b)t} = y$

55. True

SECTION 10.2 (page 641)

> **Skills Review** (page 641)
>
> 1. $\dfrac{2}{5}x^{5/2} + C$
> 2. $\dfrac{1}{4}t^4 - \dfrac{3}{4}t^{4/3} + C$
> 3. $2\ln|x - 5| + C$
> 4. $\dfrac{1}{4}\ln|2y^2 + 1| + C$
> 5. $\dfrac{1}{2}e^{2y} + C$
> 6. $-\dfrac{1}{2}e^{1-x^2} + C$
> 7. $C = -10$
> 8. $C = 5$
> 9. $k = \dfrac{\ln 5}{2} \approx 0.8047$
> 10. $k = -2\ln 3 - \ln 2 \approx -2.8904$

1. Yes; $(y+3)\,dy = x\,dx$

3. Yes; $dy = \left(\dfrac{1}{x}+1\right)dx$

5. No, the variables cannot be separated.

7. $y = x^2 + C$ 9. $r = Ce^{0.05s}$ 11. $y^2 = \tfrac{3}{2}x^2 + C$

13. $y = \sqrt[3]{x+C}$ 15. $C = 2x^2 - (y+1)^2$

17. $y = Ce^{x^2/2}$ 19. $y^2 = \tfrac{1}{2}e^t + C$

21. $y = 1 - \left(C - \dfrac{x}{2}\right)^2$ 23. $y = C(2+x)^2$

25. $y = Cx$ 27. $3y^2 + 2y^3 = 3x^2 + C$

29. $y = -e^{-x} - x + C$ 31. $y^2 = -2\cos x + C$

33. (a) $y = \tfrac{1}{2}x^2 + C$
(b)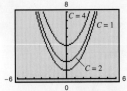

35. (a) $y = 4 - Ce^{-x}$
(b)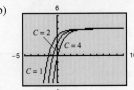

37. $y^2 = 2e^x + 14$ 39. $y = -4 - e^{-x^2/2}$

41. $P = 5e^{6t}$ 43. $y = e^{\sin x}$

45. $5y^2 = 6x^2 - 1$ or $6x^2 - 5y^2 = 1$

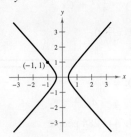

47. $v = 34.56(1 - e^{-0.1t})$ 49. $V = \left(\tfrac{1}{3}kt + C\right)^3$

51. (a) $w = 1200 - 1140e^{-0.8t}$

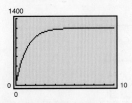

$w = 1200 - 1140e^{-0.9t}$

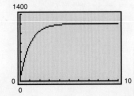

$w = 1200 - 1140e^{-t}$

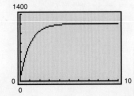

(b) ≈ 1.31 yr; ≈ 1.16 yr; ≈ 1.05 yr

(c) 1200 lb

53. 98.9%   55. Answers will vary.

MID-CHAPTER QUIZ *(page 643)*

1. $y' = -\tfrac{1}{2}Ce^{-x/2}$ and $2y' + y = 2\left(-\tfrac{1}{2}Ce^{-x/2}\right) + Ce^{-x/2} = 0$

2. $y'' = -C_1 \cos x - C_2 \sin x$ and
$y'' + y = -C_1 \cos x - C_2 \sin x + C_1 \cos x + C_2 \sin x$
$= 0$

3. $y'' = -9C_1 \sin 3x - 9C_2 \cos 3x$ and
$y'' + 9y = -9C_1 \sin 3x - 9C_2 \cos 3x$
$\qquad\quad + 9(C_1 \sin 3x + C_2 \cos 3x)$
$= 0$
$y = 2 \sin 3x - \tfrac{1}{3}\cos 3x$

4. $y' = C_1 + 3C_2 x^2$, $y'' = 6C_2 x$, and
$x^2 y'' - 3xy' + 3y = x^2(6C_2 x) - 3x(C_1 + 3C_2 x^2)$
$\qquad\qquad\qquad\quad + 3(C_1 x + C_2 x^3) = 0$
$y = -2x + \tfrac{1}{2}x^3$

5. $y = -2x^2 + 4x + C$ 6. $y = 1 + Ce^{x^2/2 + 2x}$

7. $y^2 = \ln\left|\dfrac{x-1}{x+1}\right| + C$

8. (a) $y = \pm\sqrt{\frac{1}{3}x^3 + x + C}$
   (b)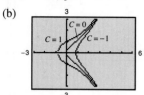

9. (a) $y = C(x - 3)$
   (b)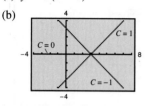

10. $y = \frac{1}{2}e^{-2x} + \frac{1}{2}$ 11. $y = -3e^{-(\cos \pi x)/\pi}$

12. $y = 2e^{x^3}$
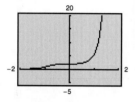

13. $v = 20 - 20e^{-1.386t}$

SECTION 10.3   *(page 647)*

Skills Review   *(page 647)*

1. $e^x + 1$ 2. $e^{3x} + 1$ 3. $\frac{1}{x^3}$ 4. $x^2 e^x$
5. $2e^{2x} + C$ 6. $\frac{1}{6}e^{3x^2} + C$ 7. $\frac{1}{2}\ln|2x + 5| + C$
8. $\frac{1}{2}\ln|x^2 + 2x + 3| + C$ 9. $\frac{1}{12}(4x - 3)^3 + C$
10. $\frac{1}{6}(x^2 - 1)^3 + C$

1. $y' + \frac{-3}{2x^2}y = \frac{x}{2}$ 3. $y' + \frac{1}{x}y = e^x$
5. $y' + \frac{1}{1-x}y = \frac{1}{x-1}$ 7. $y = 2 + Ce^{-3x}$
9. $y = e^{-x}(x + C)$ 11. $y = \frac{1}{2}x^2 + 3\ln|x| + C$
13. $y = \frac{1}{5} + Ce^{-(5/2)x^2}$ 15. $y = \frac{x^3 - 3x + C}{3(x - 1)}$
17. $y = e^{1/x^2}\left(-\frac{1}{2x^2} + C\right)$ 19. $y = Ce^{-x} + 4$

21. $y = Ce^{x^2} - 1$
23. c   24. d   25. a   26. b
27. $y = 3e^x$ 29. $xy = 4$
31. $y = 1 + 5e^{-x^3}$ 33. $y = x^2(5 - \ln|x|)$
35. $S = t + 95(1 - e^{-t/5})$

t	0	1	2	3	4	5
S	0	18.22	33.32	45.86	56.31	65.05

t	6	7	8	9	10
S	72.39	78.57	83.82	88.30	92.14

37. $v = \frac{1}{k}[9.8 + Ce^{kt}]$

39. (a) $N = 40 + Ce^{-kt}$
    (b) $N = 40 - 30.57e^{-0.0188t}$

41. (a) $y = 1 - e^{-0.25t}$
    (b) ≈ 2.77 years
    (c) $\approx 63.2\%$

SECTION 10.4   *(page 654)*

Skills Review   *(page 654)*

1. $y = \frac{3}{2}x^2 + C$ 2. $y^2 = 3x + C$
3. $y = Ce^{x^2}$ 4. $y^4 = \frac{1}{2}(x - 4)^2 + C$
5. $y = 2 + Ce^{-2x}$ 6. $y = xe^{-2x} + Ce^{-2x}$
7. $y = 1 + Ce^{-x^2/2}$ 8. $y = \frac{1}{4}x^2 + Cx^{-2}$
9. $\frac{dy}{dx} = kx^2$ 10. $\frac{dx}{dt} = k(x - t)$

1. $y = e^{(x \ln 2)/3} \approx e^{0.2310x}$ 3. $y = 4e^{-(x \ln 4)/4}$
   $\approx 4e^{-0.3466x}$
5. $y = \frac{1}{2}e^{(\ln 2)x} \approx \frac{1}{2}e^{0.6931x}$ 7. 444,284 people
9. $y = \frac{20}{1 + 19e^{-0.5889x}}$ 11. $y = \frac{5000}{1 + 19e^{-0.10156x}}$
13. (a) $N = \frac{500}{1 + 4e^{-0.2452t}}$ (b) 121 deer
    (c) ≈ 9.1 yr
15. $\frac{dP}{dn} = kP(L - P)$, $P = \frac{CL}{e^{-Lkn} + C}$

17. $y = \dfrac{360}{8 + 41t}$

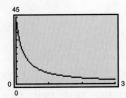

19. $\dfrac{dy}{dt} = ky$ **21.** $y = 70 + 280e^{-0.02784t}$

$y = 20e^{-0.2231t}$

23. About 74.69 min longer

25. $y = 500e^{-1.6094e^{-0.1451t}}$

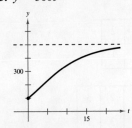

27. 34 beavers **29.** 92%

31. (a) $Q = 25e^{-0.05t}$ (b) ≈ 10.22 min

33. ≈ 3.15 h

35. $P = Ce^{kt} - \dfrac{N}{k}$ **37.** $A = \dfrac{P}{r}(e^{rt} - 1)$

39. (a) $C = C_0 e^{-Rt/V}$ (b) 0

41. (a) $C(t) = \dfrac{Q}{R}(1 - e^{-Rt/V})$ (b) $\dfrac{Q}{R}$

REVIEW EXERCISES FOR CHAPTER 10
(page 660)

1. $\dfrac{dy}{dx} = \dfrac{1}{2}Ce^{x/2}$ and $2\dfrac{dy}{dx} = 2\left(\dfrac{1}{2}Ce^{x/2}\right) = Ce^{x/2} = y$

3. $y' = C_1 e^x - C_2 e^{-x}$, $y'' = C_1 e^x + C_2 e^{-x}$, and

$y'' - y = C_1 e^x + C_2 e^{-x} - C_1 e^x - C_2 e^{-x} = 0$

5. $y = \dfrac{2}{3}x^3 + 5x + C$ **7.** $y = \dfrac{1}{2}\sin 2x + C$

9. $y = \dfrac{4}{15}(x - 7)^{3/2}(3x + 14) + C$

11. $y' = 4e^{4x}$ and $4y = 4e^{4x} = y'$

13. $y' = 2x$ and $\dfrac{1}{2}y' - \dfrac{1}{x}y = \dfrac{1}{2}(2x) - \dfrac{1}{x}(x^2) = 0$

15. Not a solution **17.** Solution **19.** $y = e^{-5x}$

21. 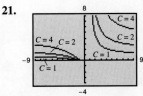 **23.** $y = \dfrac{1}{2}(x^2 + 1)$

25. Yes; $\dfrac{dy}{y} = \dfrac{dx}{x + 3}$

27. No, the variables cannot be separated.

29. $y = 2x^2 + C$ **31.** $y^4 = 5x + C$ **33.** $y = \dfrac{1}{x^2 + C}$

35. $y = Ce^{(x^2/2 + x)} - 1$ **37.** $y = Ce^{1/(4x^2)} - 2$

39. $y^2 = 2\sin x + C$

41. (a) $y = x^3 + C$

(b)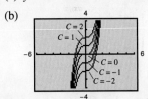

43. $y^2 = -2e^x + 6$ **45.** $y = \dfrac{1}{2}(\ln x)^2 + 2$

47. $9x^2 + 16y^2 = 25$

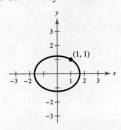

49. (a) $y = 28e^{0.6 - 0.012s}$

(b)

Speed	50	55	60	65	70
Miles per gallon	28	26.4	24.8	23.4	22.0

51. $y' - \dfrac{y}{x^2} = -\dfrac{1}{4}x^2$ **53.** $y' - y = \dfrac{1}{2x^2}$

55. $y = Ce^{4x} - 2$ **57.** $y = 2x^2 - 3x \ln|x| + Cx$

59. $y = Ce^x - 9$ **61.** $y = x^2 + 2x + \dfrac{C}{x}$

63. $y = 6 + Ce^{-x}$ **65.** $y = Ce^{-x^2} + \dfrac{1}{2}$

67. $\approx 383.298°$ F **69.** $y = Cx^{-k}$

71. (a) ≈ 22.68 in. (b) ≈ 14.38 in.

73. $\approx 88.6\%$

75. $\dfrac{dy}{dt} = ky^{1/3}$

$y = \left(\tfrac{2}{3}kt + C\right)^{3/2}$

$y = (-5t + 9)^{3/2}$

77. ≈ 13.8 gal   **79.** ≈ 50 pelicans

CHAPTER TEST   *(page 663)*

1. $y' = -2e^{-2x}$ and
$3y' + 2y = 3(-2e^{-2x}) + 2e^{-2x} = -4e^{-2x}$

2. $y'' = \dfrac{2}{(x+1)^3}$ and

$y'' - \dfrac{2y}{(x+1)^2} = \dfrac{2}{(x+1)^3} - \dfrac{2\left(\dfrac{1}{x+1}\right)}{(x+1)^2} = 0$

3. $y^2 = x^2 + C$ **4.** $y^2 = 2x^2 + C$

5. $y^3 = \dfrac{1}{\pi}\sin \pi x + C$ **6.** $y = Ce^{x^2} + 1$

7. $y = xe^{2x} + Ce^{2x}$ **8.** $y = -x - 1 + Ce^x$

9. $y = \dfrac{1}{3}x^2 + \dfrac{C}{x}$ **10.** $y = -1 + Ce^{x^3/3}$

11. $y = 1 - e^{-x^3/3}$ **12.** $y = e^{e^{x^2}}$ **13.** $y = \tfrac{1}{14}(\ln x)^2 - 2$

14. $w = 200 - 190e^{-kt}$

(a)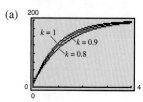

(b) ≈ 1.7 yr; ≈ 1.5 yr; ≈ 1.3 yr

(c) 200 lb

15. About 35 min longer

CHAPTER 11

SECTION 11.1   *(page 671)*

Skills Review   *(page 671)*

1. 1 **2.** 1 **3.** 2 **4.** 2 **5.** $\tfrac{1}{2}$
6. 1 **7.** 37.50% **8.** 81.82%
9. 54.17% **10.** 43.75%

1. (a) $S = \{HHH, HHT, HTH, HTT, THH,$
  $THT, TTH, TTT\}$
 (b) $A = \{HHH, HHT, HTH, THH\}$
 (c) $B = \{HTT, THT, TTH, TTT\}$

3. (a) $S = \{III, IIO, IIU, IOI, IOO, IOU, IUI, IUO, IUU,$
  $OII, OIO, OIU, OOI, OOO, OOU, OUI, OUO,$
  $OUU, UII, UIO, UIU, UOI, UOO, UOU, UUI,$
  $UUO, UUU\}$
 (b) $A = \{III, IIO, IIU, IOI, IUI, OII, UII\}$
 (c) $B = \{III, IIO, IIU, IOI, IOU, IUI, IUO, IUU, OII,$
  $OIU, OUI, OUU, UII, UIO, UIU, UOI, UOU,$
  $UUI, UUO, UUU\}$

5.

Random variable	0	1	2
Frequency	1	2	1

7.

Random variable	0	1	2	3
Frequency	1	3	3	1

9. 0.24   **11.** 0.0145   **13.** $P(3) = 0.25$

15.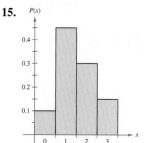

The table represents a probability distribution.

17. The table does not represent a probability distribution because the sum of the probabilities does not equal 1 and $P(4) < 0$.

19., **21.**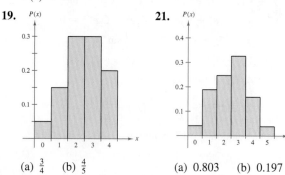

(a) $\tfrac{3}{4}$ (b) $\tfrac{4}{5}$ (a) 0.803 (b) 0.197

23. (a) $S = \{gggg, gggb, ggbg, gbgg, bggg, ggbb,$
$gbgb, gbbg, bgbg, bbgg, bggb, gbbb,$
$bgbb, bbgb, bbbg, bbbb\}$

(b)
x	0	1	2	3	4
$P(x)$	$\frac{1}{16}$	$\frac{4}{16}$	$\frac{6}{16}$	$\frac{4}{16}$	$\frac{1}{16}$

(c) 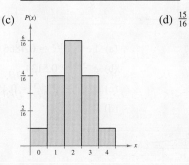   (d) $\frac{15}{16}$

25. $E(x) = 3$
$V(x) = 0.875$
$\sigma \approx 0.9354$

27. (a)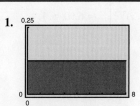

(b) 0.648   (c) 0.729

29. (a) Mean: 2.5; Variance: 1.25
(b) Mean: 5; Variance: 2.5

31. (a)
x	0	1	2	3	4
$P(x)$	$\frac{14}{50}$	$\frac{26}{50}$	$\frac{7}{50}$	$\frac{2}{50}$	$\frac{1}{50}$

(b)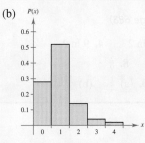

(c) $\frac{35}{50}$ (d) $E(x) = 1$, $V(x) = 0.76$, $\sigma \approx 0.87$
Answers will vary.

SECTION 11.2 (page 679)

Skills Review (page 679)

1. Yes 2. No 3. No 4. Yes 5. 1
6. $\frac{1}{2}$ 7. 1 8. $\frac{1}{4}$ 9. 1 10. 1

1.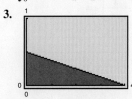

$f(x)$ is a probability density function.

$\int_0^8 \frac{1}{8} \, dx = \left[\frac{1}{8}x\right]_0^8 = 1$

3.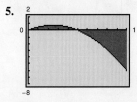

$f(x)$ is a probability density function.

$\int_0^4 \frac{4-x}{8} \, dx = \left[\frac{1}{2}x - \frac{1}{16}x^2\right]_0^4 = 1$

5.

$f(x)$ is not a probability density function because

$\int_0^1 6x(1-2x)\,dx = \left[3x^2 - 4x^3\right]_0^1 = -1 \ne 1$

and $f(x) < 0$ over

the interval $\left(\frac{1}{2}, 1\right]$.

7.

$f(x)$ is not a probability density function because

$\int_0^5 \frac{1}{5}e^{-x/5} \, dx = \left[-e^{-x/5}\right]_0^5 \approx 0.632 \ne 1$.

9.

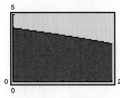

$f(x)$ is not a probability density function because

$$\int_0^2 2\sqrt{4-x}\,dx = \left[-\frac{4}{3}(4-x)^{3/2}\right]_0^2 \approx 6.90 \neq 1.$$

11.

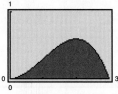

$f(x)$ is a probability density function.

$$\int_0^3 \frac{4}{27}x^2(3-x)\,dx = \frac{4}{27}\left[x^3 - \frac{x^4}{4}\right]_0^3 = 1$$

13.

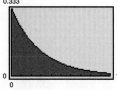

$f(x)$ is a probability density function.

$$\int_0^\infty \frac{1}{3}e^{-x/3}\,dx = \lim_{b\to\infty}\left[-e^{-x/3}\right]_0^b = 1$$

15.

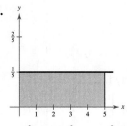

(a) $\frac{3}{5}$ (b) $\frac{2}{5}$ (c) $\frac{2}{5}$ (d) $\frac{4}{5}$

17.

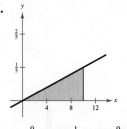

(a) $\frac{9}{25}$ (b) $\frac{1}{5}$ (c) $\frac{9}{25}$ (d) $\frac{24}{25}$

19.

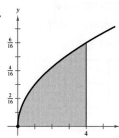

(a) $\frac{\sqrt{2}}{4} \approx 0.354$

(b) $1 - \frac{\sqrt{2}}{4} \approx 0.646$

(c) $\frac{1}{8}(3\sqrt{3}-1) \approx 0.525$

(d) $\frac{3\sqrt{3}}{8} \approx 0.650$

21.

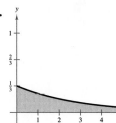

(a) $1 - e^{-2/3} \approx 0.4866$

(b) $e^{-2/3} \approx 0.5134$

(c) $e^{-1/3} - e^{-4/3} \approx 0.4529$

(d) 0

23. $\frac{2}{15}$ **25.** 1 **27.** $\frac{15}{4}$ **29.** (a) 0.87 (b) 0.34

31. (a) $1 - e^{-2/3} \approx 0.487$ (b) $e^{-2/3} - e^{-4/3} \approx 0.250$

(c) $e^{-2/3} \approx 0.513$

33. (a) $1 - e^{-6/5} \approx 0.699$ (b) $e^{-2/5} - e^{-6/5} = 0.369$

(c) $e^{-8/5} \approx 0.202$

35. (a) 0.75. There is a 75% probability that the community will receive up to 10 inches of rain.

(b) 0.25. There is a 25% probability that the community will receive 10 to 15 inches of rain.

(c) 0.25. There is a 25% probability that the community will receive up to 5 inches of rain.

(d) About 0.095. There is a probability of approximately 9.5% that the community will receive 12 to 15 inches of rain.

37. Answers will vary.

SECTION 11.3 *(page 688)*

> **Skills Review** *(page 688)*
>
> **1.** 5 **2.** 8 **3.** 3 ln 2 **4.** 9 ln 2
>
> **5.** $\frac{4}{3}$ **6.** $\frac{4}{3}$ **7.** $\frac{3}{4}$ **8.** $\frac{2}{9}$
>
> **9.** (a) $\frac{1}{4}$ (b) $\frac{1}{2}$ **10.** (a) $\frac{1}{2}$ (b) $\frac{11}{16}$

1. (a) $\frac{3}{2}$ (b) $\frac{3}{4}$ (c) $\frac{\sqrt{3}}{2}$

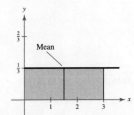

3. (a) 4 (b) 2 (c) $\sqrt{2}$

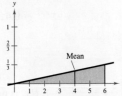

5. (a) $\frac{5}{7}$ (b) $\frac{20}{441}$ (c) $\frac{2\sqrt{5}}{21}$

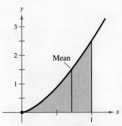

7.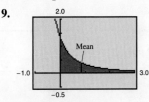

Mean: $\frac{1}{2}$

9.

Mean ≈ 0.848

11. $9 \ln 2 \approx 6.238$

13. Uniform density function
    Mean: 5
    Variance: $\frac{25}{3}$
    Standard deviation: $\frac{5\sqrt{3}}{3} \approx 2.887$

15. Exponential density function
    Mean: 8
    Variance: 64
    Standard deviation: 8

17. Normal density function
    Mean: 100
    Variance: 121
    Standard deviation: 11

19. Mean: 0
    Standard deviation: 1
    $P(0 \le x \le 0.85) \approx 0.3023$

21. Mean: 6
    Standard deviation: 6
    $P(x \ge 2.23) \approx 0.6896$

23. Mean: 8
    Standard deviation: 2
    $P(3 \le x \le 13) \approx 0.9876$

25. (a) About 0.309 (b) About 0.159
    (c) About 0.841 (d) About 0.669

27. (a) Mean: 10:05 A.M.
    Standard deviation: $\frac{5\sqrt{3}}{3} \approx 2.9$ minutes
    (b) $\frac{3}{10}$

29. (a) $f(t) = \frac{1}{2}e^{-t/2}$ (b) $1 - e^{-1/2} \approx 0.3935$

31. (a) $f(t) = \frac{1}{5}e^{-t/5}$ (b) $1 - e^{-2} \approx 0.865$

33. (a) 1.5 standard deviations (b) About 93.32%

35. About 61.5%

37. Mean: $\frac{11}{2}$; Median: $\frac{11}{2}$

39. Mean: $\frac{1}{6}$
    Median: $\frac{2 - \sqrt{2}}{4} \approx 0.1464$

41. Mean: 5
    Median: $5 \ln 2 \approx 3.4657$

43. $\frac{1}{-0.28} \ln 0.5 \approx \$2.48$

45. About 46,156 miles

47. (a) 0.6827
    (b) No, only about 0.13% of the batteries will last less than 3 years.

49. (a)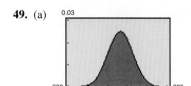

(b) About 75.7%   (c) About 19.1%

51. (a)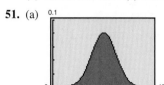

(b) About 15.7%   (c) About 29.8%

REVIEW EXERCISES FOR CHAPTER 11
(page 694)

1. $S = \{$January, February, March, April, May, June, July, August, September, October, November, December$\}$

3. If the essays are numbered 1, 2, 3, and 4, $S = \{123, 124, 134, 234\}$.

5. $S = \{0, 1, 2, 3\}$

7.

x	0	1	2	3
$n(x)$	1	3	3	1

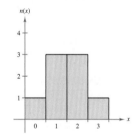

9.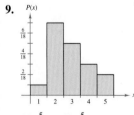

(a) $\frac{5}{6}$ (b) $\frac{5}{9}$

11. (a) $\frac{5}{36}$ (b) $\frac{5}{6}$ (c) $\frac{1}{6}$ (d) $\frac{1}{36}$

13. 19.5   **15.** $-\$2.50$

17. $V(x) \approx 440.1992$, $\sigma \approx 20.9809$

19. $V(x) \approx 0.8379$, $\sigma \approx 0.9154$

21.

$f(x)$ is a probability density function.

$$\int_0^{12} \frac{1}{12}\, dx = \left[\frac{x}{12}\right]_0^{12} = 1$$

23.

Although $\int_0^4 \frac{1}{4}(3-x)\, dx = \left[\frac{3}{4}x - \frac{x^2}{8}\right]_0^4 = 1$, $f(x)$ is not a probability density function because $f(x) < 0$ over the interval $(3, 4]$.

25.

$f(x)$ is a probability density function.

$$\int_1^9 \frac{1}{4\sqrt{x}}\, dx = \frac{1}{4}\left[2\sqrt{x}\right]_1^9 = 1$$

27. $\frac{9}{25}$ **29.** $\frac{2}{3}$ **31.** (a) $\frac{1}{2}$ (b) $\frac{1}{4}$

33. 3.5   **35.** 6

37. Variance: $\frac{9}{20}$

Standard deviation: $\frac{3\sqrt{5}}{10}$

39. Variance: 4

Standard deviation: 2

41. $\frac{1}{2}$ **43.** $4 \ln 2 \approx 2.7726$

45. (a) $1 - e^{-10/15} = 1 - e^{-2/3} \approx 0.4866$

(b) $e^{-10/15} - e^{-20/15} = e^{-2/3} - e^{-4/3} \approx 0.2498$

47. About 0.00383

49. (a)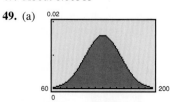

(b) 0.150   (c) 0.655

51. 0.3829

CHAPTER TEST (page 697)

1. (a) $S = \{HHHH, HHHT, HHTH, HHTT, HTHH, HTHT,$
   $HTTH, HTTT, THHH, THHT, THTH, THTT,$
   $TTHH, TTHT, TTTH, TTTT\}$; Random variable x assigns numbers to each possible outcome, depending on the number of heads that turn up.

Random variable, x	0	1	2	3	4
Frequency of x, $n(x)$	1	4	6	4	1

   (b) $P(x \geq 2) = \frac{11}{16}$

2. $P(\text{red non-face card}) = \frac{5}{13}$

3.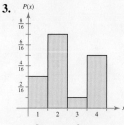

   (a) $\frac{5}{8}$ (b) $\frac{3}{8}$

4.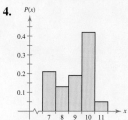

   (a) 0.95 (b) 0.66

5. $E(x) = 1.8$
   $V(x) = 1.16$
   $\sigma \approx 1.0770$

6. $E(x) = -0.056$
   $V(x) \approx 1.8649$
   $\sigma \approx 1.3656$

7.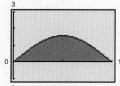

   $f(x)$ is a probability density function.
   $\int_0^1 \frac{\pi}{2} \sin \pi x \, dx = \left[-\frac{1}{2} \cos \pi x\right]_0^1 = 1$

8.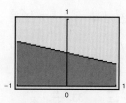

   $f(x)$ is a probability density function.
   $\int_{-1}^{1} \frac{3-x}{6} \, dx = \left[-\frac{x}{12}(x-6)\right]_{-1}^{1} = 1$

9.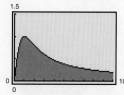

   $f(x)$ is not a probability density function because
   $\int_0^{\infty} \frac{2x}{x^2+1} \, dx = \left[\ln(x^2+1)\right]_0^{\infty} = \infty \neq 1$.

10. (a) $\frac{15}{64}$ (b) $\frac{27}{64}$ 11. (a) $\frac{7}{16}$ (b) $\frac{225}{256}$

12. (a) 0.6321 (b) 0.3679

13. $\mu = 7$
    $V(x) = \frac{49}{3}$
    $\sigma \approx 4.041$

14. $\mu = \frac{5}{8}$
    $V(x) \approx 0.0594$
    $\sigma \approx 0.2437$

15. $\mu = 1$
    $V(x) = 1$
    $\sigma = 1$

16. $P(100 \leq x \leq 120) \approx 0.683$

APPENDIX B

APPENDIX B.1 (page A8)

Skills Review (page A8)

1. $y = x^2 - 2x$ 2. $y = \dfrac{x-3}{4}$
3. $y = 1, x \neq -6$ 4. $y = -4, x \neq \pm\sqrt{3}$
5. $y = \pm\sqrt{5-x^2}$ 6. $y = \pm\sqrt{6-x^2}$ 7. $\dfrac{8}{3}$
8. $-\dfrac{1}{2}$ 9. $\dfrac{5}{7}$ 10. 1

1. $-\dfrac{y}{x}$ 3. $-\dfrac{x}{y}$ 5. $\dfrac{1-xy^2}{x^2y}$ 7. $\dfrac{y}{8y-x}$

9. $-\dfrac{1}{10y-2}$ 11. $\dfrac{1}{2}$ 13. $-\dfrac{x}{y}, 0$

A108 Answers to Selected Exercises

15. $-\dfrac{y}{x+1}, -\dfrac{1}{4}$ **17.** $\dfrac{y - 3x^2}{2y - x}, \dfrac{1}{2}$ **19.** $\dfrac{1 - 3x^2y^3}{3x^3y^2 - 1}, -1$

21. $-\sqrt{\dfrac{y}{x}}, -\dfrac{5}{4}$ **23.** $-\sqrt[3]{\dfrac{y}{x}}, -\dfrac{1}{2}$ **25.** 3

27. 0   **29.** $-\dfrac{\sqrt{5}}{3}$ **31.** $-\dfrac{x}{y}, \dfrac{4}{3}$ **33.** $\dfrac{1}{2y}, -\dfrac{1}{2}$

35. At $(8, 6)$: $y = -\dfrac{4}{3}x + \dfrac{50}{3}$
At $(-6, 8)$: $y = \dfrac{3}{4}x + \dfrac{25}{2}$

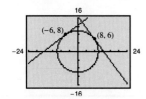

37. At $(1, \sqrt{5})$: $15x - 2\sqrt{5}y - 5 = 0$
At $(1, -\sqrt{5})$: $15x + 2\sqrt{5}y - 5 = 0$

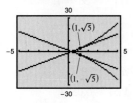

39. At $(0, 2)$: $y = 2$
At $(2, 0)$: $x = 2$

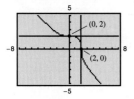

41. $-\dfrac{2}{y^2(0.00003x^2 + 0.1)}$ **43.** $-\dfrac{4xy}{2y^2 + 1}$ **45.** -2

1. (a) $\dfrac{3}{4}$ (b) 20   **3.** (a) $-\dfrac{5}{8}$ (b) $\dfrac{3}{2}$

5. Yes; If $\dfrac{dr}{dt}$ is constant, $\dfrac{dA}{dt} = 2\pi r \dfrac{dr}{dt}$ and so is proportional to r.

7. (a) $\dfrac{5}{2\pi}$ ft/min   (b) $\dfrac{5}{8\pi}$ ft/min

9. (a) 9 cm³/sec
(b) 900 cm³/sec

11. (a) $-\dfrac{7}{12}$ ft/sec   (b) $-\dfrac{3}{2}$ ft/sec
(c) $-\dfrac{48}{7}$ ft/sec

13. About 37.7 ft³/min

15. 300 mi/hr

17. $\dfrac{dv}{dt} = \dfrac{16r}{v} \sec^2\theta \dfrac{d\theta}{dt}$

$\dfrac{d\theta}{dt} = \dfrac{v}{16r} \cos^2\theta \dfrac{dv}{dt}$

19. (a) 12.5%   (b) $\dfrac{1}{144}$ m/min

21. $-V\dfrac{dp}{dt} = 1.3p\dfrac{dV}{dt}$

APPENDIX B.2   *(page A14)*

Skills Review   *(page A14)*

1. $A = \pi r^2$ **2.** $V = \dfrac{4}{3}\pi r^3$ **3.** $S = 6s^2$

4. $V = s^3$ **5.** $V = \dfrac{1}{3}\pi r^2 h$ **6.** $A = \dfrac{1}{2}bh$

7. $-\dfrac{x}{y}$ **8.** $\dfrac{2x - 3y}{3x}$ **9.** $-\dfrac{2x + y}{x + 2}$

10. $-\dfrac{y^2 - y + 1}{2xy - 2y - x}$

Answers to Checkpoints

CHAPTER 0

SECTION 0.1

Checkpoint 1 $x < 5$ or $(-\infty, 5)$

Checkpoint 2 $x < -2$ or $x > 5$; $(-\infty, -2) \cup (5, \infty)$

Checkpoint 3 Nighttime temperatures range from 20°C to 24°C.

SECTION 0.2

Checkpoint 1 $8; 8; -8$

Checkpoint 2 $2 \le x \le 10$

Checkpoint 3 (a) $0.34 \le B \le 0.40$
(b) $17{,}000 \le V \le 20{,}000$

SECTION 0.3

Checkpoint 1 $\frac{4}{9}$

Checkpoint 2 8

Checkpoint 3 (a) $3x^6$ (b) $8x^{7/2}$ (c) $4x^{4/3}$

Checkpoint 4 (a) $x(x^2 - 2)$ (b) $2x^{1/2}(1 + 4x)$

Checkpoint 5 $\dfrac{(3x - 1)^{3/2}(13x - 2)}{(x + 2)^{1/2}}$

Checkpoint 6 $\dfrac{x^2(5 + x^3)}{3}$

Checkpoint 7 (a) $[2, \infty)$ (b) $(2, \infty)$ (c) $(-\infty, \infty)$

SECTION 0.4

Checkpoint 1 (a) $\dfrac{-2 \pm \sqrt{2}}{2}$ (b) 4 (c) No real zeros

Checkpoint 2 (a) $x = -3$ and $x = 5$ (b) $x = -1$
(c) $x = \frac{3}{2}$ and $x = 2$

Checkpoint 3 $(-\infty, -2] \cup [1, \infty)$

Checkpoint 4 $-1, \frac{1}{2}, 2$

SECTION 0.5

Checkpoint 1 (a) $\dfrac{x^2 + 2}{x}$ (b) $\dfrac{3x + 1}{(x + 1)(2x + 1)}$

Checkpoint 2 (a) $\dfrac{3x + 4}{(x + 2)(x - 2)}$ (b) $-\dfrac{x + 1}{3x(x + 2)}$

Checkpoint 3

(a) $\dfrac{(A + B + C)x^2 + (A + 3B)x + (-2A + 2B - C)}{(x + 1)(x - 1)(x + 2)}$

(b) $\dfrac{(A + C)x^2 + (-A + B + 2C)x + (-2A - 2B + C)}{(x + 1)^2(x - 2)}$

Checkpoint 4 (a) $\dfrac{3x + 8}{4(x + 2)^{3/2}}$ (b) $\dfrac{1}{\sqrt{x^2 + 4}}$

Checkpoint 5 $\dfrac{\sqrt{x^2 + 4}}{x^2}$

Checkpoint 6 (a) $\dfrac{5\sqrt{2}}{4}$ (b) $\dfrac{x + 2}{4\sqrt{x + 2}}$ (c) $\dfrac{\sqrt{6} + \sqrt{3}}{3}$
(d) $\dfrac{\sqrt{x + 2} - \sqrt{x}}{2}$

CHAPTER 1

SECTION 1.1

Checkpoint 1   **Checkpoint 2**

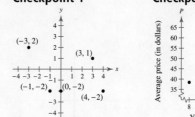

Checkpoint 3 5

Checkpoint 4 $d_1 = \sqrt{20}, d_2 = \sqrt{45}, d_3 = \sqrt{65}$
$d_1^2 + d_2^2 = 20 + 45 = 65 = d_3^2$

Checkpoint 5 0.6 mi

Checkpoint 6 $(-2, 5)$

Checkpoint 7 34.3%

Checkpoint 8 $(-1, -4), (1, -2), (1, 2), (-1, 0)$

SECTION 1.2

Checkpoint 1   **Checkpoint 2**

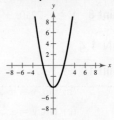

Answers to Checkpoints

Checkpoint 3 (a) x-intercepts: $(3, 0), (-1, 0)$
y-intercept: $(0, -3)$
(b) x-intercept: $(-4, 0)$
y-intercepts: $(0, 2), (0, -2)$

Checkpoint 4 $(x + 2)^2 + (y - 1)^2 = 25$

Checkpoint 5 $(x - 2)^2 + (y + 1)^2 = 4$

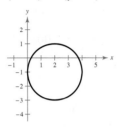

Checkpoint 6 $(-2, -1)$ and $(3, 4)$

Checkpoint 7 (a) during 2001 (b) Maryland: 5,847,000

Checkpoint 8 The projection obtained using the model is about 0.208 quadrillion Btu; this is close to the given projection.

SECTION 1.3

Checkpoint 1

(a) 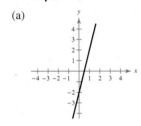 (b)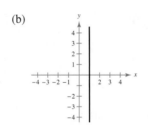

Checkpoint 2 Yes, $\frac{27}{312} \approx 0.08654 > \frac{1}{12} = 0.08\overline{3}$.

Checkpoint 3 The slope of $m = 37$ tells you that the calf gains 37 kg per month. The y-intercept $(0, 158)$ tells you that the calf's body weight at birth was 158 kg.

Checkpoint 4 (a) 2 (b) $-\frac{1}{2}$

Checkpoint 5 $y = 2x + 4$

Checkpoint 6 $y = 110t + 1575$; 2235 transplants

Checkpoint 7 (a) $y = \frac{1}{2}x$ (b) $y = -2x + 5$

Checkpoint 8 $V = -2250t + 20,000$

SECTION 1.4

Checkpoint 1 (a) Yes, $y = x - 1$.
(b) No, $y = \pm\sqrt{4 - x^2}$.
(c) No, $y = \pm\sqrt{2 - x}$.
(d) Yes, $y = x^2$.

Checkpoint 2 (a) Domain: $[-1, \infty)$; Range: $[0, \infty)$
(b) Domain: $(-\infty, \infty)$; Range: $[0, \infty)$

Checkpoint 3 $f(0) = 1, f(1) = -3, f(4) = -3$
No, f is not one-to-one.

Checkpoint 4 (a) $x^2 + 2x\,\Delta x + (\Delta x)^2 - 2x - 2\,\Delta x + 3$
(b) $2x + \Delta x - 2, \Delta x \neq 0$

Checkpoint 5 (a) $2x^2 + 5$ (b) $4x^2 + 4x + 3$

Checkpoint 6 (a) $f^{-1}(x) = 5x$ (b) $f^{-1}(x) = \frac{1}{3}(x - 2)$

Checkpoint 7 $f^{-1}(x) = \sqrt{x - 2}$

Checkpoint 8
$$f(x) = x^2 + 4$$
$$y = x^2 + 4$$
$$x = y^2 + 4$$
$$x - 4 = y^2$$
$$\pm\sqrt{x - 4} = y$$

SECTION 1.5

Checkpoint 1 6

Checkpoint 2 (a) 4 (b) Does not exist (c) 4

Checkpoint 3 5

Checkpoint 4 12

Checkpoint 5 7

Checkpoint 6 $\frac{1}{4}$

Checkpoint 7 (a) -1 (b) 1

Checkpoint 8 1

Checkpoint 9 $\lim_{x \to 3^-} f(x) = 40$ and $\lim_{x \to 3^+} f(x) = 80$
$\lim_{x \to 3^-} f(x) \neq \lim_{x \to 3^+} f(x)$

Checkpoint 10 Does not exist

SECTION 1.6

Checkpoint 1 (a) f is continuous on the entire real line.
(b) f is continuous on the entire real line.

Checkpoint 2 (a) f is continuous on $(-\infty, 1)$ and $(1, \infty)$.
(b) f is continuous on $(-\infty, 2)$ and $(2, \infty)$.
(c) f is continuous on the entire real line.

Checkpoint 3 f is continuous on $[2, \infty)$.

Checkpoint 4 f is continuous on $[-1, 5]$.

Checkpoint 5

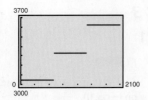

CHAPTER 2

SECTION 2.1

Checkpoint 1 3

Checkpoint 2 For the months to the left of July on the graph, the tangent lines have positive slopes. For the months to the right of July, the tangent lines have negative slopes. The average daily temperature is increasing prior to July and decreasing after July.

Checkpoint 3 4

Checkpoint 4 2

Checkpoint 5 $m = 8x$
At $(0, 1)$, $m = 0$.
At $(1, 5)$, $m = 8$.

Checkpoint 6 $2x - 5$

Checkpoint 7 $-\dfrac{4}{t^2}$

SECTION 2.2

Checkpoint 1 (a) 0 (b) 0 (c) 0 (d) 0

Checkpoint 2 (a) $4x^3$ (b) $-\dfrac{3}{x^4}$ (c) $2w$ (d) $-\dfrac{1}{t^2}$

Checkpoint 3 $f'(x) = 3x^2$
$m = f'(-1) = 3;$
$m = f'(0) = 0;$
$m = f'(1) = 3$

Checkpoint 4 (a) $8x$ (b) $\dfrac{8}{\sqrt{x}}$

Checkpoint 5 (a) $\tfrac{1}{4}$ (b) $-\tfrac{2}{5}$

Checkpoint 6 (a) $-\dfrac{9}{2x^3}$ (b) $-\dfrac{9}{8x^3}$

Checkpoint 7 (a) $\dfrac{\sqrt{5}}{2\sqrt{x}}$ (b) $\dfrac{1}{3x^{2/3}}$

Checkpoint 8 -1

Checkpoint 9 $y = -x + 2$

Checkpoint 10 Decreasing at a rate of 50,560 cases per year

SECTION 2.3

Checkpoint 1 (a) $0.5\overline{6}$ mg/ml/min
(b) 0 mg/ml/min
(c) -1.5 mg/ml/min

Checkpoint 2 (a) -16 ft/sec (b) -48 ft/sec
(c) -80 ft/sec

Checkpoint 3 When $t = 1.75$, $h'(1.75) = -56$ ft/sec
When $t = 2$, $h'(2) = -64$ ft/sec

Checkpoint 4 $h = -16t^2 + 16t + 12$
$v = h' = -32t + 16$

Checkpoint 5 (a) 50 lb/mo
(b) 53.06 lb/mo

SECTION 2.4

Checkpoint 1 $-27x^2 + 12x + 24$

Checkpoint 2 $\dfrac{2x^2 - 1}{x^2}$

Checkpoint 3 (a) $18x^2 + 30x$ (b) $12x + 15$

Checkpoint 4 $-\dfrac{22}{(5x - 2)^2}$

Checkpoint 5 $y = \tfrac{8}{25}x - \tfrac{4}{5}$;

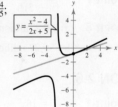

Checkpoint 6 $\dfrac{-3x^2 + 4x + 8}{x^2(x + 4)^2}$

Checkpoint 7 (a) $\tfrac{2}{5}x + \tfrac{4}{5}$ (b) $3x^3$

Checkpoint 8 $\dfrac{2x^2 - 4x}{(x - 1)^2}$

Checkpoint 9

t	0	1	2	3	4	5	6	7
$\dfrac{dP}{dt}$	0	-50	-16	-6	-2.77	-1.48	-0.88	-0.56

As t increases, the rate at which the blood pressure drops decreases.

Answers to Checkpoints

SECTION 2.5

Checkpoint 1 (a) $u = g(x) = x + 1$
$y = f(u) = \dfrac{1}{\sqrt{u}}$
(b) $u = g(x) = x^2 + 2x + 5$
$y = f(u) = u^3$

Checkpoint 2 $6x^2(x^3 + 1)$

Checkpoint 3 $4(2x + 3)(x^2 + 3x)^3$

Checkpoint 4 $y = \tfrac{1}{3}x + \tfrac{8}{3}$

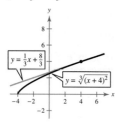

Checkpoint 5 (a) $-\dfrac{8}{(2x + 1)^2}$ (b) $-\dfrac{6}{(x - 1)^4}$

Checkpoint 6 $\dfrac{x(3x^2 + 2)}{\sqrt{x^2 + 1}}$

Checkpoint 7 $-\dfrac{12(x + 1)}{(x - 5)^3}$

Checkpoint 8 (a) $0.2 million per year
(b) $0.27 million per year

SECTION 2.6

Checkpoint 1 $f'(x) = 18x^2 - 4x$, $f''(x) = 36x - 4$,
$f'''(x) = 36$, $f^{(4)}(x) = 0$

Checkpoint 2 18

Checkpoint 3 $\dfrac{120}{x^6}$

Checkpoint 4 $s(t) = -16t^2 + 64t + 80$
$v(t) = s'(t) = -32t + 64$
$a(t) = v'(t) = s''(t) = -32$

Checkpoint 5 -9.8 m/sec^2

Checkpoint 6
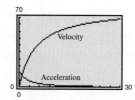
Acceleration approaches zero.

CHAPTER 3

SECTION 3.1

Checkpoint 1 $f'(x) = 4x^3$
$f'(x) < 0$ if $x < 0$; therefore, f is decreasing on $(-\infty, 0)$.
$f'(x) > 0$ if $x > 0$; therefore, f is increasing on $(0, \infty)$.

Checkpoint 2 $\dfrac{dS}{dt} = -0.022t + 0.68 > 0$ when $7 \leq t \leq 15$, which implies that the supply of rice was increasing in the years 1997 through 2005.

Checkpoint 3 Increasing on $(-\infty, -2)$ and $(2, \infty)$
Decreasing on $(-2, 2)$

Checkpoint 4 Increasing on $(0, \infty)$
Decreasing on $(-\infty, 0)$

Checkpoint 5 Because $f'(x) = -3x^2 = 0$ when $x = 0$ and because f is decreasing on $(-\infty, 0) \cup (0, \infty)$, f is decreasing on $(-\infty, \infty)$.

Checkpoint 6 $(10.6, 15)$

SECTION 3.2

Checkpoint 1 Relative maximum at $(-1, 5)$
Relative minimum at $(1, -3)$

Checkpoint 2 Relative minimum at $(3, -27)$

Checkpoint 3 Relative maximum at $(1, 1)$
Relative minimum at $(0, 0)$

Checkpoint 4 Absolute maximum at $(0, 10)$
Absolute minimum at $(4, -6)$

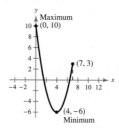

Checkpoint 5

t	0	1	2	3	4	5	6	7	8
L	1	0.5	0.6	0.7	0.76	0.81	0.84	0.86	0.88

SECTION 3.3

Checkpoint 1 (a) $f'' = -4$; because $f''(x) < 0$ for all x, f is concave downward for all x.

(b) $f''(x) = \dfrac{1}{2x^{3/2}}$; because $f''(x) > 0$ for all $x > 0$, f is concave upward for all $x > 0$.

Checkpoint 2 Because $f''(x) > 0$ for $x < -\dfrac{2\sqrt{3}}{3}$ and $x > \dfrac{2\sqrt{3}}{3}$, f is concave upward on $\left(-\infty, -\dfrac{2\sqrt{3}}{3}\right)$ and $\left(\dfrac{2\sqrt{3}}{3}, \infty\right)$.

Because $f''(x) < 0$ for $-\dfrac{2\sqrt{3}}{3} < x < \dfrac{2\sqrt{3}}{3}$, f is concave downward on $\left(-\dfrac{2\sqrt{3}}{3}, \dfrac{2\sqrt{3}}{3}\right)$.

Checkpoint 3 f is concave upward on $(-\infty, 0)$ and $(1, \infty)$.

f is concave downward on $(0, 1)$.

Points of inflection: $(0, 1)$, $(1, 0)$.

Checkpoint 4 Relative minimum: $(3, -26)$

Checkpoint 5 Relative maximum: $(2.66, 1.89)$

Relative minimum: $(12.55, 0.34)$

The death rate from HIV for 15- to 24-year-olds in the years 1990 to 2005 was highest in 1992 and lowest in 2002.

SECTION 3.4

Checkpoint 1

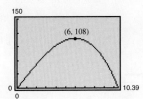

Maximum volume $= 108$ in.^{3}

Checkpoint 2 $x = 6$, $y = 12$

Checkpoint 3 $\left(\sqrt{\tfrac{1}{2}}, \tfrac{7}{2}\right)$ and $\left(-\sqrt{\tfrac{1}{2}}, \tfrac{7}{2}\right)$

Checkpoint 4 150 m by 150 m

SECTION 3.5

Checkpoint 1 (a) $\lim\limits_{x \to 2^-} \dfrac{1}{x-2} = -\infty$; $\lim\limits_{x \to 2^+} \dfrac{1}{x-2} = \infty$

(b) $\lim\limits_{x \to -3^-} \dfrac{-1}{x+3} = \infty$; $\lim\limits_{x \to -3^+} \dfrac{-1}{x+3} = -\infty$

Checkpoint 2 $x = 0$, $x = 4$

Checkpoint 3 $x = 3$

Checkpoint 4 $\lim\limits_{x \to 2^-} \dfrac{x^2 - 4x}{x - 2} = \infty$; $\lim\limits_{x \to 2^+} \dfrac{x^2 - 4x}{x - 2} = -\infty$

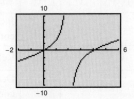

Checkpoint 5 2

Checkpoint 6 (a) $y = 0$

(b) $y = \tfrac{1}{2}$

(c) No horizontal asymptote

Checkpoint 7 No, the cost function is not defined at $p = 100$, which implies that it is not possible to remove 100% of the pollutants.

SECTION 3.6

Checkpoint 1

	$f(x)$	$f'(x)$	$f''(x)$	Shape of graph
x in $(-\infty, -1)$		$-$	$+$	Decreasing, concave upward
$x = -1$	-32	0	$+$	Relative minimum
x in $(-1, 1)$		$+$	$+$	Increasing, concave upward
$x = 1$	-16	$+$	0	Point of inflection
x in $(1, 3)$		$+$	$-$	Increasing, concave downward
$x = 3$	0	0	$-$	Relative maximum
x in $(3, \infty)$		$-$	$-$	Decreasing, concave downward

Checkpoint 2

	$f(x)$	$f'(x)$	$f''(x)$	Shape of graph
x in $(-\infty, 0)$		$-$	$+$	Decreasing, concave upward
$x = 0$	5	0	0	Point of inflection
x in $(0, 2)$		$-$	$-$	Decreasing, concave downward
$x = 2$	-11	$-$	0	Point of inflection
x in $(2, 3)$		$-$	$+$	Decreasing, concave upward
$x = 3$	-22	0	$+$	Relative minimum
x in $(3, \infty)$		$+$	$+$	Increasing, concave upward

A114 Answers to Checkpoints

Checkpoint 3

	$f(x)$	$f'(x)$	$f''(x)$	Shape of graph
x in $(-\infty, 0)$		+	−	Increasing, concave downward
$x = 0$	0	0	−	Relative maximum
x in $(0, 1)$		−	−	Decreasing, concave downward
$x = 1$	Undef.	Undef.	Undef.	Vertical asymptote
x in $(1, 2)$		−	+	Decreasing, concave upward
$x = 2$	4	0	+	Relative minimum
x in $(2, \infty)$		+	+	Increasing, concave upward

Checkpoint 4

	$f(x)$	$f'(x)$	$f''(x)$	Shape of graph
x in $(-\infty, -1)$		+	+	Increasing, concave upward
$x = -1$	Undef.	Undef.	Undef.	Vertical asymptote
x in $(-1, 0)$		+	−	Increasing, concave downward
$x = 0$	−1	0	−	Relative maximum
x in $(0, 1)$		−	−	Decreasing, concave downward
$x = 1$	Undef.	Undef.	Undef.	Vertical asymptote
x in $(1, \infty)$		−	+	Decreasing, concave upward

Checkpoint 5

	$f(x)$	$f'(x)$	$f''(x)$	Shape of graph
x in $(0, 1)$		−	+	Decreasing, concave upward
$x = 1$	−4	0	+	Relative minimum
x in $(1, \infty)$		+	+	Increasing, concave upward

SECTION 3.7

Checkpoint 1 $dy = 0.32$; $\Delta y = 0.32240801$

Checkpoint 2 0.03 m^2; Yes

Checkpoint 3 (a) $dy = 12x^2\, dx$ (b) $dy = \frac{2}{3}\, dx$
(c) $dy = (6x - 2)\, dx$ (d) $dy = -\dfrac{2}{x^3}\, dx$

Checkpoint 4 $S = 1.96\pi$ in.$^2 \approx 6.1575$ in.^{2}
$dS = \pm 0.056\pi$ in.$^2 \approx \pm 0.1759$ in.^{2}

CHAPTER 4

SECTION 4.1

Checkpoint 1 (a) 243 (b) 3 (c) 64
(d) 8 (e) $\frac{1}{2}$ (f) $\sqrt{10}$

Checkpoint 2 (a) 5.453×10^{-13} (b) 1.621×10^{-13}
(c) 2.629×10^{-14}

Checkpoint 3

x	−3	−2	−1	0	1	2	3
$f(x)$	$\frac{1}{125}$	$\frac{1}{25}$	$\frac{1}{5}$	1	5	25	125

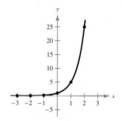

Checkpoint 4

x	−3	−2	−1	0	1	2	3
$f(x)$	9	5	3	2	$\frac{3}{2}$	$\frac{5}{4}$	$\frac{9}{8}$

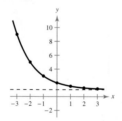

Horizontal asymptote: $y = 1$

SECTION 4.2

Checkpoint 1

x	0.00001	0.000001
$(1 + x)^{1/x}$	2.7182682	2.7182805

x	0.0000001	0.00000001
$(1 + x)^{1/x}$	2.7182817	2.7182818

Yes

Checkpoint 2

x	−2	−1	0	1	2
f(x)	$e^2 \approx 7.389$	$e \approx 2.718$	1	$\dfrac{1}{e} \approx 0.368$	$\dfrac{1}{e^2} \approx 0.135$

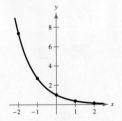

Checkpoint 3

(a)   (b)

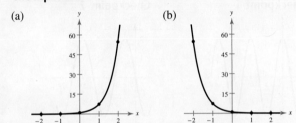

Checkpoint 4   (a)   (b) No

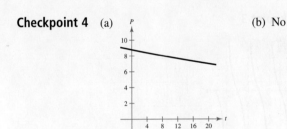

Checkpoint 5

After 0 h, $y = 1.25$ g.

After 1 h, $y \approx 1.338$ g.

After 10 h, $y \approx 1.498$ g.

$\displaystyle\lim_{t \to \infty} \frac{1.50}{1 + 0.2e^{-0.5t}} = 1.50$ g

SECTION 4.3

Checkpoint 1   At $(0, 1)$, $y = x + 1$.

At $(1, e)$, $y = ex$.

Checkpoint 2   (a) $3e^{3x}$ (b) $-\dfrac{6x^2}{e^{2x^3}}$ (c) $8xe^{x^2}$ (d) $-\dfrac{2}{e^{2x}}$

Checkpoint 3   (a) $xe^x(x + 2)$ (b) $\tfrac{1}{2}(e^x - e^{-x})$

(c) $\dfrac{e^x(x - 2)}{x^3}$ (d) $e^x(x^2 + 2x - 1)$

Checkpoint 4

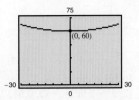

Checkpoint 5   −1.57°F/hr

Checkpoint 6

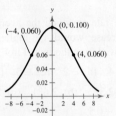

Points of inflection: $(-4, 0.060)$, $(4, 0.060)$

SECTION 4.4

Checkpoint 1

x	−1.5	−1	−0.5	0	0.5	1
f(x)	−0.693	0	0.405	0.693	0.916	1.099

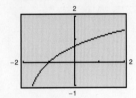

Checkpoint 2   (a) 3   (b) $x + 1$

Checkpoint 3   (a) $\ln 2 - \ln 5$ (b) $\tfrac{1}{3}\ln(x + 2)$

(c) $\ln x - \ln 5 - \ln y$

(d) $\ln x + 2\ln(x + 1)$

Checkpoint 4   (a) $\ln x^4 y^3$ (b) $\ln \dfrac{x + 1}{(x + 3)^2}$

Checkpoint 5   (a) $\ln 6$ (b) $5\ln 5$

Checkpoint 6   (a) e^4 (b) e^3

Checkpoint 7   59.6

SECTION 4.5

Checkpoint 1 $\dfrac{1}{x}$

Checkpoint 2   (a) $\dfrac{2x}{x^2 - 4}$ (b) $x(1 + 2\ln x)$

(c) $\dfrac{2\ln x - 1}{x^3}$

Checkpoint 3 $\dfrac{1}{3(x+1)}$

Checkpoint 4 $\dfrac{2}{x}+\dfrac{x}{x^2+1}$

Checkpoint 5 Relative minimum:
$(2, 2 - 2\ln 2) \approx (2, 0.6137)$

Checkpoint 6 $\dfrac{dp}{dt} = -1.3\%/\text{mo}$

The rate at which the average score would decrease would be higher than the rate derived from the model in Example 6.

Checkpoint 7 (a) 4   (b) -2 (c) -5 (d) 3

Checkpoint 8 (a) 2.322   (b) 2.631   (c) 3.161
(d) -0.5

Checkpoint 9

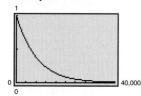

As time increases, the derivative approaches 0. The rate of change in the amount of carbon isotopes is proportional to the amount present.

SECTION 4.6

Checkpoint 1 About 2113.7 yr

Checkpoint 2 $y = 25e^{0.6931t}$

Checkpoint 3 About 47.6 hr

Checkpoint 4 After about 1.8 hr

CHAPTER 5

SECTION 5.1

Checkpoint 1 (a) 150°   (b) 30°   (c) 135°   (d) 30°

Checkpoint 2 (a) $\dfrac{5\pi}{4}$ (b) $-\dfrac{\pi}{4}$ (c) $\dfrac{4\pi}{3}$ (d) $\dfrac{5\pi}{6}$

Checkpoint 3 (a) 300°   (b) 210°   (c) 270°   (d) $-135°$

Checkpoint 4 $\tfrac{1}{2}$ ft²

SECTION 5.2

Checkpoint 1 $\sin\dfrac{\pi}{6} = \dfrac{1}{2}$

$\cos\dfrac{\pi}{6} = \dfrac{\sqrt{3}}{2}$

$\tan\dfrac{\pi}{6} = \dfrac{\sqrt{3}}{3}$

Checkpoint 2 (a) $\dfrac{1}{2}$ (b) $-\dfrac{\sqrt{2}}{2}$ (c) $-\sqrt{3}$

Checkpoint 3 (a) $-\dfrac{1}{2}$ (b) $\sqrt{2}$ (c) $\dfrac{\sqrt{6}-\sqrt{2}}{4}$
(d) 1   (e) 1   (f) 1

Checkpoint 4 About 52.5 ft

Checkpoint 5 About 245.76°

Checkpoint 6 (a) $\dfrac{\pi}{4}, \dfrac{7\pi}{4}$ (b) $\dfrac{2\pi}{3}, \dfrac{5\pi}{3}$ (c) $\dfrac{7\pi}{6}, \dfrac{11\pi}{6}$

Checkpoint 7 $\theta = 0, \dfrac{2\pi}{3}, \pi, \dfrac{4\pi}{3}, 2\pi$

SECTION 5.3

Checkpoint 1

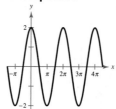

Checkpoint 2

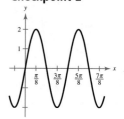

Checkpoint 3

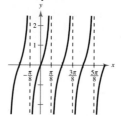

Checkpoint 4

x	-0.10	-0.05	-0.01
$\dfrac{1-\cos x}{x}$	-0.0500	-0.0250	-0.0050

x	0.01	0.05	0.10
$\dfrac{1-\cos x}{x}$	0.0050	0.0250	0.0500

$\displaystyle\lim_{x\to 0}\dfrac{1-\cos x}{x} = 0$

Checkpoint 5 Answers will vary.

Answers to Checkpoints A117

Checkpoint 6

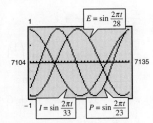

SECTION 5.4

Checkpoint 1 (a) $-4 \sin 4x$ (b) $2x \cos(x^2 - 1)$
(c) $\frac{1}{2} \sec^2 \frac{x}{2}$

Checkpoint 2 (a) $\frac{1}{2\sqrt{x}} \cos \sqrt{x}$ (b) $-6x^2 \sin x^3$

Checkpoint 3 (a) $3 \sin^2 x \cos x$ (b) $-8 \sin 2x \cos^3 2x$

Checkpoint 4 (a) $4 \sec 4x \tan 4x$ (b) $-2x \csc^2 x^2$

Checkpoint 5 (a) $-\dfrac{\sin 2x}{\sqrt{\cos 2x}}$ (b) $\dfrac{\sec^2 3x}{\sqrt[3]{\tan^2 3x}}$

Checkpoint 6 (a) $-x^2 \sin x + 2x \cos x$
(b) $2t \cos 2t + \sin 2t$

Checkpoint 7 Relative maximum: $\left(\dfrac{7\pi}{6}, \dfrac{7\pi + 6\sqrt{3}}{12}\right)$
Relative minimum: $\left(\dfrac{11\pi}{6}, \dfrac{11\pi - 6\sqrt{3}}{12}\right)$

Checkpoint 8 Relative maximum: $\left(\dfrac{\pi}{6}, \dfrac{3\sqrt{3}}{4}\right)$
Relative minimum: $\left(\dfrac{5\pi}{6}, -\dfrac{3\sqrt{3}}{4}\right)$

Checkpoint 9 About 1721 lb/day

Checkpoint 10 About $-3.9°/h$

CHAPTER 6

SECTION 6.1

Checkpoint 1 (a) $\int 3 \, dx = 3x + C$
(b) $\int 2x \, dx = x^2 + C$
(c) $\int 9t^2 \, dt = 3t^3 + C$

Checkpoint 2 (a) $5x + C$ (b) $-r + C$ (c) $2t + C$

Checkpoint 3 $\frac{5}{2}x^2 + C$

Checkpoint 4 (a) $-\dfrac{1}{x} + C$ (b) $\dfrac{3}{4}x^{4/3} + C$

Checkpoint 5 (a) $\frac{1}{2}x^2 + 4x + C$ (b) $x^4 - \frac{5}{2}x^2 + 2x + C$

Checkpoint 6 $\frac{2}{3}x^{3/2} + 4x^{1/2} + C$

Checkpoint 7 General solution: $F(x) = 2x^2 + 2x + C$
Particular solution: $F(x) = 2x^2 + 2x + 4$

Checkpoint 8 $s(t) = -16t^2 + 32t + 48$. The ball hits the ground 3 seconds after it is thrown, with a velocity of -64 feet per second.

Checkpoint 9 (a) $H = -0.00853t^2 + 0.4693t + 6.12575$
(b) About 12.3 million children

SECTION 6.2

Checkpoint 1 (a) $\dfrac{(x^3 + 6x)^3}{3} + C$ (b) $\frac{2}{3}(x^2 - 2)^{3/2} + C$

Checkpoint 2 $\frac{1}{36}(3x^4 + 1)^3 + C$

Checkpoint 3 $2x^9 + \frac{12}{5}x^5 + 2x + C$

Checkpoint 4 $\frac{5}{3}(x^2 - 1)^{3/2} + C$

Checkpoint 5 $-\frac{1}{3}(1 - 2x)^{3/2} + C$

Checkpoint 6 $\frac{1}{3}(x^2 + 4)^{3/2} + C$

Checkpoint 7 After 10 days

SECTION 6.3

Checkpoint 1 (a) $3e^x + C$ (b) $e^{5x} + C$
(c) $e^x - \dfrac{x^2}{2} + C$

Checkpoint 2 $\frac{1}{2}e^{2x+3} + C$

Checkpoint 3 $2e^{x^2} + C$

Checkpoint 4 (a) $2 \ln|x| + C$ (b) $\ln|x^3| + C$
(c) $\ln|2x + 1| + C$

Checkpoint 5 $\frac{1}{4} \ln|4x + 1| + C$

Checkpoint 6 $\frac{3}{2} \ln(x^2 + 4) + C$

Checkpoint 7 (a) $4x - 3 \ln|x| - \dfrac{2}{x} + C$
(b) $2 \ln(1 + e^x) + C$
(c) $\dfrac{x^2}{2} + x + 3 \ln|x + 1| + C$

SECTION 6.4

Checkpoint 1 $\frac{1}{2}(3)(12) = 18$

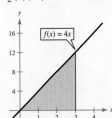

Checkpoint 2 $\frac{22}{3}$ units²

Checkpoint 3 68

Checkpoint 4 (a) $\frac{1}{4}(e^4 - 1) \approx 13.3995$
(b) $-\ln 5 + \ln 2 \approx -0.9163$

Checkpoint 5 $\frac{13}{2}$

Checkpoint 6 -144 ft

Checkpoint 7 149.9 mg/ml

Checkpoint 8 (a) $\frac{2}{5}$ (b) 0

SECTION 6.5

Checkpoint 1 $\frac{8}{3}$ units²

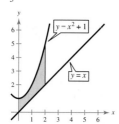

Checkpoint 2 $\frac{32}{3}$ units²

Checkpoint 3 $\frac{9}{2}$ units²

Checkpoint 4 $\frac{253}{12}$ units²

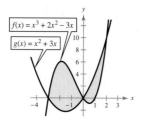

Checkpoint 5 The projection underestimated the number of households by 31,330,000.

SECTION 6.6

Checkpoint 1 $\frac{512\pi}{15} \approx 107.23$

Checkpoint 2 $\frac{832\pi}{15} \approx 174.25$

Checkpoint 3 About 151 cm³

CHAPTER 7

SECTION 7.1

Checkpoint 1 $\frac{1}{2}xe^{2x} - \frac{1}{4}e^{2x} + C$

Checkpoint 2 $\frac{x^2}{2}\ln x - \frac{1}{4}x^2 + C$

Checkpoint 3 $\frac{d}{dx}[x \ln x - x + C] = x\left(\frac{1}{x}\right) + \ln x - 1$
$= \ln x$

Checkpoint 4 $e^x(x^3 - 3x^2 + 6x - 6) + C$

Checkpoint 5 $e - 2$

Checkpoint 6 About 83.1%

SECTION 7.2

Checkpoint 1 $\frac{5}{x+3} - \frac{4}{x+4}$

Checkpoint 2 $\ln|x(x+2)^2| + \frac{1}{x+2} + C$

Checkpoint 3 $\frac{1}{2}x^2 - 2x - \frac{1}{x} + 4\ln|x+1| + C$

Checkpoint 4 $ky(1-y) = \frac{kbe^{-kt}}{(1+be^{-kt})^2}$
$y = (1 + be^{-kt})^{-1}$
$\frac{dy}{dt} = \frac{kbe^{-kt}}{(1+be^{-kt})^2}$
Therefore, $\frac{dy}{dt} = ky(1-y)$

Checkpoint 5 $y = 4$

Checkpoint 6 $y = \frac{4000}{1 + 39e^{-0.31045t}}$

SECTION 7.3

Checkpoint 1 $-5\cos x + C$

Checkpoint 2 $\sin x^4 + C$

Checkpoint 3 $\frac{1}{5}\tan 5x + C$

Checkpoint 4 $-\csc 2x + C$

Checkpoint 5 $-\frac{1}{8}\cos^4 2x + C$

Checkpoint 6 $\ln|\sin x| + C$

Checkpoint 7 1

Checkpoint 8 1

Checkpoint 9 $\frac{1}{2}\ln|\sec 2x + \tan 2x| + C$

Checkpoint 10 82.68°

SECTION 7.4

Checkpoint 1 $\frac{37}{8}$ units²

Checkpoint 2 0.436 unit²

Checkpoint 3 5.642 units²

Checkpoint 4 About 1.463

SECTION 7.5
Checkpoint 1 3.2608
Checkpoint 2 3.1956
Checkpoint 3 1.154

SECTION 7.6
Checkpoint 1 (a) Converges; $\frac{1}{2}$ (b) Diverges
Checkpoint 2 1
Checkpoint 3 $\frac{1}{2}$
Checkpoint 4 2
Checkpoint 5 Diverges
Checkpoint 6 Diverges
Checkpoint 7 0.0038 or $\approx 0.4\%$
Checkpoint 8 (a) $\int_1^\infty \frac{1}{x^{0.9}} dx = \infty$; diverges

(b) $\int_1^\infty \frac{1}{x^{1.0}} dx = \infty$; diverges

(c) $\int_1^\infty \frac{1}{x^{1.1}} dx = 10$; converges

The improper integral $\int_1^\infty \frac{1}{x^n} dx$ converges for $n > 1$.

CHAPTER 8
SECTION 8.1
Checkpoint 1 (2, 2)
Checkpoint 2 (0, 0) and (−2, 4)
Checkpoint 3 (3, 4)
Checkpoint 4 Inconsistent; no solution
Checkpoint 5 Infinitely many solutions; $\left(a, \frac{5}{6}a - \frac{1}{2}\right)$, a is a real number. (Answers will vary.)
Checkpoint 6 Answers will vary.

SECTION 8.2
Checkpoint 1 (1, −2, 4)
Checkpoint 2 (1, 2, 3)
Checkpoint 3 Inconsistent; no solution
Checkpoint 4 Infinitely many solutions; $(10 − 3a, 5a − 7, a)$, a is a real number. (Answers will vary.)
Checkpoint 5 Answers will vary; $(2a, 21a − 1, 8a)$, a is a real number
Checkpoint 6 $s = −16t^2 + 144$

SECTION 8.3
Checkpoint 1 1×2
Checkpoint 2 Multiply the first row of the original matrix by $-\frac{1}{2}$.
Checkpoint 3 (3, 2)
Checkpoint 4 Inconsistent; no solution
Checkpoint 5 (4, −3, 2)
Checkpoint 6 Answers will vary; $(2a + 1, 3a + 2, a)$, a is a real number

SECTION 8.4
Checkpoint 1 $x = −4, y = 22$

Checkpoint 2 $\begin{bmatrix} 0 \\ 0 \\ 0 \end{bmatrix}$

Checkpoint 3 (a) $\begin{bmatrix} 12 & -2 \\ 4 & 8 \\ -6 & 10 \end{bmatrix}$

(b) $\begin{bmatrix} -6 & 1 \\ -2 & -4 \\ 3 & -5 \end{bmatrix}$

(c) $\begin{bmatrix} 16 & -11 \\ 8 & 2 \\ -11 & -5 \end{bmatrix}$

Checkpoint 4 $\begin{bmatrix} 1 \\ 8 \end{bmatrix}$

Checkpoint 5 $\begin{bmatrix} 8 \\ -6 \end{bmatrix}$

Checkpoint 6 $\begin{bmatrix} -2 & -2 \\ 1 & 2 \end{bmatrix}$

Checkpoint 7 $\begin{bmatrix} -1 & 19 \\ 4 & -27 \\ 0 & 14 \end{bmatrix}$

Checkpoint 8 $\begin{bmatrix} 12 & -6 \\ 13 & -4 \end{bmatrix}$

Checkpoint 9 (a) $\begin{bmatrix} 7 & 7 & 14 \\ 8 & 8 & 16 \\ -1 & -1 & -2 \end{bmatrix}$ (b) $[13]$

Checkpoint 10 $X = \begin{bmatrix} 1 \\ -1 \\ 2 \end{bmatrix}$

SECTION 8.5

Checkpoint 1 $AB = \begin{bmatrix} 2 & 1 \\ 5 & 3 \end{bmatrix}\begin{bmatrix} 3 & -1 \\ -5 & 2 \end{bmatrix} = \begin{bmatrix} 1 & 0 \\ 0 & 1 \end{bmatrix}$

$BA = \begin{bmatrix} 3 & -1 \\ -5 & 2 \end{bmatrix}\begin{bmatrix} 2 & 1 \\ 5 & 3 \end{bmatrix} = \begin{bmatrix} 1 & 0 \\ 0 & 1 \end{bmatrix}$

Checkpoint 2 $\begin{bmatrix} -5 & 2 \\ -2 & 1 \end{bmatrix}$

Checkpoint 3 $\begin{bmatrix} 1 & 1 & -1 \\ -2 & 2 & -1 \\ 2 & -3 & 2 \end{bmatrix}$

Checkpoint 4 $\begin{bmatrix} \frac{3}{19} & \frac{2}{19} \\ -\frac{2}{19} & \frac{5}{19} \end{bmatrix}$

Checkpoint 5 $x = -1, y = 2, z = 1$

CHAPTER 9

SECTION 9.1

Checkpoint 1

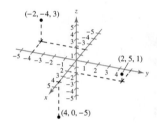

Checkpoint 2 $2\sqrt{6}$

Checkpoint 3 $\left(-\frac{5}{2}, 2, -2\right)$

Checkpoint 4 $(x - 4)^2 + (y - 3)^2 + (z - 2)^2 = 25$

Checkpoint 5 $(x - 1)^2 + (y - 3)^2 + (z - 2)^2 = 38$

Checkpoint 6 Center: $(-3, 4, -1)$; radius: 6

Checkpoint 7 $(x + 1)^2 + (y - 2)^2 = 16$

SECTION 9.2

Checkpoint 1 x-intercept: $(4, 0, 0)$;
y-intercept: $(0, 2, 0)$;
z-intercept: $(0, 0, 8)$

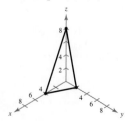

Checkpoint 2 Hyperboloid of one sheet
xy-trace: circle, $x^2 + y^2 = 1$; yz-trace: hyperbola, $y^2 - z^2 = 1$; xz-trace: hyperbola, $x^2 - z^2 = 1$; $z = 3$ trace: circle, $x^2 + y^2 = 10$

Checkpoint 3 (a) $\dfrac{x^2}{9} + \dfrac{y^2}{4} = z$; elliptic paraboloid

(b) $\dfrac{x^2}{4} + \dfrac{y^2}{9} - z^2 = 0$; elliptic cone

SECTION 9.3

Checkpoint 1 (a) 0 (b) $\frac{9}{4}$

Checkpoint 2 Domain: $x^2 + y^2 \leq 9$

Range: $0 \leq z \leq 3$

Checkpoint 3 Steep; nearly level

Checkpoint 4 Alaska contributes the products of its forest land. Alaska does not contain any manufacturing centers, but it does contain a mineral deposit.

Checkpoint 5 (a) About $-22°F$ (b) About $1°F$

SECTION 9.4

Checkpoint 1 $\dfrac{\partial z}{\partial x} = 4x - 8xy^3$

$\dfrac{\partial z}{\partial y} = -12x^2y^2 + 4y^3$

Checkpoint 2 $f_x(x, y) = 2xy^3$; $f_x(1, 2) = 16$
$f_y(x, y) = 3x^2y^2$; $f_y(1, 2) = 12$

Checkpoint 3 In the x-direction: $f_x(1, -1, 49) = 8$

In the y-direction: $f_y(1, -1, 49) = -18$

Checkpoint 4 (a) The BMI will decrease by about -0.80 for every inch that the height increases.

(b) The BMI will increase by about 0.17 for every pound that the weight increases.

Checkpoint 5 $\dfrac{\partial w}{\partial x} = xy + 2xy \ln(xz)$

$\dfrac{\partial w}{\partial y} = x^2 \ln xz$

$\dfrac{\partial w}{\partial z} = \dfrac{x^2 y}{z}$

Checkpoint 6 $f_{xx} = 8y^2$

$f_{yy} = 8x^2 + 8$

$f_{xy} = 16xy$

$f_{yx} = 16xy$

Checkpoint 7
$f_{xx} = 0$ $f_{xy} = e^y$ $f_{xz} = 2$
$f_{yx} = e^y$ $f_{yy} = xe^y + 2$ $f_{yz} = 0$
$f_{zx} = 2$ $f_{zy} = 0$ $f_{zz} = 0$

SECTION 9.5

Checkpoint 1 $f(-8, 2) = -64$: relative minimum
Checkpoint 2 $f(0, 0) = 1$: relative maximum
Checkpoint 3 $f(0, 0) = 0$: saddle point
Checkpoint 4 400 smallmouth bass, 200 largemouth bass; $T(400, 200) = 800$ lb
Checkpoint 5 $V\left(\frac{4}{3}, \frac{2}{3}, \frac{8}{3}\right) = \frac{64}{27}$ units^3

SECTION 9.6

Checkpoint 1 For $f(x)$, $S \approx 9.1$.
For $g(x)$, $S \approx 0.45715$.
The quadratic model is a better fit.
Checkpoint 2 $a = \frac{6}{5}, b = \frac{23}{10}$
Checkpoint 3 $B = 0.34w + 1306$; 4366 cm^2
Checkpoint 4 $y = -5.32t^2 + 199.3t + 1253$; 3,111,000 students

SECTION 9.7

Checkpoint 1 (a) $\frac{1}{4}x^4 + 2x^3 - 2x - \frac{1}{4}$
(b) $\ln|y^2 + y| - \ln|2y|$
Checkpoint 2 $\frac{25}{2}$
Checkpoint 3 $\int_2^4 \int_1^5 dx\,dy = 8$ units^2
Checkpoint 4 $\frac{4}{3}$ units^2
Checkpoint 5 (a)

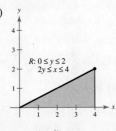

(b) $\int_0^4 \int_0^{x/2} dy\,dx$
(c) $\int_0^2 \int_{2y}^4 dx\,dy = 4 = \int_0^4 \int_0^{x/2} dy\,dx$
Checkpoint 6 $\int_{-1}^3 \int_{x^2}^{2x+3} dy\,dx = \frac{32}{3}$

SECTION 9.8

Checkpoint 1 $\frac{16}{3}$ units^3
Checkpoint 2 $e - 1$ units^3
Checkpoint 3 $\frac{176}{15}$ units^3
Checkpoint 4 Integration by parts
Checkpoint 5 3

CHAPTER 10

SECTION 10.1

Checkpoint 1 Not a solution
Checkpoint 2 Because $y' = 2Cx$, and $xy' - 2y = x(2Cx) - 2Cx^2 = 0$, $y = Cx^2$ is a solution. Particular solution: $y = \frac{1}{16}x^2$
Checkpoint 3

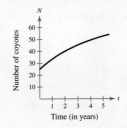

Checkpoint 4

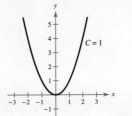
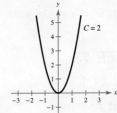

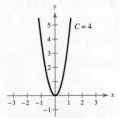

SECTION 10.2

Checkpoint 1 $y^2 = \frac{2x^3}{3} + C$
Checkpoint 2 $y^2 = (x + 1)^2 + C$

Answers to Checkpoints

Checkpoint 3 $y = \pm\sqrt{-x^2 + C}$

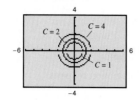

Checkpoint 4 $y^2 = 6 - 2e^x$
Checkpoint 5 $y = 10 - Ce^{-kx}$
Checkpoint 6 $y^2 = 2x^2 + 8$
Checkpoint 7 About 880,000 people

SECTION 10.3

Checkpoint 1 $y = -10 + Ce^x$
Checkpoint 2 $y = x \ln x + Cx$
Checkpoint 3 $A = 1 - e^{-0.05t}$

SECTION 10.4

Checkpoint 1 $y = \dfrac{80}{7t + 2}$

Checkpoint 2 About 14.09 more minutes
Checkpoint 3 About 76 wolves
Checkpoint 4 About 79%
Checkpoint 5 About 17.6 gallons

CHAPTER 11

SECTION 11.1

Checkpoint 1

(a) $S = \{(1, 1), (1, 2), (1, 3), (1, 4), (1, 5), (1, 6), (2, 1), (2, 2),$
$(2, 3), (2, 4), (2, 5), (2, 6), (3, 1), (3, 2), (3, 3), (3, 4),$
$(3, 5), (3, 6), (4, 1), (4, 2), (4, 3), (4, 4), (4, 5), (4, 6),$
$(5, 1), (5, 2), (5, 3), (5, 4), (5, 5), (5, 6), (6, 1), (6, 2),$
$(6, 3), (6, 4), (6, 5), (6, 6)\}$

(b) $E = \{(1, 6), (2, 5), (2, 6), (3, 4), (3, 5), (3, 6), (4, 3),$
$(4, 4), (4, 5), (4, 6), (5, 2), (5, 3), (5, 4), (5, 5), (5, 6),$
$(6, 1), (6, 2), (6, 3), (6, 4), (6, 5), (6, 6)\}$

Checkpoint 2

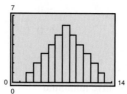

Checkpoint 3

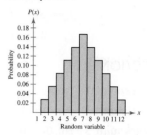

Checkpoint 4 7
Checkpoint 5 7.11
Checkpoint 6 $V(x) = 1.2$, $\sigma \approx 1.095$

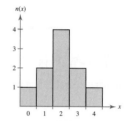

SECTION 11.2

Checkpoint 1 $\displaystyle\int_0^2 \frac{1}{2}x \, dx = 1$

Checkpoint 2 $\displaystyle\int_0^\infty 2e^{-2x} \, dx = 1$

Checkpoint 3 $\frac{3}{16}$
Checkpoint 4 0.4
Checkpoint 5 0.368
Checkpoint 6 About 0.250, or 25.0%

SECTION 11.3

Checkpoint 1 2

Checkpoint 2 $V(x) = \dfrac{4}{5}$, $\sigma \approx \sqrt{\dfrac{4}{5}} \approx 0.8944$

Checkpoint 3 84.6%
Checkpoint 4 Mean: 0.5; median: 0.35
Checkpoint 5 $\mu = 1$; $\sigma \approx 0.577$
Checkpoint 6 About 0.595, or 59.5%
Checkpoint 7 0.159

SECTION B.1

Checkpoint 1 $-\dfrac{2}{x^3}$

Checkpoint 2 (a) $12x^2$ (b) $6y\dfrac{dy}{dx}$ (c) $1 + 5\dfrac{dy}{dx}$

(d) $y^3 + 3xy^2\dfrac{dy}{dx}$

Checkpoint 3 $\dfrac{3}{4}$

Checkpoint 4 $\dfrac{dy}{dx} = -\dfrac{x-2}{y-1}$

Checkpoint 5 $\dfrac{5}{9}$

SECTION B.2

Checkpoint 1 9

Checkpoint 2 $12\pi \approx 37.7$ ft^2/sec

Checkpoint 3 $72\pi \approx 226.2$ in.^{2}/min

Index

A

Absolute
   extrema, 198
   maximum, 198
   minimum, 198
Absolute value, 8
   equation, solving, 105
   inequalities involving, 10
   model, 52
   properties of, 8
Acceleration, 169
   due to gravity, 170
   function, 169
Accuracy of a mathematical model, measuring, 594
Acute angle, 316
Addition
   of fractions, 25
   of functions, 74
   of matrices, 519
      Associative Property, 520
      Commutative Property, 520
Addition and scalar multiplication of matrices
   Distributive Property, 520
   properties, 520
Additive identity for matrices, 521
Algebra and integration techniques, 481
Algebraic equations, graphs of basic, 52
Algebraic expression(s)
   domain of, 17
   factored form of, 15
   simplifying, 174
   "unsimplifying," 418
Alternative formula for variance of a continuous random variable, 683
Amplitude, 336
Analysis, least squares regression, 51, 594
Analytic geometry, solid, 548
Angle, 316
   acute, 316
   initial ray of, 316
   obtuse, 316
   reference, 327
   right, 316
   standard position of, 316
   straight, 316
   terminal ray of, 316
   vertex of, 316
Angle measure conversion rule, 318
Angles
   coterminal, 316
   degree measure of, 316
   difference of two, 325
   radian measure of, 318
   sum of two, 325
   trigonometric values of common angles, 326
Antiderivative(s), 364
   finding, 366
   integral notation of, 365
Antidifferentiation, 364
Approximating definite integrals, 456, 462
Approximation, tangent line, 242
Arc length of a circular sector, 318
Area
   and definite integrals, 392
   finding area with a double integral, 606
   of a region bounded by two graphs, 403
Area in the plane, finding with a double integral, 606
Associative Property of Matrix Addition, 520
Associative Property of Multiplication for matrices, 524
Associative Property of Scalar Multiplication for matrices, 520, 524
Asymptote
   horizontal, 227
      of an exponential function, 262
      of a rational function, 228
   vertical, 223
      of a rational function, 224
Augmented matrix, 508
   elementary row operations on, 508
Average rate of change, 137
Average value of a function
   on a closed interval, 398
   over a region, 617
Average velocity, 139
Axis
   x-axis, 34
   y-axis, 34
   z-axis, 548
Axis of revolution, 411

B

Back-substitution, 491
   Gaussian elimination with, 510
      procedure, 511
   using, 539
Bar graph, 35
Base
   of an exponential function, 259
   of a natural logarithmic function, 280
Bases other than e, and differentiation, 294
Basic algebraic equations, graphs of, 52
Basic equation for a partial fraction, 437
Basic integration rules, 366
Behavior, unbounded, 90
Between a and b, notation for, 3
Binomial Theorem, 19
Book value, 64

C

Carrying capacity, 440
Cartesian plane, 34
Catenary, 274
Center of a circle, 47
Central tendency, measure of, 668, 684
Chain Rule for differentiation, 158
Change
   in x, 116
   in y, 116
Change-of-base formula, 293
Change of variables, integration by, 379
Characteristics of graphs of exponential functions, 262
Circle, 47
   center of, 47
   general form of the equation of, 48
   radius of, 47
   standard form of the equation of, 47
Circular function definition of the trigonometric functions, 324
Circular sector, arc length of, 318
Classifying a quadric surface, 561
Closed interval, 3
   continuous on, 97
   guidelines for finding extrema on, 199

Closed region, 586
Coefficient, correlation, 603
Coefficient matrix, 508
Column matrix, 507
Combinations, 691
Combinations of functions, 74
Common angles, trigonometric values of, 326
Common denominator, 25
Common logarithm, 280, 293
Commutative Property of Matrix Addition, 520
Completing the square, 48
Composite function, 74
   domain of, 74
Composition of two functions, 74
Concave
   downward, 203
   upward, 203
Concavity, 203
   test for, 203
      guidelines for applying, 204
Condensing logarithmic expressions, 283
Cone, elliptic, 559
Consistent system of linear equations, 493
Constant function, 184, 239
   test for, 184
Constant of integration, 365
Constant Multiple Rule
   differential form of, 245
   for differentiation, 128
   for integration, 366
Constant of proportionality, 298
Constant rate, A13
Constant Rule
   differential form of, 245
   for differentiation, 125
   for integration, 366
Continuity, 94
   on a closed interval, 97
   and differentiability, 121
   at an endpoint, 97
   from the left, 97
   on an open interval, 94
   at a point, 94
   of a polynomial function, 95
   of a rational function, 95
   from the right, 97
Continuous
   on a closed interval, 97
   at an endpoint, 97
   function, 94
   from the left, 97
   on an open interval, 94
   at a point, 94
   from the right, 97
Continuous random variable, 674
   expected value of, 681
   mean of, 681
   median of, 684
   standard deviation of, 682
   variance of, 682
      alternative formula for, 683
Contour map, 567
Convergence of an improper integral, 472, 475
Converting
   degrees to radians, 318
   radians to degrees, 318
Coordinate(s)
   of a point in a plane, 34
   of a point on the real number line, 2
   x-coordinate, 34
   y-coordinate, 34
   z-coordinate, 548
Coordinate plane, 548
   xy-plane, 548
   xz-plane, 548
   yz-plane, 548
Coordinate system
   rectangular, 34
   three-dimensional, 548
      left-handed orientation, 548
      right-handed orientation, 548
Correlation coefficient, 603
Cosecant function, 324
Cosine function, 324
Cost, depreciated, 64
Cotangent function, 324
Coterminal angles, 316
Counting principle, 691
   fundamental, 691
Critical
   number, 186
   point, 586
Cubic
   function, 239
   model, 52
Curve
   level, 567
   logistic, 440
   Lorenz, 410
   solution, 631
Curve-sketching techniques, summary of, 233

D

Decreasing function, 184
   test for, 184
Definite integral, 392, 394
   approximating, 456, 462
      Midpoint Rule, 456
      Simpson's Rule, 464
      Trapezoidal Rule, 462
   and area, 392
   as the limit of a sum, 459
   and net change, 397
   properties of, 394
Definitions of the trigonometric functions, 324
   circular function definition, 324
   right triangle definition, 324
Degree measure of angles, 316
Degrees to radians, converting, 318
Denominator, rationalizing, 31
Dependent system of linear equations, 493
Dependent variable, 69
Depreciated cost, 64
Depreciation
   linear, 64
   straight-line, 64
Derivative(s), 119
   of an exponential function with base a, 294
   of f at x, 119
   first, 167
   first partial
      notation for, 574
      with respect to x and y, 573
   of a function, 119
   higher-order, 167
      notation for, 167
      of a polynomial function, 168
   of a logarithmic function to the base a, 294
   of the natural exponential function, 271
   of the natural logarithmic function, 289
   partial, 573
      of a function of three variables, 578
      of a function of two variables, 573
      graphical interpretation of, 575
      higher-order, 579
      mixed, 579
   second, 167
   simplifying, 153, 162

third, 167
of trigonometric functions, 346
Determinant of a 2 × 2 matrix, 533
Determining area in the plane by double integrals, 606
Determining volume with double integrals, 612
Diagonal, of a matrix, main, 507
Difference
 of two angles, 325
 of two functions, 74
Difference quotient, 73, 116
Difference Rule
 differential form of, 245
 for differentiation, 131
 for integration, 366
Differentiability and continuity, 121
Differentiable, 119
Differential, 242
 of x, 242
 of y, 242
Differential equation, 369, 630
 first-order linear, 644
  standard form of, 644
 general solution of, 369, 630
 linear, guidelines for solving, 645
 logistic, 440
 particular solution of, 369, 631
 singular solutions of, 631
 solution of, 630
Differential form, 245
Differential forms of differentiation rules, 245
Differentiation, 15, 119
 and bases other than e, 294
 Chain Rule, 158
 Constant Multiple Rule, 128
 Constant Rule, 125
 Difference Rule, 131
 formulas, A2
 General Power Rule, 160
 implicit, A3, A5
 partial, 573
 Product Rule, 147
 Quotient Rule, 150
 rules, summary of, 164
 Simple Power Rule, 126
 Sum Rule, 131
Differentiation rules, differential forms of, 245
Direct substitution to evaluate a limit, 84
Directed distance on the real number line, 9

Direction
 negative, 2
 positive, 2
Discontinuity, 96
 infinite, 471
 nonremovable, 96
 removable, 96
Discrete probability, 667
Discrete random variable, 666
 expected value of, 668
 mean of, 668
 standard deviation of, 669
 variance of, 669
Disk Method for finding the volume of a solid of revolution, 411
Distance
 directed, 9
 between a point and a plane, 563
 between two points on the real number line, 9
Distance Formula, 36
 in space, 550
Distribution
 frequency, 666
 probability, 667
Distributive Property, 15
 for matrix addition and scalar multiplication, 520
 for matrix multiplication, 524
Divergence of an improper integral, 472, 475
Dividing out technique for evaluating a limit, 86
Division
 of fractions, 25
 of functions, 74
 synthetic, 22
Domain
 of a composite function, 74
 of an expression, 17
 feasible, 213
 of a function, 69
 of a function of two variables, 565
 of a function of x and y, 565
 implied, 71
 of an inverse function, 75
 of a radical expression, 21
Double angle formulas, 325
Double integral, 605
 finding area with, 606
 finding volume with, 612
Double subscript notation for matrix entries, 507
Doyle Log Rule, 181

E
e, the number, 265
 limit definition of, 265
Ebbinghaus model, 278
Elementary row operations, 508
Elimination
 Gauss-Jordan, 512
  finding inverses of matrices using, 531
 Gaussian, 499, 500
  with back-substitution, 510
  method of, 492
Ellipsoid, 560
Elliptic
 cone, 559
 paraboloid, 559
Endpoint, continuity at, 97
Endpoint of an interval, 3
Entry in a matrix, 507
 double subscript notation for, 507
Equality of matrices, 518
Equation
 absolute value, 105
 differential, 369, 630
  general solution of, 369, 630
  logistic, 440
  particular solution of, 369, 631
  singular solution of, 631
  solution of, 630
 first-order linear differential equation, 644
  standard form of, 644
 graph of, 43
 linear, 56
  general form of, 62
  point-slope form of, 61, 62
  slope-intercept form of, 56, 62
  two-point form of, 61
 linear differential equation, guidelines for solving, 645
 of a plane in space, general, 556
 primary, 213, 214
 secondary, 214
 of a sphere, standard, 551
Equation of a circle
 general form of, 48
 standard form of, 47
Equation of a line, 56
 general form of, 62
 point-slope form of, 61, 62
 slope-intercept form of, 56, 62
 two-point form of, 61
Equations, solving
 absolute value, 105

exponential, 284, 306
linear, 105
logarithmic, 284, 307
quadratic, 105
radical, 105
review, 105, 249, 538, 657
systems of (review), 620
trigonometric, 330, 355
Equations, system of, 490
  graphical interpretations of solutions, 493
  nonlinear, 492
  nonsquare, 502
  solution of, 490
  solving, 490
  with a unique solution, 534
Equations, system of linear
  consistent, 493
  dependent, 493
  inconsistent, 493, 500
  independent, 493
  number of solutions, 500
  row-echelon form, 498
Equivalent inequalities, 4
Equivalent systems of linear equations, 499
  operations that produce, 499
Error
  percentage, 246
  propagation, 246
  relative, 246
  in Simpson's Rule, 466
  in the Trapezoidal Rule, 466
Errors, sum of squared, 594, 595
Evaluating a limit
  direct substitution, 84
  dividing out technique, 86
  of a polynomial function, 85
  Replacement Theorem, 86
  of a trigonometric function, 339
Even function, 399
  integration of, 399
Event, 665
Existence of a limit, 88
Expanding logarithmic expressions, 283
Expected value, 668, 681
  of a continuous random variable, 681
  of a discrete random variable, 668
Experiment, 665
Explicit form of a function, A3
Exponential
  decay, 298
    guidelines for modeling, 300

growth, 298
  guidelines for modeling, 300
  model, 267
Exponential equations, solving, 284, 306
Exponential function(s), 259
  with base a, derivative of, 294
  base of, 259
  characteristics of graph of, 262
  graphs of, 261
  horizontal asymptotes of, 262
  integral of, 384
  natural, 265
    derivative of, 271
Exponential growth and decay, law of, 298
Exponential growth model, 267
Exponential and logarithmic form, 280
Exponential probability density function, 680, 686
Exponential Rule
  for integration (General), 384
  for integration (Simple), 384
Exponents, 13
  negative, 13
  operations with, 14
  properties of, 13, 259
  rational, 13
  zero, 13
Exponents and logarithms, inverse properties of, 282
Expression
  domain of, 17
  factored form of, 15
  logarithmic
    condensing, 283
    expanding, 283
  radical, domain of, 21
  rational, 25
    improper, 25
    proper, 25
  simplifying, 174
  "unsimplifying," 418
Extraneous solution, 105
Extrapolation, linear, 62
Extrema
  absolute, 198
  on a closed interval, guidelines for finding, 199
  relative, 193
    First-Derivative Test for, 194
    First-Partials Test for, 586
    of a function of two variables, 585, 588
  Second-Derivative Test for, 208

Second-Partials Test for, 588
  of trigonometric functions, 349
Extreme Value Theorem, 198
Extremum, relative, 193

F

Factored form of an expression, 15
Factorial, 691
Factoring by grouping, 19
Factorization techniques, 19
Family of functions, 364
Feasible domain of a function, 213
Finding
  antiderivatives, 366
  area with a double integral, 606
  extrema on a closed interval, guidelines, 199
  intercepts, 45
  inverse function, 76
  an inverse matrix, 531
  point(s) of intersection, 49
  slope of a line, 59
  volume with a double integral, 612
  volume of a solid, guidelines, 615
First derivative, 167
First-Derivative Test for relative extrema, 194
First-order linear differential equation, 644
  standard form of, 644
First partial derivative of f with respect to x and y, 573
First partial derivatives, notation for, 574
First-Partials Test for relative extrema, 586
Formula
  alternative formula for variance of a continuous random variable, 683
  change-of-base, 293
  Distance, 36
    in space, 550
  Midpoint, 38
    in space, 550
  Net Change, 397
  Quadratic, 19
  slope of a line, 59
Formulas
  differentiation, A2
  double angle, 325
  half angle, 325
  integration, A2
  trigonometric reduction formulas, 325

Fractions
  operations with, 25
  partial, 436
    basic equation, 437
    integration by, 436
Frequency, 666
Frequency distribution, 666
Function(s), 69
  acceleration, 169
  addition of, 74
  average value
    on a closed interval, 398
    over a region, 617
  combinations of, 74
  composite, 74
    domain of, 74
  composition of two, 74
  constant, 184, 239
  continuity of
    polynomial, 95
    rational, 95
  continuous, 94
  cosecant, 324
  cosine, 324
  cotangent, 324
  critical number of, 186
  cubic, 239
  decreasing, 184
  dependent variable, 69
  derivative of, 119
  difference of two, 74
  division of, 74
  domain of, 69
  even, 399
  explicit form of, A3
  exponential, 259
    base of, 259
    characteristics of graph of, 262
    graph of, 261
    horizontal asymptotes of, 262
  exponential with base a, derivative of, 294
  exponential probability density, 680, 686
  family of, 364
  feasible domain of, 213
  greatest integer, 98
  guidelines for analyzing the graph of, 233
  Horizontal Line Test for, 71
  implicit form of, A3
  implied domain of, 71
  improper rational, 439
  increasing, 184
  independent variable, 69
  inverse, 75
    domain of, 75
    finding, 76
    range of, 75
  limit of, 84
  linear, 239
  logarithmic, properties of, 280, 282
  logarithmic to the base a, derivative of, 294
  logistic growth, 267, 440
  multiplication of, 74
  natural exponential, 265
    derivative of, 271
    graph of, 266
  natural logarithmic, 280
    base of, 280
    derivative of, 289
    graph of, 280
  normal probability density, 276, 477, 686
  notation, 72
  odd, 399
  one-to-one, 71
  piecewise-defined, 71
  polynomial
    higher-order derivatives of, 168
    limit of, 85
  population density, 616
  position, 141, 169
  probability density, 433, 674
    exponential, 680, 686
    normal, 276, 477, 686
    standard normal, 686
    uniform, 685
  product of two, 74
  proper rational, 436
  quadratic, 239
  quotient of two, 74
  range of, 69
  rational
    horizontal asymptotes of, 228
    improper, 439
    proper, 436
    vertical asymptote of, 224
  secant, 324
  sine, 324
  standard normal probability density, 686
  step, 98
  subtraction of, 74
  sum of two, 74
  tangent, 324
  test for increasing and decreasing, 184
    guidelines for applying, 186
  of three variables, 565, 578
    partial derivatives of, 578
  trigonometric, 324
    circular function definition, 324
    derivatives of, 346
    graphs of, 337
    integrals of, 446, 450
    limits of, 339
    relative extrema of, 349
    right triangle definition, 324
  of two variables, 565
    domain of, 565
    graph of, 566
    partial derivatives of, 573
    range of, 565
    relative extrema, 585, 588
    relative maximum, 585, 588
    relative minimum, 585, 588
  unbounded, 90
  uniform probability density, 685
  value, 72
  velocity, 141, 169
  Vertical Line Test for, 71
  of x and y, 565
    domain of, 565
    graph of, 566
    range of, 565
Fundamental counting principle, 691
Fundamental Theorem of Algebra, 19
Fundamental Theorem of Calculus, 393
  guidelines for using, 394

G

Gabriel's horn, 478
Gauss-Jordan elimination, 512
  finding inverses of matrices using, 531
Gaussian elimination, 499, 500
  with back-substitution, 510
    procedure, 511
General equation of a plane in space, 556
General Exponential Rule for integration, 384
General form
  of the equation of a circle, 48
  of the equation of a line, 62
General Logarithmic Rule for integration, 386
General Power Rule
  for differentiation, 160
  for integration, 375
General solution of a differential equation, 369, 630
Geometry, solid analytic, 548

Gompertz growth model, 651
Graph(s)
　bar, 35
　of basic algebraic equations, 52
　of an equation, 43
　of an exponential function, 261
　of a function, guidelines for analyzing, 233
　of a function of two variables, 566
　of a function of x and y, 566
　intercept of, 45
　line, 35
　of a natural exponential function, 266
　of the natural logarithmic function, 280
　slope of, 115, 116, 137
　summary of simple polynomial graphs, 239
　tangent line to, 114
　of trigonometric functions, 337
Graphical interpretation of partial derivatives, 575
Graphical interpretations of solutions of systems of equations, 493
Graphing a linear equation, 57
Gravity, acceleration due to, 170
Greatest integer function, 98
Grouping, factoring by, 19
Guidelines
　for analyzing the graph of a function, 233
　for applying concavity test, 204
　for applying increasing/decreasing test, 186
　for finding extrema on a closed interval, 199
　for finding the volume of a solid, 615
　for integration by parts, 428
　for integration by substitution, 379
　for modeling exponential growth and decay, 300
　for solving a linear differential equation, 645
　for solving optimization problems, 214
　for solving a related-rate problem, A12
　for using the Fundamental Theorem of Calculus, 394
　for using the Midpoint Rule, 457

H

Half angle formulas, 325
Half-life, 299
Hardy-Weinberg Law, 593
Higher-order derivative, 167
　notation for, 167
　of a polynomial function, 168
Higher-order partial derivatives, 579
Horizontal asymptote, 227
　of an exponential function, 262
　of a rational function, 228
Horizontal line, 57, 62
Horizontal Line Test, 71
Hyperbola, A7
Hyperbolic paraboloid, 559
Hyperboloid
　of one sheet, 560
　of two sheets, 560

I

Identities, trigonometric, 325
　Pythagorean, 325
Identity, additive, for matrices, 521
Identity matrix, 524
　of order n, 524
Implicit differentiation, A3, A5
Implicit form of a function, A3
Implied domain of a function, 71
Improper integrals, 471
　convergence of, 472, 475
　divergence of, 472, 475
　infinite discontinuity, 471
　infinite integrand, 475
　infinite limit of integration, 472
Improper rational expression, 25
Improper rational function, 439
Inconsistent system of linear equations, 493, 500
Increasing function, 184
　test for, 184
Indefinite integral, 365
Independent system of linear equations, 493
Independent variable, 69
Inequality
　equivalent, 4
　involving absolute value, 10
　polynomial, 5
　properties of, 4
　reversal of, 4
　solution of, 4
　solution set of, 4
　solving, 4
　test intervals for, 5
　Transitive Property of, 4
Infinite
　discontinuity, 471
　integrand, 475
　interval, 3
　limit, 223
　limit of integration, 472
Infinity
　limit at, 227
　negative, 3
　positive, 3
Inflection, point of, 206
　property of, 206
Initial condition(s), 369, 631
Initial ray, 316
Initial value, 298
Inner radius, 413
Instantaneous rate of change, 140
　and velocity, 140
Integral(s)
　approximating definite
　　Midpoint Rule, 456
　　Simpson's Rule, 464
　　Trapezoidal Rule, 462
　definite, 392, 394
　　and net change, 397
　　properties of, 394
　double, 605
　　finding area with, 606
　　finding volume with, 612
　of even functions, 399
　of exponential functions, 384
　improper, 471
　　convergence of, 472, 475
　　divergence of, 472, 475
　indefinite, 365
　of logarithmic functions, 386
　notation of antiderivatives, 365
　of odd functions, 399
　partial, with respect to x, 604
　of trigonometric functions, 446, 450
Integral sign, 365
Integrand, 365
　infinite, 475
Integrating factor, 644
Integration, 365
　basic rules, 366
　by change of variables, 379
　constant of, 365
　Constant Multiple Rule, 366
　Constant Rule, 366
　Difference Rule, 366
　of even functions, 399
　of exponential functions, 384
　formulas, A2
　General Exponential Rule, 384
　General Logarithmic Rule, 386
　General Power Rule, 375

infinite limit of, 472
of logarithmic functions, 386
lower limit of, 392
numerical
   Simpson's Rule, 464
   Trapezoidal Rule, 462
of odd functions, 399
partial, with respect to x, 604
by partial fractions, 436
by parts, 428
   guidelines for, 428
   summary of common uses of, 432
Simple Exponential Rule, 384
Simple Logarithmic Rule, 386
Simple Power Rule, 366
by substitution, 379
   guidelines for, 379
Sum Rule, 366
techniques, and algebra, 481
of trigonometric functions, 446, 450
upper limit of, 392
Intercepts, 45
   finding, 45
   x-intercept, 45
   y-intercept, 45
Interpolation, linear, 62
Intersection, point of, 49
Interval on the real number line, 3
   closed, 3
   endpoint, 3
   infinite, 3
   midpoint, 11
   open, 3
Inverse
   of a 2×2 matrix, 533
   of a square matrix, 529
      using Gauss-Jordan elimination to find, 531
Inverse function, 75
   domain of, 75
   finding, 76
   range of, 75
Inverse matrix, finding, 531
Inverse properties of logarithms and exponents, 282
Invertible matrix, 530
Irrational number, 2
Irreducible quadratic, 20

L

Law of exponential growth and decay, 298
Leading "1" in a matrix, 509
Least squares regression
   analysis, 51, 594

line, 597
quadratic, 599
Left-handed orientation, three-dimensional coordinate system, 548
Level curve, 567
Limit
   direct substitution, 84
   dividing out technique, 86
   existence of, 88
   of a function, 84
   infinite, 223
   at infinity, 227
   of integration
      infinite, 472
      lower, 392
      upper, 392
   from the left, 88
   one-sided, 88
   operations with, 85
   of a polynomial function, 85
   properties of, 84
   Replacement Theorem, 86
   from the right, 88
   of trigonometric functions, 339
Limit definition of e, 265
Line
   equation of, 56
      general form of, 62
      point-slope form of, 61, 62
      slope-intercept form of, 56, 62
      two-point form of, 61
   horizontal, 57, 62
   least squares regression, 597
   parallel, 63
   perpendicular, 63
   regression, least squares, 597
   secant, 116
   slope of, 56, 59
   tangent, 114
   vertical, 57, 62
Line graph, 35
Line segment, midpoint, 38
   in space, 550
Linear
   extrapolation, 62
   interpolation, 62
Linear depreciation, 64
Linear differential equation
   first-order, 644
      standard form of, 644
   guidelines for solving, 645
Linear equation, 56
   general form of, 62
   graphing, 57
   point-slope form of, 61, 62

   slope-intercept form of, 56, 62
   solving, 105
   two-point form of, 61
Linear equations
   consistent system of, 493
   dependent system of, 493
   inconsistent system of, 493, 500
   independent system of, 493
Linear function, 239
Linear model, 52
Linear system
   consistent, 493
   dependent, 493
   inconsistent, 493
   independent, 493
   number of solutions, 500
Logarithm(s)
   to the base a, 293
   common, 280, 293
   properties of, 282
Logarithmic equations, solving, 284, 307
Logarithmic and exponential form, 280
Logarithmic expressions
   condensing, 283
   expanding, 283
Logarithmic function
   to the base a, derivative of, 294
   integral of, 386
   natural, 280
      base of, 280
      derivative of, 289
   properties of, 280, 282
Logarithmic Rule
   for integration (General), 386
   for integration (Simple), 386
Logarithms and exponents, inverse properties of, 282
Logistic
   curve, 440
   differential equation, 440
   growth
      function, 267, 440
      model, 267, 440
Lorenz curve, 410
Lower limit of integration, 392

M

Main diagonal of a matrix, 507
Map, contour, 567
Mathematical model, 51
   measuring the accuracy of, 594
Matrices, 507
   equality of, 518
   row-equivalent, 508

Matrix, 507
  addition, 519
    Associative Property, 520
    Commutative Property, 520
  additive identity, 521
  augmented, 508
    elementary row operations on, 508
  coefficient, 508
  column matrix, 507
  entry in a, 507
    double subscript notation for, 507
  identity, 524
    of order n, 524
  inverse, finding, 531
  inverse of square matrix, 529
    using Gauss-Jordan elimination to find, 531
  invertible, 530
  leading "1" in, 509
  main diagonal of, 507
  multiplication, 522
    Associative Property, 524
    Distributive Property, 524
  nonsingular, 530
  order of, 507
  properties of addition and scalar multiplication, 520
  properties of multiplication, 524
  reduced row-echelon form of, 509
  row-echelon form of, 509
  row matrix, 507
  scalar multiplication, 519
    Associative Property, 520, 524
    Scalar Identity Property, 520
  singular, 530
  square, 507
  subtraction, 519
  two-by-two
    determinant of, 533
    inverse of, 533
  zero matrix, 521
Maxima, relative, 193
Maximum
  absolute, 198
  relative, 193
    of a function of two variables, 585, 588
Mean
  of a continuous random variable, 681
  of a discrete random variable, 668
  of a probability distribution, 276
Measure of central tendency, 668, 684
Measuring the accuracy of a mathematical model, 594

Median, of a continuous random variable, 684
Method of elimination for solving a system of equations, 492
Method of substitution for solving a system of equations, 490
Midpoint
  of an interval, 11
  of a line segment, 38
  in space, 550
Midpoint Formula, 38
  in space, 550
Midpoint Rule for approximating a definite integral, 456
  guidelines for using, 457
Minima, relative, 193
Minimum
  absolute, 198
  relative, 193
    of a function of two variables, 585, 588
Mixed partial derivative, 579
Model
  absolute value, 52
  cubic, 52
  exponential growth, 267
  Gompertz growth model, 651
  linear, 52
  logistic growth, 267, 440
  mathematical, 51
    measuring the accuracy of, 594
  Newton's Law of Cooling, 650
  quadratic, 52
  rational, 52
  square root, 52
Modeling exponential growth and decay, guidelines for, 300
Multiple, scalar, 519
Multiplication
  of fractions, 25
  of functions, 74
  of matrices, 522
    Associative Property, 524
    Distributive Property, 524
    properties of matrix multiplication, 524
  scalar multiplication of matrices, 519
    Associative Property, 520, 524
    Identity Property, 520

N

Natural exponential function(s), 265
  derivative of, 271

  graphs of, 266
Natural logarithmic function, 280
  base of, 280
  derivative of, 289
Negative
  direction, 2
  exponents, 13
  infinity, 3
  number, 2
Net Change Formula, 397
Newton's Law of Cooling, 650
Nonlinear systems of equations, 492
Nonnegative number, 2
Nonremovable discontinuity, 96
Nonsingular matrix, 530
Nonsquare systems of equations, 502
Normal probability density function, 276, 477, 686
  standard normal, 686
Notation
  double subscript, for matrix entries, 507
  for first partial derivatives, 574
  for functions, 72
  for higher-order derivatives, 167
  integral, of antiderivative, 365
  for a number between a and b, 3
Number(s)
  critical, 186
  irrational, 2
  negative, 2
  nonnegative, 2
  positive, 2
  rational, 2
Number line, 2
Number of solutions of a linear system, 500
Numerator, rationalizing, 31
Numerical integration
  Simpson's Rule, 464
  Trapezoidal Rule, 462

O

Obtuse angle, 316
Occurrences of relative extrema, 193
Octants, 548
Odd function, 399
  integration of, 399
One-sided limit, 88
One-to-one correspondence, 2
One-to-one function, 71
  Horizontal Line Test, 71
Open interval, 3
  continuous on, 94

Open region, 586
Operations
    elementary row, 508
    with exponents, 14
    with fractions, 25
    with limits, 85
    order of, 104
    that produce equivalent systems, 499
    row, 499
Optimization problems
    guidelines for solving, 214
    primary equation, 213
    secondary equation, 214
    solving, 213
Order of a matrix, 507
Order of operations, 104
Order on the real number line, 3
Ordered pair, 34
Ordered triple, 498, 548
Orientation for a three-dimensional coordinate system
    left-handed, 548
    right-handed, 548
Origin
    on the real number line, 2
    in the rectangular coordinate system, 34
Outcomes, 665
Outer radius, 413

P

Parabola, 44
Paraboloid
    elliptic, 559
    hyperbolic, 559
Parallel lines, 63
Partial derivative, 573
    first, notation for, 574
    first, with respect to x and y, 573
    of a function of three variables, 578
    of a function of two variables, 573
    graphical interpretation of, 575
    higher-order, 579
    mixed, 579
Partial differentiation, 573
Partial fractions, 436
    basic equation, 437
    integration by, 436
Partial integration with respect to x, 604
Particular solution of a differential equation, 369, 631
Parts, integration by, 428
    guidelines for, 428
    summary of common uses of, 432

Percentage error, 246
Period, 336
Permutations, 691
Perpendicular lines, 63
Phase shift, 344
Piecewise-defined function, 71
Plane
    Cartesian, 34
    parallel to coordinate axes, 557
    parallel to coordinate planes, 557
    xy-plane, 548
    xz-plane, 548
    yz-plane, 548
Plane in space, general equation of, 556
Point(s)
    continuity of a function at, 94
    critical, 586
    of inflection, 206
        property of, 206
    of intersection, 49
    saddle, 586, 588
    tangent line to a graph at, 114
    translating, 39
Point-plotting method, 43
Point-slope form of the equation of a line, 61, 62
Polynomial
    factoring by grouping, 19
    inequality, 5
    rational zeros of, 23
    real zeros of, 5
    special products and factorization techniques, 19
    synthetic division for a cubic, 22
    zeros of, 5, 19
Polynomial function
    continuity of, 95
    higher-order derivative of, 168
    limit of, 85
Polynomial graphs, summary of simple, 239
Population density function, 616
Position function, 141, 169
Positive
    direction, 2
    infinity, 3
    number, 2
Power Rule
    differential form of, 245
    for differentiation (General), 160
    for differentiation (Simple), 126
    for integration (General), 375
    for integration (Simple), 366
Primary equation, 213, 214

Probability, 665
    discrete, 667
Probability density function, 433, 674
    exponential, 680, 686
    normal, 276, 477, 686
    standard normal, 686
    uniform, 685
Probability distribution, 667
Problem-solving strategies, 252
Procedure, Gaussian elimination with back-substitution, 511
Product Rule
    differential form of, 245
    for differentiation, 147
Product of two functions, 74
Proper rational expression, 25
Proper rational function, 436
Properties
    of absolute value, 8
    of definite integrals, 394
    of exponents, 13, 259
    of inequalities, 4
    inverse, of logarithms and exponents, 282
    of limits, 84
    of logarithmic functions, 280, 282
    of logarithms, 282
    of matrix addition and scalar addition, 520
    of matrix multiplication, 524
Property
    Associative
        of Matrix Addition, 520
        of Multiplication for matrices, 524
        of Scalar Multiplication for matrices, 520, 524
    Commutative, of Matrix Addition, 520
    Distributive, 15
        for matrix addition and scalar multiplication, 520
        for matrix multiplication, 524
    Scalar Identity, for matrices, 520
Property of points of inflection, 206
Proportionality, constant of, 298
Pythagorean identities, 325
Pythagorean Theorem, 36, 320

Q

Quadrants, 34
Quadratic
    equation, solving, 105
    function, 239
    irreducible, 20
    least squares regression, 599

model, 52
reducible, 20
Quadratic Formula, 19
Quadric surface, 558
    classifying, 561
Quotient Rule
    differential form of, 245
    for differentiation, 150
Quotient of two functions, 74

R

Radian measure of angles, 318
Radians to degrees, converting, 318
Radical equation, solving, 105
Radical expression, domain of, 21
Radicals, 13
Radioactive decay, 299
Radius
    inner, 413
    outer, 413
Radius of a circle, 47
Random variable, 666
    continuous, 674
        expected value of, 681
        mean of, 681
        median of, 684
        standard deviation of, 682
        variance of, 682
        variance of (alternative formula), 683
    discrete, 666
        expected value of, 668
        mean of, 668
        standard deviation of, 669
        variance of, 669
Range
    of a function, 69
    of a function of two variables, 565
    of a function of x and y, 565
    of an inverse function, 75
Rate, 58
    constant, A13
    related, A10
    variable, A13
Rate of change, 58, 137, 140
    average, 137
    instantaneous, 140
        and velocity, 140
Ratio, 58
Rational exponents, 13
Rational expressions, 25
    improper, 25
    proper, 25
Rational function
    continuity of, 95

horizontal asymptotes of, 228
improper, 439
proper, 436
vertical asymptotes of, 224
Rational model, 52
Rational number, 2
Rational Zero Theorem, 23
Rational zeros of a polynomial, 23
Rationalizing technique, 31
    for denominator, 31
    for numerator, 31
Ray
    initial, 316
    terminal, 316
Real number, 2
    irrational, 2
    rational, 2
Real number line, 2
    closed interval on, 3
    coordinate of a point on, 2
    distance between two points on, 9
    infinite interval on, 3
    interval on, 3
    negative direction, 2
    one-to-one correspondence on, 2
    open interval on, 3
    order on, 3
    origin on, 2
    positive direction, 2
Real zeros of a polynomial, 5
Rectangular coordinate system, 34
    origin in, 34
Reduced row-echelon form of a
    matrix, 509
Reducible quadratic, 20
Reduction formulas, trigonometric, 325
Reference angle, 327
Region
    average value of a function over, 617
    closed, 586
    open, 586
    solid
        guidelines for finding volume, 615
        volume of, 612
Region bounded by two graphs, area
    of, 403
Regression
    line, least squares, 597
    quadratic, least squares, 599
Regression analysis, least squares, 51, 594
Related-rate problem, guidelines for
    solving, A12
Related rates, A10
Related variables, A10

Relative error, 246
Relative extrema, 193
    First-Derivative Test for, 194
    First-Partials Test for, 586
    of a function of two variables, 585, 588
    occurrences of, 193
    Second-Derivative Test for, 208
    Second-Partials Test for, 588
    of trigonometric functions, 349
Relative extremum, 193
Relative maxima, 193
Relative maximum, 193
    of a function of two variables, 585, 588
Relative minima, 193
Relative minimum, 193
    of a function of two variables, 585, 588
Removable discontinuity, 96
Replacement Theorem, 86
Reverse the inequality, 4
Review of solving equations, 105, 249, 538, 657
Revolution
    axis of, 411
    solid of, 411
    volume of, 411
Right angle, 316
Right-handed orientation, three-dimen-
    sional coordinate system, 548
Right triangle, solving a, 329
Right triangle definition of the
    trigonometric functions, 324
Row-echelon form
    of a matrix, 509
    of a system of linear equations, 498
Row-equivalent matrices, 508
Row matrix, 507
Row operations, 499
    elementary, 508

S

Saddle point, 586, 588
Sample space, 665
Scalar, 519
    multiple, 519
Scalar Identity Property, 520
Scalar multiplication and addition of
    matrices
    Distributive Property, 520
    properties, 520
Scalar multiplication of matrices, 519
    Associative Property, 520, 524
    Identity Property, 520

Scatter plot, 35
Secant function, 324
Secant line, 116
Second derivative, 167
Second-Derivative Test, 208
Second-Partials Test for relative extrema, 588
Secondary equation, 214
Separation of variables, 636
Sign, integral, 365
Similar triangles, 320
Simple Exponential Rule for integration, 384
Simple Logarithmic Rule for integration, 386
Simple Power Rule
  for differentiation, 126
  for integration, 366
Simplifying
  algebraic expressions, 174
  derivatives, 153, 162
Simpson's Rule, 464
  error in, 466
Sine function, 324
Singular matrix, 530
Singular solutions of a differential equation, 631
Slope
  of a graph, 115, 116, 137
  and the limit process, 116
  of a line, 56, 59
    finding, 59
  in x-direction, 575
  in y-direction, 575
Slope-intercept form of the equation of a line, 56, 62
Solid analytic geometry, 548
Solid region, volume of, 612
  guidelines for finding, 615
Solid of revolution, 411
  volume of, 411
    Disk Method, 411
    Washer Method, 413
Solution, extraneous, 105
Solution, system of equations with unique, 534
Solution curves, 631
Solution of a differential equation, 630
  general, 369, 630
  particular, 369, 631
  singular, 631
Solution of an inequality, 4
  test intervals, 5
Solution set of an inequality, 4

Solution of a system of equations, 490
  graphical interpretation of, 493
Solutions of a linear system, number of, 500
Solving
  an absolute value equation, 105
  equations (review), 105, 249, 538, 657
  an exponential equation, 284, 306
  an inequality, 4
  a linear differential equation, guidelines for, 645
  a linear equation, 105
  a logarithmic equation, 284, 307
  optimization problems, 213
    guidelines, 214
  a polynomial inequality, 5
  a quadratic equation, 105
  a radical equation, 105
  a related-rate problem, guidelines for, A12
  a right triangle, 329
  a system of equations, 490
    method of elimination, 492
    method of substitution, 490
  systems of equations (review), 620
  trigonometric equations, 330, 355
Special products and factorization techniques, 19
Speed, 141
Sphere, 551
  standard equation of, 551
Square matrix, 507
  inverse of, 529
    using Gauss-Jordan elimination to find, 531
Square root, 13
  model, 52
Squared errors, sum of, 594, 595
Standard deviation
  of a continuous random variable, 682
  of a discrete random variable, 669
  of a probability distribution, 276
Standard equation of a sphere, 551
Standard form
  of the equation of a circle, 47
  of a linear first-order differential equation, 644
Standard normal probability density function, 686
Standard position of an angle, 316
Step function, 98
Straight angle, 316
Straight-line depreciation, 64
Strategies, problem-solving, 252

Substitution
  back-substitution, 491
    Gaussian elimination with, 510
    using, 539
  direct, for evaluating a limit, 84
  integration by, 379
    guidelines for, 379
  method of, for solving a system of equations, 490
Subtraction
  of fractions, 25
  of functions, 74
  of matrices, 519
Sum
  of two angles, 325
  of two functions, 74
Sum Rule
  differential form of, 245
  for differentiation, 131
  for integration, 366
Sum of the squared errors, 594, 595
Summary
  of common uses of integration by parts, 432
  of curve-sketching techniques, 233
  of differentiation rules, 164
  of rules about triangles, 320
  of simple polynomial graphs, 239
Surface
  quadric, 558
    classifying, 561
  in space, 552
  trace of, 553
Synthetic division, 22
  for a cubic polynomial, 22
System of equations, 490
  consistent linear, 493
  dependent linear, 493
  equivalent linear, 499
    operations that produce, 499
  graphical interpretations of solutions, 493
  inconsistent linear, 493, 500
  independent linear, 493
  linear
    number of solutions, 500
    row-echelon form, 498
  nonlinear, 492
  nonsquare, 502
  solution of, 490
  solving, 490
    method of elimination, 492
    method of substitution, 490
  solving (review), 620
  with a unique solution, 534

T

Tangent function, 324
Tangent line, 114
  approximation, 242
Terminal ray, 316
Test
  for concavity, 203
    guidelines for applying, 204
  First-Derivative, for relative extrema, 194
  First-Partials, for relative extrema, 586
  for increasing and decreasing functions, 184
    guidelines for applying, 186
  Second-Derivative, 208
  Second-Partials, for relative extrema, 588
Test intervals, for a polynomial inequality, 5
Theorem
  Binomial, 19
  Extreme Value, 198
  Fundamental, of Algebra, 19
  Fundamental, of Calculus, 393
    guidelines for using, 394
  Pythagorean, 36, 320
  Rational Zero, 23
  Replacement, 86
Theta, 316
Third derivative, 167
Three-dimensional coordinate system, 548
  left-handed orientation, 548
  right-handed orientation, 548
Three variables, function of, 565, 578
  partial derivatives of, 578
Trace of a surface, 553
Transitive Property of Inequality, 4
Translating points in the plane, 39
Trapezoidal Rule, 462
  error in, 466
Triangles, 320
  similar, 320
  solving a right triangle, 329
  summary of rules about, 320
Trigonometric equations, solving, 330, 355
Trigonometric functions
  cosecant, 324
  cosine, 324
  cotangent, 324
  definitions of, 324
    circular function definition, 324
    right triangle definition, 324
  derivatives of, 346
  graphs of, 337
  integrals of, 446, 450
  limits of, 339
  relative extrema of, 349
  secant, 324
  sine, 324
  tangent, 324
Trigonometric identities, 325
  Pythagorean, 325
Trigonometric reduction formulas, 325
Trigonometric values of common angles, 326
Truncating a decimal, 98
Two-by-two matrix
  determinant of, 533
  inverse of, 533
Two-point form of the equation of a line, 61
Two variables, function of, 565
  domain, 565
  graph of, 566
  partial derivatives of, 573
  range, 565
  relative extrema, 585, 588
  relative maximum, 585, 588
  relative minimum, 585, 588

U

Unbounded
  behavior, 90
  function, 90
Uniform probability density function, 685
Unique solution, system of equations with, 534
Units of measure, 177
Units of variables, 541
"Unsimplifying" an algebraic expression, 418
Upper limit of integration, 392

V

Value of a function, 72
Variable(s)
  change of, integration by, 379
  continuous random, 674
    expected value of, 681
    mean of, 681
    median of, 684
    standard deviation of, 682
    variance of, 682
    variance of (alternative formula), 683
  dependent, 69
  discrete random, 666
    expected value of, 668
    mean of, 668
    standard deviation of, 669
    variance of, 669
  independent, 69
  random, 666
  rate, A13
  related, A10
  separation of, 636
Variance
  of a continuous random variable, 682
    alternative formula for, 683
  of a discrete random variable, 669
Velocity, 140
  average, 139
  function, 141, 169
  and instantaneous rate of change, 140
Vertex of an angle, 316
Vertical asymptote, 223
  of a rational function, 224
Vertical line, 57, 62
Vertical Line Test, 71
Volume
  finding with a double integral, 612
  of a solid region, 612
    guidelines for finding, 615
  of a solid of revolution, 411
    Disk Method, 411
    Washer Method, 413

W

Washer Method for finding the volume of a solid of revolution, 413

X

x
  change in, 116
  differential of, 242
x and y
  first partial derivative of f with respect to, 573
  function of, 565
    domain, 565
    graph of, 566
    range, 565
x-axis, 34
x-coordinate, 34
x-direction, slope in, 575
x-intercept, 45
xy-plane, 548
xz-plane, 548

Y

y
   change in, 116
   differential of, 242
y-axis, 34
y-coordinate, 34
y-direction, slope in, 575
y-intercept, 45
yz-plane, 548

Z

z-axis, 548
z-coordinate, 548
Zero exponent, 13
Zero matrix, 521
Zero of a polynomial, 5, 19
   rational, 23

Roadway design, A15
Satellites, 354
Scuba diver's body (pressure), 68
Sound intensity, decibels, 296
Space launches, 178, 179
Stopping distance, 173
Surface area, 253, A14
   of an oil spill, 486
   of a swamp, 488
Surface area and volume, 256
Surveying, 469
   tree height, 329
Temperature, 454
Temperature conversion, 67
Tides, 353
Uranium concentrate, 123, 136
Velocity, 144, 180, 642
   of a bicyclist, 180
   of a booster rocket, 648
   of a diver, 141, 146
   of a racecar, 144
Velocity and acceleration, 173, 181, 461
   of an automobile, 171, 173
Vertical motion, 374, 422, 503, 505, 543, 648
Volume, 248, 417, 454, 592, A14
   of a body of water, 417
   of a box, 220
   of a funnel, 426
   of Gabriel's horn, 478
   of a hardened plastic valve part, 425
   of a jet wing tank, 417
   of an oxygen tank, 571
   of a soup bowl, 417
   of a storage tank, 593
Volume and surface area, 248
Water depth
   in a swimming pool, A15
   in a trough, A15
Water supply, 454
Water treatment facility, 80
Wire length, 41
Wooden storage crate weight, 593

Business and Economics

Advertising, 506
Advertising awareness, 640, 661
Average profit
   of a biomedical supply firm, 619
   of a medical supply company, 617
Average salary
   for associate professors at public universities, 390
   for public school nurses, 390
Biological equipment production, A9
Budget deficit, 409
Compact disc shipments, 253
Cost increases, 694
Demand, 680
Depreciation, 64, 67, 109, 166, 264, 305, 310, 313, 402

Economics, Pareto's Law, 661, 662
Economy, contour map, 568
Farming, 221
Farms
   land in, 256
   number of, 112, 305, 402
Fuel cost, 145, 409
Income, median, 602
Income distribution, 410
Inventory, 453
Investment, 648, 656
   in medical research, 163
Job offer, 409
Lead production, 545
Linear depreciation, 64, 67, 109
Lorenz curve, 410
Manufacturing, 690
Medical equipment cost, 402
National debt, 111
Productivity, 211
Quality control, 480, 672, 694
Returning phone calls, 691
Salary, public school teachers, 146
Salary contract, 103, 111
Sales
   of a biomedical syringe, 648
   of gasoline, 145
   of an orthodontic product, 485
   of pharmaceutical equipment, 648
Sales analysis, 157
Seasonal sales, 350
Social Security Trust Fund, 410
Testing for defective units, 692
Trade deficit, 143
Useful life
   of an appliance, 689
   of a battery, 689, 690
   of a hearing aid battery, 680
   of a laboratory equipment component, 696
   of a medical product, 677
   of a tire, 690
Useful lifetimes of a medical product, mean and median, 684
Wages, 690, 696
Weekly salary, 55

General

Air traffic control, A15
Athletics
   baseball, 602, 673, A15
   basketball, "Mr. Average," 690
   bicycling, 248
   diving, 181
   high school basketball attendance, 123
   parachuting, 278
   running, 136, 691
   white-water rafting, 178
Bar code selection, 694
Boating, 643, A14
Choosing a three-person group, 692

Coffee mixture, 506
Coin toss, 666, 667, 668, 671, 673, 680, 697
Computer graphics, 42
Die toss, 665, 666, 667, 668, 673, 694
Equation of a circle, 505, 543
Equation of a parabola, 505, 543
Extended application, 18, 55, 173, 241, 305, 345, 410, 435, 497, 603, 648, 673
Floral arrangements, 506
Game show, 694
Games of chance, 695
Landscaping, 12
License renewal, 689
Lottery, 694
Moving ladder, A14
Music, tuning a piano, 344
Painting a room, 592
Password, 691
Phishing, 212
Probability, 469
   ACT scores, 487
   aluminum in ore samples, 455
   American men heights, 477
   American women heights, 480
   Buffon's needle experiment, 453
   female college freshmen heights, 278
   GMAT scores, 686
   intelligence quotient (IQ), 697
   iron in ore samples, 435
   memory experiment, 433, 435
   recall in an experiment, 485
   SAT scores, 278, 487
   time between incoming calls (average), 270
Product maximum, 219
Radio station, 506
Random selection, 694
   choosing a card, 672, 694, 697
Raw materials, 536
Research Project, 81, 157, 212, 297, 354, 410, 480, 506, 583, 642, 680
Safety, 635
Security camera, A15
Sporting goods, 103
Sprinkler system, 323
Sum minimum, 215, 219, 254
Time minimum, 221
Transportation, 689
Waiting time
   for a bus, 695
   for patients arriving at a dentist's office, 696
   for patients arriving at a health clinic, 678
   for service at the checkout of a grocery store, 680
   for service in a store, 689
   for unloading Red Cross supply trucks, 680
Windshield wiper, 323

Basic Differentiation Rules

1. $\dfrac{d}{dx}[cu] = cu'$
2. $\dfrac{d}{dx}[u \pm v] = u' \pm v'$
3. $\dfrac{d}{dx}[uv] = uv' + vu'$
4. $\dfrac{d}{dx}\left[\dfrac{u}{v}\right] = \dfrac{vu' - uv'}{v^2}$
5. $\dfrac{d}{dx}[c] = 0$
6. $\dfrac{d}{dx}[u^n] = nu^{n-1}u'$
7. $\dfrac{d}{dx}[x] = 1$
8. $\dfrac{d}{dx}[\ln u] = \dfrac{u'}{u}$
9. $\dfrac{d}{dx}[e^u] = e^u u'$
10. $\dfrac{d}{dx}[\log_a u] = \dfrac{u'}{(\ln a)u}$
11. $\dfrac{d}{dx}[a^u] = (\ln a)a^u u'$
12. $\dfrac{d}{dx}[\sin u] = (\cos u)u'$
13. $\dfrac{d}{dx}[\cos u] = -(\sin u)u'$
14. $\dfrac{d}{dx}[\tan u] = (\sec^2 u)u'$
15. $\dfrac{d}{dx}[\cot u] = -(\csc^2 u)u'$
16. $\dfrac{d}{dx}[\sec u] = (\sec u \tan u)u'$
17. $\dfrac{d}{dx}[\csc u] = -(\csc u \cot u)u'$

Basic Integration Formulas

1. $\displaystyle\int kf(u)\,du = k\int f(u)\,du$
2. $\displaystyle\int [f(u) \pm g(u)]\,du = \int f(u)\,du \pm \int g(u)\,du$
3. $\displaystyle\int du = u + C$
4. $\displaystyle\int a^u\,du = \left(\dfrac{1}{\ln a}\right)a^u + C$
5. $\displaystyle\int e^u\,du = e^u + C$
6. $\displaystyle\int \ln u\,du = u(-1 + \ln u) + C$
7. $\displaystyle\int \sin u\,du = -\cos u + C$
8. $\displaystyle\int \cos u\,du = \sin u + C$
9. $\displaystyle\int \tan u\,du = -\ln|\cos u| + C$
10. $\displaystyle\int \cot u\,du = \ln|\sin u| + C$
11. $\displaystyle\int \sec u\,du = \ln|\sec u + \tan u| + C$
12. $\displaystyle\int \csc u\,du = -\ln|\csc u + \cot u| + C$
13. $\displaystyle\int \sec^2 u\,du = \tan u + C$
14. $\displaystyle\int \csc^2 u\,du = -\cot u + C$

Trigonometric Identities

Pythagorean Identities
$\sin^2\theta + \cos^2\theta = 1$
$\tan^2\theta + 1 = \sec^2\theta$
$\cot^2\theta + 1 = \csc^2\theta$

Sum or Difference of Two Angles
$\sin(\theta \pm \phi) = \sin\theta\cos\phi \pm \cos\theta\sin\phi$
$\cos(\theta \pm \phi) = \cos\theta\cos\phi \mp \sin\theta\sin\phi$
$\tan(\theta \pm \phi) = \dfrac{\tan\theta \pm \tan\phi}{1 \mp \tan\theta\tan\phi}$

Double Angle
$\sin 2\theta = 2\sin\theta\cos\theta$
$\cos 2\theta = 2\cos^2\theta - 1 = 1 - 2\sin^2\theta$

Reduction Formulas
$\sin(-\theta) = -\sin\theta$
$\cos(-\theta) = \cos\theta$
$\tan(-\theta) = -\tan\theta$
$\sin\theta = -\sin(\theta - \pi)$
$\cos\theta = -\cos(\theta - \pi)$
$\tan\theta = \tan(\theta - \pi)$

Half Angle
$\sin^2\theta = \tfrac{1}{2}(1 - \cos 2\theta)$
$\cos^2\theta = \tfrac{1}{2}(1 + \cos 2\theta)$